SOLAR WIND-MAGNETOSPHERE COUPLING

ASTROPHYSICS AND SPACE SCIENCE LIBRARY

A SERIES OF BOOKS ON THE RECENT DEVELOPMENTS
OF SPACE SCIENCE AND OF GENERAL GEOPHYSICS AND ASTROPHYSICS
PUBLISHED IN CONNECTION WITH THE JOURNAL
SPACE SCIENCE REVIEWS

VOLUME 126
PROCEEDINGS

SOLAR WIND-MAGNETOSPHERE COUPLING

Edited by

Y. KAMIDE
Kyoto Sangyo University

and

J. A. SLAVIN
Jet Propulsion Laboratory, California Institute of Technology

Terra Scientific Publishing Company / Tokyo

D. Reidel Publishing Company

A MEMBER OF THE KLUWER ACADEMIC PUBLISHERS GROUP

Dordrecht / Boston / Lancaster / Tokyo

Library of Congress Cataloging-in-Publication Data

CIP

Solar wind-magnetosphere coupling.

(Astrophysics and space science library; v. 126)
Includes bibliographies.
1. Magnetosphere—Congresses. 2. Solar wind—Congresses. I. Kamide, Y. II. Slavin, James Arthur, 1955- . III. Series.
QC809.M35S644 1986 538'.766 86-17703

ISBN-13: 978-94-010-8600-4 e-ISBN-13: 978-94-009-4722-1
DOI: 10.1007/978-94-009-4722-1

Published by Terra Scientific Publishing Company (TERRAPUB),
307 Shibuyadai-haim, 4-17 Sakuragaoka-cho, Shibuya-ku, Tokyo 150, Japan,
in co-publication with D. Reidel Publishing Company, Dordrecht, Holland

Sold and distributed in the U.S.A. and Canada
by Kluwer Academic Publishers,
101 Philip Drive, Assintppi Park, Norwell, MA 02061, U.S.A.
in Japan by Terra Scientific Publishing Company (TERRAPUB),
307 Shibuyadai-haim, 4-17 Sakuragaoka-cho, Shibuya-ku, Tokyo 150, Japan

In all other countries, sold and distributed
by Kluwer Academic Publishers Group,
P.O. Box 322, 3300 AH Dordrecht, Holland

Preface

This monograph is the outgrowth of an AGU (American Geophysical Union) Chapman Conference on Solar Wind-Magnetosphere Coupling, and it contains most of the formal papers presented at the conference including the summary panel. The conference was held on February 12–15, 1985 at the Jet Propulsion Laboratory of the California Institute of Technology in Pasadena, California, U.S.A. It was attended by over 150 scientists from most of the nations active in space research. The purpose of the conference was to bring together scientists from all areas of solar-terrestrial physics, both theoretical and experimental, in order to focus attention on the solar wind-magnetosphere coupling problem, to provide a timely forum for the exchange of ideas, and to promote interdisciplinary collaboration.

The conference agenda consisted of eight sessions: six topical, one poster, and one summary. The topical sessions were "Solar Wind and Geomagnetic Activity," "AMPTE Mission," "Solar Wind and Polar Cusp," "Auroral Zone and Polar Cap," "Solar Wind and Magnetotail," and "Theoretical Modeling and Simulations." The conference concluded with a two-hour summary panel discussion which attempted to provide an assessment of the progress that has been made as well as underscore the problems, old and new, which remain unanswered.

Sydney Chapman (1889–1970) was one of the early pioneers in the study of the influence of the sun and magnetosphere on the earth's upper atmosphere, and it was most appropriate that this conference be one in the AGU series named in his honor. As convenors of the conference, and the editors of this volume, we hope that this monograph will represent an important milestone in understanding the entire electromagnetic processes occurring in solar-terrestrial environment, and will prove useful in planning future satellite and theory programs in space physics.

January, 1986

Y. Kamide
J. A. Slavin
Editors

Acknowledgements

The cosponsors of the conference, along with the American Geophysical Union, were the Air Force Geophysics Laboratory, Kyoto Sangyo University, Los Alamos National Laboratory, the National Aeronautics and Space Administration, the National Science Foundation, and Taiyo Kikaku Co., Ltd., Japan. The publication of the volume was supported by Kyoto Sangyo University, Taiyo Kikaku Co., Ltd., the National Aeronautics and Space Administration, and the Air Force Geophysics Laboratory.

The program committee (D. N. Baker, Los Alamos National Laboratory; S. W. H. Cowley, Imperial College; D. A. Hardy, Air Force Geophysics Laboratory; Y. Kamide, Kyoto Sangyo University; L.-C. Lee, University of Alaska; R. L. McPherron, University of California, Los Angeles; P. H. Reiff, Rice University; G. Rostoker, University of Alberta; G. L. Siscoe, University of California, Los Angeles; J. A. Slavin, Caltech/Jet Propulsion Laboratory) was responsible for the organization of the sessions.

Ms. Brenda Weaver of the American Geophysical Union and Ms. Patricia McLane of Jet Propulsion Laboratory provided excellent professional assistance with the preparation of conference materials and the arrangements at Jet Propulsion Laboratory. Mr. Shoichiro Ono of Kyoto Sangyo University provided valuable guidance in administrative matters concerning the publication of this monograph. A special note of thanks is extended to Ms. Kimberly M. Malinauskas for her efficient technical assistance in the handling of the papers submitted to this monograph.

The editors would like to thank the following scientists, listed in alphabetical order, for refereeing the papers submitted to this volume:

S.-I. Akasofu
W. Baumjohann
J. L. Burch
J. R. Burrows
C. R. Clauer
N. U. Crooker
M. Dryer
D. H. Fairfield
J. C. Foster
T. A. Fritz
K.-H. Glassmeier
J. T. Gosling
T. W. Hill
W. J. Hughes
Y. Kamide
M. G. Kivelson
L. J. Lanzerotti
R. P. Lepping
K. Maezawa
I. B. McDiarmid
G. Atkinson
J. Birn
W. J. Burke
M. Candidi
S. W. H. Cowley
C. S. Deehr
T. E. Eastman
J. F. Fennell
B. J. Fraser
N. Fukushima
C. K. Goertz
M. S. Gussenhoven
J. M. Holt
F. M. Ipavich
J. R. Kan
D. M. Klumpar
K. K. Lassen
A. T. Y. Lui
K. Marubashi
R. W. McEntire
D. N. Baker
A. Brekke
L. F. Burlaga
J. K. Chao
J. D. Craven
O. de la Beaujardiere
R. C. Elphic
J. Feynman
J. W. Freeman
T. Fuller-Rowell
W. D. Gonzalez
D. A. Hardy
R. E. Holzer
T. Iyemori
J. H. King
S. Kokubun
L.-C. Lee
L. R. Lyons
H. Matsumoto
R. L. McPherron

C.-I. Meng	P. F. Mizera	F. S. Mozer
T. Murayama	A. Nishida	W. P. Olson
H. J. Opgenoorth	G. K. Parks	G. Paschmann
R. Pellinen	S. Perraut	M. I. Pudovkin
G. C. Reid	P. H. Reiff	A. D. Richmond
R. P. Rijnbeek	R. M. Robinson	G. Rostoker
C. T. Russell	J. C. Samson	H. H. Sargent III
T. Sato	M. A. Saunders	K. Schindler
M. Scholer	M. Schulz	N. Sckopke
D. G. Sibeck	G. L. Siscoe	J. A. Slavin
E. J. Smith	B. U. O. Sonnerup	D. J. Southwood
T. W. Speiser	D. P. Stern	T. Terasawa
B. T. Tsurutani	G.-H. Voigt	D. J. Williams
J. D. Winningham	R. A. Wolf	S. T. Wu
T. Yeh	K. Yumoto	L. J. Zanetti
R. D. Zwickl	—plus two anonymous referees.	

Solar Wind-Magnetosphere Coupling: Introduction

J. A. SLAVIN[1] and Y. KAMIDE[2]

[1]*Jet Propulsion Laboratory, California Institute of Technology, Pasadena, California 91109, U.S.A.*
[2]*Kyoto Sangyo University, Kyoto 603, Japan*

The continual interaction between the solar wind and the Earth's magnetic field generates a vast magnetic cavity, termed the magnetosphere. The magnetosphere is a very dynamic region of space whose three-dimensional current systems and plasma populations change rapidly in response to both conditions in the solar wind and internal instabilities. Of particular importance are magnetospheric storms and substorms which inject large amounts of energy into currents, hydromagnetic waves, charged particle acceleration, auroral lights, and heating/motion of the ionosphere and upper thermosphere. The area of space plasma physics which seeks to understand, in a synoptic sense, how the solar wind transfers energy into the magnetosphere and how efficiently that energy is deposited in the ionosphere and upper atmosphere is termed "solar wind-magnetosphere coupling." Such an understanding is a major objective of the solar-terrestrial research community and one of the chief beneficiaries of the International Magnetospheric Study (1976–79). The new observational data sets and theoretical modeling tools produced during the late 1970s and early 1980s have allowed the progress in solar wind-magnetosphere coupling which forms the basis of this monograph.

The monograph is divided into five sections covering the major regions of the magnetosphere and the techniques for their investigation. Section 1 is devoted to statistical studies of geomagnetic indices and their relation to solar wind conditions. Because much of the energy in the magnetosphere, especially during storms and substorms, goes into driving electrical currents it is possible to monitor many aspects of magnetospheric dynamics, such as the injection of particles into the ring current (i.e., the D_{st} index) and the enhancement of the auroral electrojet currents (i.e., the AE indices), using ground-based networks of magnetic observatories. Despite their limitations, including non-uniform geographic distribution of the observatories, sensitivity and linearity of the indices, statistical studies of the response of the geomagnetic indices to changing solar wind conditions have been one of the major sources of knowledge on the optimum "coupling functions" which describe the rate of energy transfer from the solar wind to the magnetosphere-ionosphere system through the inner magnetosphere. The studies of these coupling functions have provided much support, for example, in favor of magnetic reconnection at the dayside magnetopause as the dominant process for the extraction of energy from the solar wind. This result came about because of the major role the north-south component of the interplanetary magnetic field (IMF) plays in the empirically derived coupling functions, or "predictors" of geomagnetic activity. Less well understood is the influence of other parameters in solar wind plasma and in the IMF and the means by which the magnetosphere is powered during intervals of northward IMF when the contribution from dayside magnetic reconnection is at a minimum. The papers in this section contain an up-to-date description of the merits and limitations of these geomagnetic activity indices as diagnostics for the electrodynamic state of the magnetosphere. The use of correlation analyses and the linear prediction filtering techniques to search for significant relationships between solar wind conditions and these indices, and their implications for the physics of the solar wind-magnetosphere-ionosphere interactions are also reviewed. The section concludes with papers treating the operational requirements for providing real-time forecasts of conditions in the magnetosphere and ionosphere over both short and long time spans.

Section 2 considers how the conditions in the solar wind interact with the Earth's magnetic field to determine the boundary conditions on the magnetosphere. The study of the IMF and the solar wind is a major aspect of solar-terrestrial relations due to the extreme variations in plasma parameters presented to the magnetosphere over time scales of seconds to the solar cycle (11 years). In fact, the fluid-like nature of the solar wind was first inferred from the rapid increases in low-latitude magnetic field intensity just before the commencement of magnetic storms. The sharp compression of the geomagnetic field at "storm sudden commencement" suggested a shock in the interplanetary medium as a driver. However, a shock wave requires a medium with short effective mean free paths (i.e., small "Knudsen number"). We now know that shock waves are a common occurrence in the solar wind with wave-particle interactions taking the place of collisions in causing the medium to behave as a fluid despite its collisionless nature. This section reviews the effects of shocks and longer-term variations in solar wind conditions on large-scale configuration of the magnetosphere as well as on magnetospheric processes. In addition, emphasis is placed on examining the control over magnetospheric configuration exerted by the interplanetary medium through modulation of the rate of dayside reconnection. Recent developments in the proposed geometries for dayside magnetic merging along with the resulting convection in the "open" polar cap are reviewed. The possibility of solar wind plasmas entering the boundary layer even for the "closed" magnetopause is presented. Some new results from recent spacecraft observations of short-scale length "flux transfer" events are also discussed. The section concludes with an investigation of the MHD waves and turbulence generated by the global solar wind interaction and the transmission of these waves into the magnetosphere as an additional mechanism for the extraction of energy from the solar wind.

A variety of solar wind effects on electrodynamic processes in the high-latitude magnetosphere and ionosphere are addressed in Section 3. Because magnetic field lines tend to be equipotentials, there is a mapping of the high-altitude magnetospheric electric field down into the ionosphere where it produces the characteristic $\boldsymbol{E}\times\boldsymbol{B}$ drifts, or "convection." Changes in the electrodynamic state of the magnetosphere are communicated to the ionosphere through field-aligned currents. These currents themselves are carried by precipitating charged particles and the upward injection of ionospheric particles into the magnetosphere. In this way the high-latitude ionosphere acts as a dissipative load on the magnetospheric system and mirrors the changes that the solar wind and the IMF bring about in the magnetosphere. In the past, solar wind-magnetosphere coupling studies have concentrated on the north-south component of the IMF due to its dominant role in controlling reconnection at the magnetopause. However, the stresses exerted on the magnetosphere by reconnected field lines are three-dimensional and magnetospheric-ionospheric signatures associated with all three components of the IMF were reported at the conference. Accordingly, the intent of Section 3 is to present a unified discussion of the coupling between the magnetosphere and the ionosphere as well as examine the relative importance of reconnection and non-reconnection mechanisms for the supply of energy to the magnetosphere. General properties of the electric fields and currents in the polar cusp, the polar cap, and the auroral zone are reviewed. In particular, it is shown that high-latitude magnetosphere processes respond to the solar wind and the IMF in a rapid and dramatic manner. The response in terms of the cross-polar cap potential difference is perhaps the most direct and readily quantifiable indication of the efficiency of the solar wind-magnetosphere coupling. The relative importance of three separate mechanisms, magnetic merging with the IMF on the dayside, merging in the magnetotail, and viscous-like processes at the magnetopause is evaluated in view of theoretical considerations as well as of recent satellite observations. Considerable discussion of IMF B_y control of convection patterns is given on the basis of the most recent measurements by satellites and ground-based radars. The role of field-aligned currents in diagnosing plasma distribution in the magnetosphere is discussed. This section concludes with a discussion of the structure of auroral particle precipitation and auroras. In particular, variations in the spatial patterns of electron precipitation with the orientation of the IMF are extensively discussed.

The magnetotail is a critical region for the study of solar wind-magnetosphere coupling and it is the subject of Section 4. Despite the great strides in our understanding of magnetotail structure and processes over the last two decades, many aspects of tail morphology and dynamics remain poorly understood, or, at best, difficult to unify under a single model. The surface of the tail is thought to be the site for much of the energy transfer to the magnetosphere. For example, in the "open" model energy transfer from the solar wind is described in terms of a Poynting energy flux into the magnetotail produced by a non-zero magnetic field component normal to the magnetopause. Depending upon which substorm model is correct, the tail is either a reservoir for magnetic energy which is "unloaded" during substorms, a conduit channelling energy directly from the solar wind to the inner magnetosphere and ionosphere, or, perhaps on different time scales, both. In any case, the preferred mechanism for the conversion of tail lobe magnetic energy into the other forms required by the magnetosphere is reconnection at the tail midplane. However, there is considerable uncertainty about the manner in which reconnection operates. For example, how predictable is its onset, does the rate of reconnection grow "explosively" following initiation, and are "plasmoids" always created during all substorms and how many neutral lines at which locations are necessary to form these magnetic islands? In this section observations and theories of the magnetotail structure and dynamics are reviewed in some detail. A series of papers on the recent ISEE-3 observations in the translunar magnetotail relevant to the physics of reconnection, plasmoids, large-scale vortices, twisting and flattening of the tail in response to IMF changes and substorm activity are presented.

The monograph concludes in Section 5 with a summary paper by a group of panelists who were invited to review the results of the conference and present their views on the outstanding questions in solar wind-magnetosphere coupling. In contrast with past conferences in the 1960s and 1970s, there was general acceptance of magnetic reconnection as the primary mechanism enabling the transfer of solar wind energy to the magnetosphere. Experimental verifications for theoretical predictions based on "Dungey-type" magnetospheres were presented from such diverse sources as radar observations of high-latitude ionospheric convection patterns, optical auroral and particle precipitation observations by polar-orbiting spacecraft, in situ measurements at the magnetopause, and ISEE-3 observations at distances of more than 200 R_E down the magnetotail. This consensus on the importance of dayside reconnection, however, only served to underscore the many questions remaining to be answered on the properties of this fundamental plasma instability. In particular, the respective roles and controlling parameters for "steady" and "unsteady" reconnection at the magnetopause have not been resolved and the relative merits of the "component" and/or "anti-parallel" hypotheses for the dependence of reconnection rate and neutral line location on IMF orientation are still in question. Finally, there is probably less, or at least no more, known about reconnection in the magnetotail than is the case for the dayside magnetopause, and reconnection in the tail is as important to how magnetospheric energy is dissipated as dayside merging is to the input of energy from the solar wind.

The monograph papers also point out that the agreement concerning the primary role of reconnection does not diminish the need to understand the other mechanisms through which the solar wind exerts stress upon the magnetosphere. Grouped together under the heading of "quasi-viscous" interactions are all of the means by which charged particles and MHD waves enter the magnetosphere other than by simple propagation along open field lines. Despite strong evidence that such processes are operating much of the time (e.g., the existence of the low-latitude boundary layer and magnetospheric convection during periods of northward IMF), relatively little effort has gone into their study with the possible exception of the Kelvin-Helmholtz instability at the magnetopause. Finally, it is also clear that the manner and time scales over which energy from the solar wind is stored in the magnetotail and/or directly transported to the polar ionosphere and the inner magnetosphere during magnetospheric substorms is still highly controversial. Perhaps, both mechanisms are important, but there is presently no quantitative way to monitor continuously the relative importance of these two processes. A major objective of solar wind-magnetosphere coupling studies over the rest of this decade will be the resolution of these questions concerning the predictability of substorm properties and how their energy is deposited at lower altitudes.

Portions of the research described in this paper were carried out at the Jet Propulsion Laboratory of the California Institute of Technology under contract to the National Aeronautics and Space Administration.

Table of Contents

3. ELECTRODYNAMICS OF MAGNETOSPHERE-IONOSPHERE COUPLING

3.1 Electric Fields and Currents

3.2 Auroras and Auroral Precipitation

4. SOLAR WIND CONTROL OF NIGHTSIDE MAGNETOSPHERE

5. SUMMARY PANEL

1. GEOMAGNETIC INDICES AND STATISTICAL METHODS

Solar Wind-Magnetosphere Coupling, edited by Y. Kamide and J. A. Slavin, 3–15.

Merits and Limitations of the Use of Geomagnetic Indices in Solar Wind-Magnetosphere Coupling Studies

Wolfgang BAUMJOHANN

Max-Planck-Institut für Physik und Astrophysik, Institut für extraterrestrische Physik
8046 Garching bei München, W-Germany

Geomagnetic indices are meant to quantify disturbed states of geospace. While the planetary range indices like K_p describe the overall disturbance, two indices quantify the disturbance and the dissipation of energy in certain regions, namely AE (auroral oval) and D_{st} (ring current). Because of this, they are also the ones most often used in solar wind coupling studies. In this review the physical meaning of these "regional" indices will be evaluated with special emphasis placed on the merits and limitations of the use of these indices in solar wind-magnetosphere coupling studies.

1. Introduction: Solar Wind-Magnetosphere Coupling

The earth's magnetosphere is a cavity filled with hot but dilute plasma embedded in the fast-flowing denser but colder solar wind plasma. Due to the strong terrestrial magnetic field, the solar wind cannot directly penetrate the magnetopause but is deflected around the magnetosphere. Yet roughly 2% of the kinetic power of the solar wind plasma impinging on the dayside magnetosphere (about 10^4 GW) is extracted by several different coupling processes (e.g., HILL, 1983), enters the magnetosphere, and drives all dynamic and steady state plasma processes in the earth magnetosphere and ionosphere before it is finally dissipated (mainly in the form of heat).

The magnetopause coupling processes can be roughly divided into two categories. Those which imply interaction between the solar wind magnetic field and the terrestrial field at the dayside magnetopause, viz. field line merging, and other 'non-magnetic' mechanisms which are usually summarized under the heading of quasi-viscous processes. Theoretical considerations (e.g., COWLEY, 1982; HILL, 1983), direct (MOZER, 1984) and indirect observational evidence (e.g., REIFF *et al.*, 1981; BAUMJOHANN and HAERENDEL, 1985) and laboratory experiments (e.g., PODGORNY *et al.*, 1978) suggest that magnetic merging is the dominant coupling process and provides roughly 90% of the energy input into the magnetosphere.

Field line merging (or reconnection) may operate in steady state and intermittent, time-dependent fashion. The HEOS and ISEE satellite missions have provided in-situ evidence that the merging process actually takes place as both, steady-state reconnection in the vicinity of the subsolar point (e.g., SONNERUP *et al.*, 1981) and in the form of patchy, impulsive reconnection in lower latitudes (flux transfer events; e.g., SAUNDERS, 1983) and in the cusp region of the magnetopause (HAERENDEL *et al.*, 1978). However, it is at present unclear if steady-state or patchy impulsive reconnection provides the major energy input.

Quasi-viscous interaction comprises three different mechanisms: Kelvin-Helmholtz instability, impulsive penetration, and diffusive entry. The Kelvin-Helmholtz instability may operate at the dawn- and dusk-side low-latitude boundary layers and theoretically (MIURA, 1984) can provide the total 10% of the energy input attributed to quasi-viscous processes. Impulsive penetration of solar wind plasma filaments with excess momentum through the magnetopause was proposed by LEMAIRE and ROTH (1978), but it is unclear how much energy can be transferred by this process. Diffusive entry of solar wind plasma may be achieved by stochastic scattering of individual particles by resonant waves generated through micro-instabilities and HILL (1983) estimated that cross-field diffusion may account for 5–10% of the total coupling efficiency.

Once solar wind energy has entered the magnetosphere by either of the aforementioned processes, this energy is distributed, intermediately stored and eventually dissipated roughly as shown in Fig. 1. Part of the energy is directly transferred to the high-latitude ionosphere via field-aligned currents where it is dissipated by Joule heating. The other part is transferred to the magnetotail where it is intermediately stored as both plasma kinetic and thermal energy and magnetic field energy.

Part of the magnetotail energy is then transferred to the ring current via particle injection where it is dissipated by charge exchange and subsequent loss to the earth atmosphere; only a minor part of this energy is intermediately stored and transferred to the auroral ionosphere via (diffuse) precipitation. A second part of the magnetotail energy is transferred to the auroral ionosphere via particle precipitation and field-aligned currents where it is dissipated in the form of Joule heating (about 70%) and

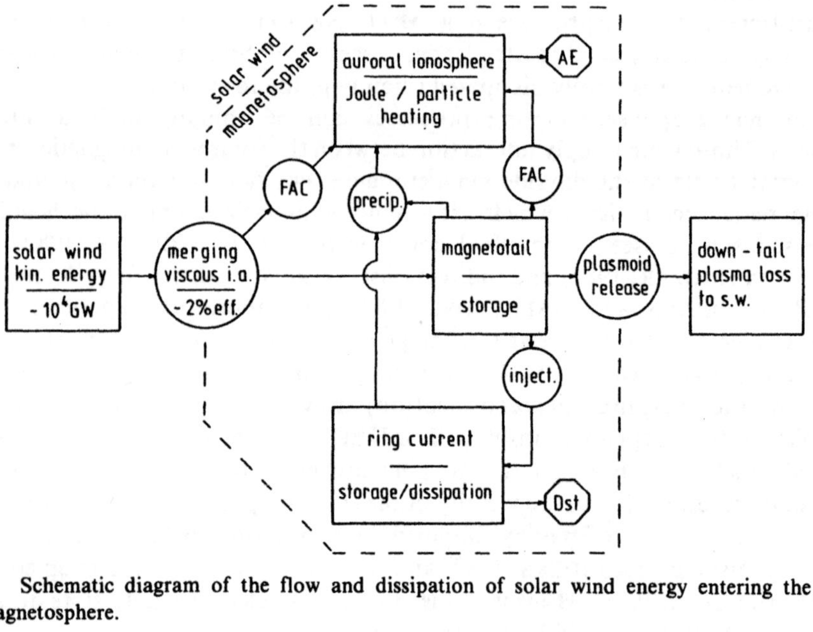

Fig. 1. Schematic diagram of the flow and dissipation of solar wind energy entering the earth's magnetosphere.

particle heating (roughly 30%; see STERN, 1984). The remaining part of the magnetotail energy (probably a substantial part; cf., PELLINEN *et al.*, 1982) is not dissipated within the magnetosphere but reenters the solar wind through the down-tail release of plasmoids.

Accordingly, there are two major sinks for solar wind energy in the magnetosphere: The probably dominant one is the auroral ionosphere (cf., HAREL *et al.*, 1981; BAUMJOHANN and KAMIDE, 1984) but a substantial part is also deposited in the ring current. The disturbances in these two regions are quantified by the *AE* and D_{st} indices, respectively. Hence, these two indices are the important ones for solar wind-magnetosphere coupling studies—and actually the ones most often used (cf., AKASOFU, 1981). The comparison of energy input into the magnetosphere (based on satellite measurements of solar wind parameters) and total magnetospheric energy dissipation (based on the *AE* and D_{st} indices)—as first done by Akasofu and co-workers (AKASOFU, 1981, 1983) and shown in Fig. 2—has substantially added to our understanding of solar wind-magnetosphere coupling.

However, for a proper use of the *AE* and D_{st} indices in solar wind-magnetosphere coupling studies, it is necessary to understand what these indices are actually describing and which problems and limitations may arise from the construction and derivation of these indices when one tries to calculate energy dissipation rates (which

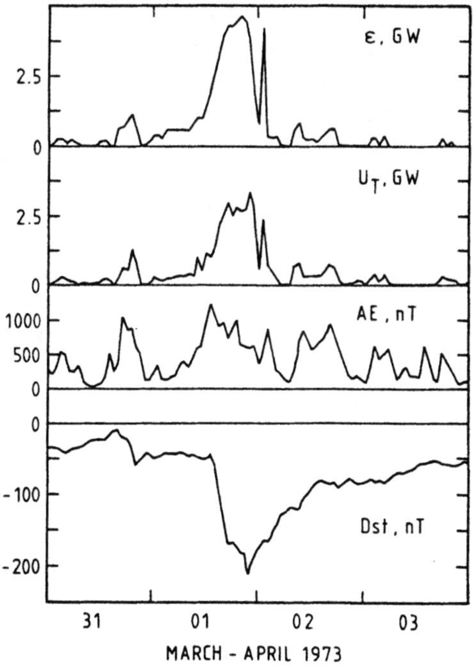

Fig. 2. An example of the solar wind energy input (ε) and the total energy dissipation rate (U_T) in the magnetosphere, together with the *AE* and D_{st} indices on which the calculation of U_T is based (after AKASOFU, 1983).

in itself introduces new problems). Hence, the remainder of this review will be devoted to a description of the construction and derivation of the *AE* and D_{st} indices and an evaluation of their merits as well as their limitations for solar wind-magnetosphere coupling studies. Since the planetary range indices like K_p are not useful for solar wind-magnetosphere coupling studies—mainly because they do not refer to certain region of the geospace but to the overall disturbance with unknown relative contributions—they will not be discussed here and the reader is referred to ROSTOKER (1972) and MAYAUD (1980).

2. The Auroral Electrojet *AE* Index

The auroral electrojet indices *AE*, *AU*, and *AL* were introduced by DAVIS and SUGIURA (1966) as a measure of global auroral electrojet activity and first used by ARNOLDY *et al.* (1971) in a solar wind-magnetosphere coupling study. The present auroral indices are based on 2.5 min readings of the *H* component trace from 12 auroral zone observatories located between about 65° and 70° magnetic latitude with a longitudinal spacing of 10–40°. For each observatory the readings are referenced to a quiet day level. The data of all 12 observatories are then plotted as a function of UT. The upper and lower envelopes are defined as *AU* and *AL* indices, respectively, and are thought to represent the maximum eastward and westward electrojet currents. The value of the *AE* index is defined as the separation between upper and lower envelope, viz. $AE=AU-AL$, and should represent the total maximum electrojet current. Note that the auroral electrojet indices actually describe maximum current density and not electrojet current strength and that, despite a good statistical correlation between those two quantities, there may be large differences in individual cases (cf., KAMIDE *et al.*, 1982; AKASOFU *et al.*, 1983).

Since it monitors the total electrojet current, the *AE* index is much more often used in solar wind-magnetosphere coupling studies than the *AU* and *AL* indices. Accordingly, we will concentrate on the former, but note that most of the following merits and limitations also refer to the latter. The main advantage in using the separate *AU* and *AL* indices instead of the combined *AE* index may be the possibility to distinguish between the convection electrojets, which show up in both *AU* and *AL* and characterize the direct dissipation of solar wind energy, and the substorm current wedge, which yields an increase mainly in the *AL* index and which is related to the sudden dissipation of energy previously stored in the magnetotail (see, for example, BAUMJOHANN, 1983; KAMIDE and BAUMJOHANN, 1985). The main disadvantage of *AU* and *AL* in comparison to *AE* is that the former are influenced by zonally uniform non-electrojet fields like D_{st}, which cancel out in *AE*. Unfortunately, the latter may be influenced by non-electrojet fields like *Sq* (see also below).

That an ideal *AE* index is indeed a good measure of the dissipation of solar wind energy in the high-latitude ionosphere can be seen in Fig. 3 (from BAUMJOHANN and KAMIDE, 1984). This figure shows a regression between the hemispherical Joule heating rate (calculated from the magnetic observations at 71 mostly auroral zone stations by using a realistic conductivity model; see KAMIDE *et al.* (1982)) and a rather ideal 71-station *AE* index. The high correlation coefficient indicates that 0.3 GW

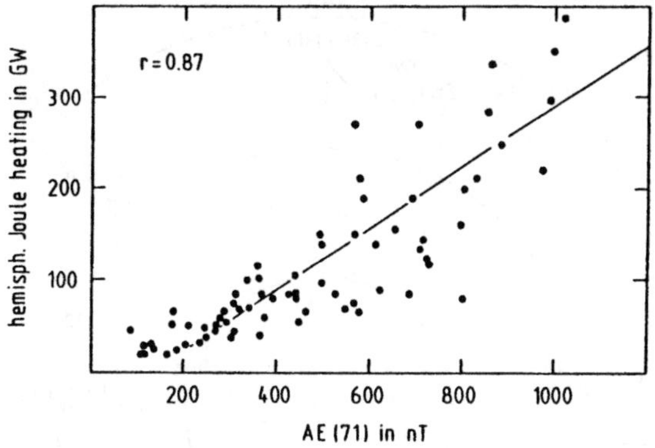

Fig. 3. Scatter plot of hourly values of hemispherical Joule heat production as a function of *AE*(71) for March 17–19, 1978 (after BAUMJOHANN and KAMIDE, 1984).

Joule heating rate per 1 nT is a good estimate. In a similar study AHN *et al.* (1983) additionally showed that the *AE*(71) index is also a good predictor for the particle heating rate.

Unfortunately, computing an "ideal" *AE* index based on the eccentric dipole north-south component observed at more than 50 stations is a very tedious and costly job and has been done for a couple of days only. However, also the standard *AE*(12) index is very useful for solar wind-magnetosphere coupling studies if one keeps in mind the differences to an "ideal" *AE* index and the resulting limitations. The main limitations of useing the standard *AE* index stem from the following (ROSTOKER, 1972; MAYAUD, 1980; KAMIDE and AKASOFU, 1983): (i) *H* component used; (ii) longitudinal gaps; and (iii) small latitudinal range. These problems are illustrated in Fig. 4 (after KAMIDE and AKASOFU, 1983). It is clearly visible that at some of the *AE* observatories the angle between the local magnetic *H* component and the eccentric dipole north–south direction is greater than 30°. Since the electrojets tend to flow along the auroral oval rather than perpendicular to the local *H* direction, these observatories tend to underestimate the electojet current in the standard *AE* index—or overestimate it if strong field-aligned currents yield strong magnetic perturbations parallel to the electrojet flow direction.

The most severe longitudinal gaps exist between Cape Wellen and Tixie Bay and Dixon Island and Murmansk, but there are also other gaps (western Canada, western and eastern Atlantic) spanning more than 2 h of local time. In particular, DP1 current wedges associated with weak or moderate substorms may cover less than 2 h of local time (e.g., BAUMJOHANN *et al.*, 1981) and can easily be missed by the 12-station *AE* netowrk.

Probably more severe is the small latitudinal range of the *AE* observatories, especially during times of very weak activity. During times of weak solar wind-magnetosphere coupling (typically associated with northward IMF) the auroral oval contracts northward and the electrojets tend to flow around or even poleward of 70° latitude (e.g., KAMIDE and AKASOFU, 1974).

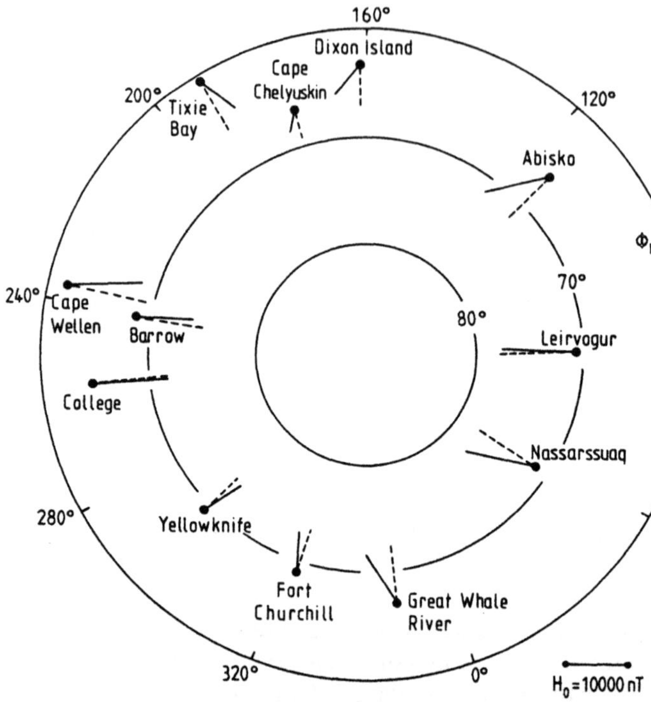

Fig. 4. Distribution of horizonal magnetic field vectors (*H* component direction) at the 12 standard *AE* observatories. The geomagnetic north direction (eccentric dipole coordinates) at each observatory is given by a dashed line (after KAMIDE and AKASOFU, 1983).

That these problems actually affect the standard *AE*(12) index is shown in Fig. 5 (from AKASOFU *et al.*, 1983) where the ratio of *AE*(12)/*AE*(71) is displayed versus *AE*(12). Besides the large scatter in the *AE*(12)/*AE*(71) ratio for a given *AE*(12) which reflects all three problems, one can see that especially below 250 nT this ratio is much less than unity. Obviously, in this range of weak disturbance the longitudinal gaps and the restriction to latitudes below 70° become so severe that solar wind-magnetosphere coupling studies using the standard *AE*(12) index may yield misleading results.

Finally, besides the above limitations which are found in the construction and derivation of *AE*(12) there is one problem which cannot be overcome even by an ideal *AE* index like *AE*(71): For regions east or west of strong net field-aligned current, a significant part of the north–south component stems from these net field-aligned currents, thus apparently enhancing or decreasing the electrojet current. This effect is most pronounced behind the head of the westward travelling surge where strong field-aligned currents flow upward (e.g., BAUMJOHANN *et al.*, 1981; OPGENOORTH *et al.*, 1983), but appreciable effects can also arise east or west of strong net field-aligned current associated with a "normal" electrojet (e.g., BAUMJOHANN *et al.*, 1980; BAUMJOHANN and KAMIDE, 1981).

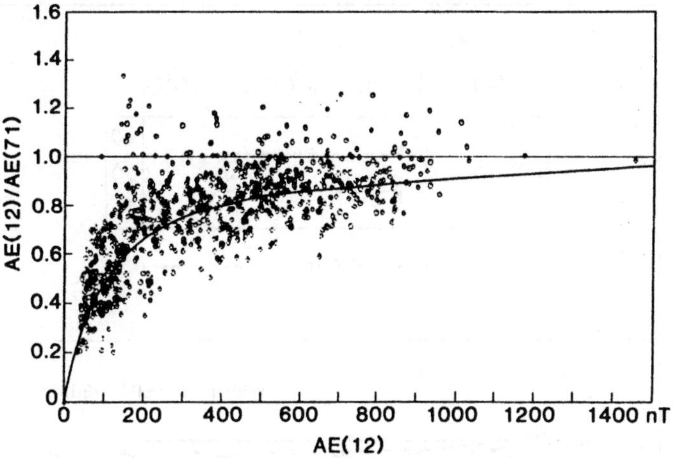

Fig. 5. The ratio $AE(12)/AE(71)$ as a function of $AE(12)$. Ratios greater than 1 are possible since $AE(12)$ is based on H components while $AE(71)$ is based on eccentric dipole north–south components (after AKASOFU *et al.*, 1983).

Figure 6 illustrates this effect for a typical substorm current wedge (1000 km east–west extent) carrying 0.6 MA. Especially under the current-carrying region the eastward ground equivalent currents (i.e., southward magnetic perturbations) reach only about 600 nT, since the southward magnetic perturbation stemming from the ionospheric part of the currnt wedge (roughly 900 nT) is reduced by about one third due to the northward magnetic perturbation (eastward equivalent current of about 300 nT) associated with the field-aligned currents.

3. The Ring Current D_{st} Index

The ring current index D_{st} was introduced by SUGIURA (1964) as a measure of the ring current magnetic field and thus its total kinetic energy and first used by BURTON *et al.* (1975) in a solar wind-magnetosphere coupling study. The present D_{st} index is based on hourly averages of the H component recorded at 4 low-latitude observatories: Honolulu, San Juan, Hermanus, and Kakioka. All four observatories are 20–30° away from the dipole equator to minimize equatorial electrojet effects and are about evenly distributed in local time. At each observatory one obtains a magnetic perturbation amplitude by subtracting an LT-dependent quiet time reference level and the Sq field from the hourly averaged H component. All four magnetic perturbation amplitudes are then averaged (to reduce local time effects) and multiplied with the average of the cosines of the observatories' dipole latitudes to obtain D_{st}, namely the value of the ring current field at the dipole equator:

$$D_{st} = \frac{1}{16} \cdot \sum_4 \cos \Lambda_i \cdot \sum_4 (H - H_0 - Sq)_i .$$

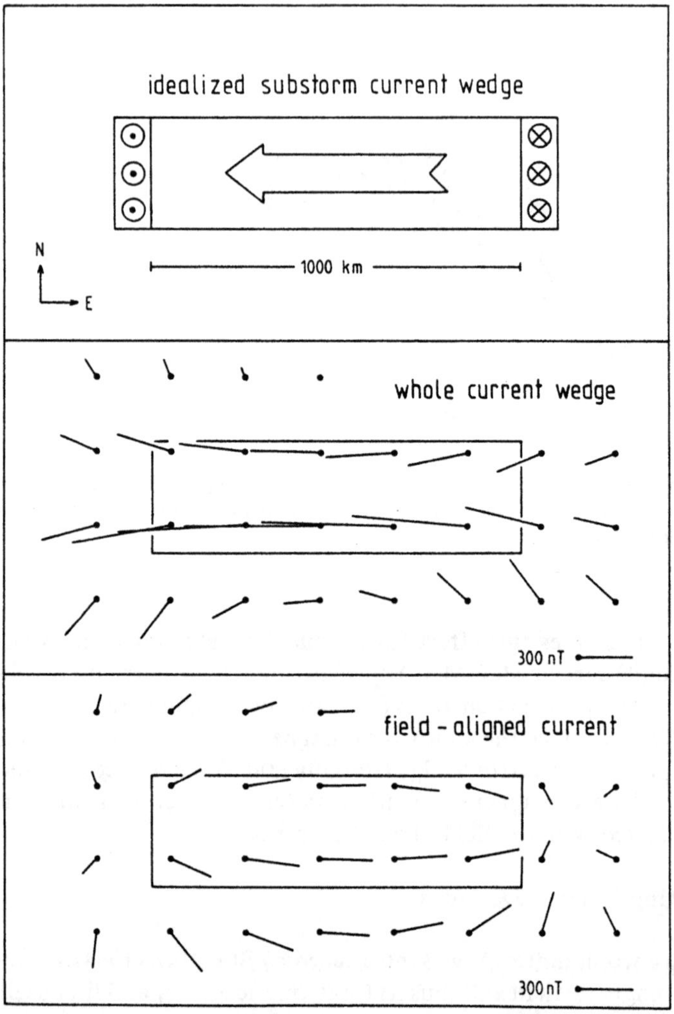

Fig. 6. Upper panel: Typical substorm current wedge carrying 0.6 MA (uniform westward ionospheric current density of 2 A/m at 100 km altitude; uniform upward and downward field-aligned current density of 20 $\mu A/m^2$ at western and eastern edge, respectively). Middle panel: Ground equivalent currents due to the whole current wedge. Lower panel: Ground equivalent current due to field-aligned part of the current wedge.

In contrast to the $AE(12)$ index, where most of the uncertainties lie in uneven and too widely spaced station network, the D_{st} uncertaintites are mainly caused by magnetic contributions of sources other than the ring current to the H component at the four D_{st} observatories, namely (e.g., ROSTOKER, 1972; CLAUER *et al.*, 1980; AKASOFU, 1981): (i) magnetopause current; (ii) asymmetric ring current; and (iii) substorm current wedge.

The magnetopause current is regulated by the solar wind pressure (e.g., NISHIDA, 1978) and its magnetic perturbation peaks around noon. A typical magnetopause contribution is, of course, included in the quiet time reference level, but dynamic variations of the solar wind pressure can yield positive or negative deviations from this average. In particular, sudden changes associated with interplanetary shock fronts (so-called sudden impulses) can change the local time average of the magnetic perturbation stemming from the magnetopause current by typically 10–40 nT, as illustrated in Fig. 7 (after NISHIDA, 1978).

The two other current systems which induce LT-asymmetric magnetic disturbances at low latitudes are illustrated in Fig. 8 (after CLAUER *et al.*, 1980), namely the asymmetric or partial ring current and the substorm current wedge. The westward partial ring current prevails in the afternoon sector and its closure via field-aligned current causes a southward magnetic disturbance around dusk. The substorm current wedge stems from short-circuiting of the cross-tail current and dominates in the midnight and early morning sector. Here, again, mainly the field-aligned currents provide a significant northward magnetic disturbance.

The combined effect of these three variable current systems yields an uncertainty of the quiet time reference level as shown in Fig. 9. The uncertainty or error has its

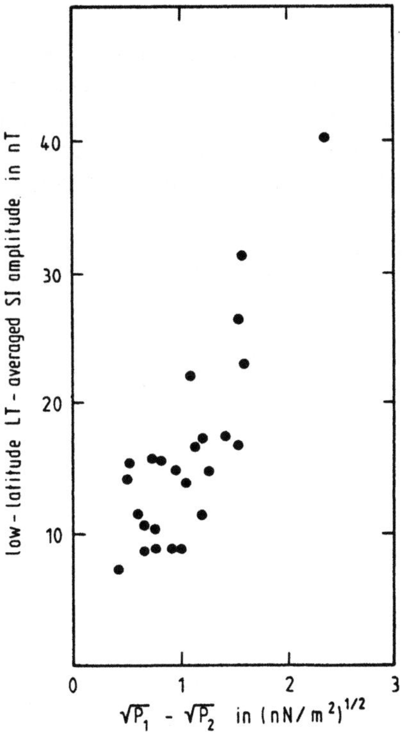

Fig. 7. Local-time averaged magnetic disturbance at low latitudes due to sudden increases in solar wind pressure (after NISHIDA, 1978).

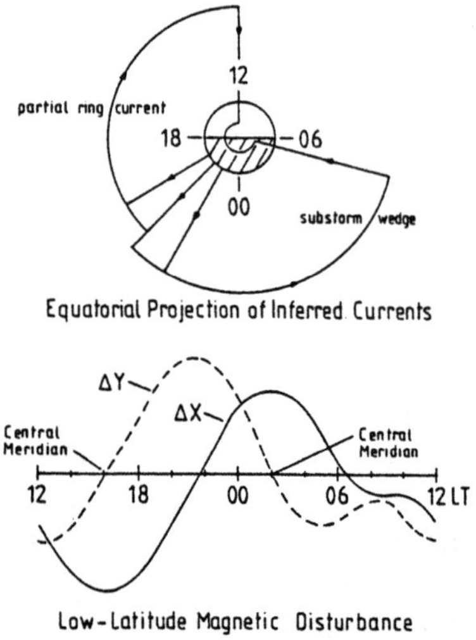

Fig. 8. Schematic equatorial projection of asymmetric partial ring current and substorm current wedge and local time distribution of associated N–S (ΔX) and E–W (ΔY) magnetic disturbances at low latitudes (after CLAUER *et al.*, 1980). Also indicated are central meridians of partial ring current (16 LT) and current wedge (02 LT).

maximum around noon, where it may reach up to 50% of a typical D_{st} of 50 nT. Hence, the influence of solar wind pressure changes seems to be dominant for errors in the quiet time level. The relatively small error around midnight probably means that partial ring current and substorm current wedge may often be closer together than indicated in Fig. 8 (both shifted towards the evening sector) so that their positive and negative ΔX disturbances tend to cancel each other.

Finally, there is one more unresolved problem which limits the use of the D_{st} index in solar wind-magnetosphere coupling studies. This problem is not so much

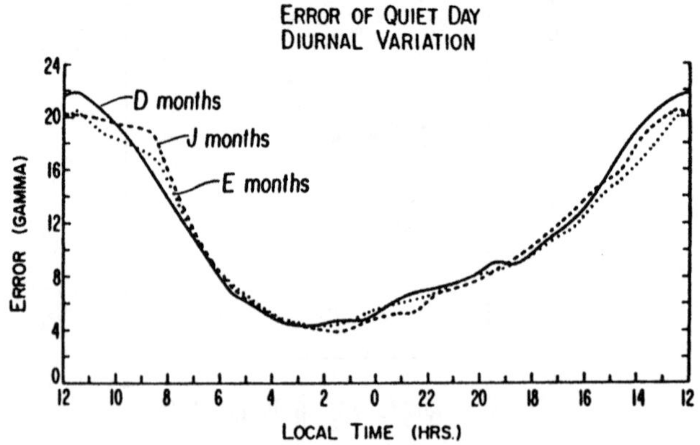

Fig. 9. Magnitude of the uncertainty of the quiet time low-latitude magnetic field vs. local time (after CLAUER *et al.*, 1980).

related to the construction and derivation of the index but rather to its use in solar wind-magnetosphere coupling studies: the D_{st} index is proportional to the total kinetic energy of the ring current (e.g., STERN, 1984), but coupling studies are interested in the rate of energy injection, since the energy of the ring currnt particles is thought of as dissipated (the actual loss occurs over hours or days through charge exchange with neutrals).

The injection into the ring current U_R can be written as (e.g., AKASOFU, 1981):

$$U_R = A\left(\frac{\partial D_{st}}{\partial t} + \frac{D_{st}}{\tau}\right)$$

where A is a constant and τ the lifetime of the ring current particles. The problem is that τ is not a constant but changes with D_{st} (AKASOFU, 1981; FELDSTEIN *et al.*, 1984) and even differs for different ion species (TINSLEY and AKASOFU, 1981). A typical range of values between 1 h and 10 hrs for τ can be found in the literature and Fig. 10 illustrates the effect of using (D_{st}-dependent) different τ's. Both models (BURTON *et al.*, 1975; FELDSTEIN *et al.*, 1984) predict D_{st} from solar wind parameters (including a correction for changing solar wind pressure) and get reasonable agreement between observed and modeled D_{st}. However, there are differences of more than 10 nT between the models which at least partially reflect differences and thus uncertainties in the value of τ.

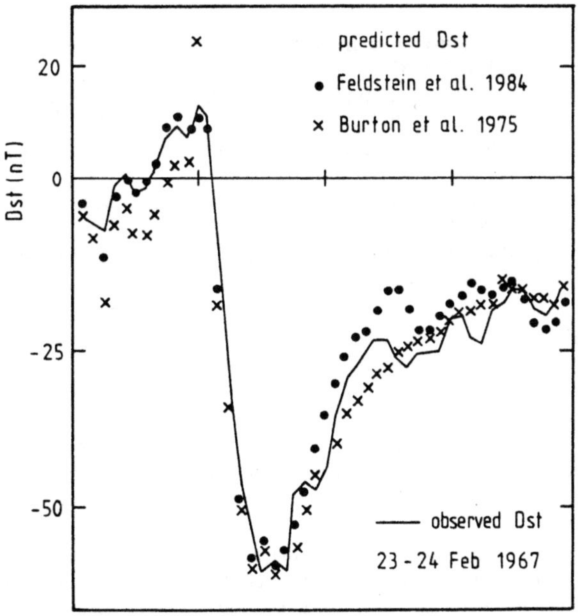

Fig. 10. Observed and predicted D_{st} values for 23–24 February 1967. The predictions given by crosses are based on the model of BURTON *et al.* (1975) while those indicated by dots use the FELDSTEIN *et al.* (1984) model (after FELDSTEIN *et al.*, 1984).

4. Conclusions

While the geomagnetic indices AE and D_{st} have proven to be very useful for solar wind-magnetosphere coupling studies, there are a couple of problems which limit their use. Some of these seem unavoidable, like too few standard AE observatories and varying contributions of field-aligned currents to AE and D_{st}, and one has to keep them in mind when interpreting correlations with solar wind parameters. Two others could be minimized by improving the indices (use of eccentric dipole coordinates instead of local H, D, Z coordinate system for AE and correction of D_{st} for solar wind pressure effects). The last problem (insufficient knowledge about the lifetime of ring current particles) remains to be solved through better understanding of the physics of the ring current.

I would like to thank J. Behrens (Universität Münster) for his help in the preparation of Fig. 6 and the Deutsche Forschungsgemeinschaft for financial support through a Heisenberg fellowship.

REFERENCES

AHN, B.-H., S.-I. AKASOFU, and Y. KAMIDE, The Joule heat production rate and the particle energy injection rate as a function of the geomagnetic indices AE and AL, *J. Geophys. Res.*, **88**, 6275–6283, 1983.

AKASOFU, S.-I., Energy coupling between the solar wind and the magnetosphere, *Space Sci. Rev.*, **28**, 121–190, 1981.

AKASOFU, S.-I., Solar wind disturbances and the solar wind-magnetosphere energy coupling function, *Space Sci. Rev.*, **34**, 173–183, 1983.

AKASOFU, S.-I., B. H. AHN, Y. KAMIDE, and J. H. ALLEN, A note on the accuracy of the auroral electrojet indices, *J. Geophys. Res.*, **88**, 5769–5772, 1983.

ARNOLDY, R. L., Signature in the interplanetary medium for substorms, *J. Geophys. Res.*, **76**, 5189–1201, 1971.

BAUMJOHANN, W., Ionospheric and field-aligned current systems in the auroral zone: A concise review, *Adv. Space Res.*, **2**, 55–62, 1983.

BAUMJOHANN, W. and G. HAERENDEL, Magnetospheric convection observed between 06 and 21 LT: Solar wind and IMF dependence, *J. Geophys. Res.*, **90**, in press, 1985.

BAUMJOHANN, W. and Y. KAMIDE, Joint two-dimensional observations of ground magnetic and ionospheric electric fields associated with auroral zone currents, 2. Three-dimensional current flow in the morning sector during substorm recovery, *J. Geomag. Geoelectr.*, 33, 297–318, 1981.

BAUMJOHANN, W. and Y. KAMIDE, Hemispherical Joule heating and the AE indices, *J. Geophys. Res.*, **89**, 383–388, 1984.

BAUMJOHANN, W., J. UNTIEDT, and R. A. GREENWALD, Joint two-dimensional observations of ground magnetic and ionospheric electric fields associated with auroral zone currents, 1. Three-dimensional current flows associated with a substorm-intensified eastward electrojet, *J. Geophys. Res.*, **85**, 1963–1978, 1980.

BAUMJOHANN, W., R. J. PELLINEN, H. J. OPGENOORTH, and E. NIELSEN, Joint two-dimensional obsevations of ground magnetic and ionospheric electric fields associated with auroral zone currents: Current systems associated with local auroral breakups., *Planet. Space Sci.*, **29**, 431–447, 1981.

BURTON, R. K., R. L. MCPHERRON, and C. T. RUSSELL, An empirical relationship between interplanetary conditions and D_{st}, *J. Geophys. Res.*, **80**, 4204–4214, 1975.

CLAUER, C. R., R. L. MCPHERRON, and M. G. KIVELSON, Uncertainty in ring current parameters due to the quiet magnetic field variability at mid-latitudes, *J. Geophys. Res.*, **85**, 633–643, 1980.

Cowley, S. W. H., The causes of convection in the earth's magnetosphere: A review of developments during the IMS, *Rev. Geophys. Space Phys.*, **20**, 531–565, 1982.

Davis, T. N. and M. Sugiura, Auroral electrojet index *AE* and its Universal Time variations, *J. Geophys. Res.*, **71**, 785–801, 1966.

Feldstein, Y. I., V. Y. Pisarsky, N. M. Rudneva, and A. Grafe, Ring current simulation in connection with interplanetary space conditions, *Planet. Space Sci.*, **32**, 975–984, 1984.

Haerendel, G., G. Paschmann, N. Sckopke, H. Rosenbauer, and P. C. Hedgecock, The frontside boundary layer of the magnetosphere and the problem of reconnection, *J. Geophys. Res.*, **83**, 3195–3216, 1978.

Harel, M., R. A. Wolf, R. W. Spiro, P. H. Reiff, C.-K. Chen, W. J. Burke, F. J. Rich, and M. Smiddy, Quantitative simulation of a magnetospheric substorm, 2. Comparison with observations, *J. Geophys. Res.*, **86**, 2242–2260, 1981.

Hill, T. W., Solar-wind magnetosphere coupling, in *Solar-Terrestrial Physics*, edited by R. L. Carovillano and J. L. Forbes, pp., 261–302, D. Reidel, Dordrecht, 1983.

Kamide, Y. and S.-I. Akasofu, Latitudinal cross sections of the auroral electrojet and its relation to the interplanetary magnetic field polarity, *J. Geophys. Res.*, **79**, 3755–3771, 1974.

Kamide, Y. and S.-I. Akasofu, Notes on the auroral electrojet indices, *Rev. Geophys. Space Phys.*, **21**, 1647–1656, 1983.

Kamide, Y. and W. Baumjohann, Estimation of electric field and currents from IMS magnetometer data for the CDAW-6 intervals: Implications for substorm dynamics, *J. Geophys. Res.*, **90**, 1305–1317, 1985.

Kamide, Y., B.-H. Ahn, S.-I. Akasofu, W. Baumjohann, E. Friis-Christensen, H. W. Kroehl, H. Maurre, A. D. Richmond, G. Rostoker, R. W. Spiro, J. K. Walker, and A. N. Zaitsev, Global distribution of ionospheric and field-aligned currents during substorms as determined from six IMS meridian chains of magnetometers: Initial results, *J. Geophys. Res.*, **87**, 8228–8240, 1982.

Lemaire, J. and M. Roth, Penetration of solar wind plasma elements into the magnetosphere, *J. Atmos. Terr. Phys.*, **40**, 331–335, 1978.

Mayaud, P. N., *Derivation, Meaning, and Use of Geomagnetic Indices*, A.G.U., Washington, 1980.

Miura, A., Anomalous transport by magnetohydrodynamic Kelvin-Helmholtz instabilities in the solar wind-magnetosphere interaction, *J. Geophys. Res.*, **89**, 801–818, 1984.

Mozer, F. S., Electric field evidence on the viscous interaction at the magnetopause, *Geophys. Res. Lett.*, **11**, 135–138, 1984.

Nishida, A., *Geomagnetic Diagnosis of the Magnetosphere*, Springer, Berlin, 1978.

Opgenoorth, H. J., R. J. Pellinen, W. Baumjohann, E. Nielsen, G. Marklund, and L. Eliasson, Three-dimensional current flow and particle precipitation in a westward travelling surge (observed during the Barium-Geos rocket experiment), *J. Geophys. Res.*, **88**, 3138–3152, 1983.

Pellinen, R. J., W. Baumjohann, W. J. Heikkila, V. A. Sergeev, A. G. Yahnin, G. Marklund, and O. A. Melnikov, Event study on presubstorm phases and their relation to the energy coupling between solar wind and magnetosphere, *Planet. Space Sci.*, **30**, 371–388, 1982.

Podgorny, I. M., E. M. Dubinin, and Y. N. Potanin, The magnetic field on the magnetospheric boundary from laboratory simulation data, *Geophys. Res. Lett.*, **4**, 207–210, 1978.

Reiff, P. H., R. W. Spiro, and T. W. Hill, Dependence of polar cap potential drop on interplanetary parameters, *J. Geophys. Res.*, **86**, 7639–7648, 1981.

Rostoker, G., Geomagnetic indices, *Rev. Geophys. Space Phys.*, **10**, 935–950, 1972.

Saunders, M. A., Recent ISEE observations of the magnetopause and low latitude boundary layer: A review, *J. Geophys.*, **52**, 190–198, 1983.

Sonnerup, B. U. Ö., G. Paschmann, I. Papamastorakis, N. Sckopke, G. Haerendel, S. J. Bame, J. R. Asbridge, J. T. Gosling, and C. T. Russell, Evidence for magnetic field reconnection at the earth's magnetopause, *J. Geophys. Res.*, **86**, 10049– 10067, 1981.

Stern, D. P., Energetics of the magnetosphere, *Space Sci. Rev.*, **39**, 193–213, 1984.

Sugiura, M., Hourly values of equatorial D_{st} for the IGY, *Ann. Int. Geophys. Year*, **35**, 1964.

Tinsley, B. A. and S.-I. Akasofu, A note on the lifetime of the ring current particles, *Planet. Space Sci.*, **30**, 733–740, 1982.

Solar Wind-Magnetosphere Coupling, edited by Y. Kamide and J. A. Slavin, 17–38.

Statistical Analyses in the Study of Solar Wind-Magnetosphere Coupling

D. N. BAKER

University of California, Los Alamos National Laboratory, Los Alamos, New Mexico 87545, U.S.A.

Statistical analyses provide a valuable method for establishing initially the existence (or lack of existence) of a relationship between diverse data sets. Statistical methods also allow one to make quantitative assessments of the strengths of observed relationships. This paper reviews the essential techniques and underlying statistical bases for the use of correlative methods in solar wind-magnetosphere coupling studies. Techniques of visual correlation and time-lagged linear cross-correlation analysis are emphasized, but methods of multiple regression, superposed epoch analysis, and linear prediction filtering are also described briefly. The long history of correlation analysis in the area of solar wind-magnetosphere coupling is reviewed with the assessments organized according to data averaging time scales (minutes to years). It is concluded that these statistical methods can be very useful first steps, but that case studies and various advanced analysis methods should be employed to understand fully the average response of the magnetosphere to solar wind input. It is clear that many workers have not always recognized underlying assumptions of statistical methods and thus the significance of correlation results can be in doubt. Long-term averages ($\gtrsim$1 hour) can reveal gross relationships, but only when dealing with high-resolution data (1–10 min) can one reach conclusions pertinent to magnetospheric response time scales and substorm onset mechanisms.

1. Introduction

It is often said that there are three forms of untruth: lies, damned lies, and statistics. Statistical methods may often be employed to cover up the truth, but they also can be most illuminating and valuable if used properly. Statistical studies have played an especially important role in the study of solar wind-magnetosphere coupling and, thus, it is important to review some of the principal results of such studies.

As noted, for example, by REIFF (1983), statistical analyses are commonly used in space physics both to search initially for patterns of relationship ("exploratory studies") and to verify an assumed theoretical relationship ("confirmatory studies"). In the early years of solar wind-magnetosphere coupling (SMC) studies, the use of statistical methods fell primarily into the category of exploratory research (e.g., SNYDER *et al.*, 1963). People continue to search for the "best" empirical SMC parameter(s) today—still using statistical methods—but theoretical ideas also are being constantly evaluated by statistical means (e.g., VASYLIUNAS *et al.*, 1982; BARGATZE *et al.*, 1985). Statistical methods, in this latter case, become the touchstone for deciding between different theoretical approaches.

In the present review, we will deal mostly with correlation analyses, although several other common methods will also be described briefly. In the course of our review, we will attempt to comment critically on the issues of "best" SMC parameters, the value (or weaknesses) of various data averaging time scales, the importance of determining confidence estimates, and the proper role of statistical methods in the overall approach to understanding magnetospheric response to solar wind variability.

2. Statistical Uncertainties in Observational Results

There are two components of the real world that make statistical analyses desirable and, indeed, crucial to our descriptions of physical relationships. The first of these is that we can not make perfect measurements of any quantity no matter how carefully we try. One may understand this by thinking of all the small, uncontrollable things that get in our way as we try to make a set of otherwise identical measurements. The result is that for any ensemble of observations, we get a range of values whose mean value (barring systematic errors) approaches the "true" value as the number of measurements goes to infinity.

A second feature of the physical world is that in complicated systems (such as the magnetosphere) there may be many parameters that affect a given "output" in a putative cause-and-effect relationship. For example, the relation between the interplanetary electric field and the cross-magnetospheric electric field may depend not only on dayside magnetic merging rates, but also on varying ionospheric conductivities, magnetospheric plasma compositions, previous ring current history, etc. Thus, even though we might feel we have excellent measures of global electric fields, the relationship determined will not be one-to-one because of covariant relationships over which there is little knowledge or control.

In general, measurement errors are well-described by the normal error (or Gaussian) distribution (BEVINGTON, 1969; WORTHING and GEFFNER, 1950). This function is an approximation to the binomial distribution as the number of observations approaches infinity. As relates to the discussion of the previous paragraphs, it is useful because it describes the distribution of random observations in most experimental situations and it provides reasonable estimations of the parameters of most probability distributions. The Gaussian curve may be expressed as

$$y = \frac{1}{\sigma\sqrt{2\pi}} \exp\left(-\frac{1}{2}\left(\frac{x-\mu}{\sigma}\right)^2\right) . \quad (1)$$

Here, y is a probability or frequency of occurrence, μ is the mean value of the parent population from which the parameter x is drawn, and σ is the standard deviation of the parent distribution. The curve is a bell-shaped function symmetric about the mean value of x.

There are a number of interesting derivations and justifications for the normal error curve (e.g., WORTHING and GEFFNER, 1950), but one may also simply appeal to plausibility arguments and to its empirically replicated form (BEVINGTON, 1969). An interesting mechanical construction that illustrates the normal error curve is Galton's quincunx (WORTHING and GEFFNER, 1950). As shown in Fig. 1, the apparatus consists of an arrangement of small pegs above a horizontal set of collection bins. Small beads or shot fall through a hole in a hopper at the top and receive a series of small deflections from the pegs as they fall downward. The sidewise deflections occur with equal likelihood to the right and the left, and are the counterpart to small random errors discussed above. The quincunx device, when set in motion, thereby produces a distribution of shot in the collection bins which approximate the normal error curve. The machine thus may be thought of as a mechanical analog of many other systems where small errors contribute to a final measurement uncertainty.

In addition to the Gaussian distribution, sometimes the distribution of possible data values will be described by the Poisson distribution (BEVINGTON, 1969). The Poisson curve is an approximation to the binomial distribution when the mean probability of a given measurement is much smaller than the total possible number (i.e., the probability is $\ll 1$ that any given value will occur). The Poisson distribution is given by

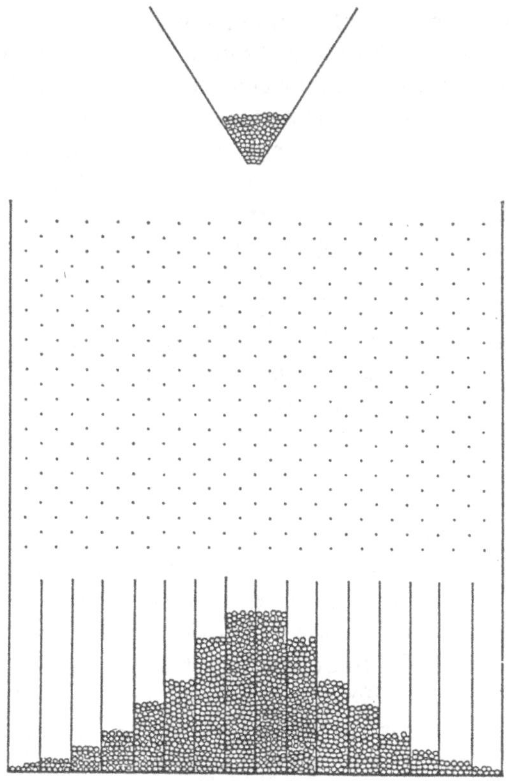

Fig. 1. Galton's quincunx, a device for illustrating the physical basis of the normal frequency distribution law (from WORTHING and GEFFNER (1950); used by permission of the publishers).

$$y=\frac{\mu^x}{x!}\, e^{-\mu} \tag{2}$$

where the parameters have the same meaning as in Eq. (1). This kind of distribution is generally appropriate in a counting experiment where data represent the number of counts per unit time.

In doing statistical analyses (e.g., least-squares regressions), one generally must take into account measurement uncertainties in the data. Either Gaussian or Poisson distributions of uncertainties will ordinarily be appropriate depending on the nature of the measurements.

3. Visual Correlation Analyses

Very often researchers will notice a seeming relationship between two or more parameters which happen to be plotted together in some way. For example, two variables may be plotted as a function of time and they may vary in a quite similar fashion or they may change in nearly opposite fashions. The human eye and brain are rather good at surveying such plotted data and spotting patterns of relationship. We term this ability to see a related behavior in parameters as "visual cross-correlation". This sort of analysis often leads to serendipitous discovery of physical relationships.

It is quite common to use visual correlation techniques in SMC studies. The time series representing various solar wind parameters (V, B, B_z, etc.) are often plotted on the same time scale as a measure of geomagnetic activity (K_p, AE, etc.) and the reader is asked to notice the similarity of behavior (perhaps with a time delay) between the time series. In this usage, visual cross-correlation is cited as evidence (or proof) of a physical relation being advocated by an author.

An example of the use of the visual correlation method is shown in Fig. 2 taken from PERREAULT and AKASOFU (1978). In the upper part of the figure are shown the values of B_z, VB_z and hourly sums of the southward B_z component ($\Sigma B_S t$) measured in the upstream solar wind. In the lower portion of the top panel are magnetospheric indices (c.f., BAUMJOHANN, 1986, this volume): AE, AL, Dst, and $ASYM$. The lower panel of Fig. 2 shows the value of estimated solar wind input energy ($\varepsilon(t)$) and the estimated magnetospheric dissipated energy ($U(t)$) on the same scales. The authors of the paper note the generally similar behavior of $\varepsilon(t)$ and $U(t)$ and conclude thereby that a "good" and meaningful relationship exists between ε and U.

Many other studies dealing with solar wind-magnetosphere coupling have used the visual correlation technique in order to assess the issue of cause-and-effect relationship, magnetospheric response time scales, and so forth. Among such studies are those of FAIRFIELD and CAHILL (1966), TSURUTANI and MENG (1972), AKASOFU and CHAO (1979), AKASOFU (1980), and AKASOFU *et al.* (1982).

In addition to the concurrent plotting of several parameter time series, visual correlation analyses may also be presented in the form of scatter plots of one parameter versus another. Without a linear or nonlinear regression fit, the reader is then asked to judge the degree of correlative behavior (or lack of it).

It may be argued that visual cross-correlation is the most primitive form of

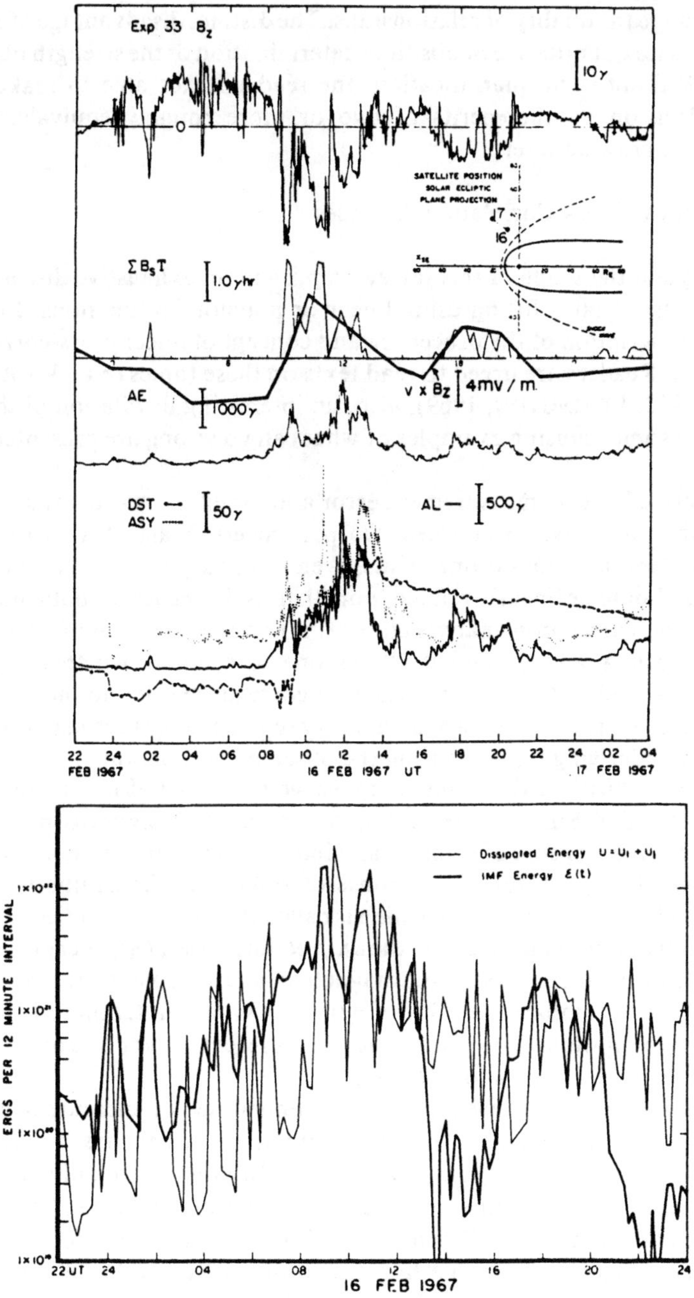

Fig. 2. Various interplanetary and geomagnetic data sets plotted on similar time scales for period in February 1967. These plots illustrate the visual correlation method discussed in the text (from PERREAULT and AKASOFU, 1978).

statistical analysis, but it is in some sense also the most "honest" since the viewer is allowed to judge the quality of relationships. The distinct disadvantage of the method is that there is no quantitative or absolute determination of the strength of correlative behavior. Without such quantification, the reader is not able to make any truly informed decision on the merits of two or more roughly equivalent apparent relationships in raw data sets.

4. Statistical Cross-Correlation Analyses

It is beyond the scope of this review to present an exhaustive discussion of the method of least-squares fitting using linear or polynomial functions. Furthermore only a brief description of the procedure and concept of linear cross-correlation will be given here. Readers are urged to read texts on these topics (e.g., WORTHING and GEFFNER, 1950; BEVINGTON, 1969); also, an interesting distillation of the essential methods and some amusing examples of what can go wrong are presented by REIFF (1983).

As described above, measurement error and/or unknown covariant factors can make the observed relation between two parameters X and Y very scattered and substantially less than one-to-one. Thus, even if we suspect on theoretical grounds that X and Y should be linearly related, nonetheless the measured data may be highly variable. Linear correlation analysis allows one to assess the strength of a correlation (if any) between X and Y, and it allows one to estimate whether the observed correlation is significant. We emphasize that correlation does not necessarily imply causality: all it says is that two parameters vary together and there may very well exist an unknown underlying factor causing the observed correlation.

In order to illustrate the point, suppose we have a set of measurements of two parameters X and Y. Since we are dealing with solar wind-magnetosphere coupling, let us assume X is a measure of some solar wind parameter and let us assume that Y is an hypothesized magnetospheric response of some sort. Generally, then, we may consider X to be a function of time which is called the *input* time series, $I(t)$, and Y is also a function of time which is called the *output* time series, $O(t)$. Very often there is a significant delay in the response of the magnetosphere to a solar wind variation and so it is appropriate to compare the value of $I(t)$ with $O(t)$ at a latter time. The *lag time* of O relative to I for optimum correlation may be represented by τ_L and one then assesses the correlation between $I(t)$ and $O(t+\tau_L)$.

As noted previously, when making a limited set of measurements, we are sampling the infinite assumed "parent population" which has a mean ($\mu \equiv \lim_{n\to\infty} (\frac{1}{n}\Sigma X_i)$) and a variance $[\sigma^2 \equiv \lim_{n\to\infty} (\frac{1}{n}(\Sigma X_i - \mu)^2)]$. In practice we only have available our limited set of measurements to estimate the experimental mean ($\bar{X} \equiv \frac{1}{n}\Sigma X_i$) and variance ($S^2 \equiv \frac{1}{n-1}\Sigma(X_i - \bar{X})^2$). When X and Y are correlated (but not related one-to-one), we can only speak of a given X having a most-probable value of Y around which other value are distributed. The closer that the observed values are to the most probable value, the more definite is the relationship of Y to X. It is this underlying assumption which is the basis for most measures of the degree of correlation.

Assuming then that the measured set of data is the most probable set, the most probable value of Y is given by least-squares fitting the values of Y versus X. This is called a regression analysis of Y on X. Conceptually, one can envision a change of variables by removal of the means: $x=(X-\bar{X})$ and $y=(Y-\bar{Y})$. Then, one has a simplified least-squares linear relation (e.g., WORTHING and GEFFNER, 1950)

$$y_c=\left(\frac{\Sigma xy}{\Sigma x^2}\right)x=bx \tag{3}$$

for the regression line. A measure of the scatter, or dispersion, of the observed values of y around the regression line is then given by

$$S_y=\sqrt{\frac{\Sigma(y-y_c)^2}{n}} \tag{4}$$

and is called the standard error of estimate. As may be seen, it is the square root of the average squared deviations from the least-squares linear fit through the data.

From this conceptual framework, one can define the correlation coefficient, r, as

$$r=\sqrt{1-\left(\frac{S_y}{\sigma_y}\right)^2} \tag{5}$$

As is clear, r is a dimensionless measure of the degree to which data points vary from the best fit line. As noted by REIFF (1983), the value of r can be thought of as the square root of the fraction of the variance (of y) that can be explained by the linear fit with x. Thus, $(1-r^2)$ is the fraction of the variance of y which *is not* explained by the linear correlation. Many correlations in SMC are in the 0.7–0.8 range. For such values of r, $(1-r^2)$ is in the range 0.4–0.5; hence, even with these high correlations, nearly half of the variance is not explained by the assumed relationship.

In practice, rather than using Eq. (4), it is more convenient to go back to the raw measured data and define the correlation coefficient as

$$r=\frac{n\Sigma X_iY_i-\Sigma X_i\Sigma Y_i}{\sqrt{(n\Sigma X_i^2-(\Sigma X_i)^2)(n\Sigma Y_i^2-(\Sigma Y_i)^2)}}\,. \tag{6}$$

We note that for simplicity we have *suppressed the weighting of the data* by which uncertainties of the X_i's and Y_i's should be taken into account explicitly (BEVINGTON, 1969). We also note that Bevington includes very convenient FORTRAN programs to do the correlation analysis. Notice that Equation (6) is quite useful in practice since only five moments are necessary to calculate r: ΣX_i, ΣY_i, ΣX_iY_i, ΣX_i^2, and ΣY_i^2.

In common usage for SMC studies, X has had values of V_{SW}, B, B_z, B_x, VB_z, VB_s, V^2B_s, ε, etc. in the solar wind. On the other hand, in such studies, Y has been K_p, AE, AL, AU, Dst, U_T, etc., and has usually been derived from magnetospheric indices or measurements. Combinations of parameters have generally been formed in order to *linearize* the assumed relationship before correlation analyses are performed.

We remind readers of the underlying assumptions of these derivations. First, it is assumed that X and Y are distributed normally about their mean (or least-squared) values. Secondly, it is assumed that the $X-Y$ relation is linear. There should be good theoretical (or other) grounds for this second assumption. In particular, if the $X-Y$ relationship is not linear, the meaning of r quickly becomes very unclear. Finally, it is normally assumed that the precision of X is much higher than that of Y; often in practice uncertainties in X can, in fact, be vested in Y.

As noted above, uncertainties in data should definitely be included in the computation of r. Many nonlinear relations (e.g., $y=ab^x$) can be linearized rather simply ($\log y=\log a+(\log b)x$). Attention should then be paid to whether the distributions of the data are log-normal or normal (in the raw, linear sense).

Let us finally consider the issue of confidence levels for a given computed value of cross correlation. Generally one can ask the question: Given a certain computed r, what is the probability that this is only spurious and the data are, in fact, uncorrelated? An answer to this question is shown in Fig. 3 (from BEVINGTON, 1969). It is a plot of r versus the number of independent observations made. The several curves P show the probabilities at the 0.1, 0.05, etc., level that variables are not correlated. For example, from the figure one can determine that with 50 observations and a value of $r \sim 0.6$, then the probability is <0.001 that the parameters are uncorrelated. Worded differently, the probability is $>99.9\%$ in such a case that a true correlation exists.

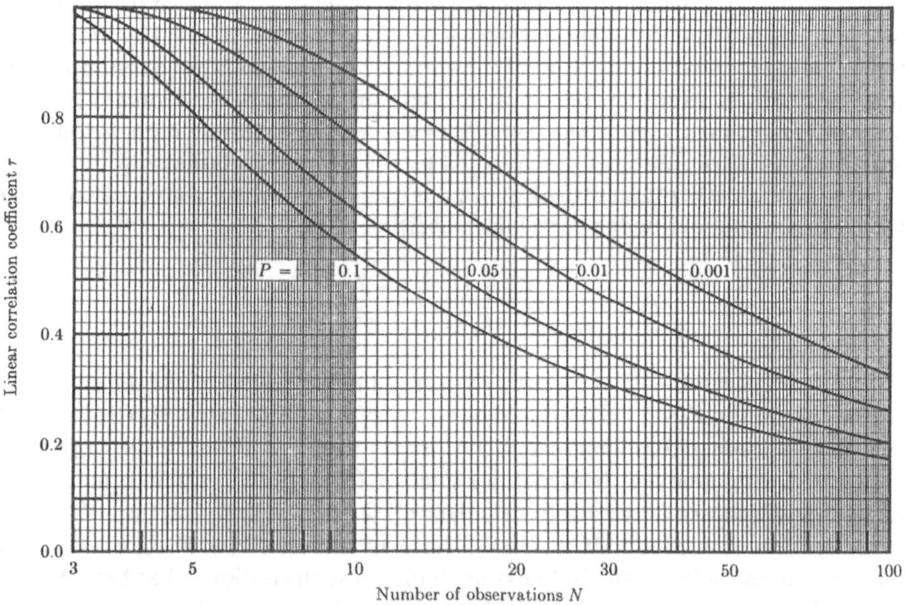

Fig. 3. The linear correlation coefficient r versus the number of observations N. The solid curves give the probability P that variables exhibiting a given correlation coefficient are not correlated (from BEVINGTON (1969); used by permission of the publishers).

4.1 Time scales for correlation analyses

Solar wind-magnetosphere coupling studies reported in the literature have utilized data averages ranging from years down to minutes. Figure 4 shows a quasilogarithmic time line marking some of the primary (and convenient) averaging intervals commonly employed in correlative analyses. Time scales greater than a solar cycle could in principle be employed but at such large times, it is not clear that one is learning much about the behavior of solar wind influence on magnetosphere behavior: Such time scales are beyond meaningful averaging ranges for these studies. At the other extreme, for data sampling at <1 minute, one is constrained by information transit times (e.g., Alfvén time scales) in the magnetosphere. Thus, uses of averages that are smaller than ~ 1 min also are probably not meaningful.

As noted in the figure, it is our opinion that data averages $\gtrsim 1$ hr make accessible information about gross coupling relationships between the solar wind and the magnetosphere, but only by using data in the $1 \sim 10$ min range can one probe the detailed coupling relationships. In the one- to multi-hour range, parcels of solar wind plasma have moved from the dayside of the magnetosphere to well beyond the explored reaches of the distant tail and one is, therefore, looking at interactions in the broadest sense. Similarly, time averages of data much less than $1 \sim 5$ min are at such high values in the frequency domain that one is probably largely studying "noise" in the input and/or output time series.

It is noted in Fig. 4 that geomagnetic storm time scales are in the many-hour to

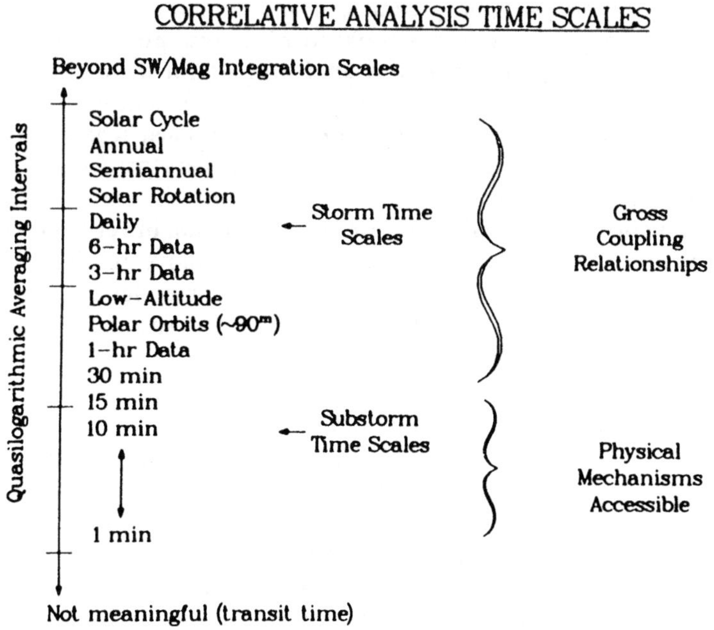

Fig. 4. A quasilogarithmic time scale illustrating commonly used data averaging intervals in solar wind magnetosphere coupling studies.

daily range. Substorm time scales, on the other hand, are in the several minute range. Since many studies of SMC are concerned with substorm processes, we would argue that correlative analyses should use relatively high resolution data. More will be said about this later.

It is a legitimate question to ask if long term solar wind coupling behavior is a superposition of a lot of short-term effects, or if different physical processes come into play at the longer time scales. If the former is true, then much information is lost using long averages, while in the latter case the long averaging is entirely appropriate. It is probably true that many features (ring current growth and decay, radiation belt diffusion, etc.) occur on slow enough time scales that longer averages are justified and, indeed, proper. On the other hand, for substorm studies one is constrained to use high time resolution data.

4.2 Correlation studies on ≳ annual time scales

There are many studies in the literature describing semiannual and annual average measurements of solar wind velocities, densities, temperatures, etc. (e.g., GOSLING *et al.*, 1971, 1976). These variations have been considered in terms of the 11-year sunspot activity cycle and, upon this basis, the relationship between solar activity and solar wind conditions at 1 AU have been well-documented (e.g., SHEELEY *et al.*, 1976).

Taking long-term averages of solar wind properties as input, CROOKER *et al.* (1977) examined the very long time scale response of the magnetosphere to such gross solar wind variations. The planetary index *Ap* averaged over yearly time scales was compared to various solar wind parameters (and combinations thereof). Figure 5 shows some of the primary results of Crooker *et al.* In the upper portion of the figure, the average solar wind speed ($\bar{V}$) is compared with $\bar{Ap}$ for the period 1962 through 1975. The authors found a relation $\bar{Ap}=10^{-5}\bar{V}^{2.3}$ with a correlation coefficient of 0.81. An improvement in the relation was found by doing a linear regression between $\bar{V}^2$ and *Ap* (see the lower part of Fig. 5). This analysis yielded $\bar{Ap}=7\times10^{-5}\bar{V}^2-1.8$ and a correlation coefficient of 0.86.

Such high correlation coefficients are not uncommon in very long-term studies (see papers in the review by FEYNMAN, 1983). As noted above, results such as those of CROOKER *et al.* (1977) are of considerable interest for understanding gross solar wind-magnetosphere relationships, but they have little to say about magnetospheric responses on substorm time scales.

4.3 Correlation studies on 3~24 hour time scales

Interplanetary magnetic field (IMF) parameters change significantly on time scales of minutes and even seconds. On the other hand, solar wind plasma parameters typically change significantly on time scales of 10's of minutes to hours. It has been quite common for workers to average solar wind, IMF, and planetary activity indices for periods of many hours to study coupling problems.

Contributing significantly to the prevalence of studies in this class has been the common availability of both interplanetary and geomagnetic activity data sets averaged over periods of 3, 6, 12 and 24 hours. For example, one of the most

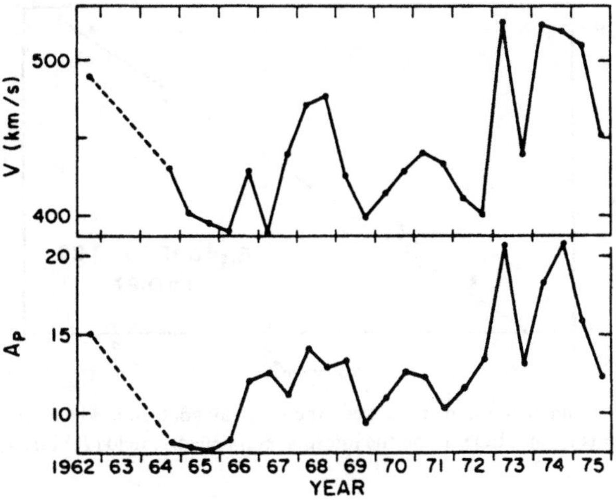

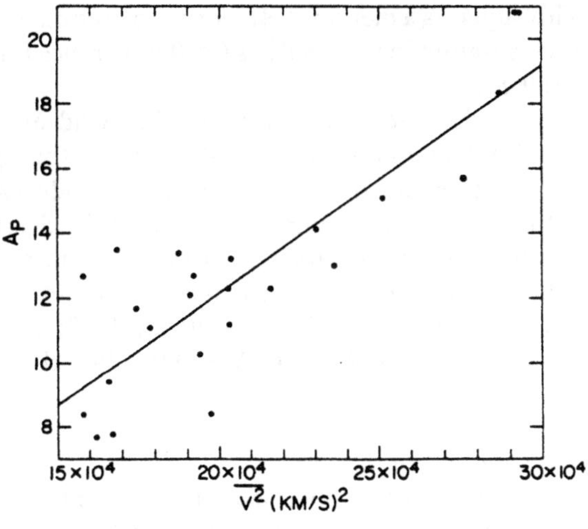

Fig. 5. Comparisons of yearly averaged values of Ap, V, V^2 showing the gross relationships between solar wind variations and geomagnetic response (from CROOKER *et al.*, 1977).

commonly used indices, K_p, is presented on 3- and 24-hour time scales. Furthermore, many spacecraft data sets are conveniently summarized on such time scales (e.g., ~100 min for low-altitude orbits, or daily for the geosynchronous orbit.

An example of the use medium scale data for SMC studies is shown in Fig. 6 (from LEI *et al.* (1981)). The solar wind electric field y-component ($E_y = VB_z$) for average periods of 3 to 6 hours is plotted versus the magnetospheric electric field (E_M) inferred from the plasmapause location as determined by GEOS-1. Such a measurement is taken by the authors to provide a direct measure of the efficiency of

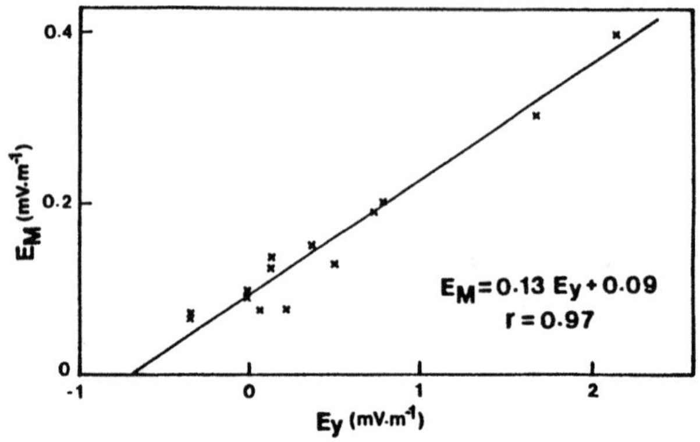

Fig. 6. A scatter plot and regression fit between the cross-magnetospheric electric field (E_M) inferred from measurements from GEOS-1 and the interplanetary electric field (E_y) inferred using VB_z (from LEI *et al.*, 1981).

dayside magnetic reconnection. As shown in Fig. 6, the result was a relation given by $E_M=0.13E_y+0.09$. This suggests a merging "efficiency" on the dayside of 13% and the high-degree of linear correlation in this analysis ($r=0.97$) argues for a close coupling relation between E_y and E_M.

Again, such relatively long-term averages of solar wind and magnetospheric properties are telling about coupling on storm-like scales (but probably not on substorm scales). A great strength of such studies is the wide variety of data on planetary-wide basis that are available going back in time for decades, or more. As illustrated by the example above, such data have been quite commonly used and very quantitative relationships have been developed in many instances (e.g., SNYDER *et al.*, 1963; OLBERT, 1968; BALLIF *et al.*, 1969; BOBROV, 1973; SVALGAARD, 1977; MURAYAMA *et al.*, 1980; GALLAGHER and D'ANGELO, 1981; HOLZER and SLAVIN, 1982).

4.4 Correlation studies on ~1-hour time scales

Some of the truly seminal work on solar wind-magnetosphere coupling has been based upon excellent data sets of hourly-averaged quantities. With such data, one is beginning to approach (c.f., Fig. 4) the time scales of substorm development.

An example of a SMC correlation study using hourly data is the classic work of ARNOLDY (1971). Figure 7 shows several of the lagged cross-correlation results obtained by Arnoldy using various parameters upstream (note the different plotting symbols) compared with the hourly-averaged *AE* index. As seen in the upper part of Fig. 7, solar wind and IMF parameters such as density (n), velocity (V), and total field strength (B) gave relatively unimpressive correlation results. In contrast, the sum of the southward (or negative) part of the IMF B_z for the 1-hour interval preceding the *AE* averaging period (denoted by $\Sigma B_s\tau$) gave an impressively high correlation (~0.8) with *AE*. This result, borne out by many subsequent studies, suggested that substorm activity (as measured by *AE*) is directly and closely related to the amount of

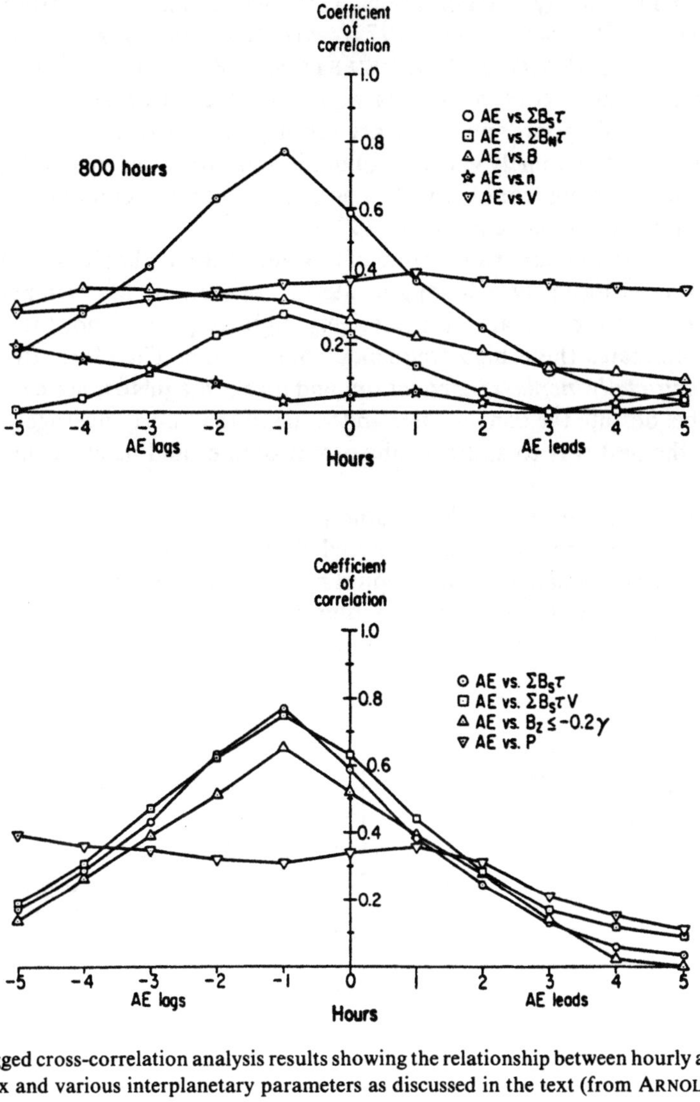

Fig. 7. Lagged cross-correlation analysis results showing the relationship between hourly averages of the *AE* index and various interplanetary parameters as discussed in the text (from ARNOLDY, 1971).

southward B_z in the interplanetary medium in the hour or so before a substorm expansive phase.

The lower part of Fig. 7 shows further results from the Arnoldy study. Convolving $\Sigma B_s\tau$ with V gave nearly identical results whereas AE was only mildly correlated with solar wind pressure. These results show the exploratory and confirmatory power of the lagged cross-correlation analysis in choosing between various theoretical possibilities.

Since the highly successful work of Arnoldy, a vast literature has grown up in this category of 1-hour analysis studies (e.g., GARRETT *et al.*, 1974; MURAYAMA and HAKAMADA, 1975; MAEZAWA, 1979; HOLZER and SLAVIN, 1978, 1979; MURAYAMA, 1979; AKASOFU, 1979; REIFF *et al.*, 1981; MELONI *et al.*, 1982). In all these instances, linear correlation analyses have been used as the discriminator between "good" and "poor" solar wind drivers of geomagnetic activity. These studies have all constituted extremely valuable guides in solar wind-magnetosphere coupling studies and have sprung from the widely accessible interplanetary and geomagnetic data sets that are available (e.g., KING, 1975).

We *conclude* from our examination of this literature set that hourly (and greater) time averaging is basically a low-pass filtering operation. Many magnetospheric processes and phenomena are of a relatively high frequency and, thus, long-term averaging eliminates these high frequency components. Therefore, this averaging procedure ***artificially increases correlations*** and masks magnetospheric response time scales. Hence, despite the convenience and value of the hourly-average studies, they still are not the best way to address substorm processes and mechanisms.

4.5 *High time resolution correlation studies*

Because of the potential problems noted above, viz., the possible masking of the details of solar wind-magnetosphere coupling mechanisms, many researchers have gone from the medium-and coarse-resolution data sets ($\gtrsim 1$ hr) to high time-resolution (1–10 min) studies. Correlation analyses employing such data hold the

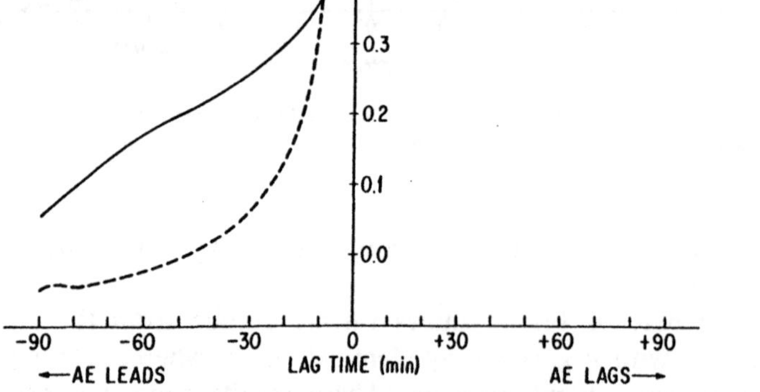

Fig. 8. **High time resolution, lagged cross-correlation analysis results comparing the efficacy of ε, VB_s, and V^2B_s as predictors of the *AE* index. (a) Results from an analysis using IMP-8 data (BAKER *et al.*, 1981). (b,c) Analyses using ISEE-3 solar wind data during both quiet and disturbed intervals (BAKER *et al.*, 1983).**

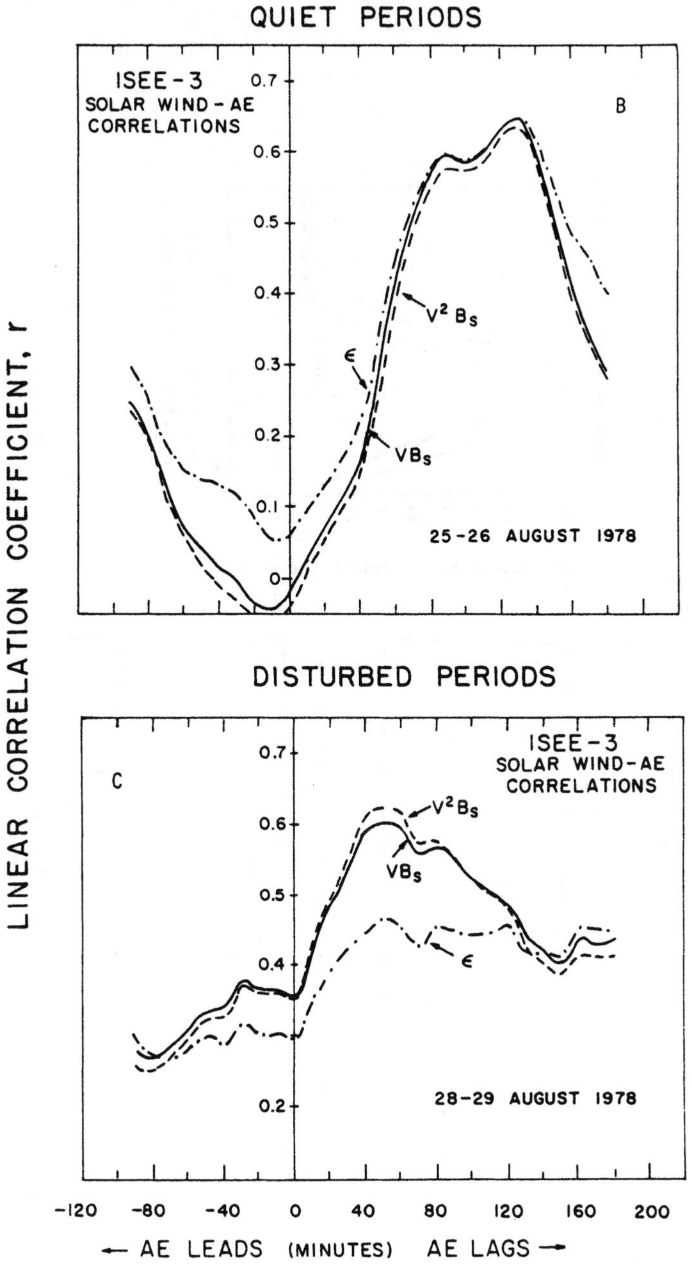

Fig. 8. (Continued).

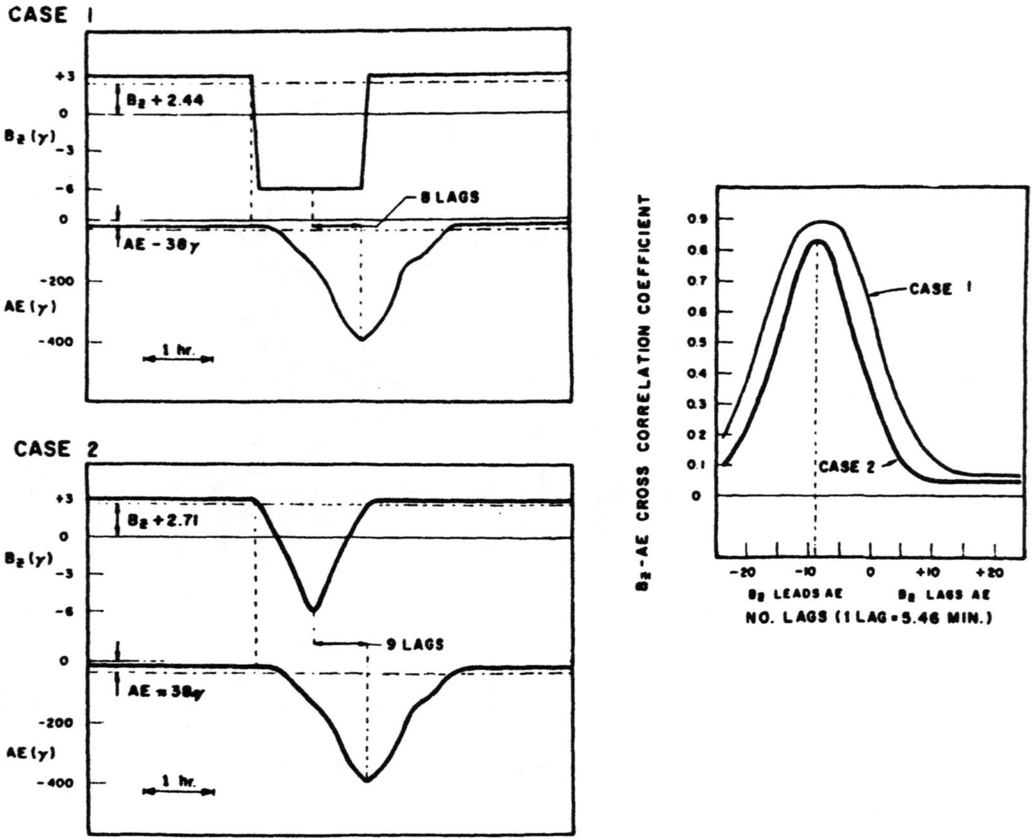

Fig. 9. Cross-correlation coefficients for two idealized sets of IMF B_z and *AE* variations (from MENG *et al.*, 1973).

possibility of addressing details of substorm relationships.

Early studies of this class tended to preselect intervals for statistical analysis. For example, MENG *et al.* (1973) chose ~30 days on which B_z (IMF) was southward to do their analysis, while other authors (e.g., BURTON *et al.*, 1975) have examined only storm intervals. Studies examining rather clear, isolated substorm activity have found that cross-correlations between VB_z (or related parameters in the solar wind) and the AE index show a maximum correlation coefficient between 0.6 and 0.7 at a lag time of ~50 min (c.f., ROSTOKER *et al.*, 1972).

In recent high-resolution correlation studies, BAKER *et al.* (1981, 1983) have emphasized evaluations of the VB_s, V^2B_s, and ε parameters. In the 1981 study they found that VB_s and V^2B_s were virtually identical in their correlative behavior with AE. Furthermore, in this nonselective study using 2.5-min time resolution data, Baker *et al.* found VB_s both qualitatively and quantitatively superior to ε in the prediction of substorms (as measured by AE). This result is illustrated in the first panel of Fig. 8.

This work was extended by BAKER *et al.* (1983) to include continuous solar wind measurements by ISEE-3 at 1-min time resolution. Figures 8b and 8c show results separated according to general levels of geomagnetic activity. The overall result found by BAKER *et al.* (1983) was that VB_s, V^2B_s, and ε had virtually identical correlation properties with AE during quiet intervals, whereas ε was substantially less well correlated with AE than VB_s during disturbed intervals. Generally speaking, all correlation levels were lowered during disturbed conditions presumably due to higher frequency, uncorrelated components both in the solar wind (input) and magnetospheric (output) time series.

Nearly all workers using high resolution data for cross-correlation analyses have interpreted the lag times between solar wind and magnetospheric parameters in terms of a "response" time of the magnetosphere. This interpretation is probably justified to an extent, but due caution should be applied. Figure 9 is taken from MENG *et al.* (1973) and shows the result of a simulation analysis of an idealized input and output time series comparison through cross-correlation. Since cross-correlation is, in effect, a convolution of the input and output functions, Meng *et al.* showed that the detailed strength and shapes of the lagged cross-correlation profile depends on the specific shape and other characteristics of the original time series (see, also, HIRSHBERG and HOLZER, 1975). Thus, one should be cautioned, in the absence of other data and analysis techniques, against interpreting peak lag times too literally as the response time of the magnetospheric systems.

We conclude, however, that analysis involving high time resolution data have revealed a great deal about substorm coupling mechanisms *including* magnetospheric response times. It is found that correlation coefficients are generally reduced from those found in longer-term average studies indicating the presence of high-frequency, internal magnetospheric instability processes. Generally, parameters involving V and B_z in the solar wind are highly "successful" in correlations with AE, AL, and other indices, but levels of correlation vary substantially with general levels of geomagnetic activity.

4.6 *Other cross-correlation methods*

Two other methods deserve mention here: multiple regression analyses and linear prediction filtering. Both of these methods rely essentially on techniques closely related to linear cross-correlation. Multiple regression is a generalization of linear correlation analysis allowing simultaneous multivariate studies (e.g., MELONI *et al.*, 1982). The method allows nonpolynomial functions and makes specific provision for parameter optimization. It is particularly useful for indicating which parameters should be included in a complete theoretical description. The method, however, greatly increases the complexity of analyses and substantial physical insight needs to be applied in order to understand results.

Linear prediction filtering uses autocorrelation and cross-correlation methods (see CLAUER, 1986, this volume) to determine the most general linear relationship between an input and an output time series. It has proven to be an excellent method for understanding magnetospheric temporal response to solar wind variations (IYEMORI *et al.*, 1979; CLAUER *et al.*, 1981; BARGATZE *et al.*, 1985). The method, described in detail by Clauer, thus allows a very general and quantitative assessment of prediction of output based upon an assumed input.

5. Selection-Based Methods

Available space does not allow a detailed treatment of the many other statistical methods that have been employed in SMC studies. Two methods deserve special mention however. The first of these we term "transfer-interval analyses" (HOLZER and SLAVIN, 1982). In this method, the AL index is taken as a measure of the rate at which magnetic flux is returned from the tail by substorm processes to the dayside magnetopause. Solar wind parameters such as VB_s or ε are taken as measures of the rate of erosion of flux from the dayside by magnetic reconnection. The time integrals of AL and VB_s (or ε, V^2B_s) are then compared quantitatively to assess the issue the "best" predicator of geomagnetic activity. Intervals of disturbed conditions bracketed at beginning and end by quiet conditions are chosen and may range in length from hours to days.

This method, involving 1-hr data resolution and quiet-to-quiet interval selection, is assessed by cross-correlation analysis. It is a careful and highly successful method (see, also, HOLZER and SLAVIN, 1979, 1981a,b). Although criticized (KAMIDE, 1983), the technique is a very valuable technique of understanding the energy transfer process from the solar wind to the magnetosphere (HOLZER and SLAVIN, 1983).

A second additional statistical method deserving special mention is superposed epoch analyses (see, e.g., REIFF, 1983). The method has been employed very profitably for many interplanetary, magnetospheric, and SMC problems (e.g., FOSTER *et al.*, 1971; CAAN *et al.*, 1978; HAKAMADA, 1980; BIEBER *et al.*, 1982). As demonstrated by these studies, the method is valuable in showing the average behavior of the system, e.g., the average AE response to a southward turning of the IMF. Given proper physical insight in the choice of parameters and in the selection of zero-epoch time, the method can be a very valuable adjunct to cross-correlation studies.

6. Summary

We conclude that a very large number of solar wind-magnetosphere studies have employed statistical methods. A partial listing of these is shown in Table 1 (BAKER *et*

Table 1. List of the correlation studies between geomagnetic indices and solar wind parameters.

Author(s)	Geomagnetic index (Time resolution)		Relation with solar wind parameters
Synder et al. (1963)	ΣKp	(24 hr)	$\Sigma Kp = (V - 330)/8.44$
Olbert (1968)	ΣKp	(24 hr)	$\Sigma Kp - (V - 262)/6.3$
Ballif et al. (1969)	Kp	(3 hr)	$Kp = 9 \left[1 - \exp\left(\frac{-\sigma B_{T} \cdot N^{-0.35}}{7.70} \right) \right]$
Arnoldy (1971)	AE	(1 hr)	$AE = -0.26(\Sigma B_s \tau)_0 - 0.91(\Sigma B_s \tau)_1$ $= -0.33(\Sigma B_s \tau)_2 + 0.12 P_0$
Rostoker et al. (1972)	AE	(5.5 min)	E_Φ
Meng et al. (1973)	AE	(5.5 min)	B_z
Bobrov (1973)	Kp	(3 hr)	$Kp = f(V, B_z, \Delta B_z)$
Garrett et al. (1974)	ap, AE	(1 hr)	$ap, AE \propto C_1 + C_2 V(B_s) + C_3 V\sigma$
Murayama and Hakamada (1975)	AE	(1 hr)	$AE = CB_s V^2$
Burton et al. (1975)	Dst	(2.5 min)	$\frac{d}{dt} Dst_0 = F(E) - a Dst_0$
Crooker et al. (1977)	Ap		$Ap = 3.5 \times 10^{-5} B_z V^2 - 1.9$
Svalgaard (1977)	am	(3 hr)	$am = 6.6 q(f, \alpha) \frac{BV_0}{21} \left[\frac{n V_0^2}{105} \right]^{1/3}$ $\times \frac{1.157}{(1 + 3\cos^2\psi)^{2/3}}$
Maezawa (1979)	Al,AU,am	(1 hr)	$AL \propto B^{0.85} V^{2.08} (\sin\theta)^{0.54}$ $AU \propto B^{0.67} B^{1.15} (\sin\theta)^{0.34}$ $am \propto B^{1.03} V^{2.34} (\sin\theta)^{0.37} n^{0.2}$
Holzer and Slavin (1979)	AL	(1 hr)	Rate of erosion $\frac{d\phi_e}{dt} = 0.2(1.9 \times 10^{10} cm) B_z V_{sw} (km\ s^{-1})$ Rate of return $\frac{d\phi_r}{dt} = 1.8 \times 10^{10} AL(\gamma)$
Murayama (1979)	AL	(1 hr)	$AL = 60(B_s + 0.5) V^2 F(\chi, B_y) n^{0.13}$
Murayama et al. (1980)	AL	(1 hr)	$AL/V^2 = K(B_s + 0.5)$
Baker et al. (1981)	AE	(2.5 min)	VB_z, V^2B, B^2V, etc.
Lei et al. (1981)	Correlation between the interplanetary and magnetospheric electric fields		
Reiff et al. (1981)	Correlation between Φ_{PC} and B_z, ε		
Holzer and Slavin (1982)	AL		BV^2, B_sV, $B_s{}^2V$
Iijima and Potemra (1982)	Correlation between field-aligned current density and VB_z, V^2B, B^2V, etc.		
Meloni et al. (1982)	AE		B_z, VB_z, ε
Baker et al. (1983)	AE, U_T	(1 min)	VB_s, V^2B_s, ε

al., 1984; AKASOFU, 1981). In many cases the confidence levels of obtained correlation coefficients is not given nor are the underlying assumptions concerning normality of distribution, etc., carefully assessed or specified. Thus, it is difficult at times to be sure of how much to trust the statistical results of many studies.

In the final analysis, we suggest in the strongest terms that researchers make every effort to familiarize themselves in great detail with the methods and bases of statistical analysis (e.g., BEVINGTON, 1969) before blindly proceeding with correlation (or other) studies.

We consider that SMC studies using hourly (or greater) averaging time scales have often obtained large correlation coefficients, but this is due to the low pass filtering effects of the temporal averaging method. This greatly downplays important internal, high frequency magnetospheric phenomena.

High time resolution (1–10 min) data sets have revealed a great deal about magnetospheric response times to solar wind variations. Further, the issue of "best" solar wind predictors has been illuminated by high resolution studies. It is clear that magnetospheric response character appears different as general levels of activity increase. We suggest that correlation studies be considered as merely part of a much broader set of methods (case studies, prediction filtering, superposed epoch analysis, etc.) in order to characterize completely the complex issue of solar wind-magnetosphere coupling.

The author wishes to express his sincere thanks to R. L. McPherron and L. F. Bargatze for very helpful discussions concerning this paper. Work at Los Alamos was performed under the auspices of the U.S. Department of Energy with support from the Los Alamos Branch of the University of California Institute for Geophysics and Planetary Physics.

REFERENCES

AKASOFU, S.-I., Interplanetary energy flux associated with magnetospheric substorms, *Planet. Space Sci.*, **27**, 425, 1979.

AKASOFU, S.-I., The solar wind-magnetosphere energy coupling and magnetospheric disturbances, *Planet. Space Sci.*, **28**, 495, 1980.

AKASOFU, S.-I., Energy coupling between the solar wind and the magnetosphere, *Space Sci. Rev.*, **28**, 121, 1981.

AKASOFU, S.-I. and J. K. CHAO, Prediction of the occurrence and intensity of magnetospheric substorms, *Geophys. Res. Lett.*, **6**, 897, 1979.

AKASOFU, S.-I., J. F. CARBARY, C.-I. MENG, J. P. SULLIVAN, and R. P. LEPPING, A high time resolution study of the solar wind-magnetosphere energy coupling function, *Planet. Space Sci.*, **30**, 537, 1982.

ARNOLDY, R. L., Signature in the interplanetary medium for substorms, *J. Geophys. Res.*, **76**, 5189, 1971.

BAKER, D. N., E. W. HONES, Jr., J. B. PAYNE, and W. C. FELDMAN, A high-time resolution study of interplanetary parameter correlations with *AE*, *Geophys. Res. Lett.*, **8**, 179, 1981.

BAKER, D. N., R. D. ZWICKL, S. J. BAME, E. W. HONES, Jr., B. T. TSURUTANI, E. J. SMITH, and S.-I. AKASOFU, An ISEE-3 high time resolution study of interplanetary parameter correlations with magnetospheric activity, *J. Geophys. Res.*, **88**, 6230, 1983.

BAKER, D. N., S.-I. AKASOFU, W. BAUMJOHANN, J. W. BIEBER, D. H. FAIRFIELD, E. W. HONES, Jr., B. H. MAUK, R. L. MCPHERRON, and T. E. MOORE, Substorms in the magnetosphere, in *Solar Terrestrial Physics—Present and Future*, chap. 8, NASA Pub. 1120, Washington, D.C., 1984.

BALLIF, J. R., D. E. JONES, and P. J. COLEMAN, Further evidence on the correlation between transverse fluctuations in the interplanetary magnetic field and *Kp*, *J. Geophys. Res.*, **74**, 2289, 1969.

BARGATZE, L. F., D. N. BAKER, R. L. MCPHERRON, and E. W. HONES, Jr., Magnetospheric response for many levels of geomagnetic activity, *J. Geophys. Res.*, **90**, 6387, 1985.

BEVINGTON, P. R., *Data Reduction and Error Analysis for the Physical Sciences*, Chap. 6–10, McGraw-Hill, New York, 1969.

BIEBER, J. W., E. C. STONE, E. W. HONES, Jr., D. N. BAKER, and S. J. BAME, Plasma behavior during energetic electron streaming events: further evidence for substorm-associated magnetic reconnection, *Geophys. Res. Lett.*, **9**, 664, 1982.

BOBROV, M. S., K_p index correlations with solar wind parameters during the first and second stages of a recurrent geomagnetic storm, *Planet. Space Sci.*, **21**, 2139, 1973.

BURTON, R. K., R. L. MCPHERRON, and C. T. RUSSELL, An empirical relationship between interplanetary condition and *Dst*, *J. Geophys. Res.*, **80**, 4204, 1975.

CAAN, M. N., R. L. MCPHERRON, and C. T. RUSSELL, The statistical magnetic signature of magnetospheric substorms, *Planet. Space Sci.*, **26**, 269, 1978.

CLAUER, C. R., R. L. MCPHERRON, C. SEARLS, and M. G. KIVELSON, Solar wind control of auroral zone geomagnetic activity, *Geophys. Res. Lett.*, **8**, 915, 1981.

CROOKER, N. U., J. FEYNMAN, and J. T. GOSLING, On the high correlation between long-term averages of solar wind speed and geomagnetic activity, *J. Geophys. Res.*, **83**, 1933, 1977.

FAIRFIELD, D. H. and L. J. CAHILL, Jr., Transition region magnetic field and polar magnetic disturbances, *J. Geophys. Res.*, **71**, 155, 1966.

FEYNMAN, J., Solar cycle and long term changes in the solar wind, *Rev. Geophys. Space Phys.*, **21**, 338, 1983.

FOSTER, J. C., D. H. FAIRFIELD, K. W. OGILVIE, and T. J. ROSENBERG, Relationship of interplanetary parameters and occurrence of magnetospheric substorms, *J. Geophys. Res.*, **76**, 6967, 1971.

GALLAGHER, D. L. and N. D'ANGELO, Correlations between solar wind parameters and auroral kilometric radiation intensity, *Geophys. Res. Lett.*, **8**, 1087, 1981.

GARRETT, H. B., A. J. DESSLER, and T. W. HILL, Influence of solar wind variability on geomagnetic activity, *J. Geophys. Res.*, **79**, 4603, 1974.

GOSLING, J. T., R. T. HANSEN, and S. J. BAME, Solar wind speed distributions: 1962–1970, *J. Geophys. Res.*, **76**, 1811–1815, 1971.

GOSLING, J. T., J. R. ASBRIDGE, S. J. BAME, and W. C. FELDMAN, Solar wind speed variations: 1962–1974, *J. Geophys. Res.*, **81**, 5061–5070, 1976.

HAKAMADA, K., Geomagnetic activity at the time of heliospheric current sheet crossings, *Geophys. Res. Lett.*, **7**, 653, 1980.

HIRSHBERG, J. and T. E. HOLZER, Relationship between the interplanetary magnetic field and isolated substorms, *J. Geophys. Res.*, **80**, 3553, 1975.

HOLZER, R. E. and J. A. SLAVIN, Magnetic flux transfer associated with expansion and contractions of the dayside magnetosphere, *J. Geophys. Res.*, **83**, 3831, 1978.

HOLZER, R. E. and J. A. SLAVIN, A correlative study of magnetic flux transfer in the magnetosphere, *J. Geophys. Res.*, **84**, 2573, 1979.

HOLZER, R. E. and J. A. SLAVIN, The effect of solar wind structure on magnetospheric energy supply during solar cycle 20, *J. Geophys. Res.*, **86**, 675, 1981a.

HOLZER, R. E. and J. A. SLAVIN, Processes influencing the diurnal variation of the AL index and its reliability, *J. Geophys. Res.*, **86**, 8977, 1981b.

HOLZER, R. E. and J. A. SLAVIN, An evaluation of three predictors of geomagnetic activity, *J. Geophys. Res.*, **87**, 2558, 1982.

HOLZER, R. E. and J. A. SLAVIN, Reply, *J. Geophys. Res.*, **88**, 1983.

IYEMORI, T., H. MAEDA, and T. KAMEI, Impulse response of geomagnetic indices to interplanetary magnetic field, *J. Geomag. Geoelectr.*, **31**, 1, 1979.

KAMIDE, Y., Comment on 'An evaluation of three predictors of geomagnetic activity' by R. E. Holzer and J. A. Slavin, *J. Geophys. Res.*, **88**, 4953, 1983.

KING, J. H., Interplanetary magnetic field data book, *Rep. NSSDC 75-04*, Goddard Space Flight Center, Greenbelt, Md., 1975.

LEI, W., R. GENDRIN, B. HIGEL, and J. BERCHEM, Relationships between the solar wind electric field and the magnetospheric convection electric field, *Geophys. Res. Lett.*, **8**, 1099, 1981.

MAEZAWA, K., Statistical study of the dependence of geomagnetic activity on solar wind parameters, *Geophys. Monograph 21*, p. 436, AGU, 1979.

MELONI, A., A. WOLFE, and L. J. LANZEROTTI, On the relationship between interplanetary quantities and the global auroral electrojet index, *J. Geophys. Res.*, **87**, 119, 1982.

MENG, C.-I., B. TSURUTANI, K. KAWASAKI, and S. -I. AKASOFU, Cross-correlation analysis of the *AE* index and the interplanetary magnetic field B_z component, *J. Geophys. Res.*, **78**, 617, 1973.

MURAYAMA, T., Principle factors controlling the development of the auroral electrojet, in *Magnetospheric Study 1979*, edited by T. Obayashi, p. 296, Tokyo, March, 1979.

MURAYAMA, T. and K. HAKAMADA, Effects of solar wind parameters on the development of magnetospheric substorms, *Planet. Space Sci.*, **23**, 75, 1975.

MURAYAMA, T., T. AOKI, H. NAKAI, and K. HAKAMADA, Empirical formula to relate the auroral electrojet intensity with interplanetary parameters, *Planet. Space Sci.*, **28**, 803, 1980.

OLBERT, S., Summary of experimental results from MIT detectors on IMP-1, in *Physics of the Magnetosphere*, edited by R. L. Caravillano, J. F. McClay, and H. R. Radoski, p. 641, Reidel, Doredrecht, 1968.

PERREAULT, P. and S.-I. AKASOFU, A study of geomagnetic storms, *Geophys. J. R. Astron. Soc.*, **54**, 547, 1978.

REIFF, P. H., The use and misuse of statistical analyses, in *Solar-Terrestrial Physics*, edited by R. L. Caravillano and J. M. Forbes, p. 493, D. Reidel, 1983.

REIFF, P. H., R. W. SPIRO, and T. W. HILL, Dependence of polar cap potential drop on interplanetary parameters, *J. Geophys. Res.*, **86**, 7639, 1981.

ROSTOKER, G., H.-L. LAM, and W. D. HUME, Response time of the magnetosphere to the interplanetary electric field, *Can. J. Phys.*, **50**, 544, 1972.

SHEELEY, N. R., Jr., J. W. HARVEY, and W. C. FELDMAN, Coronal holes, solar wind streams, and recurrent geomagnetic disturbances: 1973–1976, *Solar Phys.*, **49(2)**, 271–278, 1976.

SNYDER, C. W., M. NEUGEBAUER, and U. R. RAO, The solar wind velocity and its correlation with cosmic ray variations and with solar and geomagnetic activity, *J. Geophys. Res.*, **68**, 6361, 1963.

SVALGAARD, L., Geomagnetic activity: Dependence on solar wind parameters, in *Coronal Holes and High Speed Wind Streams*, p. 371, Colorado Assoc. Univ. Press, Boulder, 1977.

TSURUTANI, B. T. and C.-I. MENG, Interplanetary magnetic field variations and substorm activity, *J. Geophys. Res.*, **77**, 2964, 1972.

VASYLIUNAS, V. M., J. R. KAN, G. L. SISCOE, and S.-I. AKASOFU, Scaling relations governing magnetospheric energy transfer, *Planet. Space Sci.*, **30**, 359, 1982.

WORTHING, A. G. and J. GEFFNER, *Treatment of Experimental Data*, John Wiley and Sons, Inc., New York, 1950.

Solar Wind-Magnetosphere Coupling, edited by Y. Kamide and J. A. Slavin, 39–57.

The Technique of Linear Prediction Filters Applied to Studies of Solar Wind-Magnetosphere Coupling

C. Robert CLAUER

Space, Telecommunications and Radioscience Laboratory, Stanford Electronics Laboratories Stanford University, Stanford, California 94305, U.S.A.

Linear prediction filtering is a powerful empirical technique suitable for the study of stimulus-response behavior. The technique enables one to determine the most general linear relationship between multiple time-varying quantities, assuming that the physical systems relating the quantities are linear and time invariant. Several researchers have applied linear prediction analysis to investigate solar wind-magnetosphere interactions. This short review describes the method of linear prediction analysis, its application to solar wind-magnetosphere coupling studies both in terms of the assumptions and in terms of physical processes, and the results of investigations which have used this technique.

1. Introduction

A great deal of scientific effort has been invested over the past several decades to understand the causes of geomagnetic activity. It is now known that the sun through its expanding outer atmosphere, the solar wind, is the primary source of the energy which produces geomagnetic disturbances. These disturbances recorded at our geomagnetic observatories result from electrical currents which flow around the earth in space, through the ionosphere and within the conducting earth. The magnetic effects produced by these currents are characterized by various simple measures called geomagnetic indices (ROSTOKER, 1972; MAYAUD, 1980; KAMIDE and AKASOFU, 1983). Careful understanding of the methods used to create indices and the consequent limitations of the indices is required when making a physical interpretation based on their use. The benefit of the indices, however, is to synthesize a variety of complex geomagnetic disturbance measurements into a simple time series. Because of this simplification, geomagnetic indices have been used extensively in studies of magnetospheric behavior. Our understanding of the solar wind's control over geomagnetic activity has been largely acquired through studies based on correlations between solar wind quantities and geomagnetic indices. These correlations have formed the framework within which theories of solar wind-magnetosphere coupling have been developed.

Most correlative studies have used regression, superposed epoch, or cross correlation techniques to establish relationships between the solar wind and geomagnetic activity and a discussion of these techniques is given by BAKER (1986) in this volume. Recently another very powerful technique, that of linear prediction filtering, has been introduced to the field of magnetospheric physics (IYEMORI *et al.*, 1979; IYEMORI and MAEDA, 1980). This technique allows one to empirically determine the most general linear relationship between a solar wind input function and a geomagnetic disturbance output function taking into account time delays and frequency response. Such a functional relationship provides a more detailed set of constraints for models which must account for the physical mechanisms associated with the coupling of solar wind energy to magnetospheric current systems.

The technique of linear prediction filtering is based on least squares criteria and leads to filters (functions which transform the input into the output) which are linear. The method is particularly noteworthy because of the quality of the results obtained and the simplicity of the concepts involved. The theory on which linear prediction analysis is based was originally developed by the mathematician Norbert Wiener and is put forth in his book on linear least squares prediction and signal enhancement (WIENER, 1942). The technique is now commonly refered to as Wiener Filtering. While the original work by Wiener was formulated in terms of continuous time series, the important adaptation to discrete time series was developed by LEVINSON (1942) and appears as an appendix to Wiener's book. Perhaps one of the most useful books to deal with the application of Wiener filtering to geophysical problems is by ROBINSON (1967). This book provides a clear description of the technique plus Fortran computer code for the subroutines required to perform the analysis. The more recent book by ROBINSON and TREITEL (1980) also provides a good reference. In the following sections of this review, I will provide a mathematical introduction to the technique followed by a review of the applications which have used Wiener filtering in studies of solar wind-magnetosphere coupling.

2. Principles of Digital Wiener Filtering

The basic model for the filtering process considered here consists of a system having an input signal, a desired or measured output signal and a computed output signal. This is shown schematically in Fig. 1 where $I(t)$ is the input time series, $O(t)$ is the desired output time series and $\hat{O}(t)$ is the computed output. The input and the computed output are related by a filter f which must be determined from the input and the desired output time series. A basic assumption of the analysis is that the physical systems characterized by $I(t)$ and $O(t)$ can be related by a linear time invariant filter f. This can be described mathematically by the equation

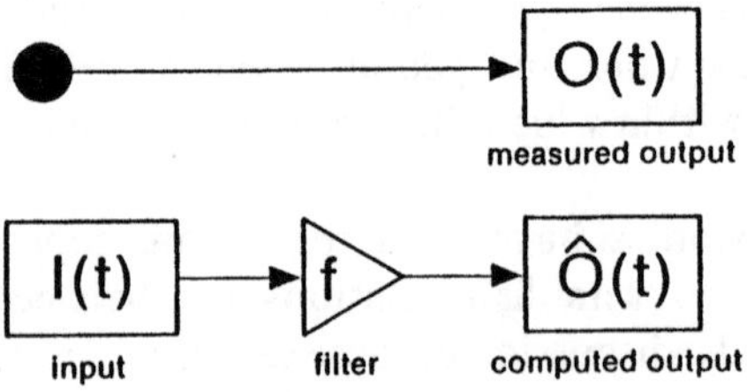

Fig. 1. The Wiener Filter model.

$$O(t) = \int_{-\infty}^{+\infty} f(\tau) I(t-\tau) d\tau. \quad (1)$$

For the case of discrete time series data, Equation (1) may be written as

$$O_t = \sum_{\tau=-\infty}^{\infty} f_\tau I_{t-\tau} = \sum_{\tau=-\infty}^{-1} f_\tau I_{t-\tau} + \sum_{\tau=0}^{\infty} f_\tau I_{t-\tau}. \quad (2)$$

Here f_τ is the prediction filter or impulse response function, t denotes time and τ denotes time lag. Note that Equation (2) can be broken into two summations, one with $\tau<0$ and the second with $\tau\geq0$. Negative values of τ indicate times in the future. Thus, applying limits to the summation can force assumptions about causality. If $\tau\geq0$, the filter will be purely causal and the output will strictly follow the input.

Given finite and noisy measurements of I_t and O_t our task is to solve for the predictive filter f. To do this, we must assume that both the input and output time series are stationary. That is, their statistical properties do not change in time. We also must assume that errors in the two time series I_t and O_t are random and uncorrelated. And finally we must assume that they are related by a linear operator as in Eq. (2). If these assumptions are not met this procedure will still produce the best linear filter which transforms the input into the output. This filter, however, will be based on the finite set of data utilized in the analysis and may not be generally valid for other subsets of the infinite data domain from which the data was sampled.

The determination of the filter coefficients ($f_0, f_1, f_2, \ldots, f_m$) is based on the least-squares criterion: one minimizes the power existing in the difference time series obtained by subtracting the actual or measured output O_t from the computed output $\hat{O}_t$. This will be illustrated through the example of a single channel filter.

2.1. *Single channel filter*

The single channel filter has one input and one output. In this example let
X_i=input,
Y_i=desired output,
$\hat{Y}_i$=computed output.
Assuming a causal relationship, the convolution Equation (2) will become

$$\hat{Y}_t = f_0 X_t + f_1 X_{t-1} + \ldots + f_m X_{t-m} = \sum_{i=0}^{m} f_i X_{t-i}. \quad (3)$$

The error at point t is given by the difference between the measured output and the computed output. This is expressed by e_t in the following equation

$$e_t = Y_t - \hat{Y}_t = Y_t - \sum_{i=0}^{m} f_i X_{t-i}. \quad (4)$$

The mean-square-error or error power is then given by

$$I = E\{e_t^2\}$$
$$= E\{[Y_t - (f_0X_t + f_1X_{t-1} + \ldots + f_mX_{t-m})]^2\} \tag{5}$$

where E is the Ensemble Average operator. This quantity is minimized by setting its partial derivatives with respect to the filter coefficients f_i equal to zero. That is $\partial I/\partial f_i=0$. Doing this produces the set of equations called the normal equations

$$\frac{\partial I}{\partial f_0} = E\{[Y_t - (f_0X_t + f_1X_{t-1} + \ldots + f_mX_{t-m})]X_t\} = 0$$
$$\frac{\partial I}{\partial f_1} = E\{[Y_t - (f_0X_t + f_1X_{t-1} + \ldots + f_mX_{t-m})]X_{t-1}\} = 0$$
$$\vdots \qquad\qquad \vdots$$
$$\frac{\partial I}{\partial f_m} = E\{[Y_t - (f_0X_t + f_1X_{t-1} + \ldots + f_mX_{t-m})]X_{t-m}\} = 0, \tag{6}$$

where both sides of the equations have been divided through by -2.

These equations can be written in terms of the autocorrelation coefficients $R_{ij}=E\{X_{t-i}X_{t-j}\}$ and the cross correlation coefficients $G_i=E\{Y_tX_{t-i}\}$ becoming

$$f_0R_{00} + f_1R_{10} + f_2R_{20} + \ldots + f_mR_{m0} = G_0$$
$$f_0R_{01} + f_1R_{11} + f_2R_{21} + \ldots + f_mR_{m1} = G_1$$
$$\vdots \qquad\qquad \vdots$$
$$f_0R_{0m} + f_1R_{1m} + f_2R_{2m} + \ldots + f_mR_{mm} = G_m. \tag{7}$$

Given the covariances R_{ij} and g_i we can solve these $m+1$ equations for the $m+1$ unknown filter coefficients f_i.

Since $R_{ij}=R_{ji}$ we can simplify the notation by letting R_τ equal the autocorrelation coefficient at lag τ. The normal equations form a set of $m+1$ linear simultaneous equations which can be written in matrix form as

$$\begin{pmatrix} R_0 & R_1 & R_2 & \cdots & R_m \\ R_1 & R_0 & R_1 & \cdots & R_{m-1} \\ \vdots & \vdots & \vdots & \ddots & \vdots \\ R_m & R_{m-1} & R_{m-2} & \cdots & R_0 \end{pmatrix} \begin{pmatrix} f_0 \\ f_1 \\ f_2 \\ \vdots \\ f_m \end{pmatrix} = \begin{pmatrix} G_0 \\ G_1 \\ G_2 \\ \vdots \\ G_m \end{pmatrix} \tag{8}$$

These equations could be solved by normal techniques, however, because of the symmetries in the matrices a much more efficient method is possible. The matrix R has the special form of a Töplitz matrix (i.e., a matrix with equal submatrices on any diagonal). Thus the matrix involves only $m+1$ distinct elements namely R_0, R_1, R_2, ..., R_m. An efficient computational solution to these equations is the Töplitz recursive method. This method is essentially the same as the Levinson algorithm (LEVINSON, 1949). A description of the recursion will not be given here but can be found in ROBINSON, (1967) and ROBINSON and TREITEL, (1980). Using the recursion technique, one usually computes filters of successively increasing length. The value of the minimum mean-square-error I_{min} decreases with each recursion. The recursion from n to $n+1$ may be stopped whenever a certain preset criterion is met. For example:

1. The filter reaches a maximum preset length.
2. The normalized mean-square-error reaches some minimum preset value.
3. The normalized mean-square-error sequence levels out and shows little likelihood of further appreciable decrease.

When the analysis is completed, we have a computed filter which transforms the input into the output time series. In addition, we have a measure of the filter's efficiency at converting the input to the output. This efficiency measure is based on the power in the residual or error time series and is given by the value of $I=I_{min}$ from Eq. (5).

2.2. *Multichannel filters*

Many systems have multiple inputs and outputs. Such systems are called multichannel systems. A multichannel system involves many time series which are related. The theory of multichannel filters is conceptually the same as the theory of the single channel filter and mathematically it represents only an extension from ordinary scalar algebra to matrix algebra. That is, multichannel theory can be obtained from single channel theory by the appropriate substitution of matrices for scalars. If the system has the same number of input and output channels, the equation will contain square matrices and vectors. If the number of input channels and output channels is different, then the matrices will be rectangular.

For example, consider a system having two input channels and two output channels where

$$\boldsymbol{X}_t = \begin{pmatrix} X_{1t} \\ X_{2t} \end{pmatrix} = \text{input},$$

and

$$\boldsymbol{Y}_t = \begin{pmatrix} Y_{1t} \\ Y_{2t} \end{pmatrix} = \text{output}.$$

The basic convolution formula is:

$$\boldsymbol{Y}_t = f_0\boldsymbol{X}_t + f_1\boldsymbol{X}_{t-1} + \ldots + f_m\boldsymbol{X}_{t-m} \tag{9}$$

where t is again an integer denoting discrete time. $\boldsymbol{Y}$ and $\boldsymbol{X}$ are 2×1 column vectors and the coefficients f_s are given by 2×2 matrices. Expanding Equation (9) yields,

$$\begin{pmatrix} Y_{1t} \\ Y_{2t} \end{pmatrix} = \begin{pmatrix} f_{11}(0) & f_{12}(0) \\ f_{21}(0) & f_{22}(0) \end{pmatrix} \begin{pmatrix} X_{1,t} \\ X_{2,t} \end{pmatrix} + \begin{pmatrix} f_{11}(1) & f_{12}(1) \\ f_{21}(1) & f_{22}(1) \end{pmatrix} \begin{pmatrix} X_{1,t-1} \\ X_{2,t-1} \end{pmatrix} + \vdots + \begin{pmatrix} f_{11}(m) & f_{12}(m) \\ f_{21}(m) & f_{22}(m) \end{pmatrix} \begin{pmatrix} X_{1,t-m} \\ X_{2,t-m} \end{pmatrix}. \quad (10)$$

In general, given N input channels and M output channels, one has:

$$\boldsymbol{X}_t \equiv X(t) = \begin{pmatrix} X_1(t) \\ X_2(t) \\ \vdots \\ X_N(t) \end{pmatrix} = \text{Input},$$

$$\boldsymbol{Y}_t \equiv Y(t) = \begin{pmatrix} Y_1(t) \\ Y_2(t) \\ \vdots \\ Y_M(t) \end{pmatrix} = \text{Desired Output},$$

$$\hat{\boldsymbol{Y}}_t \equiv \hat{Y}(t) = \begin{pmatrix} \hat{Y}_1(t) \\ \hat{Y}_2(t) \\ \vdots \\ \hat{Y}_M(t) \end{pmatrix} = \text{Computed Output},$$

and a set of $M \times N$ matrix valued filter coefficients

$$f \equiv f(s) = \begin{pmatrix} f_{11}(s) & f_{12}(s) & \dots & f_{1N}(s) \\ f_{21}(s) & f_{22}(s) & \dots & f_{2N}(s) \\ \vdots & \vdots & \ddots & \vdots \\ f_{M1}(s) & f_{M2}(s) & \dots & f_{MN}(s) \end{pmatrix}.$$

The error e_t between the desired output and the actual output is then a $M \times 1$ vector valued time series

$$e_t = \hat{\boldsymbol{Y}}_t - \boldsymbol{Y}_t = \begin{pmatrix} \hat{Y}_1(t) - Y_1(t) \\ \hat{Y}_2(t) - Y_2(t) \\ \vdots \\ \hat{Y}_M(t) - Y_M(t) \end{pmatrix} \equiv \begin{pmatrix} e_1(t) \\ e_2(t) \\ \vdots \\ e_M(t) \end{pmatrix}. \quad (11)$$

The mean-square-error matrix is defined as

$$E\{e_te_t^T\} = \begin{pmatrix} E\{e_1^2(t)\} & E\{e_1(t)e_2(t)\} & \dots & E\{e_1(t)e_M(t)\} \\ & \dots & & \\ E\{e_M(t)e_1(t)\} & E\{e_M(t)e_2(t) & \dots & E\{{}_M^2(t)\} \end{pmatrix}. \tag{12}$$

The mean-square-error or error power is given by the trace of the mean-square-error matrix which is defined to be the sum of the diagonal terms,

$$I = \mathrm{tr}E\{e_te_t^T\} = \sum_{i=1}^{m} E\{e_i^2(t)\}. \tag{13}$$

Again setting the partial derivative of I with respect to each filter coefficient equal to zero produces a set of normal equations given by:

$$\begin{aligned} f_0\phi_{xx}(0) + f_1\phi_{xx}(-1) + \dots + f_m\phi_{xx}(-m) &= \phi_{yx}(0) \\ f_0\phi_{xx}(1) + f_1\phi_{xx}(0) + \dots + f_m\phi_{xx}(1-m) &= \phi_{yx}(1) \\ \vdots \qquad\qquad\qquad &\quad \vdots \\ f_0\phi_{xx}(m) + f_1\phi_{xx}(m-1) + \dots + f_m\phi_{xx}(0) &= \phi_{yx}(m) \end{aligned} \tag{14}$$

where $\phi_{xx}(\tau)$ are the autocorrelation of the input X at lag τ and $\phi_{yx}(\tau)$ are the cross correlation of the desired output Y with the input X. This set of linear simultaneous equations can also be solved by the Töplitz recursion method discussed above. The reader is directed to texts by Robinson for further information about the calculation of the multichannel autocorrelation and cross correlation coefficients and the Töplitz recursive solution.

3. Application to Studies of Solar Wind-Magnetosphere Coupling

Wiener filtering has been applied to the problem of solar wind-magnetosphere coupling using various parameters measured in the solar wind and indices of geomagnetic activity. In these studies, solar wind parameters represent the input and geomagnetic activity the output. The computed filters or impulse response functions characterize the processes which link solar wind energy to the currents monitored by the various geomagnetic indices used in the analyses.

The technique of Wiener filtering was introduced to the solar wind-magnetosphere coupling problem by IYEMORI *et al.* (1979). In this study the authors used 300 days of hourly average data from 1967 through 1969. Filters relating the solar wind quantities VB_s and B_s to auroral magnetic indices AL, AU, AE and to the low-latitude ring current magnetic index D_{st} were computed. The D_{st} filter develops in 1 hour and decays with a time constant of about 8 hours which is consistent with the previous empirical model developed by BURTON *et al.* (1975). The AE and AL filters are a narrow pulse peaking at 1 hour lag. The AU filter is a broader peak at 1 hour lag. Since the AU filter is significantly different than the AL filter in shape, IYEMORI *et al.*, suggest that the generation for the currents monitored by AU may be different than those for AL or else the generation may be complex, consisting of multiple mechanisms. The prediction efficiency was fairly high for all of the predictions.

The Wiener Filtering technique was extended to studies of high time resolution data by a number of subsequent researchers CLAUER *et al.*, 1981, 1983; MCPHERRON *et al.*, 1984; BARGATZE *et al.*, 1985; and AKASOFU *et al.*, 1985). Figure 2 from CLAUER *et al.* (1981) shows high time resolution Wiener filters which relate 3 forms of the solar wind input function to the AL index. The three input functions have a $VB^2\sin^4(\theta/2)$, VB_s, an V^2B_s dependence respectively where

$$\theta = \begin{cases} \tan^{-1}\dfrac{|B_y|}{|B_z|}, & \text{for } B_z > 0; \\ 180° - \tan^{-1}\dfrac{|B_y|}{|B_z|}, & \text{for } B_z < 0. \end{cases}$$

and

$$B_s = \begin{cases} 0, & \text{for } B_z \geq 0 \\ -B_z, & \text{for } B_z < 0. \end{cases}$$

Two time intervals were analyzed. The filters at the top of Fig. 2 were created from an interval of moderate level goemagentic activity while the bottom panel filters were created from an interval of strong activity. The horizontal axis of the plots is time lag. A smooth curve is drawn through the filters and is shown only as an aid to the reader. The filters are very spiky. This is due to uncertainties which result from correlations based on a small number of active periods in the 10 and 12 day intervals used in the analysis. Less noisy filters result from analysis on long intervals which include many repeated correlated variations between the input and output time series. For example, compare the $VB_s \rightarrow AL$ filters in Fig. 2 with the $VB_s \rightarrow AL$ filter shown in Fig. 5 which was created using 56 days of data.

The dominant feature of the filters in Fig. 2 is a delayed negative pulse indicative of a low pass filter with a time delay. It is interesting to see that the filters for the strong activity and moderate activity intervals show different time delays suggesting some form of nonlinearity in the $VB_s \rightarrow AL$ relationship.

The variation in the $VB_s \rightarrow AL$ impulse response as a function of geomagnetic activity level has been examined in greater detail by BARGATZE *et al.* (1985). The data set examined contains 73 days of 2.5 minute resolution data. An activity level based on the AL index was assigned to each of the 34 intervals comprising the data set. These intervals were sorted by activity level to produce a time series of increasing level of geomagnetic activity. A set of running filters was determined over the entire data set by first using 5 consecutive intervals to compute a Wiener filter and then stepping ahead one interval at a time. Figure 3 shows the set of 30 $VB_s \rightarrow AL$ filters thus obtained with the weak activity filters at the bottom ranging to the strong activity filters at the top.

Two peaks can be identified in these filters, the first at 20 minutes lag and the second at 60 minutes lag. The relative importance of the peaks varies as a function of activity level, with the second peak dominant at middle activity levels, the first peak

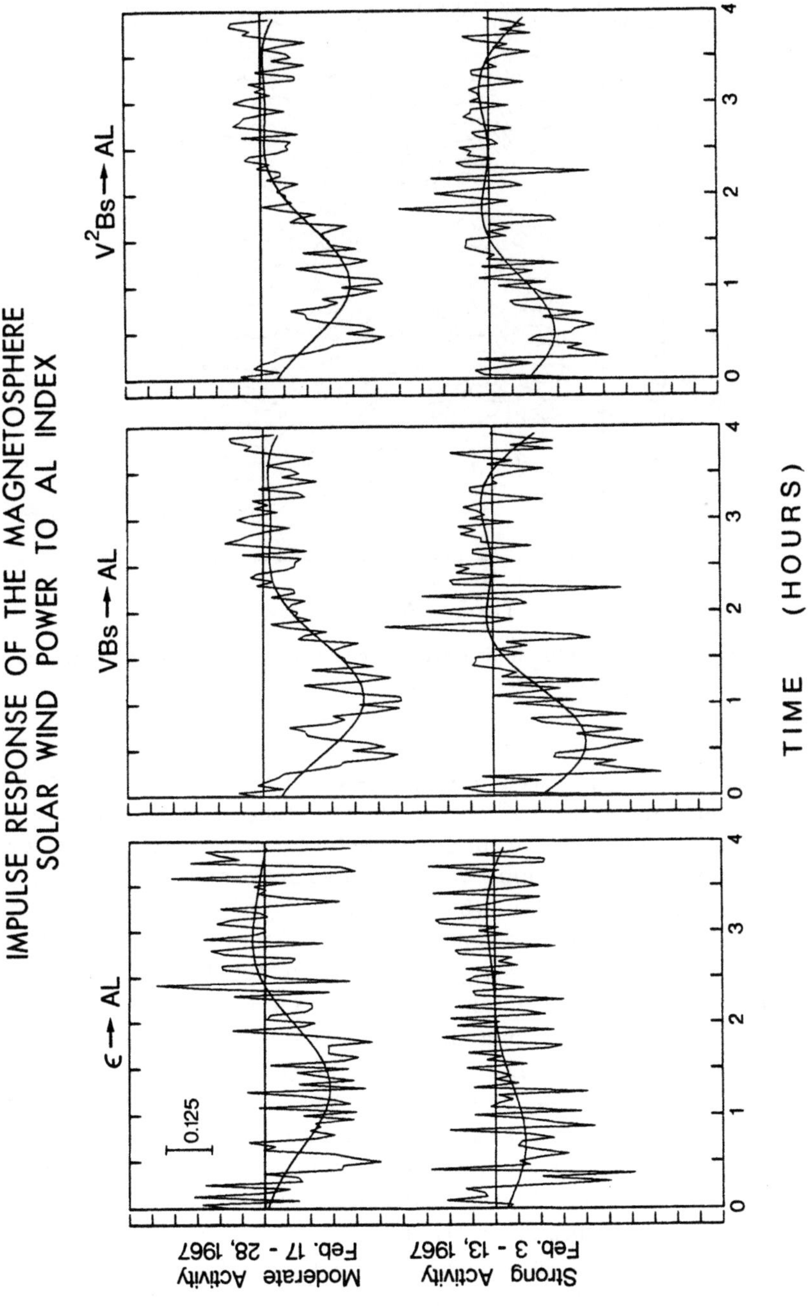

Fig. 2. Linear prediction filters relating various proposed solar wind power input parameters to the *AL* index during two intervals of moderate and severe geomagnetic activity. Time lag is plotted along the horizontal axis and the filter value is plotted along the vertical axis (Clauer *et al.*, 1981).

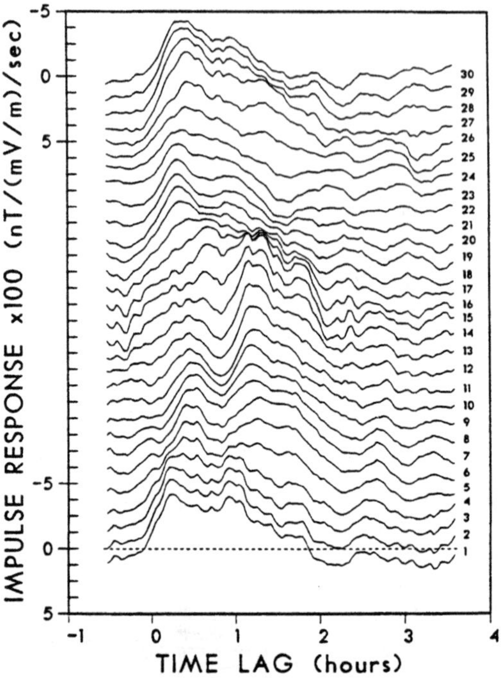

Fig. 3. Stack plot of Wiener Filters for varying levels of geomagnetic activity ranging from quiet at the bottom to severe at the top (BARGATZE *et al.*, 1985).

dominant at strong activity levels and both peaks about equal at weak activity levels.

It is suggested that the two peaks correspond to separate response modes of the magnetosphere and that the relative dominance of these modes varies with the level of activity. The physical interpretation given to the two peaks regards the first peak as corresponding to auroral activity which is directly driven by the solar wind-magnetospheric dynamo while the second peak corresponds to auroral activity which results from the release of stored energy from the magnetotail.

While auroral currents have obtained the greatest attention in linear prediction investigations, four studies have presented analysis which relate solar wind parameters to the ring current index D_{st}: IYEMORI *et al.* (1979); IYEMORI and MAEDA (1980); MCPHERRON *et al.* (1984); and indirectly AKASOFU *et al.* (1985). The results of these studies indicate that D_{st} is the best determined geomagnetic index using solar wind information alone. Using 5-minute data and a multi-channel filter with solar wind dynamic pressure and electric field inputs, MCPHERRON *et al.*, (1984) reproduced 70% of the variance in the D_{st} index.

Most of the studies using linear prediction filtering have concluded that only a portion of the output index is predictable from solar wind information alone IYEMORI *et al.*, 1979, 1980; CLAUER *et al.*, 1981, 1983; MCPHERRON *et al.*, 1984; and BARGATZE *et al.*, 1985). This is illustraited for the *AL* index in Fig. 4 which shows a

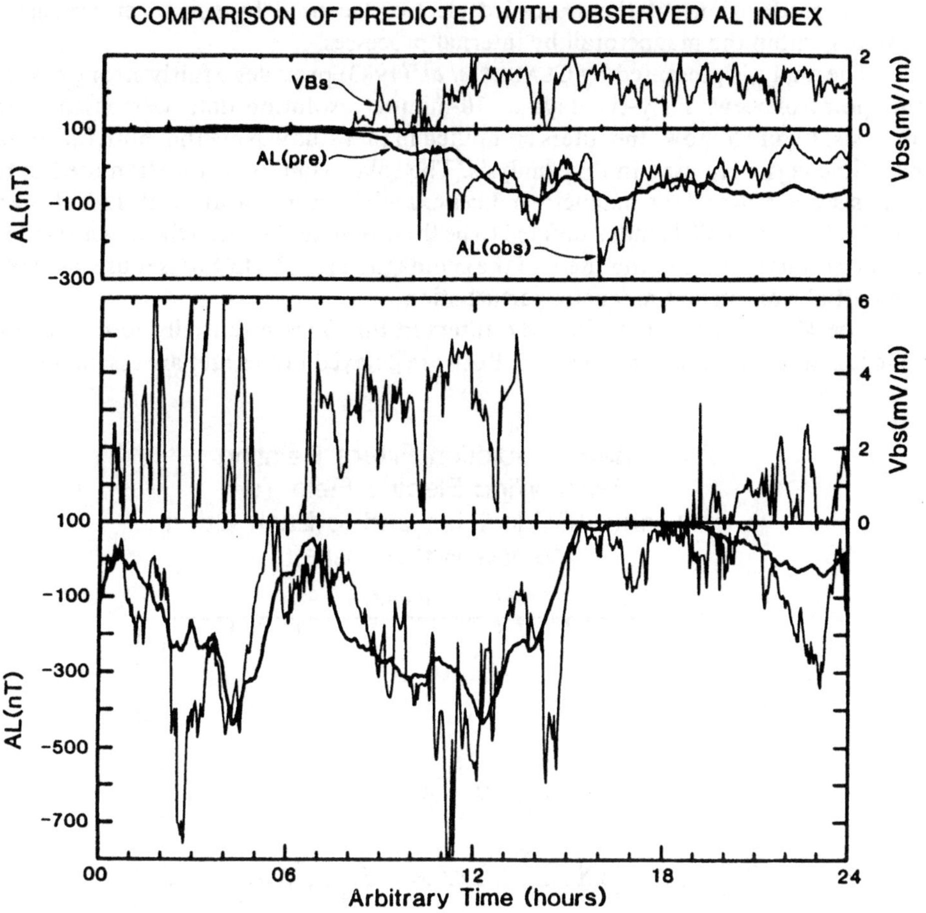

Fig. 4. Comparison of predicted and observed *AL* index for periods of moderate (top) and strong (bottom) geomagnetic activity. Each frame shows a plot of the solar wind input function VB_s at the top with the predicted (dark line) and observed (light line) *AL* index underneath, (figure courtesy of R. L. McPherron).

comparison between the measured *AL* index and the computed *AL* index. The top panel shows an interval of moderate level activity and the bottom panel shows an interval of strong activity. In each case the computed *AL* looks like a smoothed or low-pass filtered version of the observed index. It is suggested that unpredicted portions of the index can be attributed to noise in the measurements of the proper input parameters and in the indices which characterize the magnetospheric current systems, and also to real variations in the currents which are not consistently related to the solar wind driving function. That is, currents which are controlled by internal magnetospheric processes. There are several examples of large, sudden, negative excursions of the *AL* index which are not predicted and which are thought to result from intensifications of the westward electrojet caused by substorm expansions initiated within the magnetotail by internal processes.

The analysis presented by CLAUER *et al.* (1983) provides a fairly strong case for this point of view. Fifty-six days of 10-minute resolution data were used in the analysis. Figure 5 shows the filters computed in that study. Note that both future and past time lags were used in this analysis. The solar wind input function used in the analysis was VB_s and the magnetic indices examined were the auroral AL index and the low-latitude $ASYM$ index defined to be the difference between the dawn and dusk perturbations in the X component of the geomagnetic field. $ASYM$ is thus an index of the partial ring current magnetic perturbation.

The $VB_s \rightarrow ASYM$ and $VB_s \rightarrow AL$ filters in Fig. 5 are essentially zero prior to 0 time lag, thus they are purely causal. Both are delayed pulses and are essentially low

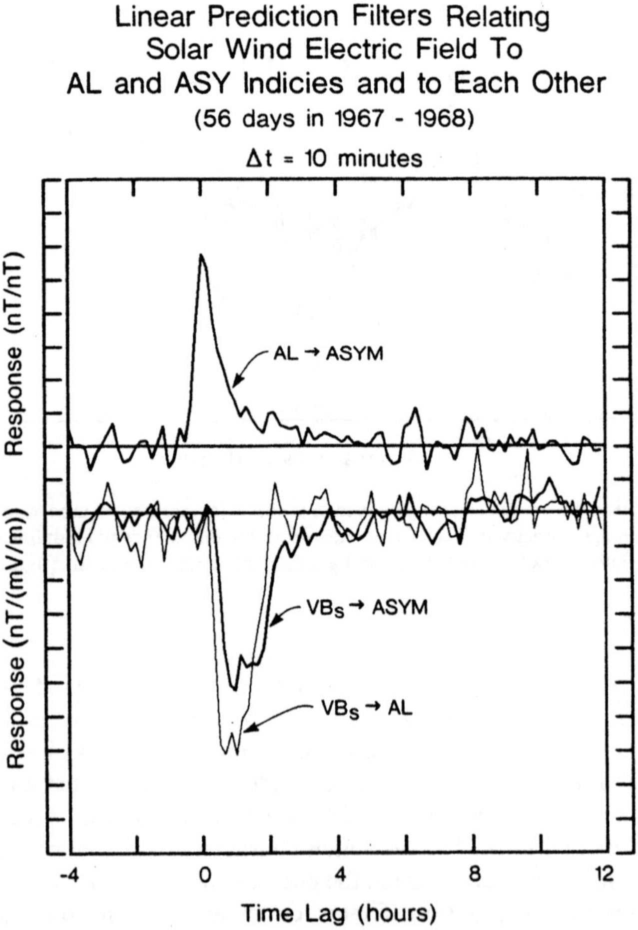

Fig. 5. Three Wiener Filters, $AL \rightarrow ASYM$, $VB_s \rightarrow ASYM$, and $VB_s \rightarrow AL$, plotted together for comparison (CLAUER, *et al.*, 1983).

pass filters with a time delay. The $AL \rightarrow ASYM$ filter is also shown in Fig. 5. This filter was created to test the notion that the partial ring current is the result of individual substorm expansions injecting bursts of energetic ions into the ring current region. These ions would drift westward to produce a partial ring current (BOGOTT and MOZER, 1973). We find, however, that the filter is nonzero at negative lags indicating that development of $ASYM$ precedes AL. This feature of the Wiener filter is consistent with the analysis of several individual events which showed the development of the partial ring current preceding the substorm expansion onset determined from both the AL index and the midlatitude positive bay signature (CLAUER and MCPHERRON, 1980).

The measures of the prediction efficiency provide some interesting information. About 45% of the variance of the $ASYM$ index is predictable from VB_s, while 56% is predictable from AL. Also, 41% of the variance of AL is predictable using VB_s. This suggests that the AL index contains information that is correlated with $ASYM$ but is unrelated to the solar wind. CLAUER *et al.* (1983) hypothesize that both AL and $ASYM$ are produced by two uncorrelated processes, one which directly involves the solar wind and another which is due to internal magnetospheric processes.

Additional analysis was performed on the error (or residual) time series. It was reasoned that the solar wind related portion of the signal is absent in the residual time series. There should, however, still remain the signal due to internal magnetospheric processes which are not directly determined by the solar wind input alone. To see if the error time series for AL and $ASYM$ both contained this internal magnetospheric signal, a filter relating the prediction residual of AL to the prediction residual of $ASYM$ was computed. It was found that 24% of the variance of the $ASYM$ residual was related to the AL residual.

Figure 6 shows a schematic model of the hypothesized system. The solar wind input VB_s is transformed to the output AL_s by filter g_1. A second component AL_x is assumed to be produced by the internal magnetospheric process through filter g_x and is added to AL_s to produce AL. The AL index in turn is transformed by filter g_2 into the $ASYM$ index. Therefore, in this model the solar wind input VB_s is related to $ASYM$ by a cascade of two filters g_1 and g_2. Provided that both systems are linear, the cascade can be represented by a single filter g_0 which is the convolution of g_1 and g_2. Thus,

$$g_{0,m} = \sum_{n-0}^{N} g_{1,n} g_{2,(m-n)}.$$

Figure 5 shows the empirically determined filters g_0 and g_2. The output of filter g_1, AL_s is not directly observed. By hypothesis, however, AL_x is uncorrelated with the solar wind so that g_1 is the filter determined from VB_s and AL. As a check of this hypothesis the $VB_s \rightarrow AL$ filter was convolved with the $AL \rightarrow ASYM$ filter for comparison with the $VB_s \rightarrow ASYM$ filter. The comparison is shown in fig. 7. A dashed line shows the empirical $VB_s \rightarrow ASYM$ filter and a solid line shows the filter produced by the convolution. The agreement between the two curves is quite good, thus supporting the hypothesis that AL is composed of both a solar wind driven

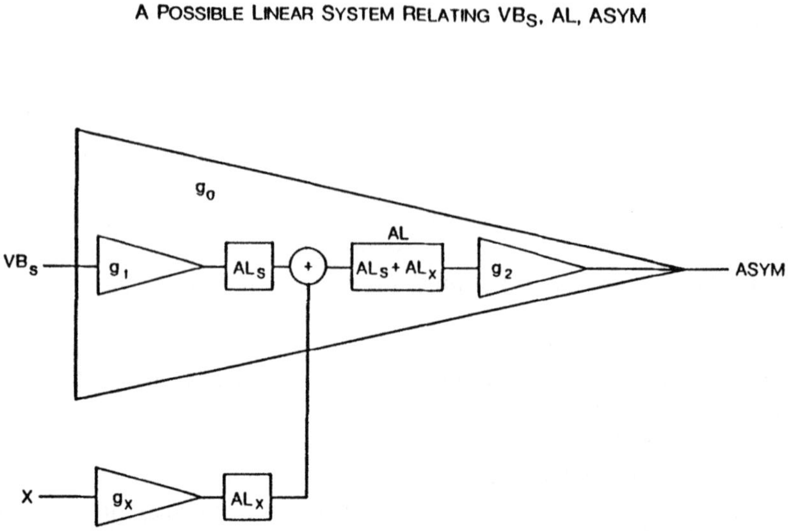

Fig. 6. Schematic diagram of a model in which the solar wind input function VB_s is transformed into the *ASYM* index by filter g_0. The *AL* index is transformed into *ASYM* by filter g_2. The *AL* index, however, is composed of two components. One component AL_s is related to the solar wind VB_s through filter g_1 while the other component AL_x is related to another process *X* by filter g_x. *AL* is the sum of these two signals (CLAUER *et al.*, 1983).

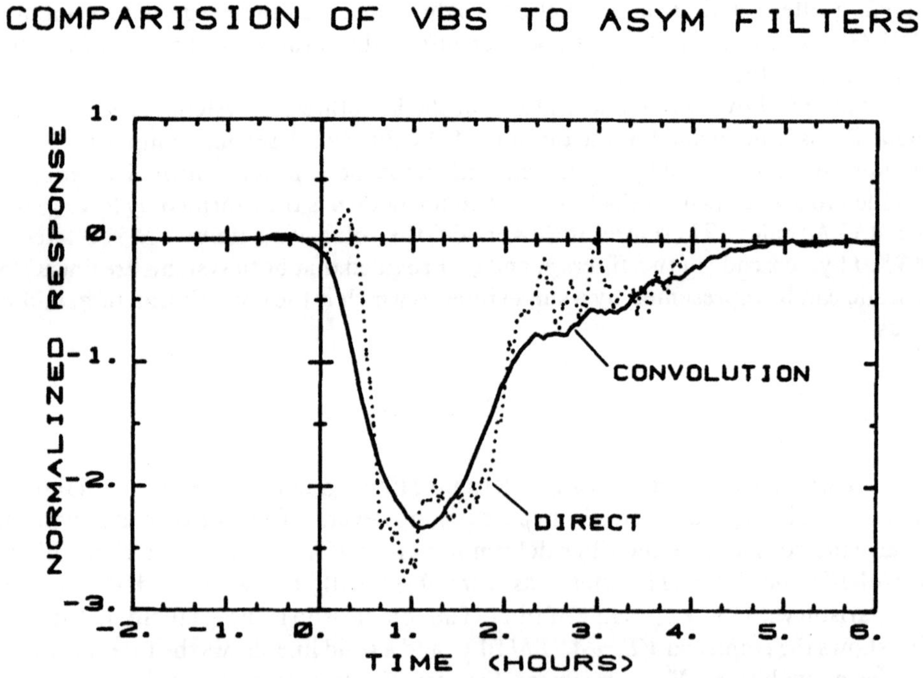

Fig. 7. Comparison of the VB_s—*ASYM* filter (g_0 in Fig. 5) with the convolution of the VB_s—*AL* and *AL*—*ASYM* filters (g_1 and g_2 in Fig. 5) (CLAUER *et al.*, 1983).

component and another component uncorrelated with the solar wind.

A somewhat different conclusion, however, was reached by AKASOFU *et al.* (1985) in an analysis which computed Wiener filters relating the ε parameter with the so-called total magnetospheric energy dissipation rate parameter, U_T. This investigation used 5-minute resolution data during magnetic storm intervals. The solar wind input parameter $\varepsilon \propto VB^2 \sin^4(\theta/2)$ was measured by the ISEE-3 satellite located 240 R_e upstream from the subsolar magnetopause. The output parameter U_T is given by:

$$U_T(\mathrm{erg/sec}) = 4 \cdot 10^{20} \left(\frac{\partial \overline{D_{st}}}{\partial t} + \frac{\overline{D_{st}}}{\tau_R} \right) + 3 \cdot 10^{15} AE$$

where $\overline{D_{st}}$ is a solar wind pressure adjusted D_{st} index and τ_R denotes a ring current decay time scale.

Figure 8 show plots from 3 of the 6 storms. Each column of plots shows data from one storm. The top panel shows the input ε parameter, the next panel shows the

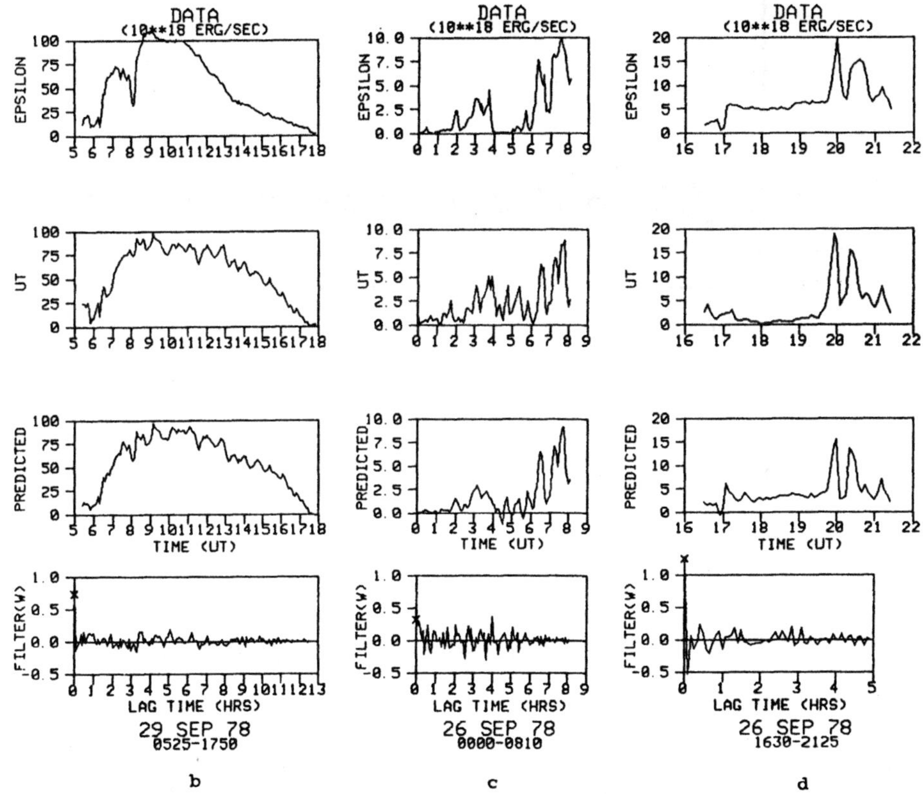

Fig. 8. Examples of linear prediction analysis for 3 geomagnetic storms on September 29, 1978. From the top are shown the input parameter ε, the output function *UT*, the prediction of *UT*, and the prediction filter (AKASOFU *et al.*, 1985).

output U_T parameter, the next panel shows the predicted $\hat{U}_T$ parameter and the bottom panel shows the computed Wiener filter. Figuer 9 shows a superposition of all six of the computed Wiener filters.

The authors note that for the 3 largest storms studied the predicted $\hat{U}_T$ is a good fit to the measured U_T while the smaller 3 storms were not well fit. The filters shown in Figs. 8 and 9 are essentially δ functions at $t=0$. These results, the zero time lag and the good prediction, are interpreted to indicate that for periods of strong activity magnetospheric energy dissipation is directly driven with little or no time lag.

The zero time lag in the Akasofu *et al.* filters, however, is troubling and seems physically implausible, particularly when the uncertainty of the transit time between ISEE-3 and the magnetopause is greater than the accuracy of the 5-minute data resolution (RUSSELL *et al.*, 1980). Further, spikes at the ends of filters are often characteristic of errors in removing the mean or trends from time series. Alternatively, it is likely that the methods used to compute U_T are cause of this artifact. The

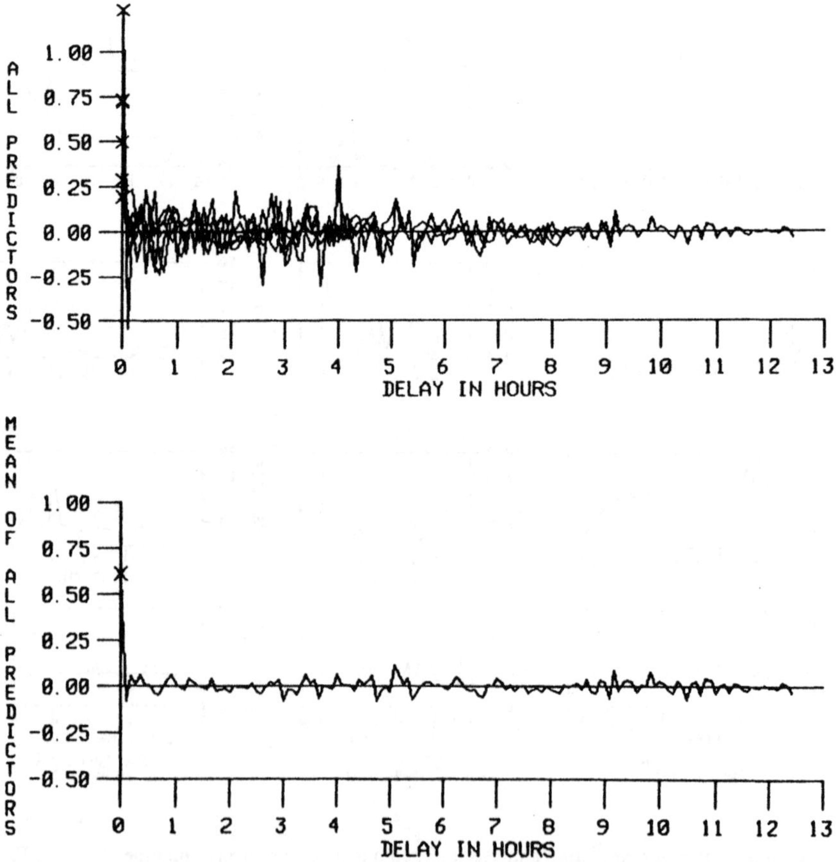

Fig. 9. Superposition (top) and average (bottom) of 6 impulse response functions computed for 6 geomagnetic storms (AKASOFU *et al.*, 1985).

derivation of U_T is based on several solar wind parameters. The value of D_{st} is adjusted for the solar wind pressure and the ring current decay "constant" τ_R is a function of the input function ε. The output parameter therefore contains terms from the measured input thereby confusing the correlation and producing a zero-lag artifact. One must also note that the application of Wiener filtering to single event studies may be inappropriate. The spikey character of the filters obtained is an indication of this. The ability to compute physically meaningful filters relies on the least squares aspects of the analysis and requires many correlated variations in the input and output time series.

The pressure effect could more properly be taken care of by using a multichannel analysis which inputs both solar wind dynamic pressure and ε to an output composed only of geomagnetic measurements. To do this, τ_R must be obtained independently from the input time series ε. For example, FELDSTEIN *et al.* (1984) estimate values for τ_R based on the magnitude of D_{st}. Such a multichannel technique would also provide a better method of determining the transit time from ISEE-3 to the magnetopause since the ρV^2—D_{st} filter was shown to be a narrow pulse at a delay time consistent with the expected propagation time between the satellite and magnetopause (MCPHERRON *et al.*, 1984).

4. Summary and Conclusions

The technique of linear prediction filtering has provided another powerful statistical tool for the data analyst to use in an attempt to resolve some of the complex relationships which exist between various geomagnetic current systems and solar wind input energy. It appears clear from all of the Wiener filtering studies cited above that a substantial portion of the geomagnetic disturbances observed on the earth can be directly related to solar wind measures by linear impulse response functions. This suggests that the responsible current systems are very directly coupled to (or driven by) the solar wind-magnetosphere MHD dynamo. In addition, however, there appears to be a substantial portion of the AL index which results from currents not strictly determined by the solar wind input.

Investigations have used Wiener analysis to compute the linear prediction filters which relate solar wind input parameters, primarily ε and VB_s, with the current systems manifested in AL, AU, D_{st} and $ASYM$. In general, these filters have low pass characteristics with linear phase and time delays. The different filters, however, quantitatively show that different indices respond differently to the solar wind electric field and hence, that different physical process produce the electric currents which influence each index. It also seems clear that the assumptions of linear and time invariant relationships between the input and output are only approximately valid.

The D_{st} index appears to be the most predictable. Solar wind inputs can reproduce over 70% of the variance in the D_{st} index (IYEMORI *et al.*, 1979; MCPHERRON *et al.*, 1984). The AL and $ASYM$ indices can be predicted relatively well with about 40% of the variance of each predicted by the solar wind electric field or ε. This seems impressive in view of the errors present in the measurements of both the input and output time series. The AU index is only poorly predicted and this is a major concern to this author. This may indicate either a major misunderstanding of the sources controlling the eastward electrojet or serious problems with the interpretation that AU in fact characterizes the eastward electrojet.

The low pass characteristics of the $VB_s \rightarrow AL$ filters indicate that the magnetosphere attenuates solar wind electric field frequencies above 10^{-4} hertz in the generation of the current systems linked with the westward auroral electrojet. Also, careful examination of the residual or error time series suggests that there are variations in the output indices which are produced by processes not directly related to the solar wind input function. It has been suggested, for example, that these separate sources of current in the westward electrojet are controlled by processes in the magnetotail which are associated with the release of stored energy.

The results of these analyses provide an empirical description of the solar wind-magnetospheric system. This, in turn, provides insight to the physical processes which operate in the system. Most of the studies reported above have utilized Wiener filtering in this way. The technique, however, may eventually be employed in the more pragmatic way of providing the ability to forecast short term geomagnetic activity if the proper inputs were provided sufficiently far upstream in the solar wind. A satellite at the Lagrangian point 240 R_e upstream, for example, might provide such short term input. If proper filters could be produced to relate observations at this location with geomagnetic activity indices, predictions could easily be computed. Based on present results, it seems reasonable to expect that such filters could be obtained and reasonably accurate predictions could be produced. Such predictions could have great benefit to a variety of commercial and scientific activities which are influenced by the level of geomagnetic disturbance.

The preparation of this review was supported by NASA under grant NAGW-235, by the National Science Foundation under grant ATM-8210562, and by the Air Force Geophysics Laboratory under Contract F19628-85-K-0001. I am indebted to Dr. R. L. McPherron for many challenging and helpful discussions and to Dr. McPherron, Mr. Lee Bargatze, and Dr. S.-I. Akasofu for assistance in obtaining figures for this paper. I very much appreciate the critical reading of an earlier version of this manuscript and helpful suggestions by Dr. R. D. Zwickl and Dr. D. N. Baker.

REFERENCES

AKASOFU, S.-I., C. OLMSTED, E. J. SMITH, B. TSURUTANI, R. OKIDA, and D. N. BAKER, Solar wind variations and geomagnetic storms: A study of individual storms based on high time resolution ISEE 3 data, *J. Geophys. Res.*, **90**, 325, 1985.

BAKER, D. N., Statistical analyses in the study of solar wind-magnetosphere coupling, this volume, 1986.

BARGATZE, L. F., D. N. BAKER, R. L. MCPHERRON, and E. W. HONES, Magnetospheric response for many levels of geomagnetic activity, *J. Geophys. Res.*, **90**, 6387, 1985.

BOGOTT, R. K. and F. S. MOZER, ATS 5 observations of energetic proton injection, *J. Geophys. Res.*, **78**, 8113, 1973.

BURTON, R. K., R. L. MCPHERRON, and C. T. RUSSELL, An emprircal relationship between interplanetary conditions and D_{st}, *J. Geophys. Res.*, **80**, 4204, 1975.

CLAUER, C. R. and R. L. MCPHERRON, The relative importance of the interplanetary electric field and magnetospheric substorms on partial ring current development, *J. Geophys. Res.*, **85**, 6747, 1980.

CLAUER, C. R., R. L. MCPHERRON, C. SEARLS, and M. G. KIVELSON, Solar wind control of auroral zone geomagnetic activity, *Geophys. Res. Lett.*, **8**, 915, 1981.

CLAUER, C. R., R. L. MCPHERRON, and C. SEARLS, Solar wind control of the low-latitude asymmetric magnetic disturbance field, *J. Geophys. Res.*, **88**, 2123, 1983.

FELDSTEIN, Y. I., V. YU. PISARSKY, N. M. RUDNEVA, and A. GRAFE, Ring current simulation in connection with interplanetary space conditions, *Planet. Space Sci.*, **32**, 975, 1984.

IYEMORI, T., H. MAEDA, and T. KAMEI, Impulse response of geomagnetic indices to interplanetary magnetic fields, *J. Geomag. Geoelectr.*, **31**, 1, 1979.

IYEMORI, T. and H. MAEDA, Prediction of geomagnetic activities from solar wind parameters based on the linear prediction theory, in *Solar-Terrestrial Predictions Proceedings, Vol. 4*, A-1, U.S. Dept. of Commerce, Boulder, CO. 1980.

KAMIDE, Y. and S.-I. AKASOFU, Notes on the auroral electrojet indices, *Rev. Geophys. Space Phys.*, **21**, 1647, 1983.

LEVINSON, N., The Wiener RMS (root mean square) error criterion in filter design and prediction, Appendix B in N. Wiener's book, 1949.

MAYAUD, P. N., *Derivation, Meaning and Use of Geomagnetic Indices*, Geophys. Monogr. 22, AGU, Washington, D.C., 1980.

MCPHERRON, R. L., R. A. FAY, C. R. GARRITY, L. F. BARGATZE, D. N. BAKER, C. R. CLAUER, and C. SEARLS, "Coupling of the solar wind to measures of magnetic activity," in *Achievements of the International Magnetospheric Study (IMS), Proceedings of an International Symposium, Graz, Austria 26–28 June*, pp. 161, ESA SP-217, European Space Agency, Paris, 1984.

ROBINSON, E. A., *Multichannel Time Series Analysis with Digital Computer Programs*, Holden-Day, San Francisco, 1967.

ROBINSON, E. A. and Sven TREITEL, *Geophysical Signal Analysis*, Prentice-Hall, Englewood Cliffs, 1980.

ROSTOKER, G., Geomagnetic indices, *Rev. Geophys. Space Phys.*, **10**, 935, 1972.

RUSSELL, C. T., G. L. SISCOE, and E. J. SMITH, Comparison of ISEE-1 and -3 interplanetary magnetic field observations, *Geophys. Res. Lett.*, **7**, 381, 1980.

WIENER, N., *Extrapolation, Interpolation, and Smoothing of Stationary Time Series with Engineering Applications*, M.I.T. Press, Cambridge, Mass. 1942.

Solar Wind-Magnetosphere Coupling, edited by Y. Kamide and J. A. Slavin, 59–83.

Solar Wind Velocity Effects on the Auroral Zone Magnetic Disturbances

K. MAEZAWA and T. MURAYAMA

Department of Physics, Nagoya University, Nagoya 464, Japan

Effects of the solar wind velocity, V, on the auroral zone magnetic disturbances are studied in order to obtain a clue to the nature of the solar wind-magnetosphere coupling mechanism. From the reviews of past works and the analyses newly performed on several aspects of the effect of V, following results were obtained: (1) V dependence of the AL index is steeper than linear; approximately proportional to V^2. The V^2 dependence is also observed for the indices *am* and D_{st}. (2) The V dependence of AL is also subject to χ, the tilt angle of the geomagnetic axis; steepest when $\chi \simeq 0$, i.e., when the geomagnetic axis is near normal to the sun-earth line. (3) The V dependence of the deviation from the quiet level of the x-component of auroral zone geomagnetic field is different in different local times; steepest in the 2200–2400 sector. Implication of these results are discussed with emphasis on the possible contribution of the viscous-like interaction in the solar wind-magnetosphere coupling.

1. Introduction

Many of the geomagnetic disturbances observed on the earth's surface are known to be the consequences of the coupling between the solar wind plasma and the terrestrial magnetosphere. Of the two types of coupling hitherto considered as the basic mechanisms, the magnetic coupling, first proposed by DUNGEY (1961), has a strong support by the observational evidences of the IMF-B_z dependence of geomagnetic disturbances. On the other hand, the role of the non-magnetic coupling, such as that first proposed by AXFORD and HINES (1961), in the magnetospheric phenomena has not been quite identified from the viewpoint of geomagnetic activity.

In the magnetic coupling, the solar wind velocity, V, and the southward component of the interplanetary magnetic field, B_s are the most important parameters of the solar wind, because the product $B_s V$ is directly related to the rate of the solar wind magnetic flux transport to the dayside magnetopause of the earth to be reconnected with geomagnetic fields. Also in the case of non-magnetic coupling, the solar wind velocity is expected to play an important role in determining the rate of energy transfer to the magnetosphere, though the way of influence is probably different from the magnetic coupling. Thus the investigation of the influence of the solar wind velocity on geomagnetic disturbances is expected to yield valuable information on the relative contributions of the two types of coupling in the solar wind-magnetosphere energy transfer.

In this paper, we deal with the effect of the solar wind velocity on the auroral zone geomagnetic disturbances. In Section 2, we are mainly concerned with the average dependence of the auroral zone index AL on the solar wind velocity. A mid-latitude index *am* and the ring current index D_{st} are also discussed to show that the three indices have nearly identical V dependence. In Section 3, the influence of the angle between the geomagnetic axis and the sun-earth line on the V dependence of the AL index is studied in relation to a possible role of the Kelvin-Helmholtz instability on the non-magnetic coupling. In Section 4, auroral zone geomagnetic disturbances are studied directly, not in terms of geomagnetic indices, to examine the change in local time of the V dependence of auroral zone geomagnetic disturbances.

2. Solar Wind Velocity Dependence of Geomagnetic Indices

2.1 *Auroral zone indices*

Auroral zone electrojet is one of the principal feature of the substorm-related magnetospheric disturbances, and its dependence on the solar wind parameters has been investigated extensively. In this subsection, we mainly deal with the AL index, which is a measure of the intensity of the westward electrojet. The AE $(=AU-AL)$ index is also widely used in the literature, but its two constituents, AU and AL, are known to respond differently to the solar wind velocity (MAEZAWA, 1979). However, as AL is the dominant contributor to AE of the two, the solar wind velocity dependence of AE almost duplicates that of AL (see, for instance, Fig. 1 and Fig. 2 of MURAYAMA (1982)).

Results of works concerning the dependence of AL (and AE) on the solar wind parameters, published before 1982 have been reviewed by MURAYAMA (1982) by the comparison of the ability of B_s, $B_s V^2$ and ε to predict AL, where B_s is the southward component of the interplanetary magnetic field (IMF), V is the solar wind velocity and the parameter ε (PERREAULT and AKASOFU, 1978) is defined by

$$\varepsilon = B^2 V \sin^4 (\theta/2)\, L_0^2 \tag{1}$$

where B is the magnitude of the IMF, θ is the colatitude of the projected vector of the IMF on the y–z plane of the GSM coordinate system, and L_0 is a constant ($=8$ R_e).

Figure 1 shows the V dependence of AL for different ranges of B_s. Three-hourly averages of B_s, V and AL are used with 1-h lag of AL from B_s and V in this result. However, as the characteristic time scale in the change of the z-component of B is usually smaller than 3 hours, B_s was calculated using hourly values of B_z and taking into account the changes within 1 hour by the use of σ_z, the standard deviation of the variability of the IMF B_z (MURAYAMA *et al.*, 1980).

It is seen in Fig. 1 that the V dependence of AL for low values of B_s is not linear, suggesting the V^2 dependence, but becomes closer to being linear as B_s increases. The V^2 dependence of AL for low values of B_s is clearly seen in Fig. 2, where average AL values are plotted against $(V/400\ \mathrm{km/s})^2$ instead of V. Curves fit very well with the lines of $AL=61\,(B_s+0.48)\,(V/400)^2$ when AL is less than 250 nT, beyond which AL values begin to separate downward from the lines. This depression was mentioned by MURAYAMA (1982) and MAEZAWA (1978) as the saturation tendency of AL for large values of functions. Origin of the saturation tendency is left to be pursued in detail, although the broadening and the shift toward low latitudes of auroral electrojet

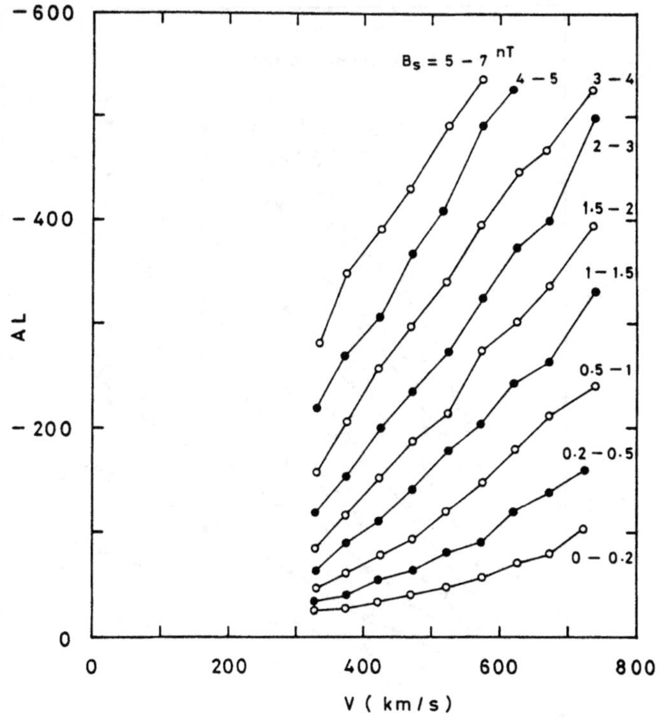

Fig. 1. Average dependence of the AL index on the solar wind velocity V, shown separately for different ranges of B_s. Data of solar wind parameters are 3-h average values obtained from their hourly values in a data set compiled by KING (1977, 1979, 1983) for the period 1967–1980. The same data set of 3-h averages are also used in the analyses for Figs. 4, 10, 11 and 12.

currents may be at least partly repsonsible for the saturation.

Figure 3 shows the average relation between AL and $B_s V$, $B_s V^2$, ε, and the fourth function $(B_s+0.5)(V/400)^2$ using 1-h average values of functions with 1-h lag of AL, shown separately for the three data sets; data with $B>10$ nT, data with $B<10$ nT and $V>600$ km/s, and others. This figure is quite similar to Fig. 1 of MURAYAMA (1982) which were derived by the same method, except that 3-h averages are used instead of 1-h averages. Degree of accordance of the curves corresponding to the three data sets in the upper three panels of Fig. 3 is a measure of the appropriateness of the values of exponents a and b of the product $B^a V^b$ which are common in three functions $B_s V$ ($a=1$, $b=1$), $B_s V^2$ ($a=1$, $b=2$) and ε ($a=2$, $b=1$), and the function $B_s V^2$ is the best of the three as the predictor of AL from this point of view. Curves for the three data sets most beautifully overlap with each other for the function $(B_s+0.5)(V/400)^2$, which is practically the same with that responsible for the lines shown in Fig. 2. Cross marks in Fig. 2 denote the average dependence of AL on $(V/400)^2$ deduced by a smoothed curve obtained from the bottom panel of Fig. 3. A good agreement between the crosses and the observed values in this figure confirms that the separation of observed

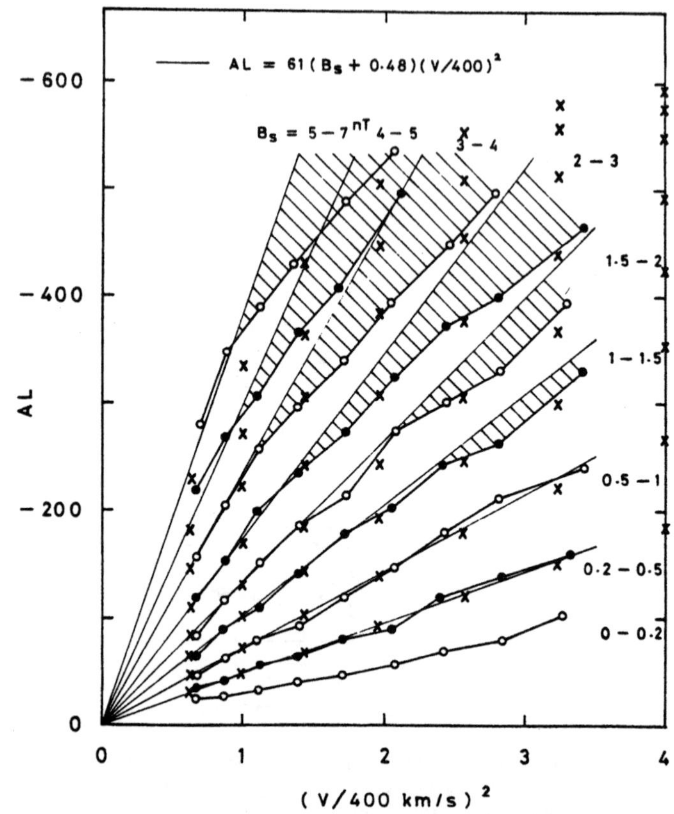

Fig. 2. Same as Fig. 1, but AL is plotted against $(V/400\ \mathrm{km/s})^2$. Straight lines denote the relation $AL = 61\,(B_s + 0.48)(V/400)^2$ with the value of B_s taken as the average for each data set of B_s. See the text for the implication of cross marks.

values from the lines in Fig. 2 originates from the saturation tendency as shown in Fig. 3.

HOLZER and SLAVIN (1982) compared $B_s V$, $B_s V^2$ and ε for the ability of the prediction of AL, and showed that $B_s V^2$ is the best of three, in agreement with our conclusion based on the results in Figs. 1–3.

MELONI *et al.* (1982) showed, in a comparison of $B_z V$ and ε using hourly data, that $B_z V$ is better than ε to predict AE when the whole data are used, whereas log ε is the better function than $B_z V$ when only data for negative B_z are used. This good correlation of log ε with AE is probably related to an approximately logarithmic shape of the saturation curve for ε in the top panel of Fig. 3. The function $B_s V^2$ is not treated in this work.

BAKER *et al.* (1983) made a fine time resolution analysis on the correlation of AE and the three functions in question for the three periods of 8 to 10 day duration, and showed that $B_s V$ and $B_s V^2$ are essentially indistinguishable in their correlation with AE,

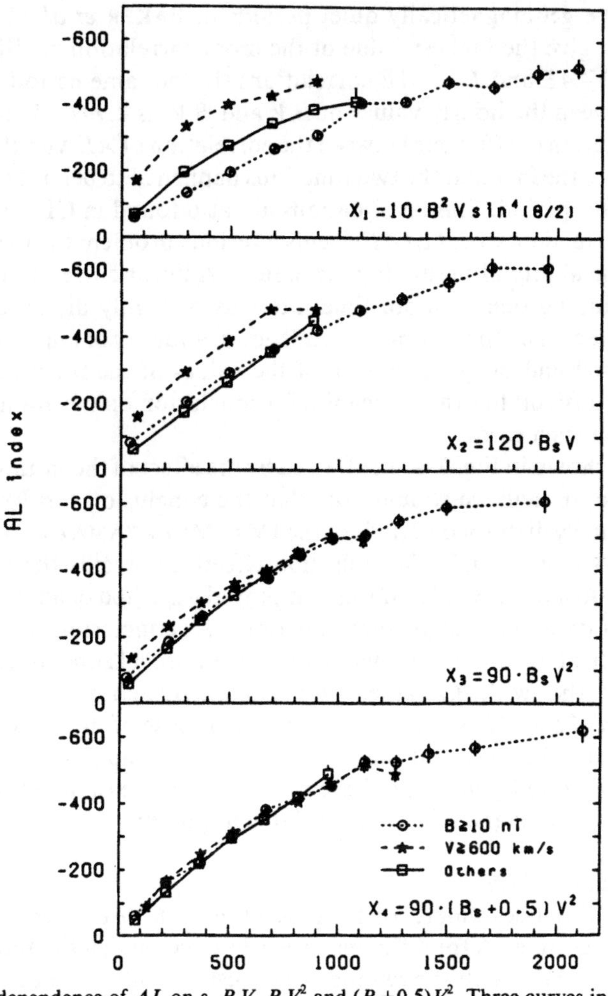

Fig. 3. Average dependence of AL on ε, B_sV, B_sV^2 and $(B_s+0.5)V^2$. Three curves in each panel are for different data sets: $B>10$ nT, $B<10$ nT and $V>600$ km/s, and $B<10$ nT and $V<600$ km/s, respectively.

but they are generally better correlated with AE than is ε. TSURUTANI *et al.* (1985) also made a fine time resolution analysis on the correlation of the three functions and AL in the two CDAW 6 periods. No significant difference in the correlation behavior for the three functions with AL was observed in these periods. It should be mentioned here that nearly identical correlations with AE of two functions do not necessarily mean the nearly identical abilities of the two functions to predict AE. This is sometimes caused by a nearly identical behavior of the two functions in a given period, which happens, for instance, between B_sV and B_sV^2 when the solar wind velocity is nearly constant in the period. A typical example is found in August 25–26,

1978, one of the geomagnetically quiet periods in BAKER *et al.* (1983). Fine time resolution data give the highest value of the cross correlation coefficients of about 0.65 for the $B_s V$–AE and $B_s V^2$–AE correlations. In the same period, the correlation coefficient between the hourly values of $B_s V$ and $B_s V^2$ is 0.997. Hence, the virtually identical correlation coefficients between the correlation of AE with the two functions are simply due to the fact that the two functions happened to behave so similarly with each other in this period. Similar situations are also found in CDAW 6 periods.

Saturation tendency of AL (AE) yields a serious problem for the comparison of functions on the ability of predicting the indices. A linear correlation coefficient for the period including one or more time intervals of highly disturbed geomagnetic conditions is affected by the saturation tendency in a variety of manners depending on the function used and on the spectrum of the values of the function in the period. Thus, it is very difficult to draw a meaningful conclusion on the abilities of functions to predict AE in such cases.

The result shown in Fig. 3 is not affected by the effect of the saturation, and hence is more reliable, though we should note that the conclusion for $B_s V^2$ has not been tested for extremely disturbed periods of the IMF. MURAYAMA *et al.* (1980) used only data with B less thatn 10 nT. This selection effectively avoids the highly disturbed periods. HOLZER and SLAVIN (1982) also employed a method of analysis to reduce the contribution of data in highly disturbed periods. As conclusions of the above three works are identical, it can be summarized that the auroral zone index AL correlates better with $B_s V^2$ that with $B_s V$, at least for moderately active condition of the IMF. The correlation of ε to AL is somewhat lower than those of the above two functions. Although it cannot be ruled out that a function other than $B_s V^2$ is more relevant in highly disturbed periods, we do not have reliable ways to examine that, by the use of the auroral zone geomagnetic indices available at present.

2.2 *Mid-latitude indices*

Mid-latitude indices widely used at present (*ap*, *am*, *aa* etc.) are derived from the values of 3-h range index K for different networks of geomagnetic stations. We will be mainly concerned here with the *am* index because the influence of the longitudinal distribution of the stations are most carefully eliminated. The K_p index is also very popular, but is not apt to our purpose because of its nonlinear dependence on the observed range of geomagnetic disturbance.

Figure 4 shows the V dependence of the *am* index for different ranges of B_s, the same representation as Fig. 1 for AL. The nonlinear dependence of *am* on the solar wind velocity is clearly observed in all of the curves. Figure 5 shows the same result but displayed differently by taking $(V/400\ \mathrm{km/s})^{1.9}$ in the abscissa. It is readily seen in the figure that the whole data fit well to the lines representing the equation 6.67 $(B_s + 1.2\ \mathrm{nT})(V/400)^{1.9}$, with the values of B_s taken to be their average value for each range of B_s. There is no indication of saturating tendency which is seen in Fig. 2 for AL.

Relatively fewer results are available for the V dependence of *am* and *ap* compared with those for AL and AE. CROOKER *et al.* (1977) compared $|B_z| V$ and $|B_z| V^2$ for the correlation with *ap* using the yearly averages from 1965 to 1974, and showed that *ap* is better and more linearly correlated with $|B_z| V^2$ than with $|B_z| V$.

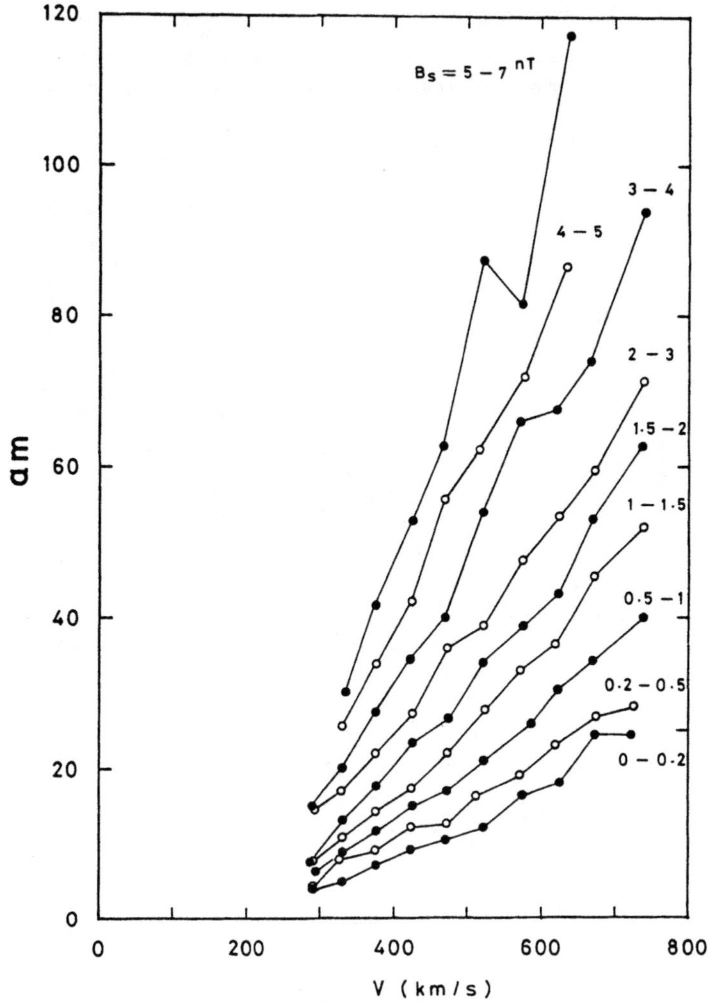

Fig. 4. Average dependence of the *am* index on *V*, shown separately for different ranges of B_s.

SVALGAARD (1977) obtained the relation $am \propto BV(DV^2)^{1/3} f(\theta)$, where θ is the latitudinal angle of the IMF direction. GARRETT *et al.* (1974) showed that *ap* is best correlated with a function which is a linear combination of $B_s V$ and σV, where σ is a square root of the total variance of the IMF. However, the correlation coefficient is not significantly different from that with $B_s V^2$. On the other hand, the result by MAEZAWA (1979), also incorporating the effect of σ, is of the form $am \propto B^{1.03}\ V^{2.34}\ D^{0.20}\ (\sigma/B)^{0.34} f(\theta)$.

There are no serious disagreements for the *V* dependence of *am* or *ap* in the results of analysis so far reported, and it can be summarized that *am* (*ap*) is approximately proportional to $B_s V^b$, where $b \cong 2.0 \pm 0.3$.

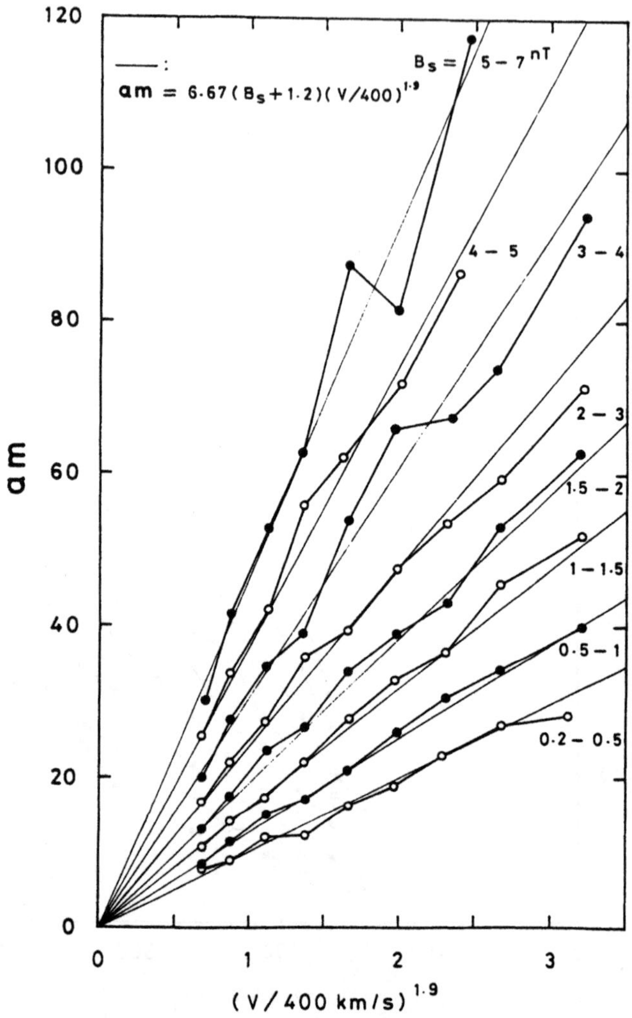

Fig. 5. Same as Fig. 4, but *am* is plotted against $(V/400\ \mathrm{km/s})^{1.9}$. Straight lines denote the relation $am = 6.67\ (B_s + 1.2)(V/400)^{1.9}$ with the value of B_s taken as the average for each data set of B_s.

2.3 D_{st} index

Analysis of the relation between D_{st} and B_sV, B_sV^2 or ε is not as simple as for *AL* or *am*, because the ring current particles injected in association with the enhancement of the values of the function survives much longer than one hour. Hence the method of direct correlation between the values of D_{st} and those of the function is not applicable but a more eleborate method such as that employed by BURTON *et al.* (1975) is needed.

Procedure of analysis in this subsection is essentially identical with that employed by MURAYAMA (1982). We tentatively assume various forms of the coupling function between the solar wind and the magnetosphere, calculate the ring current intensity predicted by each form of the coupling function, and find the function best correlated with the observed values of D_{st}. We assume that energy is injected into the ring current in proportion to the value of $F(t)$, with a time delay of one hour. The injected ring current then decays by a decay function $G(t_d)$, where t_d is time after the injection. Under these assumptions, D_{st} at the i-th hour from the beginning of the time interval of our interest is expressed as

$$D_{st}(i) = C\sum_{j=-j_0}^{i-1} F(j)G(i-1-j) + D_{CF} + K \tag{2}$$

where C and K are constants. We also define

$$D_{st}^*(i) = D_{st}(i) - D_{CF}(i) \tag{3}$$

for latter use.

The first term in the right-hand side of (2) is the sum of the contribution of the coupling function F for a (j_0+i-1) hour interval weighted by the decay function G. Introduction of j_0 means that an additional time interval of the duration j_0 hours is included before the time interval considered, which is necessary to calculate $D_{st}(i)$ near the beginning of the main time interval. In the present analysis we use $j_0>48$ hours.

The second term D_{CF} in the right-hand side of (2) denotes the contribution of Chapman-Ferraro current and the value of D_{CF} is related to the solar wind dynamic pressure DV^2 by the equation $D_{CF}=w(DV^2)^{1/2}$, where D is the solar wind proton density, with the values of $w=0.20$ nT (ev cm^{-3})$^{-1/2}$ (BURTON *et al.*, 1975).

All of the previous works of this kind (BURTON *et al.*, 1975; PERREAULT and AKASOFU, 1978; MURAYAMA, 1982; FELDSTEIN *et al.*, 1984) used a simple exponential function for the decay function G. As it was proved to be too crude, the function of the form

$$G(t) = a_1\exp(-t/\tau_1) + a_2\exp(-t/\tau_2), \quad (a_1 + a_2 = 1) \tag{4}$$

will be used in this analysis.

We hereafter denote the hourly values of D_{st} and D_{st}^* derived from observation by D_{sto} and D_{sto}^*, and the computed hourly prediction values simply by D_{st} and D_{st}^*, respectively.

Time intervals of 5 to 7 day duration were selected in 1967– 1979 on the basis of the availability of continuous data for the IMF, solar wind proton density and bulk velocity, in a data set compiled by KING (1977, 1979, 1983). The total number of the time intervals thus selected is 135. They were classified into three groups (A, B and C) according to the variance, Δ, of D_{sto}^* in the interval, i.e., $\Delta>15$ nT for A, 10 nT$<\Delta<$15 nT for B and $\Delta<10$ nT for C, respectivley. Hereafter we use the term 'event' not for the individual magnetic storm event, but for the storm activity in the individual time intervals which may contain two or more storms.

When the form of the coupling function is given, 120 to 168 values of $D_{st}^*(i)$ are calculated by (2) and (3) corresponding to 5 to 7 day duration of the time intervals. Constants C and K in (2) are determined by regression analysis between $D_{sto}*(i)$ and $D_{st}*(i)$ (see Fig. 6). The value of the correlation coefficient, r, obtained through the regression analysis, is utilized as a measure of the ability to predict D_{st} of the coupling function employed.

By the procedure as described so far, we have 135 values of r for a given form of the coupling function. Here we deal with $<r>$, the average of 90 values of r for events of groups A and B, for the comparison of the abilities of different functions to predict the time profiles of $D_{sto}*$.

We begin with three functions $f_1=B_sV$, $f_2=B_sV^2$ and $f_3=\varepsilon$, as was the case in subsection 2.1. Three curves in the lower part of Fig. 7 shows the values of $<r>$, given as a function of τ_2 for the three functions. Optimum values of other adjustable parameters τ_1 and a_1/a_2 are used to derive the curves. They are differenct for different functions, but lie around the vlaues $\tau_1=2$ h and $a_1/a_2=2$, respectively.

Next we proceed to the evaluation of the role of the solar wind proton density D in a coupling funciton. Dashed curves in the upper part of Fig. 7 shows the comparison of $f_{id}=f_iD^c$ ($i=1, 3$), where optimum values of c are used, i.e., $c=0.25$ for f_{1d}, $c=0.40$ for f_{2d} and $c=0.05$ for f_{3d}, respectively.
However, because of the small value of c for f_{3d}, the curve for f_{3d} virtually coincides with that of f_3.

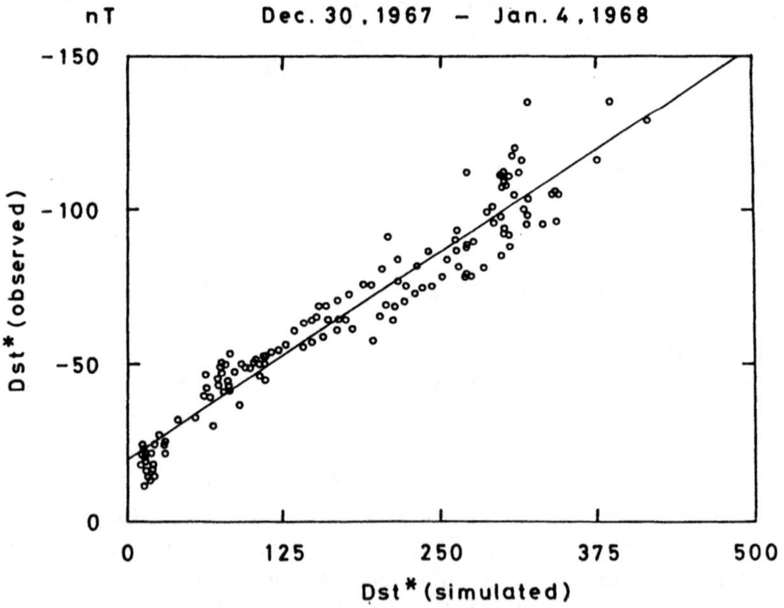

Fig. 6. A sample of the regression analysis between the observed and simulated vlaues of $D_{st}*$. The correlation coefficient for this sample is 0.967. The constant C in Eq. (4) is determined by the inclination of the regression line shown in the figure. The constant K in (4) is equal to the value of $D_{st}*$ at the intersection of the regression line with the ordinate. $F(t)=B_sV^2D^{0.4}$, $\tau_1=2$ h, $\tau_2=21$ h and $a_1/a_2=1.5$ are used to derive this result.

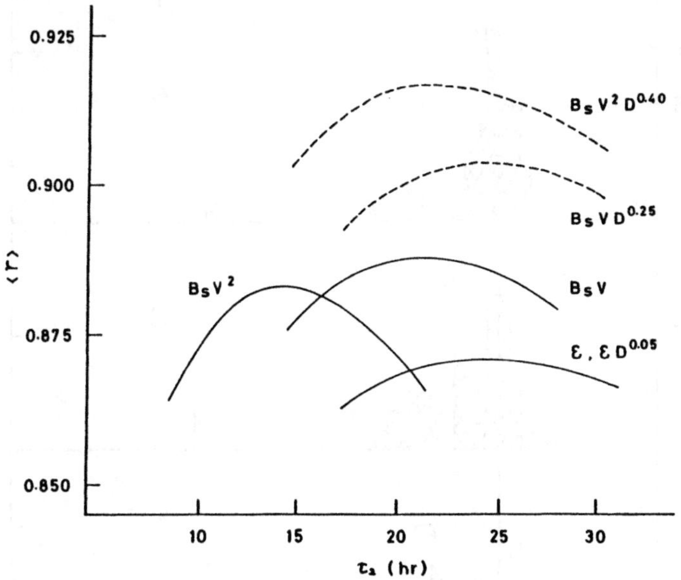

Fig. 7. Curve of $<r>$ as a function of τ_2 for each of six functions as labeled to the curves. The highest value of $<r>=0.9166$ is found for $B_sV^2D^{0.40}$ at $\tau_2=21$ hr.

Finally, we have searched for the optimum combination of the values of a, b and c for the function of the form $B_s^aV^bD^c$, independently of the forms of f_{id}'s. The result gives a new function f_4, for which $a=1.09$, $b=2.06$ and $c=0.38$, respectively. These values are close to $a=1$, $b=2$ and $c=0.4$ for f_{2d}, and hence the curves of $<r>$ for this function is only slightly higher than that of f_{2d} shown in the top part of Fig. 7.

We next evaluate the statistical significance between the maximum values of $<r>$ for different functions. Here, the significance is evaluated in the following way. Take, for example, $<r>_{max}=0.9166$ for f_{2d} and $<r>_{max}=0.9035$ for f_{1d}. We define $\delta_i=r_{2d}(i)-r_{1d}(i)$ as the difference in the values of r for the two functions for the event i, and examine the distribution of the values of δ_i for groups A, B and C separately, as shown in Fig. 8. As seen in the histogram for the group A+B in the bottom panel of the figure, δ_i's of about 60 per cent of the total events form a sharp peak centered at $\delta_i \cong 0$. This can be interpreted that the behavior of the two functions were so similar with each other in these time intervals that they are not appropriate to be used to discriminate the abilities of the two functions to predict D_{st}. However, the distribution of δ_i for the rest of the events are extremely asymmetric between the positive and negative sides, being approximately 32 to 4. Considering this asymmetry and a similar strong asymmetry in the distribution of δ_i for the group C, we conclude that the difference of 0.128 in the maximum values of $<r>$ for the functions f_{2d} and f_{1d} is statistically well significant. It is also readily proved that the statistical significance for the difference in the maximum values of $<r>$ between f_{2d} and any of f_{3d}, f_1, f_2 and f_3 is

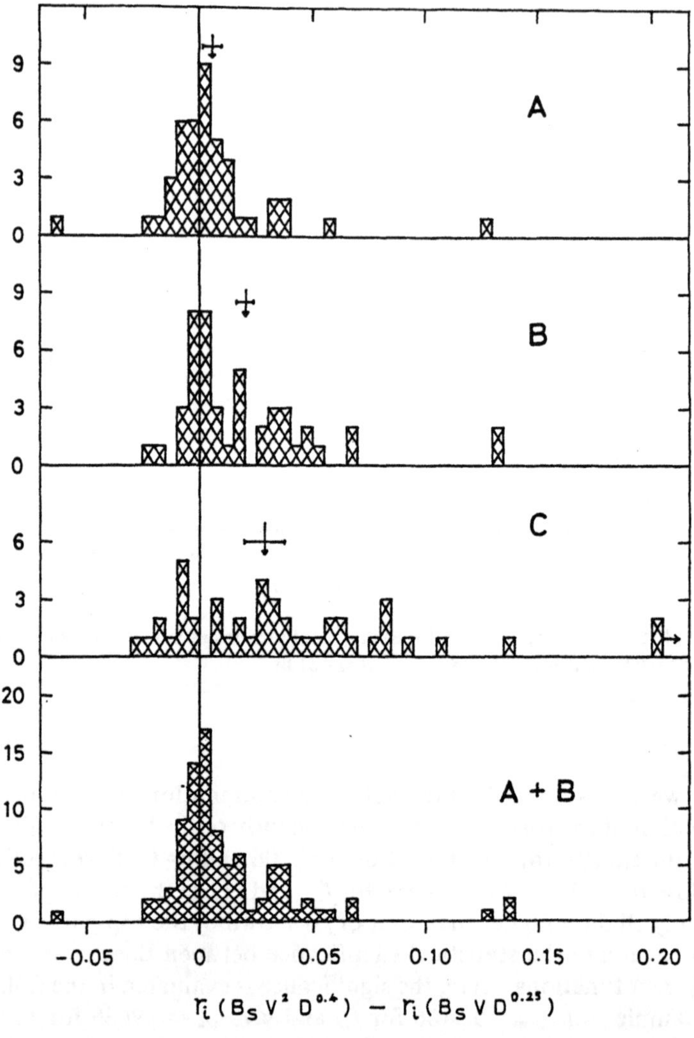

Fig. 8. Distribution of δ_i, the difference between the correlation coefficient for $B_s V^2 D^{0.40}$ and that for $B_s V D^{0.25}$ of individual events. Distributions for Group A ($\Delta > 15$ nT), Group B (10 nT$<\Delta<$15 nT), Group C ($\Delta < 10$ nT) and Group (A+B) are shown. An arrow with a horizontal bar denote the mean value and the standard error of δ_i.

much higher than that between f_{2d} and f_{1d} as discussed above.

2.4 Summary

From the results and discussions in preceding sections, the relevant functions to predict the indices *AL*, *am* and D_{st}, under the assumption on the form of $B^a V^b D^c$ for the functions, are summarized as

$$B_s^{1.0\pm0.15}\ V^{2.0\pm0.3}\ D^c \qquad c \cong \begin{cases} 0.1 & (AL) \\ 0.2 & (am) \\ 0.4 & (D_{st}). \end{cases}$$

Values of exponents for AL obtained by MAEZAWA (1979), MURAYAMA *et al.* (1980) and HOLZER and SLAVIN (1982), not referred so far, are also used to determine the values of exponents and their uncertainties. It should also be noted that owing to the saturation tendency, the above function for AL is valid only when geomagnetic activity is in a range from quiet to moderately active.

For the AU index, not discussed in this section, only one result is available; $a=0.67\pm0.08$, $b=1.15\pm0.16$ and $c=0.13\pm0.04$, respectively (MAEZAWA, 1979). AU is also subject to the saturation effect, similar to AL. The implication of the difference between AU and AL with respect to the value of the power on V will be discussed in Sections 4 and 5.

3. Influence of the Tilt Angle of the Geomagnetic Axis on the Solar Wind Velocity Dependence of Geomagnetic indices

3.1 Introduction

Among the various aspects of geomagnetic activity, the dependence on the tilt angle of the geomagnetic axis is worthy of special attention, since it suggests a possible contribution of the non-magnetic interaction to geomagnetic activity. The influence of the tilt angle was first suggested by BARTELS (1932) to interpret the equinoctial peaks in the seasonal variation of geomagnetic activity, by assuming that the activity is most enhanced when the geomagnetic axis is nearest to the normal to the direction of the sun-earth line.

The existence of the effect was later supported by the results of analyses on the universal time diurnal variation of geomagnetic activity (MCINTOSH, 1959; MAYAUD, 1970). MURAYAMA *et al.* (1980) showed that the dependence of $AL/(B_s+0.5\ \mathrm{nT})V^2$ on the tilt angle χ of the geomagnetic axis toward the sun-earth line has a clear maximum around $\chi=0$. Figure 9 shows the result of a similar analysis, in which χ dependence of the AL index is shown separately for different ranges of $B_s(V/400)^2$. It is readily observed that the equinoctial peaks become increasingly pronounced with increasing $B_s(V/400)^2$. The angle χ is defined by the geomagnetic latitude of subsolar point in this paper; hence $\chi\simeq0$ at equinoxes, $\chi>0$ in summer and $\chi<0$ in winter in the northern hemisphere of the earth.

BOLLER and STOLOV (1970) proposed that the Kelvin-Helmholtz instability is responsible for the equinoctial effect and that the instability develops when the solar wind velocity exceeds a critical value. This value decreases as ψ approaches 90°, where ψ is the angle between the direction of the geomagnetic field and that of the motion of the solar wind near the magnetopause at the flanks of the magnetosphere. Since $\psi=90°-\chi$, it is interesting to look for observational evidence for a possible coupling between the V dependence of geomagnetic activity and the dipole tilt angle χ, in order to assess the importance of non-magnetic interaction mechanism.

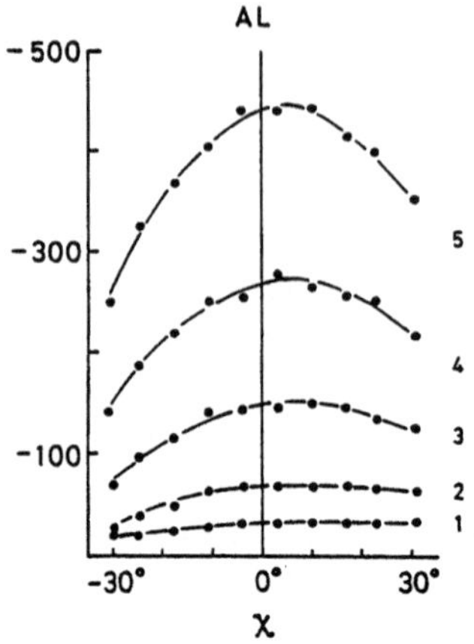

Fig. 9. Dependence of the AL index on the dipole tilt angle χ, where χ is defined as the geomagnetic latitude of the subsolar point. Curves labeled 1 through 5 correspond to the data sets for different ranges of $y=B_s(V/400)^2$ nT. 1: $y<0.3$, 2: $0.3<y<1.0$, 3: $1.0<y<2.7$, 4: $2.7<y<5.0$, 5: $5.0<y$.

3.2 Results of analysis

Analysis in this work was made with the AL index for the periods 1967–1975 and 1978–1980. Figure 10(a) shows the average dependence of AL on B_s for three different ranges of the angle χ shown separately for different ranges of V below 500 km/s. Three panels in the figure look very similar to each other in the sense that the curves for small $|\chi|$ and large positive χ values almost overlap each other, whereas the curve for large negative χ values lies considerably lower than the other two. For the ranges of V higher than 500 km/s as shown in Fig. 10(b), curves for small $|\chi|$ are clearly separated upward from those for large positive χ, and the separation increases with increasing V. The change with V of the mutual relation of curves for different values of χ can be also clearly seen in the lower panel of Fig. 11, where the same data with those for Fig. 10 are differently displayed in the form of AL versus V. The upper panel of this figure is the result of the same analysis but for AU. Three curves for different ranges of χ are approximay parallel to each other, being linearly dependent on V, and no appreciable changes in the V dependence are seen, in contrast to the case of AL.

Figure 12 shows the relation between AL and χ. Data for the two ranges of χ, $-18°<\chi<-7°$ and $7°<\chi<18°$, are added to the data used in Figs. 10 and 11. Throughout the whole ranges of B_s shown in this figure, peaks of the χ dependence of AL are obscure or found in the positive side of χ when V is low. As V increases peaks become more eminent and shifted toward $\chi=0$.

3.3 Discussion

We have shown that the dependence of AL on the solar wind velocity V varies with χ, the dependence being steeper when $|\chi|$ is small. This feature is consistent with the condition for the Kelvin-Helmholtz instability, as proposed by BOLLER and STOLOV (1970). It is also noted in Figs. 11 and 12 that the difference between the

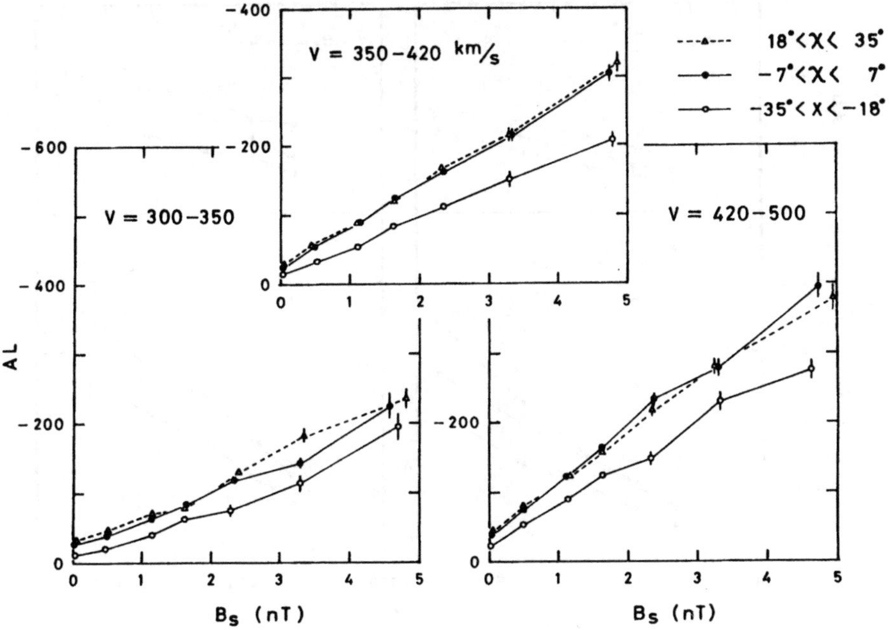

(a)

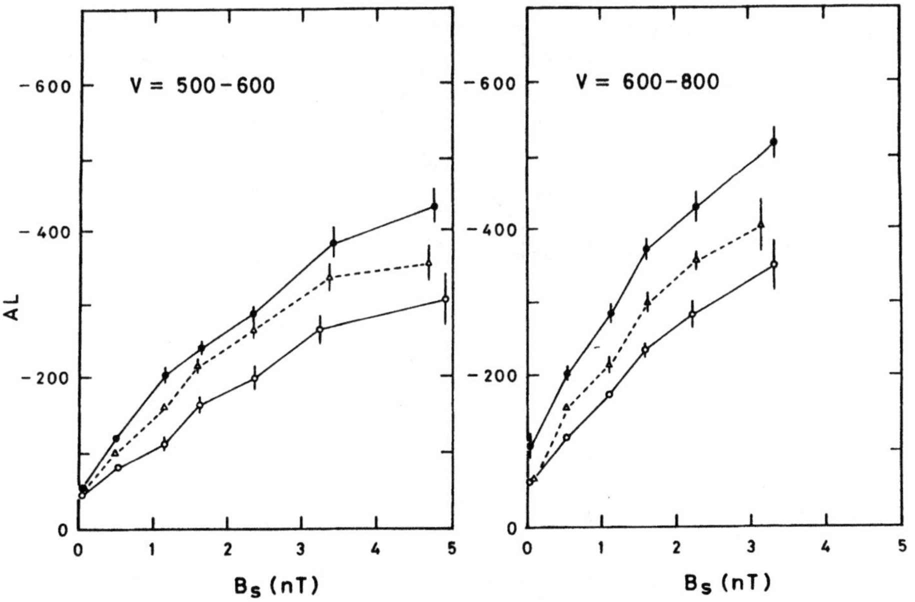

(b)

Fig. 10. (a) Average dependence of AL on B_s. Three panels are for different ranges of V, and three curves in each panel are for different ranges of χ. (b) Same as Fig. 10(a), but for the data with $V > 500$ km/s.

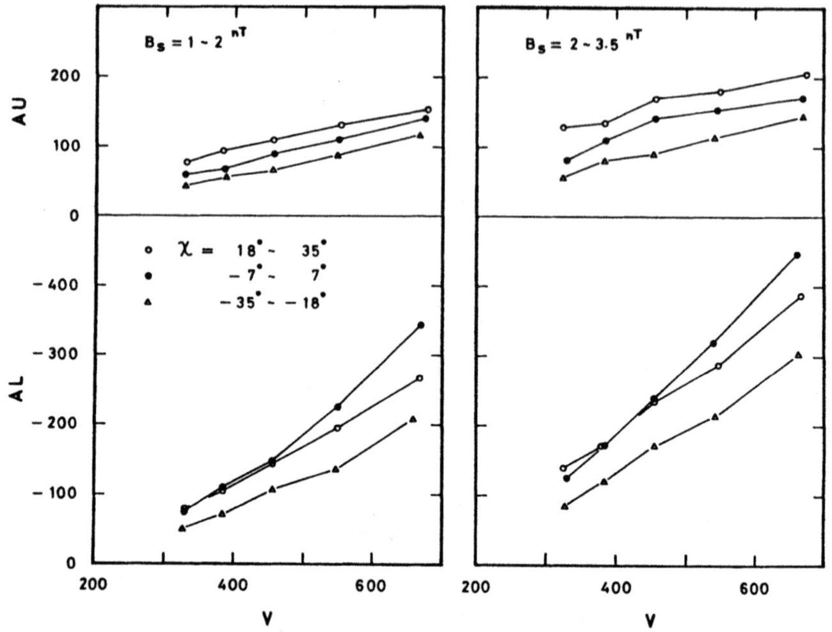

Fig. 11. Average dependence of AU (upper panel) and AL (lower panel) on V. Three curves in each panel are for different ranges of χ.

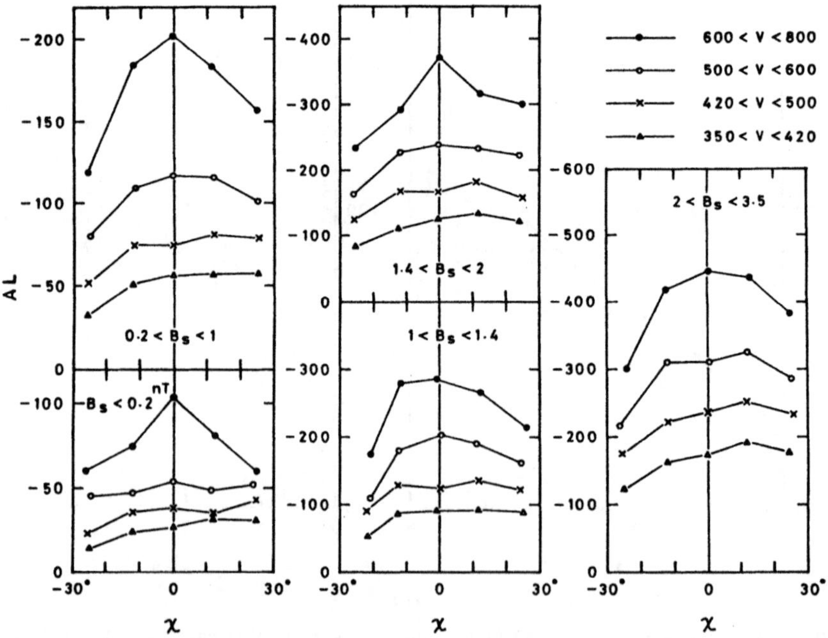

Fig. 12. Average dependence of AL on χ. Curves are shown for different ranges of B_s and V.

average values of AL for $|\chi|<7°$ and those for $|\chi|>18°$ increases with increasing B_s. This implies that the χ-modulated V effect is not a simple addition to the B_s-caused, day side reconnection-related effects; it seems that, when both effects coexist, they work cooperatively to make disturbances larger as compared to the simple addition.

Even when B_s is virtually zero, the peak of AL centered at $\chi=0$ develops, as seen in the lower left panel of Fig. 12. This effect appears to become significant when $V>600$ km/s, somewhat at a higher value of V as compared to the cases of larger B_s. The result for 0.2 nT$<B_s<$1.0 nT, shown in the upper left panel of Fig. 12, appears to be the mixture of the two tendencies.

4. Local Time Variation of the Solar Wind Velocity Effect on the Auroral Zone Magnetic Disturbances

4.1 *Introduction*

A drawback of an analysis using a geomagnetic index is that, at the price of obtaining a unified disturbance measure, we loose information that original data from individual stations possess for the spatial distribution of the magnetic disturbances. Therefore, in order to obtain further information about the nature of processes contributing to the solar wind velocity dependence of geomagnetic activity, we perform in this section an analysis using the data from individual magnetic stations at high latitudes. Our particular aim is to locate the region of strong influence of V, which may not be identical to the region of strong influence of B_s since mutually different magnetospheric processes may be concerned with each dependence.

4.2 *Analysis*

Figure 13 shows the locations in geographic polar coordinates of the 15 northern-hemisphere stations used in this analysis. As we will use the corrected geomagnetic (CGM) coordinate system to present our results, the north pole (designated as CMP), and the 70° N latitude circle (actually oval) of this system delineating roughly the poleward boundary of the auroral zone, are shown for reference. The stations are distributed from the subauroral zone to the polar cap latitudes in this coordinate system. The name and the CGM latitude of the stations are listed in Table 1.

Data from each station are processed in the following way: since we are interested in the intensity of the current along the auroral zone, hourly deviations of the X component in the CGM coordinates from the quiet day mean value (determined in the same way as in the calculation of AL and AU) are calculated for each station and used as a basic quantity to be analyzed. The hourly X deviation thus obtained (hereafter simply denoted as X) represents the local current intensity at the location determined by the UT of the hour and the latitude of the station. For example, the data for 0000–0100 UT at College represents the disturbance at 1200–1300 magnetic local time (MLT) at 65° CGM latitude. (The difference between UT and MLT has been rounded to the nearest integer.) In this way, by averaging the data for particular UT from a particular station, we can investigate the current intensity at a particular (CGM latitude-local time) location, and then synthesizing the results for different UT and different stations, we can examine the solar wind parameter dependence of the overall

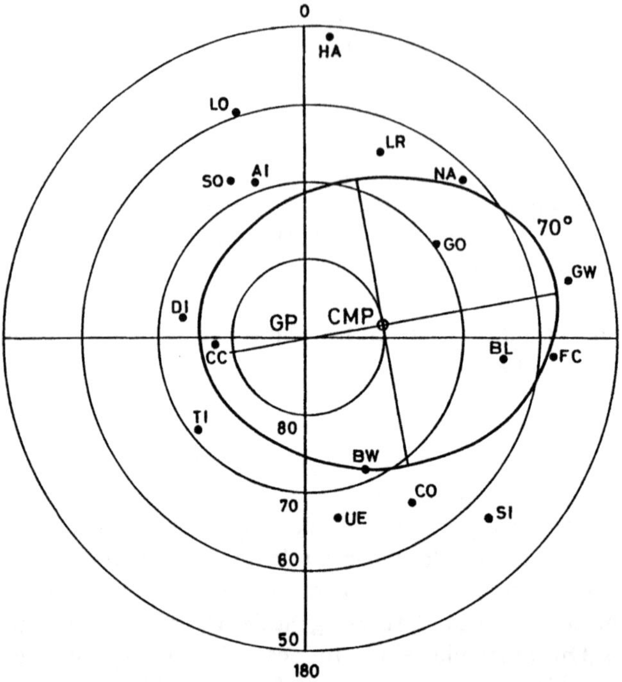

Fig. 13. Distribution of high-latitude geomagnetic stations in the northern hemisphere, used in the analysis in Section 4. GP and CMP are the geographic and the corrected geomagnetic poles, respectively. Five equi-latitude circles in geographic coordinates and equi-latitude ellipse in corrected geomagnetic coordinates are shown.

Table 1. List of stations used in this analysis.

Station name	Abbreviation	Corrected geomagnetic latitude	Data period
Baker Lake	BL	75.1	1964–1973
Cape Chelyuskin	CC	71.2	1966–1969
Fort Churchill	FC	70.0	1966–1974
Barrow	BW	69.7	1966–1974
Narssarssuaq	NA	69.0	1968–1974
Great Whale River	GW	68.1	1966–1974
Dixon Island	DI	68.0	1966–1969
Leirvogur	LR	66.6	1966–1974
Tixie Bay	TI	65.6	1966–1974
College	CO	64.8	1966–1974
Abisco	AI	64.8	1968–1974
Sodankyla	SO	63.4	1966–1967
Cape Wellen	UE	62.6	1966–1969
Sitka	SI	59.8	1964–1970
Lovo	LO	55.9	1964–1974

current distribution.

In practice, we examine the influence of the solar wind velocity, V, and the north–south component of the IMF, B_z, by calculating the mean value of X separately for different ranges of V and B_z, i.e., for a combination of four V bins, V1–V4, and four B_z bins, B_z1–B_z4, as listed in Table 2.

Figure 14 shows the latitudinal distribution of X for the 0400–0600 magnetic local time sector, in four panels corresponding to the bins B_z1–B_z4. In each panel,

Table 2. Range of B_z and V corresponding to each bin.

B_z1	$-1\ nT<B_z<0\ nT$	$V1$	$250<V<350$ km/s
B_z2	$-2\ nT<B_z<-1\ nT$	$V2$	$350<V<450$ km/s
B_z3	$-3\ nT<B_z<-2\ nT$	$V3$	$450<V<550$ km/s
B_z4	$-4\ nT<B_z<-3\ nT$	$V4$	$550<V<650$ km/s

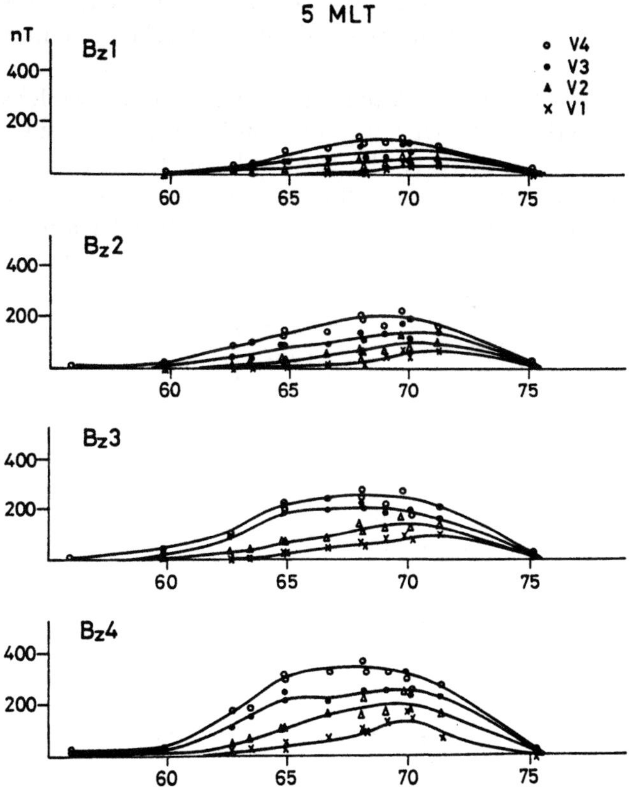

Fig. 14. Average values of X for individual stations plotted against the corrected geomagnetic latitude. Results are shown in four panels corresponding to the four ranges of the IMF-B_z as shown in Table 2. Four different marks are the four ranges of V, also shown in Table 2. Smooth curves were drawn by inspection for different ranges of V.

results for different V bins are compared using different marks. The sign of X is hereafter taken positive for the current flowing to the west, except when otherwise stated.

It is readily seen that the current is westward almost everywhere in the latitudinal range of 60°–75°, for all the ranges of B_z and V, showing the presence of the westward electrojet in this local time sector. It is also seen that, as B_z becomes more negative from the bin B_z1 to the bin B_z4, and as V increases from $V1$ to $V4$, the current intensity uniformly increases at auroral zone latitudes, though there is an indication that the location of the maximum current intensity shifts towards low latitudes with increasing $|B_z|$ and V, as is consistent with the dependence of the auroral oval latitude on geomagnetic activity. The latitudinal profile of X similar to Fig. 14 has been constructed for every two-hour local time sector. Although it is generally observed that the latitudinal location of the current at a given meridian varies with solar wind parameters, the change in the total amount of current flowing across the given meridian would be more important from the energetic point of view. We therefore use integrated values of X with respect to latitude, denoted by X_I, in order to discuss its dependence on the solar wind parameters at each meridian. (X_I is given in the unit of nT km. The numerical factor to translate X_I into the total current in A happens to be almost unity.)

Figure 15 shows the dependence of X_I on V and B_z for two local time sectors; 1600–1800 MLT in the left panel and 0200–0400 MLT in the right panel. These two local times are representative of the eastward and westward electrojet regions, respectively, and the sign of X is taken positive for the eastward current in the left panel as opposed to our convention. It is seen that the dependence of X_I for two panels

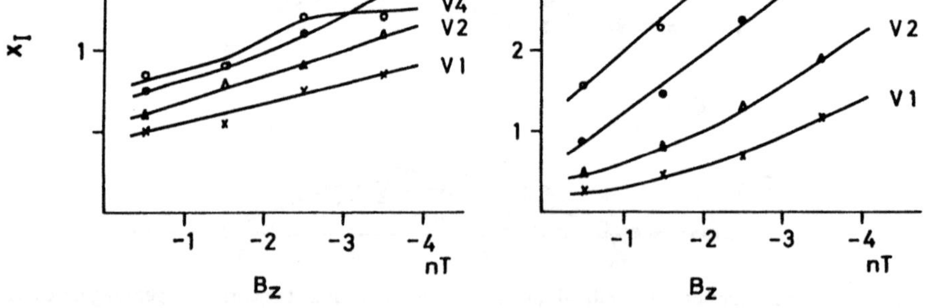

Fig. 15. Comparison of the dependence of X_I on the IMF-B_z for the magnetic local time sector 1600–1800 and that for 0200–0400. See the text for the definition of X_I. Curves are given for four ranges of V, as shown in Table 2.

is very different; in the left panel, X_I values for the $V4$ bin are almost twice those for the $V1$ bin (250–350 km/s), while in the right panel X_I values for the $V4$ bin are about four times those for the $V1$ bin, irrespective of the ranges of B_z. This result is consistent with the fact that although the AU index for the eastward electrojet is almost linear to V (MAEZAWA, 1979), the AL index for the westward electrojet shows a quadratic dependence on V as demonstrated earlier. The above finding that the latitudinally integrated current intensities of the eastward electrojet and the westward electrojet have different dependences on V have led us to suppose that the V dependence of X_I has a strong local time dependence. We have therefore examined all the local time sectors and found that the strongest dependence on V occurs in the local time sector 2200–2400 MLT.

Figure 16 shows the dependence of X_I on V and B_s for the 2200–2400 sector (denoted simply as 23 MLT in the right panel), as compared to that for the 0200–0400 MLT (denoted as 3 MLT in the left panel). Although the current intensity is generally lower in the 23 MLT sector than in the 3 MLT sector for all the ranges of V and B_z, the 23 MLT sector is distinct in that X_I values depend very strongly on V. As a matter of fact, the ratio between the X_I values for $V4$ and $V1$ bins amounts to more than 10 as compared to the value of about 4 in the case of the 3 MLT meridian.

The steep dependence of X_I on the solar wind velocity for the 23 MLT meridian is more readily seen in Fig 17, where the same data are replotted with V taken as abscissa. It is seen that the westward current at this local time sector has a very steep dependence on V for $V > 300$ km/s and almost vanishes when $V = 300$ km/s, for all ranges of B_z investigated.

It is evident why the strong dependence on V in the midnight sector is not reflected in the AL index. Since the strongest westward current usually occurs in the early morning sector, the current in the midnight sector does not contribute to the AL

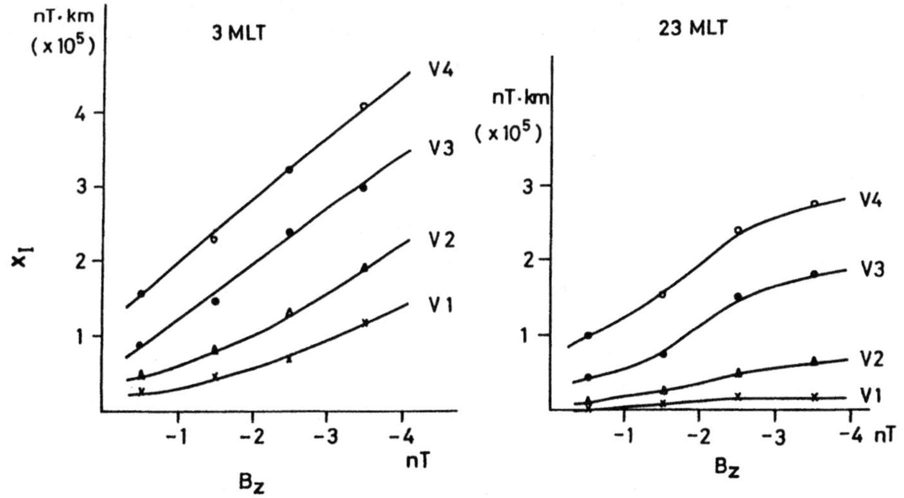

Fig. 16. Same as Fig. 16, but the comparison of sectors 0200–0400 and 2200–2400.

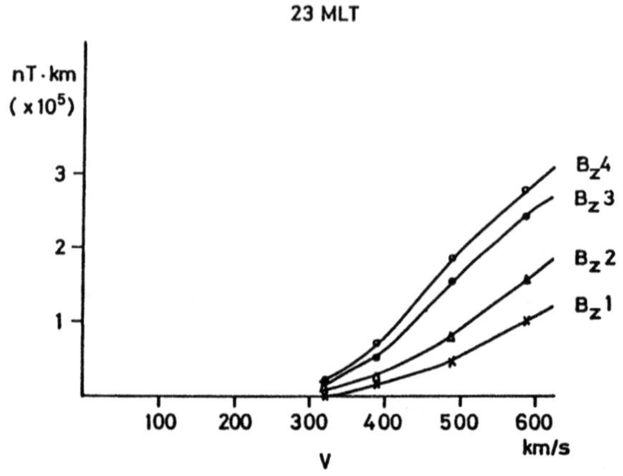

Fig. 17. Dependence of X_I on the solar wind velocity for the local time sector 2200–2400.

index, which is defined to be the maximum depression in H among the auroral zone stations. (In our data set, the maximum integrated current occurs at 3 MLT for almost all ranges of V and B_z.)

4.3 Discussion

In conclusion, although the maximum integrated current in the whole westward electrojet region is almost proportional to V^2, the degree of dependence on V varies strongly with magnetic local time. The total amount of westward current in the 2200–2400 sector is a particularly strong function of V; it almost vanishes when $V \simeq 300$ km/s and increases steeply with increasing V when $V > 300$ km/s. On the other hand, the B_z dependence of the total current does not show such a peculiarity at this local time sector. This finding supports the view that the magnetospheric processes associated with the strong V dependence operate in the night side magnetosphere and may be distinct from the processes associated with magnetic reconnection, from which the B_z dependence results.

5. Summary and Discussion

We have focused in this paper on several characteristics of the influence of the solar wind velocity on geomagnetic activity in order to obtain a clue to the solar-wind magnetosphere coupling mechanisms. Results of our analyses may be summarized as follows.

(1) The solar wind velocity dependence of the AL index is not linear; it is approximately proportional to V^2. The V^2 dependence is also observed for the midlatitude magnetic activity index *am* and the ring current index D_{st}.

(2) The degree of the V dependence of AL varies with χ, the tilt angle of the geomagnetic axis. The dependence is steepest when $\chi \simeq 0$, i.e., when the geomagnetic axis is near normal to the sun-earth line.

(3) Analysis of the hourly values from individual auroral-zone magnetic staions indicates that the influence of V appears differently in different local time sectors. The strongest effect of V is observed for the 2200–2400 MLT sector.

We first note that the origin of the solar wind velocity dependence of ground magnetic disturbances would not be as simple as it may first appear. One of the simple mechanisms discussed so far is in terms of the solar wind $V \times \boldsymbol{B}$ electric field which is transmitted to the earth's magnetosphere and ionosphere via reconnected geomagnetic field lines. However, it is clear that the V^2 dependence of geomagnetic activity in item (1) of the above summary cannot be explained by this mechanism alone.

Although it is generally agreed that the dayside reconnection process is of crucial importance for the transfer of energy to the magnetosphere, theories of reconnection still seem imperfect to predict the V dependence of the reconnection rate, because the influence of the strong external flow (solar wind in this case) has not been fully accomodated in the theory. Hence it is possible to infer that the day side reconnection process itself has a strong V dependence by some reason not yet identified theoretically.

On the other hand, it is also possible to infer that, besides reconnection, there exists another kind of solar wind-magnetosphere energy transfer mechanism, and it augments the level of geomagnetic disturbances to result in the V^2 dependence.

In relation to the second possibility, the importance of the viscous-like interaction has recently been receiving attention both observationally and theoretically. For instance, this interaction would play an important role in the momentum transport process at the low-latitude boundary layer (EASTMAN *et al.*, 1979; SCKOPKE *et al.*, 1981, MIURA, 1984), and a part of the convection system in the high-latitude region of the ionosphere was attributed to the viscous-like interaction (BURCH *et al.*, 1985; REIFF and BURCH, 1985).

Our result (item (2)) can also be included to that which may be related to the viscous interaction; the manner that χ influences the effect of V is consistent with the view that the Kelvin-Helmholtz instability is responsible for the χ dependence of the geomagnetic activity. If this view is correct, an appreciable amount of geomagnetic activity is controled by the viscous-like interaction. It is then probable that the strong V dependence of geomagnetic indices is at least partly due to the viscous-like interaction.

If the reconnection and the viscous-like interaction coexist, they likely take place in different parts of the magnetosphere and the subsequent energy transport paths may also be different. Thus we can expect that the relative contribution of the two interaction mechanisms to ground geomagnetic activity appears diffferently for different regions of the earth. In this situation, analyses using the records of individual geomagnetic stations would be needed as in our analysis in Section 4.

We have seen in Section 4 that the effect of the solar wind velocity on the auroral zone current is strongest near the midnight meridian, and this finding suggests that the near-earth plasma sheet plays a certain role in making the V dependence stronger than at other local times. Further, we can infer that the solar wind-induced convection electric field is not the direct source of this dependence, because its effect would not be localized to the midnight auroral zone.

The above consideration has led us to the picture that the particle precipitation flux from the near-earth plasma sheet is under a strong influence of the solar wind velocity and it causes the ionospheric conductivity around the midnight auroral zone to be enhanced when V is high, leading to the stronger V dependence of the local current intensity. It is worth to mention here that the flux of energetic particles in the near-earth plasma sheet does show a strong dependence on the solar wind velocity (BAKER *et al.*, 1979). The above view is also consistent with the fact that the eastward electrojet in the afternoon sector has only a linear dependence on the solar wind velocity, because the ionospheric conductivity there is mainly controled by solar radiation.

We are not in a position to discuss the relationship between the viscous-like interaction operating on the magnetopause and the particle state of the near-earth plasma sheet, although it is possible that the energy and momentum transport due to the viscous-like interaction profoundly affects the state of the plasma sheet, and this may be an ultimate cause for the formation of the plasma sheet itself.

In order to proceed futher, we may need more observational evidence for the relationships between the particle precipitation to the midnight auroral zone and the solar wind velocity. The influence of χ on this relationship is also to be confirmed from the observation. It is also desirable to make the analysis in relation to the phase of the substorm development, in view of the result by KAMIDE (1982) that the local time dependence of the auroral zone current varies with the substorm phase, suggesting the time-varying relative importance of the convection electric field and the particle precipitation.

This work is partly supported by the Space Science Data Center of the Institute of Space and Astronautical Science, Tokyo. The computational works have been made with FACOM M-362 in the Nagoya University Computation Center.

REFERENCES

AXFORD, W. I. and C. O. HINES, A unifying theory of high latitude geophysical phenomena and geomagnetic storms, *Can. J. Phys.*, **39**, 1433–1464, 1961.

BAKER, D. N., R. D. BELIAN, P. R. HIGBIE, and E. W. HONES, Jr., High-energy magnetospheric protons and their dependence on geomagnetic and interplanetary conditions, *J. Geophys. Res.*, **84**, 7138–7154, 1979.

BAKER, D. N., R. D. ZWICKL, S. J. BAME, E. W. HONES, Jr., E. J. SMITH, and S.-I. AKASOFU, An ISEE 3 high time resolution study of interplanetary parameter correlations with magnetospheric activity, *J. Geophys. Res.*, **88**, 6230–6242, 1983.

BARTELS, J., Terrestrial-magnetic activity and its relation to solar phenomena, *Terr. Magn. Atmosph. Electr.*, **37**, 1–52, 1932.

BOLLER, B. R. and H. STOLOV, Kelvin-Helmholtz instability and the semiannual variation of geomagnetic activity, *J. Geophys. Res.*, **75**, 6073–6084, 1970.

BURCH, J. L., P. H. REIFF, J. D. MENIETTI, R. A. HEELIS, W. B. HANSON, S. D. SHAWHAN, E. G. SHEELY, M. SUGIURA, D. R. WEIMER, and J. D. WINNINGHAM, IMF-B_y-dependent plasma flow and Birkeland currents in the dayside magnetosphere, 1. Dynamics explorer observations, *J. Geophys. Res.*, **90**, 1577–1593, 1985.

BURTON, R. K., R. L. MCPHERRON, and C. T. RUSSELL, An empirical relationship between interplanetary condition and D_{st}, *J. Geophys. Res.*, **80**, 4204–4214, 1975.

CROOKER, N. U., J. FEYNMAN, and J. T. GOSLING, On the high correlation between long-term averages of solar wind speed and geomagnetic activity, *J. Geophys. Res.*, **82**, 1933–1937, 1977.

DUNGEY, J. W., Interplanetary magnetic field and the auroral zones, *Phys. Rev. Lett.*, **6**, 47–48, 1961.

EASTMAN, T. E. and E. W. HONES, Jr., Characteristics of the magnetospheric boundary layer and magnetopause layer as observed by Imp-6, *J. Geophys. Res.*, **84**, 2019–2028, 1979.

FELDSTEIN, Y. I., V. Y. U. PISARSKY, N. M. RUDNEVA, and A. GRAFE, Ring current simulation in connection with interplanetary space conditions, *Planet. Space Sci.*, **32**, 975–984, 1984.

GARRETT, H. B., A. J. DESSLER, and T. W. HILL, Influence of solar wind variability on geomagnetic activity, *J. Geophys. Res.*, **79**, 4603–4610, 1974.

HOLZER, R. E. and J. A. SLAVIN, An evaluation of three predictors of geomagnetic activity, *J. Geophys. Res.*, **87**, 2558–2562, 1982.

KAMIDE, Y., The two component auroral electrojet, *Geophys. Res. Lett.*, **10**, 1175–1178, 1982.

KING, J. H., Interplanetary medium data book, Rep. NSSDC 77-04, 1977; NSSDC 79-08, 1979; NSSDC 83-01, 1983.

MAEZAWA, K., Dependence of geomagnetic activity on solar wind parameters: a statistical approach, *Solar Terr. Environ. Res. Jpn*, **2**, 103–125, 1978.

MAEZAWA, K., Statistical study of the dependence of geomagnetic activity on solar wind parameters, in *Quatitative Modelling of Magnetospheric Processes*, edited by W. P. Olson, pp. 436–447, Am. Geophys. Union, Washington, D.C., 1979.

MAYAUD, P. N., Sur quelques propriétés de l'activité magnetique, déduites de l'analyse d'une série de neuf anées des indices K_n, Ks et Km, 2. Les diverses composantes "temps universel", dans chaque Hémisphère, de la variation journaliere de l'activite, *Ann. Geophys.*, **26**, 313–336, 1970.

MCINTOSH, D. H., On the annual variation of magnetic disturbance, *Phil. Trans. R. Soc. London*, Ser. A, **251**, 525–552, 1959.

MELONI, A., A. WOLFE, and L. J. LANZEROTTI, On the relationships between interplanetary quantities and the global auroral electrojet index, *J. Geophys. Res.*, **87**, 119– 127, 1982.

MIURA, A., Anomalous transport by magnetohydrodynamic Kelvin-Helmholtz instabilities in the solar wind-magnetosphere interaction, *J. Geophys. Res.*, **89**, 801–818, 1984.

MURAYAMA, T., Coupling function between solar wind parameters and geomagnetic indices, *Rev. Geophys. Space Phys.*, **20**, 623–629, 1982.

MURAYAMA, T., T. AOKI, H. NAKAI, and K. HAKAMADA, Empirical formula to relate the auroral electrojet intensity with interplanetary parameters, *Planet. Space Sci.*, **28**, 803–813, 1980.

PERREAULT, P. and S.-I. AKASOFU, A study of geomagnetic storms, *Geophys. J. R. Astron. Soc.*, **54**, 547–573, 1978.

REIFF, P. H. and J. L. BURCH, IMF B_y-dependent plasma flow and Birkeland currents in the dayside magnetosphere, 2. A global model for northward and southward IMF, *J. Geophys. Res.*, **90**, 1595–1609, 1985.

SCKOPKE, N., G. PASCHMANN, G. HAERENDEL, B. U. O. SONNERUP, S. J. BAME, T. G. FORBES, E. W. HONES, Jr., and C. T. RUSSELL, Structure of the low-latitude boundary layer, *J. Geophys. Res.*, **86**, 2099–2110, 1981.

SVALGAARD, L., Geomagnetic activity: Dependence on solar wind parameters, in *Coronal Hole and High Speed Wind Streams*, p. 371, Colorado Associated University Press, Boulder, 1977.

TSURUTANI, B. T., J. A. SLAVIN, Y. KAMIDE, R. D. ZWICKL, J. H. KING, and C. T. RUSSELL, Coupling between the solar wind and the magnetosphere: CDAW 6, *J. Geophys. Res.*, **90**, 1191–1199, 1985.

Solar Wind-Magnetosphere Coupling, edited by Y. Kamide and J. A. Slavin, 85–92.

Linear Filters as a Method of Real Time Prediction of Geomagnetic Activity

R. L. MCPHERRON, D. N. BAKER, and L. F. BARGATZE

[1]*Department of Earth and Space Sciences, and Institute of Geophysics and Planetary Physics University of California, Los Angeles, California 90024, U.S.A.*
[2]*Los Alamos National Laboratory, Los Alamos, New Mexico 87545, U.S.A.*

Important factors controlling geomagnetic activity include the solar wind velocity, the strength of the interplanetary magnetic field (IMF), and the IMF orientation. Because these quantities change as the solar wind flows away from the sun, real-time monitoring immediately upstream of the earth is needed in any technique of real-time prediction. One such technique is linear prediction filtering which utilizes past histories of the input and output of a linear system to create a time-invariant filter characterizing the system. Problems of nonlinearity or temporal changes of the system can be handled by appropriate choice of input parameters and piecewise approximation in various ranges of the input. We have created prediction filters for all the standard magnetic indices and tested their efficiency. The filters show that the initial response of the magnetosphere to a southward turning of the IMF peaks in 20 minutes and then again in 60 minutes. After a northward turning auroral zone indices and the midlatitude *ASYM* index return to background within two hours, while *Dst* decays exponentially with a time constant of about eight hours. This paper describes a simple real-time system utilizing these filters which could predict a substantial fraction of the variation in geomagnetic activity indices 20 to 50 minutes in advance.

1. Introduction

Magnetic activity is a manifestation of "space weather." It affects many human activities including scientific, commerical and military. Just as with earth weather, the ability to forecast is important so that appropriate actions can be taken. An obvious example in scientific studies is launching balloon or rocket payloads which needs to be done some time before magnetic activity begins. A commerical example is exploration geophysics, where it is useless to fly magnetic surveys if the magnetic field is varying too rapidly. In the military context, radio communications, radar reflections, navigation, satellite orbits, and communication spacecraft are all affected by strong magnetic activity.

Historically, the level of magnetic disturbance has been characterized by magnetic indices (see article by BAUMJOHANN, this volume). Until recently, the most popular of these were *Kp* and *Ap*. These indices summarize the weighted average range of variability observed over a three-hour interval by a world-wide collection of magnetic observatories. Currently, these indices are being replaced by the *AU*, *AL*, *AE*, *D*, and *ASYM* indices. These newer indices can be determined with much higher time resolution (one minute) and are much more closely related to the physical processes which cause the activity. In particular, *AU* measures the strength of the eastward auroral electrojet, *AL* the strength of the westward electrojet, *Dst* the strength of the equatorial ring current, and *ASYM* the strength of field-aligned currents which connect the electrojets to the outer magnetosphere.

Magnetic indices are usually calculated long after the fact. The reason for this is practical. Magnetic records from a world-wide network of stations are required to produce the indices. At the present time such data are not available for as long as two years because there are no real-time data collection systems. Because of this, it is standard practice to estimate the indices on a basis of a few observatory records. Such estimates are very poor, and better techniques exist.

One such technique is linear prediction filtering (see article by CLAUER, this volume). This technique assumes that the agent causing the variations in a magnetic index is related to the index by a linear time-invariant system. Then, previous time histories of the input and output are used to calculate an impulse response characterizing the system. This response function can then be convolved with an input measured at a later time predicting the variation in a magnetic index.

2. Examples of Index Predictions

In recent years it has been well documented that the immediate cause of most geomagnetic activity is the interplanetary magnetic field (IMF). A southward orientation of the IMF at the magnetopause allows the IMF to merge with the earth's field. Subsequent transport of dayside magnetic flux to the tail and its return to the dayside are the main phenomena which drive the currents creating magnetic activity. As we show elsewhere (see article by BARGATZE *et al.*, this volume), the solar wind parameter which is most linearly related to magnetic activity indices is the rectified electric field. This parameter is able to account for more than 40% of the variance in the *AL* index and 50% of the variance in *Dst*.

A prediction filter relating the rectified solar wind electric field to the *AL* index is plotted in Fig. 1. This filter was calculated from a 52-day data set of 2.5-minute values recorded in 1967–1968. The filter has a duration of 2 hours, an initial delay of 20 minutes, and a peak between 40–60 minutes. Its rise time is more rapid than its decay, giving it the appearance of a Rayleigh function. As we show below, the filter includes an approximate 10-minute delay caused by solar wind propagation from the monitor (Explorer 33) 30–60 Re upstream of the subsolar magnetopause.

A comparison of the observed and predicted *AL* index is made in Fig. 2. In the middle panel the thin line denotes the observed index, and heavy line denotes the predicted index. The top panel shows the rectified solar wind electric field (VB_s) which was assumed to be the primary cause of auroral electrojet activity measured by the index. The bottom panel presents the prediction residual, the difference between observed and predicted *AL*. The general trends of the predicted and observed indices agree well; the predicted index appears to be a low-pass filtered version of the observed index. Some details are not well predicted as for example substorm expansion onsets (shown by vertical dashed lines), which do not appear to be directly related to the solar wind.

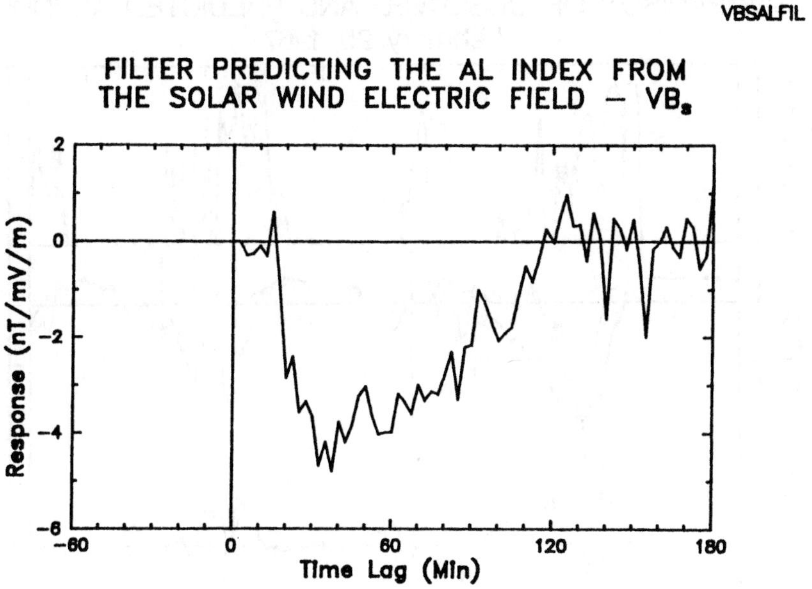

Fig. 1. Linear prediction filter relating the 2.5-minute *AL* index to VB_s, the rectified solar wind electric field (product of velocity and southward magnetic field strength).

Filters required to predict the *Dst* index are presented in Fig. 3. Since *Dst* measures effects of both the magnetopause currents and the ring current, two filters are needed to predict its variation. The first, the dynamic pressure filters, predicts a value of *Dst* proportional to the square root of solar wind dynamic pressure. The greater the pressure, the smaller the distance between the earth and the magnetopause currents causing enhanced values of *Dst*. Although not clear on the time scale of the plot, the filter is peaked at 10 minutes' delay and has a width of about 10 minutes. The delay represents the average time for the solar wind to propagate from the position of the solar wind monitor to the magnetopause. The width represents variations in this delay over the data set used to define the filters. The second filter, the solar wind electric field filter, predicts decreases in *Dst* during intervals of southward IMF. This filter represents two effects, injection of plasma into the ring current and its subsequent loss due to several processes. Because of the long time constants associated with the decay, the filter has a very long duration. Its initial behavior, however, is very similar to the *AL* filter with an initial delay of about 20 minutes and a peak at about 60 minutes.

Figure 4 illustrates predictions of *Dst* made by these filters for data from the CDAW-6 workshop (McPherron and Manka, 1985). The top two panels show the solar wind inputs, the middle two panels the variation in *Dst* predicted by the respective inputs, and the bottom panel a comparison of the observed and predicted *Dst* index. Although on average these two filters predict 70% of the *Dst* variance, there are obvious discrepancies.

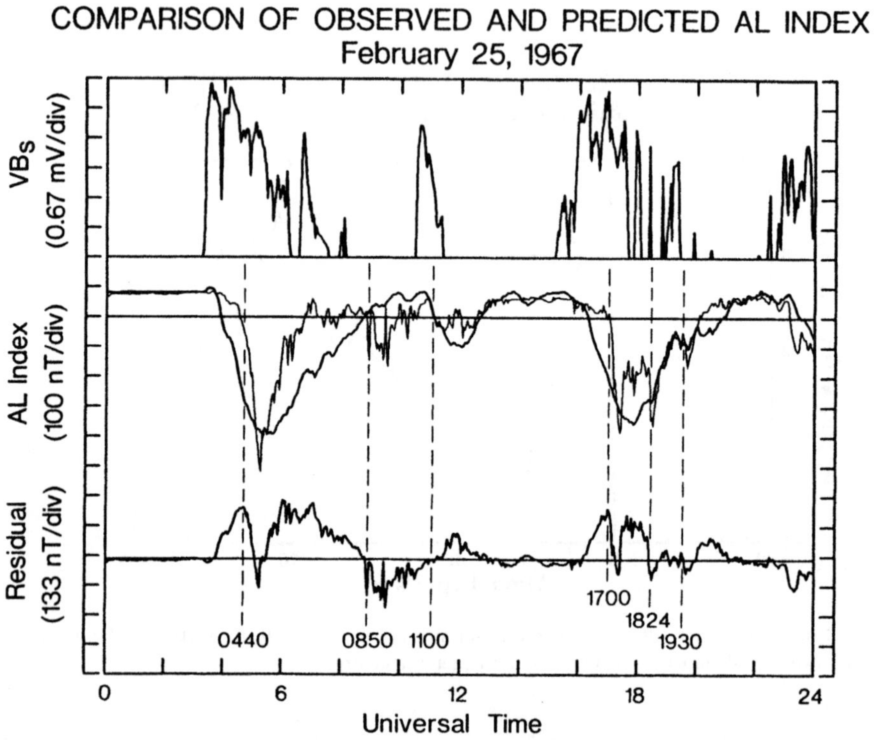

Fig. 2. A comparison of the observed *AL* index with that predicted by the rectified solar wind electric field. The top panel shows the input electric field, the middle panel compares the observed (thin line) and calculated (heavy line) *AL* index, and the bottom panel displays the prediction residual. Vertical dashed lines denote onset of substorm expansions.

3. Discussion

In the preceding section we illustrated the nature of the filters relating two of the most commonly used magnetic indices, *AL* and *Dst*, to the solar wind. Filters have also been calculated for two other indices, *ASYM*, which measures the longitudinal deviation of the midlatitude magnetic perturbation in *H* from its world-wide average *Dst*, and *AU*, which measures the strength of the eastward auroral electrojet (BAUMJOHANN, this issue). The *ASYM* filter is virtually identical to the *AL* filter and predicts a similar fraction of the index variance. In contrast, the *AU* filter is very poorly defined and predicts only about 20% of the index variance. This lack of correlation between *AU* and the solar wind is perplexing and suggests that *AU* is a poor measure, or not linearly related to the IMF, or dependent on other, unmeasured parameters.

As noted in the introduction the values of magnetic activity indices are normally calculated long after the fact from a world-wide collection of observatory records. Estimates of the indices, however, are often calculated from a collection of records

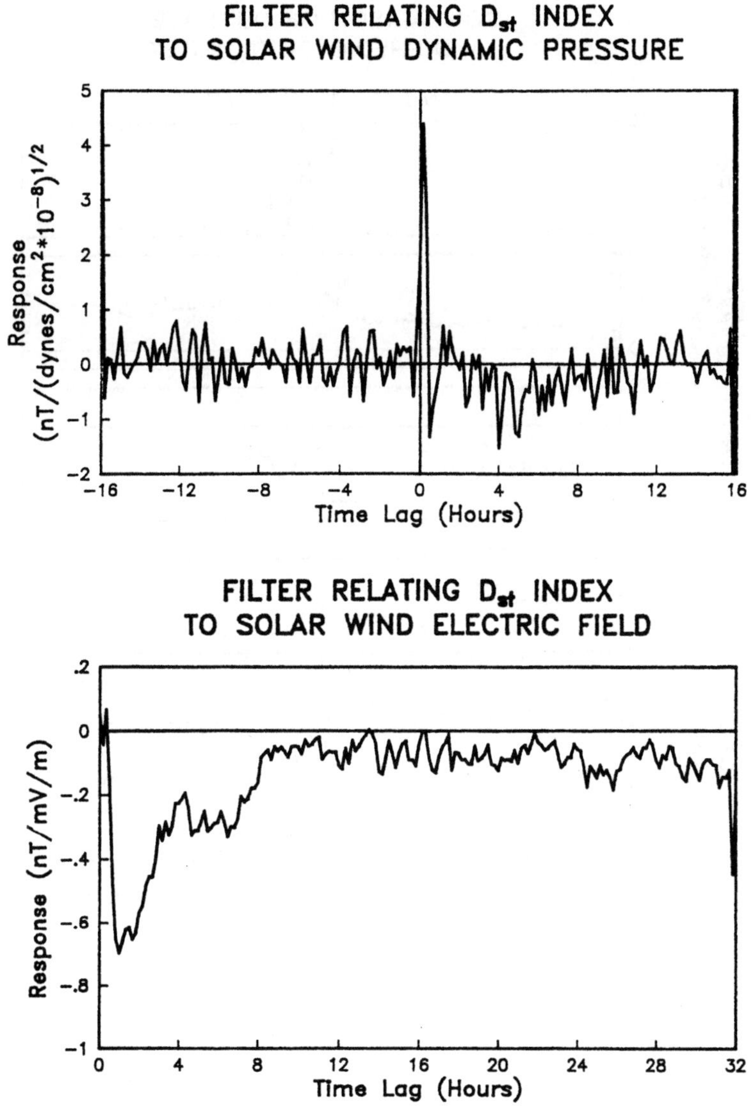

Fig. 3. Prediction filters for the *Dst* index. Top panel shows the filter relating the square root of dynamic pressure to *Dst*. Bottom panel shows the filter relating the rectified solar wind electric field to *Dst*.

available in real time, sometimes as few as one record. The technique of linear prediction filtering provides a method for improving such estimates. As we demonstrated above only two parameters, the solar wind electric field and dynamic pressure, are needed to predict a substantial fraction of the variance in standard indices. If the solar wind measurements needed to calculate these parameters are available in real time, then it is possible to make a reasonably accurate estimate of the

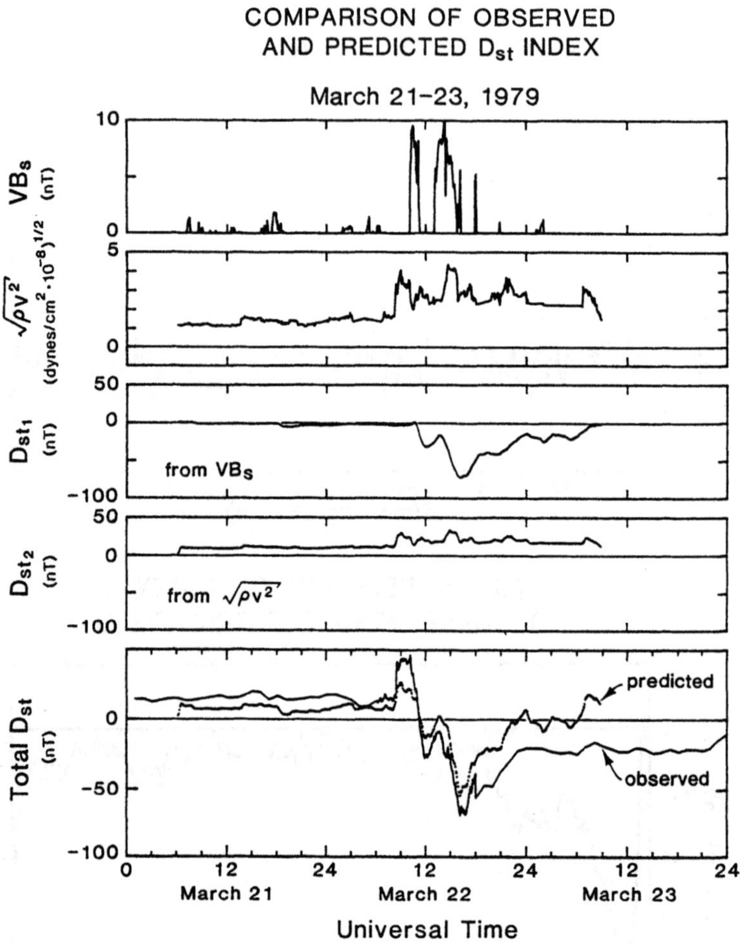

Fig. 4. An example of the prediction of the *Dst* index from solar wind measurements. Top two panels are the electric field and pressure inputs; middle panels are the separate predictions. Bottom panel compares the predicted *Dst* (dotted) to the observed *Dst* (solid).

indices.

It should be pointed out that if the solar wind measurements are made just upstream of the bow shock, there will be little warning of impending activity other than the inherent delay in the magnetospheric response functions. The situation can be improved by moving the monitor upstream of the earth to measure solar wind parameters well before they arrive at the earth. For example, at the initial location of ISEE-3 (230 Re upstream), as much as one hour advance warning was provided (e.g., TSURUTANI and BAKER, 1979). Unfortunately, as the monitor is moved upstream, the quality of the estimate is degraded. This is due to the scale of structures in the solar wind, or their evolution as they convect and propagate toward the earth. A first-order correction which assumes all parameters are convected directly to the earth at the solar wind velocity is not optimum since southward IMF is often caused by waves in the solar wind. Thus, a balance must be struck between the needs for accurate prediction of magnetic activity onsets and the needs for early warning impending activity as provided by measurements far upstream.

4. Conclusions

A real-time system to predict geomagnetic activity would consist of only a few elements. The first is a properly instrumented spacecraft monitoring the solar wind. Second is a real-time communication link between the monitor and a computer system that stores the measured quantities and computes the indices. Within this computer system, the current and preceding values of the calculated solar wind parameters are convolved with the filters to predict the indices. If the measurements are made far upstream of the earth, the instantaneous predictions apply to a significantly later time which can be determined rather accurately from the measured solar wind velocity. The predicted indices are stored in the system for examination, e.g., by remote dial-up.

This system can be improved in a variety of ways. First, it is possible that parameters other than those discussed above affect geomagnetic activity indices. Further research may determine these and may calculate corresponding filters. Inclusion of measurements of these parameters in the real-time system would then improve the estimated indices. Second, it may be possible to improve the quality of the predictions by more complex models such as piece-wise approximations of nonlinear systems by linear relations in restricted ranges or by filters whose constants change with time in predictable ways. Third, further research on the propagation of disturbances in the solar wind may improve the quality of the estimates of the solar wind at the magnetopause from measurements made upstream. Finally, there remains hope that further research on the origins of the solar wind at the sun will enable researchers to predict the existence of regions of southward IMF in the solar wind from optical measurements of the sun.

This work was supported at UCLA by grant ATM83-18200 from the National Science Foundation, grant N00014-84-C-0158 from the Office of Naval Research, and grant NGL-05-007-004 from the National Space and Aeronautics Administration. A grant from the Institute of Geophysics and Planetary Physics at Los Alamos National Laboratory also provided partial support.

REFERENCES

BAUMJOHAUN, W., Merits and limitations of the use of geomagnetic indices in solar wind: Magnetic coupling studies, this volume, 1986.

CLAUER, C. R., The technology of linear prediction filters applied to studies of solar wind-magnetosphere coupling, this volume, 1986.

MCPHERRON, R. L. and D. N. BAKER, Solar wind-magnetosphere energy input functions, this volume, 1986.

McPHERRON, R. L. and R. H. MANKA, Dynamics of the 1054 UT March 22, 1979, substorm event: CDAW-6, *J. Geophys. Res.*, **90**(A2), 1175–1190, 1985.

TSURUTANI, B. and D. N. BAKER, Substorm warnings: An ISEE-3 real time data system, *EOS Trans. AGU*, **60** (41), 702, 1979.

Solar Wind-Magnetosphere Coupling, edited by Y. Kamide and J. A. Slavin, 93–100.

Magnetospheric Response to Solar Wind Variations

L. F. BARGATZE[1,2,3], D. N. BAKER[1], and R. L. MCPHERRON[2,3]

[1]*University of California, Los Alamos National Laboratory, Los Alamos, New Mexico 87545, U.S.A.*
[2]*Institute of Geophysics and Planetary Physics, University of California, Los Angeles California 90024, U.S.A.*
[3]*Department of Earth and Space Sciences, University of California, Los Angeles California 90024, U.S.A.*

The time lagged response of the magnetosphere to solar wind variations has been determined using the linear prediction filtering method and 34 intervals of high time resolution IMP-8 solar wind data and auroral electrojet *AL* index data. The linear prediction filtering method is a powerful time series analysis technique which is utilized to produce a filter of time lagged response coefficients which estimates the most general linear relationship between magnetospheric activity and solar wind variations. This study uses the *AL* index to monitor the magnetosphere's response and VB_s to monitor the solar wind input. Before analysis, the median value of the *AL* index for each of the 34 intervals was utilized to rank the intervals according to the level of geomagnetic activity. It is found that the VB_s–AL filters are composed of two response pulses peaking at time lags of 20 minutes and 60 minutes. Our interpretation associates the 20-minute pulse with activity driven directly by solar wind-magnetosphere interaction, and it associates the 60-minute pulse with activity driven by the release of stored energy from the magnetotail. Thus, the filter results suggest that both the directly driven and the unloading models of magnetospheric response are important in describing the time lagged response of the magnetosphere to solar wind variations.

1. Introduction

An active topic of research in solar-terrestrial physics is the study of the magnetosphere's response to variations in the upstream solar wind conditions. Indeed, this is a topic of some controversy since two competing models have been proposed to describe the response of the magnetospheric system to the solar wind. One of these, the directly driven model, suggests that the magnetosphere responds directly to interactions with the solar wind. That is, one need only to prescribe the boundary conditions on the magnetopause to predict the behavior of the magnetospheric system. The other model which is often called the unloading model concludes that processes internal to the magnetosphere are important in determining when enhanced geomagnetic activity occurs. The hypothesis is that the release of stored energy from the earth's magnetotail gives rise to the energy dissipation processes commonly associated with magnetospheric substorms.

The technique of linear prediction filtering provides a means of testing the relative efficacies of the directly driven and unloading models of magnetospheric response. This technique uses a filter of time lagged response coefficients to quantify the relationship between the input and output of a linear, time-invariant system. The linear prediction filtering method has a major advantage over conventional cross-correlation methods. Once a filter is obtained, it can be convolved with the input time series to directly predict the output time series of the system. Then, comparison of the predicted and observed time series reveals valuable clues as to what signals are transferred through the system from the input to the output. For a history of the application of linear prediction filtering to the problem of solar-terrestrial physics and a self-contained description of the technique, please refer to the review by CLAUER (this volume).

In this study, we shall use VB_s which is equal to the solar wind bulk speed multiplied by the magnitude of the interplanetary magnetic field (IMF) southward component to monitor the solar wind input to the magnetosphere. We shall also use the AL index to monitor the response of the magnetosphere to the solar wind. A study by CLAUER *et al.* (1981) has already used VB_s and the AL index in a preliminary study using linear prediction filtering. They report that the VB_s–AL filters have a single response pulse which peaks after a time lag of about 40 to 60 minutes and which decays to zero after about 2.5 hours. Furthermore, CLAUER *et al.* (1981) found that the character of the filters describing the relationship between the solar wind and the AU and AL indices change depending on the magnitude of the geomagnetic disturbance in the data sets used for analysis. They suggest that the response of the magnetosphere is not completely linear; otherwise, the filter characteristics would not vary depending on the level of geomagnetic activity. Of course, the magnetosphere is a very complicated system which is not likely to behave in a completely linear fashion. This latter complication will be addressed here by organizing the data with respect to the level of geomagnetic activity before analyzing them with the filtering routines. The study presented here is a brief summary of the essential points reported in BARGATZE *et al.* (1985).

2. Data and Data Set Preparation

This study utilizes IMP-8 solar wind plasma data, IMP-8 IMF measurements, and the auroral electrojet AL index from the time interval between November 1973 and December 1974. The IMF measurements are expressed in geocentric solar magnetospheric (GSM) coordinates. All of the data are at 2.5-minute time resolution and they are the same data used by BAKER *et al.* (1981) to study the correlation between the AE index and interplanetary parameters.

A series of five steps was required to prepare a data set suitable for analysis. First, the solar wind data were time shifted to account for the transit time of the solar wind from the point of observation at IMP-8 to the subsolar point of the magnetopause. This correction takes into account both the radial and azimuthal drift of solar wind plasma in the GSM coordinate system. The time shift has the effect of setting the fiducial zero time lag for the linear filtering analysis. A more complete description of the time shifting formula is given by BAKER *et al.* (1983).

Second, the AL index which we use to model the response of the magnetospheric system and VB_s which we use to model the solar wind input to the system were scanned to identify intervals which contained at least one day's worth of nearly continuous data. The list of intervals was then modified on the basis of the following criteria: each data interval must be temporally bounded at beginning and end by 3-hour segments of weak solar wind input and weak geomagnetic activity; each data interval may contain one or more events of geomagnetic activity if and only if the events are of the same level of activity; and any data interval must be split into two separate intervals if the two new intervals independently satisfy the above criteria. These latter conditions aid in the prevention of time aliasing caused by inputting concatenated data intervals into the linear filtering routines, and they aid us later when we sort the intervals with respect to the level of geomagnetic activity.

Third, integral occurrence percentages of the AL index for each data interval were calculated to find the median value of the index in each interval. The median AL index values were then used to order the data intervals from weakest levels of geomagnetic activity (small, negative AL index medians) to strongest levels of geomagnetic activity (large, negative AL index medians). In the final two steps, the data gaps in the solar wind time series were interpolated and then, the data intervals were concatenated end-to-end in the order prescribed by activity levels. The interpolation was done since the linear filtering routines cannot accept missing data in the input or output time series; few data gaps were longer than 10 minutes long.

3. Linear Filtering Results

As mentioned above, the CLAUER *et al.* (1981) study found evidence that the response of the magnetosphere shows some nonlinear behavior. They note that the VB_s–AL and VB_s–AU filters change character depending on the level of geomagnetic disturbance in the data sets they used to calculate them. In particular, the time lag to peak response tends to shorten with increasing levels of disturbance. To study this feature, we input each set of five activity ordered data intervals into the linear filtering routines to obtain VB_s–AL filters which describe the magnetosphere's response to the solar wind over limited ranges of geomagnetic activity. In all, thirty filters were calculated using the 34 intervals in the analysis data set. These filters were numbered from 1 to 30 with #1 and #30 representing the impulse response of the magnetosphere during weakest and strongest levels of geomagnetic activity, respectively.

In Fig. 1, filter #10 and filter #27 are displayed. Both of these filters have a response pulse which rises quickly from near zero response at zero time lag, which reaches a peak value near 20-minute time lag, and which decays thereafter. In addition, filter #10 possesses a second response pulse which reaches a peak value at a time lag of about 60 minutes. If filter #27 has a corresponding pulse, it is not well resolved. Both filters decay back to near zero response at a time lag of about 2.5 hours.

Figure 2 is a stack plot of all of the VB_s–AL filters we calculated. In general, filters #1 through #18 possess the same response features which were described for filter #10 above. Likewise, filters #19 through #30 have nearly the same response character as that described for filter #27. Note, however, that there is a degree of variability in the time lag to peak magnitude of the first and second response pulses. It is likely that this variability is related to incomplete correction for solar wind transit effects. It may, however, represent some real variability in the response fo the magnetosphere to the

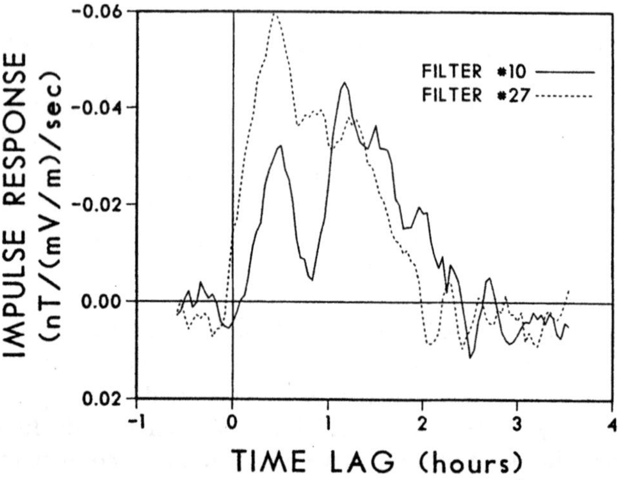

Fig. 1. A plot of filter #10 and filter #27. These filters correspond to moderately and strongly active data intervals respectively. The vertical axis is plotted using the units of the *AL* index (nT) divided by units of the solar wind electric field (mV/m) and normalized by the data sampling period (150 seconds).

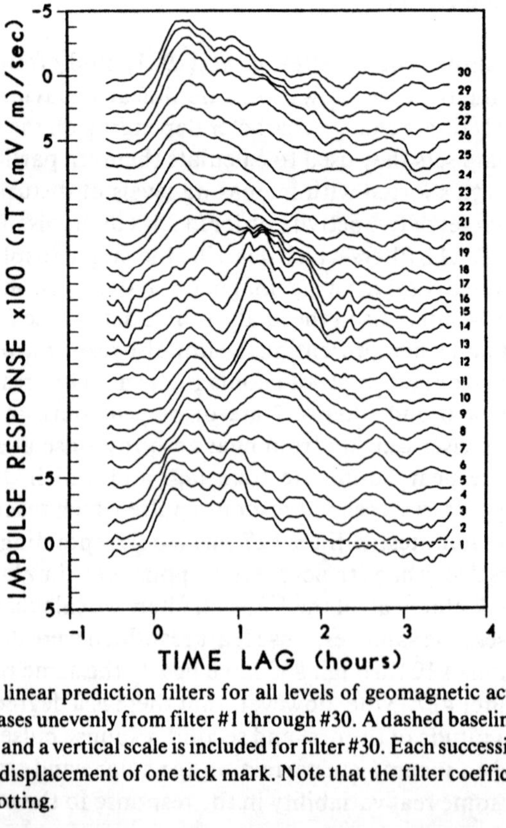

Fig. 2. A stack plot of linear prediction filters for all levels of geomagnetic activity. The geomagnetic "activity level" increases unevenly from filter #1 through #30. A dashed baseline and a vertical scale are included for filter #1 and a vertical scale is included for filter #30. Each successive filter is plotted after a cumulative, vertical displacement of one tick mark. Note that the filter coefficients were multiplied by a hundred before plotting.

solar wind.

Figure 3 is a plot of the VB_s solar wind input time series, the original AL index output time series, the predicted AL index output time series, and the residual AL index time series (=original-predicted)for the data sets used to calculate filter #10 (top three panels) and filter #27 (bottom three panels). In both cases, the predicted AL index time series appears to be a low pass averaged version of the original AL index time series. And in both cases, about 45% of the variance in the original AL index time series is modeled by the predicted AL index time series. However, there are high frequency variations in the original AL index that are not predicted using the VB_s solar wind input. Some of these variations, the sharp negative residuals, appear to be systematic response signatures of the magnetosphere. McPherron *et al.* (this volume) have correlated the occurrence of such signatures with the onset times of magnetospheric substorms. The fact that this component of the AL index is not predicted using solar wind data alone suggests that magnetospheric variables do exercise some control over the response of the magnetosphere to the solar wind.

4. Discussion and Conclusion

The VB_s–AL filter results presented here in Figs. 1 and 2 are in good agreement with the results presented by Clauer *et al.* (1981). However, by ordering our data set with respect to activity level and by correcting for the transit time of the solar wind from IMP-8 to the magnetopause, we have been able to distinguish two response pulses in our VB_s–AL filters, whereas the Clauer *et al.* study only found one pulse. These two pulses peak at time lags of about 20 minutes and 60 minutes, and their relative magnitudes depend on the level of geomagnetic activity in the data sets used to calculate them. This change in relative magnitude is responsible for the differences that Clauer *et al.* (1981) reported in their moderate and strong activity level filters.

Likewise, our filter results are in agreement with the results of a cross-correlation study performed by Baker *et al.* (1981) which used the same data set analyzed here. They found that the time lag for peak correlation between VB_s and the AL index maximizes near a time lag of about 40 minutes. This is intermediate between the time lags of the 20-minute and 60-minute pulses found using linear filtering which is understandable since they did not separate their analysis data set depending on the level of geomagnetic activity.

Our present interpretation of these results is that the 20-minute and 60-minute pulses each correspond to a separate response mode of the magnetosphere. It is possible that the 20-minute pulse corresponds to the enhancement of current flow in the ionosphere in response to magnetospheric convection driven directly by solar wind interaction. Also, it is possible that the 60-minute pulse corresponds to the enhancement of current flow in the ionosphere in response to the release of stored electromagnetic energy from the magnetotail.

There is observational support for this intepretation. For instance, Meng (1979), Holzer and Reid (1975), and Reid and Holzer (1975) reported that motions of the auroral oval respond to changes in the solar wind with an exponential time constant of about 20 minutes. Furthermore, Iyemori (1980) found that the onset of a substorm

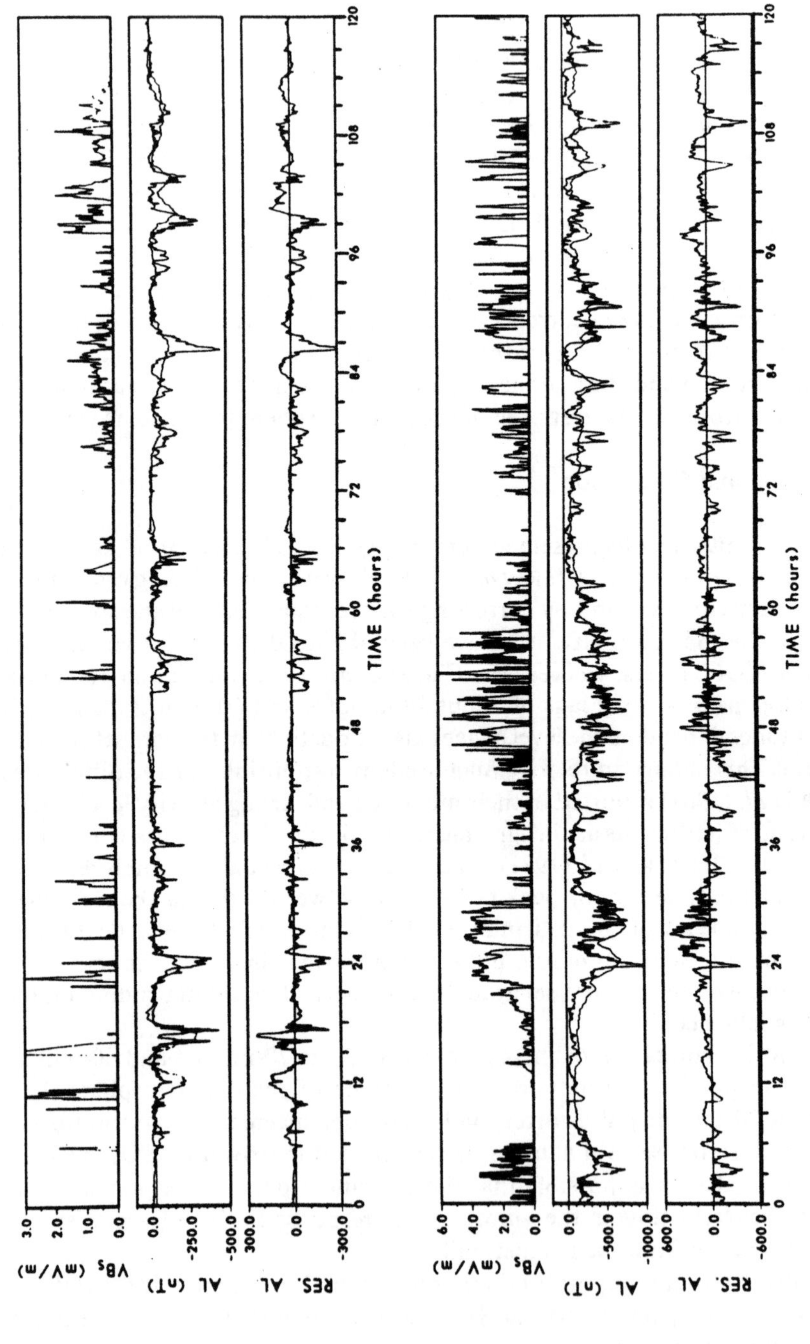

Fig. 3. A plot of the VB_s, original AL index, predicted AL index, and residual AL index time series for moderate (top three panels) and strong (bottom three panels) geomagnetic activity levels. The moderate and strong activity data shown here correspond to a 5-day subset of the data used to calculate filter #10 and filter #27. Note that the predicted AL index time series is the more smoothly varying function plotted in the middle panel of each set of panels.

is typically delayed one hour after a southward turning of the IMF. Many cross-correlation studies find a similar time lag. Also, KOKUBUN (1971) has suggested that a magnetic substorm is a two-stage process. During the first stage, which lasts about 1–2 hours, twin current vortices develop in the polar ionosphere in response to the southward turning of the IMF. Simultaneously, energy is stored in the magnetotail. MCPHERRON (1970) referred to this first stage as the growth phase of magnetospheric substorms. During the second stage or expansion phase of a substorm, auroral breakup occurs and energy is released from the tail.

If the present interpretation is correct, then the linear filtering results suggest that both the directly driven and unloading models of magnetospheric response are important in describing the complete response of the magnetosphere to solar wind variations. In the future, we hope to improve upon our capability to predict geomagnetic activity. To do this, it appears that we must identify magnetospheric variables which are readily monitored and which accurately predict the onset times of magnetospheric substorms.

The IMP-8 magnetometer data were supplied by researchers at the Goddard Space Flight Center and the *AE* indices were provided by the World Data Center, Boulder, Colorado. This work was supported at Los Alamos by the Institute of Geophysics and Planetary Physics and was done under the auspices of the United States Department of Energy. This work was supported at the University of California, Los Angeles, by the National Science Foundation, grant ATM 83-18200, by Office of Naval Research, grant N00014-84-C-0158, and by the National Aeronautics and Space Administration, grant NGL-05-007-004.

REFERENCES

BAKER, D. N., E. W. HONES, Jr., J. B. PAYNE, and W. C. FELDMAN, A high time resolution study of the interplanetary parameter correlations with *AE*, *Geophys. Res. Lett.*, **8**, 179, 1981.

BAKER, D. N., R. D. ZWICKL, S. J. BAME, E. W. HONES, Jr., B. T. TSURUTANI, E. J. SMITH, and S.-I. AKASOFU, An ISEE-3 high time resolution study of interplanetary parameter correlations with magnetospheric activity, *J. Geophys. Res.*, **88**, 6230, 1983.

BARGATZE, L. F., D. N. BAKER, and R. L. MCPHERRON, Magnetospheric response for many levels of geomagnetic activity, *J. Geophys. Res.*, **90**, 6387, 1985.

CLAUER, C. R., The technique of linear prediction filters applied to studies of solar wind-magnetosphere coupling,| this volume, 1986.

CLAUER, C. R., R. L. MCPHERRON, C. SEARLS, and M. G. KIVELSON, Solar wind control of auroral zone geomagnetic activity, *Geophys. Res. Lett.*, **88**, 915, 1981.

HOLZER, T. E. and G. C. REID, The response of the dayside magnetosphere-ionosphere system to time varying field line reconnection at the magnetopause: 1. Theoretical model, *J. Geophys. Res.*, **80**, 2041, 1975.

IYEMORI, T., Time delay of the substorm onset from the IMF southward turning, *J. Geomagn. Geoelectr.*, **32**, 267, 1980.

KOKUBUN, S., Polar substorm and the interplanetary magnetic field, *Planet. Space Sci.*, **19**, 697, 1971.

MCPHERRON, R. L., Growth phase of magnetospheric substorms, *J. Geophys. Res.*, **75**, 5592, 1970.

MCPHERRON, R. L., D. N. BAKER, and L. F. BARGATZE, Linear filters as a method of real-time prediction of geomagnetic activity,| this volume, 1986.

MENG, C.-I., Polar cap variations and the interplanetary magnetic field, in *Dynamics of the Magnetosphere*, edited by S.-I. Akasofu, D. Reidel, Dordrecht, Holland, 1979.

REID, G. C. and T. E. HOLZER, The response of the dayside magnetosphere-ionosphere system to time varying field line reconnection at the magnetopause: 2. Erosion event of March 27, 1968, *J. Geophys. Res.*, **80**, 2050, 1975.

Solar Wind-Magnetosphere Coupling, edited by Y. Kamide and J. A. Slavin, 101–109.

Solar Wind-Magnetosphere Energy Input Functions

L. F. BARGATZE[1,2,3], R. L. MCPHERRON[1,2], and D. N. BAKER[3]

[1] *Department of Earth and Space Sciences, University of California, Los Angeles California 90024, U.S.A.*

[2] *Institute of Geophysics and Planetary Physics, University of California, Los Angeles California 90024, U.S.A.*

[3] *University of California, Los Alamos National Laboratory, Los Alamos New Mexico 87545, U.S.A.*

A new formula for the solar wind-magnetosphere energy input parameter, P_i, is sought by applying the constraints imposed by dimensional analysis. Applying these constraints yields a general equation for P_i which is equal to $\rho V^3 l^2_{CF} F(M_A,\theta)$ where, ρV^3 is the solar wind kinetic energy density and l^2_{CF} is the scale size of the magnetosphere's effective energy "collection" region. The function F which depends on M_A, the Alfvén Mach number, and on θ, the interplanetary magnetic field clock angle is included in the general equation for P_i in order to model the magneto-hydrodynamic processes which are responsible for solar wind-magnetosphere energy transfer. By assuming the form of the function F, it is possible to further constrain the formula for P_i. This is accomplished by using solar wind data, geomagnetic activity indices, and simple statistical methods. It is found that P_i is proportional to $(\rho V^2)^{1/6} VBG(\theta)$ where, ρV^2 is the solar wind dynamic pressure and $VBG(\theta)$ is a rectified version of the solar wind motional electric field. Furthermore, it is found that $G(\theta)$, the gating function which modulates the energy input to the magnetosphere, is well represented by a "leaky" rectifier function such as $\sin^4(\theta/2)$. This function allows for enhanced energy input when the interplanetary magnetic field is oriented southward. This function also allows for some energy input when the interplanetary magnetic field is oriented northward.

1. Introduction

An important goal of solar-terrestrial physics is to understand how the rate of solar wind-magnetosphere energy transfer depends upon interplanetary and magnetospheric parameters. In the past, simple cross-correlation analyses have been used to establish the causal connection between enhanced geomagnetic activity and the variations of single solar wind parameters such as the solar wind plasma speed (V), the interplanetary magnetic field (IMF) strength (B), and the north–south orientation of the IMF vector (see reveiw by BAKER, 1986, in this volume, and references therein). Later, researchers began developing more complex formulas to correlate with geomagnetic indices. Such formulas include $\varepsilon = VB^2 l_0^2 \sin^4(\theta/2)$ (l_0=7 earth radii, θ is the IMF clock angle defined below), V^2B_s, and VB_s ($B_s = -B_Z$ if $B_Z \leqslant 0.0$, B_s=0.0 otherwise).

Here, we take a slightly different approach. Before cross-correlating any variables, we use the constraints imposed by dimensional analysis as a guide to develop a general formula for P_i, the rate of solar wind-magnetosphere energy transfer (VASYLIUNAS *et al.*, 1982). This general equation is listed below:

$$P_i = \rho V^3 l^2_{CF} F(M_A,\theta) \quad (1)$$

where, ρ is the solar wind plasma density, l_{CF} is the Chapman-Ferraro scale length ($l_{CF}=(M_E^2/\mu_0\rho V^2)^{1/6}$), M_A is the Alfvén Mach number ($M_A^2=(\mu_0\rho V^2 / B^2_T)^{1/2}$), θ is the IMF clock angle, and F is a function whose dependence upon M_A and θ will be discussed below. Other quantities in these formulas include M_E, the earth's magnetic dipole moment, μ_0, the permeability of free space, B_T, the magnitude of the vector sum of the IMF B_Y and B_Z components measured using geocentric solar magnetospheric (GSM) coordinates, and θ, the angle equal to the arc subtended by the GSM Z-axis and $\boldsymbol{B}_T$.

Equation 1 explicitly assumes that the amount of energy which is transferred from the solar wind to the magnetosphere is proportional to the amount of solar wind kinetic energy that is intercepted by an energy "collection" region on the magnetopause. Equation 1 includes the function F in order to model the magnetohydrodynamic interactions which are responsible for the energy transfer. Here, we assume as did VASYLIUNAS *et al.* (1982) that $F=M_A^{-2\alpha}G(\theta)$ where, α is the MHD coupling exponent, and $G(\theta)$ is a function which modulates the rate of energy input depending upon the orientation of the IMF.

Unfortunately, the energy input rate cannot be directly measured. It is customary instead to approximate P_i by using various indices of geomagnetic activity. In this study, we use the *AL* index to estimate P_i. This index is sensitive to the amount of current which flows in the westward electrojet (BAUMJOHANN, 1986; this volume). The results of BARGATZE *et al.* (1985) suggest that the amount of current which flows in the westward electrojet is controlled both by the solar wind via magnetospheric convection currents and by the release of energy from the magnetotail via the substorm current wedge.

Note that the *AL* index may not account properly for all of the energy which enters the magnetosphere. For instance, the *AL* index does not respond instantaneously to the rate of solar wind energy input. Also, the *AL* index is not directly sensitive to energy lost via plasmoids (HONES *et al.*, 1984). Hence, one must account for the time delay between solar wind energy input and the enhancement of current flow in the westward electrojet. And, one must average over a sufficiently long time scale such that the energy lost via mechanisms not monitored by *AL* is roughly proportional to the energy that is monitored by *AL*. These problems are accounted for by convolving an average *AL* index impulse response filter f with the solar wind input time series (see papers by BARGATZE *et al.*, 1986; MCPHERRON *et al.*, 1986; and CLAUER, 1986; all this volume). The filter we use is one hour long (roughly the substorm time scale) and is defined between zero and 1-hour time lag. In this light, Eq. 1 becomes:

$$P = f*(\rho V^3 l^2_{CF} M_A^{-2\alpha} kG(\theta)) \quad (2)$$

where the star represents convolution, and the normalization constant k has been introduced such that $G(\theta)$ is now defined to vary between 0.0 and 1.0. In going from Eq. 1 to Eq. 2, the i-subscript has been dropped from P_i; this is done to emphasize the fact that the AL index, a geomagnetic activity index, has been used to estimate P_i, the solar wind energy input parameter.

In the remainder of this study, three different methods are employed to investigate the dependence of P_i upon solar wind variables. Each of these methods is based on a simple statistical technique and on a modified version of Eq. 2. The analyses were completed using a data base that includes IMP 8 solar wind data from 1973–74, ISEE 3 solar wind data from 1978, and the corresponding values of the AL index. The results are presented in three separate sections below.

2. Regression Analysis Results

The formula which provides the basis for the regression analysis is easily obtained from Eq. 2 after substituting for the Chapman-Ferraro scale length:

$$\log\left(\frac{P}{\mu_0^{-1/3} M_E^{2/3} \rho^{2/3} V^{7/3}}\right) = \alpha \log(M_A^{-2}) + \log(kf^* G(\theta)) \ . \qquad (3)$$

Equation 3 is that for a straight line, $Y = \alpha X + B$, where X and Y are logarithmic variables, α is the line's slope, and $\log(kf^* G(\theta))$ is the line's Y-axis intercept. Strictly speaking, Eq. 3 is a valid representation of Eq. 2 if and only if the quantities $\rho V^3 l^2_{CF}$ and $|M_A^2$ are approximately constant since these terms have been removed from the convolution with the impulse response function, f.

To use Eq. 3, the data set was separated into twelve bins depending on the value of $G^{-1}(f^* G(\theta))$, where $G(\theta)$ is assumed to be equal to $\sin^4(\theta/2)$ and G^{-1} is its trigonometric inverse function. Each bin corresponds to a 15° range in θ starting with $0° \leqslant \theta < 15°$ (a nearly northward IMF) and ending with $165° \leqslant \theta \leqslant 180°$ (a nearly southward IMF). The data within each bin were used as input to a simple linear regression analysis routine in order to estimate the slope and intercept of the best fit line through the distribution.

Figure 1 is a scatter plot of the logarithmic variables which was made by assuming that P is equal to the AL index multiplied by an energy conversion factor of 3×10^8 ((J/s)/nT) (PERREAULT and AKASOFU, 1978). For this figure, the value of θ was restricted to lie between 135° and 150°. The best fit line has a slope equal to 0.54; the linear correlation coefficient is equal to 0.66. Figure 2 is a summary of all of the α-slope estimates plotted as a function of θ. The error in the α-estimates is shown using 2-σ error bars. For northward IMF, almost all of the slope estimates lie between 0.0 and 0.4; however, all of the corresponding regression coefficients are smaller than 0.3. For southward IMF, all but one of the slope estimates lie between 0.4 and 0.6. The regression coefficients are higher reaching values near 0.7.

3. Histogram Analysis Results

For this section only, we define a new MHD coupling exponent, α', which is equal to $1.0 - \alpha$. We do this in order to directly compare the results of this section with the results of a histogram analysis performed by KAN and AKASOFU (1982). Given

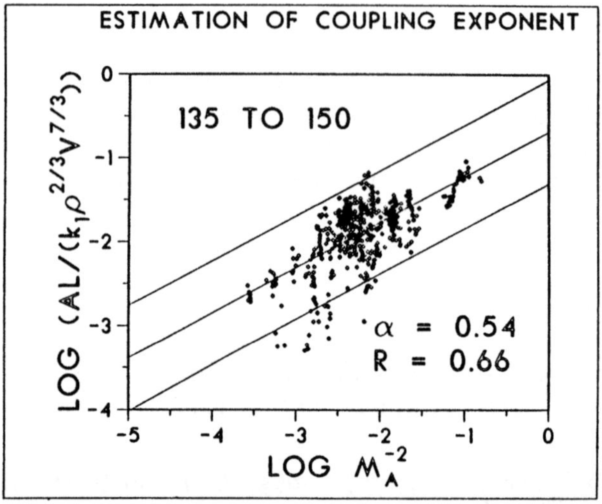

Fig. 1. The scatter plot of the logarithmic variables, $AL/k_1\rho V^3 l^2_{CF}$ and M_A^{-2}, for θ between 135° and 150° is shown. The slope of the regression line, α, is equal to 0.54; the regression coefficient, R, is equal to 0.66. The constant k_1 used in defining the ordinate is equal to $\mu_O^{-1/3} M_E^{2/3}$; k_1 should not be confused to the constant k referred to in the text.

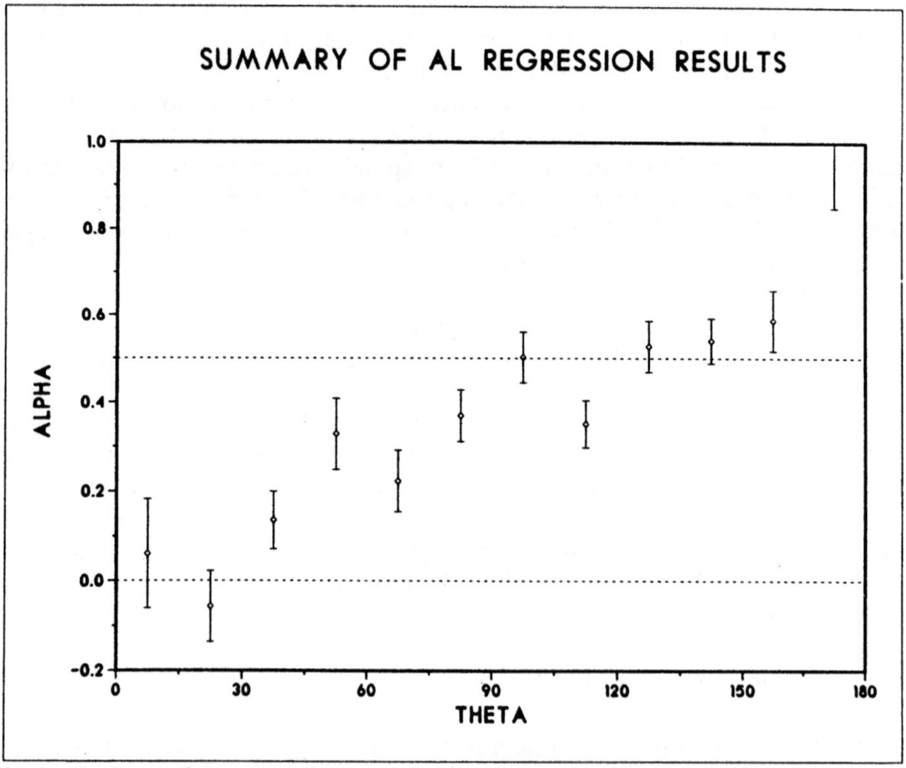

Fig. 2. Summary of the α regression results. Values of α are plotted versus the IMF clock angle, θ. 2-σ error bars are plotted for each data point.

that $\alpha'=1.0-\alpha$, one can readily manipulate Eq. 2 to yield:

$$P = f^*\Big(\frac{1}{\mu_0} V B_T^2\, k l^2_{CF}\, M_A^{2\alpha'}\, G(\theta)\Big)\,. \tag{4}$$

The histogram analysis equation is obtained directly from Eq. 4 by solving for α' explicitly:

$$\alpha' = \log\Big(P/[f^*(\frac{1}{\mu_0}\, V B_T^2\, k l^2_{CF}\, G(\theta))]\Big)/\log(M_A^2)\,. \tag{5}$$

To perform the analysis, it is assumed once again that P is proportional to the AL index and that $G(\theta)$ is equal to $\sin^4(\theta/2)$. Then, each set of data points is used as input to Eq. 5 to obtain an estimate of α'. The "best" estimate of α' is determined by histogramming all of the α'-estimates to find the most frequently occurring α'-value.

Figure 3 shows the result of the AL histogram analysis. Note that the distribution peaks between 0.5 and 0.55 and that the half-width of the distribution which provides an estimate of error is about 0.1. This result is in accord with the regression analysis results which suggest that α is equal to 0.5 (since $\alpha=1.0-\alpha'$).

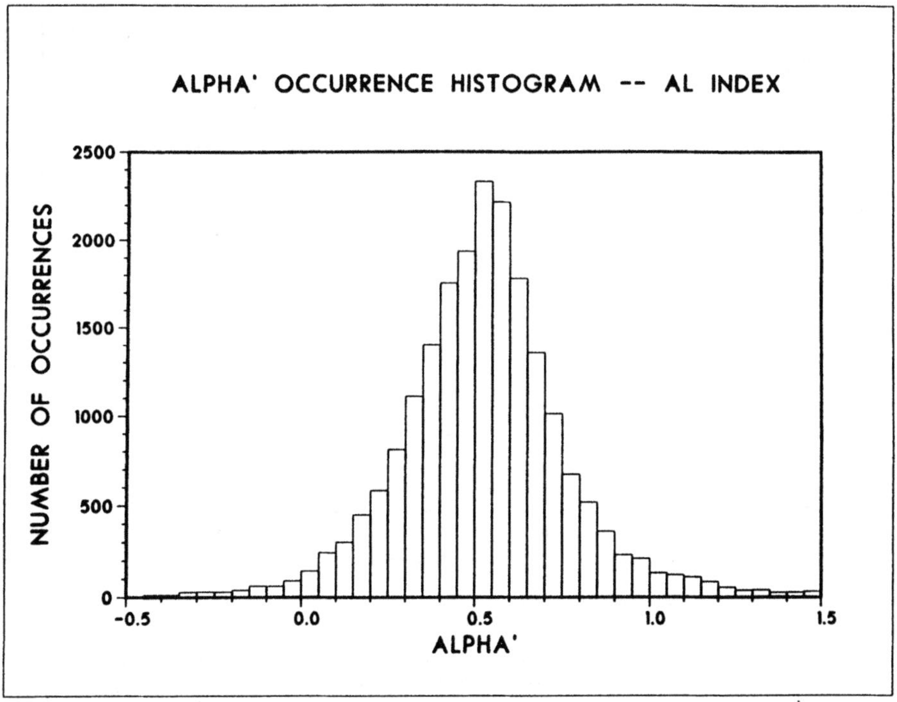

Fig. 3. The number of occurrences of α' is plotted versus the magnitude of α'. Note that α' is equal to 1.0 $-\alpha$.

4. Gating Function Analysis

Rearranging Eq. 2, one obtains the formula used to calculate the empirical dependence of $G(\theta)$ upon θ:

$$G(\theta) = \frac{P}{f^{*}(\rho V^3 k l^2_{\mathrm{CF}} M_{\mathrm{A}}^{-1})} \,. \tag{6}$$

Note that α was assumed to be equal to 0.5 to yield this equation. Before analysis, the data were separated into twelve bins depending on the average value of θ. Each data bin corresponds to a 15° range in θ. The binned data sets were then used as input to Eq. 6 in order to find the average of $G(\theta)$ within each bin. Since the $G(\theta)$ term has been removed from the convolution with f, we have limited the analysis to include only those intervals during which θ was approximately constant.

The results are displayed in Fig. 4. Also plotted are two solid lines which show the

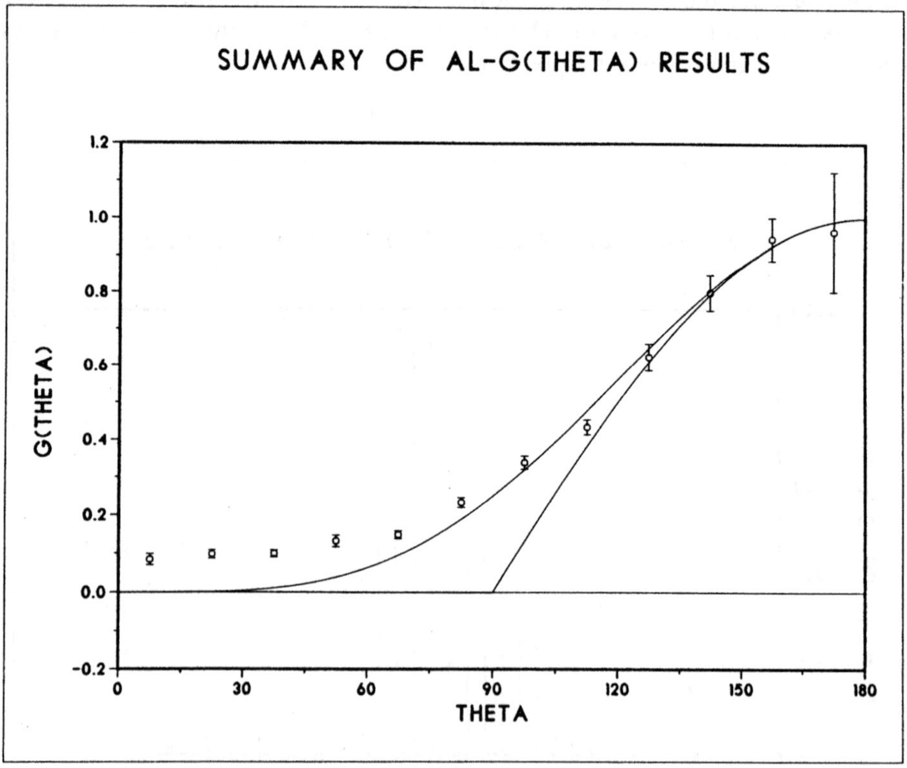

(a)

Fig. 4. Empirical estimates of the gating function, $G(\theta)$, are plotted versus the IMF clock angle, θ. Two functions, $\sin^4(\theta/2)$ (upper solid curve) and $U(\theta)\cos(\theta)$ (lower solid curve), are shown for comparison. In Fig. 4(a), the empirical $G(\theta)$-values were normalized such that they approach 1.0 for $\theta = 180°$. In Fig. 4(b), an "error" baseline was subtracted before normalizing the $G(\theta)$-values.

variation $\sin^4(\theta/2)$ and $U(\theta)\cos(\theta)$ versus θ ($U(\theta)=0.0$ if $\theta \leqslant 90°$ and $U(\theta)=-1.0$ if $\theta \geqslant 90°$). These are the gating functions used in the definitions of ε and VB_s, respectively. In Fig. 4(a), the $G(\theta)$-estimates have been normalized such that they approach 1.0 at $\theta=180°$. Note that the $G(\theta)$-estimates do not approach zero asymptotically as θ approaches 0°. This is likely due to the errors involved in the analysis or perhaps due to the effects of energy input to the magnetosphere via viscous processes (AXFORD, 1964). This "error" baseline corresponds to an AL index value of about -50 nT. Figure 4(b) shows the $G(\theta)$-estimates renormalized after the "error" baseline has been subtracted out. These estimates agree quite well with the $\sin^4(\theta/2)$ curve.

5. Conclusion

In summary, we have used the constraints imposed by dimensional analysis and three different statistical techniques to examine the relationship governing energy

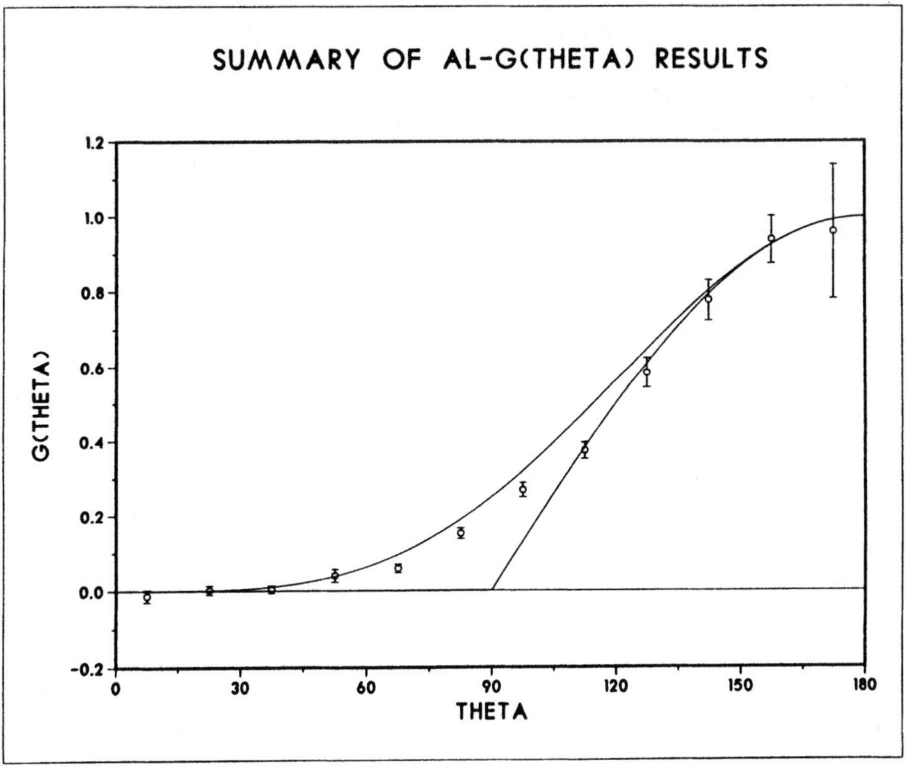

(b)

Fig. 4. (continued).

transfer from the solar wind to the magnetosphere. Two of these techniques, the regression technique and the histogram technique, have been used to find whether P_i is better represented by a parameter such as ε which is related to flow of energy described by the Poynting vector of the solar wind or by a parameter such as VB_s which is related to the solar wind motional electric field. Both the regression analysis results and the histogram analysis results suggest that the MHD coupling exponent, α, is equal to 0.5 ± 0.1. If α is equal to 0.5, the rate of energy transfer is given by $\rho V^3 k l^2_{CF} M_A^{-1} G(\theta)$. This function is proportional to $(\rho V^2)^{1/6} VBG(\theta)$ where, ρV^2 is the solar wind dynamic pressure and $VBG(\theta)$ is a rectified version of the solar wind motional electric field. Thus, the results suggest that P_i is more closely related to the strength of the solar wind motional electric field. Since P_i depends upon the solar wind dynamic pressure, the scale length of the magnetosphere which is inversely proportional to ρV^2 also plays an important role in determining the rate of energy transfer.

The above conclusions differ from the conclusions reached by KAN and AKASOFU (1982) who found that $P_i \simeq \varepsilon$. Their conclusion is based on a histogram analysis similar to the one presented here except that they modeled P_i using U_T, a measure of the total rate of magnetospheric energy dissipation (PERREAULT and AKASOFU, 1978). However, it appears that they have not performed a completely general analysis. In particular, they have implicitly chosen a value for the constant k which is about an order of magnitude larger than the value of k that we have found using the regression technique and the 1978 data base which includes U_T (the U_T results are not presented here due to lack of space; they are in general agreement with the AL index results). Referring back to Eq. 3, one finds that the constant k determines the Y-axis intercept of the line which best relates the two logarithmic variables. Hence, one cannot freely assume any value for k in order to perform a histogram analysis; it is crucial to use a value of k which is consistent with the results of regression analysis.

The results of the gating function analysis suggest that $G(\theta)$ is best represented by a "leaky" rectifier function such as $\sin^4(\theta/2)$ which is used in the definition of ε. We use the term leaky since a function like $\sin^4(\theta/2)$ allows for a small amount of energy input when the IMF is northward. In contrast, the half-wave rectifier function like that used in the definition of VB_s does not allow for any energy input when the IMF is northward. Both, of course, allow for enhanced energy input when the IMF is oriented southward.

The highest correlation coefficients of the analyses presented here lie near 0.7. If the correlation coefficients were much lower, one might question the validity of the theoretical guidelines provided by dimensional analysis which were used to obtain the original, general formula for P_i. Still, the magnitude of the correlation coefficients suggests that the model that we have used here could be improved. In fact, VASYLIUNAS et al. (1982) have suggested using a more complete model that incorporates magnetosphere-ionosphere coupling. Unfortunately, we cannot test this model until a high-time resolution data base which includes the ionospheric Pedersen conductivity becomes available.

The IMP 8 magnetometer data were supplied by researchers at the Goddard Space Flight Center and the ISEE 3 magnetometer data were supplied by researchers at the Jet Propulsion Laboratory. The AE indices were provided by the World Data Center, Boulder, Colorado. This work was supported at Los Alamos by the Institute of Geophysics and Planetary Physics and was done under the auspices of the United States Department of Energy. This work was supported at the University of California at Los Angeles by the National Science Foundation, grant ATM 83-18200, by the Office of Naval Research, grant N00014-84-C-0158, and by the National Aeronautics and Space Administration, grant NGL-05-007-004.

REFERENCES

AXFORD, W. I., Viscous interaction between the solar wind and the earth's magnetosphere, *Planet. Space Sci.*, **12**, 45, 1964.

BAKER, D. N., Statistical analyses in the study of solar wind-magnetosphere coupling, this volume, 1986.

BARGATZE, L. F., D. N. BAKER, and R. L. MCPHERRON, Magnetospheric response to solar wind variations, this volume, 1986.

BAUMJOHANN, W., Merits and limiations of the use of geomagnetic indices in solar wind-magnetosphere coupling, this volume, 1986.

CLAUER, C. R., The technique of linear prediction filters applied to studies of solar wind-magnetosphere coupling, this volume, 1986.

HONES, E. W., Jr., J. BIRN, D. N. BAKER, S. J. BAME, W. C. FELDMAN, D. J. MCCOMAS, R. D. ZWICKL, J. A. SLAVIN, E. J. SMITH, and B. T. TSURUTANI, Detailed examination of a plasmoid in the distant magnetotail with ISEE 3, *Geophys. Res. Lett.*, **11**, 1046, 1984.

KAN, J. R. and S.-I. AKASOFU, Dynamo process governing solar wind-magnetosphere energy coupling, *Planet. Space Sci.*, **30**, 367, 1982.

MCPHERRON, R. L., D. N. BAKER, and L. F. BARGATZE, Linear filters as a method of real-time prediction of geomagnetic activity, this volume, 1986.

PERREAULT, P. and S.-I. AKASOFU, A study of geomagnetic storms, *J. R. Astron. Soc.*, **54**, 547, 1978.

VASYLIUNAS, V. M., J. R. KAN, G. L. SISCOE, and S.-I. AKASOFU, Scaling relations governing magnetospheric energy transfer, *Planet. Space Sci.*, **30**, 359, 1982.

Solar Wind-Magnetosphere Coupling, edited by Y. Kamide and J. A. Slavin, 111–117.

Prediction Filters for the D_{st} Index and the Polar Cap Potential

R. A. FAY, C. R. GARRITY, R. L. MCPHERRON, and L. F. BARGATZE

Institute of Geophysics and Planetary Physics and Department of Earth and Space Sciences
University of California, Los Angeles, California 90024, U.S.A.

The technique of linear prediction filtering has been used to create filters relating solar wind parameters to the D_{st} index and to the polar cap potential. Two inputs are required to predict D_{st}, the square root of solar wind dynamic pressure and the solar wind electric field. Together these account for nearly 70% of the D_{st} variance. The dynamic pressure filter is a 10 minute wide pulse centered at 10 minutes' delay (the average time for the solar wind to travel from the monitor to the magnetopause). The electric field filter is more complicated because it represents both injection into the ring current as well as decay. This interaction is better described using a pressure and decay-corrected time derivitive. This "ring current injection" filter is a 20 minute wide pulse centered at 30 minutes' delay.

The prediction filter for the polar cap potential also required two inputs: the square of the solar wind velocity, and the solar wind electric field. The electric field filter is not well determined by the available data but appears to be a broad peak centered around one hour's delay.

These results suggest that both polar cap potential and ring current injection are primarily controlled by the solar wind, and that substorm expansions do not play a major role in ring current injection.

1. Introduction

One of the outstanding questions of magnetospheric physics is the relative importance of the solar wind as compared to internal magnetospheric processes in the control of geomagnetic activity. We have addressed this question by using the technique of linear prediction filtering. This technique is ideally suited to answering the question, is the magnetosphere a driven system? Assuming that it is, this technique provides a method for characterizing the impulse response of the system from past observations of input to, and output from the system (WEINER, 1949). If a large fraction of the output variance is unpredictable, it suggests that other parameters than those assumed as input may be involved. For example, BARGATZE *et al.* (this volume) demonstrate that only 40% of the variance in the AL index is predictable despite optimization of the input parameter, correction for solar wind propagation, and piece-wise linear approximations for different levels of activity. They conclude that internal magnetospheric processes such as unloading of stored magnetotail energy play an essential role in this type of activity.

In this paper we address the question of whether other measures of magnetic activity such as the D_{st} index and polar cap potential are directly driven by the solar wind, or whether they too exhibit significant residuals unrelated to the solar wind. Past work by BURTON *et al.* (1975), PERREAULT and AKASOFU (1978), and WYGANT *et al.* (1983) suggest these parameters should be much more closely controlled by the solar wind than is the *AL* index. In this preliminary investigation we considered only the solar wind dynamic pressure, rectified electric field, and velocity as possible coupling parameters. Further work is required to determine whether the refinements of these parameters suggested by regression analysis (e.g., MURAYAMA, this volume) will improve the prediction efficiency.

Our analysis of D_{st} makes use of data from the same 52-day data set used in our previous studies of AL and ASYM (see review by CLAUER, this volume). Our analysis of polar cap potential, however, utilizes a very short, 72-hour data set constructed from inverse modeling performed in two recent workshops (KAMIDE *et al.*, 1983). As we note in the discussion, this data set is not adequate to properly define the polar cap potential filter. However, these appear to be the only such data in existence, and until more modeling results are available we cannot do better with the prediction filter technique.

2. Experimental Results

We have examined the relation of D_{st} to the solar wind in two ways. First we calculated prediction filters directly using what are assumed to be the two major factors controlling D_{st}, the solar wind dynamic pressure, and the electric field. The results of this analysis are shown in Fig. 3 of MCPHERRON *et al.* (this volume). The pressure filter is a delta function at 10-minutes' delay, while the electric field filter is a pulse peaking at about one hour but decaying slowly over many hours. The long decay is simply the exponential recovery of D_{st} caused by loss of particles from the ring current.

A more informative way of examining the relation between D_{st} and the solar wind is to consider the ring current injection function, Q, defined as

$$Q(t) = \frac{d}{dt}D_{st} + \frac{D_{st}}{\tau}$$

As discussed by BURTON *et al.* (1975), this function is simply the rate of change of D_{st} corrected for ring current decay assuming a constant rate of recovery (8 hours). Q is approximately equal to U_r, the ring current energy dissipation rate defined by PERREAULT and AKASOFU (1978). U_r is one term in the equation for U_t, the total rate of magnetospheric energy dissipation. The two filters required to predict Q are shown in Fig. 1. The pressure filter is a wavelet, the derivitive of a delta function, which is expected since we have correlated pressure with the time rate of change of D_{st}. We point out we could equally well have corrected D_{st} using the pressure-D_{st} filter determined in the first step, and then calculated the electric field-Q filter. The results are equivalent since solar wind pressure and electric field are uncorrelated.

The electric field filter for Q is much simpler than the electric field filter for D_{st}. It is a Gaussian pulse centered at 30 minutes. Taking into account variable solar wind propagation delays of approximately 10 minutes, it is reasonable to characterize the filter as a delta function at 20-minutes' delay. We thus conclude that the rate of

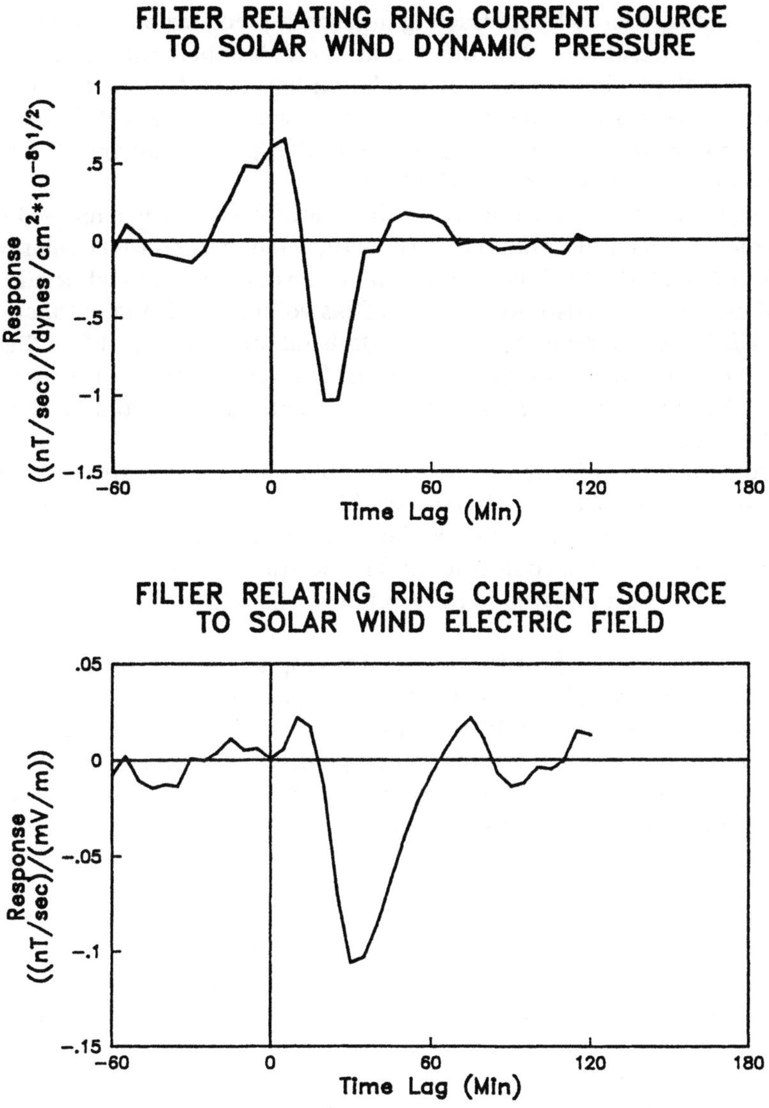

Fig. 1. Prediction filters relating the ring current injection function Q (defined in the text) to the square root of the solar wind dynamic pressure and the solar wind electric field. Top panel displays the filter for square root of dynamic pressure. Bottom panel is the electric field filter.

injection into the ring current is directly proportional to the rectified solar wind electric field delayed by 20 minutes. This is a remarkably simple result and agrees with the empirical assumptions of BURTON *et al.* (1975).

We also addressed the question of whether substorm expansions play a role in ring current injection. We did this by first eliminating solar wind electric field effects in the AL and D_{st} indices by subtracting the signals predicted by our filters, and then correlating the residuals. We found that these residuals are totally unrelated, suggesting that previous reports of a relation between AL and D_{st} are merely a reflection of their mutual dependence on the solar wind electric field. It should also be noted that the solar wind variables do a much better job of predicting D_{st} (70% of the variance) than they do in predicting AL (40%).

It is quite likely that the prediction efficiency for D_{st} can be improved by taking into account some recent results concerning the relation of D_{st} to the solar wind. For example, FELDSTEIN (1984) concluded that a constant term should be added to the rectified electric field. Also, MURAYAMA (this volume) finds that a function of the form $\rho^{.4} VE_r$ is the optimum D_{st} predictor. Both authors note in addition that a better approximation to the decay function is a sum of two exponentials with time constants of 2.5 and 8 hours. In this preliminary report, we have not investigated the effects of such refinements.

Prediction filters required to model the polar cap potential are presented in Fig. 2. The top two panels show the electric field filter and the velocity squared filter. Both filters are required to predict the potential, while it appears that only the electric field filter is required to predict Joule heating. The electric field filter is poorly defined, but is predominantly positive from 15 to 90 minutes. The velocity filter is a delta function at zero lag.

Data from which the filters were derived, and also predictions made by convolving the filters with the input data, are displayed in Fig. 3. The top two panels are the assumed inputs, the middle two panels are the predictions for each filter, and the bottom panel is a comparison of the total prediction with the observations. On the scale of the plot it is difficult to see, but the spikes in potential near 12 hours are not predicted, while the slower variations near 48 hours are. Overall, however, there is reasonable agreement.

The need for a continuous input function such as solar wind velocity is immediately apparent when it is noted that the rectified solar wind electric field is often zero, but the calculated potential is not. As a first attempt at determining an optimum function of solar wind velocity, we calculated separately filters for integral powers of velocity from 1 to 4. The square of the velocity produced the best fit to the trends present in the data.

There are clearly other models for the optimum coupling parameter between the solar wind and polar cap potential than the one we have used. For example, WYGANT *et al.* (1983) found that the potential is well predicted by a function of the form $VB_t \sin^2(\theta/2)$. Since this function is finite for northward IMF it might reduce or eliminate the need for a strong dependence on the square of the solar wind velocity. In this preliminary report, we did not investigate such possibilities since it was our intention to examine how well the rectified solar wind electric does as a predictor of geomagnetic activity.

3. Discussion

Our results demonstrate that both the D_{st} index and the polar cap potential are

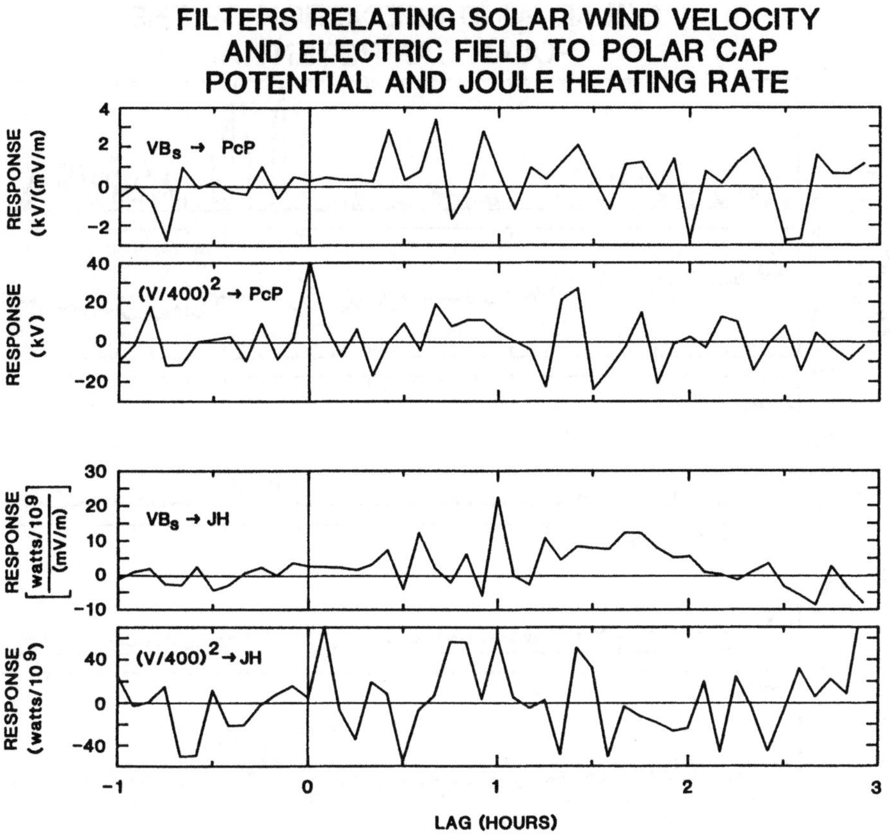

Fig 2. Top panels contain filters relating the polar cap potential to the solar wind electric field and solar wind velocity squared. Bottom panel show same for the Joule heating rate.

related to solar wind variations. For the D_{st} index, we find that 70% of the variance is predictable by the solar wind. Furthermore, the D_{st} residual is not related to the AL residual, suggesting that the substorm expansion phase does not cause detectable changes in the ring current intensity. We also find that the rate of change of D_{st} corrected for dynamic pressure and ring current decay is very simply related to the solar wind electric field by a pulse centered at 30-minutes' delay. It is this fact that has led AKASOFU (1979) to conclude that magnetic activity is primarily driven. Since the parameter, U_t, used in his analysis is dominated by the ring current during strong activity, its filter is nearly identical to that for the function, Q, defined above. As we have shown elsewhere (BARGATZE *et al.*, this volume), the auroral contribution to U_t has a significantly poorer relation to the solar wind, suggesting that other processes such as magnetotail energy storage and release also play an important role. In this regard, it is important to realize that BAUMJOHANN and KAMIDE (1984) found that the auroral contribution to U_t has been significantly underestimated, perhaps my as

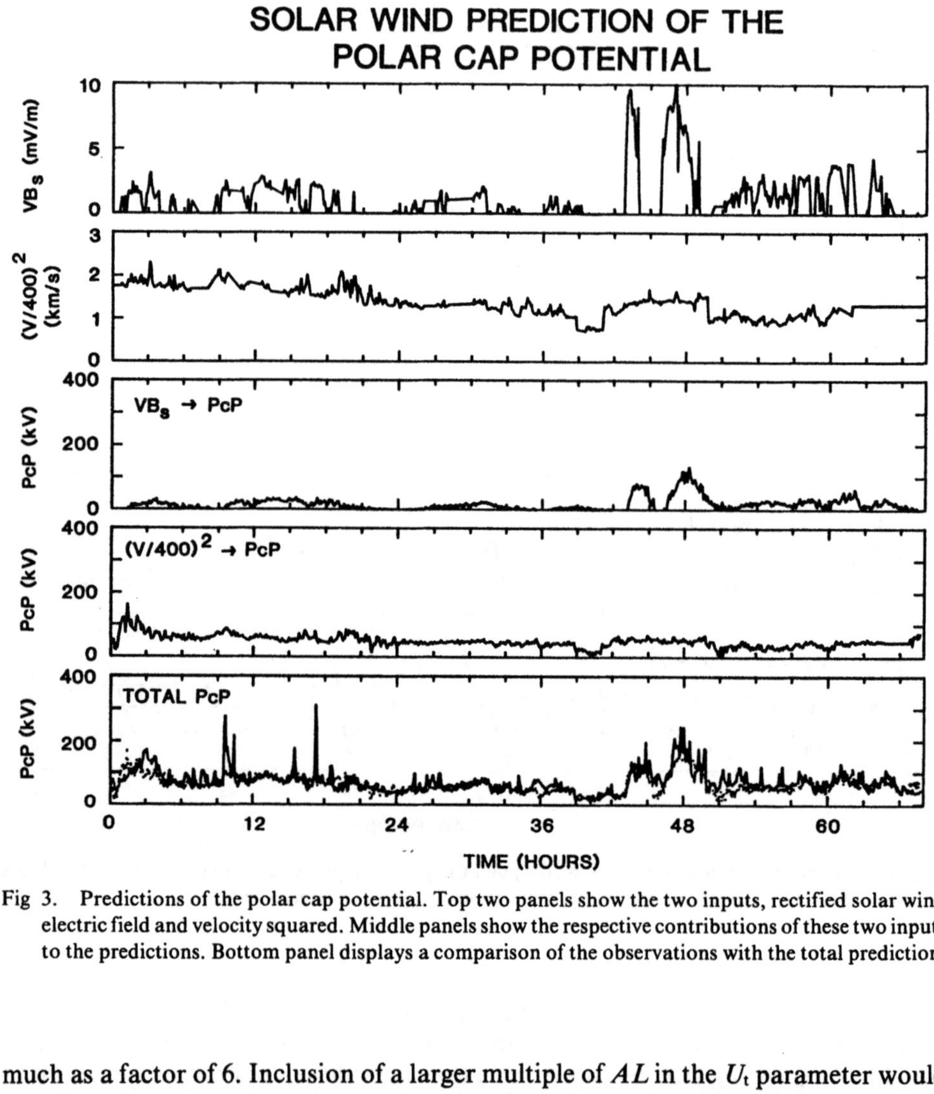

Fig 3. Predictions of the polar cap potential. Top two panels show the two inputs, rectified solar wind electric field and velocity squared. Middle panels show the respective contributions of these two inputs to the predictions. Bottom panel displays a comparison of the observations with the total prediction.

much as a factor of 6. Inclusion of a larger multiple of AL in the U_t parameter would significantly change the response function for U_t (which, with the old factors, is essentially that for Q) and would also reduce the correlation of U_t with the solar wind.

For the polar cap potential, we find that the square of the solar wind velocity is required to predict the trends in the potential. The electric field is required to predict the more rapid variations. At the peak of a major substorm, the ratio of the contributions from the velocity and the electric field is 0.4, a much larger ratio than expected from previous work. The validity of this result is, of course, dependent on whether the baseline for potential found in the inverse modeling is correct, and whether our assumed coupling function is correct. The shape of the electric field filter is not adequately defined by the available data and further inverse modeling will be required to generate a larger data set from which to determine the filters.

This work was supported at UCLA by grant ATM83-18200 from the National Science Foundation, grant N00014-84-C-0158 from the Office of Naval Research, and grant NGL-05-007-004 from the National Space and Aeronautics Administration. A grant from the Institute of Geophysics and Planetary Physics at Los Alamos also provided partial support. The authors gratefully acknowledge P. Reiff for suggesting this study and H. Kroehl and Y. Kamide for supplying some of the data.

REFERENCES

AKASOFU, S.-I., What is a magnetospheric substorm? in *Dynamics of the Magnetosphere*, edited by S.-I. Akasofu, pp. 447–460, D. Reidel, Hingam, Mass., 1979.

BARGATZE, L. F., D. N. BAKER, and R. L. MCPHERRON, Magnetospheric response for many levels of geomagnetic activity, this volume, 1986.

BAUMJOHANN, W. and Y. KAMIDE, Hemispherical Joule heating and the *AE* indices, *J. Geophys. Res.*, **89**(A1), 383–388, 1984.

BURTON, R. K., R. L. MCPHERRON, and C. T. RUSSELL, An empirical relationship between interplanetary conditions and D_{st}, *J. Geophys. Res.*, **80**(31), 4204–4214, 1975.

FELDSTEIN, Y. I., V. Yu. PISARSKY, N. M. RUDNEVA, and A. GRAFE, Ring current simulation in connection with interplanetary space conditions, *Planetary Space Sci.*, **32**(8), 975–984, 1984.

KAMIDE, Y., H. W. KROEHL, B. A. HAUSMAN, R. L. MCPHERRON, S.-I. AKASOFU, A. D. RICHMOND, P. H. REIFF, and S. MATSUSHITA, *Numerical Modeling of Ionospheric Parameters from Global IMS Magnetometer Data for the CDAW-6 Intervals*, Report UAG-88, WDC-A for Solar-Terrestrial Physics, November 1983.

MURAYAMA, T., Coupling function between the solar wind and the D_{st} index, this volume, 1986.

PERREAULT, P. D., and S.-I. AKASOFU, A study of geomagnetic storms, *Geophys. J. R. Astron. Soc.*, **54**, 547–573, 1978.

WEINER, N., *Extrapolation, Interpolation and Smoothing of Stationary Time Series with Engineering Applications*, Tech. Press of the M.I.T. and John Wiley & Son, Inc., New York, 1949.

WYGANT, J. R., R. B. TORBERT, and F. S. MOZER, Comparison of S3-3 polar cap potential drops with the interplanetary magnetic field and models of magnetopause reconnection, *J. Geophys. Res.*, **88**(A7), 5727–5735, 1983.

Solar Wind-Magnetosphere Coupling, edited by Y. Kamide and J. A. Slavin, 119–126.

Coupling Function between the Solar Wind and the D_{st} Index

Takashi MURAYAMA

Department of Physics, Nagoya University, Nagoya 464, Japan

Observed curves of the changes in time of the D_{st} index for the time intervals of 5–7 day duration were compared with the curves predicted under the assumption that the injection rate of the energy into the ring current is proportional to a function of the form $B_s^a V^b D^c$, where B_s is the southward component of the interplanetary magnetic field, V and D are the bulk velocity and the proton density of the solar wind, respectively. From the examination of functions with various combinations of the values of a, b and c for 90 time intervals, the most probable values of the exponents were found to be $a=1.09\pm0.09$, $b=2.06\pm0.18$ and $c=0.38\pm0.05$, respectively.

1. Introduction

For the investigation of the problem of energy coupling between the solar wind and the magnetosphere, the D_{st} index has an advantage over other geomagnetic indices, such as the AL index, in the sense that the D_{st} index is directly related to the energy transported into the inner magnetosphere during the course of the development of magnetospheric substorms. The depression of the D_{st} index is known to be proportional to the total amount of kinetic energies carried by the charged particles injected into the ring current (DESSLER and PARKER, 1959). On the other hand, the AL index represents the current intensity of the westward auroral electrojet which is related to the amount of energy carried by the particles precipitating into the auroral zone ionosphere. However, the intensity of the electrojet current is also subject to the electric fields in the auroral zone ionosphere, and hence the AL index relates in a less direct way to the energy transport than does the D_{st} index. The purpose of this paper is to find the most probable form of the coupling function between the solar wind and the D_{st} index, assuming that the coupling function has a form $B_s^a V^b D^c$.

2. Analysis and Results

Procedure of analysis in this work is essentially identical with that employed in MURAYAMA (1982). We assume that, in association with a positive value of a coupling function $F(t)=B_s^a V^b D^c$ as indicated by a shaded area in the upper part of Fig. 1, energy transfer takes place from the solar wind into the magnetosphere and hot plasma is injected into the inner magnetosphere to intensify the ring current with a delay of one hour. Magnitude of the abrupt decrease of D_{st} (indicated as 'O' in Fig. 1) due to the plasma injection is assumed to be proportional to the value of $F(t)$ ('I' in Fig. 1). The D_{st} then recovers by the decay of the ring current with a decay function $G(t_d)$, where t_d is time after the abrupt decrease, contributing to the total depression of D_{st} by a shaded area in the lower part of Fig. 1. From these assumptions the time

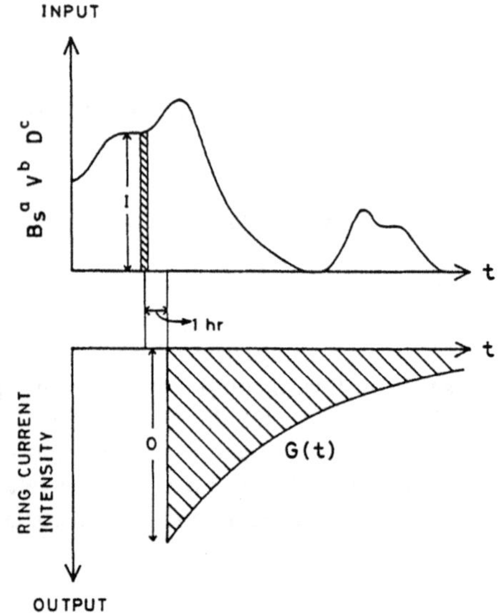

Fig. 1. Schematic representation of the method of prediction of the time variation of D_{st} by the coupling function $B_s^a V^b D^c$.

variation of D_{st} can be predicted by the superposition of saw-tooth-like areas corresponding to different times in a given time interval (see also MAEZAWA and MURAYAMA, in this volume). This curve is to be compared with the curve of $D_{sto}*(i)$ which is obtained from $D_{sto}(i)$, the hourly value of D_{st}, after eliminating the effect of the Chapman-Ferraro currents; i.e., $D_{sto}*(i)=D_{sto}(i)-w\,(DV^2)^{-1/2}$,where $w=0.20$ nT (ev cm^{-3})$^{-1/2}$ (BURTON *et al.*, 1975). For the decay function $G(t)$, we use the following formula.

$$G(t)=a_1\exp(-t/\tau_1)+a_2\exp(-t/\tau_2), \quad (a_1+a_2=1) \tag{1}$$

Time intervals of 5 to 7 day duration were selected on the basis of the availability of continuous solar wind data in a data set compiled by KING (1977, 1979, 1983). Of the 135 intervals thus selected, 90 intervals in which standard deviation of $D_{sto}*(i)$ exceeds 10 nT are used in the present analysis to exclude the time intervals of low storm activity. However, the inclusion of these intervals does not result in an appreciable change to the conclusion of this work.

From the correlation analysis between $D_{sto}*(i)$ and the predicted values $D_{st}*(i)$ in each intervals, we obtain the correlation coefficient $r_k(j)$ for the j-th interval using a coupling function F_k. Comparison of the validity of the function is made with $<r>_k$, the average value of $r_k(j)$ for the 90 intervals.

The function with the highest value of $<r>_k$ was searched for, by trying numerous combinations of adjustable parameters a, b, c, τ_1, τ_2 and a_1/a_2, and a function $F_0=B_s^{1.09}V^{2.06}D^{0.38}$ was found to give the highest value of $<r>$ of 0.9174. To estimate the error for the value of the exponent $c=0.38$, for example, we first calculate the correlation coefficient $r_\varepsilon(j)$ for each of the 90 intervals for the function $F_\varepsilon=B_s^{1.09}V^{2.06}D^{0.38+\varepsilon}$, and compare with $r_0(j)$ for the function $F_0=B_s^{1.09}V^{2.06}D^{0.38}$. Then the difference in the average value of $r_\varepsilon(j)$ and that of $r_0(j)$ is

$$\Delta_\varepsilon=<r_\varepsilon(j)>-<r_0(j)>=<\delta_j>,$$

where $<\delta_j>$ is the average value of $r_\varepsilon(j)-r_0(j)$. From the standard deviation σ_ε of δ_j; i.e., $\sigma_\varepsilon=[(\delta_j-<\delta_j>)^2/89]^{1/2}$, the error for Δ_ε can be obtained as the standard error; $s_\varepsilon=\sigma_\varepsilon/(90)^{1/2}$. The right-hand panel of Fig. 2 shows the result of the analysis. Four dots in this panel shows the values of Δ_ε, with vertical bars indicating the magnitude of s_ε, for the values of ε of -0.10, -0.05, $+0.05$ and $+0.10$, respectively, for the function F_ε. It is seen in this panel that $\Delta_{+0.05}$ and $\Delta_{-0.05}$ are very close to $s_{+0.05}$ and $s_{-0.05}$, and $\Delta_{+0.10}$ and $\Delta_{-0.10}$ are approximately two times larger than $s_{+0.10}$ and $s_{-0.10}$, respectively. From this result we estimate that the error of the exponent c is 0.05, as shown at the top of this panel. Left-hand and central panels show the results of similar analysis for the exponents a and b. The final result for F_0 thus obtained is

$$F_0=B_s^{1.09\pm0.09}V^{2.06\pm0.18}D^{0.38\pm0.05}$$

with $\tau_1=2$ h, $\tau_2=21$ h, and $a_1/a_2=1.5$, respectively.

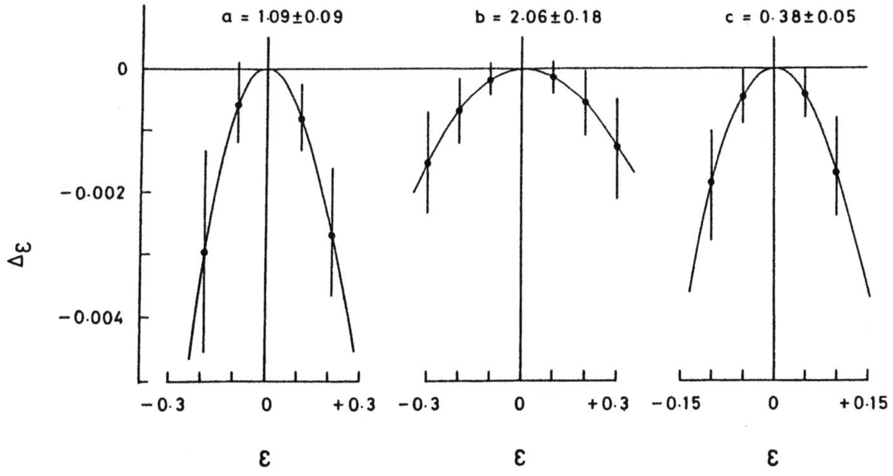

Fig. 2. Results of analysis to estimate the error of the exponents of the most probable form, $F_0=B_s^{1.09}V^{2.06}D^{0.38}$, of the coupling function obtained in this work. Black dots denote the difference Δ_ε in the average value of correlation coefficients for the function F_0 and that for a function in which one of the values of exponents of F_0 is shifted by ε. Bartical bars show the standard errors of Δ_ε. See the text for more detailed information.

3. Discussion

3.1 Decay function

The analysis in the preceding section was made with the assumption of (1) for the decay function. This choice of the form of the decay function followed the finding that significantly higher values of $<r>$ were obtained by the use of (1) than those with $G(t)$ of a single exponential function, for any coupling functions examined. It is, however, desirable to examine whether our form of $G(t)$ is good enough for our analysis or not. For this purpose the form of $G(t)$ was deduced in a different way which does not need a priori assumption on the functional form of $G(t)$. Let $G(i)$ be the value of G at i hour. Then the value of D_{st}* at the j-th hour of a time interval is written by

$$D_{st}^*(j) = K + F(j)G(0) + F(j-1)G(1) + \cdots + F(j-48)G(48) \quad (2)$$

where K is a constant, $F(j)$ is the value of the coupling function F_0 at time j, and the effects of F at $(j-49)$ h and before are neglected. For a time interval of 5 days, for instance, we have 120 equations of $D_{st}^*(j)$ and the problem reduces to the calculation of 50 unknown values of K and $G(i)$ from 120 equations of the form of (2) for different values of j by the least squares method when the values of F_0 in the 120 equations are given.

Values of $<G(i)>$, the average values of $G(i)$ thus obtained for the 90 time intervals are plotted in Fig. 3. The plot of $G(i)$'s for one time interval has a considerable random scattering of individual values of $G(i)$, but the plot of their

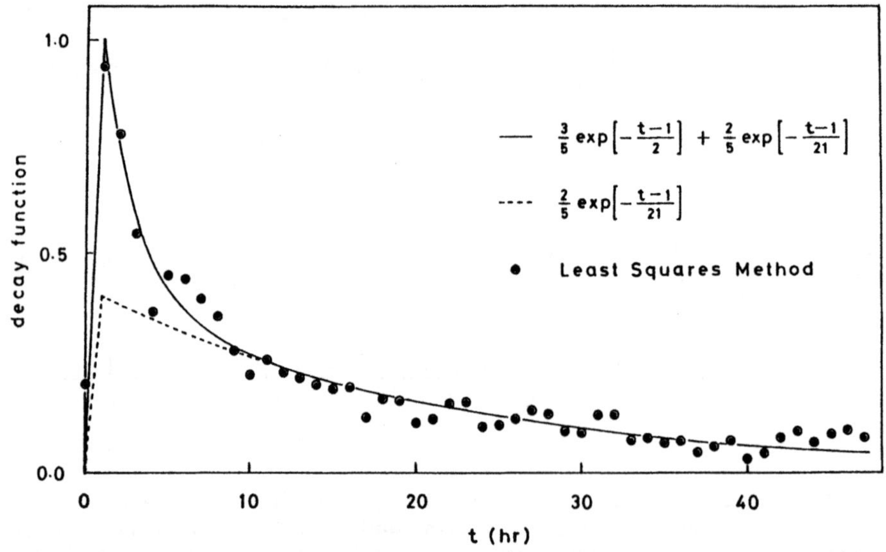

Fig. 3. Comparison of the decay function obtained in the form of Eq. (1) and that obtained by the least squares method.

average values $<G(i)>$ shows a reasonably smooth change with time as seen in Fig. 3. A good agreement between the plot of $<G(i)>$ and the form of the decay function given by (1) with $\tau_1=2$ h, $\tau_2=21$ h and $a_1/a_2=1.5$, also seen in this figure shows that our form of the decay function is a good approximation to the real form of $G(t)$.

A decay function with two values of τ was also used by FELDSTEIN *et al.* (1984) in their analysis of the D_{st} index, but unlike our function (1), their decay function is a single exponential function but the value of τ depends on the level of D_{st}; i.e., $\tau=5.8$ h when $D_{st}<-55$ nT, and $\tau=8.2$ h when $D_{st}>-55$ nT. AKASOFU (1981) also used two values of τ, which depend on the value of his coupling function ε; i.e., $\tau=1$ h when $\varepsilon>5\cdot10^{18}$ erg s^{-1}, and $\tau=20$ h when $\varepsilon<5\cdot10^{18}$ erg s^{-1}.

Our function (1) does not depend on the value of the coupling function nor the level of D_{st}, and hence the form of the decay function should be independent of the level of the magnetic storm activity. To examine this, we have calculated the values of $<G(i)>$ by the least squares method for 17 time intervals of very high storm activity (standard deviation of D_{sto}* exceeding 25 nT), and compared with those for the remaining 73 time intervals. The values of $<G(i)>$ thus obtained for the two groups of time intervals were found to agree well with each other within their uncertainties (about 10 per cent of $G(i)$'s for $i=1-10$). As all of the ranges of D_{sto}* for the 17 intervals exceed 100 nT with the average of 158 nT, this result favors our decay function (1) rather than those for other two works, although a more detailed investigation is needed to evaluate the coupling-function dependent or the D_{st}-level dependent form of $G(t)$.

The dominant mechanism responsible for the decay of the ring current is believed to be a charge exchange process between the energetic ring current ions and the thermal neutral atoms existent in the ring current region. Life times of the ions deduced from the charge exchange cross sections are given by SMITH *et al.* (1981) as a function of energy for different species of the ions. WILLIAMS (1981) deduced the energy density spectrum of the ring current under the assumption that all the ring current ions are O^+ below 50 kev and are H^+ above 50 kev. The curve of the decay function derived from this energy density spectrum and the data for the life times of O^+ and H^+ is well approximated by the function $\exp(-t/25\text{ h})$ in the range of t up to 50 hours. Thus it is reasonable to attribute the $\exp(-t/21\text{ h})$ term in our decay function to the charge exchange process.

Origin of the $\exp(-t/2\text{ h})$ component of the ring currnet is left to be clarified. Injected charged particles, not stably trapped by the geomagnetic field would be responsible for this component.

3.2 Comparison of the form of the coupling function with theory

A theoretical model for the energy coupling between the solar wind and the magnetosphere, to be compared with our result of analysis has been proposed by VASYLIUNAS *et al.* (1982). In this model, the energy transfer rate P from the solar wind into the magnetosphere is expressed as

$$P=B_T^2V/\mu_0\ L_{CF}^2\ M_A^{2\alpha}H(\theta). \qquad (3)$$

The first term $B_T^2 V/\mu_0$ in the right-hand side of this equation, where $B_T^2 = B_y^2 + B_z^2$, is the interplanetary Poynting vector and L_{CF} in the second term denotes the Chapman-Ferraro distance

$$L_{CF} = (M_E/\mu_0 D V^2)^{1/6} \tag{4}$$

where M_E is the dipole moment of the geomagnetic field. Hence, the combination of the first and the second terms implies the amount of the interplanetary electromagnetic flow per unit time through the magnetospheric boundary surface, the area of which is scaled by the square of L_{CF}. The third term is a power function of the Alfvén Mach number

$$M_A = (\mu_0 D V^2 / B_T^2)^{1/2} \tag{5}$$

with an adjustable parameter α. It is readily verified that the product of the first and the third terms in the right-hand side of (3) reduces to the kinetic energy flux DV^3 when $\alpha = 1$. $H(\theta)$ is another dimensionless function of the latitudinal angle of the IMF. Putting (4) and (5) into (3) and extracting the quantities relating to B_T, V and D, we obtain

$$P \propto B_T^{2-2\alpha} V^{2\alpha+1/3} D^{\alpha-1/3}. \tag{6}$$

From the values of exponents a, b and c of F_0 obtained by our analysis, we have three independent values of α from (6), namely, $\alpha_a = 0.455 \pm 0.045$, $\alpha_b = 0.87 \pm 0.09$ and $\alpha_c = 0.71 \pm 0.05$ corresponding to the exponents a, b and c, respectively. From the observation that any of the values of three α's are beyond the ranges of the uncertainties of the other two, and especially the value of α_a is separated from the average of α_b and α_c by 7.4 times the uncertainty of α_a, we conclude that our coupling function is not consistent with the theoretical prediction by VASYLIUNAS *et al.* (1982).

During the magnetospheric substorm, plasma sheet particles are driven into the inner magnetosphere by the collapse of the tail magnetic field and by the enhanced $\boldsymbol{E} \times \boldsymbol{B}$ drift due to the interplanetary westward electric field $B_s V$, which is penetrating into the magnetosphere, and the northward magnetic field in the low latitude region of the night-side magnetosphere. Here we try to reconcile the discrepancy between the observed and the theoretical forms of the coupling function by introducing an assumption that the efficiency of plasma sheet particles to be trapped in the ring current region is proportional to $(B_s V)^\beta$. The effect of the variability of $B_s V$ on the trapping efficiency may also be included in this term because of its close correlation with $B_s V$ ifself.

In order that our coupling function has the same form with the product of $(B_s V)^\beta$ and P in (3), the following three equations should be satisfied:

$$2 - 2\alpha + \beta = 1.09 \pm 0.09,$$

$$2\alpha + 1/3 + \beta = 2.06 \pm 0.18, \tag{7}$$

$$\alpha - 1/3 = 0.38 \pm 0.05.$$

The next task is to find a pair of the values of α and β which is consistent with all of the three equations of (7) within the uncertainties in the right-hand side of the equations. In Fig. 4, the three equations are shown in $\alpha-\beta$ diagram with the uncertainties in their locations. It is found in the figure that the shaded area with $\alpha \simeq 0.7$ and $\beta \simeq 0.4$ satisfy all of the equations within their uncertainties. Note that the positive value of β meets the requirement that $B_s V$ enhances the efficiency of plasma sheet particles to be trapped in the ring current region. Thus it is possible to interprete our coupling function between the solar wind and the D_{st} index in terms of the theoretical coupling function of VASYLIUNAS *et al.* (1982), if the efficiency of the plasma sheet particles to be trapped in the ring current region is proportional to $(B_s V)^\beta$ with $\beta \simeq 0.4$.

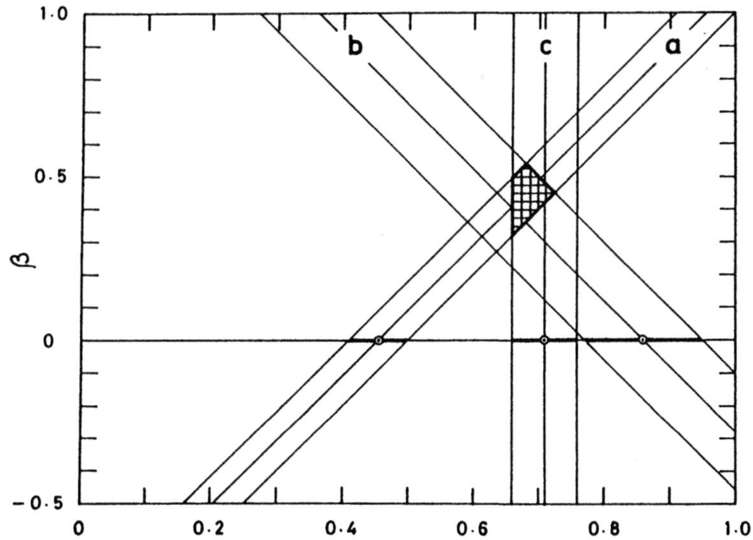

Fig. 4. Three groups of parallel lines denote the three equations of (7) in the text, shown in a α–β diagram. A shaded area shows the range of the pair of the values of α and β which is consistent with Eq. (7) within their uncertainties. Three thick horizontal bars show the values and their uncertainties of α obtained from the comparison of exponents of F_0 with those of P in (6).

REFERENCES

AKASOFU, S.-I., Energy coupling between the solar wind and the magnetosphere, *Space Sci. Rev.*, **28**, 121–190, 1981.

BURTON, R. K., R. L. MCPHERRON, and C. T. RUSSELL, An empirical relationship between interplanetary conditions and D_{st}, *J. Geophys. Res.*, **80**, 4204–4214, 1975.

DESSLER, A. J. and E. N. PARKER, Hydromagnetic theory of geomagnetic storms, *J. Geophys. Res.*, **64**, 2239- 2252, 1959.

KING, J. H., Interplanetary medium data book, NSSDC/WDC-A-R and S, 77-04, 1977; 79-08, 1979; 83-01, 1983.

FELDSTEIN, Y. I., V. Yu. PISARSKY, N. M. RUDNEVA, and A. GRAFE, Ring current simulation in connection with interplanetary space conditions, *Planet. Space Sci.*, **32**, 975–984, 1984.

MURAYAMA, T., Coupling function between solar wind parameters and geomagnetic indices, *Rev. Geophys. Space Phys.*, **20**, 623–629, 1982.

SMITH, P. H., N. K. BEWTRA, and R. A. HOFFMAN, Inference of ring current ion composition by means of charge exchange decay, *J. Geophys. Res.*, **86**, 3470–3480, 1981.

VASYLIUNAS, V. M., J. R. KAN, G. L. SISCOE, and S.-I. AKASOFU, Scaling relations governing magnetospheric energy transfer, *Planet. Space Sci.*, **30**, 359–365, 1982.

WILLIAMS, D. J., Ring current composition and sources: an update, *Planet. Space Sci.*, 1195–1203, 1981.

Solar Wind-Magnetosphere Coupling, edited by Y. Kamide and J. A. Slavin, 127–141.

Real-Time Prediction of Global Geomagnetic Activity

Jo Ann JOSELYN

NOAA/Space Environment Laboratory, 325 Broadway, Boulder, Colorado 80303, U.S.A.

The Space Environment Services Center (SESC, an around-the-clock joint operation of the Department of Commerce and the U.S. Air Force Air Weather Service) has a mandate to continually provide the best possible descriptions and predictions of the state of the geomagnetic field. Geomagnetic alerts and forecasts can be extremely important to military and civilian radiowave and satellite communication and navigation systems, electric power networks, geophysical explorers, spacecraft controllers, and scientific research campaigns. At the present time, the primary SESC methods of geomagnetic prediction are based directly on observations of solar phenomena (e.g., flares, coronal holes and filament disappearances) and recurrence. However, only a minor fraction of the observed phenomena result in geomagnetic activity and forecasters mostly rely on experience in distinguishing events expected to be geoeffective from those that may not. From 1980 through mid 1982 we learned to appreciate the predictive value of the real-time solar wind data from ISEE-3. As advances are made in the topics under consideration at this Conference, we hope to apply them to SESC forecasting efforts.

1. Introduction

A definitive test of our understanding of the coupling between the solar wind and the magnetosphere will be the successful prediction of the detailed behavior of the geomagnetic field. However, while this goal is an object of intensive scientific investigation (as demonstrated by the excellent participation in this Chapman Conference), predictions of geomagnetic activity are a present-day matter of significant practical importance to the Space Environment Services Center and the many scientific, military, and commercial concerns which it serves. The Space Environment Services Center (SESC) is an around-the-clock joint operation of the U.S. Department of Commerce and the Air Force Air Weather Service. It operates a national real-time space environment data base to accept and integrate observational data, provides operational support and services (e.g., geophysical alerts and daily forecasts) as required by Federal agencies and public users in support of the national economy, and serves as the U.S. Government focal point for international data exchange programs (NATIONAL PLAN FOR SPACE ENVIRONMENT SERVICES AND SUPPORTING RESEARCH 1983–1987,1983). While the term "space environment" includes generally the sun, interplanetary space, and the earth's magnetosphere, ionosphere, and upper atmosphere, most SESC customers are interested in the level of activity of the global geomagnetic field.

There are seven general categories of systems affected by geomagnetic activity. These categories are briefly discussed in the list below. (See also LANZEROTTI, 1979; and references therein).

1.1 The aurora

This category includes our *most* sophisticated and *least* sophisticated customers. Scientific research campaigns which may include coordinated rocket/balloon launches and ground and space-based remote observations are often keyed to the occurrence of aurora. At the same time, intense aurora at lower latitudes is an object of awe and sometimes concern to the general public, who are unaccustomed to such displays.

1.2 Radio communications and radars

In association with the particle precipitation which produces the aurora, the high and middle latitude ionosphere becomes disturbed. In particular, High Frequency (HF, 3–30 MHz) systems operating in middle latitudes may experience depressions of the maximum useable frequency of up to 50 percent, and HF systems using long (over 4000 km) eastward paths which extend poleward of 55 degrees latitude may experience periods of non-great-circle propagation. Ionospheric sporadic E-layers often form in the auroral zone (60 to 75 degrees latitude), again affecting HF propagation. HF lowest useable frequencies are elevated through the sunrise to noon sector of the auroral zone due to auroral zone absorption. LF (30–300 kHz) and VLF (3–30 kHz) systems operating across the auroral and polar regions may experience phase advances. VHF (30–300 MHz) and UHF (300 MHz–3 GHz) radars located equatorward of the auroral zone and pointing poleward may observe enhanced clutter, and VHF and UHF spacetrack radars may experience unusual signal retardation and refraction, causing ranging and pointing errors. Finally, VHF, UHF and possibly even SHF (3–30 GHz) satellite communication systems operating through the disturbed ionosphere may experience enhanced phase and amplitude scintillations. (See also DAVIES, 1981).

1.3 Spacecraft charging and control

The motion of satellites and other objects (including astronauts on extra-vehicular activity) through the upper atmosphere and ionosphere causes charges to accumulate differentially on portions of those objects. The charge can spontaneously discharge with possible detrimental effects ranging from the generation of spurious commands to permanent damage to electronic subsystems (GRARD *et al.*, 1983). The probability for spacecraft charging increases with an increase in the magnetospheric processes also responsible for geomagnetic activity. (Direct radiation damage by galactic and solar cosmic rays is a separate hazard). Some satellites use the earth's magnetic field to maintain spatial orientation. During magnetic storms operators may choose to disable control systems and use thrusters or gyros to maintain orientation. (See also GARRETT and PIKE, 1980).

1.4 Satellite drag

Particle precipitation and ionospheric currents associated with geomagnetic activity heat the neutral upper atmosphere causing it to expand. The air density at orbital altitudes up to 800 km can increase by an order of magnitude, causing serious additional frictional drag which is highly dependent on the shape and material of the satellite. Satellite drag is of interest to all agencies who must predict satellite orbital lifetimes, and is of particular concern to the Naval Astronautics Group, who must respond to existing and predicted geomagnetic activity in commanding the Transit series of navigational satellites. (See also the PROCEEDINGS OF A WORKSHOP ON SATELLITE DRAG, 1982).

1.5 Induced currents

Fluctuating geomagnetic fields induce currents in cables, pipelines and powerlines. In pipelines, the effects are short-term disruptions in flow metering and corrosion monitoring systems. In powerlines, the problems become more critical because the induced direct currents can upset the phasing of transformers (half-cycle saturation) and protective relays can trip. Once a blackout occurs, several hours are needed to return the transformer to service if it is not severely damaged. (See also LANZEROTTI, 1983).

1.6 Geophysical surveys

To prospect for oil or minerals or conduct archeological surveys, the modern exploration geophysicist maps the geomagnetic terrain. One method uses aerial surveys and it would obviously be unwise to take data on naturally disturbed days. On the other hand, magnetic storm conditions are desirable for the magnetotelluric method, which uses disturbed natural oscillations of the earth's field at frequencies below 1 Hz to explore the deep electrical resistivity at specific sites. (See also REGAN and RODRIGUEZ, 1981; KAUFMAN and KELLER, 1981; WEYMOUTH and HUGGINS, 1985).

1.7 Magnetobiology

There is a growing body of evidence that changes in the geomagnetic field affect biological systems. In particular, homing pigeons and other migratory creatures appear to use the magnetic field as a backup navigational aid. Many racing pigeon clubs now cancel events when the observed or predicted magnetic activity exceeds a specific threshold level between unsettled and storm conditions. (See also GAUTHREAUX, JR., 1980; GOULD, 1980).

This diverse list of users of geomagnetic information underscores the need for accurate descriptions and responsible predictions of geomagnetic activity, in as clear and universal a language as possible. At the present time, the SESC primarily describes and predicts geomagnetic conditions using the three-hourly K and A indices. Figures 1(a) and (b) illustrate these indices. K is a geographically local index which requires the subtraction of a quiet day curve, and is inherently not suited for real-time use. However, we manage the situation by measuring and issuing a provisional K-index every three hours using a Boulder magnetometer, while predicting (and verifying against) official U.S. Geological Survey observations of the Fredericksburg, Virginia, magnetic K and A indices. We also use 6 descriptive terms to indicate the general level of activity, as shown in Fig. 1(b) (quiet, unsettled, etc.). In a similar manner, the Air Force calculates and issues a real-time estimate of the global K_p and A_p indices, which are actually a composite of conditions at up to 12 globally

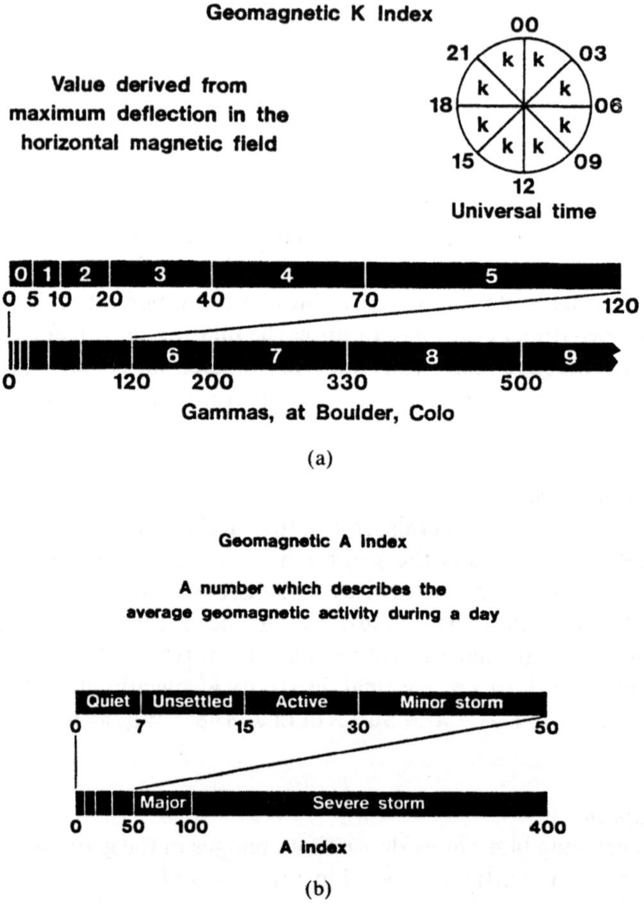

Fig. 1. (a) A schematic description of the three-hourly geomagnetic K index. The conversion scale of nanoTeslas (nT, or equivalently, gammas) to a K index is site dependent. For example, a deviation exceeding 500 nT at Boulder, Colorado is rated as a K of 9, but a deviation must be at least 2500 nT at College, Alaska to be a 9. (b) A schematic description of the daily geomagnetic A index. The A index is derived by averaging the 8 three-hourly 'a' indices obtained from the K indices according to the following pairs of (K, a): (0,0), (1,3), (2,7), (3,15), (4,27), (5,48), (6,80), (7,140), (8,240), (9,400).

spaced observatories compiled by the Institut für Geophysik at Gottingen, F.R. Germany. Because K and A indices do not represent actual and predicted geomagnetic activity in the best way for operational customers, we plan to convert to a probability format (e.g., 40% chance of quiet conditions tomorrow and 60% for unsettled conditions) in the near future (JOSELYN, 1984). However, the K and A indices are useful in that they can be used to define storm conditions (there is no internationally accepted definition of a geomagnetic storm). At the SESC, we consistently use the term 'minor storm' to describe days for which the daily A_p index was between 30 and 49, and 'major storm' to describe days for which A_p equaled or exceeded 50.

2. The Solar Origins of Geomagnetic Activity

Forecasting at the SESC is based on the premise that an understanding of solar phenomena, interplanetary propagation, and solar wind-magnetospheric coupling would lead to accurate predictions of general geomagnetic conditions on at least the time scale of hours (and perhaps even minutes). Although recurrence and persistence are ready tools which are often applied to geomagnetic forecasts, they are used in the context that all activity ultimately originates at the sun. Since most of our customers (although not all) are impacted by storm-level geomagnetic activity, we are particularly interested in the major solar phenomena which have been associated with dramatic geomagnetic effects. These are flares, coronal holes, and filament disappearances.

Beginning with Carrington's observation of a white-light flare on 1 September, 1859, and his association of the flare event with an auroral display on the following night (CHAPMAN and BARTELS, 1940), flares have been linked with geomagnetic storms (e.g., DONNELLY, 1979). However, not all flares, even big ones, cause storms. Figure 2 illustrates the frequency of storms which follow the largest of all x-ray flares. M- and X-class flares refer to the flare peak x-ray energy flux observed by the NOAA/GOES geosynchronous satellites: M-class flares identify those in the range of 10^{-2} up to 10^{-1} erg $cm^{-2}s^{-1}$ and X class flares exceed 10^{-1}. M- and X-class flares constitute about 10% of all observed x-ray flares, and X-class flares comprise less than 1%. Note that there is only a 1-in-5 chance of a major storm after a flare of the largest x-ray class, and that 3 times out of 5 during the current solar cycle there was *no* storm. This result cannot be immediately explained as a 'solar longitude' effect, as shown in Fig. 3. Here the number of X-class x-ray flares which occurred in each 15° solar

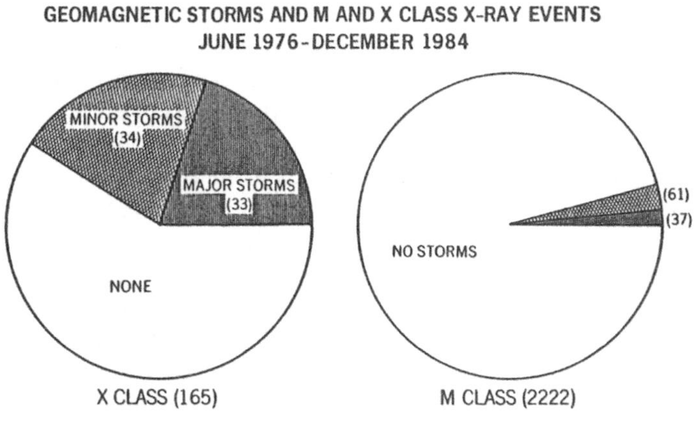

Fig. 2. Data from Solar Cycle 21 illustrating the poor correspondence between major flares and geomagnetic storms.

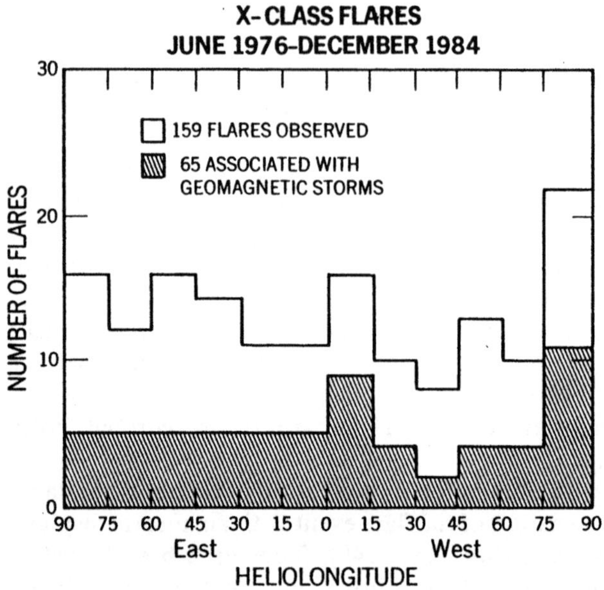

Fig. 3. A histogram showing the numbers of X-class x-ray flares which occurred in 15° bins of solar heliolongitude, and the numbers of X-class flares in each bin which can be associated with geomagnetic storms.

longitude bin across the visible disk has been plotted along with the number of geomagnetic storms which may have been caused by those flares. The extent to which the flares are not evenly distributed across the sun reveals the poorness of the statistics because with a large enough sample, no longitude bin relative to the earth should be favored. The large number of west limb events is probably an observational bias—given a huge x-ray signature and no obvious disk event, it is human nature to associate it with a region which has been in view for the past 2 weeks that just departed over the west limb. Comparing the number of geomagnetic storms which may originate with these flares, there does seem to be a peak in the number of storms in the 0– 15° West bin, but it is in proportion to the number of flares that have occurred there. Likewise, a dip in the 30– 45° West bin is a reflection of the smaller number of flares that have happened to occur in that sector. A further check using only the list of geomagnetic storms believed to have an X-class flare as a *sole* source shows that every 15° bin is represented, except the 30–45° West bin. The conclusion drawn from this analysis is that the solar longitude of a flare event is not itself a useful parameter in determining whether or not a geomagnetic storm will occur. This result is in contrast with results obtained by AKASOFU and YOSHIDA (1967) in which they plot the magnitude of the geomagnetic storm main phase decrease (maximum *Dst*) as a function of the central meridian distance of the flare assumed responsible. Of the 324 cases of their study, almost half lie between E15° and W45° solar heliolongitude, and nearly all of their major storms are associated with flares within 20° east or west of central meridian.

AKASOFU and YOSHIDA (1967) also considered the time delay between the onset of the solar flare and the associated geomagnetic sudden commencement at the earth. They found no definite relation between this time delay and the central meridian distance of solar flares. However, this result differs from recent work which studied 30 flares from this solar cycle (Cycle 21) which were determined to be the sole source of geomagnetic storms. In that study, the longitude of the flare does seem to make a difference in the time delay between flare and storm, with shorter time delays for centrally located flares (JOSELYN, 1986).

Another established source of geomagnetic disturbance are the high-speed streams which emanate from coronal holes (e.g., HUNDHAUSEN, 1977; SVALGAARD, 1977). These streams tend to be long-lived, recurring with the average period of solar rotation—27 days, thus producing recurrent geomagnetic activity. However, recurrence per se tends to be reliable only for a few years during the declining phase of the solar cycle when a pattern of streams is firmly established. But we needn't rely on recurrence—we can infer coronal holes themselves from daily He 10830 Å spectroheliograms taken by the Kitt Peak National Observatory (KPNO) near Tucson, Arizona. These coronal holes are mapped as a function of heliolongitude and published in the NOAA Solar Geophysical Data monthly publication. Figure 4 is a line drawing of the coronal holes apparent in the spectroheliograms which KPNO sends to the SESC every day for which observations are made. Additional information about potential coronal holes is obtained from green-line coronal observations from Sacramento Peak Observatory in New Mexico, the 10-cm solar radio east-west drift scan sent daily from the Canadian National Radio Observatory in Ottawa, and our own H-alpha photographs. In addition, the Fleurs Radio Astronomy Station of the University of Sidney, Australia, scans the sun daily at 692 and 1415 MHz, inferring the presence of coronal holes (FERGUSON, 1979). Coronal holes are clearly marked on SESC H-alpha synoptic charts such as shown in Fig. 5 (MCINTOSH, 1979). In a recent analysis of those coronal holes determined to be the sole source of geomagnetic storms, it was noted that geoeffective coronal holes fell almost entirely within 30° of the helioequator and were mostly within 20° of the equator (JOSELYN, 1986). Geomagnetic disturbances tended to begin when the centroid of the hole was 30° west of central meridian (2.3 days after central meridian passage). But in analogy with major flares, not every coronal hole which is observed, even equatorial ones, can be associated with geomagnetic activity. For the 3 year period of 1982–84, fewer than half of the coronal holes within 30° of the helioequator produced geomagnetic storms (Fig. 6).

A recent suggestion for a solar diagnostic of geomagnetic activity is filament disappearances (JOSELYN and MCINTOSH, 1981; WRIGHT and MCNAMARA, 1983). Solar filaments are a chromospheric feature seen in H-alpha images as dark, thread-like features snaking across the solar disk. When they are observed in emission above either photospheric limb, they are called prominences. Filament eruption events (which result in the disappearance of all or part of the filament) have been physically linked with slow coronal transients and mass ejections (SCHWENN, 1983). While filament disappearances are not subtle, neither are they rare, and only about 15% of all filament disappearances can be associated with significant magnetic activity (Fig.

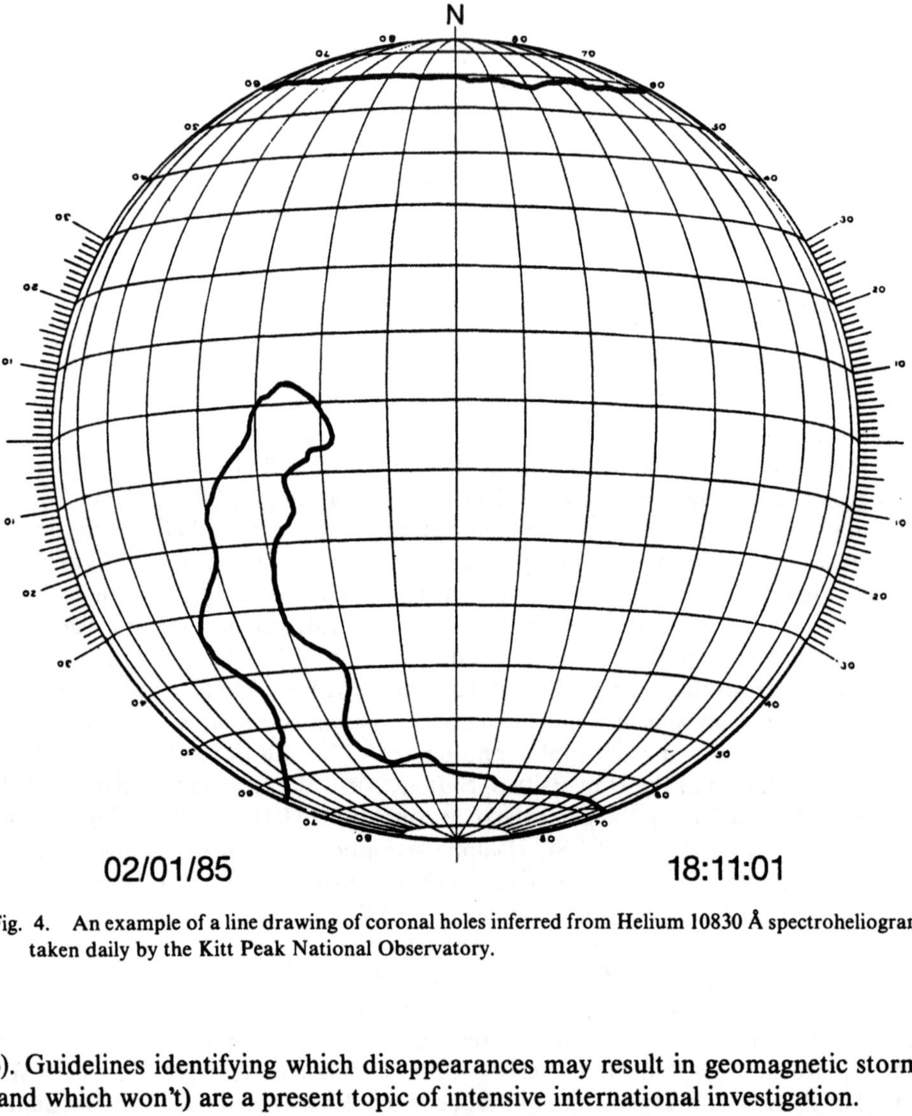

Fig. 4. An example of a line drawing of coronal holes inferred from Helium 10830 Å spectroheliograms taken daily by the Kitt Peak National Observatory.

6). Guidelines identifying which disappearances may result in geomagnetic storms (and which won't) are a present topic of intensive international investigation.

Figure 7 summarizes the author's determination of the solar origin of geomagnetic storms from the beginning of the current solar cycle through December 1984 based on the hypothesis that geomagnetic activity is 'caused' by solar flares, coronal holes, and filament disappearances. These results allow time delays between the solar source and the geomagnetic storm of 1–3 days for flares, 2–3 days for coronal holes, and 3–6 days for filament disappearances. This figure includes the results for all storms (A_p equal to or greater than 30). The results for major storms only (there have been 89 storms with A_p of 50 or more) are not significantly different. While nearly half of the cases are ambiguous (and there really may be multiple solar sources for a single geomagnetic disturbance), we find that about 20% of all storms can only be blamed on

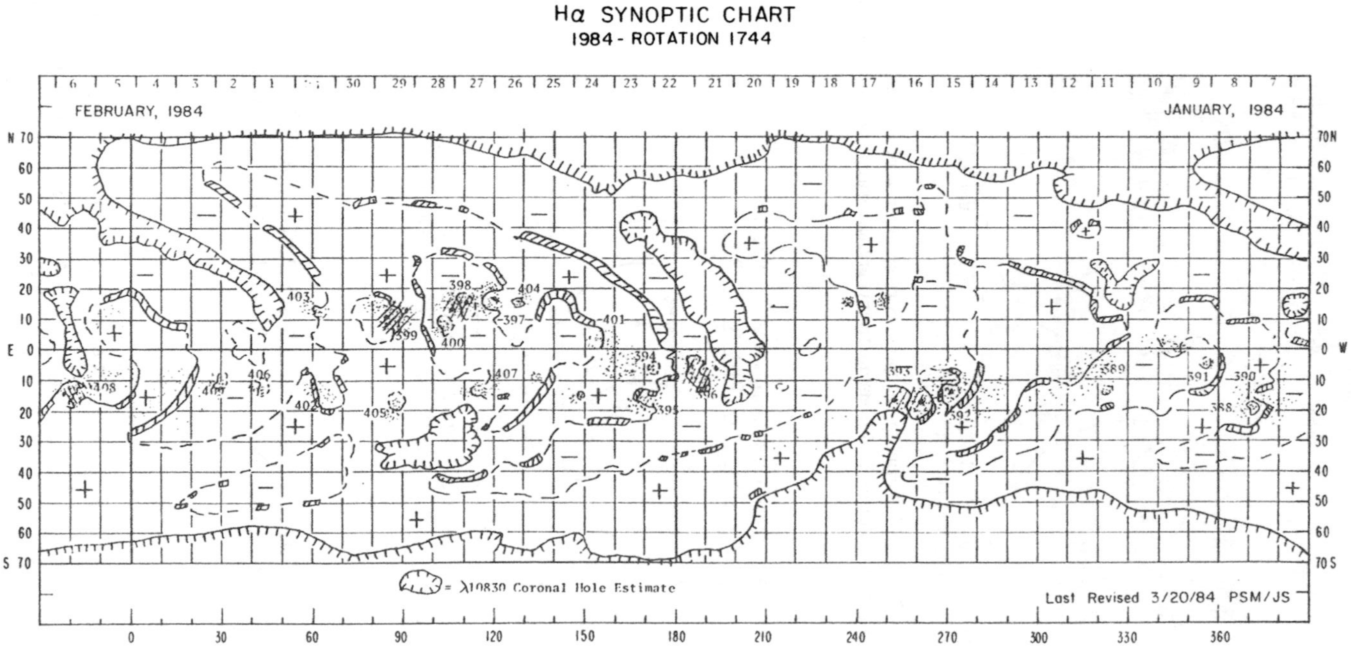

Fig. 5. Hydrogen-alpha synoptic charts in heliographic longitude and latitude. Solar active regions, filaments and coronal holes are shown in context with large-scale solar magnetic fields. Polarities are normally mapped from magnetograms transmitted from the Kitt Peak National Observatory. Dates along the top of the map are the times of central meridian passage of the longitudes appearing below the dates. The serial number of the Carrington Rotation appears at the top.

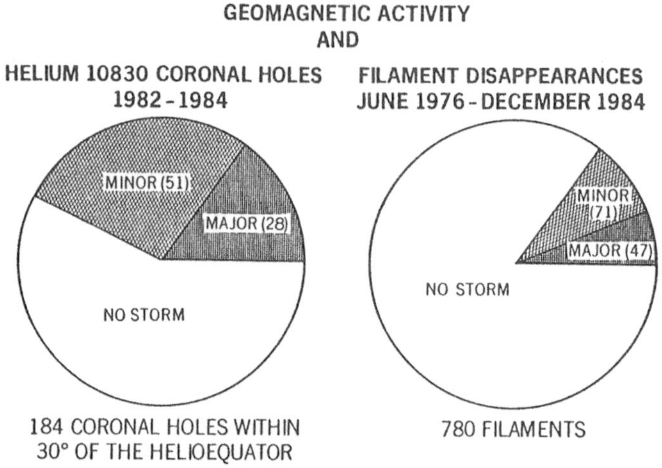

Fig. 6. Data from Solar Cycle 21 illustrating the poor correspondence of coronal holes near the helioequator with geomagnetic storms, and filament disappearances with geomagnetic storms.

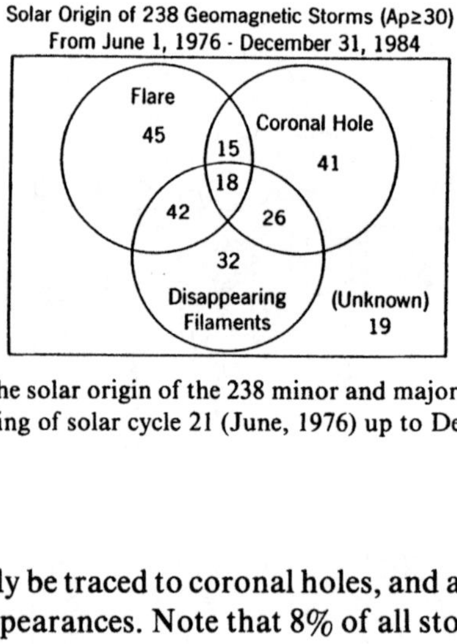

Fig. 7. A determination of the solar origin of the 238 minor and major geomagnetic storms which have occurred since the beginning of solar cycle 21 (June, 1976) up to December 31, 1984.

flares, about 17% can only be traced to coronal holes, and about 13% find their origin only with filament disappearances. Note that 8% of all storms (6% of major storms) are unexplained.

3. Real-Time Solar Wind Data

Through a special agreement with NASA, the SESC began receiving real-time data from the International Sun Earth Explorer 3 (ISEE-3) in March of 1980. This spacecraft was stationed until mid-1982 in a halo orbit about the sun-earth libration point L_1 (Fig. 8). In that orbit, it intercepted the solar wind about 30–60 minutes before the wind reached the earth and thus provided advance warning of the onset of

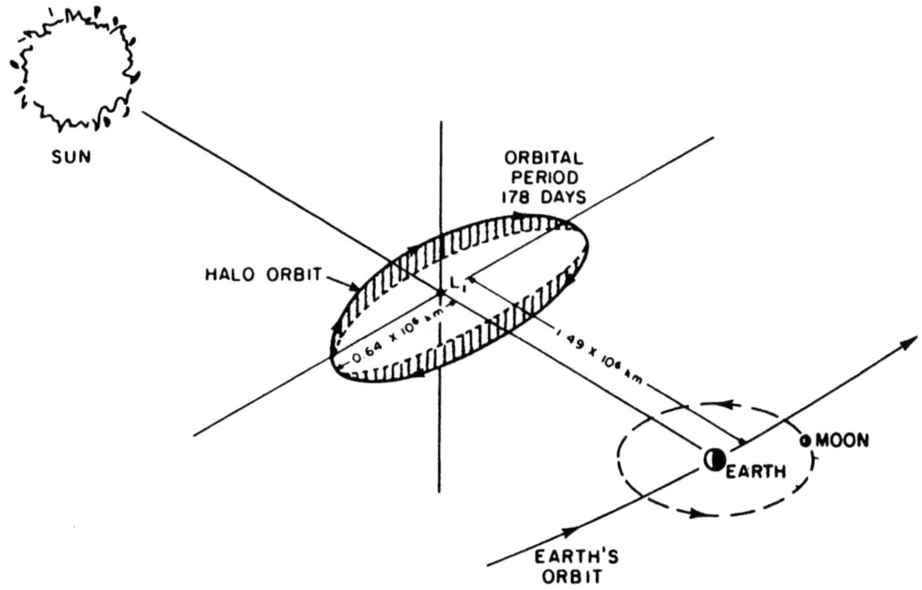

Fig. 8. The International Sun Earth Explorer-3 orbit about the sun-earth liberation point L_1 (from TSURUTANI and BAKER, 1979).

geomagnetic disturbances (TSURUTANI and BAKER, 1979; JOSELYN *et al.*, 1981). These data, primarily from experiments measuring interplanetary fields (E. J. Smith, principle investigator, JPL) and solar wind density and velocity (S. J. Bame, principal investigator, LANL), were extremely helpful in a number of ways, which can be briefly listed as follows:

1. Automatic detection of interplanetary shocks which typically preceeded sudden impulses or sudden commencements at earth. Noise on the 3 KHz plasma noise experiment (F. L. Scarf, principal investigator, TRW) often preceeded the shock signature at ISEE-3 by several hours (KENNEL *et al.*, 1982). Figure 9 illustrates a sequence of observations on September 08, 1981. The shock passed ISEE-3 at 2100 UT and was detected at the earth at 2146 UT. This is an example of a geomagnetic sudden impulse, not followed by significant activity.

2. The measurement of the north/south component of the interplanetary magnetic field. This parameter is a primary variable in algorithms which seek to express the coupling function between the solar wind and the magnetosphere (e.g., BAKER *et al.*, 1983; and many of the papers in this Chapman Conference Proceedings).

3. The measurement of the azimuthal component of the interplanetary magnetic field which identified the polarity of the interplanetary magnetic field sector. The observed velocity of the solar wind was used to estimate the heliolongitude "connected" to the earth.

4. The positive identification of high-speed streams in the solar wind. These

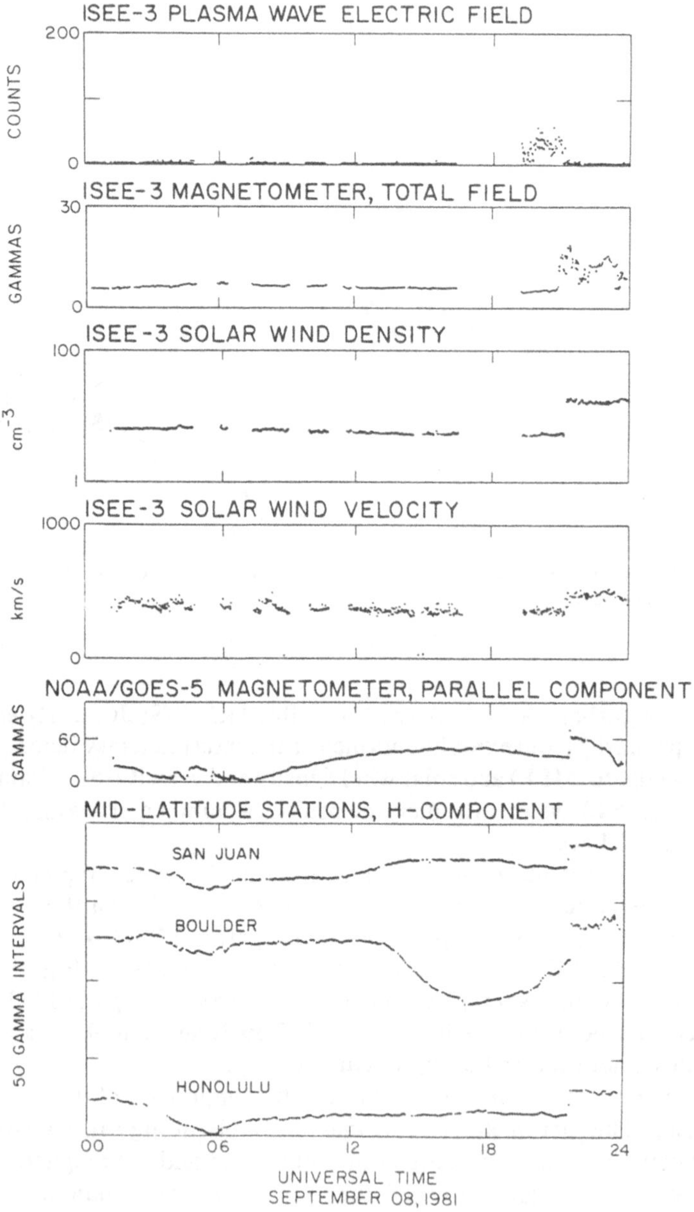

Fig. 9. Real-time data received at the SESC illustrating a shock observed at ISEE-3 at 2100UT and then at the earth at 2146UT on September 8, 1981. ISEE-3 plasma wave electric field data is from a TRW instrument (F. L. Scarf, principal investigator), ISEE-3 magnetometer data is from a JPL instrument (E. J. Smith, P. I.), and solar wind parameters are from a LANL instrument (S. J. Bame, P. I.).

features, distinguished by their very low density from flare or coronal transient events, often recurred on subsequent rotations and could sometimes be identified with specific features (coronal holes) on the sun.

Data from this spacecraft, which is now enroute to Comet Giacobini-Zinner, is still being received at the SESC, and it is still useful. For example, Figure 10 shows a week's worth of data when the spacecraft was located approximately 10° (earth-sun-probe angle) west of the earth at 1 Astronomical Unit. Even though only a few hours of data were received each day via the Deep Space Network, the increase in velocity and decrease in density between December 12 and 13 identifies the passage of a high-speed stream. We can also identify a solar sector change from a positive sector (phi angle about 135°) to a negative one between December 10 and 11.

The advantages of real-time solar wind data listed above are those particularly suited to the SESC mandate to predict daily average levels of global geomagnetic activity up to three days in advance. We have no operational requirement to predict geomagnetic substorms, which occur on time scales of hours in limited geographic sectors. However, if the capability to predict substorms were developed to a level of sufficient accuracy, many existing (and possibly several new customers) would be better served. In the absence of real-time solar wind data and/or continuous optical

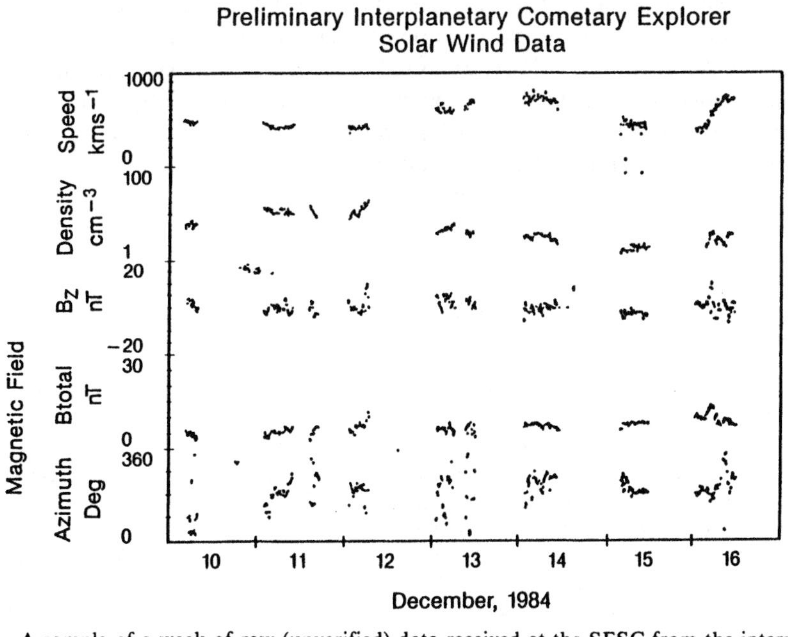

Fig. 10. A sample of a week of raw (unverified) data received at the SESC from the interplanetary Cometary Explorer spacecraft (formerly ISEE-3). From the top, the traces are solar wind velocity (0–1000 km s^{-1} linear scale), density (1–100 cm^{-3} log scale), the north/south component of the interplanetary field (B_Z, −20 to +20 nT), the total field (0–30 nT), and the azimuthal angle of the field (0–360 degrees). As in Fig. 9, solar wind plasma data is from the LANL instrument and magnetic field data is from the JPL instrument.

monitoring of the earth's polar regions, the only data available which are useful for substorm predictions are measurements of the magnetic field and energetic (E>20 keV) electrons at synchronous altitude (BAKER *et al.*,1978; KOKUBUN and MCPHERRON, 1981; NAGAI, 1982). However, there are significant technical problems preventing the accurate prediction of substorm activity using presently available data sets (KAMIDE and JOSELYN, 1986).

4. Summary

In this paper, four specific points have been made. The most important was to emphasize the need for responsible and reasonably accurate forecasts of geomagnetic activity by listing the kinds of operational customers which use SESC geomagnetic predictions. These customers range from scientific research campaigns to the general public, although the majority have military (e.g., communications and radars) or commercial interests (e.g., satellite controllers, power companies, and geophysical explorers). Next, the techniques used by the SESC to generate daily forecasts of the average level of geomagnetic activity (as opposed to substorms) were revealed. These methods are based almost entirely on observations of solar phenomena, analyses of statistical data, and forecaster experience. Third, the SESC has learned to appreciate the value of real-time solar wind data from ISEE-3, both while it was stationed at the libration point and even now at distances far from the earth. On forecast time scales (days), knowledge of solar wind parameters helps to assess the probable source of geomagnetic disturbances and the likelihood of recurrent activity. On short-time scales (minutes), real-time near-earth solar wind data could potentially be used to predict substorms and the intensity of global geomagnetic storms. Finally, as advances are made in the topics under consideration at this Chapman Conference, the SESC intends to apply them to forecasting efforts.

REFERENCES

AKASOFU, S.-I. and S. YOSHIDA, The structure of the solar flare plasma flow generated by solar flares, *Planet. Space Sci.*, **15**, 39, 1967.

BAKER, D. N., P. R. HIGBIE, E. W. HONES, Jr., and R. D. BELIAN, High-resolution energetic particle measurements at 6.6 R_E 3. Low-energy electron anisotropies and short-term substorm predictions, *J. Geophys. Res.*, **83**, 4863, 1978.

BAKER, D. N., R. D. ZWICKL, S. J. BAME, E. W. HONES, Jr., B. T. TSURUTANI, E. J. SMITH, and S.-I. AKASOFU, An ISEE 3 high time resolution study of interplanetary parameter correlations with magnetospheric activity, *J. Geophys. Res.*, **88**, 6230, 1983.

DAVIES, K., Review of recent progress in ionospheric predictions, *Radio Science*, **16**, 1407, 1981.

CHAPMAN, S. and J. BARTELS, Geomagnetic and related phenomena, in *Geomagnetism*, Vol. 1, p. 334, Oxford at the Claredom Press, London, 1940.

DONNELLY, R. F. (Ed.), *Solar-Terrestrial Predictions Proceedings*, Vols. 1–4, U.S. Government Printing Office, Washington, D.C., 1979.

FERGUSON, B. G., Coronal holes inferred from the 'Fleurs' east–west solar scans, in *Solar Terrestrial Predictions Proceedings*, Vol. 1, edited by R. F. Donnelly, p. 288, U.S. Government Printing Office, Washington, D.C., 1979.

GARRETT, H. B. and Charles P. PIKE (eds.), *Space Systems and Their Interactions with Earth's Space Environment*, American Institute of Aeronautics and Astronautics, New York, 1980.

GAUTHREAUX, Jr., S. A., *Animal Migration, Orientation, and Navigation*, Chap. 5, Academic Press, New York, 1980.

GOULD, J. L., The case for magnetic sensitivity in birds and bees (such as it is), *Am. Scientist*, **68**, 256, 1980.

GRARD, R., K. KNOTT, and A. PEDERSEN, Spacecraft charging effects, *Space Sci. Rev.*, **34**, 289, 1983.

HUNDHAUSEN, A. J., An interplanetary view of coronal holes, in *Coronal Holes and High Speed Wind Streams*, edited by J. B. Zirker, p. 225, Colorado Associated Univ. Press, Boulder, CO, 1977.

JOSELYN, J. A., Proposed major format change to geomagnetic activity reports and forecasts produced by the SESC, Boulder, Colorado, U.S.A., *Geophys. Sur.*, **6**, 419, 1984.

JOSELYN, J. A., SESC methods for short-term geomagnetic predictions, in *Proceedings of the Solar Terrestrial Predictions Workshop 18–22 June, 1984*, Observatoire de Meudon, France, in press, 1986.

JOSELYN, J. A. and P. S. MCINTOSH, Disappearing solar filaments: a useful predictor of geomagnetic activity, *J. Geophys. Res.*, **86**, 4555, 1981.

JOSELYN, J. A., J. HIRMAN, and G. R. HECKMAN, ISEE 3 in real time: an update, *EOS*, **62**, 617, 1981.

KAMIDE, Y. and J. A. JOSELYN, Toward the prediction of magnetospheric substorms from knowledge of the 'solar-wind-geomagnetic activity' relationship, in *Proceedings of the Solar Terrestrial Predictions Workshop 18–22 June, 1984*, Observatoire de Meudon, France, in press, 1986.

KAUFMAN, A. A. and G. V. KELLER, *The Magnetotelluric Method*, Elsevier Sci. Publ. Co., New York, 1981.

KENNEL, C. F., F. L. SCARF, F. V. CORONITI, E. J. SMITH, and D. A. GURNETT, Nonlocal plasma turbulence associated with interplanetary shocks, *J. Geophys. Res.*, **87**, 17, 1982.

KOKUBUN, S. and R. L. MCPHERRON, Substorm signatures at synchronous altitude, *J. Gephys. Res.*, **86**, 11265, 1981.

LANZEROTTI, L. J., (Ed.), Impacts of ionospheric/magnetospheric process on terrestrial science and technology, in *Solar System Plasma Physics*, Vol. III, edited by L. J. Lanzerotti, C. F. Kennel and E. N. Parker, North Holland Publ. Co., 1979.

LANZEROTTI, L. J., Geomagnetic induction effects in ground-based systems, *Space Sci. Rev.*, **34**, 347, 1983.

MCINTOSH, P. S., Annotated atlas of H-alpha synoptic charts for solar cycle 20 (1964–1974), Report UAG-70, World Data Center A for Solar-Terrestrial Physics, U.S. Dept. of Commerce NOAA/EDIS, Ashville, N.C. 28801, 1979.

NAGAI, T., Observed magnetic substorm signatures at synchronous altitude, *J. Geophys. Res.*, **87**, 4405, 1982.

NATIONAL PLAN FOR SPACE ENVIRONMENT SERVICES AND SUPPORTING RESEARCH 1983–1987, Federal Coordinator for Meterological Services and Supporting Research, U.S. Department of Commerce NOAA, FCM-P10-1983, Washington, D.C., July 1983.

PROCEEDINGS OF A WORKSHOP ON SATELLITE DRAG, MARCH 18–19, 1982, NOAA Environmental Research Laboratories, Space Environment Lab., Boulder, CO 80303, 1982.

REGAN, R. D. and P. RODRIGUEZ, An overview of the external magnetic field with regard to magnetic surveys, *Geophys. Surv.*, **4**, 255, 1981.

SCHWENN, R., Direct correlation between coronal transients and interplanetary disturbances, *Space Sci. Rev.*, **34**, 85, 1983.

SVALGAARD, L., Geomagnetic activity: Dependence on solar wind parameters, in *Coronal Holes and High Speed Wind Streams*, edited by J. B. Zirker, p.371, Colorado Associated Univ. Press, Boulder, CO, 1977.

TSURUTANI, B. T. and D. N. BAKER, Substorm warnings: an ISEE-3 real-time data system, *EOS*, **60**, 702, 1979.

WEYMOUTH, J. W. and R. HUGGINS, Geophysical surveying of archaeological sites, in *Archaeological Geology*, Chap.8, edited by G. Rapp, Jr., and J. Gifford, p. 191, Yale University Press, 1985.

WRIGHT, C. S. and L. F. MCNAMARA, The relationships between disappearing solar filaments, coronal mass ejections, and geomagnetic activity, *Solar Phys.*, **87**, 401, 1983.

Solar Wind-Magnetosphere Coupling, edited by Y. Kamide and J. A. Slavin, 143–148.

The 27-Day Recurrence Index

H. H. SARGENT III

Space Environment Laboratory, NOAA, Boulder, Colorado 80303, U.S.A.

The 27-day Recurrence Index is based on the geomagnetic aa-index for 1868 through 1983. The Index reveals lengthy periods when geomagnetic activity patterns are repeated every 27 days. These recurrent periods occur at the end of each sunspot cycle; they end abruptly with the beginning of a new sunspot cycle; and they last longer when associated with even-numbered sunspot cycles. The Index is thought to indicate those times when the solar wind structure remains relatively stable beyond 1 AU for periods of 1-1/2 to 5 years.

1. Introduction

The Space Environment Services Center of the National Oceanic and Atmospheric Administration (located in Boulder, Colorado) has been issuing daily forecasts of certain key solar and geophysical parameters for nearly 20 years. One element of the forecast is a projection of the 24-hour A-index of geomagnetic activity as observed at Fredericksburg, Virginia, for each of the three days following the forecast date. This mid-latitude index provides a good representation of the level of geomagnetic disturbance affecting the contiguous 48 states of the United States.

The forecasting process often begins and ends with the forecaster looking for recurrent patterns in an up-to-date 27-day rotation chart, a modification of the plotting scheme developed by BARTELS (1932). The chart (Fig. 1) is called a rotation chart because it emphasizes the association of geomagnetic activity with the 27-day synodic rotation of the Sun. The plots are stacked to allow the forecaster to quickly compare one 27-day interval with the next. In practice, the forecaster looks at the last four or five rotations to determine whether recurrent patterns exist. Patterns that drift in time (recurrence periods longer and shorter than 27 days), as well as any trends in the amplitude of disturbances, can be discerned in these charts. This information is combined with knowledge of seasonal variations in geomagnetic activity to predict both the amplitude and timing of geomagnetic disturbances. Other data, i.e., solar flare data, coronal hole observations, interplanetary scintillation data, and measurements from satellites in orbit around the Sun are used, when available, to make modifications the forecast—but each forecast originates with the Bartels charts.

Unfortunately, there are extensive periods during each sunspot cycle when recurrence simply does not work as a forecasting tool. It has been known for some time that when some major solar flares occur, shockwaves are formed, which move outward through the interplanetary medium. When many flares have occurred, the Bartels charts become virtually useless as a forecasting tool. Nevertheless, there is apparently a long and predictable period during each sunspot cycle when patterns are seen to persist in the Bartels charts for many rotations. This is evidence that well-

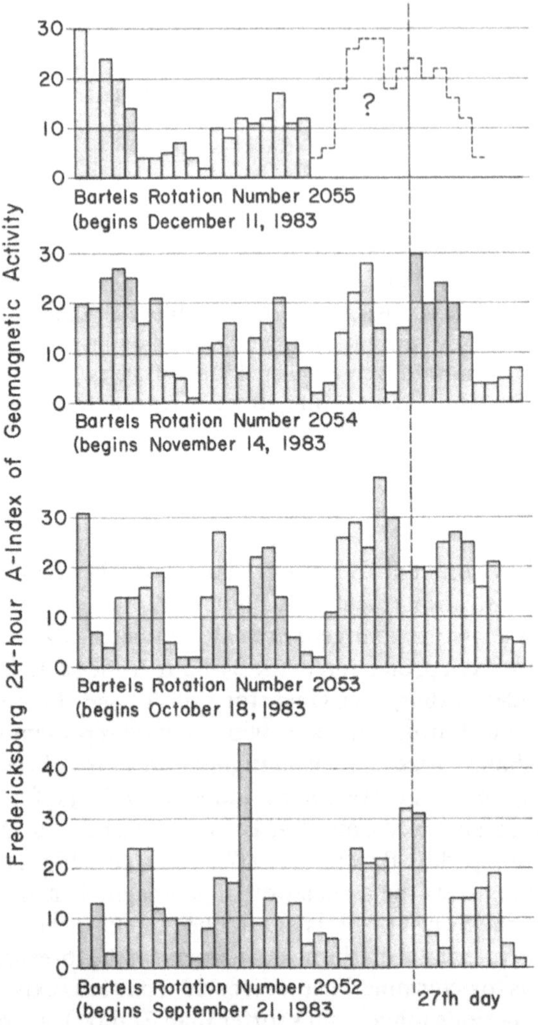

Fig. 1. Modified Bartels charts (27-day stack plots of the 24-hour *A*-index of geomagnetic activity from Fredericksburg, Virginia).

ordered stable patterns exist in the interplanetary magnetic field and velocity structure. This study was initiated to identify those periods in the past when recurrence would have served as a useful forecasting tool, and to anticipate such periods in the future. To accomplish this, a simple recurrence index was developed using a published geomagnetic data set.

2. Index Development

The data used for this study were the *aa*-index values of MAYAUD (1973), which are available from 1868 to the present. This index is published in two 12-hourly average values per day (0000–1200 and 1200–2400 UT) for the entire period. The index is particularly useful because it is derived from the average of values reported at two antipodal observatories and it can be computed quickly but remains representative of global geomagnetic conditions. Thus, 54 consecutive values of the *aa*-index can be used to describe geomagnetic activity during each 27-day Bartels rotation.

The 27-day Recurrence Index was created in the following way:

1. The 54 consecutive values of the *aa*-index for Bartels Rotation 487 (beginning on January 12, 1868) were correlated with the 54 consecutive values of Bartels Rotation 488 (beginning on February 8, 1868). The correlation coefficient (calculated to be 0.302) was assigned to February 8 (the first day of the second rotation — or, for all practical purposes, the mid-point of the 54-day period being examined). Next, Bartels Rotation 488 was correlated, in like manner, with Bartels Rotation 489 (beginning on March 6, 1868) and the correlation coefficient (0.387) was assigned to March 6 — and so on. In this way, each rotation was correlated with its immediate neighbor and a correlation coefficient was calculated for each 27 day span of data.
2. The coefficients were all multiplied by a factor of 100.
3. The entire set was smoothed using a 15-Bartels rotation running average. This smoothing function was similar in length and shape to the smoothing function traditionally applied to the monthly mean sunspot numbers. The resulting 27-day Recurrence Index is shown in Fig. 3, along with the smoothed monthly mean sunspot numbers.

3. Adjacent Period Analysis

A 27-day period was used in the original computation of the Recurrence Index. As will be discussed, strong order was found in that time series. When the analysis was performed for a 28.5-day recurrence period suggested by SVALGAARD and WILCOX (1975) similar order was not found. Eventually, this same type of analysis was performed in half-day intervals for all potential recurrence periods from 24 to 30 days. This was done to determine whether the recurrence phenomenon is centered on the 27-day period. It was also necessary to ensure that during times when the 27-day Recurrence Index is low, no adjacent period (e.g., 28.5 days) emerges. This analysis can be thought of as a sensitivity analysis (i.e., sensitivity to period). The 13 separate analyses are shown combined in Fig. 2. The inset (Fig. 2(a)) shows several fairly typical cross-sections spanning all 13 analyses at times of peak recurrence. It can be seen that the smoothed correlation does peak at the 27-day period and falls off rapidly for both longer and shorter periods.

The results of the sensitivity analysis can be summarized as follows:

1. The strongest recurrent behavior occurs, as expected, with a 27-day period.
2. The response falls off fairly rapidly for periods both longer and shorter than the 27-day period and has subsided to the general noise level within two days in either

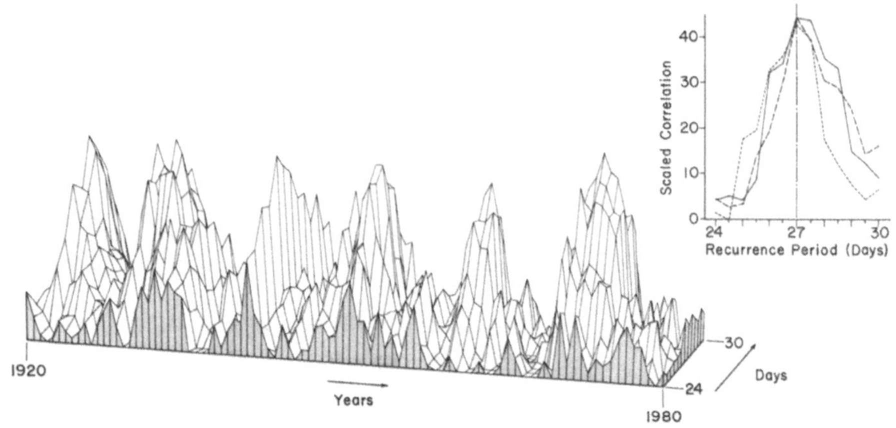

Fig. 2. Three-dimensional plot of recurrence. Transverse sections represent average relative strength of recurrence for periods of 24 to 30 days calculated for times between 1920 and 1980. Inset shows separate sections through three-dimensional plot at three different recurrence peaks.

direction, in most cases.

3. Any response at adjacent periods coincides with the 27-day response, and in no case is there any indication of a tendency for adjacent periods to emerge while the 27-day Index is at low values.

4. Discussion

It is generally accepted that 27-day recurrent behavior is directly related to the long-lived open field structures on the Sun (coronal holes) (SUESS, 1979), which are sources of high-speed streams in the solar wind. Consequently, raised values of the 27-day Recurrence Index imply both stable high-speed streams and coronal holes are present. The lowest values of the Index must occur when the stable streams are not present, at least in the ecliptic plane. One may assume that when Index values remain high for many months, the Index is evidence for long-lasting and more or less stable physical structure in interplanetary regime, at least out to 1 AU.

The reader may verify the following gross features of the Index in Fig. 3:

1. For many of the past 116 years the Index unambiguously shows the presence or absence of recurrent geomagnetic behavior. The signal, particularly during the past 60 years, is bistatic (in that values of the Index are usually in either a high state or a low state).

2. The onset of recurrent activity is generally abrupt.

3. The termination of the recurrent period is almost always as abrupt as the onset, and occurs about 1 year into the new sunspot cycle.

4. Through the entire analysis the duration of recurrent periods has varied on a 22-year basis. Recurrent periods appear to be, on average, twice as long when associated with even-numbered sunspot cycles.

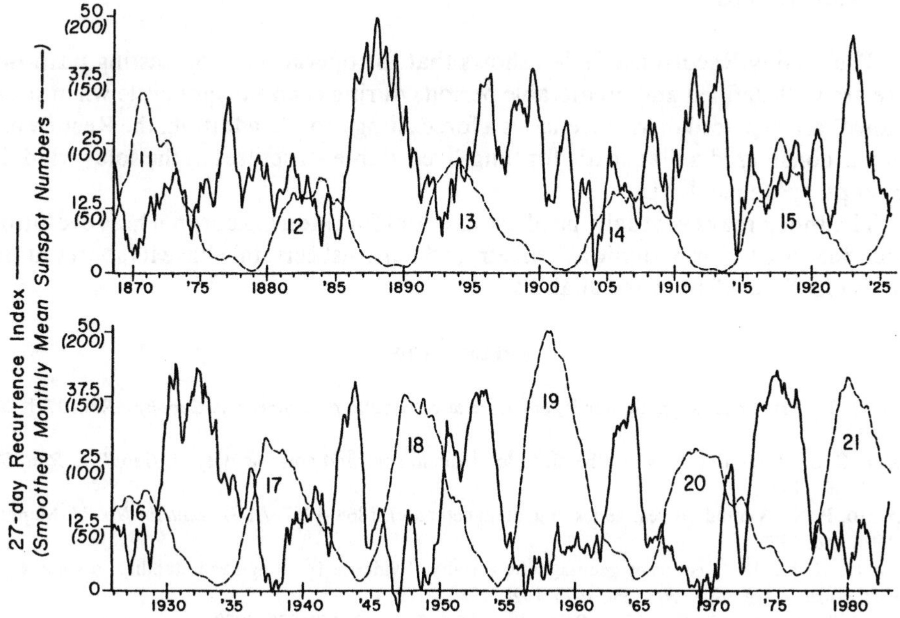

Fig. 3. 27-day Recurrence Index (solid line) and smoothed monthly mean sunspot numbers (dashed line) from 1868 to present. Sunspot cycles are numbered according to convention.

5. Recurrent periods in the Index extend over a significant portion of a sunspot cycle. During odd-numbered sunspot cycles the recurrent state lasts for 1-1/2 to 3 years. For even-numbered cycles, the recurrent state lasts for 3 to 5 years.

Many early analyses of the 27-day effect relied on identifying some regular temporal spacing between geomagnetic storms. Bartel's charting method represented an improvement over the earlier methods in that it rested on the creation of visual patterns in the charts, giving equal emphasis to geomagnetic values for each day, whether they were large or small. The Index discussed here, it should be stressed, is independent of the amplitude of the *aa*-index values used to create it, and shows only the presence or absence of stable patterns.

While SARGENT(1985) has suggested that the 27-day Recurrence Index may eventually prove useful for making inferences about coronal hole evolution at the Sun, the Index appears to be of immediate usefulness to the geomagnetic activity forecaster. Although the forecaster can never be sure when recurrence will not work, the forecaster can expect regularly occurring and well-defined periods during each sunspot cycle when recurrence will provide dependable forecasting guidance.

DEEHR (1983) has found a convincing correlation between the averaged strength of the Recurrence Index and the height of the Earth's aurora, a negative correlation. The physics of this effect is as yet unexplained. Understanding of this relation may contribute to the understanding of coupling of the high-speed streams in the solar wind to the magnetosphere.

5. Conclusions

The 27-day Recurrence Index shows that for operational forecasting purposes there are well-defined and predictable periods during each sunspot cycle when some reliance can be placed on recurrence as a forecasting tool. In addition, the Recurrence Index is considered a diagnostic for long-lived stable structures in the solar wind, in the ecliptic plane at 1 AU.

The Index may eventually be of use as a tool for inferring coronal hole evolution since 1868. A 22-year periodic asymmetry, which must certainly have its source at the Sun, is clearly evident in the analysis.

REFERENCES

BARTELS, J., Terrestrial-magnetic activity and its relation to solar phenomena, *J. Geophys. Res.*, **37**, 1–52, 1932.

DEEHR, C. S., A relation between the altitude of the aurora and solar activity, *J. Geophys. Res.*, **88**, 395–397, 1983.

MAYAUD, P. N., A hundred year series of geomagnetic data 1868–1967, *IAGA Bulletin No. 33*, Meppel, Holland, 1973.

SARGENT, H. H., III, Recurrent geomagnetic activity: Evidence for long-lived stability in solar wind structure, *J. Geophys. Res.*, **90**, 1425–1428, 1985.

SUESS, S. T., Models of coronal hole flows, *Space Sci. Rev.*, **23**, 159–200, 1979.

SVALGAARD, L. and J. M. WILCOX, Long term evolution of solar sector structure, *Solar Phys.*, **41**, 461–475, 1975.

Solar Wind-Magnetosphere Coupling, edited by Y. Kamide and J. A. Slavin, 149–160.

A Comparison of Solar and Geomagnetic Activities from the 13th to the 21st Solar Cycles

Meiqing GAO*

Cooperative Institute for Research in Environmental Sciences, University of Colorado/National Oceanic and Atmospheric Administration, Boulder, Colorado 80309, U.S.A.

This paper briefly describes several aspects of research done at the Institute of Geophysics, Academia Sinica, Beijing, P.R. China, into the relationship between solar and geomagnetic activities. Comparisons between sunspot numbers and geomagnetic indices (C_i, A_p) were made for the periods of ascending, maximum, descending, and minimum phases of solar and geomagnetic activities from the 13th to the 21st solar cycles, with emphasis on the 21st solar cycle. A prediction of time and annual average maximum A_p index for the 21st cycle made in 1979 agrees well with the observed value (the annual average maximum A_p occurred in September, 1982, and was 22.6). The study includes solar flares of importance 2 or greater, and the first, second, and third largest magnetic storms in the 17th to 21st solar cycles. For the current solar cycle the numbers and distribution of flares of importance 2 or greater and magnetic storms were compared for each year. A study of 22 spotless flares observed during the Solar Maximum Year suggests that some are associated with moderate and minor geomagnetic storms.

1. Introduction

This study uses solar and geomagnetic data since 1889 (8 full cycles of data and part of a ninth), for two purposes: to analyze the character of the magnetic cycles and their relationship with solar cycles, and, to develop a mathematical method to predict the time of occurrence and magnitude of the maximum of the 21st geomagnetic cycle. Comparisons between sunspot numbers and geomagnetic indices were made for each of the four phases of a cycle—the ascending, maximum, descending and minimum phases, and the results are given in Section 2. In Section 3, the method used to obtain a successful prediction of the maximum of the 21st geomagnetic cycle is presented. In Section 4, the relationship between solar flares (not just sunspot number) and geomagnetic activity is analyzed for the present (21th) cycle, especially noting the association of flares in spotless regions with geomagnetic storms during the Solar Maximum Year period. Section 5 contains discussion of these results, and conclusions.

2. Comparison of Solar and Geomagnetic Activity Characteristics

Sunspot numbers and geomagnetic indices (C_i or A_p) were compared for periods of ascending, maximum, descending, and minimum phases of solar and geomagnetic

*Visiting from the Institute of Geophysics, Academia Sinica, Beijing, P.R.C.

activities from the 13th to the 21st solar cycles (Table 1). One can see that solar and geomagnetic activities were highest in the 19th solar cycle. The maximum sunspot number for the 21st solar cycle is the second highest observed during the past 280 years. To determine the maximum, minimum, ascending, and descending phases for the geomagnetic cycles, the annual average values of the geomagnetic index C_i were used for the 13th through the 16th cycles. The annual average values of the A_p index (calculated from the monthly average values) were used for the 17th into the 21st cycles.

Figure 1 shows the difference between geomagnetic activity maxima and minima and sunspot maxima and minima for the various solar cycles. The average is a lag of 1.21 years during the 13th-20th solar cycles, but the average for the last four solar cycles (17th–20th) is a lag of 2.17 years.

Figure 2 shows the annual average values of either ΔC_i or ΔA_p for the odd-numbered and the even-numbered solar cycles. To get ΔC_i or ΔA_p, every cycle's beginning annual average values was subtracted from the C_i or A_p annual average values. The abscissa is the number of years from the year of geomagnetic minimum. The ordinate is $\Delta C_i \times 10$ (for the 13th–16th solar cycles) or ΔA_p (for the 17th– 21st solar cycles). The difference between the even and odd cycles points out that there is a 22-year geomagnetic cycle, the rate of increase of activity being greater in the odd-numbered cycles. This result agrees with CHERNOSKY (1966), who used annual average values of the C_i index for the 12th–19th cycles.

3. Prediction of Geomagnetic Maximum of the 21st Cycle

In 1979, the maximum value of the A_p index (annual average) and its occurrence time were predicted for the 21st cycle, using the method of extremum distribution (STATISTICS GROUP ... ACADEMIA SINICA, 1973). The method arranges a series (x_1, x_2, ... x_i, ... x_n) in order of size, using a chi-square distribution to test whether the series follows a normal distribution. If the series follows a normal distribution, its extremum distribution will be that obtained by the method outlined in the Appendix. In this way, it is possible to obtain a maximum for the present cycle. This method is used in weather forecasting.

The annual averages of A_p indices from 1932 to 1979 were chosen as the series to be treated. The series meets the criteria for a normal distribution: n=48; mathematical expectation=14.4; deviation=3.99. Application of the extremum distribution method produced a predicted maximum annual average A_p of 25.3 for the 21st solar cycle. However, the 21st solar cycle's ascending trend is slower than that of the 19th solar cycle and it is in the descending phase of an 80-year secular cycle. Hence, the maximum value should be reduced by 10– 20%. The annual average A_p maximum of the 21st solar cycle was therefore predicted to be between 20 and 23.

Two methods were used to determine the time of maximum of the 21st geomagnetic cycle. First, the relationship between the annual average A_p during the maximum and ascending phases for the 17th through the 20th cycles were analyzed, to obtain the linear correlation

$$x = .32y - 1.15 \tag{1}$$

where x=number of years in the A_p ascending phase, and y=annual average A_p maximum. When y=20, x=5.3 years. When y=23, x=6.2 years. The 21st geomagnetic cycle began in June 1977; therefore the A_p maximum should occur between September 1982 and August 1983. The second method to estimate the time of geomagnetic maximum use the correlation between sunspot ascending phase and C_i or A_p ascending phase to obtain the linear correlation

$$x = 436 - 8.33y \tag{2}$$

where x=the number of months in the ascending phase for geomagnetic activity, determined from A_p, and y=the number of months in the ascending phase, determined by the sunspot number. The 21st cycle solar maximum was predicted to occur between 1978–1982 (SARGENT, 1978). When May, 1980, is chosen, the sunspot cycle ascending period is 3.83 years. Equation (2) then yields an A_p ascending phase of 4 years and 5 months. (Because it was unlike the others, the 16th cycle was not used in Eq. (2); however, it was used to estimate a longer A_p ascending phase.) Because this prediction is shorter than the ones from Eq. (1), the middle value was chosen. Thus an annual average A_p ascending period of 5.3 years was obtained. There are two peaks of activity each year, at the equinoxes. The geomagnetic activity maximum should therefore occur in autumn 1982. The actual geomagnetic activity maximum did occur at the predicted time. The maximum annual average A_p was 22.6, and the time of maximum was September, 1982.

4. Analysis of Solar activity and Geomagnetic Activity Characteristics during the 21st Solar Cycle

Figure 3 shows the annual distributions of solar flares of optical importance ≥2 for the years of the 21st solar cycle. The shaded parts refer to the solar flares that produced magnetic storms. A delay of 1–3 days between flares and storms was considered in determining causality. The year 1981 had the largest number of ≥2 solar flares. Twenty-seven percent of ≥2 all solar flares so far produced magnetic storms, whereas 64 percent of solar flares of importance 3 have produced magnetic storms.

Figure 4 shows the annual distributions of geomagnetic storms. The data on magnetic storms were provided by Beijing Observatory; a storm was defined as having a geomagnetic K index of 5 or more and a duration of disturbance ≥1 day. The shaded parts refer to the magnetic storms produced by ≥2 solar flares. The total number of magnetic storms is 298. Of these, 183 storms were accompanied by sudden commencements; they constitute 61% of the total magnetic storms. The magnetic storms associated with ≥2 flares constitute 41% of the total magnetic storms.

Table 2 lists the total numbers of the geomagnetic storms and the largest three magnetic storms in the 17th–21st solar cycles. The magnetic storm data were collected from Chinese observatories. This list was checked with DODSON *et al.* (1974), who listed flares of importance 3 and some of importance 2 in cycles 18 through 20. A delay

Table 1. Comparison of solar and geomagnetic activities.

SOLAR CYCLE NUMBER	GEOMAGNETIC OR SOLAR PARAMETERS	TIME OF MINIMUM ACTIVITY	MAGNITUDE OF ACTIVITY AT MINIMUM	TIME OF MAXIMUM ACTIVITY	MAGNITUDE OF ACTIVITY AT MAXIMUM	DURATION OF ASCENDING PHASES	TIME OF GEOMAGNETIC MAXIMUM BEHIND SUNSPOT MAXIMUM	DURATION OF DESCENDING PHASES	TIME OF GEOMAGNETIC MINIMUM BEHIND SUNSPOT MINIMUM
13	Solar	smooth 1889.6 observed November, 1889	month smooth 5.0 annual average 6.3 month average 0.2	smooth 1894.1 observed August, 1893	month smooth 87.9 annual average 85.1 month average 129.2	smooth 4.5 years observed 3 years 9 months	-1 year	smooth 7.6 years observed 7 years 5 months	0 year
	Geomagnetic	1889	annual average 0.47 (C_1)	1892	annual average 0.82 (C_1)	3 years		8 years	
14	Solar	smooth 1901.7 observed 1901, January	month smooth 2.6 annual average 2.7 month average 0.2	smooth 1905.0 observed November 1905	month smooth 64.2 annual average 63.5 month average 107.2	smooth 4.3 years observed 3 years 10 months	1 year	smooth 6.6 years observed 7 years 6 months	-1 year
	Geomagnetic	1900	annual average 0.42 (C_1)	1906	annual average 0.65 (C_1)	6 years		5 years	
15	Solar	smooth 1913.6 observed May, 1913	month smooth 1.5 annual average 1.4 month average 0.0	smooth 1917.6 observed August, 1917	month smooth 105.4 annual average 103.9 month average 154.5	smooth 4.0 years observed 4 years 3 months	1 year	smooth 6 years observed 6 years	-1 year
	Geomagnetic	1912	annual average 0.46 (C_1)	1918	annual average 0.75 (C_1)	6 years		5.0 years	
16	Solar	smooth 1923.6 observed August, 1923	month smooth 5.6 annual average 5.8 month average 0.5	smooth 1928.4 observed February, 1928	month smooth 78.1 annual average 77.8 month average 98.0	smooth 4.8 years observed 4 years 11 months	2 years	smooth 5.4 years observed 5 years 1 month	0 year
	Geomagnetic	1923	annual average 0.48 (C_1)	1930	annual average 0.83 (C_1)	7 years		4 years	
17	Solar	smooth 1933.8 observed August, 1933	month smooth 3.4 annual average 5.7 month average 0.2	smooth 1937.4 observed July, 1937	month smooth 119.2 annual average 114.3 month average 145.1	smooth 3.6 years observed 3 years 11 months	1.75 years	smooth 6.8 years observed 6 years 9 months	0.7 year
	Geomagnetic	June, 1934	annual average 7.3 month average 5 (A_p)	April, 1939	annual average 16.5 month average 28	4 years 10 months		6 years 2 months	

18	Solar	smooth 1944.2 observed April, 1944	month smooth 7.7 annual average 9.6 month average 0.3	smooth 1947.5 observed May, 1947	month smooth 151.8 annual average 151.6 month average 201.3	smooth 3.3 years observed 3 years 1 month	4.25 years	smooth 6.8 years observed 7 years 1 month	1.3 years
	Geomagnetic	June, 1945	annual average 10.3 month average 7	September, 1951	annual average 22.3 month average 40	6 years 3 months		2 years 9 months	
19	Solar	smooth 1954.3 observed June, 1954	month smooth 3.4 annual average 4.4 month average 0.2	smooth 1957.9 observed October, 1957	month smooth 201.3 annual average 190.2 month average 235.8	smooth 3.6 years observed 3 years 4 months	2.5 years	smooth 6.9 years observed 6 years 9 months	0.2 year
	Geomagnetic	June, 1954	annual average 11.1 month average 6	April, 1960	annual average 23.5 month average 42	5 years 10 months		5 years 7 months	
20	Solar	smooth 1964.8 observed July, 1964	month smooth 9.6 annual average 10.2 month average 3.1	smooth 1968.9 observed September 1968	month smooth 110.6 annual average 105.9 month average 117.2	smooth 4.1 years observed 4 years 2 months		smooth 7.6 years observed 7 years 10 months	1.13 years
	Geomagnetic	November 1965	annual average 7.8 month average 6	November 1968	annual average 13.5 month average 17	3 years	0.17 year	8 years 7 months	
21	Solar	smooth 1976.5 observed July 1976	annual average 12.6 month average 1.9	smooth December, 1979 observed September 1979	month smooth 165.3 annual average 155.3 month average 188.4	smooth 3.5 years observed 3 years 2 months			
	Geomagnetic	June, 1977	annual average 11.9 month average 8	September 1982	annual average 22.6 month average 36	5 years 3 months	3 years		

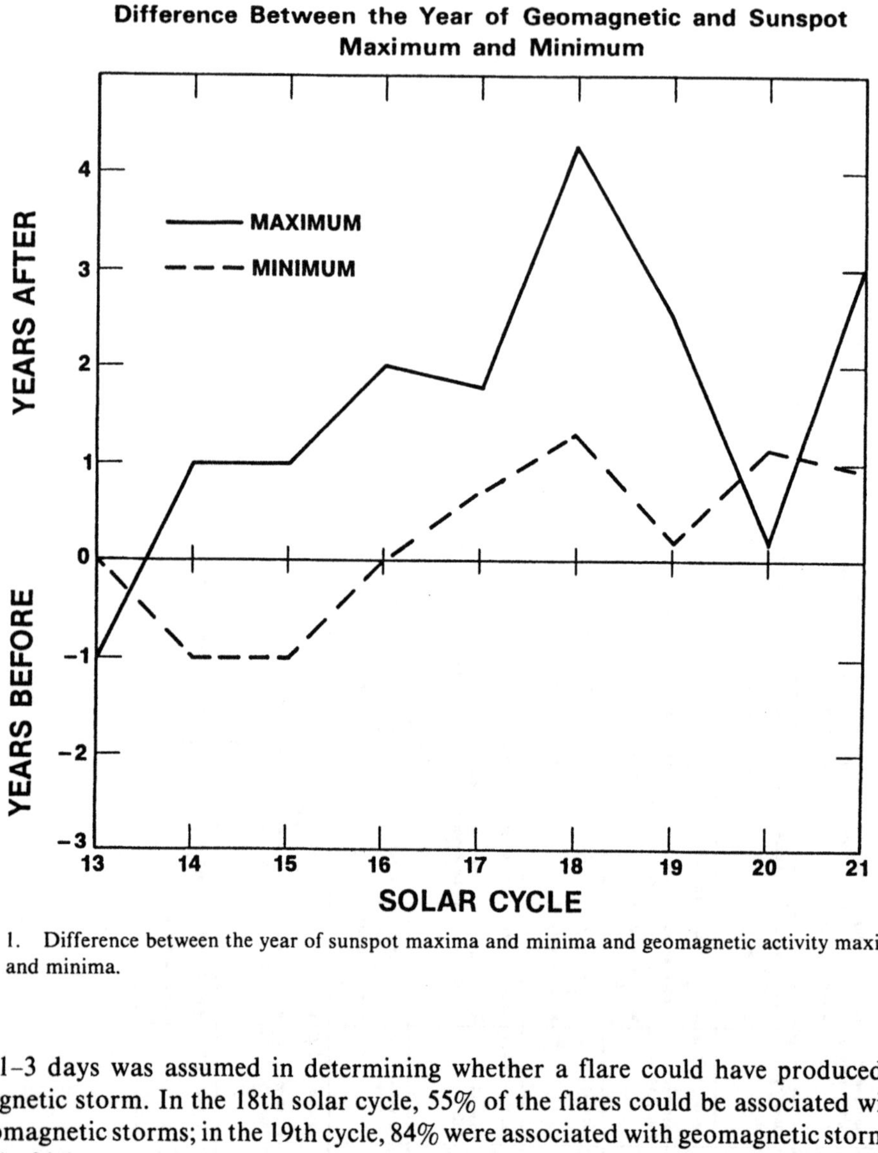

Fig 1. Difference between the year of sunspot maxima and minima and geomagnetic activity maxima and minima.

of 1–3 days was assumed in determining whether a flare could have produced a magnetic storm. In the 18th solar cycle, 55% of the flares could be associated with geomagnetic storms; in the 19th cycle, 84% were associated with geomagnetic storms; in the 20th cycle, 94% could be associated with storms. In the 21th cycle, only flares of importance 3 were considered; 64% were associated with magnetic storms.

Some special results on the relationship between flares and geomagnetic activity have been obtained for the solar Maximum Year period (August 1979–February 1981). Chinese astronomical observatories reported nine large solar flares during this time. Seven of them did not produce magnetic storms; the other two produced only small magnetic storms. Among the 22 spotless flares of importance ≥1 during SMY (from Chinese observatories) there were seven associated moderate and minor storms having short durations. Their *H* amplitudes were about 100 gammas at Beijing

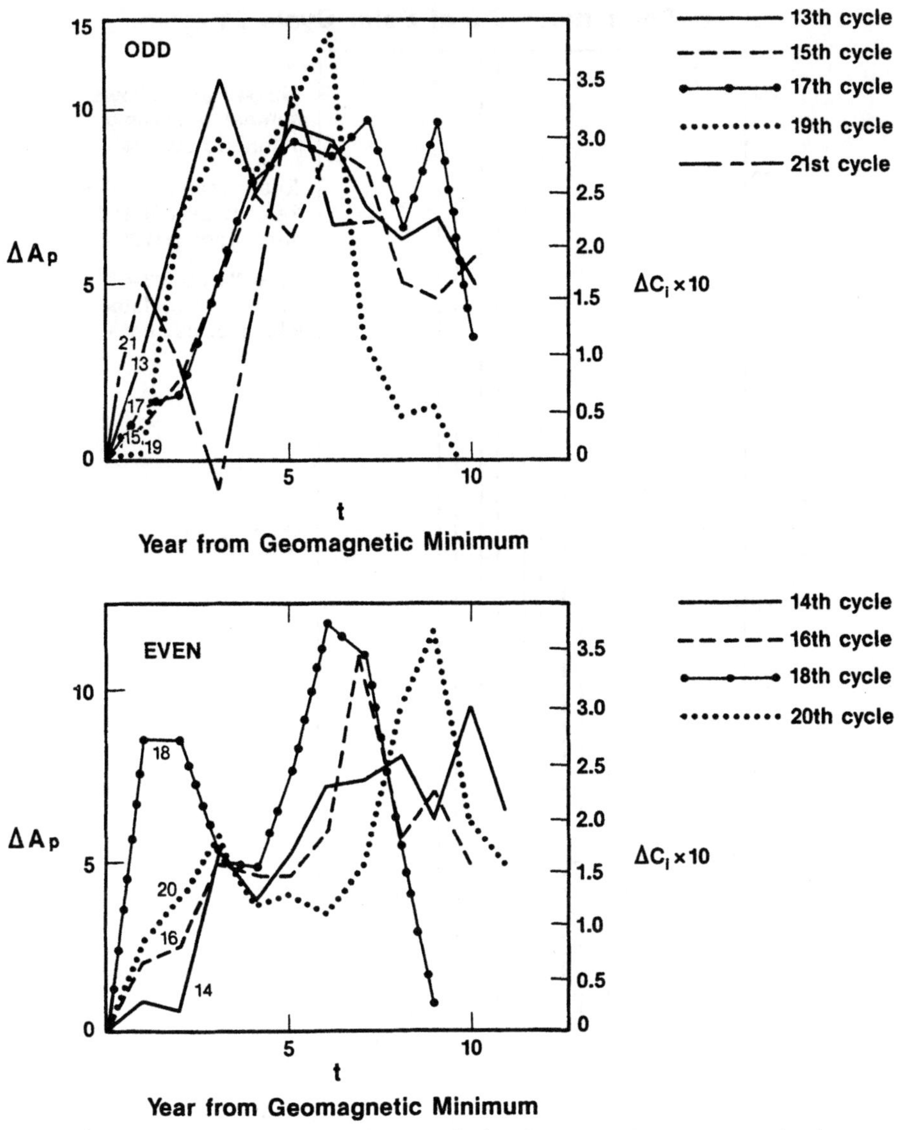

Fig. 2. Annual average values of C_i or A_p indices for odd- and even-numbered cycles.

Geomagnetic Observatory. An example of a flare in a spotless region in shown in Fig. 5.

5. Discussion and Conclusions

This paper briefly describes several aspects of research done at the Institute of Geophysics, Academia Sinica, Beijing, P.R. China, into the relationship between

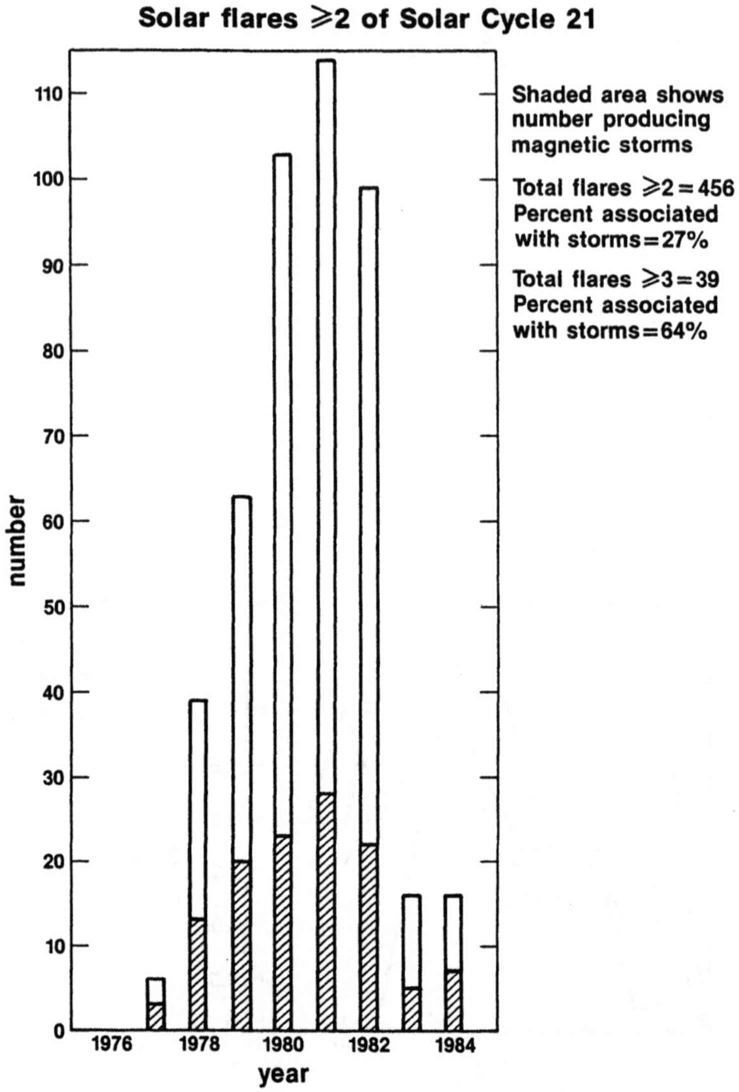

Fig. 3. Annual distributions of the number of solar flares of importance greater than 2 for the 21st solar cycle.

solar and geomagnetic activities. First the general character of the relationship was studied, using the annual average sunspot numbers and the C_i (cycles 13–16) and A_p (cycle 17 until now) geomagnetic indicies. It was noticed that the time of the beginning and maximum of the sunspot cycles did not exactly match the time of the beginning and maximum of the geomagnetic cycles. The beginning of the magnetic cycle is generally within one year of the beginning of the sunspot cycle, but of the 9 cycles studied here the geomagnetic cycle led the sunspot cycle for 2, was about the same for

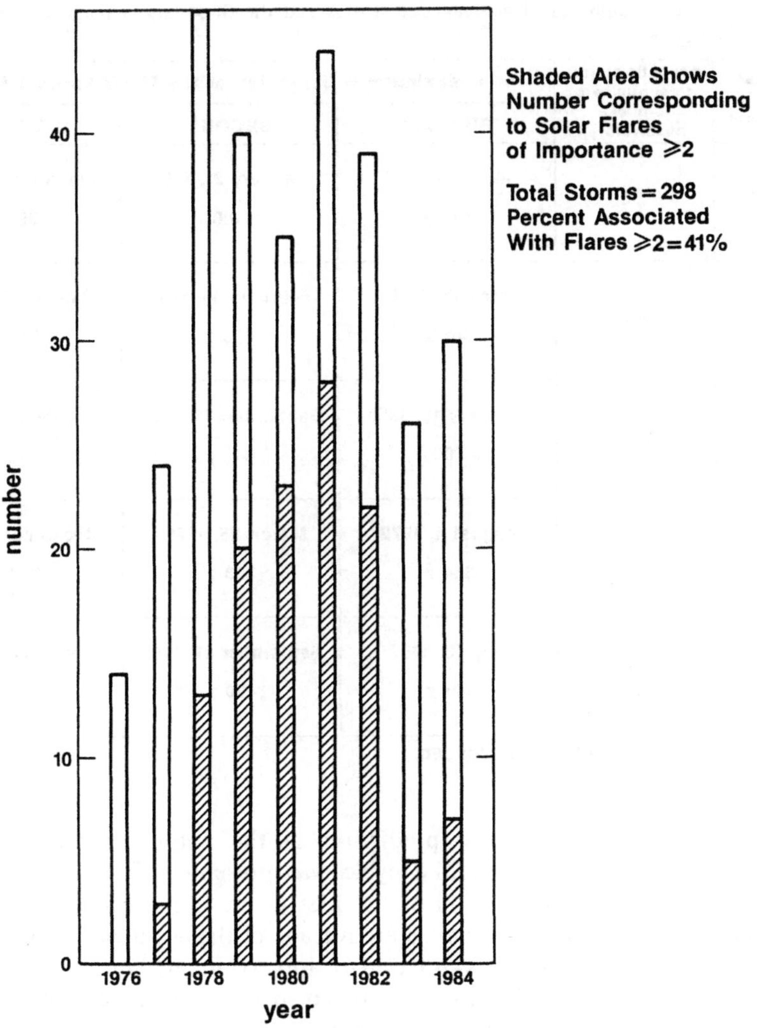

Fig. 4. Annual distributions of the number of geomagnetic storms for the 21st solar cycle.

3, and lagged the sunspot cycle for 4. A comparison of the time of maximum of the geomagnetic and solar cycles shows that nearly always (except for the 13th cycle), the geomagnetic cycle maximum has lagged the sunspot cycle. For the 18th cycle the lag was more than 4 years although for the 20th cycle the time of maximum was nearly the same. There is no known explanation for this behavior.

It was noticed that the general character of the geomagnetic cycles varies depending on whether the cycle is even or odd-numbered. Figure 2 illustrates this general character. During the odd-numbered cycles, the rise in activity from geomagnetic minimum is more rapid than for even-numbered cycles. This result agrees with work by CHERNOSKY (1966).

Table 2. The total numbers of geomagnetic storms and the three largest magnetic storms for the 17th–21st cycles.

solar cycle number	magnetic storms total numbers (Chinese observatory)	Date, Maximum H Excursion of the Three Largest Storms		
		BIGGEST	SECOND	THIRD
17	316	March 1, 1941 420 γ	January 21, 1938* > 400 γ	September 18, 1941 350 γ
18	266	May 12, 1949 461 γ	February 16, 1947 453 γ	January 24, 1949 374 γ
19	354	February 16, 1958 > 700 γ	September 13, 1957 641 γ	July 15, 1959 602 γ
20	437	August 4, 1972 354 γ	March 25, 1976* 343 γ	December 4, 1970 320 γ
21	298 (up to August 1984)	July 13, 1982 719 γ	September 21, 1982 420 γ	February 4, 1983 408 γ

*NO SUDDEN COMMENCEMENT RECORDED

In Section 3, a successful prediction of the amplitude and time of the geomagnetic maximum of the 21st cycle was presented. FEYNMAN (1980) also predicted the time and maximum of the 21st geomagnetic cycle but for the ⟨*aa*⟩ geomagnetic index. The ⟨*aa*⟩ index is an average of the *a*-indicies at two antipodal mid-latitude stations while the A_p is based on approximately 12 mid-latitude stations (including the *aa* stations). Feynman predicted an annual average *aa* of 27 to occur in 1982. The annual average of the provisional *aa* indicies for 1982 was 34.2.

In Section 4, the special relationship between flares and geomagnetic storms was discussed for the 21st solar cycle only. By comparing the numbers of flares of optical class 2 and more with geomagnetic storms occurrences, it was shown that not all large optical flares produce geomagnetic activity. During the period of the Solar Maximum Year, a special study was made of flares in spotless regions. 32% of these flares could be associated with geomagnetic storms, but these storms were only of moderate or minor intensity, and were of short duration.

Appendix: Analysis of Annual Average A_p Values by the Method of Extremum Distribution.

The series of A_p annual average values is represented by $x_1, x_2, \ldots x_i, x_m$. where the

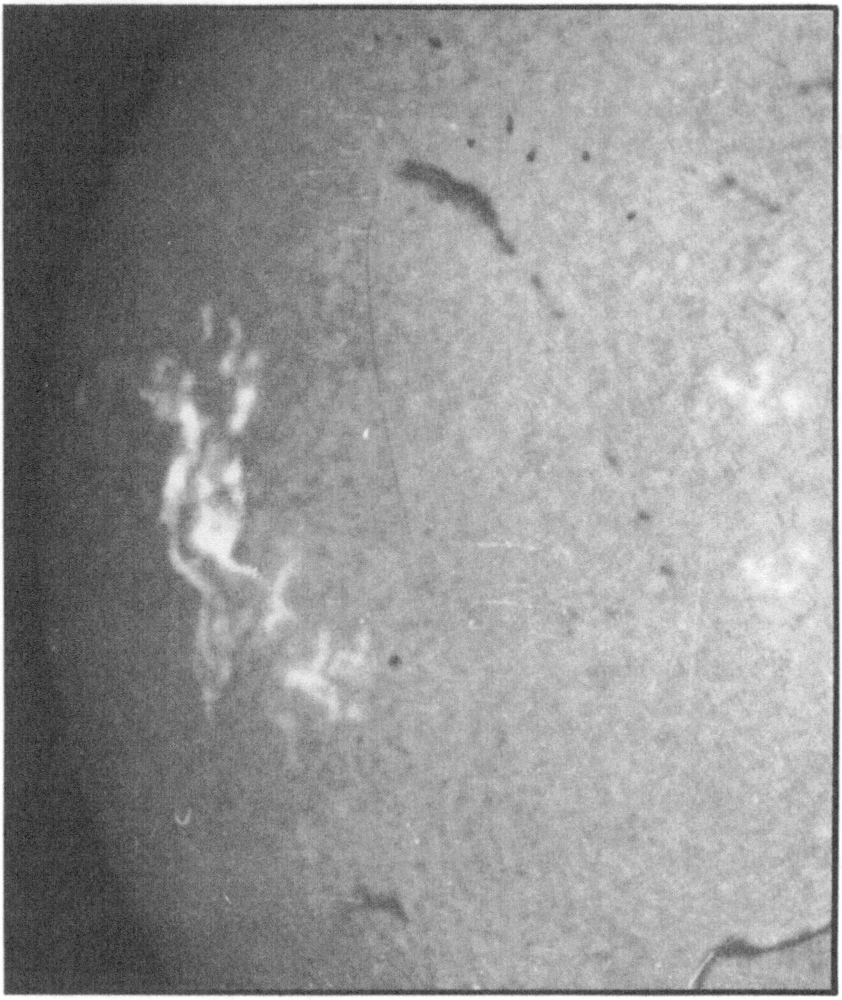

Fig. 5. A 3B spotless two-ribbon flare on October 15, 1980, observed by Beijing Observatory.

values have been arranged so that $x_1 < x_2 < x_i < x_m$. Since it has been determined that the series follows a normal distribution, the series' extremum distribution is

$$G_n(x) = \exp\left\{-e^{-y}\right\} = \exp\left\{-e\,\frac{-x+b_n}{a_n}\right\}$$

$$y = \frac{x-b_n}{a_n} \qquad = -\ln\left(-\ln G_n(x)\right).$$

It can be shown that the y_i for each x_i can be approximated by

$$y_i = -\ln\left(-\ln\frac{i}{m+1}\right).$$

where m is the total number of observations. Then, a_n and b_n are obtained using the least-squares approximation

$$\sum_{i=1}^{m} (x_i - a_n y_i - b_n)^2 = \min$$

$$a_n = \frac{\sum_{i=1}^{m} (x_i - \bar{x})(y_i - \bar{y})}{\sum_{i=1}^{m} (y_i - \bar{y})^2}$$

$$b_n = \bar{x} - a_n\bar{y}$$

where

$$\bar{x} = \frac{1}{m}\sum_{i=1}^{m} x_i \qquad \bar{y} = \frac{1}{m}\sum_{i=1}^{m} y_i.$$

For the data of this paper,

$$a_n = 3.24 \qquad b_n = 12.72$$

$$G_n(x) = 0.98 \qquad y = 3.90$$

and, therefore,

$$x = 25.3$$

where x is the expected maximum A_p.

REFERENCES

CHERNOSKY, Edwin J., Double sunspot-cycle variation in terrestrial magnetic activity. 1884–1963, *J. Geophys. Res.*, **71**, 965, 1966.

DODSON, Helen W., E. Ruth HEDEMAN, and O. C. MOHLER, Comparison of activity in solar cycles 18, 19, and 20, *Rev. Geophys. Space Phys.*, **12**, 329, 1974.

FEYNMAN, Joan, The 21st solar cycle: *aa* index of geomagnetic activity, in *Solar-Terrestrial Predictions Proceedings, Vol. 4*, U.S. Dept. of Commerce, NOAA Environmental Research Laboratories, Boulder, CO, A-8, 1979.

GAO, Meiqing, An estimation of geomagnetic activity maximum in the twenty first solar cycle, *Acta Geophysica Sinica*, **24**, 455, 1981.

SARGENT, H. H. III, A prediction for the next sunspot cycle, in *Conf. Record of the Vehicular Technology Soc., 28th IEEE Vehicular Technology Conference, Mar 22–24, 1978*, Denver, CO, 490, 1978.

STATISTICS GROUP, INSTITUTE OF MATHEMATICS, ACADEMIA SINICA, *Method of Mathematical Statistics*, Publication Bureau of Science, 1973.

2. SOLAR WIND AND MAGNETOSPHERIC PROCESSES

Solar Wind-Magnetosphere Coupling, edited by Y. Kamide and J. A. Slavin, 163–177.

Solar Wind Parameters and Magnetospheric Coupling Studies

Joseph H. KING

National Space Science Data Center, NASA/Goddard Space Flight Center, Greenbelt Maryland 20771, U.S.A.

This paper presents distributions, means, and standard deviations of the fluxes of solar wind protons, momentum, and energy as observed near Earth during the solar quiet and active years 1976 and 1979. Distributions of ratios of energies (Alfven Mach number, plasma beta) and distributions of interplanetary magnetic field orientations are also given. Finally, the uncertainties associated with the use of the libration point orbiting ISEE-3 spacecraft as a solar wind monitor are discussed.

1. Introduction

The solar wind delivers continuous but variable fluxes of plasma, momentum, and energy to the outer boundary of the Earth's magnetosphere. These fluxes couple into the magnetosphere with an effectiveness which depends on some solar wind parameters, most notably the orientation of the interplanetary magnetic field (IMF) relative to the geomagnetic field.

In order to study the solar wind-magnetosphere coupling processes, it is vital to understand how much plasma, momentum, and energy is typically incident on the magnetosphere, and what the IMF orientation typically is. It is also necessary to have time series of solar wind parameters for correlations with time series of magnetospheric response parameters (*AE*, *Dst*, etc.).

The purpose of this paper is twofold. One is to review the distributions of relevant solar wind parameters, for the solar quiet and active years 1976 and 1979, respectively. (A valuable prior review of solar wind parameter values is that of FELDMAN *et al.*, 1977.) The second is to review the availability of near-Earth solar wind data, and in particular to assess the suitability of ISEE-3 data as representative of the solar wind just outside the magnetosphere. (Recall that, for 1978–1982, ISEE-3 was 200–260 R_e upstream of the Earth and up to 100 R_e off the Earth–sun line.)

2. The Parameter Distributions

All the distributions discussed in this section are based on hourly averaged parameter values found in the NSSDC compilation (KING, 1979, 1983). The 1976 data are from IMP's 7 and 8 (magnetometer: N. F. Ness; plasma: S. J. Bame, H. S. Bridge). The 1979 data are from IMP-8 (investigators just given) and ISEE-3 (magnetometer: E. J. Smith; plasma: S.J. Bame). On the whole, distributions based on finer resolution parameter values would be insignificantly different than the distributions tabulated below.

Figure 1 shows proton density distributions. Table 1 lists the means and standard deviations of the proton densities, as well as of parameters to be discussed

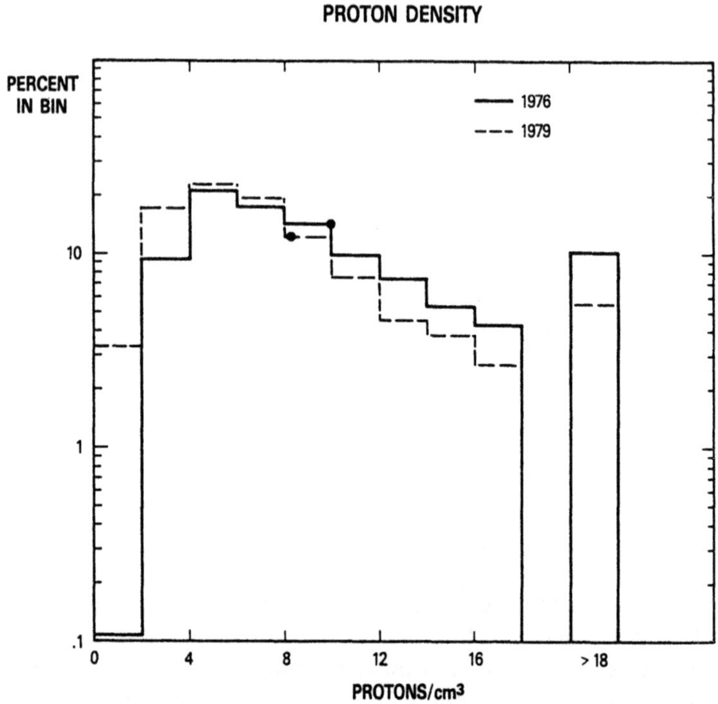

Fig. 1. Histograms of proton density values (N) for 1976 (solid lines) and 1979 (dashed lines). The heavy dots show density mean values. (These means are also listed in Table 1, along with associated standard deviations.) There are approximately 6100 and 8500 hourly values contributing to the 1976 and 1979 histograms, repectively.

Table 1. Solar wind parameter mean values.

	1971–1974[1]	1976	1979
Density (cm^{-3})	8.7±6.6	10.0±6.4	8.2±6.4
Flow speed (km/s)	468±116	445±101	419±82
Proton number flux ($cm^2\ s^{-1}$)	$(3.8\pm2.4)\cdot10^8$	$(4.1\pm2.4)\cdot10^8$	$(3.3\pm2.4)\cdot10^8$
Proton momentum flux ($g/cm\ s^2$)	$(2.9\pm1.8)\cdot10^{-8}$	$(3.0\pm1.8)\cdot10^{-8}$	$(2.3\pm1.8)\cdot10^{-8}$
Proton flow energy flux ($erg/cm^2\ s$)	.70±.52	.68±.54	.49±.48
Proton thermal energy flux ($ergs/cm^2\ s$)	8.9×10^{-3}	$(6.8\pm7.6\cdot10^{-3}$	$(4.6\pm8.4)\cdot10^{-3}$
Magnetic energy flux ($ergs/cm^2\ s$)	8.0×10^{-3}	$(6.0\pm6.2)\cdot10^{-3}$	$(11.9\pm21.3)\cdot10^{-3}$
Alfven mach number	10.7±4.8	12.5±6.1	7.5±3.5

[1] Taken from FELDMAN *et al.*, 1977.

subsequently. Table 1 also contains data from FELDMAN *et al.* (1977) which cover the 1971– 1974 period of declining solar activity. Recall that the years 1973–1975 were characterized by well established high speed solar wind streams. Note also that the solar wind parameters were normalized to the Los Alamos IMP plasma parameters in generating the NSSDC compilation so that the three columns of Table 1 are directly comparable.

Note that densities tend to be greater near solar minimum. Alpha particle densities are typically 3–5% of proton densities, and occasionally reach 10–20% in the driver gases behind interplanetary shocks (BORRINI *et al.*, 1982). Electron densities are determined by solar wind charge neutrality.

To consider the flux of solar wind protons, NV, we next present the solar wind speed distributions in Fig. 2. For both quiet and solar active times, the most probable solar wind speed is in the 350–400 km/s bin. (For 1971–74, the most probable speed was 375 km/s, according to FELDMAN *et al.*, 1977). However, the high speed streams of the solar quiet period contribute a significant high speed tail on the 1976 distribution.

Figure 3 gives the distributions of proton number fluxes. Note that there is approximately 25% more mass efflux from the sun in 1976 than in 1979, at least in the ecliptic plane where all our data were obtained.

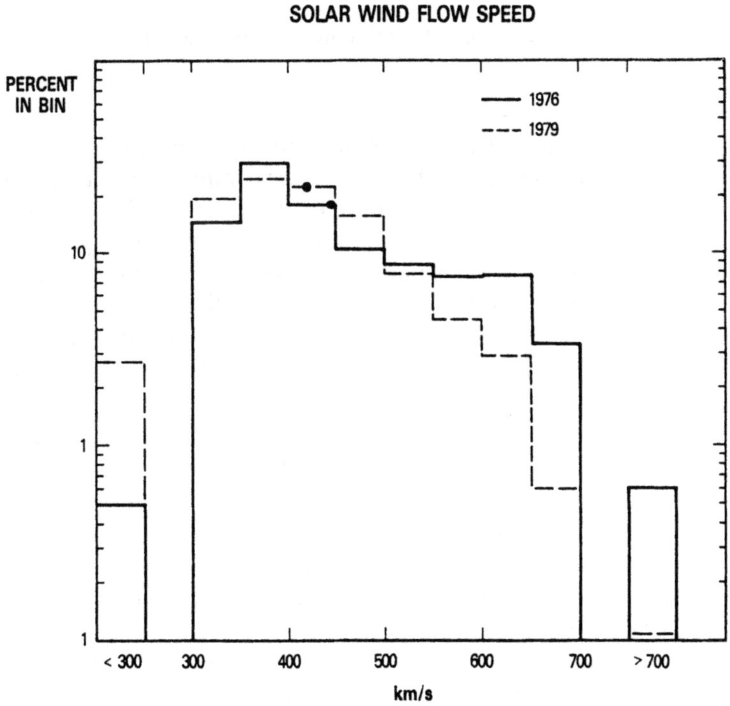

Fig. 2. Same as Fig. 1, but for flow speed (*V*).

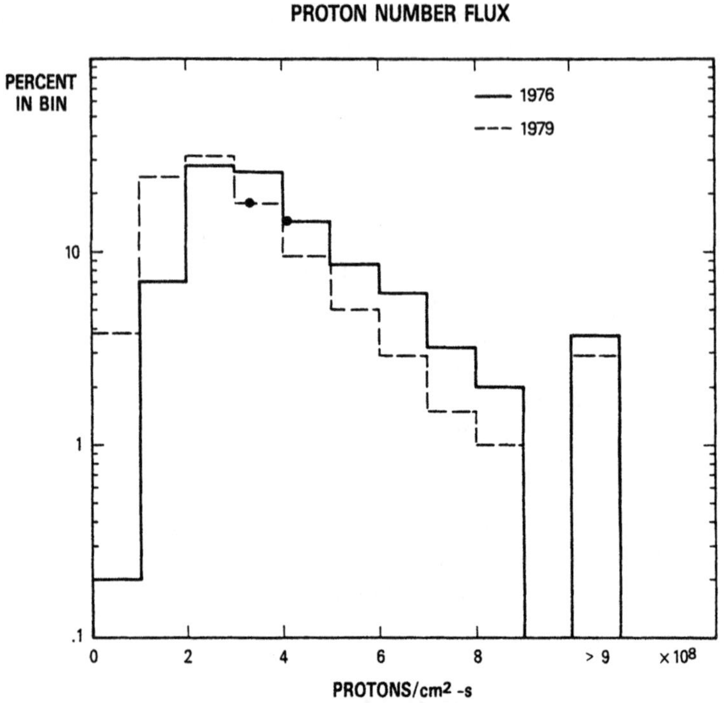

Fig. 3. Same as Fig. 1, but for proton number flux (NV).

The solar wind momentum flux, which determines the magnetosphere scale by pressure balance, is primarily given by NmV^2. The alpha particle contribution typically adds another ~15% and, in shock driver gases, ~50% or more. The distribution of solar wind (proton) momentum flux is given in Fig. 4. Note that momentum flux ((g cm/s)/(cm s)) has units of energy density (ergs/cm^3).

The solar wind energy flux is the sum of the fluxes of the bulk flow, thermal, and magnetic energies. Figures 5, 6, and 7 show the distributions of proton flow and thermal energy fluxes ($.5NmV^3$, $1.5NkTV$) and of magnetic energy flux ($VB^2/8\pi$). Note that the flow energy is typically two orders of magnitude greater than the proton thermal or magnetic energies. Note also that the proton thermal and magnetic energies are comparable to each other in 1976, but that the latter of these exceeds the former by a factor of 4 in 1979 when the magnetic field is enhanced and the density is down. Note that electron thermal energy flux, not otherwise discussed herein, is frequently greater than the corresponding proton flux. This is so because, although proton and electron densities are comparable, the electron temperature is usually in the 1–2 10^5 deg K range while proton temperatures have much greater variability and are less than 10^5 deg K 70–80% of the time. (The average proton temperature was 8.7 and $6.9*10^4$ deg K in 1976 and 1979 respectively; it was $1.2*10^5$ deg K in 1971–74, according to Feldman. *et al.*) The effect of electrons in the total plasma pressure is

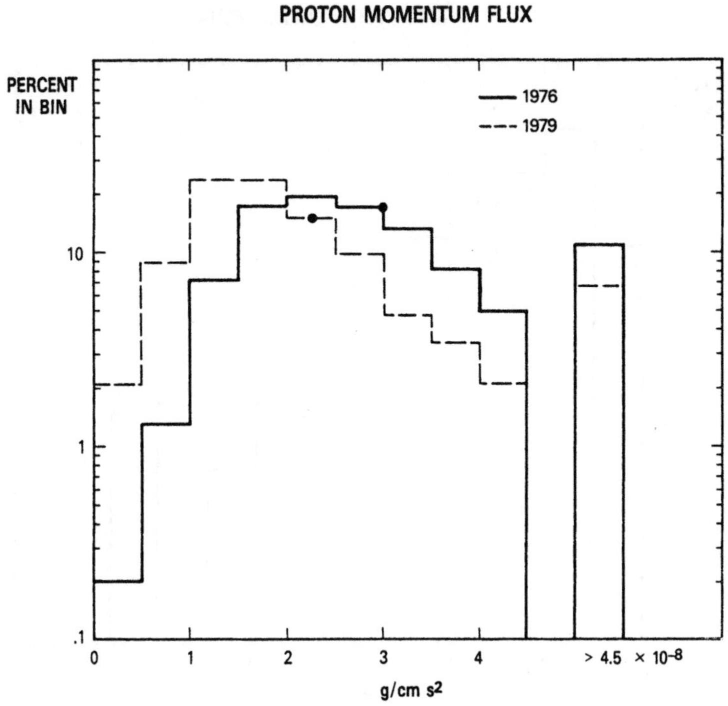

Fig. 4. Same as Fig. 1, but for proton momentum flux (NmV^2).

discussed below in considering beta.

The ratio of the flow energy density to the magnetic energy density is the square of the Alfven Mach number, M_a. The distribution of M_a values is given in Table 2. Note that higher Mach numbers are encountered during solar quiet conditions. Anomalously low M_a values, including <1.0, have been encountered during solar active times, typically in conjunction with rarefaction regions behind fast transient flows. Subalfvenic flow observations were previously discussed by GOSLING *et al.* (1982). Low densities yield large magnetopause distance. Low Mach numbers yield large bow shock standoff distances.

The ratio of plasma pressure to magnetic pressure, usually called beta, determines whether plasma or field dynamic effects will be dominant in the reference frame convecting with the solar wind. Table 3 shows the distribution of plasma beta values, where the plasma pressure has been taken to be $Nk\ (1.16T+1.08\times T_e)$, assuming $N(=N_p)/N\alpha=25$, $T(=T_p)/T\alpha=0.25$, and $T_e=1.5\times10^5$.

Note from Table 1 that the Feldman *et al.* 1971–1974 values are typically intermediate to the solar minimum 1976 values and the solar active 1979 values. The exception to this is the flow speed dominated parameters, which are greatest in the corotating stream dominated years (~1973–1975) of declining solar activity.

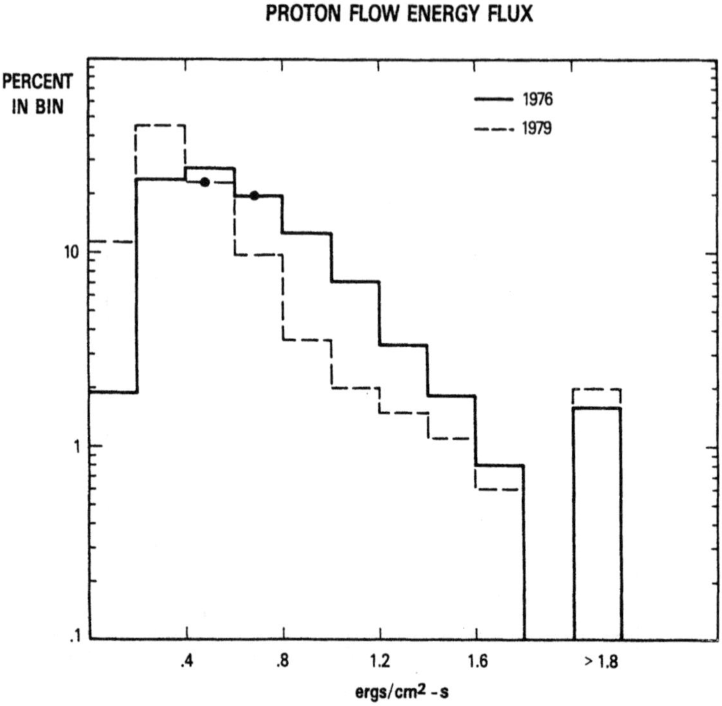

Fig. 5. Same as Fig. 1, but for flux of proton flow energy (1.5*NmV**3).

3. IMF Orientation

Plasma, momentum, and energy couple to the magnetosphere by processes which are both viscous and electrodynamic in nature. The electrodynamic processes, believed to be dominant at least for energy and plasma coupling, have an effectiveness which depends primarily on the orientation of the interplanetary magnetic field relative to the geomagnetic field.

Figure 8 shows the distribution of the projections of 22,000 1976–1982 hourly IMF vectors onto the solar ecliptic and solar magnetospheric Y–Z planes. Note that 10% (20%) of all field vectors are within 15 (30) deg of the Z axis. Given the north–south symmetry which exists, this means that 10% of all hours have field vectors which are within 30 deg of the south Z axis. Note also that only about 25% of all vectors are within 15 deg of the ecliptic plane, where the original solar wind model (PARKER, 1958) says all IMF vectors should lie.

4. ISEE-3 Data Utility

For the greater part of the past decade, the primary source of near Earth solar wind field and plasma data was IMP-8; however, this spacecraft, launched in October

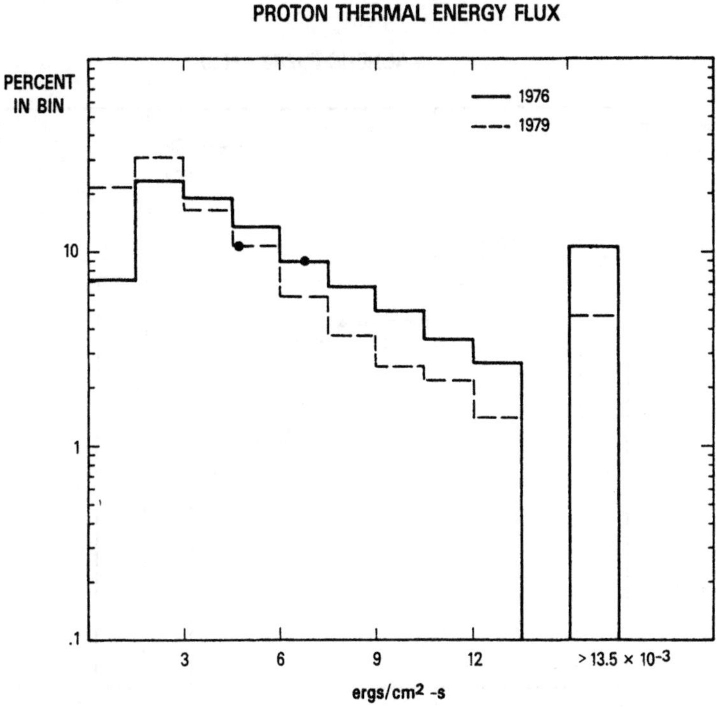

Fig. 6. Same as Fig. 1, but for flux of proton thermal energy (1.5*$NkTV$).

1973, is out of the solar wind for about 5 days of every 12.5 day orbit. Further, data coverage, which may continue for years yet, is significantly less than complete. The key alternative source of near Earth solar wind data is ISEE-3, which for most of the period between August 1978 and December 1982 was in a 6-month libration point orbit 200-260 R_e upstream of the Earth and up to 100 R_e east/west of the Earth–sun line. After this time ISEE-3 spent most of 1983 in the Earth's magnetotail, and then was renamed ICE and dispatched to a September 11, 1985, encounter with comet Giacobini-Zinner. Additional sources of near-Earth solar wind data over the past decade include IMP 7 (plasma only; operations ended in 1978) and ISEE-1 and −2 (eccentric orbits; most solar wind coverage in each August–November since 1977 launch; decay from orbit expected October 1987).

ISEE-3 provided reasonably continuous solar wind coverage for 4.5 years, and has provided the solar wind input function in many solar wind-magnetosphere coupling analyses. However, owing to its ~1-hour upstream location, its limits as a solar wind monitor merit examination. Two questions are relevant. First, to what extent will profiles seen at ISEE be repeated at Earth? This extent will be limited if evolution occurs during the ~1-hour transit time or if convecting spatial gradients and a large transverse separation between ISEE and Earth yield differing plasma domains at ISEE and Earth. Secondly, can we predict with confidence when a feature

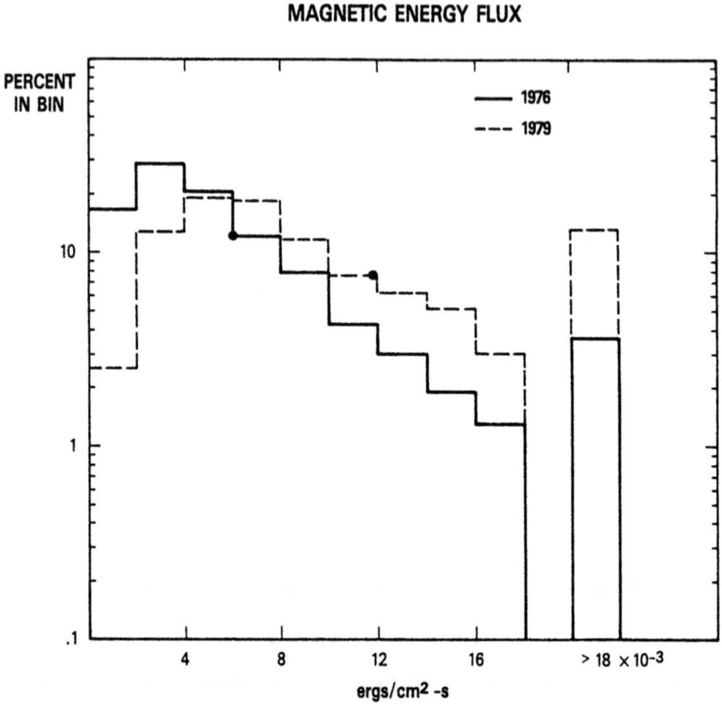

Fig. 7. Same as Fig. 1, but for magnetic energy flux ($VB^2/8\pi$). This parameter is based on about 4700 and 6500 hours for 1976 and 1979, repectively.

Table 2. Alfven Mach number.

Alfven mach No.	1976	1979
	% in bin	
0–.5	0	.02 (1 case)
.5–1	0	.06
1–2	0	.24
2–4	1.1	5.7
4–8	12.4	62.6
8–16	70.9	29.2
16–32	13.9	1.9
32–62	1.6	.23
64–128	0.1	.02
Means	12.5±6.1	7.5±3.5
No. hours	4735	6533

Table 3. Plasma beta.

Beta	1976	1979
	% in bin	
<2	0.1	1.9
.20–.30	0.3	2.9
.30–.44	0.7	6.5
.44–.67	1.6	12.5
.67–1.0	4.2	23.9
1.0–1.5	12.5	24.0
1.5–2.25	22.1	14.9
2.25–3.38	23.5	6.8
3.38–5.06	16.0	3.3
>5.06	19.1	3.0
Means	4.3±6.8	1.5±3.6
No. hours	4735	6533

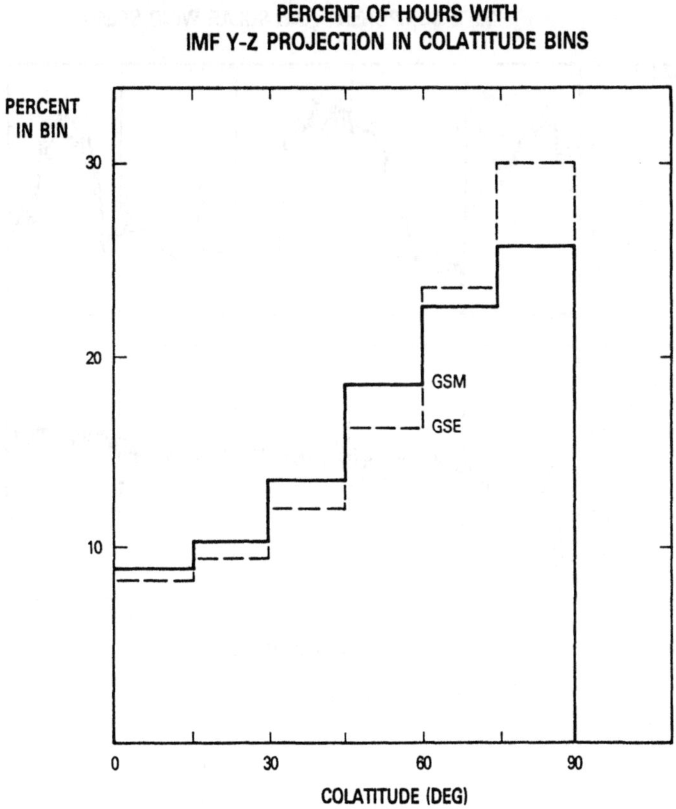

Fig. 8. Histograms of projections of 22,000 1976–1982 hourly IMF vectors onto the solar ecliptic and solar magnetospheric Y–Z planes. Using symmetry of the distributions about the Y and Z axes, all data have been folded into one quadrant.

(e.g., a discontinuity) seen at ISEE will arrive at Earth?

ISEE-3 is an effective solar wind monitor for magnetosphere studies at least some of the time. For instance, Figure 9 illustrates IMP-8 and ISEE-3 flow speed (V) and IMF Z-component for the disturbed flow whose effects were intensively studied in the CDAW-6 process (e.g., TSURUTANI *et al.*, 1984, 1985). Note that the ISEE-3 data have been time shifted by 38 minutes to align the interplanetary shock observations. The general profiles agree quite well; however, some features had ISEE-to-IMP delays longer or shorter, by up to 11 minutes, than the ISEE-to-IMP shock delay. This is a reflection of the fact that individual features have variously oriented phase fronts. Given only an ISEE-3 observation of a discontinuity, the normal must be determined from the data to predict the ISEE-to-Earth delay.

Analyses of the level of agreement between ISEE-3 and ISEE-1 (near Earth) magnetic field observations have been made by RUSSELL *et al.* (1980) and CROOKER *et al.* (1982). These authors considered levels of cross correlation between ISEE-3 and ISEE-1 time series for large numbers of 2 or 3 hour intervals. In the latter paper, it was

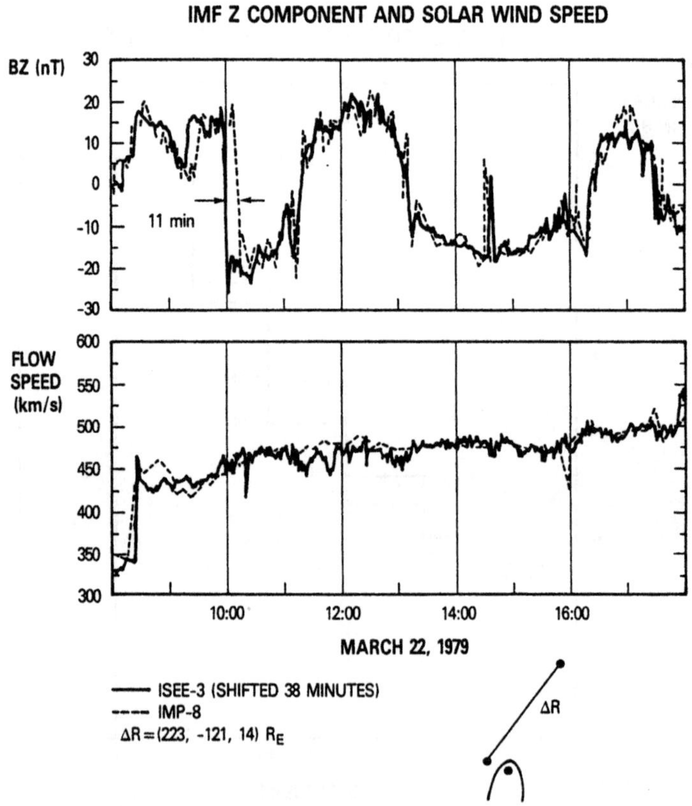

Fig. 9. ISEE-3 and IMP-8 IMF Z component (GSM) and flow speed for ten hours on March 22, 1979. The ISEE-3 data have been plotted at times 38 minutes later than observation times. The cartoon at lower right shows schematically the relative locations of the spacecraft.

shown that the value of ISEE-3 as a solar wind monitor is degraded at the $\sim 100\ R_e$ east/west extremes of the orbit.

In the generation of KING (1983), 5 min averaged values of ISEE-3 field and plasma parameters were created and were shifted to Earth by assuming corotation delay (interplanetary features aligned with ideal IMF spiral) and separately by assuming convection delay (feature normals parallel to a radius vector from sun). Such values were then compared to IMP-8 observations (5-min averages). One such comparison is shown in scatter plot form in Fig. 10. This is for the Z component of the IMF, for a 4-week period when ISEE-3 was within 50 R_e of the Earth–sun line;

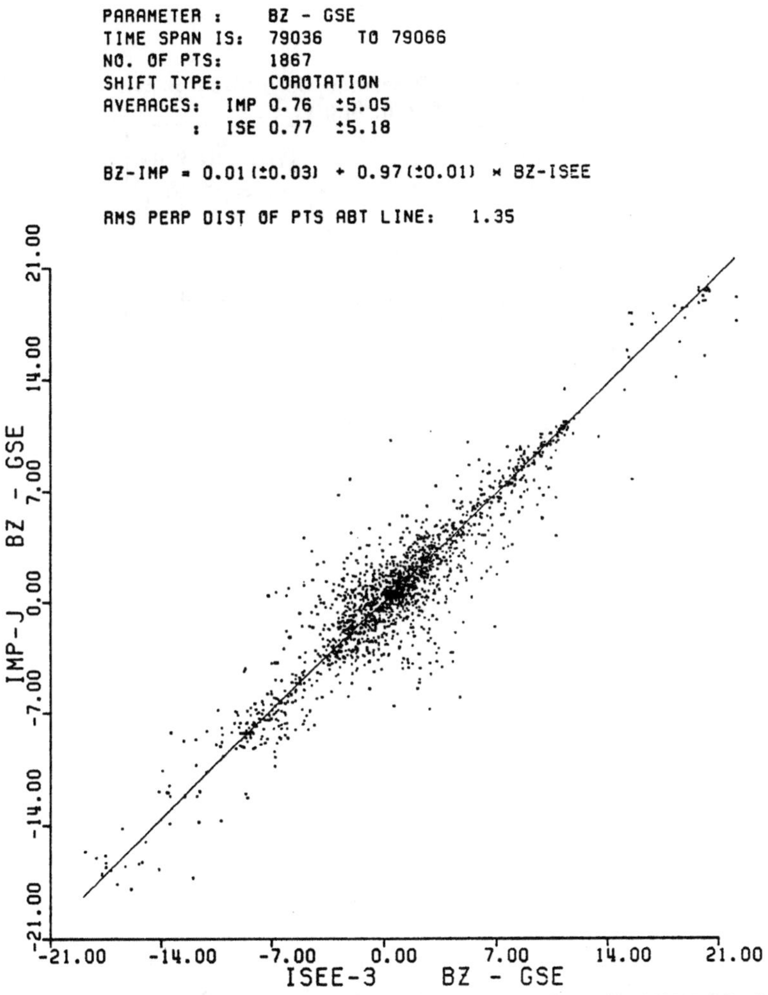

Fig. 10. Scatter plot and best fit regression line for IMP-8 and corotation-shifted ISEE-3 5-min B_z (GSE) values; 1867 points between February 6 and March 7, 1979 are given. The rms perpendicular distance between the data points and the regression line is 1.35 nT.

corotation delay was used. The RMS perpendicular distance of the data points to the best fit regression line gives a measure of the irreducible difference between the data sets, due largely to the spatial separation of the spacecraft and/or to the inappropriateness of the simple corotation shift.

Figure 11 summarizes the levels of agreement between data sets for various parameters, for cases of large and small distances of ISEE-3 from the sun–Earth line; all ISEE-3 data were corotation shifted. Note that the dependence on the location of ISEE-3 relative to the sun-Earth line is greatest in the IMF components, is apparent in the IMF intensity, and is not apparent in the plasma parameters. This reflects smaller scale size for IMF variability than for plasma parameter variability. The east/west extent of the ISEE-3 orbit does not diminish the values of ISEE-3 as a predictor of

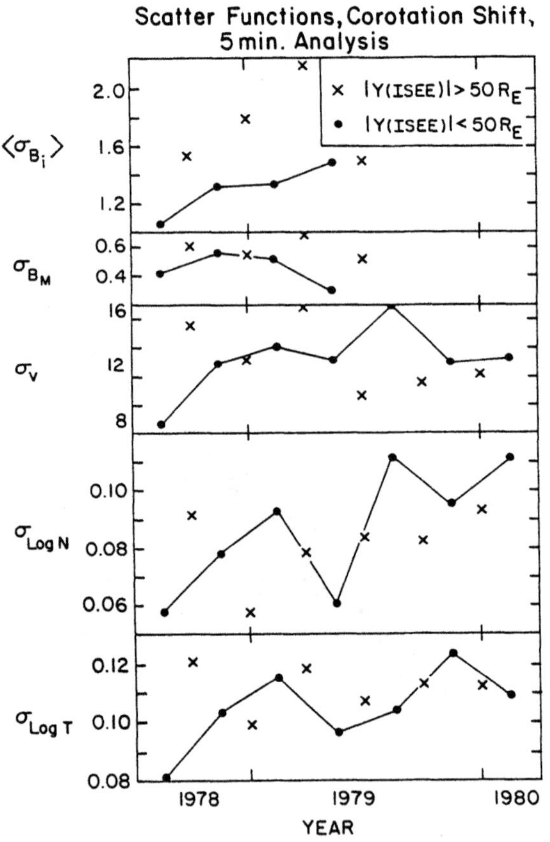

Fig. 11. Scatter functions for each of several solar wind parameters (indicated by the subscripts on the σ's at left; B_M refers to IMF intensity), for each of several time intervals characterized by the separation between ISEE-3 and the sun–Earth line. Scatter functions are defined as the rms perpendicular distances such as that discussed in the Fig. 10 caption. The points corresponding to the smaller separation have been joined by a line to guide the eye. Note that the top panel represents averages over scatter functions obtained for individual Cartesian IMF components.

plasma parameters at Earth, although it does for the IMF.

The σ values plotted in Fig. 11 give the 1-σ perpendicular distances between the data points and the regression line. These σ values must be multiplied by $\sqrt{2}$ to find the 1-σ uncertainty in a predicted Earth value (ordinate) from a given ISEE-3 observation. Thus, we assert from Fig. 11 that there is a 1-σ (~70%) probability that the 5-min averaged value of an IMF component at Earth will be within about 2 (2.5) nT of the average measured for that component at ISEE at a time earlier by the ideal corotation delay time—if ISEE is within (beyond) ~50 R_e of the Earth–sun line. For the field intensity and solar wind speed, the corresponding uncertainties are 0.7 nT (<50 R_e), 0.85 nT (>50 R_e), and 18 km/s (independent of ISEE orbit phase). Also there is a 70% probability that the 5-minute Earth value of density (temperature) will differ from the corotation shifted ISEE value by less than 30% (45%).

Figure 12 presents an even more comprehensive summary of the analysis. Here we compare the effects of corotational shifting to shifting by convection and to not shifting at all. We also consider the effects of changing the time resolution of the ISEE-IMP comparison from 5 to 60 minutes. (ISEE hourly averages were constructed for Earth hours from corotation-shifted 5-min ISEE values). Cases of large and small ISEE-3 separation are distinguished.

The first "column" of Fig. 12 reveals that there is no major, statistically significant difference between using convection vs. corotation shifting. This implies that the distribution of phase normals for interplanetary variations is not sharply peaked at right angles to the ideal IMF spiral direction nor along the helioradial

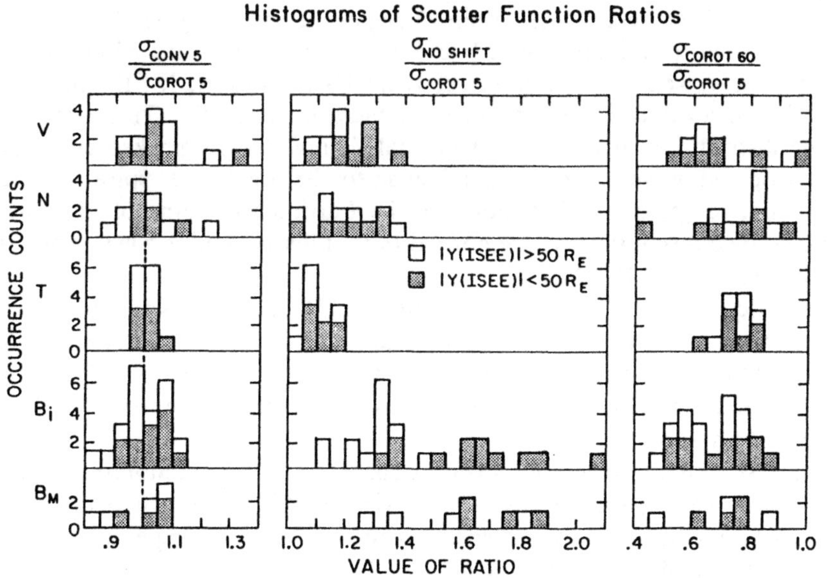

Fig. 12. Histograms of ratios of scatter functions obtained with differing approaches to ISEE-3 time-shifting and for differing time resolutions. See text for discussion.

direction.

The middle "column" of Fig. 12 shows that time shifting is meaningful, in that it always yields ISEE-IMP variances reduced relative to no-shift. Note that, particularly for the IMF parameters, time shifting makes the most difference for the cases of small ISEE-3/sun–Earth line separation.

The right "column" reflects smaller variances when considering hourly resolution data than 5-min data. This is reasonable as many errors introduced by using the wrong time shift at high resolution average out at hourly resolution. Unfortunately, this result is not relevant to many solar wind-magnetosphere studies which require data at ~5-min resolution or greater.

5. Summary

It has been shown that during the years approaching (1971–74) and at (1976) solar minimum, the proton number, momentum, and energy fluxes are comparable to each other (with values $4{\cdot}10^8/\mathrm{cm}^2\mathrm{s}$, $3{\cdot}10^{-8}\mathrm{g/cm\ s}^2$, and 0.7 ergs/$\mathrm{cm}^2$s, respectively) and are greater than corresponding solar maximum (1979) values by 20–40%. Mean values of both solar wind speed and density are lower during solar active time. On the other hand, the magnetic field intensity is greater at solar maximum; the solar-maximum magnetic energy flux (~.012 ergs/cm^2s) was twice that of solar minimum. The histograms shown in Figures 1-7 show however, that there is significant overlap in the ranges or parameters values for solar active and quiet periods.

About 10% of hourly averaged IMF vectors are within 30% of being due southward (in solar magnetospheric coordinates). Such orientation is especially conducive to transfer of solar wind plasma and energy to the magnetosphere. Only 25% of hourly IMF vectors are within 15 deg of the solar ecliptic plane.

In considering the merit of ISEE-3 as a monitor of solar wind input to the magnetosphere, we showed that on a statistical basis, shifting ISEE-3 profiles using convection or corotation yielded the same relative uncertainties. This reflects the fact that the distribution of phase front normals is not sharply peaked either along a heliocentric radius vector or perpendicular to the ideal IMF spiral direction. It follows that to predict with confidence the Earth-arrival time of any particular ISEE-observed discontinuity, the normal to that discontinuity must be determined from the detailed ISEE data.

The effectiveness of ISEE-3 as a monitor of IMF conditions degrades somewhat as ISEE-3 moves away from the Earth–sun line, as had been noted in previous analyses cited. The present analysis shows there is no difference in ISEE's ability to predict plasma parameter values according to whether ISEE is more or less than 50 R_e from the Earth–sun line.

An upstream libration point orbiting spacecraft is a good monitor of magnetospheric input. The main difficulty is that it is impossible to know, *with certainty*, that any given parameter profile observed upstream was, or will be, repeated at the Earth's magnetosphere and with what time delay.

REFERENCES

BORRINI, G., J. T. GOSLING, S. J. BAME, and W. C. FELDMAN, Helium abundance enhancements in the solar wind, *J. Geophys. Res.*, **87**, 7370, 1982.

CROOKER, N. U., G. L. SISCOE, C. T. RUSSELL, and E. J. SMITH, Factors controlling degree of correlation between ISEE-1 and ISEE-3 interplanetary magnetic field measurements, *J. Geophys. Res.*, **87**, 2224, 1982.

FELDMAN, W. C., J. R. ASBRIDGE, S. J. BAME, and J. T. GOSLING, Plasma and magnetic fields from the sun, in *The Solar Output and Its Variation*, edited by O. R. White, pp. 351, Colorado Associated University Press, 1977.

GOSLING, J. T., J. R. ASBRIDGE, S. J. BAME, W. C. FELDMAN, R. D. ZWICKL, G. PASCHMANN, N. SCKOPKE, and C. T. RUSSELL, A Sub-Alfvenic solar wind: interplanetary and magnetosheath observations, *J. Geophys. Res.*, **87**, 239, 1982.

KING, J. H., *Interplanetary Medium Data Book*, Supplement 1, NSSDC/79-08, 1979.

KING, J. H., *Interplanetary Medium Data Book*, Supplement 2, NSSDC/83-01, 1983.

PARKER, E. N., Dynamics of the interplanetary gas and magnetic fields, *Astrophys. J.*, **128**, 664, 1958.

RUSSELL, C. T., G. L. SISCOE, and E. J. SMITH, Comparison of ISEE-1 and 3 interplanetary magnetic field observations, *Geophys. Res. Lett.*, **7**, 381, 1980.

TSURUTANI, B. T., C. T. RUSSELL, J. H. KING, R. D. ZWICKL, and R. P. LIN, A kinky, heliospheric current sheet: Cause of the CDAW-6 substorms, *Geophys. Res. Lett.*, **11**, 339, 1984.

TSURUTANI, B. T., J. A. SLAVIN, Y. KAMIDE, R. D. ZWICKL, J. H. KING, and C. T. RUSSELL, Coupling between the solar wind and the magnetosphere: CDAW 6, *J. Geophys. Res.*, **90**, 1191, 1985.

Solar Wind-Magnetosphere Coupling, edited by Y. Kamide and J. A. Slavin, 179–190.

Solar Cycle Variations in the Solar Wind

John W. FREEMAN and Ramon E. LOPEZ

Department of Space Physics and Astronomy, Rice University, Houston, Texas 77251, U.S.A.

The purpose of this paper is to review the known solar cycle variations of the common solar wind parameters: solar wind velocity, temperature, density and the components of the IMF, and to look at the variations of several other parameters of particular interest to magnetospheric physicists including the IMF magnetic energy density, beta, mach number, energy flux, epsilon and the *y*-component of the interplanetary electric field. To facilitate this we employ a color graphic format that allows the visualization of six years of hourly averages of HELIOS-1 data. By way of new results, we show that high speed recurrent streams are seen with decreasing duration throughout the declining phase of solar cycle 20 and the ascending phase of cycle 21 up to solar maximum, and that the yearly average of the proton velocity continues to decrease toward solar maximum. High speed streams can clearly be seen in Akasofu's epsilon parameter and in the *y* component of the interplanetary electric field. Also, we find that beta, the ratio of the proton thermal pressure to magnetic pressure, shows a significant solar cycle variation. The individual hourly averages of beta often exceed unity with twenty cases exceeding 10 and one case as high as 25.

1. Introduction

FEYNMAN (1983), HUNDHAUSEN (1979), BRIDGE (1976), NEUGEBAUER (1975), and DIODATO (1974) have all reviewed solar cycle variations in the solar wind plasma parameters. In general, this previous work has focused on long period averages or short intervals of time-sequence data.

It is often important, however, for the magnetospheric physicist to understand the change in character of the solar wind throughout the solar cycle. This understanding involves a knowledge of the occurrence frequency of the various phenomena such as high speed streams or transient events. This can be obtained by a graphic picture of the distribution of the phenomena in time. A way is needed to compress data on a large fraction of a solar cycle on to a single frame so that the variations or the locations of different types of events are easily determined. We have attempted to develop such a format and to apply it to the HELIOS 1 data.

SHEELEY *et al.* (1976) and SHEELEY and HARVEY (1978, 1981) used color plots to study solar cycle variations in the solar wind parameters and their relations to solar surface features and geomagnetic disturbances. They used the Bartels display which assumes a 27 day solar rotation period and they used daily averages of the solar wind speed and IMF sector structure. They found solar cycle variations in high speed streams and the sector structure which usually correlated well with the existence of coronal holes when the holes were in the same hemisphere as the earth. They were also able to discern a shift in the recurrence pattern of coronal holes and their associated IMF sector structure from a 27 day pattern to a 28.5 day pattern in 1977. The one-day time resolution in their data, however, prevented them from studying solar cycle variations in shorter term features such as transient events.

KING (1979, 1981) studied solar cycle variations in the IMF and solar wind speed using yearly averages. He found a significant decrease in log B near the 1976 solar minimum. SLAVIN *et al.* (1983) have confirmed and extended this work. Their work shows a continued rise in log *B* toward solar maximum but with an interesting dip at the time of solar field polarity reversal, again based on annual averages.

KING (1981) also provided the first graph of the yearly averages of the solar wind velocities. This showed a large peak during the declining phase of solar cycle 20 associated with stable high speed streams at that time. A smaller peak was seen in 1968.

As indicated above, this report presents long duration solar wind variations with high time resolution. We have applied color graphics techniques to the hourly averages of the HELIOS 1 merged solar wind and magnetometer data sets. We have chosen to use the Carrington rotation period (27.2753 days) rather than the Bartels period. The color graphs provide a qualitative visual representation of the time period from 1974 to 1980. A more quantitative measure is provided by computing averages over each Carrington rotation.

The basic color plot scheme is similar to the Bartels display employed by SHEELEY *et al.* (1976).

2. Color Plot Format

The data used are one hour averages of the HELIOS solar wind and IMF parameters generously provided to the National Space Science Data Center by Rainer Schwenn and Fritz Neubauer.

Plate 1 illustrates the data format. The solar wind proton velocity has been divided into ten velocity ranges. Each range is represented by a color and the pixel representing each hourly average data point is colored accordingly. The color-velocity scale is shown at the top of the figure. Each horizontal line of data represents one Carrington rotation of the sun as seen from the HELIOS spacecraft with the Carrington longitude of the spacecraft shown at the bottom of the figure. This longitude is based on the time at the start of the hour average interval and in this figure has not been corrected for the solar wind travel time to the spacecraft.

The horizontal lines are labeled at the left of the figure by the Carrington rotation number as seen by an earthbased observer. Because the HELIOS spacecraft are in approximately six month heliocentric orbits, the data do not necessarily progress in time order from right to left and successively lower lines. The spacecraft and the Earth do not pass through the zero of Carrington longitude at the same time, hence the data may jump from one line to the lower one at a point other than the end of the line.

The last two digits of the year are also shown in the left margin. The 'A' and 'P's indicate aphelion and perhelion passage.

3. Solar Wind Velocity

Plate 1 displays 80 solar rotations or about six years of HELIOS 1 proton velocity data. The changing character of the solar wind can clearly be seen in this figure. Two large high speed stream regions are seen toward the end of solar cycle 20 in the longitude regions 0 to 110 degrees and 170 to 270 degrees. These, of course, are associated with the coronal holes discovered during the Skylab era (HUNDHAUSEN, 1979). No comparable high speed streams are seen from solar minimum (1976) until the end of the data period in late 1980 (solar maximum occurs in 1980). Rather, during this period of increasing solar activity, the solar wind is characterized by shorter duration streams and transient events.

There is a continuous decrease in the duration (in solar rotations) of high speed streams lasting more than one rotation. From mid 1979 through 1980 there is possibly only one high speed stream of duration longer than two solar rotations. There are, however, occasional transient events for which the solar wind velocity exceeds 700 km/s and which are seen on only one solar rotation. Enhancement of the proton temperature (Plate 2) indicates several of these may be flare associated transient events. The dominant pattern, however, is a decrease in the duration of the high speed streams across the data interval from solar minimum in 1976 through solar maximum in 1980. This may be consistent with the pattern for solar cycle 20 of high speed stream activity during the declining phase giving way to transient events during the ascending phase as summarized by HUNDHAUSEN (1979); however, we wish to stress that the recurrent streams extend well into the ascending phase. For example, a stream with duration as long as six solar rotations is seen as late as 1978 or early 1979. At this time the sunspot number is in the middle of its ascent toward maximum in 1980.

Plate 1 suggests that the solar wind velocity averaged over a solar rotation should be higher during the period of extensive high speed streams, i.e., before mid 1976. Figure 1 shows the results of computing velocity averages over each Carrington rotation during the HELIOS 1 data period. Shown for comparison are the KING (1981) annual averages and annual averages from the HELIOS rotation averages.

There are two noteworthy features in Fig. 1. First, there are excursions in individual rotation averages that are quite large. These occur rather uniformly across the solar cycle and amount to a 10% or so excursion.

The second and new result is that solar wind velocity continues to fall toward solar maximum in 1979 and 1980. This is most easily seen in the yearly averages from the HELIOS data. The rising trend started one year after solar minimum is reversed so that by solar maximum the velocity average is 15% lower than at solar minimum.

4. Temperature

Plate 2 shows the hourly averages of the proton temperatures from HELIOS 1. Comparing this with the velocities in Plate 1, we see that temperature increases are associated with velocity increases in every type of velocity event.

Particularly striking is the temperature increase associated with the high speed streams in 1974, 1975, and 1976. The streams are sufficiently well defined in the

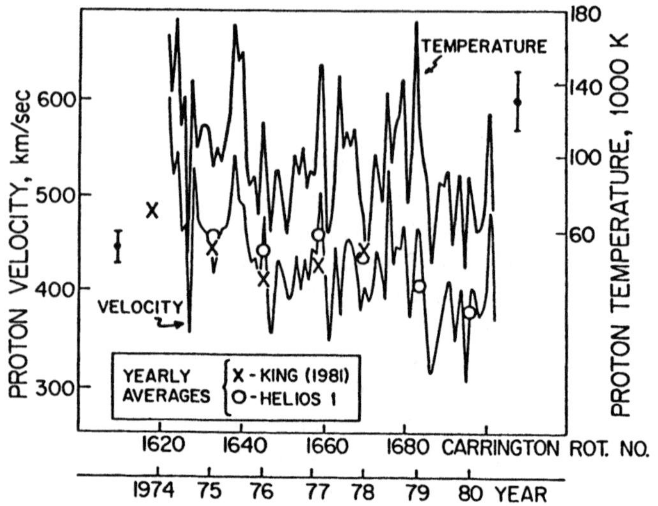

Fig. 1. Carrington rotations averages for the HELIOS-1 proton temperature and velocity data. Also shown are the yearly averages of the velocity from the HELIOS-1 Carrington rotation averages and those from the work of KING (1981). The isolated error bars represent typical errors in the temperature and velocity.

temperature data that they could be identified in temperature alone. The stream temperatures tend to range between 3×10^5 and 6×10^5 deg. K. The temperature of the general quiet or low speed solar wind ranges from 5×10^3 up to roughly 3×10^5 deg. K. FREEMAN and LOPEZ (1985) have shown that up to 10% of the time the solar wind temperature is below 15×10^3 deg. K.

Figure 1 also shows the temperature averages over each Carrington rotation. These data have been normalized to 1 AU using the velocity dependent values of the power of the *r* given by SCHWENN (1983). This correction has not been applied to the data in Plate 2 since the course temperature-color scale does not warrant it. The existence of a correlation between solar wind proton temperature and velocity is well known (NEUGEBAUER and SNYDER, 1966; HUNDHAUSEN *et al.*, 1970; BURLAGA and OGILVIE, 1970; BURLAGA and OGILVIE, 1973; PIZZO *et al.*, 1973; LOPEZ and FREEMAN, 1986). In Fig. 1, the velocity and temperature rotation average curves show the expected good correlation on a rotation by rotation basis and also over the entire data interval.

5. Density

Plate 3 shows the proton density data from HELIOS 1. Comparison with Plate 1 shows the well known general anticorrelation between proton density and velocity (NEUGEBAUER and SNYDER, 1966; HUNDHAUSEN *et al.*, 1970; PIZZO *et al.*, 1973). The high speed streams can easily be identified by the reduced number density in the midst of the streams and the enhanced density at the leading edge of the streams

(NEUGEBAUER and SNYDER, 1966; GOSLING *et al.*, 1972; PIZZO *et al.*, 1973; HUNDHAUSEN, 1979). Note that there are occasional density enhancements that do not appear to be associated with enhancements or decreases in either the proton velocity or temperature. These include the enhancements on rotations 1696, 1697, 1698, and 1699 near longitude 170° and on rotations 1694, 1695, 1697, 1698, 1699, 1670 near longitude 260°.

Figure 2 shows the rotation averages for the proton density along with rotation averages for the proton temperature. The long duration and rotation by rotation

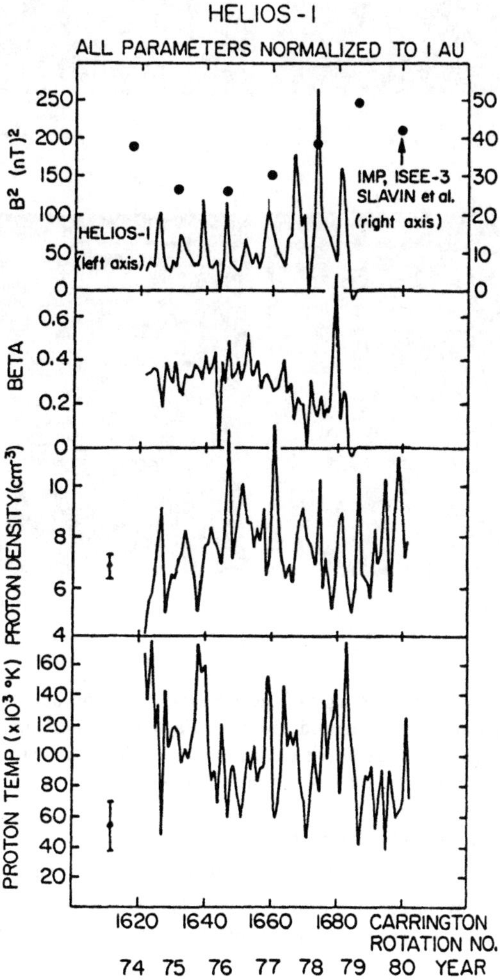

Fig. 2. Carrington rotation averages of the HELIOS-1 proton number density and temperature (both normalized to 1 AU.) and beta. Also shown for comparison with beta are the Carrington rotation averages of the square of the HELIOS-1 total magnetic field and the annual averages of *B* from IMP and ISEE-3 data from SLAVIN *et al.* (1983).

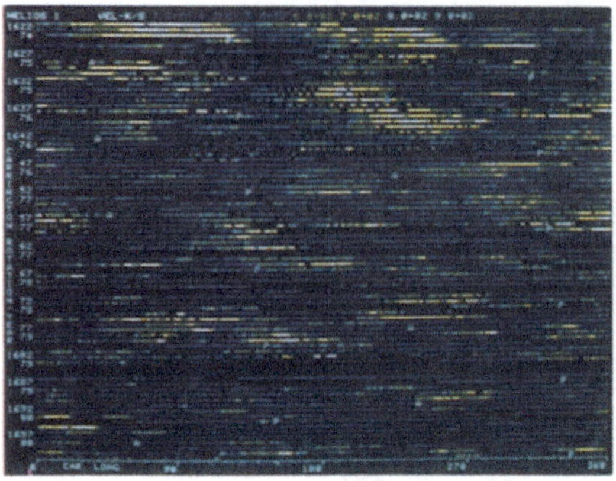

Plate 1.

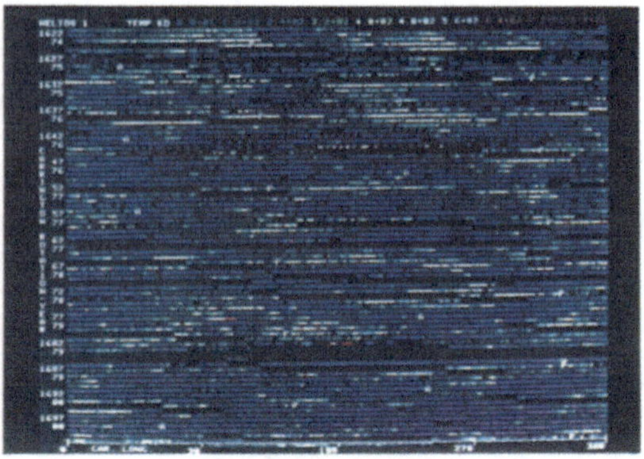

Plate 2.

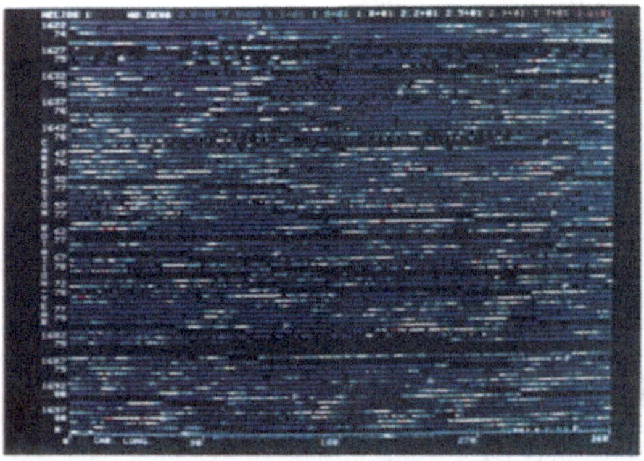

Plate 3.

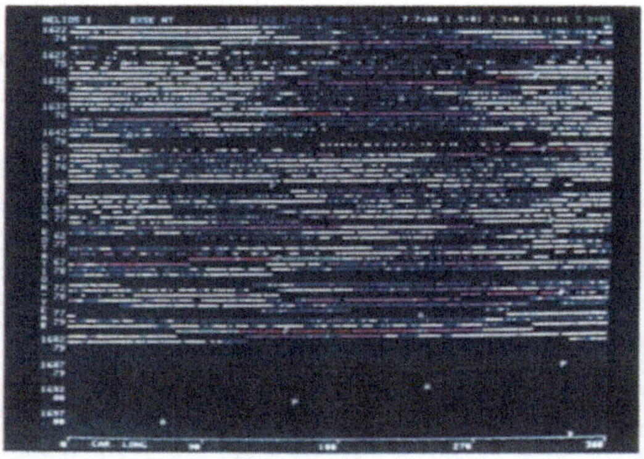

Plate 4.

Plate 1. Plate 1 represents hourly averages of the HELIOS-1 proton bulk flow velocity plotted on a Carrington rotation-Carrington longitude format. Each horizontal line of data represents one Carrington rotation. Each pixel is colored according to the velocity shown in the color code at the top of the figure. The last two digits of the year are shown below the Carrington rotation number. Further details are given in the text.

Plate 2. Same as plate 1 but showing the proton temperatures.

Plate 3. Same as plate 1 but showing the proton number density normalized to 1 AU by multiplying by r^2.

Plate 4. Same as plate 1 showing the solar ecliptic X component of the interplanetary magnetic field. The polarities are color coded with different hues. Magnetic data are not available for the full time period.

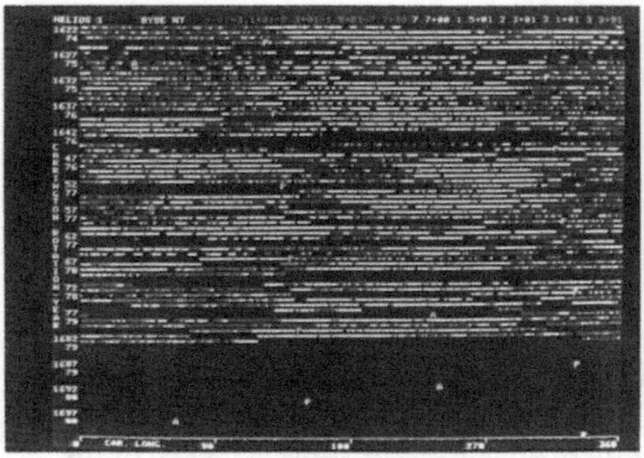

Plate 5.

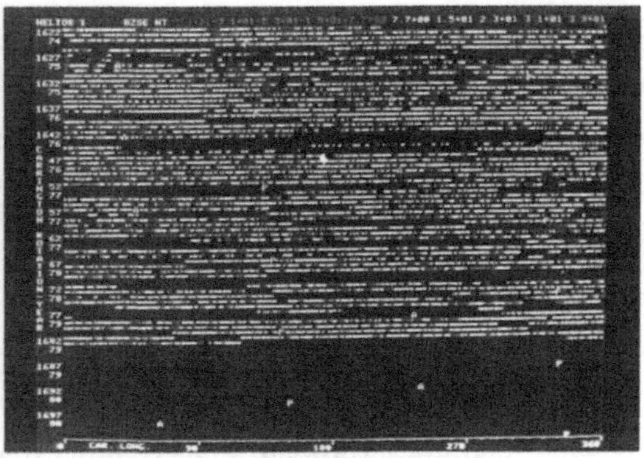

Plate 6.

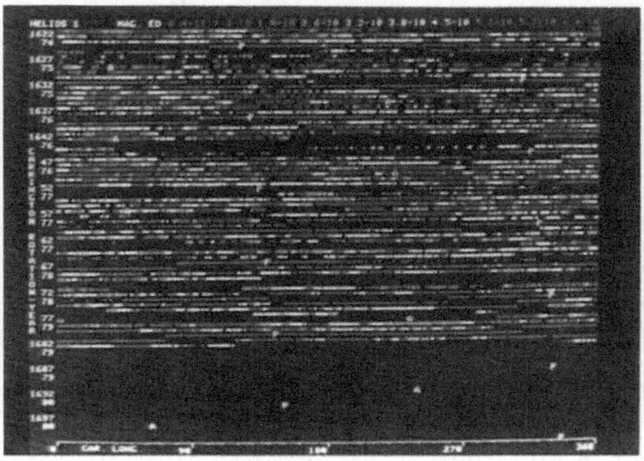

Plate 7.

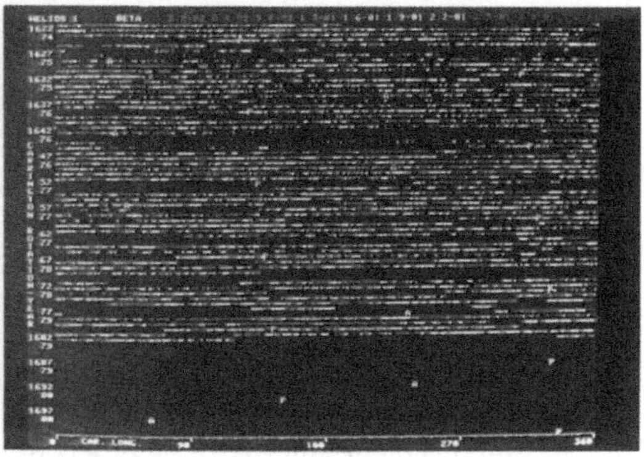

Plate 8.

Plate 5. Same as plate 4 showing the solar ecliptic *Y* component of the interplanetary magnetic field.

Plate 6. Same as plate 4 showing the solar ecliptic *Z* component of the interplanetary magnetic field.

Plate 7. Same as plate 4 showing the magnetic field energy density.

Plate 8. Same as plate 4 showing beta, the ratio of the proton thermal to magnetic pressure.

anticorrelation is sufficiently good that the thermal pressure is flat to within statistics over the entire data interval.

6. Interplanetary Magnetic Field

Plates 4, 5 and 6 show respectively the solar ecliptic *X*, *Y* and *Z* components of the interplanetary magnetic field. The plots are color coded with negative components in blue and red and positive in yellow and green. The sector structure (WILCOX and NESS, 1965) stands out clearly in B_x and B_y. We see a two-sector structure until 1976 at which time there is a switch to four sectors. A switch back to two sectors occurs in 1977 or 1978. The observation of a four sector structure is consistent with the structure inferred from photospheric magnetic field measurements by HOEKSEMA *et al.* (1982) during this time.

Careful scrutiny of Plate 6 shows that there appears to be some evidence of the sector structure in the B_z component. The role of the North–South component of the IMF in the sector structure has a controversial history (see SMITH, 1981 for a review). COLEMAN and ROSENBERG (1971) and ROSENBERG *et al.* (1971) reported a systematic behavior of the North–South component with heliographic latitude. This finding was not confirmed by a later study by HEDGECOCK (1975). HIRSHBERG and COLBURN (1973) have discussed the enhancement of B_z south following sector boundary crossing but the preference for a given polarity throughout the sector is surprising. A more careful study of the B_z component appears to be warranted.

For comparison with Plates 4, 5, and 6, Plate 7 shows the magnetic energy density in the solar wind. KING (1978, 1981) and SLAVIN *et al.* (1983) have reported the enhancement of the IMF toward solar maximum. This is confirmed by the HELIOS 1 data in the higher occurrence of red in the bottom half of the picture which indicates a stronger IMF approaching solar maximum. It is interesting to note that the enhancement begins in 1977 while the four sector structure is still present and that the enhancement appears to be independent of longitude, i.e., sector structure.

7. Derived Parameters

Beta, the ratio of the proton thermal pressure to magnetic pressure is shown in Plate 8. Since, as indicated, the thermal pressure is nearly constant, the variation in beta seen across the solar cycle is due mainly to the IMF variation already discussed.

Figure 2 also shows the Carrington rotation averages of beta. It is seen that there is a general variation of almost a factor of 2 but individual rotation averages can vary significantly from the general level. For example, rotation no. 1680 shows an exceptionally high beta.

The hourly average values of beta are often greater than 1 and in the entire HELIOS 1 dataset there are 20 cases when beta exceeds 10; the highest value being about 25.

Plate 9 shows the Alfvén mach number. This shows a solar cycle variation similar to beta, lower toward solar maximum. This variation is expected from the variation in the IMF.

Plate 10 shows the total energy flux (not including gravitational energy). The V^3 term dominates giving rise to a strong gradient over the solar cycle and permiting the transient events to stand out nicely. (By transient events we mean events appearing on a single rotation and lasting less than a few days.)

Plate 11 shows epsilon (PERREAULT and AKASOFU, 1978; AKASOFU, 1981) defined as

$$\varepsilon = VB_t^2 \, (\sin\theta/2)^4 \tag{1}$$

where

$$\theta = \arctan\,(B_h/B_z). \tag{2}$$

B_t is the total IMF, and B_h is the component in the x, y plane. The high speed streams stand out clearly but epsilon remains high toward solar maximum as the interplanetary magnetic field strengthens.

Finally, in plate 12, we look at the y-component of the interplanetary electric field, $|V \times B_z|$, another parameter of significance for magnetospheric phenomena. No attempt is made to distinguish east or west polarity in this plate because the figure becomes too confused when this is attempted. Again, the outline of the high speed streams can be seen as well as an increase toward solar maximum.

8. Summary

With this paper we have tried to give the reader a qualitative feel for the solar cycle variations of various solar wind parameters. The color graphics are useful for this purpose because they permit one to see a large fraction of solar cycle at a glance. In addition to reviewing the known variations of the proton velocity, temperature and density and the components of the IMF, we have presented the solar cycle variations of the magnetic energy density, the Alfvén mach number, beta, the energy flux, epsilon and the y-component of the interplanetary electric field.

We have shown that there is a gradual decrease in the duration of high speed streams from the declining phase of solar cycle 20 through the ascending phase of cycle 21 and a corresponding decrease in the annual average of the proton speed toward solar maximum.

Beta, the ratio of the proton thermal pressure to magnetic pressure, undergoes a significant solar cycle variation as is expected from the variation in the IMF. Individual hourly averages of beta often exceed unity with 20 cases exceeding 10 and one case as high as 25.

The Alfvén mach number shows a solar cycle variation similar to beta, lower toward solar maximum.

High speed streams can be seen clearly in epsilon and the y component of the interplanetary electric field.

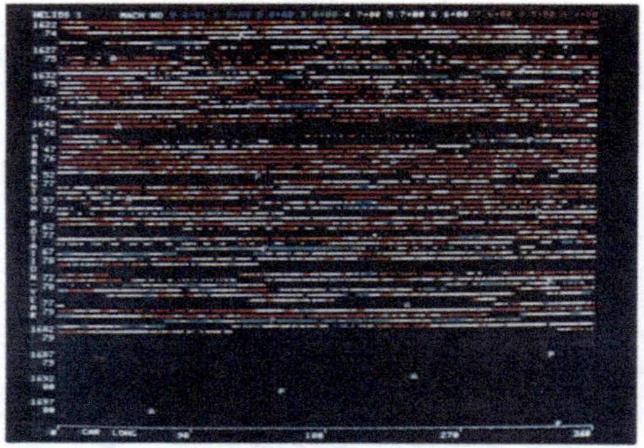

Plate 9.

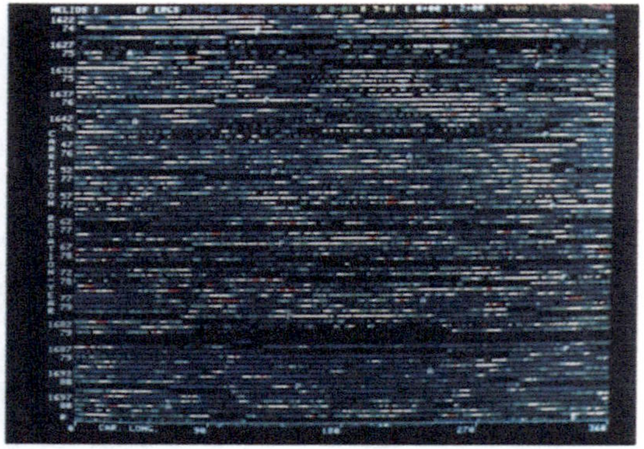

Plate 10.

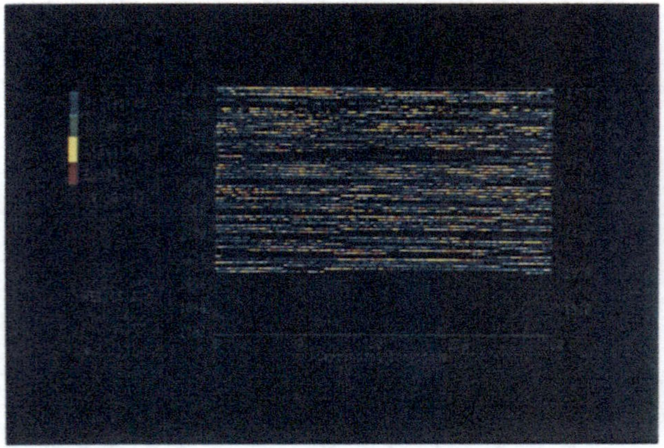

Plate 11.

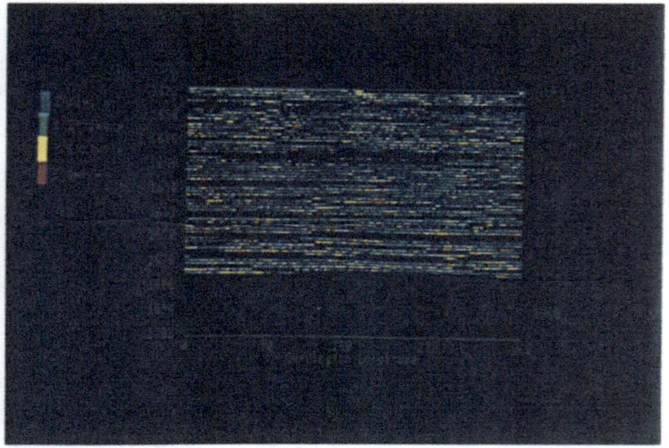

Plate 12.

Plate 9. Same as plate 4 showing the Alfvén mach number.

Plate 10. Same as plate 1 showing the total particle energy flux.

Plate 11. Epsilon plotted on a Carrington rotation-longitude plot similar to Plate 1. The computer system used for Plates 11 and 12 is an IBM-PC rather than the RAMTEK 9400 system used in Plates 1–10.

Plate 12. Same as plate 11 for the y-component of the interplanetary electric field. No attempt is made to specify the polarity of the field.

This work was begun while one of us (JWF) was at the National Space Science Data Center. We would like to acknowledge the generous help of the staff of the Center, in particular, Kent Hills, Joe King, and Jim Vette. Most the plots presented herein are available from the NSSDC by requesting dataset numbers 74-097A-01D, 74-097A-09C, 76-003A-01D, 76-003A-09C. This work was also supported by NASA grant NGL44-006-012.

REFERENCES

AKASOFU, S.-I., Energy coupling between the solar wind and the magnetosphere, *Space Science Rev.*, **28**, 121, 1981.

BRIDGE, H. S., Solar cycle manifestations in the interplanetary medium, in *Phys. of Solar Planetary Environments*, edited by D. J. Williams, AGU, Washington, D.C., 1976.

BURLAGA, L. F. and K. W. OGILVIE, Heating of the solar wind, *Astrophys. J.*, **159**, 659, 1970.

BURLAGA, L. F. and K. W. OGILVIE, Solar wind temperature and speed, *J. Geophys. Res.*, **78**, 2028–2034, 1973.

COLEMAN, P. J. and R. L. ROSENBERG, North-South component of the interplanetary magnetic field, *J. Geophys. Res.*, **76**, 2917, 1971.

DIODATO, L., G. MORENO, C. SIGNORINI, and K. W. OGILVIE, Long-term variations of the solar wind protons parameters, *J. Geophys. Res.*, **79**, 5095, 1974.

FEYNMAN, J., Solar cycle and long term changes in the solar wind, *Rev. Geophys. Space Phys.*, **21**, 338, 1983.

FREEMAN, John W. and Ramon E. LOPEZ, The cold solar wind, *J. Geophys. Res.*, **90**, 9885, 1985.

GOSLING, J. T., A. J. HUNDHAUSEN, and V. PIZZO, Compressions and rarefactions in the solar wind: Vela 3, *J. Geophys. Res.*, **77**, 5442, 1972.

HEDGECOCK, P. C., The heliographic latitude dependence and sector structure of the interplanetary magnetic field 1969–1974: Results from the HEOS satellites, *Solar Phys.*, **44**, 205, 1975.

HIRSHBERG, J. and D. S. COLBURN, Geomagnetic activity at sector boundaries, *J. Geophys. Res.*, **78**, 3952, 1973.

HOEKSEMA, J. T., J. M. WILCOX, and Philip H. SCHERRER, Structure of the heliospheric current sheet in the early portion of sunspot cycle 21, *J. Geophys. Res.*, **87**, 10, 331, 1982.

HUNDHAUSEN, A. J., Solar activity in the solar wind, *Rev. Geophys. and Space Phys.*, **17**, 2034, 1979.

HUNDHAUSEN, A. J., S. J. BAME, J. R. ASBRIDGE, and S. J. SYDORIAK, Solar wind proton properties: Vela 3 observations from July 1965 to June 1967, *J. Geophys. Res.*, **75**, 4643, 1970.

KING, H. J., Solar cycle variations in IMF intensity, *J. Geophys. Res.*, **84**, 5983, 1979.

KING, J. H., On the enhancement of the IMF during 1978–1979, *J. Geophys. Res.*, **86**, 4828, 1981.

LOPEZ, Ramon E. and John W. FREEMAN, The solar wind temperature-velocity relationship, accepted for publication, *J. Geophys. Res.*, 1986.

NEUGEBAUER, M., Large-scale and solar cycle variations of the solar wind, *Space Science Rev.*, **17**, 221, 1975.

NEUGEBAUER, M. and Conway W. SNYDER, Mariner 2 observations of the solar wind, 1, average properties, *J. Geophys. Res.*, **71**, 4469, 1966.

PERRAULT, P. and S.-I. AKASOFU, A study of geomagnetic storms, *Geophys. J. R. Astron. Soc.*, **54**, 547, 1978.

PIZZO, V., J. T. GOSLING, and A. J. HUNDHAUSEN, Large-scale dynamical effects upon the solar wind flow parameters, *J. Geophys. Res.*, **78**, 6469, 1973.

ROSENBERG, R. L., P. J. COLEMAN, Jr., and D. S. COLBURN, North-south component of interplanetary magnetic field: Explorer 33 and 35 data, *J. Geophys. Res.*, **76**, 6661, 1971.

SHEELEY, Jr., N. R. and J. W. HARVEY, Coronal holes, solar wind streams, and geomagnetic activity during the sunspot cycle, *Sol. Phys.*, **59**, 159, 1978.

SHEELEY, Jr., N. R. and J. W. HARVEY, Coronal holes, solar wind streams, and geomagnetic disturbances during 1978 and 1979, *Sol. Phys.*, **70**, 237, 1981.

SHEELEY, Jr., N. R., J. W. HARVEY, and W. C. FELDMAN, Coronal holes, solar wind streams, and recurrent geomagnetic disturbances: 1973–1976, *Sol. Phys.*, **49**, 271, 1976.

SLAVIN, J. A., E. J. SMITH, and B. T. THOMAS, Large scale temporal and radial gradients in the IMF: HELIOS 1, 2, ISEE-3, and Pioneer 10, 11, *Geophys. Res. Let.*, **11**, 279, 1983.

SCHWENN, R., The "average" solar wind in the inner heliosphere: Structures and slow variations, Solar Wind 5, NASA CP-2280, 1983.

SMITH, E. J., Solar wind magnetic field observations, in *Solar Wind Four* edited by H. Rosenbauer, Max-Planck-Institut für Aeronomie, katlenburg-Lindau, West Germany, 1981.

WILCOX, J. M. and N. F. NESS, Quasi-stationary corotating structure in the interplanetary medium, *J. Geophys, Res.*, **70**, 5793, 1965.

Solar Wind-Magnetosphere Coupling, edited by Y. Kamide and J. A. Slavin, 191–207.

MHD Simulation of the "Geoeffectiveness" of Interplanetary Disturbances

M. DRYER[1], Z. K. SMITH[1], S. T. WU[2], S. M. HAN[3], and T. YEH[1,4]

[1]*Space Environment Laboratory, NOAA, Boulder, Colorado 80303, U.S.A.*
[2]*Department of Mechanical Engineering, University of Alabama in Huntsville, Huntsville Alabama 35899, U.S.A.*
[3]*Department of Mechanical Engineering, Tennessee Technological University, Cookeville Tennessee 38505, U.S.A.*
[4]*Cooperative Institute for Research in Environmental Sciences, University of Colorado, Boulder Colorado 80307, U.S.A.*

Conceptually, if one were provided with adequate spatial and temporal information on solar (or near-solar) conditions of plasma and magnetic field, one can compute the large-scale interplanetary evolution caused by solar events such as high-speed streams, flares, and eruptive prominences. The output of this procedure can be used for the determination of the magnetospheric energy-transfer parameter and the cross-tail electric field. We will report our first series of examples that use this "geoeffectiveness" prediction technique. The first example is concerned with a simple, high-speed, corotating stream; the second, with a complex sequence of solar flares and eruptive prominences during STIP Interval VII (August 1979). These examples utilize our non-planar (2-1/2D) numerical code for time-dependent MHD flows.

We conclude with a report on our preliminary work with a fully 3D, time-dependent, interplanetary shock simulation. The results suggest that 3D work is necessary for more accurate analyses of the "geoeffectiveness" of complex solar events.

1. Introduction

During the late 1950s there was a debate, initiated amongst the community of cosmic ray physicists, concerning the modulating role of the solar magnetic field and its extension into interplanetary space. This debate took place during the early conjecture of the existence of the ever-present solar wind. The controversy is whether the solar wind is only passive in channeling solar disturbances into interplanetary space. In a recent review (DRYER, 1985) of this debate (cf., JASTROW, 1959) it was noted that the controversy continues to this day, this time within the context of the "geoeffectiveness" of solar activity. Thus, on the one hand, we note the notion of the magnetic bubble (or "tongue"; GOLD, 1959) being extended by PUDOVKIN and CHERTKOV (1976) and PUDOVKIN *et al.* (1979). On the other hand, the notion of the Parkerian solar wind with spiraling magnetic field lines IMF, one foot embedded in the sun and the other extended into interplanetary space, is being extended by HAKAMADA and AKASOFU (1982) and AKASOFU *et al.* (1983).

Several variants of these two schools have been advanced. One variant is the oblateness of interplanetary shock shapes that emanate from solar flares associated with their parent active regions (and their dominant magnetic axes) on the sun

(IVANOV *et al.*, 1974, 1980, 1984). Another variant is the statistical result that radio flux in the 10 cm and metric wave lengths are lower for large (Hα importance $\gtrsim 2$) flares with southward oriented magnetic fields (ROELOF *et al.*, 1983). Recently, however, specific flare observations with the ensuing IMF orientations (TANG *et al.*, 1985) indicate that the Pudovkin extension of the magnetic bubble proposal is not substantiated with any certainty. As to the other school, it is proposed in the variants of this idea that solar-originated disturbances propagate through the solar wind, modulating the IMF in its magnitude and polarity, until they reach the earth's magnetosphere. There, the geoeffectiveness is determined by the Poynting flux, primarily dependent on the southward IMF component and solar wind velocity, and a dynamo interaction between the solar wind and the magnetosphere.

The concept of geoeffectiveness has been examined via a kinematic approach (AKASOFU *et al.*, 1983) as well as an MHD approach (DRYER *et al.*, 1984; WU *et al.*, 1983). It has been shown that the large-scale interplanetary evolution caused by solar events can feasibly be computed (cf., DRYER *et al.*, 1978; SMITH *et al.*, 1981) if one is provided with adequate spatial and temporal information on solar (or near-solar) conditions of plasma and magnetic field. Since our knowledge of the conditions at or near the sun is not really adequate, it is often necessary to be supplemented by approximations, speculations, and even educated guesses in order to provide the input data. The output of the model codes is easily used for the determination of the magnetospheric energy-transfer parameter and the cross-tail electric field. An empirical example of this procedure, in which actual solar wind observations were used, was recently demonstrated by TSURUTANI *et al.* (1985).

The purpose of this paper is to report a series of examples that demonstrate the "geoeffectiveness" prediction technique that is based on the classical, initial boundary value problem of MHD. The reader is reminded that this report is merely the *demonstration* of the MHD technique, not its proof. Much work remains to be done in terms of improved input data as well as improved 3D code development.

2. Discussion

We will first use the nonplanar (2-1/2D) MHD code of WU *et al.* (1983) to examine a simple situation of corotating high-speed solar wind stream and a more complex situation of a sequence of solar flares and eruptive prominences. The background that provides input estimates is the epoch of STIP Interval VII (August 1979) at the beginning of the Solar Maximum Year. Then, we report on our preliminary work with a fully 3D, time-dependent code of HAN *et al.* (1984) to simulate a flare-generated shock disturbance.

2.1 Nonplanar (2-1/2D) simulation of a corotating stream

This simulation starts, firstly, with the establishment of a representative, steady-state solar wind that is invariant with respect to heliolongitude in the solar ecliptic plane. The plasma and magnetic field conditions are given by DRYER *et al.* (1984). For example, the solar wind velocity at the lower boundary (18 solar radii, 18 R_s=0.08 AU) is 250 km/sec and it is 364 km/sec at 1 AU. At t=0 hr., a high speed stream is assumed to be created by a coronal hole, about 35° in width, at the sun. This stream is modeled by a pulse that has been suggested by the interplanetary scintillation observations of TAPPIN *et al.* (1984) and a suggestion by A. Hewish (private communication, 1984). Spatially in heliolongitude, the pulse increases linearly from a steady-state velocity of 250 km/sec at the east limb to a maximum value of 320 km/sec over a 7.5° extent. The pulse remains constant for an additional 20° and then decreases linearly over another 7.5° to background. Temporally, the velocity at each grid point is increased linearly to its maximum over a 1 hr. period and is then held constant. In order to account for the solar rotation relative to Earth of 13.5°/day, the entire pulse is rotated westward in time. The density is assumed to vary inversely as the square of the velocity at each point (heliolongitudinal grid spacing was taken to be $\Delta\phi=2.5°$) within the 35° width of the assumed high speed stream's inner boundary. Note that we did not attempt to simulate the flow from the coronal hole (slightly less than 35° wide at 1 R_s) to the lower boundary of our simulation domain (18 R_s).

Figure 1 shows the spatial distribution of the solar wind speed at $t=160.4$ hr. when the earth is engulfed by the highest level of the flow. The reader is reminded that the originally undisturbed flow and magnetic field have all three components. Hence the flow at each point within the computational domain (150° in heliolongitude and nearly 1.1 AU in helioradius) exhibits the interaction of various modes (i.e., fast, Alfven, and slow) of MHD waves on the large-scale used here (grid spacing of $\Delta\phi=2.5°$, as noted above, and $\Delta R=2R_s$ in helioradius). Figure 2 shows the corresponding density change, $(n-n_0)/n_0$, at $t=160.4$ hr. in a halftone format. Note that the 25% peak *enhancement* has just passed the earth's location at nearly the

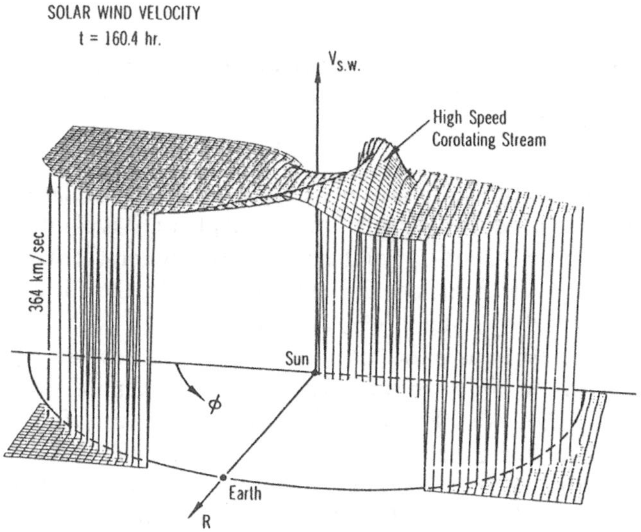

Fig. 1. Solar wind velocity in the ecliptic plane at 160.4 hr. after the emanation of a high-speed stream from a coronal hole. The initial width of the coronal hole is taken to be ~35° at $t=0$ hr. The higher-speed stream is taken to be produced over a one-hour period after which it continues unabated.

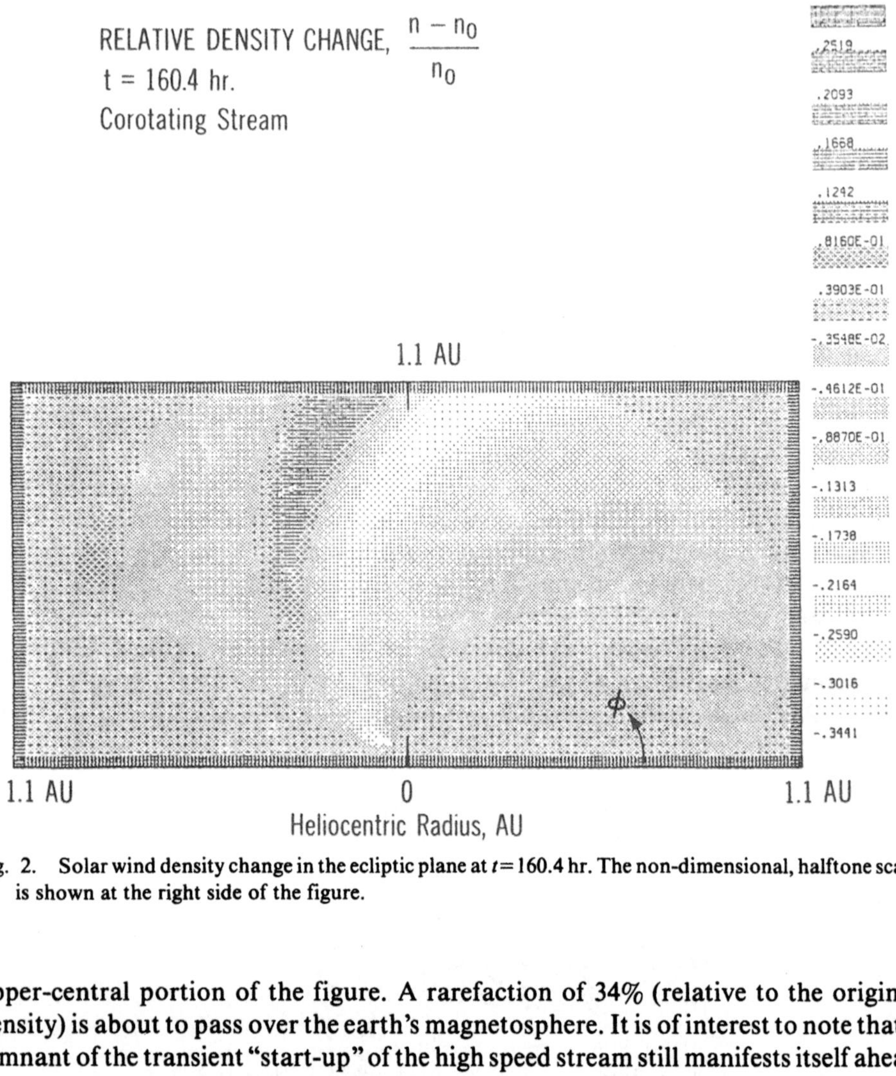

Fig. 2. Solar wind density change in the ecliptic plane at $t=160.4$ hr. The non-dimensional, halftone scale is shown at the right side of the figure.

upper-central portion of the figure. A rarefaction of 34% (relative to the original density) is about to pass over the earth's magnetosphere. It is of interest to note that a remnant of the transient "start-up" of the high speed stream still manifests itself ahead (i.e. to the west) of the peak density enhancement in the form of a slight (8%) compression, followed by a slight (4%) rareaction.

Of greater interest to us is the "geoeffectiveness" of this particular, rather mild, high speed stream. In order to suggest a temporal response within the Earth's environment, we show the Poynting flux parameter, ε, (cf., HAKAMADA and AKASOFU, 1982) in Fig. 3 and the cross-tail electric field, VB_θ ($-VB_z$) in Fig. 4, respectively. The time series of these parameters are shown at Earth's location and at two other locations (60° E and 25° W of Earth at 1 AU). At the Earth, it is seen that the peak power ($\sim 1.2 \times 10^{16}$ erg/sec) is reached at about $t=160$ hr, several hours later than the peak cross-tail field (~ 0.4 mV/meter). The fact that these peaks occur about a day after the initiation of the disturbance is a reflection of two important points: (i) it

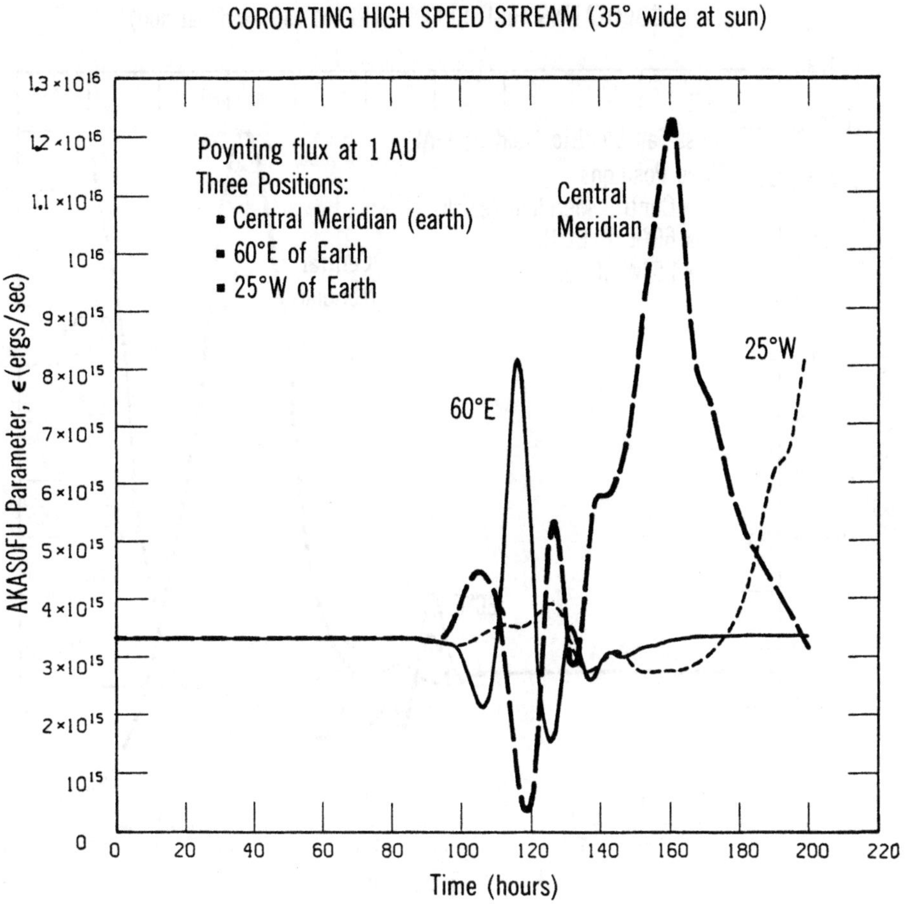

Fig. 3. Poynting flux (Akasofu's power parameter, ε) at Earth and two other locations at 1 AU positions as a result of the corotating high speed stream indicated in Figs. 1 and 2.

takes time for the quantity VB_θ to reach its peak value; and (ii) there is a delay between the time when the IMF vector swings around and the time when the southerly component is maximized. The IMF rotation has been demonstrated during the entire passage of a simulated flare-shock disturbance by Wu *et al.* (1983) and Dryer *et al.* (1984). As noted earlier, this IMF rotation reflects the passage of the MHD waves, first, the fast mode (fast shock in the case of a flare-shock), then the Alfven mode and, finally, the slow mode that returns the IMF, etc., to the pre-disturbance levels (see Wu *et al.*, 1983) of plasma and IMF conditions.

2.2 *Nonplanar (2-1/2D) simulation of a complex series of events*

As noted in the Introduction, if one were provided with spatial and temporal information of plasma and magnetic field conditions at or near the sun, it may be

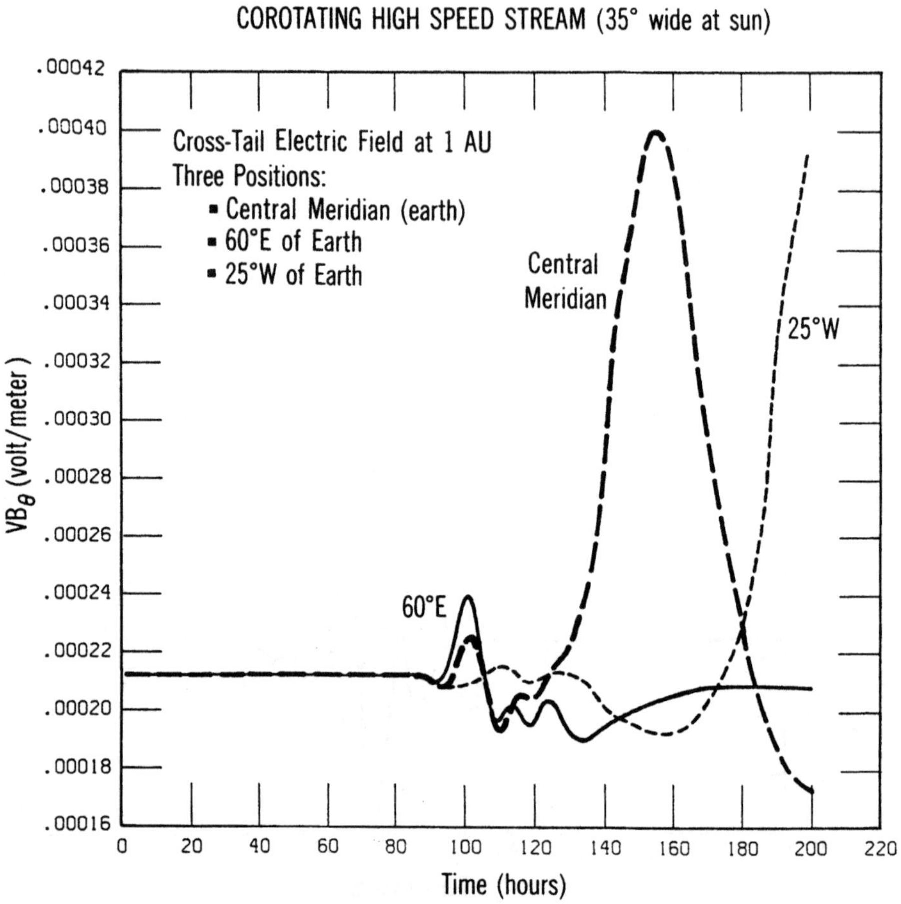

Fig. 4. Cross-tail electric field at Earth and two other locations at 1 AU positions as a result of the corotating high speed stream indicated in Figs. 1 and 2.

possible to compute the large-scale evolution caused by a series of solar events. Pioneering work along these lines has been done via the kinematic approach by AKASUFU *et al.* (1985) and SUN *et al.* (1985). Studies of this kind, as in the present MHD approach, are clearly dependent upon the input parameters that are used. The two studies just mentioned, therefore, are very successful in providing the shock arrival times at multiple spacecraft that are located in or near the ecliptic plane over a wide range of heliolongitudinal and helioradial positions. This kinematical approach has also been successful in the simulation of the solar wind plasma velocity at various spacecraft. It would be desirable, however, to incorporate additional physics, via the MHD method, within the same methodology and with the additional ingredients of observed parameters insofar as possible.

It is with this motive in mind that we now turn to a complex series of events that have been studied by many observers and analysts. This series was part of the well-studied STIP Interval VII during August 1979 (KANE *et al.*, 1984; CLIVER *et al.*, 1983; SUESS *et al.*, 1984; and others) that initiated the Solar Maximum Year. The synoptic studies not only pinpointed major events via the usual techniques (Hα, x-ray, radio bursts) but also synthesized, via white light and remote radio sensing, important parameters such as coronal shock velocities for most of the major solar flares, especially on 14 and 18 August, 1979.

Table 1 provides a time line for the MHD simulation. It should be regarded as an imperfect attempt to quantify the disturbance events. For example, the first item in Table 1 is the turn-on of a series of high and low speed streams at the lower boundary (18 R_s). They were allowed to corotate at the synodic rate of 13.5° per day. The adopted changes in magnetic polarity were such that the solenoidality of magnetic field was not satisfied. Hence errors were introduced in the form of fictitious monopoles. These errors (believed to be significant only for the most complex interactions) are presently being corrected. Thus our discussion will concentrate primarily on the plasma conditions. The second item in Table 1 refers to a filament

Table 1. Time line for major events during STIP interval VII (August 1979).

Clock time	Simulated date	Event	Reference
0 Hr.	8 Aug., 1979 0000 UT	High and low speed streans are "turned on" at 18 R_s*	SUESS *et al.* (1984)
52.0 Hr.	10 Aug., 1979 0400 UT	Eruptive prominence**	JOSELYN (1983)
154.6 Hr.	14 Aug., 1979 1240 UT	Solar Flare (1B) S 22° E 73° V_{shock}=1500 km/sec	KANE *et al.* (1984)
192.5 Hr.	16 Aug., 1979 1200 UT	Solare Flare (1B) N 18° E 25° V_{shock}=1500 km/sec	
240.6 Hr.	18 Aug., 1979 1430 UT	Solar Flare (1B) N 08° E 90° V_{shock}=2500 km/sec	KANE *et al.* (1984) WOO and ARMSTRONG (1981) MAXWELL and DRYER (1982) CANE *et al.* (1982)
319.0 Hr.	21 Aug., 1979 0700 UT	Solar Flare (2B) N 17° W 40° V_{shock}=2500 km/sec	CLIVER *et al.* (1983)

* The spatial extents and peak magnitudes of the high and low speed streams at 18 R_s are estimated on the basis of extrapolations from source surface (2.5 R_s) calculations given by SUESS *et al.* (1984).

** The eruptive prominence noted here was one of many during this epoch; it was singled out here on the speculation that it was the most "geoeffective".

that "disappeared" some time between late 9 August and 10 August. The time, 0400 UT, listed in the table, is our choice. Also assumed are sinusoidal profiles for the density, velocity, and temperature in a longitudinal width of about 24° with the peak values at the middle longitude being twice the original ambient values and zero at the edges of the 24° longitudinal extent. This total pulse was maintained for 1 hr. The original steady-state conditions were identical to those used for the previous case. The remaining four shocks in Table 1 are solar flare-generated shocks. The assumed shock velocities are also listed in Table 1 under the column labeled "Event". The downstream shock parameters were obtained by solving the Rankine-Hugoniot equations (explicitly given by HAN *et al.*, 1982) and were used for the peak values of the pulses. A sinusoidal drop-off from these peak values (at the center of the 24° heliolongitudinal extent) to background was assumed at each side of the asymmetrical, simulated shock at 18 R_s. The temporal duration of the input pulse was assumed to be 1/2 hr. at peak value for the assumed piston-driven shock. The build-up and decay were assumed to be linearly ramped over a 1 hr. period commencing at the start, and terminating at the end, of the "piston-driven" phase.

Some of the compound events are shown in Figs. 5–7. At t=60.1 hr. (10 August, 1979), for example, Figure 5 shows the solar wind velocity distribution for: (i) two high-speed streams that bracket a low-speed stream approaching the earth; and (ii) a high amplitude forward fast-mode MHD wave, caused by the eruptive prominence earlier on the same day, propagating through the low speed stream. At t=160 hr. (14 August, 1979), Figure 6 shows that the fast-mode MHD wave has passed the earth, followed by a reverse fast-mode MHD wave. Also shown are the flare-generated,

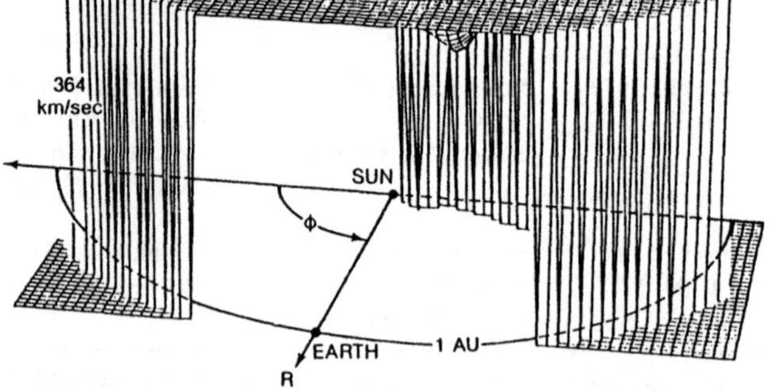

Fig. 5. Solar wind velocity in the ecliptic plane during a compound series of high/low speed streams, followed by a simulated eruptive prominence's disturbance that propagates into the low speed stream toward the earth. (Simulated date: 10 August, 1979.)

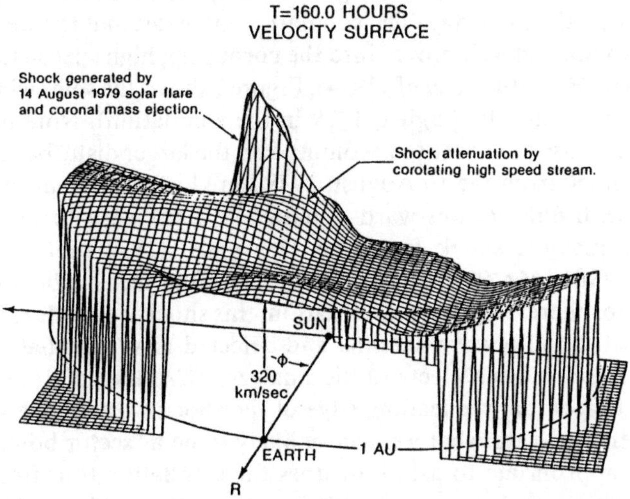

Fig. 6. Solar wind velocity in the ecliptic plane during the passage over the earth of a forward, fast-mode, MHD wave (produced by an eruptive prominence four days earlier). Also shown is a forward, fast-mode MHD shock wave from an E 73° flare. (Simulated data: 14 August, 1979.)

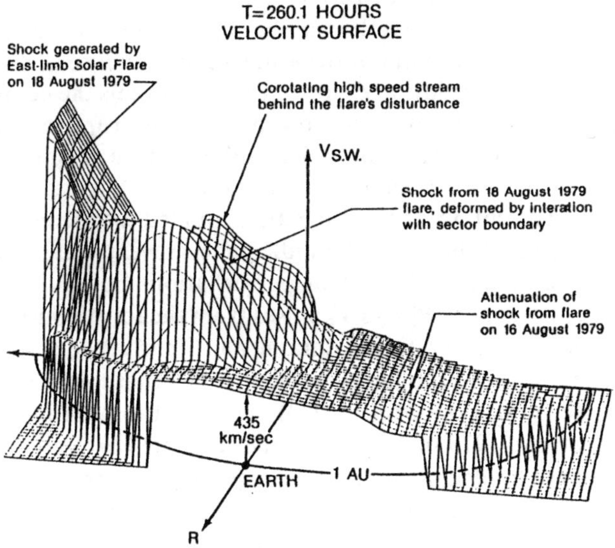

Fig. 7. Solar wind velocity in the ecliptic plane during the eastward and earthward motion of a strong interplanetary shock that was produced by an east-limb solar flare on 18 August, 1979. Reference to a sector boundary (that is suggested to be responsible for the shock's westward flank attenuation) is uncertain and remains to be studied. (Simulated date: 18 August 1979.)

forward fast-mode MHD shock wave generated by the flare and coronal mass ejecta at E73° earlier on the same day. The simulation indicates that the westward flank of this shock is attenuated as it moves into the corotating high speed stream, a process first described by HIRSHBERG *et al.* (1974). Figure 7 shows the solar wind velocity four days later (t=260.1 hr., 18 August, 1979 in the simulation). Note that the scale is substantially increased in order to accommodate the larger disturbances. Firstly, it is seen that the shock from the 16 August, 1979 flare has passed out of the domain of computation with only its westward, attenuated, flank seen to the west of earth. Secondly, the massive shock from the 18 August, 1979 limb flare (WOO and ARMSTRONG, 1981; MAXWELL and DRYER, 1982) is seen to be approaching the earth's magnetosphere. In the actual situation, this shock was, indeed, tracked by the ISEE-3 kilometric radio wave antenna and detected *in situ* at that location by the magnetometer and plasma electrostatic analyzer (CANE *et al.*, 1982). The MHD simulation indicates that the leading edge of the shock was attenuated, asymmetrically, as it interacted with what we believe likely to be a "sector boundary".

It is now appropriate to ask: how does the simulation (our first "round" of a planned interative process) compare with the observations? Figures 8 and 9 show the following information for the magnetic field strength and the radial velocity:

(i) The top panels show the hourly-averaged data of B and V_r as measured by the instruments on ISEE-3.

(ii) The middle panels show the simulation results when the disturbances include all the events listed in Table 1. The arrows denote the simulated arrivals at Earth of the leading edge of each of the five disturbances.

(iii) The bottom panels show the simulation results if the disturbance consists of the high/low speed streams without the eruptive prominence and solar flare shocks.

Comparison of the middle and bottom panels reveals the compressive effect on the magnetic flux as a result of the various pulses. The eruptive prominence (see the first arrow) has a significant effect in both B and V_r as have the shocks of 16, 18, and 21 August, 1979. On the other hand, the shock of 14 August, 1979 was relatively ineffective due, no doubt, to its E73° location coupled with a moderate initial shock velocity. Comparison of the top and middle panels reveals qualitative agreements in structure and magnitudes of the major perturbations but with important phasing shifts. The large, simulated $|B|$ and V_r pulses from the fourth flare (21 August, 1979) do not appear in the observational data at all. Clearly, the assumed initial shock velocity (2500 km/sec) was too high (despite the fact that this flare produced gamma rays as discussed by CLIVER *et al.*, 1983). The ISEE-3 data indicate that a shock arrived on 24 August, 1979 in contrast to the simulated arrival on 22 August, 1979. With the wisdom of hindsight, we would suggest that a second round in an iterative process should consider a reduction of the assumed initial shock velocity to ~1000 km/sec. Alternatively, it would be better to utilize radio observations for shock velocity determination (WOO and ARMSTRONG, 1981; MAXWELL and DRYER, 1982) for an actual case. Similar comments might be suggested vis-a-vis the other shocks in order to improve the phasing agreements.

We should also note that the large-scale nature of the MHD simulation precludes any comparison of details in the small-scale structure. Clearly, better definition of the

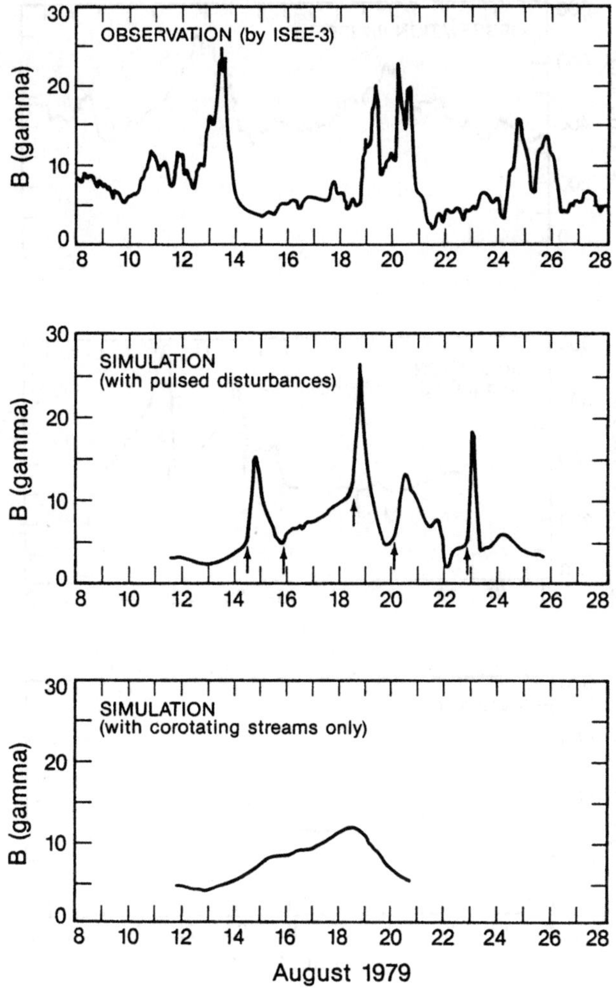

Fig. 8. Interplanetary magnetic field at Earth's location during STIP Interval VII. The upper frame, to the same scale as the others, shows the hourly-averaged measurements by ISEE-3. The middle panel incorporates the superposition in the MHD model of the five disturbances listed in Table 1. The bottom panel shows the IMF simulation if only the high speed and low speed streams existed with no additional pulses whatsoever.

input conditions are called for. Moreover, the model is still deficient in that it is constrained to the ecliptic plane, despite its inclusion of all three components of the velocity and magnetic field; and, also, in that it precludes latitudinal gradients of these quantities. Thus, a full 3D, time-dependent, MHD model is needed. We discuss some progress in this area in the next and final section.

Before moving on to the 3D discussion, we note that no attempt has been made as yet to compare the simulated ε and VB_θ with the measured values at ISEE-3 for this

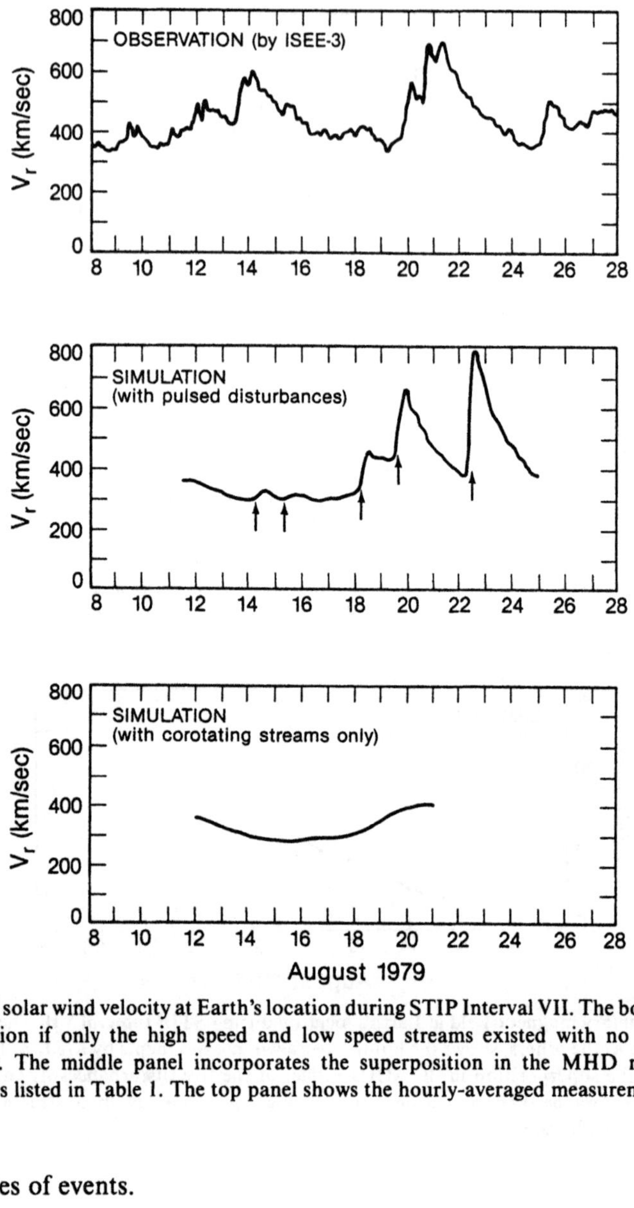

Fig. 9. Radial solar wind velocity at Earth's location during STIP Interval VII. The bottom panel shows the simulation if only the high speed and low speed streams existed with no additional pulses whatsoever. The middle panel incorporates the superposition in the MHD model of the five disturbances listed in Table 1. The top panel shows the hourly-averaged measurements by ISEE-3.

complex series of events.

2.3 *Fully 3D, time-dependent simulation of a flare-generated interplanetary shock wave*

Details of the numerical algorithm for the fully 3D, time-dependent, MHD simulation (the first such study in the field of solar/interplanetary physics to our knowledge) are given by HAN *et al.* (1984). The example chosen by those workers concerns an interplanetary shock (V_s=1000 km/sec) that is input at the lower boundary of 18 R_s. As in the 2-1/2D case, the Rankine-Hugoniot equations are solved at the center of a "bulls-eye" centered on the sun-earth axis. The pre-disturbance, steady-state, solar wind conditions are calculated with the *ad hoc* specification of a spiralling IMF within the heliosphere. In this sense, the steady-state computation is similar to that performed by PIZZO (1982) in a 3D, time-independent computation in the corotating frame. The downstream shocked conditions, as in our 2-1/2D computations, were then taken to decay sinusoidally, in both the heliolongitudinal and meridional directions, to the background levels (V, B, n, T) to the edges of the 24° width of the "bulls-eye" boundary of the assumed shock. The total computational domain was a 90° width in both angular directions. The temporal duration (again, as in the 2-1/2D studies) was taken to be 1–1/2 hr.

Figure 10 shows the radial and transverse velocities on a segment of a sphere at 32R_s five hours after the initiation of the shock pulse. The bottom panel shows the magnetic field strength. In the latter case, note that B is essentially unchanged along the central axis (which may be thought of as the sun–earth axis) because the upstream IMF at that location ($\theta=\pi/2$, $\phi=\pi/4$) is essentially radial. Thus the shock is a parallel shock at that unique position; however it is quasi-perpendicular to the west, and quasi-parallel to the east, of the central axis. Clearly, the transverse velocity is directed away from the central axis while the radial velocity displays an expected "tongue" shape.

Figure 11 shows the time series of the three IMF components at 40 R_s on the central axis ($\theta=90°$) as well as at 6° away from it ($\theta=84°$). Also shown, for comparison, are the three components of the magnetic field on the central axis for the case of 2-1/2D. Strong attenuation of the B_ϕ-component is clearly seen just behind the shock because of the diverging effect. These 3D studies are continuing.

3. Concluding Remarks

We have demonstrated, via the use of a simple, corotating high speed stream, that the MHD technique can be used in its nonplanar (2-1/2D) version to compute relevant solar wind parameters (such as ε, VB_θ) at earth's location. The situation is not so clear when a complex series of events, such as those during August 1979, are considered in a simulation that is then compared with the temporal observations (ISEE-3) of the radial velocity and total IMF at 1 AU. While qualitative agreement is found for large-scale structures, the phasing and amplitudes are not satisfactory. Several reasons for these discrepancies are offered: (i) inadequate knowledge of the spatial and temporal conditions of plasma and magnetic field near the sun, and (ii) inadequate validity of the nonplanar, 2-1/2D, time-dependent model.

Preliminary work with the 3-D, time-dependent models shows that the effect of the latitudinal variation produces attenuation of the heliolongitudinal component, B_ϕ, of the IMF. The radial velocity is similarly attenuated. Still lacking in our study, however, is the global effect of asymmetrical inputs near the sun that would very likely occur in the complex low-beta ($\beta \ll 1.0$) plasmas in the low corona.

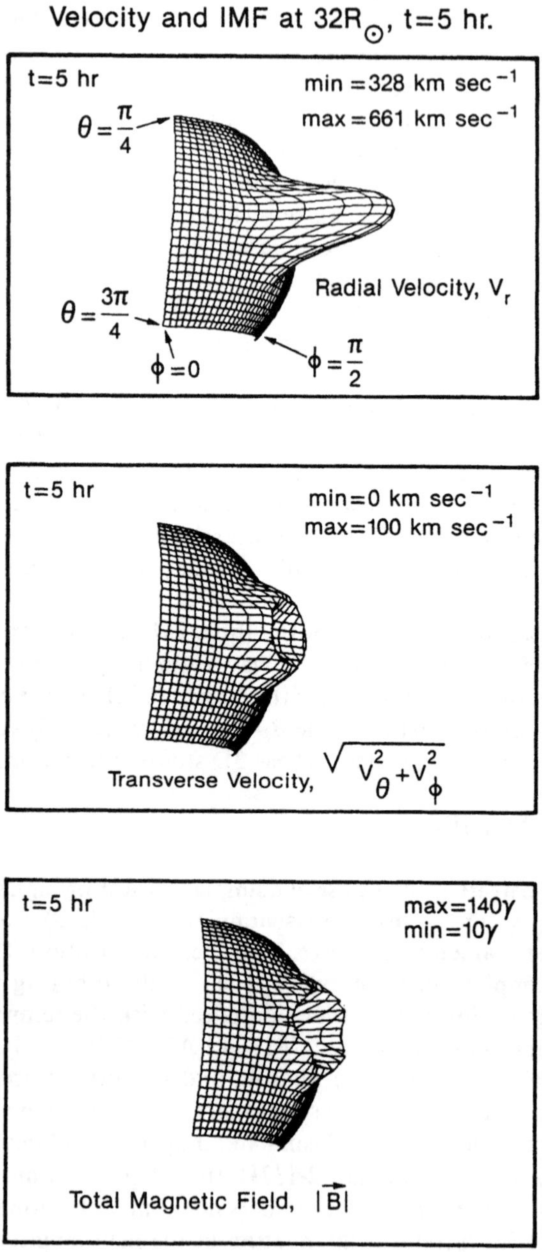

Fig. 10. Results of 3D simulation for the radal velocity, transverse velocity, and total magnetic field behind an interplanetary shock wave at 32 R_s (=32 R_o), 5 hr. after introduction of the shock pulse at 18 R_s. The magnitudes of these three quantities are shown along a 90°×90° segment of a spherical surface at 32 R_s at t=5 hr. Note that the symbol, R_o, as used in this and the next figure is identical to the symbol, R_s, used in the text.

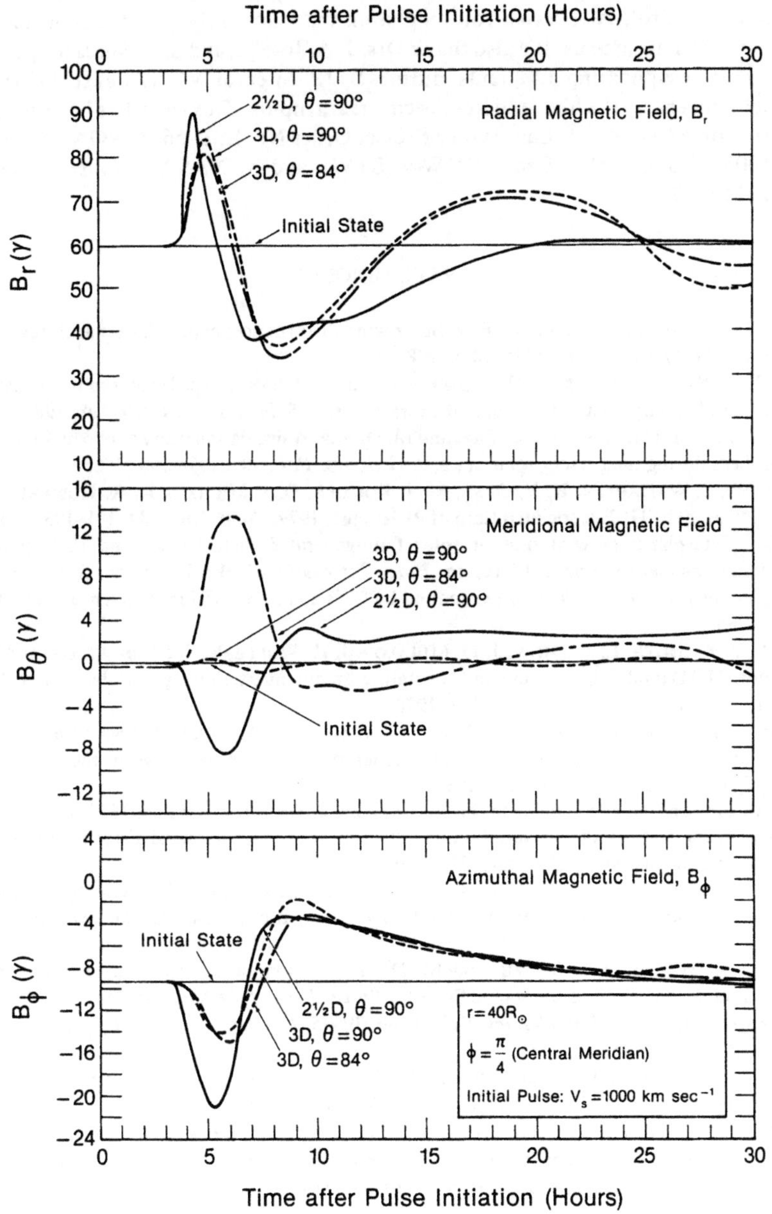

Fig. 11. Results of 3D simulation for the magnetic field components at 40 R_s (=40 R_0), one set at $\theta=90°$ and the other at $\theta=84°$, after passage of an interplanetary shock. The results for a 2-1/2D computation are also shown for $\theta=90°$ (ecliptic plane).

We wish to thank Drs. E. Smith and S. Bame for permission to use their ISEE-3 data information and Prof. A. Hewish for suggesting the transiently-initiated, corotating, high speed stream simulation to us. We also thank Drs. J. A. Joselyn and S. T. Suess for providing us with a list of disappearing filaments during STIP Interval VII (August 1979) and for suggestions concerning the high and low speed stream inputs. This work has been supported in part by Air Force Geophysics Laboratory Project Order No. ESD-5618, NOAA Contract No. NA84RA005132 and NASA Grant NHW-9 (STW), and by NASA Interagency Order No. W-15,361 (Mod. 2).

REFERENCES

AKASOFU, S.-I., K. HAKAMADA, and C. FRY, Solar wind disturbances caused by Solar flares: Equatorial plane, *Planet. Space Sci.*, **31**, 1435–1458, 1983.

CANE, H., R. G. STONE, J. FAINBERG, J. L. STEINBERG, and S. HOANG, Type II solar radio events observed in the interplanetary medium, I. General characteristics, *Solar Phys.*, **78**, 187–198, 1982.

AKASOFU, S.-I., W. FILLIUS, W. SUN, C. FRY, and M. DRYER, A simulation study of two major events in the heliosphere during the present solar cycle, *J. Geophys. Res.*, **90**, 8193–8211, 1985.

CLIVER, E. W., S. W. KAHLER, H. V. CANE, M. J. KOOMEN, D. J. MICHELS, R. A. HOWARD, and N. R. SHEELEY, Jr., The GLE-associated flare of 21 August, 1979, *Solar Phys.*, **89**, 181–193, 1983.

DRYER, M., Interplanetary evolution of solar flare-generated disturbances and their potential for producing magnetospheric activity, in *Proc. International Workshop on Solar Physics and Interplanetary Travelling Phenomena*, edited by C. Biao and C. de Jager, Science Press, Beijing, in press, 1985.

DRYER, M., Z. K. SMITH, E. J. SMITH, J. D. MIHALOV, J. H. WOLFE, R. S. STEINOLFSON, and S. T. WU, Dynamic MHD modeling of solar wind corotating stream interaction regions observed by Pioneer 10 and 11, *J. Geophys. Res.*, **83**, 4347–4352, 1978.

DRYER, M., S. T. WU, G. GISLASON, S. M. HAN, Z. K. SMITH, J. F. WANG, D. F. SMART, and M. A. SHEA, Magnetohydrodynamic modeling of interplanetary disturbances between the sun and earth, *Astrophys. Space Sci.*, **105**, 187–208, 1984.

GOLD, T., Plasma and magnetic fields in the solar system, *J. Geophys. Res.*, **64**, 1665–1674, 1959.

HAKAMADA, K. and S.-I. AKASOFU, Simulation of three-dimensional solar wind disturbances and resulting geomagnetic storms, *Space Sci. Rev.*, **31**, 3–70, 1982.

HAN, S. M., S. T. WU, and Y. NAKAGAWA, Numerical experimentation of spherically symmetric one-dimensional magnetohydrodynamic (MHD) wave propagation, *Computers and Fluids*, **10**, 159–172, 1982.

HAN, S. M., S. PANITCHOB, S. T. WU, and M. DRYER, A numerical simulation of three-dimensional transient ideal magnetohydrodynamic flows, in *Proceedings of the Southeastern 12th Conference on Theoretical and Applied Mechanics*, Vol. II, pp. 39–45, Auburn University Engineering Extension Service, Auburn, Alabama, 1984.

HIRSHBERG, J., Y. NAKAGAWA, and WELLCK, R. E., Propagation of sudden disturbances through a nonhomogeneous solar wind, *J. Geophys. Res.*, **79**, 3726–3730, 1974.

IVANOV, K. G., Observations and interpretations of interplanetary shock waves during STIP Intervals VI-XII and SMY/STIP Events 1-5, in *STIP Symposium on Solar/Interplanetary Intervals*, edited by M. A. Shea, D. F. Smart, and S. M. P. McKenna-Lawlor, pp. 89–100, Book Crafters, Inc., Chelsea, Michigan, 1984.

IVANOV, K. G., N. V. MIKERINA, and L. V. EVDOKIMOVA, Typical sequence of strong discontinuities in the head part of a non-stationary stream of interplanetary plasma, *Geomagn. Aeronom.*, **14**, 777–783, 1974 (in Russian).

IVANOV, K. G., N. V. MIKERINA, and L. V. EVDOKIMOVA, Classification and investigation of solar flare situations conformably to interplanetary and magnetospheric disturbances, in *Solar and Interplanetary Dynamics*, edited by M. Dryer and E. Tandberg-Hanssen, pp. 421–424, D. Reidel Publ. Co., Dordrecht, 1980.

JASTROW, R., Introductory remarks, in Proceedings of Symposium on the Exploration of Space, *J. Geophys. Res.*, **64**, 1647, 1959.

KANE, S. *et al.*, Energetics and interplanetary effects of the August 14 and 18, 1979, solar flares: Summary of observations made during SMY/STIP Event No. 1, in *Proc. STIP Symposium on Solar/Interplanetary Intervals*, edited by M. A. Shea, D. F. Smart, and S. M. P. McKenna-Lawlor, p. 319, Book Crafters, Inc., Chelsea, MI, USA, 1984.

MAXWELL, A. and M. DRYER, Measurements on a shock wave generated by a solar flare, *Nature*, **300**, 237–239, 1982.

PIZZO, V. J., A three-dimensional model of corotating streams in the solar wind, 3, Magnetohydrodynamic streams, *J. Geophys. Res.*, **87**, 4374, 1982.

PUDOVKIN, M. I. and A. D. CHERTKOV, The magnetic field of the solar wind, *Solar Phys.*, **50**, 213–229, 1976.

PUDOVKIN, M. I., S. A. ZAITSEVA, and E. E. BENEVOLENSKA, The structure and parameters of flare streams, *J. Geophys. Res.*, **84**, 6649–6652, 1979.

ROELOF, E. C., H. W. DODSON, and E. R. HEDEMAN, Dependence of radio emission in large Hα flares 1967–1970 upon the orientation of the local solar magnetic field, *Solar Phys.*, **85**, 339–349, 1983.

SMITH, Z. K., M. DRYER, R. W. FILLIUS, E. J. SMITH, and J. H. WOLFE, Compression of Jupiter's magnetosphere by the solar wind: Reexamination via MHD simulation of evolving corotating interaction regions, *J. Geophys. Res.*, **86**, 6773–6780, 1981.

SUESS, S. T., J. M. WILCOX, J. T. HOEKSMA, H. HENNING, and M. DRYER, Relationships between a potential field-source surface model of the coronal magnetic field and properties of the solar wind at 1 AU, *J. Geophys. Res.*, **89**, 3957–3961, 1984.

SUN, W., S.-I. AKASOFU, Z. K. SMITH, and M. DRYER, Calibration of the kinematic method of studying solar wind disturbances on the basis of a one-dimensional MHD solution and a simulation study of the heliosphere disturbances between November 22–December 6, 1977, *Planet Space Sci.*, **33**, 933–943, 1985.

TANG, F., S.-I. AKASOFU, E. SMITH, and B. TSURUTANI, Magnetic fields on the sun and the north–south component of transient variations of the interplanetary magnetic field at 1 AU, *J. Geophys. Res.*, **90**, 2703–2712, 1985.

TAPPIN, S. J., A. HEWISH, and G. R. GAPPER, Tracking a high-latitude corotating stream for more than half a solar rotation, *Planet. Space Sci.*, **32**, 1273–1281, 1984.

TSURUTANI, B. T., J. A. SLAVIN, Y. KAMIDE, R. D. ZWICKL, J. H. KING, and C. T. RUSSELL, Coupling between the solar wind and the magnetosphere: CDAW-6, *J. Geophys. Res.*, **90**, 1191–1204, 1985.

WOO, R. and J. W. ARMSTRONG, Measurements of a solar flare-generated shock wave at 13.1 R_0, *Nature*, **292**, 608–610, 1981.

WU, S. T., M. DRYER, and S. M. HAN, Non-planar MHD model for solar flare-generated disturbances in the heliospheric equatorial plane, *Solar Phys.*, **84**, 395–418, 1983.

Solar Wind-Magnetosphere Coupling, edited by Y. Kamide and J. A. Slavin, 209–231.

Solar Wind Control of Magnetospheric Configuration

C. T. RUSSELL

Department of Earth and Space Sciences, and Institute of Geophysics and Planetary Physics, University of California, Los Angeles, California 90024, U.S.A.

The solar wind exerts intimate control of the configuration of the magnetosphere by the reconnection of interplanetary magnetic field with the magnetospheric magnetic field. Thus the solar wind controls the energization of the magnetosphere proper including the ring current as well as the transport of magnetic flux to the tail and the consequent storage of energy there. The dynamics of the magnetopause are also controlled closely by the interplanetary magnetic field. The formation of flux transfer events, twisted tubes of flux across the dayside magnetopause, arise for southward interplanetary magnetic fields. Whether these flux tubes are connected to closed field lines or are completely open is not yet clearly resolved.

1. Introduction

There is intimate control of the configuration of the magnetosphere by the solar wind and the interplanetary magnetic field. This control is exerted by both normal stresses and tangential stresses on the magnetopause. Normal stresses arise from the combined dynamic pressure, thermal pressure and magnetic pressure of the solar wind. At the magnetopause, however, the post-shock flow, deflected, slowed and heated by its passage through the shock, is travelling tangentially to the magnetopause. Thus, here the normal stresses are exerted mainly through thermal and magnetic pressure.

If there were only normal stresses the magnetopause would be teardrop-shaped (SLUTZ and WINKELMAN, 1964), but, in fact, there are tangential stresses which elongate the night magnetosphere into a long tail-like configuration. The tangential stresses may be due to viscous effects, the eddy diffusion of momentum from the magnetosheath into the magnetosphere, due to fluid instabilities of the boundary layer such as the Kelvin-Helmholtz instability analogous to the action of wind over water, or due to reconnection of the interplanetary magnetic field with the terrestrial magnetic field. The magnetotail, being the end product of this tangential stress, is very sensitive to variations in this stress. The magnetic energy density integrated over the volume of the magnetotail represents a storage of energy which could be released if the normal stress were removed and the magnetosphere could assume a teardrop configuration. This reconfiguration of the night time magnetosphere occasioned by a decrease in the tangential stress on the magnetopause might be a benign event if it were not for the major fact that mechanism acting on the magnetic flux in the tail, i.e., reconnection, cuts field lines in two. Terrestrial field lines have to reconnect in the tail in order to reconfigure into a more dipole-like form. There are time delays in this process and thus the magnetotail does not appear to be immediately responsive to changes in the solar wind. Further, the tail may at times enter a metastable state, such as occurs when water is supercooled below its freezing point. In the case of water, the introduction of an ice crystal will initiate freezing but it may also occur spontaneously without any noticeable trigger. The colder the water, the smaller a perturbation necessary for the triggering of freezing, so that there is a temperature to which a container of water cannot be further cooled without the onset of freezing. So too it is likely that there is a limit to which the tail can store energy above its equilibrium state. If tail reconnection occurred in such a situation, we would judge it to be spontaneously or internally triggered.

The above discussion suggests that we must examine not only the steady-state but also the time-varying response of the magnetosphere to the solar wind if we are to understand solar wind control of the magnetospheric configuration. In this review, we will do just that. First we review the steady-state response of the magnetosphere to the solar wind. Then we examine the response to temporal variations in the solar wind input. We close with a discussion of unsteady reconnection at the magnetopause and the formation of structures that have been called Flux Transfer Events.

2. The Steady-State Response of the Magnetosphere

Nearly twenty-five years ago, DUNGEY (1961) postulated the "open" magnetosphere in which the interplanetary magnetic field reconnected with the terrestrial magnetic field. This model now forms the basis upon which the paradigm of much of our present understanding of the magnetosphere is built. Thus, it is important to review the basic features of the Dungey model. The top panel of Fig. 1 shows the steady-state configuration expected when the interplanetary magnetic field is

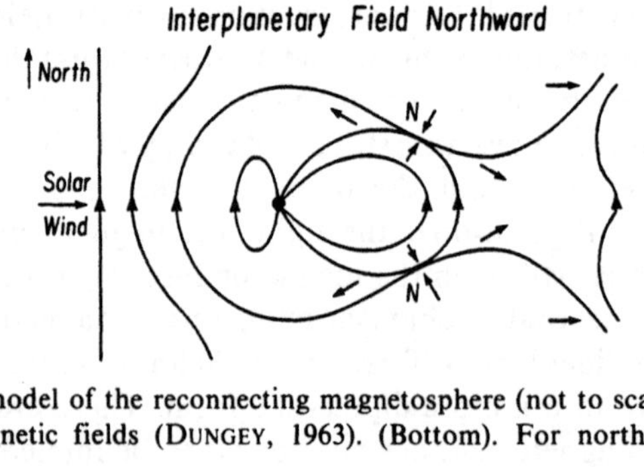

Fig. 1. The Dungey model of the reconnecting magnetosphere (not to scale). (Top). For southward interplanetary magnetic fields (DUNGEY, 1963). (Bottom). For northward interplanetary fields (DUNGEY, 1963).

southward. Moving toward the magnetosphere from the left-hand side of Fig. 1, the interplanetary field and solar wind plasma encounter the magnetic cavity. The interplanetary field line breaks, as does the terrestrial field line at the boundary, and the interplanetary and terrestrial lines reconnect to form two lines, one from the northern polar cap into the solar wind and one from the southern polar cap. The field lines continue to convect toward the right. At first the field line shortens but once the interplanetary end of the field passes the pole the field line lengthens. The field lines over the two polar caps are added to the north and south lobes of the tail. Eventually they sink to the mid-plane where they reconnect to form a field line connected to both the northern and the southern hemisphere of the earth and one connected to the solar wind at both ends. This interplanetary field line shortens as it moves tailward and the terrestrial field line shortens as it moves toward the earth. Eventually the terrestrial field line convects around to the dayside by moving out of the plane of the page and joins the reconnection process at the dayside magnetopause. In the regions in which the field lines are shortening, plasma is being accelerated and energy is flowing from the magnetic field into the plasma. In regions of magnetic field lengthening, plasma is being decelerated and energy is being transferred to the magnetic field. The entire process described is steady-state. However, if the rates of reconnection varied the system would have to respond in a time-varying manner.

As illustrated in the bottom panel of Fig. 1, DUNGEY (1963) also postulated that reconnection can occur for northward interplanetary magnetic fields. In this case the magnetic field lines of the solar wind and the magnetosphere reconnect behind the earth. This effectively transfers a closed field line on the nightside to the dayside as the field lines shorten after reconnection. There is no net transfer of flux, i.e., this is also a steady-state model. Field lines from the dayside convect out of the page and around to the nightside to replace the lost flux.

If this model of the magnetosphere is correct, the polar caps must be magnetically connected to the interplanetary medium at least when the interplanetary magnetic field is southward. There is ample evidence that this is so. First there is the response of the magnetosphere to a southward interplanetary magnetic field. The magnetopause erodes, the tail enlarges and there is an increase in geomagnetic activity (cf., BURCH, 1974; RUSSELL, 1974). There is the structure of the magnetopause, both its field and flow structure (PASCHMANN *et al.*, 1979; SONNERUP *et al.*, 1981). There is the evidence from energetic proton entry into the polar cap, for which the open model provides a tidy explanation (cf., MORFILL and SCHOLER, 1973; FENNELL, 1973). Finally, electron shadowing by the moon demonstrates the tail is open (LIN, 1968).

It is very clear that the magnetosphere can respond in a steady-state manner to steady-state reconnection. When the interplanetary magnetic field turns southward and stays southward at a nearly constant value, the magnetosphere is energized at a nearly constant rate. HONES *et al.* (1976) have observed continual earthward plasma sheet flow. CAAN *et al.* (1973) have shown that under these conditions the auroral zone and the magnetic field at synchronous orbit near midnight are continuously disturbed. At midlatitudes and in the magnetotail at such times the signatures of substorms are absent. Under such conditions KOKUBUN *et al.* (1977) have shown the existence of an *Sq*-type current system and PYTTE *et al.*, 1978) have shown that the auroral oval is continuously disturbed in what they have called a convection bay. Finally, BURTON *et al.* (1975a,b) have shown that the ring current injection depends only on the southward component of the interplanetary magnetic field and the solar wind velocity. It does not depend on the strength of auroral zone activity as measured by the *AE* index (RUSSELL *et al.*, 1974). The build up of ring current of course is the signature of the geomagnetic storm.

It is appropriate to discuss the geomagnetic storm, when we are discussing the steady-state response of the magnetosphere even though it is obvious that something has to change in the solar wind coupling in order for the storm to occur. However, once that change occurs the ring current can build up in a steady manner. When that constant input ceases, the magnetosphere appears to recover in essentially a steady manner.

A key item in the examination of the build up of the ring current is the functional form of the coupling to the solar wind input. In the late 1960's there were many correlation studies that showed that the direction of the IMF was an important factor (cf., FAIRFIELD and CAHILL, 1966; HIRSHBERG and COLBURN, 1969). In the 1970's, the work became more quantitative. First ARNOLDY (1971) proposed the half-wave rectifier model for the interaction in which reconnection occurred only for southward-directed interplanetary magnetic fields. This model was quite successful in predicting the *AE* index. Later a phenomenological model of substorms was developed based on the half-wave rectifier concept together with the observed response of the magnetosphere to southward turnings of the IMF (MCPHERRON *et al.*, 1973; RUSSELL and MCPHERRON, 1973). Since the key to explaining most of the observed phenomena was that there were different time delays in the response of different portions of the magnetosphere to a southward turning of the interplanetary magnetic field, we will defer discussion of this model to a later section. Next BURTON *et al.* (1975a,b) showed how to predict quite accurately the energy input to the magnetosphere. One of the important features of the ring current is that the energy stored in the ring current is directly determinable from the examination of the depression in ground based magnetograms. In a large storm, the order of 10^{23} ergs of particle energy are stored in the ring current. When the ring current is being energized the energy input into the ring current is a factor of 3 to 10 higher than that lost to particle precipitation or dissipated in auroral currents (PERRAULT and AKASOFU, 1978).

The energy input function found by BURTON *et al.* (1975a) is shown in Fig. 2. Here the energy injection rate is given in gammas/hour. The solar wind input is measured by the product of the solar wind velocity and the southward component of the interplanetary magnetic field. This is also the interplanetary electric field. When the interplanetary magnetic field is northward, i.e., the electric field is from dusk to dawn, there is no injection. When the interplanetary magnetic field is southward, i.e., the electric field is from dawn to dusk, the injection rate is linearly proportional to the interplanetary electric field.

This relationship was incorporated in a formula for predicting the *Dst* index by including a term proportional to the solar wind dynamic pressure to account for the changing position and strength of the magnetopause currents, by integrating the

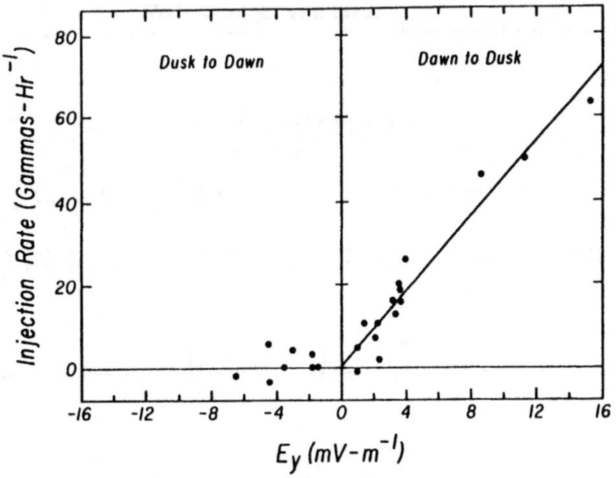

Fig. 2. The energization of the ring current measured in gammas per hour as a function of the dawn to dusk component of the interplanetary electric field (BURTON *et al.*, 1975a).

injection rate to build up the ring current and allowing a continual decay of the ring current with an 8-hour time constant. The resulting model predicted that the ring current would saturate at a constant value if the solar wind input remained steady long enough and would decay to zero eventually if the solar wind input stopped. Figure 3 shows a sample of the predictions of this model. Later, PERRAULT and AKASOFU (1978) attempted to improve on this model by altering slightly the functional form of the angular dependence of the input function on the interplanetary magnetic field. The model was still a very efficient rectifier. They also changed the dependence of the input function on the interplanetary magnetic field strength from linear to quadratic. However, these changes did not improve the correlation of the predicted magnetospheric response with that observed.

The development of the BURTON *et al.* (1975b) model had an important philosophical impact on studies of the geomagnetic storm. There had been the tendency to view geomagnetic storms as a product of substorm activity but the model treated storms independently of substorms, in accord with observations. The model also equally well explained gradual and sudden commencement storms and the long known independence of main phase minima and sudden commencement amplitudes.

3. Solar Wind Control of Magnetospheric Dynamics

The storage of energy in the ring current inflates the magnetosphere. This process is moderately slow, requiring several hours for a moderate sized storm and decaying over several days. The configuration of the magnetosphere can change much more rapidly, when only the outer reaches of the magnetosphere are involved. The magnetotail is particularly sensitive to changes in the tangential stress. To understand

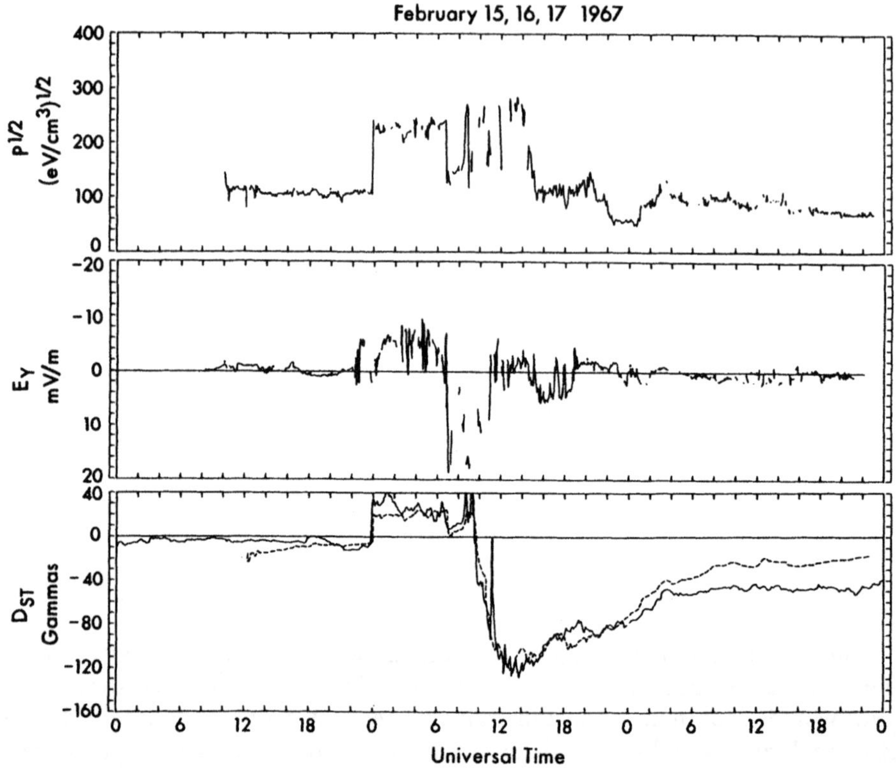

Fig. 3. The predicted response of the magnetosphere as given by the *Dst* index (dashed line) compared with the observed *Dst* index (solid line) given only the solar wind input shown in the top two panels (BURTON *et al.*, 1975b).

why this occurs we will examine a model of the sequence of dynamical changes in the magnetosphere that has been called a substorm. This model is illustrated in Figs. 4 and 5 (MCPHERRON, 1974). To treat the problem of flux transport and time delays, we arbitrarily divide the magnetosphere into three sections: the dayside magnetosphere, the tail lobes and the nightside magnetosphere and plasma sheet. When the interplanetary magnetic field turns southward the rate of reconnection or merging, M, at the dayside magnetopause increases. The newly merged flux is transported into the tail. It is removed from this region by reconnection in the tail which operates at rate, $\underline{R}$. The newly closed magnetic flux is convected to the dayside with rate, R. We do not expect instantaneous equality of all three of these rates, even though they must balance on the long term. Thus the amount of magnetic flux in the dayside magnetosphere, in the tail lobes and in the nighttime magnetosphere and plasma sheet must vary with time. This flux transport process is sketched in Fig. 4. The change in magnetic flux on the dayside is given by the time integration of the difference between the flux return rate, R, and the dayside merging rate, M. The change of magnetic flux

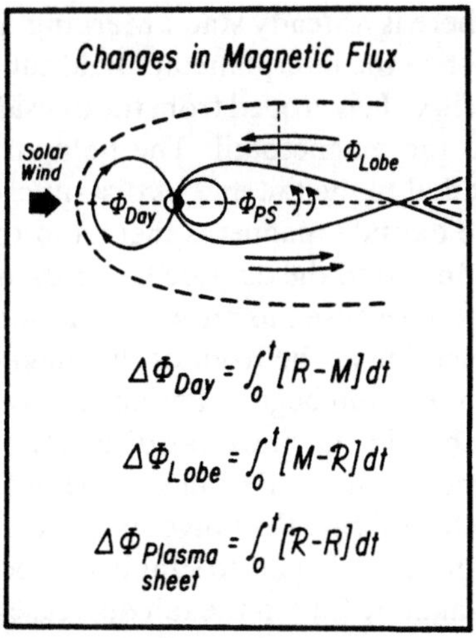

Fig. 4. Magnetic flux bookkeeping during substorms (MCPHERRON, 1974).

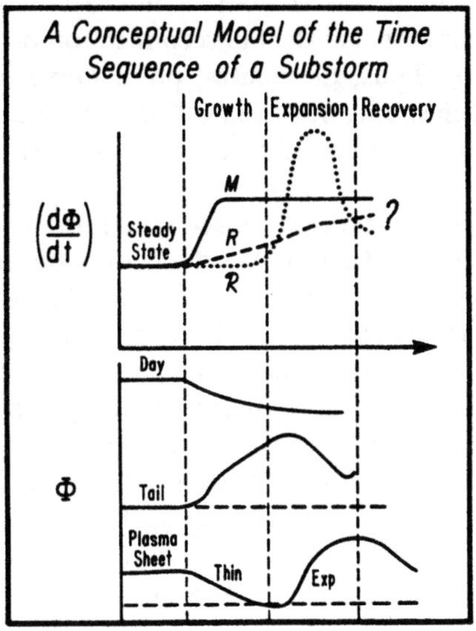

Fig. 5. Flux transport rates and resultant flux levels during substorms (MCPHERRON, 1974).

in the tail lobe is given by the time integration of the difference between the dayside merging rate, M, and the tail reconnection rate, $\mathcal{R}$. The change in magnetic flux in the plasma sheet is the time integration of the difference between the tail reconnection rate, $\mathcal{R}$, and the flux return rate to the dayside, R.

The phenomenology of a substorm can be explained by variations in the rate as shown in Fig. 5. Initially there is a steady state. Merging, tail reconnection and flux return are all in balance. When the interplanetary field turns southward, the dayside merging rate increases and flux, Φ, is eroded from the dayside magnetosphere (AUBRY *et al.*, 1970) and added to the magnetotail. The field strength in the magnetotail increases as more flux is added to the tail and its distant cross-section increases. The amount of flux in both the dayside magnetosphere and the plasma sheet decreases because flux is being transported to the dayside faster than it is being replenished by tail reconnection (if at all) and because this return rate is not sufficient to replenish the dayside flux removed by merging at the front of the magnetosphere.

At some point tail reconnection begins. Perhaps, it was simply necessary for the tail to become more stretched in the antisolar direction in order for it to become unstable to near-earth reconnection. Perhaps some interplanetary perturbation triggered tail reconnection by altering the tangential stress on the tail. Whatever the cause, the onset of tail reconnection begins to supply magnetic flux from the tail lobes to the plasma sheet. The magnetic flux in the tail decreases, the magnetic flux in the plasma sheet increases and the magnetic field in the night magnetosphere becomes more dipole-like.

If flux were convected to the dayside as fast as it were supplied by tail reconnection, the neutral point at which tail reconnection is occurring could remain stationary. However, it appears that the tail reconnection rate far exceeds that of the convection or return rate. Thus, the neutral point moves down the tail once the reconnection process reaches open field lines. For a more detailed treatment of this process the interested reader is referred to MCPHERRON *et al.* (1973) and RUSSELL and MCPHERRON (1973).

Reconnection apparently begins on closed field lines which has some important topological consequences for the tail. Reconnection on closed field lines forms magnetic islands as illustrated in Fig. 6. In the center of this island is an O-type neutral line and on the front or earthward side there is an X-type neutral line. On the earthward side of the O-type neutral line the magnetic field is southward and on the tailward side it is northward. As illustrated in the top panel of Fig. 6, the magnetic island is anchored to the earth by field lines which close around the far side of the island while reconnection is still occurring on closed field lines. When the reconnecting region reaches the open field lines, there no longer exists an anchor. Furthermore, the plasma accelerated at the reconnection point blows against the plasmoid (RUSSELL, 1974; HONES, 1977). As amply confirmed by the recent ISEE-3 measurements in the distant tail (cf., BAKER *et al.*, 1984; HONES *et al.*, 1984) the magnetic island or plasmoid as it has been termed (HONES, 1979), is ejected down the tail.

The substorm clearly has several phases. The first phase, or growth phase (MCPHERRON, 1970) occurs when dayside reconnection begins. Some energy is deposited in the auroral oval during this stage and some is stored in the tail. The expansion phase occurs at the onset of rapid tail reconnection. This may be when tail reconnection first reaches open field lines having begun earlier. It could also be triggered by sudden increases in solar wind dynamic pressure (KOKUBUN *et al.*, 1977) and by northward turnings of the IMF (CAAN *et al.*, 1977). During this stage, the

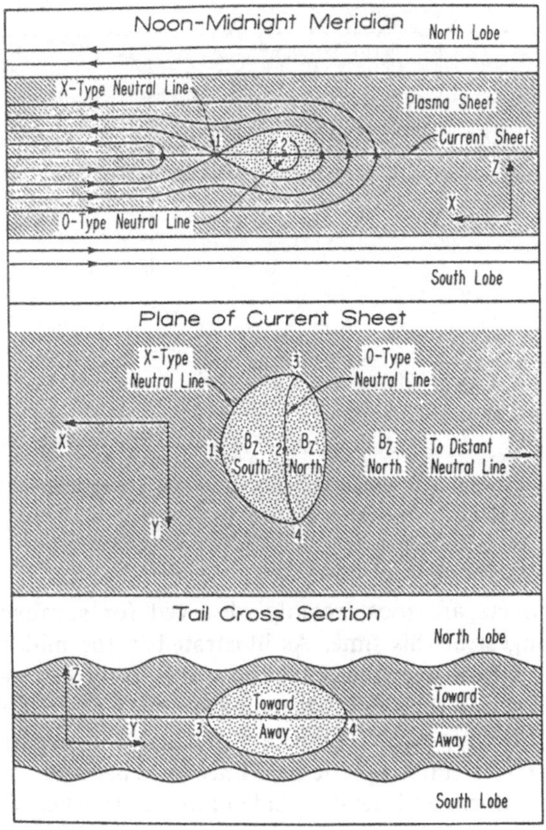

Fig. 6. A magnetic island or plasmoid formed in the tail by the initiation of reconnection on closed field lines (RUSSELL, 1974).

energy deposition is clearly dominated by the release of stored energy.

At quiet times between substorms, the neutral point lies far down the tail beyond about 100 R_e. However, there is some disagreement among the ISEE-3 experimenters as to exactly where it is (SCHOLER *et al.*, 1983; TSURUTANI *et al.*, 1984). It is important to realize that the neutral line does not extend straight across the tail (RUSSELL, 1977). As shown in Fig. 7, we expect the neutral line to be closest to the earth in the center of the tail. This is consistent with the recent deep tail ISEE-3 measurements, and resolves the conflict between conclusions based on the average north-south component of the field and the flow data. Since the plasma sheet is thinnest in the center of the tail and since the neutral point is closest to the earth in the center of the tail, we expect the center of the tail to be the region of strong flows and reconnection. This appears to be consistent with observations (HONES and SCHINDLER, 1979) but no explicit study of the cross tail location of the neutral point has been performed.

Asymmetries do occur in the convection pattern controlled by the B_y component of the IMF (cf., HEPPNER, 1972). This can occur for both northward and southward

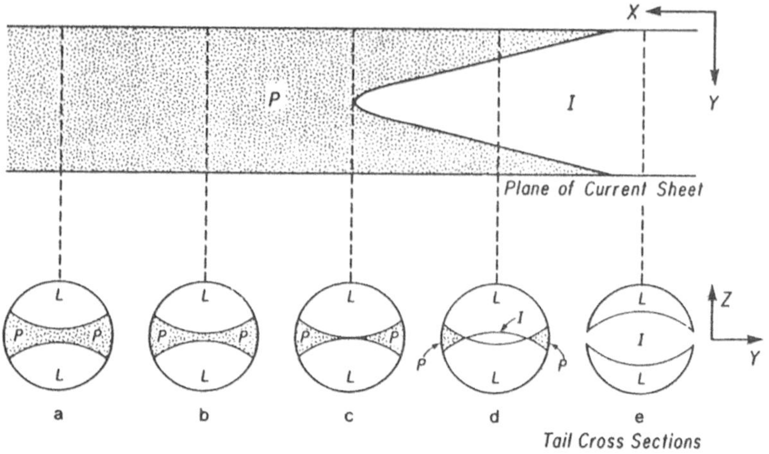

Fig. 7. The expected shape of the distant neutral line in the tail. The earth is on the left-hand side of this diagram (RUSSELL, 1977).

IMF, although effects are most readily observed for southward IMF because convection is strongest at this time. As illustrated in the middle panel of Fig. 8, (RUSSELL, 1972) the flow over the polar cap will be pulled to the dawn side in the north and the dusk side in the south when the IMF has a positive B_y component, i.e., points in the direction opposite to planetary motion. This occurs because the solar wind to which the terrestrial magnetic field lines have become connected flows over the dawn side in the north and the dusk side in the south when B_y is positive.

In the above sections, we have discussed reconnection essentially in terms of a single neutral point. Depending on the boundary conditions, formation of multiple neutral points can occur. This is called tearing of the current sheet. This has been postulated to occur in the tail by SCHINDLER and NESS (1972). This, in fact, may be true at times but seems to be irrelevant to tail dynamics except in the formation of magnetic islands which requires the existence of two X-type neutral points.

Finally, substorm analogs appear to occur in two other locations in the solar system. Two ribbon flares on the sun are thought to be caused by reconnection in a tail-like field configuration at the top of an arcade of field lines arching up from the solar surface. Comet tails are also disrupted by what may be near coma reconnection. However, both processes differ in some important aspects. In the case of solar flares, the energy probably arises from the motions at the feet of the field lines, whereas in substorms activity at high altitudes drags the feet of the field lines in the earth's ionosphere. In comets, tail reconnection will form a plasmoid or magnetic island but this island will surround the cometary coma and not be ejected down the tail.

4. Solar Wind Control of Dayside Reconnection and Tearing

Many of the same issues that occur in examining reconnection in the magnetotail

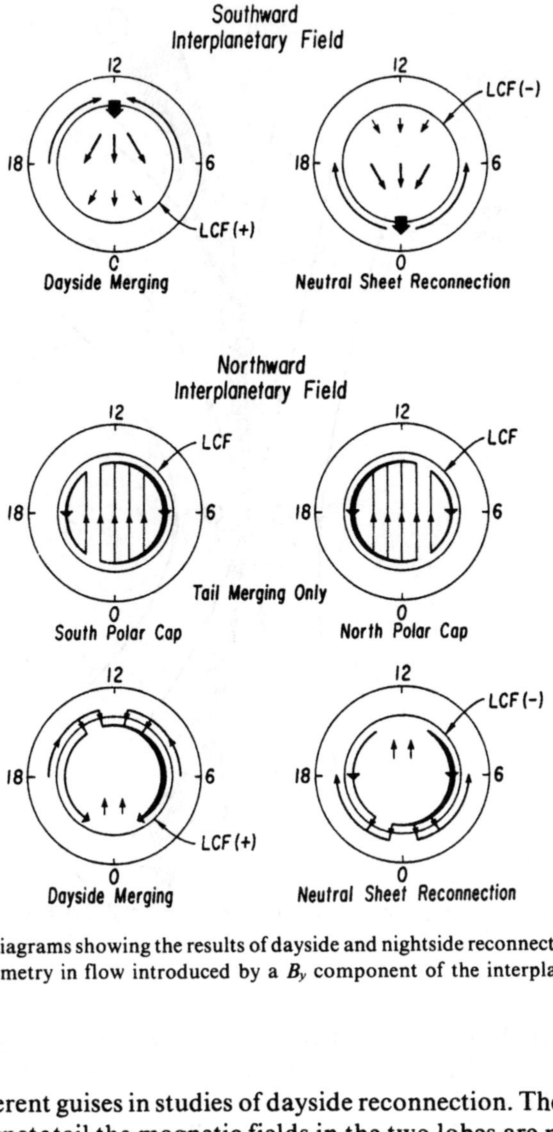

Fig. 8. Schematic diagrams showing the results of dayside and nightside reconnection on the polar cap as well as the asymmetry in flow introduced by a B_y component of the interplanetary magnetic field (RUSSELL, 1972).

also occur in different guises in studies of dayside reconnection. The major difference is that in the magnetotail the magnetic fields in the two lobes are nearly antiparallel while at the magnetopause the fields on either side of the current layer may have any relative orientation. Occasionally, the fields on either side of the boundary are, in fact, almost as antiparallel as fields in the tail. During some of these periods accelerated flows, much like the accelerated flows in the magnetotail, are observed (PASCHMANN *et al.*, 1979; SONNERUP *et al.*, 1981). However, at other times accelerated flows are not observed (PAPAMASTORAKIS *et al.*, 1984). Figure 9 shows the ISEE-1 and -2 trajectory through the reconnecting magnetopause during a particularly well documented reconnection event (PASCHMANN *et al.*, 1979). This figure illustrates as does Fig. 1 an important difference between dayside and tail reconnection. The

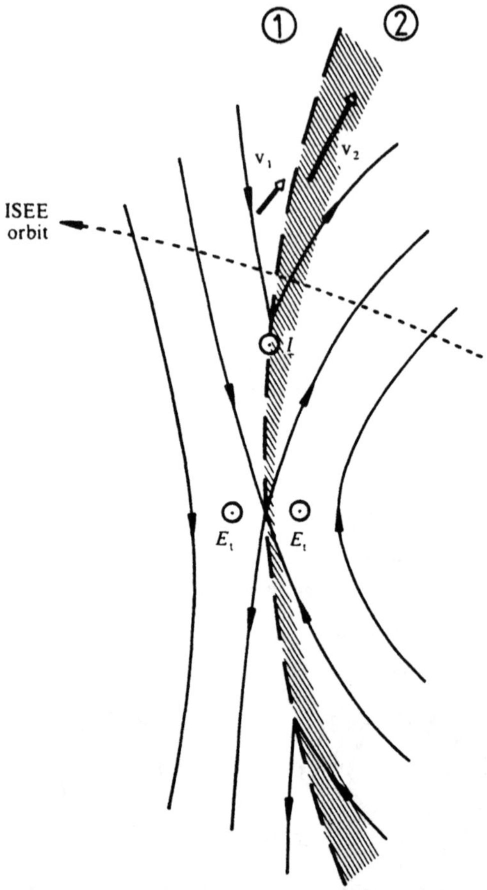

Fig. 9. The reconnecting magnetopause as detected by the ISEE-1 and -2 spacecraft (PASCHMANN *et al.*, 1979).

accelerated flows on the dayside are free to accelerate continuously, independent of the reconnection rate. Tail reconnection accelerates, at least on one side of the X-line, onto closed field lines that can fill up the available volume on the nightside and effectively push reconnection point down the tail. The X-point on the nose of the magnetopause in contrast has no such reason to move.

As mentioned above, however, the magnetic geometry is not always one of simple antiparallel fields. When the interplanetary magnetic field is not exactly due southward, the merging point might move away from the subsolar region. One postulate is that reconnection occurs on the magnetopause at those locations where the magnetospheric and magnetosheath fields are most antiparallel (CROOKER, 1979). Figure 10 shows the implication of this possible control by the IMF orientation as calculated using a magnetospheric magnetic field model and the convected-field gas dynamic model for the magnetosheath (after LUHMANN *et al.*, 1984). When the

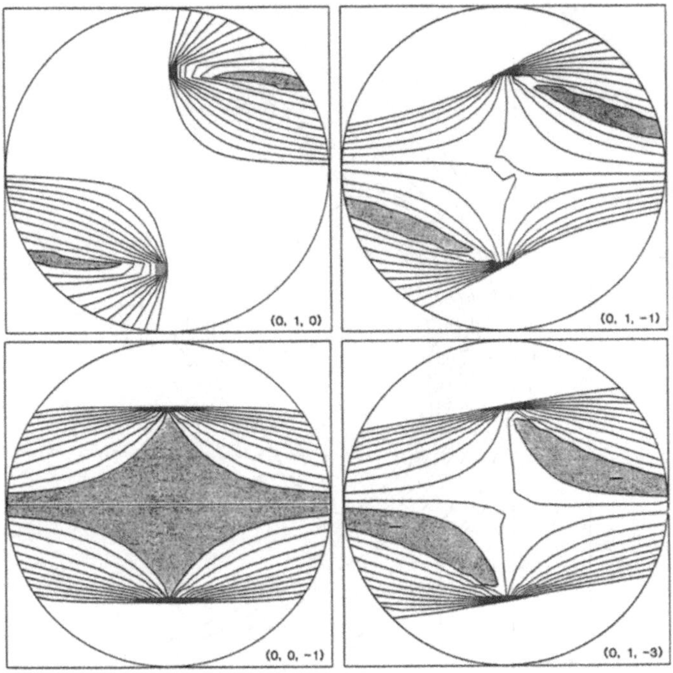

Fig. 10. Predicted sites (shaded) for merging under the anti-parallel merging hypothesis (after LUHMANN *et al.*, 1984).

magnetic field in the solar wind is due south, the reconnection region (dark shading) is along the equatorial region (see lower left panel). As the IMF turns from due south to slightly toward dusk (lower right hand panel), the reconnection region splits into two regions off the equator in the northern dusk quadrant and the southern dawn quadrant. With further rotations of the IMF toward the horizontal, the antiparallel regions become more and more separated and farther from the equator. This behavior has not been confirmed by space data for the terrestrial magnetosphere in part because of the apparent unsteady nature of reconnection.

It is very difficult to infer the behavior of the magnetopause from measurements at a single site even when the measurements come from dual spacecraft like ISEE-1 and -2 which allow the velocity of the boundary to be calculated. One way around this problem is to simulate the solar wind magnetosphere interaction. Two types of simulations have been carried out: computer simulations and laboratory simulations. Both have their limitations. The computer simulations can reproduce realistic plasma conditions but they are limited in their spatial resolution. Laboratory devices can provide adequate spatial resolution but one cannot properly scale the interaction to fit in a laboratory device. The laboratory simulations are thus limited simulations in which the relative size of plasma scale length is preserved only to the extent that they are ordered properly. Nonetheless they are instructive. Figure 11 shows the result of

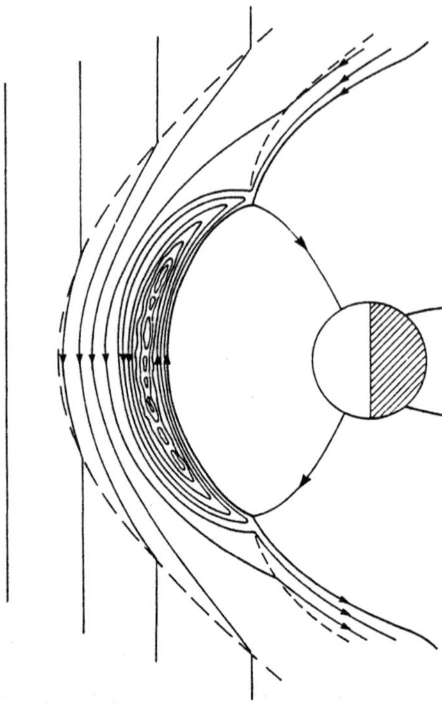

Fig. 11. Magnetopause subject to tearing at high Mach numbers in laboratory experiments (DUBININ *et al.*, 1980).

laboratory simulations of the magnetopause at an Alfven Mach number of 10 (DUBININ *et al.*, 1980). The interface between the magnetosheath and the magnetospheric magnetic fields is covered with magnetic islands, pairs of X-type and O-type neutral points. At low Mach numbers, 1 to 2.5 M_A, a pattern more resembling that of Fig. 9 is seen (DUBININ *et al.*, 1980; BAUM and BRATENAHL, 1982). Thus, it is possible that the structure of the magnetopause depends on the Alfvenic Mach number but this possible dependence has not been carefully studied with spaceflight data. One case of a sequence of magnetopause crossings with very low Mach numbers was studied by GOSLING *et al.* (1982). Figure 12 shows hodograms of the magnetic field signature through one of the magnetopause crossings at this time. The features to note are first that there is a clear rotation of the field and that there is weak evidence for an average normal component (the B_k component). Also, the crossing is slow, lasting over 18 minutes. Finally, there are knots in the hodograms. These knots which cause normal component variations suggest that tearing of the current sheet occurs even at low Mach number conditions.

A stronger perturbation of the normal component of the magnetic field at the magnetopause is associated with the flux transfer event. Figure 13 shows the magnetic field from ISEE-1 and -2 plotted versus time in boundary normal coordinates at a

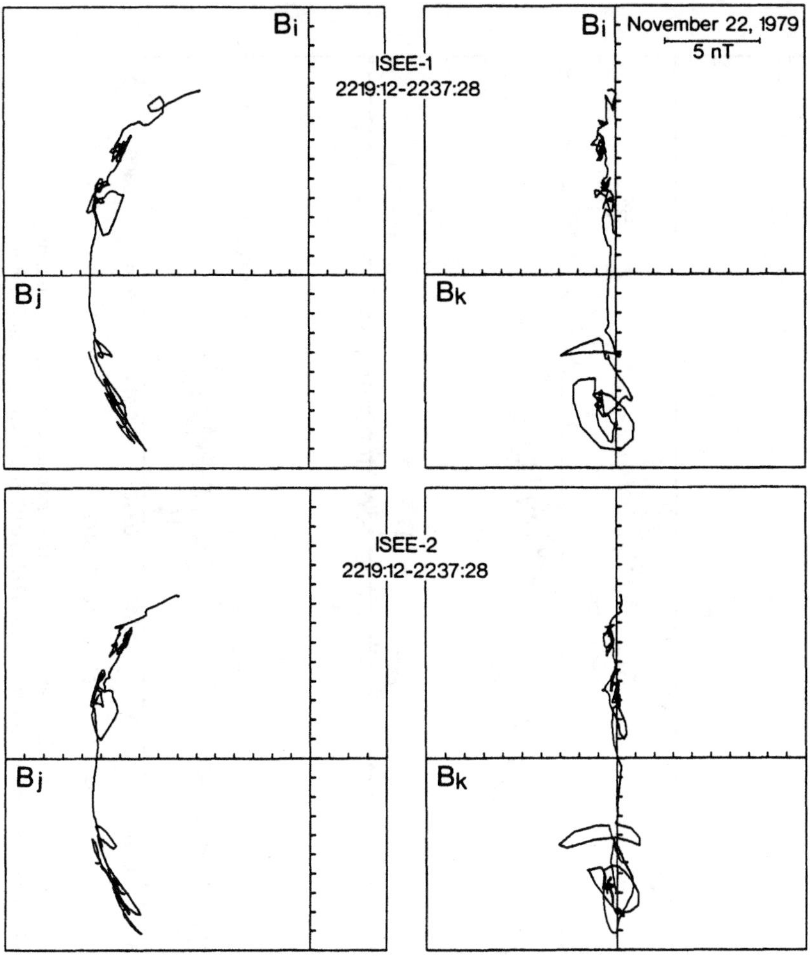

Fig. 12. Hodograms of ISEE-1 and -2 magnetic field measurements in minimum variance coordinates at time of very low Mach number, as spacecraft crossed magnetopause.

time when flux transfer events were occurring (RUSSELL and ELPHIC, 1978). Flux transfer events involve a large deviation of the normal component of the field (B_N here), a field strength enhancement, and a twist away from the direction of the magnetospheric field. Flux transfer events also contain flowing magnetosheath plasma at low energies and energetic magnetospheric particles at high energies. These properties led us to interpret these events in terms of time varying Dungey reconnection as shown in Fig. 14. In this model reconnection started and stopped convecting a flux to the magnetosphere as shown. In order to be consistent with observations this flux tube had to be twisted (SAUNDERS *et al.*, 1984) but no satisfactory mechanism has been proposed to produce this twist.

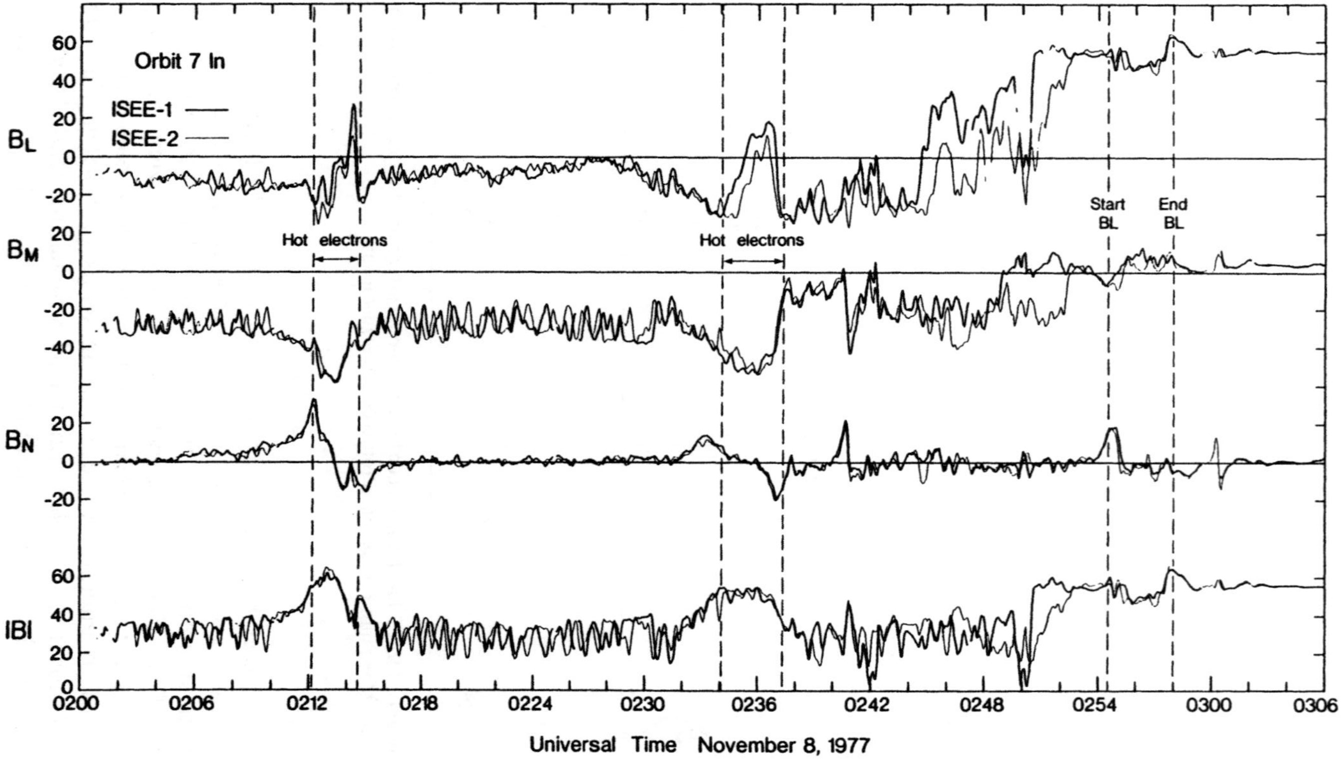

Fig. 13. Magnetic field measurements from ISEE-1 and -2 in boundary normal coordinates versus time during an interval in which flux transfer events were seen at 0213 and 0236 UT (RUSSELL and ELPHIC, 1978).

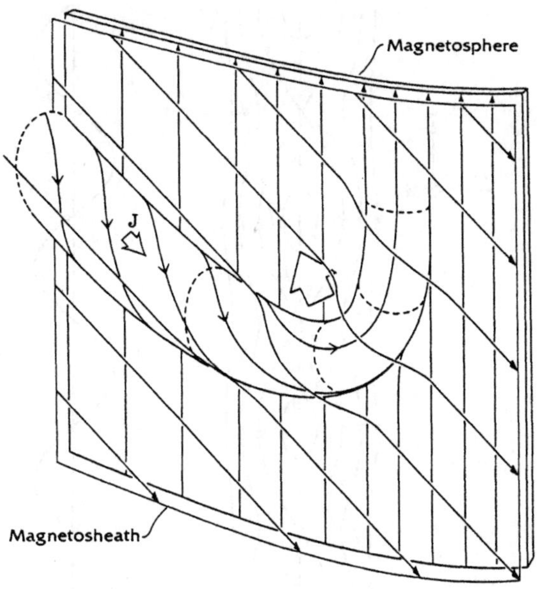

Fig. 14. Schematic of a twisted flux transfer event interpreted as time varying single neutral line reconnection (RUSSELL, 1984).

Figure 15 shows a picture of a magnetopause current sheet that has been subject to tearing together with the expected signature in the normal component. This model was rejected as a model for the flux transfer event because the signature in the normal component was always (at least in the northern hemisphere) a +/− pair. Passage through a torn current sheet would give a sequence of +/− signatures or at least the plus and minus components would be encountered first at random.

Recently, the tearing island model has been proposed to explain FTE's by LEE and FU (1985). As shown in Figure 16, they propose the occurrence of a pair of islands rather than an infinite sequence of islands. The convection of islands to the north and to the south on either side of the equator would give the observed signature of the normal component. Multiple connection of field lines along three extended X-lines on the magnetopause would produce a twisted structure. A more recent model of CROOKER (1985) extends her earlier work on antiparallel merging to multiple neutral lines and comes to similar conclusions.

Neither of these two models, however, were the first multiple neutral line models for FTE's. PODGORNY *et al.* (1980), in fact, proposed a magnetic curl across the dayside magnetopause as shown in Fig. 17. This curl was produced by reconnection at high latitudes near the north and south polar cusp. Field line 1 in Fig. 17 consists of parts of two interplanetary magnetic field lines that become connected to each other via a section of terrestrial magnetic field. The tension in the field then shrinks the north-south segment making room near the cusps for more reconnection. Field lines 2 and 3 show the further evolution of the line. This process builds up a flux bundle

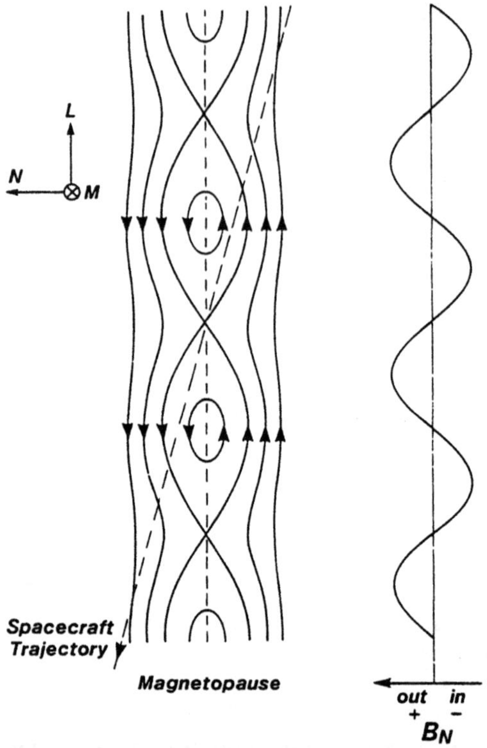

Fig. 15. Schematic of tearing at the magnetopause. Slanted spacecraft trajectory is due to upward motion of magnetic islands.

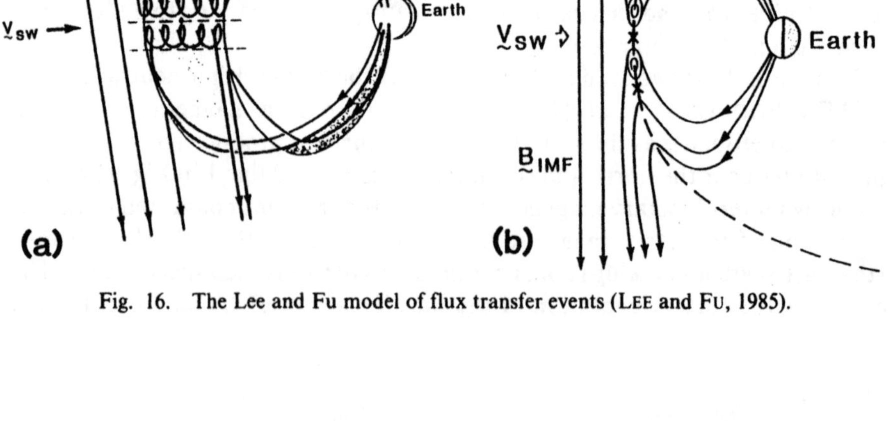

Fig. 16. The Lee and Fu model of flux transfer events (LEE and FU, 1985).

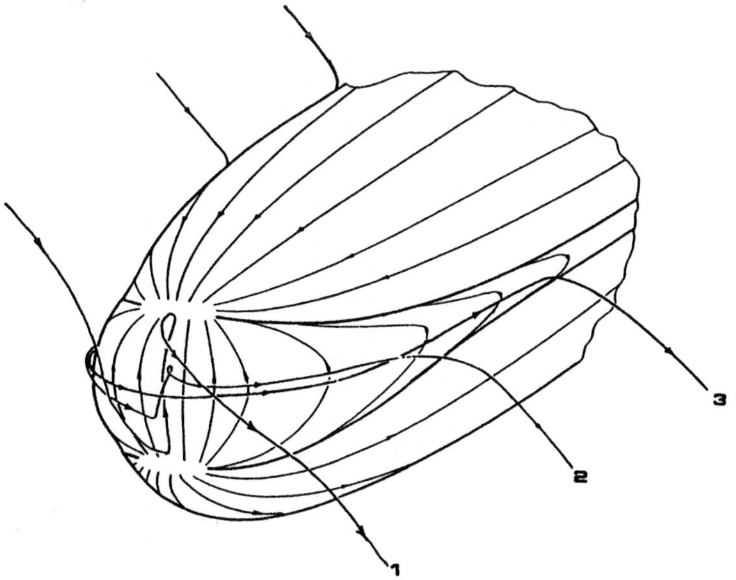

Fig. 17. The magnetic curl of PODGORNY *et al.* (1980).

across the nose of the magnetosphere. This is not a steady-state process. The bundle has to eventually roll off the magnetopause.

Figure 18 shows how Fig. 15, the tearing model, has to be modified to explain the flux transfer event. First, there must exist at least two X-lines at which reconnection can occur. In the model of PODGORNY *et al.* (1980) these two X-lines are at the polar cusp. In CROOKER's (1985) model the multiple X-lines are at lower latitudes. In LEE and FU's (1985) model there are three X-lines at lower latitudes. In any event, as long as there are at least two X-lines preferably near the equator a twisted flux tube will build up at the equator. This tube will grow only so big and have to roll off the nose of the magnetopause. If the tube rolls to the north it will produce a +/− signature as observed. If it rolls to the south it will produce a −/+ signature. This tearing model or curl formation model is attractive. It is interesting that this dayside phenomenon has many of the features of a nightside plasmoid. Thus, the dayside and nightside reconnection phenomena may not be all that different.

It is difficult to determine from present studies which, if either, model is correct. Flux transfer events occur whenever the interplanetary field is southward (RIJNBEEK *et al.*, 1984) with greatest occurrence rates for due southward fields (BERCHEM and RUSSELL, 1984). They occur all over the magnetopause when all IMF directions are considered but occur only along a diagonal swath across the nose of the magnetosphere when the IMF is aligned the diagonal (RUSSELL *et al.*, 1985). We note that we would expect some tendency to align with the IMF cross flow direction for the FTE model based on time varying reconnection (Fig. 14) and those based on multiple X-lines (Figs. 16 and 17). In fact, it is not clear why PODGORNY *et al.* (1980) drew the

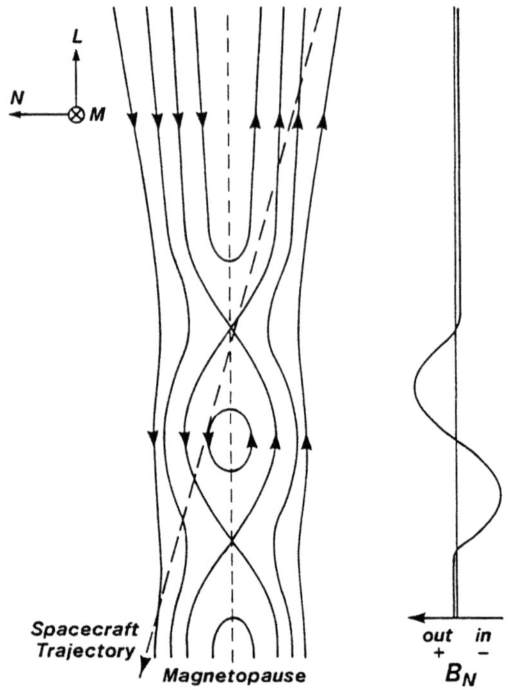

Fig. 18. Schematic of a flux transfer event, interpreted as caused by reconnection at two neutral lines near the nose. Slanted spacecraft trajectory reflects upward motion of this FTE.

magnetic curl as forming in the equatorial plane. In short then, the main fact we know about the solar wind control of FTE occurrence is that the occurrence rate maximizes for southward IMF.

5. Summary and Conclusions

The solar wind exerts intimate control over the configuration of the magnetosphere. It controls the energization and inflation of the magnetosphere during geomagnetic storms. From a knowledge of the solar wind velocity and density and the interplanetary magnetic field strength and orientation it is possible to predict the energization of the ring current and reproduce quite faithfully the magnetic measurements made by ground-based magnetometers at mid-latitudes.

The solar wind not only controls the steady-state or quasi-steady-state magnetosphere but also its dynamics. The variations in the tangential stress on the magnetosphere are mainly associated with variations in the interplanetary magnetic field rather than the solar wind velocity and density in part because the latter two quantities are less variable. When the IMF turns southward from a long period of northward orientation the magnetosphere reacts by transporting magnetic flux from the dayside magnetosphere to the tail over the polar cap. This results in a magnetic tail which is larger and more tail-like. Eventually, the tail returns this magnetic flux to the night magnetosphere through reconnection. The resulting sudden release of energy provides the expansion phase of a substorm. This sudden release may at times be associated with northward turnings and stress decreases.

Smaller scale phenomena such as the events that have been called flux transfer events are also controlled by the solar wind. It is clear that these events are associated with southward IMF. However, whether these events represent time varying Dungey-type reconnection or island formation via tearing at two or more neutral lines is not presently clear. Much more work needs to be done on the structure of flux transfer events and conditions controlling their occurrence.

This work was supported by the National Aeronautics and Space Administration under research contract NAS5-28448.

REFERENCES

ARNOLDY, R. L., Signature in the interplanetary medium for substorms, *J. Geophys. Res.*, **76**(22), 5189–5201, 1971.

AUBRY, M. P., C. T. RUSSELL, and M. G. KIVELSON, Inward motion of the magnetopause before a substorm, *J. Geophys. Res.*, **75**, 7018, 1970.

BAKER, D. N., S. J. BAME, J. BIRN, W. C. FELDMAN, J. T. GOSLING, E. W. HONES, Jr., R. D. ZWICKL, J. A. SLAVIN, E. J. SMITH, B. T. TSURUTANI, and D. G. SIBECK, Direct observations of passages of distant neutral line (80–140 R_e) following substorm onsets: ISEE-3, *Geophys. Res. Lett.*, **11**, 1042–1045, 1984.

BERCHEM, J. and C. T. RUSSELL, Flux transfer events on the magnetopause: Spatial distribution and controlling factors, *J. Geophys. Res.*, **89**, 6689–6703, 1984.

BURCH, J. L., Observations of interactions between interplanetary and geomagnetic fields, *Rev. Geophys. Space Phys.*, **12**(3), 363–378, 1974.

BAUM, P. J. and A. BRATENAHL, The laboratory magnetosphere, *Geophy. Res. Lett.*, **9**, 435, 1982.

BURTON, R. K., R. L. MCPHERRON, and C. T. RUSSELL, The terrestrial magnetosphere: A half-wave rectifier of the interplanetary electric field, *Science*, **189**, 717, 1975a.

BURTON, R. K., R. L. MCPHERRON, and C. T. RUSSELL, An empirical relationship between interplanetary conditions and *Dst*, *J. Geophys. Res.*, **80**, 4204, 1975b.

CAAN, M. N., R. L. MCPHERRON, and C. T. RUSSELL, Solar wind and substorm-related changes in the lobes of the geomagnetic tail, *J. Geophys. Res.*, **78**, 8087–8096, 1973.

CAAN, M. N., R. L. MCPHERRON, and C. T. RUSSELL, Characteristics of the association between the interplanetary magnetic field and substorms, *J. Geophys. Res.*, **82**, 4837–4842, 1977.

CROOKER, N. U., Dayside merging and cusp geometry, *J. Geophys. Res.*, **84**, 951, 1979.

CROOKER, N. U., A split separator line merging model of the dayside magnetopause, *J. Geophys. Res.*, **90**, 12104–12110, 1985.

DUBININ, E. M., I. M. PODGORNY, and Y. N. POTANIN, Structure of the magnetic field at the boundary of the magnetosphere: Analysis of a simulation experiment, *Cosmic Res.*, **18**, 77, 1980.

DUNGEY, J. W., Interplanetary field and auroral zones, *Phys. Rev. Lett.*, **6**, 47, 1961.

DUNGEY, J. W., The structure of the exosphere or adventures in velocity space, in *Geophys. The Earth's Environment*, edited by C. DeWitt, J. Hieblot, and A. Lebeau, Gordon and Breach, p. 505, New York, 1963.

FAIRFIELD, D. H. and L. J. CAHILL, Jr., Transition region magnetic field and polar magnetic disturbances, *J. Geophys. Res.*, **71**, 155–169, 1966.

FENNELL, J. F., Access of solar protons to the Earth's polar caps, *J. Geophys. Res.*, **78**, 1036–1046, 1973.

GOSLING, J. T., J. R. ASBRIDGE, S. J. BAME, W. C. FELDMAN, G. PASCHMANN, N. SCKOPKE and C. T. RUSSELL, Evidence for quasi-stationary reconnection at the dayside magnetopause, *J. Geophys. Res.*, **87**, 2147–2158, 1982.

HEPPNER, J. P., Polar cap electric field distributions related to the interplanetary magnetic field direction, *J. Geophys. Res.*, **77**, 4877, 1972.

HIRSHBERG, J. and D. S. COLBURN, Interplanetary field and geomagnetic variations: A unified view, *Planet. Space Sci*, **17**, 1183–1206, 1969.

HONES, E. W., Jr., Substorm processes in the magnetotail: Comments on 'On hot tenuous plasmas, fireballs and boundary layers in the earth's magnetotail' by L. A. Frank, K. L. Ackerson, and R. P. Lepping, *J. Geophys. Res.*, **82**, 5633, 1977.

HONES, E. W., Jr., Transient phenomena in the magnetotail and their relation to substorms, *Space Sci. Rev.*, **23**, 393, 1979.

HONES, E. W., Jr. and K. SCHINDLER, Magnetotail plasma flow during substorms: A survey with Imp 6 and Imp 8 satellites, *J. Geophys. Res.*, **84**, 7155, 1979.

HONES, E. W., Jr., S. J. BAME, and J. R. ASBRIDGE, Proton flow measurements in the magnetotail plasma sheet made with Imp 6, *J. Geophys. Res.*, **81**, 227–234, 1976.

HONES, E. W. Jr., J. BIRN, D. N. BAKER, S. J. BAME, W. C. FELDMAN, D. J. MCCOMAS, R. D. ZWICKL, J. A. SLAVIN, E. J. SMITH, and B. T. TSURUTANI, Detailed examination of a plasmoid in the distant magnetotail with ISEE-3, *Geophys. Res. Lett.*, **11**, 1046–1049, 1984.

KOKUBUN, S., R. L. MCPHERRON, and C. T. RUSSELL, Triggering of substorms by solar wind discontinuities, *J. Geophys. Res.*, **82**, 74–86, 1977.

LEE, L. C. and Z. F. FU, A theory of magnetic flux transfer at the earth's magnetopause, *Geophys. Res. Lett.*, **12**, 105–108, 1985.

LIN, R. P., Observations of lunar shadowing of energetic particles, *J. Geophys. Res.*, **73**, 3066–3071, 1968.

LUHMANN, J. G., R. J. WALKER, C. T. RUSSELL, N. U. CROOKER, J. R. SPREITER, and S. S. STAHARA, Patterns of potential magnetic field merging sites on the dayside magnetopause, *J. Geophys. Res.*, **89**, 1739–1742, 1984.

MCPHERRON, R. L., Growth phase of magnetospheric substorms, *J. Geophys. Res.*, **75**, 5592, 1970.

MCPHERRON, R. L., Current status of the growth phase controversy, (abstract), *EOS Trans. AGU*, **55**(55), 994, 1974.

MCPHERRON, R. L., C. T. RUSSELL, and M. P. AUBRY, Satellite studies of magnetospheric substorms on August 15, 1968. 9. Phenomenological model for substorms, *J. Geophys. Res.*, **78**, 3131–3149, 1973.

MORFILL, G. and M. SCHOLER, Study of the magnetosphere using energetic particles, *Space Sci. Rev.*, **15**, 267– 354, 1973.

PAPAMASTORAKIS, I., G. PASCHMANN, N. SCKOPKE, S. J. BAME, and J. BERCHEM, The magnetopause as a tangential discontinuity for large rotation angles, *J. Geophys. Res.*, **89**, 127, 1984.

PASCHMANN, G., B. U. O. SONNERUP, I. PAPAMASTORAKIS, N. SCKOPKE, G. HAERENDEL, S. J. BAME, J. R. ASBRIDGE, J. T. GOSLING, C. T. RUSSELL, and R. C. ELPHIC, Plasma acceleration at the earth's magnetopause: Evidence for reconnection, *Nature*, **282**, 243–246, 1979.

PERREAULT, P. and S-I. AKASOFU, A study of geomagnetic storms, *Geophys. J. R. Astron. Soc.*, **54**, 547–573, 1978.

PODGORNY, I. M., E. M. DUBININ, and Yu. N. POTANIN, On magnetic curl in front of the magnetopause boundry, *Geophys. Res. Lett.*, **7**, 247–250, 1980.

PYTTE, T., R. L. MCPHERRON, E. W. HONES, Jr., and H. I. WEST, Jr., Multiple-satellite studies of magnetospheric substorms: Distinction between polar magnetic substorms and convection-driven negative bays, *J. Geophys. Res.*, **83**, 663–679, 1978.

RIJNBEEK, R. P., S. W. H. COWLEY, D. J. SOUTHWOOD, and C. T. RUSSELL, A survey of dayside flux events observed by ISEE-1 and -2 magnetometers, *J. Geophys. Res.*, **89**, 786–800, 1984.

RUSSELL, C. T., The configuration of the magnetosphere, in *Critical Problems of Magnetospheric Physics*, edited by E. R. Dyer, p.1, National Academy of Sciences, Washington, D.C., 1972.

RUSSELL, C. T., The solar wind and magnetospheric dynamics, in *Proceedings of the Seventh ESLAB Symposium on Correlated Interplanetary and Magnetospheric Observations*, edited by D. E. Page, p.3, D. Reidel Publ. Co., 1974.

RUSSELL, C. T., Some comments on the topology of the geomagnetic tail, *J. Geophys. Res.*, **82**, 1625–1627, 1977.

RUSSELL, C. T., Reconnection at the earth's magnetopause: Magnetic field observations and flux transfer events, in *Magnetic Reconnection in Space and Laboratory Plasmas*, edited by E. W. Hones, Jr., pp.

124–138, American Geophysical Union, Washington, D.C., 1984.

RUSSELL, C. T. and R. C. ELPHIC, Initial ISEE magnetometer results: Magnetopause observations, *Space Sci. Rev.*, **22**, 681, 1978.

RUSSELL, C. T. and R. L. MCPHERRON, The magnetotail and substorms, *Space Sci. Rev.*, **15**, 205, 1973.

RUSSELL, C. T., R. L. MCPHERRON, and P. K. BURTON, On the cause of geomagnetic storms, *J. Geophys. Res.*, **79**, 1105, 1974.

RUSSELL, C. T., J. BERCHEM, and J. G. LUHMANN, On the source region of flux transfer events, *Adv. Space Res.*, **5**(4), 363–368, 1985.

SAUNDERS, M. A., C. T. RUSSELL, and N. SCKOPKE, Flux transfer events: Scale size and interior structure, *Geophys. Res. Lett.*, **11**, 131–134, 1984.

SCHINDLER, K. and N. F. NESS, Internal structure of the geomagnetic neutral sheet, *J. Geophys. Res.*, **77**(1), 91–100, 1972.

SCHOLER, M., G. GLOECKLER, D. HOVESTADT, F. M. IPAVICH, B. KLECKER, and C. Y. FAN, Anisotropies and flows of suprathermal particles in the distant magnetotail: ISEE-3 observations, *Geophys. Res. Lett.*, **10**, 1203, 1983.

SLUTZ, R. J. and J. R. WINKELMAN, Shape of the magnetospheric boundary under solar wind pressure, *J. Geophys. Res.*, **69**, 4933–4948, 1964.

SONNERUP, B. U. O., G. PASCHMANN, I. PAPAMASTORAKIS, N. SCKOPKE, G. HAERENDEL, S. J. BAME, J. R. ASBRIDGE, J. T. GOSLING, and C. T. RUSSELL, Evidence for magnetic field line reconnection at the earth's magnetopause, *J. Geophys. Res.*, **86**, 10,049–10,067, 1981.

TSURUTANI, B. T., J. A. SLAVIN, E. J. SMITH, R. OKIDA, and D. E. JONES, Magnetic structure of the distant geotail from −60 to −220 R_e: ISEE-3, *Geophys. Res. Lett.*, **11**, 1, 1984.

Solar Wind-Magnetosphere Coupling, edited by Y. Kamide and J. A. Slavin, 233–273.

Magnetospheric Equilibrium Configurations and Slow Adiabatic Convection

Gerd-Hannes VOIGT

Department of Space Physics and Astronomy, Rice University, P.O. Box 1892 Houston, Texas 77251, U.S.A.

This review paper describes the global magnetic field configuration of the magnetosphere and its time evolution on the convection time scale in terms of the magneto-hydrostatic equilibrium theory. This theory represents magneto-hydrodynamics (MHD) in the slow-flow approximation. In this approximation time scales are long compared to Alfvén wave travel times and flow velocities are small compared to the Alfvén speed. Then the inertial term $\rho(\mathrm{d}\boldsymbol{v}/\mathrm{d}t)$ in the MHD equation of motion is a small second-order term which can be neglected. Time derivatives and flow velocities remain first-order quantities in all the other MHD equations. Therefore one can compute slowly time-dependent processes, like magnetospheric convection, in terms of sequences of static equilibrium states. Those sequences are not arbitrary; they are constrained by the thermodynamic conditions according to which the magnetosphere evolves in time.

Under slow-flow conditions, magnetospheric convection is an adiabatic process in the course of which magnetotail flux tubes move earthward. One can show that this process requires very special assumptions about particle-loss or cooling in order to be steady in Earth's magnetosphere. Under realistic circumstances, however, convection is time-dependent and leads inevitably to the formation of a neutral line in the near-earth plasma sheet and, therefore, to the onset of a substorm. The magneto-hydrostatic theory describes the time evolution of the magnetosphere between substorm events. This description includes stretched plasma-loaded pre-substorm tail configurations as well as relaxed plasma-depleted post-substorm configurations.

In magnetospheric equilibrium models, plasma currents result from the requirement of force balance rather than being an empirical input to the models. Thus equilibrium models represent an advanced generation of magnetospheric $\boldsymbol{B}$-field models which provide physical insight into the internal structure of the plasma sheet in Earth's magnetotail.

Table of Contents

1. The Equilibrium Problem

The purpose of this review paper is to demonstrate how the magneto-hydrostatic equilibrium theory can be used to describe the large-scale magnetic field configuration of the magnetosphere and its time evolution under the influence of magnetospheric convection. The major advantage of the equilibrium theory reveals itself by the fact that magnetospheric currents result from model calculations rather than being input to the models. Thus, equilibrium models represent an advanced generation of theoretical magnetospheric **B**-field models compared to present semi-empirical models (see Section 2).

The magneto-hydrostatic theory describes magneto-hydrodynamics (MHD) in the slow-flow approximation. This approximation is applicable in the magnetosphere, whenever plasma convection velocities are small compared to the Alfvén velocity, and time scales are long compared to MHD wave travel times. To be more specific, we have to require that time derivatives $\partial/\partial t$ generate quantities of the first order in smallness, and that velocities $\boldsymbol{v}$ be first-order quantities as well, in comparison with zero-order quantities, namely, thermal pressure, currents, and magnetic field. Under these assumptions, the inertial term $\rho(\partial/\partial t + \boldsymbol{v}\cdot\nabla)\,\boldsymbol{v}$ becomes a second order term and can be neglected in the MHD equation of motion. However, time derivatives and velocities remain first-order quantities in the continuity equation, in Ohm's law, and in Maxwell's equations. Consequently, time-dependent processes, like magnetospheric convection, can be modeled by sequences of magnetospheric equilibrium configurations. We shall see below that those sequences of equilibrium states are by no means arbitrary, but are constrained by a priori assumptions about the thermodynamic evolution of the system.

A magneto-hydrostatic equilibrium is defined by a solution of the three equations

$$\boldsymbol{j} \times \boldsymbol{B} = \nabla P\,, \tag{1.1}$$

$$\nabla \times \boldsymbol{B} = \mu_0 \boldsymbol{j}\,, \tag{1.2}$$

$$\nabla \cdot \boldsymbol{B} = 0\,. \tag{1.3}$$

In general, the thermal plasma pressure is anisotropic. However, we shall see in Section 3.5 that the assumption of an isotropic pressure function can be justified in large parts of Earth's magnetosphere, especially in the magnetotail region during time periods of magnetically quiet conditions. The pressure function P is constant along magnetic field lines, according to (1.1). Therefore, P is entirely defined by its boundary values $P(\boldsymbol{x}_{eq})$ that are specified in a plane which is intersected by all field lines. A plane that meets this criterion in a "closed" magnetospheric model is, for example, the magnetospheric equatorial plane which will be marked by the subscript (eq) throughout this paper.

It is important to remember that $P(\boldsymbol{x}_{eq})$ is a free function that cannot be obtained from the equilibrium theory. Therefore, we need an additional thermodynamic condition which determines the pressure function throughout the time evolution of the system. It is reasonable to assume that the magnetospheric convection process is adiabatic. Thus the corresponding thermodynamic equation of state reads

$$P(b_i, t) \cdot [V(b_i, t)]^{\gamma} = C(b_i) . \qquad (1.4)$$

The number b_i is a field line label that in the general three-dimensional case contains two parameters such as Euler potentials α and β introduced later in Eq. (1.10). The adiabatic exponent is $\gamma = c_p / c_v = 5/3$ with the ratio of the specific heats for a monoatomic plasma. $V(b_i, t)$ represents the volume of a magnetic flux tube of unit magnetic flux (see e.g., WOLF (1983)) and is expressed by the integral

$$V(b_i, t) = \int_{x_{eq}}^{x_u} \frac{ds}{B_i(s, t)} . \qquad (1.5)$$

The integration is performed along the magnetic field line labeled b_i; $B_i(s, t)$ is the magnetic field strength along that field line. The integration extends from the equatorial intersection point $\boldsymbol{x}_{eq}(b_i, t)$ of the field line up to its ionospheric end located at $\boldsymbol{x}_u(b_i, t)$. Here we have implicitly assumed north–south symmetry in the magnetosphere. If the integral (1.5) is extended from the southern to the northern ionosphere, this symmetry assumption is unecessary.

Since we assume ideal MHD conditions, the label b_i remains attached to both the field line and the associated plasma element throughout the time development of the magnetosphere. According to (1.4) the function $C(b_i)$, once specified at the time $t=0$, does not change in time, provided there is no particle or heat loss from the field line. If there is no loss, then obviously $C(b_i)$ is constant along a drift path of a magnetospheric plasma element as it convects, for example, from the tail region toward the Earth, around the Earth, and finally to the dayside of the magnetosphere. As the magnetosphere progresses in time, the pressure function $P(\boldsymbol{x}_{eq})$ and the magnetic field configuration determine each other through the thermodynamic condition (1.4)–(1.5). We shall discuss configurational changes of the magnetosphere in response to adiabatic convection in Section 3.4.

The equilibrium theory contains information about the configuration of field aligned currents that flow between the magnetosphere and the high-latitude ionosphere. We obtain from (1.1) the current component perpendicular to $\boldsymbol{B}$,

$$j_{\perp} = \frac{\boldsymbol{B} \times \nabla P}{B^2}. \tag{1.6}$$

The parallel component follows from the requirement $\nabla \cdot \boldsymbol{j} = 0$ and reads

$$|j_{\parallel}| = -\frac{\boldsymbol{B}}{|\boldsymbol{B}|}\{\nabla P \times \nabla \int \frac{\mathrm{d}s}{B(s)}\}. \tag{1.7}$$

In (1.7) the flux-tube volume (1.5) is integrated along magnetic field lines from the southern to the northern ionosphere. For a derivation of this formula, see VASYLIUNAS (1970) or WOLF (1983).

We see that the configuration of the field aligned currents (1.7) is determined not only by the distribution of the thermal plasma $P(\boldsymbol{x}_{eq})$ in the equatorial plane, but also by the magnetic field topology via the flux tube integral $\int \mathrm{d}s/B(s)$. The value of the flux tube integral is established mainly by those parts of the flux tube where the magnetic field is weak. Along a field line, the field is weakest in the tail region, in particular in the tail plasma sheet. This means that the field geometry of the plasma sheet affects the gradient of the flux tube volume, and therefore the field aligned currents, in a very crucial and sensitive way. Here is one of the reasons why it is desirable to have a physically consistent magnetospheric $\boldsymbol{B}$-field model that satisfies the static momentum Eq. (1.1) (see also Section 2.2).

The field aligned currents (1.7) provide the physical link between the ionosphere and the magnetosphere. They are closed in the ionosphere according to assumptions about the ionospheric conductivity. The conductivity tensor is one of the inputs to the Rice Convection Model which treats magnetospheric convection in the inner magnetosphere via the process of ionosphere-magnetosphere coupling (see e.g., HAREL *et al.*, 1981; or SPIRO and WOLF, 1983).

The theoretical problem of calculating the convection electric field in the high-latitude polar-cap ionosphere leads us to the problem of specifying magnetic boundary conditions at the magnetopause. In particular, we need to specify whether the magnetosphere is open or closed. If $\hat{\boldsymbol{n}}$ denotes the surface unit vector at the magnetopause, the magnetic boundary conditions read

$$\hat{\boldsymbol{n}} \cdot \boldsymbol{B} = 0 \qquad \text{(closed)}, \tag{1.8}$$

$$\hat{\boldsymbol{n}} \cdot \boldsymbol{B} = f(\boldsymbol{x}_{mp}) \qquad \text{(open)}. \tag{1.9}$$

The function $f(\boldsymbol{x}_{mp})$ represents a given distribution of magnetic boundary values at magnetopause points $\boldsymbol{x}_{mp}$. Ultimately, this function reflects the influence of the interplanetary magnetic field (IMF) on the magnetosphere; it determines how the solar wind convection electric field maps down into the polar cap region. Thus the magnetic boundary conditions affect not only the magnetic field configuration itself, but also the polar cap convection patterns, and, through this, the whole process of magnetospheric convection.

Since we are unable at the present time to derive $f(\boldsymbol{x}_{mp})$ from plasma kinetic principles, the function for the magnetic boundary conditions must be derived from macroscopic merging theories or from observations. This topic will be discussed further by TOFFOLETTO and HILL (1986) in a separate paper in this Conference Proceedings. For our discussion, we assume that the function $f(\boldsymbol{x}_{mp})$ is given a priori. The function might be slowly time-dependent in the sense that it might be different for each magnetospheric configuration that belongs to a convection time-sequence of equilibrium states. Some aspects of the open magnetosphere will be discussed in Section 4.3.

For reasons of theoretical convenience, we formulate the equilibrium Eqs. (1.1)–(1.3) in terms of Euler potentials as well. The magnetic field can be written as

$$\boldsymbol{B} = \nabla \alpha \times \nabla \beta \tag{1.10}$$

which satisfies flux conservation (1.3). Two immediate consequences of (1.10) are the conditions

$$\boldsymbol{B} \cdot \nabla \alpha = 0 \,, \tag{1.11}$$

$$\boldsymbol{B} \cdot \nabla \beta = 0 \,. \tag{1.12}$$

In other words, the Euler potentials $\alpha(\boldsymbol{x})$ and $\beta(\boldsymbol{x})$ are constant along magnetic field lines. The lines where surfaces of constant α and β intersect represent magnetic field lines. For additional properties of Euler potentials, see STERN (1970, 1976). For isotropic equilibrium configurations, the thermal pressure is constant along magnetic field lines, according to (1.1), namely

$$\boldsymbol{B} \cdot \nabla P = 0. \tag{1.13}$$

Thus we conclude in combination with (1.11) and (1.12) that $P = P(\alpha, \beta)$ and consequently

$$\nabla P = \frac{\partial P}{\partial \alpha} \nabla \alpha + \frac{\partial P}{\partial \beta} \nabla \beta \,. \tag{1.14}$$

We combine (1.1), (1.10) and (1.14) and obtain

$$\boldsymbol{j} \times (\nabla \alpha \times \nabla \beta) = \frac{\partial P}{\partial \alpha} \nabla \alpha + \frac{\partial P}{\partial \beta} \nabla \beta \,. \tag{1.15}$$

This equation is satisfied by

$$\boldsymbol{j} \cdot \nabla \beta = + \frac{\partial P}{\partial \alpha} \,, \tag{1.16}$$

$$\boldsymbol{j} \cdot \nabla \alpha = - \frac{\partial P}{\partial \beta} \,. \tag{1.17}$$

In combination with (1.2), Equations (1.16) and (1.17) represent two equations for the two Euler potentials $\alpha(\boldsymbol{x})$ and $\beta(\boldsymbol{x})$ for any given pressure function $P(\alpha, \beta)$. Despite their esthetically appealing form, these equations have not been solved yet for general non-symmetric three-dimensinal equilibria that include a singularity like the Earth's dipole.

We note that a special class of two-dimensional equilibria can be derived from (1.16)–(1.17) by assuming that the physical quantities do not depend on the y coordinate. Under the assumption

$$\alpha(\boldsymbol{x}) = A(x, z) \tag{1.18}$$

$$\beta(\boldsymbol{x}) = y \tag{1.19}$$

the pressure function $P(\alpha, \beta)$ reduces to $P(A)$, and we find that (1.17) is satisfied identically, whereas (1.16) reduces to

$$j_y(A) = \frac{\mathrm{d}P}{\mathrm{d}A} \,. \tag{1.20}$$

The importance of (1.20) for magnetotail equilibria has been pointed out by SCHINDLER (1972). BIRN *et al.* (1975) have derived asymptotic solutions to (1.20) which lead to an integral representation for the field lines in the distant tail, where the influence of the Earth's dipole field can be neglected. The asymptotic theory will be discussed in Section 3.3.

Later, BIRN *et al.* (1977) and BIRN (1979) have combined (1.16)–(1.17) with the asymptotic theory and have derived three-dimensional models for the far-tail region. We shall discuss the characteristics of three-dimensional tail equilibria in Section 4.2.

Unfortunately, we do not have a complete three-dimensional magnetospheric magnetic field model that satisfies all the physical requirements discussed above. Equilibrium models of this type are still in an early stage of development. In Sections 3 and 4, we summarize computational results and physical conclusions which have been obtained from the magnetospheric equilibrium theory thus far.

2. Levels of *B*-Field Modeling

For quantitative modeling purposes, it is convenient to classify different regions of the magnetosphere with respect to different ranges of the plasma Beta parameter (β_{pl}) and the magnetic Mach number (M_{A}) which are defined by

$$\beta_{\mathrm{pl}} = \frac{2\mu_0 \, P(\boldsymbol{x})}{\boldsymbol{B}(\boldsymbol{x})^2} \,, \tag{2.1}$$

$$M_{\mathrm{A}}{}^2 = \frac{\mu_0 \, \rho(\boldsymbol{x}) \, \boldsymbol{v}(\boldsymbol{x})^2}{\boldsymbol{B}(\boldsymbol{x})^2} \,. \tag{2.2}$$

Equation (2.1) compares the plasma pressure with the magnetic field energy density, and (2.2) compares the kinetic and magnetic energy densities. Table 1 summarizes the theoretical levels for $\boldsymbol{B}$-field modeling that are appropriate for different ranges of β_{pl} and M_A. The models listed in Table 1 are distinguished by their basic equations they have to satisfy. It is assumed that the two differential terms in $(\partial/\partial t+\boldsymbol{v}\cdot\nabla)$ are of the same order.

2.1 *Vacuum models*

The lowest theoretical level is represented by vacuum models. These models are justified in the dipole-dominated region near the Earth up to a distance of about 5–7 R_e from the Earth. In this region of the magnetosphere, the magnetic field consists of a zero-order vacuum field $\boldsymbol{B}_0$ and a first-order perturbation field $\boldsymbol{B}_1$ which is generated by magnetospheric currents. The zero-order field consists of the dipole field $\boldsymbol{B}_d$ and of the field of the magnetopause Chapman-Ferraro currents, $\boldsymbol{B}_{cf}=-\nabla U_{cf}$, which can be derived from a scalar potential U_{cf}. The first-order field might be calculated from Biot-Savart's integral

$$\boldsymbol{B}_1(\boldsymbol{x}) = \mu_0 \int \frac{\boldsymbol{j}_1(\boldsymbol{x}') \times (\boldsymbol{x} - \boldsymbol{x}')}{|\boldsymbol{x} - \boldsymbol{x}'|^3}\, d^3x' . \tag{2.3}$$

By introducing the total field $\boldsymbol{B}=\boldsymbol{B}_0+\boldsymbol{B}_1$ into (1.1)–(1.3), we see that the zero-order equilibrium equations reduce to a linear potential problem, namely

Table 1. Applicability of magnetospheric models in various regions of the magnetosphere.

Parameter range*	Levels of $\boldsymbol{B}$-field modeling	Appropriate regions in the magnetosphere
$\beta_{pl} \ll 1$ $M_A^2 \ll 1$	Vacuum models $\boldsymbol{B} = -\nabla U$ $\nabla^2 U = 0$	The dipole dominated region up to 5–7 R_e distance from the Earth.
$\beta_{pl} \sim 1$ $M_A^2 \ll 1$	Hydromagnetic equilibrium models $\boldsymbol{j}\times\boldsymbol{B} = \nabla P$	The dipole dominated region plus the magnetotail up to about 60 R_e distance in tailward direction.
$\beta_{pl} \ll 1$ $M_A^2 \sim 1$	Flow-dominated MHD models $\rho\,(\frac{\partial}{\partial t} + \boldsymbol{v}\cdot\nabla)\,\boldsymbol{v} = \boldsymbol{j}\times\boldsymbol{B}$	The distant magnetotail between 60–250 R_e distance in tailward direction.
$\beta_{pl} \sim 1$ $M_A^2 \sim 1$	Full MHD codes $\rho\,(\frac{\partial}{\partial t} + \boldsymbol{v}\cdot\nabla)\,\boldsymbol{v} = -\nabla P + \boldsymbol{j}\times\boldsymbol{B}$	The whole magnetosphere including magnetopause and plasma sheet boundary layers.

*The quantities β_{pl} and M_A are defined in (2.1) and (2.2).

$$\boldsymbol{B}_0 = \boldsymbol{B}_d - \nabla U_{cf} , \tag{2.4}$$

$$\nabla^2 U_{cf} = 0 , \tag{2.5}$$

$$\frac{\partial U_{cf}}{\partial n} = + \hat{\boldsymbol{n}} \cdot \boldsymbol{B}_d - f(\boldsymbol{x}_{mp}) . \tag{2.6}$$

For the zero-order field $\boldsymbol{B}_0$, the magnetopause boundary conditions (2.6) satisfy (1.9). According to (1.1)–(1.3) and (1.9) the first-order equations read

$$\boldsymbol{j}_1 \times \boldsymbol{B}_0 = \nabla P_1 , \tag{2.7}$$

$$\nabla \times \boldsymbol{B}_1 = \mu_0 \boldsymbol{j}_1 , \tag{2.8}$$

$$\nabla \cdot \boldsymbol{B}_1 = 0 , \tag{2.9}$$

$$\hat{\boldsymbol{n}} \cdot \boldsymbol{B}_1(\boldsymbol{x}_{mp}) = 0 . \tag{2.10}$$

The first-order current

$$\boldsymbol{j}_1 = \boldsymbol{j}_{1\perp} + \boldsymbol{j}_{1\parallel} \tag{2.11}$$

which is needed in (2.3) to calculate $\boldsymbol{B}_1$ can be obtained from (1.6) and (1.7) by inserting the zero-order magnetic field (2.4) and the first-order pressure $P_1(\boldsymbol{x}_{eq})$ which needs to be specified in the magnetospheric equatorial plane.

The scheme (2.3)–(2.11) provides the possibility of calculating a first-order equilibrium model for the inner magnetosphere. A model of this type has been developed by SCKOPKE (1972). He used the CGL-equations (see e.g., KRALL and TRIVELPIECE, 1973) and calculated anisotropic ring currents within an axially symmetric model in order to investigate how the equatorial ring current affects the inner magnetosphere.

In the early days of magnetospheric $\boldsymbol{B}$-field modeling, theoretical efforts concentrated on solving the vacuum problem (2.4)–(2.6) for a "closed" magnetosphere (i.e., $f(\boldsymbol{x}_{mp})=0$). Some vacuum models already included a realistic magnetopause shape that meets the pressure balance condition between the solar wind and the magnetospheric $\boldsymbol{B}$-field (see BEARD (1973) for a review of the magnetopause problem and for further references). The difficulty of calculating a magnetopause shape for the "open" magnetosphere lies in the fact that the boundary conditions (1.9) are associated with $(\boldsymbol{j}\times\boldsymbol{B})$ forces parallel to the boundary that need to be balanced by a tangential flow field.

In their zero-order limitation the vacuum models had been derived from basic principles. Later, these models became semi-empirical in that they were complemented by empirical ring currents and tail currents in order to match the observed magnetic field geometry of the real magnetosphere. The reader is referred to WALKER's (1976, 1979) review papers in which he gave an assessment of the practical advantages and limitations of the semi-empirical magnetospheric models.

Unfortunately, the semi-empirical models are unsatisfactory from a theoretical point of view; they are not in agreement even with the first-order Eqs. (2.7)–(2.10). All semi-empirical models satisfy (2.8), some violate (2.9), most violate the magnetopause boundary condition (2.10), but all fail to satisfy the force balance condition (2.7). WALKER and SOUTHWOOD (1982) have tested several widely used semi-empirical magnetospheric models and have found that neither (1.1) in the magnetotail region, nor (2.7) in the dipole dominated region near the Earth, are satisfied at all.

2.2 *Quasi-static equilibrium models*

Hydromagnetic equilibrium models (explained in Section 1) represent the next higher level of **B**-field modeling. These models are not only applicable in the vacuum region of the magnetosphere ($\beta_{\text{pl}} \ll 1$), but also in the magnetotail plasma sheet up to about 60 R_e distance from the Earth. In the tail region one observes on the average an equipartition between plasma and magnetic field energy densities ($\beta_{\text{pl}} \sim 1$), and therefore the plasma distribution, the tail currents, and the magnetic field configuration determine each other via the force balance requirement (1.1). Here lies the advantage of the equilibrium models: the tail currents, for example, can be directly calculated from (1.6) and (1.7). In this respect, equilibrium models contain a higher degree of physical consistency than the semi-empirical models discussed above. Therefore they can provide a better physical insight into the internal structure of the magnetotail plasma sheet and into its connection to the near-Earth part of the magnetosphere (see Section 4.2).

Equilibrium models are especially useful in combination with the Rice Convection Model (see e.g., SPIRO and WOLF (1983) for an overview of the model characteristics). The Rice Convection Model is basically a multi-fluid code in the MHD slow-flow approximation. The total pressure consits of partial pressures P_k for a number of species (typically ~20). Each particle species k is represented by an isotropic Maxwellian energy distribution. The force-balance Equation (1.1), for example, reads

$$\boldsymbol{j} \times \boldsymbol{B} = \nabla \sum_k P_k + O(\varepsilon^2) \tag{2.12}$$

where small second-order terms $O(\varepsilon^2)$ are neglected. The combined convection velocity depends on the particle energy, i.e., on the species k, and reads

$$\boldsymbol{v}_{\perp k} = \frac{\boldsymbol{E} \times \boldsymbol{B}}{B^2} + \frac{\boldsymbol{B} \times \nabla W_k}{q \cdot B^2} . \tag{2.13}$$

The first term in (2.13) describes the ($\boldsymbol{E} \times \boldsymbol{B}$) drift velocity, the second term the bounce-averaged gradient-curvature drift velocity. For particles bouncing along a closed magnetic field line, the particle's kinetic energy

$$W_k = W_k(\boldsymbol{x}_{\text{eq}}, \mu, J) \tag{2.14}$$

is completely defined by the equatorial crossing point of the field line and the two first adiabatic invariants. For a derivation of (2.13), see WOLF (1983).

Equations (2.13)–(2.14) reveal the major advantage of the Rice Convection Model: owing to its multi-fluid formulation, it can accurately treat the energy-dependent convective drift of plasma particles around the Earth, i.e., in a region where the second term in (2.13) is comparable to, or even exceeds, the $(\boldsymbol{E}\times\boldsymbol{B})$ drift. Thus the Rice Convection Model can answer the question of how the equatorial ring current forms in the course of the time evolution of the whole magnetosphere.

Clearly, quasi-static models cannot describe processes that vary on time scales comparable to MHD travel times or that involve flows that are in the order of the Alfvén velocity. Thus, quasi-static models cannot describe the physics in flow dominated regions of the magnetosphere like the magnetopause plasma mantle or the plasma sheet boundary layer. However, they can accurately model configurational changes of the whole magnetosphere which occur on time scales that are characteristic for magnetospheric convection (see Section 3.4).

2.3 *MHD models*

The highest theoretical level of magnetospheric modeling is represented, at least conceptually, by the MHD models. MHD codes contain the inertia term $\rho(\mathrm{d}\boldsymbol{v}/\mathrm{d}t)$ and can therefore treat the whole solar-wind-magnetosphere system (BRECHT *et al.*, 1981, 1982; LYON *et al.*, 1980, 1981; OGINO and WALKER, 1984; WU *et al.*, 1981; WU, 1983, 1984). There also exist three-dimensional MHD magnetotail models which simulate magnetic reconnection and plasma acceleration owing to a tearing-mode unstable time evolution within the magnetotail region (BIRN and HONES, 1981).

Besides Ohm's law, the continuity equation and Maxwell's equations, the MHD codes essentially satisfy the equation of motion and the equation of state,

$$\rho\frac{\partial \boldsymbol{v}}{\partial t} + \rho(\boldsymbol{v}\cdot\nabla)\boldsymbol{v} = -\nabla P + (\boldsymbol{j}\times\boldsymbol{B}) , \tag{2.15}$$

$$\frac{\partial}{\partial t}\left(\frac{P}{\rho^{\gamma}}\right) + \boldsymbol{v}\cdot\nabla\left(\frac{P}{\rho^{\gamma}}\right) = 0 . \tag{2.16}$$

Note that (2.16) is completely equivalent to (1.4)–(1.5). In the slow-flow approximation, (2.15) reduces to (1.1).

At the present time, MHD models which model the whole magnetosphere are not yet well suited to a detailed study of the inner and middle magnetosphere and of the coupling between the magnetosphere and the ionosphere; they are limited by the single-fluid approximation, the model grid spacings are still too coarse, and the numerical codes need high viscosities for numerical stability. Moreover, no MHD model has reached the point yet where boundary layers can be said to be calculated accurately, even assuming the adequacy of the MHD assumptions in these regions (J. Lyon, personal communication).

Despite their present limitations, the global MHD models will eventually contribute considerably to our understanding of plasma-physical processes within the magnetosphere. In particular, they will be able to simulate the whole magnetospheric convection cycle. This will include the simulation of fluid instabilities like the tearing-mode and the accurate calculation of flow-dominated boundary layers.

The major physical insights we shall obtain from the MHD models lie in the future. Therefore, we devote the following sections of this review to the description of the magneto-hydrostatic theory. As we have seen, this theory represents the lowest theoretical level that allows us to look at the magnetospheric $\boldsymbol{B}$-field configuration and at plasma-physical phenomena from a unified point of view.

3. Results from the 2-D Theory

It is by no means trivial or useless to formulate the magnetospheric equilibrium problem in two dimensions. Two-dimensional problems are a necessary first step on the way toward understanding the principal physical processes that can be investigated by means of magneto-hydrostatic models.

In this section, we assume that all physical quantities depend on the x and z coordinates only. We use geocentric solar magnetospheric GSM coordinats. The x–z plane represents the magnetospheric noon-midnight meridian plane with x pointing toward the sun; this plane contains the Earth's dipole axis.

3.1 The Grad-Shafranov equation

In the two-dimensional approximation ($\partial/\partial y=0$), the magnetic field and the currents can be derived from a flux function $A(x, z)$ which is constant along magnetic field lines. In the special case of cartesian (x, z) coordinates, $A(x, z)$ is identical with the y-component of the magnetic vector potential. Note that the vector potential meets the Coulomb gauge $\nabla \cdot \boldsymbol{A} = 0$. We derive two magnetic field components from $\boldsymbol{B} = \nabla \times \boldsymbol{A}$ and obtain

$$B_x = -\frac{\partial A}{\partial z}, \tag{3.1}$$

$$B_z = +\frac{\partial A}{\partial x}. \tag{3.2}$$

We find that B_y is an arbitrary function of A. This functional form is required in order to satisfy the y-component of the force balance Eq. (1.1). Since $\boldsymbol{B}\cdot\nabla P = 0$ and $\boldsymbol{B}\cdot\nabla A = 0$, we conclude that the pressure function P is a function of A as well. Thus two-dimensional equilibria contain the two free functions

$$B_y = B_y(A), \tag{3.3}$$

$$P = P(A). \tag{3.4}$$

The magnetic field components (3.1)–(3.3) satisfy flux conservation (1.3).

We have to take into account that the flux function $A(x, z) = A_{\mathrm{pl}} + A_{\mathrm{d}}$ consists of a term A_{pl} which relates to the plasma currents, $\mu_0 j_{\mathrm{pl}} = -\nabla^2 A_{\mathrm{pl}}$, and a term A_{d} which represents the Earth's dipole with the dipole moment $\boldsymbol{M}_{\mathrm{D}}$. Without losing generality, we assume that $\boldsymbol{M}_{\mathrm{D}}$ points into the negative z-direction. Then we obtain the dipole magnetization current

$$\mu_0 j_{\mathrm{dipole}} = (\nabla \times \boldsymbol{M}_{\mathrm{D}})_y$$

$$= + M_{\mathrm{D}} \frac{\partial}{\partial x} \delta(x)\, \delta(z) . \qquad (3.5)$$

Therefore the plasma currents read

$$j_{\mathrm{pl}} = - \frac{1}{\mu_0} \nabla^2 A - j_{\mathrm{dipole}} . \qquad (3.6)$$

Now we combine (3.1)–(3.6) with (1.1)–(1.3) and find a non-linear partial differential equation for the flux function $A(x, z)$ which describes the 2-D equilibrium magnetosphere. This equation is valid in the whole magnetosphere, including the Earth's dipole region, and reads

$$\nabla^2 A + \frac{\mathrm{d}}{\mathrm{d}A} [\mu_0 P(A) + \frac{1}{2} B_y(A)^2] = - M_{\mathrm{D}} \frac{\partial}{\partial x} \delta(x)\, \delta(z) . \qquad (3.7)$$

Note that (3.7) is completely equivalent to the basic Eqs. (1.1)–(1.3). The homogeneous part of (3.7) is called a Grad-Shafranov equation. Equations of this type play important roles in fusion plasma physics and in the physics of solar flares. An important problem in fusion physics is to specify the two free functions (3.3) and (3.4) such that the thermal plasma evolves into a stable equilibrium. In astrophysics and magnetospheric physics, however, both stable and unstable equilibria can represent physically relevant configurations.

For convenience we introduce the two functions

$$\Phi(A) = \mathrm{d}B_y(A)/\mathrm{d}A, \qquad (3.8)$$

$$\Psi(A) = \mathrm{d}P(A)/\mathrm{d}A, \qquad (3.9)$$

and obtain the three components of the total currents from (3.1)–(3.3) via (1.2) and (3.7)

$$\mu_0\, j_x = + \Phi(A) \cdot B_x, \qquad (3.10)$$

$$\mu_0\, j_y = + \Phi(A) \cdot B_y + \mu_0\, \Psi(A) + \mu_0\, j_{\mathrm{dipole}}, \qquad (3.11)$$

$$\mu_0\, j_z = + \Phi(A) \cdot B_z. \qquad (3.12)$$

We easily identify three different extreme situations that can occur in 2-D equilibria: (a) if $\Psi(A)=0$ and $\Phi(A)=0$, the magnetic field is a vacuum field; (b) if

$\Psi(A)=0$ but $\Phi(A)\neq0$, the magnetic field is force-free and all currents and field-aligned; (c) if $\Psi(A)\neq0$ but $\Phi(A)=0$, the magnetic field is plasma pressure dominated, but the configuration has no field-aligned currents.

Since $A(x, z)$ is constant along magnetic field lines, it can serve as a field line label. Thus we re-write the thermodynamic condition (1.4),

$$P(A, t) \cdot [V(A, t)]^{\gamma} = C(A), \tag{3.13}$$

with the flux tube integral (1.5)

$$V(A, t)= \int_{A(t)} \frac{ds}{[(\nabla A)^2+B_y(A)^2]^{1/2}} \cdot \tag{3.14}$$

Finally we have to specify magnetic boundary conditions at the magnetopause according to (1.8) or (1.9). In the case of a "closed" magnetosphere, the Grad-Shafranov equation (3.7) has homogeneous boundary conditions with A_{mp}=constant representing the magnetopause field line.

General solutions to (3.7) for arbitrary functions $P(A)$ and $B_y(A)$ are still not available. There exist, however, a restricted class of linear solutions for the whole magnetosphere and a class of asymptotic solutions for the far tail region with no restrictions on $P(A)$ and $B_y(A)$. We discuss the two different aproaches in Sections 3.2 and 3.3.

3.2 *Linear equilibria*

The Grad-Shafranov equation (3.7) becomes linear if one assumes the special functional dependencies

$$P(A) = \frac{k^2}{2\mu_0} \cdot A^2 , \tag{3.15}$$

$$B_y(A) = - h \cdot A , \tag{3.16}$$

for the two free functions (3.3) and (3.4). Then (3.7) reads

$$\nabla^2 A + (k^2 + h^2) A = - M_D \cdot \frac{\partial}{\partial x} \delta(x)\, \delta(z) \quad . \tag{3.17}$$

Equation (3.17) can be solved easily by separation. Let λ_n^2 be the separation constant and

$$A_n(x, z) = F_n(x) \cdot G_n(z), \tag{3.18}$$

the eigenfunctions of the homogeneous problem. In cartesian (x, z) coordinates, the separation method requires the magnetopause to be a plane parallel to the equatorial plane. We assume the boundary condition $A(x, z)=0$ at the magnetopause $z=\pm z_0$. In addition we assume a change of the magnetic field polarity at the equatorial plane, i.e., $\partial A/\partial z=0$ at $z=0$. Then the two eigenfunctions read

$$F_n(x) \propto \exp(-\lambda_n|x|), \tag{3.19}$$

$$G_n(z) \propto \cos(\eta_n z). \tag{3.20}$$

The eigenvalues

$$\eta_n = \frac{(2n-1)}{z_0} \cdot \frac{\pi}{2}, \tag{3.21}$$

satisfy the given boundary conditions. The two free physical parameters h and k and the eigenvalues η_n and λ_n are related through

$$\eta_n^{\,2} = \lambda_n^{\,2} + h^{\,2} + k^{\,2}. \tag{3.22}$$

In the following analysis, we restrict ourselves to positive values $\lambda_n^2 \geqslant 0$. Note, however, that equilibrium configurations with $\lambda_n^2 < 0$ can satisfy the boundary conditions as well (see, for example, VOIGT and WOLF, 1985). According to (3.1)–(3.3) and (3.18)–(3.20) the equilibrium magnetic field components assume the simple form

$$B_x = +\,\eta_n \sin(\eta_n z) \cdot \exp(-\lambda_n|x|)\,, \tag{3.23}$$

$$B_y = -\,h \cos(\eta_n z) \cdot \exp(-\lambda_n|x|)\,, \tag{3.24}$$

$$B_z = +\,\lambda_n \cos(\eta_n z) \cdot \exp(-\lambda_n|x|)\,. \tag{3.25}$$

Note that the condition (3.22) determines how the magnetic field components relate to each other; if, for example, the plasma pressure (3.15) is specified by a given value k, then the strengths of the magnetic B_y and B_z components and the exponential decrease of all three magnetic components depend on each other. If, on the other hand, the exponential decrease of the magnetic field is known from observations and prescribed in terms of a given value λ_n, then the amount of the thermal plasma pressure and the strength of the magnetic B_y component determine each other. These examples demonstrate that the two physical parameters h and k in (3.15) and (3.16) are related via the boundary conditions imposed on the equilibrium Eq. (3.17). We note that COWLEY (1979) has calculated linear non-isotropic tail equilibria which include the magnetic B_y component as well.

The currents (3.10)–(3.12) follow from (3.8)–(3.9) and (3.15)–(3.16). The field aligned currents

$$\boldsymbol{j}_{\parallel} = \frac{1}{B^2}\,(\boldsymbol{j}\cdot\boldsymbol{B})\,\boldsymbol{B} \tag{3.26}$$

assume a very convenient form by introducing the plasma parameter $\beta_{pl}(x, z)$ defined in (2.1). Thus we obtain

$$j_{\|} = -\frac{h}{\mu_0}(1 + \beta_{pl}) \cdot \boldsymbol{B} . \tag{3.27}$$

We see that the strength of the field aligned currents depends not only on the strength of the magnetic B_y component but also on the amount of the thermal plasma pressure. The case of zero plasma pressure (which for $h \neq 0$ means zero plasma temperature, not vacuum!) is force-free.

The above considerations might be relevant for the interpretation of the field line rotation effect observed by the ISEE-3 spacecraft in the distant magnetotail at $x_{GSM} = -210\ R_e$ (TSURUTANI *et al.*, 1984). There were quiet time periods when the spacecraft observed a significant magnetic field component along the negative y_{GSM} direction near the neutral sheet. Note that the B_y component (3.24) reaches its maximum strength at the equatorial plane $z=0$ as well. The existence of a negative B_y component results in a gradual rotation of the $\boldsymbol{B}$-field vector when the spacecraft traverses the plasma sheet (see Fig. 4 in TSURUTANI *et al.* (1984)).

Equilibrium models of the type (3.23)–(3.25) might also be helpful to treat the theoretical question of how the dawn-dusk component of the interplanetary magnetic field penetrates into the magnetospheric tail lobes and into the plasma sheet. FAIRFIELD (1979) used IMP-6 magnetic field data and found that, averaged over the whole tail, approximately 13% of the IMF-B_y component appears as a perturbation field B_y in the tail. LUI (1983) used the same data set but restricted the analysis to data obtained near the plasma sheet. He found that approximately 50% of the IMF-B_y component is present in the plasma sheet. Thus he concluded that the IMF is shielded less in the plasma sheet than in the tail lobes. The form of the magnetic field components (3.23)–(3.25) clearly supports this conjecture. COWLEY (1981) pointed out that the B_y component of the IMF might exert a torque on the magnetotail which might result in a twist of the whole tail plasma sheet.

The inhomogeneous Equation (3.17) can be solved in terms of the eigenfunctions of the homogeneous problem (3.19)–(3.20). We simply express the flux function $A(x,z)$ as a Fourier integral,

$$A(x,z) = \sum_{n=0}^{\infty} \int_{-\infty}^{+\infty} F_n(s)\ G_n(z)\ \mathrm{e}^{-isx}\ \mathrm{d}s , \tag{3.28}$$

and remember that the δ functions can be written as

$$\begin{aligned} \delta(x) &= \frac{1}{2\pi} \int_{-\infty}^{+\infty} \mathrm{e}^{-isx}\ \mathrm{d}s \\ \delta(z) &= \sum_{n=1}^{\infty} \cos(\eta_n z) . \end{aligned} \tag{3.29}$$

We substitute (3.28)–(3.29) into (3.17) to obtain the Fourier coefficients

$$F_n(s) = + \frac{i}{2\pi} \cdot \frac{s}{[s^2 + \lambda_n^2]} \cdot M_D \tag{3.30}$$

which will be substituted into (3.28). The integral exists, and we obtain the equilibrium solution which includes the Earth's dipole,

$$A(x,z) = - \frac{M_D}{2} \sum_{n=1}^{\infty} \cos(\eta_n z) \cdot [\mathrm{sign}(x) \cdot F_n(x) - F_n(x - 2b)] \, . \tag{3.31}$$

The magnetopause of this model is approximated by a rectangular box; the tail magnetopause is located at $z = \pm z_0$, the dayside magnetopause at $x = +b$. The flux function $A(x,z)$ meets the boundary conditions $\partial A/\partial x = 0$ at $z = \pm z_0$, and $\partial A/\partial z = 0$ at $x = +b$. The boundary conditions at $x = +b$ are met by superimposing an image dipole located at $x = +2b$. The second term in (3.31) is due to the image dipole. Both functions $F_n(x)$ and $F_n(x - 2b)$ are defined according to (3.19).

Magnetospheric $\boldsymbol{B}$-field configurations that are based on (3.31) have been calculated by FUCHS and VOIGT (1979) to investigate how the magnetosphere evolves from a pure vacuum configuration toward a plasma-dominated equilibrium with stretched tail field lines. For $B_y = 0$, the field lines are maximally stretched when k approaches η_1 and therefore the first leading eignevalue λ_1 in (3.22) approaches zero. In this case the magnetic B_z component (3.25) vanishes in the far tail region, so that the magnetic field lines are almost parallel to the tail axis.

VOIGT and WOLF (1985) have calculated magnetic field configurations from (3.31) for pressure values $0 \leqslant k \leqslant \eta_1$ and for $k \geqslant \eta_1$ in order to investigate how the dayside polar cusp behaves in response to the amount of thermal plasma contained on closed dayside flux tubes. They obtained the following results: when the plasma pressure (3.15) varies between the vacuum limit ($k = 0$) and the Harris sheet limit ($k = \eta_1$) the standard polar cusp geometry results with the two neutral points on the dayside magnetopause. When, on the other hand, the plasma pressure exceeds the Harris sheet limit ($k > \eta_1$) the polar cusp flux tubes expand in response to their high amount of plasma pressure, such that the cusp field lines gradually shift to the nightside. The plasma pressure can be increased until the two magnetopause neutral points are located tailward of the dipole. Although such polar cusp configurations represent hydromagnetic equilibrium configurations, they probably do not exist in Earth's magnetosphere, owing to the lack of a powerful plasma source that could fill the dayside polar cusp flux tubes (VOIGT and WOLF, 1985).

Since the equilibrium solutions derived from (3.17) are linear, one can easily construct a magnetospheric model for any dipole orientation. A model of this type is shown in Fig. 1 for $B_y = 0$. The sequence of field line plots starts with an "earth-type" magnetosphere (i.e., the dipole is perpendicular to the solar wind) and ends with a "pole-on" configuration which has the dipole pointing into the solar wind. In the four panels of Fig. 1, the dipole is tilted by 0°, 30°, 60°, and 90° (from top to bottom). An example might be the magnetosphere of Uranus: since Uranus' dipole axis

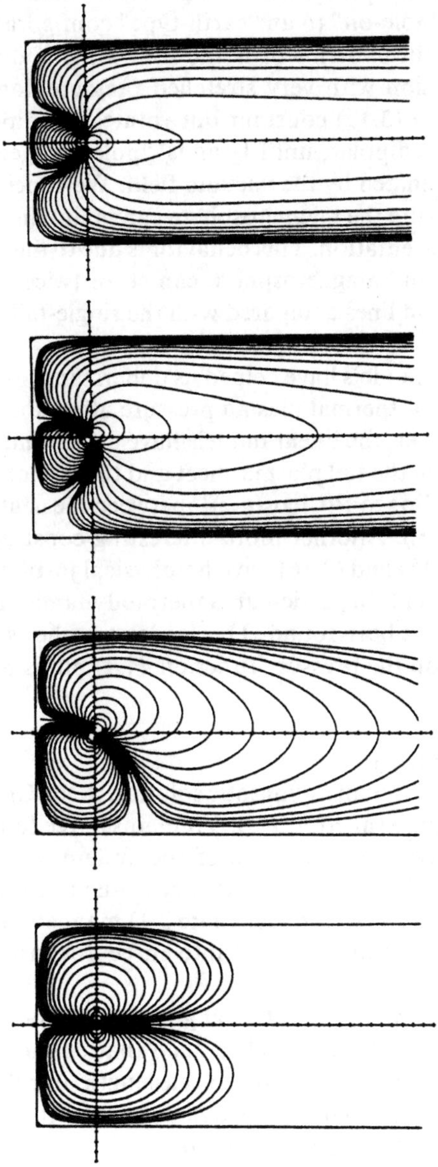

Fig. 1. Two-dimensional equilibrium magnetosphere with arbitrary dipole orientation. The shape of the magnetopause is given in this model. The dipole tilt angles are 0°, 30°, 60°, and 90° (from top to bottom). The field line plots result from a linear solution to the Grad-Shafranov equation (3.17) with $B_y(A)=0$, $P(A) \propto A^2$, and a generalized tilt-dependent dipole term. The boundary condition is $A(x,z)=0$ at the magnetopause. The top panel represents an "earth-type" magnetosphere, the bottom panel a "pole-on" magnetosphere with the dipole axis pointing into the solar wind. For a given k value in (3.15), the magnetic field turns from a plasma pressure dominated configuration (top) to a magnetic field dominated configuration (bottom).

presumably lies within this planet's orbital plane, Uranus' magnetosphere changes every 21 years from a "pole-on" to an "earth-type" configuration, and vice versa.

The result of the tilt experiment is the following: if one starts from Uranus' "earth-type" configuration with very stretched pressure-dominated tail field lines ($k=\eta_1$, $h=0$), keeps k in (3.15) constant but rotates the dipole, the tail field lines become more and more dipolar, until Uranus' "pole-on" geometry assumes a field geometry which is dominated by the vacuum field. We conclude that the maximum amount of plasma pressure the magnetosphere can maintain on closed tail field lines depends on the dipole orientation. This behavior is due to the fact that the double-tail structure of the "pole-on" magnetosphere can store twice the amount of thermal plasma on closed tail field lines compared with the single-tail structure of the "earth-type" magnetosphere.

Linear equilibrium models have helped us understand how the orientation of the dipole and the amount of thermal plasma pressure affect the magnetotail and polar cusp geometries. However, the linear models have disadvantages as well; there is no clear distinction between the tail plasma sheet and the tail lobes, because the plasma pressure behaves according to $P \propto \{\cos(\eta_n z)\}^2$ between the neutral sheet ($z=0$) and the tail magnetopause ($z=z_0$). Another more interesting constraint is the fact that the special assumptions (3.15) and (3.16) have no physical justification other than being mathematically convenient. In particular, a thermodynamic argument in favor of the assumption (3.15) has not been found. This leads us to the asymptotic theory which relaxes the constraint on the pressure function $P(A)$ but is restricted to the distant magnetotail region.

3.3 *The Asymptotic theory*

In this section, we derive an asymptotic solution to the Grad-Shafranov Eq. (3.7) and show that the pressure function $P(A,t)$ remains a free function that allows us to calculate the quasi-static time evolution of the system under the thermodynamic condition (3.13)–(3.14). The asymptotic theory is due to SCHINDLER (1972). It has been utilized by BIRN *et al.* (1975) to calculate 2-D magnetotail equilibria. Their tail models have been successfully extended into three dimensions (BIRN *et al.*, 1977; BIRN, 1979).

The asymptotic theory applies to the far geomagnetic tail region (see top panel in Fig. 2), where the field lines are stretched such that characteristic scale lengths in x-direction (sunward) are much larger than scale lengths in z-direction (perpendicular to the earth–sun line). We introduce the smallness parameter $L_z/L_x = \varepsilon \ll 1$ and find for the spatial derivatives the hierarchical order

$$\partial^2/\partial z^2 = O(1) \qquad \partial^2/\partial x^2 = O(\varepsilon^2). \tag{3.32}$$

In (3.7) we neglect quantities of the order $O(\varepsilon^2)$, take into account that the dipole term is negligible in the far tail, and assume that the magnetic B_y component be a quantity of the order $O(\varepsilon)$. Then (3.7) reduces to

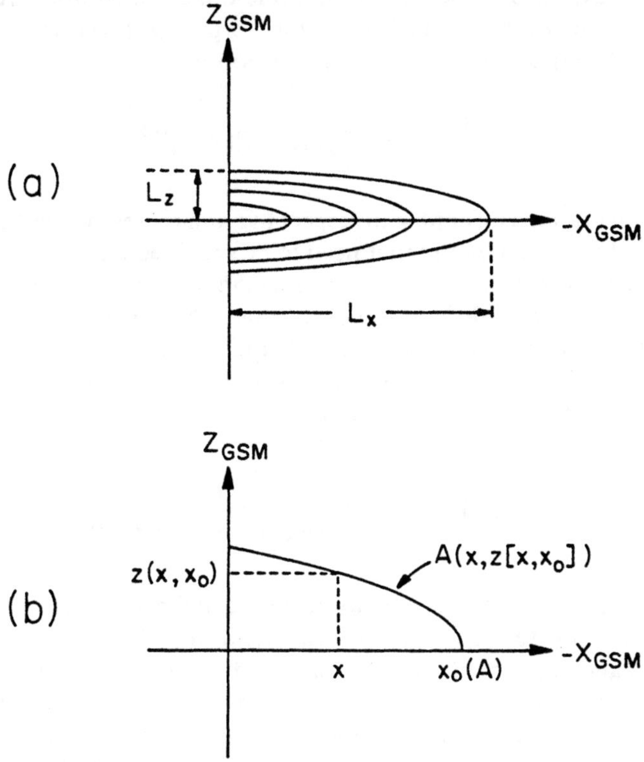

Fig. 2. Quantities defined in the asymptotic theory. (a) Due to the stretched field lines in the geomagnetic tail, characteristic scale lengths for changes of physical quantities are such that $L_x \gg L_z$. Thus spatial derivatives become $\partial/\partial x \ll \partial/\partial z$ which leads to $B_z \ll B_x$, according to (3.1)–(3.2). (b) A magnetic field line is identified by its flux function $A[x,z(x,x_0,t)]$ and crosses the equatorial plane at $x=x_0(A,t)$. For any equatorial point $0 \leqslant x \leqslant x_0$ the integral (3.37) represents this field line in a parametric form $z=z(x,x_0,t)$.

$$\frac{\partial^2 A}{\partial z^2} + \mu_0 \frac{\mathrm{d}P}{\mathrm{d}A} = 0 \,. \tag{3.33}$$

In order to visualize the geometrical implications of (3.33), we take the equivalent Eq. (1.1), neglect again terms of the order $O(\varepsilon^2)$, and obtain

$$\frac{\partial}{\partial x}\left(P + \frac{B_x^2}{2\mu_0}\right) = \frac{1}{\mu_0}(\boldsymbol{B} \cdot \nabla)\, B_x \,, \tag{3.34}$$

$$\frac{\partial}{\partial z}\left(P + \frac{B_x^2}{2\mu_0}\right) = 0 \,. \tag{3.35}$$

Therefore, the asymptotic theory assumes one-dimensional pressure balance in z-direction, but includes the magnetic curvature term in x-direction.

Equation (3.33) can be integrated, and we obtain

$$\frac{\partial A}{\partial z} = \{2\mu_0 \, [P(x,t) - P(A,t)]\}^{1/2} \; . \tag{3.36}$$

$P(x,t)$ is the integration constant with respect to z and represents the boundary values for the thermal plasma pressure that need to be specified on the tail axis at the time t=0. Note that all quantities are time-dependent in the quasi-static sense explained in Section 1. Thus the time labels $t=t_1, t_2, ..., t_i$ mark the magnetic field configurations that belong to a particular time sequence of equilibrium states.

Equation (3.36) can be integrated once more, and we find a parametric representation of the magnetic field lines

$$z(x, x_0, t) = \int\limits_{A(x,t)}^{A(x_0,t)} \frac{\mathrm{d}A}{\{2\mu_0 \, [P(x,t) - P(A,t)]\}^{1/2}} \; . \tag{3.37}$$

The bottom panel in Fig. 2 illustrates the meaning of the integral (3.37). $x_0 = x_0(A,t)$ is the equatorial intersection point of the field line labeled A. It is $A(x_0,t) = A[x,z(x,x_0,t)]$ and $A(x,t)=A(x,0,t)$. For a closed magnetosphere with the boundary condition (1.8), the magnetopause field line $z_{mp}(x,\infty,t) = z_{mp}(x,t)$ can be obtained from (3.37) by specifiying the upper integration limit $A_{mp}(\infty,t) = A_{mp}(x,z_{mp},t) = \mathrm{const}(t)$.

There are two options for calculating magnetotail models: (a) If a closed magnetosphere is assumed and $z_{mp}(t)$ is given at one location, $P(A,t)$ and $P(x,t)$ can be prescribed, and the shape of the magnetopause $z_{mp}(x,t)$ results. (b) If $P(A,t)$ and the shape of the magnetopause field line are prescribed, the distribution of the equatorial pressure $P(x,t)$ results. The magnetotail field lines plotted in Fig. 3 have been calculated according to the second option (from BIRN *et al.*, 1975).

The flux tube integral (3.14) can be carried out by utilizing the equation for the magnetic field lines. It is $\mathrm{d}s^2 = \mathrm{d}x^2 \, [1 + (B_z / B_x)^2]$. Therefore, we can re-write (3.14) and obtain

$$V(A, t) = \int\limits_{A(t)} \frac{\mathrm{d}s}{|\nabla A|} = \int\limits_{x_0}^{0} \frac{\mathrm{d}x}{B_x} \; . \tag{3.38}$$

According to (3.1), B_x can be expressed by (3.36), and we obtain from (3.38)

$$V(A, t) = \int\limits_{0}^{x_0} \frac{\mathrm{d}x}{\{2\mu_0 \, [P(x,t) - P(A,t)]\}^{1/2}} \; . \tag{3.39}$$

The times evolution of the 2-D magnetotail on the convection time scale is completely determined by a self-consistent solution of (3.13), (3.37), and (3.39). These equations reveal the major advantage of the asymptotic theory: Since $P(A,t)$ remains an

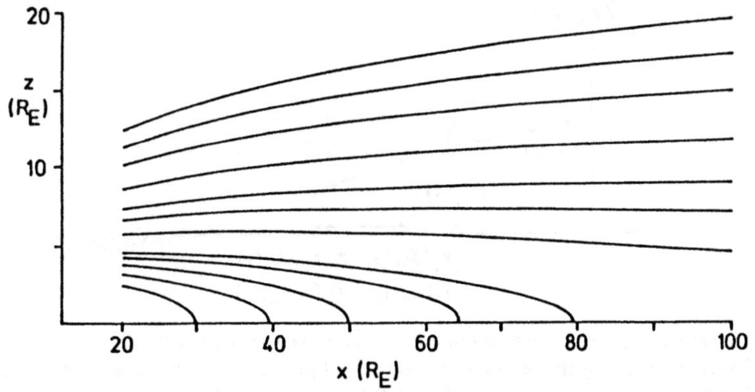

Fig. 3. Magnetic tail field lines calculated from the asymptotic theory (reproduced from BIRN *et al.* (1975)). For this configuration the field line integral (3.37) has been solved by specifying the thermal pressure $P(A)$ and the shape of the magnetopause $z_{mp}(x)$. In the static case the time label t is arbitrary.

arbitrary function in (3.37), one can calculate static magnetotail configurations for a variety of physical assumptions (BIRN *et al.*, 1975). The additional thermodynamic condition (3.13) requires that $P(A,t)$ change in time (SCHINDLER and BIRN, 1982; ERICKSON, 1984). This freedom is not available for linear models which are restricted to the special functional form $P(A) \propto A^2$ (see Eq. (3.15) in Section 3.2). We discuss time changes of the magnetospheric $\boldsymbol{B}$-field configuration in Section 3.4.

The disadvantage of the asymptotic theory lies in the fact that it cannot include the Earth's dipole in a time-dependent convection theory for the whole magnetosphere.

3.4 Magnetospheric convection and the substorm mechanism

We begin the discussion of magnetospheric convection with the following statement: Lossless magnetospheric convection must be time-dependent; steady-state convection is impossible under the conditions that exist in Earth's magnetosphere. In the remainder of this section, we review the various theoretical arguments which lead to the above statement.

Let us assume that a magnetotail flux tube convects from about $x=-60$ R_e to about $x=-20$ R_e. A small time inverval $\Delta t = t_2 - t_1$ of this process is sketched in Fig. 4. In this region of the magnetotail the asymptotic theory is applicable and we find, according to (3.35), for a system in pressure balance

$$P(\boldsymbol{x}_{eq}) \approx \frac{B_x(\boldsymbol{x}_{mp})^2}{2\mu_0}. \tag{3.40}$$

In other words, the equatorial plasma pressure should balance the magnetic field energy density at the magnetopause, where the thermal plasma pressure is negligibly small.

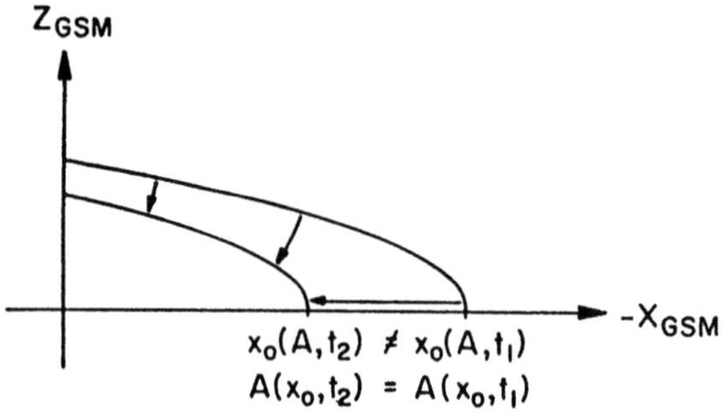

Fig. 4. A sketch of a magnetotail field line convecting earthward during the time inverval $\Delta t = t_2 - t_1$. The field line is identified by its flux function $A(x_0,t) = A\,[x,z(x,x_0,t)]$. The value of A does not change in time, but the equatorial intersection point $x_0(A,t)$ does. As the field line contracts, the flux tube volume $V = \int \mathrm{d}s/B(s)$ decreases, but the plasma pressure in this flux tube increases, according to the adiabatic condition (3.13).

Let us assume also that the above convection process be independent of time, in which case we obtain from (3.13) the thermodynamic condition

$$P(\boldsymbol{x}_{\mathrm{eq}}) \cdot [V(\boldsymbol{x}_{\mathrm{eq}})]^{\gamma} \approx C. \tag{3.41}$$

Thus steady-state convection means that the magnetic field configuration does not change in time and that therefore each flux tube that passes through the equatorial point $\boldsymbol{x}_{\mathrm{eq}}$ has the same amount of pressure and the same flux tube volume. It is obvious from Fig. 4 and from (3.41) that $V(\boldsymbol{x}_{\mathrm{eq}})$ decreases and $P(\boldsymbol{x}_{\mathrm{eq}})$ increases as the field line contracts.

ERICKSON and WOLF (1980) have asked whether (3.41) together with the constraint (3.40) is satisfied in a "typical" magnetospheric $\boldsymbol{B}$-field model. They calculated the quantities $B_x(\boldsymbol{x}_{\mathrm{mp}})^2$ and $V(\boldsymbol{x}_{\mathrm{eq}})$ in several semi-empirical 3-D magnetospheric models and in the 2-D equilibrium model of FUCHS and VOIGT (1979). They found that the pressure $P(\boldsymbol{x}_{\mathrm{eq}})$ which is contained in earthward convecting flux tubes to satisfy the thermodynamic condition (3.41) is much larger than the pressure (3.40) that would hold the system in force balance. The discrepancy becomes larger and larger for flux tubes approaching the Earth. Therefore, ERICKSON and WOLF (1980) concluded that adiabatic, lossless convection has to be time-dependent in a magnetosphere with a dipole-type magnetic main field that extends into a stretched magnetotail. The required time dependency allows the system to adjust the pressure and the volume of a convecting flux tube to the thermodynamic condition (3.13).

The basic conclusion of ERICKSON and WOLF (1980) has been confirmed by SCHINDLER and BIRN (1982). These authors calculated classes of self-consistent solutions to (3.13), (3.37), and (3.39) and demonstrated in a mathematically rigorous way the necessity for the convection process to be time dependent. They could satisfy the time-independent condition (3.41) only by constructing extremely unrealistic magnetotail magnetic field configurations, or by introducing an unrealistic cooling or particle-loss process for the thermal plasma that would reduce the pressure in earthward convecting flux tubes. BIRN and SCHINDLER (1983) extended the 2-D convection theory into three dimensions by applying the asymptotic theory to the pair of equilibrium Eqs. (1.16)–(1.17). In their latest paper, BIRN and SCHINDLER (1985) relaxed the restriction to the quasi-static approximation and calculated 2-D convection with an MHD code. They found that the system does not evolve into a steady state. Thus they confirmed the result already established by the quasi-static MHD slow-flow approximation.

Another approach to modeling magnetospheric convection is due to ERICKSON (1984). He calculated time sequences of 2-D magnetospheric models that include the Earth's dipole field. In other words, he found self-consistent solutions to the three Eqs. (3.7), (3.13), and (3.14) for $B_y(A)^2 = O(\varepsilon^2) \approx 0$. He assumes an (arbitrary) convection electric field $E_y = -\partial A/\partial t$ at the tail magnetopause that allows him to drive the convection process by changing the value of the magnetopause flux function $A_{mp}(t)$ in time.

The result of such a calculation is shown in Fig. 5. The three magnetospheric field geometries represent snapshots taken at three different times after convection has started. The time labels refer to three different values of the magnetopause flux function, namely, $A_{mp}=0$ (top), -12 (middle), and -22 (bottom). There are equal amounts of magnetic flux between adjacent tail field lines plotted. The $A=0$ field line is labeled, and the dashed field line is $A=+30$ for reference, in order to show how field

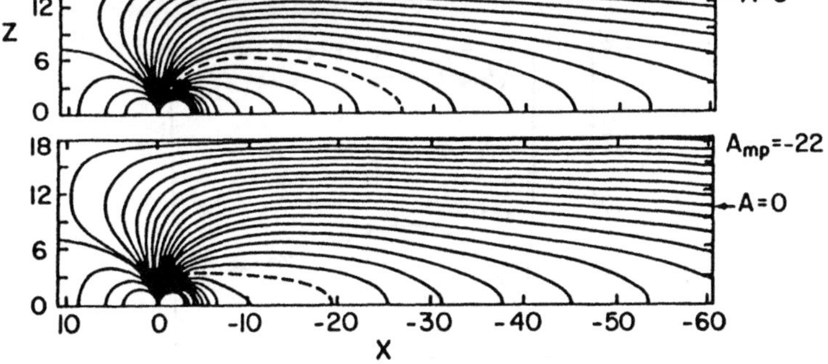

Fig. 5. Time sequence of global magnetospheric convection. Tail field lines convect due to the driving solar wind electric field $E_y = -|\partial A/\partial t|$ specified at the magnetopause. The magnetic field configuration must change in time in order to satisfy the adiabatic condition (3.13). Note that the tail field lines become more and more stretched, and that the dashed field line ($A=30$) moves earthward as convection proceeds (reproduced from ERICKSON (1984)).

lines actually move earthward during the convection process. Owing to

$$A_{mp}(t_{i+1}) = A_{mp}(t_i) - \int_{t_i}^{t_{i+1}} E_y(t)\, dt\ , \tag{3.42}$$

convection is time-dependent in the sense that for each flux tube (identified by the label A) the adiabatic condition (3.13) is maintained, provided the convection process is lossless. Note that E_y at the magnetopause may be assumed to stay constant in time!

The interesting consequence of the computer experiment shown in Fig. 5 is that the adiabatic condition (3.13) forces the system to develop a minimum in the equatorial magnetic field strength near the Earth. Figure 6 shows the equatorial B_z component plotted versus distance down the tail; the three curves correspond to the three field line plots shown in Fig. 5. Note the development of a distinct minimum in B_z which grows rapidly deeper with time. As the $\boldsymbol{B}$-field minimum develops, the tail field lines become more and more stretched, until (presumably) the region around the magnetic field minimum eventually suffers a tearing-mode instability, which results in the formation of a neutral line in the near-Earth plasma sheet. This scenario implies that the formation of a neutral line is an essential element of the magnetospheric convection cycle, and that substorms and convection are two aspects of the same global physical process. In fact, one can argue that the magnetotail does not become flute-unstable earlier in the convection process, before the neutral line has formed, because the function $Q(\boldsymbol{x}_e) = P(\boldsymbol{x}_{eq}) \cdot [V(\boldsymbol{x}_{eq})]^\gamma$ always increases with increasing distance from the Earth, which guarantees stability (Erickson, personal communication).

The time-dependent convection process illustrated in Figs. 5 and 6 can be driven by a steady solar wind. This means that the convection electric field at the magnetopause, E_y, and the magnetic boundary conditions (1.9) may remain constant in time. The time-independent boundary conditions do not prevent the system from evolving toward a configuration that eventually will meet the substorm onset criterion. Under those circumstances, the substorm mechanism is called "spontaneous."

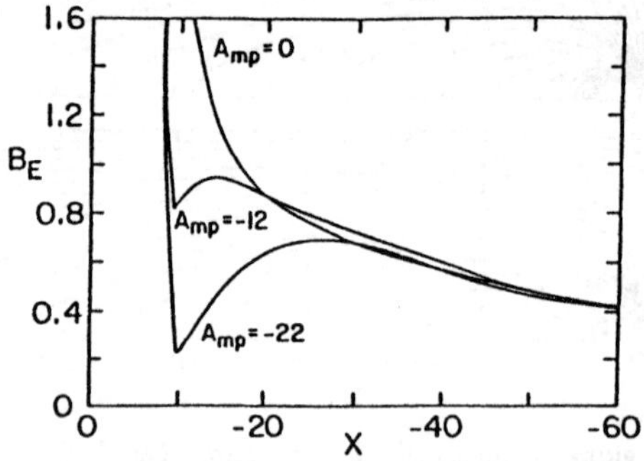

Fig. 6. The equatorial magnetic field strength (i.e., the B_z component) versus the distance down the tail. The three curves correspond to the three configurations depicted in Fig. 5. The minimum in B_z develops near $x = -10\ R_e$ and grows deeper with time as a consequence of the adiabatic condition (3.13) (reproduced from ERICKSON (1984)).

However, changes in the solar wind conditions, i.e., in the flow velocity, $\boldsymbol{v}_0$, and in the IMF strength and orientation, cause changes in the solar wind electric field,

$$\boldsymbol{E}_0(t) = -\boldsymbol{v}_0(t) \times \boldsymbol{B}_{\mathrm{IMF}}(t) \tag{3.43}$$

which lead to changes in $E_y(t)$ and in the magnetic boundary conditions (1.9), which themselves become time-dependent, namely

$$\hat{\boldsymbol{n}} \cdot \boldsymbol{B}(\boldsymbol{x}_{\mathrm{mp}}, t) = f(\boldsymbol{x}_{\mathrm{mp}}, t). \tag{3.44}$$

We shall discuss in Section 4.3 that the magnetic boundary conditions influence the equilibrium configuration of the whole magnetosphere in a very sensitive way. Therefore it is not surprising that abruptly changing boundary conditions (3.44) affect the convection process via the equilibrium conditions (1.1)–(1.3) and via the thermodynamic condition (1.4)–(1.5). Thus the occurrence of a tearing-mode instability in the tail sometimes appears to be caused directly by the solar wind. If the observational material favors a statistical correlation between changes in the solar wind and substorm-type magnetic disturbances, the substorm mechanism is called "directly driven."

Sometimes a substorm can be triggered by a northward turning of the interplanetary magnetic field. When the IMF turns northward, convection slows down or stops completely. Thus the plasma which had previously convected into the small region of minimum equatorial B_z (see Fig. 6) might stay long enough in this region to lose its adiabatic behavior, which in turn could result in an unstable evolution of the system.

The above arguments indicate that the thermodynamic conditions and the magnetopause boundary conditions, according to which the convection process evolves in time, determine how the magnetosphere develops toward a substorm. Thus the popular controversy of whether the substorm is "spontaneous" or "driven" evaporates on the level of the magneto-hydrostatic convection theory.

We summarize the results of this section in terms of a qualitative cartoon depicted in Fig. 7. The time sequence starts in the top panel where the field line B denotes an ordinary stretched tail field line. Subsequently this field line convects toward the Earth, whereby the magnetosphere goes through the configurational changes shown quantitatively in Figs. 5 and 6. Then, in the third panel, a magnetic neutral line has formed owing to the preceding development of the magnetic field minimum. The instability associated with the formation of the neutral line can, of course, not be described by the quasi-static convection theory. As a consequence of the instability a plasmoid forms which finally moves rapidly tailward (panels 4 and 5 in Fig. 7).

There is, however, an interesting alternative to the fast moving plasmoid. Zwingmann and Schindler (1980) and Zwingmann (1983) showed that after the occurrence of a substorm the magnetosphere can rest in a state of lower potential energy compared to the energy level of the pre-substorm stretched equilibrium configuration. The post-substorm low-energy configuration contains magnetic

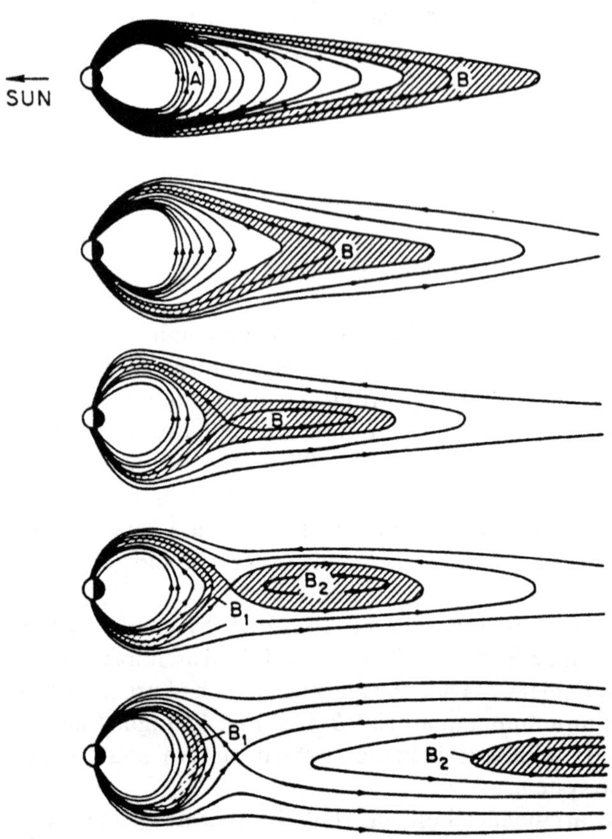

Fig. 7. Sketch of the convection process in Earth's geomagnetic tail (reproduced from WOLF and SPRIO (1983)). The time sequence starts in the top diagram with the field line B. The field line moves earthward (second diagram) and forms a neutral line (third diagram) which leads to a tearing mode instability (i.e., a substorm). As a consequence of the substorm-type instability a plasmoid forms and moves tailward (fourth and fifth diagram). This substorm scenario is based on an idea originally suggested by HONES (1977).

islands or plasmoids that are not associated with fast plasma flows. It is a presently unresolved theoretical problem, whether or not tail equilibria with plasma islands can convect adiabatically according to the thermodynamic condition (3.13)–(3.14).

We complete this section with the following conclusions:

1. The magnetospheric convection process is time-dependent; steady-state convection is not possible in Earth's magnetosphere owing to the existence of the Earth's magnetic dipole.

2. A magnetospheric substorm happens as a consequence of the convection process; the substorm is an inevitable element of the whole magnetospheric convection cycle.

3. The time evolution of the magnetosphere between substorms is determined by the thermodynamic conditions according to which the system convects, and by the (time-dependent) solar-wind-magnetosphere boundary conditions at the magnetopause.
4. Owing to the above arguments, a static "ground-state" of the magnetosphere probably does not exist. It might exist only in the limit of zero convection.

3.5 Plasma anisotropies

So far in this review, we have assumed that the thermal plasma pressure in (1.1) is isotropic. This assumption needs a physical justification. Due to a given page limit, we can only sketch the theoretical arguments which confirm that an isotropic pressure function appropriately describes the physical situation in large parts of Earth's magnetosphere, especially in the tail region during time periods of magnetically quiet conditions.

BIRD and BEARD (1972a, b) investigated two-dimensional equilibrium models of the geomagnetic tail with anisotropic and isotropic pressure distributions. They used a guiding center theory and concluded that an isotropic pitch angle distribution for the plasma particles describes the observational magnetic field data best. STILES *et al.* (1978) contributed plasma observations from the magnetotail region which confirm that plasma anisotropies are in fact very small in the quiet magnetotail.

The theoretical conclusion of Bird and Beard has been confirmed by COWLEY (1978) and by FUCHS (1980) on the basis of the double-adiabatic MHD theory. (The double-adiabatic theory is due to Chew, Goldberger, and Low and is explained in KRALL and TRIVELPIECE (1973), pp. 118–122.) Both Cowley and Fuchs derived linear solutions to the CGL equations and showed that anisotropic pressure distributions within the magnetotail give rise to the firehose ($P_{\|}>P_{\perp}$) and mirror ($P_{\perp}>P_{\|}$) instabilities.

Recently, NÖTZEL *et al.* (1985) derived a generalized Grad-Shafranov equation from the CGL theory for anisotropic two-dimensional tail equilibria. This enabled them to investigate plasma anisotropies in non-linear tail configurations. Their generalized theory largely confirms the results found by COWLEY (1978) and by FUCHS (1980).

Despite the coherent physical picture that justifies an isotropic pressure function in (1.1), we should not forget that the isotropic approximation can hardly be justified for the plasma distribution of Van Allen belt particles and for particles that form the equatorial ring current. The process of ring current formation, for example, has been modeled successfully by the Rice Convection Model (see e.g., SPIRO and WOLF, 1983). The Rice Convection Model treats the ring current plasma in terms of an isotropic multi-fluid code. It is an open theoretical question in what respect plasma anisotropies would affect the physics of magnetospheric convection and ring current formation.

4. Results from the 3-D Theory

Three-dimensional equilibrium models of Earth's magnetosphere, which satisfy

(1.1)–(1.3) in the magnetotail region and near the Earth, are not yet available. One of the reasons lies in the fact that the Grad-Shafranov equation (3.7) does not exist for three-dimensional non-symmetric configurations. Three-dimensional equilibria can be formulated in terms of the Grad-Shafranov equation only if the system is symmetric with respect to one coordinate. The anticipated "pole-on" magnetosphere of Uranus, for example, is axially symmetric with respect to the ϕ coordinate. Therefore, Equation (3.7) can be written in cylindrical (ρ,ϕ,z) coordinates and can be solved for the flux function $A(\rho,z)$. (VOIGT *et al.*, 1983; IP and VOIGT, 1985). Unfortunately, Earth's magnetosphere does not possess a symmetry of this type. The lack of axial symmetry is unfortunate from the viewpoint of magnetospheric theory, fortunate however for those engaged in agriculture (R. A. Wolf, personal communication).

In the general three-dimensional case, the Grad-Shafranov Equation (3.7), or its reduced form (1.20), must be replaced by the pair of equilibrium Eqs. (1.16)–(1.17). Thus the pressure function $P = P(A)$ becomes a function of the two Euler potentials, namely, $P = P(\alpha, \beta)$.

Magnetospheric physics benefits from computer algorithms that have been developed in fusion plasma physics to generate stable Stellarator and Tokamak equilibria. For iterative solutions to (1.1)–(1.3) on a Lagrangian grid, the reader is referred to ROBERTS (1979). Lagrangian grids which move with the magnetic field lines are particularly useful for numerical Euler potential methods because of (1.11) and (1.12). A numerical code which solves the equilibrium Eqs. (1.1)–(1.3) on a fixed Eulerian grid has been developed by CHODURA and SCHLÜTER (1981). Their method is based on the idea to decrease monotonically the potential energy of the system until the state of minimal potential energy represents a stable equlibrium.

In this section we discuss physical conclusions that have been derived from asymptotic solutions to (1.16)–(1.17) and from a model for the whole magnetosphere which includes the Earth's dipole.

4.1 *An intermediate analytical magnetospheric model*

A three-dimensional magnetospheric model which represents an intermediate step on the way toward complete equilibrium has been developed by VOIGT (1981, 1984). This model satisfies Maxwell's Eqs. (1.2) and (1.3), as well as the magnetic boundary conditions (1.8) in a mathematically rigorous way, but it satisfies only two components of the force-balance Eq. (1.1). Despite this limitation, the model reveals the advantages of the quasi-static equilibrium models referred to in Section 2.2: (a) the tail currents result from the model calculations, they are not input; (b) the magnetotail evolves in response to the amount of the thermal plasma pressure distributed in the tail plasma sheet (see Fig. 8); (c) the shape and position of the tail plasma sheet are determined by the orientation of the Earth's dipole and by the functional form of the (given) pressure function (see Section 4.2).

The model is based on a separation method for the magnetic field components and on the assumption that the field aligned currents vanish, i.e., it is $(\boldsymbol{j}\cdot\boldsymbol{B})=0$ throughout the whole magnetosphere. In a cylindrical coordinate system with the x-axis pointing toward the sun and the plane $\phi=0$ indicating the northern half of the

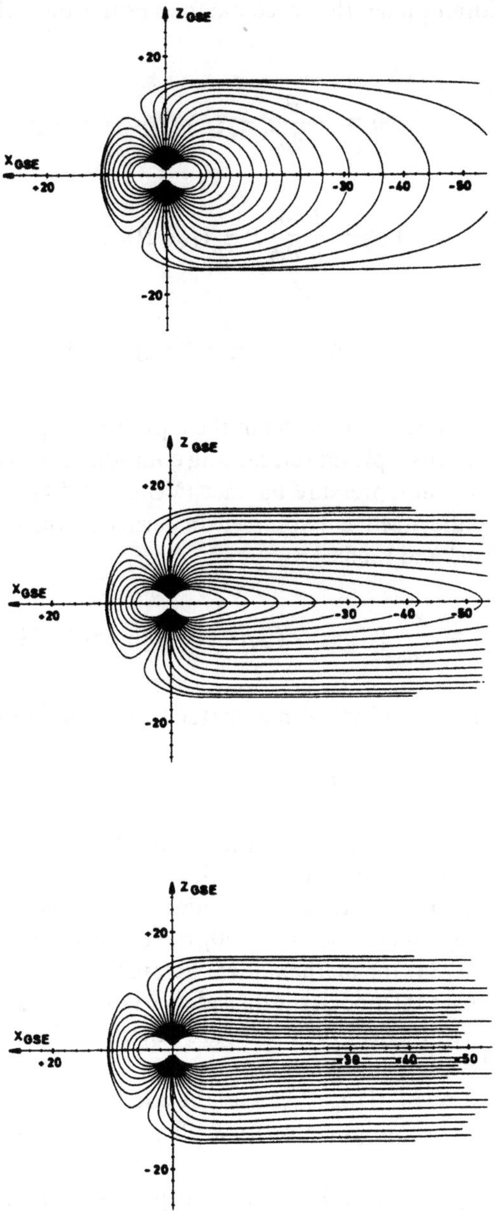

Fig. 8. Magnetic field lines of the three-dimensional magnetosphere (reproduced from VOIGT (1981)). Field lines are plotted in the noon-midnight meridian plane; they are spaced every 2° in latitude, starting at 66°. The tail currents and the whole magnetotail develop in response to the strength of the thermal plasma pressure in the tail plasma sheet. Shown are the vacuum magnetosphere without tail currents and zero plasma pressure (top), a realistic case for which the equatorial plasma pressure was chosen to fit the magnetic field in the tail lobes to observational data (middle), and the Harris sheet configuration which contains the maximum amount of thermal pressure in the magnetotail plasma sheet (bottom).

noon–midnight meridian plane, the force balance Equation (1.1) is replaced by the approximation

$$0 = -\frac{\partial}{\partial \rho} P + (\boldsymbol{j} \times \boldsymbol{B})_\rho , \tag{4.1}$$

$$0 = -\frac{1}{\rho}\frac{\partial}{\partial \phi} P + (\boldsymbol{j} \times \boldsymbol{B})_\phi , \tag{4.2}$$

$$\rho_{\mathrm{m}} \cdot (\boldsymbol{v} \cdot \nabla) v_x = -\frac{\partial}{\partial x} P + (\boldsymbol{j} \times \boldsymbol{B})_x , \tag{4.3}$$

which allows for an analytic formulation of the equilibrium problem. Equation (4.3) tells us that the model needs a plasma streaming component in x direction in order to compensate the lack of static pressure balance (VOIGT, 1984).

The plasma pressure itself is expressed in terms of the eigenfunctions of the cylindrical tail solutions and reads

$$P(\rho, \phi, x) = \frac{k^2}{2\mu_0} [P_0 - G_n(\xi_n \rho)^2 \cos^2\phi] \cdot \exp(-2\lambda_n |x|) . \tag{4.4}$$

The two eigenvalues and the physical parameter k are related by

$$k^2 = \xi_n^{\;2} - \lambda_n^{\;2}. \tag{4.5}$$

Thus the separation method employed for the 3-D model resembles the characteristics of the 2-D linear equilibria discussed in Section 3.2, including all of their physical advantages and limitations. Note the equivalence between (3.15) and (4.4), and between (3.22) and (4.5). In contrast to (3.20) the eigenfunctions $G_n(\xi_n \rho)$ are Bessel functions (see VOIGT (1984) for mathematical details).

In the three-dimensional model, the dayside part of the magnetosphere, as well as the dipole-dominated region near the Earth, are calculated in terms of the zero-order vacuum Eqs. (2.4)–(2.6).

The pressure function (4.4) together with the condition (4.5) include two extreme cases, namely, the vacuum configuration for $k=0$, and the so-called Harris sheet limit for $k=\xi_1$.

The vacuum configuration is shown in the top panel of Fig. 8. Since the thermal pressure is zero, tail currents do not exist. The asymmetry between dayside and nightside is due to the shielding Chapman-Ferraro currents which flow entirely on the magnetopause.

The other extreme is the Harris sheet configuration shown in the bottom panel of Fig. 8. Since $k=\xi_1$, the first leading eigenvalue λ_1 is zero, according to (4.5), and the pressure (4.4) does not depend on x in the far tail, where the higher eigenfunctions ($n>1$) do not contribute to the solution. Therefore the tail field lines are almost parallel to the tail axis and do not intersect the plasma sheet. The term Harris sheet refers to a special class of one-dimensional equilibrium current sheets discussed by HARRIS (1962). Obviously, the Harris sheet configuration represents the upper limit for confining plasma sheet plasma on closed tail field lines: for $k<\xi_1$ all tail field lines are closed, but they become open for $k=\xi_1$, and plasma could escape down the tail.

The middle panel in Fig. 8 depicts a realistic magnetospheric configuration with $k=0.9\ \xi_1$. This value for the plasma sheet pressure (4.4) generates a magnetic field strength in the tail lobes which is in agreement with average tail field measurements reported by MIHALOV and SONETT (1968) and by BEHANNON (1970).

The most influential physical parameters of this magnetospheric model include:

1. the magnetopause stand-off distance which can be calculated according to the actual solar wind pressure,
2. the dipole tilt angle in geocentric solar magnetospheric GSM coordinates, and
3. the constant k which in the range $0\leqslant k\leqslant\xi_1$ regulates the strength of the thermal plasma pressure in the tail plasma sheet and thus determines, whether the magnetotail configuration is more "dipole-like" or more "tail-like". Other model parameters which allow for an exact solution of the boundary value problem (1.9) will be discussed in Section 4.3.

4.2 *Magnetotail configurations*

As was mentioned in Section 2.2, one observes in the magnetotail on the average an equipartition between magnetic field and plasma energy densities. The plasma parameter is $\beta_{\text{pl}}>1$ in the plasma sheet itself, but it is $\beta_{\text{pl}}<1$ in the tail lobes, according to the definition given in (2.1). Thus the physical requirement to satisfy force balance (1.1) is especially crucial for modeling plasma, currents, and magnetic field in the tail region of Earth's magnetosphere.

The magnetospheric model discussed in Section 4.1 reveals the importance of the equilibrium concept in Fig. 9. For this figure the Earth's dipole has been tilted by 35° (summer in the northern hemisphere). Shown are circular tail cross sections taken at $x=-15$, -30, and -60 R_e. Plotted are tail current stream lines that form the tail plasma sheet (VOIGT, 1984). In this example the tail has a constant radius of 20 R_e. The thermal plasma pressure is distributed according to the series of eigenfunctions in (4.4) with the constant $k=0.9\ \xi_1$. The coefficients of the eigenfunction series are determined by the structure of the dayside and dipole-dominated parts of the magnetosphere. Thus, the tail currents and the corresponding tail magnetic field reflect changes in the dipole orientation and in the Chapman-Ferraro current system which is strongest at the dayside part of the magnetopause.

The effect of the dipole tilt on the shape of the plasma sheet is particularly striking at $x=-15$ R_e (top panel in Fig. 9). During the northern hemisphere summer, the plasma sheet is raised above the magnetospheric equatorial plane around local midnight but crosses this plane and is depressed below it near the flanks of the tail. This theoretical result is in agreement with FAIRFIELD (1980). Note also that the plasma sheet thickness increases in the y-direction toward the flanks of the tail. This phenomenon is in accordance with FAIRFIELD's (1979) data interpretation as well.

Owing to the decreasing influence of the vacuum dipole field with increasing

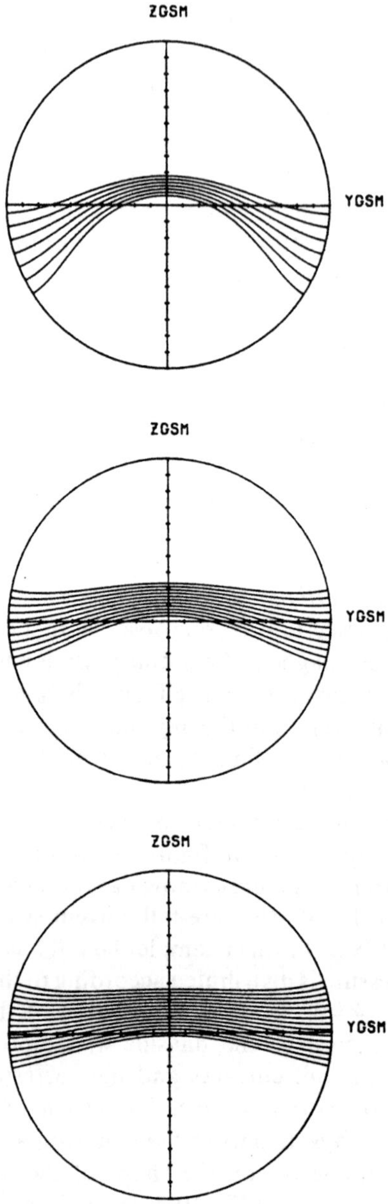

Fig. 9. Configuration of the plasma sheet in Earth's magnetotail (reproduced from VOIGT (1984)). The three tail cross sections are computed at $x_{GSM}=-15$, -30, and -60 R_e (from top to bottom). The Earth's dipole is tilted by 35° which causes the curvature of the plasma sheet. With increasing distance from the Earth, the influence of the vacuum dipole field becomes weaker, and therefore the plasma sheet becomes flatter. The middle current stream line in each panel represents the neutral sheet and carries the highest plasma density and current density.

distance from the Earth, the plasma sheet becomes flatter and thicker at greater distances (second and third panels in Fig. 9). The reason for this behavior is the fact that, mathematically speaking, the first leading eigenfunction with the lowest eigenmode tends to determine the tail configuration at greater distances from the Earth.

Despite its general agreement with the observed tail structure, VOIGT's (1981, 1984) model limited by the approximation (4.3), by the non-flaring magnetotail, and by the restriction to the given pressure function (4.4). These constraints could be relaxed by asymptotic tail models developed by BIRN *et al.* (1975) and BIRN (1979). However, despite their greater physical flexibility, the asymptotic models are restricted by the fact that they do not include the influence of the Earth's dipole field.

Under the assumptions of vanishing field aligned currents, Birn and co-workers applied the asymptotic theory (the 2-D version of which is explained in Section 3.3) to the pair of equilibrium Eqs. (1.16)–(1.17) and constructed tail models for various pressure function $P = P(\alpha, \beta)$. Regarding the magnetotail configuration, they came to the following conclusions:

1. The existence of a magnetic B_y component causes the tail magnetopause to flare in the y-direction.
2. If one assumes "closed" magnetic boundary conditions (1.8), together with other plausible physical assumptions, the tail magnetopause flares slightly more in z than in y. However, modest changes in the physical parameters result in a uniform tail flaring as well.
3. An increase in the plasma sheet thickness in y direction, toward the flanks of the tail, is associated with a reduced flaring of the tail magnetopause.
4. Not only the degree of tail flaring, but also the form of the pressure function $P(\alpha, \beta)$ influences the shape of the tail plasma sheet. If the plasma sheet has a constant thickness across the tail, it simultaneously becomes thinner in the tailward direction. In contrast, an increase of the plasma sheet thickness in the tailward direction is always associated with an increase in thickness toward dawn and dusk.

For mathematical details, especially for the hierarchy of smallness parameters in y and x upon which the asymptotic tail models are based, the reader is referred to the original papers of BIRN *et al.* (1975) and BIRN (1979).

The amount by which the plasma sheet flares in the y direction toward dawn and dusk has important consequences for the global convection process and thus for the development of substorms. The onset criterion for a tearing-mode instability will be reached first in those regions where the plasma sheet is thinnest. This argument favors the region near the Earth around local midnight (compare the three panels in Fig. 9). Owing to the onset of the instability in this region, magnetic flux tubes will release their thermal plasma first around local midnight and will become dipolar. This process leads to large gradients of both the plasma pressure and the flux tube volumes in the y-direction. Near the depleted area, the equatorial pressure $P(\mathbf{x}_{eq})$ increases toward the Earth and toward the flanks of the tail. The flux tube volumes (1.5) increase down the tail and toward the flanks of the tail. Therefore the region-1 Birkeland currents (1.7), which flow between the polar-cap ionosphere and the magnetotail, will dramatically increase such that they flow toward the ionosphere at the dawn side of the plasma sheet, and away from the ionosphere at the dusk side.

BIRN and HONES (1981) simulated the evolution of a tearing-mode instability in a segment of the far tail and demonstrated that the onset of the instability occurs in fact around local midnight in the plasma sheet.

4.3 Magnetopause boundary conditions and the influence of the IMF

Undoubtedly the magnetosphere is open, i.e., magnetospheric and interplanetary magnetic field lines are interconnected through the magnetopause. Despite this observational fact, we do not have a satisfactory theoretical model for the open magnetosphere yet. First we need a global macroscopic reconnection theory from which the magnetic boundary conditions (1.9) could be derived.

A mathematical model of the open magnetosphere has been developed by VOIGT (1978) and by VOIGT and FUCHS (1979) for a specific class of magnetic boundary conditions, namely

$$\hat{\boldsymbol{n}}\cdot\boldsymbol{B} = \hat{\boldsymbol{n}}\cdot[C_d\,\boldsymbol{B}_s(\boldsymbol{x}_{mp}) + C_{imf}\,\boldsymbol{B}_{IMF}(\boldsymbol{x}_{mp})] \tag{4.6}$$

at magnetopause points $\boldsymbol{x}_{mp}$. The functional form (4.6) is based on the hypothesis that all magnetic sources inside and outside the magnetosphere determine the boundary function $\hat{\boldsymbol{n}}\cdot\boldsymbol{B}(\boldsymbol{x}_{mp})$. Owing to two different (unspecified) shielding mechanisms inside and outside the magnetopause, the internal and external sources, $\boldsymbol{B}_s(\boldsymbol{x}_{mp})$ and $\boldsymbol{B}_{IMF}(\boldsymbol{x}_{mp})$, are scaled differently by two interconnection constants C_d and C_{imf}, respectively.

The Chapman-Ferraro current system which is consistent with the boundary condition (4.6) can be calculated in terms of a linear potential problem. For mathematical details regarding the potential problem, in particular for the derivation of (4.6), the reader is referred to VOIGT (1981). The two interconnection constants regulate the magnetic flux that penetrates the magnetopause. There are two extreme cases: for $C_d = C_{imf} = 0$ the magnetosphere is closed and satisfies the boundary condition (1.8); for $C_d = C_{imf} = 1$ the magnetopause does not exist (VOIGT, 1978).

C_d and C_{imf} are two additional model parameters of the magnetospheric model discussed in Section 4.1. Calculations regarding the latitudinal shift of the dayside polar cusp in response to the IMF B_z component (VOIGT and FUCHS, 1979) reveal that the interconnection constants satisfy the empirical relations

$$C_d^2 \ll 1, \tag{4.7}$$

$$(1 - C_{imf}^2) \ll 1. \tag{4.8}$$

These relations imply that the internal magnetospheric $\boldsymbol{B}$-field is almost completely shielded against the outside, whereas the IMF can easily penetrate the magnetopause. Thus (4.7)–(4.8) indicate that the external field $\boldsymbol{B}_{IMF}$ is the dominant factor in controlling the magnetopause normal component. A physical explanation for this empirical result can be obtained in the following way: let us consider a two-dimensional problem ($\partial/\partial y = 0$) where all physical quantities may depend on x and z

(we use again solar-magnetospheric GSM coordinates with x pointing toward the sun and z pointing north). For simplicity, we assume further that the tail magnetopause be a plane parallel to the x-axis (i.e., the tail axis) as depicted, for example, in Fig. 1.

Under the assumption of ideal MHD conditions and $B_y = v_y = 0$, Maxwell's equation $\partial \boldsymbol{B}/\partial t = \nabla \times (\boldsymbol{v} \times \boldsymbol{B})$ reduces to

$$\frac{\partial}{\partial z}(v_z B_x - v_x B_z) = -\frac{\partial}{\partial t} B_x , \tag{4.9}$$

$$\frac{\partial}{\partial x}(v_z B_x - v_x B_z) = +\frac{\partial}{\partial t} B_z . \tag{4.10}$$

In the case of time-averaged magnetic boundary conditions, time-dependencies are small. Note that at the magnetopause $B_z \ll B_x$ and that characteristic scale lengths are according to Fig. 2(a), namely $L_z/L_x = \varepsilon < 1$. Here L_z denotes the characteristic thickness d of the magnetopause. Under these conditions we obtain the following hierarchy of physical quantities,

$$\begin{aligned} v_x(x, z, t) &= O(1) \\ v_z(x, z, t) &= O(\varepsilon) \end{aligned} \tag{4.11}$$

$$\begin{aligned} B_x(x, z, t) &= O(1) \\ B_z(x, z, t) &= O(\varepsilon) \end{aligned} \tag{4.12}$$

$$\begin{aligned} \partial/\partial t &= O(\varepsilon) \\ \partial/\partial x &= O(\varepsilon) \\ \partial/\partial z &= O(1). \end{aligned} \tag{4.13}$$

If we neglect quantities of the order of $O(\varepsilon^2)$ in all MHD equations, (4.10) reduces to

$$\begin{aligned} &\frac{\partial}{\partial t} B_z = 0 \\ &\frac{\partial}{\partial x}(v_z B_x) = 0 \\ &\frac{\partial}{\partial x}(v_x B_z) = 0 . \end{aligned} \tag{4.14}$$

Thus in the asymptotic approximation, the magnetic normal component (1.9), which in our two-dimensional example is the B_z component at the magnetopause, is a small time-independent quantity which has only a very weak x-dependence along the tail magnetopause. In fact, SCHINDLER and BIRN (1978) have used the asymptotic theory (see Section 3.3 for details) to derive MHD jump conditions for a "generalized tangential discontinuity" with a finite thickness $d=L_z=\varepsilon \cdot L_x$. This type of discontinuity allows for a small magnetic field

$$B_n \approx \varepsilon \cdot B_{\text{total}} \tag{4.15}$$

perpendicular to the tail magnetopause.

Note however that time-independent boundary conditions (1.9) still allow for time-dependent magnetospheric convection and the associated change of the tail magnetic field, because time changes in (4.9) are of the order of $O(\varepsilon)$.

Another possibility is to consider the magnetopause as a pure rotational discontinuity. The very special conditions on this discontinuity, however, can usually be satisfied only with an anisotropic pressure tensor (HUDSON, 1971, 1973).

An exact solution to the boundary value problem (4.6) with the relations (4.7)–(4.8) can be used to estimate how much the stand-off position of the dayside magnetopause shifts in response to the IMF orientation. Figure 10 shows the stand-off distance, r_0, as a function of the IMF B_z component for a quiet and for a compressed state of the magnetosphere. The two states are defined in terms of the corresponding solar wind parameters, v_0 and n_0. Changes in the solar wind pressure cause changes in r_0 which are due to magnetospheric compression. In this case the field strength at the subsolar point increases with decreasing stand-off distance. On the other hand, changes in the magnitude and orientation of the IMF B_z component cause changes in r_0 which are due to an (unspecified) process of time-averaged field line erosion at the dayside magnetopause (VOIGT, 1979). In this case the subsolar magnetic field strength remains constant, although the stand-off distance shifts in response to the IMF.

We obtain from Fig. 10 that during quiet solar wind conditions the model magnetosphere is confined to subsolar distances of $r_0=10.4$ R_e and $r_0=11.2$ R_e in response to a 7 nT southward and northward directed IMF. These two values for the stand-off distance correspond qualitatively to the average values of $r_0=10.5$ R_e and $r_0=11.6$ R_e reported by FAIRFIELD (1971) with respect to the two IMF orientations.

VOIGT's (1981) model is capable of reproducing the observed shifts in the polar cusp position (VOIGT and FUCHS, 1979) and in the average stand-off distance (VOIGT, 1979) as functions of the IMF direction. However, the boundary condition (4.6) does not refer to any particular reconnection theory.

Recently, TOFFOLETTO and HILL (1986) began developing an open magnetospheric model for the generalized boundary conditions (1.9) in order to investigate how the boundary conditions affect the polar cap electric field patterns, when the solar wind electric field $\boldsymbol{E}_0=-(v_0 \times \boldsymbol{B}_{\text{IMF}})$ is mapped down into the ionosphere along open magnetic field lines. Their model consists of a vacuum superposition of three magnetic sources. The first is VOIGT's (1981) $\boldsymbol{B}$-field model with the boundary

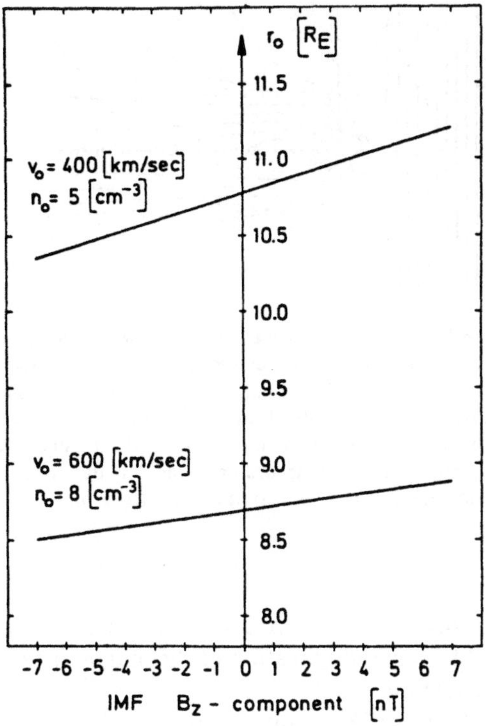

Fig. 10. Magnetopause stand-off distance r_0 as a function of the IMF B_z component for a quiet and for a compressed state of the magnetosphere. The two states are defined in terms of the solar wind velocity and particle density, v_0 and n_0. For a given B_z, large differences in the stand-off distance are caused by magnetospheric compression. For a given solar wind pressure, smaller B_z-dependent changes in the stand-off distance are caused by an assumed process of time-averaged (steady-state) field line erosion at the magnetopause (reproduced from VOIGT (1979)).

condition $C_d=0$ and $C_{imf}=0$ in (4.6). The second is the interplanetary field which in the magnetosheath region is draped around the magnetopause, because it is frozen into the magnetosheath plasma flow. The plasma flow itself results from a gas-dynamic code developed by SPREITER and STAHARA (1980).

The two combined interior and exterior magnetic fields satisfy the boundary condition (1.8). Superimposed is the third magnetic source, namely, internal and external vacuum potential fields which are due to a given distribution $f(\boldsymbol{x}_{mp})$, according to the boundary condition (1.9). The combined three sources result in a model for the open magnetosphere which is shown in Fig. 11. In this particular example, the magnetic normal component is $f(\boldsymbol{x}_{mp})=-B_0 \cos\phi$, where ϕ rotates in the magnetospheric y–z plane. The advantage of this model lies in its flexibility regarding the magnetic boundary conditions.

A quantitative model for the open magnetosphere is crucial for deriving the polar cap electric field patterns directly from the solar wind conditions. However, a model

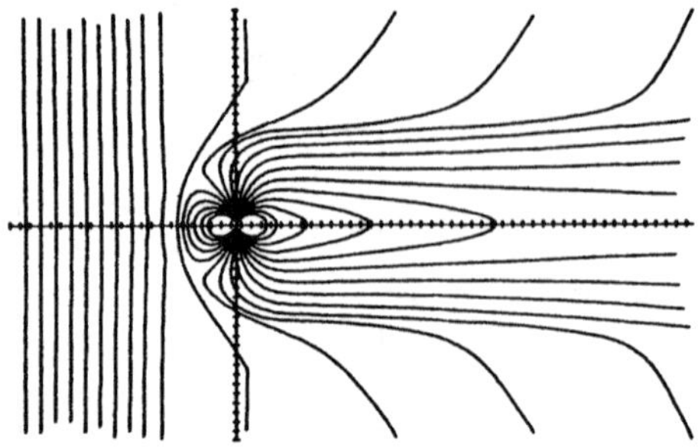

Fig. 11. Quantitative model of an open magnetosphere. The magnetic field lines result from a superposition of three magnetic sources: (a) a "closed" magnetospheric model of the type shown in the middle panel of Fig. 8 which satisfies the boundary condition (1.8); (b) the interplanetary magnetic field which is frozen into the magnetosheath plasma flow; it satisfies (1.8) as well and is draped around the magnetopause; (c) an additional vacuum potential field which is due to a given distribution $f(\boldsymbol{x}_{mp})$ of the magnetopause normal component according to the boundary condition (1.9). The numerical calculations for this figure are due to TOFFOLETTO and HILL (1985).

of this type is also important for studying the influence of the magnetic boundary conditions on the equilibrium configuration of the magnetic field and on the process of magnetospheric convection.

PILIPP and MORFILL (1978) pointed out that the $(\boldsymbol{E}\times\boldsymbol{B})$−drift (solar wind convection electric field and magnetotail magnetic field) transports plasma from the plasma mantle into the magnetotail plasma sheet. This process acts like an energy filter, such that one would expect lower-energy particles to enter the tail plasma sheet at a smaller distance from the Earth. It is important in our context that Pilipp and Morfill calculated different plasma energy distributions in the magnetospheric equatorial plane, depending on whether the magnetosheath plasma enters the plasma mantle near the dayside polar cusp or along the whole tail magnetopause. Let us assume for a moment that we were able to derive the macroscopic thermal plasma pressure $P(\boldsymbol{x}_{eq})$ in the equatorial plane from the plasma distribution function generated by the $(\boldsymbol{E}\times\boldsymbol{B})$−drift mechanism. Then the equatorial pressure function would depend sensitively on the magnetic boundary conditions at the magnetopause. In general, the function for the magnetic normal component (1.9) could favor particle entry through the polar cusps or along the tail magnetopause.

We know that the equatorial pressure function $P(\boldsymbol{x}_{eq})$ is an essential element that determines the quasi-static equilibrium configuration of the magnetotail via the equilibrium Eqs. (1.1)–(1.3). The $(\boldsymbol{E}\times\boldsymbol{B})$−drift in (2.13) determines how plasma and magnetic field convect earthward in the tail plasma sheet. Thus we conclude that the magnetospheric tail configuration and the process of earthward plasma convection depend on the mechanism of particle entry into the magnetosphere via the form of the magnetic boundary conditions (1.9) we assume at the magnetopause. Therefore, the development of a quantitative theory for the open magnetopause is one of the most relevant problems in magnetospheric physics.

I am grateful to Joachim Birn, Gary Erickson, and Dick Wolf for their comments on the manuscript.

This work was supported by the National Science Foundation under grant ATM 84-17580, by AFGL under contract F-19628-83-K-0016, and by NASA under grant NAGW-482.

REFERENCES

BEARD, D. B., The interactions of the solar wind with planetary magnetic fileds: basic principles and observations, *Planet. Space Sci.*, **21**, 1475–1496, 1973.

BEHANNON, K. W., Geometry of the geomagnetic tail, *J. Geophys. Res.*, **75**, 743–753, 1970.

BIRD, M. K. and D. B. BEARD, The self-consistent geomagnetic tail under static conditions, *Planet. Space Sci.*, **20**, 2057–2072, 1972a.

BIRD, M. K. and D. B. BEARD, Self-consistent description of the magnetotail current system, *J. Geophys. Res.*, **77**, 4864–4866, 1972b.

BIRN, J., Self-consistent magnetotail theory: general solution for the quiet tail with vanishing field-aligned currents, *J. Geophys. Res.*, **84**, 5143–5152, 1979.

BIRN, J., R. SOMMER, and K. SCHINDLER, Open and closed magnetospheric tail configurations and their stability, *Astrophys. Space Sci.*, **35**, 389–402, 1975.

BIRN, J., R. SOMMER, and K. SCHINDLER, Self-consistent theory of the quiet magnetotail in three dimensions, *J. Geophys. Res.*, **82**, 147–154, 1977.

BIRN, J. and E. W. HONES, Three-dimensional computer modeling of dynamic reconnection in the geomagnetic tail, *J. Geophys. Res.*, **86**, 6802–6808, 1981.

BIRN, J. and K. SCHINDLER, Self-consistent theory of three-dimensional convection in the geomagnetic tail, *J. Geophys. Res.*, **88**, 6969–6980, 1983.

BIRN, J. and K. SCHINDLER, Computer modeling of magnetotail convection, *J. Geophys. Res.*, **90**, 3441–3447, 1985.

BRECHT, S. H., J. G. LYON, J. A. FEDDER, and K. HAIN, A simulation study of the east-west IMF effects on the magnetosphere, *Geophys. Res. Lett.*, **8**, 397–400, 1981.

BRECHT, S. H., J. G. LYON, J. A. FEDDER, and K. HAIN, A time dependent three-dimensional simulation of the earth's magnetosphere: Reconnection events, *J. Geophys. Res.*, **87**, 6098–6108, 1982.

CHODURA, R. and A. SCHLÜTER, A 3D code for MHD equilibrium and stability, *J. Comp. Phys.*, **41**, 68–88, 1981.

COWLEY, S. W. H., The effect of pressure anisotropy on the equilibrium structure of magnetic current sheets, *Planet. Space Sci.*, **26**, 1037–1061, 1978.

COWLEY, S. W. H., On the distribution of B_y in the geomagnetic tail, *Planet. Space Sci.*, **27**, 769–793, 1979.

COWLEY, S. W. H., Magnetospheric asymmetries associated with the y-component of the IMF, *Planet. Space Sci.*, **29**, 79–96, 1981.

ERICKSON, G. M., On the cause of X-line formation in the near-earth plasma sheet: Results of adiabatic convection of plasma sheet plasma, in *Magnetic Reconnection, Geophys. Monogr. Ser., Vol. 30*, edited by E. W. Hones, pp. 296–302, AGU, Washington, D.C., 1984.

ERICKSON, G. M. and R. A. WOLF, Is steady state convection possible in the earth's magnetotail?, *Geophys. Res. Lett.*, **7**, 897–900, 1980.

FAIRFIELD, D. H., Average and unusual locations of the earth's magnetopause and bow shock, *J. Geophys. Res.*, **76**, 6700–6716, 1971.

FAIRFIELD, D. H., On the average configuration of the geomagnetic tail, *J. Geophys. Res.*, **84**, 1950–1958, 1979.

FAIRFIELD, D. H., A statistical determination of the shape and position of the geomagnetic neutral sheet, *J. Geophys. Res.*, **85**, 775–780, 1980.

FUCHS, K. and G.-H. VOIGT, Self-consistent theory of a magnetospheric ***B***-field model, in *Quantitative Modeling of Magnetospheric processes, Geophys. Monogr. Ser., Vol. 21*, edited by W. P. Olson, pp. 86–95, AGU, Washington, D.C., 1979.

FUCHS, K., Druckanisotropie in einem statischen Modell des Magnetosphärenschweifes, *Kleinheubacher Ber.*, **23**, 259–266, 1980.

HAREL, M., R. A. WOLF, P. H. REIFF, and R. W. SPIRO, Quantitative simulation of magnetospheric substorm, 1, Model logic and overview, *J. Geophys. Res.*, **86**, 2217–2241, 1981.

HARRIS, E. G., On a plasma sheet separating regions of oppositely directed magnetic fields, *Nuovo Cimento*, **23**, 115–121, 1962.

HONES, E. W., Substrom processes in the magnetotail: comments on 'on hot tenuous plasmas, fireballs, and boundary layers in earth's magnetotail' by L. A. Frank *et al.*, *J. Geophys. Res.*, **82**, 5633–5640, 1977.

HUDSON, P. D., Rotational discontinuities in an anisotropic plasma, *Planet. Space Sci.*, **19**, 1693–1699, 1971.

HUDSON, P. D., Rotational discontinuities in an anisotropic plasma II, *Planet. Space Sci.*, **21**, 475–483, 1973.

IP, A. K. and G.-H. VOIGT, Plasma-dominated magnetic field configurations for the magnetosphere of Uranus, *J. Geophys. Res.*, **90**, 6287–6293, 1985.

KRALL, N. A. and A. W. TRIVELPIECE, *Principles of Plasma Physics*, McGraw-Hill, New York, 1973.

LUI, A. T. Y., Characteristics of the cross-tail current in the earth's magnetotail, in *Magnetospheric Currents, Geophys. Monogr. Ser., Vol. 28*, edited by T. A. Potemra, pp. 158–170, AGU, Washington, DC, 1983.

LYON, J. G., S. H. BRECHT, J. A. FEDDER, and P. PALMADESSO, The effects on the earth's magnetotail from shocks in the solar wind, *Geophys. Res. Lett.*, **7**, 721–724, 1980.

LYON, J. G., S. H. BRECHT, J. D. HUBA, J. A. FEDDER, and P. J. PALMADESSO, Computer simulation of a geomagnetic substorm, *Phys. Rev. Lett.*, **46**, 1038–1041, 1981.

MIHALOV, J. D. and C. P. SONETT, The cislunar geomagnetic tail gradient in 1967, *J. Geophys. Res.*, **73**, 6837–6841, 1968.

NÖTZEL, A., K. SCHINDLER, and J. BIRN, On the cause of approximate pressure isotropy in the quiet near-earth plasma sheet, *J. Geophys. Res.*, **90**, 8293–8300, 1985.

OGINO, T. and R. J. WALKER, An MHD simulation of the bifurcation of the tail lobes during intervals with northward interplanetary magnetic field, *Geophys. Res. Lett.*, **11**, 1018–1021, 1984.

PILIPP, W. G. and G. MORFILL, The formation of the plasma sheet resulting from plasma mantle dynamics, *J. Geophys. Res.*, **83**, 5670–5678, 1978.

ROBERTS, K. V., The solution of hyperbolic equations in magneto-hydrodynamics, in *Computing Methods in Applied Sciences and Engineering, 1977, II*, edited by R. Glowinski and J. L. Lions, pp. 315–328, Springer-Verlag, NY, 1979.

SCHINDLER, K., A self-consistent theory of the tail of the magnetosphere, in *Earth's Magnetospheric Processes*, edited by B. M. McCormac, pp. 200–209, D. Reidel, Dordrecht, Netherlands, 1972.

SCHINDLER, K. and J. BIRN, Magnetospheric physics., *Physics Reports*, **47**, 109–165, 1978.

SCHINDLER, K. and J. BIRN, Self-consistent theory of time-dependent convection in the earth's magnetotail, *J. Geophys. Res.*, **87**, 2263–2275, 1982.

SCKOPKE, N., A study of self-consistent ring-current models, *Cosmic Electrodyn.*, **3**, 330–348, 1972.

SPIRO, R. W. and R. A. WOLF, Electrodynamics of convection in the inner magnetosphere, in *Magnetospheric Currents, Geophys. Monogr. Ser., Vol. 28*, edited by T. A. Potemra, pp. 247–259, AGU, Washington, D.C., 1983.

SPREITER, J. R. and S. S. STAHARA, A new predictive model for determining solar wind—terrestrial planet interactions, *J. Geophys. Res.*, **85**, 6769–6777, 1980.

STERN, D. P., Euler potentials, *American J. Phys.*, **38**, 494–501, 1970.

STERN, D. P., Representation of magnetic fields in space, *Rev. Geophys. Space Phys.*, **14**, 199–214, 1976.

STILES, G. S., E. W. HONES, S. J. BAME, and J. R. ASBRIDGE, Plasma sheet pressure anisotropies, *J. Geophys. Res.*, **83**, 3166–3172, 1978.

TOFFOLETTO, F. R. and T. W. HILL, Quantitative model of magnetic coupling between solar wind and magnetosphere, this volume, 1986.

TSURUTANI, B. T., J. A. SLAVIN, E. J. SMITH, R. OKIDA, and D. E. JONES, Magnetic structure of the distant geotail from −60 to −220 R_E: ISEE-3, *Geophys. Res. Lett.*, **11**, 1–4, 1984.

VASYLIUNAS, V. M., Mathematical models of magnetospheric convection and its coupling to the ionosphere, in *Particles and Fields in the Magnetosphere*, edited by B. M. McCormac, pp. 60–71, D. Reidel, Hingham, MA, 1970.

VOIGT, G.-H., A static-state field line reconnection model for the earth's magnetosphere, *J. Atmos. Terr. Phys.*, **40**, 355–365, 1978.

VOIGT, G.-H., Influence of the interplanetary magnetic field on the position of the dayside magnetopause, in *Proc. Magnetospheric Boundary Layers Conf.*, edited by Battrick and Mort, pp. 315–321, European Space Agency, ESA SP-148, 1979.

VOIGT, G.-H., A mathematical magnetospheric field model with independent physical parameters, *Planet. Space Sci.*, **29**, 1–20, 1981.

VOIGT, G.-H., The shape and position of the plasma sheet in the earth's magnetotail, *J. Geophys. Res.*, **89**, 2169–2179, 1984.

VOIGT, G.-H. and K. FUCHS, A macroscopic model for field line interconnection between the magnetosphere and the interplanetary space, in *Quantitative Modeling of Magnetospheric Processes, Geophys. Monogr. Ser., Vol. 21*, edited by W. P. Olson, pp. 448–459, AGU, Washington, D.C., 1979.

VOIGT, G.-H., T. W. HILL, and A. J. DESSLER, The magnetosphere of Uranus: plasma sources, convection, and field configuration, *Astrophys. J.*, **266**, 390–401, 1983.

VOIGT, G.-H. and R. A. WOLF, On the configuration of the polar cusps in earth's magnetosphere, *J. Geophys. Res.*, **90**, 4046–4054, 1985.

WALKER, R. J., An evaluation of recent quantitative magnetospheric magnetic field models, *Rev. Geophys. Space Phys.*, **14**, 411–427, 1976.

WALKER, R. J., Quantitative modeling of planetary magnetospheric magnetic fields, in *Quantitative Modeling of Magnetospheric Processes, Geophys. Monogr. Ser., Vol. 21*, edited by W. P. Olson, pp. 9–34, AGU, Washington, DC, 1979.

WALKER, R. J. and D. J. SOUTHWOOD, Momentum balance and flux conservation in model magnetospheric magnetic fields, *J. Geophys. Res.*, **87**, 7460–7466, 1982.

WOLF, R. A., The quasi-static (slow-flow) region of the magnetosphere, in *Solar Terrestrial Physics*, edited by R. L. Carovillano and J. M. Forbes, pp. 303–368, D. Reidel, Hingham, MA, 1983.

WOLF, R. A. and R. W. SPIRO, The role of the auroral ionosphere in magnetospheric substorms, in *High-Latitude Space Plasma Physics*, edited by V. Hultqvist and T. Hagfors, pp. 19–38, Plenum, 1983.

WU, C.-C., Shape of the magnetosphere, *Geophys. Res. Lett.*, **10**, 545–548, 1983.

WU, C.-C., The effects of dipole tilt on the structure of the magnetosphere, *J. Geophys. Res.*, **89**, 11,048–11,052, 1984.

WU, C.-C., R. J. WALKER, and J. M. DAWSON, A three-dimensional MHD model of the earth's magnetosphere, *Geophys. Res. Lett.*, **8**, 523–526, 1981.

ZWINGMANN, W., Self-consistent magnetotail theory: equilibrium structures including arbitrary variation along the tail axis, *J. Geophys. Res.*, **88**, 9101–9108, 1983.

ZWINGMANN, W. and K. SCHINDLER, Magnetic islands in the quiet magnetotail, *Geophys. Res. Lett.*, **7**, 909–912, 1980.

Solar Wind-Magnetosphere Coupling, edited by Y. Kamide and J. A. Slavin, 275–285.

Quantitative Model of Magnetic Coupling between Solar Wind and Magnetosphere

F. R. TOFFOLETTO and T. W. HILL

Space Physics and Astronomy Department, Rice University, Houston, Texas 77251, U.S.A.

We are engaged in an effort to develop a theoretical model of a magnetically interconnected (open) magnetosphere in sufficient quantitative detail to justify comparison with recent and anticipated observational data bases. This model incorporates a magnetopause geometry and internal geomagnetic field configuration based on the work of G.-H. Voigt and an external magnetosheath flow and magnetic-field configuration based on the work of J. R. Spreiter. The distribution of the normal magnetic-field component at the magnetopause is specified as input, from which we compute an interconnection field that links the internal and external fields. The interconnected magnetic field configuration is then used to map the solar-wind electric field onto the polar cap, thus producing a potential flow pattern suitable for comparison with observations.

Here we present preliminary results appropriate to the subsolar merging-line hypothesis, for various orientations of the interplanetary magnetic field within the GSM y–z plane. Our present specifications of the magnetopause normal component and of the magnetosheath flow field are self-consistent only on the tail magnetopause, which, however, maps to a large fraction of the polar cap. Within its region of validity, the model produces polar-cap flow patterns in agreement with observations. Future modeling of the dayside magnetopause and of the more complicated antiparallel merging hypothesis will allow a more quantitative comparison with observations.

1. Introduction

It is generally accepted that magnetic interconnection is the primary mechanism of solar-wind magnetosphere coupling. Evidence for the existence of interconnection includes direct measurements at the magnetopause (e.g., SONNERUP *et al.*, 1981) and studies of the control of polar-cap convection by the interplanetary magnetic field (IMF) (e.g., REIFF 1984a). Although many features of polar-cap convection are qualitatively understood on the basis of a mapping of the solar-wind electric field to the polar cap along interconnected magnetic-field lines (HILL, 1983, and references therein), quantitative models of this electric-field mapping have been restricted to the vacuum superposition of a uniform "IMF" with a dipole "geomagnetic field" (e.g., STERN, 1973; HILL and RASSBACH, 1975; SAUNDERS, 1976; LYONS, 1985). Such models are helpful in understanding the topology of the electric-field mapping, but they cannot simulate accurately the mapping geometry.

In this paper we present preliminary results of a second generation model of the open magnetospheric magnetic-field configuration and the corresponding electric-field mapping relation, a model intended to simulate the actual three-dimensional magnetosphere geometry as determined, to first order, by the supersonic flow of the solar wind.

One of the many unanswered questions concerning interconnection involves the location on the magnetopause where magnetic merging occurs, i.e., the position and orientation of the merging line which separates regions of inward and outward normal magnetic components at the magnetopause. One hypothesis (SONNERUP, 1974, SONNERUP *et al.*, 1981; GONZALEZ and MOZER, 1974; HILL, 1975) is that the merging line passes through the subsolar point with an orientation depending on the relative strength and orientation of the magnetosheath and magnetospheric magnetic fields. An alternate hypothesis (CROOKER, 1979, 1980) is that the merging line follows the locus of points where the internal and external magnetic fields are antiparallel. These two alternate merging-line hypotheses are further discussed by HILL (1983).

This report presents preliminary results of a quantitative three dimensional model of an open steady-state magnetosphere configuration incorporating a normal-component distribution corresponding to the subsolar merging-line hypothesis. The corresponding model for the antiparallel merging-line hypothesis remains a task for future work.

The model is a perturbation to a closed configuration consisting of an internal field from the model of VOIGT (1972, 1981) and an external field derived from the gasdynamic calculation of Spreiter and Stahara (STAHARA *et al.*, 1980, SPREITER and STAHARA, 1980), which is known to give a good approximation for the flow field in the magnetosheath. The magnetopause shape for this model is very simple, consisting of a hemispherical dayside coupled to a cylindrical tail (Fig. 1) (VOIGT, 1972, 1981). This geometry allows a semi-analytical solution to be found for the internal field as well as for the "interconnection" field (to be discussed later). Perturbing the closed configuration with a field derived from a specified distribution of the magnetopause normal component results in an open field configuration.

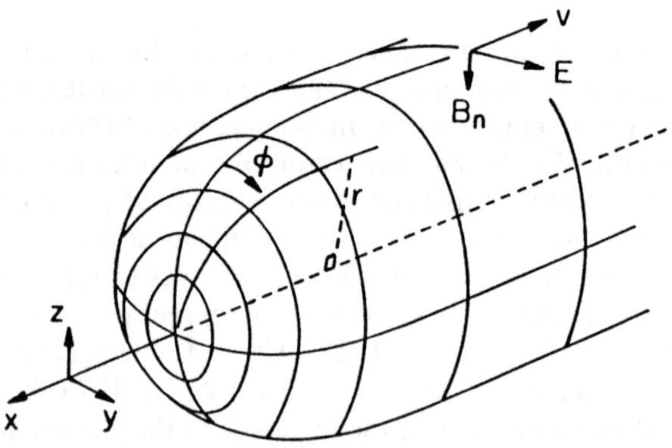

Fig. 1. Three-dimensional illustration of the magnetopause geometry consisting of a cylindrical tail ($x<0$) joined to a hemispherical day side ($x>0$). ϕ is the azimuthal angle in the y–z plane and r is the cylindrical radius which matches the spherical radius at the boundary $x=0$. The electric field E is determined by the downstream magnetopause flow (v) and the normal magnetic component B_n through Eq. (9).

2. The Closed Models

The internal field is based on the closed three-dimensional magnetospheric field model of VOIGT (1972, 1981) with a given magnetopause shape. The geometry of the magnetopause is not the result of a pressure balance calculation, but a more mathematically convenient shape with parameters that allow it to be fit to experimental data with a reasonable degree of accuracy (WALKER, 1976). By adopting a simplified magnetopause shape, one is able to solve for the internal magnetic field configuration to a high degree of accuracy through an eigenfunction expansion for the fields due to the Chapman-Ferraro and tail currents. The model parameters include a subsolar standoff distance of 10 R_E, a tail radius of 20 R_E, and a dipole tilt of 0 degrees for the present.

The model includes a tail current system with an exponentially decreasing current density with distance down the tail according to $\exp(-\alpha|x|)$, where $\alpha=0$ corresponds to the Harris sheet limit. A value $\alpha=0.05$ was used for this model, corresponding to a characteristic scale length of 20 R_E for the decay of the tail field with distance down the tail. It was found that the size and shape of the polar-cap boundary between open and closed field lines depends very sensitively on this parameter α.

A plot of the closed configuration is shown in Fig. 2 for the noon midnight meridian plane. For a further discussion of the internal field model see VOIGT (1981).

The external field was obtained from a numerical code supplied by STAHARA *et al.* (1980) which uses a gasdynamic approximation to obtain the supersonic flow of plasma around a fixed blunt body. The magnetic field was calculated by assuming that the field is embedded in the flow and that it has no significant influence on the flow. The result is the draped field shown in Fig. 2.

All the results presented herein are for a typical solar-wind Mach number of 10,

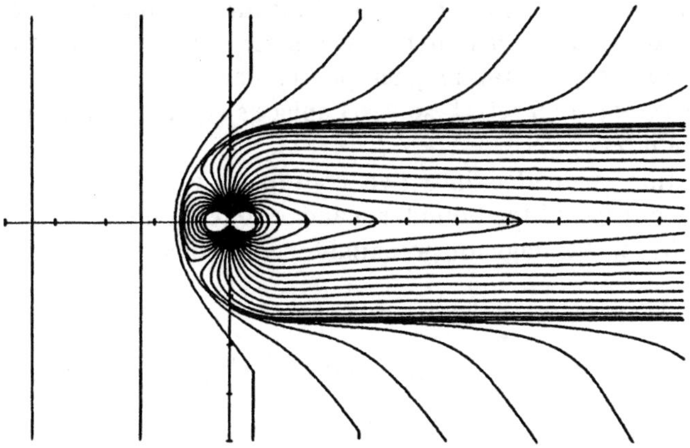

Fig. 2. Closed magnetic field configuration plotted in the symmetry plane showing the internal and external fields. The spacings on the axes correspond to a distance of 10 R_E. Note that the axes in this and subsequent figures are centered at the earth which corresponds to $x=10$ (R_E).

proton density of 5 cm^{-3}, and mean temperature of 1.2×10^5K, which corresponds to a solar-wind velocity of about 400 km/sec. The magnetopause standoff distance is taken to be 10 R_E as determined by pressure balance. The magnitude of the IMF was taken to be 6 nT.

The influence of the magnetic field on the flow is significant near the stagnation point where it leads to the development of a depletion layer (ZWAN and WOLF, 1976). However, this effect is assumed to be important only if magnetic merging is unimportant, because an equivalent effect is produced by the merging outflow away from the neutral line.

3. The Interconnection Field

In order to achieve the desired mapping of the solar-wind electric field into the magnetosphere, a perturbation field was added which is large enough to make the magnetosphere significantly open, but small enough not to noticeably change either the internal or external fields. This " interconnection field" is the solution of a Neumann boundary-value problem with a specified distribution of normal components at the magnetopause.

For the simple case discussed in this paper, the magnetopause normal component is given by:

$$B_n = -B_0 \cos\phi \tag{1}$$

where ϕ is the azimuthal angle shown in Fig. 1, and B_0 is constant. The normal component at the magnetopause is assumed to be independent of downstream distance, which is consistent with a steady state magnetosphere; for further details see VOIGT (this volume). It is also required that the interconnection field be zero at an infinite distance from the tail axis and that at infinite downstream distance the solution be independent of downstream position.

We define the "interconnection field" as the (unique) magnetic field that satisfies the given boundary conditions, that is divergence-free everywhere, and that is curl-free everywhere except at the magnetopause. This field is then the gradient of a magnetic scalar potential which solves Laplace's equation and satisfies the given boundary conditions.

The solution is divided into two coordinate regions, $x>0$ (containing the hemispherical portion of the boundary) and $x<0$ (containing the cylindrical part). The solution U_1 for the magnetic potential for $x>0$ consists of internal and external parts:

$$U_{1\,\mathrm{ext}} = -\sum_{n=1}^{\infty} [A_n/(n+1)](R^{n+2}/r^{n+1})\, P_n^1(\cos\theta)\cos\phi \tag{2}$$

$$U_{1\,\mathrm{int}} = \sum_{n=1}^{\infty} [A_n/n](r^n/R^{n-1})\, P_n^1(\cos\theta)\cos\phi \tag{3}$$

where P_n^1 are Legendre polynomials of the first kind, A_n are the coefficients to be determined, r is the (spherical) radial coordinate, R is the radius of the hemispherical magnetopause surface (taken to be 20 R_E) and θ is the angle that the radius vector makes with the dawn–dusk meridian plane.

The solution for $x<0$ consists of two parts. One part is independent of downstream distance and produces the required normal component at the magnetopause:

$$U_{2\,ext} = -B_0\,(R^2/r)\cos\phi \tag{4}$$

$$U_{2\,int} = B_0\, r\, \cos\phi \tag{5}$$

where r is now the cylindrical radial coordinate. (The cylindrical and spherical radial coordinates are identical at $x=0$.) The second part of the solution for $x<0$ corrects for the overlapping of the cylindrical and spherical fields. This potential is of the form

$$U_3 = \sum_{i=1}^{\infty} B_i \cos\phi\, J_1\,(x_{1i}r/R)e^{+x_{1i}x/R} \tag{6}$$

in which x_{1i} is the ith solution of the equation

$$J_1'\,(x_{1i})=0 \tag{7}$$

where r is distance from the tail axis, R is the radius of the tail in Earth radii (taken to be 20 R_E), and J_1 is a Bessel function of the first kind.

The method of solution follows the eigenvalue method of VOIGT (1972), which was used in the calculation of the closed model. The coefficients were obtained by numerical integration coupled with an iterative procedure to obtain convergence between the spherical and cylindrical solutions. The interconnection magnetic field is then derived by taking the negative gradient of the resulting potential.

Note that for the above solutions there is a discontinuity of the external potential at $x=0$ owing to the difference in radial dependence for $x \gtrless 0$. However, since the calculations involving the mapping of the electric field involve only the internal part of the interconnection field, a continuous external solution is not needed. Also, we are not aware of an analytical solution for the external cylinder problem; a finite element method would evidently be required. Note also that the perturbation field has the greatest influence on the overall field on or near the magnetopause where the discrepancy is smallest.

The resulting interconnection field is shown in Fig. 3, for field lines plotted in the symmetry plane $\phi=0$. For reasons discussed below only the interconnection field in the tail region was used.

4. Open Field Configuration

The magnitude of the normal component was chosen such that its average value

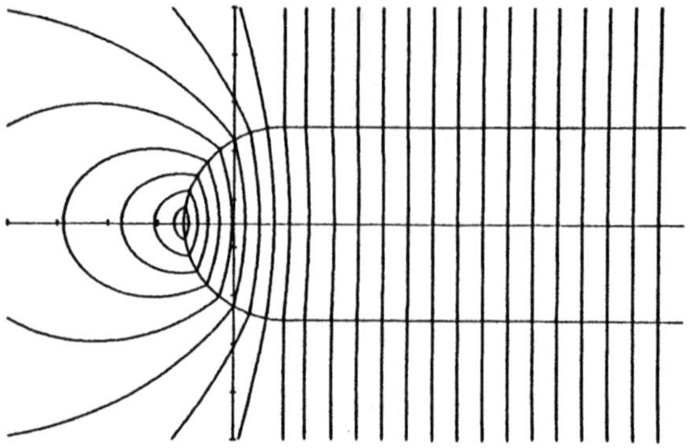

Fig. 3. Configuration of the interconnection field resulting from the solution of Laplace's equation with a normal component specified by Eq. (1). The scale is the same as Fig. 2, and the field is shown in the symmetry plane.

at the magnetopause is approximately 0.1 times the IMF magnitude of 6 nT; this results in a normal component of maximum magnitude $B_0=1$ nT. The factor 0.1 corresponds to the empirical result that the cross polar cap potential is typically ~0.1 times the total potential drop across a width equal to that of the magnetosphere during periods of southward IMF (REIFF *et al.*, 1981).

The total field is the linear combination of the internal and external fields (Fig.2) and the interconnection field (Fig. 3), and is plotted in Fig. 4. The magnetopause still contains a discontinuity of the magnetic field, although this is no longer obvious on the night side.

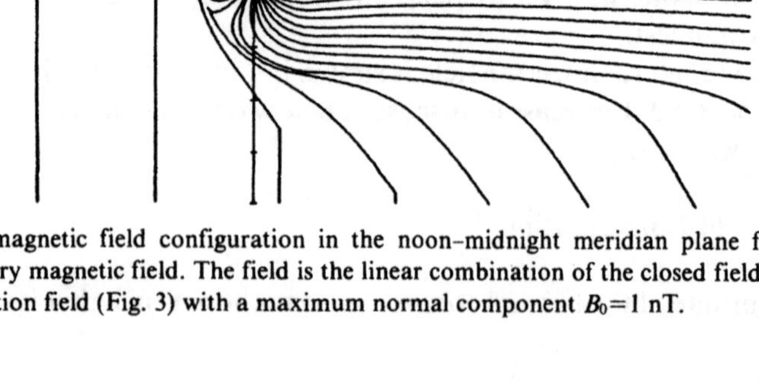

Fig. 4. Open magnetic field configuration in the noon–midnight meridian plane for a southward interplanetary magnetic field. The field is the linear combination of the closed field (Fig. 2) and the interconnection field (Fig. 3) with a maximum normal component $B_0=1$ nT.

5. Electric Field Mapping

The electric field at the magnetopause is given by the MHD condition

$$\boldsymbol{E}+\boldsymbol{v}\times\boldsymbol{B}=\boldsymbol{0} \tag{8}$$

and the azimuthal component is

$$E_\phi=vB_n\,. \tag{9}$$

With B_n given by (1) the potential at the magnetopause is then given by

$$V(\phi)=\int^{\phi} vB_0\cos\phi R\,d\phi=vB_0R\sin\phi. \tag{10}$$

Having specified the potential distribution on the magnetopause, we can map this distribution along the model magnetic field, assuming that the field lines are equipotentials, to obtain the polar-cap potential distribution. From (10) we obtain a maximum cross polar cap potential of

$$V_0=2B_0Rv\approx 90\text{ kV} \tag{11}$$

for the given input parameters $B_0=1$ nT, $R=20R_E$ and the asymptotic downstream value $v\approx 0.9\times$(upstream solar-wind speed)$\approx$360 km/s.

The potential distribution specified by (10) is only valid on the night side (cylindrical) portion of the magnetopause. The flow streamlines just outside the magnetopause are given by $\phi=$ const (by virtue of the azimuthal symmetry of the surface shape), whereas the equipotentials according to (10) are given by $vR\sin\phi=$const. Thus the requirement that streamlines also be electric equipotentials, as follows from (8), is consistent with (10) only in the regions where $vR=$const. This is clearly not the case on the day side magnetopause, although it is approximately true on the nightside portion where $R=$const by assumption and $v\approx$const$\approx 0.9\times$ (upstream solar-wind speed). The problem on the dayside is that the prescribed distribution of magnetic normal components (1) is not consistent with the prescribed (Spreiter-Stahara) flow field.

Once the normal component is specified on either side of the day side merging line, the normal component everywhere downstream is determined by the requirement that the streamlines be equipotentials (VASYLIUNAS, 1974). Moreover, the merging outflow resulting from the conversion of magnetic to kinetic energy (e.g., HILL, 1975; VASYLIUNAS, 1975) imposes a perturbation on the magnetosheath flow; this perturbation is most significant in the dayside region where the speed calculated from the gasdynamic approximation is small. The merging outflow is of the order of the local Alfvén speed which is typically about 100 km/s in the dayside magnetosheath.

For these reasons the interconnection field model, and the resultant potential mapping relations, are not reliable on polar-cap field lines that map to the dayside magnetopause. Fortunately, however, most of the polar-cap area maps to the nightside magnetopause, where the potential specified by (10) is approximately valid. The potential patterns presented here apply only to that portion of the polar cap that maps to the nightside magnetopause.

6. Polar-Cap Potential Distributions

We have modeled the polar-cap potential distribution for various orientations of the IMF within the GSM y–z plane. The simplest case is that of a strictly southward IMF, with the merging line in the equatorial (x–y) plane. The resulting polar-cap potential distribution is shown in Fig. 5(a). Note that the equipotentials seem to

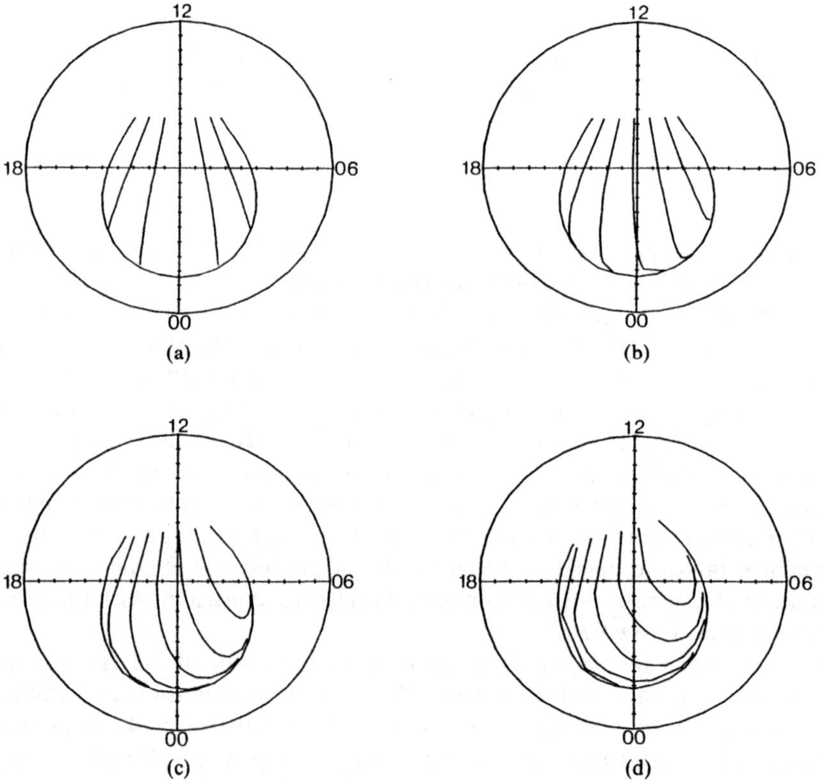

Fig. 5. Polar-cap potential distributions for various orientations of the subsolar merging line. The circle corresponds to 70 degrees geomagnetic latitude and the equipotentials are spaced at approximately 15 kV intervals. In Figs. 5(a)–5(d) the merging line is rotated from the y axis by angles of 0, 5, 22.5, and 45 degrees, respectively; i.e., 5(a) corresponds to a purely southward IMF and 5(d) corresponds approximately to a purely west–east field. The enclosing curve is the boundary between open and closed field lines. Note that the plots are incomplete, showing only results mapped down from the tail magnetopause. For reasons discussed in the text, the mapping of the equipotentials from the dayside magnetopause is not yet modeled reliably.

exhibit a divergence away from noon (a "convection throat") as is typically observed (e.g., REIFF *et al.*, 1978, and references therein).

In order to model the influence of a non-zero IMF B_y component, we imposed a rotation of the interconnection field, and hence of the merging line, about the x axis. Rotation of the interconnection field through an angle ϕ corresponds roughly to an IMF making an angle 2ϕ with the z direction in GSM coordinates, based on the assumption that the internal and external fields are roughly equal in magnitude near the subsolar point (e.g., GONZALEZ and MOZER, 1974; HILL, 1975).

The resulting potential patterns are shown in Figs. 5(b–d). The primary effect of a nonzero IMF B_y component is a rotation of the equipotentials in a manner similar to that observed (e.g., HEPPNER, 1972; HEELIS, 1984). It has long been understood qualitatively that such rotation should result from mapping of the solar-wind electric field to the polar cap along interconnected field lines (e.g., RUSSELL and ATKINSON, 1973; STERN, 1973). As the ratio $|B_y/B_z|$ increases, there is an increasing tendency for some equipotentials to close within the polar cap, as expected on the basis of qualitative arguments (e.g., CROOKER, 1979).

The input parameters B_0 and α are chosen such that the size of the polar cap and the magnitude of the potential drop across it are in accord with observations (e.g., REIFF *et al.*, 1981; MAKITA *et al.*, 1985). Note from Fig. 5 that both the polar-cap size and the total potential drop decrease with increasing $|B_y/B_z|$ even though the total amount of IMF flux intersecting the magnetopause does not change (by assumption). This is an effect of changes in the magnetosphere topology; with increasing $|B_y/B_z|$, an increasing amount of IMF flux enters one side of the magnetopause and exits the other side without intersecting the Earth.

7. Conclusion

Given the preliminary nature of the electric-field mapping model, we consider the present results encouraging rather than conclusive. Although the results presented here are qualitatively realistic, several improvements are required before quantitative comparison with observations will be justified.

For example, although most of the open polar-cap flux maps through the tail magnetopause, the mapping from the day side also needs to be considered in order to investigate the interesting "throat" region near noon. This will require specifying a merging outflow, and a distribution of normal components that is consistent with this outflow through the steady-state MHD condition $\nabla \times (\boldsymbol{v} \times \boldsymbol{B}) = \boldsymbol{0}$.

Also, our present results apply only to the simplest interconnection field geometry, corresponding to the subsolar merging hypothesis. A more complicated distribution of normal components will be needed to simulate the antiparallel merging hypothesis as well as to investigate the effect of a non-zero IMF B_x component.

Our ultimate aim is to combine our potential patterns with a model of ionospheric conductivity (e.g., REIFF, 1984b) in order to derive patterns of ionospheric and Birkeland currents for various IMF orientations. These patterns not only would be of intrinsic interest, but also would enable us to determine the feedback effect of such currents on the electric-field mapping relations.

We thank V. M. Vasyliunas, R. A. Wolf and P. H. Reiff for their helpful comments, and G. H. Voigt for many useful conversations. We are grateful to S. S. Stahara and J. R. Spreiter for supplying their computer code, and to R. W. Spiro for computing assistance. The work was supported by the Atmospheric Sciences Section of the National Science Foundation (Grant ATM-8306772) and by the Solar-Terrestrial Theory Program of the National Aeronautics and Space Administration (Grant NAGW-482.

REFERENCES

CROOKER, N. U., Dayside merging and cusp geometry, *J. Geophys. Res.*, **84**, 951, 1979.

CROOKER, N. U., The half-wave rectifier response of the magnetosphere and antiparallel merging, *J. Geophys. Res.*, **85**, 575, 1980.

GONZALEZ, W. D. and F. S. MOZER, A quantitative model for the potential resulting from reconnection with an arbitrary interplanetary magnetic field, *J. Geophys. Res.*, **79**, 4186, 1974.

HEELIS, R. A., The effects of interplanetary magnetic field orientation on dayside high-latitude ionospheric convection, *J. Geophys. Res.*, **89**, 2873, 1984.

HEPPNER, J. P., Polar cap electric field distributions related to the interplanetary magnetic field direction, *J. Geophys. Res.*, **77**, 4877, 1972.

HILL, T. W., Magnetic merging in a collisionless plasma, *J. Geophys. Res.*, **80**, 4689, 1975.

HILL, T. W., Solar-wind magnetosphere coupling, in *Solar-Terrestrial Physics*, edited by R. L. Carovillano and J. M. Forbes, pp. 262, D. Reidel, Boston 1983.

HILL, T. W. and M. E. RASSBACH, Interplanetary magnetic field direction and the configuration of the day side magnetosphere, *J. Geophys. Res.*, **80**, 1, 1975.

LYONS, L. R., A simple model for polar cap convection patterns and generation of θ auroras, *J. Geophys. Res.*, **90**, 1561, 1985.

MAKITA, K., C.-I. MENG, and S.-I. AKASOFU, Temporal and spatial variations of the polar-cap dimension inferred from precipitation boundaries, *J. Geophys. Res.*, **90**, 2744, 1985.

REIFF, P. H., Evidence of magnetic merging from low-latitude spacecraft and ground-based experiments, in *Magnetic Reconnection in Space and Laboratory Plasmas*, edited by E. W. Hones, Jr., AGU Geophysical Monograph 30, Washington, D.C., 1984a.

REIFF, P. H., Models of auroral zone conductances, in *Magnetospheric Currents*, edited by T. A. Potemra, AGU Geophysical Monograph 28, Washington, D.C., 1984b.

REIFF. P. H., J. L. BURCH, and R. A. HEELIS, Dayside auroral arcs and convection, *Geophys. Res. Lett.*, **5**, 391, 1978.

REIFF, P. H., R. W. SPIRO, and T. W. HILL, Dependence of polar cap potential drop on interplanetary parameters, *J. Geophys. Res.*, **86**, 7639, 1981.

RUSSELL, C. T. and G. ATKINSON, Comments on a paper by J. P. Heppner, "Polar cap electric field distributions related to interplanetary magnetic field directions," *J. Geophys. Res.*, **78**, 4001, 1973.

SAUNDERS, R. S., The characteristics of magnetospheric convective electric fields as mapped onto the polar caps, Ph.D. Thesis, Univ. of Denver, 1976.

SONNERUP, B. U. Ö, Magnetopause reconnection rate, *J. Geophys. Res.*, **79**, 1546, 1974.

SONNERUP, B. U. Ö., G. PASCHMANN, I. PAPAMASTORAKIS, N. SCKOPKE, G. HAERENDEL, S. J. BAME, J. R. ASBRIDGE, J. T. GOSLING, and C. T. RUSSELL, Evidence for magnetic field reconnection at the Earth's magnetopause, *J. Geophys. Res.*, **86**, 10,049, 1981.

SPREITER, J. R. and S. S. STAHARA, A new predictive model for determining solar wind-terrestrial planet interactions, *J. Geophys. Res.*, **85**, 6769, 1980.

STAHARA, S. S., D. KLENKE, B. C. TRUDDINGER, and J. R. SPREITER, Application of advanced computational procedures for modeling solar-wind interactions with Venus -theory and computer code, *NASA Contract Rep. CR 3267*, 1980.

STERN, D. P., A study of the electric field in an open magnetospheric model, *J. Geophys. Res.*, **78**, 7292,

1973.

VASYLIUNAS, V. M., Theoretical calculation of potential across a steady state open magnetosphere, *EOS, Trans. AGU*, **55**, 390, 1974.

VASYLIUNAS, V. M., Theoretical models of magnetic field-line merging, 1, *Rev. Geophys. Space Phys.*, **13**, 303, 1975.

VOIGT, G.-H., A three dimensional, analytical magnetospheric model with defined magnetopause, *Z. Geophys.*, **38**, 319, 1972.

VOIGT, G.-H., A mathematical magnetospheric field model with independent physical parameters, *Planet. Space Sci.*, **29**, 1, 1981.

VOIGT, G.-H., Magnetospheric equilibrium configurations and slow adiabatic convection, this volume, 1986.

WALKER, R. J., An evaluation of recent quantitative magnetospheric field models, *Rev. Geophys. Space Phys.*, **14**, 411, 1976.

ZWAN, B. J. and R. A. WOLF, Depletion of solar wind plasma near a planetary boundary, *J. Geophys. Res.*, **81**, 1636, 1976.

Solar Wind-Magnetosphere Coupling, edited by Y. Kamide and J. A. Slavin, 287–295.

Global Geometry of Merging at the Dayside Magnetopause

N. U. CROOKER

Department of Atmospheric Sciences, University of California, Los Angeles California 90024, U.S.A.

Models of merging at the dayside magnetopause are divided into four categories according to the location of merging site: 1) Subsolar merging lines, 2) Antiparallel field or cusp regions, 3) Across most of the dayside, and 4) Neutral points or separator lines in uniform-field-plus-dipole models. Important characteristics of all four categories are shown to be inherent in a new split merging line model, formed by the superposition of a dipole, its image, and a uniform field.

1. Introduction

A large body of evidence suggests that the central feature of solar wind-magnetosphere coupling is an interconnection between interplanetary and geomagnetic field lines. The global geometry of how this occurs is not fully understood. This paper reviews the proposed sites on the dayside magnetopause where the interplanetary magnetic field (IMF) first merges with Earth's field, and then demonstrates how a new model incorporates these sites into a unified picture.

2. Existing Models

Table 1 lists four categories of merging sites and some references to authors who have modeled them. Examples from each category are illustrated in the left column. The first merging site is along a line passing through the subsolar point. The line tilts from a horizontal position for purely southward IMF to a vertical position for purely northward IMF, while the amount of merged flux decreases from a maximum to zero. The subsolar merging line is the simplest configuration because no flow streamlines cross the merging line. Consequently, the merging line cannot be carried away with the flow, nor can it be obscured by merged flow crossing it. Further, because the magnetosheath flow is slow in the subsolar region, there is ample time for the merging process to occur.

Recent observations of merging signatures have been interpreted in terms of a subsolar merging line, although not with complete success. Flux transfer events (FTEs) flow away from the subsolar region and occur uniformly across the dayside, consistent with subsolar formation (RIJNBEEK *et al.*, 1984: BERCHEM and RUSSELL, 1984). Their energetic particle signatures also are consistent with subsolar merging (DALY *et al.*, 1984). However, FTEs occur only for southward IMF, which is not a prediction of the subsolar merging line model. Also, when their occurrence distribution is sorted according to IMF orientation, it does not fit a pattern of flow directed orthogonally away from a subsolar merging line (RUSSELL *et al.*, 1985). On the other hand, accelerated boundary layer flows are generally directed in the

Table 1. Merging models categorized by merging site.

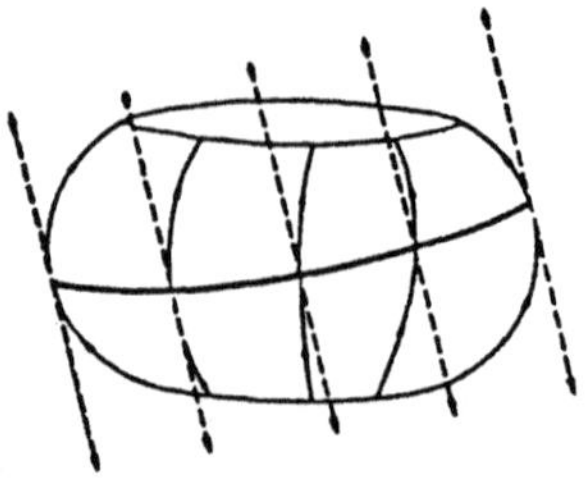	SUBSOLAR MERGING LINE Petschek, 1966 Nishida and Maezawa, 1971 Sonnerup, 1974 Gonzalez and Mozer, 1974 Yeh, 1976
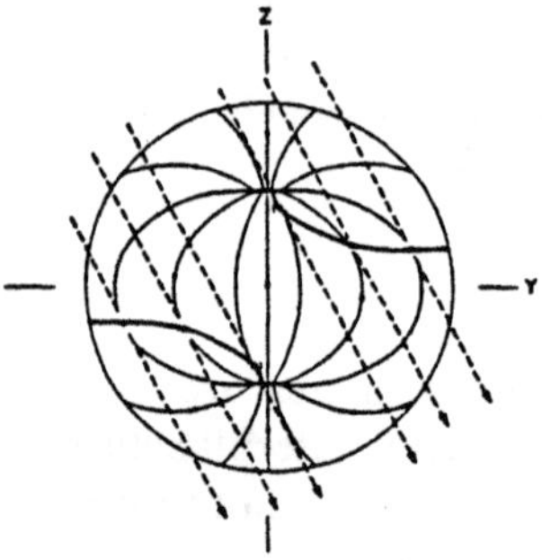 	ANTIPARALLEL FIELD OR CUSP REGIONS Haerendel et al., 1978 Podgorny et al., 1978 a Crooker, 1979 Quest and Coroniti, 1981 Luhmann et al., 1984
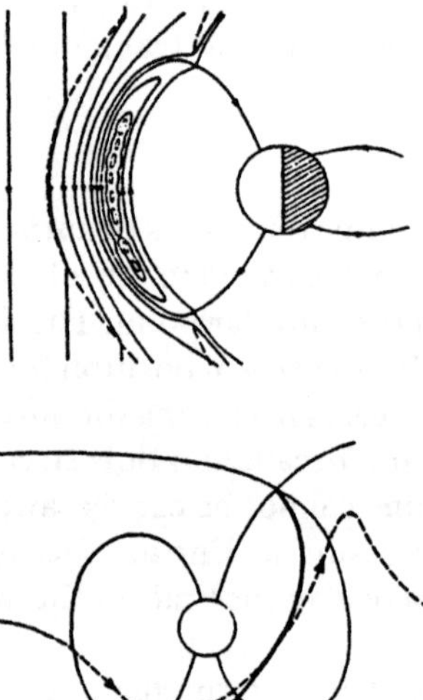	ACROSS DAYSIDE Cowley, 1976 Podgorny et al., 1980 Lee and Fu, 1985
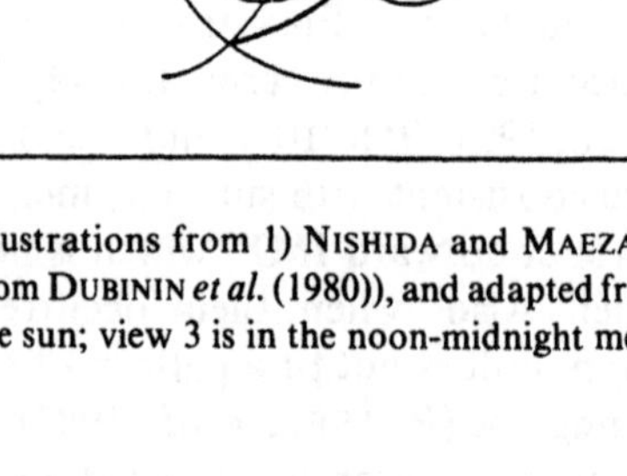	UNIFORM-FIELD-PLUS-DIPOLE NEUTRAL POINTS OR SEPARATOR LINE Dungey, 1961, 1963 Forbes and Speiser, 1971 Cowley, 1973 Stern, 1973 Yeh, 1976 Lyons, 1984

Illustrations from 1) NISHIDA and MAEZAWA (1971), 2) CROOKER (1979), 3) BAUM (1984) (adapted from DUBININ *et al.* (1980)), and adapted from 4) STERN (1973), respectively. Views 1, 2, and 4 are from the sun; view 3 is in the noon-midnight meridian plane.

predicted, orthogonal sense (SONNERUP *et al.*, 1981; CROOKER *et al.*, 1984), although the observations have not been tested rigorously for this property. The feature of accelerated boundary layer flows which seems not to agree with the subsolar merging line origin is their low rate of occurrence. Since the subsolar merging line is predicted to be present for all IMF orientations, its signature should be a common occurrence on magnetopause crossings.

The second category of merging site listed in Table 1 is predominantly away from the subsolar region, near the cusps. Several factors motivated models of this type. Before the high-resolution instruments were launched on the ISEE spacecraft, no signatures of merging had been observed in data from the subsolar region. But some signatures, probably FTEs, were observed in the vicinity of the cusps by the HEOS spacecraft (HAERENDEL *et al.*, 1978; RIJNBEEK and COWLEY, 1984). CROOKER (1979) noted that the cusps are the origins of merging lines defined as the locus of points where the IMF and geomagnetic field are antiparallel, as shown in Table 1. The antiparallel merging model, based on these merging lines, has several advantages over the subsolar merging line models. It can explain the half-wave rectifier behavior of energy transfer only for southward IMF (CROOKER, 1980), as well as asymmetries in the position of the convection throat and related seasonal patterns in geomagnetic activity (HEELIS, 1984; CROOKER *et al.*, 1985). It inspired a unification of high latitude convection patterns proposed independently for various IMF orientations (RUSSELL, 1972; HEPPNER, 1972; FRIIS-CHRISTENSEN and WILHJELM, 1975; HEELIS *et al.*, 1976; MAEZAWA, 1976). Consequently it provides a natural basis for explaining patterns of high latitude convection, field-aligned current, and auroral patterns for northward IMF (REIFF and BURCH, 1985; POTEMRA *et al.*, 1984; CHIU *et al.*, 1985). Theoretically, antiparallel merging is supported by the results of QUEST and CORONITI (1981), who found that merging by the tearing mode instability is highly favored by antiparallel fields.

For PODGORNY *et al.* (1978a), a cusp-merging model was motivated by the results of their laboratory terella experiment. They found that under simulated solar wind conditions with a normal Mach number of 10, the dayside magnetopause below the cusps is covered by a large magnetic loop or curl which prevents merging from occurring there. Instead, merging occurs in the cusp regions.

The third category of merging site in Table 1 is all across the dayside. In response to the subsolar merging line models, COWLEY (1976) argued that there is no theoretical reason why merging must be constrained to occur only along the specified line but could occur elsewhere as well. PODGORNY *et al.* (1980) refined their terella experiment and found that the large magnetic curl becomes thin and collapses to form a series of smaller loops indicative of tearing across the dayside, as pictured in the noon–midnight meridian plane in Table 1. The laboratory tearing signatures look like FTEs in ISEE data. LEE and FU (1985) also propose that tearing occurs across the dayside magnetopause. They recognized the similarity between the field tilting observed in FTEs (e.g., SAUNDERS *et al.*, 1984) and the coil structure that is predicted to result from tearing in the presence of a finite east–west (y) component of the IMF (PODGORNY *et al.*, 1978b). However, for tearing to occur all across the dayside with a finite y component, the theory of QUEST and CORONITI (1981) must be relaxed, since merging is then required to occur between fields which are not antiparallel.

The fourth category of merging site is based on models which superimpose a dipole field representing the geomagnetic field upon a uniform field representing the IMF. The first magnetospheric merging model was of this type (DUNGEY, 1961). The models give the correct global topology for an open magnetosphere. The three categories of field lines—open, closed, and interplanetary—are separated by separatrix surfaces which intersect along a circle encompassing the dipole. The circle is called a "separator line." It has the same properties as a merging line and passes through the subsolar point when the IMF has no x component. In this case the model also fits the subsolar merging line category and has been treated as such by YEH (1976). Other authors treat the two neutral points on opposite sides of the separator circle as the merging sites in these types of models. There the uniform and dipole fields are antiparallel and exactly cancel each other. An example of neutral point merging is shown in the bottom diagram in Table 1 for a purely east–west, dashed IMF line approaching the dayside portion of the heavy separator circle (after STERN, 1973). The IMF line must become highly distorted in this case before it breaks at the neutral points.

3. New Model

The available types of dayside merging models described above are all appropriate for various aspects of the problem, but none are suitable in all respects. However, a new model developed from basic principles displays features of all four categories of models and unifies their best aspects (CROOKER, 1985). The model was not specifically designed to produce this result; rather, it followed as a natural outcome.

The new model is a superposition of a uniform field with the idealized Chapman-Ferraro field shown in Fig. 1. The Chapman-Ferraro field represents the magnetic configuration produced by the surface currents which shield the solar wind from the magnetosphere (CHAPMAN and FERRARO, 1931). This configuration is the assumed magnetopause pattern in the antiparallel merging model illustrated in Table 1. In the tradition of uniform field-plus-dipole models, the new model assumes that the IMF completely penetrates the magnetopause and combines with the shielding configuration. Some degree of penetration seems reasonable, since the magnetic field is observed to rotate from its magnetosheath orientation to its magnetospheric orientation by the shortest path, consistent with a simple superposition of the two fields (BERCHEM and RUSSELL, 1982). The degree and depth of penetration may be limited, but a study of these properties is beyond the scope of the model at this time.

The results of the superposition may be anticipated by considering the nature of the antiparallel merging lines. They are the locus of points where the IMF is exactly antiparallel to the Chapman-Ferraro field. Consequently, any neutral points in the superposition must fall on these lines. They will lie wherever the magnitudes of the two fields are equal, since there the two opposing vectors will exactly cancel each other. The magnitude of the Chapman-Ferraro field along the antiparallel merging lines is zero at the cusps, rises to a maximum, and then decreases gradually outward. If

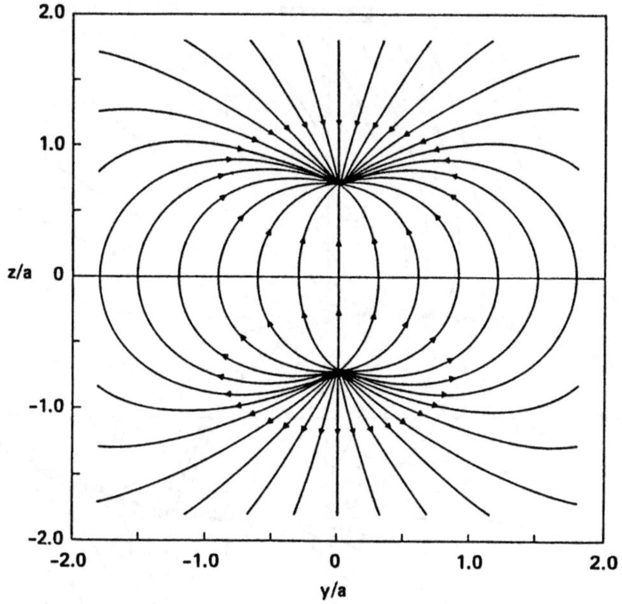

Fig. 1. Magnetopause magnetic field configuration in the idealized Chapman-Ferraro current plane, calculated from a dipole and its image, each located a distance a in back of and in front of the plane, respectively, along the x axis perpendicular to the plane.

the uniform field has a magnitude less than the maximum along the merging line, two neutral points will form, one on either side of the maximum, on both merging lines.

Figure 2 shows the results of superposing a uniform field with equal $-y$ and $-z$ components upon the Chapman-Ferraro field. As anticipated, four neutral points are present, two along each dotted antiparallel merging line. The ones nearest the z axis are displaced cusps, and the other two have X-type geometry. The neutral points also lie on the heavy separator lines, as in the uniform-field-plus-dipole models. In contrast to these models, however, the separator is not a single line across the subsolar region but has split into two lines. STERN (1973) predicted that such a split would occur if surface currents were added to the uniform-field-plus-dipole model. Since the Chapman-Ferraro configuration is obtained by superposing a dipole with its image, the new model is in effect a uniform-field-plus-dipole model with the addition of surface currents, which are represented by the image dipole.

The configuration in Fig. 2 is interpreted as the magnetic field pattern at the inner edge of the magnetopause. The split separator lines are treated as merging lines and give the new model its name of "split merging line model." The region between the merging lines corresponds directly to the magnetic curl region of PODGORNY *et al.* (1978a).

Further interpretation is illustrated in Fig. 3, where the field configuration and merging lines from Fig. 2 have been transferred. Additional merging lines represent

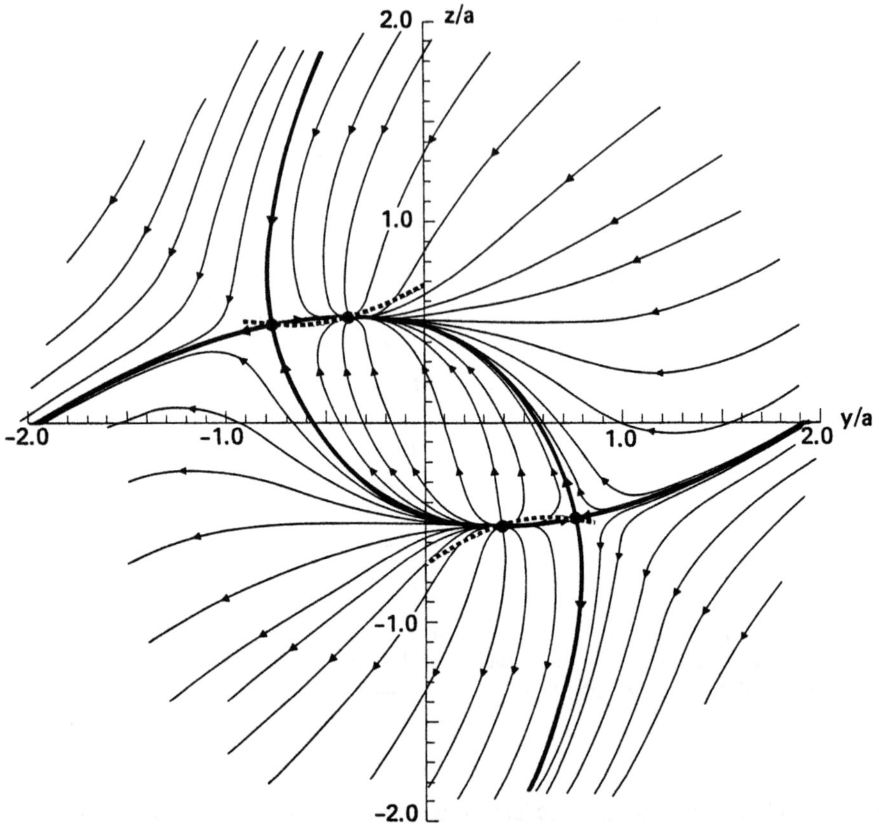

Fig. 2. Superposition of the Chapman-Ferraro field in Fig. 1 and an IMF with equal $-y$ and $-z$ components. Large dots indicate positions of the four resulting neutral points.

the collapse of the magnetic curl into FTE coils by the tearing mode instability (PODGORNY *et al.*, 1978b; LEE and FU, 1984). Heavy, straight IMF lines connect to the inner magnetopause field by normal fields through the magnetopause at the multiple merging lines. Most of the flux transfer from closed dipolar field lines to open tail lobe field lines occurs at the outermost merging lines. These should be coincident with the separator lines but are drawn inside of them to show the process of flux transfer more clearly. Accelerated boundary layer flows of the classical type are predicted to occur here. Figure 3 also illustrates how the rate of merging increases rapidly upon approach to the neutral points, where the merging fields are antiparallel.

More details of the model and its interpretation are given in CROOKER (1985). Patterns for other IMF orientations are shown, and the importance of the variation in area between the separator lines with Alfven Mach number is described. This latter feature relates to the low occurrence frequency of accelerated boundary layer flows. However, the model does not explain all of the discrepancies between observations

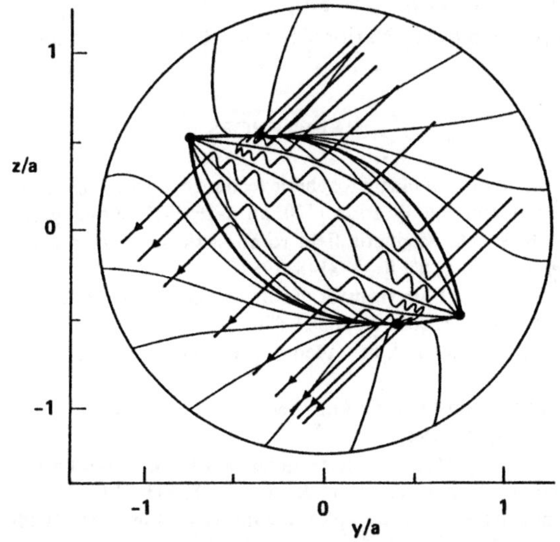

Fig. 3. Magnetic field configuration from Fig. 3 merged with straight IMF lines at multiple merging lines spanning the region between the heavy separator lines. Merging occurs in the classical geometry at the outer merging lines, producing accelerated boundary layer flows and transferring closed dayside to open tail lobe flux. The additional merging nearer to the subsolar point causes coils to form there; the coils give the signatures of FTEs.

and previous models. For example, COWLEY's (1976) objection concerning the lack of constraints on merging sites has not been addressed, and the tearing theory of QUEST and CORONITI (1981) must be relaxed to accommodate FTE coil formation.

4. Summary

The split merging line model forms a synthesis of widely diverging but useful models used in the past. In terms of the four categories listed in Table 1, the new model is a progression from:

1) The subsolar merging line models. The single merging line splits into two, and the region between the two lines is threaded with multiple X- and O-type merging lines. This geometry accommodates the observed features of both FTEs and accelerated boundary layer flows.

2) The antiparallel merging model. The antiparallel merging lines mark the paths of cusp displacement and the locations of X-type neutral points. The merging rate is highest near the cusps and neutral points, and thus the predictions of the antiparallel merging model are preserved.

3) Merging across the dayside. The area where merging may occur is explicitly determined and bound by the split merging lines.

4) Uniform-field-plus-dipole models. Adding the effects of Chapman-Ferraro surface currents produces the magnetic curl feature observed in the laboratory terella

experiments of PODGORNY *et al.* (1978a).

This work was supported by the National Science Foundation under grants ATM81-20455 and ATM85-03980.

REFERENCES

BAUM, P. J., Laboratory simulation of magnetospheric reconnection, in *Solar Terrestrial Physics—Present and Future*, NASA Reference Publication 1120, pp. 1-26–1-31, Washington, D. C., 1984.

BERCHEM, J. and C. T. RUSSELL, Magnetic field rotation through the magnetopause: ISEE 1 and 2 observations, *J. Geophys. Res.*, **87**, 8139–8148, 1982.

BERCHEM, J. and C. T. RUSSELL, Flux transfer events on the magnetopause: Spatial distribution and controlling factors, *J. Geophys. Res.*, **89**, 6689–6703, 1984.

CHAPMAN, S. and V. C. A. FERRARO, A new theory of magnetic storms, *Terr. Mag. Atmos. Elec.*, **36**, 171–186, 1931.

CHIU, Y. T., N. U. CROOKER, and D. J. GORNEY, Model of oval and polar cap arc configurations, *J. Geophys. Res.*, **90**, 5153–5157, 1985.

COWLEY, S. W. H., A qualitative study of the reconnection between the Earth's magnetic field and an interplanetary field of arbitrary orientation, *Radio Sci.*, **8**, 903–913, 1973.

COWLEY, S. W. H., Comments on the merging of nonantiparallel magnetic fields, *J. Geophys. Res.*, **81**, 3455–3458, 1976.

CROOKER, N. U., Dayside merging and cusp geometry, *J. Geophys. Res.*, **84**, 951–959, 1979.

CROOKER, N. U., The half-wave rectifier response of the magnetosphere and antiparallel merging, *J. Geophys. Res.*, **85**, 575–578, 1980.

CROOKER, N. U., A split separator line merging model of the dayside magnetopause, *J. Geophys. Res.*, **90**, 12104–12110, 1985.

CROOKER, N. U., G. L. SISCOE, T. E. EASTMAN, L. A. FRANK, and R. D. ZWICKL, Large scale flow in the dayside magnetosheath, *J. Geophys. Res.*, **89**, 9711– 9719, 1984.

CROOKER, N. U., J. G. LUHMANN, J. R. SPREITER, and S. S. STAHARA, Magnetopause merging site asymmetries, *J. Geophys. Res.*, **90**, 341–346, 1985.

DALY, P. W., M. A. SAUNDERS, R. P. RIJNBEEK, N. SCKOPKE, and C. T. RUSSELL, The distribution of reconnection geometry in flux transfer events using energetic ion, plasma, and magnetic data, *J. Geophys. Res.*, **89**, 3843–3854, 1984.

DUBININ, E. M., I. M. PODGORNY, and Y. N. POTANIN, Structure of the magnetic field at the boundary of the magnetosphere: analysis of a simulation experiment, *Cosmic Res.*, **18**, 77, 1980.

DUNGEY, J. W., Interplanetary magnetic field and the auroral zones, *Phys. Rev. Lett.*, **6**, 47–48, 1961.

DUNGEY, J. W., The structure of the exosphere or adventures in velocity space, in *Geophysics, The Earth's Environment*, edited by C. De Witt, J. Hieblot, and A. Lebean, pp. 505–550, Gordon and Breach, New York, 1963.

FORBES, T. G. and T. W. SPEISER, Mathematical models of the open magnetosphere: Application to dayside auroras, *J. Geophys. Res.*, **76**, 7542–7551, 1971.

FRIIS-CHRISTENSEN, E. and J. WILHJELM, Polar cap currents for different directions of the interplanetary magnetic field in the *Y–Z* plane, *J. Geophys. Res.*, **80**, 1248–1260, 1975.

GONZALEZ, W. D. and F. S. MOZER, A quantitative model for the potential resulting from reconnection with an arbitrary interplanetary magnetic field, *J. Geophys. Res.*, **79**, 4186–4194, 1974.

HAERENDEL, G., G. PASCHMANN, N. SCKOPKE, H. ROSENBAUER, and P. C. HEDGECOCK, The front side boundary layer of the magnetopause and the problem of reconnection, *J. Geophys. Res.*, **83**, 3195–3216, 1978.

HEELIS, R. A., The effects of interplanetary magnetic field orientation on dayside high-latitude ionospheric convection, *J. Geophys. Res.*, **89**, 2873–2880, 1984.

HEELIS, R. A., W. B. HANSON, and J. L. BURCH, Ion convection velocity reversals in the dayside cleft, *J. Geophys. Res.*, **81**, 3803–3809, 1976.

HEPPNER, J. P., Polar cap electric field distributions related to the interplanetary magnetic field direction, *J. Geophys. Res.*, **77**, 4877–4887, 1972.

LEE, L. C. and Z. F. FU, A theory of magnetic flux transfer at Earth's magnetopause, *Geophys. Res. Lett.*, **12**, 105–108, 1985.

LUHMANN, J. G., R. J. WALKER, C. T. RUSSELL, N. U. CROOKER, J. R. SPREITER, and S. S. STAHARA, Patterns of magnetic field merging sites on the magnetopause, *J. Geophys. Res.*, **89**, 1739–1742, 1984.

LYONS, L. R., A simple model for polar cap convection patterns and generation of θ-auroras, *J. Geophys. Res.*, **90**, 1561–1567, 1985.

MAEZAWA, K., Magnetospheric convection induced by the positive and negative *z* components of the interplanetary magnetic field: Quantitative analysis using polar cap magnetic records, *J. Geophys. Res.*, **81**, 2289–2303, 1976.

NISHIDA, A. and K. MAEZAWA, Two basic modes of interaction between the solar wind and the magnetosphere, *J. Geophys. Res.*, **76**, 2254–2264, 1971.

PETSCHEK, H. E., The mechanism for reconnection of geomagnetic and interplanetary field lines, in *The Solar Wind*, edited by R. J. Mackin and M. Neugebauer, pp. 257–273, Jet Propulsion Laboratory, Pasadena, California, 1966.

PODGORNY, I. M., E. M. DUBININ, and YU. N. POTANIN, The magnetic field on the magnetospheric boundary from laboratory simulation data, *Geophys. Res. Lett.*, **5**, 207–210, 1978a.

PODGORNY, I. M., E. M. DUBININ, and YU. N. POTANIN, New interpretation of changes in plasma parameters along the trajectories of Earth's satellites, *Kosmicheskiye Issledovaniya*, **16**, 522–526, 1978b.

PODGORNY, I. M., E. M. DUBININ, and YU. N. POTANIN, On magnetic curl in front of the magnetosphere boundary, *Geophys. Res. Lett.*, **7**, 247–250, 1980.

POTEMRA, T. A., L. J. ZANETTI, P. F. BYTHROW, A. T. Y. LUI, and T. IIJIMA, B_y-dependent convection patterns during northward interplanetary magnetic field, *J. Geophys. Res.*, **89**, 9753–9760, 1984.

QUEST, K. B. and F. V. CORONITI, Tearing at the dayside magnetopause, *J. Geophys. Res.*, **86**, 3289–3298, 1981.

REIFF, P. H. and J. L. BURCH, B_y-dependent dayside plasma flow and Birkeland currents in the dayside magnetosphere: 2. A global model for northward and southward IMF, *J. Geophys. Res.*, **90**, 1595–1609, 1985.

RIJNBEEK, R. P. and S. W. H. COWLEY, Magnetospheric flux erosion events are flux transfer events, *Nature*, **309**, 135–138, 1984.

RIJNBEEK, R. P., S. W. H. COWLEY, D. J. SOUTHWOOD, and C. T. RUSSELL, A survey of dayside flux transfer events observed by ISEE 1 and 2 magnetometers, *J. Geophys. Res.*, **89**, 786–800, 1984.

RUSSELL, C. T., The configuration of the magnetosphere, in *Critical Problems of Magnetospheric Physics*, edited by E. R. Dyer, pp. 1–16, Inter-Union Commission on Solar-Terrestrial Physics, Secretariat, National Academy of Sciences, Washington, D. C., 1972.

RUSSELL, C. T., J. BERCHEM, and J. G. LUHMANN, On the source region of flux transfer events, *Adv. Res.*, **5**(4), 363–368, 1985.

SAUNDERS, M. A., C. T. RUSSELL, and N. SCKOPKE, Flux transfer events: Scale size and interior structure, *Geophys. Res. Lett.*, **11**, 131–134, 1984.

SONNERUP, B. U. Ö., G. PASCHMANN, I. PAPAMASTORAKIS, N. SCKOPKE, G. HAERENDEL, S. J. BAME, J. R. ASBRIDGE, J. T. GOSLING, and C. T. RUSSELL, Evidence for magnetic field reconnection at the Earth's magnetopause, *J. Geophys. Res.*, **86**, 10049–10067, 1981.

STERN, D. P., A study of the electric field in an open magnetospheric model, *J. Geophys. Res.*, **78**, 7292–7305, 1973.

YEH, T., Day side reconnection between a dipolar geomagnetic field and a uniform interplanetary field, *J. Geophys. Res.*, **81**, 2140–2144, 1976.

Solar Wind-Magnetosphere Coupling, edited by Y. Kamide and J. A. Slavin, 297–314.

Magnetic Flux Transfer at the Earth's Magnetopause

L.-C. LEE

Geophysical Institute and Department of Physics, University of Alaska, Fairbanks Alaska 99701, U.S.A.

It is proposed that the flux transfer events (FTE's) observed by ISEE satellites are the result of multiple X-line reconnection at the dayside magnetopause. When the system length is much larger than the thickness of the current layer separating two plasma regions, magnetic reconnections will occur along several reconnection lines (X-lines). Magnetic islands (plasmoids or magnetic flux tubes)will be formed and convected out of the system intermittently and repeatedly, as a result of the multiple X-line reconnection process. Computer simulation of the multiple X-line reconnection process also confirms the theoretical predictions. In particular, impulsive magnetic reconnections associated with the formation of the magnetic flux tubes are observed. In the case of the dayside magnetopause, magnetic flux tubes with a cross-section of ~1 R_E^2 are found to be formed repeatedly every 5–10 minutes, consistent with satellite observations. The coupling of FTE's to the ionosphere is briefly discussed. Variations of the multiple X-line reconnection model are also discussed.

1. Introduction

The open magnetosphere model was first proposed by DUNGEY (1961). In this open model, the interplanetary magnetic fields and geomagnetic fields reconnect along a neutral line (X-line) near the subsolar region of the dayside magnetopause. The reconnected magnetic field lines are then convected poleward toward the cusp regions. This open model was further developed by LEVY *et al.* (1964), who predicted the existence of rotational discontinuities and high-speed plasma flows at the dayside magnetopause. However, the predicted rotational discontinuities and high-speed splama flow are not observed with the expected frequency (e.g., HEIKKILA, 1975; SONNERUP *et al.*, 1981).

The "flux transfer events" (FTE's) first reported by RUSSELL and ELPHIC (1978, 1979) also indicated that the dayside magnetic reconnection geometry is different from that in Dungey's quasi-steady state model. The properties of the flux transfer events have been further studied and discussed by PASCHMANN *et al.* (1982), SCHOLER *et al.* (1982), DALY *et al.* (1981, 1984), COWLEY (1982), RIJNBEEK *et al.* (1984), SAUNDERS *et al.* (1984), and BERCHEM and RUSSELL (1984). The "flux erosion events" observed in the dayside boundary layer by HAERENDEL *et al.* (1978) are recently identified as flux transfer events by RIJNBEEK and COWLEY (1984).

The observed properties of the flux transfer events can be summarized as follows: (1) The FTE's have a strong IMF B_z dependence and repeated large FTE's are observed to occur approximately every 5–15 minutes. (2) For the large FTE's the area of the flux tube cross-section is typically 1 R_E^2, elongated in the north–south direction. The scale size in the north–south direction is $L_z \simeq 1$–$2\ R_E$, the size in the direction normal to the magnetopause is $L_n \simeq 0.5$–$0.7\ R_E$. Smaller FTE's are also observed. (3) Twisted magnetic field lines are observed in the open flux tube, indicating the presence of field-aligned current ($J_\parallel$). (4) Alfvén waves associated with the twisted fields are observed to propagate away from the magnetopause region (SAUNDERS *et al.*, 1984). (5) The spatial distribution of the positive and negative polarity of B_n signatures indicates the FTE's originate at the low latitude regions rather than in the cusp regions.

An example of FTE's observed by ISEE 1 and 2 satellites (SAUNDERS *et al.*, 1984) is represented here as Fig. 1, in which the plasma density, velocity, magnetic field observed in the magnetosheath are plotted as a function of time. The energetic electron and proton densities ($\bar{N}_e$ and $\bar{N}_p$) are also shown. The signature in B_N (the component of the magnetic field in the direction normal to the local magnetopause) associated with three FTE's can be clearly identified. The enhanced energetic electron and proton densities indicate that the flux tubes are open.

Recently, LEE and FU (1985) proposed a multiple X-line reconnection model for the dayside magnetopause, which can explain reasonably well the occurrence and the observed properties of FTE's. The multiple X-line reconnection can be considered as a magnetic reconnection process between two approaching magnetized plasmas, which is strongly modified by the development of tearing instability. This multiple X-line reconnection process is expected to occur in a system with a characteristic length (L) much greater than the characteristic thickness (l) of the current layer separating the two plasma regions. At the dayside magnetopause, we have $L \simeq 10\ R_E$, $l \simeq 500$ km and hence $L \gg l$. The multiple X-line reconnection is not a steady process. First of all, the reconnected magnetic fluxes form tubes. As the cross-section of the magnetic flux tubes grows to a large value, the reconnection of magnetic field lines will slow down and stop, which may be caused by the nonlinear saturation of tearing instability. Subsequently, reconnection will begin again when the saturated magnetic flux tubes are convected out of the reconnection region. The magnetic flux tubes are expected to be formed intermittently and repeatedly. Therefore, the multiple X-line reconnection process is different from the steady state ones proposed by a number of workers, including SWEET (1985), PARKER (1957), PETSCHEK (1964), SONNERUP (1970), and YEH and AXFORD (1970).

Simulation study of the multiple X-line reconnection process was recently carried out by FU and LEE (1985).

In this paper, we will review the theoretical and simulation studies of FTE's by Lee and Fu. We will also discuss the coupling of FTE's to the polar ionosphere.

2. Flux Transfer Events as the Result of multiple X-line Reconnection

The multiple X-line reconnection is illustrated in Fig. 2, in which three reconnection lines (X-lines) are assumed to exist in the $X=0$ plane. The magnetic field on side 1 ($X>0$) is given by $\underline{B}=(0, B_{0y}, -B_{0z})$ and the field on side 2 ($X<0$) is $\underline{B}=(0, B_{0y}, B_{0z})$, where B_{0y} and B_{0z} are constants. We consider the case in which the magnetic fields on the two sides are not exactly antiparallel, i.e., $B_{0y} \neq 0$. As the magnetic field

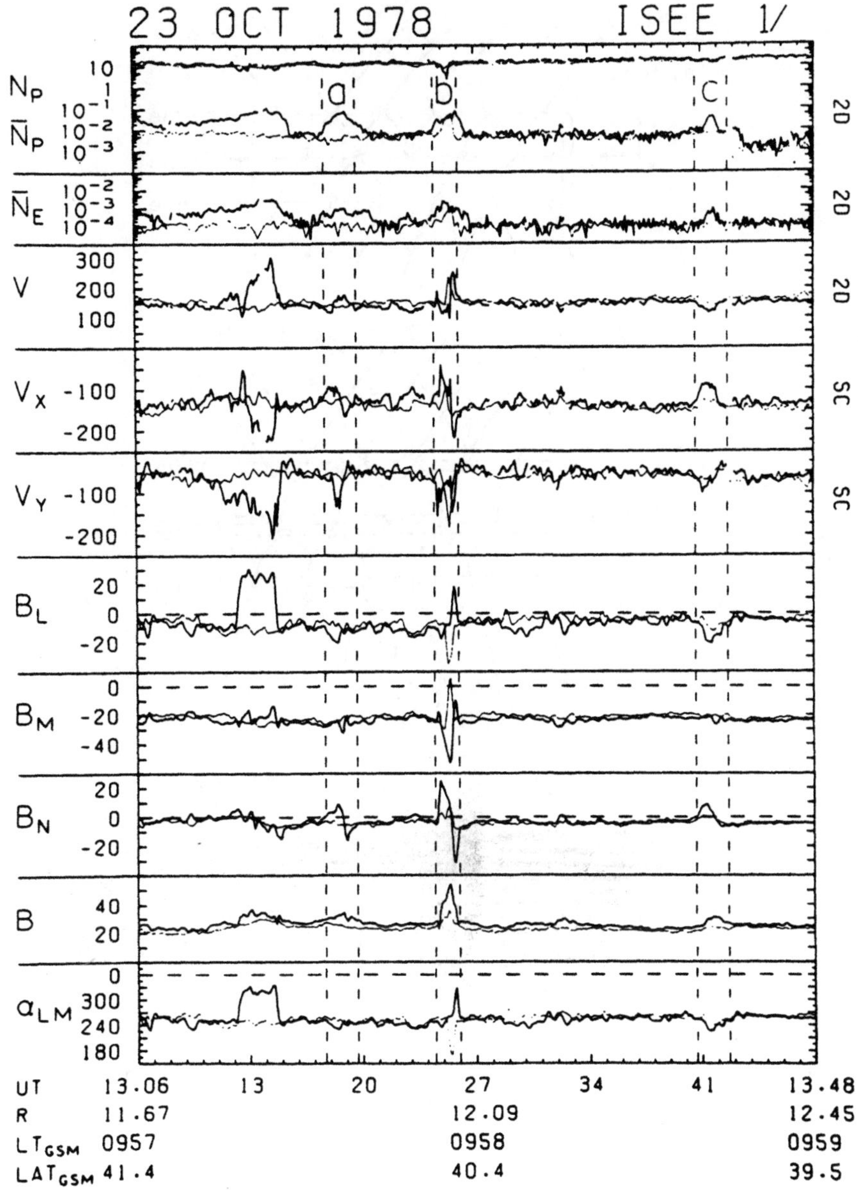

Fig. 1. ISEE-1 (heavy trace) and ISEE-2 (light trace) magnetic field and plasma measurements from the outbound magnetosheath passage on October 23, 1978. The plasma data are shown in the top five panels, and consist of various density and flow parameters. The magnetic field recordings are displayed in boundary normal coordinates where B_N is outward along the boundary normal, B_L is along the projection of the GSM Z-axis in the magnetopause plane, and B_M completes the orthogonal triad and points westward. The bottom panel shows the field angle in the LM plane, defined by $\alpha_{LM}=\text{tran}^{-1}(B_M/B_L)$. ISEE-1's position is given at the base of the Fig. in terms of geocentric radial distance (R) in Earth radii, and GSM local time (LT_{GSM}) and latitude (LAT_{GSM}). (SAUNDERS *et al.*, 1984).

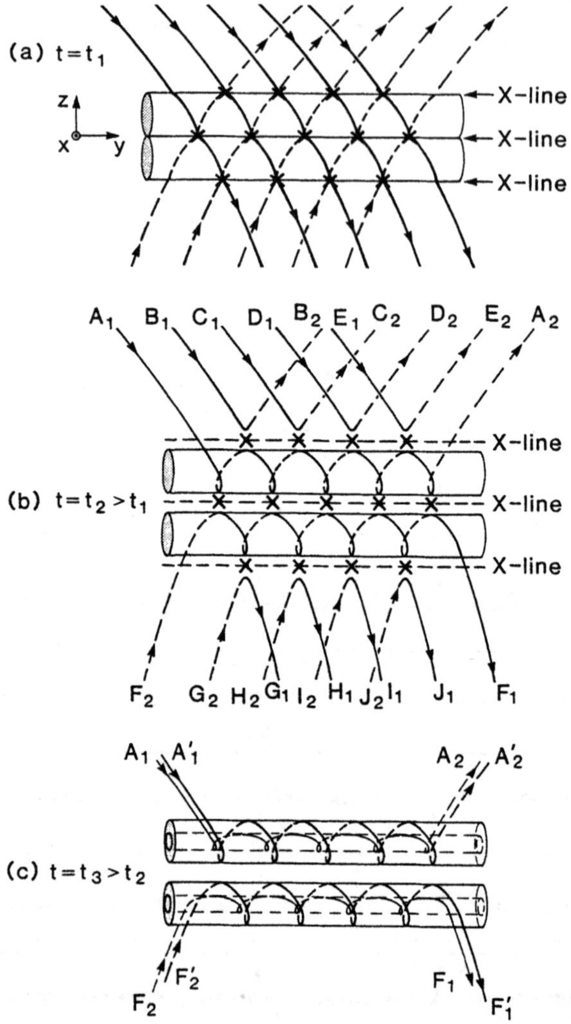

Fig. 2. Formation of magnetic flux tubes as a result of multiple X-line reconnection. (a) At $t=t_1$, the field lines reconnect at the sites marked with "X". (b) At $t=t_2$, two helical field lines A_1A_2 and F_2F_1 are formed. (c) At $t=t_3$, the helical field lines A_1A_2 and F_2F_1 are wrapped by the newly formed field lines $A_1'A_2'$ and $F_2'F_1'$.

lines on the two sides approach each other, they reconnect along the three X-lines at $t=t_1$ as shown in Fig. 2(a). The solid lines in the figure represent the magnetic field lines from side 1 and the dashed lines are for the field lines from side 2. The reconnection sites of the magnetic field lines are marked with "X".

After reconnection ($t>t_1$) the reconnected field lines will move or contract due to the field line tension as shown in Fig. 2(b). It is noted that a pair of helical field lines

A_1A_2 and F_2F_1 are formed as a consequence of the multiple X-line reconnection process. The other field lines, such as B_1B_2, C_1C_2, G_2G_1, and H_2H_1, are similar to the reconnected field lines in the single X-line reconnection case.

The reconnection lines in Fig. 2 are assumed to have finite length in the y-direction. Furthermore, the reconnection region along the central X-line is assumed to be longer than along the X-lines on the two sides. Thus, the side-1 field line marked with A_1 reconnects with the side-2 field line at the central X-line, but not at the upper X-line. This leads naturally to the interconnection of the side-1 field line and the field line in the flux tube. Similar process occurs at the ends of other tubes.

As time goes on, new magnetic field lines from sides 1 and 2 will continue to reconnect and a new pair of helical field lines $A_2'A_1'$ and $F_2'F_1'$ are formed. The old helical field lines A_1A_2 and F_2F_1 are, respectively, wrapped inside the new helical lines $A_1'A_2'$ and $F_2'F_1'$. This process will continue until the size of magnetic flux tubes containing the helical field lines are large enough to prevent further reconnection of magnetic field. The presence of a constant B_y is essential to the formation of the flux tube with helical fields. If $B_y=0$, the multiple X-line reconnection will lead only to the formation of isolated magnetic loops.

We suggest that the multiple X-line reconnection process described above may occur at the dayside magnetopause. A schematic sketch of the reconnected field lines at the magnetopause is shown in Fig. 3(a) and 3(b), in which three X-lines are assumed to exist. However, the number of X-lines is not limited to three in the general case. Figure 3(a) is a three-dimensional perspective view of the magnetopause and the reconnected magnetic field lines, while Figure 3(b) is the projection of the field lines in the noon-midnight meridian plane. It is seen that two magnetic flux tubes containing twisted field lines are formed as a natural consequence of the multiple X-line reconnection. The flux tubes northward and southward of the central X-line will be convected, respectively, to the north and south cusp regions by the magnetosheath plasma flow and the tension force of the reconnected flux tubes. The sense of field twisting as shown in Fig. 3(a) is consistent with the ISEE observations (SAUNDERS *et al.*, 1984). The northward flux tube would show a positive polarity of the observed B_n signature while the southward tube shows a negative polarity.

In the present model, there exist two types of open field lines. Field lines such B_1B_2, C_1C_2, G_2G_1, and H_2H_1 will be called the Type I open field lines while the helical field lines such as A_1A_2 and F_2F_1 will be called the Type II open field lines. Note that there are more Type I open field lines that Type II lines.

The interplanetary magnetic fields (IMF) in general have a B_y component and the X-lines are no longer in the y-direction of the solar-magnetospheric coordinates as shown in Fig. 4. Let θ be the angle between the z-axis and the projected IMF in the y–z plane. The X-lines will make an angle $\theta/2$ with the z-axis as discussed by GONZALES and MOZER (1974) and SONNERUP (1974). It can be seen from Fig. 3(a) and 4 that the flowing direction of the field-aligned current ($J_{\parallel}$) associated with the twisted field lines in the magnetopause region depends on the sign of B_y. For $B_y>0$ the current flows along the guided field, i.e., $J_{\parallel}=\underset{\sim}{J}\cdot\underset{\sim}{B}/B>0$; $J_{\parallel}<0$ for $B_y<0$.

Based on the present model the central X-line is the natural boundary between the northward and southward convected flux tubes as illustrated in Fig. 4. The X-line

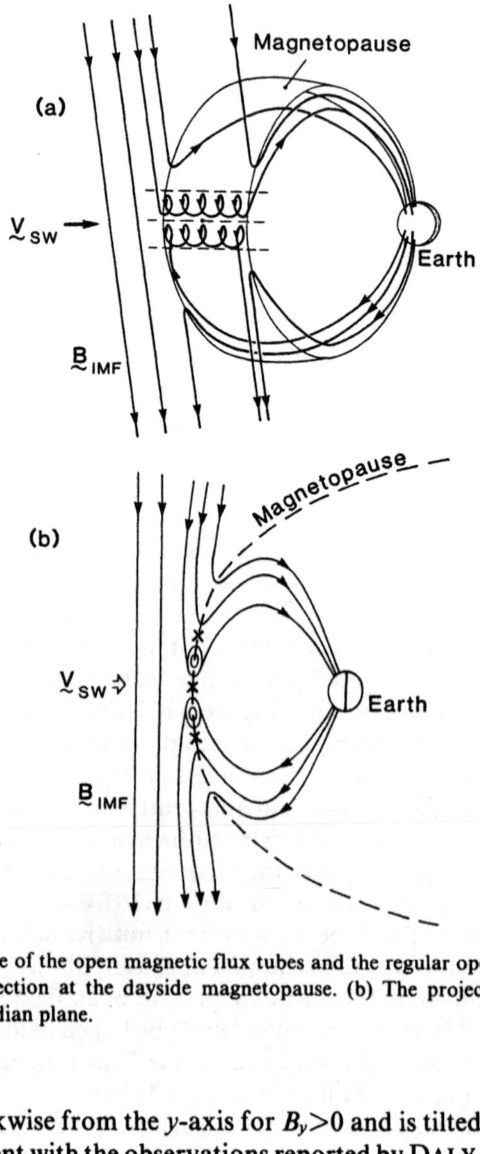

Fig. 3. (a) A perspective of the open magnetic flux tubes and the regular open field lines as a result of three X-line reconnection at the dayside magnetopause. (b) The projection of flux tubes in the noon-midnight meridian plane.

is tilted counterclockwise from the y-axis for $B_y>0$ and is tilted clockwise for $B_y<0$. This effect is consistent with the observations reported by DALY *et al.* (1984, Fig. 3). It is noted that even when the IMF turns exactly southward ($B_y=0$), the magnetic fields on the two sides of the magnetopause are not exactly antiparallel in most regions of the dayside magnetopause. The magnetic flux tubes can still be formed by the multiple X-line reconnection process.

Along each open flux tube only the portion that is embedded in the magnetopause has a transverse magnetic field component which is associated with the helical field lines. The magnetic field lines arerelatively straight outside the magnetopause. Since the Alfvén waves is the only transverse mode among the MHD waves, it is

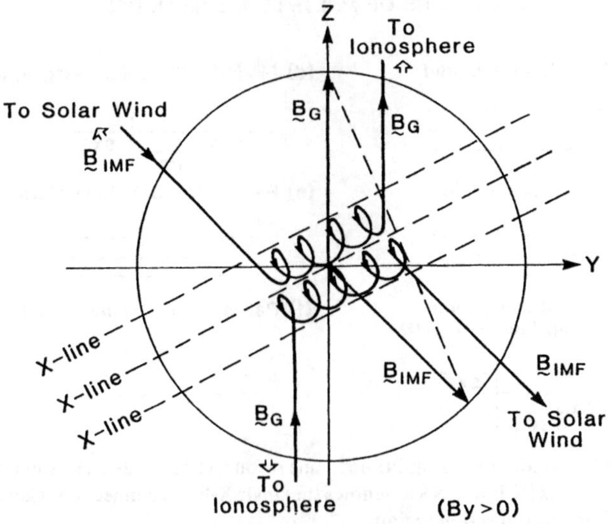

Fig. 4. Front-side view of the dayside magnetopause with open magnetic flux tubes for $B_y>0$. $\underset{\sim}{B}_G$ is the geomagnetic field.

expected that these transverse fields will propagate primarily along the magnetic field lines as Alfvén waves toward the magnetosheath and magnetosphere, as reported by SAUNDERS *et al.* (1984).

Thus, the multiple X-line reconnection process at the dayside magnetopause may lead to the formation of the observed magnetic flux tubes. The flux tubes have the following features: (1) The magnetic field lines are helical inside the tubes; (2) The flow direction of the field-aligned currents associated with the helical fields depends on B_y: *for* $B_y>0$, and $J_{\parallel}<0$ for $B_y<0$; (3) The helical fields at the magnetopause will propagate as Alfvén wave to the ionosphere and to the solar wind. These features were observed by ISEE satellites.

3. Variations of Multiple X-Line Reconnection Model

In the last section, we used three X-lines of equal length to illustrate the multiple X-line reconnection process. In fact, the number of X-lines is not limited to three (LEE and FU, 1985) and the length of the X-lines can also be unequal. Some possible variations of the multiple X-line reconnection model are illustrated in Fig. 5. The three X-line model is given Fig. 5(a), while Figure 5(b) shows the two X-line model. For the two X-line model, only one flux tube is formed at a given time. The formed flux tubes may alternatively propagate to the southern or northern cusp region.

Figure 5(c) is model in which the antiparallel merging concept is incorporated into the multiple X-line reconnection model. For IMF $B_y>0$, the reconnection will occur mostly in the duskside of the northern cusp region and in the dawnside of the southern cusp region. Single X-line reconnection (SXR) and multiple X-line

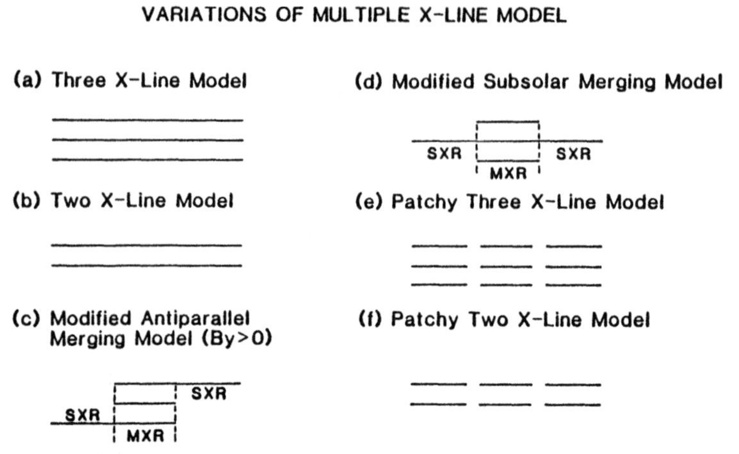

Fig. 5. Six possible variations of the multiple X-line reconnection model. The solid lines in each case are the reconnection lines (X-lines). SXR denotes the single X-line reconnection region and MXR denotes the multiple X-line reconnection region.

reconnection (MXR) occurs simultaneously at the magnetopause as shown in Fig. 5(c). The SXR region can maintain a steady-state reconnection, while the MXR region is intrinsically unsteady. This case may correspond to the split separator line merging model proposed by CROOKER (1986).

Figures 5(d) shows a modified subsolar merging model, in which the central reconnection line is longer than the other two X-lines. In this case, both the single X-line and multiple X-line reconnections may also occur simultaneously.

Figures 5(e) and 5(f) show, respectively, the patchy three X-line model and the patchy two X-line model. In these models, multiple flux-tubes with a shorter extent in the y direction can be formed. As illustrated in Fig. 5(e), three flux tubes can be formed simultaneously in each hemisphere.

As interesting feature of the multiple X-line model is that there exists a large potential difference between the field line E_2 (Type I) and the field line A_2 (Type II) in Fig. 2(b). However, the potential difference between E_2 and A_2 will be considerably reduced in the collisional conducting ionosphere. Thus, the field line A_1A_2 cannot maintain an equipotential and Alfvén waves will propagate upward from the ionosphere along the open tube A_2A_1. A large electrostatic potential difference will exist across the Alfvén wave front. The propagation of Alfvén waves tends to move the solar wind field line A_1 toward the duskside.

It may be pointed out that if the length of the flux tubes in the y direction is reduced as in the patchy multiple X-line models in Fig. 5(e) and 5(f), the potential difference between Type I open field lines and Type II open field lines will also be reduced. It remains to be seen whether the patchy multiple X-line models in Fig. 5(e) and 5(f) are favored by nature. Computer simulations and satellite observations may decide the reconnection geometry at the magnetopause.

In the next sections, we present the results of the two-dimensional MHD and particle simulations of the multiple X-line reconnection processes. Both the three X-line reconnection in Fig. 5(a) and the two X-line reconnection in Fig. 5(b) are observed.

4. MHD Simulation of Multiple X-line Reconnection

In this section, we present the simulation study of FTE's by FU and LEE (1985). In their study, it was found that magnetic flux tubes can indeed be formed intermittently and repeatedly every 5–10 min, as observed by ISEE satellites. The evolution of a current sheet between two approaching plasmas was studied based on the two-dimensional incompressible magnetohydrodynamic equations. The magnetic field $\underset{\sim}{B}$ of the current sheet at time $t=0$ is given by

$$\underset{\sim}{B} = -B_{0z} \tanh\left(\frac{x}{a}\right)\hat{z} + B_{0y}\,\hat{y} \tag{1}$$

where B_{0z} and B_{0y} are constants, $l=3a$ is the thickness of the current layer, and $\hat{y}$ and $\hat{z}$ are unit vectors in the y and z directions, respectively. All physical quantities will depend only on x and z. In the case of an incompressible fluid with a constant plasma density, the velocity $\underset{\sim}{v}$ and magnetic field $\underset{\sim}{B}$ can be expressed in terms of the velocity potential ϕ and magnetic potential A, respectively, as

$$\underset{\sim}{B} = \underset{\sim}{\nabla} A \times \hat{y} + B_y \hat{y}, \tag{2}$$

$$\underset{\sim}{v} = \hat{y} \times \underset{\sim}{\nabla}\phi + v_y \hat{y}. \tag{3}$$

It can be shown that if B_y and v_y are initially uniform as assumed in this paper, they will remain constant for $t>0$. The vorticity function Ω is related to $\underset{\sim}{v}$ and ϕ by $\Omega\hat{y} = \underset{\sim}{\nabla} \times \underset{\sim}{v} = \nabla^2\phi\hat{y}$. The dynamical equations for A and Ω are given by; (PRITCHETT and WU, 1979; FU and LEE, 1985).

$$\frac{\partial A}{\partial t} = -\underset{\sim}{v} \cdot \underset{\sim}{\nabla} A + \eta \nabla^2 A \tag{4}$$

$$\frac{\partial \Omega}{\partial t} = -\underset{\sim}{v} \cdot \nabla\Omega + \underset{\sim}{B} \cdot \underset{\sim}{\nabla} J + \nu \nabla^2 \Omega \tag{5}$$

where η is the resistivity, ν is the viscosity and $J = -\nabla^2 A$ is the current density in the y direction. Equations (4) and (5) are written in a dimensionless form. The magnetic field, length, velocity, time, resistivity and viscosity are expressed in units of B_{0z}, a, $V_A t_A = a/V_A$, $\mu_0 a V_A$, and $\rho_0 a V_A$, respectively, where V_A is the Alfvén speed based on B_{0z}, t_A is the Alfvén transit time, μ_0 is the permeability in free space, and ρ_0 is the constant plasma density. The simulation size is $2L_z \times L_x$. At $x=L_x$, a constant incoming plasma flow is maintained with a normal speed $v_n = R_0 V_A$, where R_0 can be considered as the imposed reconnection rate. At $z=\pm L_z$, a free boundary for the plasma flow and magnetic field is used. At $x=0$, a symmetric boundary condition is imposed. The magnetic fields at $x=L_x$ are maintained at the initial values as given by (1). The grid sizes Δ_x and Δ_z used are typically 0.3–1.0.

An example of the simulation is shown in Fig. (6) and (7). At $t=0$, a small magnetic field given in (1). The parameters used are: $R_0=0.1, \eta=\nu=0.03, L_z=85a$, and $L_x=11a$. The corresponding magnetic Reynolds number based on the system length ($2L_z$) and Alfvén speed V_A is $R_m=2L_z\mu_0 V_A/\nu=5100$. Note that in Fig. (6) and (7), the scales in the x and z directions are different. Figure 6 shows the magnetic field lines (contours of constant A), while Figure 7 shows the streamlines (contours of constant ϕ). A stagnation point of plasma flow at $t=0$ is located at $x=z=0$ of Fig. 6.

It can be seen from Figure 6 that magnetic islands (magnetic flux tubes) are

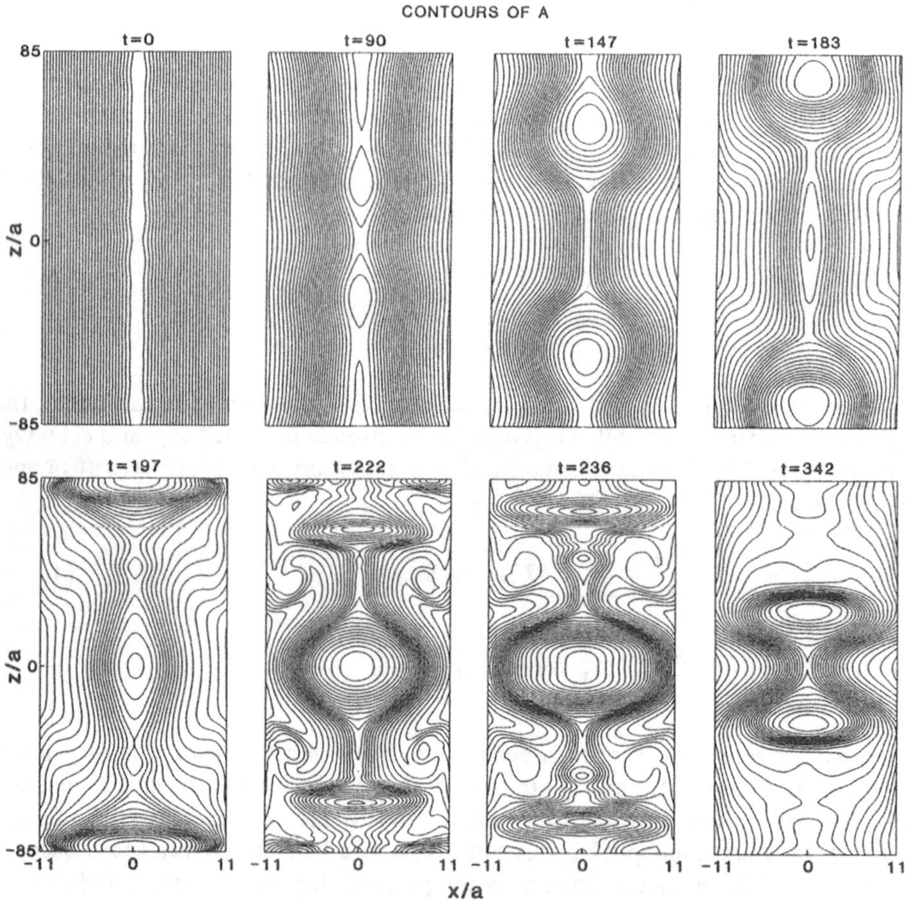

Fig. 6. Contours of constant magnetic potential A at different times. The time t is expressed in units of the Alfvén transit time t_A. The magnetic flux tubes are repeatedly formed and convected out of the boundary. At the dayside magnetopause, t_A~1–2 s.

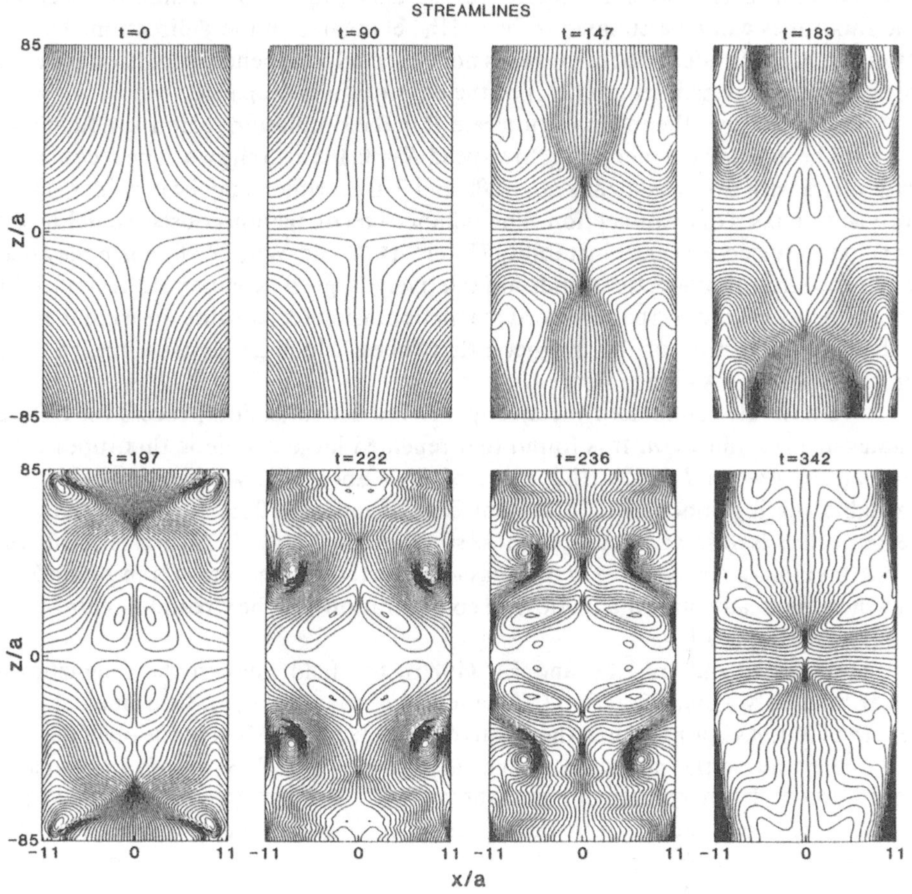

Fig. 7. Streamlines at different times for the case shown in Fig. 5.

repeatedly formed and convected out of the boundary at $z=\pm L_z$. Magnetic flux pile-up near the current sheet is often observed. The first pair of islands are formed at $t=90$, and grow to a size of $S_z \simeq 50a$ in the z direction and $S_x \simeq 15a$ in the x direction. The second pair of islands appear at $t=197$, located at $z=\pm 50a$, and reach the boundary at $t=236$. A third pair of islands are formed by splitting the large central island at $z=0$ and the splitting is nearly completed $t=342$. Note that the magnetic field configuration and the streamline pattern at $t=342$ are similar to those at $t=147$. Figure 6 shows that vortices in plasma flow are formed during the multiple X-line reconnection process.

At the dayside magnetopause near the subsolar region, the Alfvén speed is observed to be $V_A \simeq 100$–200 km/s and the thickness is $3a \simeq 450$–600 km (BERCHEM and RUSSELL, 1982). Therefore, the Alfvén transit time is $t_A \simeq 1$–2 s. The time interval that separates the magnetic flux tubes can be estimated to be 5–10 minutes, consistent with satellite observations. Furthermore, $L_z \simeq 2.5$ R_E in Fig. 6 and the cross-section of the flux tubes can be estimated to be $\sim 1R_E^2$, elongated in the z-direction. The time interval that separates two flux tubes is not expected to depend strongly on the choice of the effective system length ($2L_z$) of the dayside magnetopause.

Figure 8 shows the reconnection rate R in the simulation region as a function of time t. The normalized incoming flow speed $R_0=0.1$ is also shown. The reconnection rate varies with time and peaks at $t=200$ and 370 t_A, which may be considered as the impulsive phases of reconnection. The enhanced reconnections observed in Fig. 8 are not due to the coalescence instability (PRITCHETT and WU, 1979), since they occur when the sizes of magnetic islands are relatively large. The enhanced reconnections are probably caused by the enhanced magnetic tension force inside the islands and the poleward convection of the magnetic flux tubes, which generates a region of low pressure near the X-line region.

We have also simulated the multiple X-line reconnection process for different values of L_z/a and L_x/a. It is found that repeated large magnetic flux tubes will be formed only when $L_z \gtrsim 25a$; this value of L_z is about the wavelength of the most unstable tearing mode. For a system with length $L_z \lesssim 20a$, a quasi-steady state reconnection is observed similar to those simulated by SATO (1979) and HOSHINO and NISHIDA (1983). An example of the quasi-steady state reconnection is shown in Fig. 9, in which the steady-state magnetic field configuration and the streamlines are plotted. $\eta=0.06$, and $R_0=0.1$.

As pointed out by LEE and FU (1985), the field lines in the magnetic flux contribute only a small fraction ($\sim$one tenth) of the total open field lines, since the length of X-lines is longer than the radius of the flux tubes. Most of the open field lines are formed along the poleward X-lines as shown is Fig. 2, 3, and 6. In the case of continuously driven reconnections as shown in Fig. 6, most open field lines take place in the region between magnetic flux tubes (or magnetic islands). If the length of the

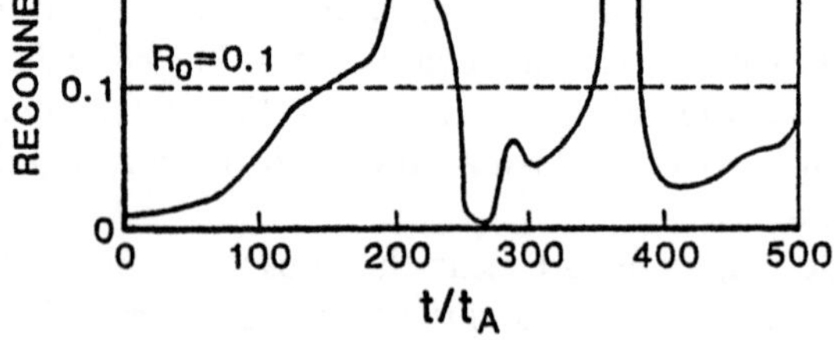

Fig. 8. The reconnection rate R as a function of time t. The dashed line shows the imposed reconnection rate $R_0=0.1$.

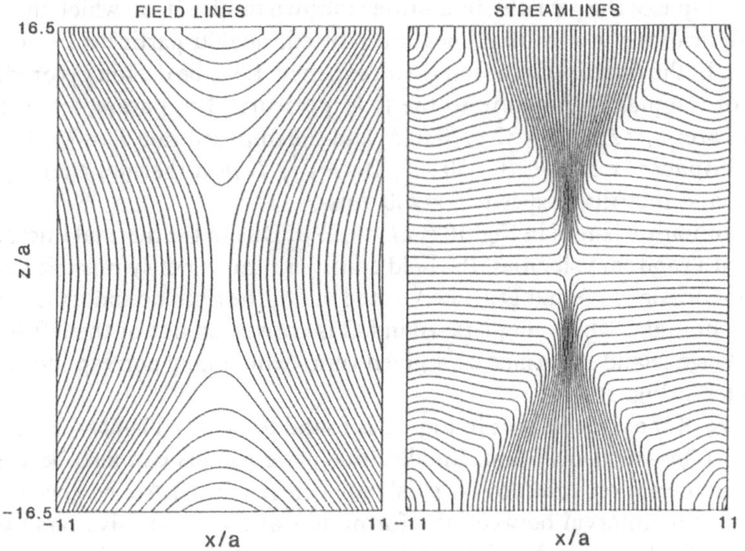

Fig. 9. The quasi-steady state configurations magnetic fields and streams for the case with $L_x = 11a$, $L_z = 16.5a$, $\eta = 0.06$, and $R_0 = 0.1$. No magnetic flux tubes are observed.

reconnection lines (X-lines) at the dayside magnetopause is assumed to be $L_y \simeq 10\ R_E$, the polar-cap potential drop associated with the multiple X-line reconnection can be estimated to be $\phi_{pc} \simeq BRL_y / t_{FTE} \simeq 50$–$150$ kV, where $B \simeq 20$–40 nT is the magnetic field at the magnetopause, $R \simeq 1 R_E$ is the radius of the flux tube and $t_{FTE} \simeq 5$–10 min is the recurrence time of the flux transfer events.

A three-dimensional MHD simulation is needed to determine if the magnetic field lines in the flux tubes are connected to the IMF and geomagnetic field lines as shown in Fig. 2 and 3.

5. Particle Simulation of the Multiple X-line Reconnection

In a collisionless plasma such as in the magnetosphere, the "anomalous" resistivity is caused either by wave-particle scattering or by the particle inertia. In the MHD simulation of magnetic reconnection, the anomalous resistivity used in the simulation has to be assumed. On the other hand, it is not necessary to assume an anomalous resistivity in a particle simulation.

Recently, DING *et al.* (1985) carried out a two-dimensional particle simulation of the multiple X-line reconnection, in which a magnetostatic ion code is used. In this simulation code, the electrostatic effect and electron currents are ignored. Non-periodic boundary conditions are employed in the simulation. A driven boundary is used in the incoming flow region ($x = \pm L_x$) while a free boundary is used in the outflow region ($z = \pm L_z$). Particles are injected from the boundary at $x = \pm L_x$ and escape through the boundary at $z = \pm L_z$.

An example of the particle simulation is shown in Fig. 10, in which the magnetic field configurations at various times are plotted. 40,000 particles are used in the simulation. In the figure, Ω is the ion gyrofrequency. Let ρ be the ion gyroradius and β the ratio of the plasma pressure to the field pressure. The parameters used in the simulation are: $L_z=2L_x$, $L_x=32\,\rho$, $\beta=0.5$, and the imposed rate $R_0=V_1/V_A=0.3$. At the magnetopause, we have $B_0=30$ nT, $\Omega=3\ s^{-1}$, T_i (ion thermal energy)$\simeq 1$ keV, $\rho=100$ km and the Alfvén speed $V_A=200$ kms^{-}.

The simulation length in Fig. 10 is $2L_z=128\,\rho$. Due to the thermal fluctuations of particles with random velocities, the field configuration is not symmetric about the z axis. Magnetic islands A and D are convected out of the right boundary ($z=L_z$) of the simulation domain, while magnetic island C is convected out of the left boundary. Only one large island is formed at a given time, which may correspond to the two x-line model in Fig. 5(b).

The size of the large islands in Fig. 10 is found to be $S_x\simeq 30$–$40\rho=3000$–4000 km and $S_z\simeq 40$–$60\rho=4000$–6000 km. Smaller magnetic islands can also be seen in the figure. The convection speed of large islands can reach the Alfvén speed ($V_A\simeq 200$ km/s). The time interval between the formation of two successive large islands is $\tau\simeq 180\,\Omega^{-1}\simeq 60$s. However, the time interval τ is found to be inversely proportional to the imposed reconnection rate R_0. Therefore, if the imposed rate $R_0\simeq 0.05$–0.1, one would expect to observe a magnetic flux tube being convected out of any one side of the boundary ever 6–12 min.

Finally, we point out that if a symmetric boundary condition is imposed at $z=0$, a splitting of magnetic islands will occur at $z=0$ as shown in Fig. 6. On the other hand, if the symmetric condition is not imposed at $z=0$, the splitting of magnetic islands is not required as shown in Fig. 10. The magnetic islands are simply convected out of the boundary at $z=L_z$ or $z=-L_z$. The asymmetric case as shown in Fig. 10 may occur more frequently than the symmetric case, since it takes a long time to split a magnetic island.

6. Ionospheric Signatures of FTE's

As mentioned in Sections 2 and 3, helical magnetic flux tubes are formed at the magnetopause through the multiple X-line reconnection process. Only the portion of the flux tubes that is embedded in the magnetopause has a transverse magnetic field component, which is associated with the field-aligned currents ($J_\parallel$) in the tubes, captured from the Chapman-Ferraro magnetopause currents (see Fig. 3(a)). It is expected that the transverse magnetic fields will propagate as Alfvén waves away from the magnetopause towards the ionosphere and magnetosheath, which may explain the observations of Alfvén waves by SAUNDERS *et al.* (1984).

Along with the Alfvén waves, the transverse plasma velocity $\underset{\sim}{v}_\perp$, electric field $\underset{\sim}{E}$ and $J_\parallel$ within the magnetic flux tubes will propagate from the magnetopause region to the ionosphere, producing footprints of FTE's in the ionosphere. The incoming Alfvén waves may be modified by the reflection processes at the polar ionosphere (e.g., TAMAO, 1984), which may reduce the electric fields and enhance the field-aligned currents in the tube. The propagation and reflection of the non-linear large-

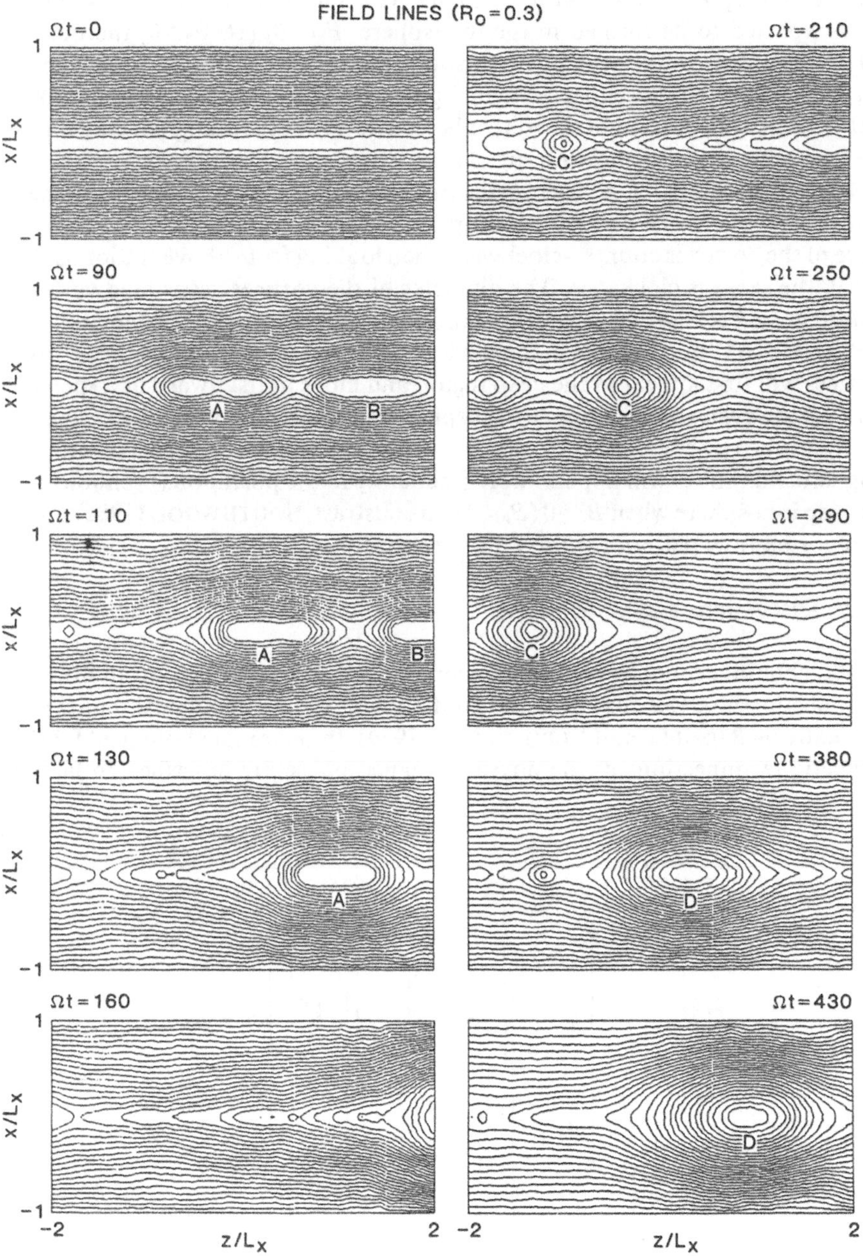

Fig. 10. Magnetic field liens at various simulation time Ωt, where Ω is the ion gyrofrequency. Magnetic islands A, B, and D are convected out of the right boundary, while the island C is convected out of the left boundary. The model is not symmetric about the z axis.

amplitude Alfvén waves associated with FTE's are complicated processes and will be studied and reported in a later paper. However, a foot print of FTE as shown in Fig. 11 is expected to be formed in the ionosphere. For B_y (IMF)>0, the field-aligned current $J_{\parallel}$ inside the tube is positive and flows toward the polar ionosphere of the northern hemisphere as expected from fig. 3(a). In figure 11, $\underset{\sim}{J}_P$ is the Pedersen current and $\underset{\sim}{J}_H$ is the Hall current in the ionosphere. The return field-aligned currents outside the flux tubes are also shown in fig. 11.

As shown in fig. 11, a large vortex of plasma flow is expected to be formed in the ionosphere as a result of the FTE coupling to the ionosphere. For B_y (IMF)>0, the sense of the vortex is counter-clockwise when looking from above the ionosphere; for $B_y<0$, the sense is clockwise. The diameter of the vortices associated with FTE's is expected to be 100–200 km. The ionosphere electric fields inside the vortex is $E\simeq 10$–100 mV/m and the associated $E\times B$ velocity is $V_E\simeq 200$–2000 ms^{-1}. The vortices will first appear in the cusp region and move antisunward at a speed of 1–10 km s^{-1}. However, the dragging of ionosphere may slow down the convection speed of the flux tube to be less than 1 km s^{-1}. From Fig. 3 and 4, it is also expected that the magnetic flux tubes will appear more frequently in the postnoon (prenoon) region of the cusp ionosphere when $B_y>0$ ($B_y<0$). In constrast, SOUTHWOOD (1985) proposed a twin vortex flow pattern for the FTE footprints.

7. Discussions and Conclusions

It should be noted that the reconnection geometry in our multiple X-line reconnection model is very different from that in the morphological model proposed originally by RUSSELL and ELPHIC (1978, 1979). In their morphological FTE model, magnetic reconnections at the dayside magnetopause are described to be "patchy"

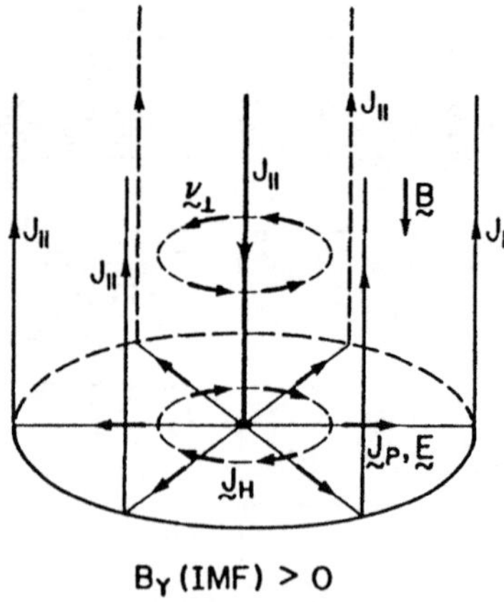

Fig. 11. The velocity ($v_\perp$), electric fields ($\underset{\sim}{E}$) and current patterns in the region of the polar ionosphere which is connected to the helical flux tubes formed at the magnetopause, for the case B_y (IMF)>0. Here, $\underset{\sim}{J}_H$ is the Hall current, $\underset{\sim}{J}_P$ is the Pedersen current density, and $J_{\parallel}$ is the field-aligned current. Inside the flux tubes, $J_{\parallel}$ is the downward; the upward return currents are located outside the flux tube.

and "intermittent". The "intermittent" nature of reconnections is supposed to lead to a finite extent of the open magnetic tubes in the z-direction, while the "patchy" property results in a finite extent of the open tubes in the y-direction. However, this morphological model requires further efforts and modifications in order to include (a) the presence of twisted field-lines and the associated field-aligned currents in the flux tubes, (b) the generation of Alfvén waves associated with FTE's, and (c) the observed B_y-dependence of the flowing direction of field-aligned currents.

In contrast, our theoretical and simulation model of multiple X-line reconnection can reproduce the above observed features of FTE's, as well as the repeated occurrence of FTE's. It is interesting to note that the reconnection models in Fig. 5(e) and 5(f) are the product of the combination of the patchy and multiple x-line models.

In conclusion, the multiple X-line reconnection model proposed by Lee and Fu can reproduce reasonably well most of the presently available observed features of FTE's. This model also provides a new reconnection configuration for the dayside magnetopause. The present theoretical and simulation studies, together with the ISEE observations of FTE's, strongly support the presence of magnetic reconnections at the dayside magnetopause.

This work is supported by NSF grant ATM 83-17454 and DOE grant to the University of Alaska. The author would like to thank Professors S.-I. Akasofu and Z. F. Fu for discussions.

REFERENCES

BERCHEM, J. and C. T. RUSSEL, Flux transfer events on the magnetopause: spatial distribution and controlling factors, *J. Geophys. Res.*, **89**, 6689, 1984.

COWLEY, S. W. H., The causes of convection in the earth's magnetosphere: A review of developments during the IMS, *Rev. Geophys. Space Phys.*, **20**, 531, 1982.

CROOKER, N. U., Global geometry of merging at the dayside magnetopause, this volume, 1986.

DALY, P. W., D. J. WILLIAMS, C. T. RUSSELL, and I. KEPPLER, Particle signature of magnetic transfer events at the magnetopause, *J. Geophys. Res.*, **86**, 1628, 1981.

DALY, P. W., M. A. SAUNDERS, R. P. RIJNBEEK, N. SCKOPKE, and C. T. RUSSELL, The distribution of reconnection geometry in flux transfer events using energetic ion plasma and magnetic data, *J. Geophys. Res.*, **89** 3843, 1984.

DING, D. Q., L. D. LEE, and Z. F. FU, A particle simulation of the multiple X-line reconnection process, in preparation, 1985.

DUNGEY, J. W., Interplanetary field and auroral zones, *Phys. Rev. Lett.*, **6**, 47, 1961.

FU, Z. F. and L. C. LEE, Simulation of multiple X-line reconnection at the dayside magnetopause, *Geophys. Res. Lett.*, May issue, 1985.

GONZALES, W. D. and F. S. MOZER, A quantitative model for the potential resulting from reconnection with an arbitrary IMF, *J. Geophys. Res.*, **79**, 4186, 1974.

HAERENDEL, G., G. PASCHMANN, N. SCKOPKE, H. ROSENBAUER, and P. C. HEDGECOCK, The frontside boundary layer of the magnetosphere and the problem of reconnection, *J. Geophys. Res.*, **83**, 3195, 1978.

HEKKILLA, W. J., Is there an electrostatic field tangential to the dayside magnetopause and neutral line? *Geophys. Res. Lett.*, **2**, 154, 1975.

HOSHINO, M. and A. NISHIDA, Numerical simulation of the dayside reconnection, *J. Geophys. Res.*, **88**, 6926, 1983.

LEE, L. C. and Z. F. FU, A theory of magnetic flux transfer at the earth's magnetopause, *Geophys. Res. Lett.*, **12**, 105, 1985.

LEVY, R. H., H. E. PETSCHEK, and G. L. SISCOE, Aerodynamic aspects of the magnetospheric flow, *AIAA J.*, **2**, 2065, 1964.
PARKER, E. N., Sweet's mechanism for merging magnetic fields in conducting fluids, *J. Geophys. Res.*, **62**, 509, 1957.
PASCHMANN, G., G. HAERENDEL, I. PAPAMASTORAKIS, N. SCKOPKE, S. J. BAME, J. T. GOSLING, and C. T. RUSSELL, Plasma and magnetic characteristics of magnetic flux transfer events, *J. Geophys. Res.*, **87**, 2159, 1982.
PETSCHEK, H. G., Magnetic annihilation, in *AAS-NASA Symposium On the Physics of Solar Flares*, edited by W. N. Hess, pp. 425, NASA Spec. Publ. SP-50, 1964.
PRITCHETT, P. L. and C. C. WU, Coalescence of magnetic islands, *Phys. Fluids*, **22**, 2140, 1979.
RIJNBEEK, R. P. and S. W. H. COWLEY, Flux erosion events are flux transfer events, *Nature*, is press, 1984.
RIJNBEEK, R. P., S. W. H. COWLEY, D. J. SOUTHWOOD, and C. T. RUSSELL, A survely of dayside flux transfer events observed by ISEE 1 and 2 magnetometers, *J. Geophys. Res.*, **89**, 786, 1984.
RUSSELL, C. T. and R. C. ELPHIC, Initial ISEE magnetometer results: magnetopause observations, *Space Sci. Rev.*, **22**, 681, 1978.
RUSSELL, C. T. and R. C. ELPHIC, ISEE observations of flux transfer events at the dayside magnetopause, *Geophys. Res. Lett.*, **6**, 33, 1979.
SATO, T., Strong plasma acceleration by slow shocks resulting from magnetic reconnection, *J. Geophys. Res.*, **84**, 7177, 1979.
SAUNDERS, M. A., C. T. RUSSELL, and N. SCKOPKE, Flux transfer events: Scale size and interior structure, *Geophys. Res. Lett.*, **11**, 131, 1984.
SCHOLER, M., D. HOVESTADT, F. M. IPAVICH, and G. GLOECKLER, Energetic protons, alpha particles, and electrons in magnetic flux transfer events, *J. Geophys. Res.*, **87**, 2169, 1982.
SONNERUP, B. U. O., Magnetic field re-connexion in a highly conducting incompressible fluid, *J. Plasma Phys.*, **4**, 161, 1970.
SONNERUP, B. U. O., G. PASCHMANN, I. PAPAMASTORAKIS, N. SCKOPKE, G. HAERENDEL, S. J. BAME, J. R. ASBRIDGE, J. T. GOSLING, and C. T. RUSSELL, Evidence for magnetic field reconnections at the Earth's magnetopause, *J. Geophys. Res.*, **86**, 10049, 1981.
SOUTHWOOD, D. J., The structure of flux transfer events, preprint, 1985.
SWEET, P. A., The neutral point theory of solar flares, in *Electromagnetic Phenomena In Cosmical Physics*, edited by B. Lehnet, p. 123, Cambridge University Press, London, 1958.
TAMAO, T., Magnetosphere-ionosphere interaction through hydromagnetic waves, in *Proc. Conf. Achievements of the IMS*, Graz, Austria, ESA SP-217, p. 427, 1984.
YEH, T. and W. I. AXFORD, On the reconnection of magnetic field lines in conducting fluids, *J. Plasma Phys.*, **4**, 207, 1970.

Solar Wind-Magnetosphere Coupling, edited by Y. Kamide and J. A. Slavin, 315–320.

Electric Field and Energy Transfer by Magnetopause Reconnection

W. D. GONZALEZ*

Instituto de Pesquisas Espaciais, Conselho Nacional de Desenvolvimento Científico e Tecnológico 12200 Sao Jose Dos Campos, SP, Brazil

Expressions related to those commonly being used in the recent literature to describe the electric field and energy transfer from the solar wind to the earth's magnetosphere are presented in this work, in a unified form, from models of large scale-magnetopause reconnection. New explicit expressions related to the component of the reconnection current, parallel and transverse to the geomagnetic field at the magnetopause, are also given. Futhermore, a new expression for the electric power related to the components of the reconnection electric field and current, that could exist parallel to the geomagnetic field, is also presented. These new derived quantities are suggested to be used in future attempts to incorporate more features of magnetopause reconnection to the study of electric field and energy transfer at the earth's magnetopause. One example of such an attempt is suggested, in which the Svalgaard-Mansurov effect could be regarded as one consequence of the presence of a finite parallel component of the reconnection current, with its direction governed by that of the dawn-dusk component of the interplanetary magnetic field.(B_y).

1. Introduction

Several studies on the electric coupling between the solar wind and the Earth's magnetosphere (PERREAULT and AKASOFU, 1978; AKASOFU, 1981; D'ANGELO and GOERTZ, 1979; KAN and LEE, 1979; GONZALEZ and GONZALEZ, 1981, 1984; REIFF *et al.*, 1981; WYGANT *et al.*, 1983; BYTHROW and POTEMRA, 1983; DOYLE and BURKE, 1983;STANISLAWSKA and PRÖLSS, 1985; TSURUTANI *et al.*, 1984, among others) have shown that quantities related to the reconnection process at the Earth's magnetopause describe well this coupling, when compared to magnetospheric parameters that represent some of the effects of such a coupling. Among those quantities, an expression called ε (PERREAULT and AKASOFU, 1978) has been successfully compared to dissipation parameters of the magnetosphere, mainly AE and U_T (AKASOFU, 1981). However, more generalized quantities, that involve ε for simple cases, have been also shown to correlate even better with U_T (GONZALEZ and GONZALEZ, 1984; STANISLAWSKA and PRÖLSS, 1985).

It is the purpose of this work to show that most, if not all, of the expressions (quantities) used in the above mentioned literature can be derived simply from large scale reconnection models, such as that given by GONZALEZ and MOZER (1974), and to propose further quantities which could be also compared to related effects in the magnetosphere. Among those effects, the Svalgaard-Mansurov effect could be

*On leave at the Jet Propulsion Lab., California Institute of Technology, Pasadena, Cal. (NASA-NRC RRA).

intimately related to an electric current-transfer between the northern and the southern hemispheres involved in asymmetric reconnection, namely when $B_y \neq 0$ (GONZALEZ and MOZER, 1974).

2. Electric Field Transfer

The expression for the total electric field due to magnetopause reconnection (GONZALEZ and MOZER, 1974) is, in gaussian units:

$$E(S,\theta)=(1/C)\ V\ B_T\ F(S,\theta); \tag{1a}$$

where V is the solar wind speed; B_T, the transverse component of the interplanetary magnetic field: $B_T \equiv (B_y^2+B_z^2)^{1/2}$ in solar-magnetospheric coordinates, and $F(S,\theta)$ is a function that describes the projection of the magnetosheath electric field to the reconnection line, given by GONZALEZ and MOZER (1974) and by GONZALEZ and GONZALEZ (1981), as:

$$F(S,\theta)=\sin(\theta-\beta)=(1-S\cos\theta)/(1+S^2-2S\cos\theta)^{1/2}, \tag{2}$$

with $S \equiv |\boldsymbol{B}_G|/|\boldsymbol{B}_M| \geqq 1$, θ the angle between the geomagnetic field $\boldsymbol{B}_G$ and the magnetosheath magnetic field $\boldsymbol{B}_M$ at the magnetopause and β the angle between the geomagnetic field $\boldsymbol{B}_G$ and the reconnection line LL′ (Fig. 1(a)). Note that $F(S,\theta)=0$ for $S\cos\theta \geqq 1$.

GONZALEZ and MOZER (1974) also presented a modified expression for $E(S,\theta)$, when geometrical effects introduced by the shape of the dayside magnetopause on the curvature of the reconnection line were taken into account and represented by the correction function $\lambda(\theta) \equiv (\sin\theta/2)^{-1}$. Thus, with this modification:

$$E_\lambda(S,\theta)=(1/C)V\ B_T\ F(S,\theta)(\sin\theta/2)^{-1}. \tag{1b}$$

However, later GONZALEZ and GONZALEZ (1981) argued that this correction factor may be basically cancelled by a similar, but opposite, geometrical effect created on the curvature of the reconnection line by the presence of the clefts.

GONZALEZ and GONZALEZ (1981) argued that some magnetospheric parameters could be better related either to the transverse (Dawn–Dusk) or to the parallel components of the reconnection electric field with respect to $\boldsymbol{B}_G$. These components (shown on Fig. 1(a), where the Dawn–Dusk direction is the line DD′) are:

$$E_D(S,\theta)=(1/C)V\ B_T(1-S\cos\theta)\times(S-\cos\theta)/(1+S^2-2S\cos\theta). \tag{3}$$

$$E_{\parallel}(S,\theta)=(1/C)V\ B_T\sin\theta(1-S\cos\theta)/(1+S^2-2S\cos\theta). \tag{4}$$

The above given expressions are general. For the limit that $S\rightarrow 1$, namely $|\boldsymbol{B}_G|=|\boldsymbol{B}_M|$ at the magnetopause, these expressions become:

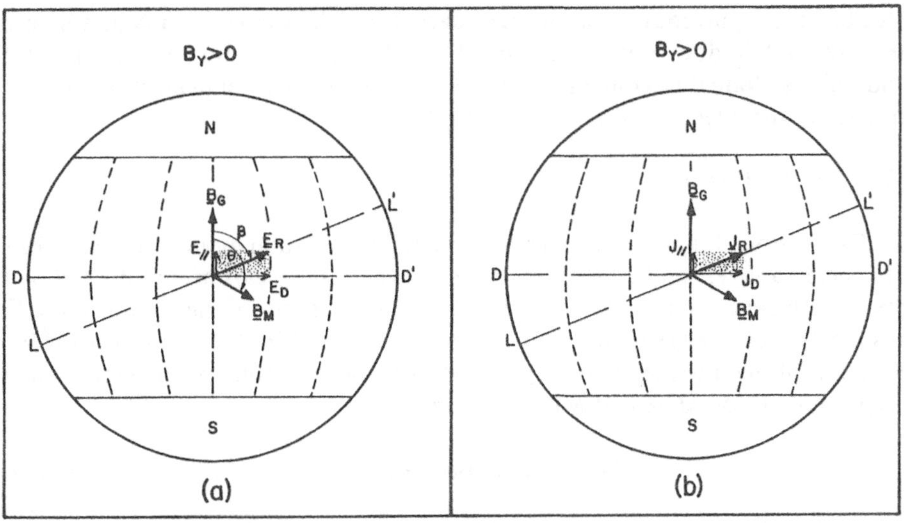

Fig. 1. (a) View from the sun of the dayside magnetopause, in which the reconnection electric field $\boldsymbol{E}_R$ and its components E_D and $E_{\parallel}$, transverse and parallel to the geomagnetic field $\boldsymbol{B}_G$, are illustrated for a case with $B_y>0$. The line LL′ is the reconnection line. The line DD′ is the Dawn–Dusk line. $\boldsymbol{B}_M$ is the magnetosheath magnetic field at the magnetopause. The angles θ and β are explained in the text. (b) Same view as in (a), where the reconnection current $\boldsymbol{J}_R$ and its components J_D and $J_{\parallel}$, transverse and parallel to the geomagnetic field $\boldsymbol{B}_G$ are illustrated for a case with $B_y>0$.

$$E(1,\theta)=(1/C)V\,B_T\sin(\theta/2), \tag{5a}$$

$$E_\lambda(1,\theta)=(1/C)V\,B_T, \tag{5b}$$

$$E_D(1,\theta)=(1/C)V\,B_T\sin^2(\theta/2), \tag{5c}$$

$$E_{\parallel}(1,\theta)=(1/C)V\,B_T\sin(\theta/2)\cos(\theta/2). \tag{5d}$$

Expression (5b) can be identified with the generalized classical VB_z function. Expression (5c) is similar to that given by KAN and LEE (1979). Expressions (1a), (1b), (5a), (5b) and (5c) have been already compared successfully to related electric field measurements (REIFF *et al.*, 1981; SONNERUP *et al.*, 1981; WYGANT *et al.*, 1983; DOYLE and BURKE, 1983; and GONZALEZ, 1984) and to electric current measurements (BYTHROW and POTEMRA, 1983). Since expression (5c) had a great success when compared to measurements, one expects that its generalized version, namely expression (3), would give interesting results as well. Expressions (4) and (5d), that refer to the component of the reconnection electric field parallel to the geomagnetic field, have not been yet compared to related magnetospheric parameters, due to the difficult task of measuring parallel electric fields at the magnetopause (Mozer, private communication). As illustrated in Fig. 1(a), this parallel electric field (with a typical value of 0.5 m V/m) could be a source of particle acceleration at the magnetopause in the case that it does not get screened out by electrons and, especially, if the conductivity along the geomagnetic field is lowered at the local magnetopause by the presence of turbulence (TSURUTANI *et al.*, 1981).

3. Electric Power

Expressions for the electric power, related to the total and component electric fields given above, can be obtained when those electric fields are multiplied by the appropriate expressions for the currents and by the length of the reconnection line. This length was approximated by $2R$ (GONZALEZ and MOZER, 1974), with R being the radius of the dayside magnetopause at the nose. The total reconnection current was given by GONZALEZ and MOZER (1974) as:

$$I=(C/2\pi)\ R|\boldsymbol{B}_{\mathrm{G}}-\boldsymbol{B}_{\mathrm{M}}|. \tag{6a}$$

In order to get the transverse and parallel components of this current with respect to the geomagnetic field at the magnetopause, one has to multiply the expression (6a), for the total current, by the transverse and parallel-projection functions $G(S,\theta)$ and $H(S,\theta)$, respectively, given by GONZALEZ and GONZALEZ (1981), namely:

$$I_{\mathrm{D}}=I\ G(S,\theta)=(C/2\pi)\ R|\boldsymbol{B}_{\mathrm{G}}-\boldsymbol{B}_{\mathrm{M}}|\times(S-\cos\theta)/(1+S^2-2S\cos\theta)^{1/2}, \tag{6b}$$

$$I_{\parallel}=I\ H(S,\theta)=(C/2\pi)\ R|\boldsymbol{B}_{\mathrm{G}}-\boldsymbol{B}_{\mathrm{M}}|\times\sin\theta/(1+S^2-2S\cos\theta)^{1/2}. \tag{6c}$$

The corresponding current densities are shown in Fig. 1(b).

Thus, the corresponding expressions for the electric power are:

$$P_{\mathrm{W}}(S,\theta)=(1/\pi)\ V\ R^2\ B_{\mathrm{T}}\ B_{\mathrm{M}}\ W(S,\theta), \tag{7}$$

with $W(S,\theta)=(1-S\cos\theta)$, for the total power,

$$P_{K}(S,\theta)=(1/\pi)\ V\ R^2\ B_{\mathrm{T}}\ B_{\mathrm{M}}\ K(S,\theta), \tag{8}$$

with $K(S,\theta)=(1-S\cos\theta)(S-\cos\theta)^2/(1+S^2-2S\cos\theta)$, for the power related to the transverse components E_{D} and I_{D}, and

$$P_{\parallel}(S,\theta)=(1/\pi)\ V\ R^2\ B_{\mathrm{T}}\ B_{\mathrm{M}}\ Y(S,\theta), \tag{9}$$

with $Y(S,\theta)=\sin^2\theta(1-S\cos\theta)/(1+S^2-2S\cos\theta)$, for the power related to the parallel components $E_{\parallel}$ and $I_{\parallel}$.

Expressions (7) and (8) have been already discussed before by GONZALEZ and GONZALEZ (1981, 1984).

For the limit $S\rightarrow 1$, and using $B_{\mathrm{T}}B_{\mathrm{M}}\sim(\pi/2)B^2$ (GONZALEZ and GONZALEZ, 1981), where B is the amplitude of the interplanetary magnetic field, expressions (7), (8) and

(9) become:

$$P_W(1.\theta)\approx V\, R^2\; B^2 \sin^2(\theta/2), \tag{10a}$$

$$P_K(1.\theta)\approx V\, R^2\; B^2 \sin^4(\theta/2), \tag{10b}$$

$$P_{//}(1,\theta)\approx V\, R^2\; B^2 \sin^2(\theta/2)\cos^2(\theta/2). \tag{10C}$$

Expressions (10a) and (10b) have been widely studied in the literature, mainly expression (10b), which is identified with the quantity $\varepsilon \equiv V l_0^2 B^2 \sin^4(\theta/2)$, given by PERREAULT and AKASOFU (1978), when $R \equiv l_0$. The quantity ε was successfully compared to dissipation parameters of the magnetosphere, mainly U_T (AKASOFU, 1981; TSURUTANI *et al.*, 1984, among others). However, expression (8), which is more general than ε, and expression (7) have been recently shown to correlate with U_T with better correlation coefficients than those obtained between ε and U_T (GONZALEZ and GONZALEZ, 1984; STANISLAWSKA and PRÖLSS, 1985). On the other hand, expressions (9) and (10c) are new and still have to be compared to related dissipation parameters of the magnetosphere. For instance, part of the energy related to dayside auroras could have an origin in processes related to these expressions, that involve electric fields and currents parallel to B_G at the magnetopause.

4. Conclusions

The general and simplified expressions given above reduce to most, if not to all, the quantities studied in the literature with respect to the electric field and power transfer at the magnetopause. Thus, one can conclude that magnetopause reconnection is an important process responsible for the coupling between the solar wind and the magnetosphere and that large scale reconnection models, like that given by GONZALEZ and MOZER (1974), represent well basic features of the reconnection process.

Besides the expressions (9) and (10c), the expression (6c) for the component of the reconnection current parallel to B_G is also given for the first time. This current is directed from the southern to the northern hemisphere when $B_y>0$, as illustrated in Fig. 1(b). It has the opposite direction when $B_y<0$. Thus, one could expect that this current would involve a net current transfer from one hemisphere to the other, probably extending its effect down to the local ionospheres. In this way, this current redistribution could be related to the well known ΔZ changes that occur at high latitudes in response to $\pm B_y$ and in association to the Svalgaard-Mansurov effect. The expected amplitudes of such effect may be related to changes in zonal ionospheric currents governed by $I_{\|}$. From expression (6c), the typical value for $I_{\|}$ is $\sim 5\times 10^5$ amperes.

This work was partially supported by the "Fundo Nacional de Desenvolvimento Científico e Tecnológico" under Contract FINEP-537/CT and by the "Fundação de Amparo a Pesquisa do Estado de São Paulo" under Project "Ciência Espacial-84/2811-1". The author appreciates the important comments made by the referees.

REFERENCES

AKASOFU, S.-I., Energy coupling between the solar wind and the magnetosphere, *Space Sci. Rev.*, **28**, 121, 1981.

BYTHROW, P. F. and T. A. POTEMRA, The relationship of total Birkeland currents to the merging electric field, *Geophys. Res. Lett.*, **10**, 573, 1983.

D'ANGELO, N. and C. K. GOERTZ, An interpretation of Akasofu's substorm parameter, *Planet. Space Sci.*, **27**, 1015, 1979.

DOYLE, M. A. and W. J. BURKE, S3-2 measurements of the polar cap potential, *J. Geophys. Res.*, **88**, 9125, 1983.

GONZALEZ, W. D. and F. S. MOZER, A quantitative model for the potential resulting from reconnection with an arbitrary interplanetary magnetic field, *J. Geophys. Res.*, **79**, 4186, 1974.

GONZALEZ, W. D. and A. L. C. GONZALEZ, Solar wind energy and electric field transfer to the Earth's magnetosphere via magnetopause reconnection, *Geophys. Res. Lett.*, **8**, 265, 1981.

GONZALEZ, W. D. and A. L. C. GONZALEZ, Energy transfer by magnetopause reconnection and the substorm parameter ε, *Plannet. Space Sci.*, **32**, 1007, 1984.

GONZALEZ, W. D., Efficiency of electric fields transfer at Earth's magnetopause and flow amplitudes related to reconnection, in *Proc. IMS Symposium, Graz-Austria, ESA, SP-217*, p.121, September 1984.

KAN, J. R. and L. C. LEE, Energy coupling functions and solar wind-magnetosphere dynamo, *Geophys. Res. Lett.*, **6**, 577, 1979.

PERREAULT, P. and S.-I. AKASOFU, A study of geomagnetic storms, *Geophys. J. R. Astron. Soc.*, **54**, 547, 1978.

REIFF, P. H., R. W. SPIRO, and T. W. HILL, Dependence of polar cap potential drop on interplanetary parameters, *J. Geophys. Res.*, **86**, 7639, 1981.

SONNERUP, B. U. Ö., G. PASCHMANN, I. PAPAMASTORAKIS, N. SCKOPKE, G. HAERENDEL, S. J. BAME, J. R. ASBRIDGE, J. T. GOSLING, and C. T. RUSSELL, Evidence for magnetic field reconnection at the Earth's magnetopause, *J. Geophys. Res.*, **86**, 10049, 1981.

STANISLAWSKA, I. M. and G. W. PRÖLSS, Correlation between the energy supplied by reconnection and the magnetospheric energy consumption, *Planet. Space Sci.*, **33**, 1091, 1985.

TSURUTANI, B. T., E. J. SMITH, R. M. THORNE, R. R. ANDERSON, D. A. GURNETT, G. K. PARKS, C. S. LIN, and C. T. RUSSELL, Wave-particle interactions at the magnetopause—Contributions to the dayside aurora, *Geophys. Res. Lett.*, **8**, 183, 1981.

TSURUTANI, B. T., J. A. SLAVIN, Y. KAMIDE, R. D. ZWICKL, J. H. KING, and C. T. RUSSELL, Coupling between the solar wind and the magnetosphere: CDAW-6, in press, *J. Geophys. Res.*, October, 1984.

WYGANT, J. R., R. B. TORBERT, and F. S. MOZER, Comparison of S3-3 polar cap potential drops with the interplanetary magnetic field and models of magnetopause reconnection, *J. Geophys. Res.*, **88**, 5727, 1983.

Solar Wind-Magnetosphere Coupling, edited by Y. Kamide and J. A. Slavin, 321–329.

Plasma and Field Observations with High Time Resolution in Flux Transfer Events

M. F. SMITH[1], D. J. RODGERS[1], R. P. RIJNBEEK[2], D. J. SOUTHWOOD[2], A. J. COATES[1], and A. D. JOHNSTONE[1]

[1]*Mullard Space Science Laboratory, University College London, Holmbury St. Mary, Dorking, Surrey RH5 6NT, U.K.*
[2]*Blackett Laboratory, Imperial College of Science and Technology, London, SW7 2BZ, U.K.*

During two of the early passes of the UKS satellite of the international AMPTE program through the dayside terrestrial magnetopause, bursts of high speed ion flow were detected by the three dimensional plasma analysers. The high time resolution measurements of the positive ion distribution have been compared with simultaneous measurements of the magnetic field in order to find out more about the structure of flux transfer events. A preliminary analysis has revealed details, previously not observable, and shown that tangential stress balance is not always satisfied by local conditions.

1. Introduction

Although DUNGEY (1961) first proposed that reconnection at the magnetopause was an important mechanism for momentum and energy transfer between the solar wind and the magnetosphere in 1961, it was not until 1978 that the ISEE spacecraft discovered evidence of fairly steady reconnection between terrestrial and interplanetary magnetic fields (SONNERUP *et al.*, 1981). Analysis of the ISEE spacecraft data also revealed magnetic perturbations near the magnetopause that were attributed to the effect of small scale, or patchy, bursts of reconnection (RUSSELL and ELPHIC, 1978). The perturbations consisted of a bipolar oscillation in the component of the field along the normal to the magnetopause often associated with substantial changes in the other components. The events, called flux transfer events (FTE's) were consistent with the passage past the spacecraft of a magnetopause flux tube that connected interior and exterior fields. Subsequently RIJNBEEK *et al.* (1984) did a statistical study of the occurrence of FTE's and found a clear correlation with southward field components in the magnetosheath. They also noted that flux transfer events were seen in conjunction with all reported examples of steady reconnection.

Although the work of RIJNBEEK *et al.* (1984) and BERCHEM and RUSSELL (1984) found a statistical link between FTE's and reconnection at the dayside magnetopause, the detailed relationship has not been established. One attempt to do so is shown in the sketch in Fig. 1 which shows the field twisting inside the connected tube. COWLEY (1982) pointed out that such twisting was necessary to be consistent with the observed field perturbations within the flux tube. SOUTHWOOD (1985) pointed out subsequently that some of the field shearing signatures reported in dual spacecraft studies of FTE's (SAUNDERS *et al.*, 1984) were likely to occur outside the connected tube.

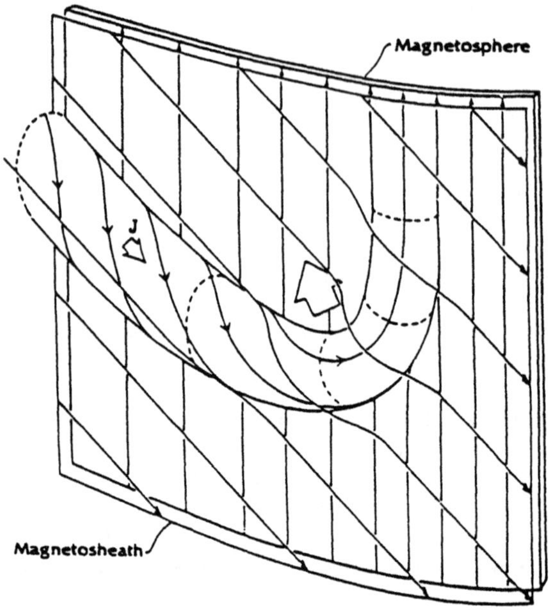

Fig. 1. Schematic of a twisted field flux transfer event.

Sketches like that shown in Fig. 1 represent concepts and must not yet be regarded as showing the proven field configuration.

In this paper we examine data from two early passes of the British satellite of the AMPTE program, UKS (United Kingdom Satellite). Both contain outbound encounters with the magnetopause on the noon meridian. In both instances the spacecraft records bursts of accelerated ions in the immediate vicinity of the magnetopause consistent with the occurrence of reconnection either at the spacecraft or elsewhere on the flux tubes connecting to the spacecraft. In both cases the external magnetic field has a southward component. Magnetosheath-like plasma was detected in bursts inside the magnetopause where the field was northward. The flow was faster than that detected in the sheath outside and thus was due to the presence of a boundary layer formed not by a simple diffusive process but by reconnection. All encounters with accelerated flows inside and outside the magnetopause are relatively shortlived (lasting less than 2–3 minutes). Their correlation with the magnetic field measured at the same time throws an interesting light on the FTE signature as defined by previous authors. It emerges that the operational definitions used in surveys like that of RIJNBEEK *et al.* (1984) and BERCHEM and RUSSELL (1984) (that the B_N amplitude is greater than 10 nT and the duration longer than 1 min in addition to showing the FTE signature) do select events containing accelerated flows symptomatic of reconnection but that the amplitude and time criteria adopted in the surveys are not adequate to identify all acceleration events detected in the ion data. It seems that there is a substantial range in the cross-sectional scale of connected flux tubes on the magnetopause.

2. Instrumentation

The AMPTE-UKS spacecraft carried a magnetometer, ion and electron plasma analysers, particle correlator, and magnetic and electric wave sensors. All instrumentation is described in a special issue of Trans. IEEE including the two instruments (magnetometer and ion analyser) whose data we call on here (SOUTHWOOD *et al.*, 1985; COATES *et al.*, 1985). The spacecraft specification was such that the magnetometer was able to determine the ambient field to an accuracy of better than 1 nT at a time resolution of 16 vectors per second. The ion detector was a three dimensional electrostatic analyser selecting E/q values in the range 10 eV to 20 keV and obtaining a full 4π angular distribution function every spin period (5 seconds).

3. Outbound Pass: 26 August 1984

On this day (Day 239) UKS was outbound in the noon meridian near but above the magnetic equator. Magnetic field and ion bulk velocity data are displayed in Fig. 2 as a function of time. There are three magnetopause crossings evident at 17:50:30, 17:51:40, and 17:54:45 UT. The coordinate system used for both vector fields is the boundary normal system introduced by RUSSELL and ELPHIC (1978). The N component is along the expected magnetopause normal direction, determined in this case by the direction in which the field variance is minimum. L and M components lie in the expected plane of the magnetopause, L being taken parallel to the Earth's dipole component in the plane. We shall refer to the L direction as northward. With such a convention the M axis then points westward. The field had a northward component until 17:50:30 UT and thus can be thought of as magnetospheric. However high speed flows were encountered in two separate bursts before this time and in each case the mean energy dropped and density rose as shown in Fig. 3.

Let us first consider the high speed flow encountered at 17:24:00 UT. The maximum velocity obtained was in excess of 200 km/s with a substantial northward component. In contrast the velocity encountered in the magnetosheath proper just before the end of the swathe of data displayed was of order 50 km/s and largely eastward (negative M direction). The flow event is embedded in a bipolar perturbation in the normal field component which extends from roughly 17:20 to 17:30 UT. The resemblance to the FTE magnetic signature is evident but the amplitude is smaller than Rijnbeek *et al.*'s 10 nT criterion. Small perturbations in all three magnetic components are evident on the same scale as that of the velocity enhancement. The M field and velocity components in particular appear to be correlated, the bipolar oscillations in each being almost exactly in antiphase. SAUNDERS *et al.* (1984) report that the perpendicular field and flow perturbations in FTE's obey the Walen relation as in an Alfven wave. Using the observed values for the amplitude of the perturbations (Δb_M=10 nT, ΔV_M=125 km/s) and the density (5 cm^{-3}) we too find a fit with such a formula.

Reminiscent also of FTE magnetic signatures is the shearing of the field both away from the exterior field direction (M component increased) and then towards it. COWLEY (1982), PASCHMANN *et al.*, (1982) and SAUNDERS (1983) have all

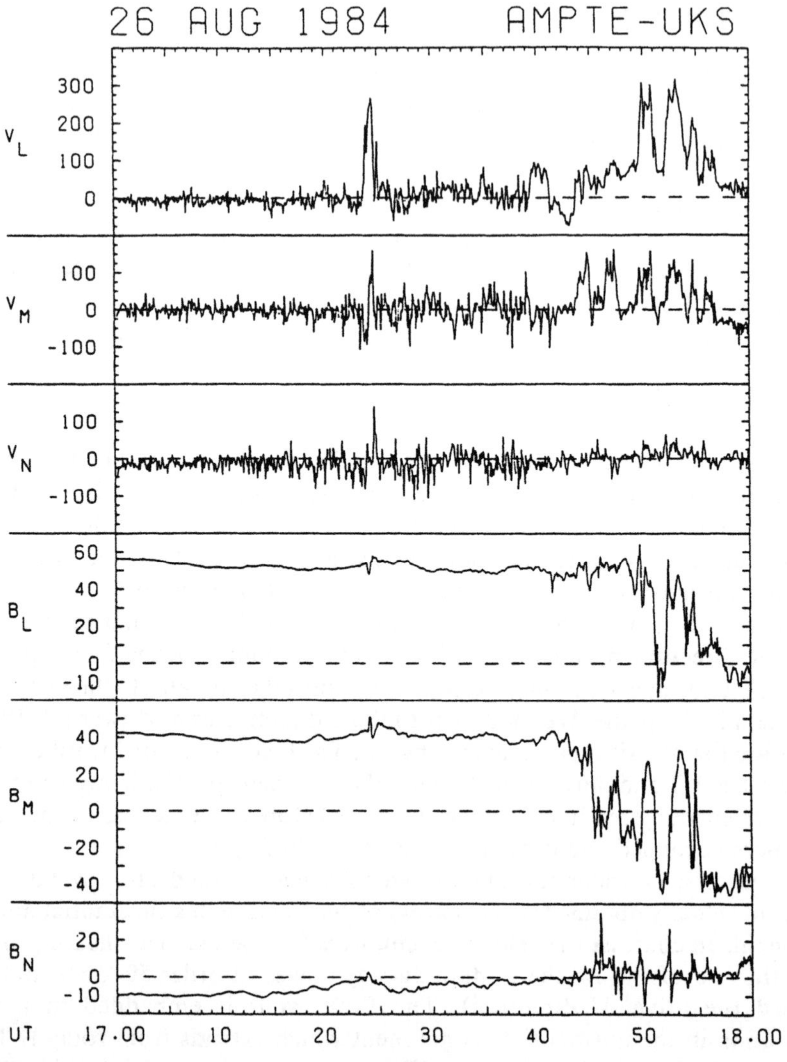

Fig. 2. Flow velocity *V*, and magnetic field *B*, in boundary normal coordinates for the magnetopause crossing of 26th August 1984.

commented on the twisting of the field apparent within FTE's. The 17:24 event thus bears resemblance in several respects to FTE's but does not pass the normal amplitude criterion.

The enhancement of the flow a factor of four above the magnetosheath level encountered later argues against the event being an encounter with a plasma that has simply diffused through the boundary. Reconnection is clearly a potential explanation of the apparent acceleration. Using the standard stress balance test for reconnection (SONNERUP *et al.*, 1981) one finds that the observed acceleration cannot

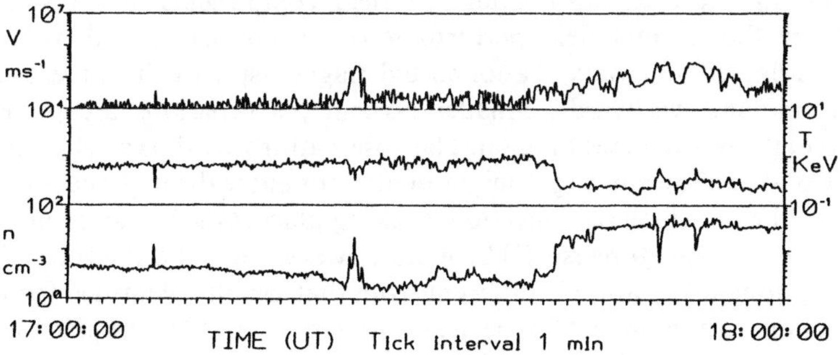

Fig. 3. Density n, temperature T and flow speed V, of the ion population for the magnetopause crossing shown in Fig. 2.

have resulted from reconnection between the locally measured terrestrial field and field in the direction of the sheath field measured later. For example, the L component of velocity (250 km/s) is larger than the M component (75 km/s), but the difference between the magnetosphere and magnetosheath magnetic fields is larger in the M component (80 nT) than the L component (60 nT). One thus concludes that any reconnection responsible cannot be local.

After the event at 17:24 up until 17:41:00 UT the distribution of ions detected is typical of the outer magnetosphere. The mean energy is high and it seems there may be a substantial fraction of the ambient distribution at energies above the range of the detector (>20 keV). The high energy ions disappear at 17:41:00 to be replaced by a magnetosheath distribution with temperature and density of order 100 eV and 30 cm^{-3} respectively. The field becomes more disturbed at this time but still has a large northward component until 17:50:30 when the magnetopause is first encountered. Two further magnetopause encounters follow. The field finally turns just southward at 17:57, after which the spacecraft is in a magnetosheath flow of about 50 km/s and a magnetic field pointing roughly eastward. The transition between regions of differing field is made in a series of jerks and it is evident that changes in L and M field and flow velocity components correlate. Indeed the flow and field changes in the L, M plane between 17:49 and the final magnetopause exit fit the stress balance relation required for local reconnection (SONNERUP *et al.*, 1981). In the earlier bursts of high speed flow the changes in the M components of flow and field correlate, and the flow velocities are substantially above sheath values. Note how the M component of flow inside the boundary is always antiparallel to that found after 17:57. Once again it is difficult to believe that any simple diffusive process could be forming the boundary layer. Reconnection is once more implicated. However there does not appear to be any clear relationship between L component changes in field and flow velocity before 17:49.

During the transition from 17:41 to 17:57 there are large amplitude fluctuations evident in the normal field component. Those at 17:49 and 17:51 have bipolar form and an amplitude comparable to FTE's. These events occur as the spacecraft enters the high speed flow region adjacent to the magnetopause, the normal component reversal taking place as both *L* and *M* field components also reverse. In this circumstance the normal field perturbations are consistent with the spacecraft crossing the rippled boundary of a connected magnetospheric flux tube, as illustrated in Fig. 1, moving westwards. Smaller normal perturbations are present as the spacecraft exits the high speed region. The observations are thus consistent with what sketches like that shown in Fig. 1 might predict for entry through the magnetopause into a flux tube on which reconnection is taking place. One has an example of what can be called a magnetopause FTE. Note however that the interpretation of the normal component signature is different from that usually identified for FTE's; the reversal in normal component has been reported in the past to take place at the centre of the connected tube without any obvious relation to changes in other components.

4. Outbound Pass: 19 September 1984

In Fig. 4 we display ion velocity and magnetic field data from UKS between 15:54 and 16:30 UT from the outbound pass on day 263. The spacecraft is once again very close to noon and just above the magnetic equator. As before boundary normal coordinates are used. When the swathe of data starts the spacecraft is clearly in the magnetosphere proper. Plasma flow velocities are small, density is low and temperature high; the field is northward and fairly regular. After about 16:18 UT the spacecraft is unambiguously in the magnetosheath and the field is mainly eastward and slightly southward. In the transition between sphere and sheath several bursts of high speed flow faster than the sheath proper are encountered. Of these the latter two at 16:14 and 16:16 UT contain velocity and field perturbations in the *L*, *M* plane tangential to the magnetopause which fit reconnection stress balance given the background sheath density of 25 cm^{-3}. Both of these events have only slight normal field perturbations associated with it. It thus seems likely that the spacecraft has encountered reconnection occurring in the magnetopause, the encounter being brought about by radial motion of a fairly planar structure.

Larger perturbations in the normal field occur earlier. At 16:00 there is a bipolar perturbation in the normal field exceeding 10 nT. There are no very evident field or flow perturbations in other components at the same time. There was no evidence of any accelerated flows when the spacecraft left the magnetopause a minute or so earlier or when it re-entered a minute or so later.

Once back in the northward field FTE signatures in the normal component are detected between 16:03:00 and 16:05:00 UT and 16:06:30 and 16:08:00 UT. Within these events very clear tangential field and plasma velocity perturbations are present. Evidently in these boundary layer plasma encounters the velocity is substantially higher than the sheath flow measured later. During the events the field adopts a direction intermediate between the interior and exterior fields. Association of the acceleration with reconnection is again supported by the stress balance test. A similar consideration applies to the burst of high speed flow detected adjacent to the magnetopause as the spacecraft exits at 16:11 UT.

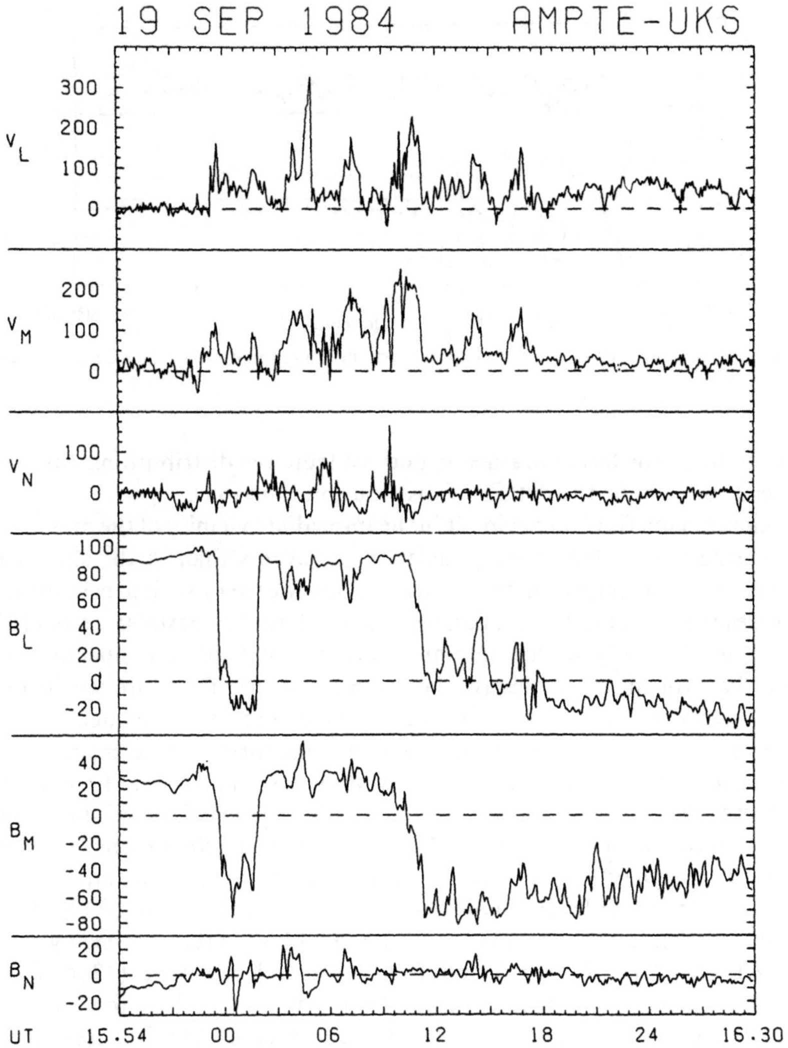

Fig. 4. Flow velocity *V*, and magnetic field ***B***, in boundary normal coordinates for the magnetopause crossing of 19th September 1984.

5. Discussion

We have reviewed UKS ion and magnetic field measurements made near the magnetopause on two outbound passes close to the noon meridian. Both passes take place when the exterior field has a southward component and thus lead one to expect to find the presence of reconnection signatures. Full three dimensional ion distribution functions are obtained in each five second spacecraft spin, and thus these data are the highest resolution set yet used for such studies. In this paper we restrict

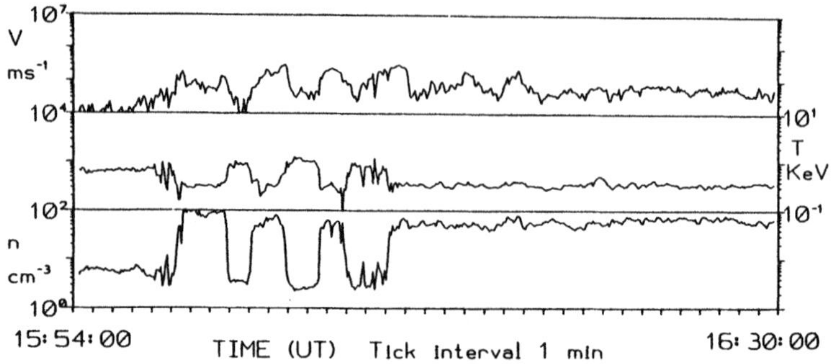

Fig. 5. Density n, temperature T and flow speed V, of the ion population for the magnetopause crossing shown in Fig. 4.

ourselves to using the bulk parameters derived from the distributions; future studies are planned to examine the details of distribution functions.

Accelerated ion flows are evident in the immediate vicinity of the magnetopause. Sheath plasma is detected moving faster in the events than in the sheath proper. Reconnection at the magnetopause is an obvious source of such acceleration and in several cases the tangential stress balance test used by PASCHMANN *et al.* (1979) and SONNERUP *et al.* (1981) fits the data and thus indicates that reconnection is taking place locally. Not all examples fit the simple test but there are no instances of boundary layers inside the magnetopause where there is no evidence of plasma gaining access to terrestrial flux tubes without some form of acceleration.

Few encounters with accelerated flows are longer than a minute or so. Nor are flows present adjacent to all magnetopause crossings detected. It seems reasonable to deduce that reconnection takes place in a sporadic or patchy manner in both space and time. The signature proposed as evidence of patchy reconnection on the magnetopause from the ISEE mission was the flux transfer event (FTE), (RUSSELL and ELPHIC, 1978). Here we have shown that the FTE signature defined by the criteria used by RIJNBEEK *et al.* (1984) and BERCHEM and RUSSELL (1984) for survey purposes does select regions where accelerated plasma is found but there are instances where smaller amplitude (<10 nT) or shorter scale events (<1 min.) are detected. We have yet to formulate any overall rule concerning the relation between ion and field signatures but, as we have described, there are clear correlations. We have shown FTE's where field and plasma data satisfy the reconnection stress balance condition as well as instances in which the relation is not satisfied using local parameters. Interestingly we have shown two examples where the spacecraft detected an FTE normal component signature as it passed through the magnetopause into or out of a region of reconnection accelerated flow. Evidently this report can only be regarded as a preliminary foray into what higher resolution measurements can provide at the magnetopause, and a more detailed analysis of these and other events should provide important information about the structure and size of reconnection regions on the magnetopause and improve our understanding of the relationship between so-called steady-state reconnection and flux transfer events.

REFERENCES

BERCHEM, J. and C. T. RUSSELL, Flux transfer events on the dayside magnetopause: spatial distribution and controlling factors, *J. Geophys. Res.*, **89**, 6689, 1984.

COATES, A. J., J. A. BOWLES, R. A. GOWEN, B. K. HANCOCK, A. D. JOHNSTONE, and S. J. KELLOCK, AMPTE UKS 3-Dimensional Ion Experiment, *IEEE Trans. Geoscience and Remote Sensing*, Vol. GE-23, 1985.

COWLEY, S. W. H., The causes of convection in the Earth's magnetosphere — a review of developments during the IMS, *Rev. Geophys. Space Phys.*, **20**, 531, 1982.

DUNGEY, J. W., Interplanetary magnetic field and auroral zones, *Phys. Rev. Lett.*, **6**, 47, 1961.

PASCHMANN, G., B. U. O. SONNERUP, I. PAPAMASTORAKIS, N. SCKOPKE, G. HAERENDEL, S. J. BAME, J. R. ASBRIDGE, J. T. GOSLING, C. T. RUSSELL, and R. C. ELPHIC, Plasma accelecration at the earth's magnetopause: Evidence for reconnection, *Nature*, **282**, 243, 1979.

PASCHMANN, G., G. HAERENDEL, I. PAPAMASTORAKIS, N. SCKOPKE, S. J. BAME, J. T. GOSLING, and C. T. RUSSELL, Plasma and magnetic field characteristics of magnetic flux transfer events, *J. Geophys. Res.*, **87**, 2159, 1982.

RIJNBEEK, R. P., S. W. H. COWLEY, D. J. SOUTHWOOD, and C. T. RUSSELL, A survey of dayside flux transfer events observed by ISEE 1 and 2 magnetometers, *J. Geophys. Res.*, **89**, 786, 1984.

RUSSELL, C. T. and R. C. ELPHIC, Initial ISEE magnetometer results: magnetopause observations, *Space Sci. Rev.*, **22**, 681, 1978.

SAUNDERS, M. A., Recent ISEE observations of the magnetopause and low latitude boundary layer: a review, *J. Geophys.*, **52**, 190, 1983.

SAUNDERS, M. A., C. T. RUSSELL, and N. SCKOPKE, Flux transfer events: scale size and interior structure, *Geophys. Res. Lett.*, **11**, 131, 1984.

SONNERUP, B. U. O., G. PASCHMANN, I. PAPAMASTORAKIS, N. SCKOPKE, G. HAERENDEL, S. J. BAME, J. R. ASHRIDGE, J. T. GOSLING, and C. T. RUSSELL, Evidence for magnetic field reconnection at the Earth's magnetopause, *J. Geophys. Res.*, **86**, 10049, 1981.

SOUTHWOOD, D. J., Theoretical aspects in ionosphere magnetosphere-solar wind coupling, in *Physics of Ionosphere Magnetosphere Connection*, edited by E. R. Schmerling, pp. Space Research Series, Pergamon, 1985.

SOUTHWOOD, D. J., W. A. C. MIER-JEDRZEJOWICZ, and C. T. RUSSELL, The fluxgate magnetometer for the AMPTE-UKS Sub Satellite, *IEEE Trans. Geoscience and Remote Sensing*, Vol. GE-23, 1985.

Solar Wind-Magnetosphere Coupling, edited by Y. Kamide and J. A. Slavin, 331–336.

Electrons at the Dayside Magnetopause: Recent Observations by AMPTE-UKS

C. P. CHALONER[1], D. S. HALL[1], R. P. RIJNBEEK[2], and D. J. SOUTHWOOD[2]

[1]*Rutherford Appleton Laboratory, Chilton, Didcot, Oxfordshire, U.K. OX11 0QX*
[2]*Blackett Laboratory, Imperial College of Science and Technology, London, U.K. SW7 2BZ*

The electron spectrometer on the AMPTE-UKS spacecraft measures the 3-D electron distribution function from 12 eV to 25 keV. The first months of the mission have produced data for many dayside magnetopause crossings. Data from one single crossing at approximately local noon is presented. It is concluded that the electrons observed during Flux Transfer Events on both sides of the magnetopause originate in the magnetosheath and, further, that the signatures of the FTEs in the magnetic field and in the electron data are not simultaneous.

1. Introduction

The AMPTE mission consists of three spacecraft: the Charge Composition Explorer (CCE), the Ion Release Module (IRM) and the UK Satellite (UKS). The UKS is a 77 kg spinning spacecraft carrying ion and electron spectrometers measuring 3-D distribution functions once per spin (~5 seconds), a 3-axis magnetometer, an electric wave experiment and a particle correlator which measures wave-particle interactions directly. The UKS and IRM are in a highly elliptical orbit with an apogee of 18.7 R_e. The UKS uses cold-gas thrusters to remain close to the IRM in order to make two-point observations in the ion-release phases of the mission. The mission has been described by BRYANT *et al.* (1985), the UKS by WARD *et al.* (1985), the UKS magnetometer by SOUTHWOOD *et al.* (1985), and the UKS electron spectrometer by SHAH *et al.* (1985).

Here we present measurements by the UKS electron instrument of the phenomena observed on a dayside magnetopause crossing. During this crossing the instrument was in a mode in which the energy range 30 eV to 18 keV was scanned once per second, so that consecutive measurements at the same energy were separated by ~74° in the GSE X–Y plane.

2. Orbit 6: 26 August 1984

Electron measurement during this outbound pass at approximately local noon show all the gross features of a dayside low-latitude magnetopause (Fig. 1). At 1700 UT the spacecraft was in the magnetosphere. It then had several encounters with the boundary layer just inside the magnetopause which (according to the magnetometer) was crossed first at ~1750 UT and with magnetosheath plasma until 1800 UT when it was well into the magnetosheath.

During this hour the magnetic field several times showed classic signatures

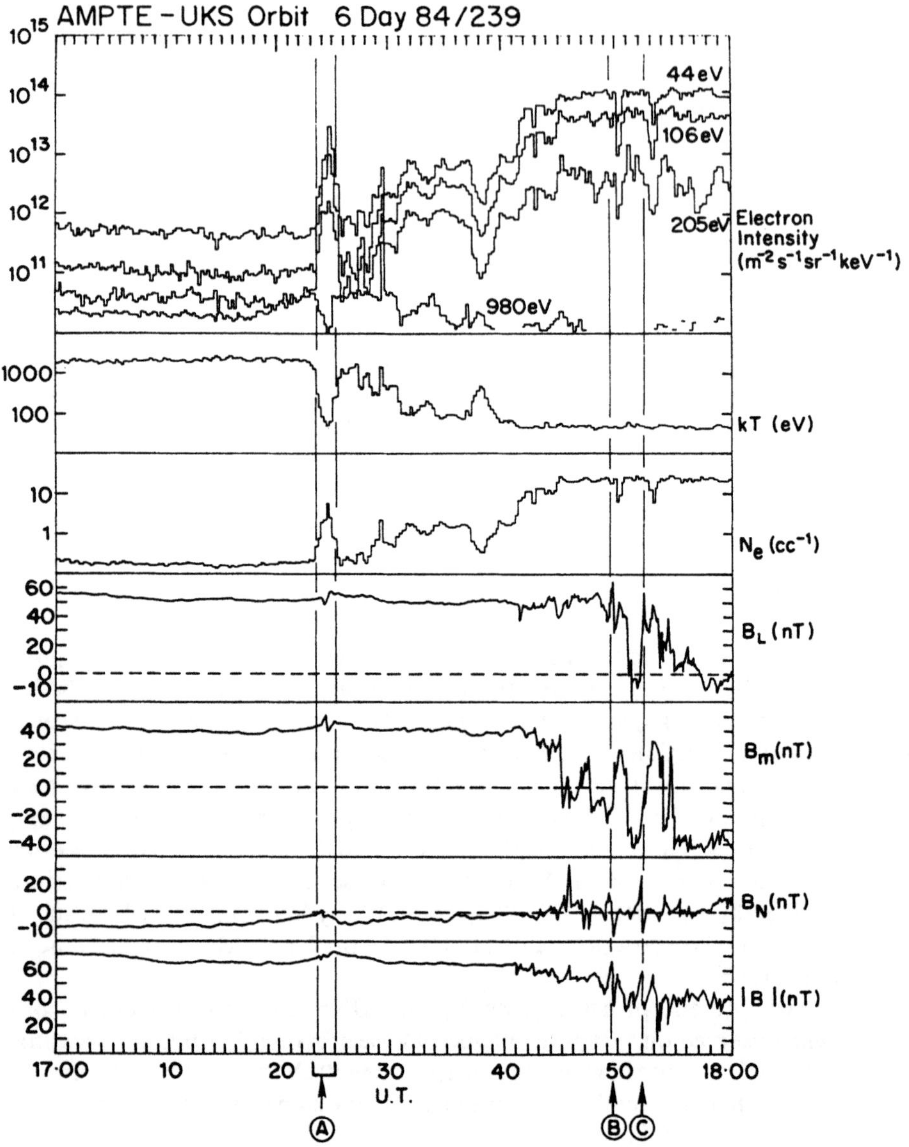

Fig. 1. Magnetic field in magnetopause-boundary-normal coordinates, together with electron density, temperature and intensity at a range of energies for the period 17:00 to 18:00 UT, showing the magnetopause crossing and FTE signatures in the B_N component for event A. Events B and C are too short to be distinguishable in this figure.

(positive-negative excursion in B_N) of magnetic reconnection events or 'FTE's—most notably at 1724 (event A), 1749 (event B) and 1752 UT (event C). Events B and C are too short to show up on the resolution of Fig. 1, but are clearly visible at the higher

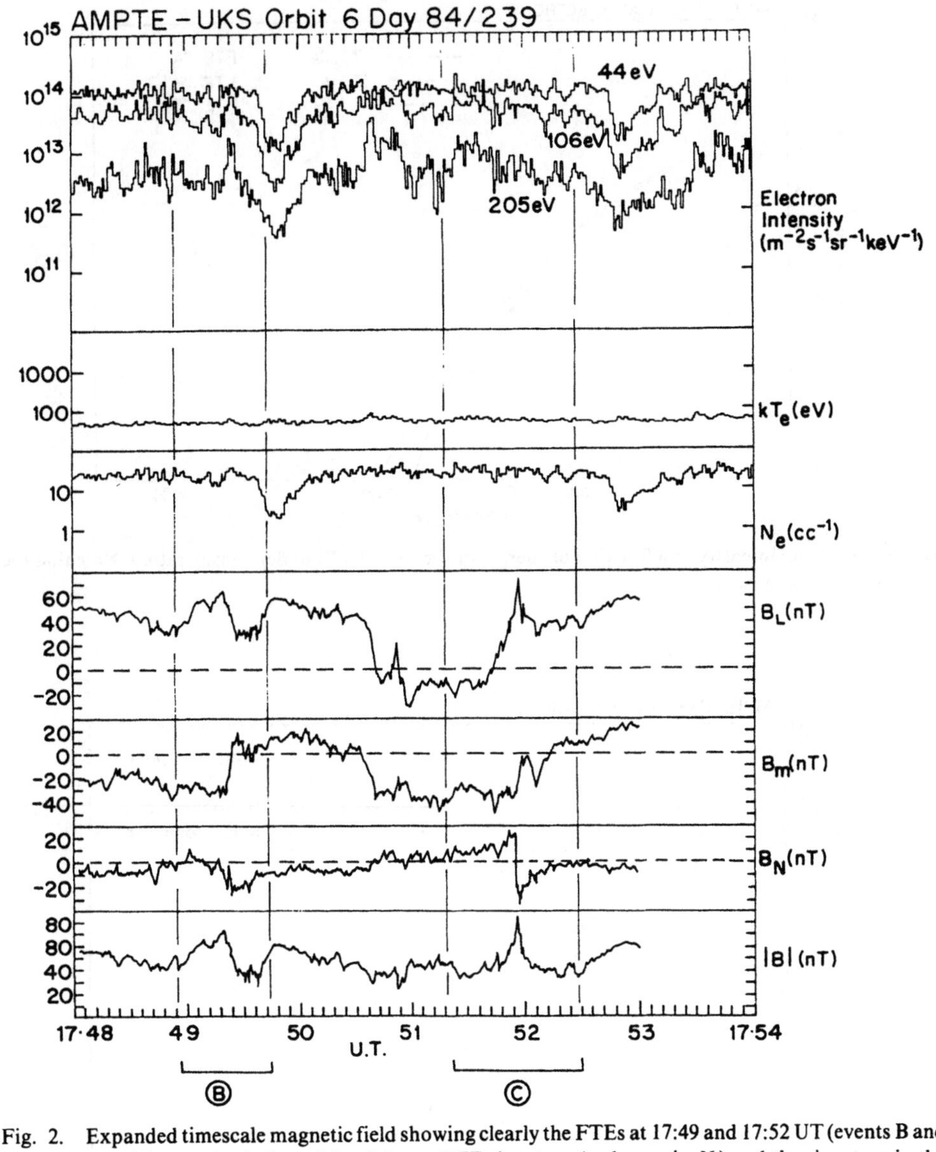

Fig. 2. Expanded timescale magnetic field showing clearly the FTEs at 17:49 and 17:52 UT (events B and C). The difference in timing of the electron FTE signature (a change in N_e) and the signature in the magnetic field (the B_N bipolar pulse) is clearly visible.

time resolution of Fig. 2. Line plots of the spectra for these events are shown in Fig. 3: it can be seen that the spectra at 1749 and 1752, which were measured in magnetosheath FTEs, are virtually identical to that at 1724, which was measured in a magnetosphere FTE.

Figure 4 shows the energy spectra for the range of plasmas encountered.

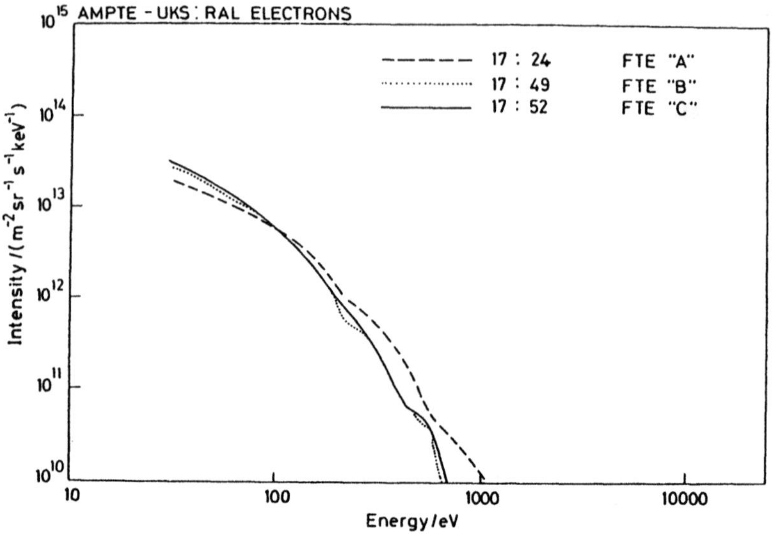

Fig. 3. Electron intensity as a function of energy for the three FTEs under consideration. Note that the spectra are almost identical.

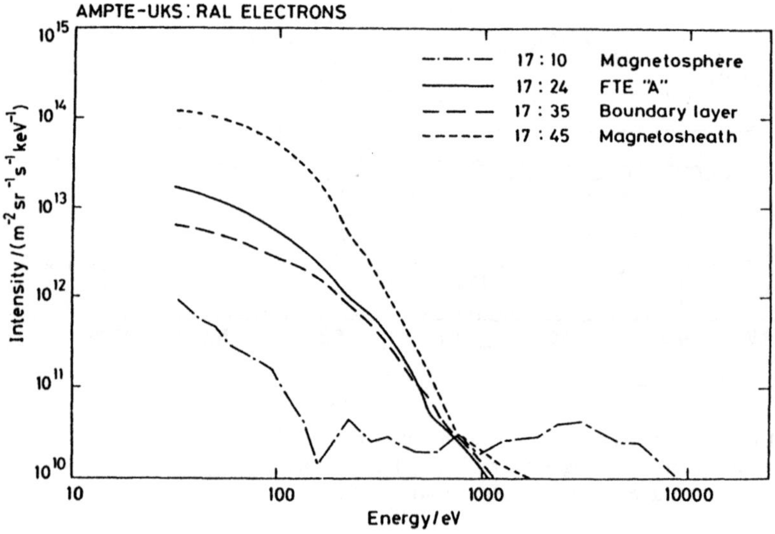

Fig. 4. Electron intensity as a function of energy at four times within the period under investigation. The magnetosphere spectrum stands out as different to the other three.

Inspection of the shape of the four spectra suggests that spectra 2 ('FTE'), 3 ('Boundary layer') and 4 ('magnetosheath') have a common source, whereas spectrum

1 ('Magnetosphere') is of a different family.

These measurements suggest that the electron signature of an FTE inside the magnetosphere (1724) is an increase in N_e accompanied by a decrease in kT, while that of a magnetosheath FTE is a decrease in N_e accompanied by virtually no change in kT. The measurements also confirm the view that the electrons seen in the boundary layer and inside FTEs on both sides of the magnetopause originate in the magnetosheath is correct (HAERENDEL *et al.*, 1978; EASTMAN and HONES, 1979).

3. Relative Timing of Magnetic and Electron FTE Signatures

It is quite clear from the 1700–1800 plots (and indeed from all other UKS magnetopause crossings we have examined so far) that there is a one-to-one correspondence between the FTE signatures in the magnetic field (the positive-negative excursion in B_N) and the electron data (the density change). On closer inspection of the exact timing of the cases shown here it becomes clear that the signatures are not always simultaneous. In the 1724 FTE (event A) the centre of the electron signature is almost exactly coincident with the B_N positive-negative excursion and yet at 1749 (event B) and 1752 (event C) follows the B_N the electron signature follows the B_N excursion by ~30 and ~50s respectively. A similar lack of simultaneity can also be seen in some of the observations of PASCHMANN *et al.* (1982), although they do not plot N_e and make no comment on this time difference. Our observations and the ion measurements of SMITH *et al.* (1986) show that on this day the FTE magnetic signature appears to herald entry of UKS into a reconnection region rather than passage through an isolated flux tube. The interpretation of FTE signatures appears to need refining.

We conclude from our preliminary assessment of AMPTE-UKS electron and magnetic measurements during the magnetopause crossing of 26 August 1984 that:

1. Our observations confirm the view that electrons observed within the boundary layer, and those associated with flux transfer events, originate in the magnetosheath.
2. Sometimes electron density changes associated with the magnetic field signature of flux transfer events occur *outside* the reconnected flux tubes that are thought to produce the magnetic field signature, and the electron density changes are not symmetric about those tubes.

REFERENCES

BRYANT, D. A., S. M. KRIMIGIS, and G. HAERENDEL, Outline of the active magnetospheric particle tracer explorers (AMPTE) mission, *IEEE Trans. on Geoscience and Remote Sensing*, **GE 23.3**, 177–181, 1985.

EASTMAN, T. E. and E. W. HONES, Jr., Characteristics of the magnetospheric boundary layer as observed by Imp 6, *J. Geophys. Res.*, **84**, 2019–2028, 1979.

HAERENDEL, G., G. PASCHMANN, N. SCKOPKE, H. ROSENBAUER, P. C. HEDGECOCK, The frontside boundary layer of the magnetosphere and the problem of reconnction, *J. Geophys. Res.*, **83**, 3195–3216, 1978.

PASCHMANN, G., G. HAERENDEL, I. PAPAMASTORAKIS, N. SCKOPKE, S. J. BAME, J. T. GOSLING, C. T. RUSSELL, Plasma and magnetic field characteristics of magnetic flux transfer events, *J. Geophys. Res.*, **87**, 2159–2168, 1982.

SHAH, H. M., D. S. HALL, and C. P. CHALONER, The electron experiment on the AMPTE UKS, *IEEE Trans. on Geoscience and Remote Sensing*, **GE 23.3**, 293–300, 1985.

SMITH, M. F., D. J. RODGERS, R. P. RIJNBEEK, D. J. SOUTHWOOD, A. J. COATES, A. D. JOHNSTONE, Small scale reconnection: AMPTE-UKS observations, this issue, 1986.

SOUTHWOOD, D. J., W. A. C. MEIR-JEDRZEJOWICZ, and C. T. RUSSELL, *IEEE Trans. on Geoscience and Remote Sensing*, **GE 23.3**, 301–304, 1985.

WARD, A. K., D. A. BRYANT, T. EDWARDS, D. J. PARKER, A. O'HEA, T. J. PATRICK, P. H. SHEATHER, K. P. BARNSDALE, and A. M. CRUISE, *IEEE Trans. on Geoscience and Remote Sensing*, **GE 23.3**, 202–211, 1985.

Solar Wind-Magnetosphere Coupling, edited by Y. Kamide and J. A. Slavin, 337–343.

Transport of Plasma across the Magnetopause

Walter J. HEIKKILA

Center for Space Sciences, The University of Texas at Dallas, Richardson Texas 75083-0688, U.S.A.

The magnetospheric boundary layer (just inside the magnetopause) appears to be partly on closed magnetic field lines, since energetic electrons often show a trapped pitch angle distribution. Solar wind plasma particles can cross the magnetopause to get into the boundary layer if the tangential electric field (in the magnetopause rest frame) changes in consort with changes in the magnetic field as $\boldsymbol{E}=-\boldsymbol{V}\times\boldsymbol{B}$ for continuous earthward flow, whether or not the field lines are closed. This has been demonstrated numerically by tracing test particle motions. Such a simultaneous change in the electric field and the magnetic field is proposed as a key feature of the so-called flux transfer events. The FTE process could more aptly be described as a transient or impulsive transfer of plasma particles on a single flux tube over a localized region of the magnetopause, ending up in general on different flux tubes after the crossing. Thus, it is proposed that FTE should denote flux "trade" event, rather than "transfer".

It is meaningless to apply the frozen field condition within the thin magnetopause current layer; in our process, the magnetopause is a sink of magnetic flux on one side and a source on the other.

1. Introduction

The magnetospheric boundary layer, first discovered by HONES *et al.* (1972) on the flanks of the magnetotail, extends to the dayside as the low latitude boundary layer LLBL (EASTMAN, 1979; EASTMAN and HONES, 1979). The frequent presence of energetic electrons with a pancake pitch angle distribution (maximum flux at 90°) indicates that it is on closed magnetic field lines. The LLBL is connected in turn to part of the entry layer, the name given to the boundary layer near the outer cusp region; another part of the latter may be on open field lines, as is the plasma mantle over the lobes of the magnetotail.

Plasma in the boundary layer is mostly solar wind in origin, and is flowing tailwards. EASTMAN (1979) estimated the total flow as $\sim 6\times 10^{26}$ ions/s near the dawn and dusk meridians. It is possible to interpret the data presented by HARDY *et al.* (1979) at lunar distances in terms of a much thicker boundary layer. The ISEE-3 observations (GOSLING *et al.*, 1984) suggest emphatically that magnetosheath plasma is able to penetrate the distant magnetopause as well.

Thus, magnetosheath plasma is apparently able to cross the magnetopause everywhere quite easily. On the dayside, at least, this penetration is onto closed field lines. By contrast, the plasma flow due to the reconnection process would be entirely on open field lines. Furthermore, in general the plasma is observed to lose to energy in the process (EASTMAN and HONES, 1979; EASTMAN *et al.*, 1985; LUNDIN and

DUBININ, 1985), rather than gaining it as it would in the case of reconnection (HEIKKILA, 1975).

LEMAIRE (1977) has proposed that the access is impulsive, localized, and transient. HEIKKILA (1982, 1983), adopting this concept, has focused on the nature of the electric field $\boldsymbol{E}$ that must be involved in the process. When there are localized changes of the magnetic field $\boldsymbol{B}$, Faraday's law tells us that curl $\boldsymbol{E}$ is finite, and is the source of a pure inductive electric field (div$\boldsymbol{E}^{\mathrm{IND}}=0$). In fact, only in this way (with curl $\boldsymbol{E}\neq 0$) is stored magnetic energy accessible, as can be realized by appealing to Poynting's theorem (HEIKKILA, 1984).

These ideas are applicable to the recently found flux transfer event (RUSSELL and ELPHIC, 1979). This feature was first recognized by magnetic measurements, and most of the past work seems to be concerned with convecting flux tubes. I prefer to focus on the plasma instead (since the principle of frozen field convection may not hold in the thin region of the magnetopause).

2. The Entry Process

Clearly, there must be some essential difference in the plasma on the magnetosheath flux tube to be involved in the FTE process when compared to the adjacent magnetosheath plasma; LEMAIRE (1977) proposed that an enhanced momentum of the flux tube would force it against the magnetopause, in what could be called a gusty wind (Fig. 1). The average flow is mostly around the magnetopause; with a southward interplanetary magnetic field the electric field in the magnetosheath would be approximately normal to the magnetopause (any tangential component near the subsolar region would be associated with reconnection). The flux tubes (or plasma clouds) with enhanced momentum will tend to deviate from the average flow, and some will strike the magnetopause, mostly toward the dawn and dusk flanks as a

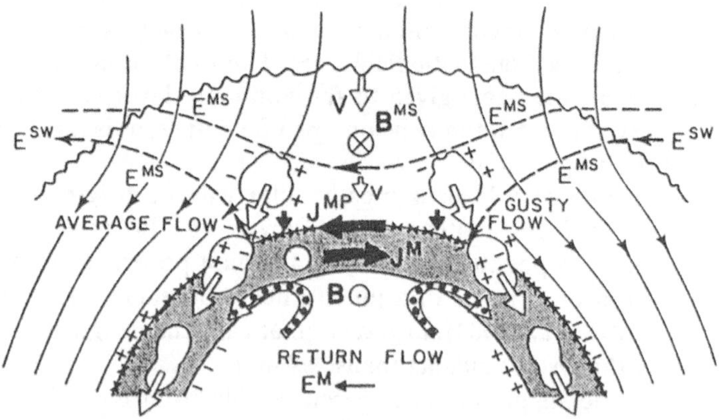

Fig. 1. Average flow and electric field lines in the magnetosheath. In a gusty flow, some plasma clouds with enhanced momentum can cross the magnetopause to form the low latitude boundary layer. The magnetospheric return flow would be mixed with solar wind in the LLBL.

deviation from the average flow. We could follow their interaction by using a sliding coordinate system; instead, for simplicity, let us consider a flux tube near the noon magnetopause (as in Figs. 4 and 5 of HEIKKILA, 1983). The magnetopause current will be pushed inwards locally, as a meander toward the earth, in the face of this enhanced pressure. We also assume that the magnetosphere is open, with a normal component B_n.

Now let us consider in detail what happens in the interaction region; as the current is pushed to the right in Fig. 2 the current density will be decreased at the trailing edge ($\Delta J<0$), and increased at the leading edge ($\Delta J>0$) as the magnetopause moves at the speed $V^{MP}=\Delta X/\Delta t$. The current perturbation produces an induction electric field (according to Lenz's law), opposing the perturbation. This field has a curl but zero divergence; the field lines have no beginning or end. We can calculate the electromotive force by using Faraday's law in integral form, as shown in Fig. 3. The electric field in general has two types of sources, (1) charge denoted by divergence, and (2) the rate of change of magnetic field marked by its curl. The total $\boldsymbol{E}$ is a combination of the induction field $\boldsymbol{E}^{IND}$ as above, and a charge-separation (or polarization) field $\boldsymbol{E}^{CS}$, produced by the plasma as in Fig. 3.

The plasma will tend to polarize when subjected to the induction electric field. The electrostatic field produced by the plasma might be able to reduce the total electric field in some portions of a closed contour, but if it does, it must enhance the field over some other part so that the line integral around the closed contour, the electromotive force, will be the same, depending only on changes in magnetic flux through the contour. The plasma can easily produce charge separation along $\boldsymbol{B}_n$ to reduce the total electric field, $\boldsymbol{E}_{\parallel}$ (parallel to $\boldsymbol{B}_n$) perhaps even to the point that $\boldsymbol{E}_{\parallel}$ may approach zero. By the previous point, it follows that over the rest of the contour, the two kinds of field will add. Thus, if anything, the tangential component will be enhanced by the plasma on either side of magnetopause (including the reversal), not reduced as one might have expected with a large isotropic conductivity. The

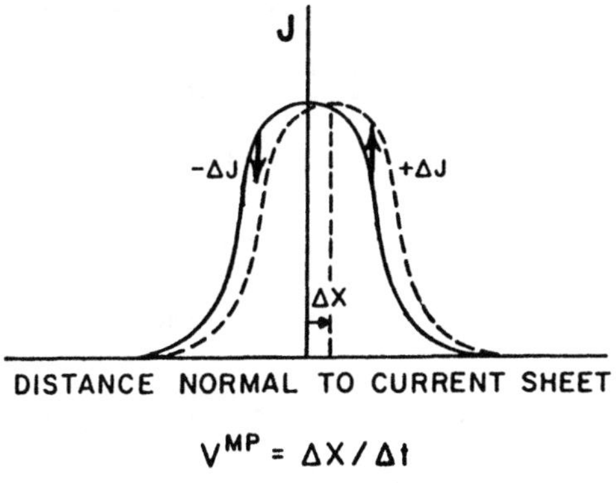

Fig. 2. Current perturbations associated with a moving current sheet would create an induction electric field according to Lenz's law, opposing the perturbation everywhere. This field reverses sense on either side of the current maximum; thus, it has a finite curl.

FLUX TUBE

$\leftarrow \underline{E}^{FT}$

(a) Electric field due to enhanced momentum

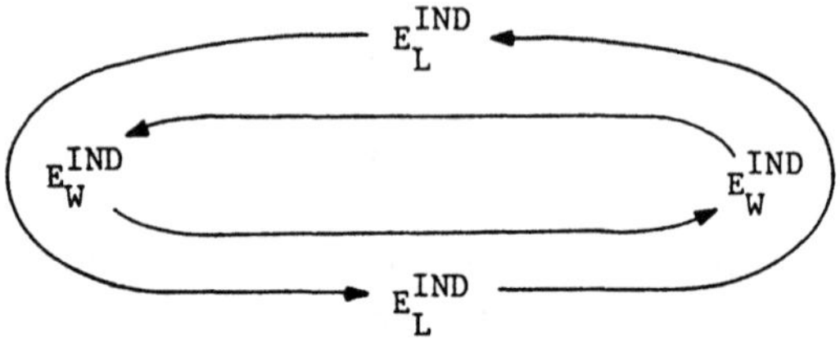

$$\oint \underline{E}\cdot\underline{dl} = -\frac{d}{dt}\iint \underline{B}\cdot\underline{dS}$$

$$2\,E_W^{IND}\cdot W + 2\,E_L^{IND}\cdot L = \frac{|\,\underline{B}_M - \underline{B}_{MS}|\cdot W\cdot L}{\Delta T}$$

$$= \Delta B\cdot V^{MP}\cdot L$$

(b) Pure induction electric (div $\underline{E}^{IND} = 0$)

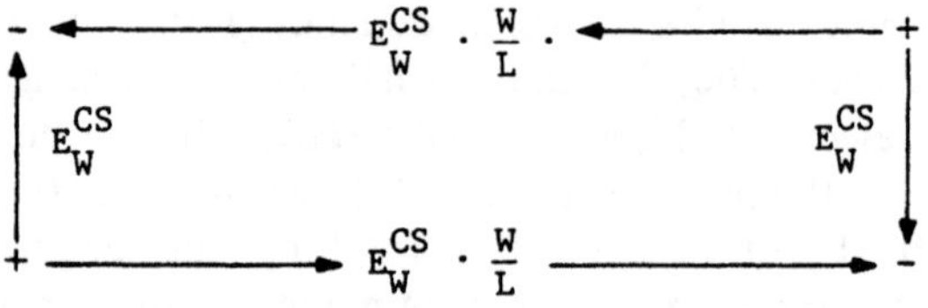

$$2\,E_W^{CS}\cdot W + 2\,E_W^{CS}\cdot\frac{W}{L}\cdot L = 0$$

(c) Charge separation field (div $\underline{E}^{CS} = \rho/Eo$)

If $|E_W^{CS}| = |E_W^{IND}|$

Then electric field in flux tube event

$$E^{FTE} = E^{FT} + \underbrace{E_L^{IND}}_{\text{(pure induction)}} + \underbrace{E_W^{CS}\cdot\frac{W}{L}}_{\text{(enhancement)}} = \frac{\Delta B\cdot V^{MP}}{2} + E^{FT} + ?\left(E^{REC.}\right)$$

(d) Total field tangent to magnetopause

Fig. 3. The electric field at the magnetopause is due to four main causes, superimposed on each other: (a) due to the flux tube, (b) pure induction, (c) charge separation, and (d) steady state reconnection.

approaching plasma cloud, with its enhanced momentum, has its own electric field, also in the same dawn-dusk sense in the magnetosheath. These fields are superimposed, with the result that the velocity of the plasma V^{PL} is greater than that of the magnetopause V^{MP}. The plasma particles can actually flow through the moving magnetopause, in a transient and patchy fachion, to create new plasma clouds inside (HEIKKILA, 1983).

In the low latitude boundary layer the electric field is now from dusk to dawn, so that the plasma can continue flowing tailward with a northward geomagnetic field.

Some model calculations shown in Fig. 4 indicate the ion trajectories of typical particles as they cross through the magnetopause. The model magnetopause uses three different sources for the magnetic field: a dipole, a constant southward field, and the field due to a current sheet; an X-line is present at the subsolar point. The current sheet is of finite width (Fig. 2). Two models of the electric field were used, a constant field as used with two dimensional theories of reconnection, and a reversal of the tangential field as in Figs. 3 and 4. With a constant electric field there is a component of the flow over the poles on open field lines, as expected on the basis of reconnection theories. This result has been used to verify that the tracing programs are correct. In contrast, with the tangential electric field reversing at the magnetopause, there is no net poleward motion, as shown in Fig. 4; the proton oscillates counter-clockwise in the magnetosheath, continues through the current sheet along B_n, and reverses its senses of gyration in the boundary layer while still drifting earthward. Thereafter, it can $\boldsymbol{E}\times\boldsymbol{B}$ drift on closed field lines, as long as the correct electric field can be maintained by the rest of the plasma in the LLBL.

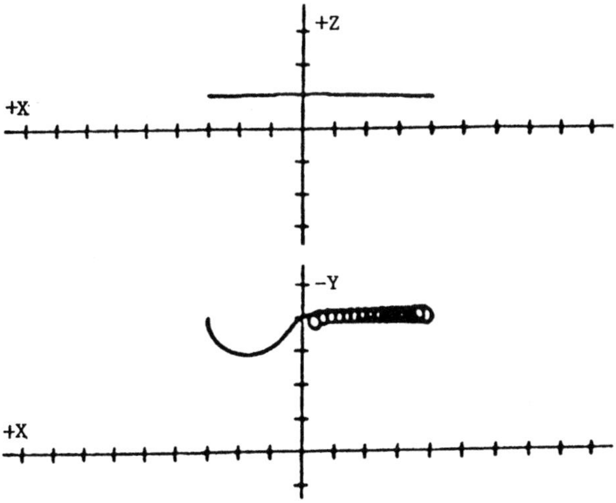

Fig. 4. Tracing the path of 1 keV proton due to a tangential electric field that reverses at the magnetopause. The proton crosses the magnetopause, going into the low latitude boundary layer on closed field lines. The assumed electric field has a finite curl, and full consideration requires a three-dimensional viewpoint.

A combination of the two electric field models gave intermediate results, as one would expect.

3. Conclusions

Several consequences follow from these ideas.

1) It is not appropriate to regard the particles within the current sheet as an ordinary plasma, conforming to the usual MHD theories. The space and time scales for a proton to cross the current sheet do not satisfy the usual criteria for MHD (HEIKKILA, 1982). It is more accurate to take the view that there is MHD fluid plasma outside, which provides particles that can penetrate through the thin magnetopause current to form a *new* MHD plasma inside. This is illustrated in Fig. 5 of HEIKKILA (1983).

2) Since the penetration is more likely toward the flanks (Fig. 1) while still continuing its tailward flow, it follows that the net effect of the continuing barrage of transient events is the charging of the sides of the boundary layer to provide the required electric field for tailward flow. Some charges would be continually lost, resulting in region 1 currents, as confirmed by BYTHROW *et al.* (1981), shown in his Fig. 7.

3) An interesting situation is brought to light when we consider a flux tube which is not quite southward, but more in the ecliptic plane, as appropriate for the garden hose angle for flux originating from the sun. Figure 5 shows that different segments of the plasma on a flux tube, which can penetrate independently through the magnetopause, would be displaced in the east-west direction: plasma on one magnetosheath flux tube would end up on different flux tubes in the boundary layer.

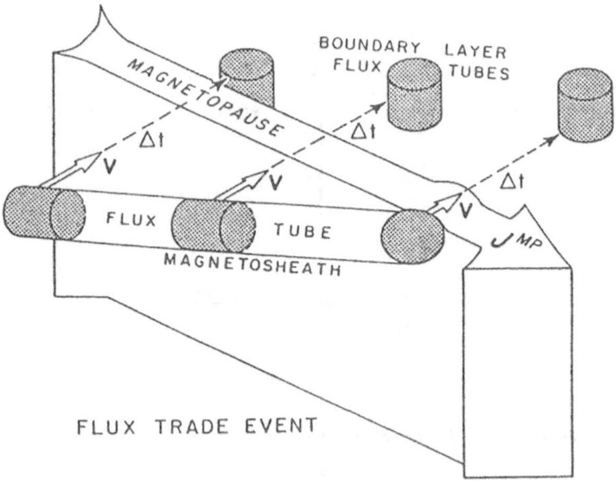

Fig. 5. All plasma elements cross independently through the magnetopause, where they end up on different flux tubes. Thus, frozen field convection is not obeyed in the thin, highly anisotropic plasma within the current sheet. The magnetopause is a sink for magnetosheath frozen field convection and a source for the boundary layer.

The term transfer, used by RUSSELL and ELPHIC (1978), to my mind implies that it should end up on one single convecting flux tube inside. I therefore propose that instead of *transfer*, we use the term *trade*.

4) This feature makes it difficult to think of a continuation of magnetic flux tubes through the magnetopause. Instead, using the idea of frozen-in-flux convecting to the magnetopause, the magnetopause now becomes a sink of magnetic flux on one side, and a source of magnetic flux on the other side. A similar sink/source combination would occur at the magnetopause in the distant magnetotail; there is no requirement for a return flow of magnetic flux.

I thank Larry Lyons for useful and incisive comments. This research was funded by NASA grant NAGW-634.

REFERENCES

BYTHROW, P. F., R. A. HEELIS, W. B. HANSON, and R. A. POWER, Observational evidence for a boundary layer source of dayside region 1 field-aligned currents, *J. Geophys. Res.*, **86**, 5577, 1981.

EASTMAN, T., The plasma boundary layer and magnetopause layer of the earth's magnetosphere, Thesis, U. of California, 1979.

EASTMAN, T. and E. W. HONES, Jr., Characteristics of the magnetospheric boundary layer and magnetopause layer as observed by imp 6, *J. Geophys. Res.*, **84**, 2019, 1979.

EASTMAN, T., L. A. FRANK, and C. Y. HUANG, The boundary layers as the primary transport regions of the earth's magnetotail, *J. Geophys. Res.*, **90**, 9541, 1985.

GOSLING, J. T., D. N. BAKER, S. J. BAME, E. W. HONES, Jr., D. J. MCCOMAS, and R. D. ZWICKL, Plasma entry into the distant tail lobes: LSEE-3, *Geophys. Res. Lett.*, **11**, 1078, 1984.

HARDY, D. A., H. K. HILLS, and J. W. FREEMAN, Occurrence of the lobe plasma at lunar distance, *J. Geophys. Res.*, **84**, 72, 1979.

HEIKKILA, W. J., Is there an electrostatic field tangential to the dayside magnetopause and neutral line?, *Geophys. Res. Lett.*, **2**, 154, 1975.

HEIKKILA, W. J., Impulsive plasma transport through the magnetopause, *Geophys. Res. Lett.*, **9**, 159, 1982.

HEIKKILA, W. J., Magnetospheric topology of fields and currents, magnetospheric currents, *Geophys. Monograph*, **28**, 208, 1983.

HEIKKILA, W. J., The electromagnetic field for an open magnetosphere, magnetic reconnection in space and laboratory plasmas, *Geophys. Monograph*, **30**, 39, 1984.

HONES, E. W., Jr., J. R. ASBRIDGE, S. J. BAME, M. D. MONTGOMERY, S. SINGER, and S.-I. AKASOFU, Measurements of magnetotail plasma flow made with Vela 4B, *J. Geophys. Res.*, **77**, 5503, 1972.

LEMAIRE, J., Impulsive penetration of filamentary plasma elements into the magnetospheres of the earth and Jupiter, *Planet. Space Sci.*, **25**, 887, 1977.

LUNDIN, R. and E. M. DUBININ, Solar wind energy transfer regions inside the dayside magnetopause: Accelerated heavy ions as tracers for MHD-processes in the dayside boundary layer, *Planet. Space Sci.*, **33**, 891, 1985.

RUSSELL, C. T. and R. C. ELPHIC, ISEE observations of flux transfer events at the dayside magnetopause, *Geophys. Res. Lett.*, **6**, 33, 1979.

Solar Wind-Magnetosphere Coupling, edited by Y. Kamide and J. A. Slavin, 345–365.

Shocks and Storm Sudden Commencements

E. J. SMITH[1], J. A. SLAVIN[1], R. D. ZWICKL[2], and S. J. BAME[2]

[1]*Jet Propulsion Laboratory, California Institute of Technology, 4800 Oak Grove Drive, Pasadena California 91125, U.S.A.*
[2]*Los Alamos National Laboratory, Los Alamos, New Mexico 87545, U.S.A.*

The ISEE-3 measurements made at the libration point upstream of Earth contain large numbers of shocks that are correlated with sudden commencement magnetic storms. A statistical survey of approximately 50 shocks and SSCs, observed between August 1978 and March 1980 (i.e., solar maximum), shows the relationship very clearly. When an SSC was seen, a shock was associated with it with an 80–90% probability. Conversely, interplanetary shocks had about 4 chances in 5 of causing an SCC. Scientific studies of SSCs involve four major areas of research each of which is reviewed. 1. The amplitude of an SSC is related to the increase in solar wind pressure from upstream to downstream of the shock which causes the magnetopause to move earth-ward. Good quantitative agreement is found between observation and theory, a result that contrasts with earlier studies which used sudden impulses as well as SSCs. 2. SSC amplitudes can also be used to infer shock properties. Preliminary analysis reveals an excellent correlation between SSC amplitude and the jump in internal pressure at the shock. It may prove possible to find a general relation to other basic shock parameters such as Mach number. Such correlations should be of use in extending our knowledge of shocks to earlier epochs for which interplanetary observations are unavailable. 3. SSCs represent the transient response of the magnetosphere to a step function input. Among the topics discussed are the dependence of SSC amplitude on local time; the velocity at which the SSC wavefront propagates through the magnetosphere; comparisons of the shock and SSC properties, including differences associated with quasi-parallel and quasi-perpendicular shocks. 4. Shocks not only cause sudden increases in the magnetospheric field but are accompanied by magnetic storms. Some tentative evidence is presented in an attempt to realte shocks to specific features of storms, in particular the development of the ring current as measured by D_{st}. Case studies indicate the latter may be caused by the simultaneous occurrence of high densities of solar wind ions (hydrogen and helium) and a significant southward component of the interplanetary magnetic field. The ions, especially the helium and perhaps heavier ions, may be accelerated at the shock or be a constituent of the solar wind driver which frequently accompanies the shock.

1. Introduction

The basic theme of this symposium (Solar Wind-Magnetosphere Coupling) addresses one of the major objectives of the ISEE Mission. ISEE-3, stationed at L1, is the upstream solar wind monitor, while ISEE 1, 2 observe the magnetospheric response to changing interplanetary conditions. Among the many features observed by ISEE-3 are large numbers of interplanetary shocks which have been identified and analyzed in detail using the JPL magnetic field and the LANL plasma data. Since shocks are known to be correlated with SSCs, such data provide an opportunity to study the physics of this relation in detail. SSCs are basically the magnetospheric reponse to a step function input and, as such, represent a response on the shortest time scales (mins).

The advantages of studying SSCs, which are unique field changes with onset times which can be determined precisely, have appealed to research workers over the years and the study of SSCs has a long history. CHAPMAN and FERRARO (1933) were the first investigators to relate the SSC to inward motion of what is now known as the magnetopause. The abruptness of the field increase allowed GOLD (1955) to infer the existence of interplanetary shocks at a time when the nature of the medium between the sun and earth was poorly understood. With spacecraft observations came confirmation of collisionless interplanetary shocks (SONETT *et al.*, 1964) and then correlation with SSCs. Numerous case studies appeared in the literature (GOSLING *et al.*, 1967; BURLAGA and OGILVIE, 1969; SUGIURA *et al.*, 1968) as well as statistical studies such as those of TAYLOR (1969) and CHAO and LEPPING (1974), the latter carried out after the solar maximum of 1968.

Unfortunately, neither these early developments nor the status of this field of research have been documented in any recent reviews. This article does not purport to fill such a need. Our purpose here is only to review briefly some recent developments with emphasis on spacecraft observations, in general, and ISEE-3 observations, in particular.

2. Correlation between Shocks and SSCs

It is appropriate to begin with a statistical examination of the shock-SSC connection. The results are presented in Table 1. The observations are analyzed here from the standpoint of the SSCs, i.e., given that an SSC has been observed, what is the probability that it was shock-associated? The data interval was restricted to the first 16.5 months of ISEE-3 operation, i.e., August 1978 through December 1979. One reason for emphasizing this interval was that most of the shocks have been analyzed using joint plasma and field data. Thus, there is little uncertainty as to whether or not a shock had actually occurred. Furthermore, many of the basic shock properties, such as Mach number and θ_B, are known.

Table 1. SSCs and shocks: 18 August 1978–31 December 1979 (16 1/2 months).

No. of SSCs	=	56
No. of data gaps	=	5
No. of test cases	=	51
No. of shocks	=	38
No. of possibles	=	7
No without shocks	=	6
		51

In this interval, the number of SSCs, tabulated in Solar Geophysical Reports month-by-month, was 56. Allowing for a propagation delay from ISEE-3 to the Earth's surface of tens of minutes, there were only 5 data gaps which made it impossible to determine whether or not a shock had occurred. The small number of SSCs which could not be investigated testifies to the value of having a spacecraft at the ISEE-3 location.

Of the remaining 51 SSCs, shocks were independently identified in 38 cases. When the field and plasma data were inspected for the other SSCs, 7 additional shocks were identified on the basis of low time resolution data. These events have not been subjected to the same thorough shock analysis as the other 38 but the probability is high that they are, in fact, shocks that were simply over looked in our initial survey. They appear to be relatively weak shocks which explains why they were missed initially.

If the basis of comparison is the number of shocks that have been unambiguously identified, a conservative criterion, the chances that a specific SSC is associated with a shock is 38/51=0.75. An upper bound is derivable if it is assumed that the number of possible shocks are truly shocks, in which case the probability becomes 45/51=0.88 or approximately 9 chances in 10 that the SSC was shock-associated.

The cases for which no shock was observed are of special interest. Three events are correlated with what appears to be a plasma piston or driver (possibly, from a flare or coronal transient) two of which (9 September 1978, 21 February 1979) are characterized by large increases in ion density and convective pressure. The magnetic field measurements for the latter are shown in Fig. 1 and involve an abrupt southward-turning of the field. The third case (18 September 1979) involves significant increase in the density of He^{++} and is also associated with an abrupt southward-turning of the interplanetary field. One of the remaining cases (19 november 1978) appears to be a large amplitude wave that is perhaps in the process of steepening into a shock.

The other two events are truly puzzling. The SSC of 17 December 1978 involves an abrupt *decrease* in ion density and convective pressure! The field has a northward component ($B_z>0$). The only distinguishing feature is the possible crossing of the heliospheric current sheet at this time. The sixth event (4 January 1979) has no obvious association with any change in the solar wind. The parameters all appear stable for several hours centered on the SSC. Both of these anomalous events appear to be bonafide SSCs with six ground stations so identifying them in each case.

The observations have also been analyzed from the point of veiw of the shocks (Table 2). For the same interval, 48 shocks have been identified without regard to whether or not they were associated with SSCs. Of these, 38 were found to be associated with an SSC while 10 shocks did not cause an SSC. Thus, given that an interplanetary shock occurs, the probability that the shock is associated with an SSC is 38/48=0.79 or about 4 chances in 5. One fifth of the shocks apparently do not lead to SSCs and are presumably associated with SIs (sudden impulses) in the magnetic field at the surface without the subsequent development of a magnetic storm. It may be speculated that such events involve a significant jump in pressure but without a change in the magnetic field to a more southward orientation (see below).

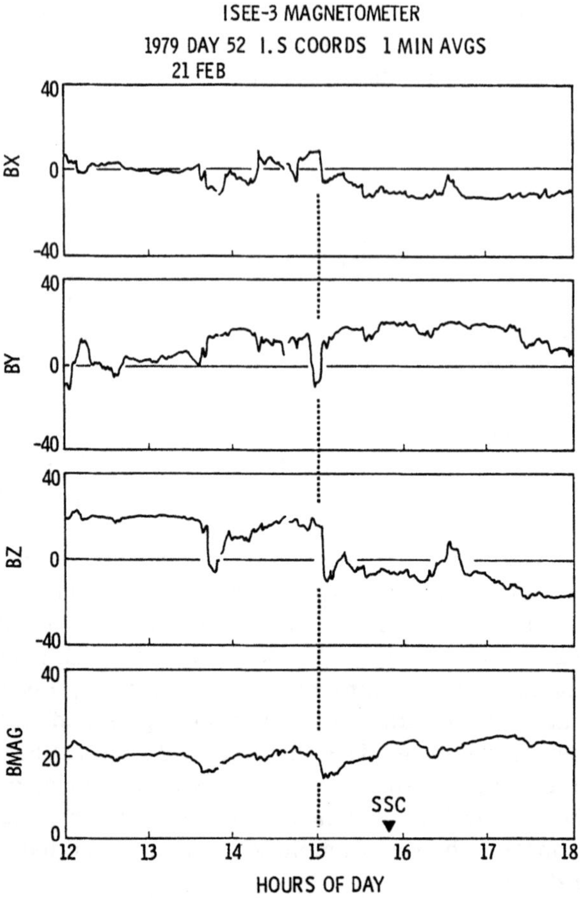

Fig. 1. Magnetic field observations for an SSC unaccompanied by a shock. The field magnitude and three components are shown prior to, and following, the SSC of 21 February 1979. The inverted filled triangle indicates the start time of the SSC. The three components are shown in a coordinate system, called IS, based on the ISEE-3 spin axis (I) and the direction to the sun(S). Since the spin axis is perpendicular to the ecliptic plane, these coordinates are the same as solar ecliptic coordinates and the B_z component is north- south. The abrupt change to a southward orientation near 15:00 is accompanied by a large increase in proton density without a corresponding change in solar wind velocity so it is not shock-related (also the field magnitude decreases). The event does involve an abrupt increase in the solar wind ram pressure which is responsible for the SSC.

Table 2. Shocks and SSCs: 18 August 1978–31 December 1979 (16 1/2 months).

No. of Shocks	=	48
No. with SSCs	=	38
No. without SSCs	=	10

3. Scientific Studies Involving SSCs

Having confirmed that a close correspondence exists between shocks and SSCs, the question may be asked: what is the scientific use of SSCs? The answer involves at least four different research areas. 1) SSCs can be used to study the dependence of the magnetopause location on the solar wind pressure. 2) In principle, they can be used to infer basic properties of interplanetary shocks. A quantitative correlation between the SSC amplitude and shock properties would extend our present knowledge of shocks to earlier epochs in the century when SSCs were being studied prior to spacecraft observations. 3) SSCs can be viewed as the transient response of the magnetosphere to a step function input. No doubt, the magnetosphere, viewed as a system whose transfer function is to be determined, is extremely complex. It involves changes in currents at the magnetopause, inside the magnetosphere, in the ionosphere and at the earth's surface. Nevertheless, progress has been made in understanding the response to an abrupt change in the solar wind and more progress is possible especially as observations are made within the magnetosphere by multiple spacecraft. 4) They can contribute to our understanding of magnetic storms. SSCs are not simply an increase in the magnetic field strength at the earth's surface. They are typically followed by large magnetic storms. At issue in one sense is the difference between SSCs and SIs. What accounts for the very different reponse of the magnetosphere in these two instances? From this point of view, we may ask whether or not the development of the ring current and the associated D_{st} is a response to certain shocks.

4. SSCs and Changes in Solar Wind Pressure

The amplitude of the SSC is the change in field strength from an initial to a final value and has customarily been interpreted in terms of the response of the magnetopause to a change in solar wind convective pressure. What physical processes are thought to be involved? Basically, an abrupt increase in solar wind pressure causes the magnetopause to move inward and launches a wave that propagates through the magnetosphere to the Earth's surface. The details are somewhat complicated as shown in the schematic (Fig. 2) taken from WILKEN *et al.* (1982).

The upper half figure shows the inward motion of the magnetopause (the solid line) from its initial position (outermost dashed line) and the accompanying wave (dashed inner line) which is compressive and propagates essentially across field lines. The motion of the field associated with this compressive wave couples into an Alfven wave, represented by the kink in the field line, which then propagates down magnetic field lines into the high latitude ionosphere. The propagation speed of this wave is more rapid than that of the compressive wave, which is slowed when it encounters the increased particle density in the plasmasphere, and the Alfven wave arrives first near Earth. The Alfven wave induces currents in the ionosphere while a component of the wave can penetrate the ionosphere and propagate quickly around the Earth in the wave guide formed by the lower ionosphere and the Earth's surface. It is the rapid speed of propagation of this electromagnetic wave that accounts for the nearly simultaneous onset of the SSC over the Earth's surface. The lower half of Fig. 2

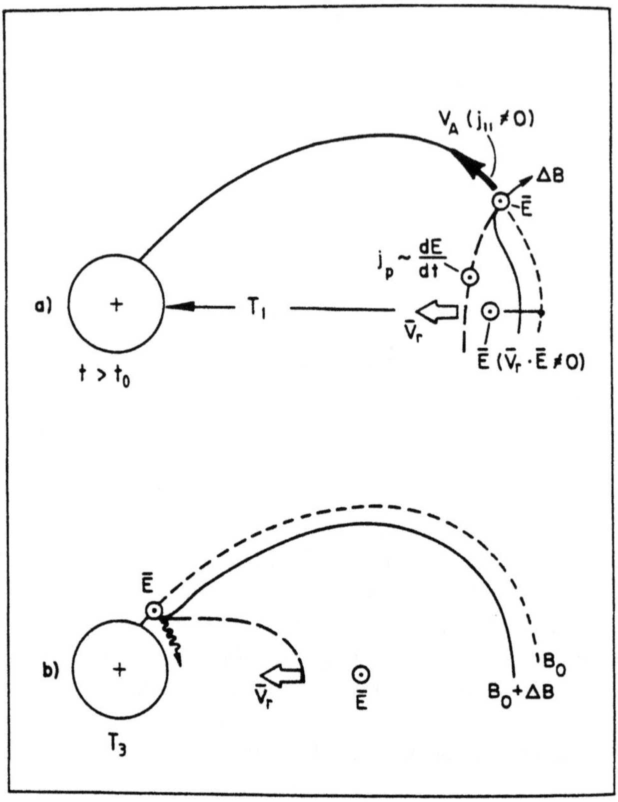

Fig. 2. Schematic response of the subsolar magnetosphere to a sudden increase in external pressure. This figure is taken from WILKEN *et al.* (1982). The upper panel shows the inward motion of the magnetopause and the two types of waves launched as a result of that motion. The lower panel shows the situation at a later stage when the last closed field line adjacent to the magnetopause has reached its new equilibrium position, the compressional wave is till propagating inward and an electromagnetic wave is travelling arond the Earth in the wave guide formed by the surface and the ionosphere. Other details of the figure are described in the text.

illustrates these latter points with the compressive wave (dashed) shown still moving Earthward when the field line representing the magnetopause (solid) has already moved to its new position and given rise to the electromagnetic wave travelling around the Earth (the oscillating line with an arrow head).

The basic physical concept has been tested quantitatively by attempting to relate the SSC amplitude, $B_2 - B_1$, to a change in solar wind convective pressure. The balance between the field pressure inside the magnetopause and the solar wind pressure outside it indicates that $B^2 \propto p$ or $B \propto p^{1/2}$ where $p = NMV_w^2$ with N, the solar wind number density, M, the mean ion mass and V_w, the solar wind speed. After the magnetopause has reached its new equilibrium position, the field at the surface will be increased as a result of the magnetopause (Chapman-Ferraro) currents being stronger and nearer the surface and the SSC will have reached its final value. Thus, the change in surface field, $B_2 - B_1$, would be expected to be proportional to $P_2^{1/2} - P_1^{1/2}$.

Figure 3, taken from SISCOE *et al.* (1968) is an attempt to test this model. The field changes were identified by inspecting ground based magnetograms and include SIs as well at least some SSCs. The interesting finding, which led to substantial discussion, was that, although there appeared to be a linear relation between $B_2 - B_1$ and $p_2^{1/2} - p_1^{1/2}$, the proportionality constant was a factor of about 2 less than predicted by theory. The latter is represented in Fig. 3 by the dashed line (theoretical) which was derived from models of the average magentopause location as a function of solar wind ram pressure. Allowance was made for an increase in $B_2 - B_1$ by a factor of 3/2 associated with the diamagnetic effect of the Earth (currents induced in the Earth tending to exclude the field change from the conducting interior). A number of effects were considered in an effort to account for the two-fold discrepancy between theory and observations.

This analysis has been repeated using the ISEE-3 plasma data and the amplitudes of the shock-associated SSCs. Because there can be significant differences in amplitude from one geomagnetic observatory to another, an average value of the field change (designated $\langle B \rangle = \langle B_2 - B_1 \rangle$, was derived. The average was based on the change in the horizontal field component reported in Solar Geophysical Reports (in the table, Principal Magnetic Storms) for stations at latitudes less than 50° (values from high

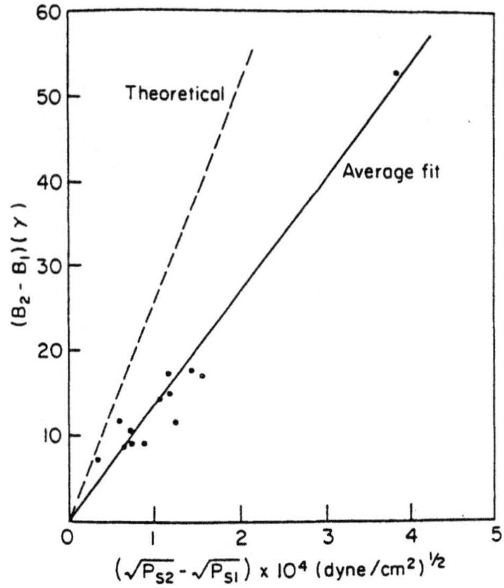

Fig. 3. Comparison between the amplitude of the magnetic impulse at the surface and the change in solar wind pressure. This figure originally appeared in SISCOE *et al.* (1968). The dashed (theoretical) line corresponds to the field change that would be expected when the magnetopause moves between the two equilibrium locations characterized by pressures P_{s1} and P_{s2}. The data exhibit a distinctly different dependence.

latitude, esentially auroral, stations were excluded). The solar wind pressures were derived from measured values of N and V_w upstream and downstream of the shock and assumed an average ion mass of 1.20 proton masses (i.e., included 5% He^{++}).

The regression line in Fig. 4 fits the observations quite well. This least squares fit to the data nearly coincides with the theoretical straight line transposed from Fig. 3. Thus, the discrepancy between theory and observation does not occur with the newer ISEE-3 data. Presumably, this difference is the consequence of using only SSCs in this case as compared to the earlier study.

5. Properties of Shocks Associated with SSCs

The second major topic, the relation between SSCs and the properties of interplanetary shocks, is addressed in Fig. 5. A log plot of the average SSC amplitude (squared in this instance so as to represent a pressure change) is shown as a function of the change in the internal pressure of the solar wind ($p=NkT$, with T the isotropic proton temperature and k the Boltzmann constant) from upstream to downstream of the shock, also plotted logarithmically. An excellent correlation can be seen corresponding to a correlation coefficient of 0.8. Correlations have also been sought and found between the SSC amplititude and other solar wind parameters but this particular correlation with Δp is judged to be the best among those tested. The existence of a good correlation of ΔB^2 with both the change in convective pressure and internal pressure presumably implies that the change in the two pressures across the shock are themselves closely correlated.

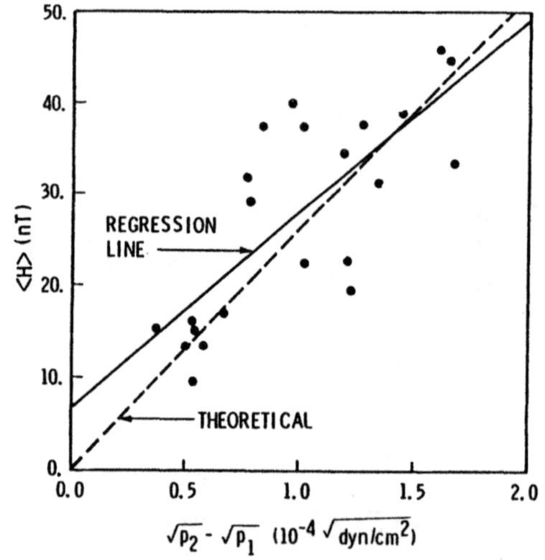

Fig. 4. Comparison between SSC amplitude and the change in solar wind pressure across the shock. The format is the same as in Fig. 3. ISEE-3 data were used to determine $P1$ and $P2$, the upstream and downstream pressures. The SSC amplitudes were obtained by averaging the changes in the horizontal component listed in Solar Geophysical Reports for observatories at geographic latitudes of less than 50°. For these data, there is a reasonable correspondence between theory and observation.

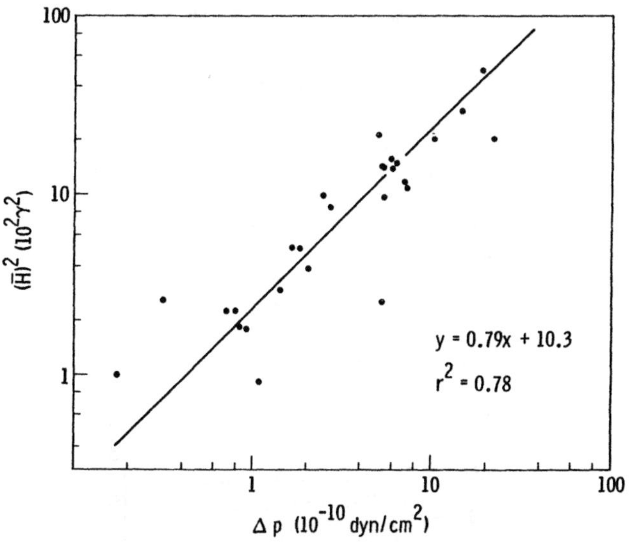

Fig. 5. Changes in thermal pressure across ISEE-3 shocks compard with the corresponding SSC amplitudes. The average values of the SSC amplitudes are the same as those used in Fig. 4 except that the field changes are squared to produce an equivalent pressure. The ordinate, Δp, is the change in the proton pressure (NkT) across the shock. The equation of the least squares straight line and the square of the correlation coefficient are shown.

A relationship of the kind demonstrated in Fig. 5 is potentially useful as a source of information about interplanetary shocks for which there are no corresponding space observations. There are substantial numbers of SSCs from 1960 onward for which there are no observations from spacecraft. Furthermore, SSC amplitudes have been detected on a regular basis since the turn of the century and it may be possible to extend our knowledge of interplanetary shocks into the past, e.g., to study the variability in shocks from one solar cycle to another.

Presumably, it would be desirable to be able to relate ΔB or $(\Delta B)^2$ to other, more fundamental shock properties such as Mach number. It might be profitable to look for a correlation between such properties and Δp or alternatively to re-derive the correlation directly in terms of such parameters. However, this has not yet been done.

6. SSCs and the Transient Response of the Magnetosphere

The third major topic is the transient response of the magnetosphere. Fig. 6 shows 18 SSCs studied by Araki and Allen (1982) as observed at Honolulu. (Incidentally, as an indication of the excellent correlation between SSCs and shocks, all of the SSCs shown in the figure are associated with shocks independently identified in the ISEE-3 data.) The waveforms shown in the figure can be considered to represent the response of the magnetosphere to a step function input. In principle, it may be possible to develop a corresponding transfer function for the magnetosphere,

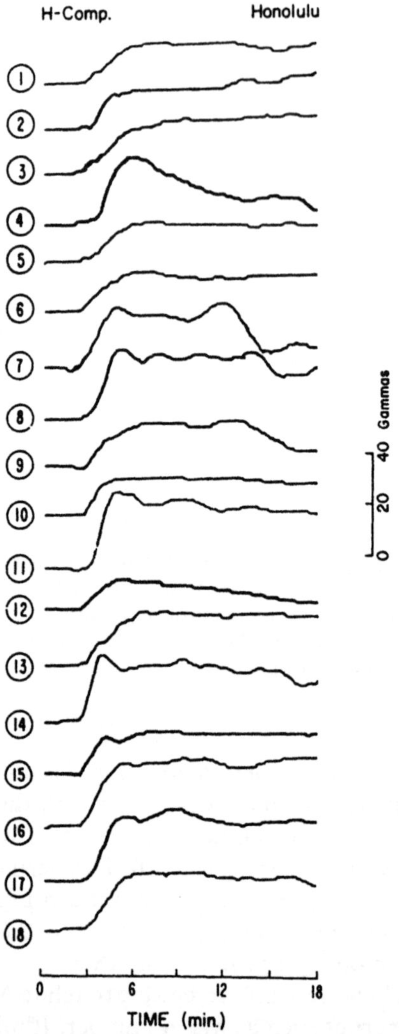

Fig. 6. SSC waveforms observed at the Earth's surface. This composite of 18 SSCs in the horizontal component of the geomagnetic field at the Honolulu observatory was assembled by ARAKI and ALLEN (1982). The events are ordered by local time with midnight at the top and noon at the bottom, although, as can be seen, this parameter is largely irrelevant. A variety of waveforms are seen, some with overshoots and other oscillations. These 18 events were checked to see whether or not they were identifiable with shocks in the ISEE-3 data. All 18 examples coincided with a shock.

taking into account the important distinction that waves are involved rather than a quasi-stationary time dependence as with electric circuits. From this conceptual point of view, the important property of each SSC is not the final value achieved (the size of the step), which has been shown to be a function of the initial and final solar wind pressures, but how the field gets from the initial to the final value (and any "ringing" or persistent fluctuations associated with the change; e.g., KAUFMANN and WALKER, 1974; and BAUMJOHANN *et al.*, 1984). As can be seen, a variety of responses are observed, some with overshoots and others without. A variety of slopes are also represented. The SSCs in the figure have been organized with respect to local time, with earliest times at the top (03:00), later times in the middle (17:00) and local midnight at the bottom, but there does not appear to be a significant dependence of waveshape on this parameter.

No doubt it is difficult to derive quantitative information from the transient response, partly because the response appears basically to be simple even though the response of the magnetosphere is complex, involving currents or current changes induced inside the magnetosphere, the ionosphere and at the Earth's surface (NISHIDA, 1964; TAMAO, 1975). The response may also depend significantly on the initial state of the magnetosphere. Another complication implicit in Fig. 6 is the limitation associated with observations at the Earth's surface which are probably dominated by ionospheric and surface currents. In order to make progress, it is probably essential to have spacecraft observations at various points in the magnetosphere. One of the features of SSCs that has been studied successfully at the surface is the polarization of the waves (ARAKI and ALLEN, 1982; and references there-in). The magnetosphere is basically anisotropic and, by analogy with propagation in anisotropic crystals, the step function resolves into components polarized in opposite senses.

Some of the limitations associated with ground based observations are illustrated in Fig. 7 which is a plot of SSC amplitude as a function of local time at two observing stations (Huancayo and Hyderabad). Clearly, there is no evidence in these data for a dependence on local time, whereas the field associated with magnetopause currents would be expected to be largest near noon. Choice of stations is important. The observations at Hyderabad are strongly influenced by the equatorial electrojet which undergoes an enhancement during SSCs. The significant variability in SSC amplititude from event to event contributes to a general scatter of the data points and tends to obscure the relationship if one exists in this data set. Careful analyses of ground based observations have, in fact, been able to resolve the expected dependence on local time, i.e., an enhancement near noon.

The local time dependence can be readily demonstrated by satellite observations such as those available at geosynchronous orbit. KOKUBUN (1983) has carried out such an analysis of GOES data with the results shown in Fig. 8. The ratio of the SSC amplitudes at the satellite and at the surface has been plotted as a function of local time. Straight line segments connect simultaneous observations at the two spacecraft (GOES 2 and GOES 3). The shading, which shows a strong dependence on local time and which reproduces the general trend of the data reasonably well, is based on the change in field when the magnetopause moves inward from 10 to 9.5 and 6.6 rE. These field values were derived from the magnetospheric model of CHOE and BEARD (1974a,b). For local times from about 18:00 to 06:00 hours, i.e., over the anti-solar hemisphere, there is an actual reduction in the field (the ratio is less than 1). This feature is attributed to the enhancement of the ring current which accompanies the intensification of the magnetopause currents and which dominates the field changes

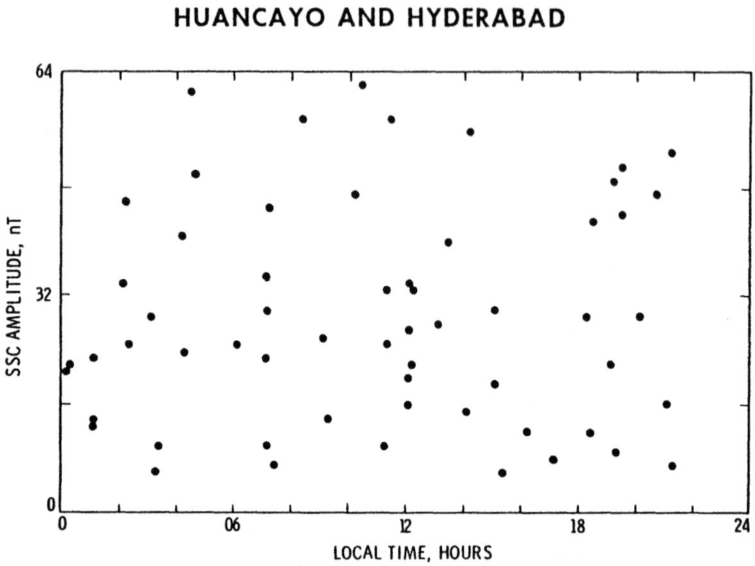

Fig. 7. An investigation of SSC amplitude with local time at two ground-based observatories. This figure shows the absence of a simple clearcut correlation between the SSC amplitude observed near the geomagnetic equator and local time. This figure is presented to contrast the surface observations with the local time dependence evident in the following figure at geosynchronous orbit. The response at the surface is strongly influenced in this case by changes in the equatorial electrojet. More careful analyses involving other stations have revealed a dependence on local time with a maximum near noon.

between the earth and the geotail.

Apart from the transient wave forms, there is another source of information implicit in observations such as those in Fig. 6. The precise time of the SSC makes it possible to study the velocity of propagation of the signal. Multi-point observations have the potential of revealing how the waves refract within the magnetosphere.

The use of magnetospheric observations to time the SSC arrival at different locations is exemplified by the schematic shown in Fig. 9, taken from WILKEN *et al.* (1982). The arrival of the SSC was studied at five satellites in geostationary orbits and at the surface, e.g., compare T_1 and T_2. Wave propagation velocities of 600 km/sec, corresponding to T_1, and 900 km/sec, corresponding to T_2, were inferred. The dashed lines in the figure show the wave front at various times as it travels through the magnetosphere. The major feature is a retardation of the wave front near the noon meridian.

Another approach which is possible using spacecraft observations is to compare the waveforms at various points in the magnetosphere with the input function (the interplanetary shock). Figure 10, also taken from KOKUBUN, (1983) shows the SSC of 6 July 1979 at three locations, GOES 2 and GOES 3 and the surface. The two GOES spacecraft were located in the subsolar hemisphere at this time and show a significant increase in field strength following the passage of the SSC. The field at both locations

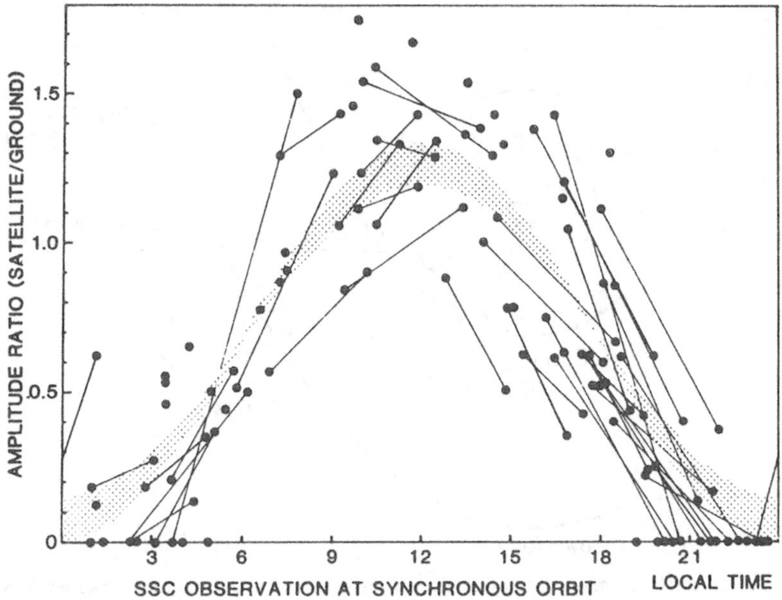

Fig. 8. Dependence of SSC amplitude at geosynchronous altitude on local time. This figure shows the results of an analysis by KOKUBUN (1983). The SSC amplitude has been normalized by taking the ratio to the corresponding amplitude at the Earth's surface, presumably eliminating the dependence on solar wind presure and reducing the scatter in the data. Simultaneous observations by pairs of satelites are connected by straight lines. Other features of the figure are discussed in the text.

is quiet before, and disturbed after, the SSC.

This event is one of many for which ISEE-3 observations of the shock are available. The change in field magnitude for the 6 July 1979 event is shown in the lower half of Fig. 11. The simultaneous field and plasma data have been analyzed successfully to obtain the shock parameters shown in the panel. The shock was quasi-parallel ($\theta_B=32°$) with a moderately large fast mode Mach number ($M_f=3.0$). Also shown in the upper half figure is a second event to be discussed below. This second shock is quasi-perpendicular ($\theta_B=83°$) and its waveshape contrasts nicely with the noisier quasi-parallel shock.

Figures 10 and 11 lead to the impression that field irregularities upstream of the quasi-parallel shock are not observed inside the magnetosphere by GOES 2 or 3, possibly because they are unable to propagate through the magnetopause. In any case, the magnetospheric field is very quiet prior to the onset of the SSC although it exhibits substantial fluctuations afterward.

The upper panel of Fig. 11 can be compared with Fig. 12 which contains GOES and surface data for the SSC of 30 Nov. 1979. By contrast, the magnetospheric field is noisy prior to the SSC, although the field upstream of the interplanetary shock is quiet as is common for quasi-perpendicular shocks. In this instance, local time effects appear to be important. The GOES observations were made in the post-midnight

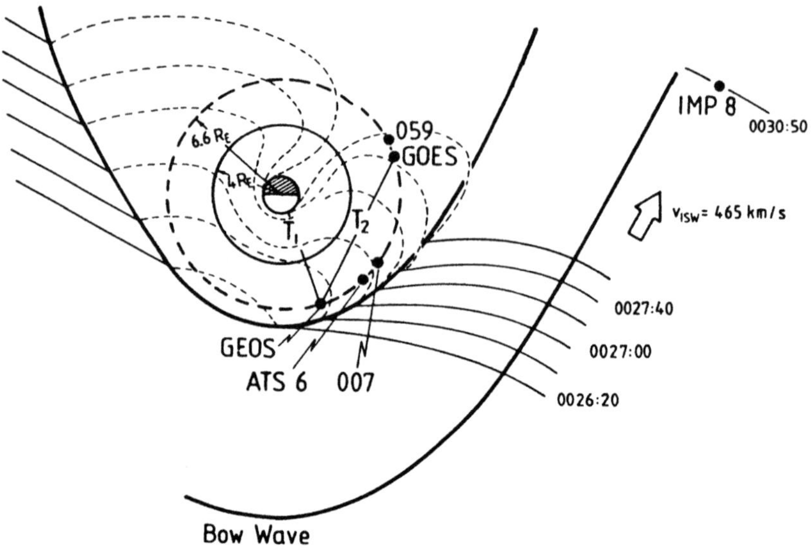

Fig. 9. Schematic of hydromagnetic wave propagation inside the magnetosphere for a specific SSC. This figure summarizes an analysis of the SSC of 29 July, 1977 by WILKEN *et al.* (1982) using multi-spacecraft observations. The bow shock, magnetopause and the locations of the spacecraft used in the study are shown. The dashed lines are a qualitative representation of the SSC wave front as it propagates tailward.

sector and show relatively slight increases in the average field, consistent with the dependence implied in Fig. 8. Furthermore, the noisly conditions preceeding the SSC onset were probably associated with auroral activity and are unrelated to the SSC.

These two examples are only indicative of the kinds of studies which are possible from a detailed comparison of the input and output functions for the large number of available cases.

7. Shocks, SSCs and Magnetic Storms

The final topic to be discussed is the possible relation of interplanetary shocks to the magnetic storms which accompany the SSCs. A close relationship need not and may not exist. Gradual commencement storms also occur whose development is at least similar to that of storms with sudden commencements. However, it is possible that the shock, the heated post-shock solar wind or the plasma ejecta which often accompanies the shock is effective in enhancing or producing distinctive features of the storm.

A feature of particular interest is the development of the ring current as indicated by the disturbance field, D_{st}. ISEE-3 plasma and field data supplemented by D_{st} and AE have been inspected for a fairly large number of magnetic storms and the following visual correlation has emerged.

There appears to be a correspondence between the intensification of the ring

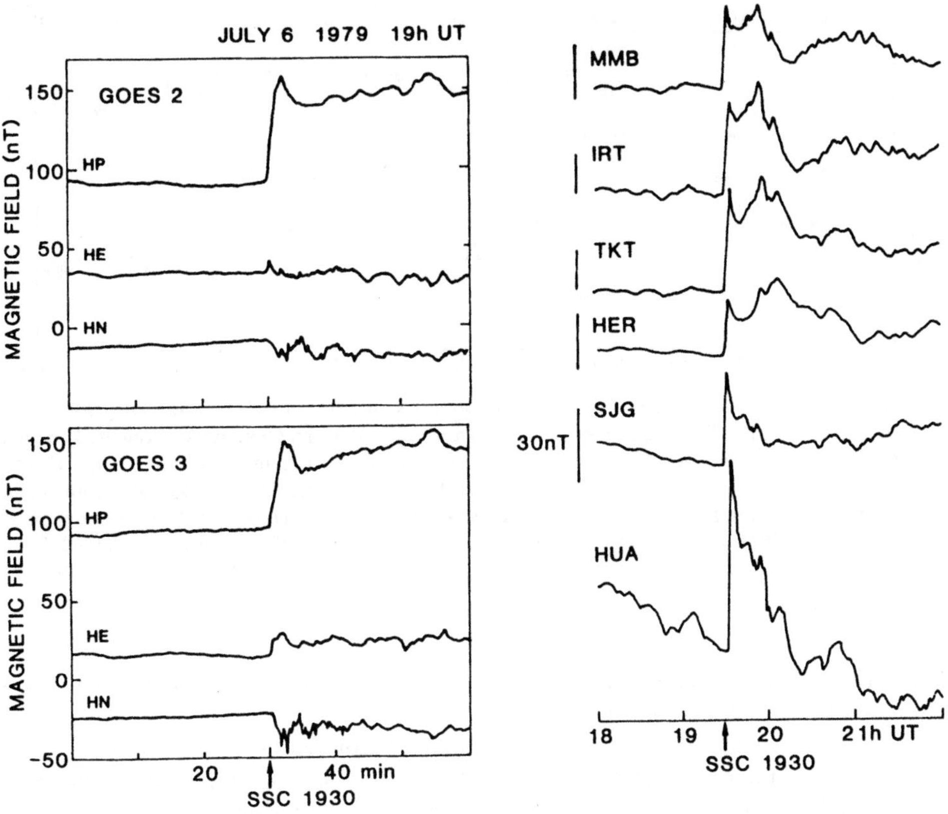

Fig. 10. SSC of 6 July 1979 observed at geosynchronous orbit and at the surface. These data were assembled by KOKUBUN (1983). The two GOES spacecraft were in the sunward hemisphere at the time and show nearly identical responses including oscillations which commence with the SSC. The field components are: *HP* is parallel to the Earth's rotation axis, *HE* is Earthward and *HN* is azimuthal and positive to the west.

current and two general aspects of the solar wind: the simultaneous presence of a large number of ions and a southward-directed magnetic field. The importance of a southward field has been noted previously by PATEL and WISKERCHEN (1975). This correlation is illustrated in Fig. 13 which contains the field magnetitude, B, the solar wind velocity, V_w, the density of He_{++}, N(He), the proton number density, N(H), the north–south component of the interplanetary field, B_z, and D_{st}. The shock was observed near 21 hours on 28 Sept. 1978 and can be identified in the plots as an increase in B, V, N(He) and N(H). There is no evidence of any enhancement of the ring current until slightly past 06 hours on 29 Sept. at which time D_{st} changes from a value of -50 nT to a value of approximately -200 nT. This intensification of the ring current is associated with increases in B, V, N(He) N(H) and a negative value of B_z, indicating the field has a strong southward component. These features are generally

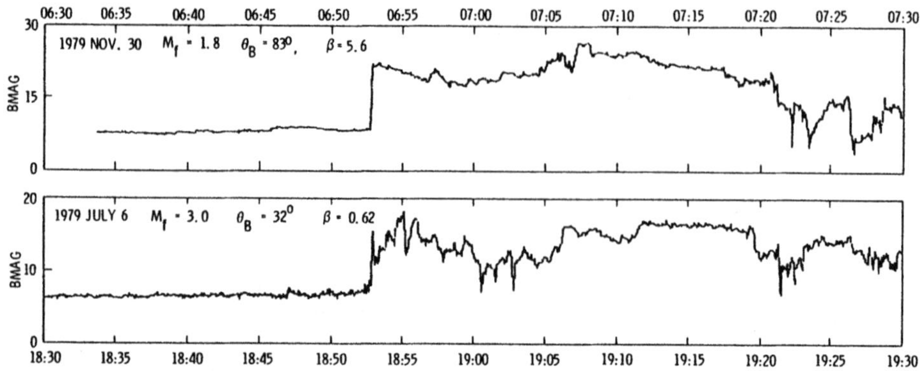

Fig. 11. Two ISEE-3 shocks associated with the two SSCs observed in 1979 on 6 July (Fig. 10) and 30 November (Fig. 12). The magnitude of the interplanetary field over a 1 hour interval centered on each shock is shown. The parameters customarily used to characterize interplanetary shocks (Mach number, M, angle between the shock normal and the upstream field direction, θ_B, and the upstream plasma beta, β) are shown. The shock in the upper panel is quasi-perpendicular with a low mach number but an unsually high beta. In the lower panel, the shock is quasi-parallel with a relatively large Mach number and a smaller than average beta. Thus, the two shocks represent an interesting contrast.

characteristic of plasma ejecta, and in fact, this event is one of several identified and discussed by ZWICKL *et al.* (1983).

The presence of large numbers of protons and alphas is not, however, a sufficient condition to enhance the ring current. Figure 14 shows another example, in the same format as Fig. 13, in which abnormally large and fluctuating values of N(H) and N(He) were observed. However, there is little, if any, effect on D_{st}. It is presumably significant that B_z is principally northward during this interval.

A southward magnetic field is also not sufficient to cause the ring current to intensify. Figure 15 shows an extended interval during which the field turns southward without any apparent effect on D_{st}. In this example, neither N(H) nor N(He) appears to be very large.

The impression gained from these and other examples is that the ring current is associated with enhanced ion densities and a southward field. A simple model that suggests itself is that large numbers of solar wind ions, including He^{++}, are injected into the magnetosphere and contribute strongly to the ring current. Their injection or penetration into the magnetosphere, which may involve their temporary storage in or passage through the magnetotail, apparently depends on the presence of the southward field, as expected if reconnection is important.

These few examples cannot be claimed to prove a general relation. More careful study will be required to determine whether these impressions are valid or not and whether or not the suggestions stemming from them have any merit. Even if the observations prove to be of general validity, futher analysis will be required to determine the role played by the shock. Although shocks are unlikely to favor the production of southward fields, they could be responseible for the heating and/or

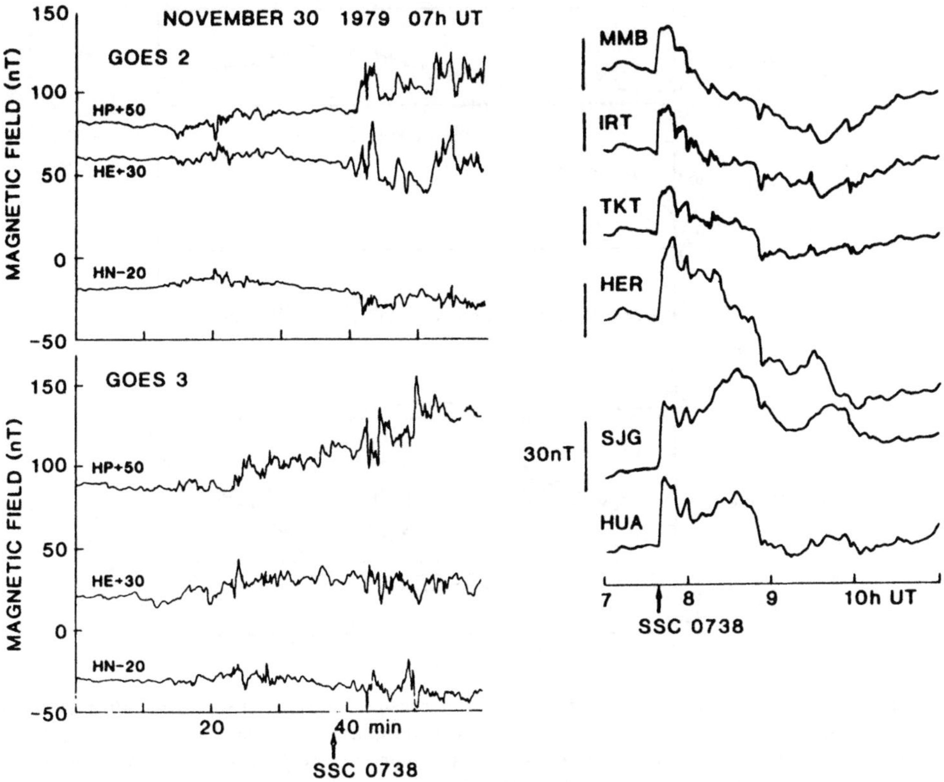

Fig. 12. SSC of 30 November 1979 as observed at Geosynchronous orbit and at the surface. This second example also appeared in KOKUBUN (1983). The magnetospheric response to the quasi-perpendicular shock (Fig. 11) is distinctly different than in Fig. 10 and is different at the two spacecraft. The observations were made in the anti-solar hemisphere during a noisy interval apparently associated with auroral activity. The three field directions are defined in the caption to Fig. 10. We infer that designations such as $HP+50$ indicate that 50 nT have been added to the true value of HP, etc.

acceleration of the ions which contribute to the ring current. This possibility could contribute to the rapid development of D_{st}. In many storms, the onset of D_{st} follows the SSC in a matter of minutes rather than of hours as in Fig. 13.

8. Summary

The ISEE mission is well suited to studies of the phenomena involved in interplanetary shocks and SSCs. Our reexamination of the statistics of this relationship has benefitted from nearly continuous monitoring of the solar wind throughout the recent solar maximum. In addition to ISEE- 1 and 2, a number of other spacecraft were making magnetospheric observations during the interval that the shocks identified by ISEE-3 impacted the magnetosphere.

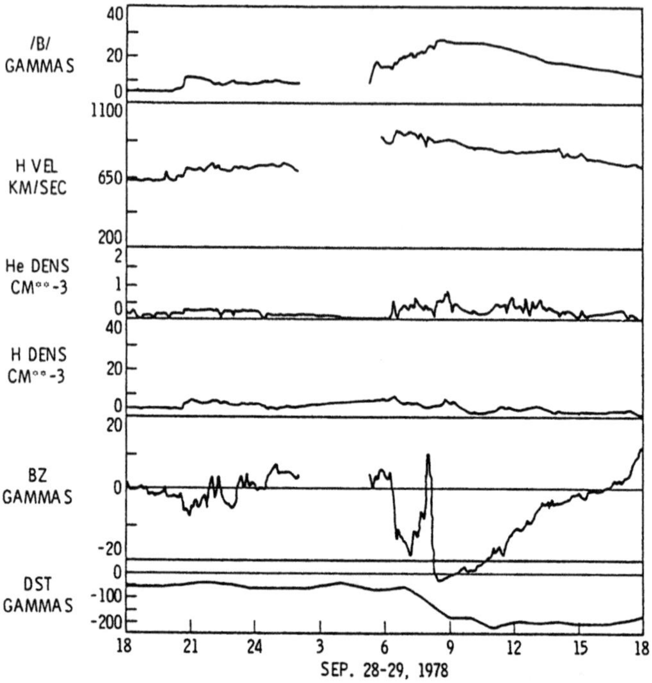

Fig. 13. Solar wind conditions during the sudden commencement storm of 29 September, 1978. The field and plasma parameters in the upper five panels are defined in the text. The lowest panel contains D_{st} which shows the development of the storm. The storm onset coincides with an enhanced density of protons N(H), and helium, N(He), as well as a strong southward-turning of the field.

The high degree of correlation between shocks and SSCs at the 80–90% level basically confirms earlier studies. At the conference, there was some discussion of the difficulties encountered by ground-based ovservers in deciding whether an SSC had occurred or not. The high degree of correlation shows that the "consensus approach" (i.e., a positive identification at several observatories) must be working reasonably well. No doubt, there is a twilight zone involving weak shocks and weak SSCs or noisy initial conditions in which ambiguities may lurk.

Our analysis has resolved one issue that was in doubt. The ISEE data confirm the expected relation between the amplitude of an SSC and the change in solar wind pressure across the shock. An earlier study had cast doubt on the validity of this relation but presumably suffered from limited numbers of cases and a mixture of SSCs and SIs. It would be interesting, and possibly informative, to examine sudden impulses and their interplanetary signatures during the ISEE epoch.

A good correlation has also been found between the jump in the internal plasma pressure (NkT) across the shock and SSC amplitude. It is not clear how best to interpret this result. Presumably, the different measures of shock strength are intercorrelated and it may be a matter of indifference as to which one is selected, but such

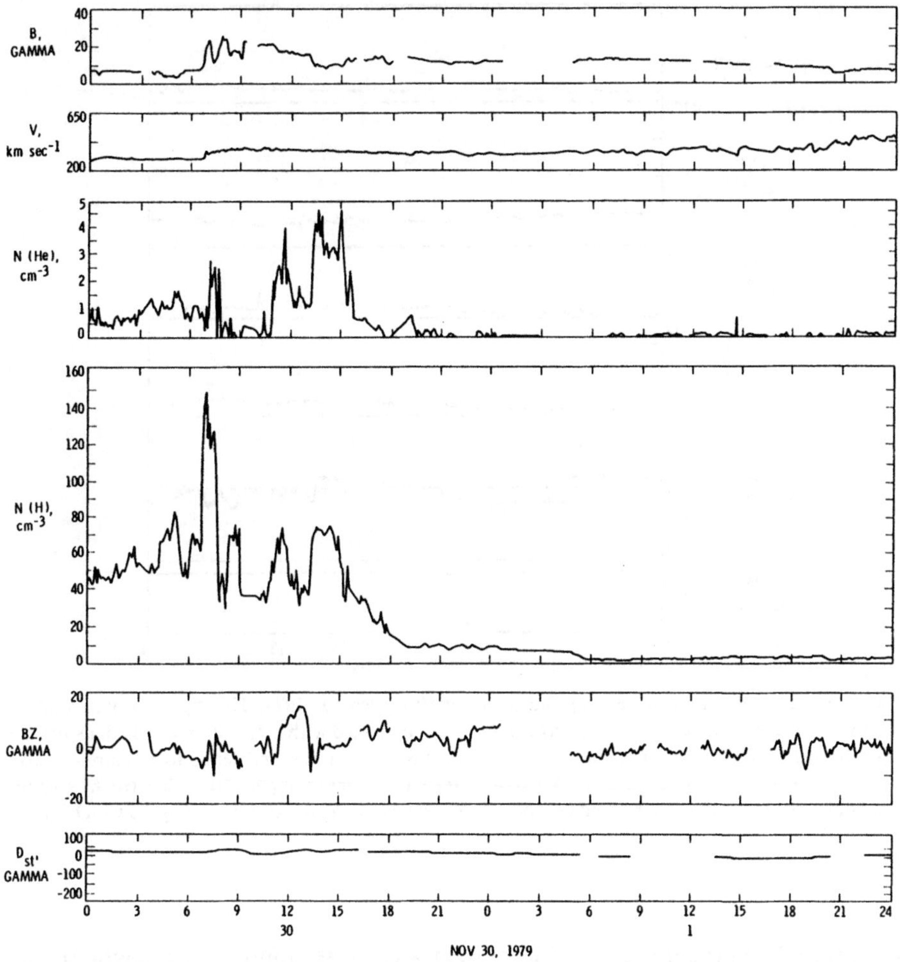

Fig. 14. Solar wind observations during the sudden commencement storm of 30 November, 1979. This event corresponds to the SSC shown in Fig. 12. It is not characterized by a significant D_{st}, implying little build-up of the mangetospheric ring current. In spite of this, abnormally large densities of protons and helium were present in the solar wind. The B_z component, however, shows the field to be northward almost continuously throughout the storm.

an expectation has not yet been established. Furthermore, it would be desirable, if possible, to derive a single basic shock parameter, such as Mach number, that is well correlated with SSC amplitude. Such a relation would enable studies of shocks during earlier epochs including many previous soalr cycles.

We have also reviewed a number of interesting recent studies of SSCs using earth-orbiting satellites. There is, however, a large amount of research involving SSCs that is still waiting to be done. SSCs provide a unique opportunity to conduct

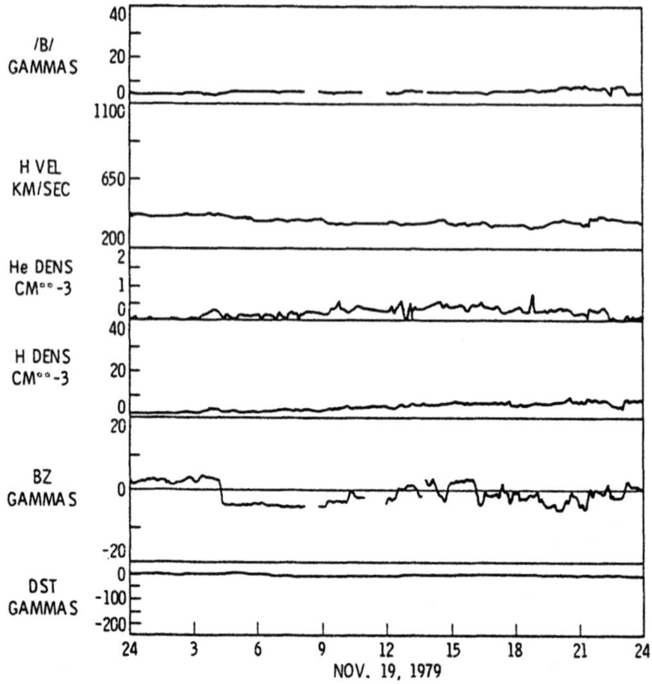

Fig. 15. Solar wind conditions during the storm of 19 November 1979. This figure provides another counter-example to the tendency for strong D_{st} to be correlated with both enhanced ion densities and a southward field. In this case, the field turns southward abruptly and tends to remain southward throughout the following interval. No significant enhancement appears in either the hydrogen or helium densities. There is a marked absence of ring current build-up as indicated by D_{st} which is almost featureless.

observational and theoretical studies of the transient response of the magnetosphere to a step function input. Simultaneous observations by ISEE-1, 2, GEOS and other magnetosphere satellites are available for study which could provide much new information about the propagation of large amplitude hydromagnetic waves inside the magnetosphere as well as the coupling between changes in the magnetopause and changes in the ionosphere and magnetotail.

Finally, we have suggested that the storms which accompany SSCs may have unique aspects which have been largely overlooked in the present strong emphasis on substorms.

S. Akasofu prepared the original plots from which Figures 13, 14, 15 were selected. He also pointed out the principal reasons why no correlation is to be expected in Fig. 7.

This article reports work carried out by the Jet Propulsion Laboratory, California Institute of Technology, under contract with the National Aeronautics and Space Administration.

REFERENCES

ARAKI, T. and J. H. ALLEN, Latitudinal reversal of polarization of the geomagnetic sudden commencement, *J. Geophys. Res.*, **87**, 5207, 1982.

BAUMJOHANN, W., H. JUNGINGER, G. HAERENDEL, and O. H. BAUER, Resonant Alfven waves excited by a sudden impulse, *J. Geophys. Res.*, **89**, 2765, 1984.

BURLAGA, L. F. and K. W. OGILVIE, Causes of sudden commencements and sudden impulses, *J. Geophys. Res.*, **74**, 2815, 1969.

CHAO, J. K. and R. P. LEPPING, A correlative study of SSC's, interplanetary shocks, and solar activity, *J. Geophys. Res.*, **79**, 1799, 1974.

CHAPMAN, S. and V. C. A. FERRARO, A new theory of geomagnetic storms, *Terrest. Magn. Atmosph. Elect.*, **38**, 79–96, 1933.

CHOE, J. Y. and D. B. BEARD, The compressed geomagnetic field as a function of dipole tilt, *Planet. Space Sci.*, **22**, 595–608, 1974a.

CHOE, J. Y. and D. B. BEARD, The near magnetic field of the magnetotail current, *Planet Space Sci.*, **22**, 609–615, 1974b.

GOLD, T., *Gas Dynamics of Cosmic Clouds*, Chapter 17, p. 103, North-Holland Publishing Company, Amsterdam, 1955.

GOSLING, J. T., J. R. ASBRIDGE, S. J. BAME, A. J. HUNDHAUSEN, and I. B. STRONG, Discontinuities in the solar wind associated with sudden geomagnetic impulses and storm commencements, *J. Geophys. Res.*, **72**, 3357, 1967.

KAUFMANN, R. L. and D. N. WALKER, Hydromagnetic waves excited during an SSC, *J. Geophys. Res.*, **79**, 5187, 1974.

KOKUBUN, S., Characteristics of storm sudden commencement at geostationary orbit, *J. Geophys. Res.*, **88**, 10,025, 1983.

NISHIDA, A., Ionospheric screening effect and storm sudden commencements, *J. Geophys. Res.*, **69**, 1861, 1964.

PATEL, V. L. and M. J. WISKERCHEN, Interplanetary field and plasma during initial phase of geomagnetic storms, *J. Geomag. Geoelectr.*, **27**, 363–382, 1975.

SISCOE, G. L., V. FORMISANO, and A. J. LAZARUS, Relation between geomagnetic sudden impulses and solar wind pressure changes—an experimental investigation, *J. Geophys. Res.*, **73**, 4869, 1968.

SONETT, C. P., D. S. COLBURN, L. DAVIS, Jr., E. J. SMITH, and P. J. COLEMAN, Jr., Evidence for a collision-free magnetohydrodynamic shock in interplanetary space, *Phys. Rev. Lett.*, **13**, 153, 1964.

SUGIURA, M., T. L. SKILLMAN, B. G. LEDLEY, and J. P. HEPPNER, Propagation of the sudden commencement of July 8, 1966, to the magnetotail, *J. Geophys. Res.*, **73**, 6699, 1968.

TAMAO, T., Unsteady interactions of solar wind disturbances with the magnetosphere, *J. Geophys. Res.*, **80**, 4230, 1975.

TAYLOR, H. E., Sudden commencement associated discontinuities in the interplanetary magnetic field observed by Imp 3, *Solar Phys.*, **6**, 320, 1969.

WILKEN, B., C. K. GOERTZ, D. N. BAKER, P. R. HIBGIE, and T. A. FRITZ, The SSC on July 29, 1977 and its propagation within the magnetosphere, *J. Geophys. Res.*, **87**, 5901, 1982.

ZWICKL, R. D., J. R. ASBRIDGE, S. J. BAME, W.D C. FELDMAN, J. T. GOSLING, and E. J. SMITH, Plasma properties of driver gas following interplanetary shocks observed by ISEE-3, in *Solar Wind Five*, NASA Conference publication 2280, pp. 711–717, 1983.

Solar Wind-Magnetosphere Coupling, edited by Y. Kamide and J. A. Slavin, 367–380.

Global Auroral Responses to Magnetospheric Compressions by Shocks in the Solar Wind: Two Case Studies

J. D. CRAVEN[1], L. A. FRANK[1], C. T. RUSSELL[2], E. J. SMITH[3], and R. P. LEPPING[4]

[1]*Department of Physics and Astronomy, The University of Iowa, Iowa City, Iowa 52242, U.S.A.*
[2]*Institute of Geophysics and Planetary Physics, University of California at Los Angeles, Los Angeles California 90024, U.S.A.*
[3]*Jet Propulsion Laboratory, California Institute of Technology, Pasadena, California 91109, U.S.A.*
[4]*Laboratory for Extraterrestrial Physics, NASA/Goddard Space Flight Center, Greenbelt Maryland 20771, U.S.A.*

Two case studies are presented of the global auroral responses to shocks in the solar wind at Earth. The z component of the interplanetary magnetic field, B_z, is negative ahead and behind the first shock and positive for the second case. A sudden-commencement geomagnetic storm develops in each case, with maximum $D_{st} \lesssim 190$ nT. An immediate auroral response is detected at all longitudes around the auroral oval, in which auroral luminosities increase by a factor of 2–3 with the first samples after each sudden commencement. The time delay in obtaining the first sample varies with local time from ~1 to 18 minutes. No other significant variations in the aurora are associated with the immediate response. Beginning ~30 minutes after each sudden commencement, the aurora becomes active and displays significant variations in its luminosity and spatial distribution. For $B_z<0$ an intense substorm develops. A sun-aligned transpolar arc forms when $B_z>0$, appearing first at local midnight as a polar arc and then lengthening sunward from the auroral oval across the polar cap to noon at an average speed of ~1 km/sec.

1. Introduction

Variations in solar-wind dynamic pressure and orientation of the interplanetary magnetic field (IMF) alter the magnitudes and distributions of currents around and within the magnetospheric cavity. One well-established signature of changes in these currents is an increase or decrease in the horizontal component of the magnetic field detected with magnetometers at low-latitude ground stations. For example, the magnetic field at all local times begins to increase abruptly, in ≲1 minute, in response to the collision of an interplanetary shock with the magnetosphere (e.g., NISHIDA, 1978). The magnitudes of these increases are typically tens of nanoteslas. If no further changes occur in the solar wind or within the magnetosphere, the increase in field strength is identified as a sudden (positive) impulse. Of further significance is a rapid compression followed by a geomagnetic storm, in which the ring current undergoes a rapid (hours) main phase enhancement and auroral activity increases dramatically at higher latitudes. In this case, the low-latitude increase in the magnetic field is identified as the sudden commencement, or SC, of a geomagnetic storm. Unfortunately, the detection of changes in the surface magnetic field associated with interplanetary discontinuities can be ambiguous (e.g., BURLAGA and OGILVIE, 1969), and a unique determination of the type of discontinuity can not be obtained from a clearly observed ground signature (e.g., TAYLOR, 1969; CHAO and LEPPING, 1974).

There is also an immediate response to a shock impact in the form of enhanced energetic (tens of keV) electron precipitation along the auroral oval. These increases are observed with riometers and balloon-borne x-ray detectors (e.g., BROWN *et al.*, 1961; MATSUSHITA, 1961; ORTNER *et al.*, 1962; ULLALAND *et al.*, 1970). Increases in auroral luminosities are also reported (VOROB'YEV, 1974). The enhancements can begin simultaneously ($\lesssim 1$ min) with SC onsets and last for ~3–10 minutes. The magnitudes of the enhancements vary from ~10% to factors of >2, depending on the method of detection, and also vary from event to event. These immediate, brief increases are not associated with auroral substorms.

Many investigators have studied the response of the auroral electrojet to impacts by interplanetary discontinuities on the magnetosphere (WILSON and SUGIURA, 1961; KAWASAKI *et al.*, 1971; BURCH, 1972; KOKUBUN *et al.*, 1977), with an emphasis on the triggering of auroral substorms. In brief summary, it appears that a substorm can be initiated following a shock impact for which the magnitude of the SC exceeds ~10 nT, and for which the IMF z-component, B_z, is negative for ~30 minutes before the SC (BURCH, 1972). However, the magnitude of the SC for triggering a substorm may depend on other parameters of the solar cycle (KAWASAKI *et al.*, 1971) and the amplitude of the IMF (BURCH, 1972). Extended periods of B_z positive prior to a SC will in general not lead to substorms (BURCH, 1972). BURCH (1972) and KOKUBUN *et al.* (1977) interpret their observations to indicate that the state of the magnetosphere at the time of arrival of a shock is important in determining the magnetospheric response: pre-conditioning, such as in a growth phase, is important. Kokubun and colleagues conclude that substorm initiation remains an internal magnetospheric process accelerated by the shock-induced compression of the magnetosphere. In contrast, AKASOFU and CHAO (1980) argue that substorms which can follow shock impacts and SCs are the result of increases in the rate of direct energy transfer by the solar wind-magnetospheric dynamo. For no change in IMF direction, the rate of energy transfer increases with increasing V and B across the shock. The more southward B, the greater the coupling and energy dissipation.

The maximum time delays from SCs to auroral substorm onsets used in the various investigations have not been identical. For example, KAWASAKI *et al.* (1971) appear to use ±3 minutes, while KOKUBUN *et al.* (1977) allow delays of up to 15 minutes in defining simultaneity. No unique relation is known for the time of substorm onset after a SC.

We wish to pursue further the auroral responses to the arrival of interplanetary shocks using time sequences of global auroral images routinely provided by the imaging photometers on board Dynamics Explorer 1 (DE 1) (FRANK *et al.*, 1981a). Our objective is to monitor auroral luminosities prior to and following SCs for which the sign of B_z is the same on each side of the shock associated with the SC. For the first case, the IMF is directed well southward, and in the second case it is northward. In each case there is an immediate response followed in ~30 minutes by the onset of significant variations in luminosities and spacial distributions. For $B_z<0$, an auroral substorm is observed. For the case of $B_z>0$, sun-aligned, polar-cap auroral forms brighten, and a transpolar arc of a theta aurora is seen to develop initially at local midnight.

2. Observations

2.1. Magnetic activity

The time histories of D_{st} and K_p (Solar-Geophysical Data, NOAA) for 20–22 October 1981 (Fig. 1) reveal the presence of two small geomagnetic storms with peak D_{st} value ≤ 190 nT. The first storm is preceeded by an 11-nT SC reported at 1309 UT on 20 October only by the Honolulu ground magnetometer. A rapid main-phase decrease follows in the surface magnetic field. Auroral activity in the three-hour interval from 1200 to 1500 UT which includes the SC is characterized by $K_p=7_-$. The K_p values increase to 7_0 in the next two three-hour intervals, and then decrease with the recovery phase of the geomagnetic storm. As shown in Fig. 2(a), the *AE* index is >300 nT after 0600 UT, and after ~1100 UT semi-periodic, large increases to $\gtrsim 1000$ nT are detected at ~two-hour intervals until ~2000 UT. It will be shown in the next section that the IMF is predominantly southward for this event (See Fig. 3). The downward direction of the arrow locating the SC at 1309 UT in Fig. 1 denotes this predominant direction.

The second geomagnetic storm is preceded by a significant SC detected at more than 16 stations on 22 October 1981 (see Fig. 1). Mean onset time for the SC is 0524:45 UT. A maximum SC amplitude of 88 nT is reported by Toolangi (magnetic latitude of

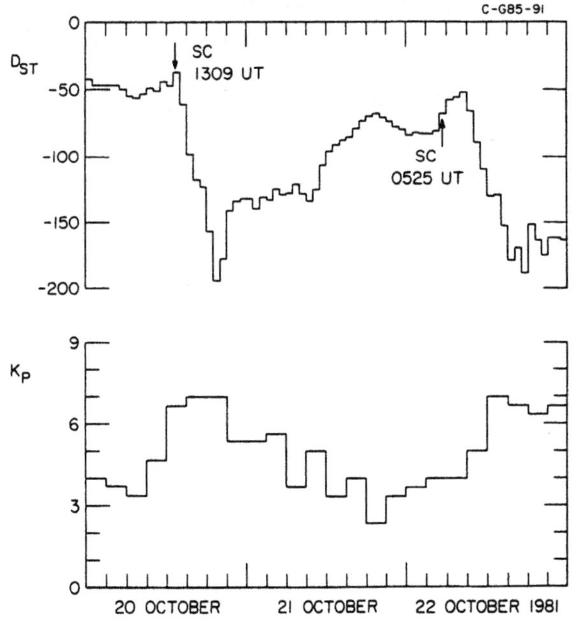

Fig. 1. Hourly averages of D_{st} and the three-hour values of K_p for 20–22 October 1981. The onset time for each sudden commencement (SC) of interest here is identified.

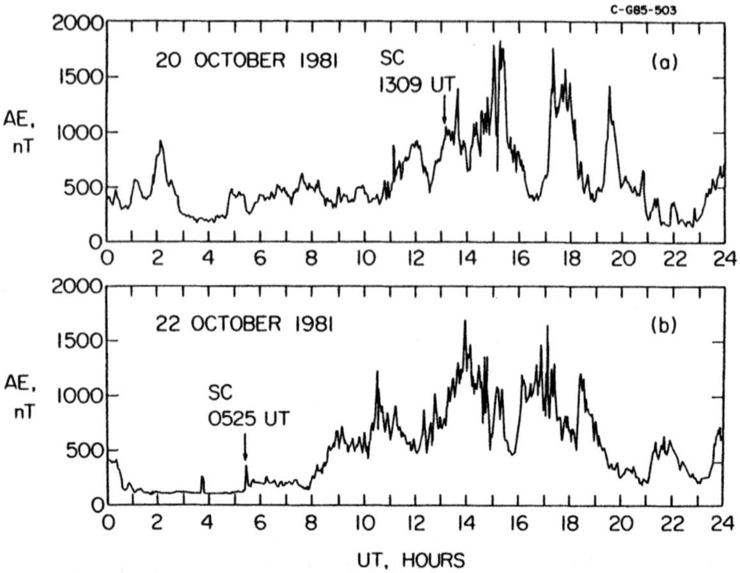

Fig. 2. (a) The *AE* index for 20 October 1985. (b) The *AE* index for 22 October 1985. Times for the SCs of interest here are identified.

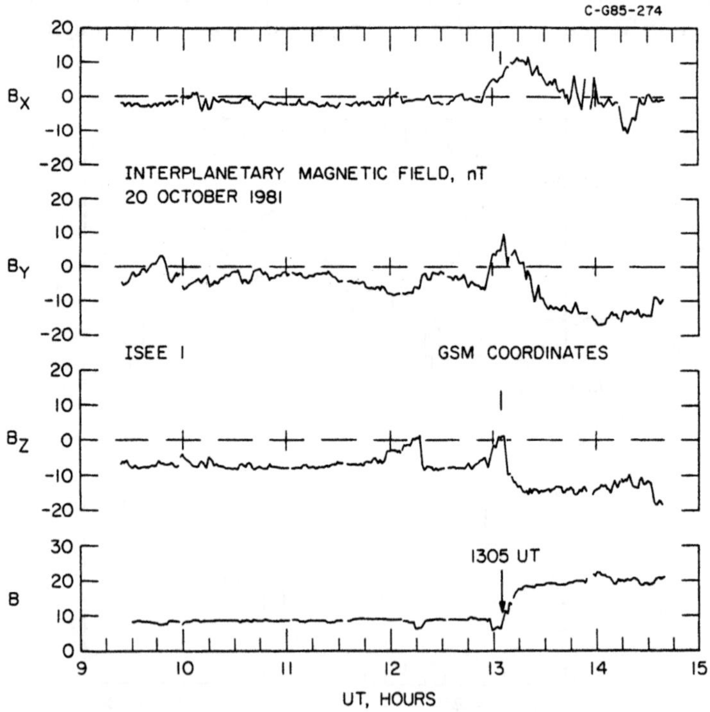

Fig. 3. Magnitude of the IMF and its components in GSM coordinates as observed with ISEE 1 on 20 October 1981 from 0920 to 1440 UT.

$-46.7°$) for the horizontal component of the magnetic field. At Honolulu the reported amplitude is 22 nT. The K_p index is 4_0 from 0300 to 0900 UT, and then increases with the delayed onset of the main-phase decrease. Maximum K_p values of 7_0 are reached, identical to those for the earlier storm. The *AE* index is $\lesssim$125 nT in the 4 hours before the SC, and increases to ~190 nT for ~2 hours after the SC (see Fig. 2(b)). A brief (~3 min) increase to ~360 nT is initiated within one minute of the SC. Significant magnetic activity with $AE>1000$ nT is present after ~1000 UT, at about the same time that the main-phase decrease develops. It will be shown that the IMF is predominantly northward for this event (see Fig. 5). The upward direction of the arrow locationg the SC at 0525 UT in Fig. 1 denotes this predominant direction.

2.2 *Event of 20 October*

Observations with the ISEE-1 and -3 spacecraft demonstrate that B_z is predominantly negative for more than 12 hours prior to and for more than 6 hours after the SC of 20 October 1981. A portion of these observations with ISEE 1 is presented in Fig. 3 for the period 0920–1440 UT to show that $|B_z| \sim B$ in the several hours prior to arrival of the solar-wind discontinuity at ~1300 UT, and that B_z is large and negative after passage of the discontinuity. Geocentric solar magnetospheric (GSM) coordinates are used. The time history of ground magnetic activity described previously is in full accord with these variations in B and B_z. Two points of note which make this a less than ideal case study for $B_z<0$ are (1) the brief ~10 minute, depression in B and concomitant rotation in IMF direction beginning at ~1258 UT, during which B_z is positive, and (2) the slow, ~10 minute, ramp in magnitude from ~6 nT to ~18 nT after 1305 UT. The IMF direction returns southward at ~1307 UT.

Analysis of the solar-wind plasma for the time interval around 1300 UT reveals abrupt increases in bulk speed, density and temperature at 1305:30 UT (±30 sec), coincident with the initial rapid increase in the magnitude of the IMF. The bulk flow speed and ion temperature increase from ~600 km/sec and $\sim 2\times 10^5$ K to ~660 km/sec and $\sim 3\times 10^5$ K, respectively, and the relative density increases by a factor of ~1.6 (J. T. Gosling, private communication). We identify the discontinuity at 1305 UT as an interplanetary shock, which when detected at ISEE 1 is embedded within a weaker solar-wind discontinuity for which the most prominant signature is a rotation of the IMF direction. The ~3.5-minute delay between detection of the shock with ISEE 1 at GSM coordinates (20.5, −7.5, −0.10 R_E) and the SC at Honolulu is consistent with the ~660 km/sec (6.2 R_E/min) speed of the shock in the solar wind, for a radially directed shock normal, and the greater Alfven speeds of propagation within the magnetosphere.

Auroral imaging of northern polar latitudes in the time intervals 0226–0718, 0917–1054 and 1200–1425 UT on 20 October 1981 shows that auroral activity is also consistent with the time history of AE in this period of $B_z<0$. The 12 vacuum-ultraviolet (VUV) images of the latter time interval are presented in Plate 1 using a false-color format in which weaker luminosities of several kilorayleighs (kR) are coded in red and the greater luminosities of the active aurora and from the sunlit hemisphere (~20 kR) are coded gold to yellow. Below each image is the UT for the beginning of the 12-minute telemetering period for the image. During this observing period, the VUV-wavelength photometer operates in a mode for which an image is obtained with one filter and the next image is scanned with a companion filter of different passband. The two-filter sequence repeats cyclically. The first image, at upper left in the plate, has been gained using a passband which extends from 120 to 155 nm. Above Earth's limb, the photometer responds solely to solar Lyman-α radiation resonantly scattered by exospheric hydrogen (RAIRDEN *et al.*, 1983). At auroral latitudes the principal emissions are due to atomic oxygen (OI) at 130.4 and 135.6 nm. The passband for the second image is 123–155 nm, which removes contributions to the photometer response by Lyman-α radiation while maintaining high sensitivity to emissions from OI. Slightly reduced apparent brightness of the sunlit hemisphere for images of the first and fourth columns in the plate arises from the photographic processing.

During the 66-minute break in telemetry from 1054 to 1200 UT an auroral substorm onset and associated expansion are detected by ground magnetometers, beginning at ~1100 UT. Maximum $AE \simeq 900$ nT is recorded at 1200 UT as imaging is resumed. The first images of Plate 1 record the late recovery phase of this substorm. A discrete auroral form is clearly visible in the evening-to-midnight sector at the poleward edge of the auroral oval, and weaker, more diffuse auroral distributions are observed to extend around the auroral oval at lower latitudes in the night sector. Auroral luminosities are enhanced at the latitude of the polar cusp in the local-noon sector.

Careful comparisons of the first six images (1200 to 1313 UT) with the next two images (1313 to 1337 UT) show that auroral luminosities are greater around the auroral oval after the SC at 1309 UT (see frame at 1313 UT). This increase is seen more clearly in Fig. 4(a), where maximum absolute intensities are plotted for the local noon, dusk and midnight sectors as a function of the sample time at each local time. In the dusk sector, these maxima in emissions arise from the discrete auroral forms at the poleward edge of the auroral oval while at local midnight the emissions arise from diffuse emissions at more equatorward latitudes. Contributions from solar Lyman-α radiation and dayglow have been removed in determining the absolute luminosities. Prior to 1308 UT, luminosities around the oval vary between ~3 and 6 kR. The last sample at local noon (at 1308 UT) before the SC does not indicate an increase in luminosities. One minute later, at 1309 UT, samples in the dusk sector provide the first indication of an auroral response, with luminosities increasing from ~3.5 to 6 kR. From 1317–1335 UT luminosities of ~8.5–12 kR are observed in the dusk and midnight sectors. At local noon, luminosities increase to ~20 kR before the polar cusp blends into the dayglow near the limb. We refer to this increase as the immediate auroral response to magnetospheric compression. With the ninth image of the sequence in Plate 1, beginning at 1337 UT, an intense and rapidly developing substorm appears near the equatorward edge of the oval at midnight. Maximum luminosities exceed 40 kR. The brief 500-nT increase in AE after 1330 UT (Fig. 2(a)) is associated with the expansive phase of this substorm.

2.3. *Event of 22 October*

Observations of the IMF prior to and during the onset of the geomagnetic storm

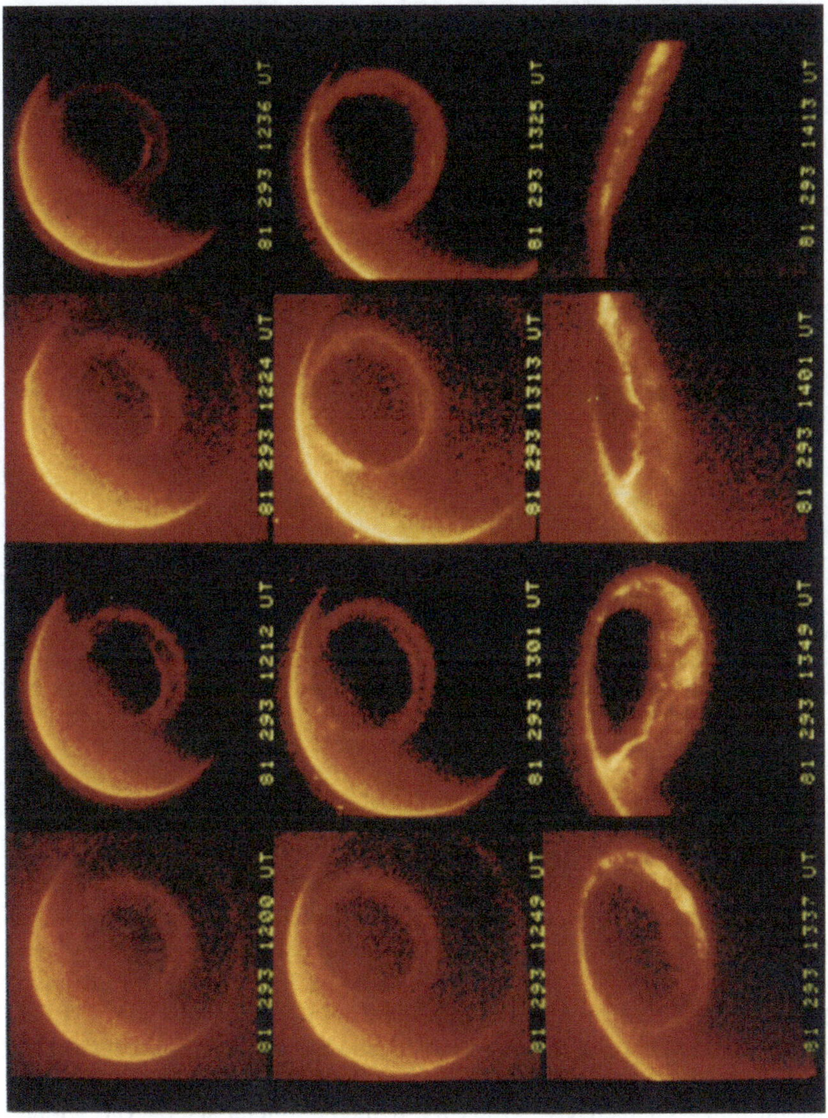

Plate 1. Sequence of 12 consecutive false-color auroral images at ultraviolet wavelengths in the time interval 1200 through 1425 UT on 20 October 1981. Below each image is the UT for the beginning of the 12-minute telemetering period for the image. Increased luminosities following the SC at 1309 UT are seen with the seventh image at 1313 UT. Substorm onset is first detected with the ninth image at 1337 UT. Predominant direction of the IMF B_z component is southward. See the text for a discussion of ultraviolet filters used.

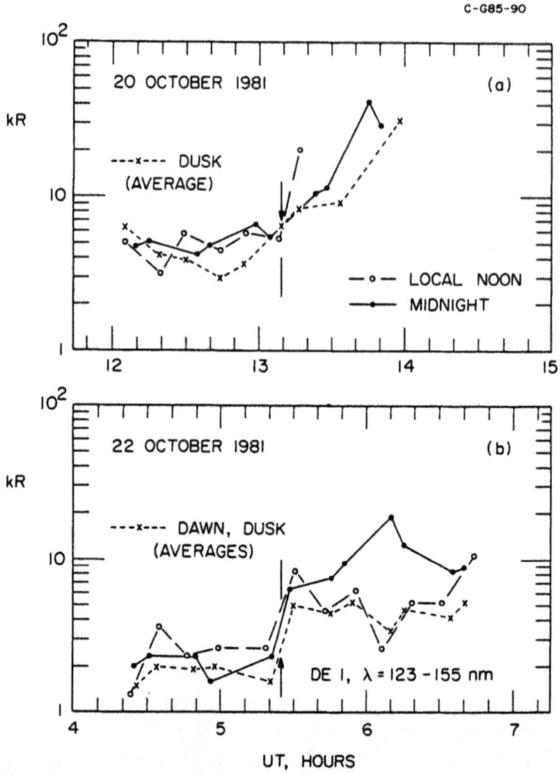

Fig. 4. (a) Maximum emission intensities of atomic oxygen at 130.4 and 135.6 nm along the auroral oval in the noon, dusk and midnight sectors for the ~2-hour time interval surrounding the SC at 1309 UT on 20 October 1981. (b) Continuation of Fig. 4(a) for 22 October 1981 and the SC at 0525 UT. Averages are given for the intensities in the dawn and dusk sectors.

on 22 October 1981 are complicated by the absence of ISEE-1 measurements before the SC and a gap in telemetry from ISEE 3 during passage of the discontinuity. Direct observations near Earth coincident with the SC are available with IMP 8, but again, telemetry is brief in duration. Observations of the IMF available in the period 0215 through 0750 UT are presented in Fig. 5. Telemetry transmission times (UT) are given for ISEE 1 and IMP 8. For ISEE 3, times are advanced by 36 minutes as determined by matching time profiles of the magnetic components of ISEEs 1 and 3 after ISEE 1 re-enters the solar wind at ~0640 UT. Using this time base, B_z turns northward at ~2354 UT on 21 October, more than 5 hours before the SC. The leading edge of the solar-wind discontinuity is detected with the IMP-8 magnetometer at 0528 UT, which is ~3 minutes after onset of the SC. The magnetic field turns southward after ~0830 UT, and is followed by the onset of the main-phase decrease for the geomagnetic storm and increasing values for K_p in the next two 3-hour intervals (Fig. 1). The amplitude of AE begins to increase after ~0800 UT in response to the decreasing positive amplitude of the B_z component of the IMF (see Fig. 2(b)).

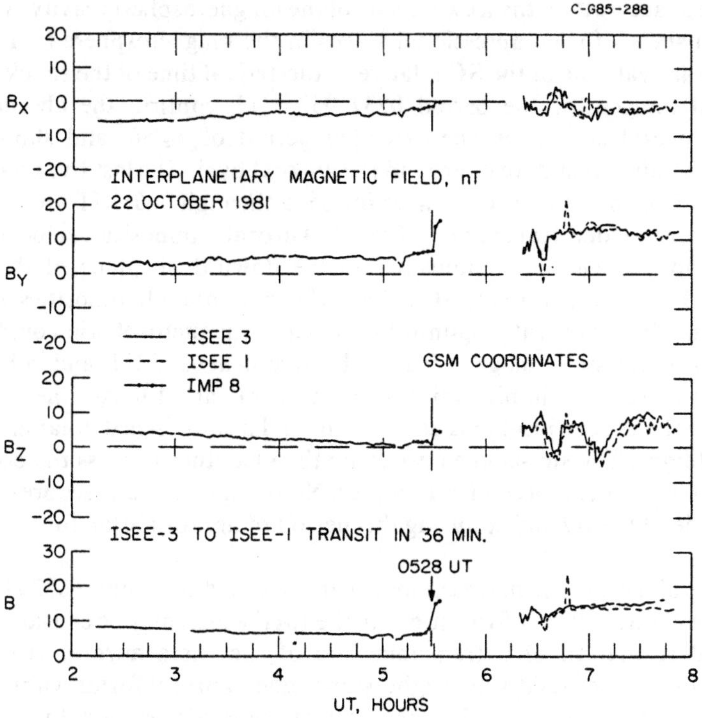

Fig. 5. Continuation of Fig. 3 for 22 October 1981 from 0215 to 0750 UT, using observations with ISEEs 1 and 3 and IMP 8. For ISEE 3, time is advanced by 36 minutes as determined by matching time profiles of the magnetic components of ISEEs 1 and 3 after ISEE 1 re-enters the solar wind at ~0640 UT.

Plasma parameters for the solar wind measured on either side of the discontinuity are consistent with an identification of the discontinuity as a shock: bulk velocity increases from 455 to 575 km/s, ion density from 5.5. to ~14 cm^{-3} and ion temperature from ~5.5×10^4 to ~3.0×10^5 K (A. J. Lazarus, private communication). The solar-wind speed of 575 km/s behind the shock and a distance of 198 R_E to ISEE 3 within ~6° of the sun-earth line yields a time delay of 37 minutes, in agreement with that determined above. In the slower solar wind ahead of the shock, the time delay for radial propagation is 46 minutes. The approximate 3-minute delay between SC onset and detection of the shock at IMP 8 is believed to be due in part to the orientation of the shock normal relative to the location of the spacecraft at GSM coordinates (−2.4, 22.8, −30.8 R_E). Using averages of the IMF magnetic field measurements obtained just before and after the shock, and applying the coplanarity theorem, the shock normal direction is (−0.937, 0.293, −0.192) in GSM coordinates: the shock normal is oriented ~10° downward and ~20° toward dusk away from the antisolar direction. The angle between the shock normal and geocentric spacecraft position is 67°. Hence, the shock interacts with the magnetospheric cavity before reaching the spacecraft on the leeward side of the magnetospheric cavity. The greater propagation speeds for compressional waves in the magnetosphere (~1000 km/s) advance the arrival time of the SC relative to the arrival time of the shock at IMP 8.

Auroral images for the period 0237–0525 UT confirm the absence of any important auroral activity in the extended period of $B_z>0$, and show that the emissions are uniform and weak around the auroral oval. The last five images in this time interval and the next seven images for 0525 through 0651 UT are displayed in Plate 2, in a format identical to that of Plate 1. Auroral luminosities at local noon and midnight, and averages of luminosities in the dawn-dusk plane of the oval are presented in Fig. 4(b). Before the SC at 0525 UT, maximum luminosities are 1.5 to 3 kR. The immediate auroral response to the rapid magnetospheric compression is clearly visible in the sixth image of Plate 1, beginning at 0525 UT, and in Fig. 4(b) by the step-like increase in luminosities to ~5 to 8 kR at all local times. About 30-minutes later the aurora brightens at midnight and a sun-aligned polar arc begins to form at midnight. The sun-aligned polar arc then lengthens across the polar cap to local noon at an average speed of ~1 km/sec. No significant increases are observed in the magnitude of the *AE* index during this period of arc formation following the SC (Fig. 2(b)).

The lack of measurements in the interplanetary medium from ~0535 to 0620 UT prohibits the identification of variations in the IMF which may be causally related to this unusual formation of a transpolar arc of the theta aurora. However, the continuing presence of relatively bright, sun-aligned auroral forms within the polar cap throughout the interval of the gap in telemetry is consistent with B_z remaining positive. A southward turning of the IMF is expected to cause a decrease in luminosities within the polar cap (FRANK *et al.*, 1985).

3. Discussion

Two case studies of the auroral responses to magnetospheric compressions by shocks in the solar wind identify an immediate response within 1–10 minutes of the SC. This response is independent of the sign of B_z and is observed at all local times around the auroral oval. There seems little doubt that this global response is related to the 'direct' auroral response observed from the ground over a wide range of longitudes by VOROB'YEV (1974) within ≲2 minutes of SCs. However, durations and amplitudes for the increases in auroral luminosities observed by Vorob'yev are ~3–8 minutes and ~25%, respectively. This short duration is similar to the duration of initial response to SCs observed in auroral x-rays and with riometers (e.g., BROWN *et al.*, 1961; MATSUSHITA, 1961; ORTNER *et al.*, 1962; ULLALAND *et al.*, 1970), and to the ~3-minute, ~230-nT increase in *AE* observed here for the SC of 22 October 1981 ($B_z>0$ case). In contrast, the immediate auroral responses presented here are increases by factors ≳2 in auroral luminosities over durations of greater than 10–20 minutes ($B_z>0$). For the case of $B_z<0$, no decrease in luminosities is observed before substorm onset (see Fig. 4(a)). We have not investigated here, using simultaneous auroral observations with DE at visible wavelengths, any dependence of the immediate auroral response on the energy of the precipitating electrons.

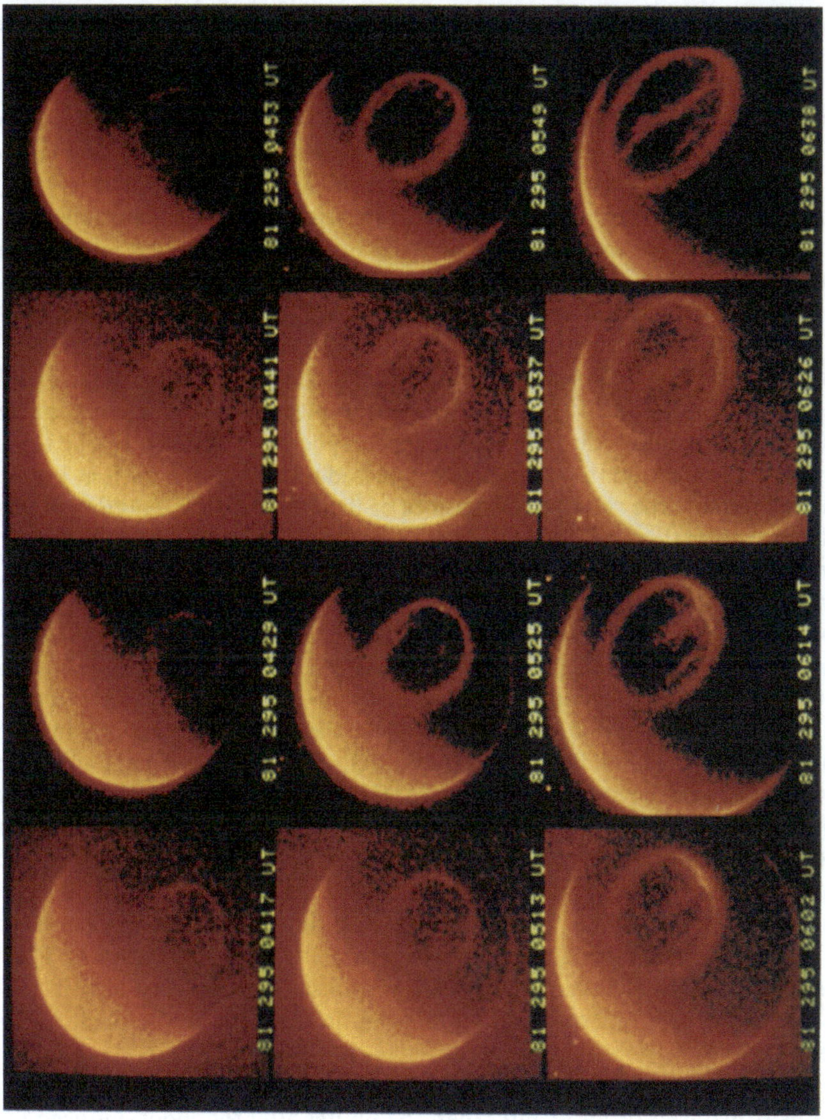

Plate 2. Continuation of Plate 1 for the time interval 0417 through 0650 UT on 22 October 1981. The SC occurs at 0525 UT with the beginning of the sixth frame. Predominant direction of the IMF B_z component is northward.

The amplitude-time profile for the immediate auroral response is probably related to the speed and orientation of the IMF discontinuity and the magnitude of the increase in solar-wind momentum flux. For example, at 600 km/sec (5.6 R_E/min), a discontinuity propagating in the antisunward direction will envelop the magnetosphere from $X=10$ to -30 R_e in ~7 minutes. Faster propagation at non-radial orientations may speed the global onset and shorten the interaction time. For the relatively slow shocks discussed here, the higher Alfven speeds within the magnetospheric cavity (~1000 km/s) may also influence the onset times for magnetic and auroral activity. The increases in solar-wind momentum flux for the two shocks are by factors of 1.8 ($B_z<0$ on 20 October) and 4.1 ($B_z>0$ on 22 October). The second, larger increase is associated with the more widely reported SC of greater amplitude.

While the latitudinal response of the auroral emissions has not been studied here quantitatively, the auroral images show that the pre-existing discrete and diffuse auroral forms both brighten after the SC. Surveys of the DE-1 auroral images and the IMP-8 and ISEE-1 and -3 measurements of the IMF show that the immediate increases in auroral luminosities around the auroral oval are routinely detected in association with abrupt increases in IMF magnitude. Auroras in the local-noon sector appear to be particularly sensitive. We conclude that plasmas of the magnetopause boundary layer, and the dayside and outer magnetosphere are affected by the rapid magnetospheric compression that stimulates increased particle precipitation. Future analysis may determine the relative importance of increases in IMF magnitude and solar-wind momentum flux in determining the amplitude of the immediate increase in auroral luminosities.

For the case of $B_z>0$, the immediate response is more stiking than for $B_z<0$ due to the lower emission rates before the SC. The later, delayed onset of a localized brightening at midnight followed by the ~1 km/sec sunward lengthening of a sun-aligned, polar arc of the theta aurora has only one counterpart in the DE data base of auroral images reviewed to date. However, the other example of sunward lengthening of a polar arc from the oval near midnight occurs in a period of apparent quiescent conditions in the solar wind. The IMF B_z component is positive. These observations suggest that in response to magnetospheric compressions for which $B_z>0$, plasma flows in the plasma-sheet boundary layer can be directed out of the plane of the geomagnetic tail and into the tail lobes. The direction of plasma flow and width of this plasma region would be guided and determined by pre-existing sunward flow patterns established during the lengthy period of $B_z>0$ (MAEZAWA, 1976; BURKE *et al.*, 1979; 1982). The magnetospheric response time observed here is ~30 minutes for the onset of initial or enhanced plasma flow normal to the plane of the plasma sheet.

Research cited earlier is concerned with the triggering of substorms by shocks, where the substorm onsets occur no later than 15 minutes after the shock-induced SC onsets. From this earlier work it is seen that under certain favorable conditions a substorm onset, as determined by ground magnetometers, can occur within minutes of the shock impact. For the case of $B_z<0$ studied here, these favorable conditions are not present.

We note from Figs. 1 and 2 that the maximum amplitudes and overall temporal variations of the auroral and ring-current indices are very similar for the two geomagnetic-storm periods, where the storm of 22 October begins only after the IMF turns southward ~3 hours after the shock crossing. We conclude that the gross energetics of the storm periods are similar, and that the only significant variable which differentiates physical conditions external to the magnetosphere for the two cases is the sign of B_z. In each case, significant auroral activity begins within ~30 minutes of the SC. For the case of $B_z<0$ an auroral substorm is observed along the oval in the midnight sector. For the case of $B_z>0$, the dynamic evaluation of the aurora takes place within the polar cap: a polar substorm. The sunward edge of the lengthening polar arc represents, for the polar substorm, an analog to the westward-traveling surge of an auroral-oval substorm. The 1-km/sec speed of the sunward-traveling surge is typical of westward-traveling surges which travel along the pre-existing discrete auroral forms. This suggests that faint, discrete, sun-aligned auroral forms undetected with DE 1 may have been present within the polar cap in advance of the shock impact as part of the established polar-cap convection pattern.

We conclude that the ~30-minute time delay in the onset of auroral activity is independent of the sign of B_z and is related to the time constant for reconfiguration of the geomagnetic tail in response to changes in the IMF and solar-wind plasma behind the shock. In the solar wind, the shocks of 20 and 22 October travel to $X \simeq -186$ and -162 R_E, respectively, in 30 minutes. As typical Alfven speeds within the plasma sheet and its boundary layer ($n \sim 0.1$–1 cm^{-3}, $B \sim 25$ nT; $V_A \sim 545$–1723 km/s) are similar to or greater than the speeds of these relatively slow shocks (~660 and 575 km/sec), the compressional waves will travel at least as far down the geomagnetic tail in 30 minutes. Propagation speeds within the tail lobes exceed those in the plasma sheet. If a localized critical region within the geomagnetic tail is dominant in controlling the substorm process, then the time from first signal arrival to substorm onset may be a measure of the time required for reconfiguration of the geomagnetic tail.

For the plasmoid model of Hones (e.g., HONES, 1984) the critical-region distance is $X \sim -15$ R_E, so >25 minutes are available for the reconfiguration. For a model of magnetotail dynamics which emphasizes the plasma-sheet boundary layer (FRANK *et al.*, 1981b; EASTMAN *et al.*, 1984) and a more distant acceleration region associated by WILLIAMS (1981) with energetic ions in the plasma-sheet boundary layer, the distance is $X \sim -100$ R_E and the available time for reconfiguration is >10 minutes.

The *AE* data have been kindly provided through the efforts of Dr. S.-I. Akasofu at the University of Alaska. We wish to thank Dr. A. J. Lazarus of the Massachusetts Institute of Technology and Dr. J. T. Gosling of the Los Alamos National Laboratory for analyzing and providing solar-wind plasma parameters associated with the shocks. This research at the University of Iowa has been supported in part by the National Aeronautics and Space Administration under contract NAS5-25689 and grants NGL-16-001-002 and NAG5-483, and by the Office of Naval Research under grant N00014-76-C-0016. At the University of California at Los Angeles this research has been supported by the National Aeronautics and Space Administration under contract NAS5-28448. The research conducted at the Jet Propulsion Laboratory of the California Institute of Technology was carried out under contract to the National Aeronautics and Space Administration.

REFERENCES

AKASOFU, S.-I. and J. K. CHAO, Interplanetary shock waves and magnetospheric substorms, *Planet. Space Sci.*, **28**, 381–385, 1980.

BROWN, R. R., T. R. HARTZ, B. LANDMARK, H. LEINBACH, and J. ORTNER, Large-scale electron bombardment of the atmosphere at the sudden commencement of a geomagnetic storm, *J. Geophys. Res.*, **66**, 1035–1041, 1961.

BURCH, J. L., Preconditions for the triggering of polar magnetic substorms by storm sudden commencements, *J. Geophys. Res.*, **77**, 5629–5632, 1972.

BURKE, W. J., M. C. KELLEY, R. C. SAGALYN, M. SMIDDY, and S. T. LAI, Polar cap electric field structures with a northward interplanetary magnetic field, *Geophys. Res. Lett.*, **6**, 21–24, 1979.

BURKE, W. J., M. S. GUSSENHOVEN, M. C. KELLEY, D. A. HARDY, and F. J. RICH, Electric and magnetic field characteristics of discrete arcs in the polar cap, *J. Geophys. Res.*, **87**, 2431–2443, 1982.

BURLAGA, L. F. and K. W. OGILVIE, Causes of sudden commencements and sudden impulses, *J. Geophys. Res.*, **74**, 2815–2825, 1969.

CHAO, J. K. and R. P. LEPPING, A correlative study of SSC's, interplanetary shocks, and solar activity, *J. Geophys. Res.*, **79**, 1799–1807, 1974.

EASTMAN, T. E., L. A. FRANK, W. K. PETERSON, and W. LENNARTSSON, The plasma sheet boundary layer, *J. Geophys. Res.*, **89**, 1553–1572, 1984.

FRANK, L. A., J. D. CRAVEN, K. L. ACKERSON, M. R. ENGLISH, R. H. EATHER and R. L. CAROVILLANO, Global auroral imaging instrumentation for the Dynamics Explorer mission, *Sp. Sci. Instr.*, **5**, 369–393, 1981a.

FRANK, L. A., R. L. MCPHERRON, R. J. DECOSTER, B. G. BUREK, K. L. ACKERSON, and C. T. RUSSELL, Field-aligned currents in the earth's magnetotail, *J. Geophys. Res.*, **86**, 687–700, 1981b.

FRANK, L. A., J. D. CRAVEN, and R. L. RAIRDEN, Images of the earth's aurora and geocorona from the Dynamics Explorer mission, *Adv. Space Res.*, **5**, No. 4, 53–68, 1985.

HONES, E. W., Jr., Plasma sheet behavior during substorms, in *Magnetic Reconnection in Space and Laboratory Plamas*, edited by E. W. Hones, Jr., p. 178, AGU Geophys. Monograph No. 30, 1984.

KAWASAKI, K., S.-I. AKASOFU, F. YASUHARA, and C.-I. MENG, Storm sudden commencements and polar magnetic substorms, *J. Geophys. Res.*, **76**, 6781–6789, 1971.

KOKUBUN, S., R. L. MCPHERRON, and C. T. RUSSELL, Triggering of substorms by solar wind discontinuities, *J. Geophys. Res.*, **82**, 76–86, 1977.

MAEZAWA, K., Magnetospheric convection induced by the positive and negative Z components of the interplanetary magnetic field: quantitative analysis using polar cap magnetic records, *J. Geophys. Res.*, **81**, 2289–2303, 1976.

MATSUSHITA, S., Increase of ionization associated with geomagnetic sudden commencements, *J. Geophys. Res.*, **66**, 3958–3961, 1961.

NISHIDA, A., *Geomagnetic Diagnosis of the Magnetosphere*, Springer-Verlag, New York, 1978.

ORTNER, J., B. HULTQVIST, R. R. BROWN, T. R. HARTZ, O. HOLT, B. LANDMARK, J. L. HOOK, and H. LEINBACH, Cosmic noise absorption accompanying geomagnetic storm sudden commencements, *J. Geophys. Res.*, **67**, 4169–4186, 1962.

RAIRDEN, R. L., L. A. FRANK, and J. D. CRAVEN, Geocoronal imaging with Dynamics Explorer: A first look, *Geophys. Res. Lett.*, **10**, 533–536, 1983.

TAYLOR, H. E., Sudden commencement associated discontinuities in the interplanetary magnetic field observed by IMP 3, *Solar Phys.*, **6**, 320–334, 1969.

ULLALAND, S. L., K. WILHEM, J. KANGES, and W. RIEDLER, Electron precipitation associated with a sudden commencement of a geomagnetic storm, *J. Atmos. and Terr. Phys.*, **32**, 1545–1553, 1970.

VOROB'YEV, V. G., Sc-associated effects in auroras, *Geomagn. Aeron.*, **14**, 72–74, 1974.

WILLIAMS, D. J., Energetic ion beams at the edge of the plasma sheet: ISEE 1 observations plus a simple explanatory model, *J. Geophys. Res.*, **86**, 5507–5518, 1981.

WILSON, C. R. and M. SUGIURA, Hydromagnetic interpretation of sudden commencements of magnetic storms, *J. Geophys. Res.*, **66**, 4097–4111, 1961.

Solar Wind-Magnetosphere Coupling, edited by Y. Kamide and J. A. Slavin, 381–390.

Triggering of Magnetospheric Particle Bursts by SSC Events

S. M. THOLEN and T. P. ARMSTRONG

Dept. of Physics and Astronomy, University of Kansas, Lawrence, Kansas 66045, U.S.A.

A comprehensive study of energetic proton and electron count rates (E_p>0.29 MeV; E_e>0.22 MeV) as recorded by the Charged Particle Measurement Experiment on board the IMP 7 and 8 satellites has been performed for the periods of time surrounding times of geomagnetic storm sudden commencements (SSC's) for a ten year period from 1973 to 1983. The results indicate that SSC events significantly increase the probability of magnetospheric particle bursts and that triggering probability depends on satellite location in the magnetosphere. The highest incidence of triggered bursts occurs in the magnetotail (60.0%), followed by the magnetosheath region (44.4%). About half of the magnetotail triggered bursts occur within 20 minutes of the SSC time. In the interplanetary medium, the incidence of triggered bursts differs insignificantly from chance occurrence. These results provide a quantitative and statistically complete study of the energetic particle responses of the Earth's magnetosphere to an impulsive perturbation. We hope this study will provide the basis for developing specific, global, quantitative models of the electric and magnetic fields initiated by the SSC event which produce, in turn, the observed charged particle energization.

1. Introduction

Energetic electron (>40 keV) and proton (>125 keV) bursts have been reported (ANDERSON, 1965; KONRADI, 1966) in the terrestrial magnetotail, in the magnetosheath and bow shock region (FAN *et al.*, 1964), and upstream in the solar wind (LIN *et al.*, 1974). Later studies have shown that particle bursts occur frequently in the magnetosphere (SARRIS *et al.*, 1976) and that some electron bursts occurred following SSC's (DECKER and ARMSTRONG, 1977). A study in 1971 indicated a high probability of SSC events triggering magnetospheric substorms (AKASOFU and KAMIDE, 1980). Since the occurrence of energetic particle bursts in the magnetosphere is a probable indicator of the recent energization of the particles, the possibility that such bursts can be specifically triggered by the arrival at the magnetosphere of the interplanetary disturbances is clearly of importance to the description of magnetospheric particle acceleration.

In the present report we have investigated the correlation between SSC occurrences and particle bursts. Energetic proton (E_p>0.29 MeV) and electron (E_e>0.22 MeV) intensities observed with IMP 7 and 8 during 338 SSC events which occurred between 1973 and 1983 were examined. Our principal finding is that there is a high probability of an SSC event triggering a particle burst. This is most likely to be observed within the magnetotail. We present here a summary of the statistical results as well as specific cases.

2. Description of Experiment

The data used in this analysis were obtained from the Johns Hopkins University Applied Physics Laboratory Charged Particle Measurement Experiment on board the IMP 7 and 8 satellites. Both have near-circular orbits (IMP 7–32 by 38 Re; IMP 8–34 by 36 Re) with small inclinations to the ecliptic plane. The spacecraft have approximately 13-day orbits, spending about 7.5 days in the interplanetary medium, 2.5 days in the magnetosheath region, and 3 days in the magnetotail. Particle counts are measured in 30 channels corresponding to protons, electrons, and alpha particles of various energies. Data was available from IMP 7 from 1973 to 1978 and from IMP 8 from 1974 to 1983, and this study examined the 5.5 minute (327.68 sec) averaged count rates for two proton channels ($P1$=0.29–0.50 MeV; $P2$=0.50–0.97 MeV) and three electron channels ($E4$=0.22–2.5 MeV; $E5$=0.50–0.97 MeV; $E6$=0.80–2.5 MeV). For every report of an SSC in the Solar-Geophysical Data (COFFEY, 1983), the particle count rates were plotted from 3 hours before the SSC time to 6 hours after and individually classified into 7 groups:

1. "Yes" events when correlated particle bursts were observed to begin within the time period of $-30<t<60$ minutes from the SSC.
2. "No" events when observing conditions were good but no particle bursts were detected.
3. Events in which bursts of particles were seen both before and after the SSC.
4. Times when the data was missing or incomplete.
5. Events when particle counts were obscured by solar activity.
6. Times when proton bursts were observed but no electron bursts.
7. Events in which bursts did occur within the 9-hour period examined but which began at least 100 minutes after the SSC occurrence.

3. Observations

A typical "yes" event was required to have normal or quiet observing conditions before the SSC; in addition, the particle burst must have begun within a period from 30 minutes before to 60 minutes after the SSC. It has been found that storm sudden commencements can propagate from the magnetotail to the ground as well as from the ground to the magnetotail. A study of 13 events (PATEL, 1972) showed that 9 SSC's were witnessed on the ground first and 4 were seen first in the magnetotail. Delay times ranged up to 12 minutes for these 13 events. The 30 minutes before the SSC time was allowed to account for such cases where an interplanetary shock may have passed the satellite before any indication of same was noted by ground-based magnetometers. The 60 minute time limit was chosen primarily because a definite majority of cases fell within these limits or occurred much later, after 100 minutes or more. Of 26 events with clear onset times between -30 and 100 minutes, all occur between -10 and 70 minutes, with a fairly strong concentration from 0 to 20 minutes after the SSC passage. Figure 1 shows the distribution of burst onset times with reference to the SSC times for all of the "yes" events. The dominant feature of this distribution is the peak at 0 to 20 minutes delay. A broader, second peak at 30 to 60

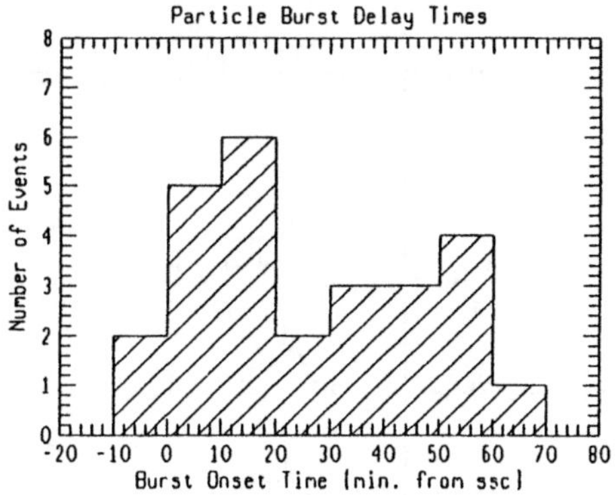

Fig. 1. Particle burst delay times for all "yes" events observed.

minutes delay is also present. The expectation of a burst onset for randomly chosen times is about 0.2 onsets per 10-minute bin for all times sampled. For the periods from -3 to $+6$ hours around an SSC, the expectation of a burst onset is slightly higher, namely, about 0.3 events per 10-minute bin. Figure 1 shows an effect far beyond any reasonable chance occurrence. We may safely conclude that there is a prompt triggering, with 0 to 20 minutes delay, as well as significantly increased likelihood of burst onsets out to delays of 60 minutes. The longer delays may correspond to bursts occurring as a result of a substorm triggered by the SSC while the promptly triggered bursts may result directly from magnetotail processes in response to the SSC impulse. Figure 2 shows the 5.5 minute-averaged counting rates for electrons (a) and protons (b) for a 9 hour period on day 299 of 1977 acquired with IMP 8. The sudden particle increase here is typical of the "yes" events. The vertical line indicates the time of the SSC passage; the horizontal axis is time measured in minutes from the SSC passage and the vertical axis is the log of the averaged count rate. The channels plotted are labelled to the right. This event has close to the maximum allowed delay between the SSC and the burst onset but has a clear burst signature in both electrons and ions.

The typical "no" event was also required to have fairly quiet observing conditions before the SSC passage so that a burst could have been observed if one had occurred. No significant increase in count rates was observed over the entire nine-hour period examined. Another class of events consisted of times when bursts were already in progress before the SSC passage and may have continued afterward as well. These were not considered in the triggering statistics since it is not clear whether or not the bursts were correlated with the SSC passage. Solar events were characterized by very high particle count rates over the entire period, generally averaging around 1000 counts per second or more. Solar and interplanetary fluxes are also distinguishable from magnetospheric bursts by having much harder energy spectra in the 300 keV to 4 MeV interval. Any particle burst that may have occurred would probably have been masked by these high counts. In another category, proton bursts were observed but no change in the electron count rates was seen. The final group of events consisted of

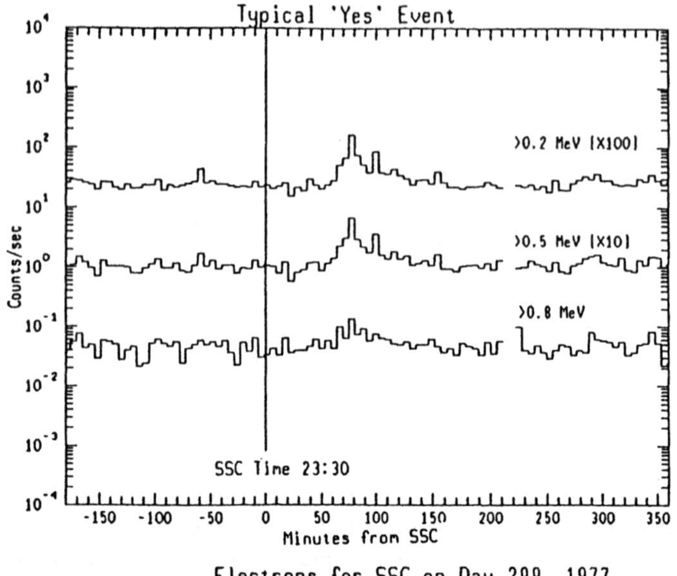

(a)

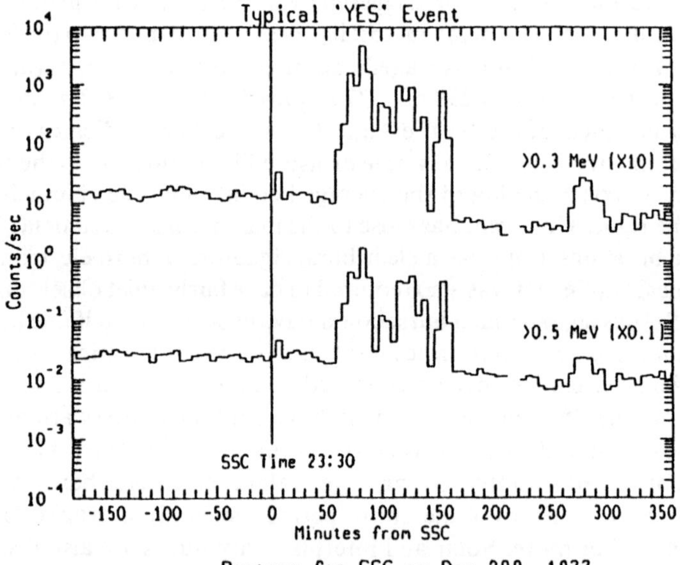

(b)

Fig. 2. Electron and proton count rates surrounding an SSC time for a typical "yes" event.

those exhibiting particle bursts within the nine-hour time span but beginning at least 100 minutes after the time reported for the SSC passage. This time limit was chosen to clearly differentiate between this group and the "yes" group. There were very few cases in which a possible burst was seen between 60 and 100 minutes after the SSC passage.

After all the SSC times from 1973 to 1983 were classified according to the scheme outlined above, 100 random times were selected from this same time period. The same classification procedure was carried out for these times to provide an idea of what particle count rates would be expected from a random distribution of times. Table 1 lists the number of events placed in each category for both the SSC times and the random times examined.

While "yes" events occurred only 5.2% of the random times, the SSC events produced a 12.4% probability of a "yes" event. Likewise for the "no" events, 48% of the random times selected showed no particle bursts at all, while the same was true for only 21.4% of the SSC times. The probability of solar events is markedly higher for the SSC times than the random times, due to the fact that the passage of an SSC often indicates some solar activity and the SSC times would be preferentially selecting times of increased solar activity.

For each event of the SSC times as well as the random times, the spacecraft coordinates were obtained and plotted. Figure 3 shows the coordinates of the spacecraft recording the data (there were only a few events in which both spacecraft provided data for the same SSC occurrence) for all the "yes" and "no" events. The top three graphs are the x–y, x–z, and y–z (Earth radii) plots for all the "yes" events and the bottom three for all the "no" events. On the x–y and x–z plots are nominal magnetopause and bow shock locations, as obtained from an empirical formula by FAIRFIELD (1971); the coordinates are GSE. The average locations of these boundaries were used since there was no data available for particle and field parameters. This unavoidably introduces some uncertainty into our classification of regions. The "yes" events show a strong concentration in the magnetotail region and some in the magnetosheath. The "no" events show no such concentration and are evenly distributed throughout the orbit of the satellite, with the ratio of events occurring in each region corresponding to the amount of observing time spent there.

Table 1. Number of events in each category.

Number of events	SSC times	Random times
Total events	338	100
Events with enough data	210	77
Yes events	27 (12.4%)	4 (5.2%)
No events	45 (21.4%)	37 (48.0%)
Bursts before and after	17 (8.1%)	15 (19.4%)
Solar activity	96 (45.7%)	16 (20.8%)
Protons but no electrons	18 (8.6%)	4 (5.2%)
Late bursts	8 (3.8%)	1 (1.3%)

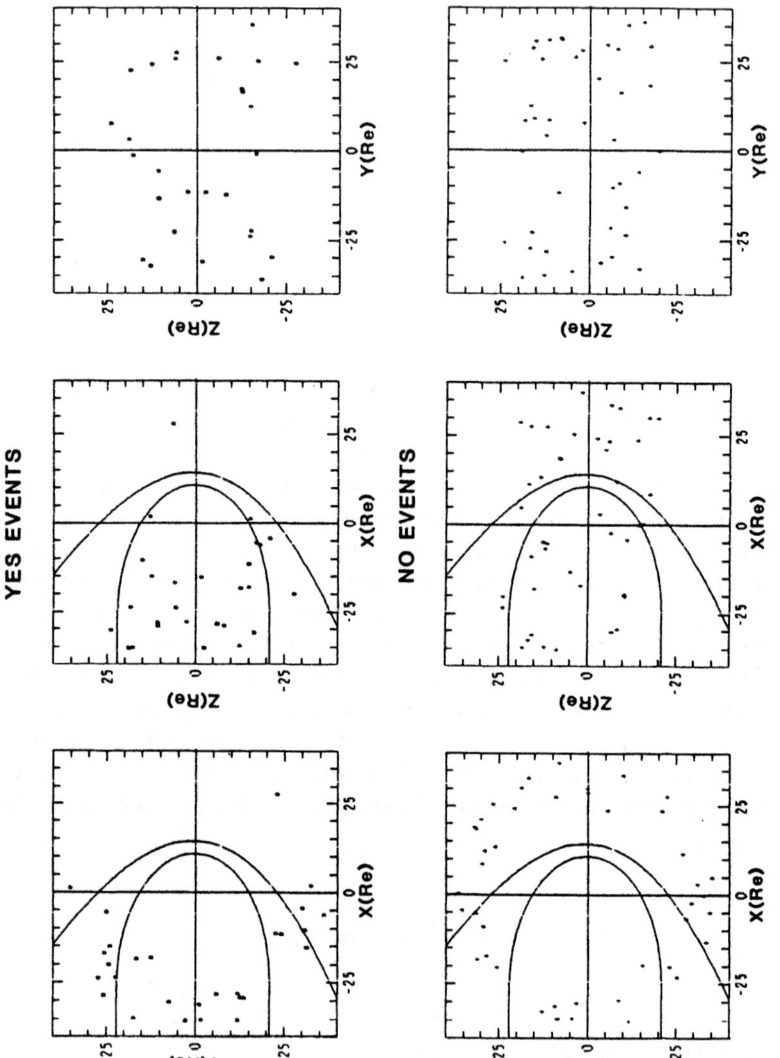

Fig. 3. Satellite coordinate plots during particle observations of times surrounding SSC events.

Figure 4 shows the same plots for the random time "yes" and "no" events. There were only 4 "yes" events, so a probabilistic distribution is not appropriate. However, in direct contrast to the SSC times when a clear majority of the events occurred in the magnetotail, none of these "yes" events were observed when the satellite was in the magnetotail and only one in the magnetosheath. The plots of the "no" events are essentially the same as for the SSC times. The events are again distributed around the orbit of the satellite with the same distribution as the observing time of the satellite in each region.

In Table 2 we have tabulated the exact numbers of "yes" events in each region and "no" events in each region for the SSC times and random times. If we consider only events classified as either "yes" or "no" we see a definite difference in the SSC times and random times. Of the "yes" and "no" events for the SSC times, 36.6% are "yes" events and 63.4% are "no" events. For the random times, only 9.8% are "yes" and 90.2% are "no". The "no" events are spatially distributed about the same for the SSC times and random times as was seen on the coordinate plots. Over half the "no" events occurred while the satellite was in the interplanetary medium; the remainder were distributed about equally between the magnetotail and magnetosheath regions. The "yes" events for the random times are too few to analyze, but just like the "no" events, most occurred in the interplanetary medium where the satellite spends most of its time. The "yes" events for the SSC times are markedly different from these. Of all the "yes" events recorded following an SSC, 46.2% occurred when the satellite was in the magnetotail, 30.8% in the magnetosheath, and 23.1% in the interplanetary medium.

When the results are organized by location, the SSC enhancement of burst probability in the magnetotail is readily apparent. Table 3 shows the statistics according solely to location. When the satellite was in the magnetotail and an SSC was reported, a particle burst followed on 12 out of 20 occasions, 60.0% of the time. The random observation of a "yes" event in the magnetotail was, in direct contrast, 0%. When the satellite was located in the magnetosheath, an SSC produced a particle burst 44.4% of the time while only 16.7% were observed in the random times. In the interplanetary medium, "yes" events occurred with approximately the same probability for the SSC times and random times. For any location, bursts occur 9.8% of the times (4 out of 41) randomly selected; bursts occur 36.6% of the times (26 out of 71) an SSC was reported. The "yes" distributions in the interplanetary medium are virtually the same for the SSC times and random times, indicating little or no incidence of triggered bursts in this region.

4. Conclusions

The results indicate that SSC's do significantly increase the probability of magnetospheric particle bursts; both proton and electron bursts can be triggered by SSC occurrences. In addition, the triggering probability depends largely on satellite location in the magnetosphere. The highest incidence of triggered bursts occurs in the magnetotail, where bursts followed SSC's 60.0% of the time and never occurred in the random times examined. In the magnetosheath, triggered bursts also occurred but with less frequency. Bursts followed SSC's on 44.4% of the occasions observed, as

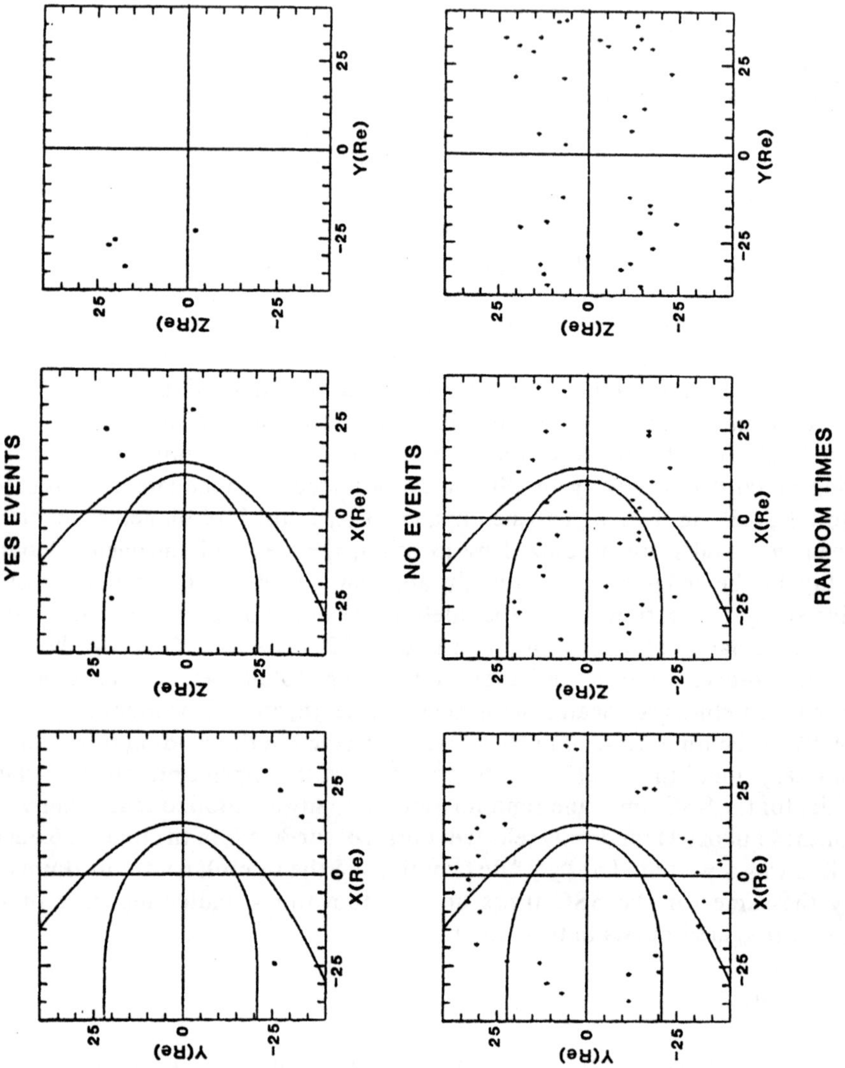

Fig. 4. Satellite coordinate plots during particle observations of randomly selected times.

Table 2. Results broken down according to location.

Number of events	SSC times	Random times
Yes and no events	71	41
Yes events	26 (36.6%)	4 (9.8%)
Magnetotail	12 (46.2%)	0 (0%)
Magnetosheath	8 (30.8%)	1 (25%)
Interplanetary medium	6 (23.1%)	3 (75%)
No events	45 (63.4%)	37 (90.2%)
Magnetotail	8 (17.8%)	4 (10.8%)
Magnetosheath	10 (22.2%)	5 (13.5%)
Interplanetary medium	27 (60.0%)	28 (75.7%)

Table 3. Results broken down into yes and no events.

Number of events	SSC times	Random times
Magnetotail	20	4
Yes events	12 (60.0%)	0 (0%)
No events	8 (40.0%)	4 (100%)
Magnetosheath	18	6
Yes events	8 (44.4%)	1 (16.7%)
No events	10 (55.6%)	5 (83.3%)
Interplanetary medium	33	31
Yes events	6 (18.2%)	3 (9.7%)
No events	27 (81.8%)	28 (90.3%)

opposed to a random expectation of 16.7%. Implicit in this investigation was the assumption that an SSC report was indicative of the passage of an interplanetary shock which is not universally true. While the occurrence of SSC's without interplanetary shocks and the occurrence of interplanetary shocks without SSC's is small, taking this into account would undoubtedly alter our statistics somewhat. In general, however, we can say that the impact of an interplanetary shock on the earth's magnetosphere significantly increases the probability of particle bursts, with the greatest triggering probability in the magnetotail region.

This work was supported in part by the National Science Foundation under grant ATM 83-05537.

REFERENCES

AKASOFU, S.-I. and Y. KAMIDE, Recent progress in studies of magnetospheric storms and substorms, *J. Geomag. Geoelectricity*, **32**, 585–615, 1980.

ANDERSON, Kinsey A., Radial dependence of energetic electron fluxes in the tail of the earth's magnetic field, *Phys. Rev. Lett.*, **14**, 888–890, 1965.

COFFEY, H. E. (ed.), *Solar Geophysical Data*, US Dept. of Commerce, National Oceanic and Atmospheric Administration, Boulder, Colorado, 343–460, 1983.

DECKER, Robert B. and T. P. ARMSTRONG, SSC-associated effects on >0.2 MeV solar electrons in 1972–1976, *EOS Trans. AGU*, **58**, 466, 1977.

FAIRFIELD, D., Average and unusual locations of the earth's magnetopause and bow shock, *J. Geophys. Res.*, **76**, 6700–6716, 1971.

FAN, C. Y., G. GLOECKLER, and J. A. SIMPSON, Evidence for >30 keV electrons accelerated in the shock transition region beyond the earth's magnetospheric boundary, *Phys. Rev. Lett.*, **13**, 149, 1964.

KONRADI, A., Electron and proton fluxes in the tail of the magnetosphere, *J. Geophys. Res.*, **71**, 2317–2325, 1966.

LIN, R. P., C.-I. MENG, and K. A. ANDERSON, 30- to 100-keV protons upstream from the earth's bow shock, *J. Geophys. Res.*, **79**, 489, 1974.

PATEL, V. L., Sudden impulses in the geomagnetotail and the vicinity, *J. Geophys. Res.*, **20**, 1127–1136, 1972.

SARRIS, E. T., S. M. KRIMIGIS and T. P. ARMSTRONG, Observations of magnetospheric bursts of high-energy protons and electrons at 35Re with IMP 7, *J. Geophys. Res.*, **81**, 2341–2355, 1976.

Solar Wind-Magnetosphere Coupling, edited by Y. Kamide and J. A. Slavin, 391–399.

Observations of Energetic Electrons and VLF Waves at Geostationary Orbit during Storm Sudden Commencements (SSC)

A. KORTH[1], G. KREMSER[1], N. CORNILLEAU-WEHRLIN[2], and J. SOLOMON[2]

[1]*Max-Planck-Institut für Aeronomie, D-3411 Katlenburg-Lindau, F.R.G.*
[2]*Centre de Recherches en Physique de l'Environnement Terrestre et Planétaire, CNET/CRPE F-92131 Issy-les-Moulineaux, France*

Simultaneous observations of energetic electrons (16–300 keV) and ELF/VLF waves (200 Hz–3 kHz) have been analysed. They were obtained on board GEOS-2 during storm sudden commencements (ssc) in 1978 and 1979. A strong local time asymmetry of the ssc signatures was found. On the dayside the electron distribution exhibits a large anisotropy so that wave particle interaction could take place. The interaction was demonstrated by calculations of the anisotropy and the wave growth rate for one selected ssc event. On the nightside the wave amplitude was influenced by the ssc only slightly or not at all.

1. Introduction

Storm sudden commencements (ssc) have been studied already since a few decades. They are associated with the passage of a solar wind structure and changes in solar wind parameters (GOSLING *et al.*, 1967; BURLAGA and OGILVIE, 1969). Characteristics of ssc's in the magnetosphere were examined by NISHIDA and CAHILL (1964), PATEL (1968), and SUGIURA *et al.* (1968).

At geosynchronous altitude ssc's have been studied by PATEL and COLEMAN (1970), BAUMJOHANN *et al.* (1984), and WILKEN *et al.* (1982). PATEL and COLEMAN (1970) reported on ssc associated magnetic field measurements in the dusk sector, BAUMJOHANN *et al.* (1984) on Alfvén waves excited by ssc's, and WILKEN *et al.* (1982) on propagation of ssc associated magnetohydrodynamic waves based on multi-satellite observations. A statistical analysis was done on magnetic field measurements at various local times in association with ssc's by KOKOBUN (1983). He found that the variation of the ssc amplitude tends to become very small near midnight.

In association with ssc's there have also been observed particle precipitation (ULLALAND *et al.*, 1984) and ULF and VLF wave stimulations (OLSEN and LEE, 1983; KOKOBUN and OGUTI, 1968). It is agreed that a sudden compression of the magnetospheric plasma could trigger ULF and VLF emissions. One expects that as a result of such a compression the particle distribution function would become increasingly anisotropic. Particles with such anisotropic distribution have been shown to be unstable to the growth of ion-cyclotron waves (CORNWALL, 1965; GENDRIN *et al.*, 1971; KORTH *et al.*, 1984) and VLF or whistler mode waves (CORNILLEAU-WEHRLIN *et al.*, 1985).

In this paper we report on coordinated observations of energetic electrons and ELF/VLF waves from the European geostationary satellite GEOS-2. About 35 ssc events occurred between August 1978 and March 1979. We selected a representative set of six events at different local times to investigate the local time dependence of ssc's and analyzed one event in detail by calculating the frequency dependent anisotropy A and the growth rate γ.

2. Data

The electron data have been obtained with the energetic particle spectrometer S321 on GEOS-2 (KORTH *et al.*, 1978; KORTH and WILKEN, 1978). For the present paper we use the integral and differential data. The energy distribution of the electrons is determined in 15 energy channels in the range from 16 keV to 300 keV for adjacent pitch angle intervals of 10 ° width. The differential data are sampled for 0.344 s every 5.5 s. Integral data ($E>22$ keV) are measured continuously with a time resolution of 0.172 s.

The wave data have been obtained from the wave experiment S300 (JONES, 1978; S300 EXPERIMENTERS, 1979). For the purpose of studying whistler mode waves (frequency f is smaller than the electron gyrofrequency f_{ce}), we only consider the B_x magnetic component of the wave, which is about perpendicular to the static magnetic field. The wave data are continuously recorded from the onboard filterbank in the frequency range 0.2–5 kHz. These filterbank data are well suited to study the time evolution of whistler mode waves whose frequencies are some tenths of the electron gyrofrequency f_{ce} ($f_{ce} \cong 2$ to 4 kHz in the equatorial region for $L=6.6$). Nevertheless those filters have not sharp cut-offs. They only give rough information on the wave frequency range. The accurate frequency spectra are provided by the onboard correlator. The time resolution of the correlator is 20 ms in the high speed telemetry rate, but most of the time we have only access to compressed data with a time resolution of 5.5 s. This is sufficient for the present work, as we will integrate over 5 minutes for the calculation of the anisotropy A and the growth rate γ.

3. Observations

Plate 1 shows the intensity changes of energetic electrons associated with storm sudden commencements, ssc, for six selected different local times in colour coded diagrams. The onset times of the ssc's—taken from the reports of Solar Geophysical Data—of the six events are marked in the corresponding wave data in Fig. 1. During the period from August 1978 to March 1979, the satellite GEOS-2 was located at a geographic longitude of 37.3° E—that means LT=UT+2.5 hrs—and a geomagnetic latitude of about −3°.

The measurements show that in the first five coloured diagrams of Plate 1 a clear electron flux increase could be observed. However, there are also cases in the midnight sector indicating a flux decrease—one example is given on Plate 1 for 3 Feb. 1979. From the observations we also found that near local noon the electron intensity increase is restricted to pitch angles very close to 90° (see ssc's on 11 Sept. 1978 and 29 Oct. 1978 in Plate 1). Hence it follows that the anisotropy of the electrons after the ssc is largest near local noon, where the increase in the magnetic field is strongest.

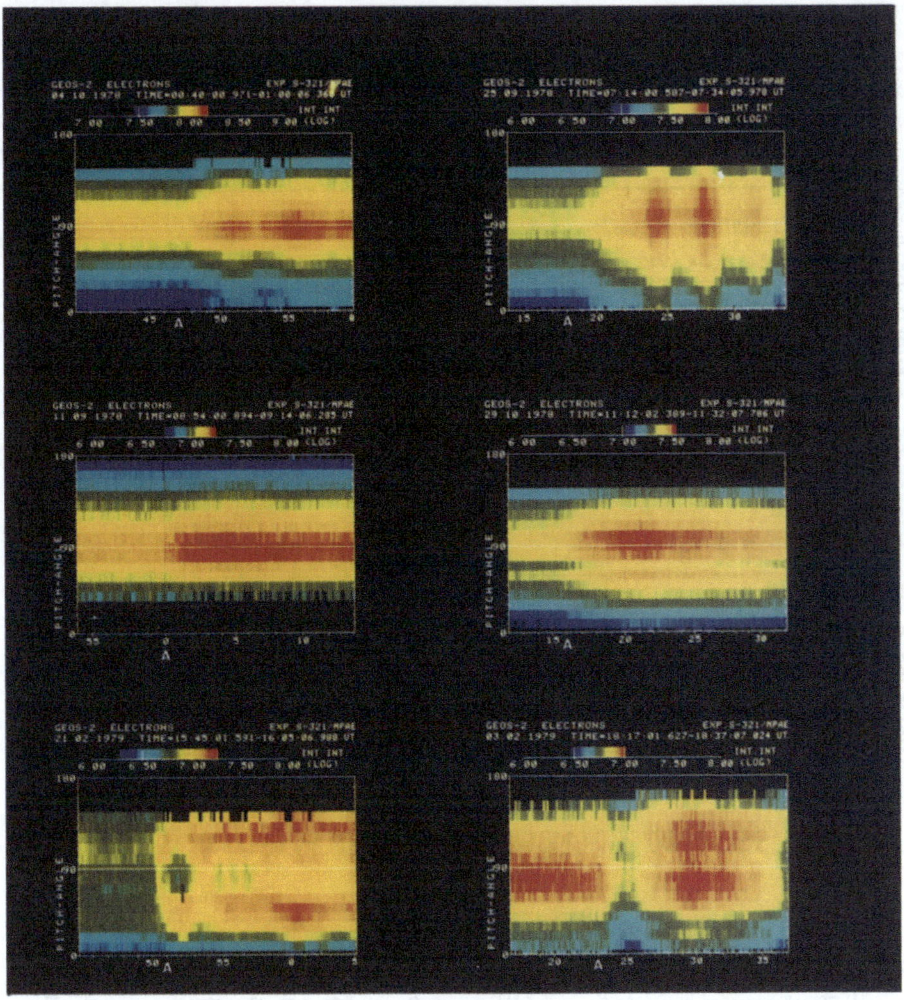

Plate 1. Integral electron intensities ($E > 16$ kev) are displayed in a time versus pitch angle diagram. The time intervals include the occurrence of ssc's at different local times. The intensity is given in a colour coded logarithmic scale. The time scale is indicated in minutes and the time interval is shown on top of each panel in universal time. The data are averaged over 5.5 s. The onset times of the ssc for the six events are indicated by triangles.

Figure 1 shows the ELF/VLF wave data for the same six events as on Plate 1. The amplitude of the magnetic component B_x is plotted versus time for four frequency bands. On the morning side of the magnetosphere (4 Oct. 1978 and 25 Sept. 1978), the influence of the ssc on the wave amplitude level is small and the wave amplitude is only slightly reinforced. Around noon (11 Sept. 1978 and 29 Oct. 1978) the wave amplitude increases considerably, whereas there is no influence on the wave amplitude in the

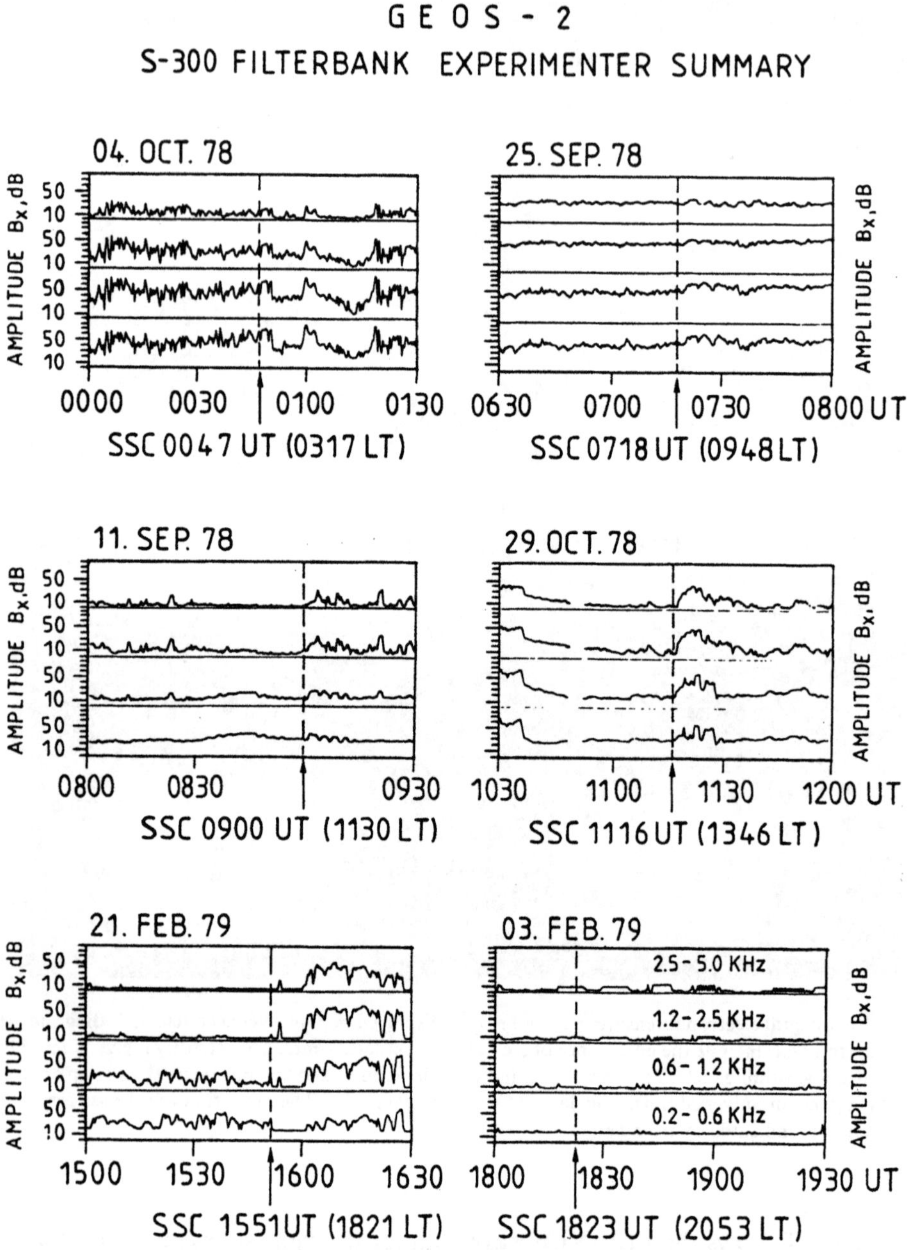

Fig. 1. Wave data for six ssc's presented in Plate 1. For each ssc event 90 minutes of integrated wave magnetic field data are plotted in four frequency channels. The four frequency ranges are given in this Figure for event 6 (3 Feb. 1979). The time resolution is 22 s. The amplitude is given in decibels with a 70 dB dynamic.

evening sector (21 Feb. 1979 and 3 Feb. 1979).

In the following we will concentrate on one ssc observed on the dayside of the magnetosphere near local noon. We selected the 29 Oct. 1978 event as an example out of many, where a change in the electron anisotropy and an amplification of the waves was found. On this day the magnetometer showed an increase by 10 nT only in the *H* component (parallel to the earth rotation axis), whereas the *V* and *D* component (radial and azimuthal) remained constant. This case gives a clear indication of a compression of the magnetosphere. The response of this compression is seen in Fig. 2 in the energetic electron data. All electron channels in the energy range between 16 keV and 214 keV showed a clear simultaneous increase of the electron intensity at pitch angles near 90° following the ssc. This is in accordance with the conservation of the first and second adiabatic invariants (ASHOUR-ABDALLA and COWLEY, 1974). The other time variations of the electron intensities in Fig. 2 are not simultaneous in all energy channels. They result either from the plasma sheet boundary crossing (0100–0530 UT) or from drift effects.

4. Wave Particle Interaction

ELF/VLF waves in the whistler mode can be amplified by anisotropic electron distributions. A necessary condition for this process is that the anisotropy *A* is larger

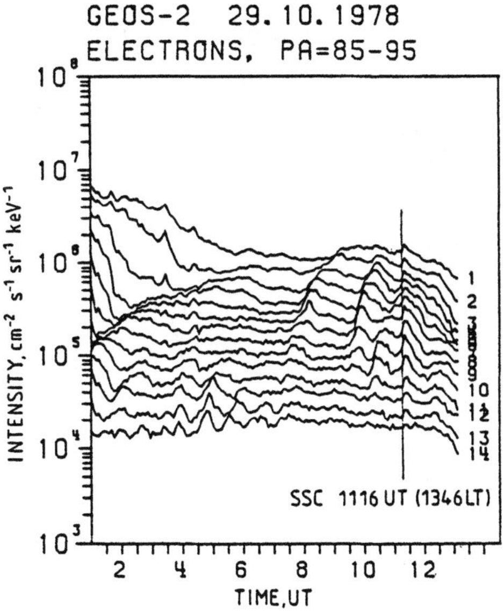

Fig. 2. Differential electron intensity variations versus universal time. The ssc is indicated by a vertical line. The numbers at the curves correspond to the energy channels. PA is the pitch angle range. Energy channels 1: 16–24 keV, 2: 24–30 keV, 3: 30–38 keV, 4: 38–45 keV, 5: 45–52 keV, 6: 52–61 keV, 7: 61–71 keV, 8: 71–83 keV, 9: 83–96 keV, 10: 96–114 keV, 11: 114–133 keV, 12: 133–154 keV, 13: 154–180 keV, 14: 180–214 keV.

than the critical anisotropy A_c (KENNEL and PETSCHEK, 1966). CORNILLEAU-WEHRLIN *et al.* (1985) performed a detailed experimental study using energetic electrons and ELF/VLF waves and gave a method to calculate the particle anisotropy A and the temporal growth rate γ of the waves from the data using the exact formulas given by KENNEL and PETSCHEK (1966). A and γ are frequency dependent.

In the following we calculated the anisotropy A and the normalized wave growth rate γ/ω_{ce} for the 29 Oct. 1978 ssc and compared their frequency dependence to that of the observed wave spectrum. The electron gyrofrequency f_{ce} was determined from the measurements of the magnetic field. The electron plasma frequency f_{pe} was provided by the relaxation sounder experiment (before the ssc: $f_{ce}=3.1$ kHz, $f_{pe}=20$ kHz; after the ssc: $f_{ce}=3.3$ kHz, $f_{pe}=23$ kHz).

The results are shown in Figs. 3 and 4. Figure 3 indicates that before and after the ssc the calculated values of the anisotropy $A(x)$ are larger than the critical anisotropy $A_c=x/(1-x)$, for $0.05 \leq x \leq 0.45$ ($x=f/f_{ce}$, reduced wave frequency). The reason that the waves are not amplified before the ssc although the anisotropy is high enough will be explained from the results in Fig. 4.

In Fig. 4(a) we have computed the reduced temporal growth rate γ/ω_{ce} of the whistler mode instability for two five minute time intervals preceding and following the

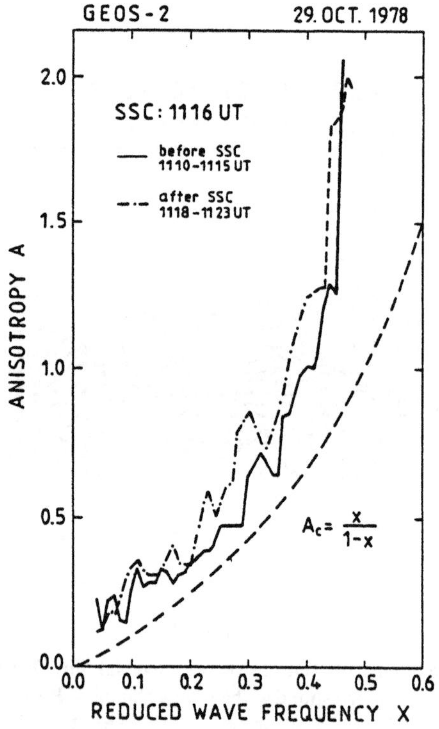

Fig. 3. Numerical calculations of the electron anisotropy A versus reduced wave frequency $x=f/f_{ce}$ for a five minute time interval before and after the ssc. The critical anisotropy A_c is plotted for comparison.

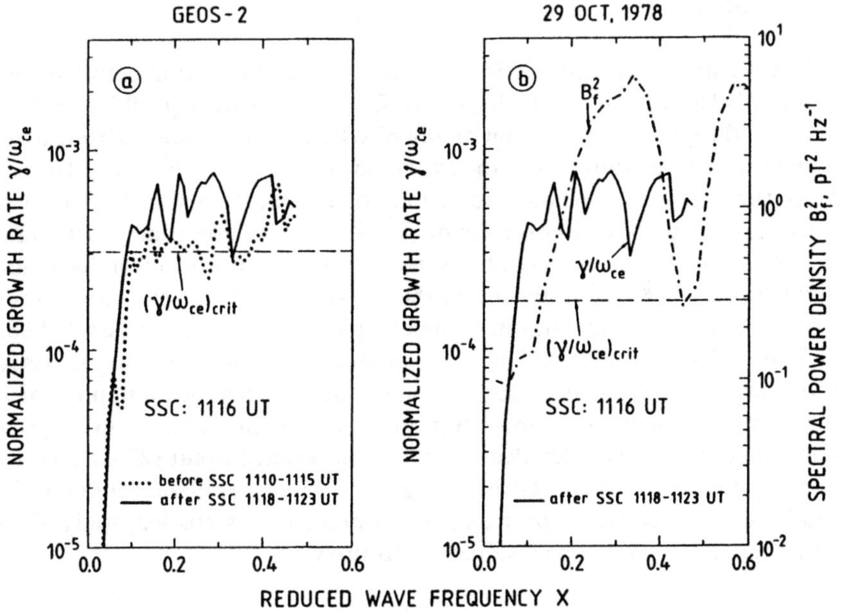

Fig. 4. (a) The temporal growth rates (γ/ω_{ce}) of the whistler instability versus reduced wave frequency $x=f/f_{ce}$ for the same time intervals as in Fig. 3. Also shown is the critical value $(\gamma/\omega_{ce})_{crit}$, below which the wave cannot be sustained in the medium. (b) Temporal growth rate of the waves after the ssc together with the emitted wave spectrum. Note that the growth rate exceeds by far the critical limit for continuous wave emission.

ssc. The spatial growth rate γ/V_g, where V_g is the wave group velocity, integrated along the line of force has at least to compensate for the losses of wave energy at the ionosphere in order that waves can be emitted continuously around the magnetic equator. Therefore γ/V_g must exceed some critical limit that CORNILLEAU-WEHRLIN *et al.* (1985) estimated. Expressed for the reduced temporal growth rate, this critical limit appears roughly as $(\gamma/\omega_{ce})_{crit}\simeq(6\times10^{-3})/f_{pe}$, where f_{pe} is the electron plasma frequency expressed in kHz. In the first time interval 1110–1115 UT before the ssc, where $f_{pe}\simeq20$ kHz, we get $(\gamma/\omega_{ce})_{crit}\simeq3\times10^{-4}$. In Fig. 4(a) we observe that the value of (γ/ω_{ce}) that we have obtained before the ssc is found near the critical limit (horizontal dashed line), whereas after the ssc (γ/ω_{ce}) clearly exceeds this critical limit. Moreover, after the ssc the critical limit has been slightly decreased, because at this time $f_{pe}\simeq23$ kHz (Fig. 4(b)). In agreement with the above results the wave field intensity $B^2{}_f$ before the ssc (not shown) amounts to not more than 5×10^{-3} $(\mathrm{pT})^2/\mathrm{Hz}$, which results in a very weak diffusion in pitch angle ($y_0{}^2\simeq1.5\times10^2$, see KENNEL and PETSCHEK (1966) and CORNILLEAU-WEHRLIN *et al.* (1985) for definition and discussion of the diffusion parameter $y_0{}^2$). On the contrary, as shown in Fig. 4(b) after the ssc, where (γ/ω_{ce}) largely exceeds the critical value, $B^2{}_f$ has increased up to 6 $(\mathrm{pT})^2/\mathrm{Hz}$, a value corresponding to a strong diffusion regime ($y_0{}^2\simeq3\times10^{-2}$).

5. Conclusion

We have studied in conjunction with the ssc the interaction between energetic electrons ($E>16$ keV) and whistler mode ELF/VLF waves ($f\sim0.2$–5 kHz), both measured in the geostationary orbit on board GEOS-2. We found a strong local time asymmetry of ssc signatures from the examination of the two data sets. The analysis indicates that the ssc associated variations of the electron flux lead to a higher intensity and anisotropy after the ssc on the dayside, in accordance with adiabatic effects. We obtain the largest variations in intensity and anisotropy at local noon. In the night sector (1800 LT to 0600 LT) the ssc associated electron variations could show intensity decreases and negative anisotropies (as examples ssc's on 21 Feb. 1979 and 3 Feb. 1979 on Plate 1). The wave observations indicate that ssc associated emissions are most frequently observed on the dayside near local noon, which is in agreement with the increase in anisotropy and electron flux resulting in the growth of whistler mode waves. The detailed study of one selected event (29 Oct. 1978) gave evidence for the relationship between the growth rate γ calculated from the electron data and the observed wave spectrum. A comprehensive statistical study of ssc and wave associated events will be a subject of future work.

REFERENCES

ASHOUR-ABDALLA, M. and S. W. H. COWLEY, Wave-particle interactions near the geostationary orbit, in *Magnetospheric Physics*, edited by B. M. McCormac, pp. 241–270, D. Reidel, Hingham, Mass., 1974.

BAUMJOHANN, W., H. JUNGINGER, G. HAERENDEL, and O. H. BAUER, Resonant Alfvén waves excited by a sudden impulse, *J. Geophys. Res.*, **89**, 2765–2769, 1984.

BURLAGA, L. F. and K. W. OGILVIE, Causes of sudden commencements and impulses, *J. Geophys. Res.*, **74**, 2815–2825, 1969.

CORNILLEAU-WEHRLIN, N., J. SOLOMON, A. KORTH, and G. KREMSER, Experimental study of the relationship between energetic electrons and ELF waves observed onboard GEOS: A support to quasilinear theory, *J. Geophys. Res.*, **90**, 4141–4154, 1985.

CORNWALL, J. M., Cyclotron instabilities and electromagnetic emission in the ultra low frequency and very low frequency ranges, *J. Geophys. Res.*, **70**, 61–69, 1965.

GENDRIN, R., S. LACOURLY, A. ROUX, J. SOLOMON, F. Z. FERGIN, M. B. GOKHBERG, V. A. TROITSKAYA, and V. L. YAKIMENKO, Wave packet propagation in an amplifying medium and its application to the dispersion characteristics and to the generation mechanisms of Pc1 events, *Planet. Space Sci.*, **19**, 165–194,1971.

GOSLING, J. T., J. R. ASBRIDGE, S. J. BAME, A. J. HUNDHAUSEN, and I. B. STRONG, Discontinuities in the solar wind associated with sudden geomagnetic impulses and storm commencements, *J. Geophys. Res.*, **72**, 3357–3363, 1967.

JONES, D., Introduction to the S-300 wave experiment onboard GEOS, *Space Sci. Rev.*, **22**, 327–332, 1978.

KENNEL, C. F. and H. E. PETSCHEK, Limit on stably trapped particle fluxes, *J. Geophys. Res.*, **71**, 1–28, 1966.

KOKUBUN, S., Characteristics of storm sudden commencement at geostationary orbit, *J. Geophys. Res.*, **88**, 10,025–10,033, 1983.

KOKUBUN, S. and T. OGUTI, Hydromagnetic emissions associated with storm sudden commencements, *Rep. Ionos. Space Res. Jpn.*, **22**, 45–59, 1968.

KORTH, A. and B. WILKEN, New magnetic electron spectrometer with directional sensitivity for a satellite application, *Rev. Sci. Instrum.*, **49**, 1435–1442, 1978.

KORTH, A., G. KREMSER, and B. WILKEN, Observations of substorm associated particle-flux variations at $6<L<8$ with GEOS-1, *Space Sci. Rev.*, **22**, 501–509, 1978.

KORTH, A., G. KREMSER, S. PERRAUT, and A. ROUX, Interaction of particles with ion cyclotron waves and magnetosonic waves. Observations from GEOS 1 and GEOS 2, *Planet. Space Sci.*, **32**, 1393–1406, 1984.

NISHIDA, A. and L. J. CAHILL, Jr., Sudden impulses in the magnetosphere, observed by Explorer 12, *J. Geophys. Res.*, **69**, 2243–2256, 1964.

OLSON, J. V. and L. C. LEE, Pc1 wave generation by sudden impulses, *Planet. Space Sci.*, **31**, 295–302, 1983.

PATEL, V. L., Sudden impulses in the geomagnetotail, *J. Geophys. Res.*, **73**, 3407–3419, 1968.

PATEL, V. L. and P. J. COLEMAN, Jr., Sudden impulses in the magnetosphere, observed at synchronous orbit, *J. Geophys. Res.*, **75**, 7255–7260, 1970.

S-300 EXPERIMENTERS, Measurements of electric and magnetic wave fields and of cold plasma parameters onboard GEOS-1. Preliminary results, *Planet. Space Sci.*, **27**, 317–339, 1979.

SUGIURA, M., T. L. SKILLMAN, B. B. LEDLEY, and J. P. HEPPNER, Propagation of the sudden commencement of July 8, 1966, to the magnetotail, *J. Geophys. Res.*, **73**, 6699–6709, 1968.

ULLALAND, S., J. STADSNES, M. HAVAG, J. BJORDAL, K. BRONSTAD, G. KREMSER, A. KORTH, L. P. BLOCK, I. B. IVERSEN, K. M. TORKAR, W. RIEDLER, J. KANGAS, P. TANSKANEN, P. STAUNING, and E. AMATA, Observations during a magnetospheric substorm triggered by an SSC. a case study, in *Proc. Conf. Achievements of the IMS*, 26–28 June 1984, Graz, Austria, ESA SP-217, pp. 309–314, 1984.

WILKEN, B., C. K. GOERTZ, D. N. BAKER, P. R. HIGBIE, and T. A. FRITZ, The SSC on July 29, 1977 and its propagation within the magnetosphere, *J. Geophys. Res.*, **87**, 5901–5910, 1982.

Solar Wind-Magnetosphere Coupling, edited by Y. Kamide and J. A. Slavin, 401–411.

Observations of Hydromagnetic-Band Energy in the Magnetosphere near $L \sim 2$ to $L \sim 3$ and Solar Wind Wave and Field Conditions

D. VENKATESAN[1,2], A. WOLFE[3,4], A. MELONI[5], L. J. LANZEROTTI[4], C. G. MACLENNAN[4], and J. F. BAMBER[1]

[1]*University of Calgary, Alberta, Canada T2N 1N4*
[2]*Applied Physics Laboratory, The Johns Hopkins University, Laurel, Maryland 20707, U.S.A.*
[3]*New York City Technical College, City University of New York, Brooklyn, New York 11201, U.S.A.*
[4]*AT&T Bell Laboratories, Murray Hill, New Jersey 07974, U.S.A.*
[5]*Instituto Nazionale di Geofisica, Osservatorio Geofisica Monte Porzio, 00040 Rome, Italy*

An anlysis of the *H*-component of the magnetometer data from AT&T Bell Laboratories and AFGL observatories near $L \sim 2$ to $L \sim 3$, across the USA, during the interval of moderate geomagnetic activity, Jan. 27-30, 1979, reveals a high correlation between hourly values of the magnetic energy densities in the Pc 3 frequency range and the direction of the interplanetary magnetic field. The statistical results support theories of magnetohydrodynamic wave generation in the solar wind on the sunward side of the bow shock and the transmission of the wave energy through the bow shock, magnetosheath, and magnetopause. The results of dynamic power spectral analyses of the ground data reveal a number of sudden power enhancements for field variations with periods in the range 15–50 seconds. The observation of large compressional power in the solar wind at the time of two of these events are cited to be specific examples of magneto-acoustic wave transmission in the magnetosphere.

1. Introduction

The solar wind interacts with the earth producing a geomagnetic cavity; a primary question is 'how is energy transferred from the solar wind to the geophysical environment?' The manifestation of various phenomena naturally involves different modes of transfer. Here we are concerned with one such study. We specifically address the question 'what is the relation between solar wind parameters and geomagnetic field fluctuations measured in the magnetosphere and on the terrestrial surface?' A number of investigations in the past (e.g., YUMOTO *et al.*, 1984; TAKAHASHI *et al.*, 1984; YUMOTO and SAITO, 1983; TAKAHASHI and MCPHERRON, 1982; TAKAHASHI *et al.*, 1981 and references therein) have dealt with this topic; from a statistical point of view, they have considered the association between geomagnetic fluctuations observed at ground level observatories at mid- to sub-auroral latitudes and solar wind parameters. Only a few other studies have dealt with observations at low L values (e.g., RUSSELL *et al.*, 1983 at $L=2.4$).

We cite here examples of two specific events. The results of our extensive study dealing with events on an individual as well as on a statistical basis have been discussed in detail elsewhere (WOLFE *et al.*, 1985). Thus the present paper has the limited objective of focusing attention on the general problem of hydromagnetic energy transfer from the solar wind to deep within the magnetosphere, with which our group has been involved over the past few years.

LANZEROTTI *et al.* (1981) have reported rapid changes (~ few minutes) in the phase and polarization of hydromagnetic waves in the Pc 3 frequency range at latitudinally closely spaced stations ($\Delta L{\sim}0.5$) near $L{\sim}2$. MEDFORD *et al.* (1982) from a further analysis of the data, have reported the existence of substantial differences in the hydromagnetic power measured at periods $\lesssim$100s amongst the stations. Such differences over small spatial scales naturally suggest the ionosphere as an important element as well as the necessity for the study of low latitude magnetic field variations with interplanetary conditions as the energy source. In this investigation, we study the interplanetary magnetic field data and the observations of magnetometers at ground stations across the USA, near $L{\sim}2$ and $L{\sim}3$, to examine potential sources of hydromagnetic energy input to the low-latitude magnetosphere.

2. Data Analysis

Power spectra of ground magnetic data and their connection with solar wind parameters have been examined in our previous studies using linear regression and multiple regression methods (refer to WOLFE, 1980, 1982; LANZEROTTI *et al.*, 1981; WOLFE and MELONI, 1981; WOLFE *et al.*, 1980). The magnetometer data from three sets of ground stations (see Table 1) have been investigated in this study over the interval January 27–30, 1979 during local daytime hours when good interplanetary magnetic field (IMF) data were available from the IMP-J spacecraft. We have performed analysis on the data of the 28th when particularly strong evidence of a latitudinal dependence of spectral power has been observed.

Dynamic (or moving) power spectral analysis technique has been adopted in our study. This involves 385 minute data sets from each station (1400–2025 UT, January 28, 1979); for each station, the power spectra are computed over segments of successive 30 minutes, each segment being shifted from the previous by 5 minutes. Thus 72 power spectra are obtained. Note the sampling intervals for the AT&T Bell Labs and AFGL Stations (see Table 1) are respectively 2s and 1s. After decimating the number of points, each spectrum is calculated using a 512 point Fast Fourier Transform with data through a prolate spheroidal window (see, THOMSON *et al.*, 1976), smoothing subsequently over frequency. The spectra are plotted sequentially, with frequency on the vertical axis, time on the horizontal axis and power logarithmically on a color scale. It is relevant to point out that a spectrum plotted between, say, 1700 and 1705 UT, refers to the calculation over the interval 1700 to 1730 UT; thus an increase in power due to a short term fluctuation (order 15 min) with above average amplitude appears on the dynamic spectra approximately 15 minutes "earlier" than the actual fluctuations appear on an intensity-time plot.

2.1. *Results of the power spectral analysis for January 28, 1979*

The power spectra of the *H* component magnetic data from stations Duckwater, Burma Road and Barstow (stations at $L{\sim}2$) for hour 11-12 L.T., January 28, 1979, are shown in Fig. 1. Since considerable power exists at higher frequencies, the bands under study have been extended to include the ~15–30 range. Note the similarity in

Table 1. Recording station locations.

Station	Geographic latitude (°N)	Geographic longitude (°W)	Geomagnetic latitude (°N)	Geomagnetic longitude (°W)	≈ L values
symbol/name					
SI					
Siple, Antarctica	76.00 (°S)	84.00	64.76 (°S)	8.02	4.0
PB					
Pittsburg, NH	45.03	71.44	56.85	2.30	3.5
DU					
Durham, NH	43.10	70.90	54.43	1.43	3.2
DW					
Duckwater, NE	38.92	115.62	46.17	52.67	2.0
BR					
Burma Road, NE	36.70	115.97	43.95	52.45	1.9
BT					
Barstow, CA	34.93	116.75	42.09	52.84	1.8
LOC					
Lompoc, CA	34.7	120.6	40.2	59.4	1.7
TPA					
Tampa, FL	27.8	82.5	40.7	15.1	1.7
NEW					
Newport, WA	48.3	117.1	55.2	60.4	3.0
MCL					
Mt. Clemens, MI	42.6	82.9	55.8	15.2	3.1
SUD					
Sudbury, MA	42.2	71.3	55.8	1.9	3.1

the spectra at all stations at the lower frequencies ($<10^{-2}$ Hz); this confirms the earlier study of our group (MEDFORD *et al.*, 1982). However significant differences exist at the higher frequencies.

Figure 2 shows a plot of power in the four frequency bands corresponding to 15–30, 30–60, 60–120 and 120–240s as a function of time; L.T. is shown at the top and UT at the bottom, for the same day, for the same three stations. The following points can be made:

(a) The local time (L.T.) dependence of power at the three stations is reasonably

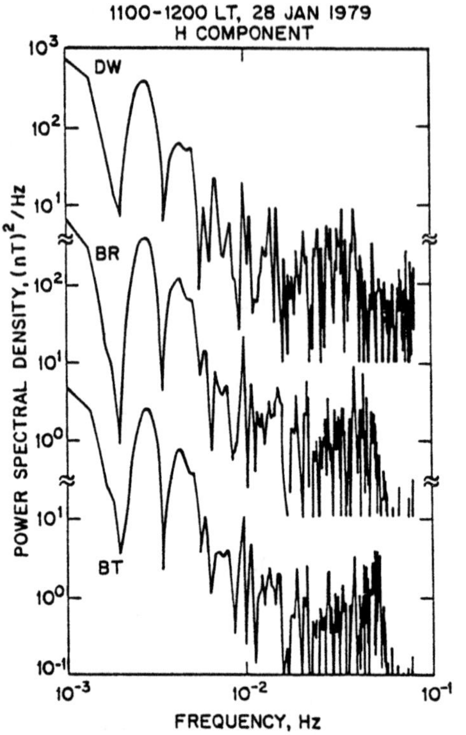

Fig. 1. Hourly power spectral densities near $L=2$. Note the similarities in structure from low frequencies to approximately 60 mhz. Enhancements in Pc 3 wave power are seen.

similar in all bands (15–30, 30–60, 60–120, and 120–240s).

(b) Enhanced power levels are particularly evident near local noon for Pc 3 (15–30s).

(c) The lower frequency band (120–240s) tends to have the highest power levels throughout the local daytime; this is consistent with a spectrum falling off as an inverse power of frequency.

(d) Further, this band tends to be enhanced during local morning hours.

(e) On the other hand, the other three frequency bands tend to have enhanced power at local noon.

2.2. *Correlation with interplanetary magnetic field (IMF)*

Here we focus our attention on two specific occurences of interplanetary disturbances and/or waves and simultaneous change in hydromagnetic energy in the terrestrial magnetosphere. The following provides an appropriate background for this study; for details see WOLFE *et al.* (1985). The power spectra computed for magnetometer data for all the stations listed in Table 1 and integrated power in all the four frequency bands is correlated with the angle which the interplanetary magnetic field makes with the earth-sun line. The obtained linear correlation coefficients are found to be significant in terms of the geomagnetic power to this angle of IMF;

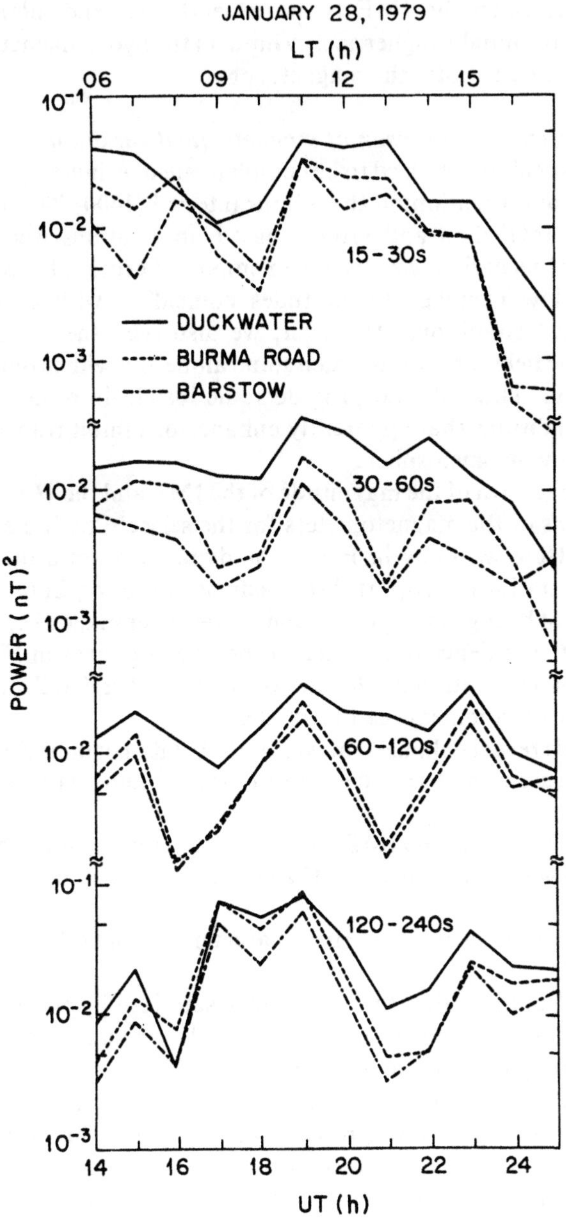

Fig. 2. Local daytime variation of integrated power in four frequency bands at three stations. Relatively high power levels in the shortest period band (top panel) are calculated for morning and early afternoon hours.

substantially higher power can be expected in the Pc 3 range as the angle of IMF decreases. Briefly, when the IMF is more nearly perpendicular to the dayside magnetopause, substantially higher energy input in the hydromagnetic frequency Pc 3 range can be expected to enter the magnetosphere.

2.3. Dynamic spectral comparison of magnetic field variations

Overall statistical studies need to be complemetned by investigation of individual events. We focus our attention on the 6-hour interval (1400–2000 UT), January 28, 1979 when good IMF data and ground-based observations were simultaneously available. During this period, we observe examples of ground-based hydromagnetic wave energy at low geomagnetic latitudes coinciding with events observed in interplanetary field conditions. However, we also see other events in the interplanetary magnetic field conditions which stand alone and with no coincident features in the ground data. These should provide some insight into the characteristics of interplanetary conditions that apparently enhance or inhibit transfer of solar wind energy deep into the magnetosphere.

The dynamic spectra of the magnitude of the IMF and the H and D components of the observations of the magnetometers for the selected period at all the stations were computed. Here we show in Fig. 3 the dynamic spectra of IMF and the D component at six stations, Newport, Mt. Clemens, Sudbury, Lompoc, Barstow and Tampa. The level of background power tends to be lower in the D component than in the H component, and hence our choice of the D component in Fig. 3. For results relating to the enhancements in the H component and other details, see WOLFE *et al.*, (1985). The following comments can be made:

(1) Two intervals reveal enhanced power in IMF data beginning at ~1720 UT and ~1855 UT. These are also observed at most of the ground stations with a lag of ~5 minutes.

(2) Both events persists in the IMF data for ~15 minutes while the ground events last somewhat longer: ≳20 minutes at 1725 UT and up to 50 minutes at 1900 UT at several stations.

(3) One observes a similarity in spectral characteristics in both the IMF and ground magnetic field data.

(4) The 1720 UT event in the IMF data is characterized by power enhancement at frequencies in a narrow band around 0.04 Hz; the ground event enhancement is around the same frequency but in a wider band.

(5) The 1720 UT event shows this sort of correspondence with the IMF in a narrow band near 0.05 Hz; but only some ground stations exhibit this correspondence (the eastern stations do not; the significance of this is not obvious).

(6) Similar enhancement correspondence is seen for the 1900 UT event ~0.04 to 0.05 Hz.

(7) Note that during both, and particularly for the 1900 UT event, the ground dynamic spectra show significantly higher absolute levels of power in the bands mentioned than exist in the spectrum of the IMF magnitude.

(8) It is appropriate to mention some of the non-correlated features as well, to complete the perspective. The enhancements in the IMF data during 1550-1600 UT

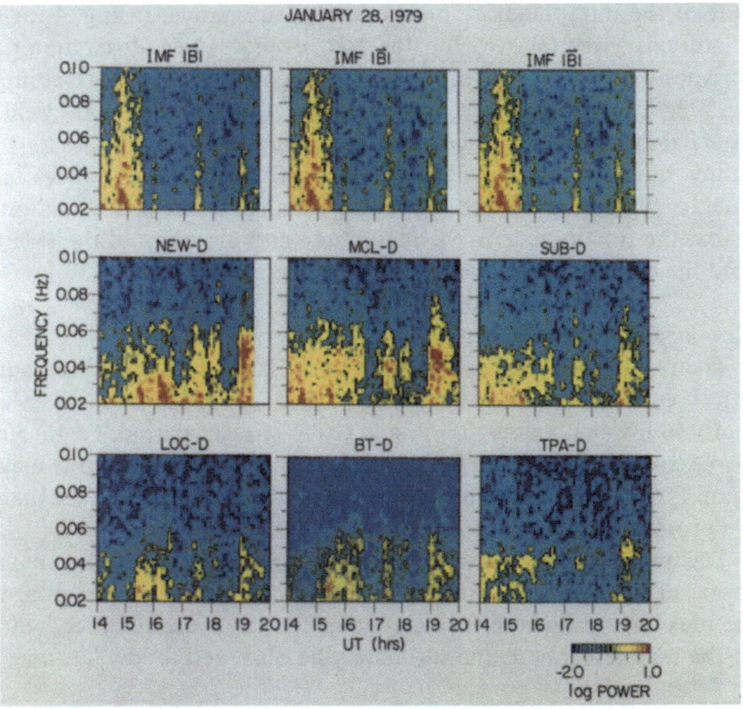

Fig. 3. Comparison of dynamic power spectra of magnetic field data measured in the interplanetary medium and at ground stations (D component) near $L=2$ and $L=3$ across the U.S. Enhancements is ground-measured Pc 3 wave power at widely separated stations occur simultaneously with enhancements for these same frequencies in the magnetic field power in interplanetary space beginning at ~1730 UT and at ~1900 UT.

reveal no correspondence in the ground data enhancements; other periods of enhancements seen during 1400–1530 UT at some ground stations do not reveal any clearcut correspondence with IMF data. Again several events seen on the ground at 1545, 1625 and 1810 UT are not observed in the IMF data.

(9) In conclusion, a good question to pose is "what are the IMF conditions that lead to corresponding enhancements in the ground magnetic field data?"

Comparison of stations separated in latitude (for example, Siple (L~4), Pittsburg (L~3.5) and Durham (L~3.2)) has also been carried out; in general, power decreases with decreasing latitude (or equivalently, with decreasing values of L). Further details of the study are reported by WOLFE *et al.*, (1985).

2.4. *Detailed examination of IMF data*

The IMF data were examined in detail during the two intervals of time 1725–1745 UT and 1905–1925 UT. The objective was to explore the existence of any qualitative association between changes in IMF and those in the characteristics of the low latitude hydromagnetic wave phenomena. The X, Y, Z components and the magnitude of the interplanetary magnetic field are shown in Fig. 4 for the two 20 minute intervals cited above, for the two events when discontinuities in field characteristics were seen. Note the sudden change in the three IMF components (*X*, *Y*, *Z*) at ~1736 UT, resulting in a consequent directional change in the field. The slight drop in the field magnitude can be noticed; of course, it recovered to its ~7 nT value within a few minutes. Again, the initial part of the 1905 UT interval also reveals interplanetary field disturbances with a large change in the field direction. It is to be stressed that the fluctuations in the magnitude and direction of the IMF take place just before the enhanced fluctuations seen on the ground.

2.5. *Summary and conclusions*

1. The results from this study of two specific events extend previous studies of our group (LANZEROTTI *et al.*, 1981; MEDFORD *et al.*, 1982) to (a) lower latitudes (or equivalently to lower L values, $L\sim2$) and (b) to shorter periods (~15–30s).

2. We can visualize that a tangential discontinuity model of the magnetopause can account for the transmission of compressional waves through the magnetopause and wave propagation inward from the outer magnetosphere.

3. The question is, how do such waves propagate deep into the magnetosphere to be observed at such low latitudes? These wide-band compressional waves may reach the plasmapause where they can couple with or excite other hydromagnetic waves in the magnetosphere and ultimately be observed at low latitudes near local noon.

4. In conclusion, solar wind conditions, particularly the two directional changes in IMF appear to control the generation and propagation of MHD waves (Pc 3 band) in the solar wind which are then later observed on the dayside at low latitudes (near $L\sim2$).

The results of this study are consistent with the proposed mechanism in Fig. 5 to consider the transfer of hydromagnetic energy from the solar wind source to the observations at ground level. However, rather than being generated "locally" as upstream waves, the waves penetrating into the magnetosphere could also be generated elsewhere upstream (E. W. Greenstadt, private communication, 1985) and propagate to the magnetopause. Obvious alternative locales for wave production include upstream interplanetary shock waves. The study of GREENSTADT *et al.* (1983) regarding the transfer of pulsation-related wave-activity across the magnetopause is also relevant in this context.

This study clarifies further the problem raised by LANZEROTTI *et al.* (1981) concerning the propagation of hydromagnetic energy deep into the magnetosphere. In that paper it was noted that strong damping of the surface waves from a Kelvin-Helmholtz source would be expected inside the magnetosphere; no substantial energy would be expected at such low *L*-values. However, this present work, as well as that by YUMOTO *et al.* (1985) strongly suggest that compressional waves can penetrate deeply, to low *L*-values, providing energy for the signals observed. This study does not answer completely the observations by LANZEROTTI *et al.* (1981) of the rapid changes in wave polarizations over a short latitudinal baseline at low latitudes nor the changes in spectral content of the signals (MEDFORD *et al.*, 1982). Perhaps interferences

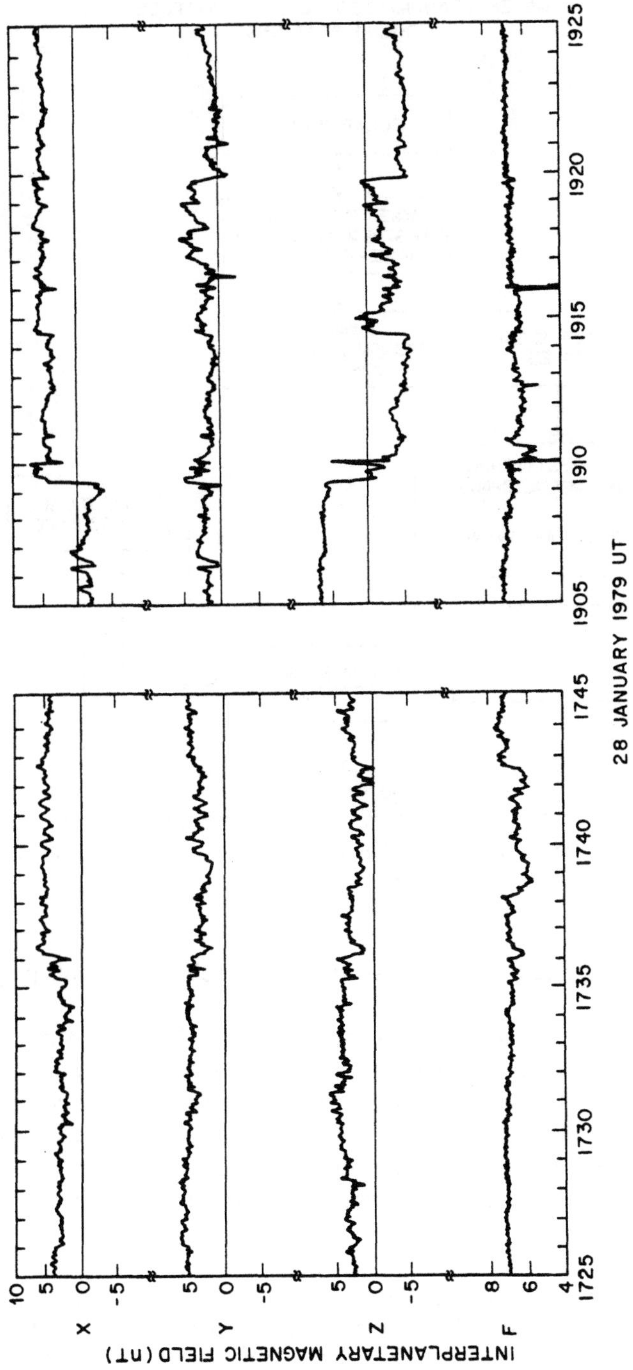

Fig. 4. Interplanetary magnetic field data in solar-magnetospheric coordinates and total field magnitude for two time intervals. Left panel: abrupt changes in the field occurred near 1735 UT. Right panel: changes in the field begin near 1910 UT.

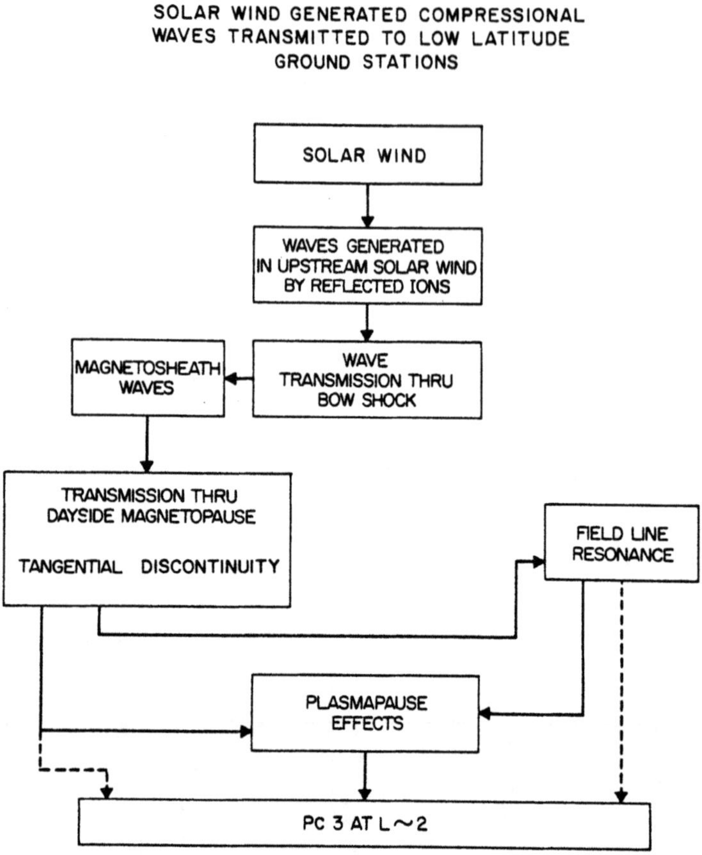

Fig. 5. Flowchart to show MHD wave generation mechanisms and transfer processes from the solar wind to produce dayside Pc 3 at low latitudes.

between multiple compressional waves, as well as ionospheric currents play a role in this regime.

Ideally, an experiment coordinating simultaneous plasma and magnetic field measurements in the solar wind, magnetosheath, outer and inner magnetospheres, and at various terrestrial locations could test out the ideas of wave propagation and transmission through various regions of geospace more convincingly. In spite of practical difficulties in doing such an experiment, progress is still possible by piecing together the results from individual experiments studying different parts of the overall problem.

A portion of the work by A. Wolfe at New York City Tachnical College was supported by the US Air Force Office of Scientific Research grant AF78-3707. A portion of the work by A. Meloni was supported by a grant from Consiglio Nazionale delle Ricerche and Instituto Nazionale di Geofisica. The work by D. Venkatesan and J. Bamber was supported by the

University of Calgary Research Grants Special Award, and NSERC, Canada Research grant 69-1565 to Venkatesan and a summer research fellowship to J. Bamber by the National Science and Engineering Research Council of Canada (NSERC). We would like to thank J. King and L. Brenton of NSSDC for providing the composite omnitape containing IMP-J hourly averages of solar wind parameters. We are also greateful to D. Knecht and P. Fougére of AFGL for providing a tape containing magnetometer data from their network of ground stations. We are especially grateful to D. Norwood and N. Harris for their help with the data reduction and analysis.

REFERENCES

GREENSTADT, E. W., W. W. MELLOTT, R. L. MCPHERRON, C. T. RUSSELL, H. J. SINGER, and D. J. KNECHT, Transfer of pulsation related wave activity across the magnetopause: Observations of corresponding spectra by ISEE-1 and ISEE-2, *Geophys. Res. Lett.*, **10**, 659, 1983.

LANZEROTTI, L. J., L. V. MEDFORD, C. G. MACLENNAN, T. HASEGAWA, M. H. ACUNA, and S. R. DOLCE, Polarization characteristics of hydromagnetic waves at low geomagnetic latitudes, *J. Geophys. Res.*, **86**, 5500, 1981.

MEDFORD, L. V., L. J. LANZEROTTI, C. G. MACLENNAN, A. MELONI, and A. WOLFE, Variations in hydromagnetic wave frequencies at low geomagnetic latitudes, *AGU Trans. Eos*, **63**, 18, abstract SM42B-3, 1982.

RUSSELL, C. T., J. G. LUHMANN, T. J. ODERA, and W. F. STUART, The rate of occurrence of dayside Pc 3, 4 pulsations: The *L*-value dependence of the IMF cone angle effect, *Geophys. Res. Lett.*, **10**, 663, 1983.

TAKAHASHI, K. and R. L. MCPHERRON, Harmonic structure of Pc 3–4 pulsations, *J. Geophys. Res.*, **81**, 1504–1516, 1982.

TAKAHASHI, K., R. L. MCPHERRON, E. W. GREENSTADT, and Arthur NEELEY, Factors controlling the occurence of the Pc 3 magnetic pulsations at synchronous orbit, *J. Geophys. Res.*, **86**, 5472, 1981.

TAKAHASHI, K., R. L. MCPHERRON, and T. TERASAWA, Dependence of spectrum of Pc 3–4 pulsation on the interplanetary magnetic field, *J. Geophys. Res.*, **83**, 2770–2780, 1984.

THOMSON, D. J., M. F. ROBBINS, C. G. MACLENNAN, and L. J. LANZEROTTI, Spectral and windowing techniques in power spectral analysis of geomagnetic data, *Phys. Earth Planet. Inter.*, **12**, 217–231, 1976.

WOLFE, A., L. J. LANZEROTTI, and C. G. MACLENNAN, Dependence of hydromagnetic energy spectra on solar wind velocity and interplanetary magnetic field direction, *J. Geophys. Res.*, **85**, 114–118, 1980.

WOLFE, A., ULF geomagnetic power near L=4,7. A conjugate area study of power controlled by solar wind quantities, *J. Geophys. Res.*, **87**, 10425–10431, 1982.

WOLFE, A. and A. MELONI, ULF geomagnetic power near L=4,6. Relationship to upstream solar wind quantities, *J. Geophys. Res.*, **86**, 7507–7512, 1981.

WOLFE, A., Dependence of mid-latitude hydromagnetic energy spectra on solar wind speed and interplanetary magnetic field direction, *J. Geophys. Res.*, **85**, 5977–5982, 1980.

WOLFE, A., A. MELONI, L. J. LANZEROTTI, C. G. MACLENNAN, J. BAMBER, and D. VENKATESAN, Dependence of hydromagnetic energy spectra near L=2 and L=3 on upstream solar wind parameters, *J. Geophys. Res.*, **90**, 5117, 1985.

YUKMOTO, K. and T. SAITO, Relation of compressional HM waves at GOES 2 to low latitude Pc 3 magnetic pulsations, *J. Geophys. Res.*, **88**, 10041, 1983.

YUMOTO, K., T. SAITO, B. T. TSURUTANI, E. J. SMITH, and S.-I. AKASOFU, Relationship between the IMF magnitude and Pc 3 magnetic pulsations in the magnetosphere, *J. Geophys. Res.*, **89**, 9731, 1984.

YUMOTO, K., T. SAITO, S.-I. AKASOFU, B. T. TSURUTANI, and E. J. SMITH, Propagation mechanism of daytime Pc 3–4 pulsations observed at synchronous orbit and multiple ground-based stations, *J. Geophys. Res.*, **90**, 6439, 1985.

3. ELECTRODYNAMICS OF MAGNETOSPHERE-IONOSPHERE COUPLING

Solar Wind-Magnetosphere Coupling, edited by Y. Kamide and J. A. Slavin, 415–421.

Dayside Convection, Viscous Interaction and Magnetic Merging

Wolfgang BAUMJOHANN and Gerhard HAERENDEL

Max-Planck-Institut für Physik und Astrophysik, Institut für extraterrestrische Physik 8046 Garching b. München, W-Germany

Using data from the GEOS-2 electron gun experiment we have analyzed the dependence of the dayside convection electric field at $L=6.6$ on solar wind and IMF conditions. The convection electric field does not correlate at all with the solar wind momentum flux density ($r<0.25$). Hence, viscous interaction plays only a minor role for equatorial magnetospheric convection at $L=6.6$. The correlation coefficients for convection electric field vs. merging electric field are of the order of 0.5–0.6, thus indicating that dayside convection is predominantly driven by dayside merging. The LT-dependence yields a dawn-dusk asymmetry with the dusk sector electric fields being about twice as strong as those in the dawn sector. The regression coefficients are consistent with a merging efficiency of about 20%. The regression constants (of the order of 0.05 mV/m) apparently describe an electric field stemming from the ionospheric wind dynamo. During very quiet times ($Kp=0$–1) solar wind and ionospheric dynamo are of about equal importance for magnetospheric convection at synchronous orbit.

1. Introduction

The question, if viscous interaction or dayside merging is the dominant agent of magnetospheric convection, was a matter of debate for some 20 years, and it was only recently, when correlation studies using measurements of the polar cap potential drop by low-altitude satellites gave firm evidence that dayside merging is the dominant agent of convection (REIFF *et al.*, 1981; WYGANT *et al.*, 1983; DOYLE and BURKE, 1983). However, since these studies were based on electric field measurements in the polar cap which is threaded by open lobe field lines, they could not rule out the possibility that "viscous-like" interaction in the low-latitude boundary layer (e.g., SONNERUP, 1980) may account for some part of the convection on closed auroral oval field lines.

Accordingly, it seemed sensible to combine the data set of about 1500 hourly averaged electric field vectors from the electron gun measurements aboard the synchronous orbit GEOS-2 satellite (between August 1978 and April 1979) presented by BAUMJOHANN *et al.* (1985) with solar wind plasma and IMF data from KING (1983) and study the relative importance of the above two mechanisms for dayside magnetospheric convection via a regression analysis. Here, we summarize the main results of this study. The material is reported in full in BAUMJOHANN and HAERENDEL (1985).

2. Regression Analysis

For the regression analysis we had to make a compromise between local time resolution and enough data samples to do a meaningful regression and we have thus performed separate regressions on data falling into the following three-hour LT-bins: 05–07 LT, 08–10 LT, 11–13 LT, 14–16 LT, 17–19 LT, and 20–22 LT. In all regressions, GEOS-2 data was correlated with IMF data from the same UT-hour.

Let us first describe the regression analysis for the viscous interaction hypothesis: A good parameter to describe viscous-like interaction between the solar wind plasma and the magnetosphere should be the solar wind momentum flux density (WYGANT *et al.*, 1983) which can be approximated by

$$F = m_p n_{sw} v_{sw}^2 \tag{1}$$

where m_p represents the proton mass and n_{sw} and v_{sw} denote solar wind proton density and bulk velocity, respectively.

Figure 1(a) shows scatter plots of the two components of the convection electric field (in a corotating frame) for one of the six three-hour LT-bins, namely 15 LT, versus the solar wind momentum flux density given by Eq. (1). It can clearly be seen that the convection electric field components display no dependence whatsoever on the solar wind momentum flux density. This is also supported by the regression analysis which yields correlation coefficients of 0.21 (radial component) and 0.12 (azimuthal component). The regression analysis for the remaining three-hour LT-bins gave correlation coefficients which were mostly even below 0.1 and never above 0.25. Hence, there exists no linear correlation between solar wind momentum flux density and dayside magnetospheric convection.

For the correlation between dayside merging and dayside convection we have chosen the simplest form of the merging electric field (see, for example, KAN and LEE, 1979), namely

$$E_{merg} = v_{sw}\ (B_y^2 + B_z^2)^{1/2} \sin^2\left(\frac{\theta}{2}\right), \tag{2}$$

where B_y and B_z denote GSM IMF components, and θ is the angle between the transverse component of the IMF and the B_z-axis.

Similarly to Fig. 1(a), Figure 1(b) gives scatter plots of the convection electric field components versus the merging electric field given by Eq. (2) for the 15 LT interval. In contrast to Fig. 1(a), here a dependence of the convection electric field on the merging electric field is visible and the regression analysis confirms this by yielding correlation coefficients of 0.64 and 0.62 for radial and azimuthal components, respectively. The solid line gives the regression line with regression coefficients of 0.15 and 0.09 and regression constants of −0.03 mV/m and −0.06 mV/m for radial and azimuthal components, respectively. The correlation coefficients for the other LT sectors were of similar magnitude (unless the convection electric field component was close to zero) thus confirming that merging at the dayside magnetopause is an important agent for magnetospheric convection.

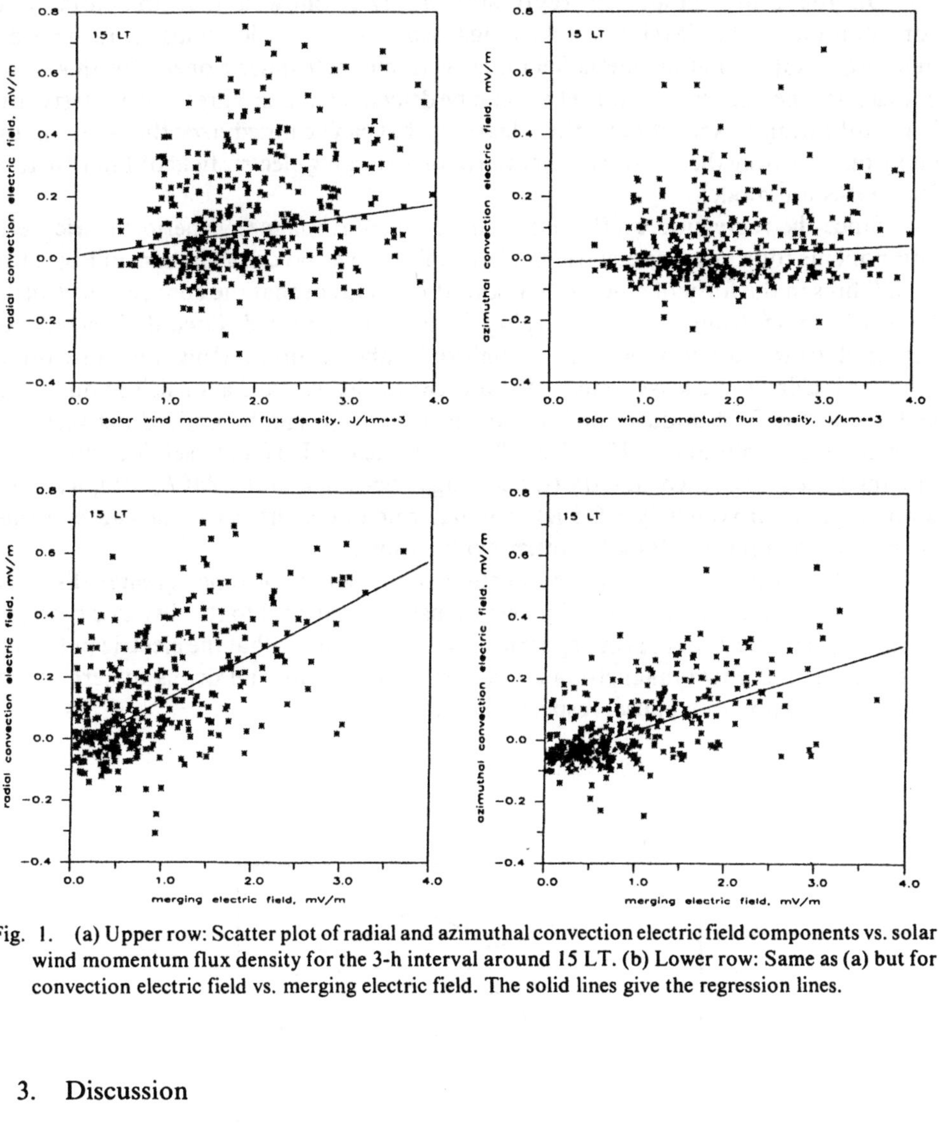

Fig. 1. (a) Upper row: Scatter plot of radial and azimuthal convection electric field components vs. solar wind momentum flux density for the 3-h interval around 15 LT. (b) Lower row: Same as (a) but for convection electric field vs. merging electric field. The solid lines give the regression lines.

3. Discussion

Our regression analysis showed that there is close to no correlation between the solar wind momentum flux density and the synchronous orbit electric field. In this we corroborate the result of WYGANT *et al.* (1983) who concluded that $|n_{sw}v_{sw}^2$ does not correlate with the cross polar cap potential. Our result is, moreover, consistent with a more direct search for evidence on viscous interaction at the magnetopause by MOZER (1984). Mozer looked for dusk-to-dawn electric fields caused by viscous-like interaction just inside the magnetopause and found that those fields are indeed often observed, but that the viscous interaction potentials were typically very low, only a few kV, and constituted less than 10% of typical cross-tail potentials.

On the other hand, our regression analysis showed that the dominant contribution to the dayside synchronous orbit electric field stems from dayside merging of solar wind magnetic field lines with magnetospheric ones. But this is, of course, not the only conclusion which can be drawn from our regression analysis and in the following we shall discuss three features that can be noted from this analysis, or come out by fitting the electric field pattern for a merging electric field of 1 mV/m to a Volland-Stern model.

Since the regression coefficients can be taken as magnetospheric electric field components for a merging electric field of 1 mV/m, we have drawn them in Fig. 2 for all LT-bins in the form of a polar diagram. It is apparent that the convection electric field stemming from dayside merging is predominantly duskward directed, as expected from theory as well as low-altitude observations. However, also quite obvious is a dawn-dusk asymmetry: the duskward electric field at dusk is about twice as large as the field at dawn and about the same ratio can be noticed between the electric field amplitudes at 15 LT and 09 LT. Since our IMF data set does not show any intrinsic (net) B_y asymmetry (the average over the about 1500 B_y samples used amounts to approximetely −0.5 nT) we may rule out B_y effects as the cause of this asymmetry and have to look for other explanations.

Part of this asymmetry is obviously caused by the 1.4 times greater distance between dawn magnetopause and plasmapause as compared to the distance between dusk magnetopause and plasmaspheric bulge. If we assume that the electric potential at the dusk and dawn magnetopause decreases uniformly and drops to zero at the dusk and dawn plasmapause, respectively, we may get

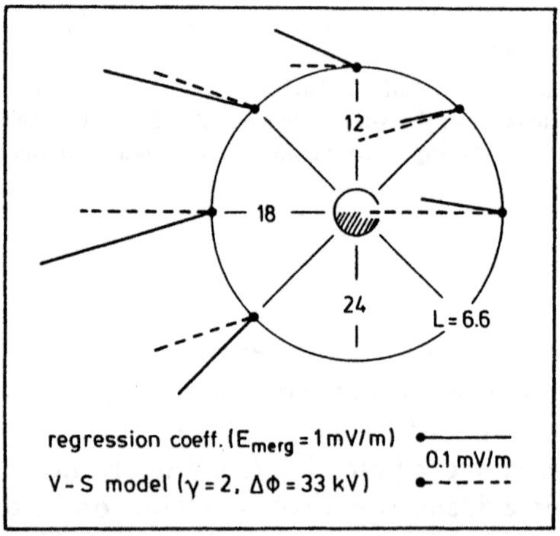

Fig. 2. Comparison of electric field vectors at 6.6 R_E. The solid vectors give the predominantly dawn-to-dusk electric field caused by dayside merging as predicted by our regression analysis (regression coefficients for a merging electric field of 1 mV/m). The dotted arrows give a Volland-Stern-type electric field (for a shielding factor $\gamma=2$ and a cross-tail potential $\Delta\Phi=33$ kV).

$$\Phi_{dusk} = \frac{2}{1.4} \Phi_{dawn}. \tag{3}$$

Hence, roughly 60% of the cross-tail potential drop off in the dusk sector and only 40% in the dawn sector. This estimate corroborates low-altitude satellite observations (P. H. Reiff, pers. comm., 1985).

Additionally, we have drawn into Fig. 2 electric field vectors given by a Volland-Stern-Type potential (VOLLAND, 1973; STERN, 1975) of the form:

$$\Phi_y = -\frac{1}{2} \left(\frac{r}{\Delta y} \right)^{\gamma} \Delta\Phi \sin \phi \tag{4}$$

with $\gamma = 2$ denoting the shielding factor, $r = 6.6\ R_E$ radial distance, ϕ azimuth ($\phi = 0$ at 12 LT), $\Delta\Phi = 33$ kV the cross-tail potential, and $\Delta y = 15\ R_E$ half the distance between dawn and dusk magnetopause. The fit is not extraordinarily good, but it shows that our electric field values represent a cross-tail potential drop of about 33 kV. If we compare this with the cross-tail potential difference of about 190 kV which would result if a merging electric field of 1 mV/m would actually exist along a merging line covering the whole frontside magnetopause (roughly 30 R_E) we get a merging efficiency of about 0.17. This merging efficiency is somewhat lower than an early estimate of 0.35 by GONZALES and MOZER (1974) but agrees nicely with more recent values given by WU *et al.* (1981) and WYGANT *et al.* (1983) which range between 0.13 and 0.18.

The low merging efficiency indicates that only about 20% of the solar wind flow impinging on the dayside magnetopause cross-section is in "ideal contact", i.e. ideal for reconnection, with the magnetopause. Actually, most of the solar wind entering the magnetosheath is deflected around the magnetosphere and never comes into contact with the earth's magnetopause so that the effective scale-length of the interplanetary region in actual contact with the dayside magnetopause is much less than 30 R_E (cf., GONZALES and MOZER, 1974). Moreover, the detection of patchy reconnection and flux transfer events (HAERENDEL *et al.*, 1978; RUSSELL and ELPHIC, 1979) indicates that the "ideal contact" surface is not smooth but rather patchy and transient, thus reducing the effective length of the merging line even further.

The synchronous orbit electric field of about 0.05 mV/m given by the regression constants obviously cannot stem from dayside merging. Since this electric field is not dawn-to-dusk directed but rather towards 09 LT (see Fig. 3 where the regression constants are plotted as electric field vectors) it cannot stem from some weak background viscous interaction either. Hence, we have to follow the same line of arguments as given by BAUMJOHANN *et al.* (1985) and attribute the electric field described by the regression constants to the only source of large-scale electric fields other than the solar wind dynamo, namely the ionospheric wind dynamo (e.g., RICHMOND, 1979). Unfortunately, no realistic models for the ionospheric dynamo fields at auroral latitudes are available with which we could compare our data. But we may, at least, note that direction (outward in the morning and inward in the afternoon and evening hours) as well as amplitude (1–2 mV/m if mapped to the ionosphere

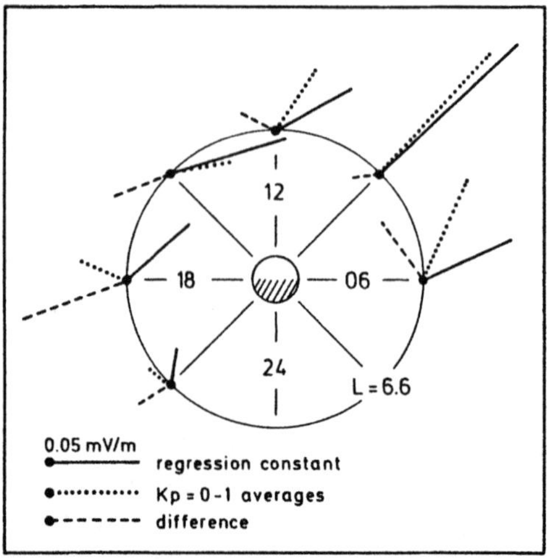

Fig. 3. Comparison of electric field vectors at 6.6 R_E. The solid vectors give the electric field independent from dayside merging as predicted by our regression analysis (i.e., the regression constants). The dotted arrows give *Kp*=0–1 electric field averages. The dashed arrows denote the differences between the regression constants and the *Kp*=0–1 averages.

along dipolar field lines) roughly agree with theoretical predictions (cf., RICHMOND, 1979). However, even if we cannot compare our regression constants with theoretical values for the ionospheric wind dynamo field, we may compare them with empirical models of the quiet day electric field. Of course, the most obvious choice is to compare the regression constants with the *Kp*=0–1 averages of BAUMJOHANN *et al.* (1985). This comparison is shown in Fig. 3 where we also show the difference vectors (i.e., *Kp*=0–1 averages minus regression constants). As can be seen regression constants and *Kp*=0–1 averages have about similar magnitudes (of the order of 0.02–0.1 mV/m) but differ in direction thus yielding difference vectors of 0.02–0.05 mV/m and consistently duskward direction. This duskward electric field would require a cross-tail potential drop of 10 kV for a shielding factor $\gamma=2$. Since this cross-tail potential must be due to merging (otherwise it would also be included in the regression constants) we may conclude that weak merging electric fields of the order of 0.1–0.3 mV/m do exist at the dayside magnetopause even during very quiet times.

REFERENCES

BAUMJOHANN, W. and G. HAERENDEL, Magnetospheric convection observed between 06 and 21 LT: Solar wind and IMF dependence, *J. Geophys. Res.*, **90**, 6370–6378, 1985.

BAUMJOHANN, W., G. HAERENDEL, and F. MELZNER, Magnetospheric convection observed between 06 and 21 LT: Variations with *Kp*, *J. Geophys. Res.*, **90**, 393–398, 1985.

DOYLE, M. A. and W. J. BURKE, S3-2 measurements of the polar cap potential, *J. Geophys. Res.*, **88**, 9125–9133, 1983.

GONZALES, W. D. and F. S. MOZER, A quantitative model for the potential resulting from reconnection with an arbitrary interplanetary magnetic field, *J. Geophys. Res.*, **79**, 4186–4194, 1974.

HAERENDEL, G., G. PASCHMANN, N. SCKOPKE, H. ROSENBAUER, and P. C. HEDGECOCK, The frontside boundary layer of the magnetosphere and the problem of reconnection, *J. Geophys. Res.*, **83**, 3195–3216, 1978.

KAN, J. R. and L. C. LEE, Energy coupling function and solar wind-magnetosphere dynamo, *Geophys. Res., Lett.*, **6**, 577–580, 1979.

KING, J. H., Interplanetary medium data book—Supplement 2: 1978–1982, NSSDC/WDC-A-R&S 83-01, Greenbelt, MD, 1983.

MOZER, F. S., Electric field evidence on the viscous interaction at the magnetopause, *Geophys. Res. Lett.*, **11**, 135–138, 1984.

REIFF, P. H., R. W. SPIRO, and T. W. HILL, Dependence of polar cap potential drop on interplanetary parameters, *J. Geophys. Res.*, **86**, 7639–7648, 1981.

RICHMOND, A. D., Ionospheric wind dynamo theory: A review, *J. Geomag. Geoelectr.*, **31**, 287–310, 1979.

RUSSELL, C. T. and R. C. ELPHIC, ISEE observations of flux transfer events at the dayside magnetopause, *Geophys. Res. Lett.*, **6**, 33–36, 1979.

SONNERUP, B. U. Ö., Theory of the low-latitude boundary layer, *J. Geophys. Res.*, **85**, 2071-2026, 1980.

STERN, D. P., The motion of a proton in the equatorial magnetosphere, *J. Geophys. Res.*, **80**, 595–599, 1975.

VOLLAND, H., A semiempirical model of large-scale magnetospheric electric fields, *J. Geophys. Res.*, **78**, 171–180, 1973.

WU, L., R. GENDRIN, B. HIGEL, and J. BERCHEM, Relationship between the solar wind electric field and the magnetospheric convection electric field, *Geophys. Res. Lett.*, **8**, 1099–1102, 1981.

WYGANT, J. R., R. B. TORBERT, and F. S. MOZER, Comparison of S3-3 polar cap potential drops with the interplanetary magnetic field and models of magnetopause reconnection, *J. Geophys. Res.*, **88**, 5727–5735, 1983.

Solar Wind-Magnetosphere Coupling, edited by Y. Kamide and J. A. Slavin, 423–440.

Solar Wind Control of the Polar Cusp

Eigil FRIIS-CHRISTENSEN

Division of Geophysics, Danish Meteorological Institute, Lyngbyvej 100, DK-2100 Copenhagen Denmark

The polar cusps are characterized by a number of physical processes which are closely related to the solar wind. A number of studies, statistical as well as sevent studies, have been carried out in order to describe and explain these relations. This paper reviews briefly some of the observational information regarding the solar wind control of the location of the cusp, its particle distribution, and its electric fields and currents. Particular emphasis has been put on comparisons between contradictory interpretations of the observations by applying new ground-based observational evidence to some of the important and controversial physics issues like the displacement of the "throat" in local time, the relation between simultaneously existing convection cells of different origin, and the field-aligned currrent regions for northward IMF.

1. Introduction

The original definition of the polar cusps was purely topological and based upon magnetic field models including a cusp-like feature of the earth's magnetic field at high latitudes. On account of the topological characteristica it was expected that the polar cusps would be places where direct entry of solar wind plasma, or rather magnetosheath plasma, could occur.

A historical review of the observational evidences for the existence of the cusp may be found in for example WINNINGHAM and HEELIS (1983) and will not be repeated here. The identification of the cusp in terms of regions of soft electron fluxes has however changed the definition of the cusp from being purely topological, and hence difficult to measure directly, to a definition being based upon characteristic features of paticle precipitation.

Although such a definition is more practical and easier to use observationally its value is dependent on a good physical model of the access mechanism of solar wind plasma into the magnetospheric cusps. However, even today no single theory of this has gained universal acceptance and therefore there still exists quite a lot of controversy about the proper way of defining the cusp.

The difficulties with a proper definition of the cusps are particularly important when comparing different data sets. One of the difficulties with polar cusp observations is that the cusp is a very dynamic region with large spatial gradients and, because of the varying solar wind, also with large temporal variations.

The time and space variations are furthermore coupled in a complicated way because the position of the cusp itself is not fixed, but varies according to solar wind and magnetospheric variations. It is therefore natural that the region of the magnetospheric cusp has been difficult to describe, although the recent years have provided new and interesting results based on observations from spacecrafts designed particularly for polar cusp studies like for example the Dynamics Explorer satellites.

Ground based observations of the projection of the cusp into the ionosphere have also increased and they have the advantage of being able to monitor the timevariations at specific locations. The projection to ionospheric altitudes of the cusp corresponds to large areas of the magnetopause boundary layers, which means that large spatial gradients in the observed physical parameters are expected.

2. Cusp Particle Distributions

Today the most generally used characterization of the cusp is that it is a region with very high electron fluxes with very low average energy, i.e., from 10 ev to 500 ev. One of the problems with this definition is that particle populations of this kind may be found at noon as well as in the dawn and dusk sectors, although the average energy tends to increase when moving more than two hours away from local magnetic noon. Using spacecraft data taken during meridional passes across the cusp the latitudinal boundaries of this region are relatively well defined. The extension of the cusp in magnetic local time seems harder to define since similar sharp longitudinal boundaries have not been found. This is one of the reasons for the alternative name of the cusp namely the "cleft".

One of the few studies which have tried to identify the average location of the cusp in latitude as well as in longitude on a statistical basis is a study by CANDIDI *et al.* (1983)who used a large number of DMSP F-2 satellite observations of very low energy electrons. The particle distribution will of course be influenced by the effect of the varying position of the cusp which will tend to smear out the normally very sharp latitudinal boundaries of the cusp. Nevertheless, a maximum in the intensity of very low energy electrons is seen around noon, and in particular for northward IMF a difference in response to the azimuthal component B_y is seen. This is demonstrated in Fig. 1, taken from CANDIDI *et al.* (1983), showing maximum contour lines of average spatial intensity of 50 eV electrons over the Southern Hemisphere for $B_y>0$ and $B_y<0$. The distribution consists of two parts, a narrow part, not particularly B_y dependent, which is supposed to be the dayside auroral oval and hence populated with particles from the dayside plasma sheet boundary layer. A second part comprises a much broader (statistically) region extending to very high latitudes and which displays a pronounced B_y dependence.

If the solar wind is the direct source of the cusp particles the intensity of precipitating low energy electron flux should vary according to the density variations in the solar wind. CANDIDI and MENG (1984) used this starting point in a study and found that the intensity of precipitating electron flux increases with increasing solar wind density, although not linearly. Furthermore, they found that a southward IMF increases the intensity of the electron flux wiht a factor of 2 relative to the positive B_z condition. The observed features agree with models by which electrons enter into the magnetosphere in association with field-line merging although the possibility of non-linear diffusion processes cannot be eliminated (CANDIDI and MENG, 1984).

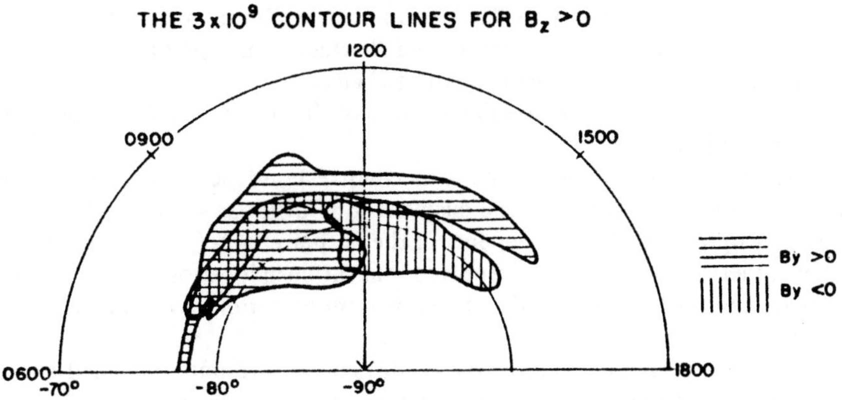

Fig. 1. Regions of maximum precipitation of 50 eV electron fluxes recorded when B_z was positive. (After CANDIDI *et al.*, 1983).

Since a definition of the polar cusps from the electron energy spectrum alone seems ambiguous, other particle characteristics have been used in an attempt to identify the cusp. BURCH *et al.* (1982) have found characteristic V-shaped energy/pitch-angle injection signatures in the positive ions observed on DE 1, which have been explained as caused by convection into the polar cap of cusp field-lines through a spatially restricted region of plasma injection. This signature of the cups has been used with success to identify the location of the cusp relative to plasma flow observations by BURCH *et al.* (1985).

3. Location of the Cusp

The latitudinal position of the polar cusp is known to be very variable. A number of factors of solar wind or magnetospheric origin have been proposed to be responsible for the movements of the cusp but also here there still exists fundamental disagreement on the relative importance of these factors. The solution of this problem is important because it is related to the question of how the solar wind plasma enters the magnetosphere and, in particular, the question of the importance of magnetic merging in this process.

A consequence of the merging hypothesis is that the cusp moves towards the equator following an increased southward component of the IMF which increases the erosion of closed field lines at the front of the magnetosphere. Such an equatorward movement of the cusp due to B_z was demonstrated by BURCH (1973) and was also concluded in a study of storm time variations of the polar cusp by MENG (1983). The poleward movement during the recovery is, hoever, not directly related to B_z according to MENG (1983).

MAKITA and MENG (1984) studied the relation of the different boundaries of polar cusp precipitation zones near the cusp to the northward component of the IMF. They concluded that only the poleward boundary of the cusp (<500 ev electrons) does show a positive correlation with B_z. The equatorward boundary of the cusp even has an equatorward movement with increasing B_z which is interpretated as a broadening of the soft zone normally associated with the cusp.

So it seems difficult to relate the location of the cusp to the B_z component of the IMF in a simple way.

This lack of an obvious one-to-one correspondence between the IMF B_z component and the latitude of the cusp was one of the reasons for EATHER *et al.* (1979) to propose that the movements of the cusp are largely controlled by internal magnetospheric processes. They reported that the cusp location, as determined from the position of the 6300Å dayside aurora, was much better ordered by the *AE* index than by B_z.

SANDHOLT *et al.* (1983) also studied dayside aurora but concluded a far better correlation with B_z than with *AE*. This interpretation has been questioned by EATHER (1985) claiming that the data show just as good support for *AE* correlation.

Since the auroral data are not immediately comparable with the soft electron precipitation data one explanation of the apparent discrepancies could be that the auroral data are not exclusively due to cusp particle precipitation. In the paper by EATHER *et al.* (1979) 12 hours of data from dawn to dusk were considered, but no obvious difference in the behaviour was noted when only ±4.5 hours around local noon were considered (EATHER, 1985). The question therefore arises whether or not "cusp" precipitation, either directly measured electron fluxes or aurora, is in fact all originating in the cusp or if part of the "cusp" precipitation is due to other sources like the dayside boundary plasma sheet.

Another parameter which is related to the cusp is the ionospheric convection current DPY, which is very closely related to the B_y component of the IMF. This current has been shown to be located poleward of the region of enhanced *F*-region electron temperatures which are supposed to be related to cusp particle precipitation. Figure 2 after VENNERSTRØM *et al.* (1984) shows how the equatorward boundary of the DPY current follows the time variations of the latitude of the region of maximum T_e and the convection reversal.

In Fig. 3 is shown the locations of the DPY current observed from the Greenland chain of magnetometers during summer months of 1972 and 1973. The positions are plotted against B_z as well as against *AE* but with no obvious difference in the correlation.

In conclusion the various data do neither significantly support an exclusive B_z nor an *AE* dependence of the polar cusp location. On the contrary, examples do show that both parameters may be simultaneously operating.

4. Plasma Convection and Birkeland Currents: Models

A second major solar wind contorl of the cusp is the control of the plasma convection and of the Birkeland or field-aligned currents.

HEELIS (1984) studied a large data set from the *AE* satellites and presented models for the ionospheric convection in and around the cusp and its relation to the orientation of hte interplanetary magnetic field. The models, all for southward IMF,

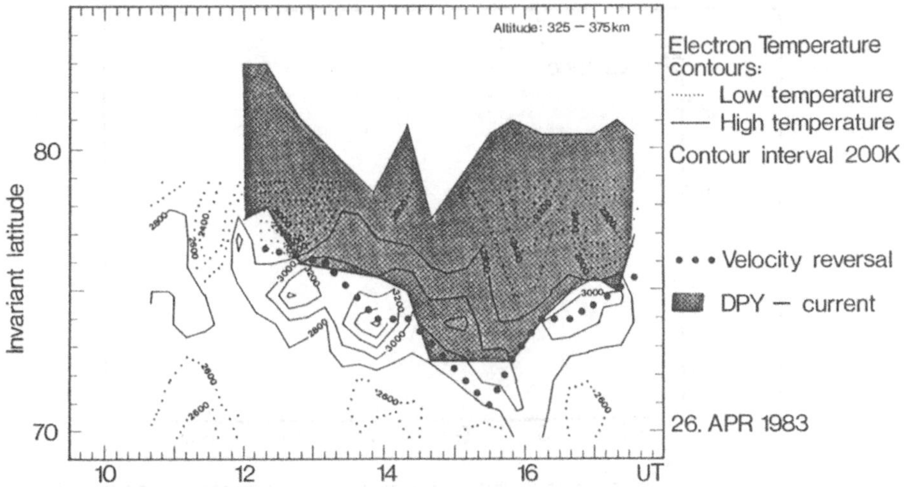

Fig. 2. Comparison of dayside *F*-region electron temperature maxima with the ionspheric DPY current and the convection reversal. (After VENNERSTRØM *et al.*, 1984).

but for different azimuthal components, are shown in the Fig. 4. A consistent tendency in the observations of convection is a "throat-like" feature displaced to the morning side. This was proposed to be caused by the preferred garden hose orientation of the solar wind magnetic field which causes merging to take place predominantly in the prenoon sector for both signs of B_y as illustrated in Fig. 5. The displacement is proposed to grow with the relation between $|B_y|$ and $|B_x|$. The model also predicts a displacement to the postnoon side for the orto-garden hose orientation of the IMF, but examples of this kind were not given by HEELIS (1984), probably because of missing spacecraft data at the right place and time for this non-typical orientation of the IMF.

Using ground based data, however, it should be possible to test the hypothesis. Figuere 6 shows Greenland magnetometer data during a consistent non-typical relation between the signs of B_x and B_y for 3–4 hours around local magnetic noon. The expected displacement of the throat in the equivalent convection velocities towards the afternoon is not seen in this example nor during a similar example for opposite B_y and B_x (not shown here). Although the convection velocities shown here are inferred from the magnetometer measurements assuming a homogeneous ionospheric conductivity (summer season) and zero neutral wind, it is not anticipated that direct electric field measurement would have changed the interpretation since radar measurement of convection velocities normally agree well with the magnetometer data in the sunlit part of the ionospheric polar cap and cusp as will be demonstrated later on in this paper.

Another recent model of convection patterns and field-aligned currents is the model by BURCH *et al.* (1985) and REIFF and BURCH (1985) based on DE 2 obsrvations of plasma flow and Birkeland currents. Three aspects of this model which

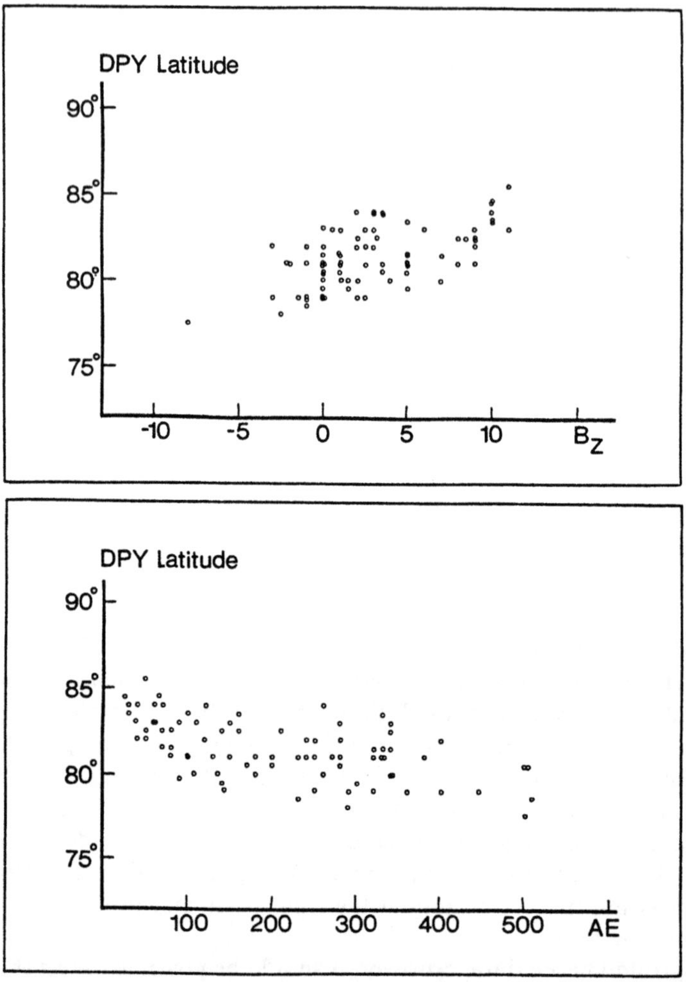

Fig. 3. Observed positions (invariant latitudes) of the DPY current as a function of the IMF B_z component and as a function of the AE index.

is shown in Fig. 7 will be discussed. First for moderately negative B_z, i.e., $B_z \simeq -|B_y|$ the model predicts simultaneous existence of a merging convection cell connecting open and closed field lines and a lobe cell, completely on open field lines. Secondly is noted the existence of permanent "viscous" cells in the dawn and dusk sector. Finally for northward IMF the model has a double cell structure in the polar cap for $B_y <> 0$ with both cells having the same orientation.

The last aspect is in contradiction to the model shown in Fig. 8 which has been proposed by POTEMRA *et al.* (1984) and which is based on an analysis of MAGSAT magnetometer data by ZANETTI *et al.* (1984). Here a double cell structure in the polar cap for $B_y <> 0$ consists of two cells of different orientation like the case: $B_y = 0$. Only

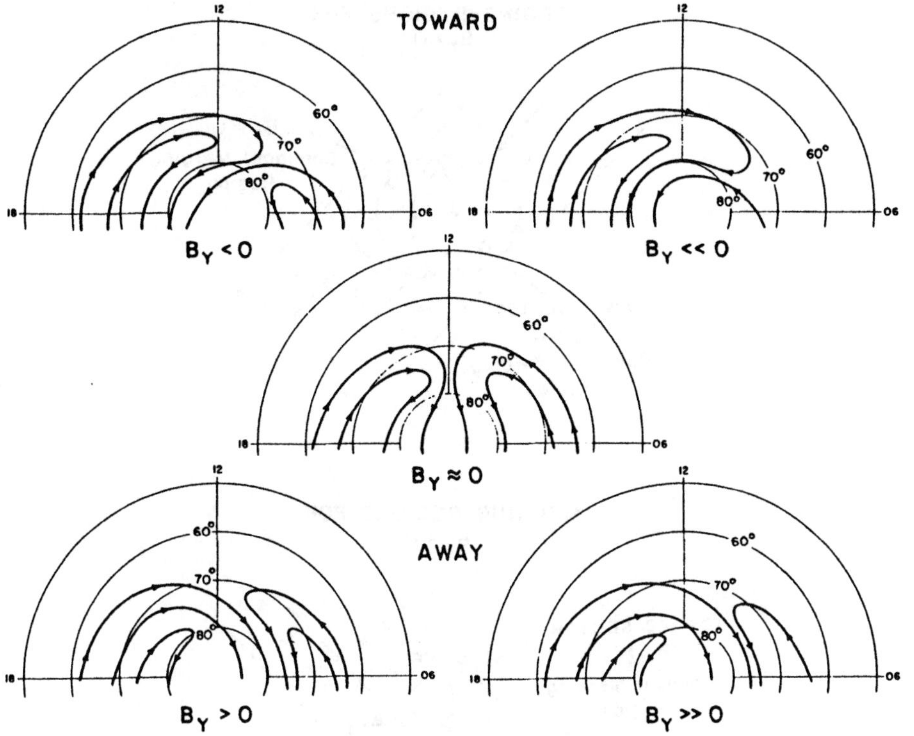

Fig. 4. Schematic representation of flow geometries on the dayside northern hemisphere for different magnitudes of B_y. B_z is negative and $|B_x|$ is constant in all cases. (After HEELIS, 1984).

the relative sizes of the cells are influenced by B_y.

A common feature of all these convection models is that they are qualitative ones, based upon theoretical considerations and single event studies which support the predictions of the models. Quantitative empirical models of the convection around the cusp have not bee reported due to the difficulty in using the rather inhomogeneous set of satellite observatoins.

Ground-based observations, however, are more homogeneous and are suitable to be used for an empirical model. A model of this kind, using normal magnetic observatory data has been described by FELDSTEIN *et al.* (1984). FRIIS-CHRISTENSEN *et al.* (1985), created an empirical model based on a more dense network of stations, namely the meridional magnetometer chain in Greenland. The model consists of calculated electric fields and currents for various direction of the IMF based upon empirical linear relations between the IMF B_y and B_z components using the algorithm as described in KAMIDE *et al.* (1981).

The model correspond to a typical magnitude of the IMF-vector, 5 nT, which means that large IMF magnitudes are not really considered by FRIIS-CHRISTENSEN (1985). Since the IMF independent part of the convection models, which was also

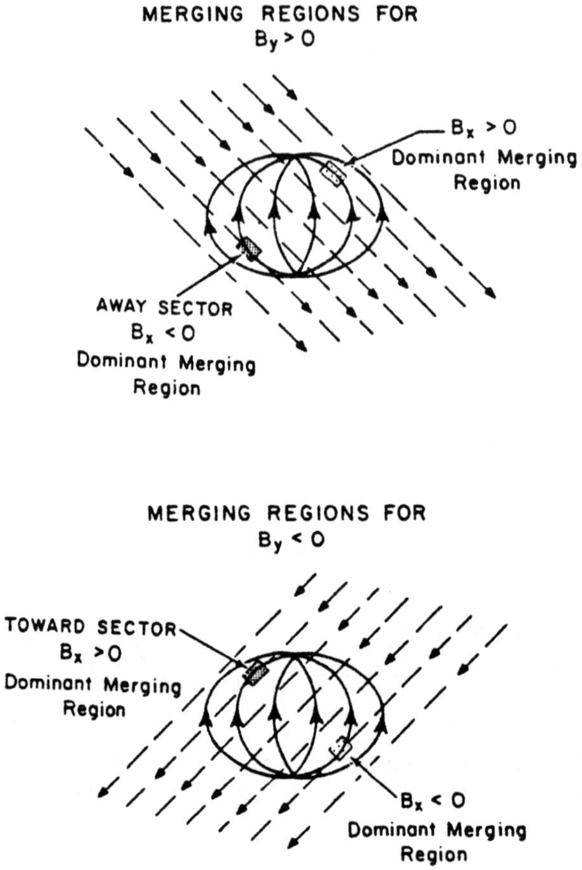

Fig. 5. Schematic view of the IMF and geomagnetic field geometries viewed from the sun. Regions of preferred merging for different orientations of the IMF are indicated by the shaded boxes. (After HEELIS, 1984).

calculated, does contribute considerably to the total convection for typical IMF values, it is not immediately clear which parts do arise from IMF dependent part of the convection.

In order to give an estimate of the IMF-dependent part of the high-latitude convection pattern Figure 9 shows the residual electric potential corresponding to the patterns for $B_z>0$ and $B_z<0$ in Fig. 4 of FRIIS-CHRISTENSEN *et al.* (1985) but with the IMF independent part subtracted. The residual patterns have the advantage of a more pronounced display of the different influences of the IMF B_z and B_y components. The patterns have been derived using the assumption of linear relationship between the geomagnetic variations and the IMF B_y and B_z compoents which means that increased IMF magnitudes will affect the pattern linearly.

For northward IMF the effect of the IMF is the existence of two convection cells

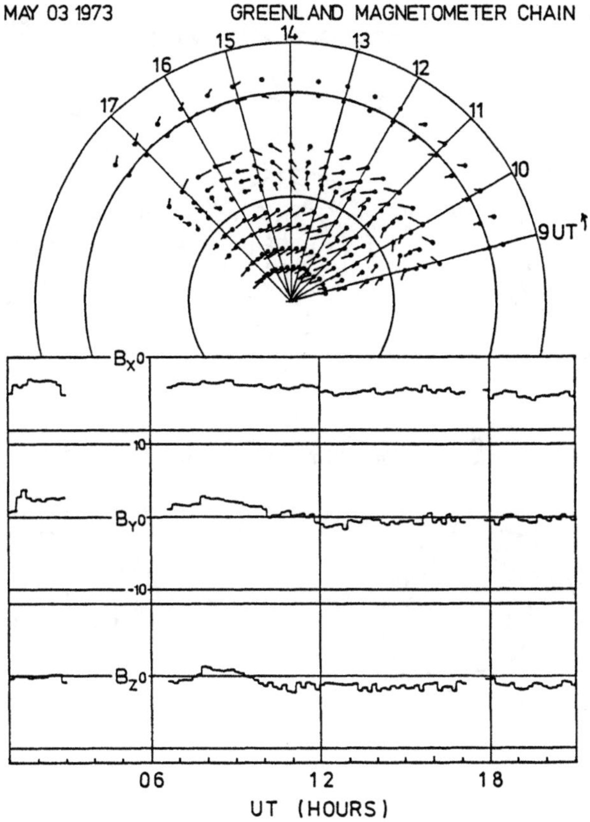

Fig. 6. Convection velocity vectors in the dayside northern hemisphere inferred from the Greenland magnetometer chain comprising an interval from 1200 to 1530 UT of orto-garden hose orientation of the IMF ($B_x<0$ and $B_y<$). Magnetic local time is at 1400 UT.

of opposite orientation and also oppositely directed to the basic two-cell convection pattern. The two cells are displaced towards the dayside of the polar cap and they are of similar size for $B_y=0$ implying sunward convection in the dayside polar cap around magnetic noon. These features are consistent with earlier works by for example MAEZAWA (1976) and BURKE *et al.* (1979). For $B_y<0$ the afternoon cell is largest and for $B_y>0$ the morning cell is dominating. These features are not in agreement with the convection model of REIFF and BURCH (1985) for $B_z>0$ (shown in Fig. 7) but are rather in agreement with the model in Fig. 8 by POTEMRA *et al.* (1984). The two cells in the residual patterns of Fig. 9 are not very prominent in the total convection patterns of FRIIS-CHRISTENSEN *et al.* (1985) because only when the IMF vector is sufficently large these features dominate the total convection patterns. A direct comparison between the residual flow patterns in Fig. 9 with the complete flow patterns in Figs. 7 and 8 might seem inlegitimate. Since, however, the flow pattern in Figs. 7 and 8 have been sketched qualitatively on the basis of selected cases with clear, but not

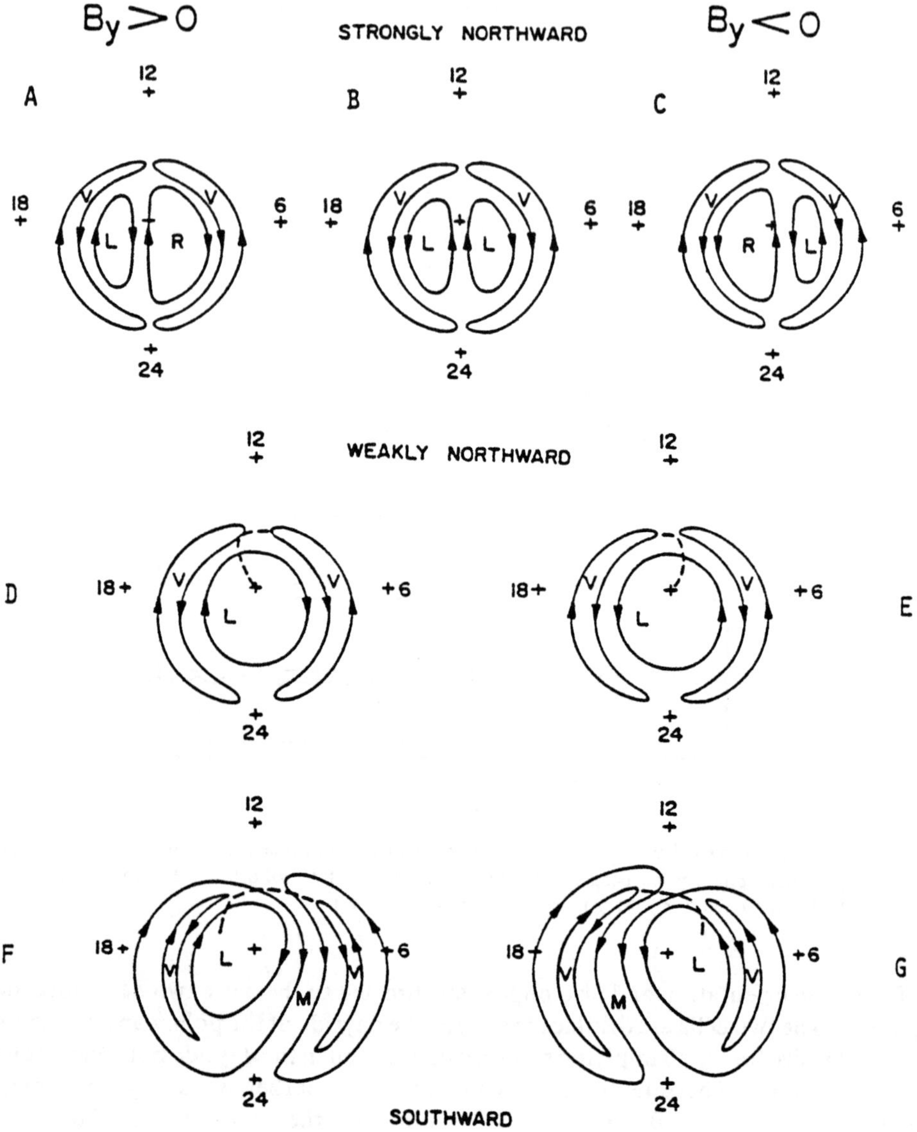

Fig. 7. Schematic northern hemisphere polar-cap convection patterns for various orientations of the IMF. (After REIFF and BURCH, 1985).

necessarily average IMF effects it seemed more appropriate to compare with statistical patterns where IMF effects have been analogously emphasized by removing the IMF-independent part of the convection.

Corresponding residual patterns of the distribution of field-aligned currents are shown in Fig. 10. They show that the field-aligned currents for $B_z>0$ are also

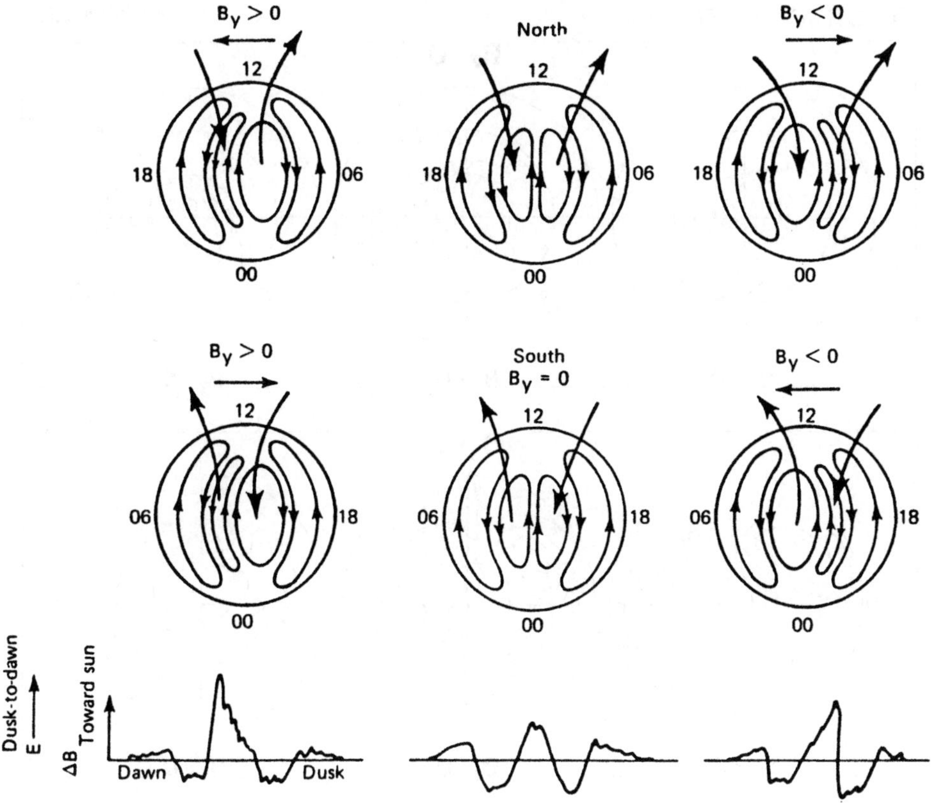

Fig. 8. Schematic diagrams of convective flow and Birkeland current patterns in the northern and southern polar regions for positive B_z. (After POTEMRA *et al.*, 1984).

displaced towards the dayside.

For southward IMF the lower panel of Fig. 9 shows the IMF B_y and B_z effect on the convection pattern. We notice for $B_y>0$ a single convection reversal in the plasma flow in the prenoon sector and this feature is in agreement with the model of BURCH *et al.* (1985). For $B_y<0$ a satellite passing the convection pattern towards the magnetic pole along the meridian at 09 magnetic lcoal time would encounter eastward convection in the auroral zone. Poleward of this is situated a reversal to westward convection in a narrow region before finally eastward convection follows in the polar cap. This is in agreement with the model of BURCH *et al.* (1985) and their observations on the DE2 satellite during large IMF magnitudes (15–20 nT). It should be noticed that in the electric potential patterns in Fig. 9 the meandering of the potential contours seen in the prenoon sector for $B_y<0$ is probably the statistical effect of two cells, a narrow one corresponding to a crescent shaped one and a circular one in the polar cap as described in the model of REIFF and BURCH (1985).

The model of REIFF and BURCH (1985) is antisymmetric with respect to the midnight–noon meridian regarding the B_y effects. The observations of the DE 2

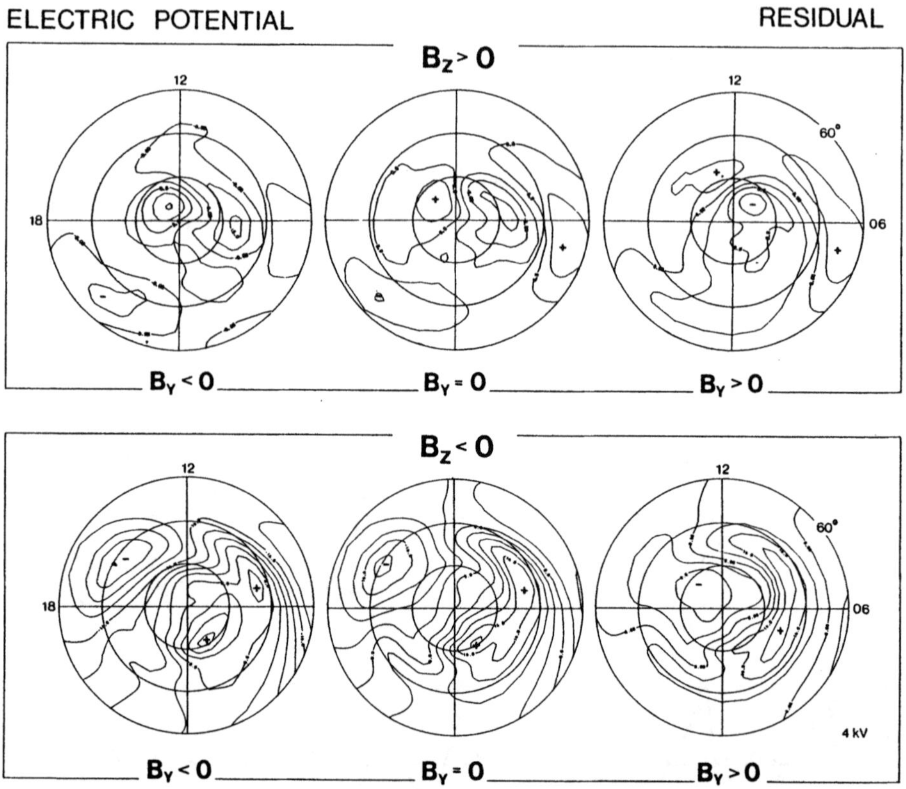

Fig. 9. Residual electric potential patterns in the northern hemisphere for various IMF orientations calculated on the basis of an empirical model of IMF effects on ground-based magnetic variations. The IMF independent part of the pattern has been subtracted.

satellite however, only support the prenoon part of the convection pattern since the spacecraft orbits shown in the paper by BURCH *et al.* (1985) are more or less along the 09 to 21 magnetic local time meridian and thus are unable to give information along the 15 magnetic local time meridian. Although REIFF and BURCH (1985) have constructed the convection models assuming strict antisymmetry with respect to the noon–midnight meridian the empirical model in Fig. 9 does not support this since the prenoon cell is considerably stronger than the postnoon cell under all B_y conditions. Correspondingly a double reversal in the convection is not seen in the afternoon for $B_y>0$ as predicted in the model of REIFF and BURCH (1985).

For the prenoon sector the model of REIFF and BURCH (1985) does compare relatively well with the empirical model in Fig. 9. The corresponding model of field-aligned currents in Fig. 10 which shows three current sheets for $B_y<0$ and four current sheets for $B_y>0$ are also consistent with DE 2 observations and interpretations of BURCH *et al.* (1985).

However, one difference does exist in the interpretation. While Figures 9 and 10

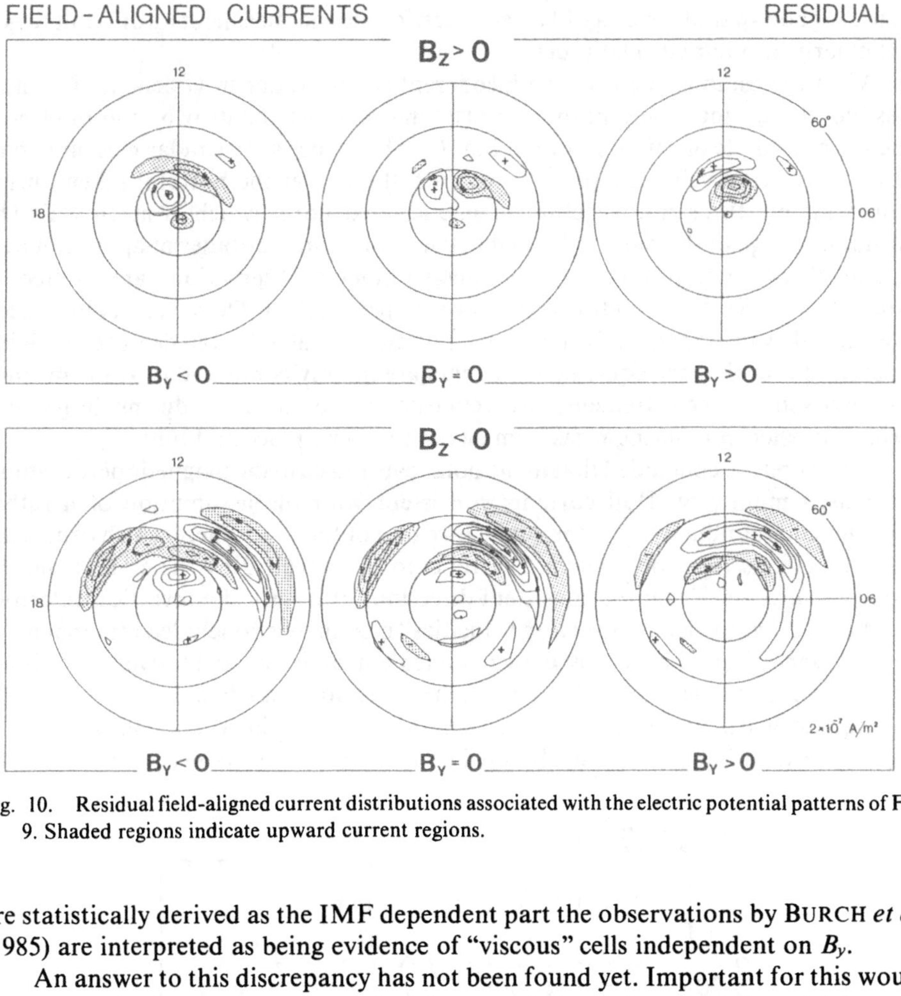

Fig. 10. Residual field-aligned current distributions associated with the electric potential patterns of Fig. 9. Shaded regions indicate upward current regions.

are statistically derived as the IMF dependent part the observations by BURCH *et al.* (1985) are interpreted as being evidence of "viscous" cells independent on B_y.

An answer to this discrepancy has not been found yet. Important for this would be spacecraft observations along the 15 magnetic local time meridian which could confirm the antisymmetric behaviour of the convection anticipated by REIFF and BURCH (1985).

Another possibility could be that the "viscous'" cell is not completely independent on B_z which could mean that the asymmetric throat for $B_y=0$ in Fig. 9 might be associated, at least partly, with viscous interaction.

5. Plasma Convection in the Ionospheric Cusp: Observations

Observations are normally not so simple as the presented models of high-latitude convection and field-aligned currents. The global study of the Coordinated Data Analysis Workshop (CDAW) 6 have shown the importance of the time and spatial variability (KROEHL and KAMIDE, 1985). Even from regional studies we know that large spatial differences in the response of the convection to changes in the IMF may occur. A good spatial coverage of direct electric field measurements is difficult to get, particularly at polar cusp latitudes.

With the establishment of the incoherent scatter radar in Greenland it is now possible to look into the ionospheric cusp and measure the distribution of electric fields. An example of this is shown in Fig. 11. This figure shows radar measurements of the convection velocities (arrows with a full circle at the bottom) taken during approximately three hours before magnetic local noon which is at 14 UT. The latitudinal range of the radar is limited between invariant latitudes of approximately 70 and 80 degrees. For comparison magnetometer observations are plotted as "equivalent convection" vectors (arrows with open circles). These vectors have been constructed as overhead equivalent current vectors rotated by 180 degrees, implying the assumption that the equivalent currents are mainly composed of Hall currents. The two sets of measurements are remarkably similar even during large time variations when comparing measurements at the same place and time.

It is therefore concluded that in the polar cap and cusp the magnetic pertubations are caused mainly by Hall currents consistent with the assumption of a rather homogeneous conductivity in the sunlit *E*-region of the ionosphere which causes the magnetic effects from field-aligned currents to be cancelled by the effect from the Pedersen currents. From the consistent directions of observed convection velocities and inferred "equivalent convection velocities" it is further concluded that magnetometer observations may be used to complement at least qualitatively the direct measurements at places or at times when these are not available.

Equatorward of the convection reversal at about invariant latitudes of 76 degrees, eastward convection is observed consistent with the convection models for

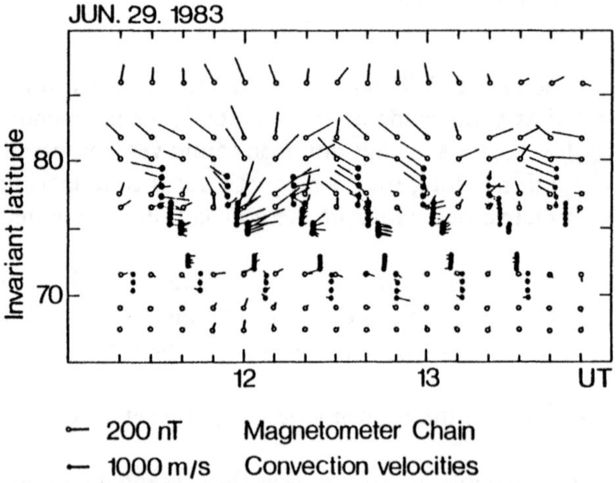

Fig. 11. Simultaneous observations of convection velocity vectors obtained from the ISR radar in Søndre Strømfjord (vectors with filled circles at the base) and inferred from magnetometer observations along the West Coast of Greenland three hours before local magnetic noon at 14 UT on June 29, 1983.

the prenoon sector. Poleward of the reversal convection velocities are more variable due to changes in the interplanetary B_y and B_z components. Before 1150 UT the convection is westward corresponding to a positive B_y. The change in the B_y component takes place at about 1140 UT and is seen about 20 minutes later in the convection velocities which turn eastward for a short time interval until about 1230 UT when a change is caused by a positive B_y occurring at the satellite position at 1200 UT. The time resolution of the radar measurements of the convection velocity vectors for selected points along a whole meridian is about 20 minutes and hardly sufficient to give an exact estimate of the time delay from the satellite position to the ionosphere. The chain of magnetometers, however, indicate that the change in the convection is starting shortly after 1220 UT, most clearly on the East Coast of Greenland, which is located in postnoon sector.

This is further illustrated in Fig. 12 which shows equivalent convection velocity vectors as observed on the West Coast of Greenland together with the measurements on the East Coast. Using two chains spatially separated it is possible to show spatial distribution of the time variations of the polar cap and cusp convection pattern. The

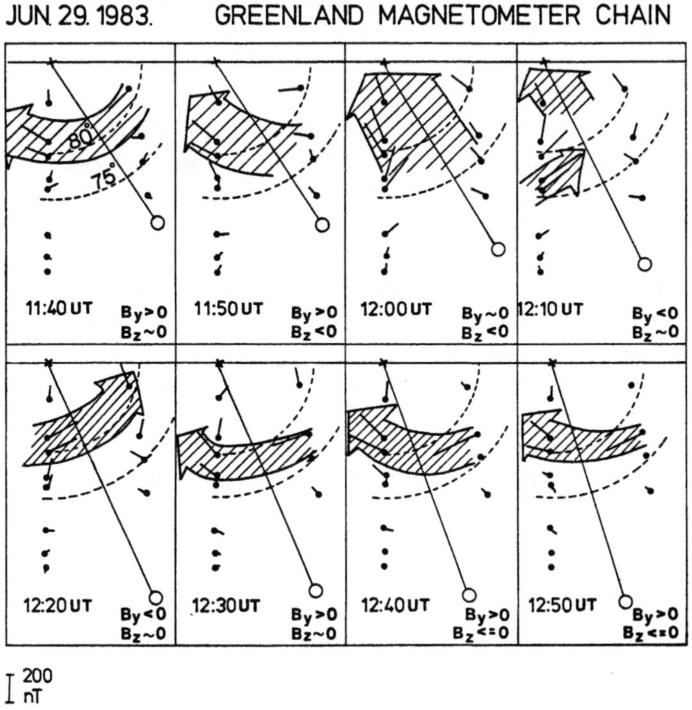

Fig. 12. Time development of the spatial distribution of convection velocity vectors inferred from the magnetometer chains in Greenland. The magnetic pole is indicated by a cross and the meridional magnetometer chain on the Wes Coast of Greenland is placed on a vertical line. The East Coast stations are placed according to the position in an invariant latitude-magnetic local time diagram where magnetic local noon is indicated by a line connecting the pole and the position of the sun. The main areas of convection are emphasized by hatching.

Figure shows observations at all Greenland stations taken with 10 minutes intervals. A cross indicates the position of the pole and the invariant latitudes of 70 and 80 degrees are given as well as the longitude of the sun (magnetic local noon) which is located between the West Coast chain in the prenoon sector and the East Coast stations in the postnoon sector. To facilitate the interpretation of the data the main convection regions are indicated as hatched areas. For each interval is given the signs of the IMF B_y and B_z components taking into account a typical time delay from the position of the satellite to the ionosphere of about 20 to 30 minutes.

At 1140 UT B_y is positive while B_z is around zero. A nearly circular convection cell is seen, probably associated with the lobe cell (L-cell) in the model of REIFF and BURCH (1985) (see Fig. 7). From 1150 UT to 1200 UT the IMF is southward while B_y is decreasing to zero, illustrated by the convection direction turning from westward to poleward. The influence of negative B_y is seen first in the prenoon sector corresponding to a convection from lower to higher latitudes and probably from closed to open field-lines in agreement with the merging cell (M-cell) of REIFF and BURCH (1985) in Fig. 7. Analogously the change to $B_y>0$ is seen most pronounced in the postnoon sector, which is also consistent with a merging cell now crossing from closed field-lines in the postnoon sector to open field lines in the prenoon polar cap.

6. Summary

During the recent years the number of relevant high-latitude observations have increased considerably due to the extended ground-based observations during the International Magnetospheric Study (IMS) and a number of satellite experiments specifically designed in order to provide good measurements of cusp particles and fields.

In this paper some of these observations have been discussed with particular emphasis on those aspects of these studies which enable a comparison between ground-based and space-based observations.

The conclusion of the paper may be summarized in the following:

1) The cusp is characterized by intense fluxes of very low energy electrons but a definition of the cusp based upon this characteristic seems not to be sufficient to exclude other sources like the dayside extension of the plasma-sheet boundary layer. Additional criteria for the cusp are the existence of injection signatures of positive ions reported by BURCH *et al.* (1982).

2) The location of the cusp seems to be controlled, at least partly, by B_z although a number of observations indicate an important contribution to the cusp movements from magnetospheric sources.

3) For northward IMF most observations support a polar cusp and polar cap convection pattern dominated by an extra two-cell pattern (reversed relative to the normal pattern) for all B_y. The relative sizes of the two cells, however, is related to the sign and magnitude of B_y.

4) For southward IMF the prenoon sector shows a double convection reversal for $B_y<0$ compared to a single reversal for $B_y>0$. A similar behaviour of the postnoon

sector for opposite B_y has been proposed but has not been verified observatonally. In contrary, some observations indicate a fundamental asymmetri between the prenoon and the postnoon sector regardless of the sign of B_y.

5) Space- and ground-based observations support the models by BURCH and REIFF (1985) and CROOKER (1979) of two IMF controlled convection cells, of which one, the lobe cell, is confined exclusively on open field-lines, while the other, the merging cell, consists of streamlines passing from the closed field-line region on the prenoon (postnoon) side for $B_y<0$ ($B_y>0$) into the open field-line region of the northern hemisphere.

REFERENCES

BURCH, J. L., Rate of erosion of dayside magnetic flux based on quantitative study of the dependence of polar cusp latitude on the interplanetary magnetic field, *Radio Sci.*, **8**, 955, 1973.

BURCH, J. L., P. H. REIFF, R. A. HEELIS, J. D. WINNINGHAM, W. B. HANSON, C. GURGIOLO, J. D. MENIETTI, R. A. HOFFMAN, and J. N. BARFIELD, Plasma injection and transport in the mid-altitude polar cusp, *Geophys. Res. Lett.*, **9**, 921, 1982.

BURCH, J. L., P. H. REIFF, J. D. MENIETTI, R. A. HEELIS, W. B. HANSON, S. D. SHAWHAN, E. G. SHELLEY, M. SUGIURA, D. R. WEIMER, and J. D. WINNINGHAM, IMF B_y-dependent plasma flow and Birkeland currents in the dayside magnetosphere. 1. Dynamics Explorer Observations, *J. Geophys. Res.*, **90**, 1577, 1985.

BURKE, W. J., M. C. KELLEY, R. C. SAGALYN, M. SMIDDY, and S. T. LAI, Polar cap electric field structures with a northward interplanetary magnetic field, *Geophys. Res. Lett.*, **6**, 21, 1979.

CANDIDI, M. and C.-I. MENG, The relation of the cusp precipitating flux to the solar wind and interplanetary magnetic field, *J. Geophys. Res.*, **89**, 9741, 1984.

CANDIDI, M., H. W. KROEHL, and C.-I. MENG, Intensity distribution of dayside polar soft electron precipitation and the IMF, *Planet. Space Sci.*, **31**, 489, 1983.

CROOKER, N. U., Dayside merging and cusp geometry, *J. Geophys. Res.*, **84**, 951, 1979.

EATHER, R. H., Polar cusp dynamics, *J. Geophys. Res.*, **90**, 1569, 1985.

EATHER, R. H., S. B. MENDE, and E. J. WEBER, Dayside aurora and relevance to substorm current systems and dayside merging, *J. Geophys. Res.*, **84**, 3339, 1979.

FELDSTEIN, Y. I., A. E. LEVITIN, D. S. FAERMARK, R. G. AFONINA, and B. A. BELOV, Electric fields and potential patterns in the high-latitude ionosphere for different situations in the interplanetary space, *Planet. Space Sci.*, **32**, 907, 1984.

FRIIS-CHRISTENSEN, E., Y. KAMIDE, A. D. RICHMOND, and S. MATSUSHITA, Interplanetary magnetic field control of high-latitude electric fields and currents determined from Greenland magnetometer data, *J. Geophys. Res.*, **90**, 1325, 1985.

HEELIS, R. A., The effect of interplanetary magnetic field orientation on dayside high-latitude ionospheric convection, *J. Geophys. Res.*, **89**, 2873, 1984.

KAMIDE, Y., A. D. RICHMOND, and S. MATSUSHITA, Estimation of ionospheric electric fields, ionospheric currents, and field-aligned currents from ground magnetic records, *J. Geophys. Res.*, **86**, 801, 1981.

KROEHL, H. W. and Y. KAMIDE, High-latitude indices of electric and magnetic variability during the CDAW 6 intervals, *J. Geophys. Res.*, **90**, 1367, 1985.

MAEZAWA, K., Magnetic Convection induced by the positive and negative *z*-components of the interplanetary magnetic field: Quantitative analysis using polar cap magnetic records, *J. Geophys. Res.*, **81**, 2289, 1976.

MAKITA, K. and C.-I. MENG, Average electron Precipitation patterns and visual aurora characteristics during geomagnetic quiescence, *J. Geophys. Res.*, **89**, 2861, 1984.

MENG, C.-I., Case studies of the storm time variatn of the polar cusp, *J. Geophys. Res.*, **88**, 137, 1983.

POTEMRA, T. A., L. J. ZANETTI, P. F. BYTHROW, A. T. Y. LUI, and T. IIJIMA, B_y-dependent convection patterns during northward interplanetary magnetic field, *J. Geophys. Res.*, **89**, 9753, 1984.

REIFF, P. H. and J. L. BURCH, IMF B_y-dependent plasma flow and Birkeland currents in the dayside magnetosphere. 2. A global model for northward snad southward IMF, *J. Geophys. Res.*, **90**, 1595, 1985.

SANDHOLT, P. E., A. EGELAND, B. LYBEKK, C. S. DEEHR, G. G. SIVJEE, and G. J. ROMICK, Effects of interplanetary magnetic field and magnetospheric substorm variations on the dayside aurora, *Planet. Space Sci.*, **31**, 1345, 1983.

VENNERSTRØM, S., E. FRIIS-CHRISTENSEN, T. S. JØRGENSEN, O. RASMUSSEN, C. R. CLAUER, and V. B. WICKWAR, Ionospheric currents and *F*-region plasma boundaries near the dayside cusp, *Geophys. Res. Lett.*, **11**, 903, 1984.

WINNINGHAM, J. D. and R. A. HEELIS, The low altitude cleft: Plasma entry and magnetospheric topology, in *High-Latitude Space Plasma Physics*, edited by B. Hultqvist and T. Hagfors, p. 335, Plenum Publishing Co., 1983.

ZANETTI, L. J., T. A. POTEMRA, T. IIJIMA, W. BAUMJOHANN, and P. F. BYTHROW, Ionospheric and Birkeland current distributions for northward interplanetary magnetic field: Inferred polar convection, *J. Geophys. Res.*, **89**, 7453, 1984.

Solar Wind-Magnetosphere Coupling, edited by Y. Kamide and J. A. Slavin, 441–452.

DE-1 Observations of Solar Wind-Magnetosphere Coupling Processes in the Polar Cusp

J. L. BURCH, J. D. MENIETTI, and J. N. BARFIELD

Department of Space Sciences, Southwest Research Institute, San Antonio, Texas 78284, U.S.A.

Measurements of hot plasma distribution functions in the mid-altitude polar cusp by the Dynamics Explorer-1 spacecraft have provided new information on the processes of magnetosheath plasma injection and transport and the dependence of these phenomena on solar-wind parameters. Results on magnetosheath plasma injection have included: (1) Localization of the region of plasma injection to the high-latitude cusp-magnetosheath interface, (2) movement of this region toward lower latitudes as B_z becomes more southward, (3) maintenance of quasi-neutrality of suprathermal plasmas within the cusp, and (4) support for merging-type injection at the high-latitude cusp boundary for northward IMF. Plasma flow measurements from Dynamics Explorer-1 have led to a global IMF-dependent convection model which contains both merging and viscous driving forces.

1. Introduction

The early orbital configuration of Dynamics Explorer 1 (DE-1) resulted in many traversals of the mid-altitude (~to 3.5 earth radii) polar cusp as the spacecraft climbed toward apogee over the northern polar cap. The unique orbit, along with the comprehensive plasma and field instrumentation on board the spacecraft, made possible several new findings on the dynamics of polar-cusp plasmas and their dependence on solar-wind conditions. As observed by BURCH *et al.* (1982), and explained by their model, the velocity-space distributions of polar-cusp plasmas in this mid-altitude regime can be used to infer the locations where magnetosheath plasma is injected into the cusp. It was found that plasma injection occurs at the high-latitude intersection of the cusp with the magnetosheath, i.e., near the neutral points, and not near the subpolar point. A subsequent study of the DE-1 and DE-2 plasma and field data by BURCH *et al.* (1985) led to the development of an IMF B_y-dependent convection model which elucidated the role of merging and viscous interactions for negative IMF B_z. REIFF and BURCH (1985) extended this model to all IMF orientations, offering explanations for several previously known phenomena, including sunward convection in the polar cap for positive B_z and the dependence of theta aurora development on B_y, while making some predictions for future observations.

Recently we have been able to confirm the prediction by REIFF *et al.* (1977) of quasi-neutrality in the polar cusp (BURCH, 1985). This approximate equality of suprathermal electron and positive-ion densities throughout the cusp is consistent with a negative potential barrier at the magnetopause produced by ambipolar diffusion. Such a potential barrier would limit the entry of magnetosheath electrons into the cusp to the level needed to neutralize the charge carried by the injected positive ions.

Another important source of polar-cusp plasma is the ionosphere. A strong outflow of ions, coupled with perpendicular heating that produces conical ion distributions, was observed in the cusp by GURGIOLO and BURCH (1982). Later LOCKWOOD *et al.* (1984) identified these ions as primarily O^+ and showed evidence that, as they move away from the earth, they convect toward the night side, filling large portions of the polar cap. Further information on the outflow of ions in the cusp, and the dependence of this phenomenon on the interplanetary magnetic field (IMF) is presented in this volume by WAITE *et al.* (1986).

In this paper these recent results on polar-cusp plasmas are reviewed, and one new study is reported. The new study involved the determination of the IMF dependence of the location of cusp ion injection by tracing the mid-altitude ion distributions backward to their sources in the Olson-Pfitzer magnetic field model.

2. Cusp Plasma Injection

One of the first new observations made by the High Altitude Plasma Instrument (HAPI) on DE-1 was a unique energy/pitch-angle signature, which we refer to as an ion 'V' structure (BURCH *et al.*, 1982). An example of this phenomenon appears in Fig. 1. First seen in the cusp, and later on the night-side, the V structure was so distinct and was observed so often that it demanded a rather simple interpretation. Our interpretation is that the V structure represents those energies and pitch angles that have equal travel times from a localized source or injection region. In order that only a small range of transit times are accessible to the spacecraft at any time, the source must either turn on only occasionally and briefly (producing a temporal effect) or the plasma must be convected away from the injection point (producing a spatial effect). Although temporal effects surely occur, they are observed in the HAPI data to be superimposed on the larger-scale spatial effect. The simple model of BURCH *et al.* (1982) showed why the V structures are not observed at low altitudes and showed further that their shape is independent of particle mass.

2.1 *IMF B_z effects*

We know that the cusp is found at lower latitudes during periods of southward IMF. However, the question of the dependence of the cusp plasma injection altitude on the IMF has not previously been explored. Analysis of the energy-angle V structures has provided a tool for the investigation of this question, and some initial results have been obtained using the DE-1 HAPI data.

In this study theoretical energy-angle V structures were computed for each of nine DE-1 cusp passes in an Olson-Pfitzer magnetic-field model for various injection altitudes. The equatorward boundary of injected ions was determined and identified as the injection field line. A magnetic-field-line tracing routine was then used to map that Olson-Pfitzer field line upward to the selected injection altitude. Then the guiding centers of ions with various injection pitch angles were traced downward to the DE-1 altitude and then to their mirror points and back up to the DE-1 altitude with

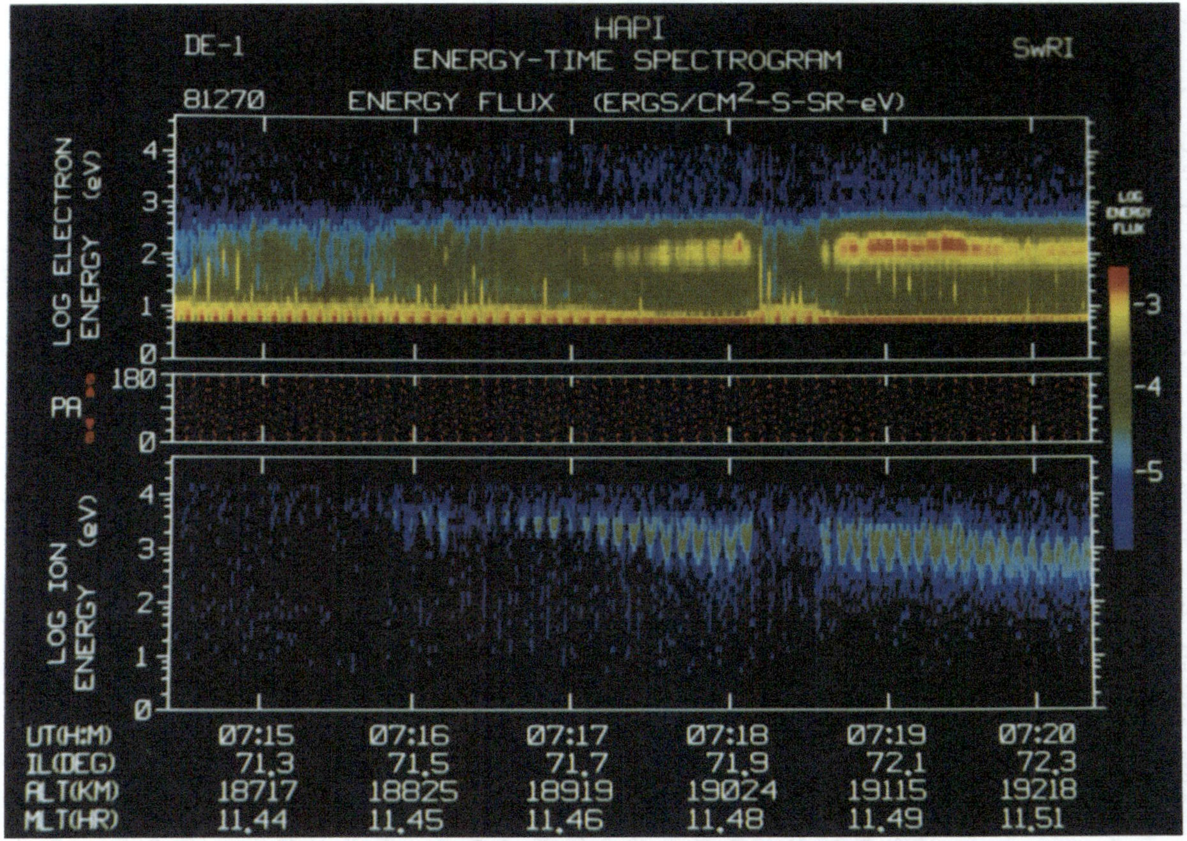

Fig. 1. Energy-time spectrogram of electron and ion energy flux measured on day 81270 (September 27, 1981) during a traversal of the mid-altitude polar cusp by DE-1, showing the energy-angle V structures resulting from cusp ion injection. The $\log_{10}$ of the energy flux in ergs $cm^{-2} s^{-1} sr^{-1} eV^{-1}$ is encoded according to the color bar at the right-hand side of the figure. The pitch angle of the measured particles is shown by the red trace in the middle panel. Spacecraft ephemeris data are listed along the horizontal axis.

atmospheric absorption assumed for ions that reached altitudes below 400 km. The theoretical energy-angle V structures represented those ions with a selected constant travel time between the injection point and the DE-1 spacecraft. Inspection of the equation representing particle energy, E, as a function of the observed pitch angle, α_o, and the transit time, t, shows how the computation of $E(\alpha_o)$ for a selected value of t was carried out. This relationship has the form

$$E(\alpha_o, t) = \frac{M}{2t^2}\left[\int_{s_i}^{s_o} ds/\sqrt{1 - \sin^2\alpha_o(B(s)/B_o)}\right]^2 \qquad (1)$$

where,

ds is arc length along an Olson-Pfitzer magnetic field line,
s_o, s_i are the observation and injection points,
M is the particle mass,
$B(s)$ is the magnetic induction along the field line,
B_o is the magnetic induction at the observation point,
α_o is the observed pitch angle, and
t is the transit time.

Unlike the case of a dipole magnetic field (BURCH *et al.*, 1982), the differential arc length of the field line, ds, is not expressible analytically for the Olson-Pfitzer model. Nevertheless, numerical integration is required in either case, and the results, as far as the shape of the V structures for given injection altitudes is concerned, are virtually the same for each model. The ion trajectory calculations were carried out for the magnetic local time where DE-1 actually crossed each ion injection boundary, and the MLT's were generally between 10 and 12 hrs. However, for comparison purposes the injection points derived from the best fits of the HAPI data to the theoretical energy-angle V structures are plotted in the noon meridian in Fig. 2. Since calculations were performed for every 1 R_E along the field line, the data points in Fig. 2 are located with an accuracy of at least ± 1 R_E.

As shown in Fig. 2, the previous result (BURCH *et al.*, 1982) that injection occurs near the high-latitude cusp-magnetosheath interface holds for both B_z polarities and for a wide range of cusp latitudes. As expected, negative B_z periods are associated with lower (and sometimes much lower) cusp latitudes.

2.2 *Quasi-neutrality*

A recent study (BURCH, 1985) has shown a remarkably close correspondence between the densities of suprathermal electrons and positive ions at altitudes between 16,000 and 23,000 km in the cusp. We suggest that this quasi-neutrality results from a process of ambipolar diffusion at the magnetopause, which allows the entry of magnetosheath electrons only to the extent required to balance the charge carried by the positive ions.

That quasi-neutrality should exist in the cusp was predicted by REIFF *et al.* (1977), who suggested that a negative potential barrier exists at the magnetopause to limit the entry of magnetosheath electrons to the fluxes required to neutralize the positive charge carried by the injected ions. In a study of 19 DE-1 cusp passes, BURCH

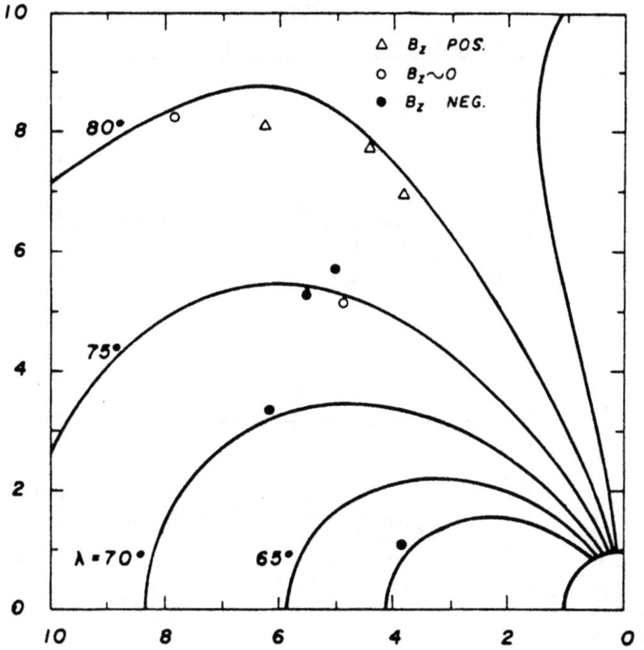

Fig. 2. Olson-Pfitzer magnetic field lines in the noon meridian along with computed cusp ion injection locations based on comparisons with DE-1 data from nine orbits. Triangles denote times with IMF B_z positive, open circles with B_z near 0 or fluctuating, and solid circles for B_z negative.

(1985) found such quasi-neutrality to hold, within instrumental accuracy, everywhere that the v-shaped cusp ion injection signature was detected. Figure 1, which is taken from BURCH (1985), shows in energy-time spectrogram format the qualitative correspondence between electron and positive-ion energy flux that is readily observed for all DE-1 cusp passes analyzed. Figure 3 contains plots of electron and ion density, downward electron number flux, and upward electron field-aligned current for the same time period as shown in Fig. 1. The electron moments plotted in Fig. 3 were computed for energies above 18 eV to avoid contamination by spacecraft photo-electrons, while the ion moments extended down to the instrumental limit of 5 eV. Since magnetosheath electron and ion energies are well above these limits, very little inaccuracy should result from them. The largest uncertainty in the measurement is the counting efficiency of the channel electron multipliers as a function of particle mass and specie, since the detector geometry is identical for electrons and ions. The counting efficiencies determined by BORDONI (1971) and FIELDS *et al.* (1977) were used for electrons and ions (assumed H^+), respectively.

The small quantitative differences between the electron and ion densities in Fig. 3 are well within the experimental error due to counting-efficiency uncertainties. The close correspondence of the spatial variation of the electron and ion densities for all cusp passes is, however, quite remarkable and in agreement with the quasi-neutrality

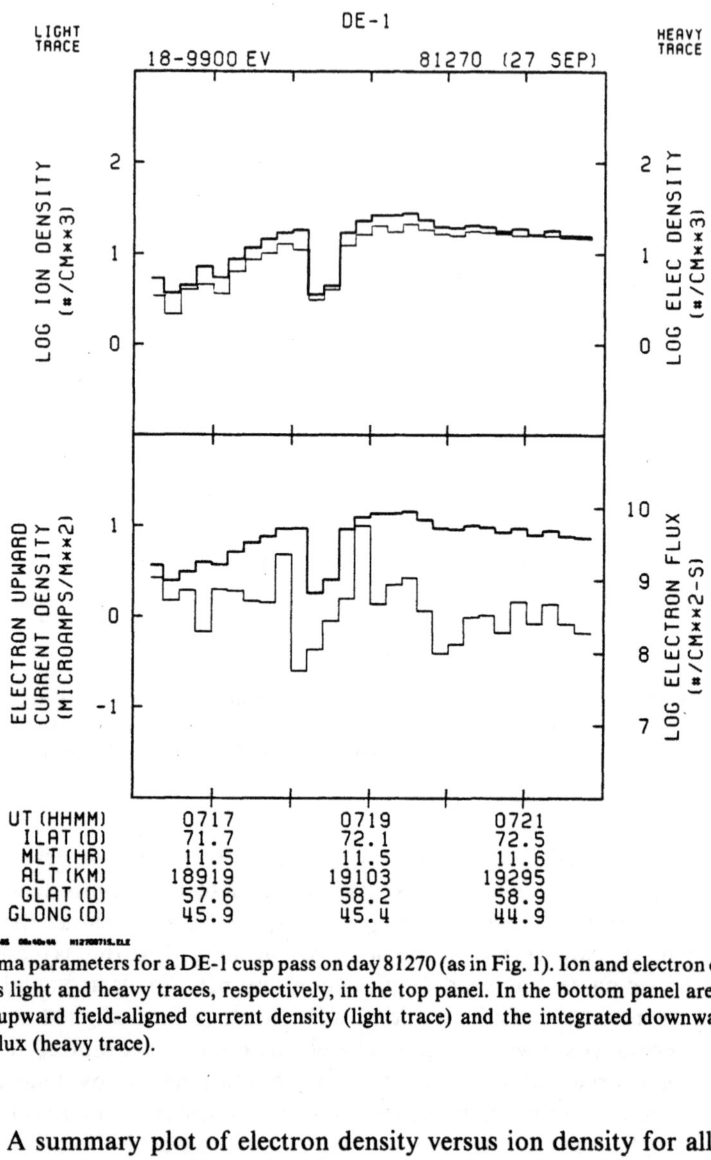

Fig. 3. Plasma parameters for a DE-1 cusp pass on day 81270 (as in Fig. 1). Ion and electron densities are plotted as light and heavy traces, respectively, in the top panel. In the bottom panel are plotted the electron upward field-aligned current density (light trace) and the integrated downward electron number flux (heavy trace).

prediction. A summary plot of electron density versus ion density for all two-spin samples (12 seconds) throughout the 19 cusp passes investigated is shown in Fig. 4. The upward deviation of the data points from the diagonal line representing equal electron and ion densities in the range below 2 cm^{-3} is caused by the poor counting statistics that occur near the detection threshold of the ion sensor.

The following interpretation for the observed quasi-neutrality was suggested by BURCH (1985). Magnetosheath positive ions are injected into the cusp through a spatially restricted region and are convected large distances as they travel toward the earth and as some of them mirror to flow upward, forming the plasma mantle. Magnetosheath electrons, which have thermal speeds much higher than the magneto-

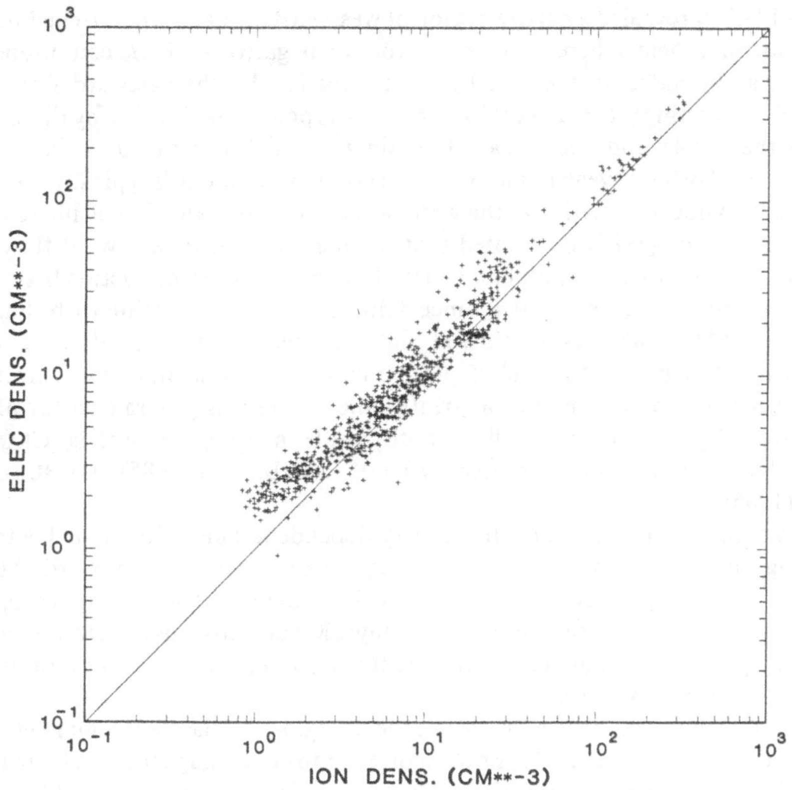

Fig. 4. Summary plot of electron density vs. ion density for 19 DE-1 cusp passes. The diagonal line is the locus of equal electron and ion densities. (Taken from BURCH, 1985).

sheath flow velocity, can enter the magnetosphere essentially anywhere along open field lines. However, ambipolar diffusion limits the entry of electrons to just the flux that is needed to maintain charge neutrality. Since the cusp electrons have velocities more than an order of magnitude higher than the cusp ions, they are either absorbed by the atmosphere, or mirror and leave the magnetosphere, while convecting less than one-tenth the way through the cusp. Hence, magnetosheath electrons must be able to enter the cusp along its full latitudinal width. However, as noted by REIFF *et al.* (1977) the supersonic nature of the positive ion distributions in the downstream magnetosheath would severely limit the ability of ions to enter the cusp except near the equatorward boundary. Finally we note that the significance of the observed quasi-neutrality lies in the fact that it holds for suprathermal plasma without any need for the neutralizing effect of thermal ionospheric plasma.

3. IMF-Dependent Convection

As reported by BURCH *et al.* (1985), plasma convection measurements made by the DE-1 HAPI revealed a narrow region of westward flow at the polar-cap boundary in the northern-hemisphere morning sector for negative IMF B_y conditions. This feature was bounded in the equatorward direction by the eastward flow of the sunward-convecting plasma-sheet ions, and in the poleward direction by the eastward flow of the low-latitude polar cap. Examination of drift-meter data from several hundred DE-2 orbits revealed that this convection morphology is typically present for negative B_y, while for positive B_y the westward convection extends well into the polar cap. BURCH *et al.* (1985) postulated that the narrow zone of westward flow at the morning-side polar-cap boundary is caused by a viscous-type interaction between the solar wind and the magnetosphere, since it does not reverse direction with changes in the sign of IMF B_y as does the flow at higher latitudes. These results were used to construct a B_y-dependent model of plasma convection and Birkeland currents for southward IMF conditions. Subsequently, the model was generalized for all IMF orientations by combining it with the anti-parallel merging hypothesis (CROOKER, 1979). The full model was presented by REIFF and BURCH (1985) and BURCH and REIFF (1985).

A distinguishing feature of this new B_y-dependent convection model is that the following three types of convection cells may coexist, even for southward IMF:

(1) Two merging cells, wherein closed magnetospheric flux tubes open by merging with solar-wind field lines at the dayside magnetopause, flow antisunward across the polar cap, reconnect on the night side, and return at lower latitudes (the traditional two-cell system).

(2) A lobe cell, driven by magnetic merging at the polar magnetopause, tailward of the cusp. This cell does not involve a transfer of magnetic flux from the day side to the night side, but only a stirring and shortening of magnetotail field lines. The sense of rotation of the vortex is dependent on B_y: clockwise in the northern hemisphere for $B_y>0$ and vice versa.

(3) Two narrow viscous cells, one on each side of the polar cap, driven by quasi-viscous processes on the flanks of the magnetosphere. These cells do not contain a large potential difference (<10 kV). They are, therefore, much more easily observed (in the northern hemisphere) on the dawn side for $B_y<0$ and on the dusk side for $B_y>0$, where their flow direction is opposite to that of the lobe cell.

A summary of the B_y-dependent global convection patterns of the REIFF and BURCH (1985) model is shown in Fig. 5. The lobe cells (L) in Fig. 5B are driven by northward-IMF tail merging. The cells marked "R" in Fig. 5A and 5C are "reclosure" cells, with the sunward-flowing portion on closed field lines. The dashed contours in Fig. 5A through 5G are postulated merging lines.

For the polar cusp the relevance of the B_y-convection model is mainly its ability to predict the motion of magnetosheath ions after they are injected into the cusp. The model was based on DE-1 cusp plasma observations for southward IMF conditions. Therefore, it can easily explain, for example, the observation of ion energy-angle V structures in the morning sector for $B_y>0$ and in the afternoon sector for $B_y<0$ in the northern hemisphere. Injection for southward IMF is presumably along the dashed merging contours in Figs. 5D through 5G. The question of the plasma injection mechanism for northward IMF has been an outstanding one. Although the model

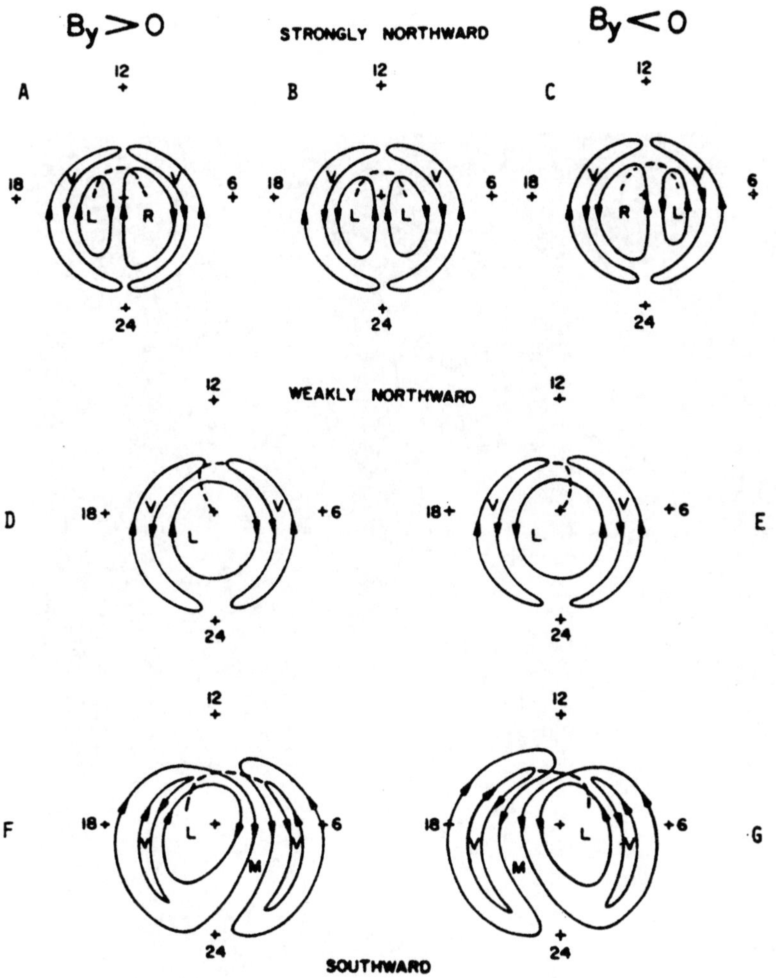

Fig. 5. Schematic northern-hemisphere polar-cap convection patterns for various IMF orientations. Viscous cells are marked with a "V", merging cells with an "M", and lobe cells with and "L". Reclosure cells are marked with an "R".

incorporates tail merging with northward IMF, it is not clear whether magnetosheath ions would be injected at these merging sites or sunward of them by neutral-point injection.

Data of the type shown in Fig. 6 may help answer this question. This cusp pass occurred on day 81285 (Oct. 12, 1981), more than 2.5 hours after the IMF had turned from southward to strongly northward. Clearly evident in Fig. 6 are the positive-ion energy-angle V structures characteristic of injection through a spatially-restricted region and subsequent strong convection. Also evident is an energy-latitude velocity dispersion, but in this case the ion energy increased slowly with increasing latitude as

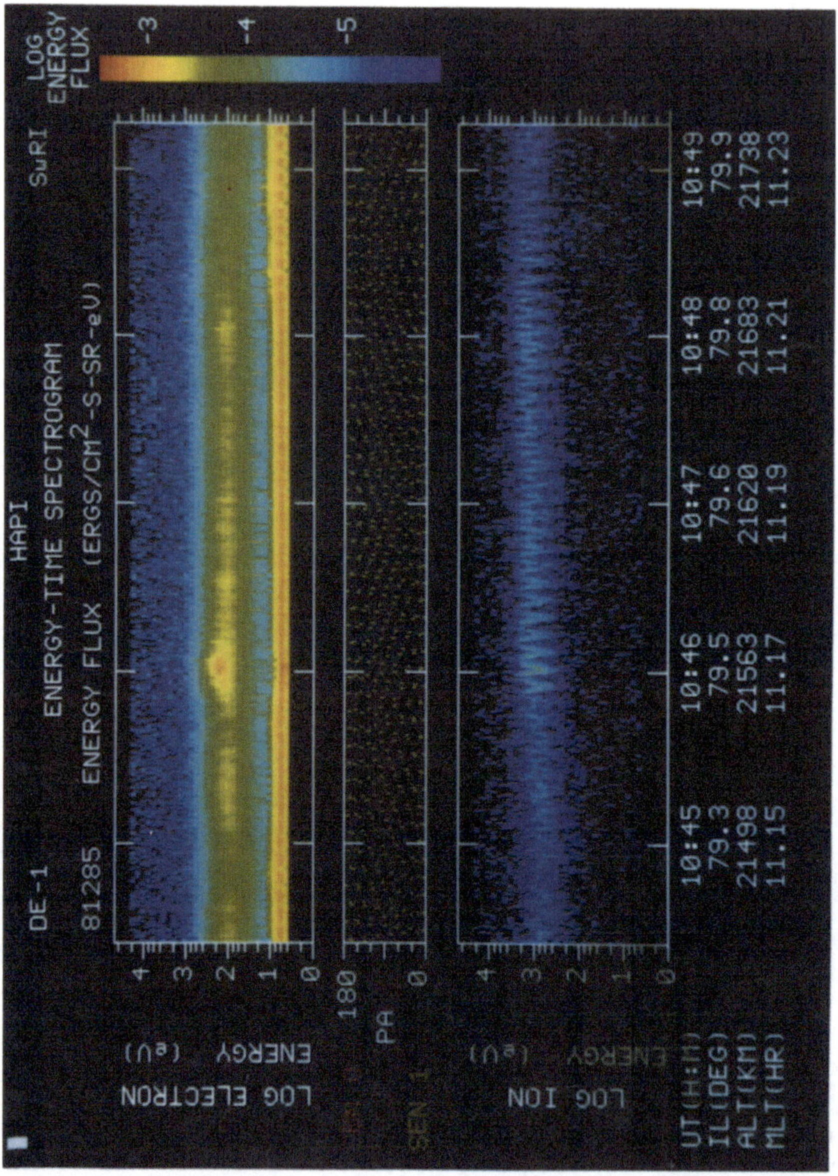

Fig. 6. Energy-time spectrogram of electron and positive-ion energy flux measured on day 81285 (Oct. 12, 1981) during a traversal of the mid-altitude polar cusp by DE-1 following an extended period of northward IMF. Plot format is the same as in Fig. 1.

observed previously with low-altitude spacecraft during periods of northward IMF and sunward convection (BURCH *et al.*, 1980). Taken together, the ion cusp injection and sunward convection signatures are consistent with plasma injection at a northward tail merging site at the poleward edge of the cusp as opposed to a neutral-point diffusive entry process.

4. Summary and Conclusions

Measurements of hot plasma distribution functions in the mid-altitude polar cusp by the DE-1 spacecraft have provided new information on the processes of magnetosheath plasma injection and transport and the dependence of these phenomena on solar-wind parameters. Results on magnetosheath plasma injection have included: (1) localization of the region of plasma injection to the high-latitude cusp-magnetosheath interface; (2) movement of this region toward lower latitudes as B_z becomes more southward; (3) maintenance of quasi-neutrality of suprathermal plasmas within the cusp; and (4) support for merging-type injection at the high-latitude cusp boundary for northward IMF.

Plasma flow measurements from DE-1 have led to a global IMF-dependent convection model (REIFF and BURCH, 1985) which contains both merging and viscous driving forces. Although the model was initially based on measurements for southward IMF, use of the antiparallel merging hypothesis allowed its extension to all IMF directions. The complete model can explain a number of observed phenomena, including the tail-merging entry of magnetosheath plasma for northward IMF, for which additional experimental evidence has now been obtained from DE-1 (Fig. 6).

This research was supported by NASA Contract No. NAS5-28711 at Southwest Research Institute and NSF Grant No. ATM 83-13422 at the University of Texas at San Antonio.

REFERENCES

BORDONI, F., Channel electron multiplier efficiency for 10-1000 eV electrons, *Nucl. Instr. and Methods*, **97**, 405, 1971.

BURCH, J. L., Quasi-neutrality in the polar cusp, *Geophys. Res. Lett.*, **12**, 469–472, 1985.

BURCH, J. L. and P. H. REIFF, Field-aligned currents and ion convection at high altitudes, *Advances in Space Research*, **5**, 23–40, 1985.

BURCH, J. L., P. H. REIFF, R. A. HEELIS, R. W. SPIRO, and S. A. FIELDS, Cusp region particle precipitation and ion convection for northward interplanetary magnetic field, *Geophys. Res. Lett.*, **7**, 393, 1980.

BURCH, J. L., P. H. REIFF, R. A. HEELIS, J. D. WINNINGHAM, W. B. HANSON, C. GURGIOLO, J. D. MENIETTI, R. A. HOFFMAN, and J. N. BARFIELD, Plasma injection and transport in the mid-altitude polar cusp, *Geophys. Res. Lett.*, **9**, 921, 1982.

BURCH, J. L., P. H. REIFF, J. D. MENIETTI, R. A. HEELIS, W. B. HANSON, S. D. SHAWHAN, E. G. SHELLEY, M. SUGIURA, and J. D. WINNINGHAM, IMF B_y-dependent plasma flow and Birkeland currents in the dayside magnetosphere: 1. Dynamics Explorer observations, *J. Geophys. Res.*, **90**, 1577, 1985.

CROOKER, N. U., Dayside merging and cusp geometry, *J. Geophys. Res.*, **84**, 951, 1979.

FIELDS, S. A., J. L. BURCH, and W. A. ORAN, Mass-dependent channel electron multiplier operation, *Rev. Sci. Instrum.*, **48**, 1076, 1977.

GORNEY, D. J., A. CLARKE, D. CROLEY, J. FENNELL, J. LUHMANN, and P. MIZERA, The distribution of ion beams and conics below 8000 km, *J. Geophys. Res.*, **86**, 83, 1981.

GURGIOLO, C. and J. L. BURCH, DE-1 observations of the polar wind—a heated and an unheated component, *Geophys. Res. Lett.*, **9**, 945, 1982.

HEELIS, R. A., J. D. WINNINGHAM, M. SUGIURA and N. C. MAYNARD, Particle acceleration parallel and perpendicular to the magnetic field observed by DE-2, *J. Geophys. Res.*, **89**, 3893, 1984.

LOCKWOOD, M., J. H. WAITE, Jr., T. E. MOORE, J. F. E. JOHNSON, and C. R. CHAPPELL, A new source of suprathermal O^+ ions near the dayside polar cap boundary, in press, *J. Geophys. Res.*, 1985.

REIFF, P. H. and J. L. BURCH, IMF B_y-dependent plasma flow and Birkeland currents in the dayside magnetosphere: 2. A global model for northward and southward IMF, *J. Geophys. Res.*, **90**, 1595, 1985.

REIFF, P. H., T. W. HILL, and J. L. BURCH, Solar-wind plasma injection at the dayside magnetospheric cusp, *J. Geophys. Res.*, **82**, 479, 1977.

SHARP, R. D., R. G. JOHNSON, and E. G. SHELLEY, Observation of an ionospheric acceleration mechanism producing energetic (keV) ions primarily normal to the geomagnetic field direction, *J. Geophys. Res.*, **82**, 3324, 1977.

WAITE, J. H., Jr., M. LOCKWOOD, T. E. MOORE, M. O. CHANDLER, J. L. HORWITZ, and C. R. CHAPPELL, Solar-wind control of the geomagnetic mass spectrometer, this volume, 1986.

WHALEN, B. A., W. BERNSTEIN, and P. W. DALY, Low altitude acceleration of ionospheric ions, *Geophys. Res. Lett.*, **5**, 55, 1978.

WINNINGHAM, J. D. and J. L. BURCH, Observations of large scale ion conic generation with DE-1, in *Proceedings of the 1982-4 MIT Symposium on Physics of Space Plasmas*, edited by J. Belcher, H. Bridge, T. Chang, B. Coppi, and J. Jasperse, pp. 137–158, MIT Press, 1984.

Solar Wind-Magnetosphere Coupling, edited by Y. Kamide and J. A. Slavin, 453–476.

Solar Wind Control of the Polar-Cap Voltage

Patricia H. REIFF[1] and Janet G. LUHMANN[2]

[1]*Center for Space Physics, Rice University, Houston, Texas 77251, U.S.A.*
[2]*Institute for Geophysics and Planetary Physics, University of California, Los Angeles California 90024, U.S.A.*

The voltage across the polar cap is the most easily measured indicator of the strength of the solar-wind magnetosphere interaction. The polar-cap voltage can be produced in three ways: by merging of closed dayside magnetospheric field lines with the interplanetary magnetic field, by merging of magnetotail field lines with the IMF, and by quasi-viscous transfer processes at the magnetopause. It is possible to estimate observationally the variuous separate contributions to the total polar-cap voltage. When the IMF is southward, dayside merging contributes 80–95% of the total voltage, while the contributions from viscous processes and tail-lobe merging are quite variable, each typically less than 10 kV. When the IMF is northward, the voltage due to dayside merging diminishes while that due to tail-lobe merging contributes as much as 35 kV.

Two new empirical formulas for the dependence of the asymptotic polar cap potential on the IMF are presented; the first is given by Φ (kV)=22+0.069 $vBQ1$, where $Q1$ is the Sonnerup-Hill theoretical merging formula, and the second is Φ (kV)=6.7+ 0.047 $vB \sin^3(\theta/2)$ with the solar wind velocity expressed in km/s, the B-field in nanotesla, and the polar angle θ in GSM coordinates. These formulae are determined from data taken during times when the IMF, if northward, had been northward for 3 hours or more; if the IMF suddenly turns northward, the potential drop decays slowly towards these values. Alternate formulas are presented for times when the IMF is large.

Predictions of the dependence of dayside and lobe merging voltages on the IMF are derived from the antiparallel merging hypothesis.

1. Introduction

The cross-polar-cap electrostatic voltage is a direct measure of the instantaneous rate of plasma flow through the magnetospheric convection system. Ranging in magnitude from as high as 240 kV (GURNETT and FRANK, 1973) to as low as 10–12 kV (WYGANT *et al.*, 1983; DOYLE and BURKE, 1983), its typical value is 50–60 kV. This value implies that the magnetosphere intercepts, on average, about 10– 15% of the solar-wind voltage that would occur across the diameter of the magnetosphere. (For an interplanetary magnetic field (IMF) strength of 5 nT perpendicular to the earth-sun line and a 350 km/s solar-wind flow, a voltage of 467 kV would be induced across 40 R_E diameter.) This means that most of the magnetosheath flow diverges and flows around the magnetopause, much as in the gas-dynamic flow calculations of SPREITER and ALKSNE (1969).

The efficiency of penetration of the solar-wind electric field is sensitive to the direction of the IMF, particularly its north–south component B_z. This dependence was anticipated on the basis of DUNGEY's (1961) open model of the magnetosphere, and was confirmed indirectly when MOZER *et al.* (1974) found a substorm dependence of polar cap electric fields in balloon data, and HEPPNER (1973, 1977) found a *Kp* dependence of the polar-cap voltage inferred from satellite data. Figure 1 (from COWLEY, 1984) combines the AE-C and -D polar-cap voltage measurements from REIFF *et al.* (1981) (filled squares), S3-2 data from DOYLE and BURKE (1983) (open circles) and S3-3 data from WYGANT *et al.* (1983) (filled circles). Three effects are immediately obvious: first, the voltage is not simply proportional to the solar-wind dawn–dusk electric field. For an interplanetary dawn–dusk electric field $|VB_z|$ of 1 mV/m, the typical polar-cap voltage is about 25 kV if $B_z>0$ (corresponding to an efficiency of 5%), whereas if $B_z<0$, the typical voltage is about 100 kV (corresponding to an efficiency of 20%). Second, the voltage does not fall to zero as the IMF turns northward (the right-hand side), but rather appears to decay to a value of 10–30 kV. Third, as B_z becomes more negative, the voltage does not continue rising but instead appears to saturate in the range 100–150 kV.

In an ideal-MHD approximation (no viscosity, no resistivity, no heat conductivity), the magnetopause is an impenetrable barrier to the solar wind, and no transverse momentum transfer across the magnetopause is possible, except through waves. Since the efficiency is observed to be non-negligible (and for southward IMF's is substantial), the ideal-MHD approximation must be violated somewhere on the magnetopause. Two classes of mechanisms have been offered to locally violate the MHD condition and therefore allow such transverse momentum transfer: magnetic merging, in which interplanetary magnetic field lines become attached to magnetopheric field lines at the magnetopause (the violation of frozen-in-flux occurs in a diffusion region surrounding the X-line) (DUNGEY, 1961; see review by VASYLIUNAS,

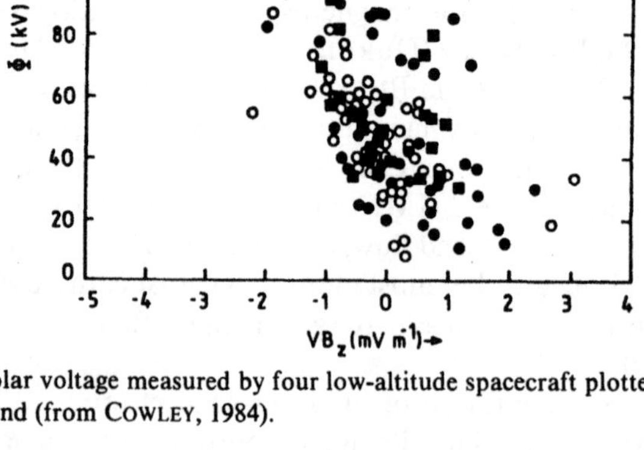

Fig. 1. Total transpolar voltage measured by four low-altitude spacecraft plotted versus simultaneous vB_z in the solar wind (from COWLEY, 1984).

1975), and "viscous interaction", by which momentum is transferred without magnetic interconnection (AXFORD and HINES, 1961). Several mechanisms have been suggested for the source of the anomalous 'viscosity': single-particle diffusion (enhanced by any of several micro-instabilities); large amplitude (non-linear) Kelvin-Helmholtz waves; gradient-drift injection; and "impulsive injection" (for a recent review, see HILL, 1979, 1983). (Note: as pointed out by Vasyliunas in this meeting, "viscosity" is defined as transferring momentum without transferring particles, and thus does not violate "frozen-in flux" (ALFVÉN and FÄLTHAMMER, 1963). Most of these mechanisms also transfer particles, and therefore are more correctly called "diffusion," which does violate frozen-in-flux.)

Most dayside magnetic merging models also fall into two general categories: subsolar merging models, in which the merging line crosses the subsolar point (e.g., SONNERUP, 1974; GONZALES and MOZER, 1974; HILL, 1975), and antiparallel merging models, in which the merging line occurs at high latitudes in the vicinity of the polar cusps (e.g., HAERENDEL *et al.*, 1978; CROOKER, 1979, 1980; LUHMANN *et al.*, 1984). In addition, recent tearing-mode models predict that several merging lines can occur on the dayside magnetopause (LEE and FU, 1985; see CROOKER, this volume, for a review).

2. Predictions of Subsolar Merging Models

The first quantitative predictions of the merging rate (and hence of the polar-cap voltage) were based on the subsolar merging models. These predictions were first subjected to direct experimental tests using the AE-C and AE-D convection data (REIFF *et al.*, 1981) and subsequently using S3-3 (WYGANT *et al.*, 1983) and S3-2 (DOYLE and BURKE, 1983) electric field measurements (Fig. 1). These papers typically correlated the polar-cap voltage Φ with various functions of the IMF, in particular $-vB_z$ (the solar-wind dawn–dusk electric field); vB_s, where $B_s=-B_z$ if $B_z<0$, and $=0$ if $B_z>0$ (the "half-wave rectifier": ROSTOKER *et al.*, 1972; BURTON, 1973); $vB\sin^2(\theta/2)$, where θ is the polar angle of the IMF ($=\cos^{-1}(B_z/|B|)$) (SONNERUP, 1974; KAN and LEE, 1979); as well as more complicated functions of the IMF (SONNERUP, 1974; GONZALES and MOZER, 1974; HILL, 1975; ATKINSON, 1978; PERRAULT and AKASOFU, 1978).

It can be shown from dimensional analysis (VASYLIUNAS *et al.*, 1982) that any quantitative estimate of the polar-cap voltage must be of the form

$$\Phi_{pc}=vB_TL_{CF}Q(\theta_T,M_A{}^2) \tag{1}$$

where Φ_{pc} is the polar-cap voltage, v is the solar-wind speed, B_T is the projection of the IMF on the solar magnetospheric y–z plane, L_{CF} is the Chapman-Ferraro scale length, and $Q(\theta_T,M_A{}^2)$ is a dimensionless function of two variables: θ_T, the "clock-angle," is defined by $\cos^{-1}(B_z/B_T)$ (the influence of B_x is assumed to be negligible) and $M_A{}^2=\mu_0\rho v^2/B_T{}^2$. VASYLIUNAS *et al.* argue that only two other dimensionless ratios can occur in the formula: $H=\mu_0\Sigma_p v$, (where Σ_p is the height-integrated Pedersen conductivity), which measures the relative importance of ionospheric line tying compared to inertial effects, and R, the Reynolds number $= M_A(2)^{1/2}\, eB_T\, L_{CF}/fmv$, (where f is a dimensionless coefficient), which measures the relative importance of inertial vs. viscous effects. We will suggest below that each of these ratios may have second-order effects on the determination of the polar-cap voltage.

3. Tests of Dayside Merging Models

REIFF *et al.* (1981), WYGANT *et al.* (1983) and DOYLE and BURKE (1983) (subsequently referred to as papers 1, 2 and 3, respectively) analyzed linear correlations $\Phi = \Phi_0 + \Phi_1$, where Φ_0 is a constant and Φ_1 is a formula of the type of Eq. (1). Included in the formula Φ_1 were various angular dependences, including the theoretical results

$$Q1 = \frac{\alpha(\alpha - \cos\theta)}{(\alpha^2 + 1 - 2\alpha\cos\theta)^{1/2}} \qquad \text{[SONNERUP, 1974; HILL, 1975]}, \qquad (2)$$

$$Q2 = \frac{\alpha(\alpha - \cos\theta)^2}{(\alpha^2 + 1 - 2\alpha\cos\theta)} \qquad \text{[HILL, 1975]}, \qquad (3)$$

$$Q3 = Q1/(\alpha\sin(\theta/2)) \qquad \text{[GONZALES and MOZER, 1974]}, \qquad (4)$$

where $\alpha = |B_T|/B_E$, B_E being the magnetospheric field just inside the magnetopause, and θ can be either the polar angle θ or the clock angle θ_T.

Considering the limited number of data points in each study, the hourly-averaged IMF data used, and uncertainties as to whether the spacecraft orbit completely traversed the polar-cap convection pattern, the results obtained in the three papers were remarkably similar (see paper 3 for a summary). The three fundamental conclusions reached by these three papers are:

(1) Almost any function that emphasizes the southward component of the IMF correlates reasonably well with the polar-cap voltage, but the functions which correlate best are those that do not cut off sharply for weakly northward IMF's (e.g., $Q1$, $Q2$, $Q3$, and KAN and LEE, 1979).

(2) Each correlation, of the form $\Phi_0 + \Phi_1$, indicated a substantial "residual potential" Φ_0, which varied from 16–40 kV. These residual potentials, measured during northward IMF, were suggested to be caused by 'viscous' processes in paper 1; however, paper 2 showed that the voltage measured during times of northward IMF was influenced by how long the IMF had been northward (Fig. 2, from WYGANT *et al.*, 1983). From this observation they suggested that only 10–20 kV of the voltage can actually be ascribed to viscous processes. A value of 10–20 kV for the viscous potential is consistent with both theoretical expectations (EVIATAR and WOLF, 1968; HILL, 1979; SONNERUP, 1980; TSURUTANI and THORNE, 1982; PU and KIVELSON, 1983; MIURA, 1984; see HILL, 1979; 1983 for a review) and independent experimental evidence (EASTMAN and HONES, 1979; SANDERS *et al.*, 1980;MOZER, 1984; see COWLEY, 1984 for a review).

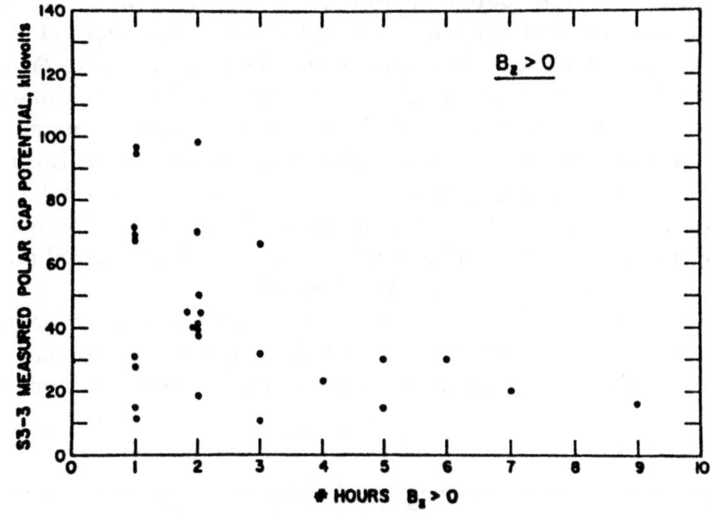

Fig. 2. Decay of the polar-cap voltage as a function of time after a northward turning of the IMF (from WYGANT *et al.*, 1983).

(3) As mentioned earlier, there appears to be a "saturation" effect, whereby the voltage, which is nearly linearly related to IMF functions when the IMF magnitude $|B|$ is small ($<\sim 5$ nT), saturates at a value of 100–160 kV even when the IMF magnitude is very large (30 nT or more). Paper 1 suggested that this was caused by the amplification of the magnetic field strength in the magnetosheath to values ~8 times the IMF value, but limited by the condition that the ratio of external to internal field strength never exceeded unity (if it ever did, the magnetopause would move inwards until pressure balance were reestablished). The actual best-fit value of the average amplification factor determined in each paper was strongly influenced by the scarcity of measurements at high field strengths, and this is no doubt why the results of the three papers differ most on that parameter. Subsequent measurements of the magnetic field compression at the magnetopause (CROOKER *et al.*, 1982) have shown that the sheath field just outside the magnetopause is approximated by $B_{sh} \simeq (4 B_T B_E)^{1/2}$, where B_T and B_E are defined as above. This implies an amplification factor which is largest (10–15) for small B_T's (<2nT), moderate (7–8) for moderate B_T's (4–6 nT), and continues to decrease for larger B_T's, again in accord with the range of inferred values.

It should be noted that there is an independent alternate argument that the polar-cap voltage should not exceed a maximum value. HILL *et al.* (1976) (following RASSBACH *et al.*, 1974) argued that ionospheric line-tying should imply a maximum potential difference such that the currents which are induced in the ionosphere, closing via Birkeland currents which flow between the ionosphere and the magnetopause, should not be sufficiently large to cause major alterations in the magnetic field near the magnetopause. The relevant dimensionless parameter that determines the maximum potential drop is Σ_p/Σ_0, where Σ_p is the ionospheric conductivity as before, and Σ_0 is a critical conductivity $= B_{sh}/(B_E \mu_0 v_A) \sim 20$ mho for typical values (B_{sh} is the sheath field just outside the magnetopause, B_E is the inside value and v_A the Alfvén speed as before). Line-tying yields a maximum potential difference of about 155 kV (HILL *et al.*, 1976); since this value is comparable to the maximum potentials observed, it appears that this mechanism may also be important in determining the saturation potential. We note that Σ_p/Σ_0 is similar in form to Vasyliunas' dimensionless H parameter above ($\Sigma_p/\Sigma_0 = H/M_A$) (although the Alfvén velocity appears instead of the solar wind speed); thus H may indeed play a role of limiting the voltage in times of large IMF field values.

These three effects are best seen in Fig. 3. In this figure, we have updated the results of paper 1. In order to better establish the "baseline" or "viscous" potentials, we have eliminated all northward IMF data unless the IMF was northward for 3

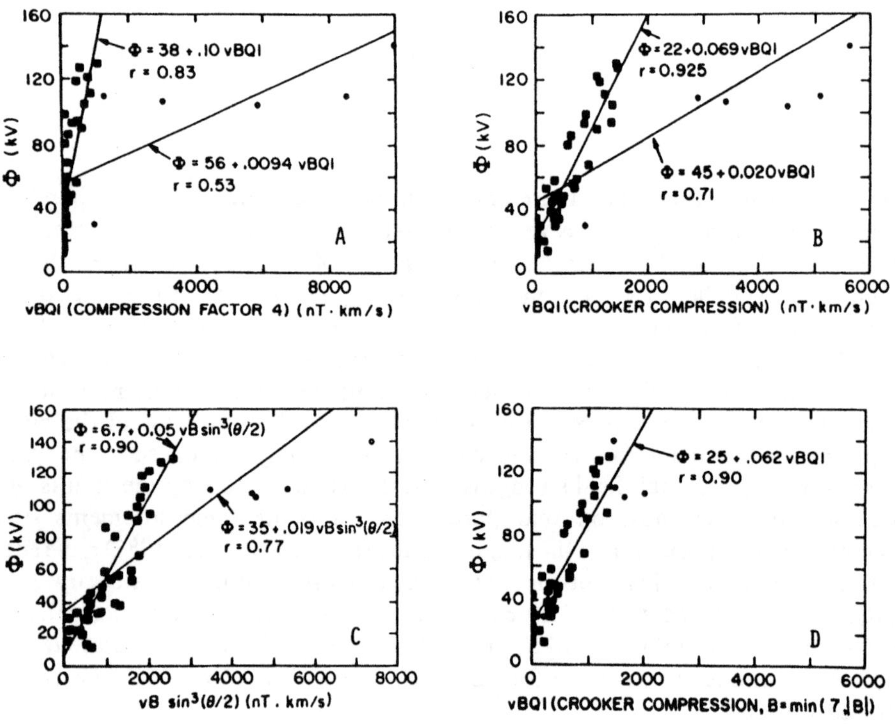

Fig. 3. Correlation of the observed cross-polar-cap voltage Φ with the function vBQI predicted by a particular magnetic merging theory, (a) with compression/limitation (i.e., $\alpha = \min(g|B|/60\,\text{nT}\,1)$, with $g=4$); and (b) using the magnetosheath compression formula given by CROOKER *et al.* (1982) i.e., $\alpha = \min((4B_T B_E)^{1/2}, 1)$. (c) Best-fit formula of the type $\Phi = \Phi_0 + A_0 vB\sin^n(\theta/2)$. Each plot is shown both including and not including the 6 data points for which $|B| > 10$ nT.(For each, the data set includes AE-C and -D data from REIFF *et al.* (1981), S3-3 data during northward IMF from WYGANT *et al.* (1983), and S3-2 data from northward IMF plus passes for which $\Phi > 90$ kV from DOYLE and BURKE (1983). (d) Same as (b), but now setting $|B| = \min(|B|, 7\text{ nT})$.

hours or more (following the suggestion of paper 2). In addition, we have included the potential values from papers 2 and 3 for times when the IMF had been northward for 3 hours or more (S3-3 data courtesy of J. Wygant, personal communication, 1985; S3-2 data from DOYLE and BURKE, 1983). Lastly, we added the data points from paper 3 for which $\Phi \geqslant 90$ kV, in order to examine the saturation effect. Figure 3(a) shows the best-fit line to function $vBQ1$ (in all of these, if $\{\alpha - \cos(\theta)\} < 0$, the entire function $vBQ1$ is set to zero), using the amplification/limitation procedure of paper 1. For this case, the amplification/limitation procedure redefines $\alpha = \max(g|B|/B_E, 1)$, where g is the optimal amplification value, in this case, 4. (The angle θ and total field $|B|$ gave better fits than θ_T and B_T.) The points fit very nicely to a line at low values of $vBQ1$, but the "saturation potential" observed at the five large values of $vBQ1$ make the fitted line to the entire set ($\Phi = 56 + 0.01\ vBQ1$) less meaningful, with a relatively low ($r = 0.53$) correlation coefficient. This amplification/limitation formula, however, is still "ad hoc", and so we have tested separately the "Crooker compression" (Fig. 3(b)), in which now $\alpha = (4\ B_T/B_E)^{1/2}$. The best-fit line ($\Phi = 45 + 0.20\ vBQ1$) has a correlation coefficient of 0.71, which is significantly better than with the simple amplification/limitation procedure of paper 1, but still shows the saturation effect in five of the six points with high field strengths (dots). Therefore we are led to conclude that another additional saturation mechanism, possibly the ionospheric conductivity mechanism discussed above, is at work.

It is notable that all five saturated values occurred for IMF strengths > 10 nT. In the interest of determining a function that may be used routinely to estimate the polar-cap voltage, we have redone all of the correlations, eliminating all orbits for which $|B| > 10$ nT. These are shown as the lefthand lines in Figs. 3(a) and 3(b). In addition, we have tested formulae of the type $\Phi = \Phi_0 + A_0\ vB\sin^n(\theta/2)$. The best example of that type is shown in Fig. 3(c), which is $\Phi = 6.7 + 0.047(vB\sin^3(\theta/2))$, with a correlation coefficient of 0.90. For comparison, the correlation including the six high-field points is also shown. The results for several functions vBQ are shown in Table 1. The numbers without parentheses are the values calculated by eliminating all points with $|B| > 10$ nT, e.g. the left-hand plots of Figs. 3(a)-(c); the numbers in parentheses indicate the best fits from setting the maximum $|B|$ magnitude to a saturated value, i.e., Figure 3(d).

The previous best-fit formula from paper 1 ($\Phi = 38 + 122\,\alpha Q2$, compression factor 7) was tested against a completely independent technique for inferring the polar-cap voltage: the magnetogram-inversion (KRM) algorithm (KAMIDE *et al.*, 1981). Figure 4(a) (from REIFF *et al.*, 1985) shows the results of that comparison. The agreement is quite impressive (a correlation coefficient of 0.74). We have now redone this comparison, using our new improved formulae (Fig. 4(b)). The general agreement is very good, particularly in the steady quiet times (6–10 UT) and the times of high voltage (10–12 UT and 14–16 UT).

We see that, for times such as this, when the IMF turns from being strongly southward to strongly northward (12–14 UT and 16–18 UT), the new formula underestimates the potential drop. Because the magnetotail is $\gtrsim 200\ R_E$ long, it requires 1–2 hours for a given field line to convect through the system (faster for larger voltages). Thus, if dayside merging is suddenly stopped, polar cap convection can

Table 1. Best-fit values of $\Phi=\Phi_0+A_0X(v, B, \theta)$, $|B_{sw}|<10$ nT.

X	(saturation $\|B\|$ value)	r	ϕ_0 (kV)	A_0 (kV·s/km·nT)
vB_s	(B_s = min(B_s,4))	0.82 (0.87)	35 (34)	0.049 (0.051)
vBsin(θ/2)	(B = min(B,10))	0.80 (0.84)	-7.1 (-3.3)	0.034 (0.032)
$vB\sin^2(\theta/2)$	(9)	0.87 (0.88)	-0.52 (3.9)	0.041 (0.038)
$vB\sin^3(\theta/2)$	(9)	0.90 (0.89)	6.7 (12.7)	0.047 (0.041)
$vB\sin^4(\theta/2)$	(8)	0.90 (0.88)	13 (18)	0.053 (0.048)
vBQ1 (compression 4)		0.83	38	0.102
vBQ1 (Crooker compression)	(7)	0.92 (0.90)	22 (25)	0.069 (0.062)
vBQ2 (compression 6)		0.81	40	0.076
vBQ2 (Crooker compression)	(7)	0.89 (0.84)	29 (34)	0.079 (0.063)
vBQ3 (compression 4.5)		0.91	29	0.041
vBQ3 (Crooker compression)	(21)	0.85 (0.82)	20 (22)	0.024 (0.022)

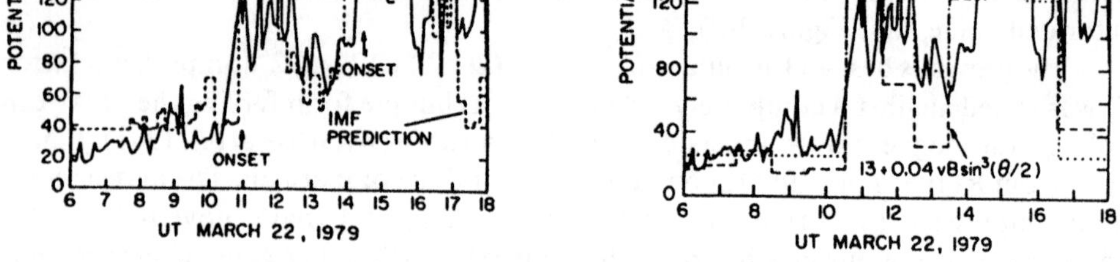

Fig. 4. (a) Cross-polar-cap voltage estimated from the IMF using $\Phi=38+122\,\alpha Q2$ (compression factor 7) (dashed), and estimated from inversion of magnetograms (solid). Considering the various limitations and sources of error in the two techniques, the agreement is quite gratifying (from REIFF *et al.*, 1985). (b) Same as (a), but now using our new best-fit functions $vBQ1$ (Crooker compression) (light line) and $vB\sin^3(\theta/2)$ (dashed). These formulas are asymptotic (steady state)formulas, and so are better at predicting the voltage during slowly-changing times, (i.e. 6–10 UT), but not when the IMF B_z is changing from large and negative to large and positive, as it does from 12–13 and 16–17 UT. The B-field is limited to 9 and 7 nT, respectively (see figures in parentheses in Table 1).

persist for 2–3 hours. The new formulae were derived from data when the IMF, if northward, had been northward for 3 hours or more; therefore, is an underestimate in these dynamic times.

4. Predictions of Cusp Merging Models

The other major type of merging model, those with merging lines at or near the dayside cusp, has not yet matured enough to make quantitative predictions of the polar-cap voltage. Part of the difficulty lies in that the magnetosheath at locations removed from the subsolar point has a significant flow parallel to the boundary. Nevertheless, the antiparallel merging models have made several qualitative predictions which have been borne out by experimental work:

(1) Magnetic merging can occur between the IMF and the magnetotail field lines poleward of the cusp, resulting in sunward polar cap convection during times when the IMF is nearly directly northward (DUNGEY, 1963; RUSSELL, 1972; confirmed by MAEZAWA, 1976; BURKE *et al.*, 1979; REIFF, 1982; ZANETTI *et al.*, 1984; see REIFF, 1984 for a general review of evidence for magnetic merging).

(2) For IMF vectors which are northward but have a strong y-component, the predominant flow in the polar cap is nearly circular (called a "lobe cell" by BURCH and REIFF, 1984), with the direction of rotation determined by the sign of the y-component (for the Northern hemisphere, $+Y$ is associated with clockwise flow around the pole, with the opposite for the Southern hemisphere) (CROOKER, 1979; first discovered experimentally by SVALGAARD, 1973; also observed by HEELIS, 1984; POTEMRA *et al.*, 1984; BURCH and REIFF, 1984).

(3) For IMF directions with $|B_y| \gtrsim |B_z|$, the IMF can merge both with closed dayside magnetic field lines and open tail lines simultaneously (CROOKER, 1979; REIFF and BURCH, 1985; observed by BURCH *et al.*, 1985). This results in a "dayside merging cell" and a "lobe cell" occurring together in the polar cap. Figures 5(b) and 5(g), from CROOKER (1979) show two flow lines circulating entirely poleward of the open-closed boundary (shown dashed), i.e. our "lobe cell", and two flow lines that cross the open-closed boundary, i.e. our "merging cell."

Figure 6(a), from BURCH *et al.* (1985) shows Dynamics Explorer data containing both a dayside merging cell, a lobe cell, and two flanking viscous cells. Although it appears that the viscous cell is identified by but a single data point, it is in fact observed in four instrument cycles (only every fourth point is plotted on this polar plot). This type of viscous cell (or viscous "spike") is observed on roughly 30% of the DE orbits (this orbit occurred when the IMF was southward). Since these viscous spikes occur at the location where the particle signature changes from a trapped plasma sheet distribution to a polar rain or polar cusp distribution, we gain confidence in their significance. Figure 6(b) shows their model of the convection pattern for southward B_z and $B_y<0$. Figure 6(c) shows the convection observed for $B_y>0$ using two nearly-simultaneous passes of DE 2 and Intercosmos Bulgaria 1300, also showing lobe, viscous and merging cells, but now the lobe cell occurs on the dusk side, as predicted. One gains more confidence in assigning the spike at ~68 ILAT in the DE 2 data with a viscous "cell", since it is also observed at nearly the same latitude

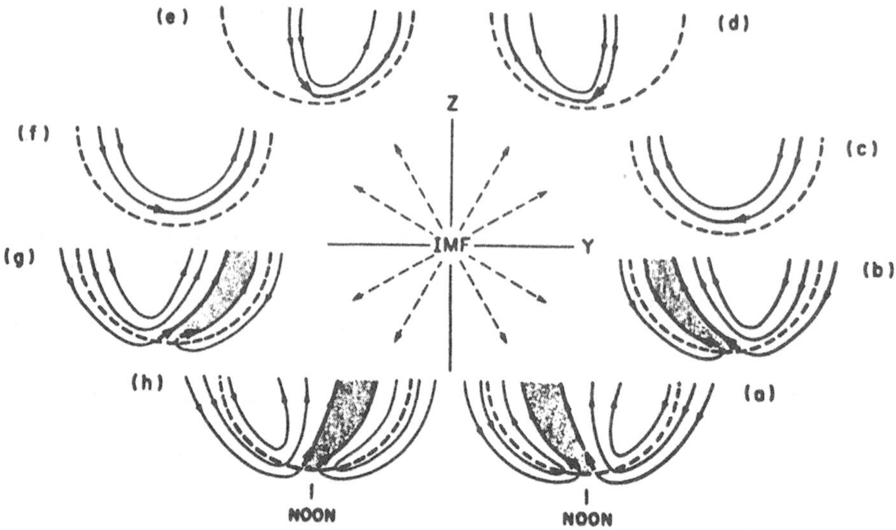

Fig. 5. Predicted polar cap convection patterns in the dayside northern hemisphere, viewed from above, for various orientations of the interplanetary magnetic field (IMF) in the y–z plane, perpendicular to the earth–sun line. The short heavy arrows represent the direction of flow initiated by dayside merging near the cusp. Note the pattern of rotation of the arrows around the figure in response to the rotation of the IMF. The heavy streamline in each pattern indicates the position of maximum convection. Those streamlines which cross the polar cap boundary (the heavy dashed line) effect a transfer of flux from closed dayside field lines to open tail lobe field lines. The shaded regions are convection gaps, predicted by Crooker; however, these should map to closed portions of the magnetopause and have no net width in the ionosphere. Note the coexistence of merging cells and lobe cells in Figs. (b) and (g) (from CROOKER, 1979).

by ICB-1300. These viscous cells do not exhibit a large potential difference across them (a few kV), but the independent measurements of viscous potentials yield values of that order (see COWLEY, 1984 for a review). Figure 6(b) differs from Fig. 5(b) (Crooker's models) only by the addition of viscous cells which are narrow and contain only a small potential difference; thus Crooker's model may be considered a very good approximation to the typical observations.

(4) If the IMF has a significant x-component, the merging location and strength is significantly different between the two hemispheres. For example, if the IMF has only a $-x$-component, merging occurs between the IMF and the lobe field lines in the northern hemisphere, and between the IMF and closed field lines in the southern hemisphere (3rd row left of Fig. 7). In this case, the "merging potential" should be equal in the two hemispheres, but the "lobe potential" should be considerably different. (See also COWLEY, 1981 for other asymmetries expected in relation to the x-component of the IMF.)

We present here the first estimate of the predicted polar-cap voltage for the antiparallel merging model. We estimate the polar-cap voltage by the following a

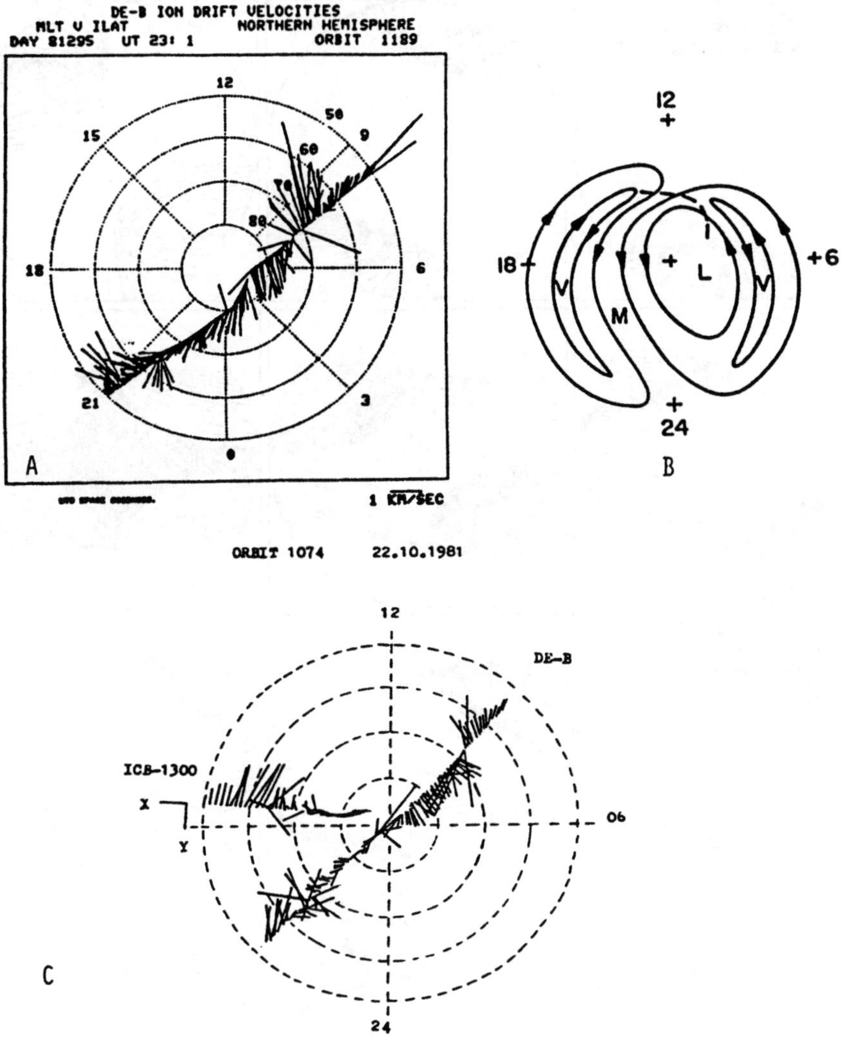

Fig. 6. (a) Polar cap convection pattern for orbit 1189. A small reversed flow spike near 8 MLT is the signature of the viscous cell. The plasma signature shows a sudden change there from plasma sheet to boundary layer conditions. (b) Schematic flow pattern corresponding to (a) for $B_y < 0$ ((a) and (b) from BURCH *et al.*, 1985). (c) Convection pattern for $B_y > 0$ observed nearly simultaneously by DE-B (courtesy R. A. Heelis) and Intercosmos Bulgaria-1300 (courtesy E. Dubinen). Here the viscous spike is observed by both satellites near $\Lambda = 65°$, and the lobe cell is observed on the dusk side, as predicted by BURCH *et al.* (1985).

four-step procedure: (1) We determine the locus of points along the magnetopause at which the internal and external fields are antiparallel. We use the model of LUHMANN *et al.* (1984), and choose as our operational definition of "antiparallel" that the cosine

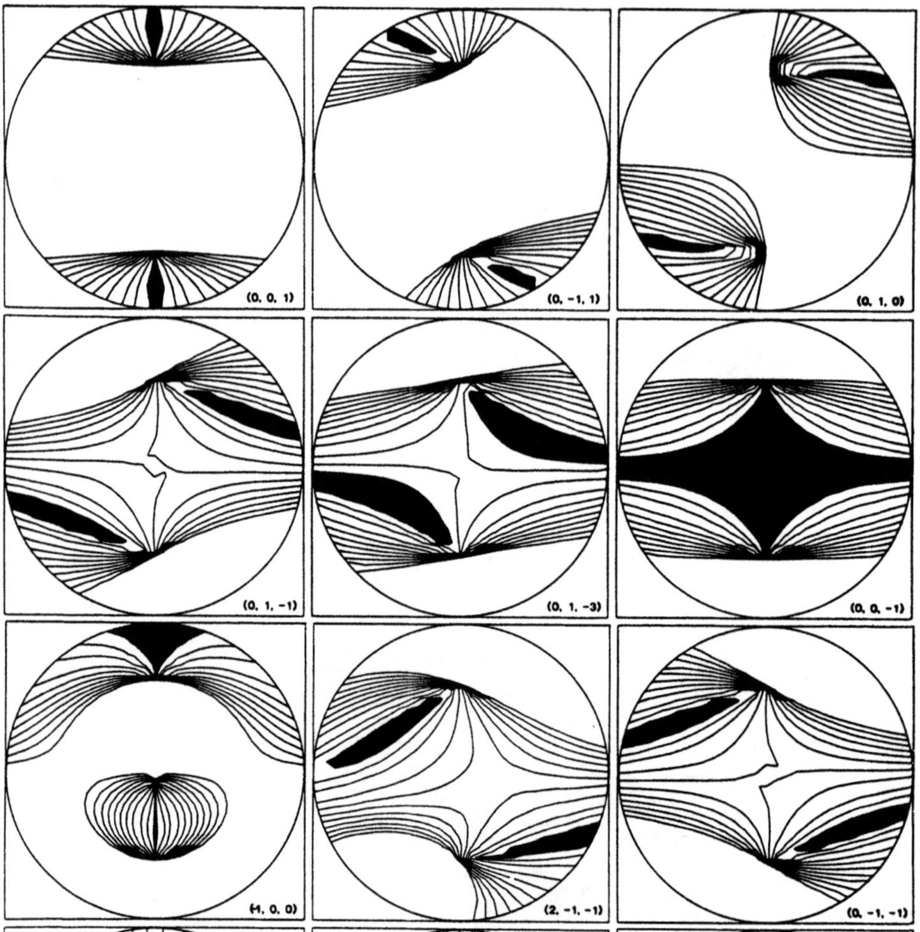

Fig. 7. Contours on the dayside magnetopause (viewed from the sun, i.e., the GSE *YZ* plane projection) of equal value of the cosine of the angle between the magnetospheric and magnetosheath model fields. Only contours with negative values, implying some antiparallel component, are shown. Values at the contours, starting with the contour filled with shading are −0.98, −0.95, −0.9, −0.8, −0.7, −0.6, −0.5, −0.4, −0.3, −0.2, −0.1. The interplanetary field orientations are indicated in vector notation (B_x, B_y, B_z) in the lower right corners. In particular, the patterns for northward and southward fields occupy the first position in the left column and the second position in the right column, respectively. The pattern for a 45° cone angle is shown in the third position of the middle column (from LUHMANN *et al.*, 1984).

of the angle between the internal field (the Hedgecock and Thomas model (WALKER and SOUTHWOOD, 1982) and the draped magnetosheath external field, taken at a distance 0.5 R_E outside the magnetopause (from the gas dynamic model of SPREITER and STAHARA [1981]) be less than −0.98. (This definition of antiparallel corresponds to the area within the dark shading of Fig. 7 (LUHMANN *et al.*, 1984)). (2) Using the gasdynamic magnetosheath model, we follow the flux tube of draped field lines which map this area of the magnetopause out into the solar wind (Fig. 8). (3) We measure the width W of the flux tube in the solar wind perpendicular to both the magnetic field direction and the solar wind flow direction (for these cases, we choose $B_x=0$, so that the solar-wind electric field is in the solar magnetospheric y–z plane, and $\theta=\theta_T$). Since there typically is a merging region in each hemisphere, we add the widths $w1$ and $w2$ to get W. (4) We normalize these widths by dividing the width W_0 of the packet that

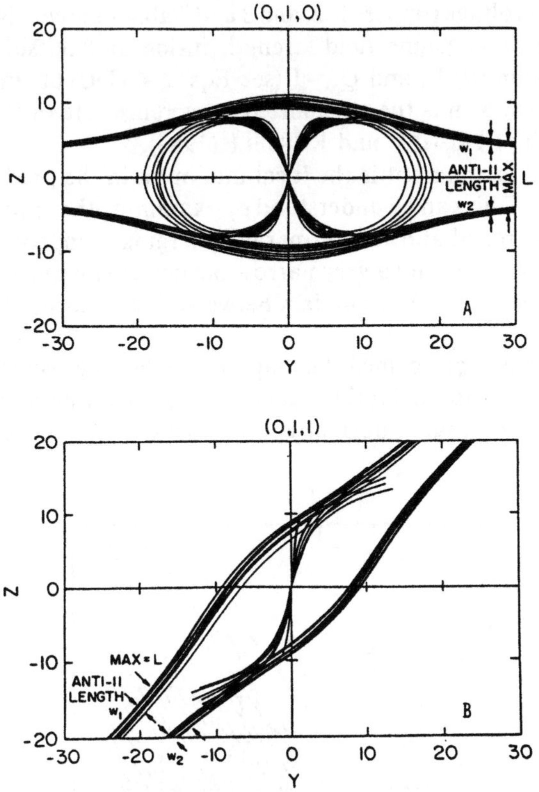

Fig. 8. Estimate of the polar cap potential drop from the antiparallel merging hypothesis. Field lines which touch the merging area (the area with $\cos(B_{sh}, B_E) < -0.98$) are mapped both inside the magnetosphere to the earth (interior lines) and from the magnetosheath into the solar wind (exterior lines) (view from the sun). Since for these cases $B_x=0$, the potential drop should be proportional to the width $W=w1+w2$. The maximum potential (if merging occurs at the subsolar point as well) is proportional to L. (a) $B_x=0$, $B_y=1$, $B_z=1$ ($\theta=45°$), (b) $B_x=0$, $B_y=1$, $B_z=0$ ($\theta=90°$).

intersects the neutral line for a directly southward magnetic field (note that for such a field, the neutral line is in the equatorial plane, passing through the subsolar point: Fig. 7).

The results are shown in Fig. 9, where we show the normalized width W/W_0 (see Fig. 8) in the solar wind, which should be directly proportional to the polar-cap voltage for a given solar speed, as a function of θ. Also shown is the "maximum merging length", which is the distance L (see Fig. 8), also normalized by dividing by W_0. This length would correspond to the maximum voltage in this model if the merging line crossed the subsolar point as well (i.e., if multiple neutral lines or flux transfer events allowed merging to occur at the subsolar point at the same time that antiparallel merging was occurring near the cusps). We see that both W/W_0 and L/W_0 are strongly varying functions of θ_T, and non-zero for northward IMF's, as required by the polar cap potential data (conclusions 1 and 2 above).

Also shown for comparison are two angular functions, $\sin^2(\theta/2)$ and $\sin^4(\theta/2)$ (PERRAULT and AKASOFU, 1978), both of which have been shown to give reasonable fits to the polar-cap voltage (papers 1, 2, and 3 and Table 1 above. (Note: in the limit of large compression, $\alpha=1$ (same field strength inside and outside magnetopause), $Q1=\sin(\theta/2)$, $Q2=\sin^2(\theta/2)$, and $Q3=1$ (see Eqs. 2–4 (DOYLE and BURKE, 1983)). The histogram also shown is the normalized occurrence rate of flux transfer events (FTE's) presented by BERCHEM and RUSSELL (1984).

One result that is apparent is the local minimum in the predicted anti-parallel potential for $\theta=90°$. This can be understood by examining the appropriate panel (top right) of Fig. 7: the area of allowed antiparallel merging is approximately parallel to the IMF direction, so that only a very narrow packet can merge in each hemisphere. This direction is also close to the boundary between closed dayside fields and open tail fields.

We have separately performed the mapping to the solar wind for the field lines that connect to internal closed dayside field lines and those that map to open tail-lobe field lines. (We assume that a normal component exists, inward in the Northern

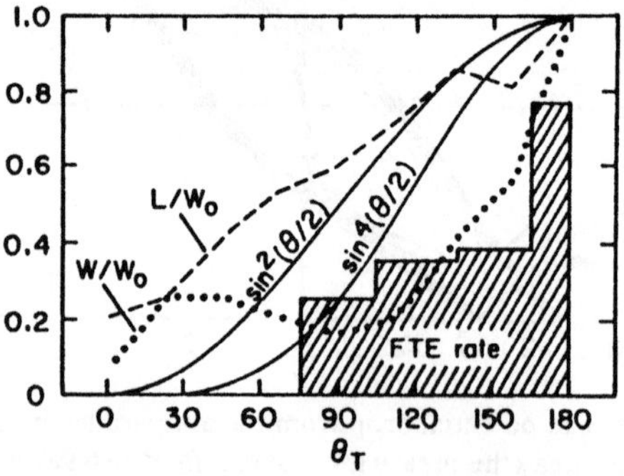

Fig. 9. Normalized antiparallel merging length W/W_0 (dotted), and maximum merging length L/W_0 (dashed), compared to $\sin^2(\theta/2)$, $\sin^4(\theta/2)$ and the flux transfer event rate (histogram), as a function of polar angle θ (B_x is chosen to be zero in these cases). W/W_0 is proportional to the predicted polar cap potential drop for the antiparallel hypothesis.

hemisphere and outward in the Southern hemisphere, that will connect the internal fields to the magnetosheath fields; unlike TOFFOLETTO and HILL (this volume), we do not calculate these normal fields explicitly.) We thus obtain the normalized widths W/W_0 and L/W_0 separately for the potential drops associated with merging of closed dayside field lines with the IMF (the "merging cell," shown as X's in Fig. 10) and with merging of open tail field lines with the IMF (the "lobe cell," shown as O's in Fig. 10). We see that, in the antiparallel merging model (top half of Fig. 10), both kinds of merging can occur simultaneously for weakly northward fields ($45° < \theta < 90°$), and the maximum lobe potential is predicted to be considerably (~4×) smaller than the maximum merging potential.

We have also performed a similar tracing exercise for the case of "component merging" (bottom half of Fig. 10), again normalizing the widths W/W_0 and L/W_0 as before. Unlike the case of antiparallel merging, when we mapped into the solar wind only the sheath fields which overlay the dark shaded areas on Fig. 7, we now map out into the solar wind the sheath fields that have *any* antiparallel component with the model internal fields (i.e. the outermost contours on Fig. 7). However, when we map

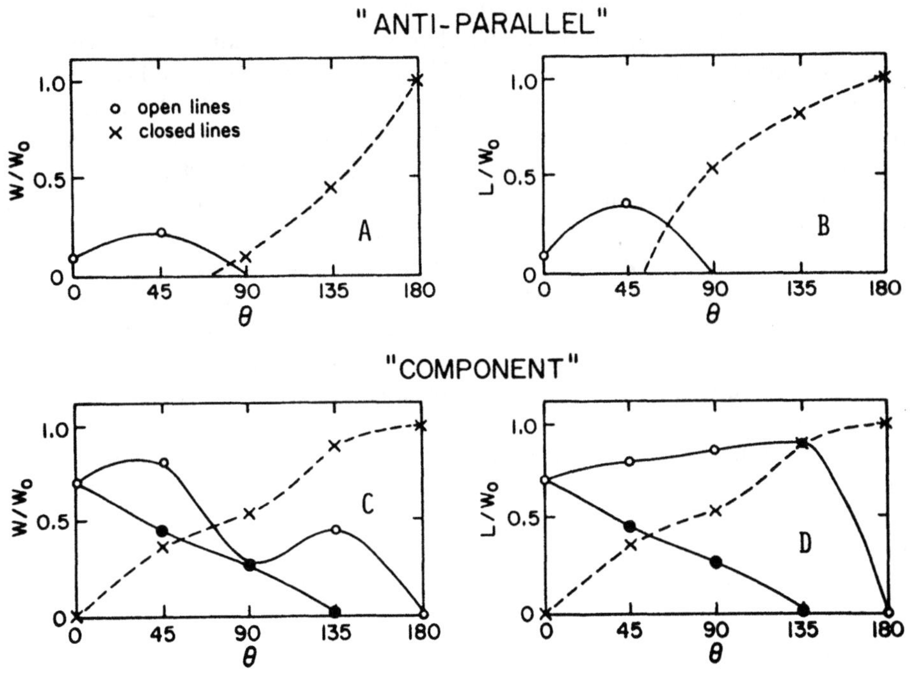

Fig. 10. Normalized merging lengths W/W_0 (top left) and L/W_0 (top right) for the "antiparallel merging" hypothesis (only field lines which map to the shaded areas of Fig. 7 are traced), and W/W_0 (bottom left) and L/W_0 (bottom right) for the "component merging" hypothesis (all field lines with any antiparallel component are traced). The length associated with field lines which were initially closed are shown by X's, with field lines initially open by O's. For the lower two panels, some overlap between the two tubes occurs; the width of the open flux tube exclusive of the overlap is shown by solid circles.

the flux tubes that overlap open field lines into the solar wind (open circles in the bottom of Fig. 10), there is a considerable overlap with the flux tubes that should connect to closed field lines (X's). Since a given IMF flux tube will impact the closed portion of the magnetopause first, it is reasonable to reduce the prediction for the open-field merging to only that region without overlap (filled circles in bottom of Fig. 10). Here the predicted maximum lobe potential is nearly two thirds of the maximum merging potential, which is not supported by the available data.

5. Testing of the Predictions

Determining separately the potential drop associated with merging with closed field lines (the "merging cell") and with merging with open field lines (the "lobe cell") is a difficult task, since the outflow from each is antisunward. However, in certain cases, a narrow "viscous cell" is observed near the polar cap boundary. When this narrow channel of antisunward flow lies at the trapping boundary for 40 keV electrons, it separates the sunward flow in the plasma sheet from the sunward flow in the lobe cell, allowing a separate determination of the potential in each cell (Fig. 11(a), from REIFF *et al.*, 1981). One such case is shown in Fig. 6(a), from BURCH *et al.* (1985), where the antisunward flow spike at 70 degrees invariant, 8 MLT, separates sunward-flowing plasma-sheet-like plasma from sunward-flowing polar-rain-type plasma. The integrated voltage for that case is shown in Fig. 11(b), from HEELIS and REIFF (1984). We see 2 kV in the dawn viscous cell, 14 kV in the lobe cell, and 95 kV in the merging cell. Examining the ISEE-3 IMF data for that time period (C. T. Russell amd E. J. Smith, personal communication, 1984), we find that the IMF was strongly southward (~9 nT) and the *y*-component strongly negative (~ −9 nT) for at least 10 hours before the orbit. However, the *y*-component changed sign during the one hour prior to the pass, yielding an average value of approximately −3 nT. For this case, the clock angle θ_T varied from ~130 to ~160 degrees, and strict antiparallel merging (top half of Fig. 10) would predict no lobe potential, whereas component merging (bottom left of Fig. 10) would predict a lobe potential of 0–40% of the merging potential. The measured ratio, 14%, certainly falls in the appropriate range. However, considering the large time variation, this one case cannot be considered a definitive test.

It is hoped that cases like this may ultimately be used to test the predictions of the polar-cap voltage from the various models. However, this is a difficult task because it requires the coexistence of the viscous, lobe, and merging cells, which apparently occurs only about 30% of the time (BURCH *et al.*, 1985). Such a study is currently underway (R. A. Heelis, personal communication, 1985).

6. Dependence of the Viscous Potential on the IMF

No good model yet exists which predicts how magnitude of the viscous potential should depend on the IMF. If the viscous potential is generated by large-amplitude Kelvin-Helmholtz waves (e.g., EVIATAR and WOLF, 1968; and other references above), we would expect their growth rate to be largest when the external field is either parallel or antiparallel to the internal field. If the viscous potential is generated by

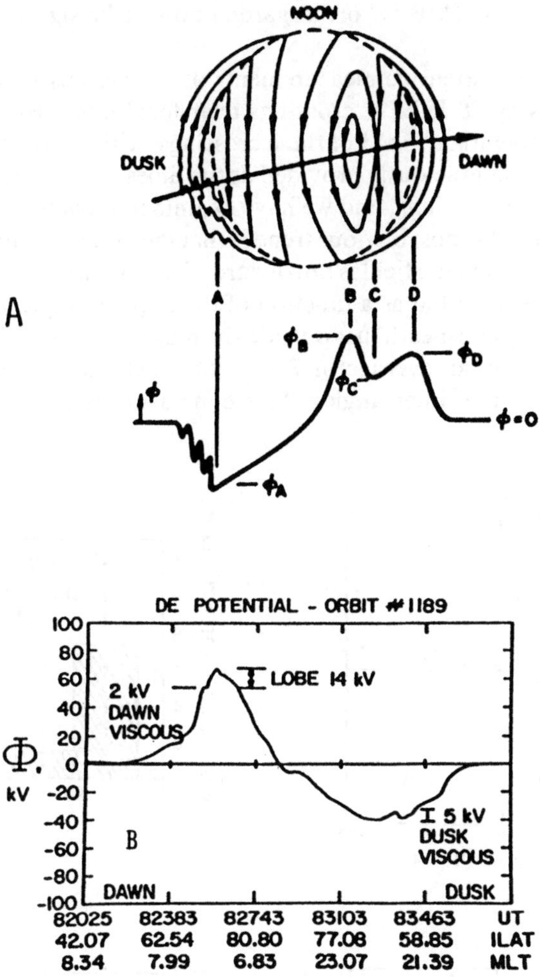

Fig. 11. (a) Identification of the lobe potential $\Phi_B-\Phi_C$, the viscous potential $\Phi_D-\Phi_C$ and the merging potential $\Phi_C-\Phi_A$ for a case where the lobe cell occurs on the dawn side ($B_y<0$) (from REIFF *et al.*, 1981). (b) Example of the potentials so calculated from the data in Fig. 6(a) (from HEELIS and REIFF, 1984).

cross-field diffusion, then its value should be larger on the flank of the magnetosphere where gyroresonant waves have the larger amplitudes. It is known that the bow shock is more turbulent when the IMF is nearly parallel to the shock normal (see RUSSELL and GREENSTADT, 1983, for a review). For example, if the IMF is nearly aligned with the x-direction, the turbulence is large near the nose of the magnetopause (LUHMANN *et al.*, 1985). Therefore, we might expect that the level of turbulence convected to the magnetopause (and hence viscous potential) would be larger on the dawn flank when the IMF has a "garden hose" orientation (the sign of B_x and B_y are opposite), and on the dusk side when the IMF is "ortho-garden hose" (the sign of B_x and B_y are the same).

So far, the most extensive measurements of the viscous potential are those of MOZER (1984), using 28 ISEE-1 crossings near local dusk. He found evidence of non-zero viscous potentials for 23 of these crossings, with an average potential of 2.5 kV (Fig. 12(a)). He has graciously provided us the times and magnetosheath magnetic field values for these crossings, and we have examined them to see if any significant trends are apparent. The most obvious trend is that the potential appears to be larger if the magnetosheath magnetic field is northward. We have used these data to calculate the average viscous potential as a function of the popular angle $\theta_{sh}=\cos^{-1}(B_z/|B_{sh}|)$, measured in the magnetosheath just outside the magnetopause (Fig. 12(b)). We see a significant trend, from an average of 7.5 kV for angles less than 45°, to a nearly-constant 1.5 kV for the other angles. This conclusion is compatible with Kelvin-

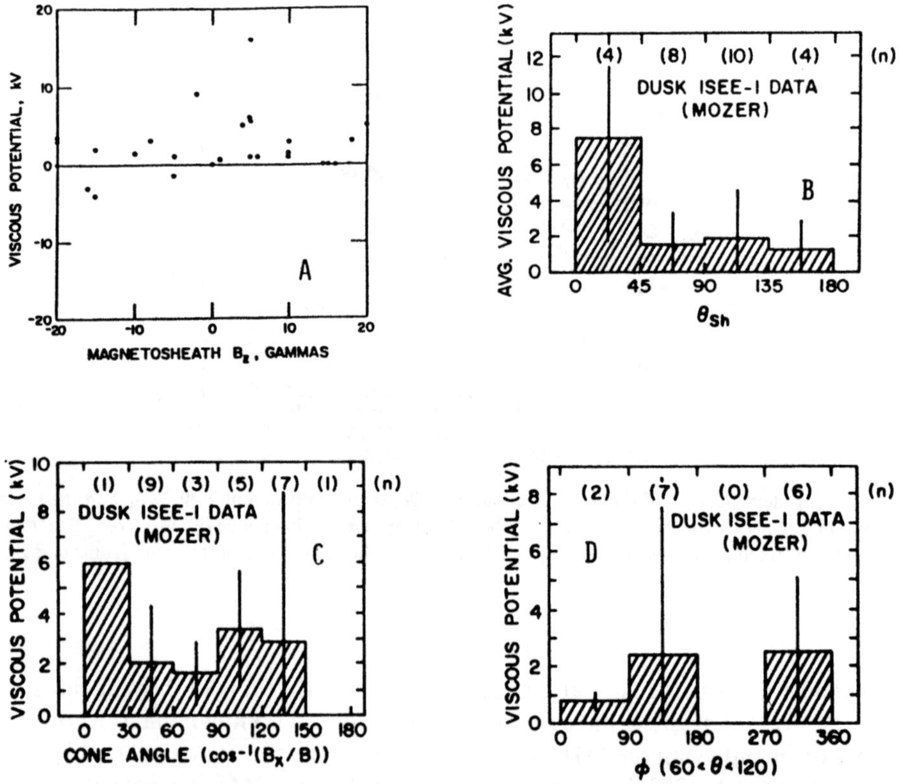

Fig. 12. (a) Viscous potential measured from ISEE near the dawn-dusk terminator as a function of the magnetosheath field (from MOZER, 1984). (B) The same data, averaged (with negative values replaced by zero) and plotted as a function of theta measured in the magnetosheath (data courtesy of F. S. Mozer). Average viscous potential measured as a function of IMF cone angle ($\cos^{-1} B_x/B$) (c) and of angle ϕ in the equatorial plane (d).

Helmholtz models, when one recalls that for strongly southward fields, the dayside magnetopause is open (CROOKER, 1977) and thus not subject to the instability. (Magnetic merging "eats up" the viscous boundary layer on the dayside).

We have also examined Mozer's data set to look for other trends, for example the cone angle effect of LUHMANN *et al.* (1985) (Fig. 12(c)) and the gardenhose-vs-orthogardenhose effect (Fig. 12(d)). Unfortunately, there is only one pass with a cone angle ($\cos^{-1} B_x/|B|$) less than 30°; that pass showed a viscous potential of 6 kV for a cone angle of 19°, which was larger than the average potentials observed at other cone angles (Fig. 12(c)); however, that large single data point is balanced by a null(0 kV) reading for a cone angle of 155°, which should also allow large turbulence according to the model of LUHMANN *et al.* (1985). Therefore the cone angle test is inconclusive. A similar test of gardenhose ($90° < \Phi < 180°$ or $270° < \Phi < 360°$) versus orthogardenhose IMF conditions is also inconclusive, because of the scarcity of orthogardenhose events (Fig. 12(d)), although the trend is negative. It may well be that the turbulence associated with parallel shocks is of too low a frequency to cause gyroresonant diffusion at the magnetopause.

7. Dependence of the Viscous Potential on the Solar Wind Velocity

SERGEEV and KUZNETSOV (1981) have, in effect, determined a simple dependence of the viscous potential on interplanetary parameters. They used polar-cap magnetic perturbations under summer conditions to infer the polar cap potential drop (the summertime restriction insured reasonably constant conductivity). They found Φ (kV)$=6(v/300\text{ km/s})^2-9B_z$, for $B_z<1$ nT. They attributed the B_z term to magnetic merging and the v^2 term to boundary layer effects. Thus the viscous potential should be ~6 kV for $v=300$ km/s and ~11 kV for $v=400$ km/s. A viscous potential of this size is certainly consistent with our best-fit residuals (~6 kV for $vB\sin^3(\theta/2)$, 13 kV for $vB\sin^4(\theta/2)$. This is certainly an interesting question for future study; the above interpretation suggests that the Reynolds number may also play a role in determining the potential. A second-order effect, certainly, but perhaps not negligible as suggested by VASYLIUNAS *et al.* (1982).

8. Conclusions

The asymptotic (steady-state) polar-cap voltage Φ can be shown to be nearly linearly proportional to $v_{sw}B\sin^n(\theta/2)$, where $\theta=\cos^{-1}(B_z/B)$ and $n=3$ or 4, so long as $|B|\leq 10$ nT. For $|B|>10$ nT, a strong saturation effect occurs. If one includes northward IMF data when only the IMF has remained northward for three hours or more, the residual potential (presumably generated by viscosity or diffusion at the magnetopause) reduces to $\leq$6–13 kV, which is comparable to direct measurements of the viscous potential. The saturation effect may be due to line-tying in the ionosphere (HILL *et al.*, 1976; HILL, 1984) although it should be noted that an alternate model has recently been proposed (PUDOVKIN *et al.*, 1985) whereby the length of the merging line depends on the solar wind velocity and magnetic field strength.

We have presented the first estimate of the magnitude of the voltage associated with the lobe cell, for cases with $B_x=0$, on the basis of the antiparallel merging model. We find that the lobe voltage should typically be considerably smaller (~4×) than typical merging voltages and should maximize, not for directly northward fields, but for fields with $|B_y| \sim B_z$ (Fig. 10). If $B_x=0$, no lobe convection should occur for IMF's with a southward component. Since the observations indicate that lobe cells can occur if the IMF is southward, some merging of fields which are not exactly antiparallel may occur. However, since the largest potential drops associated with lobe cells are ≲35 kV, the results are in general closer to the "antiparallel merging" predictions (top left of Fig. 10), than to the "component merging" predictions (bottom left of Fig. 10).

The authors gratefully acknowledge the following scientists for use of their data and/or models: J. Spreiter and S. Stahara, for the external magnetic field model; R. Walker for the internal magnetic field model; F. Mozer for viscous potential drop data; J. Wygant for S3-3 potential drop data; W. Burke for S3-2 potential drop data; E. Smith and C. Russell for ISEE'3 magnetometer data, and E. Dubinen for Intercosmos Bulgaria 1300 convection data.

This research was supported in part by the National Aeronautics and Space Administration (grant NAS2-9491 at UCLA, and grant NGR44-006-137 at Rice) and by the National Science Foundation (grant ATM-8518710).

APPENDIX

In order to minimize confusion as to how exactly to use the Crooker compression formulas to estimate the polar cap potential drop, the following BASIC subroutine is offered. This subroutine uses the formula $\Phi=25+0.062vBQ$ with $|B|=\min(|B|, 7\text{nT})$. Changes necessary in order to calculate $vBQ2$ or $vBQ3$ are noted.

```
3000  REM SUBROUTINE TO COMPUTE FITTING FUNCTION
3010  PRINT "VBQ1 (HILL EQN.65), CROOKER COMPRESSION"
3020  SET RADIANS
3030  P2=1.5708
3040  REM MU NOUGHT FOR D IN #/CM3, V IN KM/SEC, B IN
      NANOTESLAS
3050  MO=0.002098
3060  FOR I=1 TO N1
3070  B4=B(I)
3080  IF B(I)<7  THEN 3100
3090  B(I)=7
3100  BT=SQR(B1(I)*B1(I)+B2(I)*B2(I))
3110  REM    B1=By; B2=Bz; B=|B|; BT=Btangential; BE=Bearth;
      BS=Bsheath
3120  BE=SQR (2*M0*D1(I)*V(I)*V(I)+BT*BT)
3130  REM    D1=Density(#/cm3); V=Velocity (km/s)
3140  BS=2*SQR(BT*BE)
3150  IF BS<BE  THEN 3170
3160  BS=BE
```

```
3170  X(I)=0
3180  C8=BS/BE
3190  C9=B2(I)/BT
3200  REM    C8=alpha; C9=cos (thetatangential)=cos (clock angle)
3210  IF C8-C9<0  THEN   3260
3220  X(I)=V(I)*B(I)*C8*(C8-C9)/SQR(1+C8*C8-2*C8*C9)
3230  REM    for vBQ2 use    X(I)=V(I)*B(I)*C8*(C8-C9)*(C8-C9)/
      (1+C8*C8-2*C8*C9)
3240  REM    for vBQ3 add next line (with 3220)
3250  REM    X(I)=X(I)/(C8*SIN(ACS(C9)/2))
3260  PHI(I)=25+0.062*X(I)
3270  REM    for vBQ2 use    PHI(I)=34+0.063*X(I)
3280  REM    for vBQ3 use    PHI(I)=22+0.022*X(I)
3290  NEXT I
3300  RETURN
```

REFERENCES

ALFVÉN, H. and C.-G. FÄLTHAMMAR, *Cosmical Electrodynamics*, Oxford Univ. Press, London, 1963.

ATKINSON, G., Energy flow and closure of current systems in the magnetosphere, *J. Geophys. Res.*, **83**, 1089–1103, 1978.

AXFORD, W. I. and C. O. HINES, A unifying theory of high-latitude geophysical phenomena and geomagnetic storms, *Can. J. Phys.*, **39**, 1433–1464, 1961

BERCHEM, J. and C. T. RUSSELL, Flux transfer events on the magnetopause: spatial distributions and controlling factors, *J. Geophys. Res.*, **89**, 6689–6703, 1984.

BINGHAM, R., D. A. BRYANT, and D. S. HALL, A wave model for the aurora, *Geophys. Res. Lett.*, **11**, 327–330, 1984.

BURCH, J. L. and P. H. REIFF, Field-aligned currents and ion convection at high altitudes, *Space Research*, (in press) 1984.

BURCH, J. L., J. D. WINNINGHAM, V. A. BLEVINS, N. EAKER, W. C. GIBSON, and R. A. HOFFMAN, High-altitude plasma instrument for Dynamics Explorer-A, *Space Sci. Instrum.*, **5**, 455–463, 1981.

BURCH, J. L., P. H. REIFF, J. D. MENIETTI, R. A. HEELIS, W. B. HANSON, S. D. SHAWHAN, E. G. SHELLEY, M. SUGIURA, D. R. WEIMER, and J. D. WINNINGHAM, IMF B_y-dependent plasma flow and Birkeland currents in the dayside magnetosphere: 1. Dynamics Explorer Observations, *J. Geophys. Res.*, **90**, 1577– 1593, 1985.

BURKE, W. J., M. C. KELLEY, R. C. SAGALYN, M. SMIDDY, and S. T. LAI, Polar cap electric field structures with a northward interplanetary magnetic field, *Geophys. Res. Lett.*, **6**, 21–24, 1979.

BURTON, R. K., An empirical relation between interplanetary conditions and *Dst*, *J. Geophys. Res.*, **80**, 4024, 1975.

COLLIN, H. L. and R. D. SHARP, The relationship between the energies of upflowing ions and the measured accelerating potential, *EOS Trans. AGU*, **65**, 257, 1984.

COWLEY, S. W. H., Asymmetry effects associated with the *X*-component of the IMF in a magnetically open magnetosphere, *Planet. Space Sci.*, **29**, 809–818, 1981.

COWLEY, S. W. H., Solar wind control of magnetospheric convection, Proc. Conf. Achievements of the IMS, 26–28 June, 1984, Graz, Austria, *ESA SP-271*, 1984.

CROOKER, N. U., The magnetospheric boundary layers: a geometrically explicit model, *J. Geophys. Res.*, **82**, 3629–3633, 1977.

CROOKER, N. U., Dayside merging and cusp geometry, *J. Geophys. Res.*, **84**, 951–959, 1979.

CROOKER, N. U., The half-wave rectifier response of the magnetosphere and antiparallel merging, *J. Geophys. Res.*, **85**, 575–578, 1980.

CROOKER, N. U., Global geometry of merging at the dayside magnetopause, this volume, 1986.

CROOKER, N. U., G. L. SISCOE, P. R. MULLEN, C. T. RUSSELL, and E. J. SMITH, Magnetic field compression at the dayside magnetosphere, *J. Geophys. Res.*, **87**, 10407–10412, 1982.

DOYLE, M. A. and W. J. BURKE, S3-2 measurements of the polar cap potential, *J. Geophys. Res.*, **88**, 9125–9134, 1983.

DUNGEY, J. W., Interplanetary magnetic field and the auroral zones, *Phys. Rev. Lett.*, **6**, 47, 1961.

DUNGEY, J. W., The structure of the ionosphere, or adventures in velocity space, in *Geophysics: The Earth's Environment*, edited by C. DeWitt, J. Hiebolt, and A. Lebeau, pp. 526–536, Gordon and Breach, New York, 1963.

EASTMAN, T. E. and E. W. HONES, Jr., Characteristics of the magnetospheric boundary layer as observed by IMP 6, *J. Geophys. Res.*, **84**, 2019–2028, 1979.

EVANS, D. S., Precipitating electron fluxes formed by a magnetic field aligned potential difference, *J. Geophys. Res.*, **79**, 2853–2858, 1974.

EVIATAR, A. and R. A. WOLF, Transfer process in the magnetopause, *J. Geophys. Res.*, **73**, 5561, 1968.

GONZALES, W. D. and F. S. MOZER, A quantitative model for the potential resulting from reconnection with an arbitrary interplanetary magnetic field, *J. Geophys. Res.*, **79**, 4186–4194, 1974.

GURNETT, D. A. and L. A. FRANK, Observed relationships between electric fields and auroral particle precipitation, *J. Geophys. Res.*, **78**, 145–170, 1973.

HAERENDEL, G., G. PASCHMANN, N. SCKOPKE, H. ROSENBAUER, and P. C. HEDGECOCK, The frontside boundary layer of the magnetosphere and the problem of reconnection, *J. Geophys. Res.*, **83**, 3195–3216, 1978.

HEELIS, R. A., The effects of interplanetary magnetic field orientation on dayside high-latitude ionospheric convection, *J. Geophys. Res.*, **89**, 2873–2880, 1984.

HEELIS, R. A. and P. H. REIFF, Observations of magnetospheric convection from low altitudes, *Space Research*, in press, 1984.

HEPPNER, J. P., High latitude electric fields and the modulations related to interplanetary magnetic field parameters, *Radio Sci.*, **8**, 933–948, 1973.

HEPPNER, J. P., Empirical models of high-latitude electric fields, *J. Geophys. Res.*, **82**, 1115–1125, 1977.

HILL, T. W., Magnetic merging in a collisionless plasma, *J. Geophys. Res.*, **80**, 4689–4699, 1975.

HILL, T. W., Rates of mass, momentum, and energy transfer at the magnetopause, in *Magnetospheric Boundary Layers, ESA SP-148*, pp. 325–332, 1979.

HILL, T. W., Solar-wind magnetosphere coupling, in *Solar-Terrestrial Physics*, edited by R. L. Carovillano and J. M. Forbes, pp. 261–302, D. Reidel, Hingham, MA, 1983.

HILL, T. W., A. J. DESSLER, and R. A. WOLF, Mercury and Mars: The role of ionospheric conductivity in the acceleration of magnetospheric particles, *Geophys. Res. Lett.*, **3**, 429–432, 1976.

KAMIDE, Y., A. D. RICHMOND, and S. MASUSHITA, Estimation of ionospheric electric fields, ionospheric currents, and field aligned currents from ground magnetic records, *J. Geophys. Res.*, **86**, 801–813, 1981.

KAN, J. R. and L. C. LEE, Energy coupling function and solar wind-magnetosphere dynamo, *Geophys. Res. Lett.*, **6**, 577–580, 1979.

LEE, L. C. and Z. F. FU, A theory of magnetic flux transfer at the Earth's magnetopause, *Geophys. Res. Lett.*, **12**, 105–108, 1985.

LUHMANN, J. G., R. J. WALKER, C. T. RUSSELL, N. U. CROOKER, J. R. SPREITER, and S. S. STAHARA, Patterns of potential magnetic field merging sites on the dayside magnetopause, *J. Geophys. Res.*, **89**, 1739–1742, 1984.

MAEZAWA, K., Magnetic convection induced by the positive and negative z-components of the interplanetary magnetic field: Quantitative analysis using polar cap magnetic records, *J. Geophys. Res.*, **81**, 2289–2303, 1976.

MIURA, A., Anomalous transport by magnetohydrodynamic Kelvin-Helmholtz instabilities in the solar wind-magnetosphere interaction, *J. Geophys. Res.*, **89**, 801–818, 1984.

MOZER, F. S., Electric field evidence on the viscous interaction at the magnetopause, *Geophys. Res. Lett.*, **11**, 135–138, 1984.

MOZER, F. S., M. D. GONZALES, F. BOGOTT, M. C. KELLEY, and S. SCHUTZ, High-latitude electric fields and the three-dimensional interaction between interplanetary and terrestrial magnetic fields, *J. Geophys. Res.*, **79**, 56–63, 1974.

PERRAULT, P. and S.-I. AKASOFU, A study of geomagnetic storms, *Geophys. J. R. Astron. Soc.*, **54**, 547, 1978.

POTEMRA, T. A., L. J. ZANETTI, P. F. BYTHROW, A. T. Y. LUI, and T. IIJIMA, B_y□dependent convection patterns during northward IMF, *J. Geophys. Res.*, **89**, 9753–9760, 1984.

PU, Z.-Y., and M. G. KIVELSON, Kelvin-Helmholtz instability at the magnetopause: energy flux into the magnetosphere, *J. Geophys. Res.*, **88**, 853–861, 1983.

PUDOVKIN, M. I., S. A. ZAITSEVA, T. A. BAZHENOVA, and V. G. ANDREZEN, Electric fields and currents in the Earth's polar caps, *Planet. Space Sci.*, **33**, 407–414, 1985.

RASSBACH, M. E., R. A. WOLF, and R. E. DANIEL, Jr., Convection in a Martian magnetosphere, *J. Geophys. Res.*, **79**, 1125–1127, 1974.

REIFF, P. H., Sunward convection in both polar caps, *J. Geophys. Res.*, **87**, 5976–5980, 1982.

REIFF, P. H., Evidence of magnetic merging from low-altitude spacecraft and ground-based experiments, in *Magnetic Reconnection in Space and Laboratory Plasmas*, edited by E. W. Hones, Jr., pp. 104–113, AGU Press, Washington, D.C., 1984.

REIFF, P. H. and J. L. BURCH, B_y-dependent plasma flow and Birkeland currents in the dayside magnetosphere: A global model for northward and southward IMF, *J. Geophys. Res.*, **90**, 1595–1609, 1985.

REIFF, P. H., R. W. SPIRO, and T. W. HILL, Dependence of polar cap potential drop on interplanetary parameters, *J. Geophys. Res.*, **86**, 7639–7648, 1981.

REIFF, P. H., R. A. WOLF, R. W. SPIRO, Y. KAMIDE, and J. H. KING, Comparison of polar cap potential drops estimated from solar wind and ground magnetometer data: CDAW-6, *J. Geophys. Res.*, **90**, 1318–1324, 1985.

ROSTOKER, G., H.-L. LAM, and W. D. HUME, Response time of the magnetosphere to the interplanetary electric field, *Can. J. Phys.*, **5**, 544, 1972.

RUSSELL, C. T., The configuration of the magnetosphere, in *Critical Problems of Magnetospheric Physics*, edited by E. R. Dyer, Jr., pp. 1–16, National Academy of Science, Washington, D.C., 1972.

RUSSELL, C. T., and E. W. GREENSTADT, Plasma boundaries and shocks, *Rev. Geophys. Space Phys.*, **21**, 449–462, 1983.

SANDERS, G. D., L. J. MAHER, and J. W. FREEMAN, Observation of the plasma boundary layers at lunar distances: direct injection of plasma into the plasma sheet, *J. Geophys. Res.*, **85**, 4607–4615, 1980.

SERGEEV, V. A. and B. M. KUZNETSOV, Quantitative dependence of the polar cap electric field on the IMF B_z-component and solar wind velocity, *Planet. Space Sci.*, **29**, 205–213, 1981.

SISCOE, G. L., Polar cap size and potential: A predicted relationship, *Geophys. Res. Lett.*, **9**, 672-675, 1982.

SONNERUP, B. U. Ö., Magnetopause reconnection rate, *J. Geophys. Res.*, **79**, 1546–1549, 1974.

SONNERUP, B. U. Ö., Theory of the low latitude boundary layer, *J. Geophys. Res.*, **85**, 2017–2026, 1980.

SPREITER, J. R. and A. Y. ALKSNE, Plasma flow around the magnetosphere, *Rev. Geophys.*, **7**, 11, 1969.

SPREITER, J. R. and S. S. STAHARA, A new predictive model for determining polar wind-terrestrial planet interactions, *J. Geophys. Res.*, **85**, 6769–6777, 1980.

SVALGAARD, L., Polar cap magnetic variations and their relationship with the interplanetary magnetic sector structure, *J. Geophys. Res.*, **78**, 2064–2078, 1973.

TOFFOLETTO, F. R. and T. W. HILL, Quatitative model of magnetic coupling between solar wind and magnetosphere, this volume, 1986.

TSURUTANI, B. T. and R. M. THORNE-, Diffusion processes in the magnetopause boundary, *Geophys. Res. Lett.*, **9**, 1247–1250, 1982.

VASYLIUNAS, V. M., Theoretical models of magnetic field-line merging, 1, *Rev. Geophys. Space Phys.*, **13**, 303–336, 1975.

VASYLIUNAS, V. M., J. R. KAN, G. L. SISCOE, and S.-I. AKASOFU, Scaling relations governing magnetospheric energy transfer, *Planet. Space, Sci.*, **30**, 359–365, 1982.

WALKER, R. J. and D. S. SOUTHWOOD, Momentum balance and flux conservation in model magnetospheric magnetic fields, *J. Geophys. Res.*, **87**, 7460–7466, 1982.

WYGANT, J. R., R. B. TORBERT, and F. S. MOZER, Comparison of S3-3 polar cap potential drops with the interplanetary magnetic field and models of magnetopause reconnection, *J. Geophys. Res.*, **85**, 5727–5735, 1983.

ZANETTI, L. J., T. A. POTEMRA, T. IIJIMA, W. BAUMJOHANN, and P. R. BYTHROW, Ionospheric and Birkeland current distributions for northward interplanetary magnetic field: inferred polar convection, *J. Geophys. Res*, **89**, 7453–7458, 1984.

Solar Wind-Magnetosphere Coupling, edited by Y. Kamide and J. A. Slavin, 477–494.

Solar Wind Dependencies of High-Latitude Convection and Precipitation

J. C. FOSTER[1], J. M. HOLT[1], R. G. MUSGROVE[1], and D. S. EVANS[2]

[1]*MIT Haystack Observatory, Westford, Massachusetts 01886, U.S.A.*
[2]*NOAA/ERL, Boulder, Colorado 80302, U.S.A.*

The solar wind plasma and interplanetary magnetic field (IMF) dependencies of the patterns of plasma convection and particle precipitation in the high-latitude ionosphere are investigated by constructing average patterns of the ionospheric electric field and the precipitating energy influx for different solar wind conditions. A data base of over 100 days' observations of high-latitude plasma convection with the Millstone Hill radar has been sorted and binned according to the one-hour averaged values of IMF B_z and B_y. The individual line of sight velocities observed by the radar have been combined in each magnetic latitude-local time bin to derive statistically averaged convection patterns between 50 and 73 degrees latitude. A 25 mV/m southward electric field is found at auroral latitudes in the midnight sector for IMF away sectors ($B_y>0$) while a southward directed IMF increases the cross polar cap potential difference from 15 kV (for $B_z>+3$ nT) to 75 kV (for $B_z<-3$ nT). For geomagnetically disturbed conditions (K_p=3-to 5-), the additional requirement that B_z be negative results in a 30 kV potential difference across an enhanced dayside region of plasma entry into the polar cap. Average patterns of the energy influx to the ionosphere by precipitating electrons have been produced by a similar sorting and averaging technique applied to a data set consisting of six satellite years' data from the NOAA/TIROS satellites. Average patterns are presented for six levels of geomagnetic activity and for data selection criteria based on the IMF orientation. It is found that the energy influx carried by 300 eV to 20 keV ions and electrons reaches an asymptotic limit at and above a latitude which decreases from 80 to 64 degrees as K_p increases from 0 to 5. The average total energy influx into a single hemisphere increases from 14 GW for $B_z>+3$ nT to 32 GW when $B_z<-3$ nT and is enhanced on the dusk (dawn) half of the northern hemisphere auroral and polar region for IMF away (toward) sectors.

1. Introduction

The high-latitude ionosphere responds to changes in the near-earth interplanetary medium due to the electric fields, currents, and particle precipitation which couple the low altitude regime to the magnetosphere and its source of energy in the solar wind. The dynamics of the ionosphere, principally driven by magnetospheric electric fields, is a direct consequence of the solar wind-magnetosphere coupling. Although ionospheric currents and magnetospheric substorm activity respond in a repeatable fashion to changes in the near-earth interplanetary magnetic field, IMF (e.g., FOSTER *et al.*, 1971), the ionospheric effects are filtered from the solar wind by the coupling through the magnetosphere and so the details and timing of their response to solar wind variability are difficult to predict. Case studies examine the synchronism and repeatability of the ionospheric response and correlations and characteristic signatures are sought. Alternately, the dependencies can be determined in a statistical sense if a large, unbiased data set is available. Satellite observations of precipitating electrons have been used to address the solar wind dependencies of particle precipitation patterns in the polar cap (GUSSENHOVEN *et al.*, 1984) and across auroral latitudes (HARDY *et al.*, 1985). Qualitative models of the convective circulation of the magnetosphere and ionosphere have been derived from satellite observations of plasma drifts and currents (e.g., HEELIS, 1984; POTEMRA *et al.*, 1984; REIFF and BURCH, 1985). Empirical models of the plasma convection in the ionosphere have been developed from incoherent scatter radar observations (FOSTER, 1983; OLIVER *et al.*, 1983; HOLT *et al.*, 1985a) and have been related to the magnetospheric circulation by mapping along magnetic field lines (SOJKA *et al.*, 1983; FOSTER, 1984). For this study we have assembled a compressed computer accessible data base detailing the auroral and polar latitude distribution of particle precipitation and convection electric field at ionospheric heights and have merged this with hourly average values of solar wind parameters in order to determine how these two important sources of energy in the high-latitude ionosphere respond to the condition of the solar wind.

2. Precipitation Patterns

The NOAA/TIROS weather satellites carry precipitating ion and electron detectors in circular sun-synchronous dawn–dusk orbits at 850 km altitude. Normally there are two such systems aloft at any given time. By integrating the precipitating energy flux along the satellite track the average energy flux carried by ions and electrons between 300 eV and 20 keV is determined at each degree of magnetic latitude poleward of 45 degrees. Although the satellites are in dawn–dusk orbits, the diurnal motion of the magnetic pole about the geographic pole enables them to survey a large portion of magnetic latitude-local time space each day. Combining data acquired by satellites in different orbits, and data from both hemispheres results in almost complete coverage of the high latitude region. Data are not available for the region below 60 degrees at local times immediately before midnight. A data base consisting of six satellite years spanning 1979–1981 was used.

For the purpose of this study each one-degree averaged energy flux sample was tagged with the invariant latitude, dipole magnetic local time, K_p index and solar wind plasma and interplanetary magnetic field parameters appropriate to its measurement. Figure 1 presents average patterns of energy deposition by particle precipitation for six levels of the planetary geomagnetic activity index K_p. These patterns were derived by bin averaging the individual samples in 2 degrees invariant latitude bins at each 0.25 hour of dipole magnetic local time. The resultant average values of the energy flux were color coded in 21 steps between 0.01 and 14.48 ergs/cm^2/ster/sec. Intensity levels are separated by a factor of the square root of 2. Patterns were generated for K_p between 0–1, 1–2, 2–3, 3–4, 4–5, and >5. The number of one-degree averaged samples (in thousands) included in the six patterns was 329, 743, 791, 542, 232, and 111, respectively. These patterns clearly show the progressive expansion and increase in

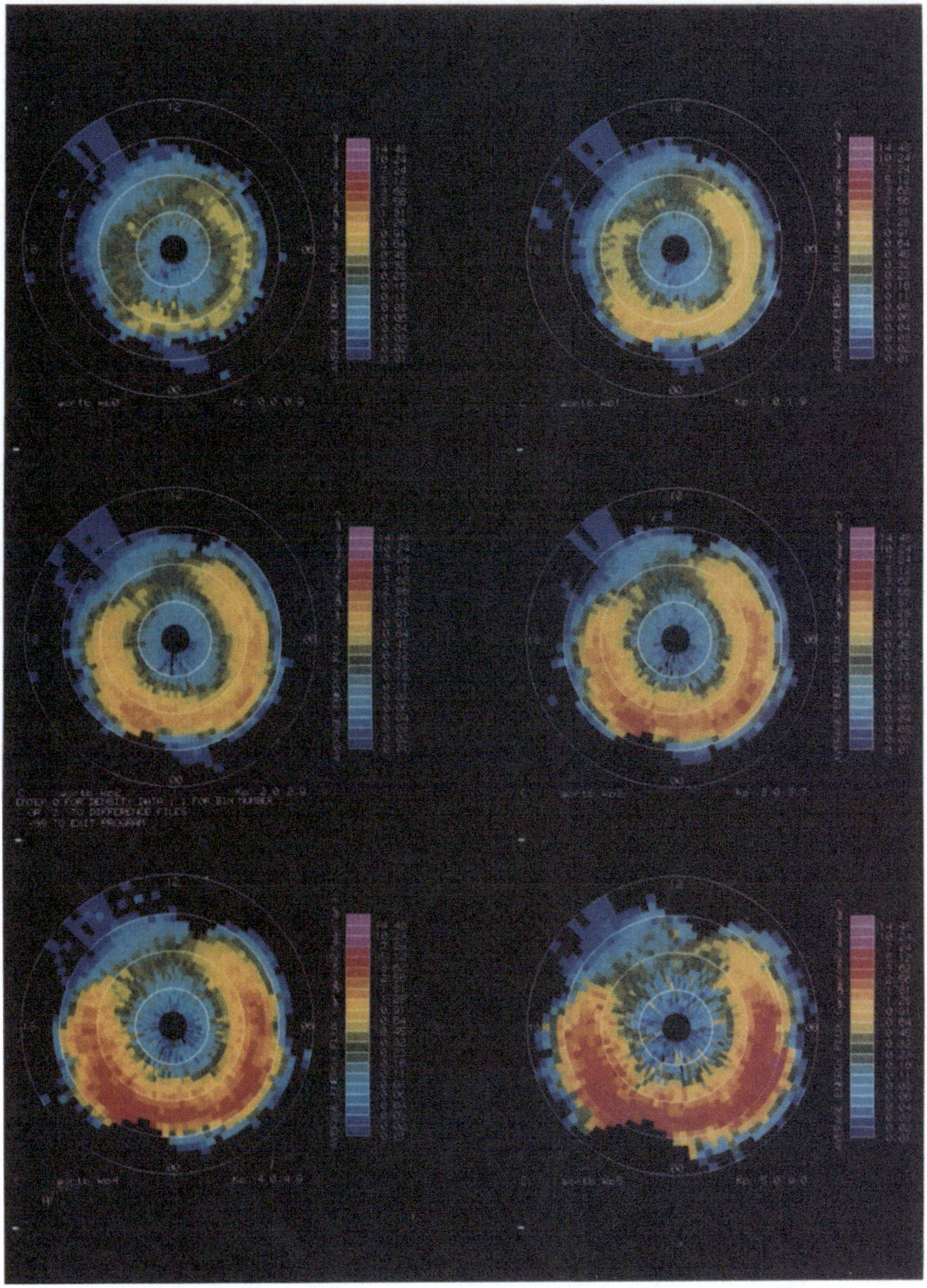

Fig. 1. Average patterns of energy flux carried by precipitating electrons with energies from 200 eV to 20 keV for six ranges of the geomagnetic activity index K_p. For each range of K_p 0–1, 1–2, etc. the average energy influx has been calculated for two degree bins in invariant latitude between 50 and 86 degrees and at each 15 min of dipole magnetic local time.

average intensity of the precipation pattern with increasing magnetic activity. The energy input patterns have the shape of a horseshoe opening toward 13 hours local time and are nearly circular except on the dayside where they are somewhat blunted across the early post-noon hours. Whereas the nightside precipation energy input increases and expands progressively with increasing K_p, precipitation in the dayside region between 13–16 LT is most intense for the lowest activity levels. Similar patterns of average number flux, integral energy flux, and average energy have been presented in the work of HARDY *et al.* (1985) whose statistical patterns were derived from Defense Meteorological Satellite Program (DMSP) detectors sensitive to precipitating electrons in the energy range from 20 eV to 20 keV. The HARDY *et al.* (1985) patterns show local maxima in the integral energy flux across the noon sector due to the influx of electrons below the energy threshold used in this study. The further DMSP study of GUSSENHOVEN *et al.* (1984), concentrating on polar latitudes, examined the activity and IMF B_z dependencies of polar rain electrons over the central polar cap. The statistical patterns derived in that study largely pertain to energies below the detector threshold used in the current study and describe the access to the ionosphere of plasma mantle electrons.

Figure 2 quantifies the six patterns of Fig. 1 by integrating around all local times at each latitude in order to determine the latitudinal distribution of the precipitating energy input. Curves for each K_p level are displayed in units of gigawatts (GW) per 2 degree latitude bin, and the total hemisphere-averaged energy input (in GW) is indicated. Electron precipitation accounts for 5 to 50 GW as K_p increases from 0 to >5. Whereas the total power input increases exponentially and the peak energy flux

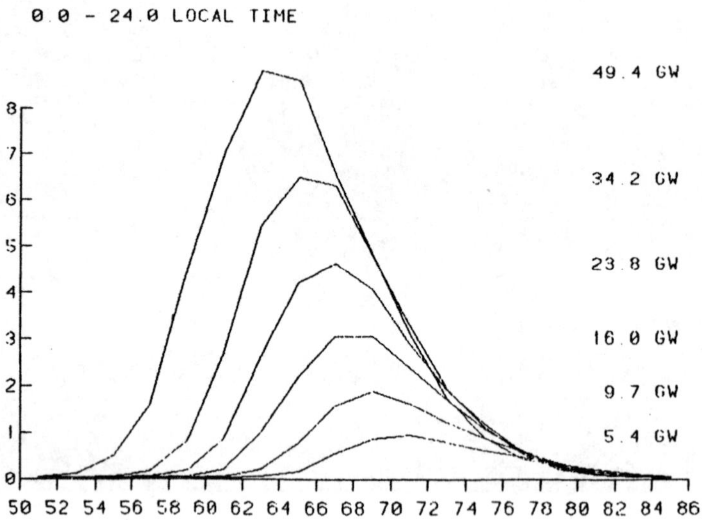

Fig. 2. The distribution of precipitating energy influx as a function of invariant latitude for the six K_p-dependent patterns of Fig. 1 are displayed in units of gigawatts (GW) per 2 degree latitude. The total energy input in a single hemisphere above 50 degrees is indicated for each K_p level.

moves steadily to lower latitude with increasing K_p, the distribution of the energy input poleward of the peak remains remarkably constant. The latitudinal distribution of the energy input reaches an asymptotic limit at progressively lower latitudes as a function of increasing K_p. For example, the total energy input poleward of 72 degrees, and its distribution above that latitude, is constant for any value of K_p greater than 2. The asymptotic limiting curve indicated by this study is presented in Fig. 3. This distribution is not that which would result from uniform precipitation in the region poleward of the peak, but rather indicates that at energies between 300 eV and 20 keV the energy input at polar and high auroral latitudes is independent of the K_p index of ionospheric currents.

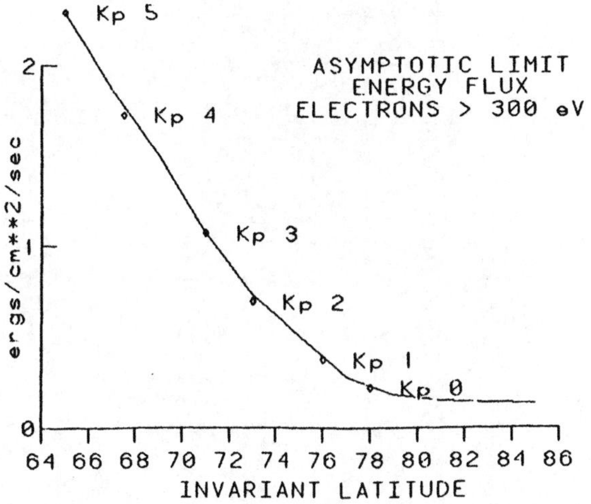

Fig. 3. The K_p-dependent curves of Fig. 2 indicate that the precipitating electron energy input at each latitude approaches an asymptotic limit. The energy influx remains constant at and above a latitude which decreases as indicated as K_p increases.

The satellite data base has been merged with the NASA 1-hour averaged interplanetary parameter data set which includes solar wind plasma and magnetic field data (KING, 1983). The bin sorting program used to derive the K_p dependent patterns was used to determine average precipitation patterns for different configurations of the IMF in solar magnetospheric coordinates. Several points are worth mentioning in regards to this study. The total number of precipitation data samples for which IMF data exist is considerably less than the total data set available for a K_p sort. When sorting by IMF sector (B_y), data for B_y positive in the northern hemisphere were combined with data for B_y negative from the southern hemisphere (and the converse). Patterns derived from simultaneously sorting on B_z (solar magnetospheric north–south component) and B_y (negative for toward sectors and positive for away sectors) are presented in Fig. 4. In order to accentuate the

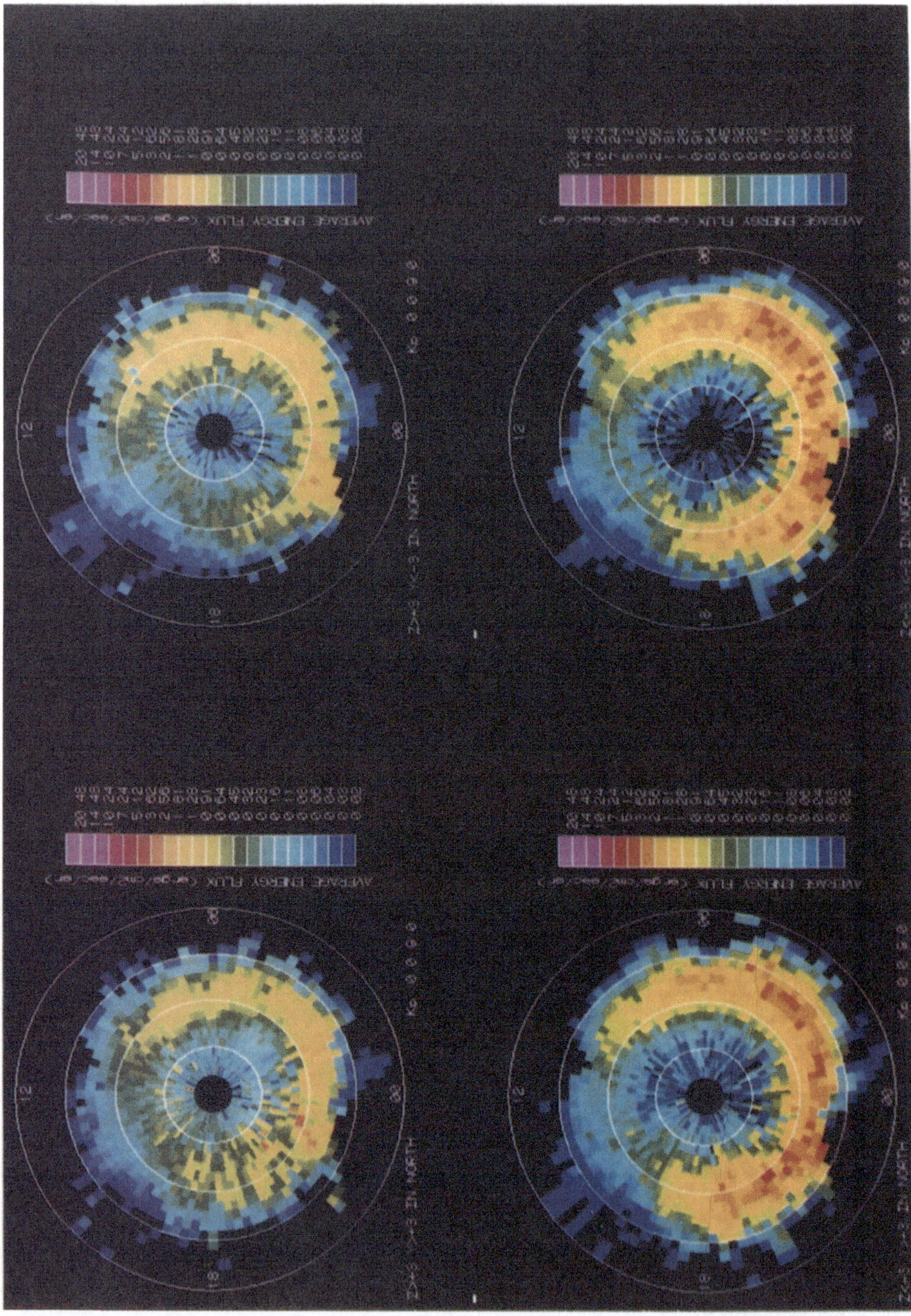

Fig. 4. Average patterns of energy flux determined for data selection criteria based solely on the IMF orientation. $B_z > +3$ nT at the top and $B_z < -3$ nT below. $B_y > +3$ nT at the left; $B_y < -3$ nT at the right.

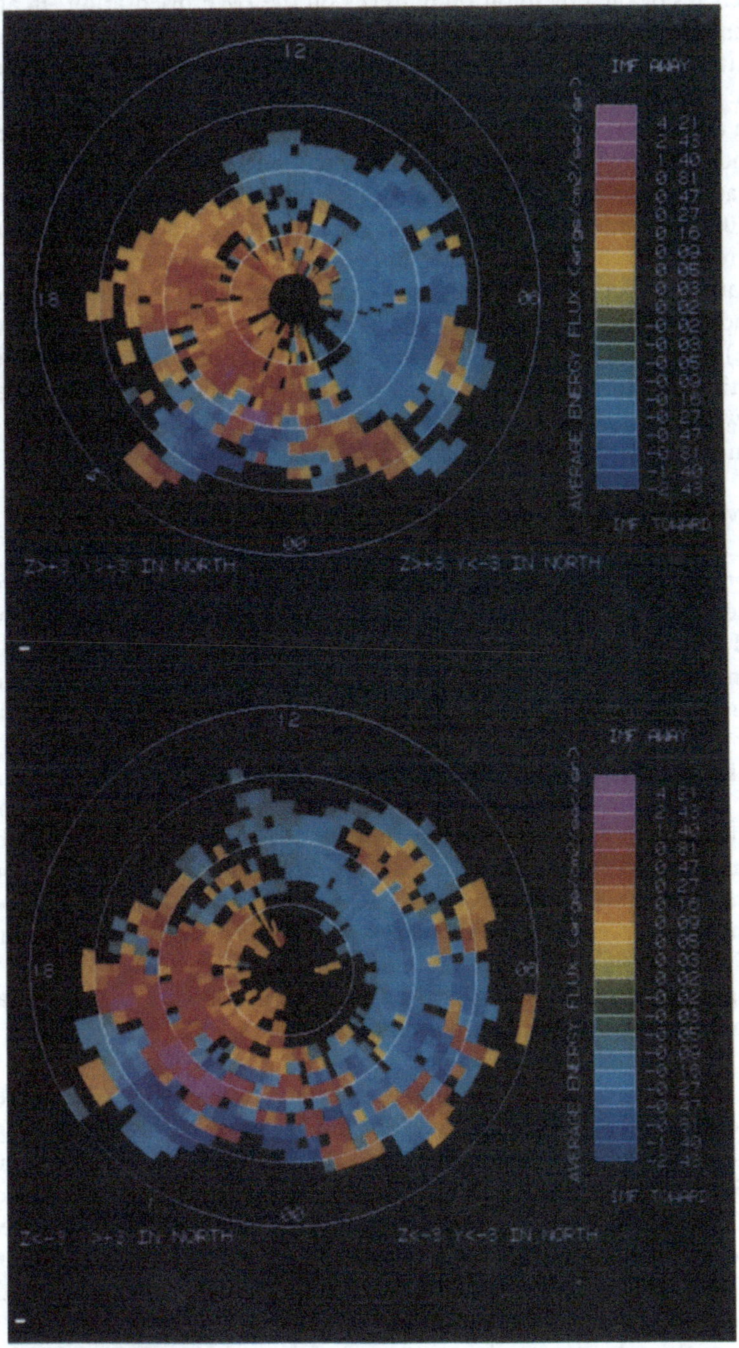

Fig. 5. Subtracting the $B_y < -3$ nT (toward sector) patterns of Fig. 4 from the $B_y > +3$ nT (away sector) patterns reveals that the precipitating energy input is greater on the evening (morning) side of the 02–14 MLT meridian for IMF away (toward) sectors.

dependencies of the precipitation patterns on the IMF orientation, B_z and B_y were constrained to each have absolute value in excess of 3 nT in this study. The resulting four patterns each consisted of the average of approximately 60 thousand data samples. The energy input is clearly less for positive B_z (top) but the B_y dependence between away (left) and toward (right) sector data are not so easily discerned in the color-coded figure. In Fig. 5 we have subtracted the toward sector data from the away sector patterns and have color coded the difference values (differences with absolute value <0.1 erg/cm^2/ster/sec have been suppressed). In the resultant figure regions shaded red and yellow denote that portion of the high-latitude region where the precipitating energy influx is enhanced for IMF away sectors (B_y positive) while in blue shaded regions the energy input is greater during toward sector conditions (B_y negative). For both signs of B_z it is seen that the evening sector at both auroral and polar latitudes experiences enhanced precipitation energy input during away IMF sectors while precipitation in the morning sector is greater for toward sectors. The meridian from 14–02 magnetic LT roughly separates these regimes.

3. IMF-Dependent Convection Patterns

From its location in eastern Massachusetts the Millstone Hill incoherent scatter radar uses its 46 meter steerable antenna and 2.5 MW UHF transmitters to probe the mid and auroral latitude ionosphere. By scanning the antenna in azimuth at low elevation angle Doppler returns indicating the line of sight component of the *F* region plasma motion are observed over a large range of latitudes and longitudes. Each Doppler velocity sample is taken at a specified latitude and longitude and local time and at a known direction from the radar. The line of sight data may be combined in a number of ways to derive the vector convection velocities over the spatial area surveyed. When the data from a large number of experiments are binned by latitude and local time, it is found that each bin has been sampled from a number of look directions. We have developed a sequential least squares fitting procedure to determine the average vector velocity in each such bin directly from the composite set of the line of sight velocity measurements in each bin (HOLT *et al.*, 1985b). By including only those samples satisfying a given set of selection criteria we can derive a statistical best-fit convection pattern under any set of external conditions for which sufficient data samples are available.

For this study we have created an on-line data base of all wide-latitude coverage experiments performed at Millstone Hill from 1978 to early 1982. This range of dates was chosen since Millstone Hill's steerable antenna went into operation in 1978 and the digitized solar wind data base available to us ended in May 1982. These dates span the most recent solar cycle maximum as well as the data coverage of the precipitation data base described above. In this data base there are over 1 million separate velocity observations for which solar wind magnetic field data are available from the NASA 1-hour averaged solar wind parameter data set. Figure 6 presents the statistical coverage of the velocity data base in terms of K_p and IMF B_z and B_y components. Histograms showing the distribution of the samples with respect to these parameters are shown. Unbiased distributions exist for each of these parameters. Subsequent

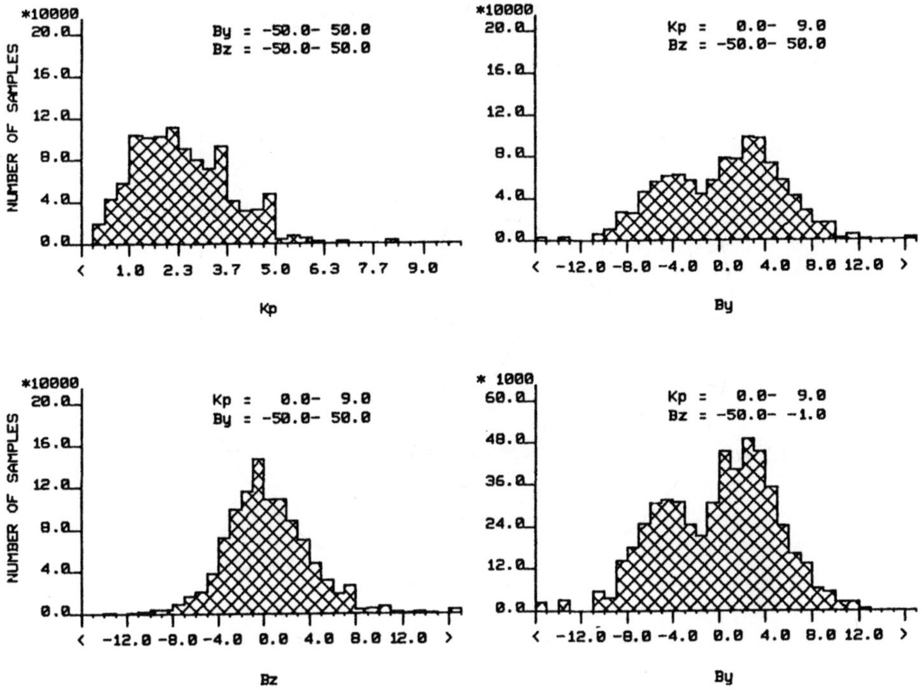

Fig. 6. K_p, B_z, and B_y distributions of the over 1 million Millstone Hill velocity observations for which IMF data are available. In the lower right is shown the distribution with B_y for velocity samples observed when $B_z < -1$ nT.

data sorting and averaging was done by specifying the ranges of K_p, B_z, and B_y to accept in building the average patterns. Data were then sorted into bins of 2 degrees apex magnetic latitude and 0.5 hour of magnetic local time. In general, 60,000 to 100,000 velocity samples were included in determining each average convection pattern. Bins for which there were fewer than 15 samples were excluded. Since only Millstone Hill velocity data were used, the patterns derived cover the range of apex magnetic latitudes between 55 and 73 degrees. Our fitting procedures are applicable to any data set of convection velocity observations. Future studies will extend these patterns to polar cap latitudes by including Sondrestrom radar and satellite observations.

The IMF sector dependence of the high-latitude ionospheric convection pattern was investigated by separating data taken with $B_y > +2$ nT (AWAY sectors) from those taken when $B_y < -2$ nT (TOWARD sectors). The convection patterns are most easily visualized as plots of electrostatic equipotential contours calculated from the average velocity data as has been discussed by FOSTER (1983). The potential contour patterns for $B_z > +1.25$ nT are presented in Fig 7(a) while those for $B_z < -1.25$ nT are shown in Fig. 7(b). The contours have been connected across latitudes above 73

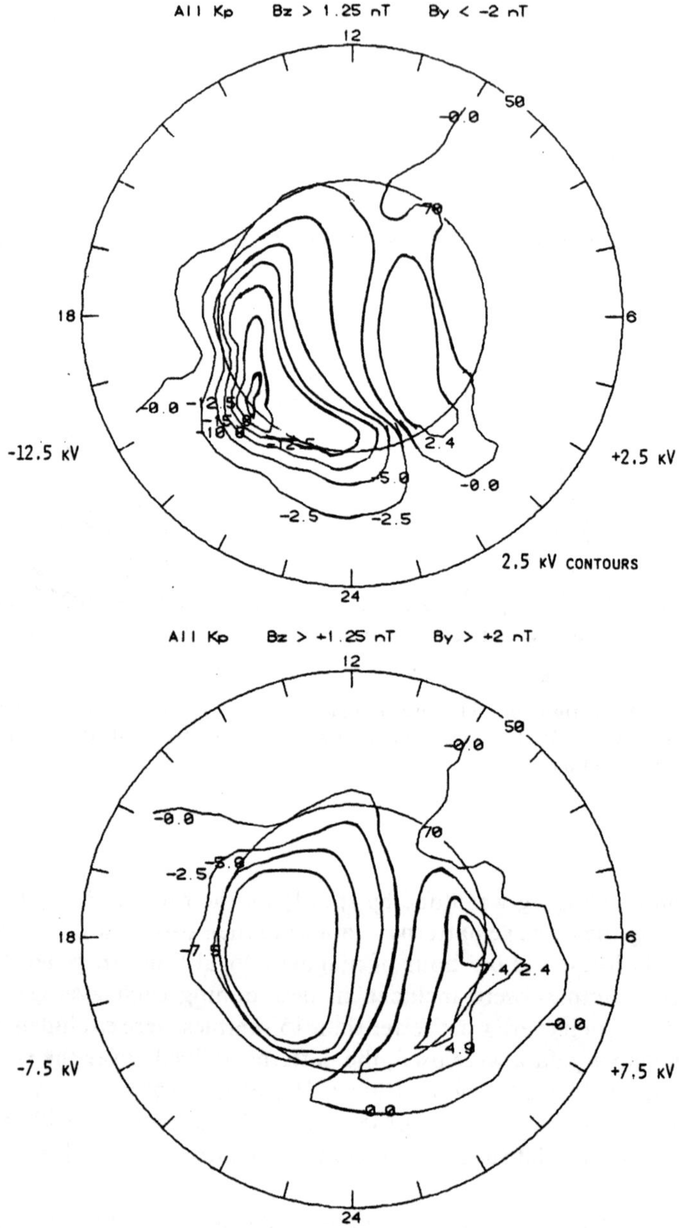
All Kp Bz > 1.25 nT By < -2 nT
12
18
6
24
50
70
-0.0
-12.5
-15.0
-10.0
2.4
-5.0
-2.5
-12.5 kV
+2.5 kV
2.5 kV CONTOURS
All Kp Bz > +1.25 nT By > +2 nT
12
18
6
24
50
70
-0.0
-2.5
-5.0
-7.5
2.4
4.9
-7.5 kV
+7.5 kV

(a)

Fig. 7.

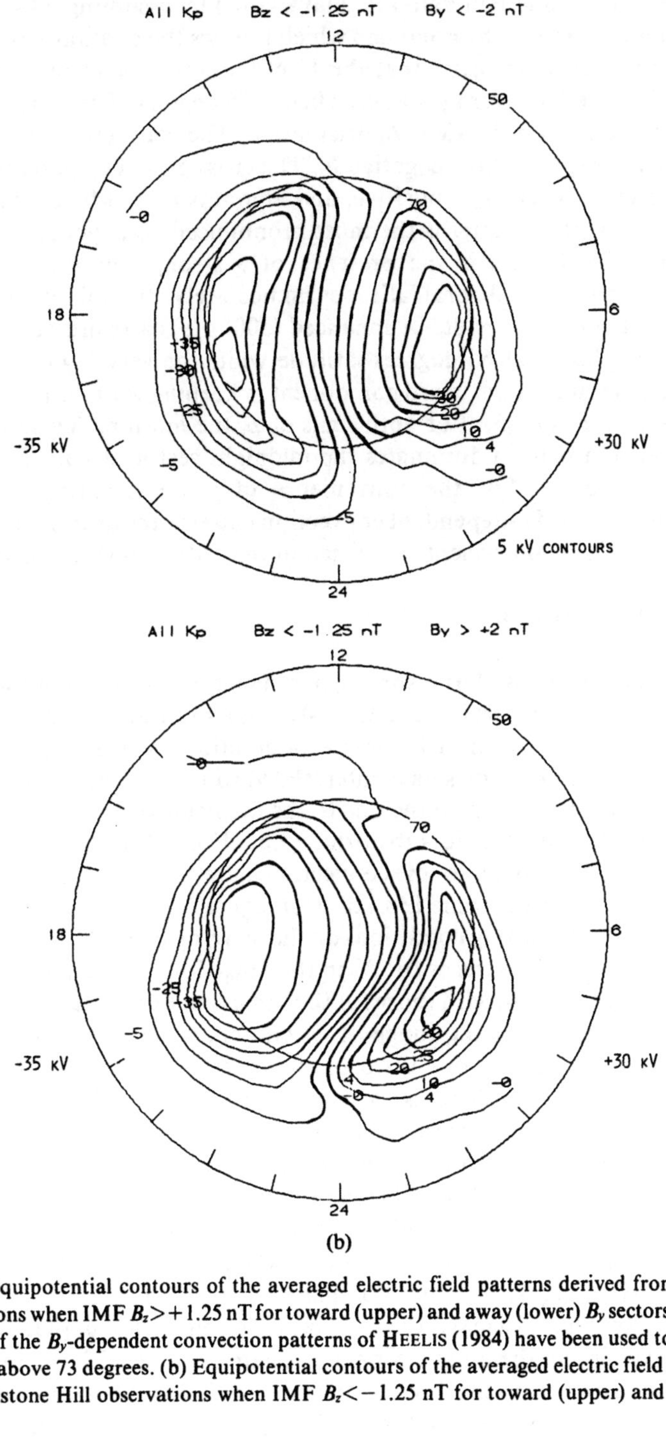

(b)

Fig. 7. (a) Equipotential contours of the averaged electric field patterns derived from Millstone Hill observations when IMF $B_z > +1.25$ nT for toward (upper) and away (lower) B_y sectors. The qualitative features of the B_y-dependent convection patterns of HEELIS (1984) have been used to extrapolate the contours above 73 degrees. (b) Equipotential contours of the averaged electric field patterns derived from Millstone Hill observations when IMF $B_z < -1.25$ nT for toward (upper) and away (lower) B_y sectors.

degrees (where Millstone Hill data are unavailable) by assuming a two-cell pattern with an asymmetry across the polar cap which follows the qualitative description of HEELIS (1984). It is noteworthy that the Heelis sector-dependent patterns closely match the Millstone Hill average patterns below 75 degrees. The convection electric field depends strongly on the IMF B_z orientation. The convection pattern is greatly strengthened and expanded for negative B_z. The cross polar cap potential drop is 15 kV for both sectors when B_z is positive, and amounts to 75 kV for the negative B_z cases. In our convection patterns the most pronounced B_y dependent difference at latitudes below 75 degrees is the presence of a strong southward electric field component (25 mV/m) at auroral latitudes immediately after midnight during away sectors (B_y positive). This results in enhanced (500 m/s) eastward convection of the midnight-sector plasma at 65 degrees latitude while the very-high latitude plasma would have a pronounced westward component in keeping with the findings of DE LA BEAUJARDIERE *et al.* (1985). For both signs of B_z the morning (evening) cell of the two-cell convection pattern dominates the midnight sector at auroral latitudes for away (toward) sectors. For the convenience of those requiring a quantitative representation of the IMF-dependent convection patterns the average northward and eastward electric field components are listed in the Tables in the appendix.

4. Effect of Negative B_z

The K_p index when used as a sorting parameter is a measure of the strength of ionospheric currents in general. When the IMF has a southward component the solar and geomagnetic fields are in a favorable orientation for merging at the dayside magnetopause. Previous studies have interrelated southward B_z to an enhanced rate of transfer of magnetic flux from the day to the nightside (CAAN *et al.*, 1975) and to the occurrence of geomagnetic substorms (e.g., FOSTER *et al.*, 1971). We have investigated the B_z dependence of the convection patterns by intercomparing average convection patterns for the same level of geomagnetic activity (as based on the K_p index) but for different sorting conditions on the merging index B_z. For both B_y sector orientations we compare patterns for K_p in the range 3-to 5- and for which B_z is less than -1.25 nT (negative B_z) with those for the same range of K_p but with no selection criterion on B_z (all-B_z). For the range of K_p chosen, corresponding to moderately disturbed conditions, nearly one third of the cases have positive B_z. This is not surprising since the K_p index denotes the maximum disturbance level observed during a three hour interval. Figure 8 presents convection equipotential contours for toward (top) and away (bottom) sectors for the negative B_z cases (left) and for the data with no selection criterion on B_z (right). Whereas the negative B_z patterns have the same overall qualitative features as the all-B_z cases, the total cross polar cap potential is increased by 30 kV, to in excess of 100 kV, when B_z is negative. The qualitative features of the HEELIS (1984) convection patterns are prominent.

FOSTER (1984) has discussed the high-latitude convection pattern and how it is related to convection in the magnetospheric equatorial plane. That paper identified regions of plasma entry into the dayside polar cap with flux tubes involved in dayside merging at the magnetopause and discussed the possible temporal variability of that portion of the ionospheric pattern as controlled by magnetospheric merging and antisunward flux transfer rates. If this is the case we expect to find a greater potential drop across the dayside polar cap entry region for the negative B_z patterns which reflect the effects of enhanced merging. Based on direct radar observations FOSTER and DOUPNIK (1984) reported that this region is approximately three hours of local time wide and accounts for about 25 kV potential difference in disturbed times.

In order to accentuate the enhancement of the convection patterns which accompanies negative B_z we have differenced the potential patterns of Fig. 8, subtracting the all-B_z values from the negative B_z values bin by bin. In Fig. 9 we

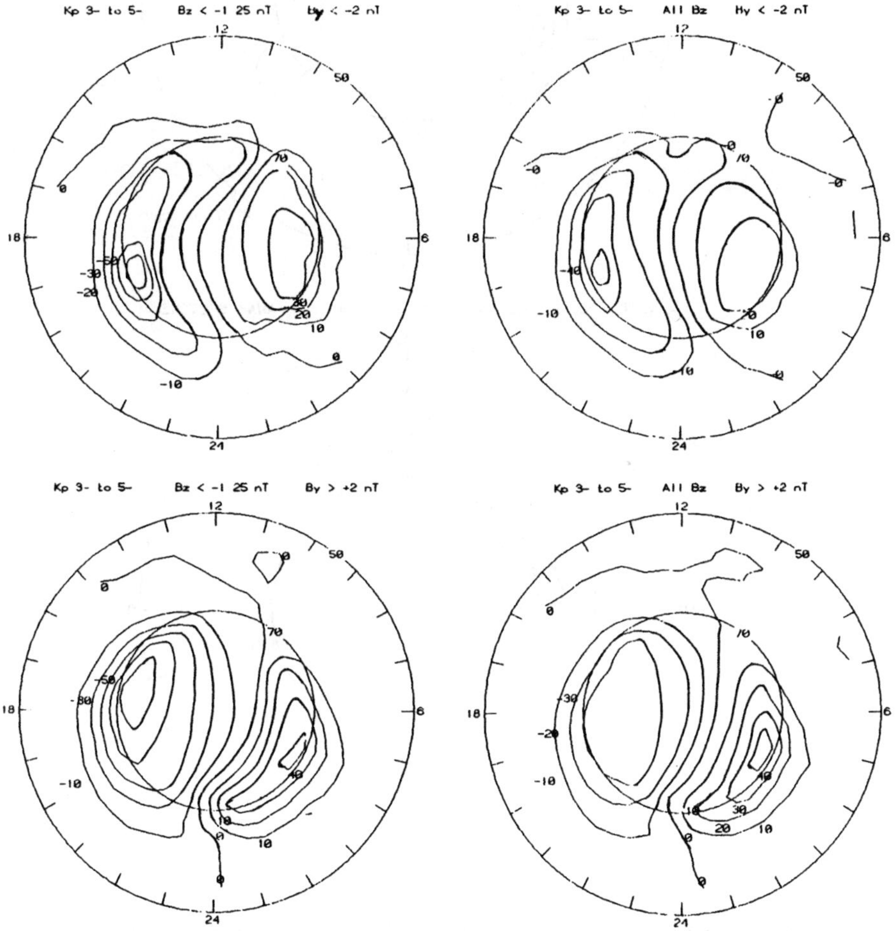

Fig. 8. Equipotential contours of the average convection patterns for IMF toward sector data (top) and away sector data (bottom) for disturbed geomagnetic conditions with K_p 3-to 5-. The negative B_z patterns of Fig. 8 are shown to the left while patterns derived with no restriction on B_z are shown to the right. For either B_y sector the requirement that B_z be negative results in an increase of the cross polar cap potential difference by 30 kV.

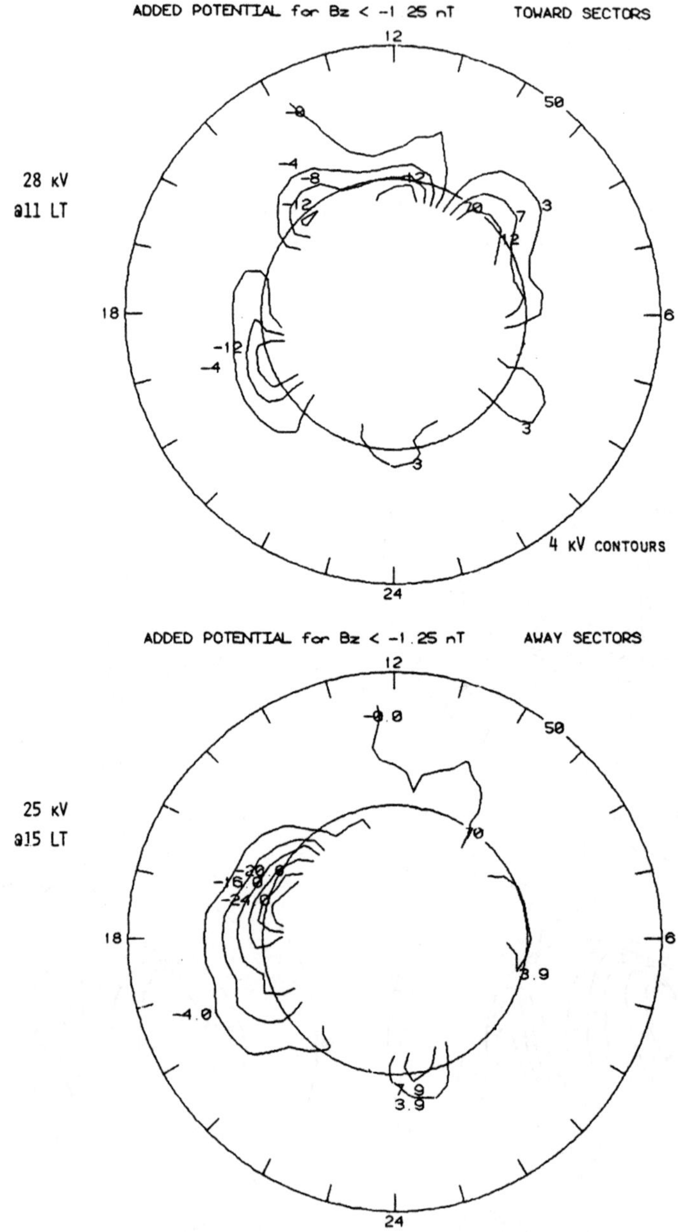

Fig. 9. Contours of the additional potential associated with negative B_z, derived by subtracting the all-B_z patterns of Fig. 9 from those with negative B_z. For toward sector data a 25 kV potential difference is associated with a region of plasma entry into the polar cap centered at 1030 MLT, while for away sectors 25 kV appears across a region at 1400 MLT.

Table 1.

Bz > +1.25 nT EASTWARD ELECTRIC FIELD (mV/m) By > +2 nT

MLT / ALAT	1	2	3	4	5	6	7	8	9	10	11	12	13	14	15	16	17	18	19	20	21	22	23	24
73	-3	-1	1	3	4	7	7	4	2	1	0	0	1	2	0	0	0	0	1	-1	-3	-4	-10	-9
71	-2	0	1	2	3	6	5	2	1	0	0	0	0	1	0	0	0	0	0	0	-2	-2	-6	-6
69	0	0	0	1	1	4	3	1	0	0	0	0	0	0	0	0	0	0	0	0	-1	-1	-2	-2
67	0	0	1	0	0	2	2	1	0	0	0	0	0	0	0	0	0	0	0	0	0	0	0	0
65	0	0	1	0	0	1	1	0	0	1	1	0	0	0	0	0	0	0	0	0	0	0	0	0
63	0	0	0	0	0	1	0	0	0	0	0	0	0	0	0	0	0	0	0	0	0	0	0	0
61	0	0	1	0	0	2	1	0	0	0	0	0	0	0	0	0	0	0	0	0	0	0	0	0
59	0	0	1	0	0	1	0	0	0	0	0	0	0	0	0	0	0	0	0	0	0	0	0	0
57	0	0	0	-1	0	1	0	0	0	0	0	0	0	0	0	0	0	0	0	0	0	0	0	0
55	-1	-1	0	0	-1	0	0	0	0	0	0	0	0	0	0	0	0	0	0	0	-1	-1	0	0

Bz > +1.25 nT NORTHWARD ELECTRIC FIELD (mV/m) By > +2 nT

MLT / ALAT	1	2	3	4	5	6	7	8	9	10	11	12	13	14	15	16	17	18	19	20	21	22	23	24
73	8	2	7	-11	-4	-2	-13	-12	-11	-1	0	2	11	7	6	8	12	16	14	13	15	18	12	13
71	0	-3	-2	-7	-10	-11	-10	-3	0	0	1	1	5	4	3	4	8	9	9	12	14	14	7	3
69	-6	-6	-5	-4	-3	-8	-5	-1	0	0	0	1	2	2	1	3	5	5	5	5	5	7	5	0
67	-4	-3	-4	-2	0	-2	-2	0	0	0	0	0	1	0	0	3	4	2	3	2	1	0	0	0
65	-3	-3	-4	-2	0	-1	0	0	0	0	-1	0	0	0	0	1	2	2	2	1	1	0	-1	-2
63	-2	-2	-2	-2	-2	-3	0	0	0	0	0	0	0	-1	0	1	1	1	2	1	1	0	0	-1
61	0	0	-1	-3	-3	-3	-1	0	0	0	0	0	0	0	0	0	0	0	1	0	0	1	0	0
59	0	0	0	0	-1	-1	0	0	0	0	0	0	0	0	0	0	0	0	1	1	1	0	0	0
57	1	0	0	1	0	0	0	0	0	0	0	0	0	0	0	0	0	0	0	0	0	0	0	1
55	1	1	1	2	1	0	0	0	0	0	0	0	0	0	0	0	0	0	0	0	0	0	0	1

Bz < -1.25 nT EASTWARD ELECTRIC FIELD (mV/m) By > +2 nT

MLT / ALAT	1	2	3	4	5	6	7	8	9	10	11	12	13	14	15	16	17	18	19	20	21	22	23	24
73	-22	-12	-6	-5	-1	7	12	13	8	5	12	11	8	4	5	6	3	-1	-3	-3	-12	-21	-22	-24
71	-19	-10	-3	1	4	8	10	9	5	3	6	5	4	2	2	5	2	0	-2	0	-8	-18	-19	-19
69	-12	-3	4	6	4	6	5	4	3	2	4	4	3	1	0	2	1	1	-1	-2	-7	-14	-15	-14
67	-5	-2	2	3	4	4	3	3	2	2	3	2	1	0	0	1	1	1	-1	-6	-8	-12	-10	-8
65	-3	0	1	2	2	1	1	1	1	2	2	1	1	0	0	1	0	0	-1	-5	-5	-4	-4	-4
63	-1	0	0	0	0	0	0	0	0	0	0	0	0	0	0	0	0	0	-1	-2	-1	-1	-1	-2
61	-3	-1	0	1	1	0	0	0	0	0	0	0	0	0	0	0	0	0	-2	-6	-3	-1	-1	-2
59	-2	-1	-1	0	0	0	-1	0	0	0	0	0	0	0	0	0	0	0	-1	-2	-1	-1	-1	-1
57	-3	-2	-2	0	0	0	-1	0	0	0	0	0	0	0	0	0	0	0	-1	-3	-2	0	0	-1
55	-3	-3	-4	-2	-1	-1	-1	0	0	0	0	0	0	0	0	0	0	0	0	1	0	-1	-1	-1

Bz < -1.25 nT NORTHWARD ELECTRIC FIELD (mV/m) By > +2 nT

MLT / ALAT	1	2	3	4	5	6	7	8	9	10	11	12	13	14	15	16	17	18	19	20	21	22	23	24
73	16	16	5	7	7	-8	-24	-27	-20	0	-6	-1	11	22	39	32	29	29	32	17	-3	8	15	12
71	3	-8	-4	-20	-24	-27	-26	-18	-7	0	0	5	10	18	28	25	33	35	30	18	8	12	10	4
69	-10	-17	-27	-32	-26	-22	-15	-9	-3	0	0	1	5	12	20	20	27	31	22	19	19	18	8	-2
67	-21	-20	-27	-25	-17	-13	-8	-5	-2	-1	0	0	2	5	10	11	16	18	17	28	27	19	7	-12
65	-8	-14	-19	-16	-11	-7	-4	-3	-1	-1	0	0	1	3	4	4	7	9	11	25	23	14	6	0
63	-4	-10	-11	-8	-8	-5	-1	-2	-2	-1	0	0	0	1	1	1	3	5	8	17	16	8	8	6
61	-1	-5	-8	-9	-7	-3	0	-1	-1	0	0	0	0	0	0	0	1	2	4	11	14	12	8	5
59	0	-2	-4	-5	-4	-1	0	0	0	0	0	0	0	0	0	0	1	2	4	7	8	6	5	4
57	0	0	-1	-3	-2	0	0	0	0	0	0	0	0	0	0	0	1	1	3	5	6	5	4	3
55	1	1	0	0	0	1	1	0	0	0	0	0	0	0	0	0	0	1	2	4	4	3	3	2

Table 2.

Bz > +1.25 nT EASTWARD ELECTRIC FIELD (mV/m) By < -2 nT

MLT / ALAT	1	2	3	4	5	6	7	8	9	10	11	12	13	14	15	16	17	18	19	20	21	22	23	24
73	-8	-2	5	1	1	3	3	4	4	1	2	1	0	2	2	1	1	0	-2	-99	13	8	-7	-9
71	-7	-5	-1	0	2	2	2	1	1	1	1	0	0	0	1	0	0	0	0	1	1	1	-7	-8
69	-5	-5	-2	0	1	1	1	0	0	0	0	0	0	0	0	0	0	-1	-1	-1	-7	-7	-3	-6
67	-5	-3	0	0	0	1	1	0	1	0	0	0	0	0	0	0	0	-1	0	-3	-7	-5	-3	-3
65	-3	-1	0	0	0	1	1	0	1	0	0	0	0	0	0	0	0	-1	0	-2	-3	-3	0	0
63	-1	0	0	0	0	0	0	0	0	0	0	0	0	0	0	0	0	0	0	0	-2	0	2	0
61	-3	0	0	0	2	0	0	0	0	0	0	0	0	0	0	0	0	-1	0	1	0	0	0	-1
59	-1	0	0	0	0	0	0	0	0	0	0	0	0	0	0	0	0	0	0	0	0	0	1	0
57	0	0	0	0	0	0	0	0	0	0	0	0	0	0	0	0	0	-1	-1	0	0	0	0	-1
55	1	1	1	0	0	0	0	0	0	0	0	0	0	0	0	0	0	-1	-2	-1	0	0	0	0

Bz > +1.25 nT NORTHWARD ELECTRIC FIELD (mV/m) By < -2 nT

MLT / ALAT	1	2	3	4	5	6	7	8	9	10	11	12	13	14	15	16	17	18	19	20	21	22	23	24
73	0	-1	-14	0	3	-3	-1	-6	-4	1	0	0	6	9	5	9	12	12	17	99	-22	-30	-8	-1
71	6	2	0	0	-3	-2	-1	0	2	2	1	1	3	6	5	6	8	12	18	16	3	-4	10	8
69	3	0	-4	-4	-1	0	1	2	1	2	2	1	1	4	3	3	3	7	9	8	18	16	10	10
67	6	0	-4	-1	1	1	0	1	0	2	1	0	0	1	1	1	1	4	4	5	14	12	9	7
65	3	0	0	0	0	0	0	0	0	0	0	0	0	0	0	0	1	3	0	1	7	8	4	3
63	3	0	-1	0	0	0	0	0	0	0	0	0	-1	-1	-1	0	0	1	0	0	3	2	1	3
61	3	0	-4	-1	0	0	0	0	0	0	0	0	0	-1	-1	-1	0	0	0	1	3	2	2	3
59	2	1	0	0	1	1	1	0	0	0	0	0	-1	-1	-1	-1	0	0	0	0	2	2	2	2
57	2	2	1	0	1	1	1	0	0	0	0	0	-1	-1	-1	-1	0	0	0	0	1	1	2	2
55	1	2	2	1	1	1	1	0	0	0	0	0	-1	-1	-1	-1	0	0	0	0	1	1	2	2

Bz < -1.25 nT EASTWARD ELECTRIC FIELD (mV/m) By < -2 nT

MLT / ALAT	1	2	3	4	5	6	7	8	9	10	11	12	13	14	15	16	17	18	19	20	21	22	23	24
73	-18	-16	-9	-4	-1	10	11	11	16	9	7	7	9	11	12	7	7	5	-3	-10	-16	-19	-19	-18
71	-17	-14	-8	-4	0	14	14	9	9	6	4	4	6	8	8	5	5	3	-4	-8	-12	-16	-17	-17
69	-15	-9	-5	-3	0	6	6	4	5	3	2	3	3	4	3	3	1	0	-4	-6	-8	-12	-16	-18
67	-11	-7	-4	-1	0	2	1	2	3	2	2	3	2	2	2	1	1	0	-1	-4	-7	-8	-11	-12
65	-6	-5	-2	0	2	0	0	1	3	2	2	2	2	1	1	1	0	0	0	0	-3	-3	-3	-5
63	-3	-3	-2	0	0	0	0	0	1	0	1	3	1	0	0	0	0	0	0	0	-1	-2	-1	-2
61	-3	-3	-2	0	0	0	1	0	1	1	1	1	0	0	0	0	0	0	0	0	-1	-1	-1	-2
59	-3	-3	-2	0	0	0	1	0	0	0	1	1	1	0	0	0	0	0	0	0	0	-1	0	-2
57	-3	-3	-2	-1	-1	0	0	0	0	1	1	1	0	0	0	0	0	0	0	1	0	0	0	-2
55	-2	-2	-2	-2	-2	0	0	0	0	0	1	1	0	0	0	0	0	0	0	1	0	0	0	0

Bz < -1.25 nT NORTHWARD ELECTRIC FIELD (mV/m) By < -2 nT

MLT / ALAT	1	2	3	4	5	6	7	8	9	10	11	12	13	14	15	16	17	18	19	20	21	22	23	24
73	-11	-21	-18	-21	-23	-32	-31	-26	-34	-15	3	9	16	27	20	9	15	10	5	9	-2	-14	-7	-5
71	-2	-10	-13	-19	-20	-33	-40	-24	-20	-14	3	10	15	24	31	29	24	19	23	29	11	-9	-9	-6
69	-2	-8	-9	-25	-24	-20	-19	-9	-9	-8	1	4	6	14	21	25	27	29	33	35	20	8	1	0
67	1	-3	-7	-19	-16	-13	-9	-5	-6	-5	-1	-1	2	8	12	15	19	25	23	28	27	24	18	11
65	0	-3	-9	-15	-13	-7	-6	-4	-5	-5	-2	-3	0	3	7	8	11	14	13	14	19	22	19	10
63	0	-1	-5	-9	-7	-5	-4	-4	-3	-2	-3	-5	-1	1	3	4	5	9	8	9	16	15	14	8
61	3	0	-3	-9	-7	-6	-3	-2	-1	-1	-2	-2	0	0	1	1	3	4	4	8	12	12	11	9
59	2	0	-2	-7	-6	-4	-2	-1	0	0	-1	-1	0	0	0	0	1	3	3	5	8	8	9	8
57	3	2	0	-5	-5	-1	0	0	0	0	-1	-1	0	0	0	0	0	1	2	4	5	5	5	6
55	3	2	1	-1	-2	0	0	0	0	0	-1	-1	0	0	0	0	0	1	1	3	4	3	4	5

present contours of the additional potential which appears when B_z is southward for the two B_y sector orientations. It is found that a 25–30 kV potential difference is associated with a region of enhanced poleward plasma convection on the dayside. For IMF toward sectors this potential difference spans 3 hours of pre-noon local time centered at 1030, while for away IMF sector data the center of the enhanced region is near 1500 LT. Similar to the conclusions of FOSTER and DOUPNIK (1984), we find that for equal levels of geomagnetic activity (as determined by the K_p index) the further condition that B_z be southward results in a 30 kV enhancement of the potential difference across the dayside. The IMF sector dependence is such that enhanced plasma entry into the polar cap occurs in the pre-noon (post-noon) quadrant for toward (away) sectors. Following the qualitative arguments and ionosphere-magnetosphere mapping presented by FOSTER (1984) we propose that these enhancements in the high-latitude convection pattern when B_z is negative constitute the signature, somewhat smoothed by averaging, of enhanced field line merging at the dayside magnetopause.

We thank J. R. Doupnik for assistance in processing the NOAA/TIROS precipitation data at Utah State Univ. This work was supported by National Science Foundation grants ATM-81-19477 to Utah State University and ATM-81-19812 to the Massachusetts Institute of Technology.

APPENDIX

Electric field components, east and north in apex magnetic coordinates, at apex latitudes between 55 and 73 degrees are presented at each hour of magnetic local time in the following tables. These average values are derived from a data base of over 1 million observations of the ionospheric convection velocity made at Millstone Hill between 1978 and 1982. Figs. 7(a) and 7(b) display the equipotential contour patterns corresponding to the four IMF-dependent models shown. Table 1 presents the IMF away sector model components while Table 2 presents electric field components for toward sector conditions. In both cases, model components for positive B_z are presented in the upper half of the table and values for negative B_z below.

REFERENCES

CAAN, M. N., R. L. MCPHERRON, and C. T. RUSSELL, Substorm and interplanetary magnetic field effects on the geomagnetic tail lobes, *J. Geophys. Res.*, **80**, 191–194, 1975.

DE LA BEAUJARDIERE, O., V. B. WICKWAR, J. D. KELLY, and J. H. KING, Effect of the interplanetary magnetic field *Y* component on the high-latitude nightside convection, *Geophys. Res. Lett.*, **12**, 461–464, 1985.

FOSTER, J. C., An empirical electric field model derived from Chatanika radar data, *J. Geophys. Res.*, **88**, 981–987, 1983.

FOSTER, J. C., Ionospheric signatures of magnetospheric convection, *J. Geophys. Res.*, **89**, 855–865, 1984.

FOSTER, J. C. and J. R. DOUPNIK, Plasma convection in the vicinity of the dayside cleft, *J. Geophys. Res.*, **89**, 9107–9113, 1984.

FOSTER, J. C., D. H. FAIRFIELD, K. W. OGILVIE, and T. J. ROSENBERG, Relationship of interplanetary parameters and occurrence of magnetospheric substorms, *J. Geophys. Res.*, **76**, 6971–6975, 1971.

GUSSENHOVEN, M. S., D. A. HARDY, N. HEINEMANN, and R. K. BURKHARDT, Morphology of polar rain, *J. Geophys. Res.*, **89**, 9785–9800, 1984.

HARDY, D. A., M. S. GUSSENHOVEN, and E. HOLEMAN, A statistical model of auroral electron precipitation, *J. Geophys. Res.*, **90**, 4229–4248, 1985.

HEELIS, R. A., The effects of interplanetary magnetic field orientation on dayside high-latitude ionospheric convection, *J. Geophys. Res.*, **89**, 2873–2880, 1984.

HOLT, J. M., R. H. WAND, J. V. EVANS, and W. L. OLIVER, Models for the plasma convection at high latitudes from Millstone Hill observations, *J. Geophys. Res.*, in press, 1985a.

HOLT, J. M., J. V. EVANS, W. L. OLIVER, and R. H. WAND, Millstone Hill observations of ionospheric convection, in *Physics of Space Plasmas*, pp. 53–72, Scientific Pubs., Cambridge MA, 1985b.

KING, J. H., Interplanetary medium data book—supplement 2, 1978–1982, Nat. Space Sci. Data Center publication # NSSDC/WDC-A-R&S 83-01, Greenbelt, Maryland 20771, 1983.

OLIVER, W. L., J. M. HOLT, R. H. WAND, and J. V. EVANS, Millstone Hill incoherent scatter observations of auroral convection over 60<N<75. 3. average patterns versus K_p, *J. Geophys. Res.*, **88**, 5505–5516, 1983.

POTEMRA, T. A., L. J. ZANETTI, P. F. BYTHROW, and A. T. Y. LUI, B_y-dependent convection patterns during northward interplanetary magnetic field, *J. Geophys. Res.*, **89**, 9753, 1984.

REIFF, P. H. and J. L. BURCH, IMF By-dependent plasma flow and Birkeland currents in the dayside magnetosphere. 2. a global model for northward and southward IMF, *J. Geophys. Res.*, **90**, 1595, 1985.

SOJKA, J. J., J. C. FOSTER, P. M. BANKS, and J. R. DOUPNIK, Mapping electrostatic potentials from the ionosphere to the magnetosphere, *Planet. Space Sci.*, **31**, 1329–1388, 1983.

Solar Wind-Magnetosphere Coupling, edited by Y. Kamide and J. A. Slavin, 495–505.

Sondrestrom Radar Observations of the Effect of the IMF B_y Component on Polar Cap Convection

O. DE LA BEAUJARDIERE[1], V. B. WICKWAR[1], and J. H. KING[2]

[1]*SRI International, Menlo Park, California 94025, U.S.A.*
[2]*NASA, National Space Science Data Center, Greenbelt, Maryland 20771, U.S.A.*

Average patterns of convection, derived from Sondrestrom radar observations, reveal that the interplanetary magnetic field dawn-dusk component (IMF B_y) strongly influences the nighttime polar convection. The convection for one orientation of B_y is not the mirror image of the other orientation. A positive B_y seems to organize the velocities such that, at all local times, they are predominantly westward within the radar latitude range. On a case-by-case basis, auroral oval boundaries can be determined by coincident DMSP particle data and by radar-measured *E*-region densities. On one occasion of positive B_y, sunward flow (i.e., westward flow) is observed in the polar cap between dusk and midnight. For large negative B_y, large southward velocities are observed about three hours before midnight. These are the only times when the predominant velocity component is clearly southward. When B_y is negative, in the midnight and dawn sectors, the plasma velocities appear random. However, the average drifts are mostly southward. The radar average patterns are compared with theoretical predictions and recently proposed convection patterns. Although none of the models seem to reproduce the observed convections well, some features present in the models are also seen in the radar data.

1. Introduction

There is little doubt that the interplanetary magnetic field strongly influences the magnetospheric and ionospheric convection (SVALGAARD, 1968, 1973; MANSUROV, 1969; FRIIS-CHRISTENSEN *et al.*, 1985 and references therein). However, the exact processes involved in the solar-wind/magnetosphere coupling are not well known, nor is the exact response of the large-scale convection. Our inability to measure an instantaneous snapshot of the global convection pattern has made it difficult to establish what electric field configuration corresponds to each IMF orientation. The range of possibilities is exemplified by the variety of convection patterns proposed in recent publications (POTEMRA *et al.*, 1984; REIFF and BURCH, 1985; CHIU *et al.*, 1985; FRIIS-CHRISTENSEN *et al.*, 1985). The main focus of these papers was on the noon sector, because this is where the connection between the solar wind and the ionospheric plasma is the most direct (e.g., JORGENSEN *et al.*, 1972; JORGENSEN, 1981; CROOKER, 1979). In contrast, the nightside of the polar convection has not been studied very extensively. However, early works by HEPPNER (1972, 1977) and MOZER *et al.* (1974) indicate that the B_y component has an effect on the nightside convection as well.

The purpose of this paper is to examine the effect of the IMF B_y component in the dusk, midnight, and dawn sectors. Sondrestrom incoherent-scatter radar measurements of plasma drift in the auroral oval and the polar cap are used. In a previous paper (DE LA BEAUJARDIERE *et al.*, 1985a) the effect of the B_y component was derived from a case-by-case examination of the data. In this paper, we briefly review the individual radar observations and then show average convection patterns obtained by binning the drift data for the two orientations of the IMF B_y component. These patterns are compared to the various models of convection that have been proposed.

Ground-based observations of plasma convection have an advantage over space-borne observations in that they are continuous over long periods of time—several hours or days. Furthermore, the incoherent-scatter radars have an advantage over other ground-based instruments such as magnetometers in that they measure the ion velocity directly rather than infer it using mathematical inversion techniques that rely heavily on conductivities (KAMIDE and RICHMOND, 1982).

2. Observations

The radar data were obtained during 24-hr continuous observations. The operating mode for these runs is described in WICKWAR *et al.* (1984). Electric fields and profiles of ionospheric parameters were measured between 68° and 82° invariant latitude with a cycle time of about 20 minutes.

There is a total of 14 day-long experiments for which the IMF data are available from the IMP-J satellite. The data for four fairly typical days are shown in Fig. 1 and 2. The IMF components in GSM coordinates are plotted on Fig. 1, and the radar-measured plasma drifts are in Fig. 2. The velocity vectors are color coded to show more clearly the regions of sunward or antisunward flow. The blue vectors are for eastward drifts, and the red vectors for westward drifts. The IMF vectors in the *YZ* plane are reproduced along the inside circle. Positive B_z point towards the center of the circle, and positive B_y towards the west direction (see Fig. 2(d)). The vectors are red and blue for positive and negative B_y, respectively. DMSP particle data were used to show the position of the auroral oval (GUSSENHOVEN *et al.*, 1983). The oval boundaries were determined from coincident passes and are indicated by wide blue arrows (these arrows tend to disappear among the convection vectors; therefore, the times at which the satellite passes occurred are marked by a thinner arrow on the outside periphery).

These four examples show the characteristics in the convection patterns that have been seen in the other Sondrestrom 24-hr runs. Figure 2 is taken from DE LA BEAUJARDIERE *et al.* (1985a) where these observations are described in some detail. The available data were averaged and the results are shown in Figs. 3(a) and 3(b) for B_y positive and negative, respectively. The data were averaged in bins of one degree by one hour. The drift components were weighted according to their statistical uncertainty. The selection was based on 10-min IMF averages, and the times considered were offset by 10 min to take into consideration the propagation times.

The conclusions reached on a case-by-case basis by DE LA BEAUJARDIERE *et al.* (1985a) are still valid when examining the averaged data. The main features of the convection pattern illustrated both by the four examples of Fig. 2, and by Fig. 3 can be briefly described as follows.

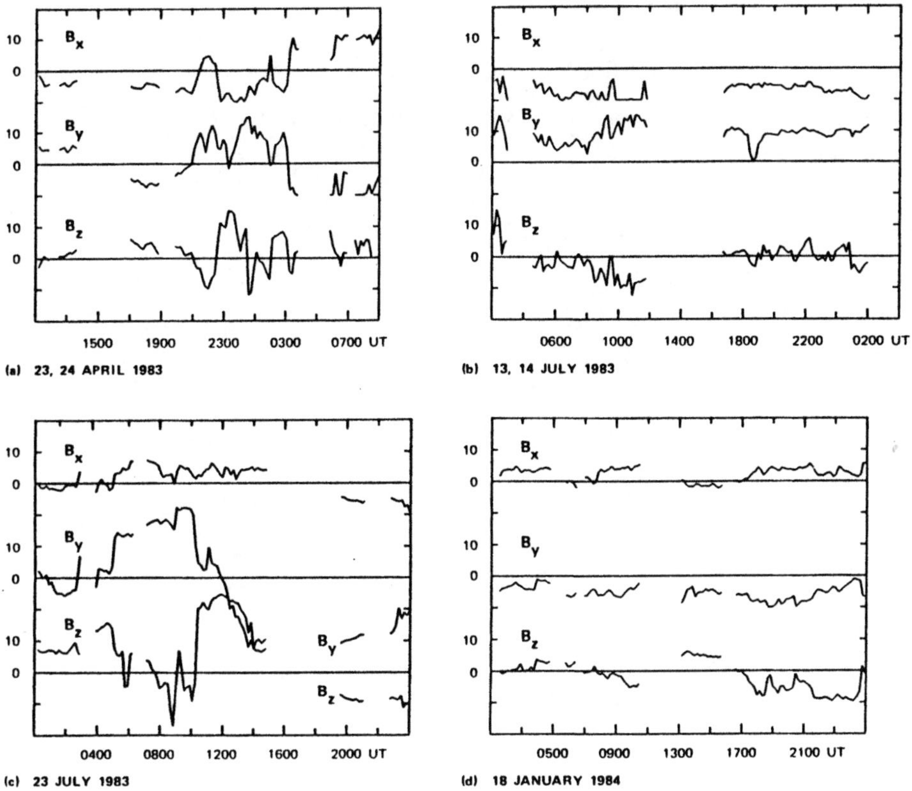

Fig. 1. IMF parameters (GSM coordinates) from the IMP-J spacecraft for the four periods of Fig. 2.

2.1 *Case $B_y>0$.*

When B_y is positive, the velocities are predominantly westward within the radar field of view at all local times (Fig. 3(a)). Fig. 2(b) offers a clear individual example: B_y remained positive throughout the observation period, and the radar appears to remain under a large westward vortex for 24 hours.

A shear convection reversal is observed in the morning cell. See, for example Figs. 2(b), 2(c), and 3(a).

The observations of 23–24 April are discussed by DE LA BEAUJARDIERE *et al.* (1985b) who show that between dusk and midnight on 23–24 April 1983, sunward (westward) flow is observed within the polar cap (Fig. 2(a)). Following GUSSENHOVEN *et al.* (1981), the polar cap is poleward of the auroral oval which is taken as the region where the integral number flux and the energy flux rise noticeably above background. When considering Sondrestrom data alone, the oval is defined as the region where discernible E-region precipitation is observed (i.e., where the density is above $\sim 3\times 10^4$ el/cm^3, in the absence of solar EUV production).

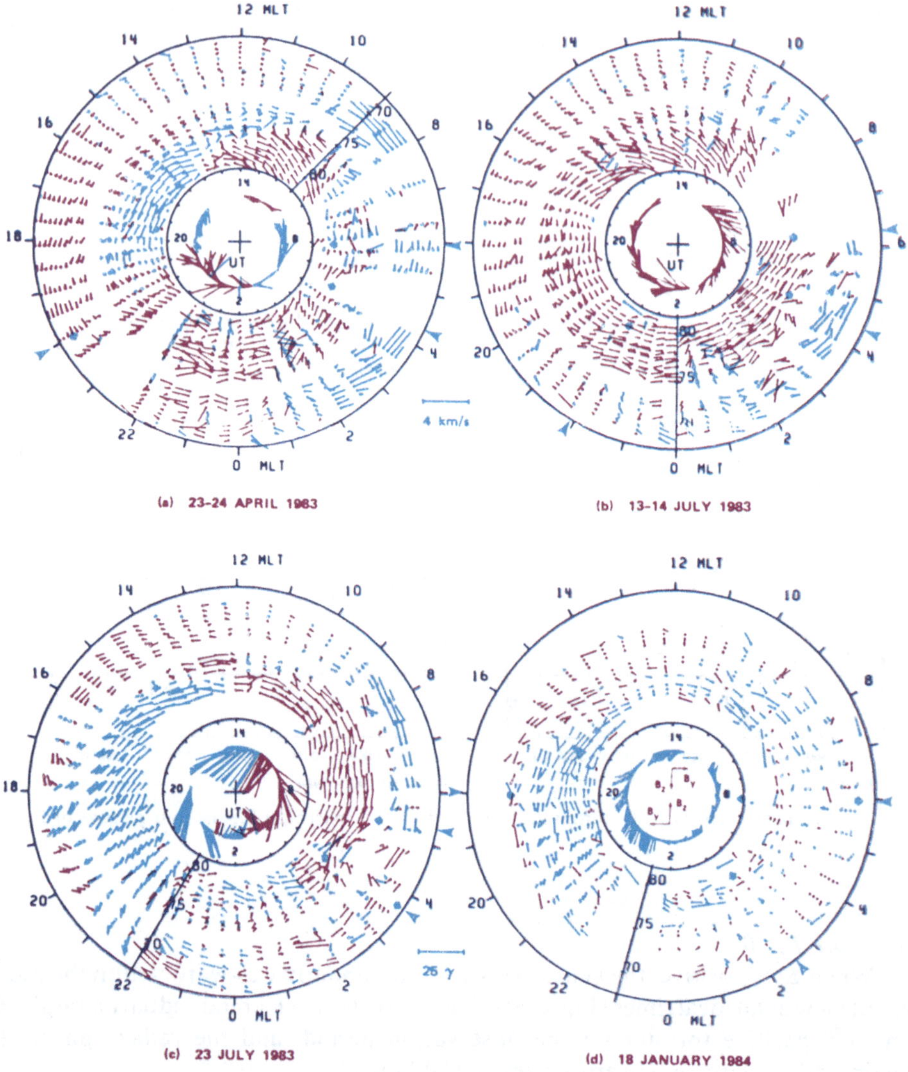

Fig. 2. Plasma drift and IMF B_y and B_z vectors for four typical examples. Red (blue) is for west (east) velocity, and for positive (negative) B_y. The thick blue arrows indicate the poleward or equatorward boundaries of the auroral oval, as determined from coincident DMSP passes.

2.2 *Case* $B_y<0$.

In the midnight and dawn sectors, when B_y is negative, the plasma velocities are small and random in the individual observations (see 18 January, Fig. 2(d)). The velocities also appear quite irregular in the midnight and dawn sector of the average pattern (Fig. 3(b)) but, in the midnight sector, the dominant averaged component is often southward. For large negative B_y, the afternoon cell appears shifted toward

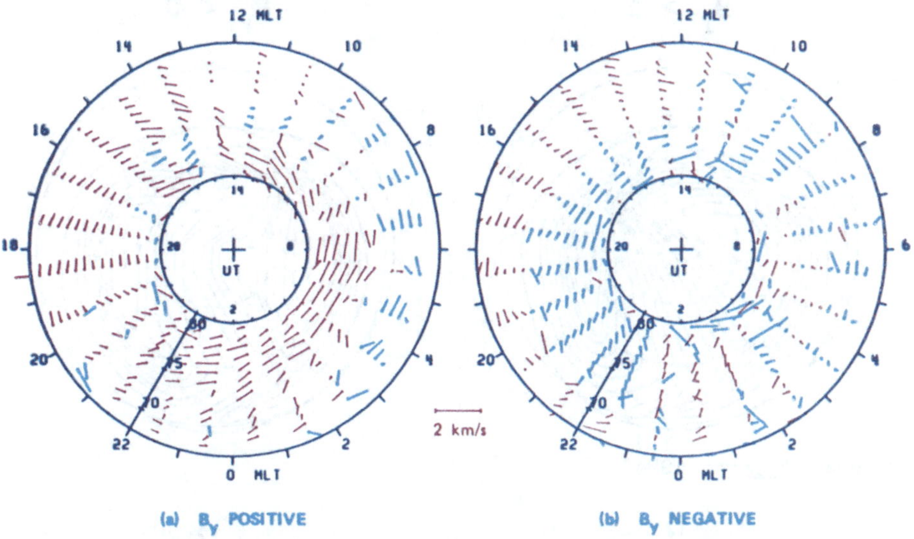

Fig. 3. Average of ion drifts measured by the Sondrestrom radar, for B_y positive and negative.

early hours such that large southward velocities are observed about 3 hours before midnight, as in the case of 24 July (Fig. 2(c)). Large southward velocities are only observed in these premidnight hours, when B_y is large and negative.

A shear convection reversal is observed in the afternoon cell (Fig. 2(a), 2(c), 2(d) and 3(c)), symmetrical to that seen in the morning cell when $B_y>0$.

In conclusion, the nightside convection is strongly influenced by the B_y component. This influence is quite complex. The color convention for the IMF and drift vectors in Fig. 2 reveals clearly that a "red" (positive) B_y corresponds to a "red" (westward) convection, and a "blue" (negative) B_y corresponds to a "blue" (eastward) velocity. In other words, in the poleward edge of the oval and in the polar cap, the drifts are mainly westward for positive B_y, and eastward for negative B_y. However, the convection for one orientation of B_y is not the mirror image of that of the other orientation.

3. Discussion

We now compare the radar observations with the various empirical models or schematic convection patterns that have been published. Examples of some recent models are shown in Figs. 4, 5, and 6, in two columns with positive B_y patterns on the left, and negative B_y on the right. The two pairs in Fig. 4 are from POTEMRA *et al.* (1979) and MAYNARD (1985). In these works, the B_z component was not taken into consideration. The patterns on Figs. 5 and 6 are for negative and positive B_y, respectively. They are from REIFF and BURCH (1985), FRIIS-CHRISTENSEN *et al.* (1985), and POTEMRA *et al.* (1984).

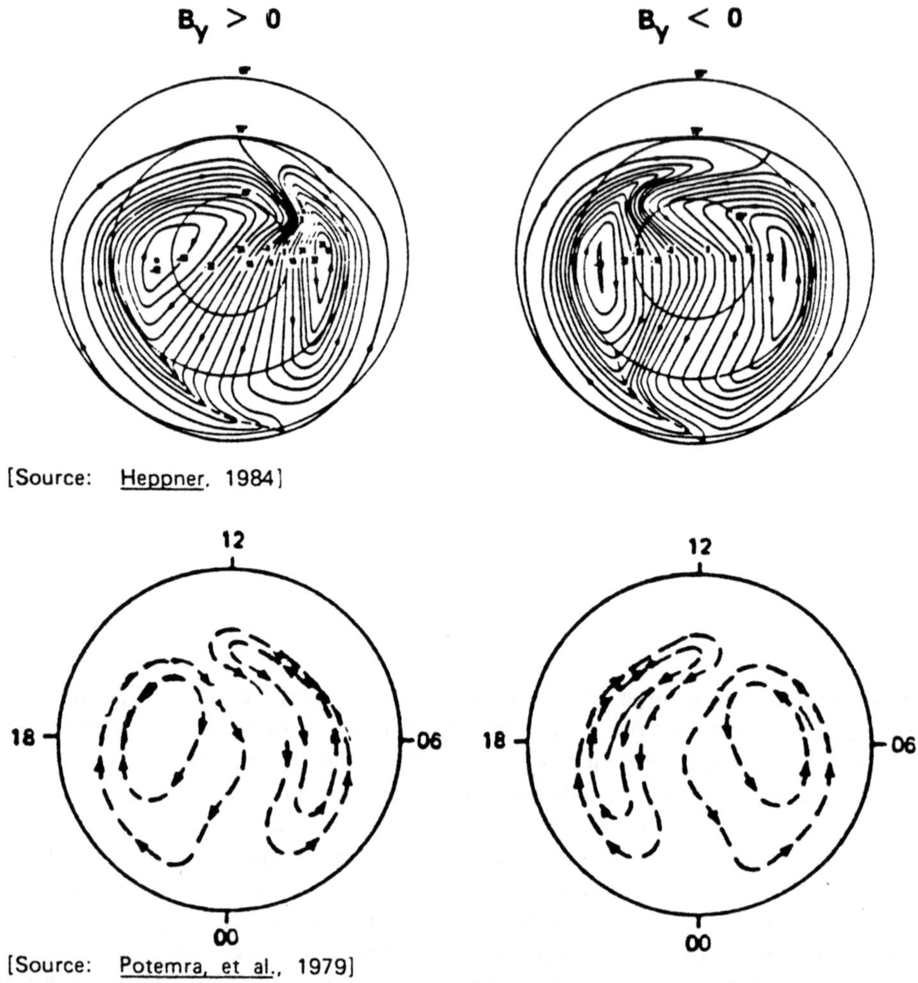

Fig. 4. Convection patterns for the two orientations of B_y, derived independently of B_z.

The first impression one has when comparing the observations with these various models, is that in the nightside, none of the models match the observations well. A possible reason for this poor match is that most of these models are based on observations close to noon. Even the electrostatic potential configuration obtained by FRIIS-CHRISTENSEN *et al.* (1985) (Figs. 5 and 6, lowest panels), are mostly valid in the dayside, because the potential function derived from magnetograms depends greatly on the conductivity model adopted. On the dayside, this model is known with reasonable accuracy, but not on the nightside where the position and strength of auroral precipitation are not well known. In the Heppner model shown by MAYNARD (1985), the dayside convection has been updated from the HEPPNER (1977) model with recent DE-2 data but not the nightside convection. These nightside patterns were

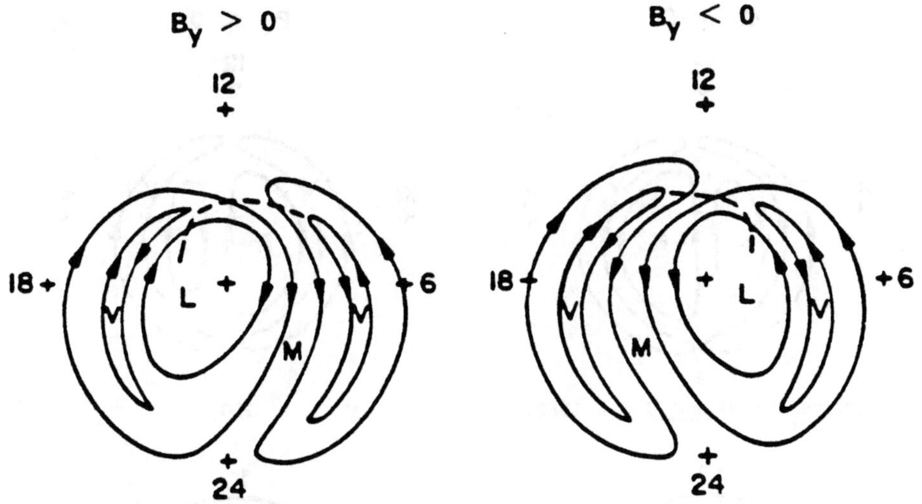

[Source: Reiff and Burch, 1985]

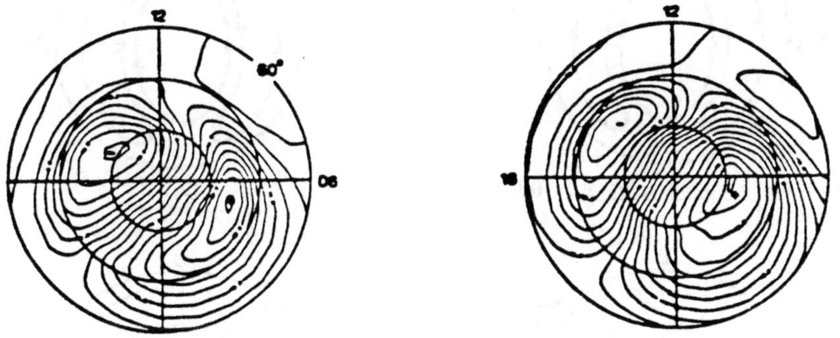

[Source: Friis-Christensen, et al., 1985]

Fig. 5. Convection patterns for B_z negative.

based on Barium releases and on OGO-6 measurements. OGO-6 only yielded one component of the velocity, and, in addition, the OGO-6 noon-midnight passes suffered from spacecraft attitude problems (MAYNARD, 1974). However, the Heppner and Friis-Christensen models, which are based in part on nightside observation, do not indicate that the pattern for one sign of B_y is symmetric to that for the opposite B_y. This behavior is in disagreement with all the other schematic patterns, but the Sondrestrom observation confirm this lack of symmetry. Interestingly, however, the radar observations seem to indicate that the influence of the B_y component is stronger than what is implied by Heppner's and Friis-Christensen's models.

When considering only the dawn or dusk local times, the simple Potemra patterns of Figs. 4(c) and 4(d) correspond well to the observations. This is also

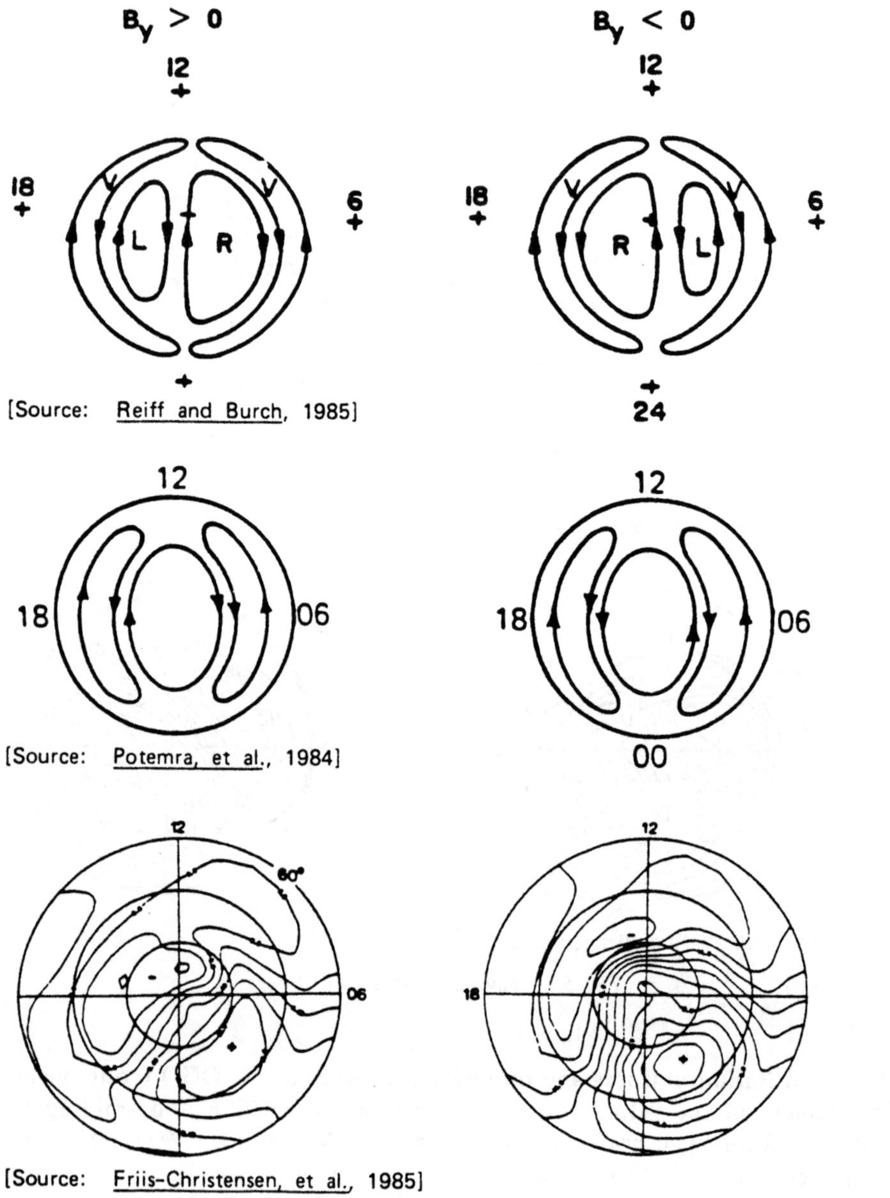

Fig. 6. Convection patterns for B_z positive.

apparent in the Sondrestrom data shown by ROBINSON *et al.* (1986). A narrow convex cell with a sharp convection reversal is located on the afternoon side when B_y is positive. A similar topology was also observed by HEELIS *et al.* (1983) using MITHRAS data.

The convection pattern is also dependent on B_z and, possibly, B_x, as well as season and substorm disturbances. In particular, to examine how the B_z component affects the average patterns, the radar observations were binned according to the sign of B_z as well as of B_y. The patterns appeared more noisy and the number of vectors in several bins were too small to allow conclusive determination of the dependence on B_z.

However, these patterns appeared very similar to the averages shown in Fig. 3. The main characteristics of the observations summarized above remain valid. The only noticeable difference was that the latitudes of the reversals were lower for negative B_z, as can be expected.

One difficulty in comparing the observations with the models is that the radar field of view can be either in the auroral oval, or in the polar cap, depending on the geomagnetic activity or, equivalently, on the $-B_z$ component. It is to alleviate this difficulty that oval boundaries are indicated in Fig. 2. However the definition of polar cap is highly dependent on the physical parameters that are measured and on the set of instruments used (TORBERT *et al.*, 1981).

Sunward flow in the polar cap has been observed experimentally (MAEZAWA, 1976; HEPPNER, 1977; BURKE *et al.*, 1979; BURCH *et al.*, 1985), but most such observations were in the dayside of the polar cap. The Sondrestrom data seem to indicate that sunward (westward) flows can occur around magnetic midnight in the polar cap, where the polar cap is determined by the DMSP and radar observations.

Sunward drifts in the polar cap were also predicted theoretically by STERN (1983), LYONS (1985), REIFF ad BURCH (1985), and CHIU *et al.* (1985). In these models, the flow in the polar cap is mostly clockwise for positive B_y, and counterclockwise for negative B_y. The Sondrestrom observations seem to be consistent with the former, but not the latter case.

The fact that the convection pattern appears disorderly when B_y is negative, may be explained in part by Lyons' model. The polar cap convection that he infers for $B_y<0$ is very complex. Because the actual IMF components are not constant in time but vary continuously, this pattern is highly variable. As a result, as the various elements of this intricate pattern move around, the observed drifts appear disorderly.

The Potemra and Reiff patterns of Figs. 5 and 6 show a complex cell topology. The fact that these multiple cells are not readily apparent in the radar data should not be taken as conclusive evidence that they do not exist. The radar latitude is often too low to see some of the reversals predicted in the polar cap. Also, as pointed out by BURCH *et al.* (1985), the viscous cell (marked V in Figs. 5 and 6) can be extremely narrow and, thus, could have been missed by the radar whose spatial resolution is about 100 km at the farthest latitudes.

4. Conclusion

In conclusion, these Sondrestrom observations reveal that the IMF B_y component strongly influences the nightside polar convection. Various models were compared with the radar observations. Although none of them offer a very good match, some features from each model are found in the observations:

- When $B_y>0$, the velocities are predominantly westward, at all local times. This is in agreement with schematic diagrams (POTEMRA *et al.*, 1979, 1984; REIFF and BURCH, 1985), and with theoretical predictions (STERN, 1973; LYONS, 1985; CHIU *et al.*, 1985).
- When $B_y<0$, the convection is small and appears random. This is akin to Birkeland-current observations by ZANETTI *et al.* (1984), and may be explained in part by LYONS (1985).
- There is a tendency to observe velocity shears in the morning (evening) convection cells for B_y positive (negative). This matches well the POTEMRA *et al.* (1979) sketch.
- In the nightside, the convection for one orientation of B_y is not symmetric to the pattern for the other orientation. This asymmetry is in disagreement with most schematic patterns proposed, but it is present in the empirical models of HEPPNER (1977) and FRIIS-CHRISTENSEN *et al.* (1985).

This work was supported by NSF cooperative agreement ATM8121671 and AFOSR contract F49620-83-K-0005. Our thanks go to D. Hardy, S. Gussenhoven, and N. Heinnman who provided the DMSP data.

REFERENCES

BURCH, J. L., P. H. REIFF, J. D. MENIETTI, R. A. HEELIS, W. B. HANSON, S. D. SHAWHAN, E. G. SHELLEY, M. SUGIURA, D. R. WEIMER, and J. D. WINNINGHAM, IMF B_y-dependent plasma flow and Birkeland currents in the dayside magnetosphere. 1. Dynamics Explorer observations, *J. Geophys. Res.*, **90**, 1577, 1985.

BURKE, W. J., M. C. KELLEY, R. C. SAGALYN, M. SMIDDY, and S. T. LAI, Polar cap electric field structures with a northward interplanetary magnetic field, *Geophys. Res. Lett.*, **6**, 21, 1979.

CHIU, Y. T., N. U. CROOKER, and D. J. GORNEY, Model of oval and polar cap arc configurations, *J. Geophys. Res.*, **90**, 5133, 1985.

CROOKER, N. U., Dayside merging and cusp geometry, *J. Geophys. Res.*, **84**, 951, 1979.

DE LA BEAUJARDIERE, O., V. B. WICKAWAR, J. D. KELLY, and J. H. KING, IMF-B_y effect on the high-latitude nightside convection, *Geophys. Res. Lett.*, **12**, 461, 1985a.

DE LA BEAUJARDIERE, O., V. B. WICKWAR, and J. D. KELLY, Observations of polar-cap ionospheric signatures of solar wind/magnetosphere coupling, in *Results of THE ARCAD 3 PROJECT and of the Recent Programmes in Magnetospheric and Ionospheric Physics, Toulouse 84*, p. 353, Cepadues-Editions, France, 1985b.

FRIIS-CHRISTENSEN, E., Y. KAMIDE, A. D. RICHMOND, and S. MATSUSHITA, Interplanetary magnetic field control of high-latitude electric fields and currents determined from Greenland magnetometer data, *J. Geophys. Res.*, **90**, 1325, 1985.

GUSSENHOVEN, M. S., D. A. HARDY, and W. J. BURKE, DMSP/F2 electron observations of equatorward auroral boundaries and their relationship to magnetospheric electric fields, *J. Geophys. Res.*, **86**, 768, 1981.

GUSSENHOVEN, M. S., D. A. HARDY, and N. HEINEMANN, Systematics of the equatorward diffuse auroral boundary, *J. Geophys. Res.*, **88**, 5692, 1983.

HEELIS, R. A., J. C. FOSTER, O. DE LA BEAUJARDIERE, and J. HOLT, Multistation measurements of high-latitude ionospheric convection, *J. Geophys. Res.*, **88**, 10111, 1983.

HEPPNER, J. P., Polar-cap electric field distributions related to the interplanetary magnetic field direction, *J. Geophys. Res.*, **77**, 4877, 1972.

HEPPNER, J. P., Empirical models of high-latitude electric fields, *J. Geophys. Res.*, **82**, 1115, 1977.

JORGENSEN, T. S., E. FRIIS-CHRISTENSEN, and J. WILHELM, Interplanetary magnetic field direction and high-latitude ionospheric currents, *J. Geophys. Res.*, **77**, 1976, 1972.

JORGENSEN, T. S., Electric fields in the dayside auroral region, in *Exploration of the Polar Upper Atmosphere*, edited by C. S. Deehr and J. A. Holtet, p. 267, D. Reidel Publishing Company, Boston, Mass., 1981.

KAMIDE, Y. and A. D. RICHMOND, Ionospheric conductivity dependence of electric fields and currents estimated from ground magnetic observations, *J. Geophys. Res.*, **87**, 8331, 1982.

LYONS, L. R., A simple model for polar cap convection patterns and generation of θ auroras, *J. Geophys. Res.*, **90**, 1561, 1985.

MAEZAWA, K., Magnetospheric convection induced by the positive and negative Z components of the interplanetary field: Quantitative analysis using polar cap magnetic records, *J. Geophys. Res.*, **81**, 2289, 1976.

MANSUROV, S. M., New evidence of a relationship between magnetic fields in space and on Earth, *Geomagn. Aeron.*, Engl Transl., **9**, 662, 1969.

MAYNARD, N. C., Electric field measurements across the Harang discontinuity, *J. Geophys. Res.*, **79**, 4620, 1974.

MAYNARD, N. C., Structure in the DC and AC electric fields associated with the dayside cusp region, in *The Polar Cusp*, edited by J. A. Holtet and A. Egeland, pp. 305–322, D. Reidel Publishing Company, 1985.

MOZER, F. S., W. D. GONZALEZ, F. BOGOTT, M. C. KELLEY, and S. SCHUTZ, High-latitude electric field and the three-dimensional interaction between the interplanetary and terrestrial magnetic fields, *J. Geophys. Res.*, **70**, 56, 1974.

POTEMRA, T. A., T. IIJIMA, and N. A. SAFLEKOS, Large-scale characteristics of Birkeland currents, in *Dynamics of the Magnetosphere*, pp. 165, D. Reidel, Hingham, Mass., 1979.

POTEMRA, T. A., L. J. ZANETTI, P. F. BYTHROW, and A. T. Y. LUI, B_y-dependent convection patterns during northward interplanetary magnetic field, *J. Geophys. Res.*, **89**, 9753, 1984.

REIFF, P. H. and J. L. BURCH, IMF B_y-dependent plasma flow and Birkeland currents in the dayside magnetosphere: 2. A global model for northward and southward IMF, *J. Geophys. Res.*, **90**, 1595, 1985.

ROBINSON, R. M., C. R. CLAUER, O. DE LA BEAUJARDIERE, J. D. KELLY, and D. S. EVANS, IMF B_y control of ionization and electric fields measured by the Sondrestrom radar, this volume, 1986.

STERN, D. P., A study of the electric field in an open magnetospheric model, *J. Geophys. Res.*, **78**, 7292, 1973.

SVALGAARD, L., Sector structure of the interplanetary magnetic field and daily variation of geomagnetic field at high latitudes, *Geophys. Pap. R-6*, Dan. Meteorol. Inst., Copenhagen, Denmark, 1968.

TORBERT, R. B., C. A. CATTELL, F. S. MOZER, and C.-I. MENG, The boundary of the polar cap and its relation to electric fields, field-aligned currents, and auroral particle precipitation, in *Physics of Auroral Arc Formation*, edited by S.-I. Akasofu and J. R. Kan, p. 143, American Geophysical Union, Washington, D. C., 1981.

WICKWAR, V. B., J. D. KELLY, O. DE LA BEAUJARDIERE, C. A. LEGER, F. STEENSTRUP, and C. H. DAWSON, Sondrestrom overview, *Geophys. Res. Lett.*, **11**, 883, 1984.

ZANETTI, L. J., T. A. POTEMRA, T. IIJIMA, W. BAUMJOHANN, and P. F. BYTHROW, Ionospheric and Birkeland current distributions for northward interplanetary magnetic field: Inferred polar convection, *J. Geophys. Res.*, **89**, 7453, 1984.

Solar Wind-Magnetosphere Coupling, edited by Y. Kamide and J. A. Slavin, 507–518.

IMF B_y Control of Ionization and Electric Fields Measured by the Sondrestrom Radar

R. M. ROBINSON[1], C. R. CLAUER[2], O. DE LA BEAUJARDIERE[3], J. D. KELLY[3], and D. S. EVANS[1,*]

[1]*Space Sciences Laboratory, Lockheed Palo Alto Research Laboratory, Palo Alto California 94304, U.S.A.*
[2]*STAR Laboratory, Stanford University, Stanford, California 94305, U.S.A.*
[3]*Radio Physics Laboratory, SRI International, Menlo Park, California 94025, U.S.A.*

The pattern of ionization and electric fields measured in the dayside auroral zone by the Sondrestrom incoherent scatter radar has been compared with interplanetary magnetic field (IMF) data from the IMP-8 spacecraft. The electric field configuration is most sensitive to the B_y component of the IMF. For $B_y<0$ there is a strong reversal in convective flow between 1300 and 1600 MLT with a region of enhanced F region ionization associated with low energy particle precipitation at the reversal. For $B_y>0$ the convection reversal is most apparent on the morning side. Ionization produced by particle precipitation for $B_y>0$ is much less conspicuous than that observed on the afternoon side for $B_y<0$. We show radar measurements of ionization, electric fields and electron temperatures associated with two distinct types of particle precipitation in the pre-noon local time sector for $B_y>0$. The patterns of precipitation for $B_y<0$ and $B_y>0$ are consistent with the distribution of field-aligned currents, in that precipitation occurs primarily in the upward current regions. With this interpretation, only the precipitation associated with the cusp field-aligned currents on the dawn side may be intense enough to produce ionospheric signatures observable from the ground.

1. Introduction

The large-scale pattern of electric fields and field-aligned currents in the auroral zone and its response to changes in the interplanetary magnetic field have been studied extensively in past years (see, for example, HEPPNER, 1972; FRIIS-CHRISTENSEN *et al.*, 1972; IIJIMA and POTEMRA, 1978; HEELIS, 1984). BURCH *et al.* (1985) have shown that the electric field pattern can be understood in terms of the superposition of several cells referred to as either merging, viscous or lobe cells. Each cell responds differently to changes in the interplanetary magnetic field. Precipitation is expected in various parts of the convection pattern depending on the connection of these field lines to various plasma regimes.

Since operations began in Greenland in February, 1983, numerous experiments were performed in which the Sondrestrom radar was used to measure the distribution of ionization and electric fields at very high latitudes. The radar data provide a

*Permanent address: Space Environment Laboratory, NOAA, Boulder, Colorado, U.S.A.

complimentary view to that obtained with polar orbiting satellites, because patterns of ionization and electric fields can be constructed from measurements made at different local times. However, the interpretation of these observed patterns must be done carefully because electric fields may change substantially during the time over which the measurements are made. Interplanetary magnetic field data are presently available for many of the radar experiments so that variations of electric fields in response to IMF conditions can be identified.

In this paper we describe the results of a study in which we examined radar measurements in the dayside auroral oval and polar cap made during 24 experiments. After describing the data used in the study, we present a short, qualitative description of the observed electric field patterns and their dependence on the IMF. We then discuss the ionization that occurs in association with these patterns and look in detail at two examples of ionization enhancements produced by particle precipitation in the morning sector.

2. Data Base

The radar data used in this study were obtained during experiments on 24 days chosen on the basis of the availability of IMF data. Some of these experiments were performed as part of the World Day observations; others were performed as part of campaigns to study electrodynamics in the polar cusp. Preliminary results of these experiments have been described in a number of papers (see, for example, JORGENSEN *et al.*, 1984; CLAUER *et al.*, 1984; KELLY and VICKREY, 1984). For the present study, we examined the data set for the specific purpose of identifying IMF B_y-dependent patterns of particle precipitation and electric fields in the prenoon and postnoon local time sectors. The radar operating mode for most of the 24 days consisted of some combination of fixed-position measurements and scans. The fixed-position measurements at a sequence of elevation angles and azimuths provided electric field data over a broad latitudinal range, while the scans yielded high time-and spatial-resolution measurements of ionization and drifts over a more limited region.

Table 1 shows the number of times different orientations of B_y occurred during radar measurements in the morning and afternoon sectors. All IMF components given in this paper refer to the GSM coordinate system. To be counted as an occurrence, B_y had to maintain the same sign for longer than an hour, which is greater than the 15 minutes to one-half hour required for electric fields on the dayside to

Table 1. Occurrence frequency of various orientations of IMF B_y in morning and afternoon sectors for the data set used in this study.

	$B_y>0$	$B_y<0$
Morning	14	8
Afternoon	9	11

respond to changes in the IMF at the location of IMP-8 (CLAUER *et al.*, 1984). No sorting according to the magnitude of B_y was performed and the data include cases of both positive and negative values of B_z. Although B_y was highly variable, there were many days when this component of the IMF remained fairly constant over long time intervals. Out of the 24 days there were eight in which B_y was positive throughout the morning and afternoon sectors, while there were five in which B_y remained negative during this local time interval.

3. Large-Scale Electric Fields and Their Relation to IMF B_y

In general, the electric fields measured by the radar during the local times between 0600 and 1800 are quite consistent with the results of previous studies based on satellite data. Of particular note is the strong dependence of the electric field on the y component of the IMF (DE LA BEAUJARDIERE *et al.*, this volume). For $B_y<0$ the radar measures predominantly eastward drifts in the morning sector, while in the evening sector both eastward and westward convecting plasma is observed. The westward (sunward) drifting plasma is confined to the equatorward part of the oval, with the eastward (antisunward) drifting plasma poleward of it. The antisunward drifting plasma is typically quite strong and confined to the region immediately poleward of the convection reversal. As pointed out by JORGENSEN *et al.* (1984) the convection reversal is typically rotational; however, the change in the east-west component of the velocity is particularly large and is thus an easily identifiable feature in latitudinal plots of electric field in the post-noon sector (ROBINSON *et al.*, 1984).

For $B_y>0$ the convection reversal in the afternoon sector disappears and westward drifts dominate at all latitudes. A corresponding reversal appears, however, on the morningside with, again, sunward drifts on the equatorward side and antisunward drifts on the poleward side. The reversal in velocity on the morning side is not as dramatic as on the afternoon side, but is still quite obvious in the radar data.

The convection observed near noon is highly variable, but tends to be dominated by westward drifts for $B_y>0$ and eastward drifts for $B_y<0$. Occasionally, antisunward drifting plasma at noon can be observed, especially when B_z is large and negative.

These general statements about the patterns of electric fields and their dependence on B_y underscore the importance of IMF data in the interpretation of radar measurements. Under conditions when B_y is changing rapidly, radar measurements through the prenoon to postnoon local time sector reveal multiple reversals in the zonal convection of plasma. Fortunately, there are enough periods of fairly stable IMF conditions to allow unambiguous mapping of the local time and latitudinal distribution of electric fields from radar measurements. These studies are currently in progress.

4. Ionization Produced by Particle Precipitation and Its Dependence on IMF B_y

Because much of the electron precipitation that occurs in the dayside oval and polar cap is very soft, ionization is predominantly produced at altitudes above 200 km. At these altitudes, ionization has lifetimes from several tens of minutes to hours and can be transported great distances from the location at which it is produced. Thus, the appearance of ionization enhancements does not necessarily indicate the presence of local production by particle precipitation. Although there are other measureable indicators of local production that can be used, we show here only ionization that can be definitely attributed to particle precipitation, either because the precipitation was hard enough to ionize at altitudes below 200 km, or because simultaneous satellite data were available to confirm the association. Because we limited ourselves to these obvious indicators of particle precipitation, we may have overlooked regions of precipitation in which the energy of the particles was too low to produce measureable enhancements in ionization. KOFMAN and WICKWAR (1984) and WICKWAR (this volume) describe a method based on solving the electron heat balance equation by which regions of local production can be identified regardless of the accompanying ionization. This technique may prove useful for future studies.

The most obvious region of particle precipitation observed with the radar at Sondrestrom is that associated with the convection reversal in the afternoon sector when $B_y<0$ (FOSTER, *et al.*, 1983; CLAUER *et al.*, 1983). ROBINSON *et al.* (1984) presented radar and satellite measurements in this precipitation region to show that it is associated with the region 1 Birkeland current and precipitating electrons accelerated through parallel potential drops of about 1 kV from a source of density 3.5 cm^{-3} and temperature 350 eV. EVANS (1984) studied precipitation in this local time sector using data from the NOAA satellites and confirmed the persistence of this feature even in periods of very low magnetic acitivity. The ionospheric signature and induced effects of this precipitation have been examined by WICKWAR (this volume). For $B_y<0$ on the morning side no localized region of enhanced ionization was identified in the radar data that could be attributed to particle precipitation.

For $B_y>0$, a reversal in the convective flow of plasma appears on the morning side. In examining the fourteen periods of measurements during which B_y was positive in the morning hours we were able to identify two examples of ionization produced by particle precipitation. Figure 1 shows data obtained on 13 July 1983 while the radar was performing an elevation scan in the plane of the magnetic meridian. At the time of these measurements, B_y was $+10$ nT and B_z was -10 nT. The top panel in Fig. 1 is a contour plot of electron density in the meridian plane constructed from the radar measurements. The center of the plot corresponds to the location of the radar which is at 74 degrees invariant latitude. Directly above the radar a region of ionization was observed with a peak density of about 3×10^5 cm^{-3} at an altitude of 200 km. Because this region of ionization extends downward to altitudes well below 200 km, we can be reasonably sure that local production by precipitating particles was occurring at this location. The second panel in Fig. 1 shows the latitudinal profile of electric field measured during the scan. The eastward electric field is small to the south and increases gradually to the north. The meridional electric field is southward with a peak value of about 75 mV/m just poleward of the ionization enhancement. Assuming no significant gradients in the east-west electric field in the zonal direction and uniform ionospheric conductivity, the north-south gradients in the meridional electric field are consistent with the presence of upward field-aligned currents within the ionization enhancement and downward field-aligned currents poleward of the

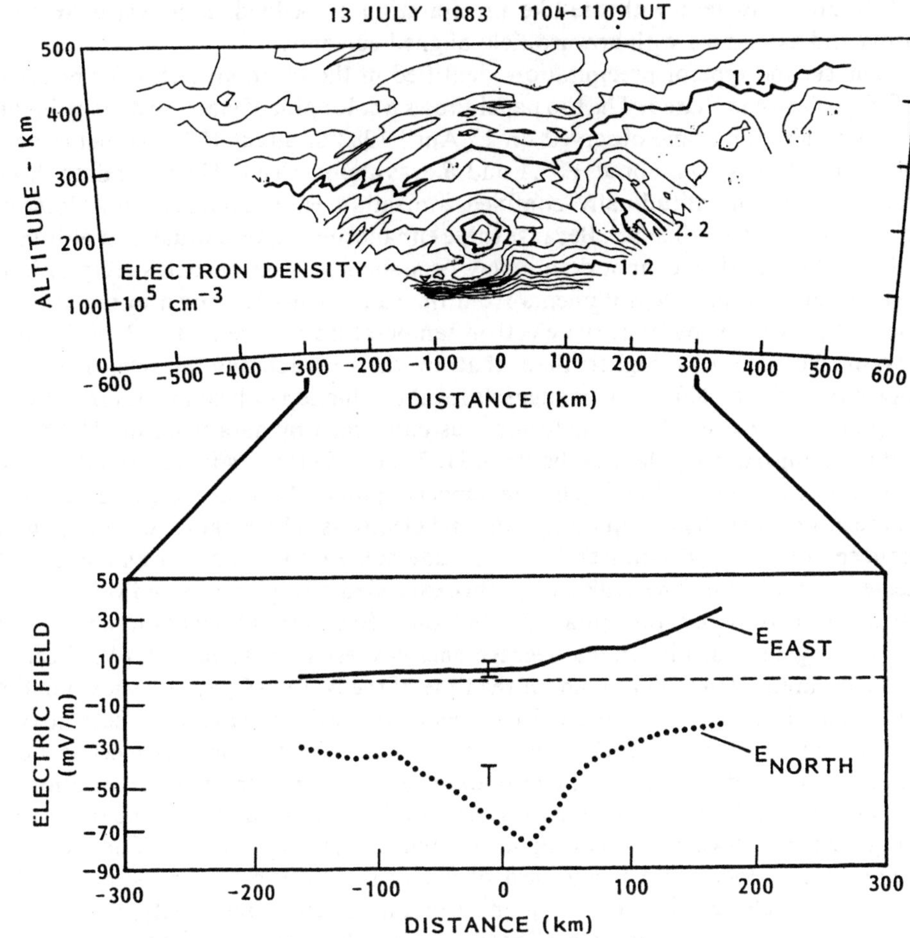

Fig. 1. Radar data obtained during a 5-minute scan in the magnetic meridian on 13 July, 1983. Distances are in km magnetically north and south from the radar.

enhancement. The sense of these currents is consistent with that of the region 1 and region 2 Birkeland currents (IIJIMA and POTEMRA, 1976). The ionization feature at 74 degrees latitude was apparent at the same location in the data from the subsequent radar scan (not shown) which was made about a half hour later. Thus, the precipitation producing this ionization was probably extended at constant invariant latitude along the auroral oval over about one-half hour in local time and persisted for at least this long in Universal Time. The latitudinal structure of electric field observed during the second scan was somewhat different. The precipitation was associated with a local perturbation in a wide region of fairly uniform southward fields. However, the gradient in electric field associated with the perturbation was, as before, consistent with the presence of upward field-aligned currents. In summary, the observations on

13 July are consistent with arc-like precipitation embedded in sunward drifting plasma and associated with upward field aligned currents.

The second type of precipitation identified in the morning sector for positive IMF B_y is shown in Plate 1. The top panel shows a color plot of electron density in the magnetic meridian plane observed on 13 April 1983 at about 1000 magnetic local time. At this time B_y was about +8 nT and B_z was about −5 nT. The radar data show several regions of latitudinally confined ionization enhancements with electron densities of about 3×10^5 cm^{-3}. Because these enhancements were situated at altitudes of about 400 km, they cannot definitively be associated with particle precipitation. Electron temperature measurements from the same scan are shown in the second panel. These data show that the electron temperature between 73 and 74 degrees invariant latitude was higher than that in the surrounding plasma, reaching tempertures of over 4000 K. This suggests local production of heat and ionization by precipitating particles. This conclusion was confirmed by data from the NOAA-6 satellite during a north polar pass between 1137 and 1154 UT. Radar and satellite data at this time are shown in Fig. 2. The top panel is a plot of the total energy flux carried into the atmosphere by precipitating ions and electrons. These data show a region of enhanced precipitation within the same latitude range as the ionization enhancement detected by the radar. The peak energy flux exceeded 10 ergs/cm^2-s and was carried almost exclusively by precipitating electrons. Spectral information about the precipitating electrons indicated average energies less than about 300 eV. Electric fields measured by the radar at about the time of the NOAA-6 pass are shown in the bottom panel of Fig. 2. Latitudinal profiles of the northward electric field at two different times are shown. The similarity between the two sets of electric field measurements confirms the constancy of the electric field pattern over the duration of the observations. The electric field over most of the region scanned by the radar was strong and northward indicating antisunward drifting plasma. The particle precipitation is located just poleward of where the electric field maximizes. The gradient in the electric field assuming uniform ionospheric conductivity is consistent with the presence of upward field-aligned currents. The location of these currents is well within the polar cap as determined by the direction of the plasma drift. Thus, the ionization shown in Plate 1 was probably the ionospheric signature of a polar cap arc. Note that, as pointed out by WICKWAR and KOFMAN (1984), the high electron temperatures along with the enhanced ionization can lead to substantial thermal excitation of the O^1D state of atomic oxygen, so that 630.0 nm emissions probably exceeded a few kR in intensity from this source alone.

5. Discussion

By examining Sondrestrom radar measurements made during the morning and afternoon hours, we have attempted to relate ionization produced by particle precipitation to boundaries of the convective flow and to the *y* component of the IMF. The most obvious signature of electron precipitation is the *F* region ionization enhancement associated with the afternoon sector convection reversal during times when IMF B_y is negative. EVANS (1984) showed that this precipitation probably

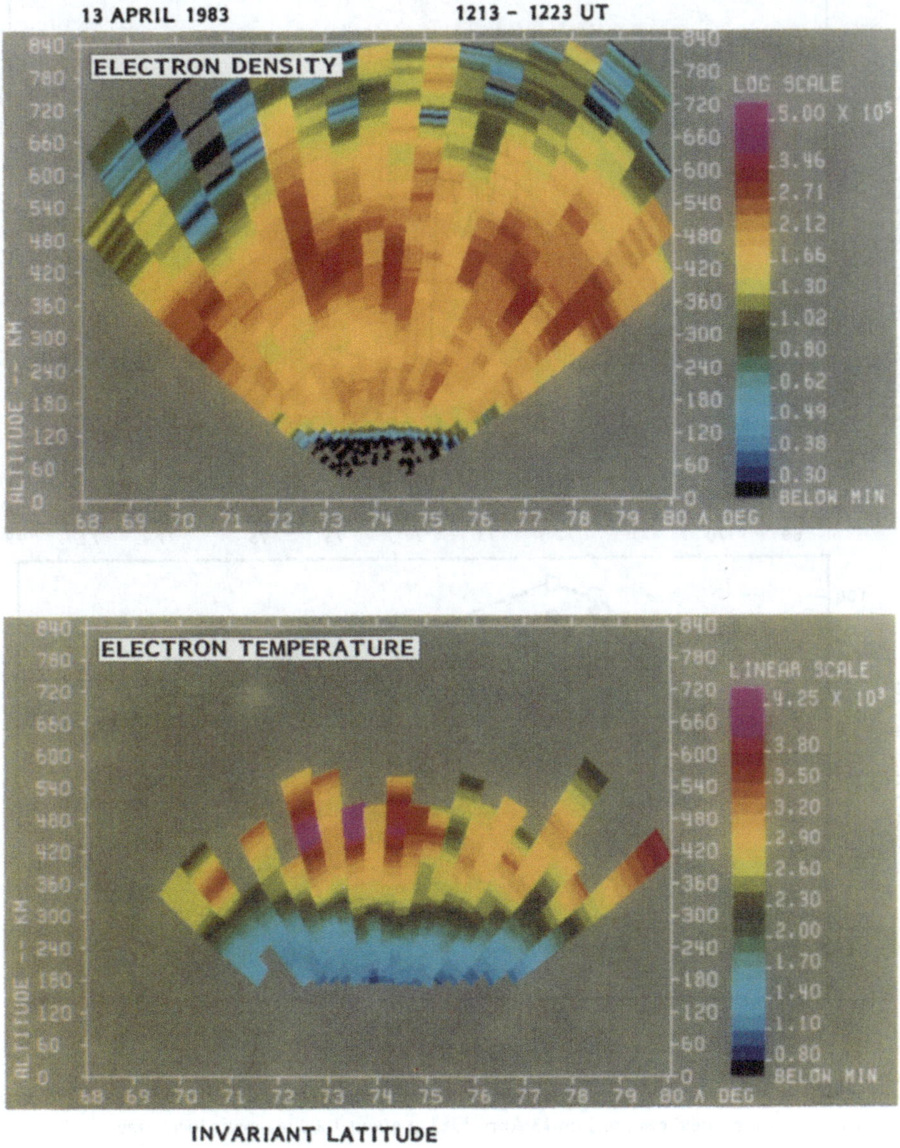

Plate 1. Color plots of electron density and electron temperature measured during an elevation scan on 13 April, 1983, at about 1000 magnetic local time.

consists of electrons from the low latitude boundary layer accelerated through parallel potential differences of about a kV.

In contrast to the afternoon sector, ionization enhancements in the morning

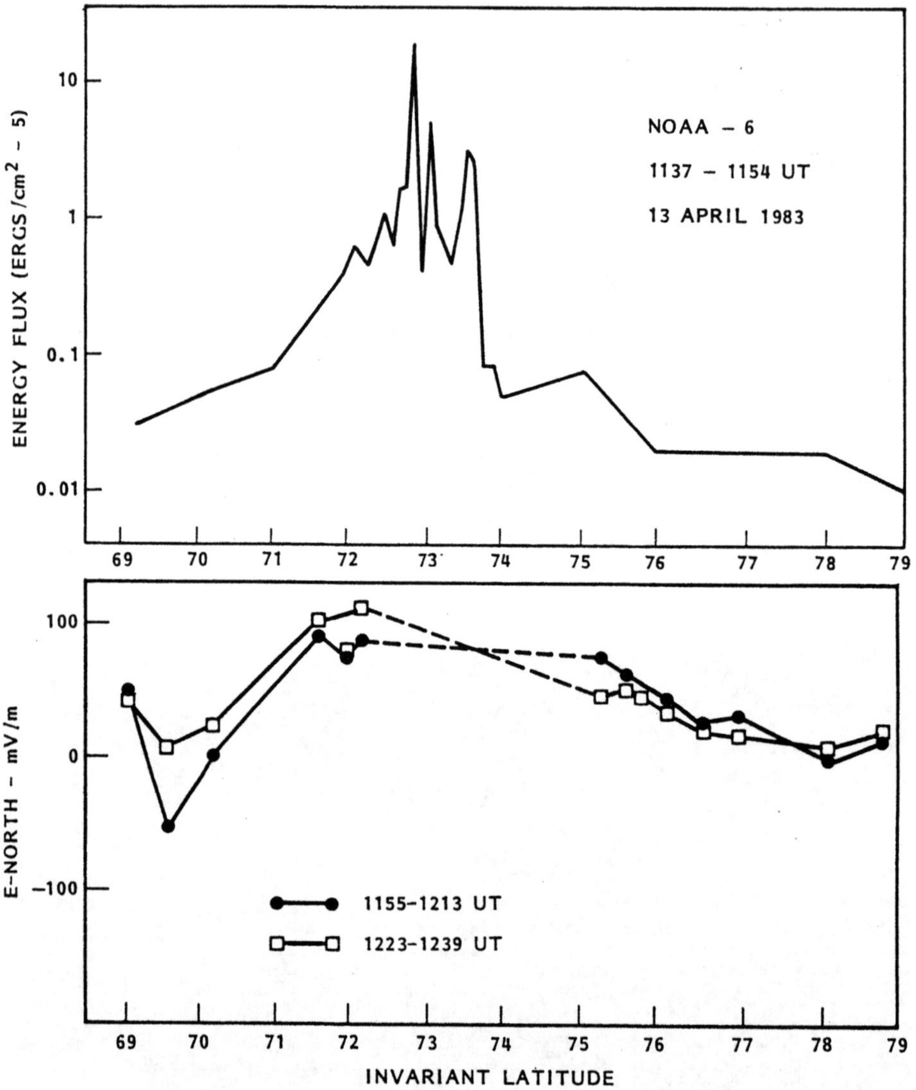

Fig. 2. NOAA-6 measurements of total energy flux and Sondrestrom radar measurements of the northward electric field obtained on 13 April 1983 at about 1000 magnetic local time.

sector that are obviously produced by discrete particle precipitation are rare. In examining the radar data for the 22 mornings during which IMF data were available, we found only two examples of ionization produced by particle precipitation. The absence of such ionospheric enhancements at this local time may be due to several effects. First, as noted above, the precipitating electrons may be very low in energy, thus depositing most of their energy at high altitudes where ionization production is slow. Second, the precipitation may be weak and diffuse so that it is indistinguishable from ionization produced by solar illumination which is a significant source at these local times. Finally, the discrete precipitation at these local times may occur outside of the radar field of view.

In this paper we have shown two cases of ionization in the morning sector traceable to discrete particle precipitation. Both occurred between 1000 and 1200 MLT during times when B_y was positive and B_z was negative. One was observed in the region of sunward convecting plasma while the other was observed in the region of antisunward convecting plasma.

While the event in the sunward convecting plasma can probably be identified as a morning sector auroral arc, the precipitation in the region of antisunward moving plasma must have been situated within the polar cap. The presence of discrete aurora at very high latitudes in the morning sector has been noted by LASSEN and DANIELSEN (1978) and GUSSENHOVEN (1982). That these features are not observed more often from Sondrestrom may be a result of the size of the auroral oval relative to the latitude range within which the Sondrestrom radar can make measurements. On April 13 the K_p index for the three-hour period from 0900 to 1200 UT was 6 so that the oval was probably greatly expanded at this time. For most of the other mornings included in our data set, K_p was between 2 and 4. When a convection reversal was present on these days ($B_y>0$) it was usually located between 74° and 76° invariant latitude. Thus, the radar field of view only extended a few hundred km into the polar cap. On 13 April, the convection reversal was located at 71° invariant latitude and a greater portion of the polar cap was observed.

In trying to account for the precipitation features that have been observed from Sondrestrom, it is significant to note that all are associated with gradients in the electric field that are consistent with upward field-aligned currents. In the case of the afternoon precipitation, this corresponded to the region 1 Birkeland current. For the precipitation occurring in the sunward drifting plasma on the morning side, we associated the upward current with region 2 Birkeland current. The precipitation observed on antisunward drifting field lines probably coincided with the polar cusp currents. These currents have been studied by IIJIMA *et al.* (1978) and MCDIARMID *et al.* (1978) and were found to depend on the B_y component of the IMF.

With these considerations we can predict the locations at which intense particle precipitation is likely to occur based on the convection patterns that have been observed. Figure 3 shows the general pattern of convection for $B_y<0$ and $B_y>0$. Upward field-aligned currents occur where the divergence of the electric field is negative (LYONS, 1978). These regions on the dayside are shown by shading in Fig. 3. Each of the regions of particle precipitation thus far identified from the radar measurements can be associated with one of these shaded areas. In the nomenclature of BURCH *et al.* (1985), the precipitation at the convection reversal in the afternoon for $B_y<0$ and equatorward of the convection reversal in the morning for $B_y>0$ are both associated with the viscous cell. The poleward region of precipitation for $B_y>0$ on the morning side is associated with the merging cell.

It should be noted that although we did not sort our data by B_z, this component of the IMF was never large and positive for a sustained period for any of the data sets

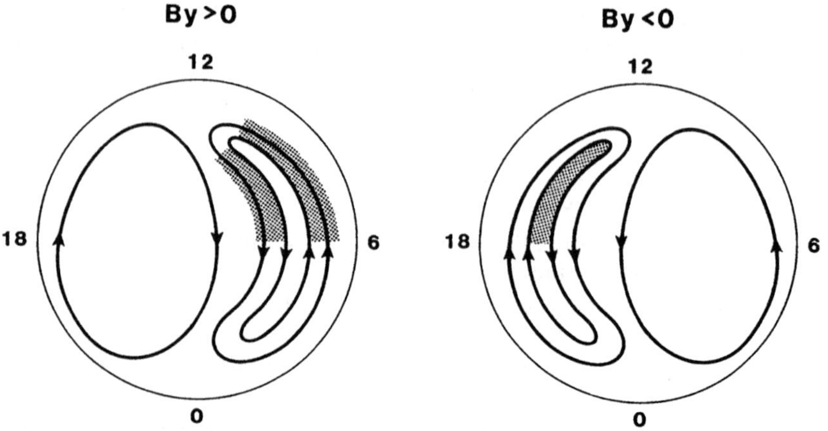

Fig. 3. Relationship between observed precipitation and the large-scale convective flow of plasma. The shaded regions show the locations on the dayside where ionospheric effects of particle precipitation were observed.

and there were no examples in which the convection at very high latitude was sunward. Thus, the type of polar cap arcs that appear under sustained B_z north conditions were not observed on these days. On the other hand, the data shown for April 13 clearly indicates precipitation in the polar cap. The distinction between these two types of polar cap arcs may be that the former is associated with the lobe convection cell, while the latter is associated with the merging cell (BURCH *et al.*, 1985).

6. Conclusion

Many claims have been made about identification of polar cusp precipitation from ground observations. An identification of a precipitation feature as originating from the polar cusp cannot be made solely on the proximity of the feature to local noon. One should instead consider the possible sources of the precipitation. The precipitation associated with the post-noon convection reversal boundary appears to originate from the low latitude boundary layer, and thus is not polar cusp precipitation. However, the precipitation observed on 13 April in a region of antisunward convecting plasma is probably associated with the merging cell (BURCH *et al.*, 1985). For $B_y<0$, a corresponding region of antisunward drifting plasma exists on the afternoon side of the polar cap. However, as the polar cusp currents are downward in this region, we should not expect the precipitation to be as intense as on the dawnside for $B_y>0$. Because the polar cusp field-aligned currents change direction on opposite sides of the polar cap, it follows that the fluxes of precipitating electrons associated with the polar cusp must exhibit a corresponding asymmetry. This may account for the results of studies of the distribution of polar cap arcs which show that these precipitation features appear more often on the dawn side in the northern hemisphere (see, for example, LASSEN and DANIELSEN, 1978).

We thank N. Ness and J. King of NASA-Goddard space flight Center for providing IMP-8 data. The site crew of the Sondrestrom radar maintained and operated the radar during the experiments. This work was supported at Stanford by AFGL contract F19628-85-K-0001 and National Science Foundation grant ATM-82-10562, at SRI by National Science Foundation cooperative agreement ATM-81-17792 and at Lockheed by the Lockheed Independent Research Program.

REFERENCES

BURCH, J. L., P. H. REIFF, J. D. MENIETTI, R. A. HEELIS, W. B. HANSON, S. D. SHAWHAN, E. G. SHELLEY, M. SUGIURA, D. R. WEIMAR, and J. D. WINNINGHAM, IMF B_y-dependent plasma flow and Birkeland currents in the Dayside magnetosphere, 1. Dynamics Explorer observations, *J. Geophys. Res.*, **90**, 1577, 1985.

CLAUER, C. R., P. M. BANKS, E. FRIIS-CHRISTENSEN, T. S. JORGENSEN, and V. B. WICKWAR, Ion convection in the dayside cleft measured by the Sondre Stromfjord radar, *EOS Transactions, American Geophysical Union*, **64**, 820, 1983.

CLAUER, C. R., P. M. BANKS, A. Q. SMITH, T. S. JORGENSEN, E. FRIIS-CHRISTENSEN, S. VENNERSTROM, V. B. WICKWAR, J. D. KELLY, and J. DOUPNIK, Observation of interplanetary magnetic field and of ionospheric plasma convection in the vicinity of the dayside polar cleft, *Geophys. Res. Lett.*, **11**, 891, 1984.

DE LA BEAUJARDERE, O., V. B. WICKWAR, and J. H. KING, Sondrestrom radar observations of the effect of the IMF B_y component on polar cap convection, this volume, 1986.

EVANS, D. S., The characteristics of a persistent auroral arc at high latitude in the 1400 MLT sector, *Proceedings from NATO Conference on the Polar Cusp in Lillehammer, Norway, 7–11 May, 1984*, 1984.

FOSTER, J. C., V. B. WICKWAR, and J. D. KELLY, Dayside patterns of plasma convection and density, *EOS Transactions, American Geophysical Union*, **64** 819, 1983.

FRIIS-CHRISTENSEN, E. K. LASSEN, J. WILHJELM, J. M. WILCOX, W. GONZALEZ, and D. S. COTBURN, Critical component of the interplanetary magnetic field responsible for large geomagnetic effects in the polar cap, *J. Geophys. Res.*, **77**, 3371, 1972.

GUSSENHOVEN, M. S., Extremely high latitude auroras, *J. Geophys. Res.*, **87**, 2401, 1982.

HEELIS, R. A., The effects of interplanetary magnetic field orientation on dayside high-latitude ionospheric convection, *J. Geophys. Res.*, **89**, 2873, 1984.

HEPPNER, J. P., Polar cap electric field distributions related to the interplanetary magnetic field direction, *J. Geophys. Res.*, **77**, 4877, 1972.

IIJIMA, T. and T. A. POTEMRA, The amplitude distribution of field-aligned currents at northern high latitudes observed by Triad, *J. Geophys. Res.*, **81**, 2165, 1976.

IIJIMA, T., R. FUJII, T. A. POTEMRA, and N. A. SAFLEKOS, Field-aligned currents in the south polar cusp and their relationship to the interplanetary magnetic field, *J. Geophys. Res.*, **83**, 5595, 1978.

JORGENSEN, T. S., E. FRIIS-CHRISTENSEN, V. B. WICKWAR, J. D. KELLY, C. R. CLAUER, and P. M. BANKS, On the reversal from "sunward" to "antisunward" plasma convection in the dayside high latitude ionosphere, *Geophys. Res. Lett.*, **11**, 887, 1984.

KELLY, J. D. and J. F. VICKREY, *F*-region ionospheric structure associated with antisunward flow near the dayside polar cusp, *Geophys. Res. Lett.*, **11**, 907, 1984.

KOFMAN, W. and V. B. WICKWAR, Very high electron temperatures in the daytime *F*-region at Sondrestrom, *Geophs. Res. Lett.*, **77**, 919, 1984

LASSEN, K. and C. DANIELSEN, Quiet-time pattern of auroral arcs for different directions of the interplanetary magnetic field in the *y–z* plane, *J. Geophys. Res.*, **83**, 5277, 1978.

LYONS, L. R., Generation of large-scale regions of auroral currents, electric potentials and precipitation by the divergence of the convection electric field, *J. Geophys. Res.*, **85**, 17, 1980.

MCDIARMID, I. B., J. R. BURROWS, and M. D. WILSON, Magnetic field perturbations in the dayside cleft and their relationship to the IMF, *J. Geophys. Res.*, **83**, 5753, 1978.
ROBINSON, R. M., D. S. EVANS, T. A. POTEMRA, and J. D. KELLY, Radar and satellite measurements of an *F*-region ionization enhancement in the post-noon sector, *Geophys. Res. Lett.*, **11**, 899, 1984.
WICKWAR, V. B. and W. KOFMAN, Dayside red auroras at very high latitudes: The importance of thermal excitation, *Geophys. Res. Lett.*, **1**, 923, 1984.
WICKWAR, V. B., The polar cap boundary observed in the early afternoon, this volume, 1986.

Solar Wind-Magnetosphere Coupling, edited by Y. Kamide and J. A. Slavin, 519–545.

Equatorial Ionospheric Electric Fields Associated with Magnetospheric Disturbances

B. G. FEJER

School of Electrical Engineering, Cornell University, Ithaca, New York 14853, U.S.A.

Radar measurements have been used extensively to determine the low latitude electric field pattern during magnetically disturbed conditions. *F*-region plasma drift measurements at the Jicamarca Radar Observatory have shown that the equatorial zonal electric field is strongly affected by magnetic activity, but the vertical electric field is not. Large electric field perturbations lasting from a few minutes to a few hours have been observed to occur simultaneously at all longitudes. Electric field measurements at auroral and equatorial latitudes showed that some of these events are associated with rapid increases and decreases in magnetospheric convection. In this case, large decreases in the equatorial zonal electric fields, relative to the quiet time values, are usually associated with northward turnings of the interplanetary magnetic field. Southward interplanetary magnetic field changes are usually less effective in disturbing the equatorial electric fields. There are also large perturbations during periods of high substorm activity and large asymmetric ring currents. The magnitude of the electric field perturbations is strongly dependent on local time, with the largest effects in the midnight-sunrise period. These results will be compared with other observations and with recent numerical simulations. Our present understanding of the high latitude processes responsible for low latitude electric field disturbances will also be discussed.

1. Introduction

The low latitude ionospheric dynamics is controlled, to a large extent, by ionospheric electric fields. In the equatorial lower *E*-region the zonal electric field gives rise to a large vertical electric field which drives the equatorial electrojet (e.g., FORBES, 1981, REDDY, 1981). The *F*-region $\boldsymbol{E}\times\boldsymbol{B}$ plasma drifts affect the latitudinal distribution of *F*-region electron density and the thermal balance of the equatorial ionosphere and thermosphere. The understanding of the equatorial electric fields during quiet and disturbed times is also essential for the study of plasma instabilities in the *E*- and *F*-regions (FEJER and KELLEY, 1980), and for modelling the low latitude ionosphere (e.g., ANDERSON, 1981).

There are several sources of electric fields in the Earth's ionosphere. During magnetically quiet periods the two main sources of ionospheric electric fields, in the reference frame moving with the Earth, are the magnetospheric circulation due to the interaction of the solar wind with the Earth's magnetic field, and the neutral wind dynamo. The magnetospheric convection dominates at geomagnetic latitudes above about 60°, whereas the ionospheric dynamo electric fields are most important at lower latitudes (MOZER, 1973; RICHMOND *et al.*, 1976). Incoherent scatter radar measurements have shown that the quiet time ionospheric electric fields from L shells less than about 3 can usually be explained by ionospheric dynamo theories, at least during daytime (e.g., GONZALES, 1979).

The existence of an electrical coupling between the high and low latitude ionospheres, at least during geomagnetically disturbed periods, has been known for servaral years from magnetometer studies (e.g., ONWUMECHILLI and OGBUEHI, 1962; ONWUMECHILLI *et al.*, 1973). Large fluctuations in the intensity of the equatorial electrojet were shown to be often closely correlated with similar variations in the auroral current system. This correlation was attributed to a current system generated during substorms and closing in the low latitude daytime ionosphere. NISHIDA (1968) showed the existence of coherent worldwide magnetic fluctuations, largely or entirely due to ionospheric currents, with periods of 0.5 to 2 h. These so called DP-2 fluctuations are positively correlated with southward variations of the interplanetary magnetic field (IMF) (NISHIDA, 1968; NISHIDA and MAEKAWA, 1971). There are other considerably more complex perturbations in the equatorial current system during substorms and magnetic storms (ONWUMECILLI *et al.*, 1973; KANE, 1974).

The coupling between polar and equatorial electric fields and currents cannot be established easily using magnetic records alone since their baselines are often displaced by magnetic effects from the ring current. In addition, nighttime low latitude ionospheric effects cannot be observed with magnetograms due to the reduced ionospheric conductivities. Several studies using electric field measurements have confirmed the existence of this coupling during disturbed times (e.g., TESTUD *et al.*, 1975; CARPENTER and KIRCHOFF, 1975; GONZALES *et al.*, 1979). The fluctuations and reversals of the equatorial east-west electric field can be related to perturbations in the auroral current system (MATSUSHITA and BALSLEY, 1972; FEJER *et al.*, 1976; CARTER *et al.*, 1976; REDDY, 1981). In addition, RASTOGI and PATEL (1975) suggested that large and rapid northward turnings of the IMF cause large perturbations in the equatorial current system. Recently, the relationship between the high and low latitude current systems was studied in more detail using electrical and magnetic field measurements at equatorial, middle and high latitudes and also interplanetary magneitc field data (KELLEY *et al.*, 1979; FEJER *et al.*, 1979b; GONZALES *et al.*, 1979, 1983; BLANC, 1978, 1983b; BLANC *et al.*, 1983). Several numerical models have been used to determine the low latitude currents and electric fields produced by given field aligned current distributions (MAEKAWA and MAEDA, 1978; MAEKAWA, 1980; NOPPER and CAROVILLANO, 1978; KAMIDE and MATSUSHITA, 1981; MATSUSHITA and KAMIDE, 1981), or by sudden increases in the polar cap potential drop (BLANC, 1983b; SENIOR and BLANC, 1984), for given conductivity models.

There are two possible processes which seem to account for most of the middle and low latitude electric field perturbations during disturbed times. One is the direct penetration of electric fields and currents from the magnetosphere and high latitude ionosphere to lower latitudes through the conducting ionosphere. This penetration does not always occur (BLANC, 1978; FEJER *et al.*, 1979b; GONZALES *et al.*, 1979), and it is controlled by several electrodynamical processes (MAEKAWA, 1980; KAMIDE and MATSUSHITA, 1981; SENIOR and BLANC, 1984). The second mechanism is due to the dynamic action of storm time thermospheric winds produced by auroral heating. These winds alter the global circulation pattern and generate disturbance dynamo electric fields at middle and low latitudes (BLANC and RICHMOND, 1980) and storm time changes in thermospheric density and composition (RISHBETH, 1975). Recent radar observations have shown clear evidence for the occurrence of disturbance dynamo drifts at mid-latitudes (BLANC *et al.*, 1983; GANGULY, 1984) and at the equator (FEJER *et al.*, 1983).

The direct penetration and the disturbance dynamo electric fields operate at considerably different time scales. The former seems to propagate to the low latitude ionosphere almost instantaneously. KIKUCHI and ARAKI (1979) suggested that this occurs through a TM_0 mode excited by polar electric fields which propagates in the north-south direction in the Earth-ionosphere waveguide. In contrast, the disturbance dynamo electric field has a latitude dependent time delay of a few hours or longer (BLANC and RICHMOND, 1980). At low latitudes these effects can be separated easily for isolated substorms but not for long lasting magnetic storms. These two processes are also effective in transfering part of the energy stored in the magnetosphere to the middle and low latitude ionosphere and thermosphere.

In the following section we describe initially the radar measurement techniques used for equatorial ionospheric electric field observations. We will then discuss in detail the typical electric field signatures observed during disturbed periods and their relationship with the high latitude current system and with the interplanetary magnetic field. Some recent electric field measurements obtained with incoherent scatter radars at middle and high latitudes will also be discussed but only when necessary for the understanding of the low latitude processes. Finally, we will discuss briefly some recent numerical modelling efforts. Here also we will focus only on the numerical results that can be compared directly with the equatorial electric field data.

2. Measurement Techniques

Most of the equatorial electric field data were obtained from radar measurements of plasma drifts at the Jicamarca Radio Observatory near Lima, Peru (11°57'S, 76°52'W, magnetic dip 2°N). Initially only the E-region east-west plasma drifts were measured using coherent radar observations of plasma irregularities in the equatorial electrojet. Over Jicamarca, and E-region electron drift velocity of 100 m/s corresponds to an east-west electric field of about 0.1 mV/m. This technique is not accurate when the electron drift velocity is smaller than about 50 m/s or somewhat larger than the nominal ion acoustic speed (about 360 m/s). E-region coherent radar studies of the equatorial zonal electric field have also been made in Thumba (8.5°N, 77°E, magnetic dip 0.9°S), India (e.g., REDDY, 1981).

Since the late 1960s, incoherent scatter radar observations have been used extensively to measure the equatorial F-region vertical and east-west plasma drifts at Jicamarca. These measurements are directly related to the F-region east-west and vertical electric field, since $\boldsymbol{V}=\boldsymbol{E}\times\boldsymbol{B}/B^2$. Over Jicamarca a vertical (eastward) F-region drift velocity of 40 m/s corresponds to an eastward (downward) electric field of 1 mV/m. The typical integration time is about 5 min, and the accuracy of the measurements is about 1–2 m/s for the vertical drifts and about 15–20 m/s for the zonal drifts, depending on the signal to noise ratio (WOODMAN, 1970, 1972). Simultaneous electrojet and *F*-region drift measurements have shown that the inferred east-west electric fields are usually well correlated (BALSLEY and WOODMAN, 1969). Since the electric fields are obtained directly from the *F*-region drift measurements, we will refer mostly to the drift data.

3. Low Latitude Electric Fields

The *F*-region zonal drifts are driven by vertical electric fields which map as north-south electric fields to the *E*-region at slightly higher latitudes. These drifts are generated by *E*- and *F*-region neutral winds, and show no apparent effect with magnetic activity (WOODMAN, 1972; FEJER *et al.*, 1981). On the other hand, the equatorial *F*-region vertical drifts (east-west electric fields) are strongly affected by magnetic activity (e.g., FEJER *et al.*, 1976, 1979b; GONZALES *et al.*, 1979, 1983). Furthermore, the Jicamarca vertical drift measurements are considerably more accurate than the *F*-region zonal drifts. Therefore, for the remainder of this study we will discuss only the vertical drift (east–west electric field) data.

The equatorial *F*-region east-west electric fields (i.e., vertical drifts) show considerable day to day variations even during magnetically quiet periods. There are also substantial seasonal and solar cycle effects (FEJER *et al.*, 1979a; FEJER, 1981). Figure 1 shows a composite of the drift data taken during the 1968–1969 winter. The daytime upward drift velocity is about 20 m/s, corresponding to an eastward electric field of 0.5 mV/m. The nighttime downward drifts are somewhat larger, and the reversals occur near sunrise and sunset. The data in Fig. 1 correspond to relatively quiet geomagnetic periods except for the July 3 nighttime data. In this case there was a

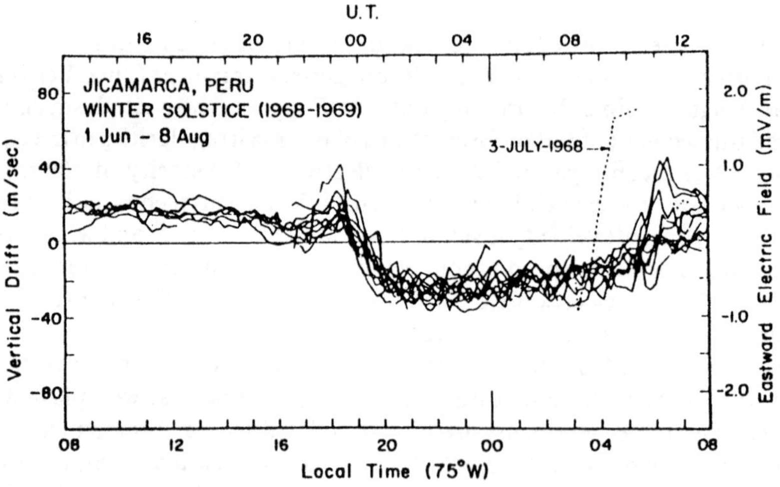

Fig. 1. *F*-region vertical plasma drifts (eastward electric fields) during the 1968–1969 winter solstice (after, WOODMAN, 1970).

large and sudden departure from the quiet time profiles.

During substorms and geomagnetic storms, the east-west electric fields often show significant departures from the quiet day values. Figure 2 illustrates a number of such variations during an extended disturbed period. The dashed curve indicates the quiet time average pattern. The periods of occurrence of equatorial spread-*F* echoes, which can sometimes affect severely the incoherent scatter drift measurements, are also shown. Large perturbations in the vertical velocity (zonal electric field) are present during both daytime and nighttime, but are most frequent and largest in the midnight-early morning period. They usually decrease and sometimes even reverse drift direction. Figure 2 shows both rapid fluctuations as well as slower departures from the quiet curve. The fast perturbations are associated with large changes in the high latitude current system. The slower perturbations (for example, in the post-midnight period on May 14) are due to disturbance dynamo electric field effects (FEJER *et al.*, 1983), which will not be discussed here.

Magnetometers have detected large disturbances and even anomalous reversals of the daytime zonal electric field for many years (e.g., GOUIN and MAYAUD, 1967; HUTTON and OYINLOYE, 1970). These reversals, commonly referred to as counter-electrojet events, can occur during both magnetically quiet and disturbed days (REDDY, 1981; FORBES, 1981). On quiet days they take place in the early morning and afternoon hours, are characterized by slow and smooth changes in the electrojet current intensity, and occur progressively at all longitudes at about the same local time. These events seem to be closely related to large lunar and solar tidal effects

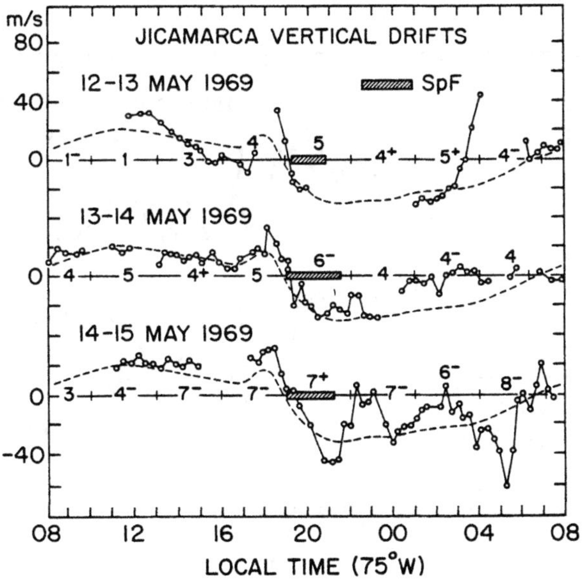

Fig. 2. Examples of the large variability of the vertical plasma drifts during an extended disturbed period.

(FORBES, 1981). On the other hand, the fluctuations and reversals during geomagnetically disturbed periods are considerably more sudden and occur at all longitudes at the same universal time (GONZALES *et al.*, 1979, 1983). Figure 3 shows an example of the close correlation between rapid fluctuations in the nighttime zonal electric field measured at Jicamarca and simultaneous daytime magnetic field perturbations at Addis Ababa, Ethiopia. Additional examples of simultaneous electric field perturbations on the dayside and nightside will be presented later.

The frequency of occurrence and magnitude of vertical drift disturbances have significant local time dependence. A number of daytime velocity profiles during disturbed conditions are plotted in Fig. 4. The velocity and IMF data on 29 December 1970 were discussed in detail by RASTOGI *et al.* (1978). On May 9, 1968 there were both positive and negative pertubations. Positive daytime drift perturbations are relatively uncommon, however, and are usually restricted to the midnight-noon period, except during extremely disturbed times. That is, the drift perturbations are predominantly downward during the day and upward at night, corresponding to westward and eastward electric field perturbations, respectively. The next three profiles show more typical daytime drift perturbations during moderately disturbed times. The daytime downward drift perturbations have relatively fast recovery times in the morning, whereas in the afternoon they are longer lasting and more pronounced. We will show later that these disturbances are often associated with sudden changes in the IMF B_z component.

Near solar minimum large and sudden drift perturbations can often occur for considerably smaller levels of magnetic activity (as indicated by the K_p index) than during the maximum. As a result the drifts are considerably more variable at solar

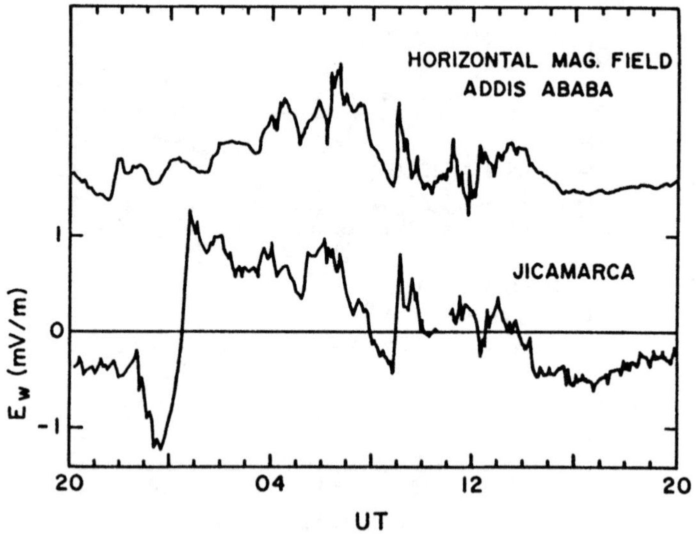

Fig. 3. Westward electric field at Jicamarca and horizontal component of the magnetic field at Addis Abbaba on October 10–11, 1980 (after GONZALES *et al.*, 1983).

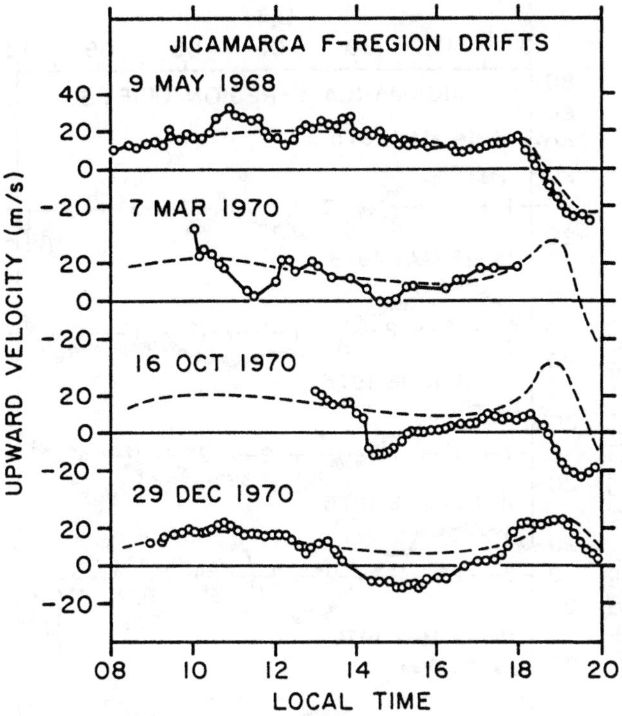

Fig. 4. Typical *F*-region vertical drifts at Jicamarca during magnetically disturbed periods. The dashed lines indicate the corresponding quiet time variations.

minimum than during solar maximum. Figure 5 shows the drift data during three 48 hour periods during solar minimum. In this case, in the evening and nighttime periods there are large upward velocity perturbations relative to the quiet days. During solar maximum large drift perturbations in the sunset-midnight sector are observed only during very disturbed conditions. The velocity profile during 19–20 May 1976 show small amplitude DP-2 like fluctuations. Near solar minimum these coherent perturbations seem to occur more frequently and are present over a larger local time sector than during the maximum.

4. Disturbed Equatorial Zonal Electric Fields

GONZALES *et al.* (1979) pointed out four possible coupling processes between the magnetospheric and the low latitude current systems: (1) the penetration of electric fields to low latitudes during rapid changes in magnetospheric convection; (2) the partial closure of the auroral and ring current systems at low latitudes during substorms; (3) storm sudden commencements and sudden impulses; (4) disturbance dynamo electric fields. Here we will examine only the first two processes, which seem to be associated with the largest and most common equatorial electric field

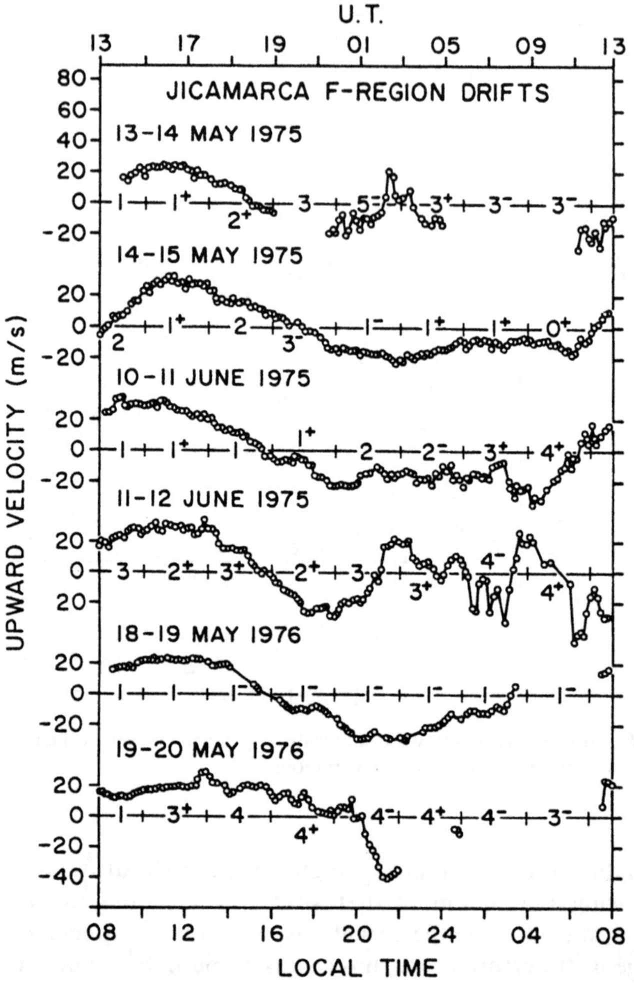

Fig. 5. *F*-region vertical drifts and K_p values during three 48 hour periods at solar minimum. Note the large variablity of the nighttime drifts.

perturbations.

4.1 The effect of the IMF

MATSUSHITA (1977) has reviewed the earlier results on IMF polarity effects in the low latitude thermosphere. Most of the past work has dealt with the relationship between sudden northward IMF turnings and the occurrence of anomalous equatorial zonal electric fields. Some recent work has suggested, however, that IMF sector or B_y effects can also produce noticeable changes in the equatorial electric fields. In-situ satellite measurements using the Cosmos 184 satellite during October–November 1967 indicated an increase in the equatorial upward and eastward plasma drifts when B_y reversed from negative to positive (GALPERIN *et al.*, 1978). Ionosonde data from a 3-year period suggested larger *F*-region vertical drifts in the dusk sector when the IMF polarity was positive during September–October and negative during March–April (SASTRI, 1984).

The relationship between equatorial electric fields and the north–south component of the IMF has been discussed by several authors. RASTOGI and PATEL (1975) suggested that large and sudden northward B_z turnings were associated with large perturbations in the equatorial ionospheric east–west electric field. Several examples of east-west electric field reversals well correlated in time with sudden northward turnings of the IMF were presented in a number of papers (RASTOGI and PATEL, 1975; PATEL, 1978; FEJER *et al.*, 1979b; GONZALES *et al.*, 1979). Figure 6 shows one such example together with the auroral electrojet indices. A positive *E*-region drift velocity of 100 m/s corresponds to an eastward electric field of about 0.1 mV/m. The rapid northward B_z change was well correlated with a large westward electron drift velocity (i.e., eastward electric field) perturbation and with the onset of a substorm recovery phase, as is typically the case. Most of the equatorial electric field perturbations seem to occur during periods of large and sudden northward B_z

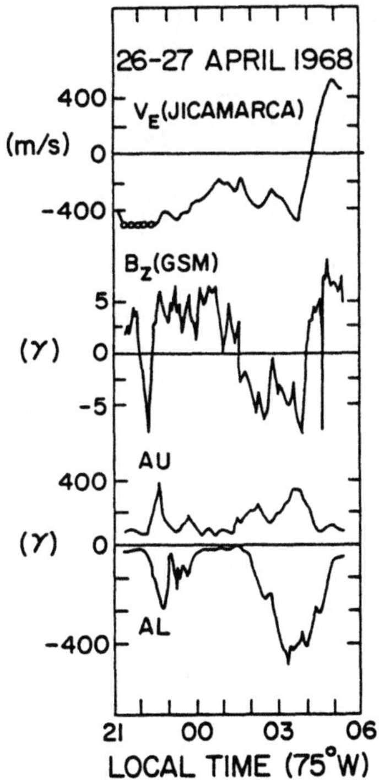

Fig. 6. Example of a nighttime reversal of the equatorial electrojet westward electron drifts (i.e., zonal electric field) associated with a large northward B_z turning and with a substorm recovery. The dots indicate a lower bound on the velocity (after FEJER *et al.*, 1979a).

turnings. However, not every such IMF change is associated with equatorial electric field perturbations. The large northward change at about 2200 LT in Fig. 6, for example, did not produce any detectable velocity perturbation. There are large anomalous electric field reversals also when the IMF has a large southward value.

Figure 7 shows an example of large daytime drift variations during a period of large changes in the IMF and in the auroral current system. In this case, no drift

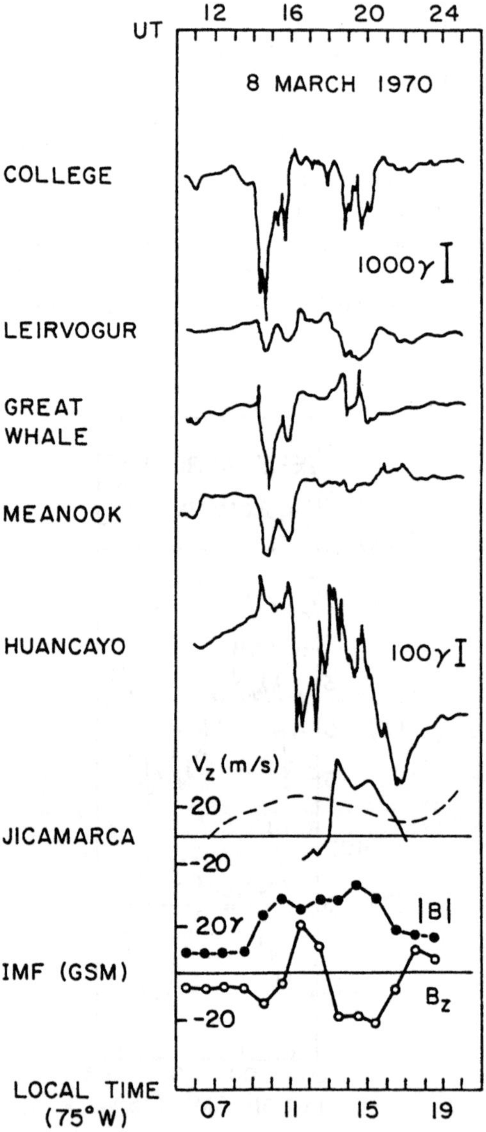

Fig. 7. Example of large daytime vertical drift (electric field) variations associated with high latitude current and IMF pertubations. The dashed curve indicates the quiet time variation (after FEJER *et al.*, 1979b).

measurements were made before 12 LT, but using also the Huancayo magnetic field data, we can estimate that the downward drift perturbation (relative to the time quiet time value) associated with the large northward B_z turning at 1600 UT (1100 LT) was about −45 m/s (westward electric field of about 1 mV/m). There was also an upward drift perturbation of the same order associated with the large southward B_z turning. The disturbance electric field associated with the northward B_z turning is of opposite sign but of comparable amplitude to that shown in Fig. 6 although the latter was associated with considerably smaller high latitude currents (indicated by the auroral indices). Electric field measurements at Jicamarca indicate that the electric field perturbations are usually larger in the nightside than in the dayside (see also GONZALES *et al.*, 1979). When high resolution data are used, it is clear that the equatorial electric field perturbations occur nearly at the same time as the rapid northward IMF turnings. However, as mentioned above, although every sudden equatorial electric field perturbation occurs during periods of significant perturbations in the high latitude current system, the converse is not true.

The frequency of occurrence and the amplitude of the equatorial electric field perturbations associated with northward B_z turnings change considerably with local time. Figure 8 shows the time dependence of the eastward electric field perturbations (relative to the quiet time values) associated with hourly northward B_z turnings of 2 gammas or greater. We have omitted the data when the hourly B_z polarity changed back and forth, and near sunrise and sunset when the drifts vary considerably even during quiet times. Therefore, this pattern should reflect the perturbation amplitudes associated with large northward B_z changes from steady southward values. The hourly values are certainly not ideal for the study of the high and low electrical coupling, but nevertheless, Figure 8 illustrates some important points. The daytime electric field perturabations are westward with slightly larger values in the afternoon sector. The nighttime eastward perturbations are largest in the 02–05 local time sector and smallest prior to midnight. The occurrence of larger electric field disturbances at

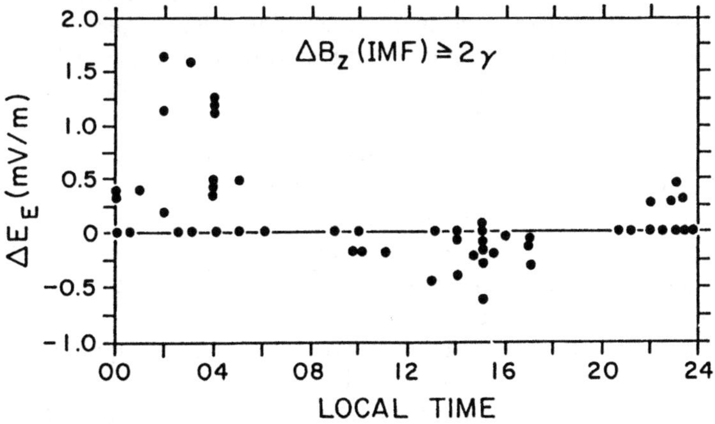

Fig. 8. Local time dependence of equatorial eastward electric field perturbations associated with northward IMF turnings of 2 gammas or greater.

night is consistent with the results of numerical simulations. (JAGGI and WOLF, 1973; NOPPER and CAROVILLANO, 1978; SENIOR and BLANC, 1984).

RASTOGI and PATEL (1975) suggested that the zonal equatorial electric field perturbations were due to the direct penetration of the interplanetary electric field through the magnetosphere and into the low latitude ionosphere. When the IMF is northward the ecliptic interplanetary electric field is directed from dusk to dawn which corresponds to a westward electric field in the dayside ionosphere and an eastward field in the nightside, in agreement with the observations. KELLEY *et al.* (1979) pointed out, however, that the direct penetration of the interplanetary electric field into the low latitude ionosphere is not consistent with the electrical coupling properties of the magnetosphere.

KELLEY *et al.* (1979) proposed a mechanism which has some similarity to the model of Rastogi and Patel but with the important difference that the zonal electric field perturbation is ultimately driven by the convection electric field with an indirect control from the IMF. They suggested that sudden IMF northward turnings from a steady southward direction cause a temporary imbalance between the convection related charge density and the charge in the inner edge of the ring current which normally shields the effect of the high latitude electric field from middle and low latitudes. Since the time constant for the decay or build up of shielding charges in this Alfven layer is greater than several minutes (VASYLIUNAS, 1972), this imbalance produces a dusk to dawn electric field perturbation which can penetrate the plasmasphere. This disturbance electric field is westward during the day and eastward at night. Since the field aligned currents couple the high latitude ionosphere and the magnetosphere, this process is equivalent to the partial closure of the region 2 field aligned currents through the middle and low latitude ionosphere when the region 1 currents are turned off. The efficiency of this process should depend also on other parameters such as the ionospheric conductivities, the level of geomagnetic activity, and the local time.

Rapid increases in the convection electric field also should penetrate the plasmasphere during the time required to set up the Alfven layer shielding charges. This is equivalent to the partial closure of the region 1 field aligned current system in the middle and low latitude ionosphere. Large southward changes in the IMF increase the dawn to dusk magnetospheric dynamo electric field, corresponding to an eastward electric field on the dayside and westward on the nightside; i.e., with the same polarity as the quiet time equatorial electric fields. Our data suggest, however, that this process is considerably less effective than that associated with sudden northward B_z turnings. These anomalous electric field perturbations cause upward daytime and downward nighttime drifts, are most common on the midnight-noon period, and have the largest amplitudes between about 02 and 05 local time as shown, for example, in Fig. 2. No IMF data were available for this period, however.

Figure 9 shows examples of pronounced downward and upward perturbations associated with large southward and northward turnings of the IMF, respectively. The nighttime downward drift perturbations sometimes coincide with the onset of magnetospheric substorms. GONZALES *et al.* (1979, 1983) presented several other examples of large velocity increases associated with southward B_z changes. They all

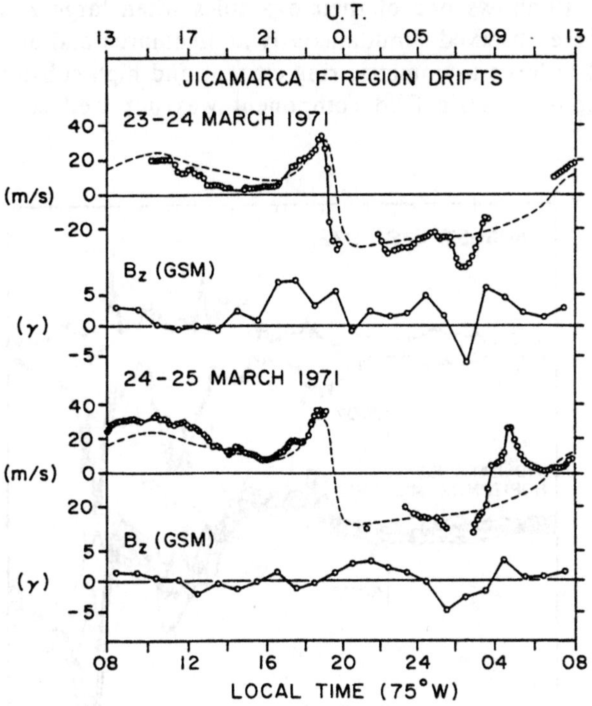

Fig. 9. *F*-region vertical drifts and IMF data. Note the large nighttime drift perturbations well correlated with southward and northward IMF changes.

occurred in the midnight-noon period and had largest amplitudes shortly before sunrise. The March 23 daytime data also show an example of a large northward B_z change at about 16 LT without any detectable effect on the drifts. In this case, however, the IMF was either zero or northward for an extended period prior to the northward change. This supports the idea that some sort of preconditioning is required for IMF effects to extend into low latitudes as suggested by MATSUSHITA (1979).

Statistical studies do not show significant variations of the average equatorial electric fields with magnetic activity. This is probably due to the transient and sporadic dependence of the electric fields on B_z. FEJER *et al.* (1979b) showed that the hourly average vertical drifts between 1000 and 1400 LT change only slightly with B_z. It is possible that there is a more significant drift variation with B_z in the midnight-dawn period when the high latitude electric fields can penetrate more easily to lower latitudes.

4.2 *Electric field perturbations associated with large substorm and ring current activity*

GONZALES *et al.* (1979) used detailed worldwide electric and magnetic field measurements to study the origin of the equatorial perturbations during disturbed periods. Figure 10 shows one of their examples when large zonal electric field perturbations were observed simultaneously at Jicamarca and at College, Alaska, during a period of large asymmetric ring current and high substorm activity. The auroral north-south electric field component was not well correlated with the

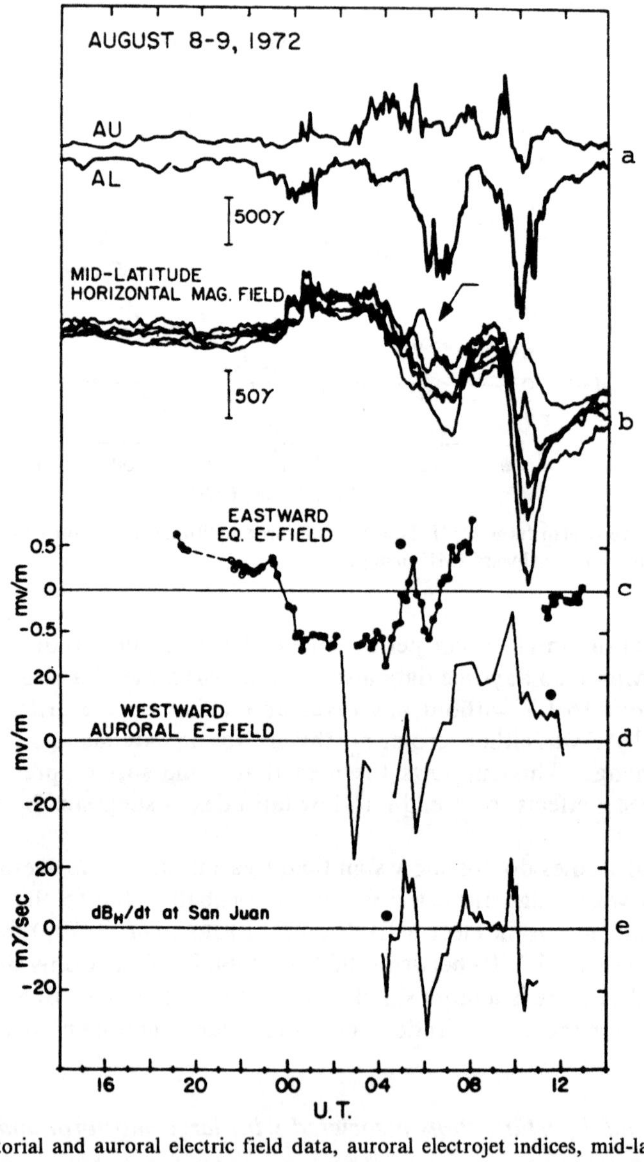

Fig. 10. Equatorial and auroral electric field data, auroral electrojet indices, mid-latitude magnetograms, and time derivative of the horizontal component of the magnetic field at San Juan during a strongly disturbed period. The arrow indicates the San Juan magnetogram, and the dots indicate local magnetic midnight (after GONZALES *et al.*, 1979).

equatorial data. The time derivative of the horizontal component of the magnetic field measured in San Juan, Puerto Rico was well correlated with the electric field perturbations, as is often the case. Figure 11 shows nighttime electric field data along with daytime magnetic field data from two low latitude stations in India, Kodaikanal near the dip equator, and Alibag outside the equatorial electrojet region. The intensity of the equatorial electrojet at Kodaikanal can be estimated by subtracting the horizontal components of the magnetic field at these two stations. The westward electric field fluctuations in the daytime electrojet are well correlated with the eastward nighttime perturbations. GONZALES *et al.* (1979) showed also that the electric field perturbations on the nightside have larger amplitudes than on the dayside.

Figure 12 shows the variation of the IMF B_z component, *AE* index, and equatorial electric field during another strongly disturbed period. These data illustrate the difficulty in associating the equatorial electric field perturbations with changes in the auroral indices and in the IMF during highly disturbed periods. The onset of the first large electric field perturbation at about 03 UT ocurred at the same

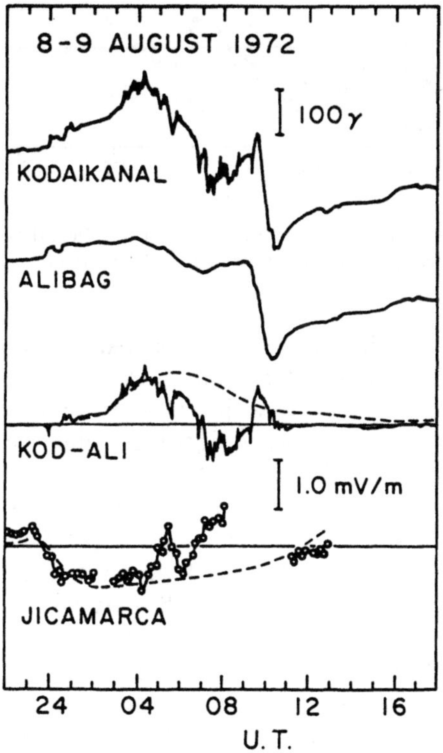

Fig. 11. Horizontal component of the magnetometer at Kodaikanal and Alibag, India, electrojet intensity at Kodaikanal obtained by subtracting Alibag from Kodaikanal, and equatorial electric field at Jicamarca. The dashed lines indicate the quiet time patterns (after GONZALES *et al.*, 1979).

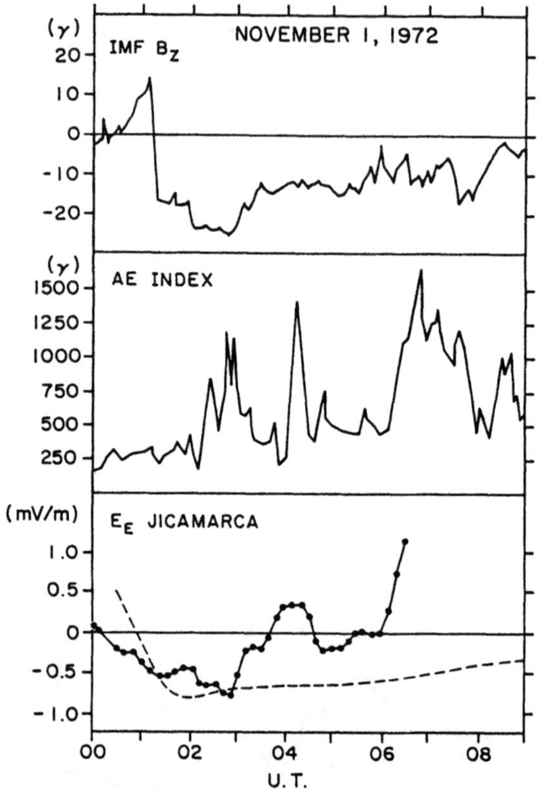

Fig. 12. IMF B_z component, AE index, and equatorial electric field data during a strongly disturbed period. The dashed line indicates the quiet time zonal electric field.

time as a sudden decrease in the AE index related, perhaps, to the simultaneous northward IMF change from a very large southward value. However, the second large electric field perturbation coincided with a large increase in the AE index but no important change in B_z, and there was no significant change associated with the large variations in AE at about 04 UT. AKASOFU and LEPPING (1977) examined in detail the auroral data during this period and concluded there was no systematic relationship between changes in B_z and the occurrence of individual substorms.

GONZALES *et al.* (1979) suggested that large electric field perturbations during periods of high magnetic activity were due to the partial closure of high latitude field aligned currents in the dayside low latitude ionosphere. This process is not fundamentally different, however, from that proposed by KELLEY *et al.* (1979) since the latter mechanism can be interpreted as due to the penetration of region 2 field aligned currents into the low latitude ionosphere. Large pertubations in the mid-latitude north-south electric fields (east–west drifts) are often observed during periods of large southward B_z and large rates of energy injection into the ring current system (derived from the D_{st} index) (BLANC, 1978, 1983). The effect on the mid-latitude zonal electric field is usually smaller than on the meridional field. The largest mid-latitude electric field disturbances are observed during periods of large excursions in the injection rate. This relationship is similar to that between the equatorial zonal electric field perturbations and the time derivative of the horizontal component of the mid-latitude magnetic field found by GONZALES *et al.* (1979). The Saint-Santin electric field perturbations were attributed to short term extensions of the magnetospheric convection into middle latitudes due to the increase in the dawn-dusk potential drop caused by B_z, and with a corresponding increase in plasma injection rate into the ring current system. This process leads to an eastward daytime and westward nighttime electric field perturbation, which does not agree with the sign of the dominant equatorial electric field perturbations. Clearly the variation of the ionospheric electric fields during periods of strong substorm activity is considerably more complex than during isolated substorms.

4.3 Fast electric field fluctuations

The electric field perturbations discussed so far consisted mostly of sudden departures from the quiet time level with usual recovery times of 1–3 hours. However, DP-2 like electric field fluctuations with time scales of about 1 hour are also observed. Figure 13 shows an example of large coherent fluctuations in the equatorial current system well correlated with similar variations in the auroral zonal electric field and in B_z. The phase relationship between B_z and the ionospheric electric and magnetic field fluctuations is not clear, however, due to the uncertainty of propagation time delays. According to NISHIDA (1968) the daytime DP-2 fluctuations are due to eastward electric field perturbations associated with southward turnings of the IMF. NISHIDA (1973) also suggested that it is difficult to determine the sign of the electric field perturbations from a single station due to the possible occurrence of additional electric field disturbances which usually accompany the DP-2. On the other hand, GONZALES *et al.* (1983) argued that equatorial electric field fluctuations of both signs are possible. In this case, southward B_z changes would be associated with eastward daytime (westward nighttime) electric field perturbations and northward B_z turnings would correspond to perturbations of opposite sign. They also suggested that the ring current shielding charges play an important role in determining the sign of the electric field perturbations. In the absence of an additional quasi-dc electric field perturbation, the Jicamarca data shown in Fig. 13 would suggest that the fluctuations are associated with eastward electric field fluctuations on the nightside and westward on the dayside.

Figure 14 shows several examples of daytime periodic drift fluctuations near solar minimum. These data show predominantly upward velocity perturbations with largest amplitudes before about 12 LT. Recently, PAPAMASTORAKIS *et al.* (1984) showed evidence for pronounced local time asymmetries in the DP-2 and SI (sudden impulse) disturbances, with maximum electric field around 10 LT. This is consistent with the data in Fig. 14, and with the local time dependence of the equatorial electric field perturbations associated with southward B_z changes, as discussed previously.

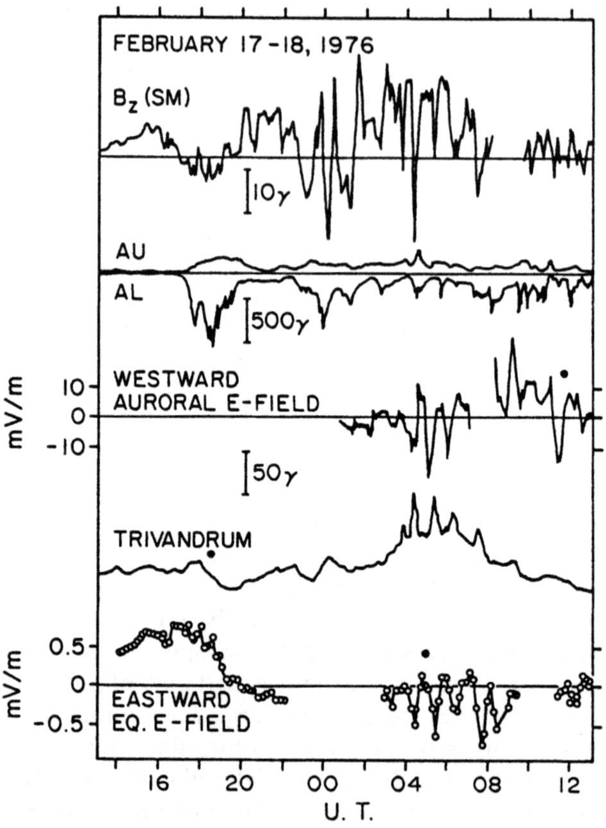

Fig. 13. IMF B_z component, auroral indices, zonal electric field at Chatanika, Alaska, horizontal component of the magnetic field at Trivandrum, India, and zonal electric field at Jicamarca during a period of large electric and magnetic field fluctuations. The dots indicate local magnetic midnight (after GONZALES *et al.*, 1979).

5. Global Electric Field Measurements

Recently several radar experiments have been conducted to determine the latitudinal variation of the ionospheric electric fields during disturbed periods. The measurements during the first of these campaigns, discussed in detail by KELLEY *et al.* (1979), are presented in Fig. 15. Note the different scales used for the auroral and equatorial electric fields. The large fluctuations in B_z did not affect the zonal field at Millstone Hill and Jicamarca prior to 0600 UT (0100 LT at these two stations). However, the large northward B_z change at about 0730 was correlated with large and nearly identical eastward electric field perturbations observed by the three radars. In addition, the large southward IMF turning after about 0930 UT was coincident with a large increase in the equatorial westward electric field. These results agree with our conclusion that the largest equatorial electric field perturbations occur between about

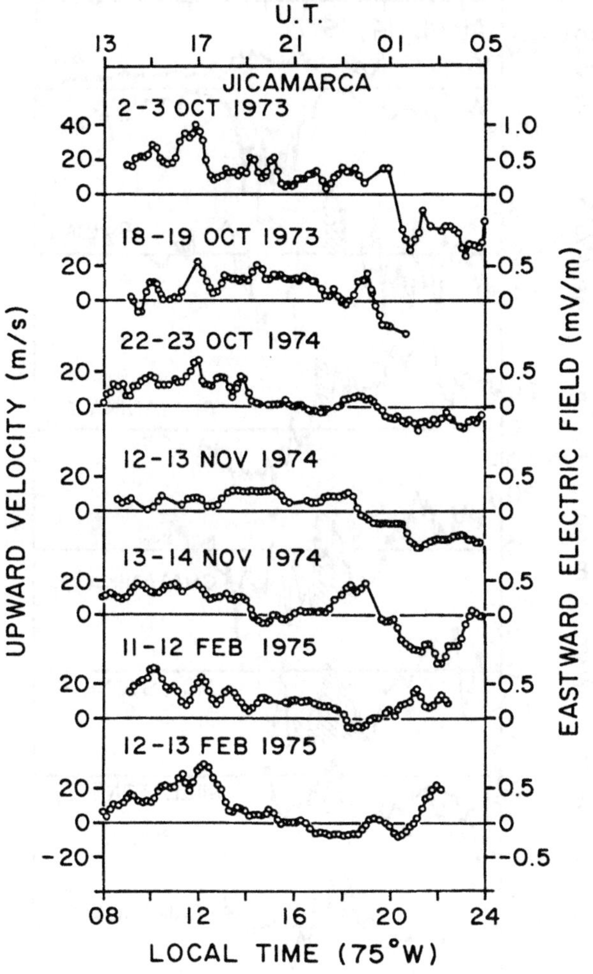

Fig. 14. Examples of daytime drift fluctuations near solar minimum.

0800–1200 UT (0300–0700 local time). Additional examples of simultaneous electric field observations at high and low latitudes were discused by GONZALES *et al.* (1983). They also showed, in particular, the occurrence of large auroral and equatorial zonal electric field disturbances when the IMF had a large northward change but did not reverse.

GONZALES *et al.* (1983) showed that the amplitude ratios between the Arecibo ($L=1.46$) and the Jicamarca electric field perturbations during two active nighttime periods were 2.4 and 3.5. This is larger than the ratio of $(1.46)^{3/2}=1.7$, which corresponds to the mapping of a constant electric field in the equatorial plane (MOZER, 1970). The identification of electric field perturbations at Arecibo during disturbed days is considerably more difficult than at Jicamarca due to the higher

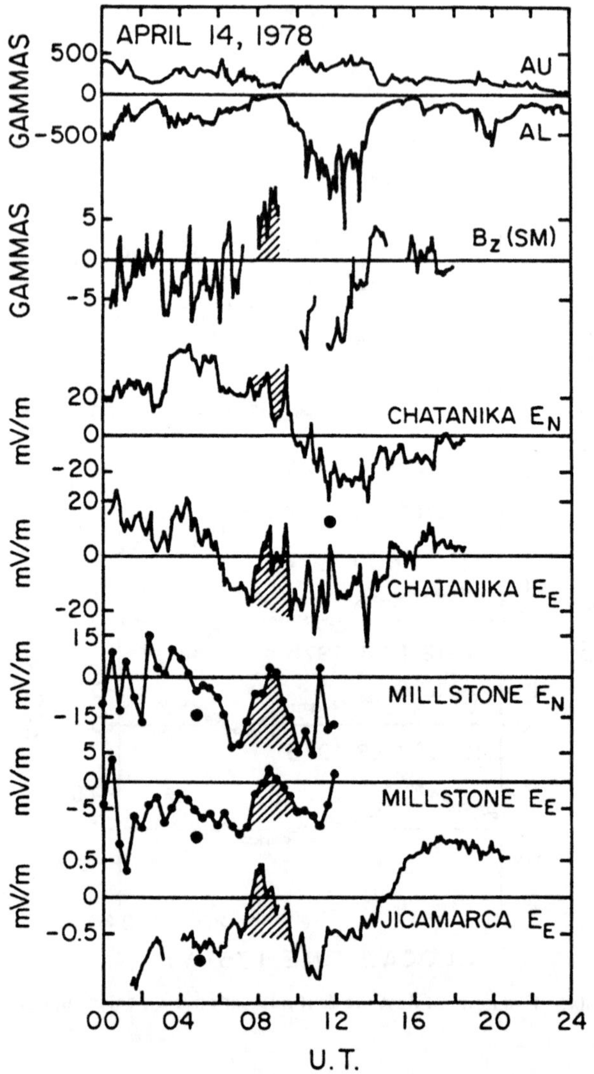

Fig. 15. Equatorial and auroral electric fields, along with auroral indices and IMF data during a strongly disturbed period. The dots indicate local magnetic midnight (after KELLEY *et al.*, 1979).

uncertainty in the electric field measurements at Arecibo. The difference between the two measured ratios might not be entirely due to experimental errors, however. KAMIDE and MATSUSHITA (1981) showed that the shielding factor is larger in the evening than in the morning sector, and changes with the intensity of the region 2 field aligned currents during both quiet and substorm times.

6. Summary of the Experimental Results

The zonal equatorial electric fields often show large and sudden perturbations during geomagnetically disturbed periods, whereas the *F*-region vertical electric field does not seem to change significantly with magnetic activity. The zonal electric field perturbations occur approximately simultaneously at all longitudes and latitudes and are always associated with sudden commencements, sudden impulses, or with large perturbations in the high latitude electric fields and currents. The high latitude ionosphere can also affect the lower latitudes by the generation of disturbance dynamo electric fields which have time delays of a few hours and longer.

Large equatorial electric field perturbations often coincide with sudden changes in the IMF. The zonal electric field disturbances associated with large northward B_z turnings are westward during the day and eastward at night. The electric field perturbations related to southward B_z changes are eastward on the dayside and westward on the nightside, but are less frequent and have smaller amplitudes. These electric field perturbations are not driven directly by the interplanetary magnetic field but are the result of changes in magnetospheric convection and substorm phenomena. Southward B_z changes seem less effective in disturbing the equatorial electric fields. Large equatorial electric field perturbations are also observed during periods of asymmetric ring current development and high substorm activity. To first order, perturbation electric fields are westward during the day and eastward at night. The electric field perturbations associated with northward B_z turnings are often associated with substorm recovery onsets but, in general, there is no clear relationship between the auroral indices and the equatorial electric field disturbances. There is also some evidence that IMF sector or B_y variations can affect the low latitude electric fields.

There is a strong local time dependence on the amplitude of the electric field perturbations related with both northward and southward IMF changes. In the first case they are observed at essentially all local times, with largest and smallest amplitudes in the 02–05 and 20–24 local time sectors, respectively. The electric field perturbations related to southward B_z changes are observed predominantly in the midnight-noon local time sector with largest amplitudes shortly before sunrise.

Large amplitude coherent DP-2 like electric field fluctuations with a time scale of about one hour are sometimes observed in the sunrise-noon sector. These fluctuations are most common near solar minimum when they are often present also in the evening and nighttime sectors. The amplitude of DP-2 fluctuations seems to have the same local time dependence as the sudden eastward electric field disturbances associated with southward B_z turnings.

7. Comparison with Recent Numerical Models

Extensive numerical models have been developed to study the penetration of high latitude electric fields into the middle and low latitude ionosphere (e.g., NOPPER and CAROVILLANO, 1978; NISBET *et al.*, 1978; MAEKAWA and MAEDA, 1978; MAEKAWA, 1980; KAMIDE and MATSUSHITA, 1981; MATSUSHITA and KAMIDE, 1981; BLANC, 1983; SENIOR and BLANC, 1984). Here will compare briefly the radar data with the low latitude results from these models.

There are several parameters which seem to affect the response of the middle and low latitude electric fields to high latitude pertubations during active times. MAEKAWA (1980) showed that the penetration of region 1 field aligned currents into lower latitudes decreases with the increase in the ratio of the region 2 to region 1 current intensity and with their spatial separation. KAMIDE and MATSUSHITA (1981) have examined in detail the shielding efficiency process assuming anisotropic conductivities. Their study shows that the north-south electric field component is larger than the east-west component for a latitude of 30°, in agreement with electric field measurements at Arecibo (e.g., GANGULY *et al.*, 1984). They also showed that the shielding efficiency is higher in the evening than in the morning sector, as suggested by the radar measurements. NISBET *et al.* (1978) calculated the worldwide electric field and current distribution using empirical field aligned current models for $4^{+}>K_{p}>2^{-}$, and the conductivities in KIRCHHOFF and CARPENTER (1975). The calculated zonal electric fields decreased from their values at 69° latitude by factors of 14 at 45°, 140 at 20°, and 200 at the equator. Other studies suggest considerably smaller degrees of shielding.

NOPPER and CAROVILLANO (1978) calculated the worldwide ionospheric potential due to high latitude field aligned currents using realistic conductivities, particularly at low latitudes. This model did not include the effect of low latitude winds which produce *E*- and *F*-region dynamo electric fields. Figure 16 shows the

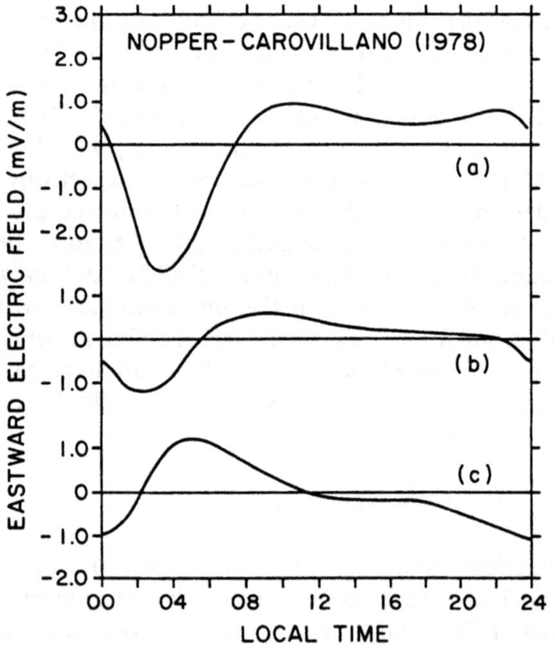

Fig. 16. Equatorial ionospheric electric field patterns calculated for three given field aligned currents distributions. Ionospheric dynamo effects were not included (after NOPPER and CAROVILLANO, 1978).

equatorial electric fields obtained for three different field aligned current distributions. The first pattern (a), obtained assuming the presence of only a pair of region 1 field aligned currents of 10^6 A at 72° latitude, has the same polarity, except for the sunset-midnight sector, as the quiet time equatorial electric fields due to the ionospheric dynamo. The calculated quiet time pattern, shown in (b), was obtained by adding region 2 currents of 5×10^5 A at 66° to the current system used in (a). The inclusion of the equatorward field aligned currents reduces significantly the penetration of the region 1 current system into low latitudes. When the equatorward field aligned current intensity was increased to 10^6 A, to represent active conditions, the perturbed equatorial electric fields reversed from their quiet time direction, as shown in (c).

The quiet time low latitude electric fields are mostly due to ionospheric dynamo effects which indicates considerably more effective shielding between the high and low latitude current systems than suggested in (b). The electric field variations shown in Fig. 16 do reproduce, however, several features of the experimental data. The pattern in (b) is in reasonably good agreement with the local time dependence of Jicamarca electric field perturbations associated with southward B_z changes, and with the local time dependence of the DP-2 and SI magnetic field perturbations reported by PAPAMASTORAKIS *et al.* (1984). The pattern for active periods, shown in curve (c), agrees well (except for the dusk-midnight sector) with the disturbance electric fields associated with large northward B_z changes, illustrated in Fig. 8. This was expected since the increase of the region 2 current intensity is equivalent to a sudden decrease in the convection electric field caused by a northward turning of B_z from a steady southward value. NOPPER and CAROVILLANO (1978) suggested that for impulsive fluctuations of region 1 and region 2 currents, the electric field perturbation would alternate between the second and third patterns.

SENIOR and BLANC (1984) have presented a detailed model for magnetospheric convection, with latitudinal and local time dependent ionospheric conductivities, as well as a time dependent ring current. They have also calculated middle latitude and equatorial ionospheric electric fields generated by sudden increases in the dawn to dusk potential drop. Electric fields calculated for mid-latitudes are in good agreement with the experimental curves. Figure 17 shows the initial and steady state equatorial electric field perturbations calculated for an increase of 100 kV in the dawn to dusk potential drop. The initial time pattern with eastward electric fields during the day and westward at night is similar to Nopper and Carovillano's curve (b), except for considerably larger evening drifts. The largest westward electric perturbation fields occur in the postmidnight period, a result obtained also in previous studies (e.g., JAGGI and WOLF, 1973; BLANC, 1983b; and others). These results are in general agreement with the electric field pertubations associated with southward B_z turnings, but with higher amplitudes than suggested by our data. The disturbance electric field pattern in Fig. 8 is in good agreement with the local time dependence of the inital state pattern of Fig. 17, but taken with reversed polarity (to correspond to a sudden decrease of the dawn to dusk potential drop). The electric fields associated with northward B_z changes have larger amplitudes than in the previous case, indicating weaker shielding for dusk to dawn electric field perturbations than in the opposite

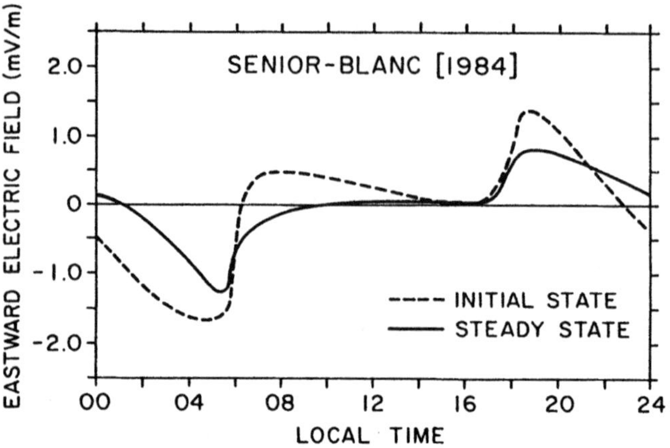

Fig. 17. Initial and steady state ionospheric equatorial electric field patterns calculated for a sudden increase of 100 kV across the polar cap (after SENIOR and BLANC, 1984).

case. The steady state electric fields have very small daytime amplitudes, in good agreement with statistical studies which do not show any significant variation of the average daytime electric fields between quiet and disturbed periods.

8. Conclusions

Radar studies of ionospheric electric fields have provided extensive evidence for the existence of an effective coupling between the high and low latitude ionospheres during magnetically active times. The main characteristics of the equatorial electric field perturbations associated with high latitude processes seem well established. Numerical simulations have determined the effect of a number of important parameters on the shielding between the high and low latitude current systems, and have reproduced some features of the equatorial disturbance electric field data. Further progress will require extensive simultaneous measurements from low to high latitudes and considerably more detailed simulation studies. Hopefully, progress is being made on both fronts. Extensive measurements with incoherent scatter radars, ground based and in-situ magnetometers, and with other probes have recently been started. Highly sophisticated self consistent numerical models are being developed and expanded to include the low latitude ionosphere. These studies should soon provide important new information on the complex magnetosphere-ionosphere coupling processes and on their effects in the low and equatorial ionosphere.

I thank D. Farley and M. Kelley for helpful comments and the staff of the Jicamarca Radio Observatory for the extensive electric field measurements. The Jicamarca Observatory is operated by the Geophysical Institute of Peru, with support from the National Science Foundation. This work was supported by the Aeronomy Program, Division of Atmospheric Sciences, of the National Science Foundation, through grants ATM-8309662 and ATM-8305500.

REFERENCES

ANDERSON, D. N., Modelling the ambient, low latitude *F*-region ionosphere, *J. Atmos. Terr. Phys.*, **43**, 753, 1981.

AKASOFU, S.-I. and R. P. LEPPING, Interplanetary magnetic field and magnetospheric substorms, *Planet. Space Sci.*, **25**, 895, 1977.

BALSLEY, B. B. and R. F. WOODMAN, On the control of the *F*-region drift velocity by the *E*-region electric field: Experimental evidence, *J. Atmos. Terr. Phys.*, **31**, 865, 1969.

BLANC, M., Mid-latitude convection electric fields and their relation to ring current development, *Geophys. Res. Lett.*, **5**, 203, 1978.

BLANC, M., Magnetospheric convection effects at midlatitudes, 1, Saint-Santin observations, *J. Geophys. Res.*, **88**, 211, 1983a.

BLANC, M., Magnetospheric convection effects at midlatitudes, 3, Theoretical derivation of the disturbance convection pattern in the plasmasphere, *J. Geophys. Res.*, **88**, 235, 1983b.

BLANC, M. and A. D. RICHMOND, The ionospheric disturbance dynamo, *J. Geophys. Res.*, **85**, 1669, 1980.

BLANC, M., D. ALCAYDE, and J. D. KELLY, Magnetospheric convection effects at midlatitudes, 2, A coordinated Chatanika/Saint-Santin study of the April 10–14, 1978, magnetic storm, *J. Geophys. Res.*, **88**, 224, 1983.

CARPENTER, L. and V. W. KIRCHOFF, Daytime three-dimensional drifts at Millstone Hill Observatory, *Radio Sci.*, **9**, 217, 1975.

CARTER, D. A., B. B. BALSLEY, and W. L. ECKLUND, VHF Doppler radar observations of the African equatorial electrojet, *J. Geophys. Res.*, **81**, 2786, 1976.

FEJER, B. G., The equatorial ionospheric electric fields. A review, *J. Atmos. Terr. Phys.*, **43**, 377, 1981.

FEJER, B. G. and M. C. KELLEY, Ionospheric irregularities, *Rev. Geophys. Space Physics*, **18**, 401, 1980.

FEJER, B. G., D. T. FARLEY, B. B. BALSLEY, and R. F. WOODMAN, Radar studies of anomalous velocity reversals in the equatorial ionosphere, *J. Geophys. Res.*, **81**, 4621, 1976.

FEJER, B. G., D. T. FARLEY, R. F. WOODMAN, and C. CALDERON, Dependence of equatorial *F*-region vertical drifts on season and solar cycle, *J. Geophys. Res.*, **84**, 5792, 1979a.

FEJER, B. G., C. A. GONZALES, D. T. FARLEY, M. C. KELLEY, and R. F. WOODMAN, Equatorial electric fields during magnetically disturbed periods, 1, The effect of the interplanetary magnetic field, *J. Geophys. Res.*, **84**, 5797, 1979b.

FEJER, B. G., D. T. FARLEY, C. A. GONZALES, R. F. WOODMAN, and C. CALDERON, *F*-region east-west drifts at Jicamarca, *J. Geophys. Res.*, **86**, 215, 1981.

FEJER, B. G., M. F. LARSEN, and D. T. FARLEY, Equatorial disturbance dynamo electric fields, *Geophys. Res. Lett.*, **10**, 537, 1983.

FORBES, J. M., The equatorial electrojet, *Rev. Geophys. Space Phys.*, **19**, 469, 1981.

GALPERIN, Yu. I., V. N. POMOMAREV, and A. G. ZOSIMOVA, Equatorial ionospheric anomal and interplanetary magnetic field, *J. Geophys. Res.*, **83**, 4265, 1978.

GANGULY, S., R. A. BEHNKE, and B. A. EMERY, Average electric field behavior above Arecibo, submitted to *J. Geophys. Res.*, 1984.

GONZALES, C. A., Electric fields in the low latitude ionosphere and their relationship to magnetospheric and interplanetary phenomena, PhD Thesis, Cornell Univ., Ithaca, N.Y., 1979.

GONZALES, C. A., M. C. KELLEY, B. G. FEJER, J. F. VICKREY, and R. F. WOODMAN, Equatorial electric fields during magnetically disturbed periods, 2, Implications of simultaneous auroral and equatorial measurements, *J. Geophys. Res.*, **84**, 5803, 1979.

GONZALES, C. A., M. C. KELLEY, R. A. BEHNKE, J. F. VICKREY, R. WAND, and J. HOLT, On the latitudinal variations of the ionospheric electric field during magnetospheric disturbances, *J. Geophys. Res.*, **88**, 9135, 1983.

GOUIN, P. and P. N. MAYAUD, A propos de l'existence d'un contre-electrojet aux latitudes magnetiques equatoriales, *Ann. Geophys.*, **26**, 41, 1967.

HUTTON, R. and J. OYINLOYE, The counter-electrojet in Nigeria, *Ann. Geophys.*, **26**, 921, 1970.

JAGGI, R. K. and R. A. WOLF, Self-consistent calculation of the motion of a sheet of ions in the magnetosphere, *J. Geophys. Res.*, **78**, 2852, 1973.

KAMIDE, Y. and S. MATSUSHITA, Penetration of high-latitude electric fields into low latitudes, *J. Atmos. Terr. Phys.*, **43**, 411, 1981.

KANE, R. P., Relationship between interplanetary plasma parameters and geomagnetic D_{st}, *J. Geophys. Res.*, **79**, 64, 1974.

KELLEY, M. C., B. G. FEJER, and C. A. GONZALES, Anomalous low latitude electric fields associated with a northward interplanetary magnetic field, *Geophys. Res. Lett.*, **6**, 301, 1979.

KIKUCHI, T., T. ARAKI, H. MAEDA, and K. MAEKAWA, Transmission of ionospheric electric fields to the equator, *Nature*, **273**, 650, 1978.

KIRCHHOFF, V. W. J. H., and L. A. CARPENTER, Calculations of conductivities and currents in the ionosphere, *PSU-IRL-SCI-439*, Penn. State Univ., University Park, 1975.

MAEKAWA, K., Global distribution of electric fields and currents in the ionosphere produced by field aligned currents of magnetospheric origin-I, *J. Atmos. Terr. Phys.*, **42**, 673, 1980.

MAEKAWA, K. and H. MAEDA, Electric fields in the ionosphere produced by polar field aligned currents, *Nature*, **273**, 649, 1978.

MATSUSHITA, S., IMFP effects on the equatorial geomagnetic field and ionosphere- a review, *J. Atmos. Terr. Phys.*, **39**, 1207, 1977.

MATSUSHITA, S., IMF an lower thermospheric currents and motions: A review, *J. Geomag. Geoelectr.*, **31**, 225, 1979.

MATSUSHITA, S. and B. B. BALSLEY, A question of DP-2, *Planet. Space Sci.*, **20**, 1259, 1972.

MATSUSHITA, S and Y. KAMIDE, Electromagnetic interaction between high and low latitudes shown by computer simulation movie, *J. Atmos. Terr. Phys.*, **43**, 403, 1981.

MOZER, F. S., Electric field mapping in the ionosphere at the equatorial plane, *Planet Space Sci.*, **18**, 259, 1970.

MOZER, F. S., Electric fields and plasma convection in the plasmasphere, *Rev. Geophys. Space Phys.*, **11**, 755, 1973.

NISBET, J. S., M. J. MILLER, and L. A. CARPENTER, Currents and electric fields in the ionosphere due to field aligned auroral currents, *J. Geophys. Res.*, **83**, 2647, 1978.

NISHIDA, A., Geomagnetic DP-2 fluctuations and associated magnetospheric phenomena, *J. Geophys. Res.*, **73**, 1975, 1968.

NISHIDA, A., Reply to "A question of DP-2" by S. Matsushita and B. B. Balsley, *Planet. Space Sci.*, **21**, 1255, 1973.

NISHIDA, A., Coherence of geomagnetic DP-2 fluctuations with interplanetary magnetic variations, *J. Geophys. Res.*, **73**, 5549, 1968.

NISHIDA, A. and K. MAEKAWA, Two basic modes of interaction in the solar wind and the magnetosphere, *J. Geophys. Res.*, **76**, 2254, 1971.

NOPPER, R. W. and R. L. CAROVILANO, Polar equatorial coupling during magnetically active periods, *Geophys. Res. Lett.*, **5**, 699, 1978.

ONWUMECHILLI, A. and P. O. OGBUEHI, Fluctuations in the geomagnetic horizontal field, *J. Atmos. Terr. Phys.*, **24**, 173, 1962.

ONWUMECHILLI, A., K. KAWASAKI, and S.-I. AKASOFU, Relationship between the equatorial electrojet and polar magnetic variations, *Planet. Space Sci.*, **21**, 1, 1973.

PAPAMASTORAKIS, I., G. HAERENDEL, and W. BAUMJOHANN, Local time dependence of the response of the equatorial electrojet to DP-2 and SI disturbances, *J. Geophy.*, **54**, 213, 1984.

RASTOGI, R. G. and V. L. PATEL, Effect of interplanetary magnetic field on the ionosphere over the equator, *Proc. Indian Acad. Sci., Sect. A*, **82**, 121, 1975.

RASTOGI, R. G., R. F. WOODMAN, and P. C. HEDGECOCK, Correlated changes in the equatorial electrojet and in the interplanetary magnetic field during a geomagnetic storm, *J. Atmos. Terr. Phys.*, **40**, 189, 1978.

REDDY, C. A., The equatorial electrojet: a review of the ionospheric and geomagnetic effect, *J. Atmos. Terr. Phys.*, **43**, 557, 1981.

RICHMOND, A. D., S. MATSUSHITA, and J. D. TARPLEY, On the production mechanism of electric currents and fields in the ionosphere, *J. Geophys. Res.*, **81**, 547, 1976.

RISHBETH, *F*-region storms and thermospheric circulation, *J. Atmos. Terr. Phys.*, **37**, 1055, 1975.

SASTRI, J. H., IMF polarity effects on the equatorial ionospheric *F*-region, Paper presented at Physics of the Magnetosphere-Ionosphere connection, Graz, 1984.

SENIOR, C. and M. BLANC, On the control of magnetospheric convection by the spatial distribution of ionospheric conductivities, *J. Geophys. Res.*, **89**, 261, 1984.

TESTUD, J., P. AMAYENC, and M. BLANC, Middle and low latitude effects of auroral disturbances from incoherent scatter, *J. Atmos. Terr. Phys.*, **37**, 989, 1975.

VASYLIUNAS, V. M., Mathematical models of magnetospheric convection and its coupling to the ionosphere, in *Particles and Fields in the Magnetosphere*, edited by B. M. McCormck, 60, D. Reidel, Dordrecht, Netherlands, 1970.

VASYLIUNAS, V. M., The interrelationship of magnetospheric processes, in *Earth's Magnetospheric Processes*, edited by B. M. McCormac, 29, D. Reidel, Dordrecht, Holland, 1972.

WOODMAN, R. F., Vertical velocities and east-west electric fields at the magnetic equator, *J. Geophys. Res.*, **75**, 6249, 1970.

WOODMAN, R. F., East–west ionospheric drifts at the magnetic equator, *Space Res.*, **12**, 969, 1972.

Solar Wind-Magnetosphere Coupling, edited by Y. Kamide and J. A. Slavin, 547–562.

The Relationship of Birkeland and Ionospheric Current Systems to the Interplanetary Magnetic Field

L. J. ZANETTI and T. A. POTEMRA[1]

Johns Hopkins University, Applied Physics Laboratory, Laurel, Maryland 20707, U.S.A.

Comparisons are made between Birkeland and ionospheric current systems as a function of the interplanetary magnetic field (IMF). A southward IMF produces the Region 1 and Region 2 Birkeland and electrojet current systems. The effects on these magnetospheric-ionospheric current systems when the IMF has a large northward component (≥5 nT) have been considered in recent studies. Stable, large-scale Birkeland current systmes, referred to as the NBZ (for "northward B_z") currents, have been inferred from the MAGSAT magnetometer data to exist during long periods (~1/2 day) of northward IMF. This system includes the Region 1 and Region 2 system but at a much lower intensity. Adjoining the Region 1 system both at dawn and at dusk, and covering the entire polar cap is another pair of large-scale currents oppositely directed to the Region 1 system. Antisunward Hall current accompanies the NBZ Birkeland currents, and asymmetries in these current systems due to the IMF B_y component are also found. These B_y-dependent patterns are compared to those observed during southward IMF periods. Significant structure is evident in the NBZ system on the tailward side of the dawn–dusk magnetic meridian, whether this region is illuminated or not. It has been suggested that these complicated patterns are related to competing merging processes in the geomagnetic tail during periods of northward IMF.

1. Introduction

The Birkeland and ionospheric current systems at auroral zone and high latitudes are familiar manifestations of solar wind-magnetospheric coupling, depositing a significant amount of the energy transferred from the solar wind to the near earth environment. It has been quite well known that such energy transfer is a function of the orientation of the interplanetary magnetic field (IMF) although traditional concentration has been on the characteristics and details of the coupling to the auroral region during southward IMF. Specifically, the auroral zone has received the major concentration of effort, generally from statistical data studies (e.g., FELDSTEIN, 1963; HOFFMAN and BURCH, 1973; EATHER *et al.*, 1976; WINNINGHAM *et al.*, 1975; IIJIMA and POTEMRA, 1976b, 1978; MCDIARMID *et al.*, 1979). Ionospheric current system analysis has primarily been performed from well-distributed ground magnetometer chains in North America (e.g., AKASOFU *et al.*, 1980), in Scandinavia (e.g., KÜPPERS *et al.*, 1979; MAURER and THEILE, 1978; BAUMJOHANN *et al.*, 1978), and Greenland (WILHELM and FRIIS-CHRISTENSEN,

[1]Now at Office of Science and Technology Policy, Executive Office of the President, Washington, D.C. 20506, U.S.A.

1976). Extensive modeling of the equivalent ionospheric current systems has been developed with the input of reasonable conductivity models (e.g., HUGHES and ROSTOKER, 1977; KISABETH and ROSTOKER, 1971; KISABETH, 1979; AKASOFU *et al.*, 1981; KAMIDE *et al.*, 1981) or by using simultaneous electric field data (e.g., BAUMJOHANN *et al.*, 1980, 1981; BAUMJOHANN and KAMIDE, 1981). Again the primary area investigated has been the general auroral zone except for the IMF B_y influence on the DPY current system (FRIIS-CHRISTENSEN and WILHJELM, 1975; MANSUROV, 1969; SVALGAARD, 1973). Magnetic perturbation studies from satellite data have also generally concentrated on the auroral zone with the exception of some high-latitude studies (MCDIARMID *et al.*, 1977, 1978, 1980).

The purpose of this paper is to review characteristics of large-scale Birkeland and ionospheric current systems and their association with the interplanetary magnetic field. These studies have primarily used low altitude satellite data to assess signatures and the distribution of Birkeland currents; see POTEMRA (1979, 1983) for a comprehensive review of other characteristics of Birkeland currents.

The figures that follow summarize our understanding of the low altitude Birkeland and ionospheric current systems and how they are influenced by the IMF. The concentration will be on large scale and stable features, defined as those structures with scale lengths greater than 0.5° in latitude and stability on the order of hours, dependent on the stability of the IMF of course. An interval of 0.5° in latitude maps to about half an earth radius in the dawn-dusk equatorial region. The entire 5°–7° latitudinal cross section would thus map to 10–15 R_e in this area of space. These current systems have a close association to the large-scale convection and electric field patterns in the high-latitude region. The patterns shown in the following figures were determined primarily from magnetometer data acquired by the TRIAD and MAGSAT satellite experiments.

2. Southward IMF Current Systems

Figure 1(a) shows the familiar patterns of Birkeland currents determined by IIJIMA and POTEMRA (1978, their Fig. 13) which categorize the transverse magnetic field disturbances in the TRIAD magnetometer data (ZMUDA and ARMSTRONG, 1974). These regions were arbitrarily labelled Region 1, for the poleward currents that flow toward the ionosphere in the dawn sector and away from the ionosphere in the dusk sector, and Region 2 Birkeland currents which encircle Region 1, being more distant from the poles, and flow oppositely to the Region 1 currents at any given local time. Given the similar current densities (few μA/m) of these regions from dusk through midnight to dawn, they are most likely connected in the ionosphere by Pedersen currents. The dayside current densities are not as balanced (IIJIMA and POTEMRA, 1978; SUGIURA and POTEMRA, 1976), the Region 1 current densities being a factor of two to four higher than the Region 2 densities depending on local time and magnetic activity. Some of the excess Birkeland current on the dayside must be connected in a different manner, e.g., across the dayside high latitude region (WILHELM and FRIIS-CHRISTENSEN, 1976; AKASOFU, 1982). Some have suggested a direct coupling of these Birkeland currents into the electrojet currents (HUGHES and

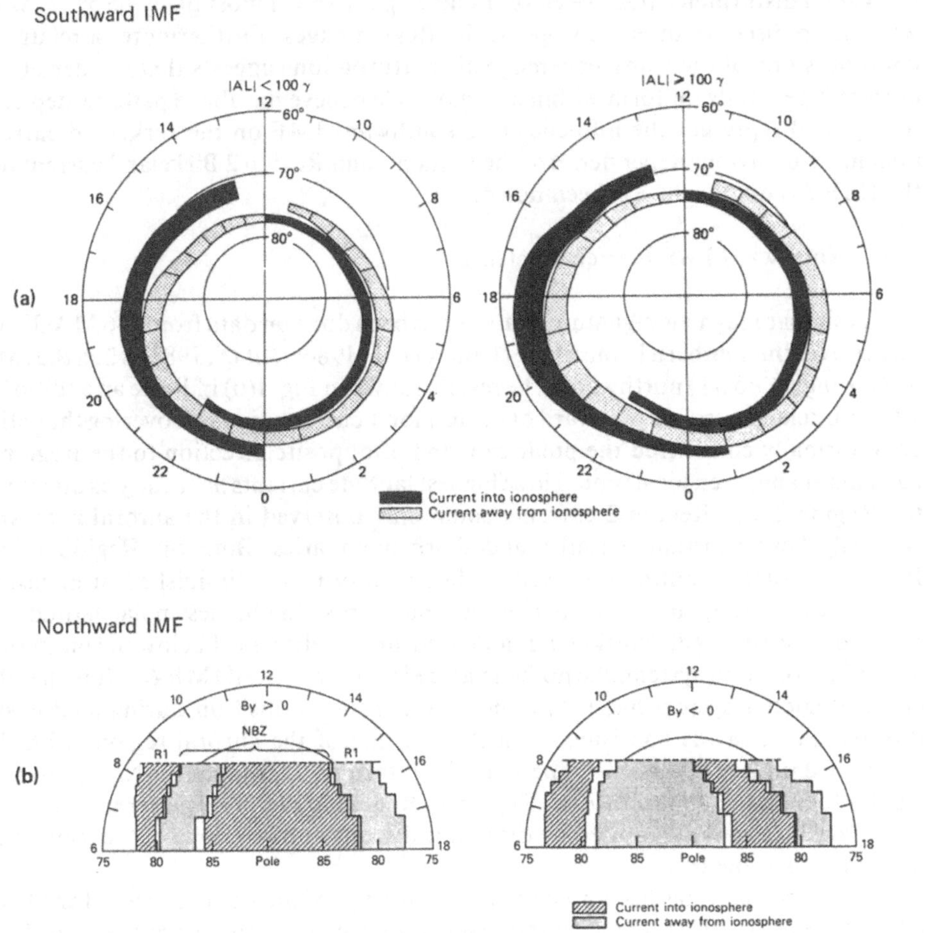

Fig. 1. (a) A summary of the distributions and flow directions of large-scale field-aligned currents determined from data obtained from the 439 Triad passes during weakly disturbed conditions ($|AL| \leqslant 100\gamma$, left panel) and from data obtained from 366 Triad passes during active periods ($|AL| \geqslant 100\gamma$, right panel). (b) The spatial distribution of the flow directions of the large-scale NBZ Birkeland currents in the south polar region during $B_y > 0$ (the left panel) and during $B_y < 0$ (the right panel). This statistical plot was determined from an analysis of 146 MAGSAT orbits.

ROSTOKER, 1977; 1979).

The second panel of Fig. 1(a) shows an expansion of the individual Regions and their circumferences as magnetic activity increases. These results are consistent with the expansion of the auroral oval with an increase of magnetic activity. The Hall electrojet currents are generally located at these same latitudes. We would like to make a note on the method of this statistical accumulation of data; the boundaries shown are average boundary locations from individual signatures. Thus any individual disturbance trace (Region 1 and Region 2) will most likely be 5°–7° wide with sharp discontinuities in slope at the Region edges. Furthermore, a relatively constant slope of the transverse magnetic perturbation suggests that the density of current is generally uniform within a Region. We believe the above patterns depicted in Fig. 1(a) represent the influence of a southward IMF on the Birkeland current systems. No obvious dependence of the Region 1 and Region 2 Birkeland currents on the IMF B_y component has been noted.

3. Northward IMF Current Systems

A similar magnetic signature analysis has been done on data from 146 MAGSAT passes over the southern hemisphere (IIJIMA *et al.*, 1984; IIJIMA, 1983) when the IMF was strongly (⩾5 nT) northward. The results shown in Fig. 1(b) indicate an additional set of Birkeland currents poleward of the Region 1 currents, in fact covering the entire area normally considered the polar cap, and of opposite direction to the Region 1 currents at any given local time. These highest latitude currents are nearly as intense as the Region 1 and Region 2 currents commonly observed in the auroral zone and generally have constant densities and sharp boundaries. Both the Region 1 and Region 2 systems continue to exist (although very much diminished in intensity, particularly in Region 2). The pair of currents across the highest polar latitudes is referred to as the "NBZ" Birkeland (and ionospheric, discussed below in connection with Fig. 2) current system, the nomenclature is for northward IMF B_z. Note that the entire Region 1 system has moved poleward from its position during southward directed IMF; this is consistent with the quieting of the auroral regions with the northward turning of the IMF. The analysis performed to produce the NBZ pattern is similar to that used to produce the Region 1– Region 2 (Fig. 1(a)) pattern; an average boundary location is identified by the major discontinuity in slope of the transverse magnetic disturbance.

An asymmetry has been noted that is a function of the *y* component of the IMF (POTEMRA *et al.*, 1984). These patterns were produced for southern hemisphere data; one may switch the dawn and dusk labels in order to view conditions appropriate for the northern hemisphere, since the asymmetry is the opposite for opposite hemispheres. The NBZ and Region 1 Birkeland currents are so labelled. The more intense and spatially smaller Birkeland currents are due to preferred anti-parallel merging on the tail lobe magnetopause boundary. Thus, for example, positive B_y for the IMF will be antiparallel to the earth's field on the dusk side of the magnetotail, related to the higher density NBZ Birkeland currents. This asymmetry will be discussed more fully in the description of Fig. 3 (see below).

Figure 2 (ZANETTI *et al.*, 1983b; IIJIMA *et al.*, 1984) illustrates the difference in the transverse magnetic disturbance field observed during periods of southward (2a) and northward IMF (2b) by the magnetometer on the MAGSAT satellite as it traverses the auroral oval approximately along the dawn–dusk meridian. The orbits are summarized to the right of each panel and are plotted against magnetic local time and latitude. The statistical locations of Birkeland currents (Region 1/Region 2, IIJIMA and POTEMRA, 1976) are included on the right side of this figure for general

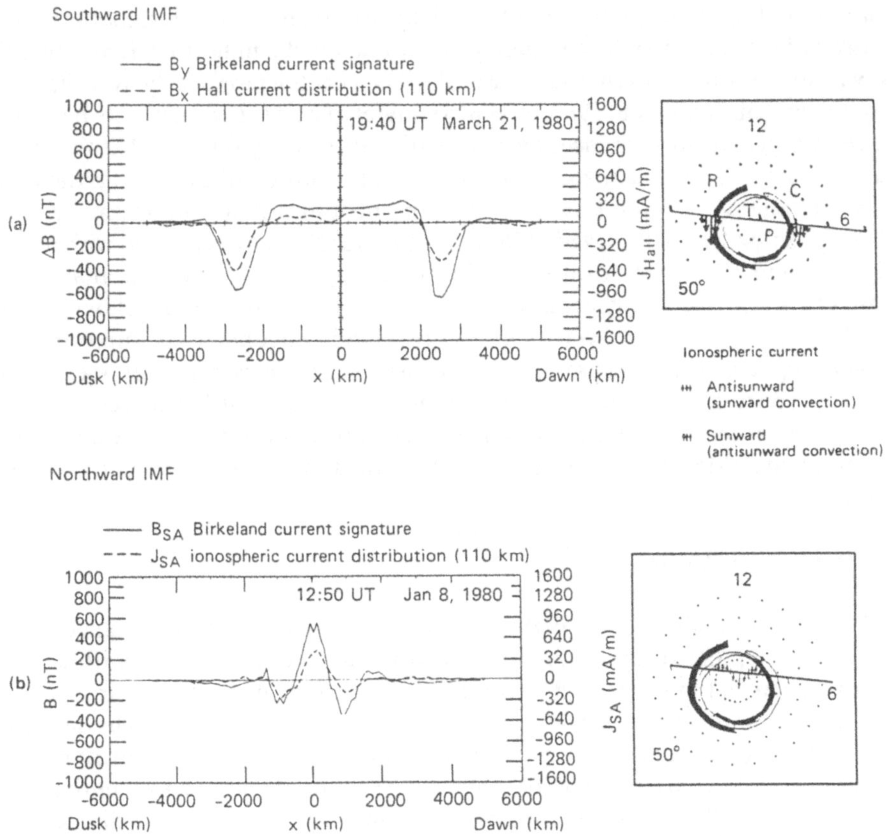

Fig. 2. (a) The one-dimensional current analysis procedure on March 21, 1980, at 19:40 UT. The right panel is the location of the orbit relative to geomagnetic coordinates and the statistical pattern of Birkeland currents (IIJIMA and POTEMRA, 1978). The traces are of magnetic field disturbances versus Cartesian coordinate x centered on the magnetic pole. Reproduced is the Birkeland current signature as a solid line and the dotted line represents the ΔB in the along track direction at ~110 km due to a Hall current distribution. (b) Ionospheric current analysis example for January 8, 1980, at 12:52 UT. The ionospheric current distribution (dotted line, J_s) plotted against distance (km) from the pole. The right hand scale (mA/m) applies to the dotted trace with positive values indicating antisunward-directed horizontal current. The solid trace is the B_{SA} disturbance component in nT plotted vs. distance (km). On the right is a magnetic latitude vs. local time plot of the satellite dawn-to-dusk track with arrows schematically indicating the horizontal current directions.

auroral zone reference. The solid trace in Fig. 2(a) is the transverse magnetic disturbance signature ΔB (nT) in approximately a geomagnetic east-west direction. The triangular signatures are the Region 1 (poleward) and Region 2 (equatorward) Birkeland current signatures in the nominal auroral zone at dawn and dusk. Figure 2(a) is an example of MAGSAT data on March 21, 1980 at approximately 19:40 UT, the abscissa having been mapped to a Cartesian coordinate system centered (0 km) on the noon–midnight magnetic meridian. This Birkeland current signature for a southward IMF case may be compared to data taken at similar locations when the IMF was strongly northward. Figure 2(b) shows an orbital pass nearly over the south magnetic pole on January 8, 1980, at approximately 12:50 UT from which data show transverse magnetic disturbances over the entire auroral region as well as the region usually considered the polar cap. The disturbance signature (solid trace) has the shape of a "W'" (Region 2/Region 1/NBZ) in the sun-antisunward (hereupon referred to as the SA) direction (see IIJIMA *et al.*, 1984). Such a coordinate system fixed with respect to the sun and the magnetosphere is the proper system for such an analysis, as opposed to an earth-fixed system, which was appropriate for the case shown in Fig. 2(a). The inward and outward NBZ Birkeland currents that surround the noon–midnight meridian cause as large a disturbance as is generally seen in the auroral oval during southward IMF. Surrounding these highest-latitude currents, the Region 1 and Region 2 currents occupy similar latitudinal extents as are occupied with southward IMF although intensities are quite diminished, particularly those of Region 2.

4. Ionospheric Currents

A detailed analysis was performed on the component of magnetic field disturbances parallel to the main field in order to determine the distribution of ionospheric current intensity and location. Magnetic field disturbances arising from distant current sources are curl-free, allowing the disturbance field $\boldsymbol{B}$ to be described by the gradient of a magnetic potential (Φ_m). Since the divergence of $\boldsymbol{B}$ must be zero, we must solve $\nabla^2\Phi_m=0$ in order to describe the magnetic disturbance potential for all current-free space. The data are conformally mapped to a Cartesian coordinate frame in which the solutions to Laplace's equation are oscillatory in two directions and exponential in the third (altitude in this case). Using a single orbit of data, we perform a two-dimensional analysis (height and distance approximately along the track being two of the standard Cartesian coordinates) while assuming all is constant in the third dimension (distance approximately across the track). A Fourier transform of the component of $\boldsymbol{B}$ parallel to the main field ($B_\parallel$) will automatically give the Φ_m oscillatory coefficients for various wavelengths along the horizontal analysis direction. The exponential dependence on altitude is imposed; it is a function of wavelength, and the disturbances in the "along track" field component are calculated at 110 km. These calculations are the dotted traces in Fig. 2, assumed to be directly outside the ionospheric current sheet source and therefore convertible to current intensity by expressing $J=(2/\mu_0)B$ in units of milliamperes per meter; see MERSMANN *et al.*, (1979), KÜPPERS *et al.* (1979), BAUMJOHANN *et al.* (1980), and/or ZANETTI *et al.* (1983a) for details.

Figure 2(a) contains the results of the Hall current analysis as described above for the MAGSAT orbit starting at 19:40 UT. The solid lines represent the disturbances resulting from Birkeland currents and are plotted at the same resolution as the Hall current analysis (250 km). The dotted line is a plot of the magnetic disturbance at 110 km calculated from the Hall current analysis by assuming that this location is just outside the source current distribution. This quantity is thus a measure of Hall current density $J_{Hall}=(2/\mu_0)B_x$. Note the general coincidence of the Birkeland and Hall electrojet current distributions, particularly the coincidence of their edges. Equal amplitudes of ΔB would indicate (assuming that Birkeland currents are closed meridionally by Pedersen currents) a Hall to Pedersen conductivity ratio of 2.0. This ratio is about 1.0 at dawn and 1.5 at dusk in this example. The sign of ΔB in Fig. 2(a) indicates the direction of the Hall current; the dotted trace at dusk indicates the eastward electrojet current distribution whereas the dotted trace at dawn indicates the westward electrojet. From −2000 to +2000 km, there appear to be some small Hall currents across the polar cap, and one may assume that Pedersen currents connect the net poleward Region 1 currents. The integrated Hall current is 0.86×10^6 amps for the combined electrojets and 0.33×10^6 amps for the sunward Hall current in the polar cap. The electrojet Hall distributions from the Fourier technique are shown schematically to the right of Fig. 2(a) for comparison with Fig. 2(b). The left and right scales of Fig. 2(b) both describe the dotted trace that indicates antisunward horizontal current in the center surrounded by sunward current on either side. The abscissa is measured in kilometers from the center of the Cartesian coordinate frame, the center being the south magnetic pole. The solid line is the transverse magnetic disturbance vector in the SA direction (B_{SA}) showing signatures of Birkeland currents assumed to be in sheets aligned in that same direction. This trace is for comparison and is essentially a low time resolution copy of the B_{SA} component data. The right side of Fig. 2(b) schematically places the inferred ionospheric current system within magnetic coordinates and relative to the statistical pattern of Birkeland currents in the auroral oval. An antisunward current exists directly over the magnetic pole and is skirted by two sunward currents of lesser intensity. Interpreting these currents in terms of convection directions presupposes the currents to be ionospheric Hall currents.

These current patterns during northward IMF are consistent with the "four-celled" convection flow proposed previously (MAEZAWA, 1976; BURKE *et al.*, 1979, see ZANETTI *et al.*, 1984 for further refrences). The "viscous" convection cells at the auroral zone locations still exist but are diminished in magnitude and have been neglected in the NBZ schematic for clarity. As in the southward IMF case, there is a remarkable similarity between the NBZ Birkeland current signature and the distribution of NBZ ionospheric Hall current intensity. This fact allows the same conclusion for the Hall to Pedersen conductivity ratio as was made for the Region 1/Region 2/electrojet case, i.e., that the ratio must be a constant with Pedersen currents joining the Birkeland currents and Hall currents present at a constant ratio. Note that these data were taken over a sunlit ionosphere, which is to say that photoionization should provide fairly even conductivity. Note also that the northward IMF case existed during an unusual time when the IMF B_y component was equal to zero.

5. B_y-Dependent Patterns

Figure 3(a) is a schematic diagram of convective flow patterns in the north polar regions during periods of positive and negative IMF B_y, which incorporate the

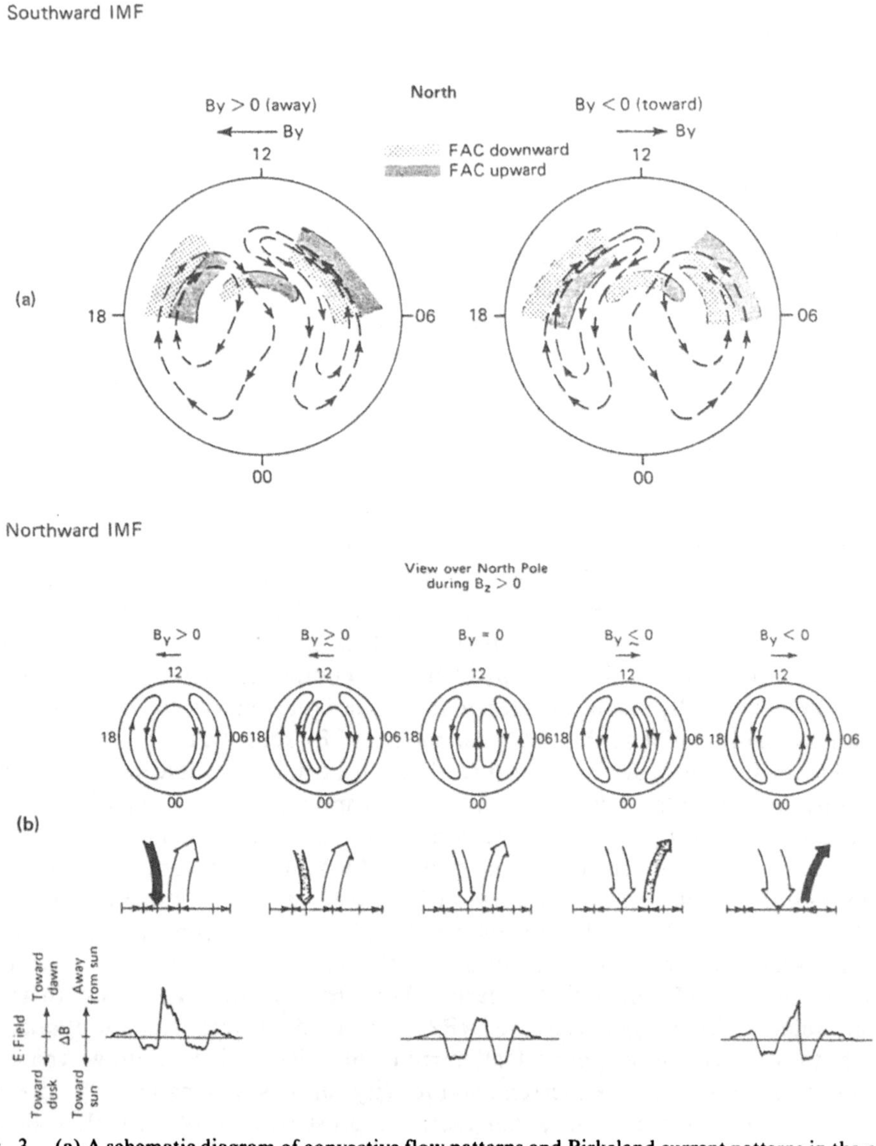

Fig. 3. (a) A schematic diagram of convective flow patterns and Birkeland current patterns in the north and south polar regions during conditions of positive and negative B_y. The most important feature is the shift of the "throat" to morning or afternoon depending upon the polarity of B_y. (b) A schematic diagram of convection flow, electric fields, and Birkeland currents in the north polar regions during positive B_z with B_y ranging from strongly positive (on the left) to zero (the center panel) to strongly negative (in the right panel). The traces of transverse magnetic disturbance ΔB or electric field observed by dusk–dawn orbiting satellite over the north polar regions are shown schematically along the bottom. The Region 1 and 2 Birkeland current systems have been deleted to avoid confusion. Notice that the flow directions of the NBZ Birkeland currents associated with the dusk-to-dawn electric field in the polar cap remains the same for all conditions of B_y. The density of the currents (denoted by the shading, with "black" indicating the most dense and no shading indicating the least dense currents) does depend upon B_y as shown.

characteristics of Birkeland currents and earlier suggestions (e.g., BERTHELIER *et al.*, 1974; FAIRFIELD, 1977) about the effects of B_y on the location of the region of maximum convection and the "throat". The convective flow streamlines are shown as the dashed lines in each panel of Fig. 3(a) and comprise the basic "two-cell" pattern. The patterns in Fig. 3(a) include the important relationship to the IMF azimuthal direction; namely, that during periods of positive B_y (an away IMF), the flow reversal region (i.e., the "throat") is located in the afternoon sector of the north polar cap. The flow is directed to the opposite side of the polar region so as to produce regions of maximum convection (and thereby regions of maximum convective electric fields) on the dawn side of the north polar cap. These patterns systematically reverse for negative B_y.

The dayside location and flow direction of the Regions 1 and 2 Birkeland current system have been drawn in each panel of Fig. 3(a). These regions of Birkeland current are confined within regions of sunward convective flow and consequently on closed geomagnetic field lines. The spatial patterns and flow direction of the Regions 1 and 2 Birkeland currents do not depend upon B_y, but are closely correlated with *Kp* and substorm activity and are believed to be associated with processes confined inside the magnetosphere. The associated convective electric fields in these regions are directed in the proper direction to drive Pedersen currents which could connect the Birkeland currents in the ionosphere. In the dayside region where the Region 1 currents are often larger than the Region 2 currents, they cannot, however, be coupled by ionospheric Pedersen currents alone.

The location of the Birkeland current system of the cusp (as defined by IIJIMA and POTEMRA, 1976; IIJIMA and POTEMRA, 1978) is shown in Fig. 3(a) as the crescent-shaped region between 1000 MLT and 1400 MLT, poleward of the Region 1 system. The currents of this system are believed to flow on the outermost boundary of the magnetosphere, where interactions with the solar wind and interplanetary magnetic field may occur. The flow of these currents is generally upward before noon and downward in the afternoon, but one direction dominates over the other (depending on the polarity of the IMF B_y). For each sign of the IMF B_y, one of the auroral zone convection cells dominates the polar cap region, producing higher shear (for example) at dawn during positive IMF B_y. For both southward and northward IMF the asymmetry due to a given sign of the IMF B_y is opposite in opposite hemispheres. The orientation of the "throat" and the resultant convection flow is consistent with the DPY currents (FRIIS-CHRISTENSEN and WILHJELM, 1975). The field aligned cusp currents have the same flow directions as the NBZ Birkeland currents, but important differences include the following. The alignment of the magnetic disturbance associated with the cusp Birkeland currents is east–west; it is sunward for the NBZ system. The NBZ current system extends across the polar region up to the pole, whereas the cusp Birkeland system is confined to a small crescent-shaped region below about 81° (invariant) latitude near noon. The NBZ currents increase in intensity as the IMF becomes stronger in the northward direction, whereas the cusp currents increase in magnitude as the IMF increases in the southward direction.

The basic feature of the patterns shown in Fig. 3(b) is the existence of two convection cells in the polar region during periods of northward IMF (in addition to the viscous cells at auroral latitudes, which are also shown). The flow directions of these cells are clockwise on the dawn side and counterclockwise on the dusk side, in the northern hemisphere. During periods of weak B_y, the two cells are symmetrically positioned with respect to the noon-midnight meridian, thereby producing sunward plasma flow over each pole. The electric fields associated with these cells are directed from dusk to dawn from the center of one cell to the center of the other (as shown in the center panel of Fig. 3(b). The Birkeland current flow consistent with the dusk-to-dawn electric field is downward on the dusk side and upward on the dawn side (identical to that of the NBZ system). The current densities of the upward and downward NBZ currents are equal, resulting in symmetric W patterns for the magnetic field disturbances and electric fields (as shown below the center panel).

As B_y takes on stronger positive values, the clockwise dawn cell in the northern polar cap expands and displaces the counterclockwise dusk cell toward the afternoon side of the polar cap. As the dusk cell collapses, the flow reversal at its center intensifies, resulting in stronger electric fields and higher density Birkeland currents on the afternoon side. Recall that the high shear for southward IMF and positive B_y was at dawn. The spatial region of Birkeland current associated with this cell becomes smaller as well (darker Birkeland current patterns in the center panel). The spatial region of the Birkeland current associated with the morning cell expands and the density of this current diminishes. The total (integrated) downward and total (integrated) upward NBZ currents remain equal. The afternoon cell may collapse to such a degree that it cannot be observed by satellite or ground observations, and it may be assumed that the single cell model prevails.

The pattern of transverse magnetic field disturbance and electric field for a dawn-dusk satellite pass is shown below the far left panel in Fig. 3(b) for positive B_y. The largest gradient in this pattern is shifted toward the dusk side of the northern hemisphere. As B_y diminishes in magnitude from positive to negative values (as B_z remains strongly positive), the dusk cell "inflates" and displaces the dawn cell to the morning side as shown in the sequence of panels in Fig. 3(b), proceeding from left to right. For larger negative values of B_y the dawn cell may collapse to a degree such that a single counterclockwise cell dominates in the polar region.

It would most likely be quite rare for the IMF to have a zero B_y component and a strong northward component as well. Moreover, it is very difficult to distinguish a collapse cell from shear in the auroral zone. Only recently (FRIIS-CHRISTENSEN, 1983, 1986) have ground magnetometer data inversion techniques been able to decipher this double cell within the polar cap. The observations are of counter rotating cells adding to sunward convection over the magnetic pole. It is not surprising, therefore, that the majority of the observations and models have concentrated on "single" cells rotating in opposite directions depending on the sign of the IMF B_y. REIFF and BURCH (1985) have extended Dynamics Explorer (DE) 1 and 2 observations of southward IMF patterns (consistent with Fig. 3(a) if an additional viscous cell is included within the spatially dominant cell) to all orientations of the IMF. This model imposes cells over the polar cap that rotate in the same direction assuming there is a significant IMF B_y component. These patterns could be consistent with Fig. 3(b) only with the addition of "eddies" upon the basic patterns.

There is evidence of irregular patterns superimposed on the NBZ system past the dawn–dusk magnetic meridian (nightward); see BYTHROW *et al.* (1985). These additional shears may well be associated with the polar aurora. The collapsed cell most likely reaches from the dayside auroral region to the nightside auroral region; this also would be consistent with DE observations (HEELIS, 1984). The shift of the spatially dominant cell from dusk to dawn with increasing IMF B_y (northern hemisphere) is consistent with the motion of the theta aurora (FRANK, 1985), which is proposed to lie in this vicinity and be due to plasma sheet populations filling into a bifurcated tail. Preferred antiparallel merging is expected on the tail magnetopause; this would result in the high convection shear and intense Birkeland currents (POTEMRA *et al.*, 1984).

6. Conclusions

The statistical dependence of the current density (and associated magnetic field disturbance amplitude) on the orientation and amplitude of the IMF component normal to the earth-sun line, i.e., on $(B_y^2+B_z^2)^{1/2}$, has been studied with the use of "transmission functions". Figure 4(a) is from Fig. 2 of IIJIMA and POTEMRA (1982) and Figure 4(b) is from Fig. 10 of IIJIMA *et al.* (1984). A good relationship exists between the strength and orientation of the interplanetary magnetic field and the strength of the Region 1/Region 2 system for southward IMF and the NBZ system for northward IMF. The "transmission functions" reflect the antiparallel alignment of the IMF and the earth's magnetic field. Specifically the transmission function $B_T=(B_y^2+B_z^2)^{1/2}\sin(\theta/2)$ illustrated in Fig. 4(a) attains its maximum, for a given transverse (perpendicular to the earth-sun line) IMF magnitude, when the IMF is directed southward (i.e., when $\theta=180°$, where θ is the polar angle measured from the $+z$ axis). This implies that the solar wind electric field may be transferred into the magnetosphere at the dayside magnetopause and could drive the Region 1 Birkeland current system. In contrast, the "complementary" transmission function $\varepsilon^*=(B_y^2+B_z^2)^{1/2}\cos(\theta/2)$ attains its maximum when the IMF is directed exactly northward ($\theta=0°$) for a given transverse IMF magnitude. This fact, along with the IMF B_y asymmetry obervations discussed previously, support antiparallel merging models (DUNGEY, 1963; RUSSELL, 1972; CROOKER, 1979; and others) and suggest that the NBZ Birkeland current system is driven by a polarization electric field due to the solar wind velocity imposed upon the magnetotail during northward IMF. In both cases there is residual Birkeland current for vanishing values of the transmission functions, indicating the presence of other sources for producing Birkeland currents as well.

Recent simulation efforts (e.g., OGINO *et al.*, 1984) have produced the NBZ current system from a three dimensional magnetohydrodynamic model using a northward IMF. An idealized analytic model (CORNWALL, 1985) produces both the NBZ current system and the convection patterns. Also confirmed are the irregular patterns in the nightside hemisphere as the plasma sheet protrudes "filaments" into the high latitudes of the magnetotail. The well-developed "cells" of NBZ currents tend

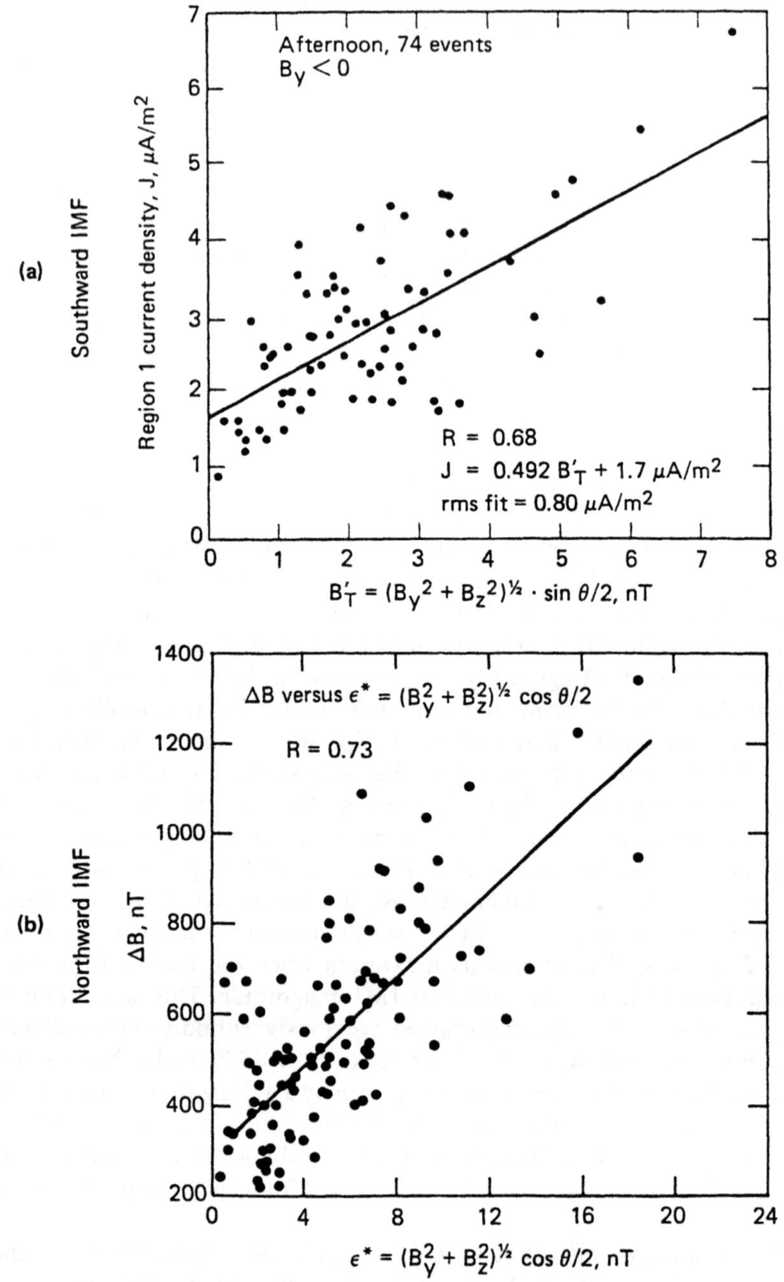

Fig. 4. (a) Region 1 current density versus the intensity and orientation of the IMF in the y–z plane, i.e., versus $(B_y^2+B_z^2)^{1/2}\sin(\theta/2)$, for southward IMF. (b) The magnitude of peak triangular disturbance (B_{SA}) plotted versus the "complementary transmission function" $\varepsilon^*=(B_y^2+B_z^2)^{1/2}\cos(\theta/2)$ for all 146 events, northward IMF.

to appear only on the day side of the dawn–dusk geomagnetic meridian (BYTHROW *et al.*, 1985), a result shared by the work of FRIIS-CHRISTENSEN (1985) using the Greenland magnetometer chain. Bythrow *et al.* have acknowledged that the "W" pattern is difficult to recognize on the nightside; however, filtering of the fine structure still reveals an average (but lower intensity) "W" pattern in the magnetic disturbances.

An important conclusion of these studies is that the magnetosphere is not "closed" to the solar wind with regard to the Birkeland and ionospheric currents when the IMF turns northward. Figure 5 summarizes these results showing in schematic form the integrated Birkeland currents as a function of the B_z component of the IMF. The effects of the B_x and B_y components of the IMF have been ignored in this diagram. The actual relationship between current amplitudes and the IMF is more complicated, as has been discussed with regard to Figs. 4(a) and 4(b) earlier. the total Region 1 input current intensity decreases from 5×10^6 amps as the IMF increases from -10 nT (i.e., from a very strong southward value) to a base level of approximately 1×10^4 amps. The maximum value was estimated by integrating the input current density of 7 μA/m^2 (from Fig. 4(a) for $B_z=-10$ nT, $B_y=0$, and $\theta=180°$) over one-half the area of the Region 1 system in one hemishere (taken as $\sim1.5\times10^6$ km^2). The minimum value was estimated from values of Region 1 current density given by BYTHROW and POTEMRA (1983). As the B_z component of the IMF increases from 0 to $+10$ nT, the integrated NBZ Birkeland current reaches a value of about 5×10^5 amps (approximately 10% of the total current in the Region 1 system for $B_z=-10$ nT). The value of 5×10^5 amps was estimated by integrating the input NBZ current intensity of approximately 0.6 A/m (corresponding to the magnetic disturbance of 800 nT for $B_z=10$ nT, $B_y=0$, and $\theta=0°$ from Fig. 4(b), over one-half of one polar region (taken to be the area above $\sim80°$ invariant latitude.

7. Summary

(1) The Region 1 and Region 2 Birkeland current systems predominate when the IMF has a southward component. These systems expand in latitude with increased geomagnetic activity, and the current density of the Region 1 system is directly

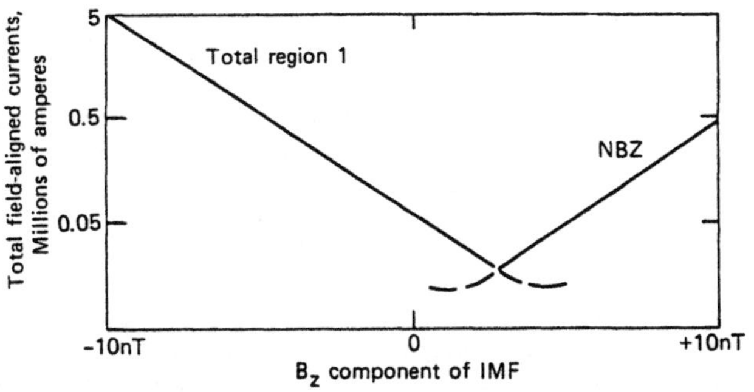

Fig. 5. Summary schematic of the functional relationship between the integrated current intensity of the Region 1 and NBZ Birkeland current systems to the B_z component of the IMF.

correlated with the negative IMF B_z component (i.e., the Region 1 current densities on the day side increase as B_z becomes more negative).

(2) The NBZ Birkeland current system appears over the magnetic poles and dominates the Region 1/Region 2 system when the IMF has a northward configuration. The Birkeland current amplitudes are approximately linearly proportional to the complementary transmission function (ε^*) which is a function of the positive IMF B_z component.

(3) In both the southward and northward IMF cases, the Birkeland currents are collocated with ionospheric Hall currents: the electrojet currents in the case of Region 1 and Region 2 currents, and the NBZ polar cap Hall current in the case of the NBZ Birkeland currents. A good fraction of the electrojet current may be returned to the day side by way of the polar cap. The electrojet and NBZ Hall currents have a distribution peaked at the center, the boundary between an upward and a downward directed Birkeland current. If these Birkeland currents are connected in the ionosphere by Pedersen currents, the Pedersen currents must also peak at this boundary and the Hall to Pedersen conductivity ratio will be a constant across the region.

(4) Asymmetry in the current systems results from a significant B_y component of the IMF. The asymmetries are consistent with antiparallel merging between the IMF and the earth's magnetic field.

The data used here have resulted from the efforts of the TRIAD and MAGSAT project teams at the Johns Hopkins University Applied Physics Laboratory, the Geophysical Research Laboratory of the University of Tokyo, and the NASA Goddard Space Flight Center. This work was supported by the National Science Foundation and the Office of Naval Research.

REFERENCES

AKASOFU, S.-I., Hall current as a source of the cross-tail current interruption, in *The Asymmetric Main Phase Field and the Poleward Expanding Auroral Bulge*, Preprint, 1982.

AKASOFU, S.-I., J. L. KISABETH, G. J. ROMICK, H. W. KROEHL, and B.-H. AHN, Day-to-day and average magnetic variations along the IMF Alaska meridian chain of observatories and modeling of a three-dimensional current system, *J. Geophys. Res.*, **83**, 2065, 1980.

AKASOFU, S.-I., Y. KAMIDE, and J. KISABETH, Comparison of two modeling methods for three-dimensional current systems, *J. Geophys. Res.*, **86**, 3389, 1981.

BAUMJOHANN, W. and Y. KAMIDE, Joint two-dimensional observation of ground magnetic and ionospheric electric fields associated with auroral zone currents. 2. Three-dimensional current flow in the morning sector during substorm recovery, *J. Geomag. Geoelectr.*, **33**, 297, 1981.

BAUMJOHANN, W., R. A. GREENWALD, and F. KÜPPERS, Joint magnetometer array and radar backscatter observations of auroral currents in northern Scandinavia, *J. Geophys.*, **44**, 373, 1978.

BAUMJOHANN, W., J. UNTIEDT, and R. A. GREENWALD, Joint two-dimensional observations of ground magnetic and ionospheric electric fields associated with auroral zone currents. 1. Three-dimensional current flows associated with a substorm-intensified eastward electrojet, *J. Geophys. Res.*, **85**, 1963–1978, 1980.

BAUMJOHANN, W., R. J. PELLINEN, H. J. OPGENOORTH, and E. NIELSON, Joint two-dimensional observations of ground magnetic and ionospheric electric fields associated with auroral zone currents. 4. Current systems associated with local auroral breakups, *Planet Space Sci.*, **29**, 431–447, 1981.

BERTHELIER, A., J. J. BERTHELIER, and C. GUERIN, The effect of the east–west component of the interplanetary magnetic field on magnetospheric convection as deduced from magnetic perturbations at high latitudes, *J. Geophys. Res.*, **79**, 3187, 1974.

BURKE, W. J., M. C. KELLEY, R. C. SAGALYN, M. SMIDDY, and S. T. LAI, Polar cap electric field structures with a northward interplanetary magnetic field, *Geophys. Res. Lett.*, **6**, 21, 1979.

BYTHROW, P. F. and T. A. POTEMRA, The relationship of total Birkeland currents to the merging electric field, *Geophys. Res. Lett.*, **10**, 573, 1983.

BYTHROW, P. F., W. J. BURKE, T. A. POTEMRA, L. J. ZANETTI, and A. T. Y. LUI, Ionospheric evidence for turbulent reconnection and plasma flows in the magnetic tail during periods of northward IMF, *J. Geophys. Res.*, **90**, 5319, 1985.

CORNWALL, J. M., Idealized model of polar cap currents, fields, and aurora, *J. Geophys. Res.*, **90**, 3541, 1985.

CROOKER, N. U., Dayside merging and cusp geometry, *J. Geophys. Res.*, **84**, 951, 1979.

DUNGEY, J. W., The structure of the ionosphere, or adventures in velocity space, in *Geophysics: The Earth's Environment*, edited by C. DeWitt, J. Hiebolt and A. Lebeau, pp. 526–537, Gordon and Breach, New York, 1963.

EATHER, R. H., S. B. MENDE, and R. J. R. JUDGE, Plasma injection at synchronous orbit and spatial temporal auroral morphology, *J. Geophys. Res.*, **81**, 2805, 1976.

FAIRFIELD, D. H., Electric and magnetic fields in the high-latitude magnetosphere, *Revs. Geophys. Space Phys.*, **15**, 285, 1977.

FELDSTEIN, Y. I., Some problems concerning the morphology of auroras and magnetic disturbances at high latitudes, *Geomag. Aeron.*, **3**, 183–192, 1963.

FRANK, L. A., Solar wind-magnetospheric coupling as inferred from global auroral imaging, AGU Chapman Conference on Solar Wind-Magnetosphere Coupling, Pasadena, California, 1985.

FRIIS-CHRISTENSEN, E., Polar cap current systems, in *Magnetospheric Currents*, edited by T. A. Potemra, pp. 86–95, Geophysical Monograph 28, 1983.

FRIIS-CHRISTENSEN, E., Solar wind control of the polar cusp, this volume, 1986.

FRIIS-CHRISTENSEN, E. and J. WILHJELM, Polar cap currents for different directions of the interplanetary magnetic field in the *Y–Z* plane, *J. Geophys. Res.*, **80**, 1248, 1975.

HEELIS, R. A., Ionospheric convection during times of northward interplanetary magnetic field, *EOS*, **65**, 1065, 1984.

HOFFMAN, R. A. and J. L. BURCH, Electron precipitation patterns and substorm morphology, *J. Geophys. Res.*, **78**, 2867, 1973.

HUGHES, T. J. and G. ROSTOKER, Current flow in the magnetosphere and ionosphere during periods of moderate activity, *J. Geophys. Res.*, **82**, 2271–2282, 1977.

HUGHES, T. J. and G. ROSTOKER, A comprehensive model current system for high-latitude magnetic activity. 1. The steady state system, *Geophys. R. J. Astron. Soc.*, **58**, 525– 569, 1979.

IIJIMA, T. and T. A. POTEMRA, Field-aligned currents in the dayside cusp oberved by triad, *J. Geophys. Res.*, **81**, 5971, 1976.

IIJIMA, T. and T. A. POTEMRA, Large-scale characteristics of field-aligned currents associated with substorms, *J. Geophys. Res.*, **83**, 599, 1978.

IIJIMA, T. and T. A. POTEMRA, The relationship between interplanetary quantities and Birkeland current densities, *Geophys. Res. Lett.*, **9**, 442–445, 1982.

IIJIMA, T., Field-aligned currents during northward IMF, in *Magnetospheric Currents*, edited by T. A. Potemra, pp. 115–122, Geophysical Monograph 28, 1983.

IIJIMA, T., T. A. POTEMRA, L. J. ZANETTI, and P. F. BYTHROW, Large-scale birkeland currents in the dayside polar region during strongly northward IMF—A new Birkeland current system, *J. Geophys. Res.*, **89**, 7441–7452, 1984.

KAMIDE Y., A. D. RICHMOND, and S. MATSUSHITA, Estimation of ionospheric electric fields, ionospheric currents, and field-aligned currents from ground magnetic records, *J. Geophys. Res.*, **86**, 801, 1981.

KISABETH, J. L., On calculating magnetic and vector potential fields due to large-scale magnetospheric current systems and induced currents in an infinitely conducting earth, in *Quantitative Modeling Magnetospheric Processes*, Geophys. Monogr. Ser., Vol. 21, edited by W. P. Olson, AGU, Washington, D.C., 1979.

KISABETH, J. L. and G. ROSTOKER, Development of the polar electrojet during polar magnetic substorms, *J. Geophys. Res.*, **76**, 6815, 1971.

KÜPPERS, F., J. UNTIEDT, W. BAUMJOHANN, K. LANGE, and A. G. JONES, A two-dimensional magnetometer array for ground-based observations of auroral zone electric currents during the international magnetospheric study (IMS), *J. Geophys. Res.*, **46**, 429, 1979.

MAEZAWA, K., Magnetospheric convection induced by the positive and negative *Z* components of the interplanetary magnetic field: Quantitative analysis using polar cap magnetic records, *J. Geophys. Res.*, **81**, 2289, 1976.

MANSUROV, S. M., New evidences of the connection between the magnetic fields of cosmic space and the earth, *Geomag. Aeron.*, **9**, 622, 1969.

MAURER, H. and B. THEILE, Parameters of the auroral electrojet from magnetic variations along a meridian, *J. Geophys.*, **44**, 415, 1978.

MCDIAMID, I. B., E. E. BUDZINSKI, M. D. WILSON, and J. R. BURROWS, Reverse polarity field-aligned currents at high latitudes, *J. Geophys. Res.*, **82**, 1977.

MCDIARMID, I. B., J. R. BURROWS, and M. D. WILSON, Large-scale magnetic field peturbations and particle measurements at 1400 km on the dayside, *J. Geophys. Res.*, **84**, 1431, 1979.

MCDIARMID, I. B., J. R. BURROWS, and M. D. WILSON, Comparison of magnetic field perturbations at high latitudes with charged particle and IMF measurements, *J. Geophys. Res.*, **83**, 681, 1978.

MCDIARMID, I. B., J. R. BURROWS, and M. D. WILSON, Comparison of magnetic field perturbations and solar electron profiles in the polar cap, *J. Geophys. Res.*, **85**, 1163, 1980.

MERSMANN, U., W. BAUMJOHANN, F. KÜPPERS, and K. LANGE, Analysis of an eastward electrojet by means of upward continuation of ground-based magnetometer data, *J. Geophys.*, **45**, 281, 1979.

OGINO, T., R. J. WALKER, M. ASHOUR-ABDALLA, and J. M. DAWSON, An MHD Simulation of B_y-dependent magnetospheric convection and field-aligned currents during northward IMF, *J. Geophys. Res.*, **90**, 10,835, 1985.

POTEMRA, T. A., Current systems in the earth's magnetosphere, IUGG Qadrennial Report, *Revs. Geophys. Space Phys.*, **17**, 640–656, 1979.

POTEMRA, T. A., Birkeland currents: Present understanding and some remaining questions, in *High Latitude Space Plasma Physics*, edited by Bengt Hultqvist and Tor Hagfors, pp. 335- 353, Plenum Pub. Corp., 1983.

POTEMRA, T. A., L. J. ZANETTI, P. F. BYTHROW, A. T. Y. LUI, and T. IIJIMA, B_y dependent covection patterns during northward interplanetary magnetic field, *J. Geophys. Res.*, **89**, 9753–9760, 1984.

REIFF, P. H. and J. L. BURCH, IMF B_y-dependent plasma flow and Birkeland currents in the dayside magnetosphere. 2. Global model for northward and southward IMF, *J. Geophys. Res.*, **90**, 1595, 1985.

RUSSELL, C. T., The configuration of the magnetosphere, in *Critical Problems of Magnetospheric Physics*, edited by E. R. Dyer, pp. 1–16, National Academy of Science, Washington, DC, 1972.

SUGIURA, M. and T. A. POTEMRA, Net field-aligned currents observed by Triad, *J. Geophys. Res.*, **81**, 2155, 1976.

SVALGAARD, L., Polar cap magnetic variations and their relationship with the interplanetary magnetic sector structure, *J. Geophys. Res.*, **78**, 2064, 1973.

WILHELM, J. and E. FRIIS-CHRISTENSEN, Magnetometer chain in Greenland, *Geophys. Pap.*, R-48, Dan Meteorol. Inst., Copenhagen, 1976.

WINNINGHAM, J. D., F. YASUHARA, S.-I. AKASOFU, and W. J. HEIKKILA, The latitudinal morphology of 10 eV to 10 keV electron fluxes during magnetically quiet and disturbed times in the 2100-0300 MLT sector, *J. Geophys. Res.*, **80**, 3148, 1975.

ZANETTI, L. J., W. BAUMJOHANN, and T. A. POTEMRA, Ionospheric and Birkeland current distributions inferred from the MAGSAT magnetometer data, *J. Geophys. Res.*, **88**, 4875, 1983a.

ZANETTI, L. J., W. BAUMJOHANN, T. A. POTEMRA, and P. F. BYTHROW, Three-dimensional Birkeland-ionospheric current system, determined from MAGSAT in *Magnetospheric Currents*, edited by T. A. Potemra, pp. 123–130, Geophysical Monograph 28, 1983b.

ZANETTI, L. J., T. A. POTEMRA, T. IIJIMA, W. BAUMJOHANN, and P. F. BYTHROW, Ionospheric and Birkeland current distributions for northward interplanetary magnetic field; Inferred polar convection, *J. Geophys. Res.*, **89**, 7453–7458, 1984.

ZMUDA, A. J. and J. C. ARMSTRONG, The diürnal-variation of the region with vector magetic field changes assciated with field-aligned currents, *J. Geophys. Res.*, **79**, 2501, 1974.

Solar Wind-Magnetosphere Coupling, edited by Y. Kamide and J. A. Slavin, 563–573.

Magnetospheric Models from Field-Aligned Currents

G. ATKINSON

Herzberg Institute of Astrophysics, National Research Council of Canada, 100 Sussex Drive, Ottawa Ontario K1A 0R6, Canada

The interaction of the magnetosphere with the solar wind causes the convection of magnetic flux tubes into the distant tail creating the necessity for a sunward return convection inside the magnetosphere. The nature and location of this sunward flow in the plasma sheet is controlled by the distribution of plasma. Many properties of this sunward convection can be obtained from a study of the field-aligned currents (Birkeland currents). Upward (downward) Birkeland current indicates that plasma depletion (enhancement) is occurring on flux tubes. From the observed Birkeland current and convection patterns at low altitudes, the variation of the plasma content of flux tubes is obtained. This is mapped to the equatorial plane, producing a model of magnetospheric plasma distribution and convection, including among other things the Harang Discontinuity. An explanation for the shape of the plasma sheet is offered. In the model, flux tubes are added to the tail and plasma sheet by dayside merging and removed by the plasma depletion (and collapse of flux tubes at the near-earth edge) that accompanies the Birkeland currents. If the first dominates, the plasma sheet advances towards the earth. If the latter dominates, it recedes.

However, the recession appears to be unstable. Quantitative evidence is discussed that supports: (1) the Birkeland-current-associated plasma depletion and enhancement picture, (2) the time-dependent picture above, (3) the idea that the substorm expansion is an instability of the recession of the plasma sheet. The substorm expansion is in many ways a miniature version of the large scale convection. The formation of a near-eath neutral line may not be necessary for expansion, but it can be fit into the picture. An explanation of the initial brightening of arcs arises from the model.

1. Introduction

This paper is a summary of the features and logic of the steady-state and time-dependent model of the magnetosphere that has been developed by the author in a series of four papers (ATKINSON, 1983, 1984a, 1984b, 1985, henceforth referred to as papers I, II, III and IV). This model is based on deductions from the observed distribution of Birkeland currents and hence is empirical rather than theoretical. The relationship between Birkeland currents and the plasma content of flux tubes is implicit in the theoretical models of FEJER (1964), VASYLIUNAS (1972), SOUTHWOOD (1977) and also in the computer modelling work of the Rice University group (e.g., HAREL and WOLF, 1976).

First it is necessary to define the meanings of some of the terms as they are used in this paper.

1. The term "convection" is used to describe the motion of the zero electric field reference frame, that is the $\boldsymbol{E}\times\boldsymbol{B}/B^2$ component of the plasma drift.

2. Correspondingly a "flux tube" will be defined as moving with this convection and hence is defined in terms of the cold electrons rather than the hot protons (the model assumed hot protons and cold electrons).

3. "Gradient drift" will be used to refer to the other components of proton motion, namely field-gradient and curvature drift and also non-adiabatic drift due to magnetic-field gradients.

4. "Protons" is to be understood to include positive ions.

The choice of the above definitions is intended to simplify the dicussions of the physical principles in this paper. The choice of alternative definitions should lead to the same conclusions about the physics.

2. Control of Convection

It is generally believed that flux tubes are convected from the dayside magnetosphere to the nightside magnetosphere by two types of interaction with the solar wind: dayside merging with the interplanetary field and some kind of cross-magnetopause momentum transfer causing a boundary-layer convection in the anti-solar direction. The former appears to dominate.

The above convection to the nightside creates a necessity for a return sunward convection internal to the magnetosphere from the distant tail, through the plasma sheet toward the dayside. However the nature and location of the return convection is determined to a great extent by the hot plasma in the plasma sheet since this plasma is compressed in the process. Is the convection primarily on the dawnside or the duskside? Most current models assume it is uniformly distributed. A full theoretical solution requires consideration of the ionosphere, hot plasma and self-consistent magnetic fields. Fortunately insight can be gained by the semi-empirical approach used by the author, which being based on observation avoids confrontation with the above.

3. Relationship between Birkeland Currents and Plasma

Since the energy of protons in the plasma sheet is observed to be greater than the energy of electrons, a first approximation to the behaviour can be obtained by assuming the plasma consists of cold electrons and hot protons. The electrons follow the convection whereas the protons undergo gradient drift in addition to convection.

It is generally believed that Birkeland currents are carried by electrons at low altitudes and that they close in the outer magnetosphere to cross-field currents carried by the gradient-drift of protons (magnetization currents can be neglected because they are divergence-free). The flux tube is shown as a "black box" in Fig. 1 with electrons entering or leaving the ionospheric end of the tube and protons entering or leaving the outer magnetospheric part of the tube. Consider upward current (left side of Fig. 1). For each electron leaving the ionospheric end of the flux tube, current continuity requires the loss of a proton from the outer magnetospheric part of the flux tube. The loss of protons occurs as a divergence of their gradient drift. Thus the intensity of upward Birkeland current at low altitudes gives the rate of "depletion" of total plasma

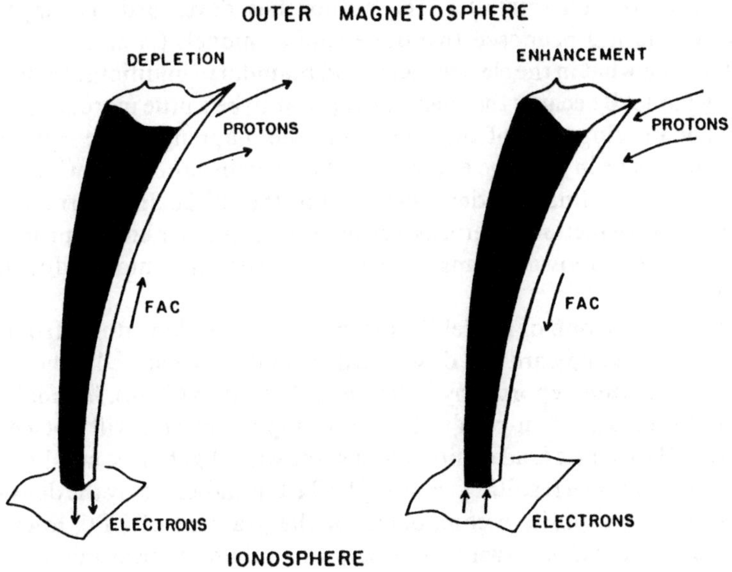

Fig. 1. Schematic of flux tubes carrying Birkeland current.

content (TPC) of the flux tube. Similarly, for downward current (right hand side of Fig. 1) the low altitude current gives the rate of "enhancement" of TPC on that flux tube. In addition, since the outer magnetospheric currents are carried by energetic protons, the low-altitude current indicates that substantial depletion or enhancement of the plasma energy content of the flux tubes is occurring in association with the changes in TPC.

The above description is true in the convecting flux-tube reference frame. In this frame the Birkeland current indicates the time rate of change of the TPC. In a geostationary frame this transforms to include spatial variations ($\mathrm{d}/\mathrm{d}t = \partial/\partial t + \boldsymbol{V}\cdot\nabla$). Thus in the case of a steady-state magnetosphere the Birkeland current represents convection of the flux tubes through stationary gradients of the TPC.

4. Ionospheric Altitudes—Steady State

From the foregoing discussion, it is apparent that the time integral of the Birkeland current through the ionospheric foot of a particular flux tube will give the net decrease or increase of the TPC of that flux tube. Thus if we knew the TPC of flux tubes at the polar cap boundary, and the intensity of convection and Birkeland currents we could calculate numerical values of the TPC everywhere south of the boundary. At ionospheric heights we have some models of the convection electric fields and the Birkeland currents. These are probably not of sufficient accuracy for a detailed calculation. An even greater difficulty is that we have to use informed guesses at the boundary conditions.

Because of the uncertainties, the author has developed the approach only qualitatively, and has proposed two quite similar models (A and B, see paper IV) which differ somewhat in the electric fields and boundary conditions. In this paper, we discuss only model B because the energetics appear to be a little more satisfactory than model A. The development of the model and the mapping to the equatorial plane (next section) has been to some extent a trial and error procedure which will not be repeated here. The final models agree with the Birkeland current associated enhancement and depletion criteria as well as with most other known properties of the magnetosphere. The allowable range of variations from these models does not appear to be large.

Figure 2 is a cartoon of model B as seen in a view looking down from above the polar cap. Regions of upward and downward current are separated by heavy lines and are similar to the sytem reported by IIJIMA and POTEMRA (1978). (A small difference near midnight, discussed in paper III, is not important and will not be discussed further here.) The electric field equipotentials, shown as light lines, are similar to those in HEPPNER (1977). The resulting value of the TPC is indicated by the density of dots. In general there is a positive component of the gradient of TPC towards higher latitudes and also towards dawn and dusk with a minimum near midnight.

5. Equatorial Plane—Steady State

Figure 3 is a cartoon showing the mapping along field lines of Fig. 2 to the

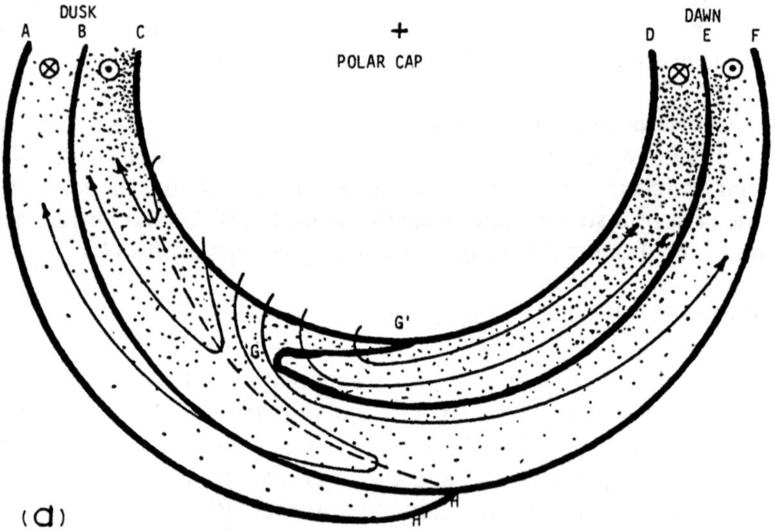

Fig. 2. Polar-cap view of model B at ionospheric altitudes. Heavy lines are boundaries of regions of FAC with upward and downward denoted by dots and crosses in circles. Light lines are electric equipotentials. The density of dots indicates TPC of flux tubes. The dashed line is the Harang Discontinuity.

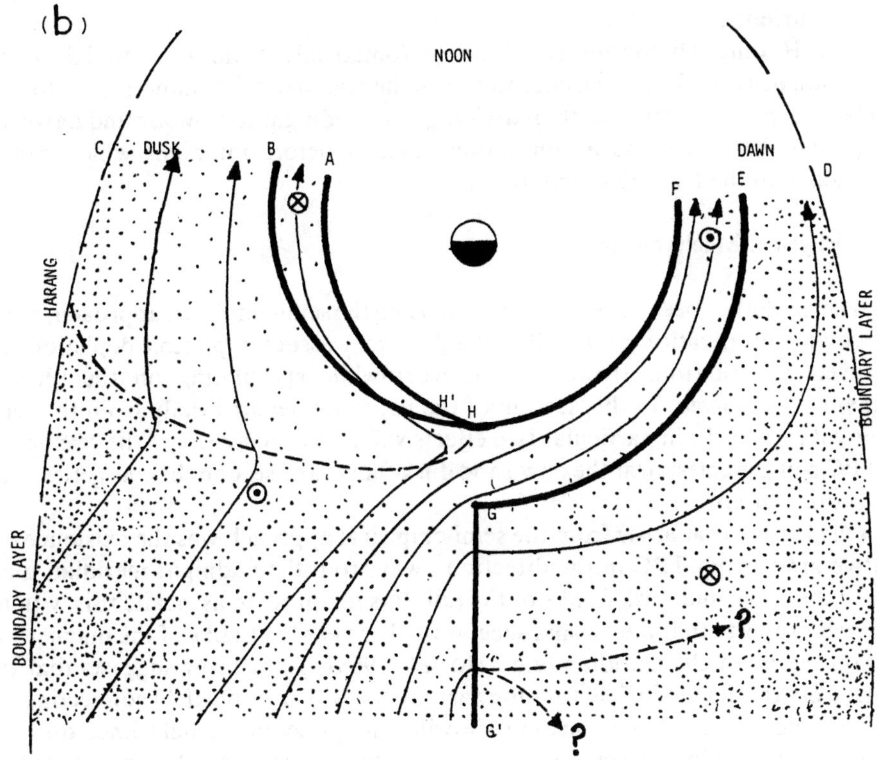

Fig. 3. Mapping of Fig. 2 along magnetic field lines to the equatorial plane. Symbolism is the same as Fig. 2. The magnetospheric boundary layer is not included.

equatorial plane. This mapping has been discussed in papers III and IV and will not be pursued in detail here. The figure shows: the regions of upward (from the ionosphere) and downward Birkeland currents separated by heavy lines, the convection equipotentials indicated by light lines, and the TPC indicated by the density of dots. The gradient of TPC has a positive component radially away from the earth and also from the midnight meridian toward both the dawn and dusk flanks of the tail. A region of "tail symmetry" exists further down the tail and is defined as the region of the tail where the plasma sheet and the region I currents are symmetric about the midnight meridian.

The dusk to dawn component of the convection in the region of tail symmetry maps to the eastward convection north of the Harang Discontinuity. A further argument for the existence of this component comes from the tail itself. Observationally the plasma sheet exhibits "tail symmetry". If the convection were also symmetric about the midnight meridian, the region I currents would both have the same sign, that is upwards or downwards according to the principles of FAC-associated plasma enhancement and depletion. Therefore a major assymetry to the convection must exist. In particular the dusk to dawn component shown in Fig. 3 produces the correct region I currents.

The Harang Discontinuity results automatically from this model. If the convection has a dusk to dawn component in the region of tail symmetry, but follows the classical picture nearer earth of dividing near midnight to flow around dawn and dusk, then there must be a convection reversal before midnight which can be identified with the Harang Discontinuity.

6. Physical Mechanisms

The model has been developed thus far using the semi-empirical approach, and in particular consideration of the Birkeland-current-assocted plasma depletion and enhancement criterion. However it is worthwhile speculating on the physical mechanisms involved. In this section of the paper we depart briefly from the semi-empirical approach. In particular two effects will be considered: first the reasons for the TPC being greatest near the dawn and dusk flanks and second the energetics of the convection pattern.

One feature that arose from the semi-empirical approach was the existence of a positive gradient in TPC in the direction away from the midnight meridian plane toward the dawn and dusk flanks of the tail. It is tempting to assocaite this with the observed property that the plasma sheet is thicker near the flanks of the magnetotail than on the midnight meridian (e.g., HONES, 1968). The author suggests that this results from the magnetic field topology associated with a dipole field connecting to a tail-like field. Figure 4 is a cartoon showing the prjection of field lines onto the equatorial plane. Since the curvature is toward the midnight meridian plane the field should be stronger near the midnight meridian plane. From energy considerations, plasma-rich flux tubes tend to occupy weaker-field regions because the required compression of the plasma is less. Hence it is reasonable to expect a higher TPC (and a thicker plasma sheet) near the flanks in an equilibrium situation. This interpretation

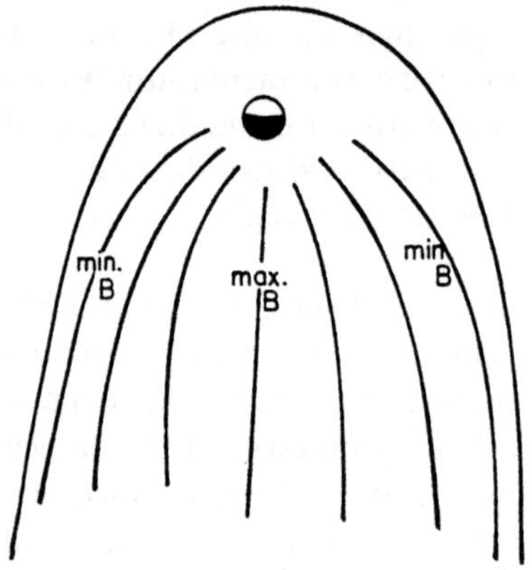

Fig. 4. Equatorial section of the magnetosphere showing projections of magnetic field lines onto the equatorial plane.

implies that plasma pressure forces are more important than inertial forces in determining the interior configuration of the magnetotail.

Some insight can also be gained into the form of the convection pattern. Given a magnetotail and plasma sheet near equilibrium, it is clear that a convection pattern with inflow in regions of flux tube plasma depletion (upward Birkeland current) and perhaps some outflow in regions of plasma enhancement (downward current) is an energetically favoured convection. The cross-tail component provides continuity of the flow. The reader should be very careful here *not* to think of enhancement and depletion as causing convection. They are merely a measurable indicator of the convection pattern and of TPC and plasma energy loss and gain of flux tubes.

7. Time-Dependent Convection

The remainder of this paper is concerned with time-dependent convection. The plasma depletion associated with the upward Birkeland currents (see Figs. 2 and 3) allows flux tubes to convect earthwards and collapse to a dipolar form, thus removing them from the plasma sheet. Dayside merging and boundary layer convection, on the other hand, adds flux tubes to the magnetotial and ultimately to the plasma sheet. In the steady-state, the rates of the two processes are equal. In time-dependent flows, the two rates are unequal.

First, consider the case where the rate of addition of flux tubes to the tail due to solar wind interaction (dayside merging, boundary layer) exceeds the rate of collapse associated with Birkeland currents. Then the inner edge of the plasma sheet, the Harang discontinuity and the region of tail symmetry would all be expected to move earthwards. This would be the type of large-scale two-celled convection that has been given the names convection bay, DP2 and growth phase of substorms. It is consistent with the observed southward drift of the Harang Discontinuity between substorm expansions (NIELSEN and GREENWALD, 1979) and in the earthward displacement of the inner edge of the plasma sheet in active times (FAIRFIELD and VINOS, 1984). Indications are that the Birkeland current increases in strength as the region of tail symmetry, Harang Discontinuity and plasma sheet move earthwards. In fact, for constant dayside merging rate a steady-state might be produced when the Birkeland current become sufficiently strong that the two rates are equal.

At some stage the region of tail symmetry, etc., must retreat. In this process, the rate of flux tube collapse due to plasma depletion should be greater than the rate of flux tube addition to the tail due to dayside merging. A northward turning of the interplanetary field is one way to produce this condition and these are known to produce substorm expansions (CAAN *et al.*, 1977; ROSTOKER, 1983). However, the observed suddenness of substorm expansions and their longitudinally limited extent indicate that this recession of the region of tail symmetry is unstable. We shall see further proof for this in the next section of the paper, and then subsequently see what the observations of Birkeland currents can tell us about the nature of the instability.

8. Quantitative Prediction and Verification

For a plane field-aligned current sheet, a simple prediction relating velocities and Birkeland currents at ionospheric heights can be made:

$$V_{HD} = V_{CONV} - kI_{//} \tag{1a}$$

where

$$K = 1/(\alpha e N B_i) \approx 600 \text{ mks units.} \tag{1b}$$

The physics of Eq. 1a is that the velocity of the current sheet V_{HD} (in this case the Harang Discontinuity will be considered), is equal to the convection velocity, V_{CONV}, minus the rate at which flux tubes pass through the current sheet due to plasma depletion, $kI_{//}$. $I_{//}$ is the Birkeland current (positive upwards), $I_{//}/e$ the rate at which electron-ion pairs are depleted, and $N\alpha$ the number of electron ion pairs that must be removed from a unit flux tube to allow it to collapse through the current sheet (N is the TPC, α is the fraction to be removed). B_i, the ionospheric magnetic field, enters because N is defined as TPC per unit flux. The constant k would be expected to stay reasonably constant over times when the plasma-sheet plasma does not change properties very much. From previous work on the subject, for one hemisphere $N\alpha \approx 2 \times 10^{20}$ (see papers III and IV). If V_{HD} and V_{CONV} are measured at ionospheric heights in m/sec and $I_{//}$ in amps/m, then k has a value of approximately 600.

In paper IV, an equation equivalent to equation 1a was applied to an event reported by PELLINEN *et al.* (1982). The event consisted on a convection bay leading into a substorm expansion. Motions of the westward electrojet and hence the Harang Discontinuity were reported. The results during the convection bay were surprisingly consistent indicating little variation in k during this period. This must be considered strong support for the Birkeland-current-associated enhancement and depletion model. Similar studies of other events are needed. At substorm onset k increased suddenly by an order of magnitude. Thus either $N\alpha$ decreased by a factor of 10, or the two-dimensional sheet approximation was no longer valid. Both these possibilities will be considered in the next section on substorm expansions. A further point of note in the PELLINEN *et al.* (1982) event is that the substorm expansion onset occurred suddenly at the time when the last term of Eq. 1a became equal to V_{CONV} supporting the view that it is the recession from the earth of the region of tail symmetry that is unstable and is the cause of the substorm expansion.

9. Substorm Expansion

The substorm expansion (intensification) is generally associated with the onset of the current system that has in recent years come to be called the current wedge. The equatorial projection of this current wedge is illustrated in Fig. 5. The location of depletion and enhancement has been indicated. One of the most striking things about the system is that in many of its features it is a miniature version of the large scale

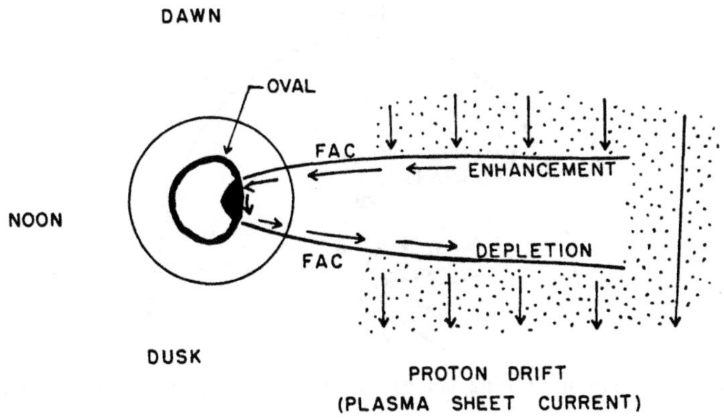

Fig. 5. Equatorial projection of the substorm current wedge showing location of depletion and enhancement.

region I current system discussed earlier. These features include:

1. gradients of TPC from the centre toward the dawn and dusk flanks;
2. depletion and earthward collapse (earthward convection) of flux tubes on the westward edge and enhancement and possible inflation (outward convection) on the eastward edge;
3. eastward convection (the westward electrojet) providing continuity of convection.

However a major difference from the large-scale system is that the convection is not equipotential. Time scales are too short for the ionospheric feet of flux tubes to follow the convection in the outer magnetosphere, and hence there is a topology change (curl $E \neq 0$).

In view of the similarity of this small scale system to the larger one, it is reasonable to believe that physical processes must be quite similar. As in the case of the large scale system, the convection is favoured by the energetics associated with the gradient drifting plasma: earthward collapse of flux tubes in regions of plasma depletion (upward Birkeland current) and possibly some outward expansion in regions of plasma enhancement.

The brightening of the arcs and the westward propagation of the brightening at high velocities can be explained by this model. The initial collapse near midnight will cause the type of convection indicated in Fig. 6. Since, as pointed out earlier, the time scales are too short for the field line feet to follow, this represents a distortion of field lines and hence increased upward current and arc brightening. The onset of this non-equipotential convection propagates outward from midnight at magnetohydrodynamic velocities in the outer magnetosphere. This is consistent with the velocities associated with the brightening (SAMSON, 1985). The westward travelling surge, which is the indication of the plasma-depletion associated with the upward current travels much more slowly.

In fact this initial brightening and enhancement of upward current may give

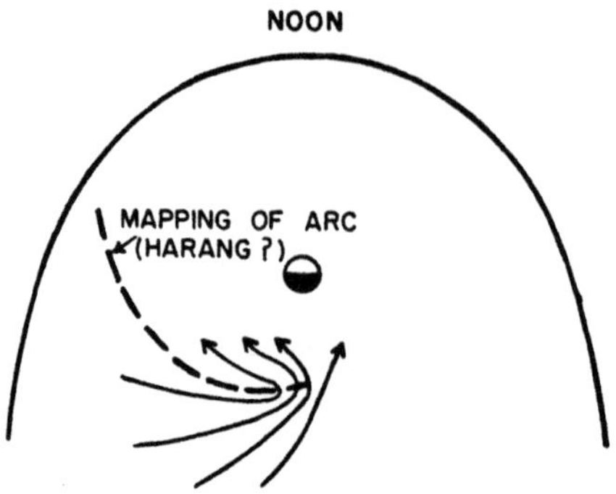

Fig. 6. Equatoral section illustrating convection stream-lines (not equipotential) in a substorm expansion.

some idea as to why the recession of the tail is unstable. Given a current wedge, the Birkeland current and magnetic stress *increase* to the west of the wedge as the stress and Birkeland current decrease within the wedge. Under these circumstances the recession of the region of tail symmetry is unlikely to occur uniformly cross the tail. However the currents are only an indicator of processes and a solution involving hot plasma and self-consistent fields is necessary. We note that since the current wedge is essentially three-dimensional, equation 1a may break down and hence this is one possible reason for the apparent large increases in k at expansion onset.

Now let us turn our attention to the possiblity of the existence or formation of a near-earth neutral line during substorm expansion. The formation of such a neutral line also offers a ready explanation of the large increase in k at expansion phase onset. That is, plasma tailward of the neutral line is disconnected from the earth and hence the TPC (N in Eq. 1b) is suddenly reduced.

The existing theory of near-earth neutral lines is that the tail lobe flux increases, thereby increasing the forces trying to push flux tubes into the neutral sheet ultimately causing an X type neutral line. In this model, the onset of tail reconnection leads to the substorm expansion. This model explains many features of the expansion, but not the westward propagation or why the onset occurs at time when $V_{HD}=0$.

A more satisfactory theory may be that the neutral line formation is the result of another process. In particular there is a second way that flow through a neutral line may be increased. (The first was increasing the forces driving inflow.) It is to remove forces blocking outflow. In particular, the onset of rapid earthward collapse of flux tubes may allow the onset of nightside near-earth reconnection. If the earthward collapse of flux tubes is associated with plasma depletion, then the westward propagation of the neutral line would be associated with the westward spreading of the current wedge.

We now summarize this section. The basic suggestion is that the recession of the region of tail symmetry is unstable, and that a full solution which considers hot plasma, self-consistent magnetic fields and the ionosphere would reveal the instability. The observed flow appears to be favoured by energy considerations and is a miniature version of the large-scale flow. The formation of a near-earth neutral line may not be a necessary part of the process. However, if it is, then the neutral line probably results when the above instability starts an outflow from the site of the onset of the X-type neutral line. A model in which an X-type neutral line is formed due to the onset of tearing-mode or similar instabilities due to increased taillobe flux is not as satisfactory.

REFERENCES

ATKINSON, G., Time dependent flows in the renovated model of the magnetosphere, *J. Geophys. Res.*, **90**, 10843–10850, 1985.

ATKINSON, G., The role of currents in plasma redistribution, in *Magnetospheric Currents*, edited by T. Potemra, pp. 325–330, Geophysical Monograph 28, American Geophysical Union, Washington, D.C., 1983.

ATKINSON, G., Field-aligned currents as a diagnostic tool. Result: A renovated model of the magnetosphere, *J. Geophys. Res.*, **89**, 217–226, 1984a.

ATKINSON, G., Thick current sheets in the renovated model of the magnetosphere, *J. Geophys. Res.*, **89**, 8949–8955, 1984b.

CAAN, M. N., R. L. MCPHERRON, and C. T. RUSSELL, Characteristics of the association between the interplanetary magnetic field and substorms, *J. Geophys. Res.*, **82**, 4837–4842, 1977.

FAIRFIELD, D. H. and A. F. VINAS, The inner edge of the plasma sheet and the diffuse aurora, *J. Geophys. Res.*, **89**, 841–854, 1984.

FEJER, J. A., Theory of the geomagnetic dialy disturbence variations, *J. Geophys. Res.*, **69**, 123–135, 1964.

HAREL, M. and R. A. WOLF, convection, in *Physics of Solar-Planetary Environments*, Vol. II, edited by D. J. Williams, pp. 617–629, Am. Geophys. Un., Washington, D.C., 1976.

HEPPNER, J. P., Empirical models of high-latitude electric fields, *J. Geophys. Res.*, **83**, 1115, 1977.

HONES, E. W., Jr., Review and interpretation of particle measurements made by the VELA satellites in the magnetotail, in *Physics of the Magnetosphere*, edited by R. L. Carovillano, pp. 392–408, D. Reidel, Hingham, Mass., 1968.

IIJIMA, T. and T. A. POTEMRA, Large-scale characteristics of field-aligned currents assocaited with substorms, *J. Geophys. Res.*, **83**, 599–615, 1978.

NIELSEN, E. and R. A. GREENWALD, Electron flow and visual aurora at the Harang discontinuity, *J. Geophys. Res.*, **84**, 4189–4200, 1979.

PELLINEN, R. J., W. BAUMJOHANN, W. J. HEIKKILA, V. A. SERGEEV, A. G. YAHNIN, G. MARKLUND, and A. O. MELNIKOV, Event study on pre-substorm phases and their relation to the energy coupling between solar wind and magnetosphere, *Planet. Space Sci.*, **30**, 371–388, 1982.

ROSTOKER, G., Triggering of expansive phase intensifications of magnetospheric substorms by northward turnings of the interplanetary magnetic field, *J. Geophys. Res.*, **88**, 6981–6993, 1983.

SAMSON, J. C., Large scale studies of Pi 2's associated with auroral breakups, *J. Geophys.*, **56**, 133–145, 1985.

SOUTHWOOD, D. J., The role of hot plasma in magnetospheric convection, *J. Geophys. Res.*, **82**, 5512, 1977.

VASYLIUNAS, V. M., The Interrelationship of Magnetospheric processes, in *Earth's magnetospheric Process*, edited by B. M. McCormac, pp.29–38, D. Reidel, 1972.

Solar Wind-Magnetosphere Coupling, edited by Y. Kamide and J. A. Slavin, 575–604.

The Characteristics of Polar Cap Precipitation and Their Dependence on the Interplanetary Magnetic Field and the Solar Wind

D. A. HARDY[1], M. S. GUSSENHOVEN[1], K. RIEHL[1], R. BURKHARDT[2], N. HEINEMANN[3], and T. SCHUMAKER[3]

[1]*Air Force Geophysics Laboratory, Hanscom Air Force Base, Massachusetts 01731, U.S.A.*
[2]*Emmanuel College, Boston, Massachusetts, U.S.A.*
[3]*Boston College, Chestnut Hill, Massachusetts, U.S.A.*

Work completed at the Air Force Geophysics Laboratory over the last several years on the nature of the precipitation occurring in the polar cap and its relationship to the state of the solar wind and the Interplanetary Magnetic Field is reviewed. This work has shown that there are two states of polar cap precipitation; an active state occurring when the IMF B_z component is positive in which "polar showers" or "polar squalls" occur over much of the cap and a quiet state occurring when the IMF B_z is negative in which only the unperturbed polar rain is observed. The two states occur with approximately equal frequency. The characteristics of both the precipitating electrons and ions for the two states and the variation in these characteristics with the orientation of the IMF and the state of the solar wind plasma are discussed along with the implications of these observations as to the mechanisms leading to these two states.

1. Introduction

A considerable body of literature exists on the characteristics of precipitation observed in the polar cap (BURCH, 1968; WHALEN *et al.*, 1971; WINNINGHAM and HEIKKILA, 1974; WINNINGHAM *et al.*, 1975; FENNELL *et al.*, 1975; YEAGER and FRANK, 1976; FOSTER and BURROWS, 1976, 1977; MENG and KROEHL, 1977; MENG *et al.*, 1977; MIZERA and FENNELL, 1978; MENG, 1981a and b; BURKE *et al.*, 1982; HARDY *et al.*, 1982; FRANK *et al.*, 1982; WINNINGHAM and GURGIOLO, 1982; GUSSENHOVEN *et al.*, 1984; PETTERSON and SHELLEY, 1984). These papers have shown that there are two basic classes of precipitation observed in the polar cap. The first is the polar rain. This consists of a weak flux of electrons with average energies of approximately 100 eV and an energy flux of between 10^{-2} and 10^{-3} ergs/cm^2 sec. The intensity of the polar rain varies slowly across the polar region often exhibiting clear gradients in the integral number flux along the satellite trajectory. The overall intensity of the polar rain is dependent on the sector structure of the IMF. For away (toward) sectors in the north (south) pole the intensity is approximately a factor of two higher than for the opposite case. From these observations a picture has been built up of the polar rain as occurring on the open field lines that map to the lobes of the geomagnetic tail and that the polar rain consists of solar wind particles that gain access to these open field lines.

The second type of precipitation are the so called "polar showers" and "polar squalls" (WINNINGHAM and HEIKKILA, 1974). These events exhibit much higher levels of integral number flux than the polar rain. Although the electrons in many such events have average energies of approximately 100 eV there are cases where field aligned acceleration of several kilovolts is observed. Typically these events are found embedded in the polar rain, i.e., as observations are made along the satellite trajectory electron fluxes at the level of the polar rain are observed both before and after the observation of the region of enhanced flux. Research has shown that the "polar showers" and "polar squalls" can produce both visual and subvisual sun-aligned arcs and are therefore intimately related to the occurrence and dynamics of polar cap arcs and theta aurora (EATHER and AKASOFU, 1969; BERKEY *et al.*, 1976; MENG and AKASOFU, 1976; ISMAIL *et al.*, 1977; LASSEN and DANIELSON, 1978; WEBER and BUCHAU, 1981; MENG, 1981a; GUSSENHOVEN, 1982; FRANK *et al.*, 1982).

Polar cap precipitation can therefore be thought of as having a quiet state corresponding to the occurrence of the unperturbed polar rain and an active state where "polar showers" or "polar squalls" occur. For the purpose of this paper we will use the term "active polar cap precipitation" to refer to the occurrence of "polar showers" or "polar squalls".

Over the past five years a series of research projects have been completed at the Air Force Geophysics Laboratory dealing with observations of the polar cap precipitation and dynamics. (HARDY *et al.*, 1982; BURKE *et al.*, 1982; HARDY, 1984; GUSSENHOVEN *et al.*, 1984; RIEHL and HARDY, 1985). This work was motivated by the availability of large data sets taken continuously in time with instruments of sufficient sensitivity to measure both the precipitating electrons and ions in the polar regions. These characteristics of the data sets permitted the completion of studies that better quantified the average characteristics of the polar particle environment for both electron and ions and contributed to the determination of the way in which the environment is affected by the solar wind and the IMF.

This paper is a review of this work. The paper is divided into five sections. Following the Introduction, Section 2 describes the data sets used in the studies. Section 3 summarizes the work on the low altitude two-dimensional pattern of polar rain precipitation. The spectral characteristics of the polar rain ions and electrons are discussed. Section 4 covers the research on the occurrence of "polar squalls" and "polar showers" and their relationship to visual polar cap arc and theta aurora. Section 5 is a summary and discussion.

2. Instrumentation

The data used in the studies reviewed here are primarily from the SSJ sensors flown on the satellites of the Defense Meteorological Satellite Program. The SSJ/3 sensors were flown on the DMSP F2 and F4 satellites. Each SSJ/3 consists of two cylindrical electrostatic analyzers that together return a sixteen point spectrum once per second over the energy range from 50 eV to 20 keV. The sensors are oriented on the satellite with their look direction towards the local vertical. The F2 satellite was placed initially in an 840 kilometer, circular, sun synchronous orbit in the dawn-dusk meridian but was subject to a very slow drift towards later local times. The F4 satellite was placed in a similar sun synchronous orbit in the 1030–2230 meridian (For more details see HARDY *et al.*, 1979).

Starting with the launch of the DMSP F6 satellite the SSJ/3 sensors were replaced by the improved SSJ/4 sensor. The SSJ/4 sensor consists of four cylindrical electrostatic analyzers arranged into two pairs; one to measure electrons and one to measure ions. Each pair returns a 20 point spectrum once per second over the energy range from 30 eV to 30 keV. The ion analyzers have very large geometric factors which permit measurements of auroral phenomena with good count rate statistics. There are presently two SSJ/4 detectors operational on the DMSP F6 and F7 satellites. The F6 satellite is in a 840 kilometer, circular, sun synchronous orbit in the dawn-dusk meridian. The F7 satellite is in a similar orbit in the 1030–2230 meridian. (For more details see HARDY *et al.*, 1984)

3. Polar Rain Measurements

A survey of individual passes of the DMSP satellites across the polar cap shows that there are two general patterns of electron precipitation. Examples of these two patterns are shown in Figs. 1 and 2 where we have plotted in the three panels from top to bottom, the average energy, the integral energy flux and the integral number flux of electrons over the SSJ/3 energy range. In Fig. 1 there is a clear region of enhanced flux in the auroral oval. At latitudes above this region the electrons of the polar rain are observed. The electron flux is uniformly low and slowly varying across the cap with an average energy on the order of 100 eV. In Fig. 2, by contrast, even though there is still a region of auroral precipitation, there are intense fluxes of low energy electrons observed up to geomagnetic latitudes exceeding 85 degrees. In this pass these enhancements, which we identify as "polar showers" or "polar squalls", are seen to be embedded in the polar rain, i.e., electrons with the characteristics of the polar rain are observed both before and after each of the enhancements. This is a point that will be discussed in more detail in Section 4.

In order to determine the statistical characteristics of the polar rain two approaches were used. In the first of these the polar cap region was divided into a series of spatial elements, and using one year of data from the SSJ/3 detector on the DMSP F2 satellite, the average polar rain spectrum in each spatial element was determined as a function of activity, season, and the orientation of the IMF (GUSSENHOVEN *et al.*, 1984). In the second approach, individual passes were analyzed for which solar wind and IMF data were available (RIEHL and HARDY, 1985). For these passes the average polar rain electron spectra were calculated and fit to Maxwellian distributions. From these fits the distribution of spectral parameters and the relationship of the polar rain to the IMF and solar wind were investigated.

The key results of the two dimensional study are shown in Figs. 3 and 4. In the first of these figures, the two dimensional maps of the integral number flux and average energy taken over the average spectra are shown in a color spectrogram format for the low and high activity cases. In Fig. 4, maps are shown for the same quantities for the high activity case when the data are separated according to the sign of the IMF B_y component. In Fig. 4 cases in which B_y is positive (negative) for north

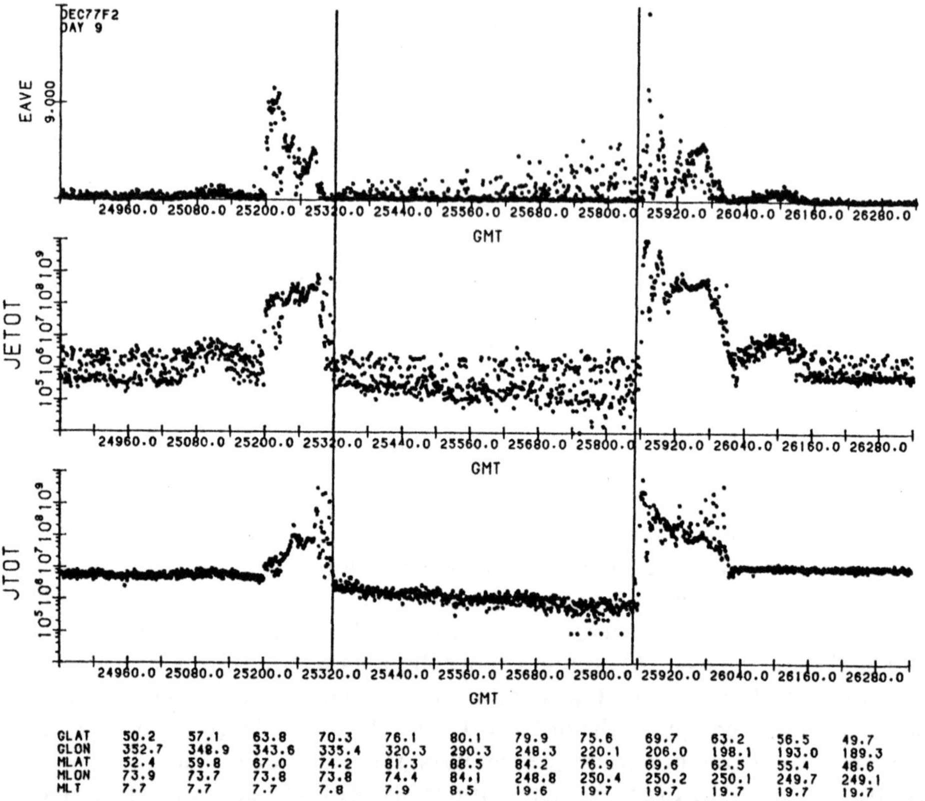

Fig. 1. Integral quantities over the electron spectrum measured by the SSJ/3 sensor are plotted for the northern hemisphere pass between 0654 and 0719 on December 9, 1977. Plotted in the three panels from top to bottom are the average energy of the electrons in units of keV on a linear scale, the integral energy flux in units of keV/cm^2 sec ster on a logarithmic scale and the integral number flux in units of electrons/cm^2 sec ster on a logarithmic scale. The bottom of the figure is annotated with the geographic latitude and longitude (GLAT and GLON); and the geomagnetic latitude and longitude (MLAT and MLON); and the magnetic local time (MLT) of the satellite all mapped down the field line to 110 kilometers. The universal time is given in seconds.

pole have been combined with cases in which B_y is negative (positive) in the south pole.

The figures illustrate the following points: First, the polar rain occurs within a circular region whose center is offset towards premidnight and whose radius increases with increasing activity. Second, there is a large scale trend for the integral number flux to increase and the average energy to decrease across the cap, symmetric about a line running from pre-midnight to pre-noon. The integral flux increases by a factor of fifteen while the average energy decreases by a factor of three along this line. This behavior is maintained in all separations of the data, by activity and by the sign of the IMF B_x and B_y components. Third, in the separation of the data by the IMF B_y component the integral number flux is clearly higher for the case of B_y positive in the

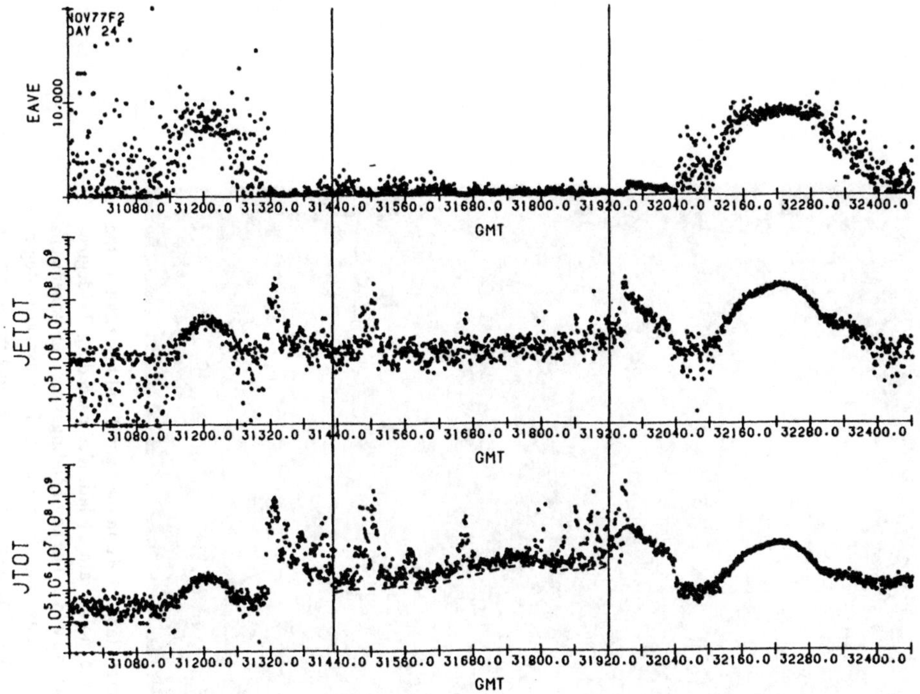

GLAT	-42.1	-49.0	-55.7	-62.4	-68.8	-74.7	-79.3	-80.5	-77.3	-71.8	-65.6	-59.0
GLON	149.2	146.4	142.9	138.1	130.9	118.4	93.8	53.6	20.2	2.9	353.6	347.7
MLAT	-55.1	-62.4	-69.5	-76.8	-83.7	-88.2	-82.2	-76.0	-70.0	-64.2	-58.6	-53.2
MLON	226.0	225.0	223.7	221.2	214.3	109.9	55.9	49.9	46.9	44.7	42.9	41.3
MLT	19.6	19.5	19.5	19.3	18.9	12.0	8.4	8.1	7.9	7.8	7.7	7.6

Fig. 2. Integral quantities in the same format as Figure 1 for the southern hemisphere pass between 0836 and 0901 UT on November 24, 1977.

north pole and negative in the south pole than for the opposite case; in agreement with the earlier work of YEAGER and FRANK (1976) and MENG and KROEHL (1977). Besides the difference in the intensity, there is a rotation in the overall pattern with the sign of the B_y component. For the case of B_y positive (negative) in the north (south) pole the pattern and gradient in the integral number flux and average energy are roughly symmetric about a line running premidnight to pre-noon. For the opposite case the entire pattern appears to have been rotated towards later local times such that the quantities are more nearly symmetric about the noon-midnight meridian. The same behavior is seen for lower activity and when the data are separated by the sign of the B_x component. The lack of any difference in the behavior when the separation of the data is made by the sign of the B_x and B_y components is the result of the high correlation between the signs of these two components due to the sector structure of the IMF.

These results on the response of the polar rain pattern to the sector structure of the IMF are in disagreement with the earlier work of MENG and coworkers (1977). They reported that for B_y positive (negative) in the north (south) pole the gradient in

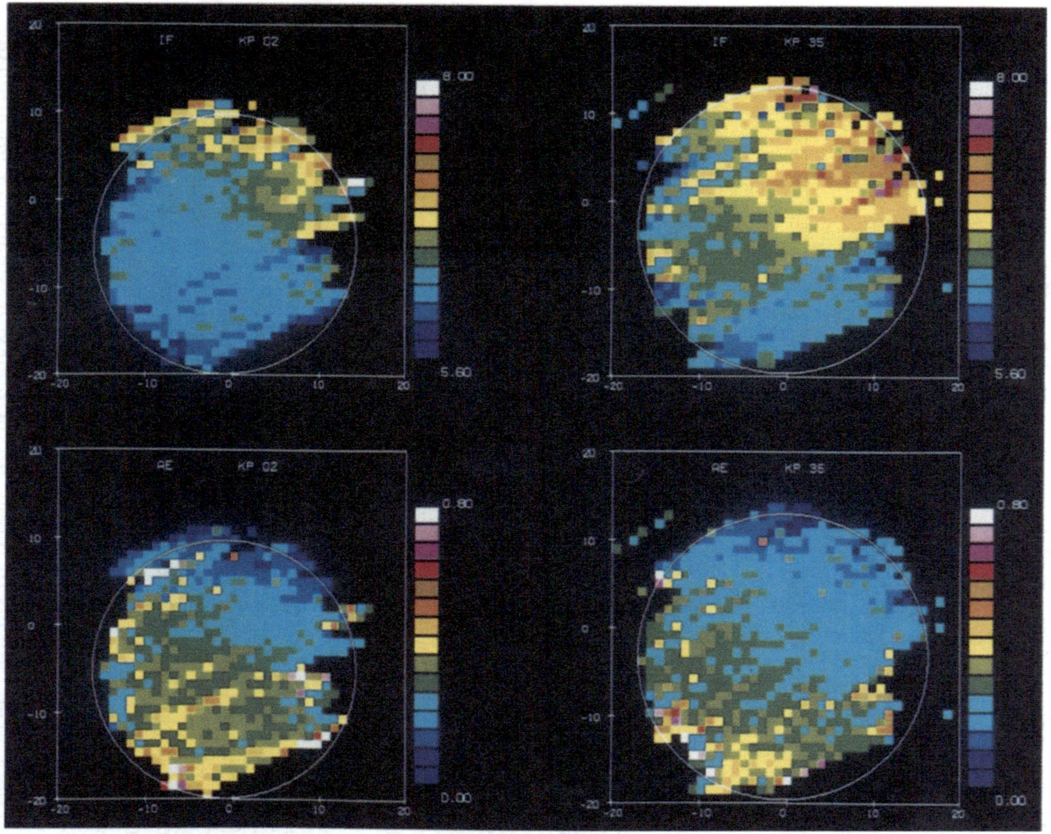

Fig. 3. Polar color spectrograms of the average energy (labeled AE in the figure) and logarithm of the average integral number flux (labeled IF) of the polar rain at latitudes exceeding 70 degrees geomagnetic for KP between 0 and 2 and KP between 3 and 5. The color coding is in equal linear steps of the average energy between 0 and 800 eV and in equal logarithmic steps of the average integral number fluy between 5×10^5 and 10^8 e/cm^2 sec ster. The axes are labeled in degrees along the noon-midnight (top to bottom) and along the dusk-dawn (left to right) meridians.

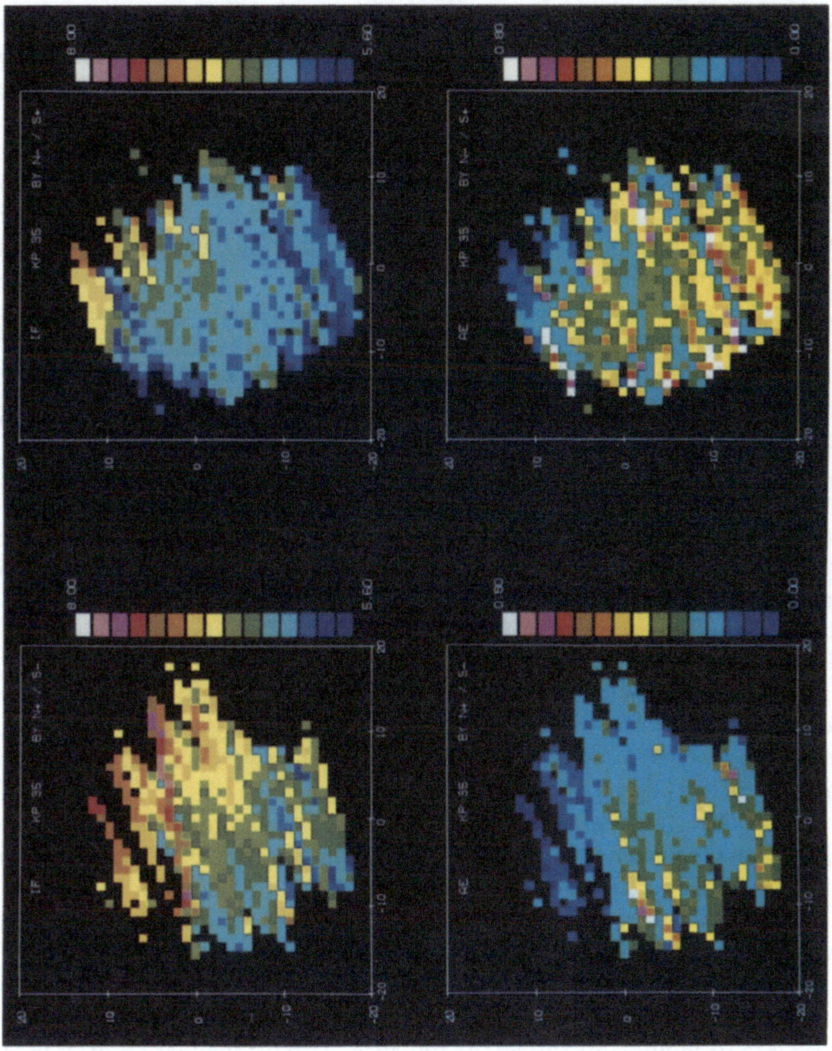

Fig. 4. A plot in the same format as Fig. 3 where the data in the range KP 3 to 5 have been seperated by the sign of the IMF B_y component. Passes over the north pole with B_y positive (negative) have been combined with passes over the south pole for B_y negative (positive).

the integral number flux is positive from dawn to dusk and that for the opposite orientations the gradients reverse. We see no such reversal in the large scale average pattern. To examine this in more detail a set of 262 passes from the DMSP F2 satellite, for which IMF data were available, were studied (RIEHL and HARDY, 1985). We found that the pattern of the integral number flux could be separated into five classes as shown in Fig. 5; a clear positive gradient in the integral number flux from dawn to dusk along the satellite track, a clear negative gradient, a curved contour with the maximum in the middle of the cap, a curved contour with a minimum in the middle of the cap, and a contour with no discernible gradient. Of the 262 cases studied, 10.9% showed a positive dawn to dusk gradient, 48.7% a positive dusk to dawn gradient, 25.8% no gradient, 12.7% a maximum in the center of the cap, and 1.9% a minimum in the center of the cap. Since these passes are approximately equally distributed between poles, if the results stated by MENG and coworkers (1977) were correct, there should be approximately an equal number of passes with gradients dawn to dusk and dusk to dawn. Our results show there are 7 times as many passes with no gradient or a positive gradient dusk to dawn as for a positive gradient dawn to dusk.

Additionally we looked at the dependence of the occurrence of the five classes of polar rain patterns on the sign of the B_y component and on the pole. Percentage histograms for the five classes as a function of B_y for the north and south pole are shown in Figs. 6 and 7. In the north pole the effect reported by MENG and coworkers (1977) is weakly present. Cases of positive gradients from dawn to dusk are confined primarily to times with B_y negative and positive gradients dusk to dawn, while not occurring exclusively for B_y positive, are strongly skewed in that direction. It is important to note, however, that for B_y negative there are more cases with no measurable gradient than for a positive gradient dawn to dusk. In the south pole by contrast there is little evidence of the effect reported by MENG and coworkers (1977). The vast preponderance of cases show either no gradient or a gradient dusk to dawn. For those cases with a positive gradient dusk to dawn there is only a weak skew towards negative B_y (57% $B_y<0$, and 43% $B_y>0$). The cases with no gradient show a significant skew towards positive values of B_y.

In summary, the polar rain tends to display a positive gradient in the integral number flux from the evening towards the morning side of the cap. This behavior is most prominent for B_y positive (negative) in the north (south) pole. For the opposite case the tendency is for the gradient to remain in the same direction or to flatten out. Only in a small number of cases is the gradient seen to reverse and appear positive from the morning side to the evening side of the cap. This is consistent with the two dimensional analysis of the polar rain pattern, since there are too few cases with the gradient reversed to significantly affect the statistics. Also for the DMSP orbit in the 0800–2000 meridian and for the rotation of the two dimensional pattern as seen in Fig. 4 for B_y negative (positive) in the north (south) pole, individual passes would tend to display little if any gradient as we found.

In Fig. 8 we show the relationship between the average energy and the integral number flux obtained from the study of the two dimensional pattern of the polar rain. The same approximately linear relationship between the logarithm of the integral

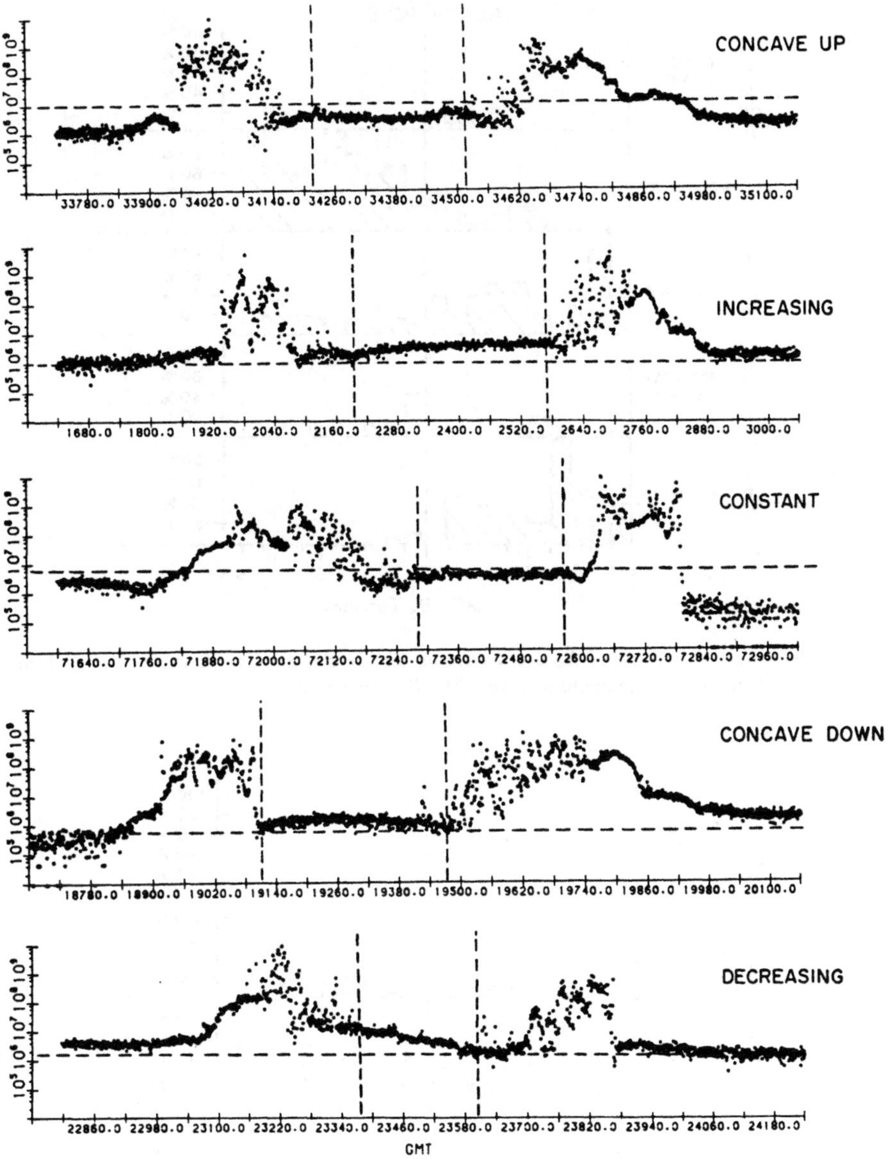

Fig. 5. Plots of the logarithm of the integral number flux of electrons across the polar region for the five classes of gradients of the polar rain observed with the SSJ/3 detector on the DMSP F2 satellite. In each panel dawn is to the left and dusk to the right.

number flux and the average energy is seen to hold for all separations of the data by activity and by the IMF orientation.

In order to examine the source of this relationship it is necessary to consider the detailed spectral characteristics of the polar rain electrons. For this purpose we used

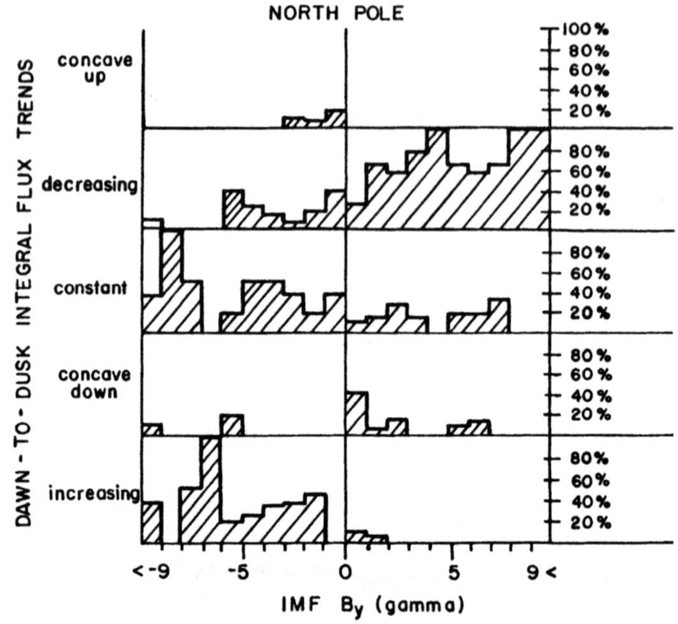

Fig. 6. Percentage histogram of the occurrence of the five classes of gradients in the north pole as a function of the sign and magnitude of the IMF B_y component.

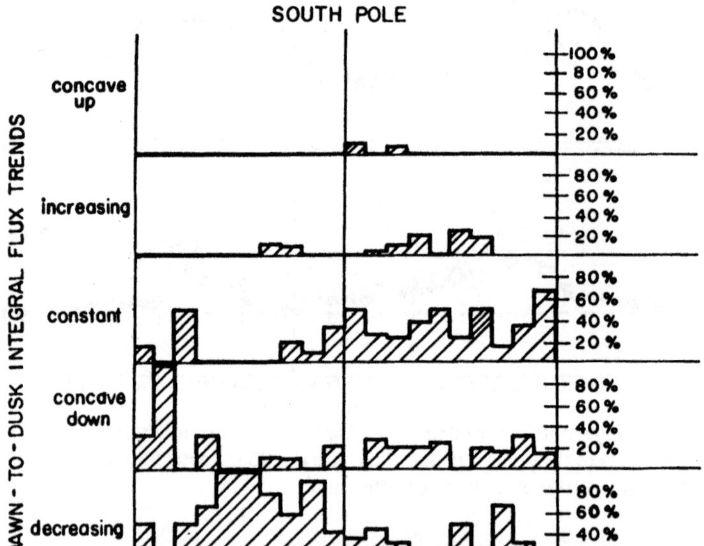

Fig. 7. Same as Fig. 6 for south pole passes.

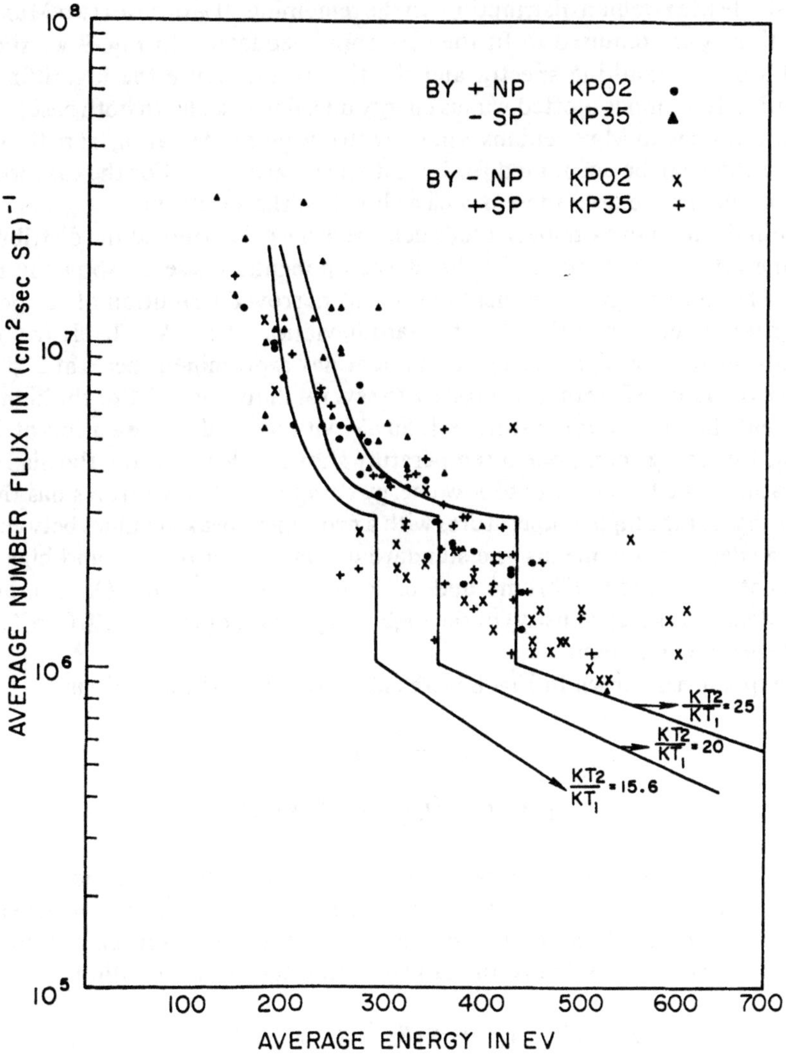

Fig. 8. The average energy of the polar rain versus its integral number flux as calculated from the two dimensional study. Different symbols are used to indicate points obtained from different seperation of the data by the orientation of the IMF and activity as measured by *KP*. The three superimposed curves represent the variation in the average energy with the integral number flux predicted based on the measured relationship between the density of the low and high energy components of the polar rain and for three different average ratios of the temperature of the high and low energy components.

the 262 passes of the DMSP data cited above for which the solar wind and IMF data were available. For these passes 45 second averages of the polar rain spectrum were calculated and the resulting distributions fit to either one or two Maxwellian distributions. In 70 percent of the cases, we found that the distribution could be well fit by a single Maxwellian distribution. In the remaining 30% of cases two Maxwellian distributions were required to fit the spectrum adequately. In Fig. 9 we show two examples of the resulting spectra and the fits. In the figure the logarithm of the distribution function is plotted versus energy on a linear scale. In both cases the fit of the distribution with Maxwellians is good with few points deviating from the straight line more than can be accounted for by statistical variations. For the case where the two Maxwellians are used there is a clear break in the spectrum.

From fits to approximately 1400 such spectra we determined the distribution of the temperatures and densities for the two components. These are shown in Figs. 10 and 11. The low energy component has a very narrow distribution of temperatures with a mean value of 80 eV and a standard deviation of 13 eV. The distribution of temperatures for the high energy component has a prominent peak at 525 eV but extends over a range from several 100 eV to several thousand eV. For the high energy component the mean temperature is 1250 eV with a standard deviation of 750 eV. Since the low energy component temperature shows little variation, the distribution of the ratio of the high energy to low energy component temperatures has the same shape as that for the high temperatures with a prominent peak at values between 5 and 8. For the densities the means and standard deviations for the low and high energy components are $.055 \pm .038/\text{cm}^3$ and $6.1 \times 10^{-4} \pm 3.6 \times 10^{-4}/\text{cm}^3$. On average, the number density (energy density) of the high energy component is 1.1% (6.6%) that of the low energy component.

The quantities shown in Fig. 8 were calculated from the equations

$$IF = \int j(E)\mathrm{d}E$$

and

$$EAVE = \int Ej(E)\mathrm{d}E / \int j(E)\mathrm{d}E,$$

where $j(E)$ is the differential number flux of electrons at energy, E, and the integrals are taken over the detector energy range from 50 eV to 20 keV. For the general case where the distribution function for the electrons in the polar rain can be represented by the sum of two Maxwellians one can show that the above equations reduce to

$$EAVE = 2((1 + R_1 R_2^{1.5})/(1 + R_1 R_2^{.5}))kT_1$$

and

$$IF = 5.5 \times 10^6 \, (n_1(kT_1)^{.5} + n_2(kT_2)^{.5})$$

where the subscripts 1 and 2 refer to the values of the high and low energy components respectively and $R_1 = (n_2/n_1)$ and $R_2 = (kT_2/kT_1)$. In deriving these equations the limits of the integral were taken to be zero and infinity whereas the detector measures only from 50 eV to 20 keV. It is straightforward to show that for typical values of the fitted parameters, taking the integral over the detector energy range makes only a 10 percent difference in the calculated values of $EAVE$ and IF.

In order for the behavior shown in Fig. 8 to hold either R_1 or R_2 or both must increase as n_1 decreases. The required relationship is found only for R_1 and is shown in Fig. 12. In the figure the mean and standard deviation for n_2 is plotted as a function of

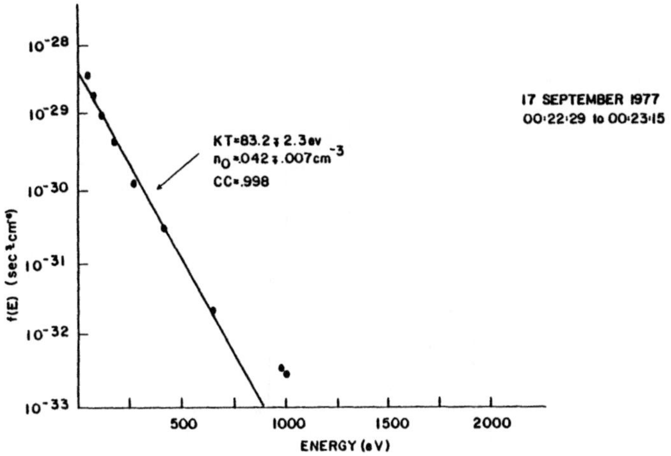

a

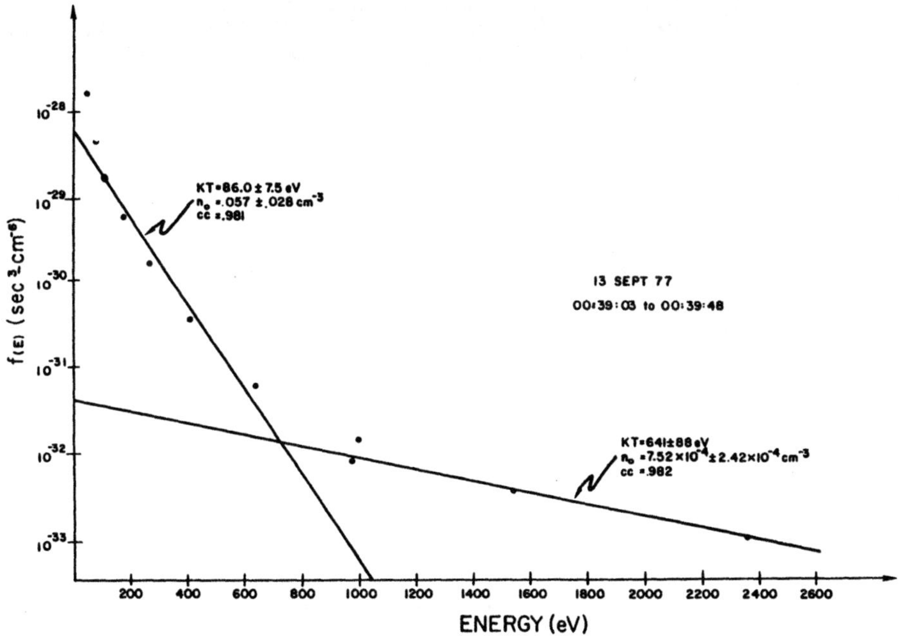

b

Fig. 9. Two examples of the polar rain spectrum where the logarithm of the distribution function is plotted versus energy on a linear scale. The straight lines represent the linear fits to the data and the temperatures and number densities are those derived from the fits.

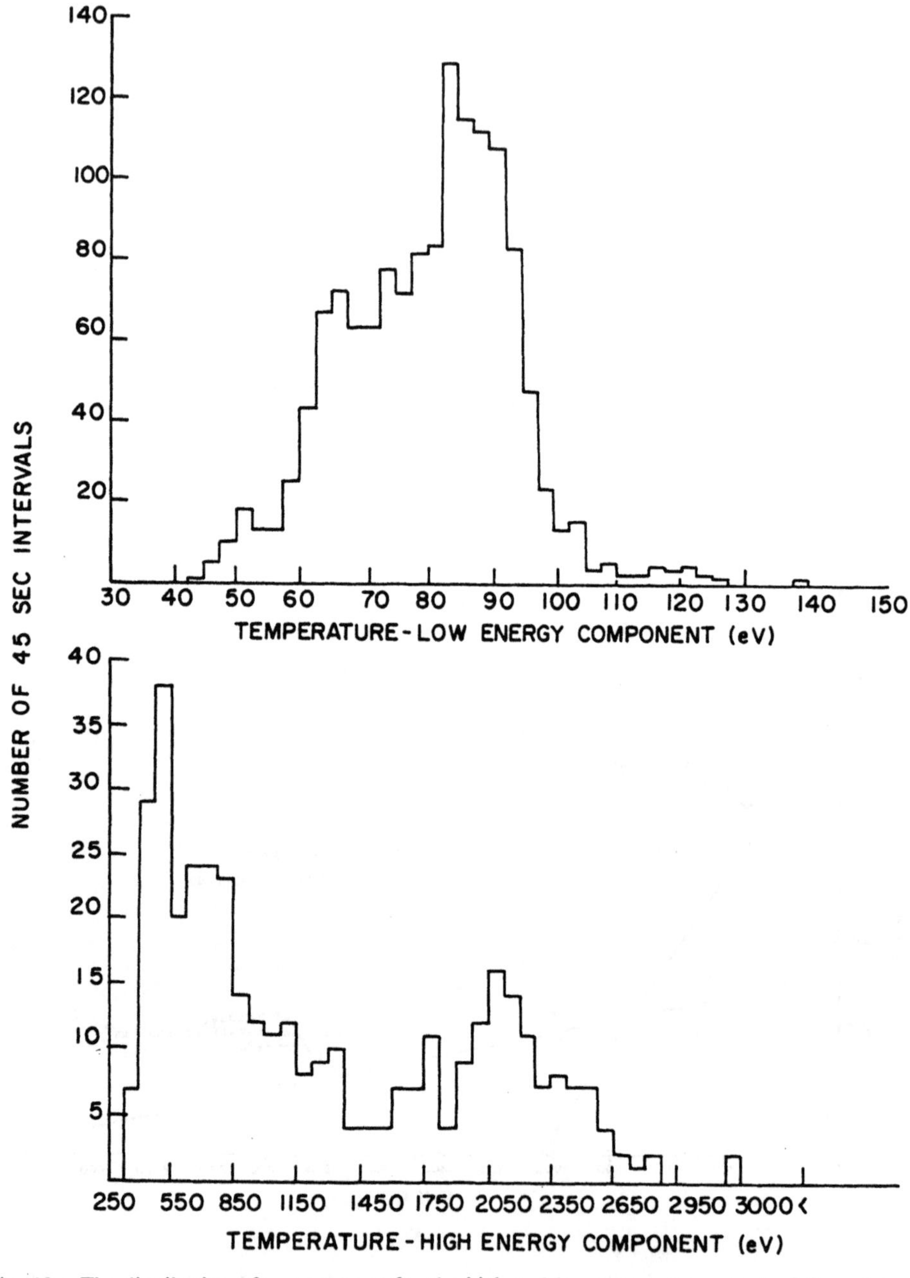

Fig. 10. The distribution of temperatures for the high and low energy component of the polar rain spectrum.

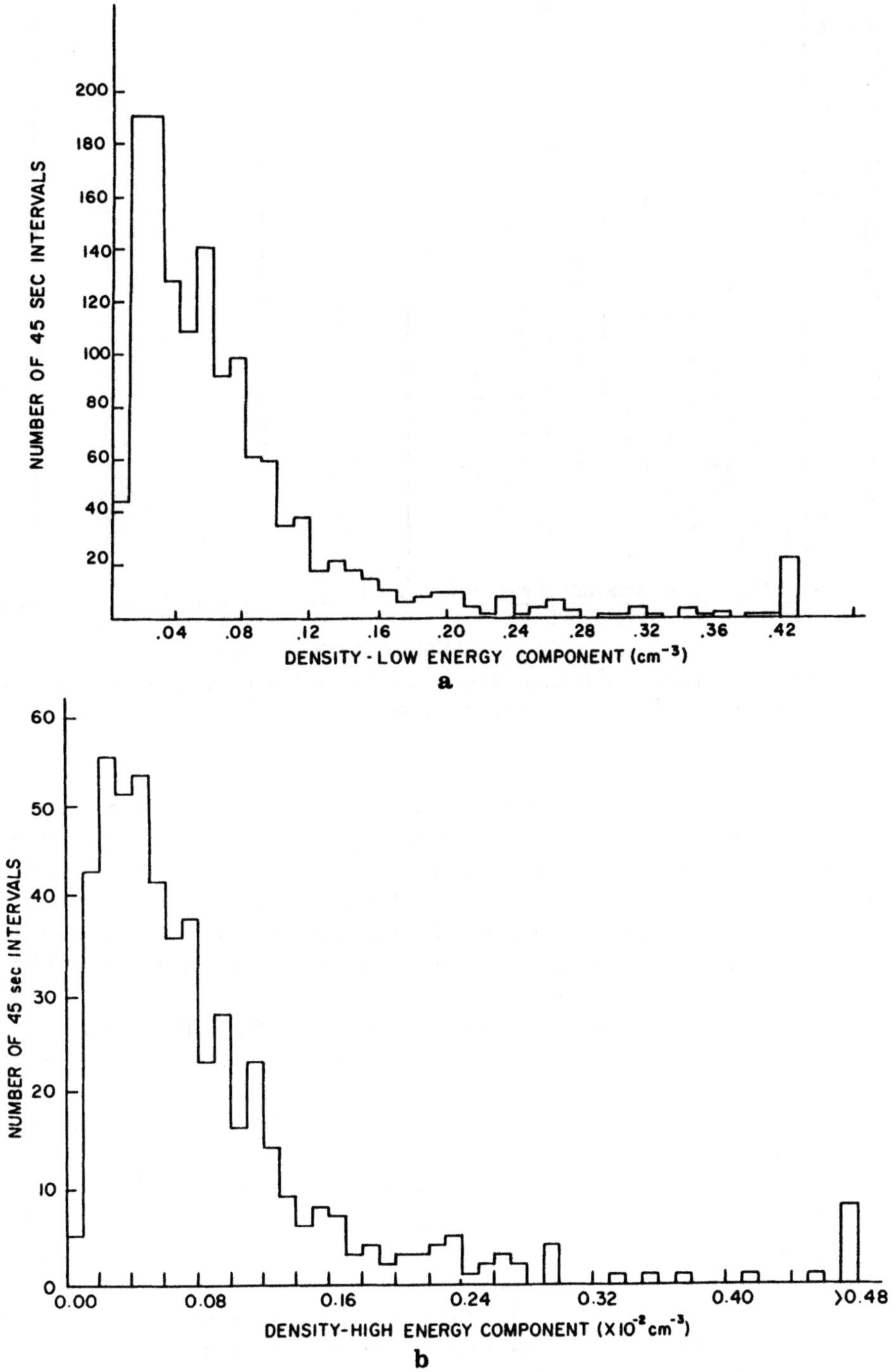

Fig. 11. The distribution of number densities for the high and low energy components of the polar rain spectrum.

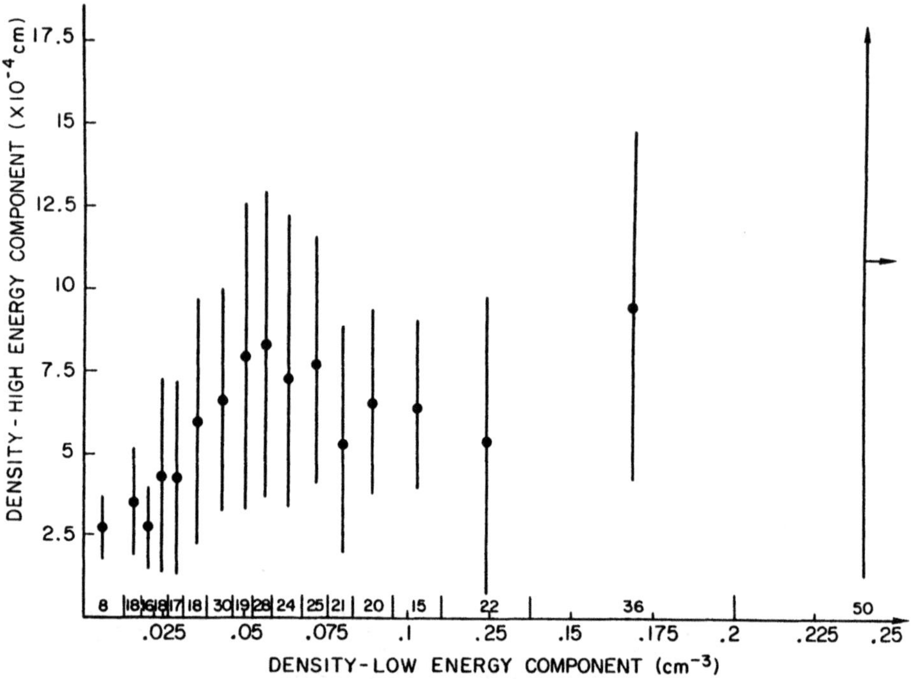

Fig. 12. The mean and standard deviation of the density of the high energy component of the polar rain plotted as a function of the low energy component density.

a series of bins in n_l. One notes that at the smallest values of n_l, R_l is largest with a value of .043 in the first density bin. In the density range from .018 to .058/cm^3 R_l is approximately constant and above this value R_l decreases as n_l increases. In Fig. 8 we show the results of using the curve from Fig. 12 for three values of R_2. The lower curve uses the mean value for the temperatures of the two components. The other curves are for higher values of R_2. It appears that these curves provide a reasonable fit to the results from the two dimensional study.

This result implies that there is a spatial variation in the relative intensity of the high energy component across the polar cap. This is shown in Fig. 13 where we have plotted the average spectrum from the two dimensional study at noon and at midnight. The spectrum clearly has a more prominent high energy component near midnight than at noon.

As a last step in our analysis we sought to measure the ion component in the polar rain. For this analysis we selected passes from the DMSP F6 satellite when the SSJ/4 detector showed a wide region of polar rain electron precipitation as in Fig. 1. Because of the extremely low level of ion flux and even though the SSJ/4 detector has a very large geometric factor we had to sum the ion counts across the entire polar cap in order to produce a measurable spectrum. In Fig. 14 we have plotted the count rate spectrum for one such case where the counts have been summed over 300 seconds. The figure shows that there was a clear signal in the first five channels of the detector. In

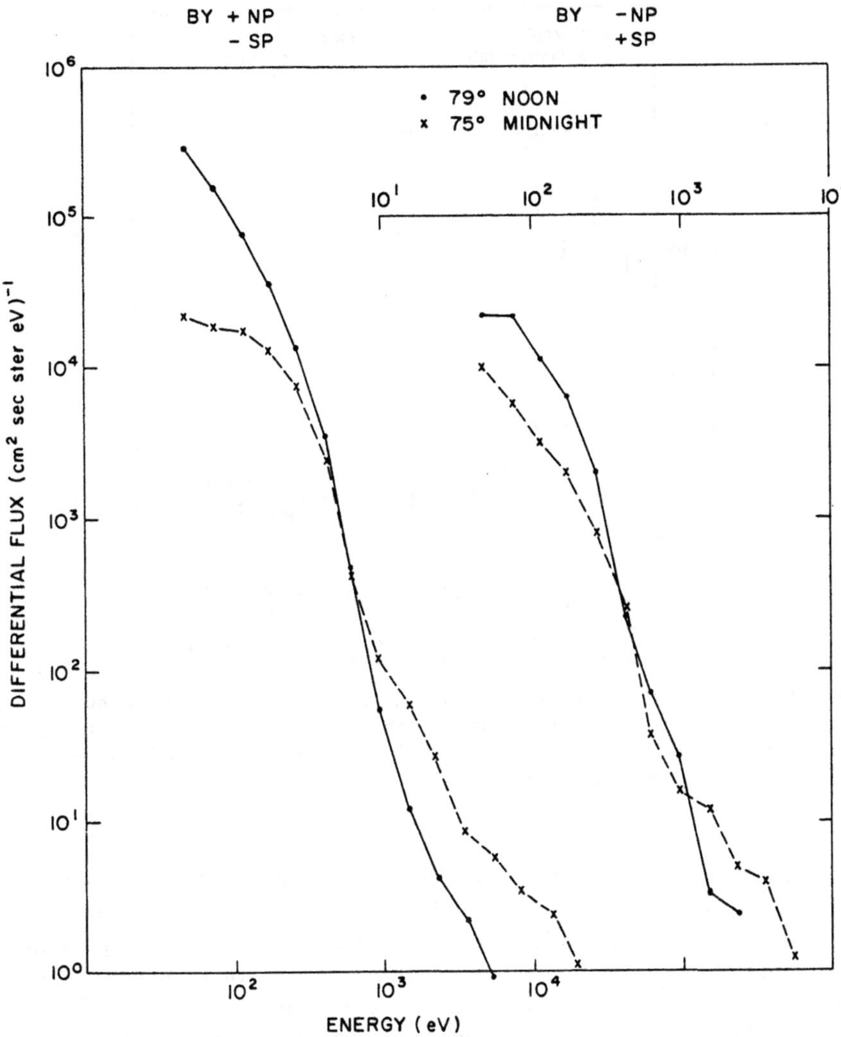

Fig. 13. Average polar rain spectra at 79 degrees CGL, noon (circles) and 75 degrees CGL, midnight (crosses) for B_y positive (negative) in the north (south) pole (left hand side) and for B_y negative (positive) in the north (south) pole (right hand side). *KP* is 3 to 5.

Fig. 15 the distribution function values calculated from these counts are plotted as is the linear fit to the data. The values of the density and the temperature from the linear regression are much lower than the typical values obtained for electrons.

A total of 54 passes were analyzed. In these passes we found that the resulting spectra could be divided into three classes. The first class included thermal spectra as in Fig. 15. For these cases the density was found to range from 1.4×10^{-3} to 7.2×10^{-3} $/cm^3$ and the temperatures from 9 to 41 eV. The second class consisted of spectra

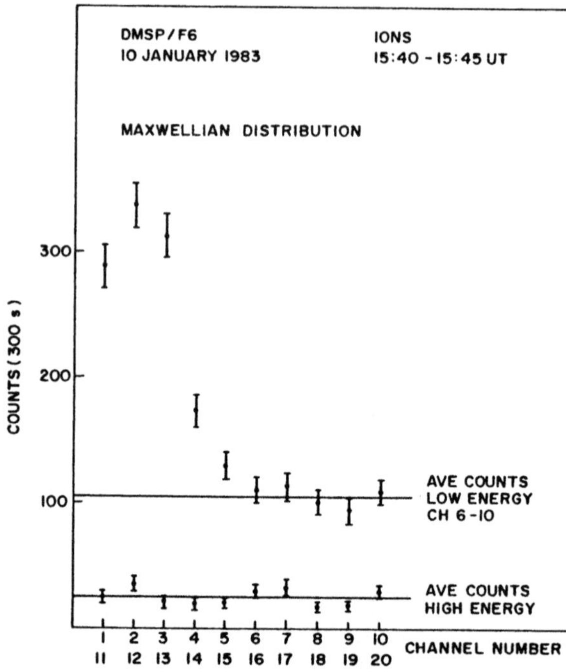

Fig. 14. The summed counts in the 20 ion channels of the SSJ/4 detector for a three hundred second interval over the polar rain. The channels labeled 1 through 10 are for the channels of the low energy detector between 30 eV and 1 keV. The remaining channels labeled 11 through 20 are the high energy channels in the range from 1 keV to 30 keV.

whose shape could best be fit by a Maxwellian distribution with a bulk flow component. Whether the distribution had a bulk flow cannot be determined from a detector such as the SSJ/4 that looks only in a fixed direction. By fitting the spectra, bulk velocities of 100 to 200 kilometers/second were inferred along with densities in the range from 3.8×10^{-5} to 4.6×10^{-4} cm^{-3} and temperatures of 30 to 75 eV. The last class consisted of passes where the ion spectrum displayed both a thermal component and a component with a inferred bulk velocity. The two components in such cases, when fit, had temperatures, densities, and bulk velocities in the same ranges as those for passes where either the thermal spectrum or the flowing distribution were seen separately.

4. Active Polar Cap Precipitation (Polar Showers and Polar Squalls)

The fact has been well established in the literature that fluxes of low energy electrons with intensities above the level of the polar rain are commonly seen in the polar cap (WINNINGHAM and HEIKKILA, 1974) and that these fluxes can be associated with both visual and subvisual sun aligned arcs (ISMAIL *et al.*, 1975; WEBER and BUCHAU, 1978; HARDY *et al.*, 1982). In our work, we undertook to better

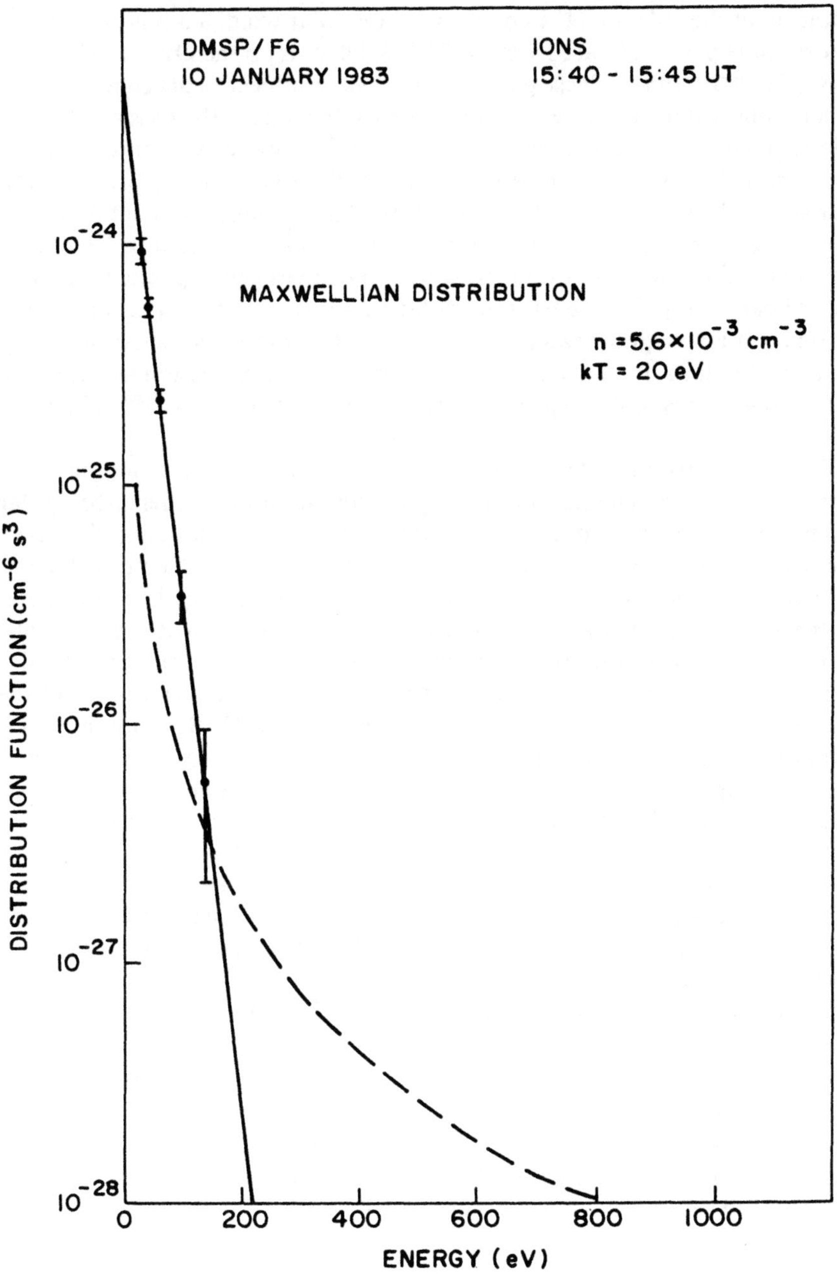

Fig. 15. The logarithm of the distribution function for a case of precipitating polar rain ions plotted versus energy on a linear scale. The straight line is a linear fit to the data and the number density and temperature are from the linear fit.

quantify the frequency with which such events occur and their relationship to the orientation of the IMF and the condition in the solar wind. To this end, we selected 452 north pole passes of data from the SSJ/3 detector on the DMSP F2 satellite for analysis (HARDY, 1984). These passes were selected subject to the conditions that the IMF and solar wind data were available both for the time of the pass and for the hour previous and that the orbit passed to a geomagnetic latitude exceeding 85 degrees. As a first step, each pass was examined to determine if active polar cap precipitation, i.e., "polar shower" or "polar squall", was observed at a geomagnetic latitude above 85 degrees. The key result of this analysis was the determination that in 229 cases active polar cap precipitation was observed above 85 degrees geomagnetic latitude. In the other 233 cases only the unperturbed polar rain was observed. This means that the occurrence of active polar cap precipitation is as common as the occurrence of the unperturbed polar rain and is not a rarely occurring phenomenon as has often been maintained for visual polar cap arcs or theta aurora (ISMAIL *et al.*, 1977; FRANK *et al.*, 1982).

We next determined the relationship between the IMF B_z component and the occurrence of the two classes of polar cap precipitation i.e., unperturbed polar rain and active polar cap precipitation at geomagnetic latitudes exceeding 85 degrees. The occurrence of the passes in the two classes was binned in 1 gamma intervals from -12 to $+12$ gamma. The results are shown in Fig. 16. The three panels in the figure give, from top to bottom, the distribution of the number of cases where active polar cap precipitation was observed, the distribution of cases for the unperturbed polar rain and the total number of cases irrespective of the class. The occurrence of the two classes is strongly skewed according to the sign of the IMF B_z component. For the observations of active polar cap precipitation, 80% occur for B_z positive and for the cases of the unperturbed polar rain, 73% occur for B_z negative. This is in agreement with the studies of the polar rain cited earlier in this paper (GUSSENHOVEN *et al.*, 1984; RIEHL and HARDY, 1985). For the passes chosen for the two dimensional polar rain study approximately 70 percent occurred for B_z negative and a similar percentage was found for those passes used in the study of the spectral characteristics of the polar rain. This skew in the occurrence is a lower bound since the use of hourly averages of the IMF would tend to smear out the results.

Binning in a similar manner we also looked for a dependence of the occurrence of the two classes on the IMF B_x and B_y components. In all such separations no relationship was found. This is to be expected, since the occurrence can be predicted with approximately an 80 percent confidence based just on the sign of the IMF B_z component. Similarly we looked for relationships between occurrence of active polar cap precipitation and the solar wind velocity and number density. Binning the two quantities both separately and together failed to produce any significant ordering of the data.

In the final part of the analysis we scaled off, for each pass exhibiting active polar cap precipitation, the maximum intensity event seen above 85 degrees geomagnetic latitude. In Fig. 17 we have plotted the integral energy flux for these events versus their integral number flux. Diagonal lines have been drawn for fixed values of the average energy. One notes that the "polar showers" occur over three orders of

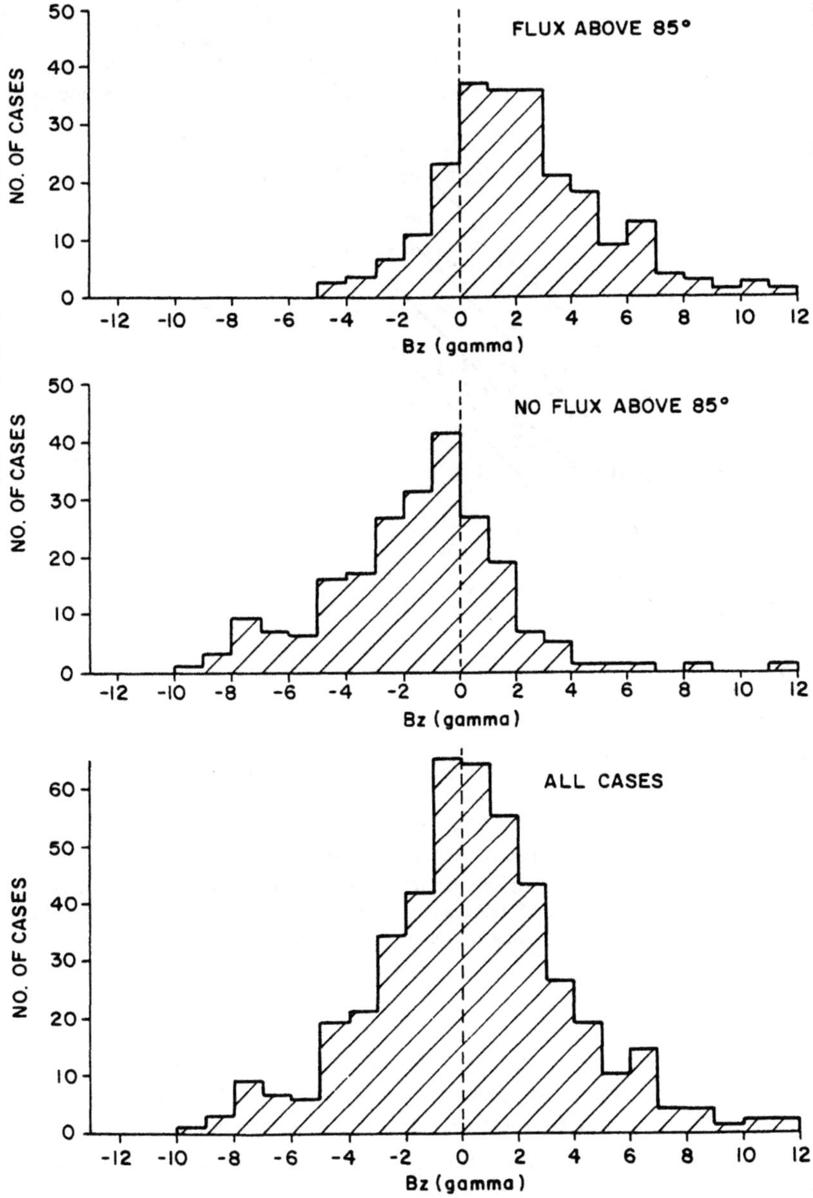

Fig. 16. The three panels show from top to bottom the distribution of passes with respect to the IMF B_z for which active polar cap precipitation was seen above 85 degrees, for which only polar rain was observed, and the distribution irrespective of the polar cap precipitation pattern.

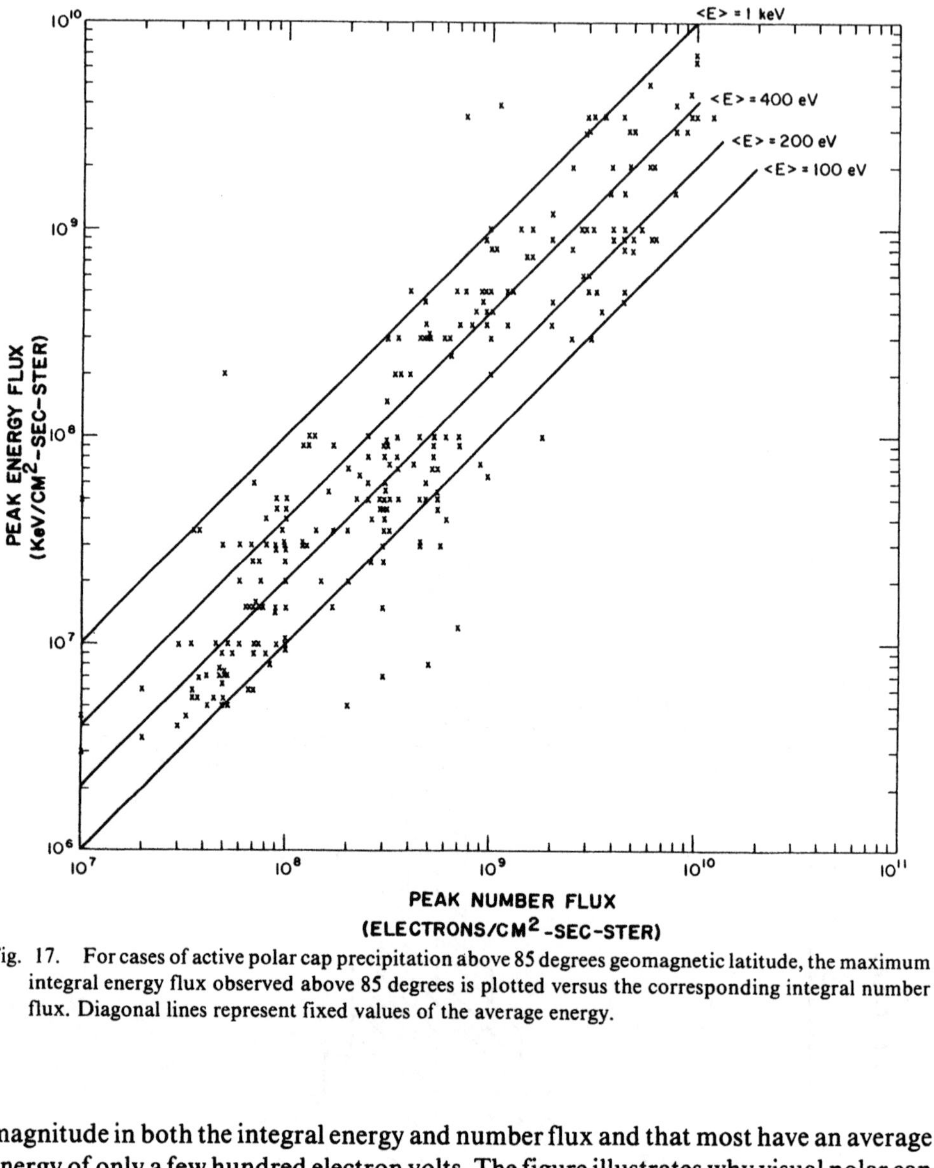

Fig. 17. For cases of active polar cap precipitation above 85 degrees geomagnetic latitude, the maximum integral energy flux observed above 85 degrees is plotted versus the corresponding integral number flux. Diagonal lines represent fixed values of the average energy.

magnitude in both the integral energy and number flux and that most have an average energy of only a few hundred electron volts. The figure illustrates why visual polar cap arcs are reported to occur with relatively low frequency (ISMAIL *et al.*, 1977) to moderately high frequency (GUSSENHOVEN, 1982). Which of the events in Fig. 17 can be seen with optical systems will depend on the sensitivity of that instrumentation; with the number increasing as the sensitivity of the optical system increases. Bright polar cap arcs and "theta aurora" therefore represent merely the most intense cases of a general class of events whose intensity varies over this wide range.

We also looked for relationships between the magnitude of the IMF B_z positive component and the most intense active polar cap precipitation observed above 85 degrees geomagnetic latitude. In Fig. 18 we have plotted the integral energy flux of these events versus the IMF B_z component. The integral energy flux is seen to occur above a threshold that increases for increasing values of the IMF B_z component. A similar plot of the integral number flux shows the same behavior. Taking the equations for the straight lines representing these thresholds, the corresponding equation for the average energy is,

$$EAVE = 125 \times 10^{1.2(B_z/14.7)} \text{ (eV)}$$

where B_z is given in gamma. For typical values of B_z the average energy is on the order of 100 to 200 eV, i.e., typical magnetosheath temperatures. We take the thresholds to represent the level at which electrons are getting into the magnetosphere from the magnetosheath. Values above the threshold result from processes, such as acceleration, that the electrons experience once they are within the magnetosphere.

With the availability of the high sensitivity ion data from the SSJ/4 detectors, we have been able to look at both the ion and electron signatures for passes where intense fluxes of particles are seen at high latitudes and above the level of the polar rain. Based upon examination of a number of passes it appears that such cases can be separated into two additional classes. The first of these classes is illustrated in a color spectrogram format in Fig. 19. This is a pass of southern hemisphere data from the

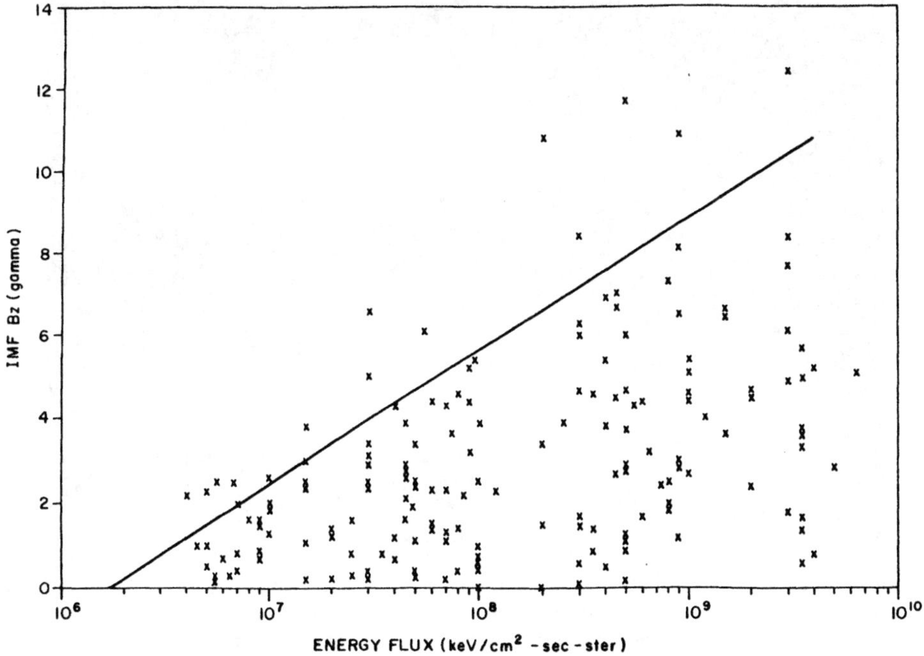

Fig. 18. For cases of active polar cap precipitation above 85 degrees geomagnetic latitude the maximum integral energy flux is plotted versus the magnitude of the IMF B_z for the hour in which the measurement was made.

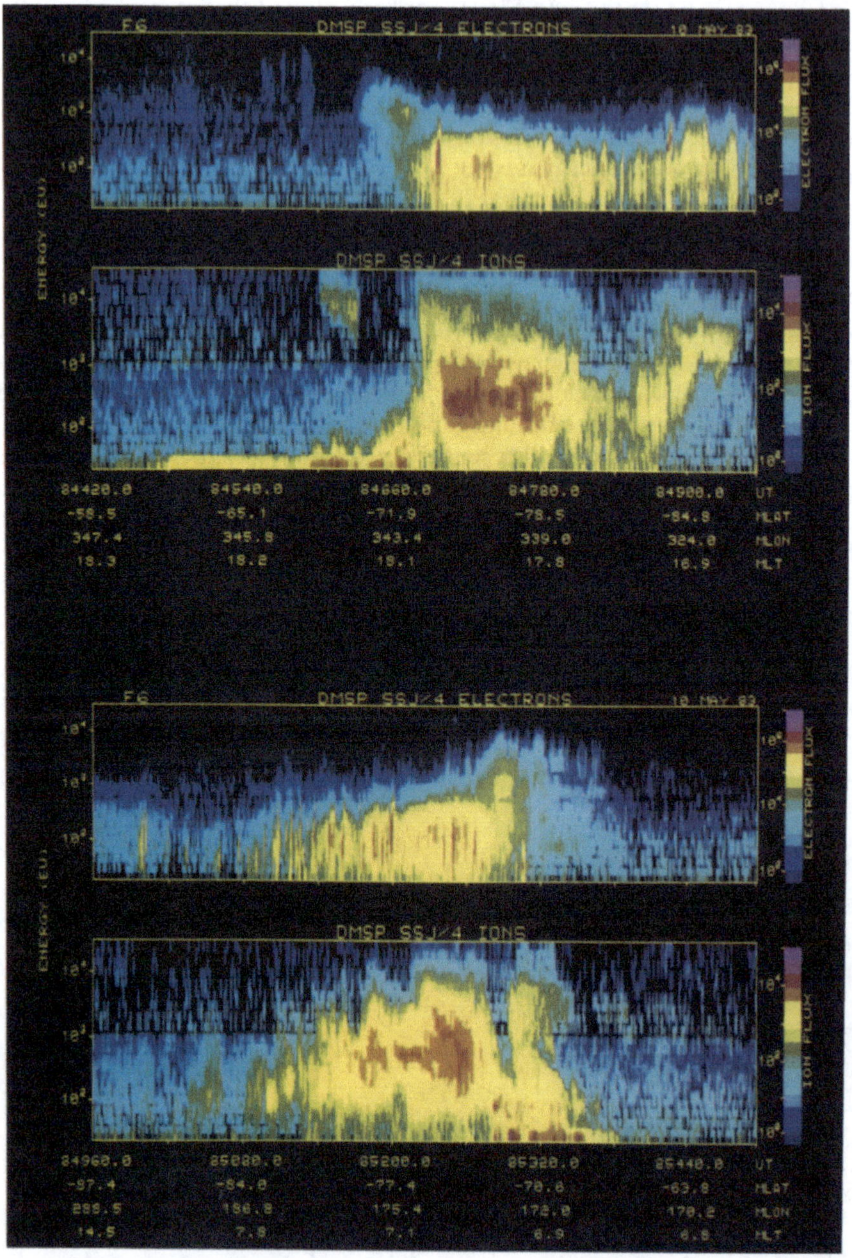

Fig. 19. The electron and ion data from the SSJ/4 plotted in a color spectrogram format for a south pole pass of the F6 satellite on 10 May, 1983. The data are plotted in units of electrons/cm^2 sec ster eV. The electron and ion color scales are offset by a factor of one hundred. The bottom of the figure is annotated with the universal time in seconds and the magnetic latitude (MLAT), magnetic longitude (MLON) and magnetic local time (MLT) of the satellite, all mapped down the magnetic field to 110 kilometers.

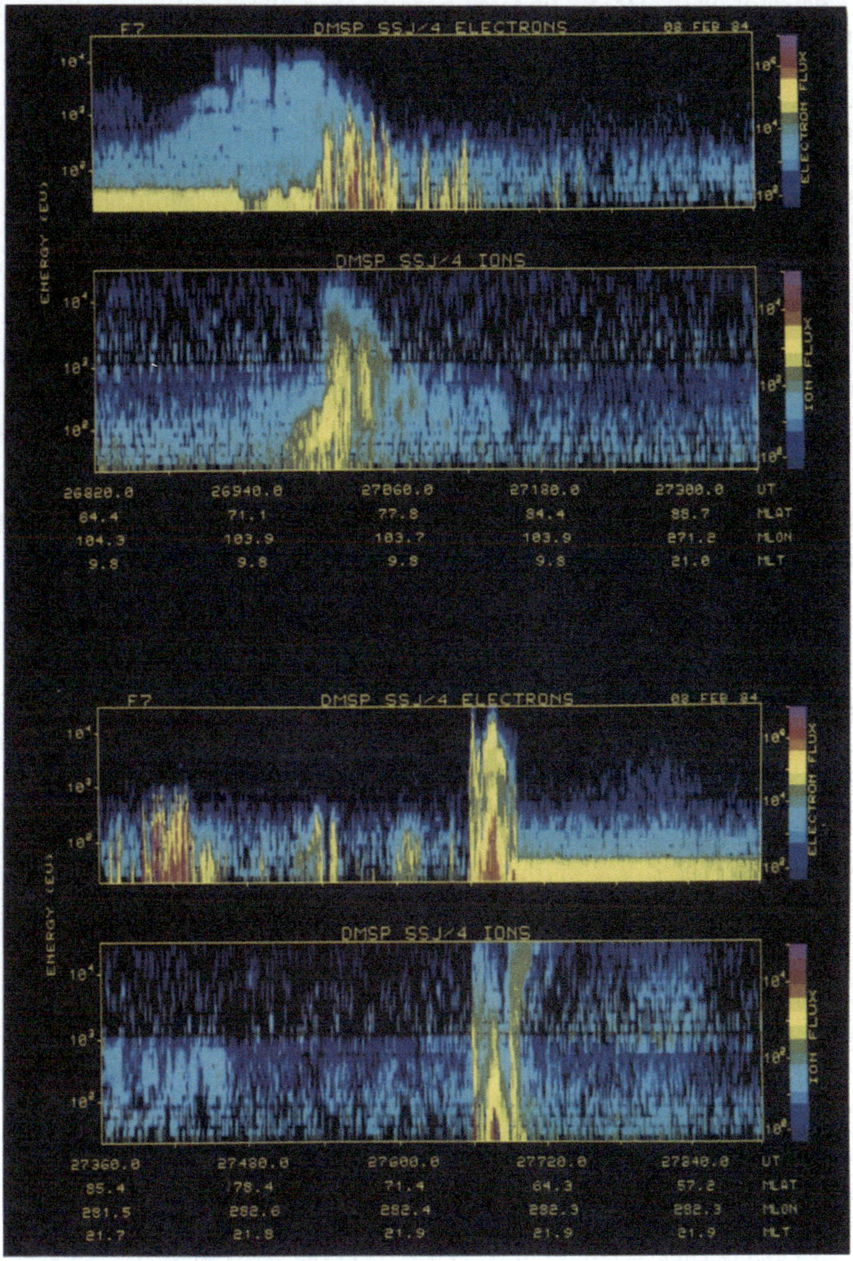

Fig. 20. The same format as Fig. 19 for a north pole pass of the F7 satellite on 8 February, 1984.

SSJ/4 detector on the DMSP F6 satellite on 10 May, 1983. In this pass the satellite reached a maximum geomagnetic latitude of −88 degrees. In the pass there is a clear region of diffuse auroral or central plasma sheet precipitation seen on both the evening and morning sides of the oval. In this region the level of electron and ion precipitation is relatively weak with fluxes observed to energies exceeding 1 keV. On both sides of the oval there is the same very sharp transition between the central plasma sheet and the boundary plasma sheet (at approximately 84670 seconds on the evening side and 85300 seconds on the morning side). For the electrons the intensity of the high energy electrons decreases while that of the low energy electrons increases dramatically. For the ions there is a similar increase in intensity and a general tendency for the ions to be observed at higher energies. In this class of events what is of note is that the electron and ion precipitation appears to extend up to 87 degrees geomagnetic latitude without ever falling to the level of the polar rain. In such passes it appears that the boundary plasma sheet and/or cleft has simply expanded poleward. For such cases this expansion is seen in both poles.

The second class is illustrated in Figure 20. Here we have plotted a northern hemisphere pass of SSJ/4 data from the DSMP F7 satellite on 8 February, 1984. The satellite passed to a maximum geomagnetic latitude of approximately 89 degrees. In this pass a very narrow auroral oval is seen on the evening side poleward of which the level of electron and ion precipitation falls to the level of the polar rain. On the morning side there is a clear region of diffuse auroral precipitation poleward of which electrons and ions of the cusp or cleft are seen. Above the cleft the precipitation again drops to the level of the polar rain. At approximately 81 degrees a very intense case of "polar showers" is observed along with a very weak enhancement in the ion flux. For this case the enhancement is clearly unassociated with the oval precipitation of the boundary plasma sheet and/or cleft. The enhancement is clearly embedded in the polar rain, i.e., the level of the electron flux observed both before and after the event exhibits the characteristics of the polar rain. For such cases the level of ion precipitation associated with such electron enhancements varies from not measurable to moderately intense.

5. Discussion

In summary, there is an active and a quiet state of the precipitation in the polar cap. The active state occurs when the IMF B_z component is positive and is characterized by highly structured fluxes of electrons with typical average energies of 100 eV and with relatively high number densities. These structured fluxes often show evidence of field aligned accelerations which, when sufficiently intense, produce visual polar cap arcs. The quiet state occurs when the IMF B_z component is southward and the oval is more active. This state of the cap is characterized by polar rain electron precipitation that is uniformly low in intensity. Since the IMFB_z has equal probability of being positive or negative, the two states occur with approximately equal probability.

In general, when the IMF B_z component is northward, active polar cap precipitation is seen up to geomagnetic latitudes exceeding 85 degrees. The maximum intensities of these events are found to cover a continuum of values with bright polar cap arcs and theta aurora representing only the most intense members of a general class. Active polar cap precipitation events appear to be either the result of the expansion of the boundary plasma sheet and/or cleft to high latitudes or the occurrence of isolated enhancements of the electron and ion flux clearly embedded in the polar rain. For both such cases the enhancements occur in both poles simultaneously.

For the quiet state of the polar cap, the unperturbed polar rain is observed within a circular region, centered towards pre-midnight, and with a radius that increases with activity. There are large scale spatial variations in the integral number flux and average energy across the circular region. These variations are approximately symmetric about a meridian slightly pre-midnight and pre-noon along which the integral number flux (average energy) increases (decreases) from the nightside to the dayside of the cap. The general tendency is for the gradients in these quantities to be maintained in this sense irrespective of the sign of the IMF B_y component. For B_y positive (negative) in the north (south) pole the gradients are clearly present in the sense stated above. For B_y negative (positive) in the north (south) pole the gradients remain in the same sense or disappear. Only in a small minority of cases do the gradients reverse with the integral number flux and average energy observed to decrease (increase) from nightside to the dayside of the cap. In the two dimensional statistical model this change in behavior is seen as a rotation of the pattern of integral number flux and average energy to be more symmetric about the noon-midnight meridian.

The electron spectrum of the polar rain is well fit by either one or two Maxwellian distributions. The colder of the two Maxwellians has a temperature of 80 eV $\pm$ 13 eV and a number density of $.055/\mathrm{cm}^3 \pm .038/\mathrm{cm}^3$. The high energy Maxwellian, that could be fit in 30% of the spectra studied, has an average temperature of 1250 eV $\pm$ 750 eV and a number density of $6.1\times10^{-4}\pm3.6\times10^{-4}/\mathrm{cm}^3$. The relationship between the average energy and the integral number flux in the polar rain results from a spatial variation across the cap in the relative intensity of the high and low energy components. The precipitating ions in the polar rain are found to have much smaller densities than the polar rain electrons and can be fit either to a purely thermal or flowing Maxwellian distribution.

Our observations of the active cap place constraints on any theory which attempts to account for these phenomena. First, since the intensity of the events appears to occur within a continuum from very low energy depositions that would produce sub-visual arcs as reported by WEBER and BUCHAU (1981) up to very large energy depositions that would correspond to bright polar cap arcs and theta aurora (FRANK *et al.*, 1982), any theory must be able to account for the entire class of events. This is especially true since a necessary and sufficient condition for these events to occur, irrespective of intensity, is that the IMF B_z component be northward. Second, any theory of high latitude dynamics must take into account the fact that in the typical active cap numerous enhancements are seen across the cap. One cannot limit the explanation to that of the "theta bar". If a theory is to invoke bifurcation of the tail lobes such that these events appear on closed field lines then in a typical event the tail lobe must have been split many times by regions of closed field lines. Third, these events cannot be looked upon as exclusively resulting from the expansion of the auroral oval (MENG, 1981a). Although there are numerous cases where the precipitation at high latitudes is contiguous with the precipitation in the oval, there are an equal if not greater number of cases where the enhancements appear clearly embedded in the polar rain. Even for those cases where the precipitation is contiguous it appears as an extension of the boundary plasma sheet and/or cleft. Thus, based on the pattern of particles alone we cannot say definitively whether the precipitation is occurring on open or closed field lines.

Even the observation of the polar rain is no assurance that the field lines on which it occurs are open. Flux tube volume increases with increasing latitude. If at very high latitudes a flux tube were to go from open to closed due to processes deep in the tail, the total plasma content on the flux tube would be great enough to maintain precipitation in the cap so long as sufficient wave particle interactions were taking place to keep the loss cone filled.

Our observations of the quiet polar cap provide some support for the theory that the polar rain originates from the strahl component of the solar wind (ROSENBAUER *et al.*, 1976; FELDMAN *et al.*, 1978; OGILVIE and SCUDDER, 1981) as suggested by FAIRFIELD and SCUDDER (1985). The strahl component of the solar wind is a highly field aligned population of electrons that originates in the solar corona and reaches the vicinity of the earth without having been scattered. The temperature we determined for the low energy component of the polar rain electrons is in good agreement with the temperatures of 65 eV reported by OGILVIE and SCUDDER (1981) for the strahl and 83 eV reported by FELDMAN *et al.* (1978). In addition the average density of the low energy component of the polar rain is close to the value of $8.6\times10^{-2}/cm^3$ reported by FAIRFIELD and SCUDDER (1985) in the lobes.

In addition the high energy population in the polar rain has the proper characteristics for the high energy tail of the strahl distribution. SCUDDER and OLBERT (1979 a and b) have shown for conditions in the solar corona leading to the strahl the high energy tail should appear at approximately 7 times the local most probable electron energy. The fact that the distribution of the ratios of the high energy component to low energy component temperatures showed a clear peak between 5 and 8 clearly supports the identification of the high energy component as the tail of the strahl.

Thus our results offer some support for the theory that the polar rain originates in the solar wind and gains access to the caps along open field lines connecting through the lobes to the solar wind. Such a theory, however, leaves several unanswered questions. First, it does not explain the presence of the gradients in the polar rain or their response to the IMF. Second, it does not explain why there is a variation in the relative intensity of the two components across the polar cap. Lastly, it leaves the question of flux balance unanswered. As we have shown the flux of electrons in the polar cap far exceeds the flux of ions. In order for charge balance to be maintained there must be a net outflow of electrons from the polar ionosphere. Whether this return flux is made up by the polar wind or from photoelectrons and how the various plasma regions interact remain open questions.

REFERENCES

Berkey, F. T., L. L. Cogger, S. Ismail, and Y. Kamide, Evidence for a correlation between sun aligned arcs and the interplanetary magnetic field direction, *Geophys. Res. Lett.*, **3**, 145, 1976.

Burch, J. L., Low energy electron fluxes at latitudes above the auroral zone, *J. Geophys. Res.*, **73**, 3585, 1968.

Burch, J. L., S. A. Fields, and R. A. Heelis, Polar cap acceleration regions, *J. Geophys. Res.*, **84**, 5863, 1979.

Burke, W. J., M. S. Gussenhoven, M. C. Kelley, D. A. Hardy, and F. J. Rich, Electric and magnetic field characteristics of discrete arcs in the polar cap, *J. Geophys. Res.*, **87**, 2431, 1982.

Davis, T. N., The morphology of the polar auroras, *J. Geophys. Res.*, **65**, 3497, 1960.

Eather, R. H. and S.-I. Akasofu, Characteristics of polar cap auroras, *J. Geophys. Res.*, **74**, 4794, 1969.

Fairfield, D. H. and J. D. Scudder, Polar rain: Solar coronal electrons in the Earth's magnetosphere, *J. Geophys. Res.*, **89**, 4655, 1985.

Feldman, W. C., J. R. Asbridge, S. J. Bame, J. T. Gosling, and D. S. Lemous, Characteristic electric field variations across simple high-speed solar wind streams, *J. Geophys. Res.*, **83**, 5285, 1978.

Fennell, J. F., P. F. Mizera, and D. R. Croley, Jr., Low energy polar cap electrons during quiet times, in *Proc. Int. Conf. Cosmic Rays 14th*, **MG8-3**, 1267, 1975.

Foster, J. C. and J. R. Burrows, Electron fluxes over the polar cap, 1, Intense keV fluxes during poststorm quieting, *J. Geophys. Res.*, **81**, 6016, 1976.

Foster, J. C. and J. R. Burrows, Electron fluxes over the polar cap, 2, Electron trapping and energization on open field lines, *J. Geophys. Res.*, **82**, 5165, 1977.

Frank, L. A., J. D. Craven, J. L. Burch, and J. D. Winningham, Polar views of the earth's aurora with Dynamics Explorer, *Geophys. Res. Lett.*, **9**, 1001, 1982.

Gurgiolo, C. and J. L. Burch, DE-1 observations of the polar wind—A heated and an unheated component, *Geophys. Res. Lett.*, **9**, 945, 1982.

Gussenhoven, M. S., Extremely high latitude auroras, *J. Geophys. Res.*, **87**, 2401, 1982.

Gussenhoven, M. S., D. A. Hardy, N. Heinemann, and R. K. Burkhardt, Morphology of the polar rain, *J. Geophys. Res.*, **89**, 9785, 1984.

Hardy, D. A., Intense fluxes of low-energy electrons at geomagnetic latitudes above 85 degrees, *J. Geophys. Res.*, **89**, 3883, 1984.

Hardy, D. A., M. S. Gussenhoven, and A. Huber, The precipitating electron detectors (SSJ/3) for the block 5D/flights 2-5 DMSP satellites, calibration and data presentation, Rep. AFGL-TR-79-0210, Air Force Geophysics Lab., Hanscom Air Force Base, Bedford, Mass., 1979.

Hardy, D. A., W. J. Burke, and M. S. Gussenhoven, DMSP optical and electron measurements in the vicinity of polar cap arcs, *J. Geophys. Res.*, **87**, 2412, 1982.

Hardy, D. A., L. K. Schmitt, M. S. Gussenhoven, F. J. Marshall, H. C. Yeh, T. L., Schumaker, A. Huber, and J. Pantazis, Precipitating Electron and Ion Detectors (SSJ/4) for the block 5D/Flights 6-10 DMSP satellites: calibration and data presentation, Rep. AFGL-TR-84-0317, Air Force Geophysics Laboratory, Hanscom AFB, Mass., 1984.

Ismail, S., D. D. Wallis, and L. L. Cogger, Characteristics of the polar cap sun-aligned arcs, *J. Geophys. Res.*, **82**, 4741, 1977.

Lassen, K. and C. Danielsen, The quiet time pattern of auroral arcs for different directions of the interplanetary field in the *Y–Z* plane, *J. Geophys. Res.*, **83**, 5277, 1978.

Meng, C.-I., Polar cap arcs and the plasma sheet, *Geophys. Res. Lett.*, **8**, 273, 1981a.

Meng, C.-I., The auroral electron precipitation during extremely quiet geomagnetic conditions, *J. Geophys. Res.*, **86**, 4607, 1981b.

Meng, C.-I. and S.-I. Akasofu, the relationship between the polar cap auroral arc and the auroral oval arc, *J. Geophys. Res.*, **81**, 4004, 1976.

Meng, C.-I. and H. W. Kroehl, Intense uniform precipitation of low energy electrons over the polar cap, *J. Geophys. Res.*, **82**, 2305, 1977.

Meng, C.-I., S.-I. Akasofu, and K. A. Anderson, Dawn-dusk gradient in the precipitation of low energy electrons over the polar cap and its relationship to the interplanetary magnetic field, *J. Geophys. Res.*,

82, 5271, 1977.

MIZERA, P. F. and J. F. FENNELL, Satellite observations of polar, magnetotail lobe, and interplanetary electrons at low energies, *Rev. Geophys. Space Phys.*, **16**, 147, 1978.

MONTGOMERY, M. D., S. J. BAME, and A. J. HUNDHAUSEN, Solar wind electrons: Relay measurements, *J. Geophys. Res.*, **73**, 4999, 1968.

OGILVIE, K. W. and J. D. SCUDDER, Observations of the "strahl" by the solar wind spectrometer on Mariner 10, in *Solar Wind Four*, edited by H. Rosenbauer, 226, the Max Planck Institute fur Extraterrestriche Physik, Garching, West Germany, 1981.

OLBERT, S., Inference about the solar wind dynamics from observed distributions of electrons and ions, in *Proc. International School and Workshop on Plasma Astrophysics, in ESA SP-161*, 1981.

PETERSON, W. K. and E. G. SHELLEY, Origin of the plasma in the cross polar cap auroral feature (theta aurora), *J. Geophys. Res.*, **89**, 6729, 1984.

RIEHL, K. and D. A. HARDY, Average characteristics of the polar rain and their relationship to the solar wind and the interplanetary magnetic field, submitted to *J. Geophys. Res.*, 1985.

ROSENBAUER, H., H. MIGGENRIEDER, M. MONTGOMERY, and R. SCHWENN, Preliminary results of the Helios plasma measurement, in *Physics of the Solar Planetary Environment*, edited by D. J. Williams, 319, American Geophysical Union, 1976.

SCUDDER, J. D. and S. OLBERT, A theory of local and global processes which affect solar wind electrons 1, The origin of typical 1 AU velocity distribution functions—steady state theory, *J. Geophys. Res.*, **84**, 2755, 1979a.

SCUDDER, J. D. and S. OLBERT, A theory of local and global processes which affect solar wind electrons 2, experimental support, *J. Geophys. Res.*, **84**, 1979b.

WEBER, E. J. and J. BUCHAU, Polar cap F-layer auroras, *Geophys. Res. Lett.*, **8**, 125, 1981.

WHALEN, B. A., J. R. MILLER, and I. B. MCDIARMID, Sounding rocket observations of particle precipitation in the polar cap electron aurora, *J. Geophys. Res.*, **76**, 6847, 1971.

WINNINGHAM, J. D. and C. GURGIOLO, DE-2 photoelectron measurements consistent with a large scale electric field over the polar cap. *Geophys. Res. Lett.*, **9**, 977, 1982.

WINNINGHAM, J. D. and W. J. HEIKKILA, Polar cap auroral electron fluxes observed with ISIS 1, *J. Geophys. Res.*, **79**, 949, 1974.

WINNINGHAM, J. D., F. YASUHARA, S.-I. AKASOFU, and W. J. HEIKKILA, The latitudinal morphology of 10-eV to 10-keV electron fluxes during magnetically quiet and disturbed times in the 2100–0300 MLT sector, *J. Geophys. Res.*, **80**, 314, 1975.

YEAGER, D. M. and L. A. FRANK, Low-energy electron intensities at large distances over the earth's polar cap, *J. Geophys. Res.*, **81**, 3966, 1976.

Solar Wind-Magnetosphere Coupling, edited by Y. Kamide and J. A. Slavin, 605–631.

Dynamic Variations of the Polar Cap

Ching-I. MENG and Kazuo MAKITA*

Applied Physics Laboratory, The Johns Hopkins University, Laurel, Maryland 20707, U.S.A.

The morphology and dynamics of the polar cap, defined as the region poleward of auroral electron precipitations, are discussed, based on observations made by the low-altitude polar-orbiting satellites of the U.S. Defense Meteorological Satellite Program (DMSP). Particular attention is given to polar cap variations with the interplanetary magnetic field (IMF) and substorm activity. Their implications on the magnetospheric configuration and the solar-terrestrial interaction will also be presented.

It is found that during geomagnetically quiet times the average location of the electron precipitation poleward boundary is at particularly high geomagnetic latitudes: about 82° to 84° in the morning, noon, and evening sectors and at about 81° to 82° in the midnight sector. This indicates a very small polar cap. The equatorward edge of the auroral oval is at about 70° in the morning, noon, and evening sectors and at about 69° in the midnight sector. The latitudinal width of the region of auroral electron precipitation is unusually large, indicating a widening of the auroral oval in periods of geomagnetic quiescence. Faint, stable auroral arcs sometimes are seen in the precipitation region, associated with the so-called polar cap arcs. The location of the poleward boundary (i.e., the polar cap size) is affected by the magnitude of the northward interplanetary magnetic field (IMF) component; however, the location of the auroral precipitation equatorward boundary is not related to the positive IMF B_z magnitude. These results indicate that the unusually high latitude of the auroral oval poleward boundary is the consequence of a very small number of geomagnetic flux lines interconnected with the IMF during a northward IMF condition. They also imply that the size of the polar cap is inversely controlled by the magnitude of the northward B_z component. The insensitivity of the equatorward auroral precipitation boundary to the northward IMF B_z can be used to infer that its dynamics is dominated by internal magnetospheric processes and not by direct solar-magnetospheric interactions.

It is also found that both the dawn–dusk and noon–midnight dimensions of the polar cap tend to vary in harmony with the *AE* index. The increase of the polar cap size begins after the southward turning of the IMF with an increase of the energy input ε parameters but prior to the corresponding increase of the *AE* index. The period of the maximum dimension of the polar cap approximately coincides with the period of the maximum *AE* value. The polar cap size decrease begins at about the same time that the substorm activity begins to subside, but it continues well after the *AE* index is reduced to very small values. If the size of the polar cap is considered as a measure of the open geomagnetic flux and thus of the magnetic energy stored in the magnetotail, these results suggest that the magnetic energy in the magnetotail increases and decreases in harmony with the growth and decay of substorm activity, respectively.

*Permanent Adress: Takushoku University, Tokyo, Japan.

1. Introduction

In the past 20 years, various types of satellite observations have revealed that the earth's magnetosphere is open: namely, that geomagnetic field lines from the polar cap regions extend into interplanetary space and connect with the magnetic field lines of solar origin. Variations of the solar wind and the interplanetary magnetic field (IMF) will undoubtedly affect the configuration of the terrestrial magnetosphere. Associated with an increase of the solar wind dynamic pressure, compression of the dayside magnetosphere and the magnetotail has been detected. The location of the magnetopause can be accurately determined from magnetohydrodynamics by balancing the solar wind dynamic pressure in the interplanetary space with the magnetic pressure of the geomagnetic field inside the magnetosphere. However, responses of the terrestrial magnetosphere to variations of the IMF are far from clear, and the importance of the IMF in the solar-terrestrial interaction, especially in the energy coupling, has been recognized.

One of the possible processes of the energy transfer from the solar wind into the magnetosphere to produce the magnetospheric substorm is the field line merging between the IMF and the geomagnetic field. The dayside northward-directed closed geomagnetic field lines near the magnetopause can be eroded away by a southward-directed IMF (i.e., the geomagnetic field lines are opened by and connected with the interplanetary field lines); consequently, the dayside magnetopause moves earthward. Due to the "frozen-in" condition between the interplanetary field and the solar wind, the antisunward motion of the solar wind carries the newly merged geomagnetic field lines from the dayside magnetosphere (or the magnetosheath) across the polar cap into the magnetotail. Associated with this so-called "magnetic flux transfer" process, one of the expected configurational responses of the magnetosphere is the change in the polar cap size, which is controlled by the flux of open geomagnetic field lines. The polar cap is defined as the area bounded by the auroral oval, and the bundle of the geomagnetic field lines originating in the polar cap is connected (i.e., open) to the IMF of the solar origin. The purpose of this paper is to report progress in the observation and understanding of the polar cap size variation with changes of the IMF; some new results on configurational changes of the auroral oval (i.e., the polar cap) with the different orientation of the IMF are also discussed based on the Defense Meteroro-logical Satellite Program (DMSP) observations. The polar cap size is defined here by using the poleward boundary of the instantaneous auroral electron precipitations.

Usual observations of the auroral electron precipitations have corresponded generally to moderately active geomagnetic conditions (during magnetospheric substorms) when the auroral optical display and electron precipitations are intense and dynamic. Therefore, previous studies of auroral electron precipitations primarily represent the active or moderately active magnetosphere. During periods of relative magnetic quiescence, the global auroral display observed by imagers onboard ISIS-2 and DMSP satellites showed that quiet auroral arcs extend along the contracted poleward edge of the auroral oval and, at times, the luminosity of the auroral oval is even below the detector threshold of these satellites (AKASOFU, 1974). Due to the rarity of extremely quiet magnetospheric conditions and the nondynamic nature of the auroral display, the morphology of the very quiet auroral oval has not yet drawn much of the attention of magnetospheric and auroral physicists (HOFFMAN and BURCH, 1973; WINNINGHAM *et al.*, 1975; LUI *et al.*, 1976; MENG, 1981a, b; MURPHREE *et al.*, 1982). The purpose of this paper is to examine the dynamics of the polar cap inferred from the auroral electron precipitations observed by polar-orbiting USAF DMSP satellites. The main questions of importance are: (1) what is the electron precipitation at very quiet conditions, (2) how small is the polar cap, and (3) what are the distinct variations of the polar cap region from quiet to disturbed geomagnetic conditions?

The DMSP is a continuous U.S. Department of Defense operational system for global weather monitoring. Normally, the program has two satellites in ~840-km sun-synchronous circular polar orbits with a period of 101 min and 98.75° inclination (PIKE, 1975). One satellite is in the dawn–dusk meridian orbit, and the other is in the noon–midnight meridian orbit. Depending on launch dates and system lifetimes, there can be as many as three satellites or as few as one satellite in operation at a given time. Radiometers onboard DMSP satellites produce the familiar images of the global auroral display in the dark hemisphere. In addition to the principal meteorological instruments, since 1974 some of the satellites have carried low-energy electron detectors to monitor auroral electron precipitation. These DMSP satellites provide continual soft electron ($E_e \leqslant 20$ keV) measurements from both northern and southern polar regions, without any systematic data gaps, except from 1980 to 1982. Improvements of particle detectors were made initially form only 8-channel electron measurement between 200 and 20 keV to the present 20-channel measurements of both electrons and ions from 30 eV to 30 keV.

In the following, the auroral electron precipitation and the polar cap size of the quiet magnetosphere will be discussed first, followed by an investigation of dynamic variation of the polar cap with substorm development and IMF. Detailed reports of these results are published elsewhere (MAKITA and MENG, 1984; MAKITA *et al.*, 1983; 1985).

2. Polar Cap of Geomagnetic Quiescence

A typical example of the polar-region domains inferred from electron precipitations observed by a dawn–dusk satellite during the quiet condition is shown in Fig. 1 (October 6, 1978). The top panel illustrates the auroral electron integral number flux, energy flux, and the electron average energy. The bottom left diagram shows the satellite trajectory, given at the top of this diagram, in corrected geomagnetic-latitude local-time coordinates. As the satellite traveled from the evening to the morning sector, it passed through the approximate center of the polar cap, at about 85° geomagnetic latitude along the midnight meridian (HOLZWORTH and MENG, 1975). The bottom right panel shows 24 hours of *AE* index, with arrows indicating this and three other passes used in the later statistical analysis to demonstrate the data selection and the condition of the defined geomagnetic quiescence. In this event, the hourly value of the *AE* index was less than 50 nT from 2000 UT on October 5 to 0300 UT on October 6. A moderate magnetic disturbance (~300 nT) occurred at about

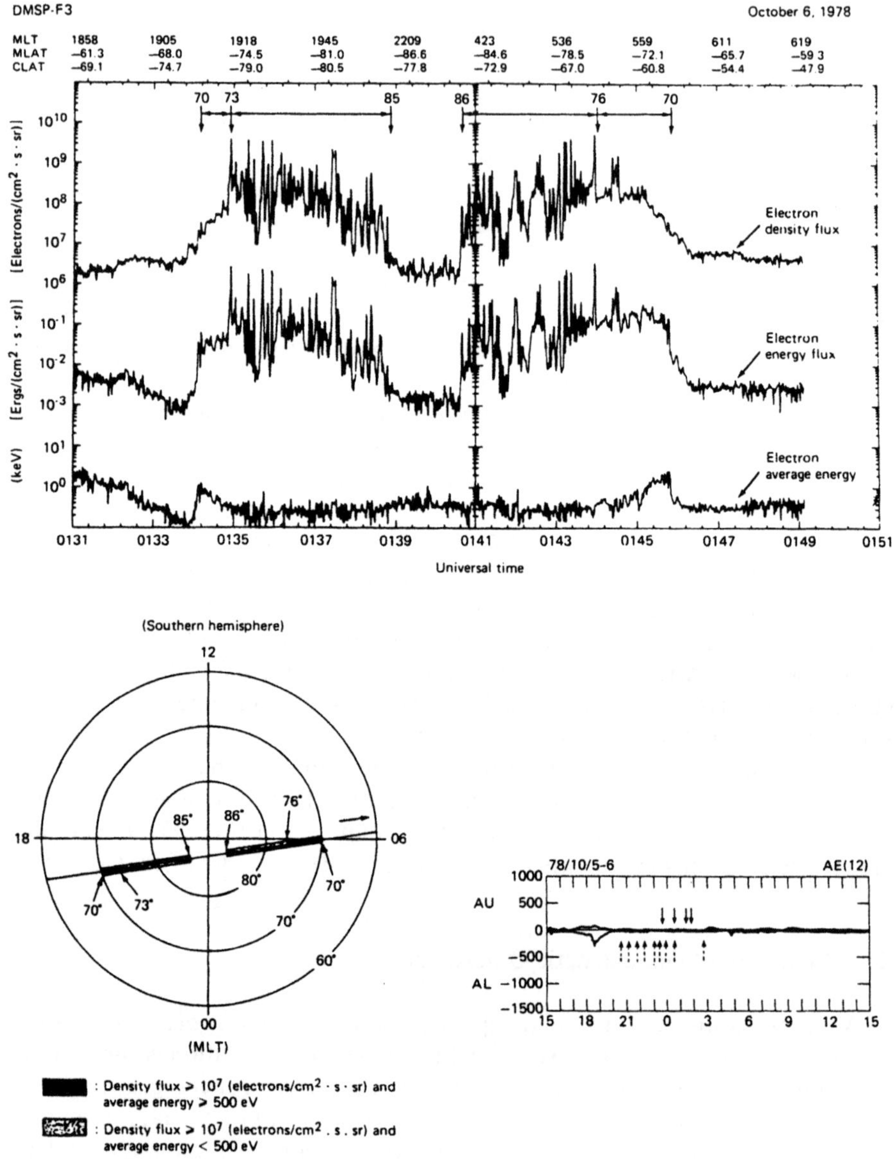

Fig. 1. An example of the auroral electron precipitation in the dawn–dusk sector during a quiet period. Top panel consists of the electron total number flux, energy flux, and average energy observed by the DMSP-F3 satellite. Bottom left panel shows the satellite trajectory in the corrected geomagnetic latitude-local time coordinates as given at te top of the figure. The solid and dotted regions indicate the hard and soft electron precipitation regions, respectively, as defined in the text. The bottom right panel shows 24 hours of AE index with the solid arrows indicating this and three other passes used in the statistical analysis of the study. Dashed arrows indicate available DMSP passes that do not satisfy the quiescence and trajectory selection criteria and that are deleted from this analysis.

1830 UT on October 5 and ended at about 2000 UT. Four orbits of DMSP made observations of electron precipitation that were within 3 hours after the activity (shown by dashed arrows) from 2000 UT to 2300 UT were deleted from the data base. Between 2300 UT and 0300 UT, there were nine DMSP polar passes, but the maximum geomagnetic latitudes of two passes were below 80° and may not traverse the polar cap region. Three others were partial polar-crossing observations. Therefore, out of 13 DMSP passes in this interval, we can use only the four passes shown by solid arrows to study the polar cap morphology in quiet condition. Figure 1 represents one of the four good passes within the time span indicated by the *AE* plot. The enhanced auroral electron number flux and energy flux were detected over most of the polar region above 70° geomagnetic latitude. The equatorward and poleward boundaries of the region of auroral electron precipitation were determined at the edge of the region extended precipitation, where the electron number flux and energy flux rise and/or drop noticeably with respect to the background level. In this example, the region of enhanced electron precipitation extends from 70° to 85° geomagnetic latitude in the dusk sector and 70° to 86° in the dawn sector. Thus, the polar cap extends only to ~85° gm latitude in the evening side and to ~86° gm latitude in the morningside; the dawn–dusk diameter of the polar cap is only ~10° wide. Based on characteristics of electron precipitations, two distinctly different precipitation regions over the auroral oval can be recognized on both the dawn and the dusk sectors. The average energy in the lower latitude part is generally higher than 500 eV, corresponding to a region of hard electron precipitation that maps to the central plasma sheet. The poleward part of the precipitation region has an average electron energy lower than 500 eV, corresponding to a region of soft precipitation or the boundary plasma sheet as defined by WINNINGHAM *et al.* (1975). The boundary between these hard and soft regions is defined here as the transition boundary. The selection of 500 eV is somewhat arbitrary since it merely represent as change in precipitation characteristics and is used in this analysis for the sake of consistency. The latitudinal characteristics of the precipitation also change near the transition boundary from somewhat smooth and continuous in the lower latitude to highly fluctuating burst type in the higher latitude. In this example, the region of high-average-energy electron precipitation is rather narrow, about 3° in the dusk sector and about 6° in the dawn sector, compared with the width of the entire auroral oval. The region of low-average-energy electron precipitation is very wide, with widths of about 12° in the dusk sector and about 10° in the dawn sector.

The electron precipitations in the noon and midnight sectors are illustrated by the next example (Fig. 2). The geomagnetic condition is again quiet as shown by the *AE* index. The hourly value of the *AE* index was continuously low from 0000 UT to 2400 UT on December 25. From high-resolution *AE* data, a disturbance of ~200 nT occurred between 2100 UT and 2300 UT on December 24. Thus, the electron precipitation observations from 0000 UT to 0200 UT shown by dashed arrows are not included in the statistical analysis, which will be discussed in the next section. From 0200 UT to 2400 UT, there were 43 DMSP polar passes from two DMSP satellites, of which 18 passes were the low-latitude polar passes below 80° gm latitude and the other 14 passes were only partial polar crossings. Therefore, only 11 passes, shown by

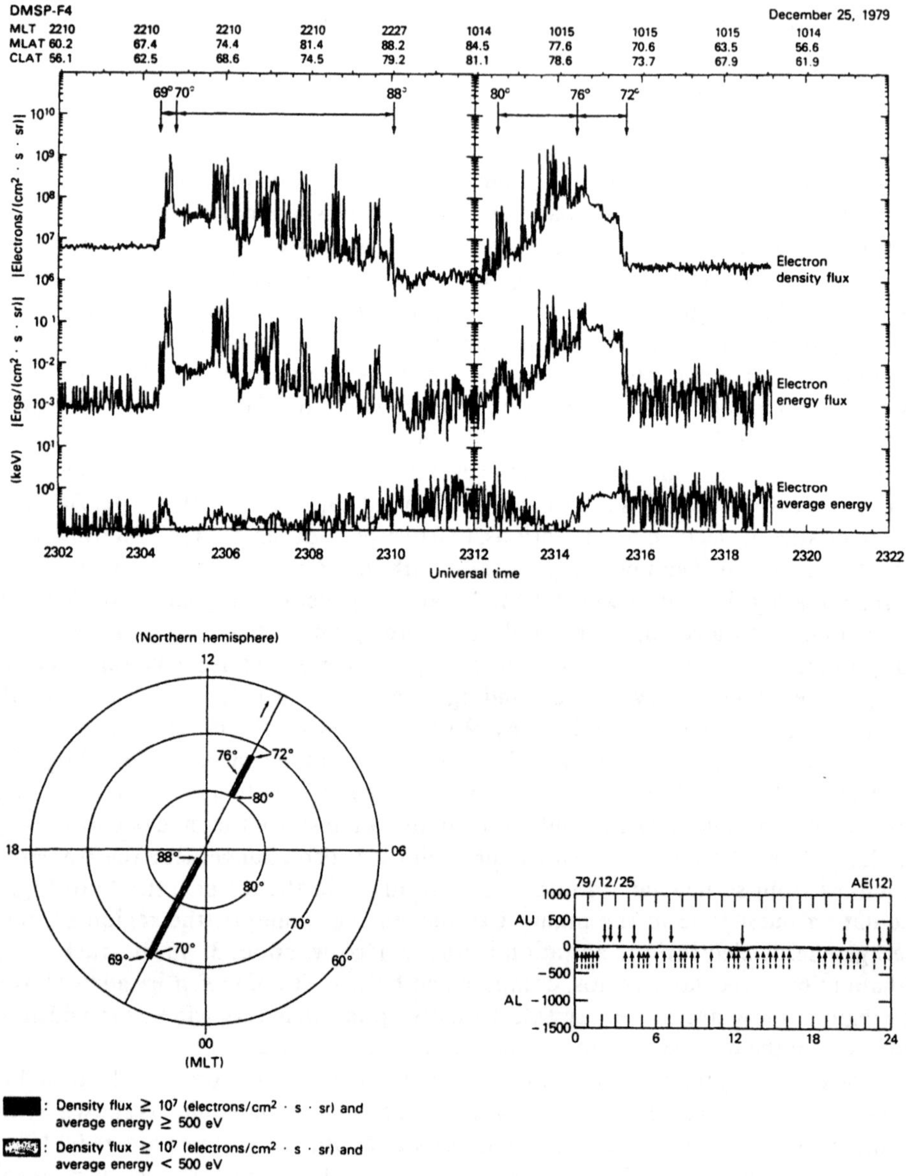

Fig. 2. An example of the auroral electron precipitation in the noon–midnight sectors during quiet periods observed by the DMSP-F4 satellite. The format of the presentation is the same as that of Fig. 1 but for December 25, 1979, ~2310 UT.

solid arrows, out of 49 passes in this interval are used in understanding the morphology of quiet-time auroral precipitations. Figure 2 is one of the selected passes observed from 2302 UT to 2322 UT along the noon–midnight meridian. In this example, the region of high-average-energy electron precipitation extends from 69° to 70° gm latitude in the midnight sector and 72° to 76° in the noon sector. The region of low-average-energy electron precipitation extends from 70° to 88° in the midnight sector and 76° to 80° in the noon sector. The polar cap is only ~12° in dimension in the noon–midnight orientation. It again indicates a very small polar cap in the quiet times as shown by the trajectory plot at lower left.

It is important to know whether the polar cap region defined by using the poleward boundary of the soft extended auroral electron precipitations is a reasonable one. One way to check is to examine the polar cap locations over both the northern and southern hemispheres. Based on the definition of the polar cap as the area of opened geomagnetic field lines, a near symmetry is expected. Figure 3 is an example of nearly simultaneous conjugate observations made by DMSP-F2 and -F3 satellites at ~0930 UT on August 14, 1978. The boundary of the polar cap was at ~82°–83° in the dusk sector and at ~84°–85° in the dawn sector over both polar regions. Therefore, the polar cap dimension as determined is a resonable one; a small polar cap region was detected over both the northern and southern hemispheres, similar to previous results reported in MENG (1981b).

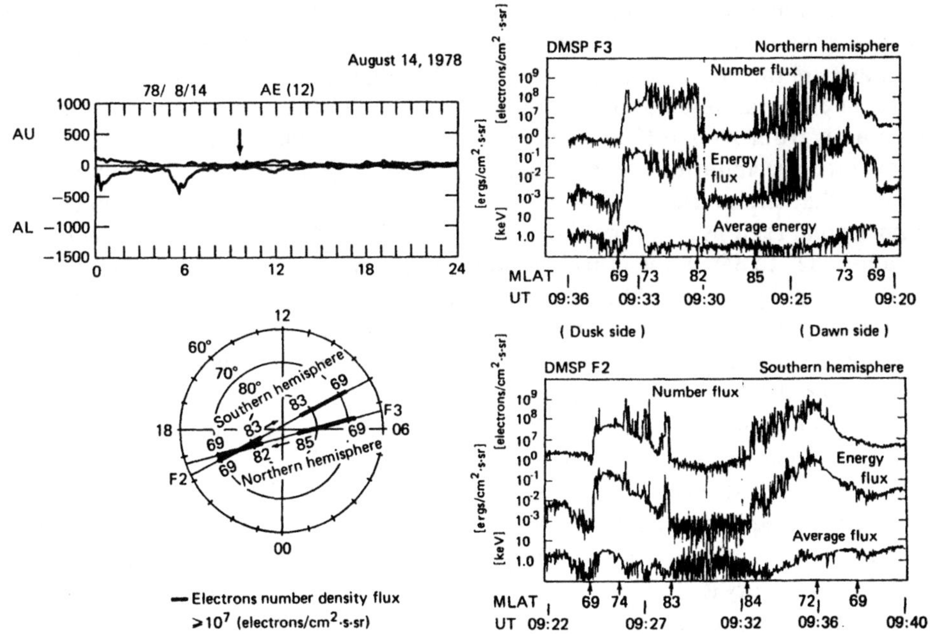

Fig. 3. Auroral electron precipitations observed at both northern and southern polar regions, near simultaneously at about 0930 UT on August 14, 1978, by two DMSP satellites. Note the similarity between two hemispheres, especially in the defined polar cap configuration.

3. Average Location of Quiet-Time Polar Cap and Precipitation Boundaries

From the above illustrations, it is clear that the region of polar cap is rather small in all of the local time sectors and also that, in the quiet time, the soft electron precipitation region is extremely wide in latitude. In this section, we examine statistically the spatial distribution of the polar cap and the auroral electron precipitation regions based on the latitudes of the poleward, transition, and equatorward boundaries during quiet geomagnetic conditions from 434 polar pass observations. We have examined the average boundary locations in the northern and southern hemispheres separately to search for possible seasonal variations. Over the northern hemisphere, there were 114 winter observations (October to March) and only 41 summer ones (April to September). For the southern hemisphere observations, there were 62 winter passes (April to September) and 217 summer passes (October to March). Since the northern hemispheric electron precipitation data were mostly obtained in the winter season and the southern ones were mostly obtained in the summer season, the average electron precipitation patterns are thus grouped only by the local season, irrespective of the hemisphere. These data were subgrouped into four magnetic local time (MT) sectors, namely, from 0300 to 0900, 0900 to 1500, 1500 to 2100, and 2100 to 0300 MLT.

Figure 4 illustrates the average locations of the poleward, transition, and equatorward boundaries of the precipitation region in the winter and summer hemispheres during geomagnetically quiet periods. Because of the limited coverage of DMSP satellite orbits, observations from 1200 to 1800 and 0000 to 0300 MLT zones were not abundant compared with other sectors. Therefore, the average location of boundaries in the noon sector (0900 to 1500 MLT), the evening sector (1500 to 2100 MLT), and the midnight sector (2100 to 0300 MLT) were determined by using the data mainly from 0900 to 1200, 1800 to 2100, and 2100 to 0000 MLT, respectively. In this average precipitation pattern during quiet times (Fig. 4), one of the interesting results is that the averaged location of the poleward boundary in both winter and summer polar regions is at very high geomagnetic latitudes in all local time sectors. It is at about 82° to 84° in the morning, noon, and evening sectors and at about 81° to 82° in the midnight sector. The standard deviation of the poleward boundary location in the midnight sector is larger than that in other magnetic local time sectors because of greater scattering. The average size of the polar cap of the very quiet magnetosphere is about 12° to 14° in diameter.

It is important to point out that the average locations of the auroral equatorward and transition boundaries between the local winter and the local summer hemisphere are nearly identical, except in the noon sector (0900 to 1500 MLT). The average location of the equatorward boundary is at about 70.0° to 70.2°, 68.6° to 68.9°, and 69.5° to 69.7° in the evening, midnight, and morning sectors, respectively. The transition boundary is at about 73.4° to 73.5°, 71.3° to 71.6°, and 75.4° to 75.5° in the evening, midnight, and morning sectors. In the noon sector, there is a slight difference between the winter and the summer hemisphere in the averaged equatorward and transition boundary locations. The locations are at 70.0° and 76.7° in the winter hemisphere and 70.8° and 77.5° in the summer hemisphere, perhaps indicating that,

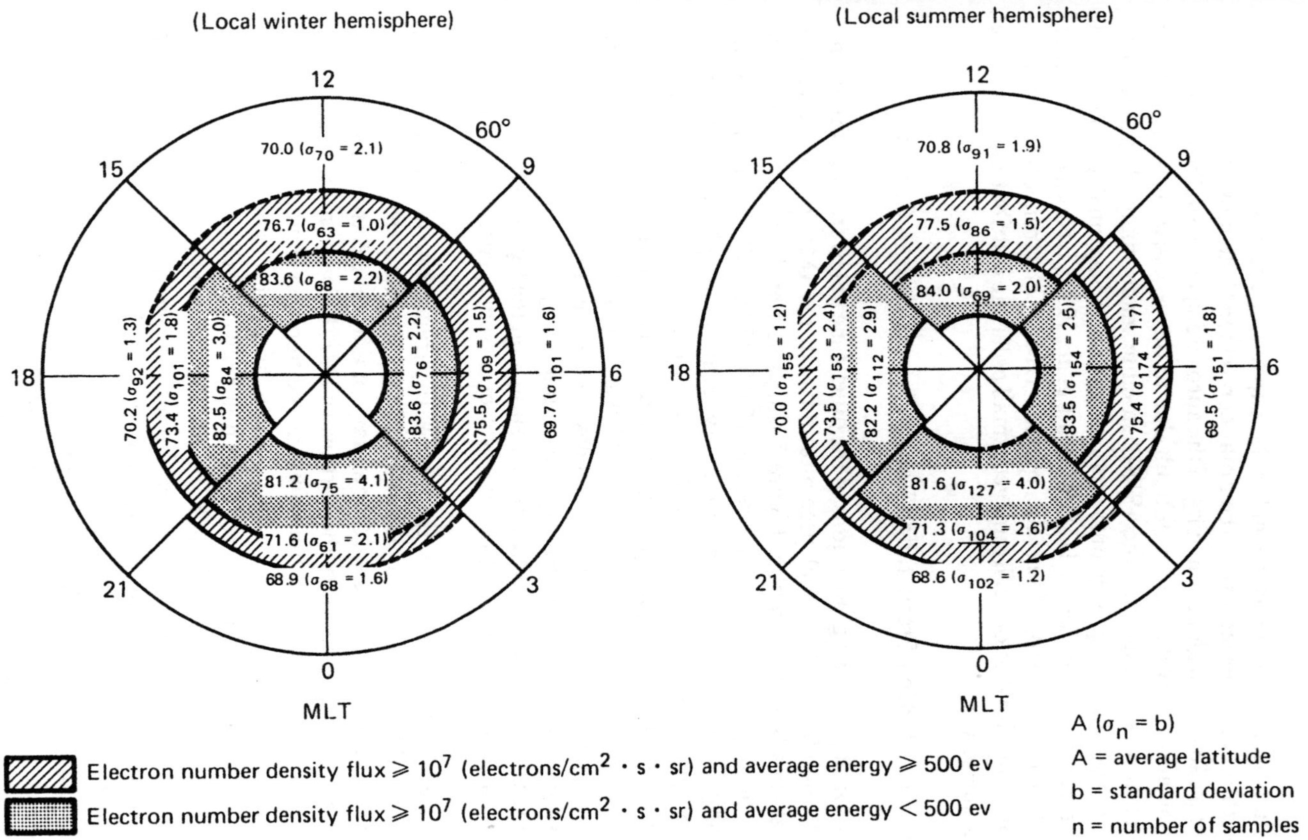

Fig. 4. Average distribution of auroral electron precipitation on the local winter and the local summer hemispheres during geomagnetic quiescence. The lightly shadowed region corresponds to the poleward low-average-energy auroral electron precipitation region, and the heavily shadowed region corresponds to the equatorward high-average-energy auroral electron precipitation. Note that the boundaries of poleward precipitation are statistically located at very high latitude, above 80° in all local time sectors during very quiet periods.

in the noon sector, the region of high-average-energy electron precipitation in the summer hemisphere is located about 1 degree higher than it is in the winter hemisphere.

Is this statistical result of a very small polar cap during geomagnetic quiescence consistent with other observations of the polar cap region? The most direct way to define the polar cap is from global imaging of the auroral display. Usually imaging the very quiet time auroral oval is rather difficult due to the very low auroral emission intensity. The auroral scanner on the ISIS-2 satellite made many global auroral distributions during the very quiet geomagnetic condition. A very small polar cap and a widened auroral oval were observed (MURPHREE *et al.*, 1982); it confirms the earlier suggestion of MENG (1981b). Thus the present DMSP electron observation is certainly consistent with that optically imaged, small polar cap distribution during the geomagnetic quiescence.

4. Quiet-Time Electron Precipitation Boundaries and IMF B_z Component

The relationship of average locations of the quiet-time poleward, transition, and equatorward boundaries of auroral precipitation with the IMF B_z component is examined statistically in this section. Figure 5 shows scatter diagrams of the three boundaries subgrouped into four MLT sectors (0900–1500, 1500–2100, 2100–0300, and 0300–0900). Within each diagram, boundary locations are plotted according to the concurrent hourly average of the IMF B_z value. Since DMSP electron data used here were selected during the quiet period, the concurrent hourly average value of IMF B_z was mostly northward (>80%). The poleward, transition, and equatorward boundaries are represented by circles, triangles, and solid dots, respectively. The linear regression was calculated for each type of boundary by using the least-squares-fit method, and the correlation coefficients are shown on the figure. There are large scatters with respect to the linear regressions; however, some general trends of these boundaries can be stated.

It is noticed that responses of the poleward, transition, and equatorward boundaries to the IMF B_z fluctuation are similar in all local time sectors. Generally, the latitudinal position of the poleward boundary of auroral electron precipitation was proportional to the magnitude of the northward IMF B_z component. The average displacement rate of the poleward boundary in the noon, evening, midnight, and morning sector was about 0.14°, 0.37°, 0.43°, and 0.26° per 1 nT change of the IMF B_z, respectively. This indicates an increase of the polar cap radius by ~0.3° per 1 nT of southward IMF B_z. The effect of IMF B_z on the transition and equatorward boundaries during quiet times was quite different from that of the poleward boundary. The transition boundary moves only slightly equatorward with the increasing northward IMF component in all local time sectors, whereas the equatorward boundary did not show any significant variation with the northward change of IMF B_z magnitude. These results indicate that, in all local time sectors, the width of the region of low-average-energy electron precipitation (between the poleward and transition boundaries) expanded both poleward and equatorward with increasing northward IMF B_z intensity; however, the width of the region of high-

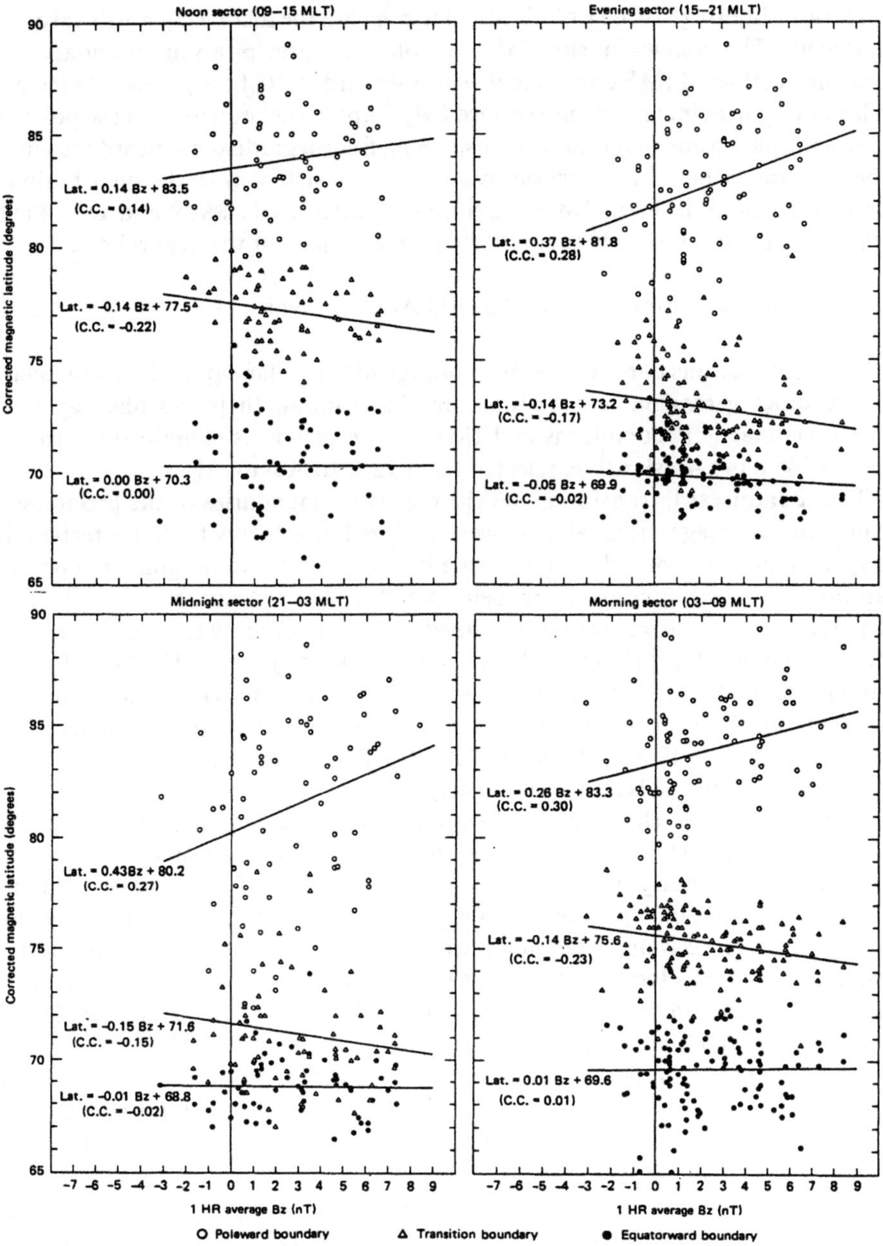

Fig. 5. The scatter plot of the poleward, transition, and equatorward boundaries of the auroral electron precipitation in the four MLT sectors (09–15, 15–21, 21–03, and 03–09). The poleward, transition, and equatorward boundaries are represented by circles, triangles, and solid dots, respectively. The linear regression was calculated for each type of boundary by using the least-squares-fit method; the correlation coefficients (c.c.) are also shown.

average-energy electron precipitation (between the transition and equatorward boundaries) slightly reduced its latitudinal width with the increasing northward IMF component. The poleward shift of the poleward precipitation boundary with increasing northward IMF component. The poleward shift of the poleward precipitation boundary with increasing northward IMF implies the reduction of the polar cap size with the increasing magnitude of positive B_z. Such a gradual poleward movement of the poleward edge of the auroral oval is very consistent with the distribution of discrete auroras in different IMF conditions reported by LASSEN and DANIELSEN (1978), since the discrete arcs occur in the poleward part of the auroral oval.

5. General Variations of Polar Cap and Auroral Oval

Previous sections concern the morphology of the polar cap and auroral oval in the very quiet geomagnetic condition. In this section, their morphology under different geomagnetic conditions and IMF B_z component is examined statistically based on 246 dawn–dusk orbits selected from June to August 1978.

The equatorward, transition, and the poleward boundaries of the precipitation region at various magnetic local times were analyzed, together with the corresponding hourly average AE index during the satellite traverse over the polar region. The locations of these boundaries are determined with an accuracy of 0.1°. These dawn–dusk DMSP passes were subgrouped according to three ranges of the hourly AE index, namely, $0 \leqslant AE \leqslant 150$ (nT), $150 < AE \leqslant 400$ (nT), and $AE > 400$ (nT); there are 100 passes in $0 \leqslant AE < 150$ (nT), 85 passes in $150 < AE \leqslant 400$ (nT), and 61 passes in $AE > 400$ (nT). We also subgrouped the data into four dawn– dusk magnetic local time sectors, 1700 to 1900 and 1900 to 2100 MLT in the evening sector and 0600 to 0800 and 0800 to 1000 MLT in the morning sector.

Figure 6 shows the average locations of the poleward, transition, and equatorward boundaries of the precipitation region during geomagnetically quiet periods when the hourly average AE index was less than 150 nT. The light- and heavy-hatched areas indicate the low-energy and the high-energy electron precipitation regions, respectively. The poleward boundary of the low-energy electron precipitation region is located at about 80° to 82° in both the evening and morning sectors. The transition boundary is seen at about 73° to 75° in both the evening and morning sectors, and the equatorward boundary is located at 69° to 71° in the evening sector and 67° to 69° in the morning sector. The latitudinal extent of the low-energy electron precipitation region is about 7° to 8° in both the evening and morning sectors, while the width of the high-energy electron precipitation region is about 4° in the evening sector and 6° in the morning sector. The polar cap size and locations of various boundaries are somewhat different from those in Fig. 4, indicating a larger polar cap and the equatorial shift of the auroral oval. It is contributed by including a slightly disturbed condition and using a less stringent definition of geomagnetic quietness.

Figure 7 shows the average width of the electron precipitation region during disturbed periods ($AE > 400$ nT). The poleward boundary is located at about 73° to 75° in the evening sector and 76° to 77° in the morning sector. The transition and the

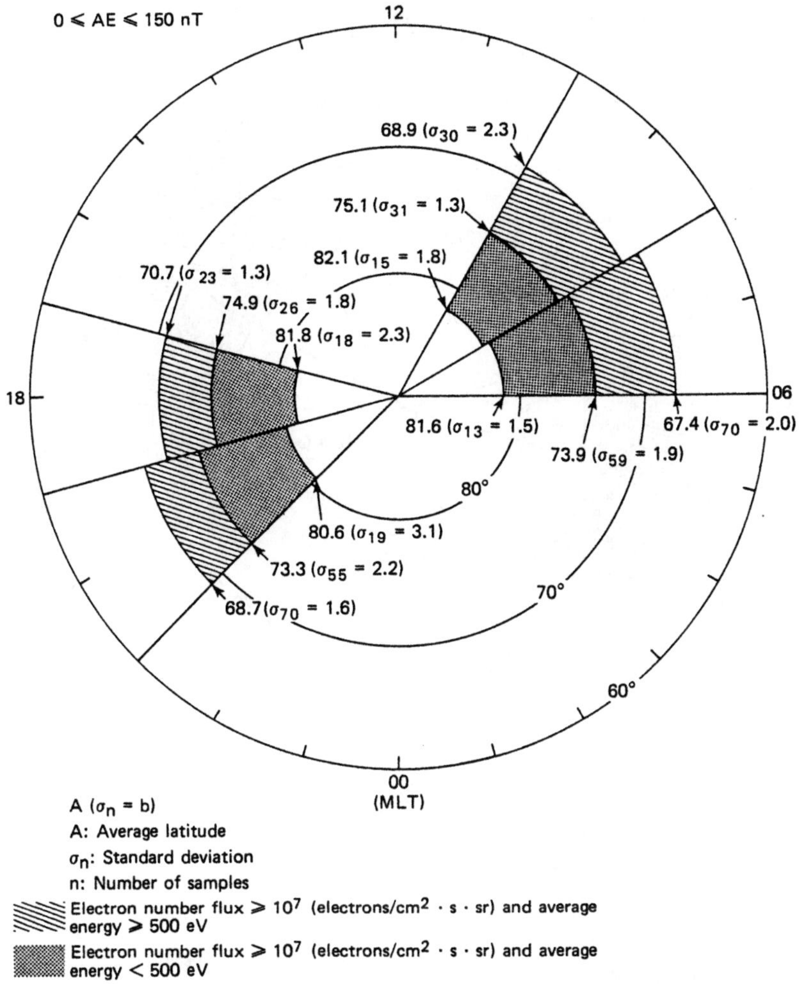

Fig. 6. Averaged auroral electron precipitation distribution on the dawn–dusk side during quiet geomagnetic conditions ($0 \leqslant AE \leqslant 150$ nT). The poleward light-shadowed region corresponds to the low-energy auroral electron precipitation region, and the equatorward heavy-shadowed region corresponds to the high-energy auroral electron precipitation region. Note that the poleward precipitation boundary is statistically above 80° geomagnetic latitude during moderately quiet times.

equatorward boundaries are located at about 70° to 72° and 64° to 66°, respectively, in both the evening and morning sectors. The polar cap size is much larger than that of the quiet condition with the diameter increased from only ~12° to 14° of very quiet to ~16° to 18° of moderately quiet and finally to ~30° wide in latitude during this moderately active time. During magnetic storms, the polar cap size is even larger to >50° wide in latitude (MENG, 1984). The average width of the low-energy electron precipitation region during disturbed time is about 1° to 4° in the evening sector and

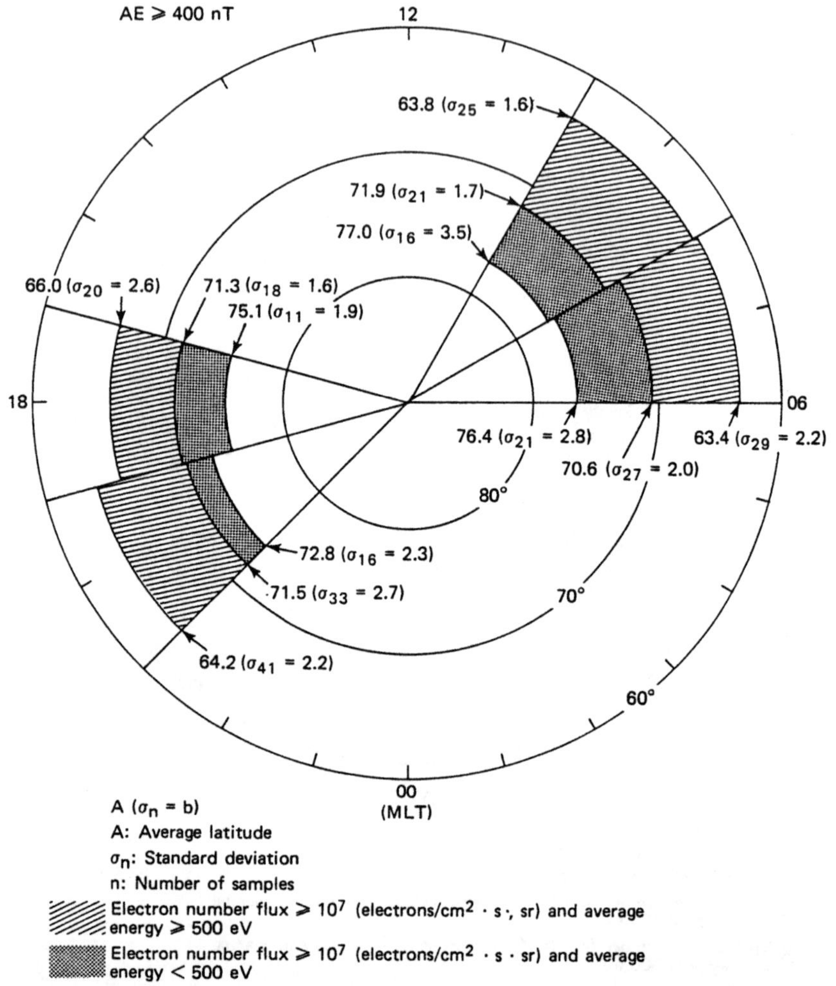

Fig. 7. The same as Fig. 6, but for the geomagnetically active time (AE>400 nT). Note the equatorward shift of the averaged poleward boundary of 5° in the morning sector (0600–1000 MLT) and of 7° to 8° in the evening sector (1700–2100 MLT). The reduction of the auroral oval width is about 2° to 4° in the evening and less than 1° in the morning sectors from the average quiet conditions.

about 5° to 6° in the morning sector, while the width of the high-energy electron precipitation region is about 5° to 7° in the evening sector and about 7° to 8° in the morning sector. The location of the poleward, transition, and equatorward boundaries for moderately active periods (150 nT $<AE\leqslant 400$ nT) falls between the above two geomagnetic conditions. Generally, as the geometric activity increases, the polar cap size increases, the width of the low-energy auroral electron precipitation region decrease, and the width of the high-energy auroral electron precipitation region increases.

The locations of the poleward, transition, and equatorward boundaries and their relationships to the IMF B_z component were also statistically examined. The three boundaries were subgrouped for two magnetic local time sectors (1700–2100 and 0600–1000) and then were compared with the concurrent hourly IMF B_z components and also at 1 hour preceding the satellite crossing. Figures 8(a) and 8(b) show the scatter plot of the poleward, transition, and equatorward boundaries and the corresponding hourly average values of the IMF B_z component at the hour of and 1 hour before the satellite pass, respectively. The poleward, transition, and equatorward boundaries are represented by open circles, triangles, and dots, respectively. The left and right panels show variations in the evening and morning sectors. The correlation coefficient and the linear regression by the least-squares method were calculated for the entire B_z range and for positive and negative ranges separately, as is illustrated in the diagram.

The responses of the equatorward boundary of the precipitation region to the positive and negative IMF B_z were established by HARDY *et al.* (1981). From our analysis, it is clear that a linear correlation is present between decreasing B_z values at the hour and 1 hour before, and the equatorial displacement of the poleward, transition, and equatorward boundaries, when B_z is negative.

Figure 8(a) shows that the average poleward, transition, and equatorward boundaries move toward lower latitudes by about 1.1°, 0.37°, and 0.66° per 1 nT, respectively, in the evening sector and about 0.77°, 0.54°, and 0.55° per 1 nT, respectively, in the morning sector.

Similar tendencies can be recognized in Fig. 8(b). It shows that the average poleward, transition, and equatorward boundaries move toward lower latitudes by about 0.98°, 0.42°, and 0.85° per 1 nT in the evening sector and about 0.96°, 0.52°, and 0.65° per 1 nT in the morning sector. Both figures show that the correlation can also be seen between the increasing values of the positive B_z component and the poleward displacement of the poleward boundary of the extended precipitation region.

The above analyses reveal that both the poleward and equatorward auroral precipitation boundaries are located at higher latitudes during quiet periods than during disturbed periods, indicating a drastic variation of the polar cap size. It was also found that there is a widening of the precipitation region during quiet periods that coincides with periods of positive IMF B_z. This results from the fact that the poleward boundary shifts poleward considerably more than the equatorward boundary. It is also shown that the poleward boundary moves toward lower latitudes with a decreasing IMF B_z value. Further, the shift of the equatorward boundary can be seen only when IMF B_z is negative. However, there is no distinct correlation between the positive B_z magnitude and the displacement of the equatorward and transition boundaries in both the dawn and dusk sectors.

6. Dynamic Variation of the Auroral Oval

The above observations indicate that the unusually high latitude of the precipitation poleward boundary (i.e., a small polar cap) is the consequence of a very

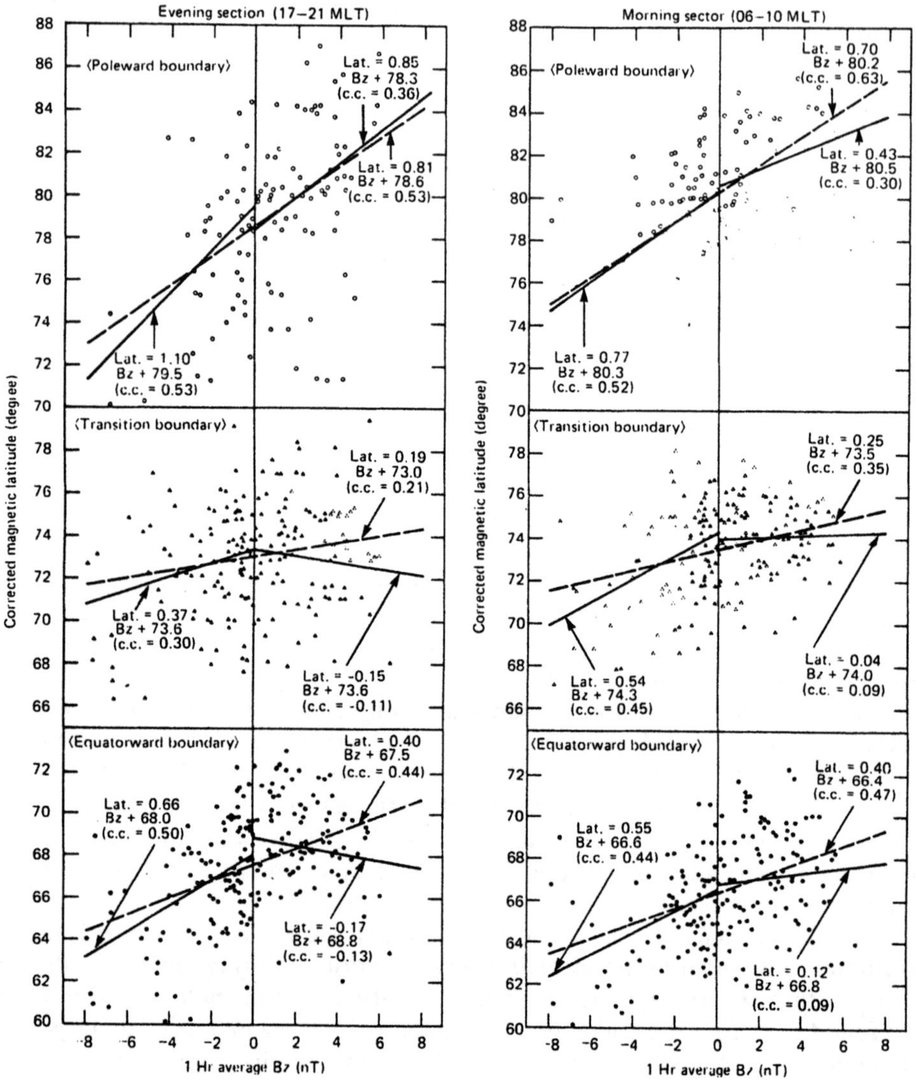

(a)

Fig. 8. (a) The scatter plot of the poleward transition and equatorward boundaries of the auroral electron precipitation in the magnetic local time 1700–2100 MLT and 0600–1000 MLT and their relationship to the corresponding hourly average values of the IMF B_z component at the hour of the polar pass. The poleward transition and equatorward boundaries of the auroral electron precipitation are represented by open circles, triangles, and dots, respectively. The solid lines give the least-squares fit to the data points with IMF B_z>0 nT and B_z>0 nT separately. The dotted lines give the least-squares fit for points without B_z separation. The correlation coefficients (c.c.) are also shown in this figure. (b) The scatter plot of the poleward transition and equatorward boundaries of the auroral electron precipitation in the magnetic local time 1700–2100 MLT and 0600–1000 MLT and their relationship to the corresponding hourly average values of the IMF B_z at 1 hour preceding the polar pass. The format of the presentation is the same as that of Fig. 8(a).

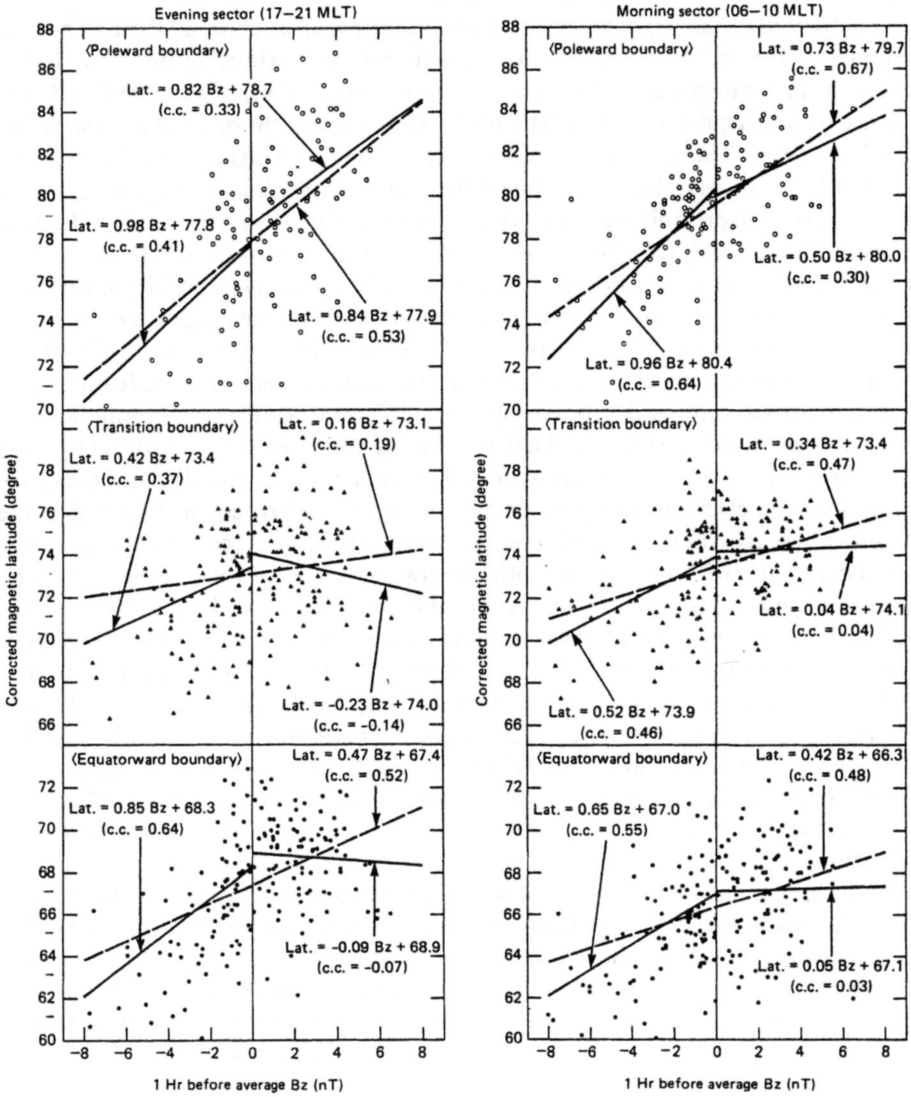

(b)

small number of geomagnetic flux lines interconnected with the IMF during a northward IMF condition. They also imply that the size of the polar cap is inversely controlled by the magnitude of the northward B_z component. The insensitivity of the transition and equatorward precipitation boundaries to the northward IMF B_z can be used to infer that their dynamics are dominated by internal magnetospheric processes and not by direct solar-magnetospheric interactions. In this section, we will present the evolution of the polar cap size with the IMF and substorm variations based on observations from a pair of DMSP satellites.

Figure 9 illustrates typical electron precipitation profiles along the dawn–dusk meridian for three consecutive northern polar region crossings during the development of a magnetospheric substorm. The upper panel from 0231 to 0311 UT was obtained during a quiet period (*AE*(hourly)=32 nT) before the onset of a substorm on Octorber 8, 1978. The *AE* index at the time of the top panel is shown by the arrow (a) in Fig. 10. From the electron number flux and energy flux data, the region of enhanced electron precipitation extended from gm latitude ~69° to ~80° in the dusk sector and from ~71° to ~84° in the dawn sector, respectively. The high-average-energy electron precipitation region was located between 69° and 71° in the dusk sector and between 71° and 75° in the dawn sector. The low-average-energy electron precipitation region was located between 71° and 80° in the dusk sector and between 75° and 84° in the dawn sector. The latter can be distinguished from the former by its highly structured features. The polar cap was ~16° wide in latitude. The middle panel was observed between 0432 and 0450 UT during the moderately disturbed period (*AE*(hourly)=200 nT); see the arrow (b) in Fig. 10. From the middle panel data, one can see that the electron precipitation region clearly shifted equatorward after the previous polar crossing. The high-average-energy electron precipitation region was located between ~69° and ~67° in both the dawn and dusk sectors. The low-average-energy electron precipitation region extended from ~69° to ~71° in the dusk sector and from ~73° to ~81° in the dawn sector with a polar cap width of ~28°. The lower panel was observed from 0613 to 0630 UT during a disturbed period (*AE*(hourly)=250 nT) in the recovery of a substorm; see the arrow (c) in Fig. 10. The high-average-energy electron precipitation region was detected from ~65° to ~72° in the dusk sector and from 65° to 74° in the dawn sector. The low-average-energy electron precipitation region cannot be identified in the dusk sector and were from 74° to 78° in the dawn sector. The polar cap was about 30° wide in the dawn–dusk orientation.

If the precipitation boundary can be defined by the poleward edge of the elevated level of the low-average-energy electron precipitation region, this sequence of DMSP polar crossings suggests that the poleward boundary of the low-average-energy electron precipitation region shifted drastically equatorward and thus the dawn–dusk distance between the precipitation boundaries (i.e., the polar cap) distinctly increased during this disturbed period. Note that there were a few bursts of electron precipitation in the polar cap thus defined during the quiet and moderately disturbed periods, but they disappeared during the highly disturbed periods. By using consecutive observations from two identical DMSP satellites made over both polar regions, the dynamic variation in the polar cap dimension can be inferred. In this analysis, Akasofu's solar wind input ε parameter ($\varepsilon = VB^2\sin^4(\theta/2)l_0^2$) is calculated using the ISEE-3 IMF data by assuming that the solar wind velocity is constant (400 km/sec). We examined in detail 15 examples, but, for presentation purposes, we selected two typical examples in the dawn–dusk sectors and one in the noon–midnight sectors.

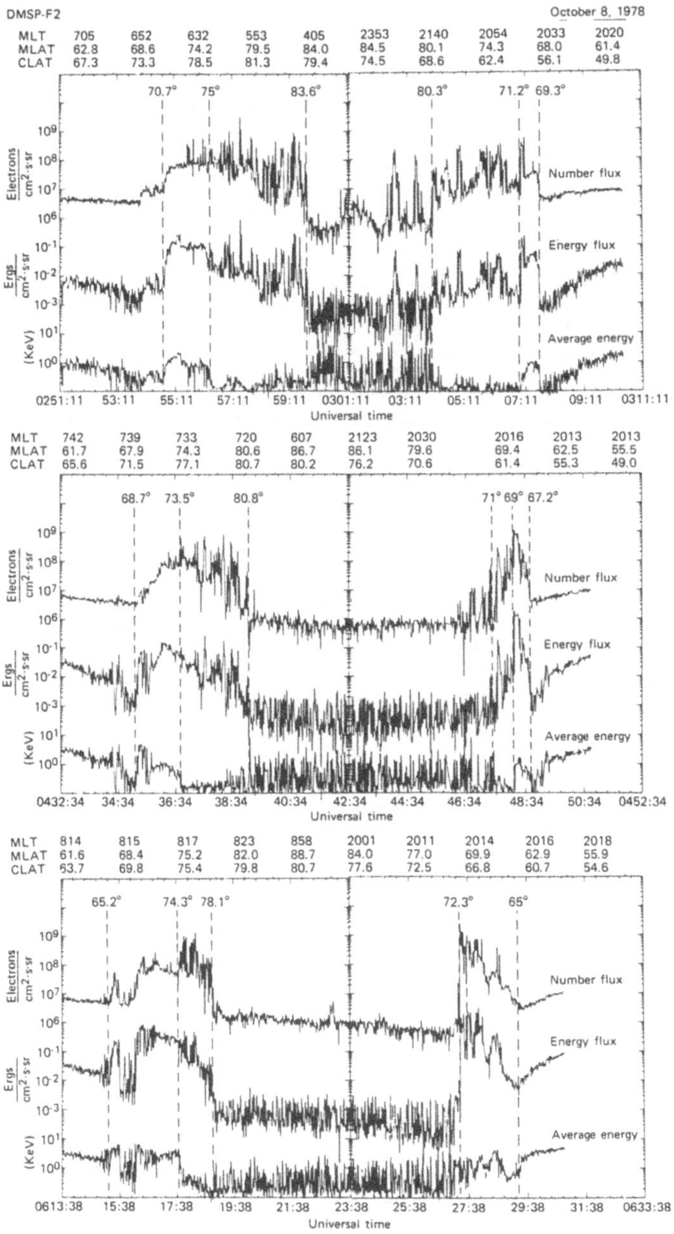

Fig. 9. Typical electron precipitation profile along the dawn- dusk meridian for three consecutive northern polar region crossings. The upper panel was obtained during a quiet period (AE=32 nT), the middle panel was observed during the moderate disturbed period (AE=200 nT) and the lower panel was observed during a disturbed period (AE=250 nT).

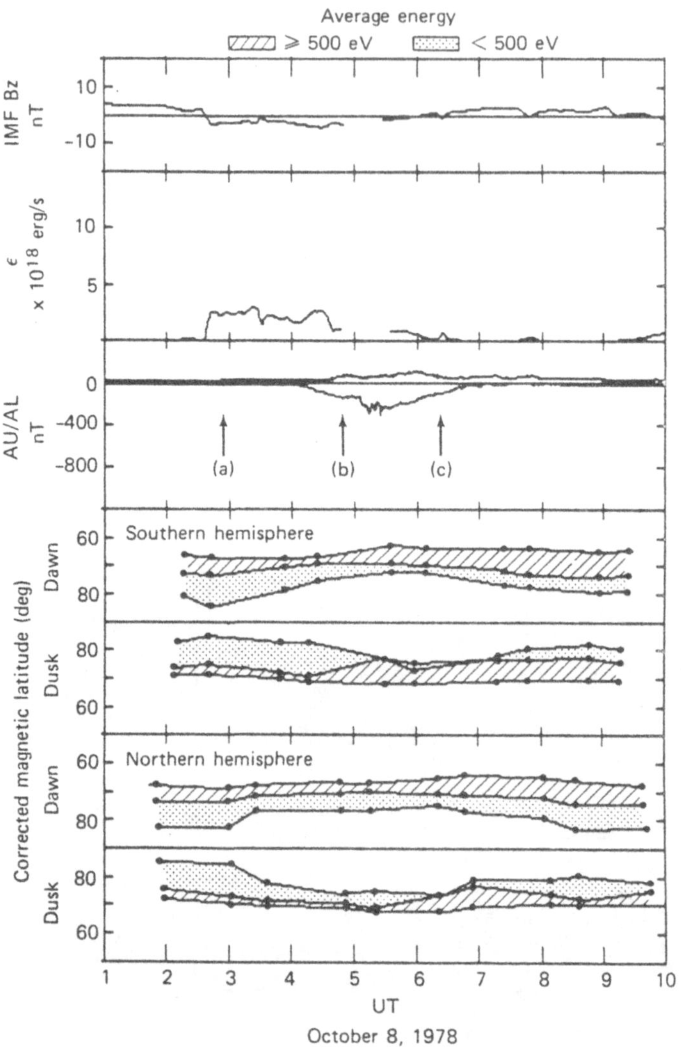

Fig. 10. Temporal and spatial variations of the polar cap size and of the electron precipitation region in the dawn–dusk sector with the IMF B_z, ε, and AE index. The lightly shaded region (poleward) corresponds to the low-average energy electron precipitation region (E_{ave}<500 eV), and the hatched region (equatorward) corresponds to the region of high-average-energy electron precipitation (E_{ave}<500 eV). Note that there is about a 1 hour time delay for the solar wind to propagate between the ISEE-3 location and the magnetosphere. Thus, both the IMF B_z and ε data should be shifted to the right by about 1 hour in order to examine them in conjunction with the DMSP observation and the AE index.

6.1 October 8, 1978

In Fig. 10, temporal and spatial variations of the polar cap size and of the electron precipitation region are plotted, together with the IMF B_z, ε, and *AE* index. From the *AE* index, one can see that an isolated substorm occurred during this period. The electron precipitation regions along the dawn–dusk meridian observed from both the southern and northern hemispheres are shown in the two bottom panels. The electron precipitation data are obtained by combining observations from the DMSP-F2 and -F3 satellites, enabling us to determine the boundaries with time resolution of ~30 min on the average. Note that the observed locations of these electron precipitation boundaries were normalized to those of the 1800 MLT in the dusk sector and the 0600 MLT in the dawn sector by assuming a circular distribution with the center shifted by 4.2° from the geomagnetic pole toward the 0015 MLT (MENG *et al.*, 1977).

As shown in this figure, during the period of the northward IMF and low ε values (namely, before 0230 UT), the poleward boundaries of the low-average-energy electron precipitation region were located at 80° in gm latitude or higher. Thus, the dimension of the polar cap in the dawn–dusk direction during the quiet period was about 15° or less. Note that there is about a 1 hour time interval for the solar wind to propagate from the ISEE-3 location to the magnetosphere. Therefore, both the IMF B_z and ε data should be shifted to the right by about 1 hour in order to examine them in conjunction with the DMSP observations and the *AE* index.

Associated with a negative excursion of the IMF B_z component and the associated increase of the ε parameter after 0230 UT, the poleward boundary of the electron precipitation region began to shift equatorward. By combining both the northern and southern polar data, one can infer that the shift began between 0300 and 0320 UT (at about the time when the southward-directed IMF reached the magnetosphere). The geomagnetic activity represented by the *AE* index began gradually at about 0400 UT and reached the maximum values at about 0515 UT. The polar cap became largest at about 0530 UT; the maximum dimension of the polar cap was 33° in the southern hemisphere and 31° in the northern hemisphere. On the other hand, the equatorward shifts of both the equatorial boundary of the high-energy precipitation region and the transition boundary between the high- and low-energy precipitation regions were much less.

After 0430 UT, the ε parameter began to decrease. Subsequently, the substorm activity began to decline after 0520 UT. The poleward boundary of the electron precipitation region gradually shifted poleward after 0600 UT. As a result, the distance between the precipitation boundaries was reduced. Note that this reduction lasted well after the substorm activity (represented by the *AE* index) ceased and the ε parameter became less than 10^{18} erg/sec.

6.2 August 18–19, 1978

Figure 11 shows an example that covers the recovery phase of one substorm and the development of a new substorm activity with a fairly quiet period between them. The recovery of the first substorm was associated with the northward turning of the IMF. During this interval, the poleward boundary of the auroral electron precipita-

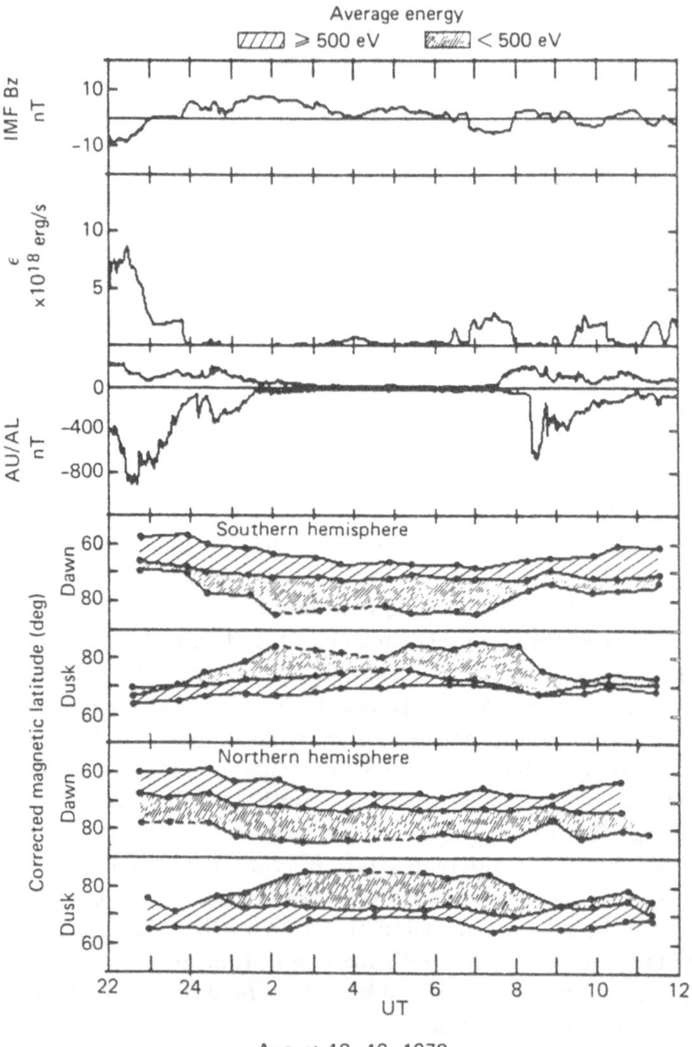

Fig. 11. The format in the figure is the same as that for Fig. 10. The poleward boundary of the electron precipitation region in both hemispheres began to shift poleward (equatorward) after the northward (southward) turning of the IMF and the corresponding decrease (increase) of the ε parameter at ISEE-3 at 2300 UT (0600 UT).

tion region gradually shifted poleward. After 0200 UT on August 19, the poleward boundary was located at latitudes higher than 80° in both hemispheres. Between ~0300 UT and ~0700 UT, the ε parameter was very low, and the poleward boundary of the auroral electron precipitation region remained persistently at magnetic latitudes above 80°. It should be noted that the equatorward boundary of auroral electron precipitation regions also moved slightly poleward during this quiet period. With an increase of the ε parameter after 0650 UT at the ISEE-3 location, a new substorm activity began at about 0730 UT, first a significant increase of the AU and a sharp increase of AL at ~0815 UT. Concurrently, the poleward boundary of the electron precipitation region gradually moved equatorward and then sharply after ~0800 UT.

6.3 December 3–4, 1979

The previous two examples illustrate the dynamics of the precipitation region observed in the dawn–dusk sector. This example illustrates the auroral electron precipitation region in the noon–midnight sector (Fig. 12). We normalize the observed locations of these electron precipitation boundaries to the 00 MLT in the midnight sector and to the 1200 MLT sector in the noon sector as described previously. One can see that the IMF B_z component was positive until about the end of December 3, and both the ε parameter and the AE index were low. The poleward boundary of the auroral electron precipitation region was located at latitudes higher than 80° on the dayside and 71° on the nightside. First, the overall trend of the equatorward and poleward shifts and their relationship to the AE index (as well as to the IMF B_z component and the ε parameter) are quite similar to those for the dawn–dusk sector, although the shift is a little more pronounced in the midday sector than in the midnight sector. Thus, there is no doubt that the noon–midnight dimension of the polar cap is also significantly greater during a substorm than during a quiet period. The equatorward shift of the midday part of the auroral oval (the cusp) during a substorm was studied earlier by AKASOFU (1972) and EATHER *et al.* (1979). It is quite likely that the nightside shift is somewhat obscured by both gradual and rapid (~0630, 1212 UT) poleward expansion of the aurora. Note that the nightside auroral oval consisted almost wholly of the high-average energy electron precipitation region, since the low-energy precipitation region disappeared. When the ε parameter began to decrease after 1500 UT, the polar cap boundaries began to shift poleward.

7. Interpretations of the Results

It may be worthwhile to summarize some of the findings before attempting to interpret the results in terms of changes of the open flux and of magnetic energy in the magnetotail as a consequence of the solar wind-magnetosphere interaction. It is quite clear that the distance between the precipitation boundaries along both the dawn–dusk and noon–midnight meridians increases during substorm activity. This feature can be considered as a response of the magnetosphere to the IMF southward turning and the resulting increase of the ε parameter. The increase is more prominent along the dawn–dusk meridian than in the noon–midnight meridian. In the noon–midnight meridian, the increase is more evident on the dayside than on the nightside. At times, the nightside dimension decreases for brief periods; this feature is due to the poleward expansion of the auroral bulge during a substorm.

The poleward boundary of the instantaneous auroral electron precipitation can

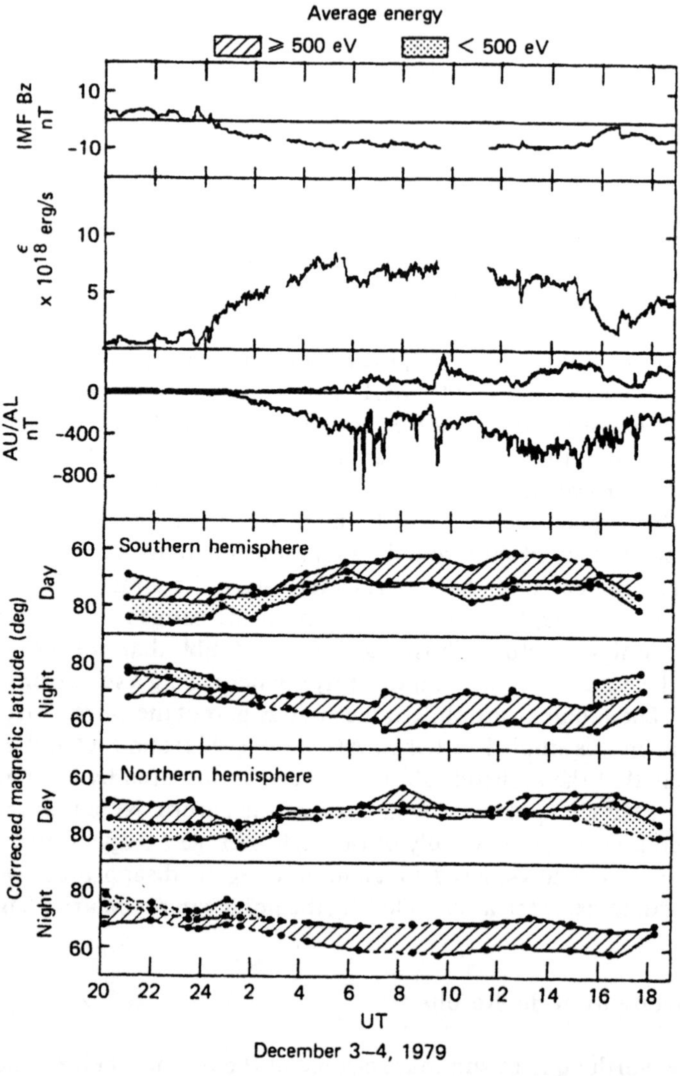

Fig. 12. Temporal and spatial variations of the polar cap size and of the electron precipitation region in the noon–midnight sector. The format in this figure is the same as that for Fig. 10. Note that the overall trend of the equatorward and poleward shifts and their relationship to the AE index (as well as to the IMF B_z component and the ε parameter) are quite similar to those for the dawn–dusk sector.

be used to define the dimension of the polar cap. Then the polar cap begins to shift poleward slowly at about the time when the AE index begins to decline. The polar cap dimension is largest at about the period when the AE index becomes largest. However, the shift continues well after the AE index decreases to the noise level. The size of the polar cap is very small corresponding to a quiet magnetosphere. The equatorward boundary of the precipitation region appears to respond in a similar way, but the extent of the shift is much less than that of the poleward boundary. Thus, the resulting short distance between the precipitating boundaries is characterized mainly by the poleward shift (or advance) of the poleward precipitation boundaries. Further, this poleward shift is due to widening of the low-energy precipitation region as previously suggested (MENG, 1981b). On the other hand, the low-energy precipitation region during an intense substorm becomes very narrow, particularly in the night sector. Note also that the electron precipitation pattern in the latitude range between 60° and 70° (where most *AE* stations are located) is quite insensitive to the IMF changes during the initial epoch. We schematically illustrate these main features in Fig. 13. Note that these findings are consistent with earlier results by MENG (1981b), MURPHREE *et al.* (1982), and MAKITA *et al.* (1983).

We now attempt to interpret some of these findings in terms of changes of the open geomagnetic flux and of magnetic energy in the magnetotail. The basic assumptions in this interpretation are that the poleward boundaries of the electron precipitation belt can be identified as the polar cap boundary and that the dimension of the polar cap thus is a defined measure of the amount of the open magnetic flux.

The amount of the open flux begins to increase soon after the southward turning of the IMF B_z reaching the front of the magnetosphere. The subsequent overall shifts of the polar cap boundary follow changes of ε fairly well except that the decrease continues gradually well after ε becomes less than $\sim 10^{18}$ erg/sec.

The accompanying substorm activity expressed by the *AE* index begins often about ~30 min to 1 hr after the beginning of the open flux increase. This is possibly due to the fact that the *AE* stations cannot monitor geomagnetic changes that take place at latitudes higher than 70°; thus we must be cautious about the limitation of the *AE* index in monitoring polar cap changes. The subsequent overall changes of the *AE* index and the amount of the open flux are similar. The peak periods of the *AE* and of the open flux occur at about the same time. The open flux decreases gradually well after the *AE* index is reduced to the noise level.

It is quite likely that the open flux (i.e., the size of the polar cap) is a qualitative measure of the total magnetic energy in the magnetotail. If this would indeed be the case, the above observation suggests that the total amount of magnetic energy in the magnetotail tends to vary similarly with the ε parameter and the *AE* index. In particular, there is no significant time shift for the period of maximum values of the polar cap dimension and the *AE* index. Therefore, substorms cannot be described as a manifestation of explosive conversion of the magnetotail magnetic energy that is accumulated prior to substorm onset. If this were the case, the polar cap dimension should decrease rapidly during the expansive phase. It may be added that AKASOFU (1972) and EATHER *et al.* (1979) found that the equatorward shift of the dayside cusp during the expansive phase is inconsistent with the hypothesis of explosive return of the magnetic flux from the magnetotail to the dayside magnetosphere. It is also interesting to note that the magnetic energy in the magnetotail can decrease gradually without significant magnetic effects in the auroral zone.

It is also interesting to speculate that the observed changes of the total open flux may be directly related to the so-called "thinning" and "thickening" of the plasma

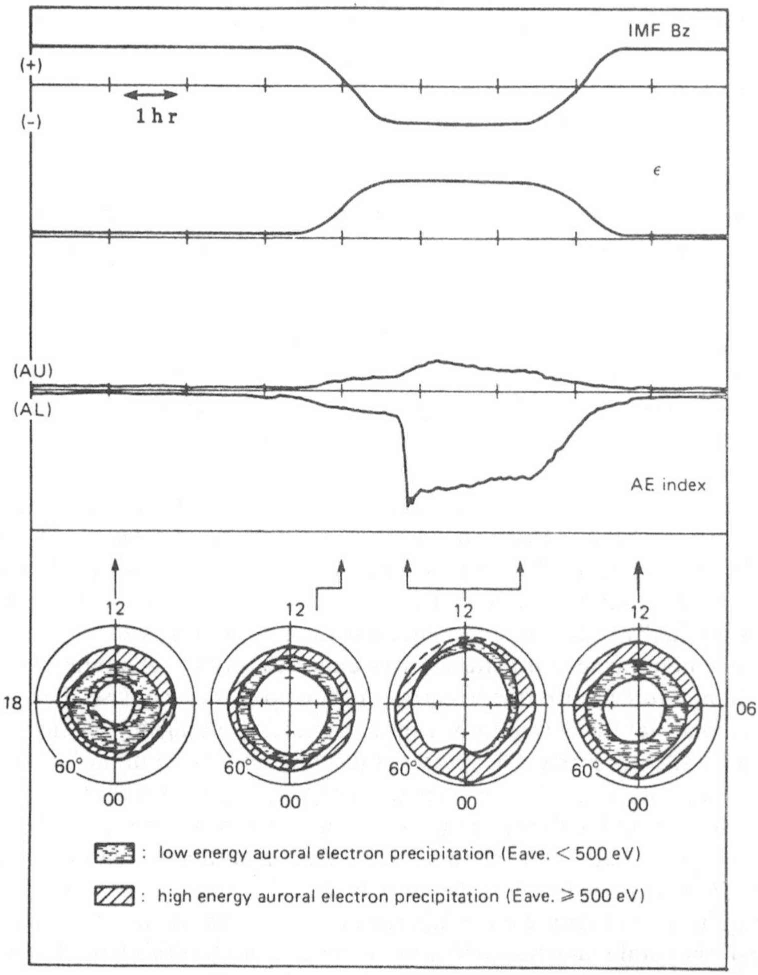

Fig. 13. Schematic presentation of the results. The IMF period. The size of the polar cap begins to i B_z at the front of the magnetosphere. The peak p same. The polar cap begins to contract graduall

sheet during a substorm. If the magnetotail during a substorm, an increased open flu sheet, resulting in thinning of the plasma sh lobe cross section begin to decrease durin plasma finds its way into the plasma she contracting high-latitude lobe.

This research is supported in part by the

Grant AFOSR 84-0049 and by the Division of Atmospheric Sciences, National Science Foundation, under Grant ATM-83-15041 to The Johns Hopkins University, Applied Physics Laboratory.

REFERENCES

AKASOFU, S.-I., Midday auroras and magnetospheric substorms, *J. Geophys. Res.*, **77**, 244, 1972.

AKASOFU, S.-I., A study of auroral displays photographed from the DMSP-2 satellite and from the Alaska meridian chain of stations, *Space Sci. Rev.*, **16**, 617, 1974.

EATHER, R. H., S. B. MENDE, and E. J. WEBER, Dayside aurora and relevance to substorm current systems and dayside merging, *J. Geophys. Res.*, **84**, 3339, 1979.

HARDY, D. A., W. J. BURKE, M. S. GUSSENHOVEN, N. HEINEMANN, and E. HOLEMAN, DMSP/F2 electron observations of equatorward auroral boundaries and their relationship to the solar wind velocity and the north-south component of the interplanetary magnetic field, *J. Geophys. Res.*, **86**, 9961, 1981.

HOFFMAN, R. A. and J. L. BURCH, Electron precipitation patterns and substorm morphology, *J. Geophys. Res.*, **78**, 2867, 1973.

HOLZWORTH, R. H. and C.-I. MENG, Mathematical representation of the auroral oval, *Geophys. Res. Lett.*, **2**, 377, 1975.

LASSEN, K. and C. DANIELSEN, Quiet time pattern of auroral arcs for different directions of the interplanetary magnetic field in the *Y–Z* plane, *J. Geophys. Res.*, **83**, 5277, 1978.

LUI, A. T. Y., S.-I. AKASOFU, E. W. HONES, Jr., S. J. BAME, and C. E. MCILWAIN, Observations of the plasma sheet during a contracted oval substorm in a prolonged quiet period, *J. Geophys. Res.*, **81**, 1415, 1976.

MAKITA, K. and C.-I. MENG, Average electron precipitation patterns and visual auroral characteristics during geomagnetic quiescence, *J. Geophys. Res.*, **89**, 2861, 1984.

MAKITA, K., C. -I. MENG, and S.-I. AKASOFU, The shift of the auroral electron precipitation boundaries in the dawn–dusk sector in association with geomagnetic activity and interplanetary magnetic field, *J. Geophys. Res.*, **88**, 7967, 1983.

MAKITA, K., C.-I. MENG, and S.-I. AKASOFU, Temporal and spatial variations of the polar cap dimension and its relation to the energy input rate ε and *AE* index, *J. Geophys. Res.*, **90**, 2744, 1985.

MENG, C.-I., The auroral electron precipitation during extremely quiet geomagnetic conditions, *J. Geophys. Res.*, **86**, 4607, 1981a.

MENG, C.-I., Polar cap arcs and the plasma sheet, *Geophys. Res. Lett.*, **8**, 273, 1981b.

MENG, C.-I., Dynamic variation of the auroral oval during intense magnetic storms, *J. Geophys. Res.*, **89**, 227, 1984.

MENG, C.-I., R. H. HOLZWORTH, and S.-I. AKASOFU, Auroral circle-delineating the poleward boundary of the quiet auroral belt, *J. Geophys. Res.*, **82**, 164, 1977.

MURPHREE, J. S., C. D. ANGER, and L. L. COGGER, The instantaneous relationship between polar cap and oval auroras at times of northward interplanetary magnetic field, *Can. J. Phys.*, **60**, 349, 1982.

PIKE, C. P. (Ed.), Defense meteorological satellite program auroral-ionospheric interpretation guide, *Air Force Surv. Geophys.*, **306**, 1975.

WINNINGHAM, J. D., F. YASUHARA, S.-I. AKASOFU, and W. J. HEIKKILA, The latitudinal morphology of 10-eV to 10-keV electron fluxes during magnetically quiet and disturbed times in the 2100–0300 MLT sector, *J. Geophys. Res.*, **80**, 3148, 1975.

Solar Wind-Magnetosphere Coupling, edited by Y. Kamide and J. A. Slavin, 633–641.

The Dynamics of the Equatorward Boundary of the Auroral Oval

H. NAKAI[1], Y. KAMIDE[2], D. A. HARDY[3], and M. S. GUSSENHOVEN[3]

[1]*Makino High School/Physics, Hirakata, Osaka 573, Japan*
[2]*Kyoto Sangyo University, Kyoto 603, Japan*
[3]*AF Geophysics Laboratory, Hanscom Air Force Base, Bedford, Massachusetts 01731, U.S.A.*

The Auroral Boundary Index determined by GUSSENHOVEN *et al.* (1983) on the basis of DMSP precipitating electron data is used to examine the expansion and contraction of the auroral oval in the midnight sector. The time scales of the expansion and the contraction of the auroral oval are found to be about 45 minutes and 8 hours, respectively. An empirical formula including these two time scales to predict the oval location from variations in the interplanetary magnetic field (IMF) is devised. The correlation coefficient between the measured and the predicted boundaris is 0.82.

1. Introduction

Numerous studies have shown that the equatorward boundary of the diffuse aurora maps along field lines to the inner edge of the plasma sheet (e.g., LASSEN, 1974; LUI *et al.*, 1977; SLATER *et al.*, 1980). Thus the location of the equatorward boundary of the nightside auroral oval can provide us with useful information on electromagnetic processes in the magnetosphere-ionosphere system. In the last decade, extensive investigations have shown that the equatorward boundary moves in a systematic manner in response to geomagnetic activity, the north–south component of the IMF, and the velocity of the solar wind (e.g., KAMIDE and WINNINGHAM, 1981). In particular, the southward component of the IMF has been shown to be the main contributor in controlling the position of the auroral oval (e.g., HARDY *et al.*, 1981; NAKAI and KAMIDE, 1983; MAKITA and MENG, 1984). Since each satellite crosses the auroral boundary at a different local time, previous studies have been limited to average behavior of the response of the auroral oval to the IMF. GUSSENHOVEN *et al.* (1981), using DMSP F/2 satellite data of precipitating soft electrons, devised the so-called 'equivalent' boundary, which is called the Auroral Boundary Index. Since this index gives the approximate location of the equatorward boundary of the auroral oval in corrected geomagnetic latitude at the fixed magnetic local time, the use of this index should be quite useful in discussing the large-scale dynamics of the auroral oval and associated processes in the magnetotail.

In this report, first we utilize the auroral boundary index to obtain this characteristic response time to the IMF, and then attempt to find to what degree the observed boundary index can be reproduced from IMF variations and the obtained response time.

2. The Auroral Boundary Index

Although the concept and the practical procedure of the auroral boundary index is described in detail by GUSSENHOVEN *et al.* (1982, 1983), it will be briefly outlined here. The corrected geomagnetic latitude of the equatorward boundary of the auroral oval in the *i*-th magnetic local time bin can be expressed by a linear regression function of K_p (GUSSENHOVEN *et al.*, 1981):

$$\Lambda_i = \Lambda_{0i} + \alpha_i K_p. \tag{1}$$

They showed that the boundary in the evening sector generally correlates better with K_p than that in the morning sector. Because of this, only the boundary in the evening sector is used in calculating the auroral boundary index. Figure 1 shows the local time dependence of the auroral oval boundary sorted in terms of K_p. The equivalent boundary at midnight, the auroral boundary index, is derived from the measured boundary by projecting it along these curves of the 'average' auroral oval to midnight. Thus, the auroral boundary

$$\Lambda = (\Lambda_{0m} - (\alpha_m/\alpha_i)\ \Lambda_{0i}) + (\alpha_m/\alpha_i)\ \Lambda_{ob} \tag{2}$$

where Λ_{ob} is a measured boundary in the *i*-th MLT, and the subscript m stands for midnight, that is, the local time sector between 2300–2400 MLT. It should be noted that (2) does not include K_p explicitly.

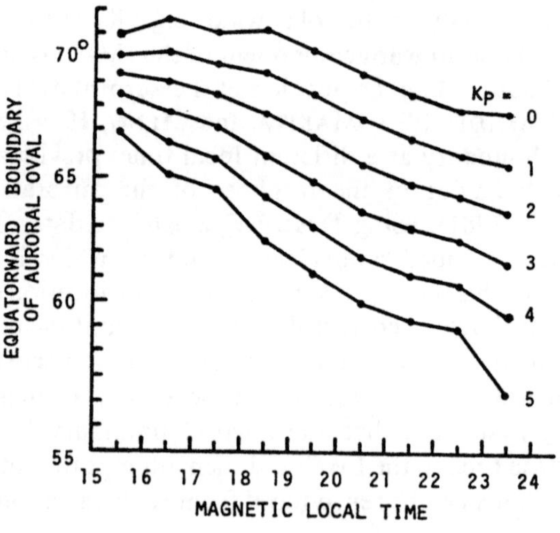

Fig. 1. Local time dependence of the equatorward boundary of the auroral oval for various values of K_p. Each point is obtained from the empirical linear relationship between the oval boundary and K_p in each local time sector (GUSSENHOVEN *et al.*, 1981).

3. Response of the Equatorward Boundary to IMF Variations

In the present paper the auroral boundary index is transformed to the location of the plasma sheet inner edge in the midnight sector. This transformation is performed by mapping the boundary location of the auroral oval to the equatorial plane of the magnetosphere along dipole field lines. Then we attempt to quantify the location of the plasma sheet inner edge as a function of the electric field in the solar wind. It is assumed in this process that with changes in the polar cap potential drop, which reflects the solar wind electric field, the plasma sheet moves earthward or tailward until an equilibrium condition is reached (see, for example, HAREL *et al.*, 1981). Here, we define "equilibrium" as the state when the inner boundary of the plasma sheet becomes coincident with the zero energy Alfven layer. It is assumed that the zero energy Alfven layer at midnight (L_A) can be expressed in the form:

$$L_A = \alpha \Phi_{pc}^{-1/\delta} \tag{3}$$

where Φ_{pc} refers to the polar cap potential drop. To relate Φ_{pc} to the solar wind electric field $B_s V$, the following form is adopted:

$$\Phi_{pc} = a_0 + a_1 \, B_s V \eta \tag{4}$$

where $B_s = -B_z$ when B_z is negative and $B_s = 0$ when B_z is positive, and V is the solar wind speed. The factor η, which indicates the compression effect of the magnetic field in the geomagnetic sheath, is defined here as 12 nT/B when the magnitude of the IMF (B) exceeds 12 nT, and is replaced by $\eta = 1$ when the ratio exceeds unity (REIFF *et al.*, 1981). In (3) and (4) α, δ, a_0, and a_1 are constants. Before the equilibrium condition is reached, the inner edge of the plasma sheet moves toward the zero energy Alfven layer given by (3) and (4) through variations of $B_z V$. We also assume that the rate of the inner edge movement is proportional to the distance between the inner edge and the zero energy Alfven layer as expressed by

$$\mathrm{d}\, L(t)/\mathrm{d}t = (1/\tau)(L_A - L(t)).$$

Then we obtain

$$L(t) = (1/\tau) \int_0^t L_A \, e^{(t'-t)/\tau} \, \mathrm{d}t' + L_i \, e^{-t/\tau} \tag{5}$$

where τ is a time scale of the inner edge movement, and L_i is a radial distance of the inner edge at $t=0$. Since it has been shown that the auroral oval generally expands more rapidly than it contracts (PIKE and DANDEKAR, 1979), two kinds of time scale, namely the expansion time scale and the contraction time scale in (5), are to be determined from a bulk data set.

Using formulas (4) and (5), we can predict the movement of the L value of the plasma sheet inner edge whenever the solar wind parameter in (4) is observed. A complete description of the practical procedure to obtain the constants in (4) and (5) is given in NAKAI *et al.* (1986).

4. Prediction of the Auroral Oval Location

We use the data set of the solar wind parameters (5 min. values of B_z and 1 hour values of the solar wind velocity from the IMP-J satellite) and the auroral boundary index during the period from January 1 to May 1, 1978, to find the values of the unknown constants in Eqs. (3)–(5). The practical procedures can be divided into two parts. The first is to obtain the expression of the radial distance of the Alfven layer as a function of the solar wind parameters. There we assume that the Alfven layer is approximately coincident with the lowest latitude in each of the auroral expansion cases. The second procedure is to obtain the suitable time scales for the auroral oval expansion and the contraction, separately. By means of the least squares method, we have found the following results: (a) Combining (3) and (4), the radial distance of the Alfven layer at midnight is represented by

$$L_A^{-2.5}=4.9\times10^{-3}\ B_s V\eta/400+1.25\times10^{-2}. \qquad (6)$$

(b) The most probable values for the time scales for the expansion and the contraction of the auroral oval are 45 minutes and 8 hours, respectively (NAKAI *et al.*, 1986). Formula (6) is shown in Fig. 2, where the abscissa is normalized by $B_s V\eta/400$ and the ordinate shows L_A.

By using (5) and (6), the dynamics of the auroral boundary can be predicted from variations of the solar wind parameters. The accuracy of this prediction can be seen in Fig. 3 where the predicted boundary is plotted as a function of the 'observed' auroral boundary index. The correlation coefficient between the two boundaries is found to be 0.82. Since the equatorward boundary during extremely quiet periods is assumed

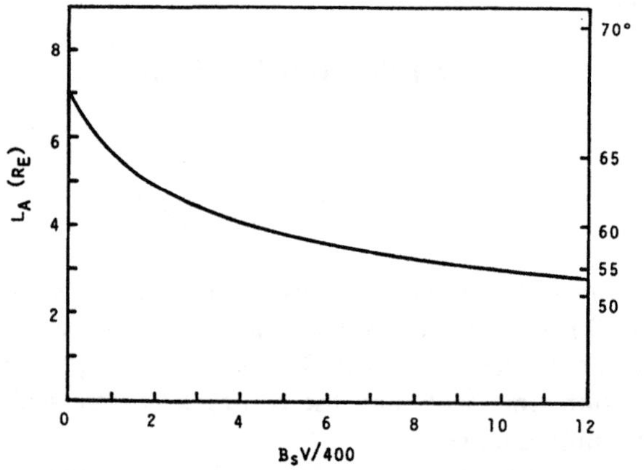

Fig. 2. Radial distance of the zero energy Alfven layer in the midnight sector as a function of the solar wind parameter $B_s V\eta/400$ (nT km/s). The scales on the right-hand side of the figure indicate the latitudes to which the zero energy Alfven layer is mapped along dipole field lines.

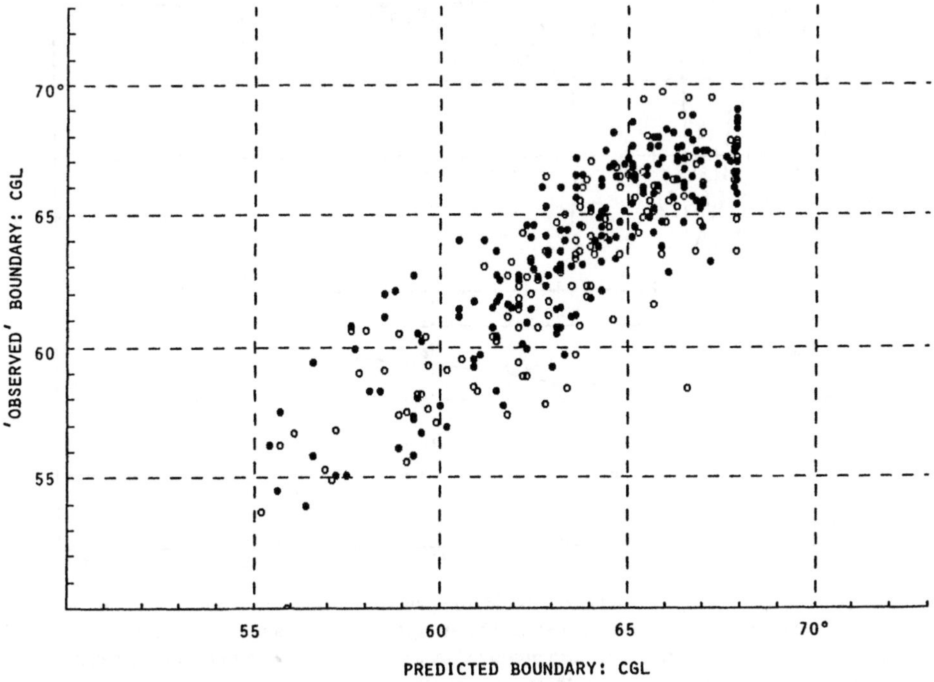

Fig. 3. Scatter plot of the auroral boundary index values versus the predicted values. Open and closed circles indicate the auroral boundary index obtained from the oval in the southern and northern hemispheres, respectively.

to be settled at 68 degrees, the distribution of the predicted latitudes has an upper limit of 68 degrees.

Figure 4 shows a typical example of the auroral oval expansion and the subsequent contraction on April 18, 1978. The top panel shows variations of $B_z V/400$ (nT km/s/400) with interpolations of missing data over no more than three 5-min. averages by a linear function. The middle panel shows the location of the zero energy Alfven layer predicted by using (6). The bottom panel shows variations of the auroral boundary index, where closed circles show the index obtained from the oval in the northern hemisphere. The solid curve in the bottom panel indicates the predicted boundary. In Fig. 4, the IMF turned south at 0155 UT, and showed some sharp negative peaks before the time it turned northward at 0425 UT. The predicted zero energy Alfven layer shifted earthward with decreasing of $B_z V$. Although $B_z V$ exceeded 10 (nT km/s), the zero energy Alfven layer did not become less than 3 R_E. This was due to the fact that the magnitude of the IMF (B) was greater than 12 nT during the period concerned. Both the expansion and the contraction of the auroral oval were well represented by the predicted boundary. The maximum deflection of the predicted boundary agreed reasonably well with that of the measured one.

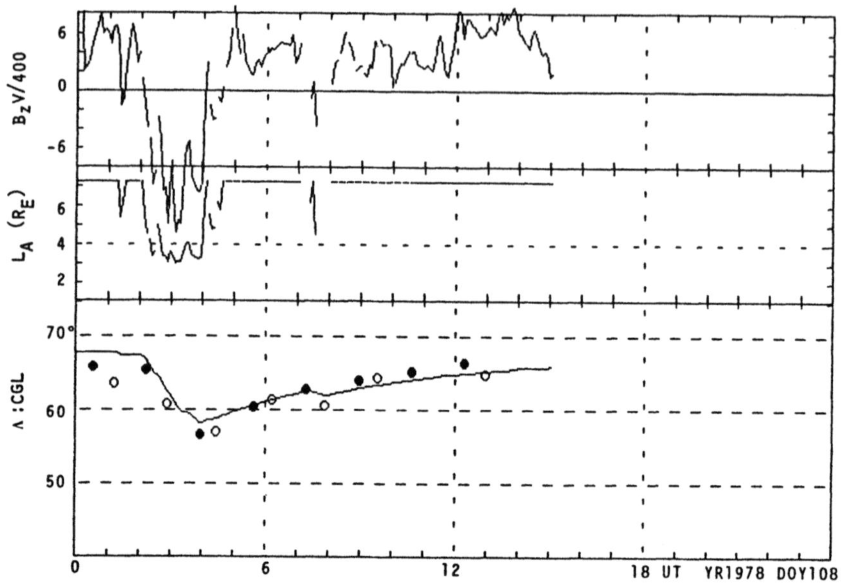

Fig. 4. Variations of the solar wind electric field (top panel) and the corresponding locations of the zero energy Alfven layer (middle panel) predicted by using (6), and of the auroral boundary index (bottom panel). The predicted boundary is also presented by a solid line in the bottom panel. Open and closed circles are for the southern and northern hemispheres, respectively.

5. Discussion

In the present paper, we have used the Auroral Boundary Index in an attempt to predict the auroral oval location and the corresponding plasma sheet boundary in the magnetosphere on the basis of solar wind electric fields. As shown in Section 2, the index was derived by using the average boundary of the auroral oval sorted by K_p. The shape of the auroral oval boundary, however, is expected to change in association with the development of magnetic disturbances. Such distortions of the boundary would cause misestimations of the auroral boundary index. This is also responsible for some portion of the residual scatter of (1) distributed with a standard deviation of about 2 degrees (GUSSENHOVEN *et al.*, 1982). There is also the possibility that the equatorward boundary of precipitating electrons at low altitudes is not identical to the inner edge of the plasma sheet in the equatorial plane of the magnetosphere. Recent simultaneous measurements by the DE satellites at two altitudes should be able to unveil the validity of this caveat. The range of errors in estimating the auroral boundary index is probably within about 2 degrees.

In Sections 3 and 4 the auroral boundary index has been utilized to examine the auroral oval dynamics. That confirmed that the north-south component of the IMF, or the dawn–dusk component of the solar wind electric field, plays a primary role in determining the location of the auroral oval boundary. The agreement between the measured and the predicted boundary derived by using (5) and (6) assures us that the

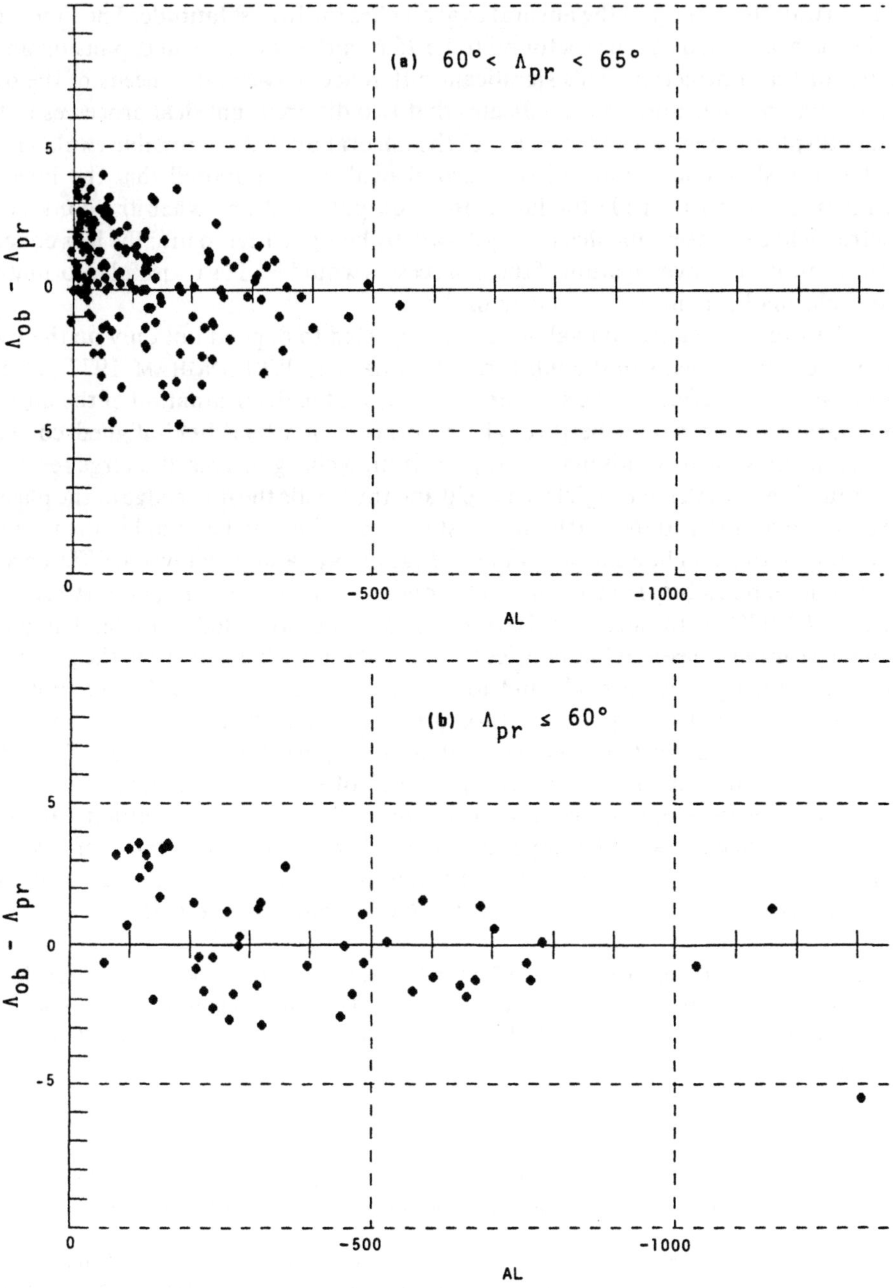

Fig. 5. (a) Difference between the measured and the predicted boundary for 60° (CGL)$<\Lambda_{pr}\leqslant$65° (CGL) plotted on the corresponding *AL* index averaged over 1 hour preceding the observation of the auroral oval boundary. (b) Same as Fig. 5(a) but for $\Lambda_{pr}\leqslant$60° (CGL).

inner edge of the plasma sheet is coincident with the zero energy Alfven layer when the equatorward boundary of the auroral oval reaches the lowest latitude. The time scale of the auroral oval dynamics is found to be 45 minutes for the oval expansion and 8 hours for the contraction. This significant difference between the speeds of the oval expansion and the contraction indicates that two different physical processes in the magnetosphere cause each movement of the auroral oval. As a possible explanation of the very slow contraction of the auroral oval, it is supposed that the injected particles are trapped inside the last closed equipotential line when the convection electric field decreases, and decay very slowly by being scattered into the loss cone. A more quantitative investigation of these processes would lead us to useful information about plasma behavior in the magnetosphere.

The size of the auroral oval has been suggested to depend not only on the IMF but also on the intensity of the substorm (KAMIDE and WINNINGHAM, 1977). One of the causes for the effect of the substorm activity on the determination of the auroral boundary position would be through the process in which field-aligned currents intensified by substorms discharge the polarization charges. This discharge results in the reduction of the 'shielding' electric field and thus leads the inner edge of the plasma sheet to move closer to the earth. Such a substorm effect can be found in the residual scattering of the oval boundaries in Fig. 3. Figures 5(a) and (b) show the difference of the measured boundary from the predicted one as a function of AL for 60° (CGL) and for $\Lambda_{pr} \leqslant 60$ (CGL), respectively. Here Λ_{pr} means the predicted latitude. There is a tendency for the measured boundaries to locate at higher latitudes than the predicted boundary during quiet periods and to locate at lower latitudes during substorm periods. Using the least squares method it is noted that the auroral oval boundary shifts equatorward by 0.53 degrees and 0.37 degrees for every 100 nT of AL, respectively. Three data points exceeding 1000 nT of $-AL$ in Fig. 5(b) are excluded in calculations of the slope to avoid a too strong influence on the resultant values. It appears that the values of the slopes are generally consistent with those of KAMIDE and WINNINGHAM (1977), but the influence of substorms on the auroral oval boundary is much smaller than that of the B_z component of the IMF.

Data of the interplanetary magnetic field and the solar wind from IMP-J were provided by N. F. Ness and others through the WDC-A for Rockets and Satellites in NASA and the Interplanetary Medium Data Book. K. Maezawa and T. Iyemori assisted us in rearranging the interplanetary magnetic field records and the AL indices, respectively.

REFERENCES

GUSSENHOVEN, M. S., D. A. HARDY, and W. J. BURKE, DMSP/F2 electron observations of equatorward auroral boundaries and their relationship to magnetospheric electric fields, *J. Geophys. Res.*, **86**, 768, 1981.

GUSSENHOVEN, M. S., D. A. HARDY, N. HEINEMANN, and E. HOLEMAN, 1978 diffuse auroral boundaries and a derived auroral boundary index, *AFGL-TR-82-C398*, Air Force Geophysics Laboratory, Hanscom Air Force Base, Mass., December, 1982.

HARDY, D. A., W. J. BURKE, M. S. GUSSENHOVEN, N. HEINEMANN, and E. HOLEMAN, DMSP/F2 electron observations of equatorward auroral boundaries and their relationship to the solar wind velocity and the north-south component of the interplanetary magnetic field, *J. Geophys. Res.*, **86**, 9961, 1981.

HAREL, M., R. A. WOLF, P. H. REIFF, R. W. SPIRO, W. J. BURKE, and F. J. RICH, Quantitative simulation of a magnetospheric substorm; 1. Model logic and overview, *J. Geophys. Res.*, **86**, 2217, 1981.

KAMIDE, Y. and J. D. WINNINGHAM, A statistical study of the instantaneous nightside auroral oval: The equatorward boundary of electron participation as observed by the ISIS 1 and 2 satellites, *J. Geophys. Res.*, **82**, 5573, 1977.

LASSEN, K., Relation of the plasma sheet to the nightside auroral oval, *J. Geophys. Res.*, **79**, 3857, 1974.

LUI, A. T. Y., C. D. ANGER, and S.-I. AKASOFU, The equatorward boundary of the diffuse aurora and auroral substorms as seen by the ISIS 2 auroral scanning photometer, *J. Geophys. Res.*, **80**, 3603, 1975.

MAKITA, K. and C.-I. MENG, Average electron precipitation patterns and visual auroral characteristics during quiescence, *J. Geophys. Res.*, **89**, 2861, 1984.

NAKAI, H. and Y. KAMIDE, Response of nightside auroral oval boundaries to the interplanetary magnetic field, *J. Geophys. Res.*, **88**, 4005, 1983.

NAKAI, H., Y. KAMIDE, D. A. HARDY, and M. S. GUSSENHOVEN, Time scales of expansion and contraction of the auroral oval, *J. Geophys. Res.*, **91**, in press, 1986.

PIKE, C. P. and B. S. DANDEKAR, Evening sector auroral oval dynamics from DMSP photographs, *J. Geophys. Res.*, **84**, 3389, 1979.

REIFF, P. H., R. W. SPIRO, and T. W. HILL, Dependence of polar cap potential drop on interplanetary parameters, *J. Geophys. Res.*, **86**, 7639, 1981.

SHEEHAN, R. E. and R. L. CAROVILLANO, Characteristics of the equatorward auroral boundary near midnight determined from DMSP images, *J. Geophys. Res.*, **83**, 4749, 1978.

SLATER, D. W., L. L. SMITH, and E. W. KLECKNESS, Correlated observations of the equatorward diffuse auroral boundary, *J. Geophys. Res.*, **85**, 531, 1980.

Solar Wind-Magnetosphere Coupling, edited by Y. Kamide and J. A. Slavin, 643–654.

Annual Variation of Aurora and Solar Wind Coupling

S. M. SILVERMAN

Physics Department, Boston College, Chestnut Hill, Massachusetts 02167, U.S.A.

An annual variation of the frequency of occurrence of auroras and of magnetic activity, with maxima in the spring and fall, has been recognized for over two centuries. The most common explanations for this variation involve either a heliographic latitude dependence, varying with the position of the earth in its orbit (the axial hypothesis), or an effect depending on the earth-sun geometry with maxima at the equinoxes. Some hypotheses combine one of these with an additional factor. The majority of investigators conclude that the equinoctial hypothesis is more nearly in accord with the facts.

A study of auroral data from Sweden for a period of 220 years, and of a New England data set for the period 1840–1906, confirms the existence of maxima in occurrence frequency in the spring and fall. These maxima do not occur consistently over the entire period. During the prolonged solar activity minimum at the beginning of the nineteenth century, for example, there is no definite maximum. Data for New England, where the summer days are not as long as in Sweden, follow the solar declination closely. The results indicate that an equinoctial origin for the variation is preferred. Power spectral studies of the Swedish data confirm that the amplitude of the six months peak, that is, the equinoctial variation, is less marked during the prolonged sunspot minima at the beginnings of the eighteenth and nineteenth centuries.

A consequence of the Russell-McPherron hypothesis is that the spring maximum is associated, on the average, with radial fields towards the sun, and the fall maximum with fields away from the sun. Other observations tend to support this, at least in part. Consequently the spring/fall ratio may be used to estimate at least the trends in the relative importance of these fields over extended periods. The ratios show an increase from the earliest data, about 1720, to a maximum in the period around 1780, followed by a decline to a minimum about 1830. This pattern is then repeated, with maximum around 1870, and a subsequent minimum for the cycle beginning in 1923. The variation of these fields is not in phase with that of the intensity of southward component/solar wind velocity as indicated by the amplitude of the 6-months peak.

1. Introduction

A variation of the frequency of auroral occurrence in the course of the year has been recognized since 1736 (KRAFFT, 1737; cited in LOVERING, 1866). Krafft, on the basis of 141 auroras observed at St. Petersburg, concluded that the aurora was more frequent at the equinoxes than at the solstices. This conclusion was firmly established by the middle of the nineteenth century (see, for example, HANSTEEN, 1847; BROUN, 1848; LOVERING, loc. cit. at 205, and references cited therein). It was also recognized by the mid-nineteenth century that magnetic disturbances behaved in a similar manner. Thus BROUN (1848), basing his conclusions on several years of observations at Makerstoun, Scotland, noted that "the law of the frequency of aurorae is the same as that of the intensity of magnetic disturbances; the latter have their greatest value near the equinoxes, and their least value near the solstices." LOVERING (loc. cit.), based on his tables and a harmonic analysis of the data, gave the dates of the maxima as March 16 and September 24, and the minima as January 7 and June 23. Wolf, based on an analysis of 5764 auroras, derived maxima near March 20 and October 15, and minima near June 22 and December 25. Wolf also compared the auroral curve with the yearly changes of magnetic inclination as given by Hansteen, the magnetic storms given by Sabine, and his own data on sunspots and magnetic declination, and concluded that the agreement between these curves was as exact as could be expected from the nature of the observations. Hansteen, cited in QUETELET (1854), found maxima in the changes of magnetic inclination at Christiania (now Oslo), Norway, on April 1 and September 30, and minima on July 1 and December 31. Attempts at explanation of these variations focussed on potential correlations with such other periodic phenomena as meteor showers, the age of the moon, and possible connections with meteorological parameters.

At the turn of the century the question was again raised, using the much more extensive data sets then available. ELLIS' work (1899, 1901, 1904, 1905a, 1905b), in particular, covering a range of fifty years of magnetic data at Greenwich, became the basis of several proposed explanations of the equinoctial maxima. ELLIS (1899, 1901) initially proposed that the explanation resided in the fact that at the equinoxes the whole surface of the earth was presented to the sun, while at the solstices one or the other pole is continually in shadow. LOCKYER and LOCKYER (1904) proposed that either polar prominences or activity in the solar regions was responsible for magnetic disturbances. Consequently they concluded that the seasonal variation in frequency depended on the positions of the sun's axis in relation to the earth, and that the epochs of greatest inclinations of the sun's axis towards the earth corresponded to those of greatest magnetic and auroral frequency, that is, around March 6 and September 5, with intermediate positions on June 5 and December 6. ELLIS (1905) subsequently found maximum disturbances at about March 3 and October 9. MAUNDER (1905), noting the lack of symmetry about the equinoxes, suggested that the inequality was not due to a single cause alone, but was a combination of two or more causes. CORTIE (1912, 1915) noted that the dates of the maxima and minima corresponded to the positions of maximum or minimum heliographic latitude. Since sunspots are almost never on the equator, and sunspots are correlated with magnetic activity, these results indicated that the equinoxes are favored because the earth is more nearly in a line with more sunspots then.

The distinction between a heliographic latitude dependence (the axial hypothesis) and a more purely equinoctial hypothesis is the earlier date, by about two weeks, for maximum disturbance predicted by the axial hypothesis. Since, as recognized early on, the actual dates of maxima and minima are quite variable from one period to another, tests of which hypothesis is correct depend heavily on statistical treatments. Although some authors (PRIESTER and CATTANI, 1962) have favored the axial hypothesis, the majority of investigators, using a variety of magnetic indices and tests, have concluded that the equinoctial hypothesis is more nearly in accord with observation (BARTELS, 1932, 1963; MCINTOSH, 1959; ROOSEN, 1966; SHAPIRO, 1969; FRASER-SMITH, 1972). Some authors have proposed a combination of an equinoctial effect and some other factor, such as the variation of sunspot latitude over the solar cycle (MEYER, 1966; SAITO, 1972; MURAYAMA, 1974). CURRIE (1966) proposed a ring current origin, with its seat in the quiet-time proton belt. BOLLER and STOLOV (1970), in a much discussed paper, proposed a mechanism relying on Kelvin-Helmholtz instabilities and resulting in maxima at the equinoxes.

A new dimension to the argument was added by LEWIS and MCINTOSH (1953), who pointed out that differences in the diurnal variation of magnetic activity could serve as a criterion for distinguishing between different hypotheses. This work was subsequently extended by MCINTOSH (1959) who concluded that the equinoctial hypothesis was strongly supported. The use of the diurnal variation as a test has since become a standard feature of proposed explanations.

With the advent of in situ space data discussion has tended to focus on the relationship of magnetic activity to properties of the solar wind and the interplanetary field. Attention has largely focussed on the proposal of RUSSELL and MCPHERRON (1973), based on the interaction of the magnetosphere with a variable southward component of the interplanetary field resulting from the relative orientations of the solar magnetospheric and solar equatorial coordinate systems. Their theory is thus both axial and equinoctial, with maxima, later than those predicted by either, occurring in early April and early October. Their proposal has been modified or criticized, in whole or in part, by several authors. MAYAUD (1974) found good agreement between predicted and observed behavior using only the variation of angle between the earth's dipole axis and the earth-sun direction. BERTHELIER (1976) found that the predicted amplitude of the diurnal variations was much lower than would be expected from the Russell-McPherron hypothesis. HOLZER and SLAVIN (1982) found that while the basic Russell-McPherron postulate was supported, there was nevertheless an unexplained deficiency in the predicted amplitudes. GREEN (1984) used *a* indices for 1868–1980. His results clearly favor an equinoctial generation mechanism, and do not support the Russell-McPherron proposal as a major cause of the semiannual variation.

2. General Behavior of the Annual Variation

The data set used in this research is that of RUBENSON (1879, 1882) for the period prior to 1877 and the extension of this catalog to 1943 by Silverman (unpublished work, 1982). The data set is discussed in FEYNMAN and SILVERMAN (1980), SILVERMAN and FEYNMAN (1981), SILVERMAN and SHAPIRO(1983), and SILVERMAN (1983). A smaller data set from New England (1840–1906) (Silverman, unpublished, 1985) complements the Swedish data set.

The data used to determine the annual variation over the period 1723 to 1943 are the number of days on which aurora occurred for latitudes less than or equal to 65 degrees. This restriction eliminates biasing due to a large number of auroras which may be observed at stations near the auroral oval. This effect is especially noticeable in the later years when data from a station such as Karesuando is reported. To provide sufficient numbers of auroras for analysis the data was broken up into periods corresponding to each sunspot cycle, as measured from minimum to minimum. The use of such long periods also tends to average out spikes due to more conscientious observation by individual observers. Such sampling fluctuations are nevertheless still possible, and this must be kept in mind in considering the results.

The number of days with aurora varies considerably for different cycles. The minimum number is 104 for the period 1810–1822, and the maximum number is 744 for the period 1867–1877. The total number of days with aurora for the entire period 1723–1943 is 7870. The numbers for the period after 1877 are not strictly comparable to those prior to 1877, since the later data base is drawn from a smaller number of stations. While it is possible to approximate a correction for this effect (SILVERMAN, 1983) it is not necessary to do so here since we are interested only in relative values over the course of the year. The data used are summarized in Table 1.

The procedure used was to divide the number of days with auroras in each month by the total for the cycle, resulting in a frequency distribution. This allows for intercomparison of different time periods on a comparable basis. Auroral occurrence is rare in June and infrequent in July. To bring out the overall behavior more clearly, taking account of the summer minimum, the results have been plotted from July to the following June.

The results are plotted in Fig. 1. The frequencies for the entire period, 1723–1943, as well as for each sunspot cycle separately, are shown in the Figure. Auroral notations for the summer months in Sweden are rare, and the Figures have therefore been drawn for July through June. Considering the curve for the entire period, 1723–1943, we notice first the well known spring and fall maxima. The fall maximum is in October, though the value for September is only a little bit lower. The spring maximum is in March, with a relatively rapid decline thereafter. The maxima for individual cycles are, generally speaking, in March and October, with occasional excursions to neighboring months. The relative importance of the Spring and Fall maxima varies from one cycle to another. For the period 1810–1842 we no longer have two maxima. For 1823–1842 there is essentially only one maximum, in October. The period 1810–1822 has an apparent maximum in January, but this may be an artifact of the small number of auroras observed during this period, which corresponds also to a part of the prolonged solar activity minimum at the beginning of the nineteenth century.

The trends in these curves are probably real, though the values themselves appear to be strongly influenced by the longer days in the summer months at these latitudes. This conclusion results from a comparison of the frequencies in Sweden with those calculated from a sample of 2980 days with aurora for the period 1840–1906 in New England, based primarily on observations in Maine. (The sample used is part of a more complete study of New England auroras, not yet completed, of which about 6,000 records were used in this study.) Figure 2 compares frequencies for the two regions. The frequency curve from the New England observations is much more symmetrical than the Swedish one, with the equinoctial maxima and solstitial minima being approximately equal.

ROOSEN (1966) has used a comparison of the annual variation of magnetic

Table 1. (a) Annual variation of aurora Sweden Lat. ⩽65°00′.

Period	Jan.	Feb.	March	April	May	June	July	Aug.	Sept.	Oct.	Nov.	Dec.	Total
1723-1732	36	47	65	29	5	1	1	20	50	73	63	49	439
1733-1743	45	61	62	29	7	0	4	38	61	79	44	39	469
1744-1754	73	83	94	45	2	0	1	38	70	94	70	70	640
1775-1765	55	74	78	69	13	0	0	42	91	73	53	60	608
1766-1774	40	44	62	33	7	1	3	24	36	67	41	39	397
1775-1783	34	35	52	43	12	0	4	24	66	42	19	30	361
1784-1797	61	81	97	100	16	1	5	46	97	90	69	63	726
1798-1809	35	45	53	17	2	0	0	15	48	54	40	34	343
1810-1822	19	16	12	8	0	0	0	0	11	13	13	11	104
1823-1832	20	10	16	10	1	0	0	9	25	18	23	16	148
1833-1842	34	35	25	27	0	0	1	10	45	41	38	32	288
1843-1855	37	55	49	42	4	0	0	18	50	46	39	40	380
1856-1866	45	64	71	55	6	0	1	29	70	79	50	64	534
1867-1877	53	77	113	99	10	3	3	39	109	106	86	46	744
1878-1888	42	47	65	24	7	3	4	9	42	53	55	39	390
1889-1900	65	76	71	26	7	0	1	11	66	73	64	43	503
1901-1912	27	29	38	15	2	0	0	6	30	32	27	19	225
1913-1922	21	18	25	18	4	0	1	13	27	37	20	11	195
1923-1932	12	24	32	7	2	0	0	3	32	39	21	16	188
1933-1943	21	14	31	23	8	0	1	6	27	29	19	9	188
1723-1943	775	935	1112	719	115	9	30	400	1053	1138	854	730	7870

Table 1. (b) Annual variation: Sweden ⩽ 65°00′, frequencies.

Period	Jan.	Feb.	March	April	May	June	July	Aug.	Sept.	Oct.	Nov.	Dec.
1723-1732	.082	.107	.148	.066	.011	.002	.002	.046	.114	.166	.144	.112
1733-1743	.096	.130	.132	.062	.015	0	.009	.081	.130	.168	.094	.083
1744-1754	.114	.130	.147	.070	.003	0	.002	.059	.109	.147	.109	.109
1775-1765	.090	.122	.128	.113	.021	0	0	.069	.150	.120	.087	.099
1766-1774	.101	.111	.156	.083	.018	.003	.008	.060	.091	.169	.103	.098
1775-1783	.094	.097	.144	.119	.033	0	.011	.066	.183	.116	.053	.083
1784-1797	.084	.112	.134	.138	.022	.001	.007	.063	.134	.124	.095	.087
1798-1809	.102	.131	.155	.050	.006	0	0	.044	.140	.157	.117	.099
1810-1822	.183	.154	.125	.007	0	0	0	0	.106	.125	.125	.106
1823-1832	.135	.068	.108	.068	.007	0	0	.061	.169	.122	.155	.108
1833-1842	.118	.122	.087	.094	0	0	.003	.035	.156	.142	.132	.111
1843-1855	.097	.145	.129	.111	.011	0	0	.047	.132	.211	.103	.105
1856-1866	.084	.120	.133	.103	.011	0	.002	.054	.131	.148	.094	.120
1867-1877	.071	.103	.152	.133	.013	.004	.004	.052	.147	.142	.116	.062
1878-1888	.108	.121	.167	.062	.018	.008	.010	.023	.108	.136	.141	.100
1889-1900	.129	.151	.141	.052	.014	0	.002	.022	.131	.145	.127	.085
1901-1912	.120	.129	.169	.067	.009	0	0	.027	.133	.142	.120	.084
1913-1922	.108	.092	.128	.092	.021	0	.005	.067	.138	.190	.103	.056
1923-1932	.064	.112	.170	.037	.011	0	0	.016	.170	.207	.112	.085
1933-1943	.112	.074	.165	.122	.043	0	.005	.032	.144	.154	.101	.048
1723-1943	.099	.119	.141	.091	.015	.001	.004	.051	.134	.145	.109	.093

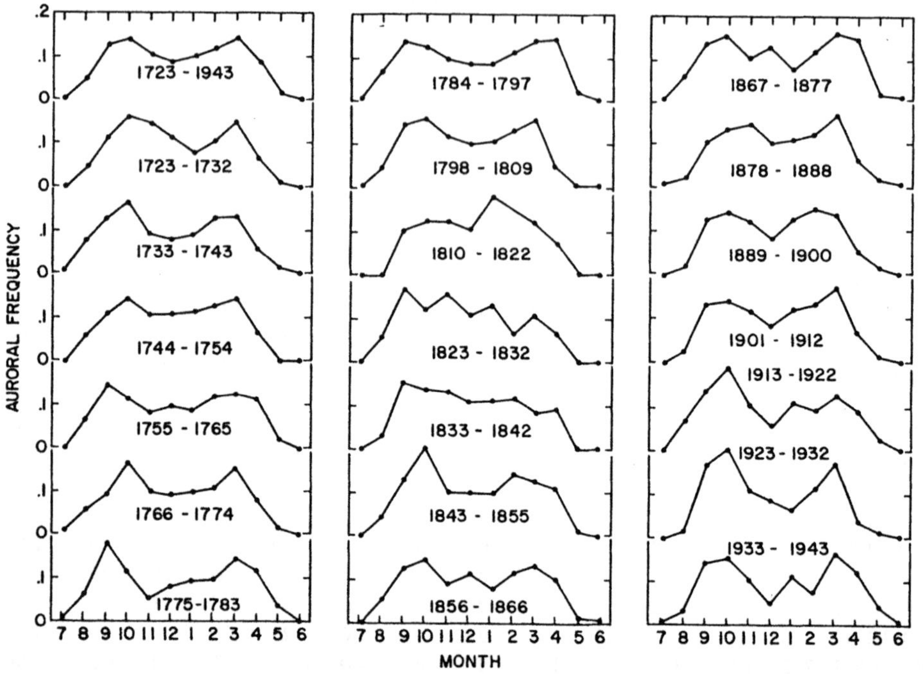

Fig. 1. Annual variation of occurrence frequency of Swedish Auroras, 1723–1943.

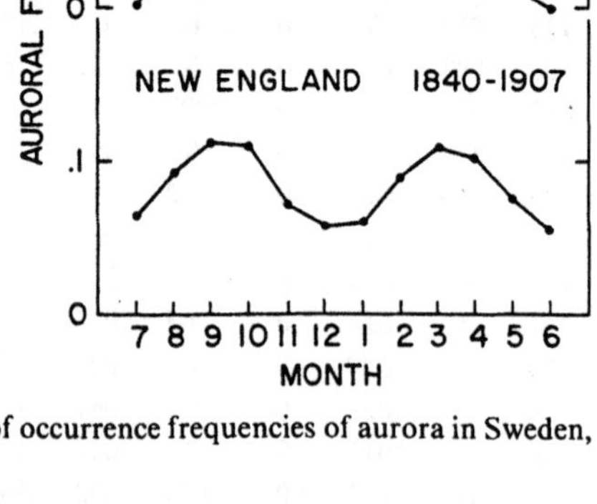

Fig. 2. Annual variation of occurrence frequencies of aurora in Sweden, 1723–1943, and New England, 1840–1907.

activity and the solar declination angle as a test of the equinoctial vs. axial hypotheses. In Fig. 3 a similar comparison of the annual variation of the auroral frequencies for both Sweden and New England is shown. The scales have been chosen to approximately equalize the minima and maxima. The solar declination used was that calculated by Roosen, five days before the terrestrial observation date, to take account of a presumed average travel time between sun and earth. The monthly value of frequency has been plotted at the middle of the month.

For Sweden the trend of the two is very similar, though the relative values are much different during the winter months. This is likely due to the distorting effects of the small number of summer observations. For New England the correspondence is fairly close, indicating that for auroral activity, as for magnetic activity, an equinoctial origin for the annual variation is preferred. The time resolution of the auroral data as used is not sufficient to establish with any precision the dates of maxima and minima.

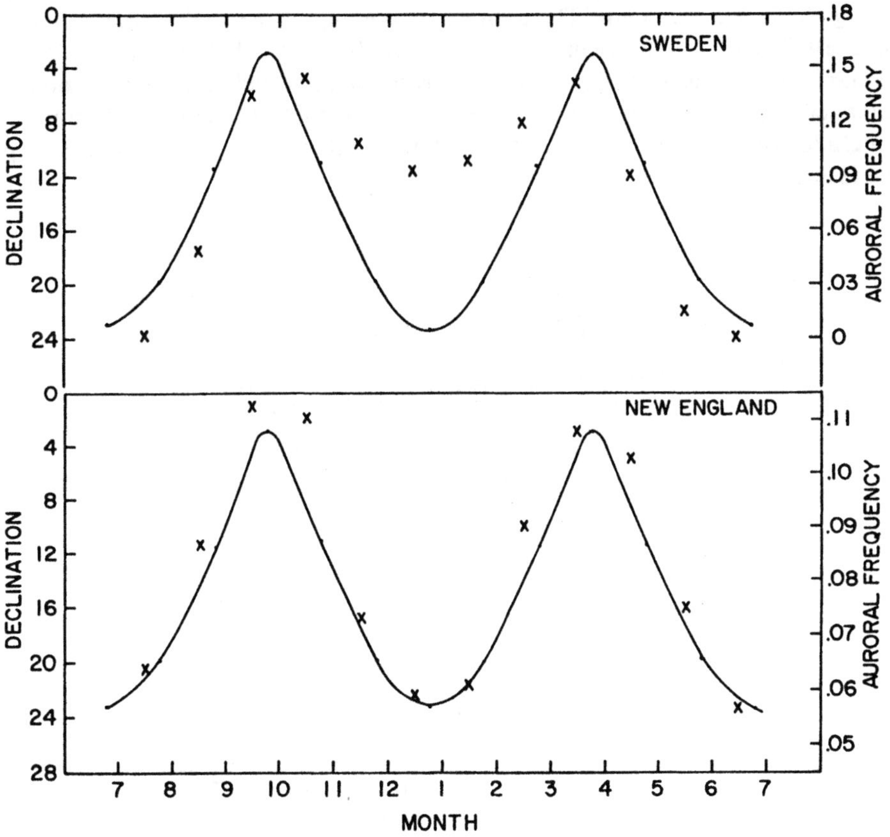

Fig. 3. Comparison of annual variation of occurrence frequency of aurora in Sweden and New England with solar declination.

3. Amplitude of the Semiannual Variation

In a previous paper (SILVERMAN and SHAPIRO, 1983) power spectral analyses were carried out on the Swedish data set. Power spectra were calculated for a sequence of consecutive, overlapping, approximately 22-year data segments. This procedure allowed variations over time of individual peaks to be clearly seen. In that paper it was assumed that the most prominent peak, that at twelve months, would change only if some terrestrial parameter changed. The most likely factors, the magnitude of the earth's magnetic field and the position of the dipole, were not considered likely to change significantly. This peak could then be used as a reference peak. The ratio of the amplitudes of the 6-months peak to the 12-months peak was then plotted. A pronounced decline in the magnitude of the 6-month peak during the period from 1797–1843, and in particular from 1810–1833, was noted. A smaller, similar, effect was noted for the period from 1878 to 1912.

In Fig. 4 the amplitude of the 6-months peak, using the same data as in the previous paper, is plotted directly, rather than as a ratio. The near-disappearance of this peak during the period from 1810–1833, which can also be inferred from the plots in Fig. 1, is clearly evident. The definite weakening of the peak in the period from 1878–1912 can also be seen. The two periods in which the magnitude of the semiannual peak declines significantly correspond to periods of prolonged minima in solar activity around the beginning of each century. SILVERMAN and SHAPIRO (1983) concluded that a lessening of the solar radial field during these periods could explain the results.

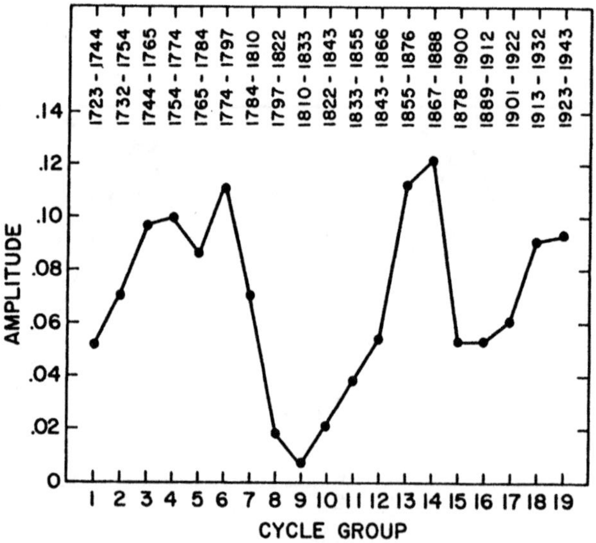

Fig. 4. Amplitude of six-month peak in power spectrum of auroral occurrence in Sweden, 1723–1943.

4. The Spring/ Fall Ratio

The more recent studies of the interaction between the solar wind and magnetic activity have related magnetic disturbances to some function of the velocity of the solar wind and the southward component of the interplanetary magnetic field (see, for example, FEYNMAN and CROOKER, 1978; HOLZER and SLAVIN, 1982, and references cited therein). Historical data on auroral occurrence frequency has also been used to determine characteristics of the solar wind over the past few centuries (FEYNMAN and SILVERMAN, 1980; SILVERMAN and FEYNMAN, 1981). As a result of these studies and others listed in the introduction, it seems reasonable to make the following assumptions:

(1) The intensity and occurrence frequency of magnetic disturbances and auroras are a function of $B_s^m V^n$, where B_s is the magnitude of the southward component of the interplanetary field, and V is the solar wind velocity. Auroras occur at a particular latitude only when the magnetic field is disturbed above some threshold. The value of the exponent for B_s has been proposed as between 0.85 and 2 by different authors.

(2) While the solar wind velocity may show considerable variation over a period of one or more solar cycles, we will assume here that over a period of six months to a year the average velocity is the same.

(3) A consequence of the Russell-McPherron hypothesis is that the spring maximum is associated, on the average, with fields towards the sun, and the fall maximum with fields away from the sun. While there is disagreement as to the extent of this association, the observations do appear to support such a connection (BERTHELIER, 1976; HOLZER and SLAVIN, 1982).

The frequency of occurrence of aurora is not a measure of the intensity of disturbances. It is, rather, essentially a measure of the number of days on which the disturbance level exceeds some threshold value of disturbance. We will assume here that this measure is also a function of $B_s^n V^m$, as is the magnetic activity. With the assumptions above, the relative frequency of the spring compared to the fall will then be equal to the spring/fall ratio of the southward component of the interplanetary field, to the power n. This in turn will reflect the ratio of occurrence of field towards the sun to that of the field away from the sun. Since the annual variation, if similar to that of magnetic activity, is only in part due to this factor, we can deduce, at best, only trends from the present analysis.

In order to improve the reliability of the analysis we may take the ratio of the sum of the number of days on which auroras are observed in March and April as that for the spring, and the corresponding sum for September and October as that for the fall. Furthermore, to increase the numbers we may take the sum over an entire sunspot cycle. This will weight the results towards the period around sunspot maximum. For Sweden, in addition, the skewness of the annual variation because of the summer effect, and the sharp decline in April, means that the results should be treated as indicating changes only, and not be used in an absolute sense. Comparison of these ratios with a New England sample (Silverman, unpublished, 1985) indicates at least the approximate correctness of these ratios. The New England sample consists of 2813 days on which aurora occurred, for the period 1843 to 1900. About ninety percent of the observations are from Maine. Table 2 presents a comparison of the ratios of the sums for March and April to those of September and October for both Sweden and New England. For four of the five solar cycles the agreement is good or reasonable. The ratio for Sweden for the cycle from 1889–1900 appears to be low. The analogous ratio for Sweden south of 61.5 degrees is 0.9, indicating that the ratio for the region south of 65 degrees appears to be too low and probably represents a sampling fluctuation. The results for Sweden for the period 1723–1943 are plotted in Fig. 5.

Despite the uncertainties in the data there appear to be clear indications of trends, with an increase in ratio from the earliest data, about 1720, to a maximum in the period around 1780, followed by a decline to a minimum about 1830. This pattern is then repeated, with maximum around 1870, and a subsequent minimum for the cycle beginning in 1923.

5. Discussion and Conclusions

The 6-months peak reflects an increase in the disturbance levels at or near the

Table 2. Spring/fall ratios for Sweden and New England.

Period	Ratio	
	New England	Sweden
1843-1855	0.92	0.95
1856-1866	0.86	0.85
1867-1877	1.07	0.99
1878-1888	0.86	0.94
1889-1900	0.86	0.70

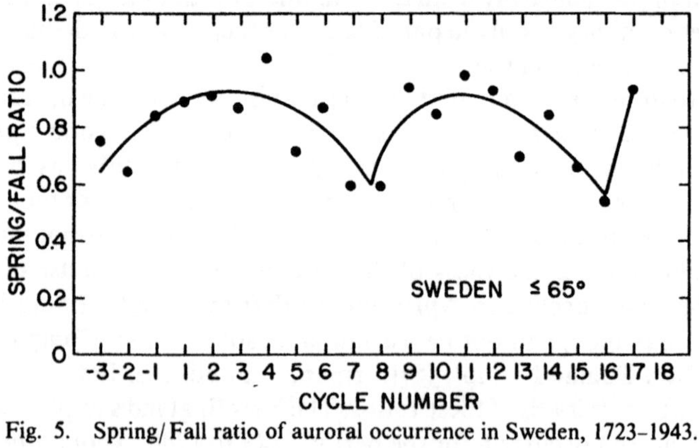

Fig. 5. Spring/Fall ratio of auroral occurrence in Sweden, 1723–1943.

equinoxes. This increase appears to be, at least in part, a result of the relative orientation of the coordinate system dependent on the earth's dipole axis with that of a coordinate system centered on the sun, producing a southward component of the interplanetary field allowing for energy transfer from the solar wind into the magnetosphere. The disturbance level is dependent not only on this southward component but also on the solar wind velocity. The auroral data for the amplitude of the 6-months peak indicates that one or both of these parameters declined considerably during the prolonged solar minima around the beginnings of the nineteenth and twentieth ccenturies. This result is consistent with other evidence (cf., FEYNMAN and SILVERMAN, 1980; SILVERMAN and FEYNMAN, 1981). The spring/fall ratio, to a first approximation, can be taken as reflective of the relative importance of the radial fields pointing towards and away from the sun. The auroral evidence indicates that these fields are not, on the average, equal from one solar cycle to the next. The variation of these fields, furthermore, is not in phase with that of the intensity of southward component/solar wind velocity as indicated by the amplitude of the 6-months peak. The field away from the sun is strongest somewhat after the prolonged solar minima, during which the southward component/solar wind velocity is weakest. Parenthetically it may be noted that the minimum spring/fall ratio for the cycle beginning in 1923 coincides with the decrease in corrrespondence between modulation of sunspots and that of magnetic activity noted by GREEN (1984).

Models for the explanation of the annual variation of either magnetic activity or auroral occurrence are far from definitive at this time. The discussion of the auroral results will therefore not be taken beyond the narrow limits above.

This material is based upon work supported by the National Science Foundation under Grant No. ATM-83-11139.

REFERENCES

BARTELS, J., Terrestrial-magnetic activity and its relations to solar phenomena, *Terr. Mag. Atm. Elect.*, **37**, 1-52, 1932.

BARTELS, J., Discussion of time-variations of geomagnetic activity, indices K_p and A_p, 1932–1961, *Ann. Geophys.*, **19**, 1-20, 1963.

BERTHELIER, A., Influence of the polarity of the interplanetary magnetic field on the annual and the diurnal variations of magnetic activity, *J. Geophys. Res.*, **81**, 4546–4552, 1976.

BOLLER, B. R. and H. L. STOLOV, Kelvin-Helmholtz instability and the semiannual variation of geomagnetic activity, *J. Geophys. Res.*, **75**, 6073, 1970.

BROUN, J. A., Observations in magnetism and meteorology made at Makerstoun in Scotland, *Trans. R. Soc. Edinburgh*, **18**, 401-402, 1848.

CORTIE, A. L., Sun-spots and terrestrial magnetic phenomena, 1898–1911: the cause of the annual variation in magnetic disturbances, *M.N.R.A.S.*, **73**, 52–60, 1912.

CORTIE, A. L., The efficiency of sun-spots in relation to terrestrial magnetic disturbances, *M.N.R.A.S.*, **76**, 15- 18, 1915.

CURRIE, R. G., The geomagnetic spectrum — 40 days to 5.5 years, *J. Geophys. Res.*, **71**, 4579–4598, 1966.

ELLIS, W., On the relation between magnetic disturbance and the period of solar spot frequency, *M.N.R.A.S.*, **60**, 142–157, 1899.

ELLIS, W., Sun-spots and magnetic disturbance, *M.N.R.A.S.*, **61**, 537–541, 1901.

ELLIS, W., The aurora and magnetic disturbance, *M.N.R.A.S.*, **64**, 228–236, 1904; **65**, 520–524, 720, 1905.

ELLIS, W., The seasonal variation in magnetic disturbance, with other remarks, *M.N.R.A.S.*, **65**, 520–524, 1905.

ELLIS, W., Note on the annual inequality in the frequency of magnetic disturbance, *M.N.R.A.S.*, **65**, 720, 1905.

FEYNMAN, J. and N. U. CROOKER, The solar wind at the turn of the century, *Nature*, **275**, 626, 1978.

FEYNMAN, J. and S. M. SILVERMAN, Auroral changes during the 18th and 19th centuries and their implications for the solar wind and long term variation of sunspot activity, *J. Geophys. Res.*, **85**, 2991–2997, 1980.

FRASER-SMITH, A. C., Spectrum of the geomagnetic activity index A_p, *J. Geophys. Res.*, **77**, 4209–4220, 1972.

GREEN, C. A., The semiannual variation in the magnetic activity indices A_a and A_p, *Planet. Space Sci.*, **32**, 297–305, 1984.

HANSTEEN, C., Sur les aurores boréales observées a Christiania, de 1837 a 1846, *Mem. Acad. R. Belgique*, **20**, 103–120, 1847.

HOLZER, R. E. and J. A. SLAVIN, A quantitative model of geomagnetic activity, *J. Geophys. Res.*, **87**, 9054–9058, 1982.

LEWIS, R. P. W. and D. H. MCINTOSH, A universal time component in geomagnetic disturbance, *J. Atm. Terr. Phys.*, **4**, 78–80, 1953.

LOCKYER, N. and W. J. S. LOCKYER, A probable cause of the yearly variation of magnetic storms and aurorae, *Proc. R. Soc. London*, **74**, 90–95, 1904.

LOVERING, J., On the periodicity of the aurora borealis, *Mem. Am. Acad. Arts Sci.*, **10**, 9–351, 1866-1871.

MAUNDER, E. W., Magnetic disturbances as recorded at the Royal Observatory, Greenwich, and their association with sun-spots, *M.N.R.A.S.*, **65**, 666–681, 1905.

MAYAUD, P.-N., Comment on 'Semiannual variation of geomagnetic activity' by C. T. Russell and R. L. McPherron, *J. Geophys. Res.*, **79**, 1131, 1974.

MCINTOSH, D. H., On the annual variation of magnetic disturbance, *Phil. Trans. R. Soc. London*, **A251**, 525- 552, 1959.

MEYER, J., A semiannual wave in the recurrence tendency of geomagnetic activity, *J. Geophys. Res.*, **71**, 2397–2400, 1966.

MURAYAMA, T., Origin of the semiannual variation of geomagnetic K_p indices, *J. Geophys. Res.*, **79**, 297–300, 1974.

PRIESTER, W. and D. CATTANI, On the semiannual variation of geomagnetic activity and its relation to the solar corpuscular radiation, *J. Atm. Sci.*, **19**, 121–126, 1962.

QUETELET, A., Sur les aurores boréales, *Bull. Acad. R. Belgique*, **21**, Part I, 282, 1854.

ROOSEN, J., The seasonal variation of geomagnetic disturbance amplitudes, *Bull. Astron. Inst. Netherlands*, **18**, 295–305, 1966.

RUBENSON, R., Catalogue des aurores boréales observées en Suède, *Kgl. Sv. Vetenskapakad. Handl.*, **15**(5), 1879; **18**(1), 1882.

RUSSELL, C. T. and R. L. MCPHERRON, Semiannual variation of geomagnetic activity, *J. Geophys. Res.*, **78**, 92–108, 1973.

SAITO, T., Recurrent magnetic storms in relation to the structure of solar and interplanetary magnetic fields, *Rept. Ionos, Space Res. Japan*, **26**, 245–266, 1972.

SHAPIRO, R., Semiannual variation of geomagnetic disturbance and its modulation of shorter period variations, *J. Geophys. Res.*, **74**, 2356–2367, 1969.

SILVERMAN, S. M., The visual aurora as a predictor of solar activity, *J. Geophys. Res.*, **88**, 8123–8128, 1983.

SILVERMAN, S. M. and J. FEYNMAN, The changing aurora of the past three centuries, in *Exploration of the Polar Upper Atmosphere*, edited by C. S. Deehr and J. A. Holtet, p. 407, 1981.

SILVERMAN, S. M. and R. SHAPIRO, Power spectral analysis of auroral occurrence frequency, *J. Geophys. Res.*, **88**, 6310–6316, 1983.

4. SOLAR WIND CONTROL OF NIGHTSIDE MAGNETOSPHERE

Solar Wind-Magnetosphere Coupling, edited by Y. Kamide and J. A. Slavin, 657–670.

Solar Wind Control of the Magnetotail

D. H. FAIRFIELD

Laboratory for Extraterrestrial Physics NASA Goddard Space Flight Center, Greenbelt Maryland 20771, U.S.A.

As the solar wind impinges on the geomagnetic dipole, high latitude field lines are drawn out behind the earth to form the geomagnetic tail. Reconnection between interplanetary and geomagnetic field lines is generally thought to be the primary mechanism causing this tail formation. The presence of a southward interplanetary field enhances reconnection on the dayside and increases the amount of flux transported to the magnetotail. Increased tail flux increases the flaring of tail field lines inside of about 100 R_e which increases the component of solar wind pressure normal to the boundary. This increased normal pressure creates a larger tail field strength. Beyond 100 R_e the tail ceases to flare and external thermal and magnetic pressure alone balance the tail pressure. Flux return to the dayside occurs either in concert with tailward transfer (resulting in directly driven geomagnetic activity) or impulsively and sporadically (magnetospheric substorms). This impulsive release of energy seems to accompany the release of concentric closed loops of magnetic flux containing hot plasma (plasmoids) which are frequently detected by ISEE-3 at about 200 R_e down the tail. The interplanetary B_y component seems to control the location where flux crosses the magnetotail boundary and where additional plasma can enter the distant magnetotail.

1. Introduction

After two decades of magnetospheric research most investigators seem to agree that the primary mechanism involved in forming the geomagnetic tail is the dayside reconnection of geomagnetic and interplanetary field lines and their subsequent transport to the nightside hemisphere. Additional plasma may cross the dayside magnetopause via diffusive processes and transport closed field lines to the tail, but this process appears to be less important than reconnection except under northward interplanetary field conditions when reconnection is of minimal importance (e.g., COWLEY, 1982). This paper will review knowledge of the configuration of the geomagnetic tail and the ways it changes in response to the solar wind plasma and magnetic field.

2. Interplanetary Control of Tail Energy

The existence of the geomagnetic tail is ultimately due to a dawn-dusk current near the equatorial plane and its closure around the northern and southern halves of the tail magnetopause, but it is useful to consider the problem from the hydromagnetic viewpoint where geomagnetic flux tubes of the tail are confined by the pressure of the external solar wind. Assuming pressure balance at the magnetopause we may equate the external pressure of the solar wind to the internal pressure in the magnetosphere:

$$Kn_{sw}m_p v^2 \sin^2\theta + n_{sw}kT_{sw} + B_{sw}/8\pi = nkT + B^2/8\pi \quad (1)$$

where K is a constant which is a measure of how efficiently solar wind particles transfer their momentum to the magnetosphere ($K=1$ for inelastic collisions which we assume, $K=2$ for elastic collisions, etc. (e.g., SCHIELD, 1969)), T_{sw} is the sum of the solar wind electron and proton temperatures, n_{sw}, B_{sw}, and v are solar wind density, field strength and velocity, m_p is the proton mass (heavy ions are neglected), θ is the angle betweend the solar wind and the magnetopause, B is the magnetosphere field strength, n is the magnetosphere plasma density, T is the sum of the magnetosphere electron and proton temperatures, and k is the Boltzmann constant. On the dayside of the earth θ is large and the first term in Eq. 1 is the dominant term on the left hand side. On the nightside the flaring angle θ becomes smaller at larger distances down the tail and eventually the other two terms on the left hand side become of comparable magnitude. The tail pressure on the right hand side (which over most of the boundary is that contributed by the tail lobe field since plasma pressure is unimportant in the tail lobe) decreases down the tail as θ decreases on the left hand side. At some point the tail will cease flaring; θ will become zero, the tail radius will remain constant, and B will be constant and balanced by only the external thermal and magnetic pressure. Isee-3 measurements have shown that on average this situation occurs at about $X_{sm} \cong 120$ R$_e$ with a tail radius of about 30 R$_e$ (SLAVIN *et al.*, 1985). Beyond this point the 9.2 nT average lobe field that is observed is nearly that which would be balanced by solar wind field and thermal pressure very close to long term average values.

A specific example of solar wind control of tail pressure is shown in Fig. 1 (FAIRFIELD, 1985). The solid line in Fig. 1(a) shows the total field plus plasma pressure measured by Isee 2 in the tail about 14 R$_e$ behind the earth on March 22, 1979. The triangles indicate the total solar wind pressure incident on the tail as measured by Imp-8. A flaring angle, θ, of 20° has been assumed. The solar wind data are shifted to the right by 10 minutes to account for the solar wind transit time from Imp-8 to Isee-2. Plasma β ($=8\pi nkT/B^2$) is shown in Fig. 1(b) to indicate the relative importance of the field and plasma contributions to the total pressure. At 0830 UT a solar wind pressure increase associated with an interplanetary shock wave caused an increase in the tail pressure by a comparable factor. There is no reason to expect that θ should remain 20° and alternatively we can assume the validity of equation 1 and use the measurements to determine an experimental flaring angle, θ_f, as a function of time. This procedure produces the trace in Fig. 1(c). The angle θ_f decreases at the time of the interplanetary shock wave indicating the tail is compressed. This angle then remains relatively constant near 18° until 1020 when it begins a 30 minute increase to 32°. The apparent reason for this behavior is an abrupt southward turning of the interplanetary field at 1020 UT (Fig. 1(d)) which previous work (e.g., SLAVIN and HOLZER, 1979) has shown enhances dayside reconnection and increases flux in the magnetotail.

Two spacecraft measurements such as those shown in Fig. 2 have demonstrated

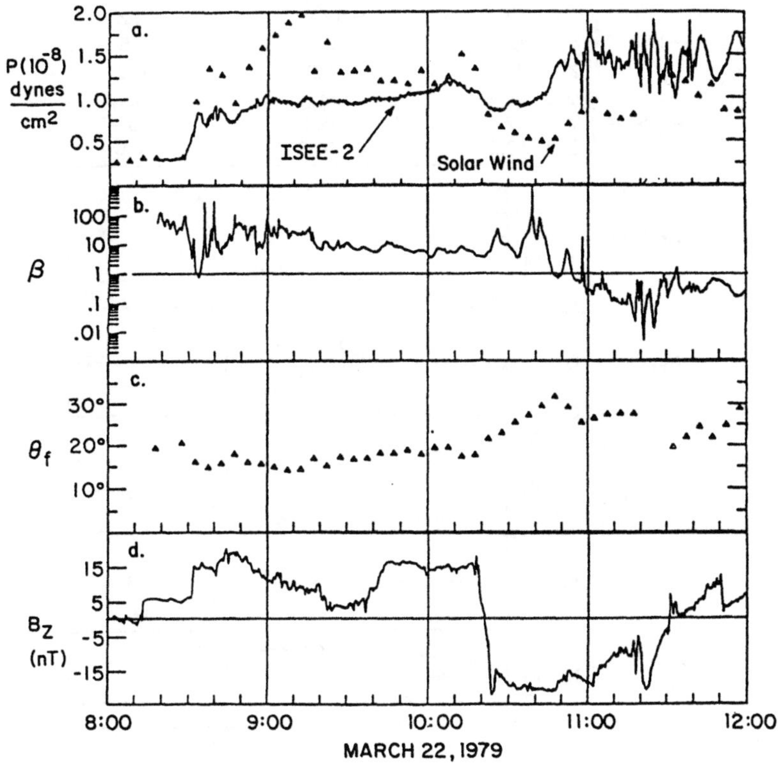

Fig. 1. (a) The magnetotail pressure and the component of the total solar wind pressure normally incident on the tail magnetopause assuming a 20° flaring angle. (b) The plasma β in the magnetotail measured by Isee-2. (c) The flaring angle of the tail calculated from Eq. 1 which balances the measured solar wind and magnetotail pressures. (d) The B_z component of the interplanetary magnetic field. An icnrease in the tail flaring angle appears to be caused by the southward IMF and terminated by the 1055 substorm onset.

that increasing tail flux before large substorm intensifications occurs frequently and throughout the whole tail (FAIRFIELD *et al.*, 1981). During an interval of southward IMF (southward interplanetary magnetic field θ_{IP}) from 0245–0424, B at two widely separated spacecraft in the geomagnetic tail lobe is seen to increase. A pair of substorm intensifications were apparent in ground magnetograms at 0419 and 0439 and a corresponding pair of tail field decreases were seen by the two Imp spacecraft (vertical dashed lines). FAIRFIELD *et al.* (1981) estimate a tail energy loss of 6×10^{21} ergs between 0430 and 0600 which gives a loss rate of 10^{18} ergs/s. This value is large enough to account for the energy dissipated in the substorm.

Enhanced tail flux and pressure lead to more tail-like fields (fields that are more aligned with the solar wind direction and have a smaller north–south component) which are detected particularly easily at the 6.6 R_e synchronous orbit. Figure 3 shows 30 keV to 300 keV electron measurements from a spacecraft at this location (BAKER *et*

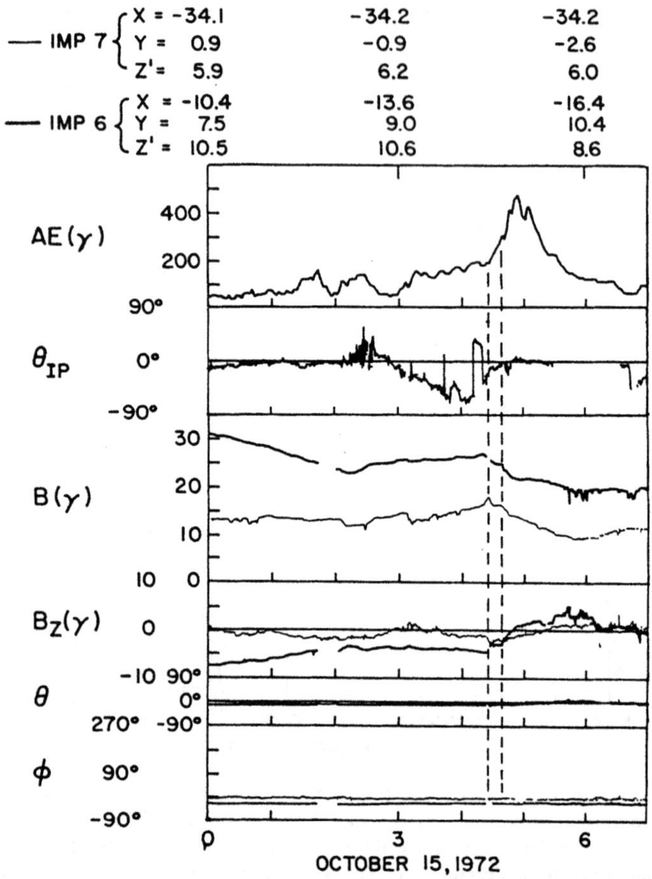

Fig. 2. Magnetic field observations from two IMP spacecraft at widely separated locations in the lobes of the magnetotail. The field magnitude B and solar magnetospheric latitude and longitude angles θ and ϕ are shown along with the tail B_z component, the IMF latitude angle θ_{IP} and the AE index. The vertical dashed lines denote the times of B decreases and B_z increases associated with substorms intensifications.

al., 1984). The C_2 parameter is an indicator of the electron pitch angle anisotropy where intervals of positive C_2 have been shown to correspond to field aligned electrons (termed cigar phase regions after the shape of the distribution function contours) and tail-like fields. BAKER *et al.*, (1978) have shown that these cigar phases are reliable predictors of magnetospheric substorms whose onsets are indicated by the abrupt enhancement of plasma fluxes such as occur at 0950 UT in Fig. 3. The accompanying decrease in C_2 at substorm onset in Fig. 3 is known to be associated with a reconfiguration of the tail magnetic field and a diversion of cross tail current through the ionosphere where it contributes to the east–west electrojet responsible for large decreases on auroral zone magnetograms (BAKER *et al.*, 1978). Earthward

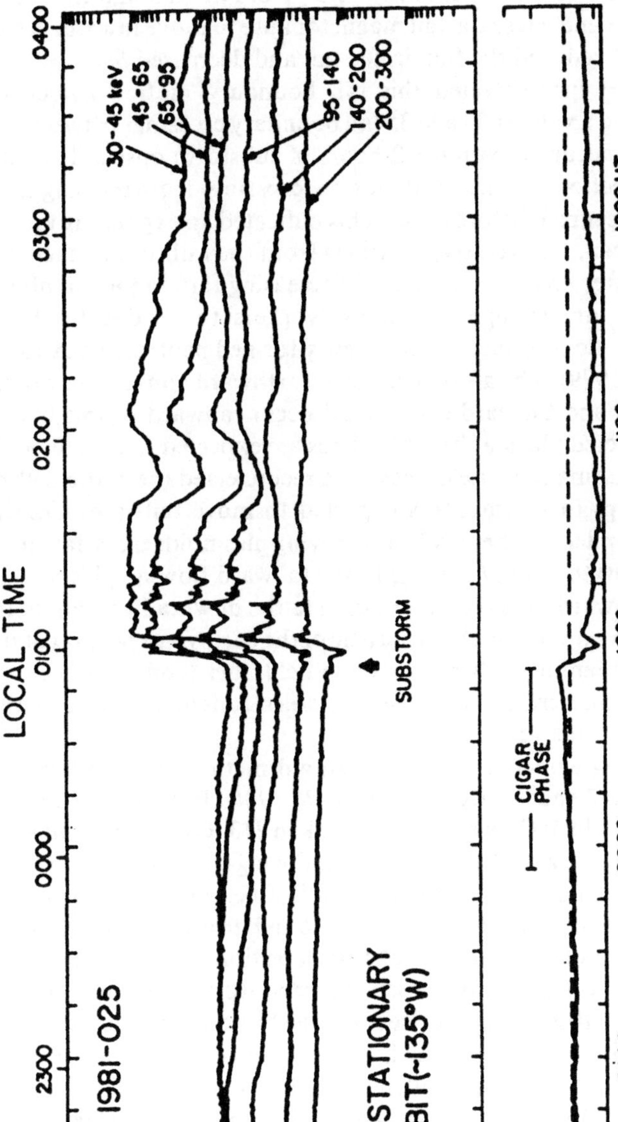

Fig. 3. Differential electron fluxes and >30 keV electron anisotropy amplitudes C_2 measured on spacecraft 1981–025 at geostationary orbit near local midnight. Flux increases at substorm onset near 0950 separate an interval of positive C_2 before onset from an interval of negative C_2 after onset (BAKER *et al.*, 1984).

moving flux tubes at substorm onset remove flux from the tail and thus reduce the pressure within the tail and cause the tail magnetopause to move inward. A substorm onset at 1055 in Fig. 1 reduced the flux in the tail and decreased θ_f.

MAEZAWA (1975) first detected this tail boundary motion associated with substorms by collecting Explorer 33 and 35 tail boundary crossings between 18 and 70 R_e down the tail that occurred within ± 2 hours of substorm onsets. He found that irrespective of the direction of spacecraft motion, crossings from the magnetosheath to the tail invariably occurred before onset. This indicated an expanding tail causd by enhanced flux in the tail. Conversely, crossings from the tail to the magnetosheath invariably occurred after onset and indicated decreasing flux in the tail after onset.

Substorm onsets are thought by many workers to be due to the sudden enhancement of magnetic reconnection at a newly formed neutral line at 10–20 R_e in the tail (e.g., HONES, 1979). The aforementioned earthward convection would occur earthward of this line and tailward flow would occur tailward of this line. On the tailward side, reconnected plasma sheet field lines are expected to form closed loops. When all the closed plasma sheet field lines have reconnected the entire collection of concentric closed loops (a plasmoid) is expected to move tailward. Considerable evidence has been compiled for these tailward moving plasmoids from data in cislunar space (HONES, 1979; BIEBER *et al.*, 1982). In the tailward flowing plasma seen after substorm onset, northward fields corresponding to the downtail portion of the loop typically precede southward fields characteristic of the earthward portion of the loop. Energetic electrons streaming tailward on open field lines tend to be seen after the loops have passed the spacecraft whereas isotropic electrons tend to be seen within the closed loop regions (BIEBER *et al.*, 1982).

Further convincing evidence for this plasmoid picture of substorms has been obtained from the Isee-3 spacecraft (HONES *et al.*, 1984a,b; SCHOLER *et al.*, 1984). Figure 4 (BAKER *et al.*, 1984) shows Isee-3 data taken 213 R_e down the tail at the time of the synchronous orbit data of Fig. 3. During the cigar phase detected at 6.6 R_e, Isee-3 moves from the magnetosheath to the tail lobe. This motion indicates an expanding magnetotail before onset like that found nearer earth by MAEZAWA (1975). Baker *et al.* found 30 examples of this behavior which corresponds to an appreciable fraction of the substorm onsets detected during this period. In Fig. 4, some 20 minutes after substorm onset, Isee-3 moved into the plasma sheet where it observed plasma flowing rapidly tailward. This is also typical behavior as 42 of 75 onsets were found to be followed (25 ± 10 minutes) by isolated episodes of high speed (>750 km/sec) plasma flow. Apparently the plasma arriving at ≈ 200 R_e occupies considerably more volume than the pre-existing plasma sheet so that it frequently engulfs Isee-3. Greater volume is expected since an approximately constant amount of plasma within the plasmoid will be confined by increasingly weaker field at greater distances down the tail. The tail radius probably increases as the plasmoid moves past at 200 R_e whereas nearer earth the radius decreases due to the plasmoid departure.

The above picture is further substantiated by Isee-3 observations of traveling compression regions seen in the distant tail lobes (SLAVIN *et al.*, 1984). These events are enhancements in B that are typically 10–30% of the field strength. They occur 10–30 minutes after substorm onsets, with the longer delays occurring further down

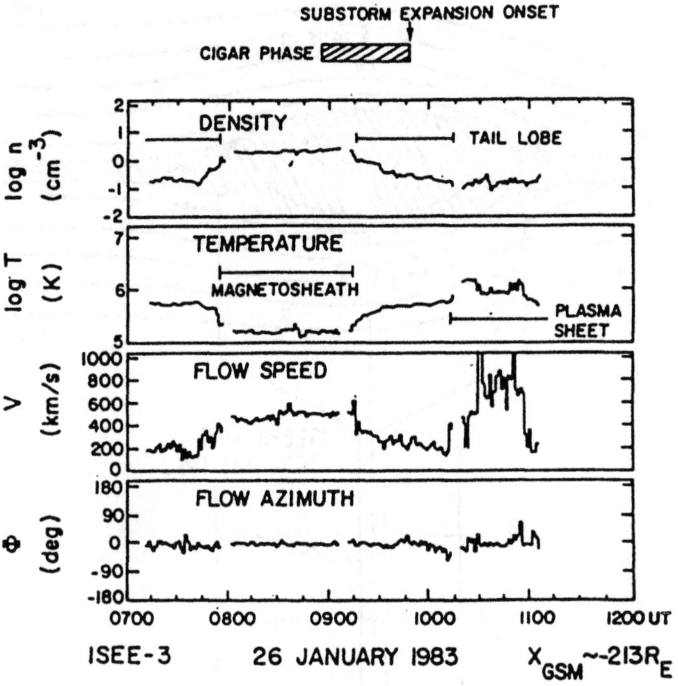

Fig. 4. ISEE-3 electron data taken 213 R_e behind the earth at the time the synchronous orbit data in Fig. 3. The various electron parameters indicate that ISEE-3 moved from the magnetosheath to the tail lobe during the substorm growth phase and from the lobe to the plasma sheet some twenty minutes after the near-earth substorm onset (BAKER *et al.*, 1984).

the tail. The B perturbations are invariably accompanied by a northward, followed by a southward, deflection of the field with the point of inflection corresponding to the peak in the magnitude enhancement. An example of such B and B_z pertubations is shown in Fig. 5 along with an explanatory sketch (SLAVIN *et al.*, 1984). Isee-3 is shown in the tail lobe with the passing plasmoid compressing the lobe field between the plasmoid and the magnetosheath. The plasmoid is moving from left to right so the positive B_z perturbation precedes the negaive B_z perturbation with $B_z=0$ corresponding to maximum compression. MAEZAWA (1975) had also noted such tail compressions in data nearer the earth and he gave a similar explanation. A similar B compression event can be seen in the Imp-7 data at 0425 in Fig. 2.

3. Interplanetary B_y Effects on the Magnetotail

Although the IMF B_z component clearly controls the input of solar wind energy to the magnetotail, the B_y component exerts control over other processes. A number of observed effects can be explained by considering the location where tail field lines are apt to cross the magnetopause and connect to the IMF. Figure 6 (TSURUTANI *et*

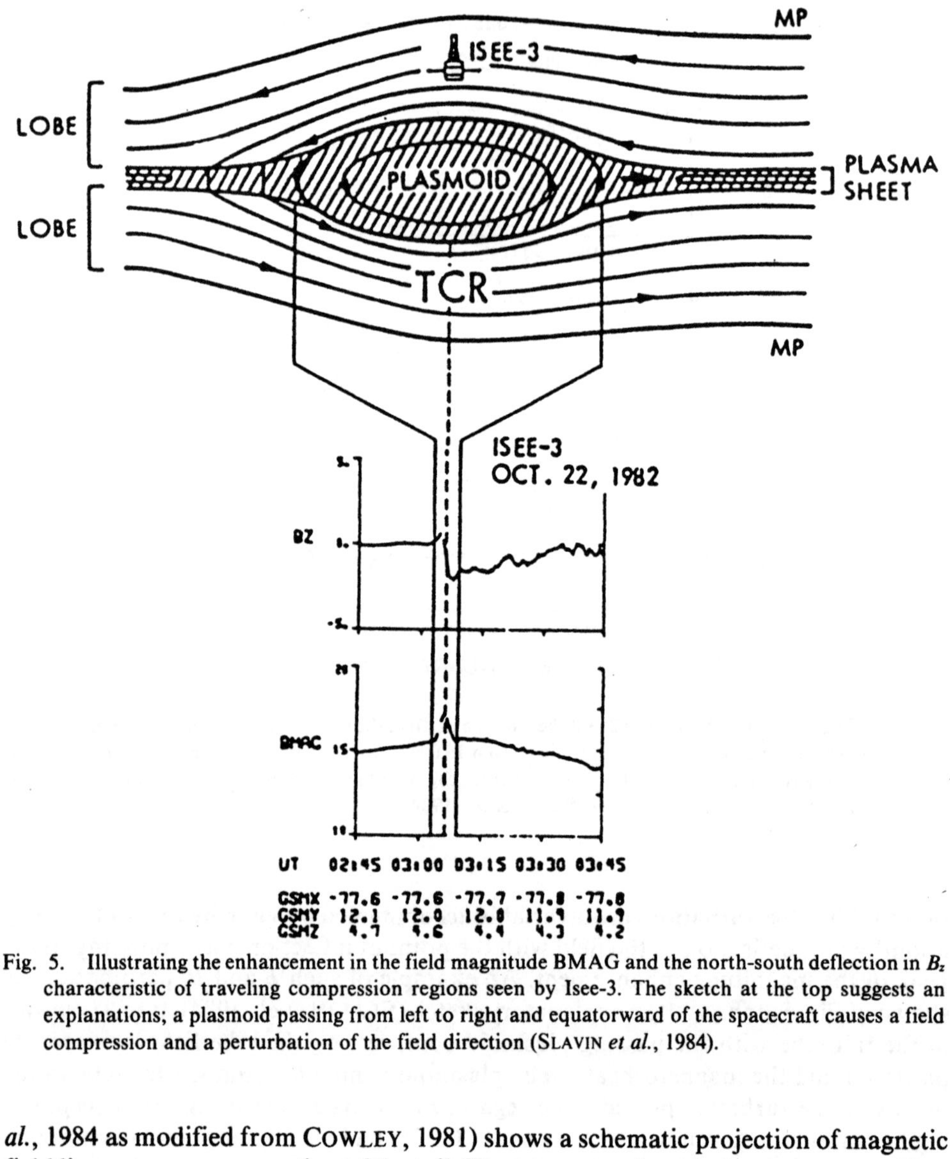

Fig. 5. Illustrating the enhancement in the field magnitude BMAG and the north-south deflection in B_z characteristic of traveling compression regions seen by Isee-3. The sketch at the top suggests an explanations; a plasmoid passing from left to right and equatorward of the spacecraft causes a field compression and a perturbation of the field direction (SLAVIN *et al.*, 1984).

al., 1984 as modified from COWLEY, 1981) shows a schematic projection of magnetic field lines on a *yz* cross section of the tail. The magnetosphere is shown immersed in a positive sector (IMF B_y positive) where reconnected field lines should preferentially enter the tail in the northern dawn sector and preferentially leave the tail in the southern dusk sector. This geometry leads to a cross tail component which is small compared to the large B_x component in the tail but which has been detected by FAIRFIELD (1979) between 20 and 40 R_e, by COWLEY and HUGHES (1983) at synchronous orbit, and by TSURUTANI *et al.* (1984) in the deep tail. In a negative sector the geometry reverses and the field lines enter and leave in the northern dusk and southern dawn sectors respectively, adding a negative B_y component to the tail.

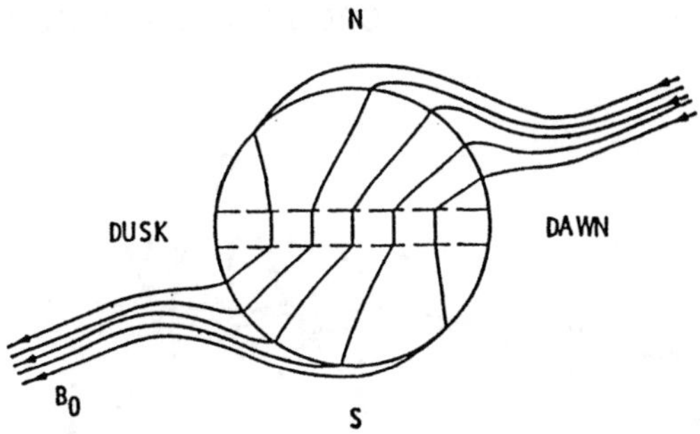

Fig. 6. A schematic view of a magnetotail cross section as viewed looking toward the sun. For the positive interplanetary sector shown, the field lines preferentially enter the tail in the northern dawn quadrant and leave the tail in the southern dusk quadrant thus producing a small dawn–dusk field component (TSURUTANI *et al.*, 1984 as modified from COWLEY, 1981).

Plasma can pass through magnetopause rotational discontinuities in these preferred quadrants and enhanced plasma has indeed been seen in these quadrants at lunar orbit (HARDY *et al.*, 1979) and in the deep tail (GOSLING *et al.*, 1985). The absence of such plasma inside the tail magnetopause is sometimes seen by Isee-3 and is interpreted as evidence for a locally closed magnetopause (GOSLING *et al.*, 1984). Nearer earth the plasma mantle appears to be thicker when B_z is southward (SCKOPKE *et al.*, 1976). The field line projections in Fig. 6 are also electric equipotentials (COWLEY, 1981) which implies that stronger electric fields exist in the preferred quadrants. This fact is consistent with many additional low altitude observations (e.g., BURKE, 1982).

SIBECK and SISCOE (1984) have studied the time evolution of a reconnected flux tube as it moves from the dayside to the deep tail. They find that the magnetosheath portion of the reconnected tube loses plasma to the magnetotail portion. This may leave a magnetotail-like flux tube extending out into the magnetosheath. These authors suggest that previously reported multiple tail crossings may in fact be the magnetosheath portion of reconnected flux tubes passing the spacecraft. This suggestion provides an alternative explanation for multiple crossings previously interpreted as waves on the tail boundary. The suggestion may also explain the fact that tail boundary crossings are seen over a surprisingly large range of distance from the tail axis. This range is illustrated in Fig. 7 (HONE and BINSACK, 1972) which shows a collection of Explorer 33 tail crossings. Beyond 40 R_e the range of distances over which the boundary may be seen is almost as large as the tail radius.

CROOKER and SISCOE (1979) have studied one Explorer 33 pass in detail whose trajectory segment is indicatd in Fig. 7. They find that the multiple boundary crossings on this pass were phased relative to the *AE* index in the same manner as found by MAEZAWA (1975) but they also point out that the distance ragne over which crossings were found is considerably greater than can be accounted for by the known movement of dayside flux to the magnetotail (SLAVIN and HOLZER, 1979). They proposed a tailward moving compression wave (and a violation of local magnetotail

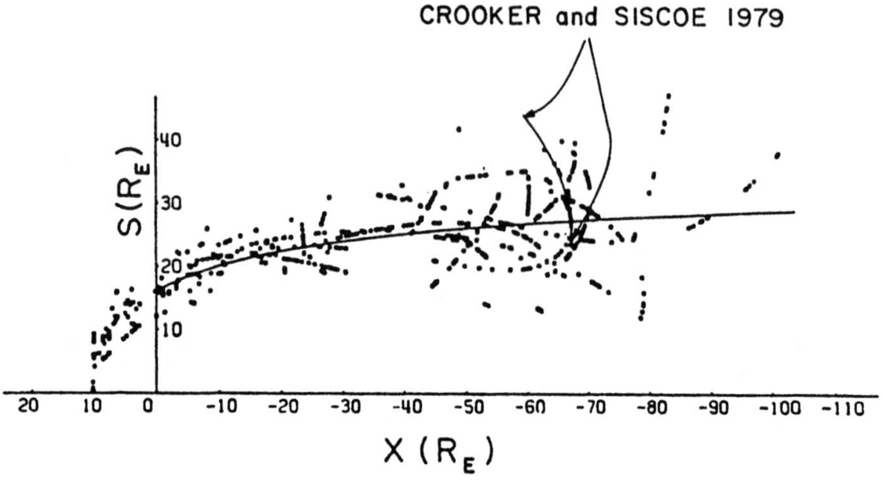

Fig. 7. Explorer 33 determinations of the distances of the magnetotail boundary from the tail axis are plotted at their observed distances down the tail (HONE and BINSACK, 1972). The range of distances beyond 40 R_e seems to be larger than could be explained by flux transfer to the tail.

pressure balance) as an explanation, but now the open flux tubes (SIBECK and SISCOE, 1984) provide an alternative explanation.

4. Tail Unloading Versus Driven Substorms

Although it is clear that the IMF B_z component closely controls the occurrence of geomagnetic substorms, there remains some controversy over whether substorm energy is directly supplied by the solar wind (AKASOFU, 1980) or whether it is stored in the magnetotail for subsequent release (e.g., NISHIDA, 1983; and references therein). The latter position has been developed largely through the analyses of spacecraft data in the magnetosphere and this position has been well documented for many years. The former position has been developed more recently by comparing solar wind conditions with estimates of energy input to the magnetosphere made from ground observations, with little input of in situ magnetospheric data.

To illustrate this controversy we focus on an event that occurred on March 19, 1978. AHN *et al.* (1984) have conducted an extensive analysis of this event utilizing data from 71 magnetic observations plus numerous all-sky camera pictures and riometer records. The authors conclude that a substorm began at about 1020 UT with the appearance of an auroral arc which brightened and moved poleward 10 minutes later. Weak currnts gradually began flowing and a weak absorption event occurred at 1030 UT. The authors apparently view this event as a normal, but small, substorm and they explicitly warn against "misidentifying" the interval subsequent to 1030 UT as a growth phase prior to a larger intensification that followed at 1137 UT.

Isee-1 magnetic field data taken about 14 R_e behind the earth have been obtained from C. T. Russell and synchronous orbit electron data have been obtained from D.

N. Baker. These data are displayed in Fig. 8. A time near 1030 marks the beginning of a tail field increase and B_z decrease that are indeed the magnetospheric characteristics of a growth phase. Also electron intensities begin decreasing at synchronous orbit and the C_2 parameter also begins increasing as it does prior to large substorm intensifications. The intensification at 1137 is clearly seen in both data sets in terms of the field reconfiguration (an ISEE B_z increase and synchronous orbit electron increase) and tail energy loss.

Irrespective of semantics, the situation seems relatively clear. At 1020 UT a southward IMF B_z began driving weak auroral zone currents and also initiated energy input to the magnetotail. By 1137 UT the tail had accumulted considerable excess energy and was in a highly stressed, tail-like state such that a large substorm intensification occurred cuasing this energy to be released. Thus it seems quite reasonable to consider the 1030–1137 period as being directly driven by the solar

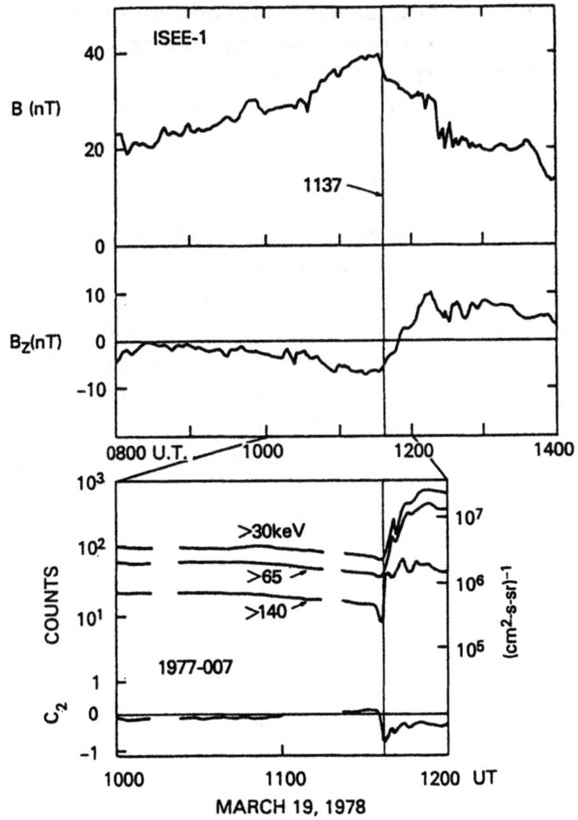

Fig. 8. Magnetic field data from Isee-1 14 R_e behind the earth and electron data from spacecraft 1977-007 in geostationary orbit. The data show typical growth phase behavior prior to a large 1137 substorm intensification even though ground data had earlier been used to argue against a growth phase.

wind. After 1137 it seems highly likely that energy "unloaded" from the tail was being supplied to the magnetosphere. It also seems reasonable to consider the period before 1137 as a growth phase since energy was being stored for eventual and inevitable loss during a large intensification. Energy storage and loss takes place on a time scale of an hour or less whereas on longer time scales the energy transfer is reliably controlled by the IMF B_z and can be considered to be directly driven. A breakdown in the magnetic activity-B_z correlation is in fact seen on short time scales by correlation analyses using linear prediction filters (e.g., BAKER, 1986).

A directly driven component of geomagnetic activity was described by PYTTE *et al.*, (1978) before the current controversy developed. In fact today many researchers believe that there is both a driven and an unloading component to most substorms. The important point is that once differing energy input and output rates are accepted, there is a need for an energy storage reservoir. This is the role played by the magnetotail. If a controvery exists, it must be between those who believe the magnetospheric response has both unloading and directly driven components versus those who advocate only a directly driven component. Those who advocate directly driven substorms downplay the importance of sudden substorm onsets and argue that onsets are spatially localized and of minimal importance from the global energy view point. On the other hand those who claim a role for tail unloading are impressed by the clear tail morphology and the wide ranging spatial effects of these tail changes.

5. Summary

In the above discussion it has been emphasized how solar wind pressure controls the magnetotail. It has also been emphasized how the energetically insignificant interplanetary B_z field component controls the degree to which energy is added to the magnetosphere and stored in the tail. Although not the main subject of the paper, the manner in which internal processes such as tail reconnection release this energy have been briefly reviewed. It is concluded that the magnetotail plays an important role in magnetospheric substorms, although a directly driven component also makes a contribution, particularly prior to intensifications that are associated with tail reconfigurations.

The role of the interplanetary B_y component has also been discussed. This component apparently controls the location of field lines passing through the tail boundary and thereby the location of local plasma entry. The large variations in the cross-sectional dimensions of the distant tail have been emphasized and this is pointed out as a presently unsolved problem.

The author thanks C. T. Russell and D. N. Baker for providing the ISEE-1 and 1977-007 spacecraft data included in Fig. 8. The comments of D. N. Baker and R. P. Lepping and the referees are also greatfully acknowledged.

REFERENCES

AHN, B.-H., Y. KAMIDE, and S.-I. AKASOFU, Global ionospheric current distribution during substorms, *J.*

Geophys. Res., **89**, 1613–1625, 1984.
AKASOFU, S.-I., The solar wind-magnetosphere energy coupling and magnetospheric disturbances, *Planet. Space Sci.*, **28**, 495–509, 1980.
BAKER, D. N., Correlation analyses in the study of solar wind-magnetosphere coupling, this volume, 1986.
BAKER, D. N., P. R. HIGBIE, E. W. HONES, Jr., and R. D. BELIAN, High resolution energetic particle measurements at 6.6 R_E 3. Low energy electron anisotropies and short-term substorm predictions, *J. Geophys. Res.*, **85**, 4863–4868, 1978.
BAKER, D. N., S. J. BAME, R. D. BELIAN, W. C. FELDMAN, J. T. GOSLING, P. R. HIGBIE, E. W. HONES, Jr., D. J. MCCOMAS, and R. D. ZWICKL, Correlated dynamical changes in the near-earth and distant magnetotail regions: ISEE-3, *J. Geophys. Res.*, **89**, 3855–3864, 1984.
BIEBER, J. W., E. C. STONE, E. W. HONES, Jr., D. N. BAKER, and S. J. BAME, Plasma behavior during energetic electron streaming events; further evidence for substorm-associated magnetic reconnection, *Geophys. Res. Lett.*, **9**, 664–667, 1982.
BURKE, W. J., Magnetosphere-ionosphere coupling: Contributions from IMS satellite observations, *Rev. Geophys. Space Phys.*, **20**, 685–708, 1982.
COWLEY, S. W. H., Magnetospheric asymmetries associated with the *Y*-component of the IMF, *Planet. Space Sci.*, **29**, 79–96, 1981.
COWLEY, S. W. H., The causes of convection in the earth's magnetosphere: A review of developments during the IMS, *Rev. Geophys. Space Phys.*, **20**, 531–565, 1982.
COWLEY, S. W. H. and J. W. HUGHES, Observations of an IMF sector effect in the *Y*-magnetic field component at geostationary orbit, *Planet. Space Sci.*, **31**, 73–90, 1983.
CROOKER, N. U. and G. L. SISCOE, Large amplitude substorm motion of the magnetotail boundary, *Geophys. Res. Lett.*, **6**, 105–108, 1979.
FAIRFIELD, D. H., On the average configuration of the geomagnetic tail, *J. Geophys. Res.*, **84**, 1950–1958, 1979.
FAIRFIELD, D. H., Solar wind control of magnetospheric pressure, *J. Geophys. Res.*, **90**, 1201–1204, 1985.
FAIRFIELD, D. H., R. P. LEPPING, E. W. HONES, Jr., S. J. BAME, and J. R. ASBRIDGE, Simultaneous measurements of magnetotail dynamics by IMP spacecraft, *J. Geophys. Res.*, **86**, 1396–1414, 1981.
GOSLING, J. T., D. N. BAKER, S. J. BAME, E. W. HONES, Jr., D. J. MCCOMAS, R. D. ZWICKL, J. A. SLAVIN, E. J. SMITH, and B. T. TSURUTANI, Plasma entry into the distant tail lobes: ISEE-3, *Geophys. Res. Lett.*, **11**, 1078–1081, 1984.
GOSLING, J. T., D. N. BAKER, S. J. BAME, W. C. FELDMAN, R. D. ZWICKL, and E. J. SMITH, North–south and dawn–dusk asymmetries in the distant tail lobe: ISEE-3, *J. Geophys. Res.*, **90**, 6354–6360, 1985.
HARDY, D. A., H. K. HILLS, and J. W. FREEMAN, Occurrence of the lobe plasma at lunar diatance, *J. Geophys. Res.*, **84**, 72–78, 1979.
HONES, E. W., Jr., Transient phenomena in the magneotail and their relation to substorms, *Space Sci. Rev.*, **23**, 393–410, 1979.
HONES, E. W., Jr. and J. H. BINSACK, Explorer 33 and 35 plasma observations of magnetosheath flow, *J. Geophys. Res.*, **77**, 3334–3344, 1972.
HONES, E. W., Jr., D. N. BAKER, S. J. BAME, W. C. FELDMAN, J. T. GOSLING, D. J. MCCOMAS, R. D. ZWICKL, J. A. SLAVIN, E. J. SMITH, and B. T. TSURUTANI, Structure of the magnetotail at 220 R_E and its response to geomagnetic activity, *Geophys. Res. Lett.*, **11**, 5–7, 1984a.
HONES, E. W., J. BIRN, Jr., D. N. BAKER, S. J. BAME, W. C. FELDMAN, D. J. MCCOMAS, R. D. ZWICKL, J. A. SLAVIN, E. J. SMITH, and B. T. TSURUTANI, Detailed examination of a plasmoid in the distant magnetotail with ISEE-3, *Geophys. Res. Lett.*, **11**, 1046–1049, 1984b.
MAEZAWA, K., Magnetotail boundary motion associated with geomagnetic substorms, *J. Geophys. Res.*, **80**, 3543– 3548, 1975.
NISHIDA, A., IMF control of the earth's magnetosphere, *Space Sci. Rev.*, **34**, 185–200, 1983.
PYTTE, T., R. L. MCPHERRON, E. W. HONES, Jr., and H. I. WEST, Jr., Multiple satellite studies of magnetospheric substorms: distinction between polar magnetic substorms and convection driven magnetic bays, *J. Geophys. Res.*, **83**, 663–679, 1978.
SCHIELD, M. A., Pressure balance between solar wind and magnetosphere, *J. Geophys. Res.*, **74**, 1275–1286, 1969.

SCHOLER, M., G. GLOECKLER, B. KLECKER, F. M. IPAVICH, D. HOVESTADT, and E. J. SMITH, Fast moving plasma structures in the distant magnetotail, *J. Geophys. Res.*, **87**, 6717–6727, 1984.

SCKOPKE, N., G. PASCHMANN, H. ROSENBAUER, and D. H. FAIRFIELD, Influence on the interplanetary magnetic field on the occurrence and the thickness of the plasma mantle, *J. Geophys. Res.*, **81**, 2687–2691, 1976.

SIBECK, D. G. and G. L. SISCOE, Downstream properties of magnetic flux transfer events, *J. Geophys. Res.*, **89**, 10709–10715, 1984.

SLAVIN, J. A. and R. E. HOLZER, Empirical relationships between interplanetary conditions, magnetic flux transfer and the *AL* index, in *Quantitative Modeling of Magnetospheric Processes*, edited by W. P. Olson, pp. 423–436, AGU, Washington, D.C., 1979.

SLAVIN, J. A., E. J. SMITH, B. T. TSURUTANI, D. G. SIBECK, H. J. SINGER, D. N. BAKER, J. T. GOSLING, E. W. HONES, Jr., and F. L. SCARF, Substorm associated traveling compression regions in the distant tail: ISEE-3 geotail observations, *Geophys. Res. Lett.*, **11**, 657–660, 1984.

SLAVIN, J. A., E. J. SMITH, D. G. SIBECK, D. N. BAKER, R. D. ZWICKL, and S.-I. AKASOFU, An ISEE-3 study of average substorm conditions in the magnetotail, *J. Geophys. Res.*, **90**, 10875–10895, 1985.

TSURUTANI, B. T., D. E. JONES, R. P. LEPPING, E. J. SMITH, and D. G. SIBECK, The relationship between the IMF B_Y and the distant tail (150–238 R_E) lobe and plasmasheet B_Y fields, *Geophys. Res. Lett.*, **11**, 1082–1085, 1984.

Solar Wind-Magnetosphere Coupling, edited by Y. Kamide and J. A. Slavin, 671–690.

Solar Wind Influence on Magnetotail Configuration and Dynamics

A. T. Y. Lui

Applied Physics Laboratory, Johns Hopkins University, Laurel, Maryland 20707, U.S.A.

The relationships between solar wind parameters and magnetotail phenomena are briefly reviewed. Manifestations of the external solar wind condition are evident in a diversity of effects in the tail, which include tail axis orientation, flapping motion, cross-sectional area variation, tail twisting, and asymmetries between dawn–dusk sectoral as well as north–south hemispheric differences. Solar wind control of the tail length is also inferred here for the first time. In addition, recent findings suggest a possible drastic tail topological change during periods of strong and sustained northward interplanetary magnetic field. There are also newly proposed tail features such as vortices in the tail and reconnected magnetic flux tubes, which are presumably tied to the solar wind condition in some form. Finally, what appears to be still under much debate is the issue of directly-driven versus unloading as the prime nature of the magnetospheric substorm process. However, there is an indication that perhaps both the directly-driven and the unloading aspects are present in magnetospheric substorms.

1. Introduction

Research on the earth's magnetotail has progressed from theoretical predictions (Piddington, 1960; Johnson, 1960), to inference from near-earth magnetospheric observations (Smith, 1962; Heppner *et al.*, 1963), and then to investigation with in situ measurements (Ness, 1965; Singer *et al.*, 1965; Bame *et al.*, 1966) all within the early part of the 1960's. Extensive surveys of the tail within ~80 R_e downstream have now been made with many satellites such as the IMP and Vela series, OGO-5, Explorers 33 and 35, as well as the more recent ISEE series (Fairfield and Ness, 1967; Mihalov *et al.*, 1968; Behannon, 1970; Hones *et al.*, 1972; Aubry *et al.*, 1972; Meng and Anderson 1974; Lui *et al.*, 1976, 1978, 1984; Parks *et al.*, 1979; Lui, 1980; Huang *et al.*, 1984; Lui and Krimigis, 1984). The intermediate range of the tail from ~80 to 230 R_e has been explored to some extent by ISEE-3 (Slavin *et al.*, 1983; Bame *et al.*, 1983; Gloecker *et al.*, 1984; Feldman *et al.*, 1984; Daly *et al.*, 1984; Zwickl *et al*, 1984; Klecker *et al.*, 1984). Possible encounters of the far distant tail at ~500 to 3100 R_e downstream have also been reported with Pioneer 7 and 8 observations (Ness *et al.*, 1967; Wolfe *et al.*, 1967; Fairfield, 1968; Mariani and Ness, 1969; Intriligator *et al.*, 1969, 1979; Walker *et al.*, 1975; Villante, 1977).

With more than two decades of research, there is little doubt that the magnetotail results from the coupling of the solar wind and the earth's magnetic field. Furthermore, both the configuration and dynamics of the magnetotail are influenced by the solar wind condition in many ways with varying degrees. The intent of this paper is to review briefly our current knowledge of the solar wind's control of the magnetotail based primarily on observational information and prudent analyses from the various satellite missions mentioned above. Controversial issues which often exist in active research disciplines will be discussed. The first part of this paper addresses the non-substorm related aspects of the tail configuration which are considered to be influenced by the solar wind. The second part deals only with those aspects of the tail dynamics which can be more or less directly relatable to the external solar wind condition. No attempt will be made to include tail phenomena attributed to intrinsic tail structures or linked clearly to substorm activities. However, we will comment on the question of whether or not the magnetospheric substorm is a process directly driven by the solar wind.

2. Configuration

2.1 *Tail orientation*

Since the interaction between the solar wind and the geomagnetic field creates the magnetotail by reconnection process or viscous-like interaction (Dungey, 1961; Axford and Hines, 1961; Eastman *et al.*, 1976; 1985; Paschmann *et al.*, 1979; Heikkila, 1978, 1982, 1983), the tail orientation is naturally controlled to a large extent by the solar wind. Both the solar wind speed and direction cause the tail axis to deviate from the sun-earth line analogous to a wind sock. Tilting of the tail axis away from the ecliptic plane occurs when the solar wind flows at an angle to the ecliptic plane. Because the earth moves perpendicular to the solar wind as it orbits the sun, the magnetotail lags the earth, resulting in a duskward aberration of the tail axis by 3 to 4 degrees from the earth-sun line. As the earth's magnetic dipole is not aligned with the earth's spin axis, there is a diurnal wobbling of the earth's dipole, which is one of the factors giving rise to the flapping motion of the tail axis in the north–south as well as east–west direction (Speiser and Ness, 1967). Studies have also shown that flapping motion of the tail is frequently detected during periods when the interplanetary magnetic field (IMF) is strong (>10 nT) and has predominantly a southward component (Toichi and Miyazaki, 1976). It is suggested that the tail flapping motions during these occasions are caused by the penetration of Alfvenic fluctuations in the solar wind into the tail. The dipole tilt also causes the neutral sheet to become an arched surface (Russell and Brody, 1967; Fairfield, 1980). This effect has been reproduced rather well in MHD modelling of the magnetospheric structure (Wu, 1984). The various displacements of the tail axis and neutral sheet away from the sun-earth line or equatorial plane introduces variations of the tail boundaries and plasma regions with respect to a given spacecraft location. However, the variations due to tilting, aberration, and wobbling of the dipole axis can often be corrected for by a proper choice of the reference frame.

2.2 *Tail cross-section*

The amount of magnetic flux in the tail is determined primarily by magnetospheric currents, in particular, currents in the magnetopause boundary closing in the neutral sheet, which in turn are related to the solar wind parameters. One of the parameters in evaluating the total magnetic flux in the tail is the size of the tail cross-section bounded by the tail magnetopause on a plane parallel to the solar magnetospheric *YZ*-plane at some fixed downstream distance. Figure 1 shows a study on the relationship between the radial dimension of the near-earth tail ($-15>X>-20$ R_e) and the impinging solar wind momentum flux. The tail size is characterized for simplicity by one parameter R_T which is the radial distance of the magnetopause from the aberrated central tail axis. This one-parameter measure is resonably good since the tail within ~30 R_e does not deviate appreciably from a cylindrical shape (see, e.g., MENG *et al.*, 1981). Both the tail radius and the simultaneously observed solar wind momentum flux (represented by the product of the solar wind density *n* times the square of the solar wind velocity *v*) are plotted in logarithmic scale. The magnetopause location is determined with IMP-6 plasma measurements in the downstream distance of 15 to 20 R_e. Data are obtained in the 1971 to 1973 period and IMP-6 crosses the dawn magnetopause in August to September and the dusk magnetopause in November to December of each year. Unfortunately, the statistics are rather limited by the lack of corresponding solar wind plasma parameters in the compilation

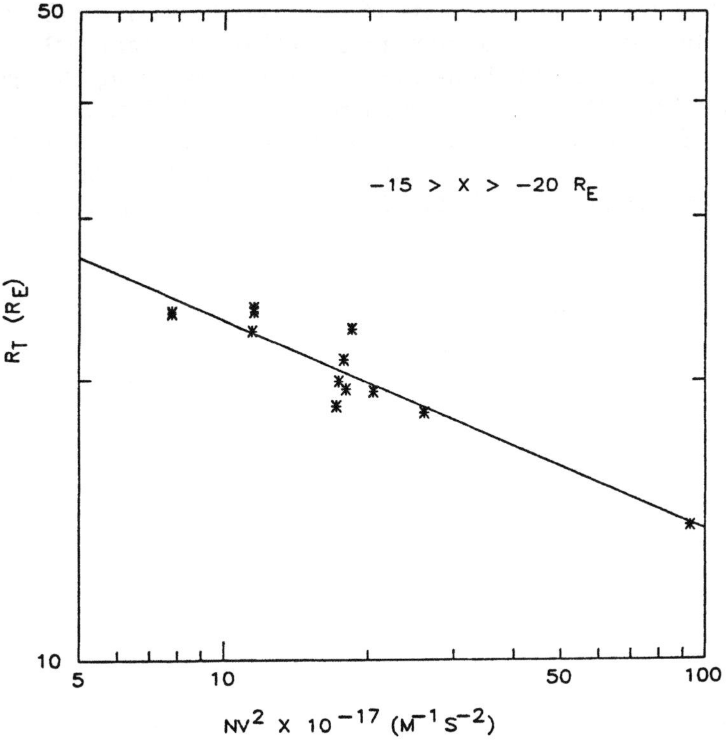

Fig. 1. A plot of the radial dimension of the tail R_T versus solar wind number density *n* multiplied by the square of the solar wind speed *v*. The straight line is obtained from a least square fit to the points, suggesting that the tail radius is roughly proportional to one-fourth to one-fifth power of the solar wind momentum flux.

from KING (1979) for many of the determined magnetopause crossing times. Nevertheless, it is apparent that the larger the solar wind momentum flux, the smaller is the dimension of the tail. The least square linear fit to the data points yields a slope of -0.227 ± 0.026. The correlation coefficient of the fit is -0.934. Since one point at $nv^2 \simeq 92 \times 10^{-17} \mathrm{m}^{-1}\mathrm{s}^{-2}$ is separated from the main distribution and may dominate the trend, we have also carried out the fit without that point. The resulting slope is -0.221 ± 0.047, with a correlation coefficient of -0.827. This exercise indicates that the slope found in Fig. 1 is not affected appreciably by the single point at the high value of solar wind momentum flux. Judging from the derived slope, we find the radial dimension of the tail to be inversely scaled as one-fourth to one-fifth power of the solar wind momentum flux. This scaling is rather similar to that of the magnetospheric nose location which is inversely proportional to one-sixth power of the incident solar wind momentum flux (SPREITER and ALKSNE, 1969). Comparison of the two dependences suggests that the tail radial dimension is slightly more sensitive to the solar wind momentum flux than the dayside magnetopause location. Note that the derived slope in Fig. 1 deviates from -0.167 (one-sixth power dependence) by about three times its standard deviation.

The north–south component of the interplanetary magnetic field (IMF B_z) plays an important role in many aspects of the magnetospheric configuration and dynamics. In particular, it has been reported that southward IMF tends to be associated with earthward displacement of the dayside magnetopause loction (FAIRFIELD, 1971). Therefore, it is of interest to examine the dependence of the radial dimension of the tail on the IMF B_z. Figure 2 shows a scatter plot of these two quantities based on magnetopause locations determined from the study in Fig. 1. More data points are present in Fig. 2 than in Fig. 1 because the interplanetary

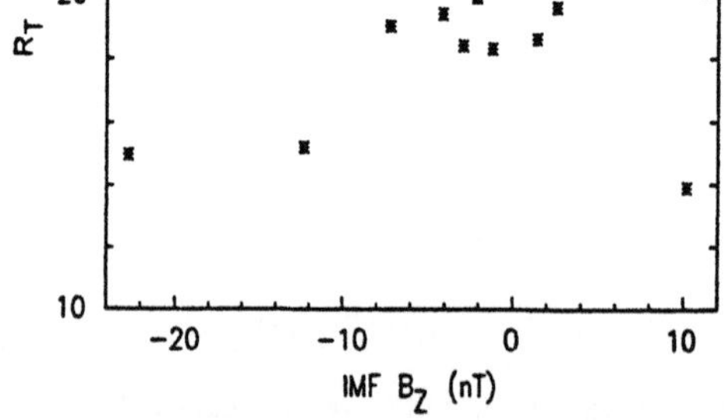

Fig. 2. A plot of the radial dimension of the tail R_T versus the IMF B_z. It is seen that the tail radius tends to be large when the absolute value of the IMF B_z is small. The tail dimension appears to be reduced when the IMF B_z takes a large positive or negative value.

magnetic field information is more often available than the solar wind plasma parameters (KING, 1979). Figure 2 shows that the tail size appears to be large when the IMF B_z is small in absolute magnitude. However, for large absolute value of the IMF B_z, the tail dimension tends to be small. There is therefore no obvious linear relationship present. A more subtle relationship cannot be ruled out though, since this negative result may well be due to other more dominating factors such as magnetospheric substorms masking a possibly less prominent correlation. Indeed, MAEZAWA (1975) has reported that the tail radius increases before the onset of mid-latitude positive bays (taken to denote the substorm expansion phase onset) and decreases subsequently.

2.3 *Tail length*

The tail length is a poorly known quantity due to the limited survey capability from spacecraft. DESSLER (1964) has suggested theoretically that the tail may extend to about 20 to 50 A.U., i.e. $\sim 10^6$ R_e. On the other hand, DUNGEY (1965) has argued that the tail length is about 10^3 R_e on the basis of polar cap convection speed and the latitude of the last closed field line. In this context, the tail length is equated to the downstream distance of the nightside neutral line.

It is possible to use Dungey's method to infer the dependence of the tail length on the solar wind. MAKITA *et al.* (1983) have examined the poleward boundary of the auroral electron precipitation region from DMSP data to determine empirical relationships between the IMF B_z and the poleward boundaries in the evening as well as morning sectors. Hourly averaged IMF parameters from KING (1979) are used in their study. Using the Makita *et al.* results and taking the boundary difference between the evening and morning sectors to denote the size of the polar cap D, we obtain

$$D\,(m) = \begin{cases} (21.8 - 0.97\, B_z\,(\mathrm{nT})) \times 10^5 & B_z < 0, \\ (21.3 - 0.66\, B_z\,(\mathrm{nT})) \times 10^5 & B_z > 0. \end{cases}$$

The factor 10^5 is the conversion factor from degrees in latitude to meters. There is a discontinuous jump at $B_z=0$ because northward IMF intervals and southward IMF intervals are differentiated in obtaining the least square fits. The dependence of cross-polar cap electric potential Φ on IMF B_z has been obtained by REIFF *et al.* (1981) as

$$\Phi\,(\mathrm{kV}) = 62 - 8.9\, B_z\,(\mathrm{nT}) \qquad -9\,\mathrm{nT} < B_z < 6\,\mathrm{nT}.$$

With these two relationships, we can infer the tail length variations with respect to the IMF B_z using Dungey's technique. The average convection speed v in the polar ionosphere is $\Phi / B_e D$, where B_e (=0.5 gauss) is the earth's magnetic field in the polar ionosphere. The time required to convect from the dayside poleward boundary to nightside poleward boundary is $D/v = B_e D^2 / \Phi$. If we assume the average solar wind flow in the magnetosheath to be a constant, then the tail length L varies with IMF B_z

as $L \sim D^2/\Phi$. More explicitly,

$$L\,(\mathrm{R_e}) = \begin{cases} 180\,[1 - 0.044\,B_z\,(\mathrm{nT})]^2/[1 - 0.144\,B_z\,(\mathrm{nT})] & -9\,\mathrm{nT} < B_z < 0, \\ 172\,[1 - 0.031\,B_z\,(\mathrm{nT})]^2/[1 - 0.144\,B_z\,(\mathrm{nT})] & 6\,\mathrm{nT} > B_z > 0, \end{cases}$$

and for small B_z, it can be further simplified to

$$L\,(\mathrm{r_e}) = \begin{cases} 180\,[1 + 0.056\,B_z\,(\mathrm{nT})] & B_z < 0, \\ 172[1 + 0.082\,B_z\,(\mathrm{nT})] & B_z > 0. \end{cases}$$

In the above derivation, the average magnetosheath downstream flow speed is taken to be 300 km/s. The resulting expression indicates that the tail length decreses as the IMF becomes more southward. The discontinuity of L at $B_z=0$ arises from that in the polar cap size D obtained from the results of MAKITA *et al.* (1983).

It is interesting to compare the actual value of L from this expression with the location of the nightside neutral line which marks the downstream extent of the last closed field line. The statistical location of the neutral line may be estimated through an examination of the averaged values of the north–south component of the magnetic field B_z in the tail region. Recent ISEE-3 observations show that the distribution of B_z is shifted progressively towards negative values with increasing tail distanes (SLAVIN *et al.*, 1983, 1985). The averaged B_z is 1.50 ± 1.26 nT in the tail region within 100 $\mathrm{R_e}$ from earth, 0.43 ± 1.28 nT at the 100 to 210 $\mathrm{R_e}$ tail region, and -0.22 ± 1.29 nT further downstream than 210 $\mathrm{R_e}$. This set of statistics suggests that the average location of the last closed field line to be about 100 to 200 $\mathrm{R_e}$ in the tail, in resonable agreement with the values in the formula for the dependence of tail length on IMF B_z. It should be noted that the nightside neutral line location deduced above does not include the possible formation of a near-earth neutral line during substorms.

2.4 *Tail asymmetries*

Asymmetric distribution of tail plasma are often indicated by tail observations. Differences can exist between the dawn and dusk sections of the tail and/or between the northern and southern hemispheres. These asymmetries are related to the IMF orientation, confirming expectations of an open magnetosphere in which the IMF interconnects with the geomagnetic field. Observations indicate that the B_x and B_y component of the IMF are the key components in the development of tail asymmetries.

Entry of energetic solar particles to the magnetotail is found to be controlled by the sense of the interplanetary sector. In the 'towards' sector (when IMF $B_x>0$, $B_y<0$), the energetic solar particles are observed in the southern hemisphere tail lobe but not in the northern hemisphere. Likewise, entry into the northern hemisphere is seen in the 'away' sector when IMF $B_x<0$ and $B_y>0$ (REID and SAUER, 1967; VAN ALLEN *et al.*, 1971). Similar north–south entry asymmetry is found for electrons at energies below a few keV (YEAGER and FRANK, 1976).

From simple theoretical considerations that there should be a partial penetration of the IMF in an open magnetosphere, COWLEY (1981) has reasoned that the magnetic flux in the north and south tail lobes should be unequal when IMF B_x is non zero. The magnetic flux is predicted to be larger in the northern lobe than in the southern lobe for periods with IMF $B_x>0$ (and alternatively larger in the southern lobe when IMF $B_x<0$). This asymmetry can be accomplished with a north–south displacement of neutral sheet location. Figure 3 shows the result from his study searching for this displacement. The location of the observed neutral sheet crossings relative to the estimated neutral sheet location is plotted versus the IMF B_x. There appears to be a hint of a trend as suggested by the theory.

The interconnection of the geomagnetic field with an IMF of non-zero B_y component introduces an IMF-correlated B_y component within the tail. This effect has been observed and is illustrated in Fig. 4, where the B_y component at the neutral sheet is plotted against the hourly averaged IMF B_y component. Neutral sheet crossings are determined from IMP-6 within the downstream distance of 30 R_e using 15.36 sec averaged magnetic field measurements in 1971 to 1972. A linear relationship between these two quantities is apparent and is given by a linear regression fit as

$$B_y\ (\text{tail}) = (0.5 \pm 0.1)\ B_y\ (\text{IMF}) - 0.5 \pm 0.2$$

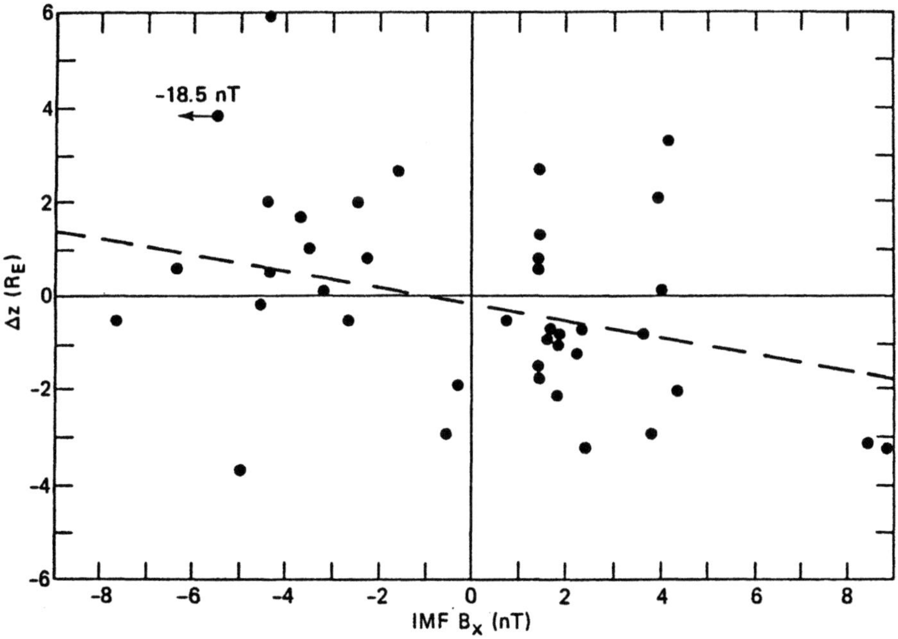

Fig. 3. The location of the actual observed neutral sheet crossings relative to its estimated location as a function of IMF B_x. There is a hint that the neutral sheet is progressively further below its estimated location as the IMF B_x increases positively and progressively further above its estimated position as the IMF B_x decreases negatively (after COWLEY, 1981).

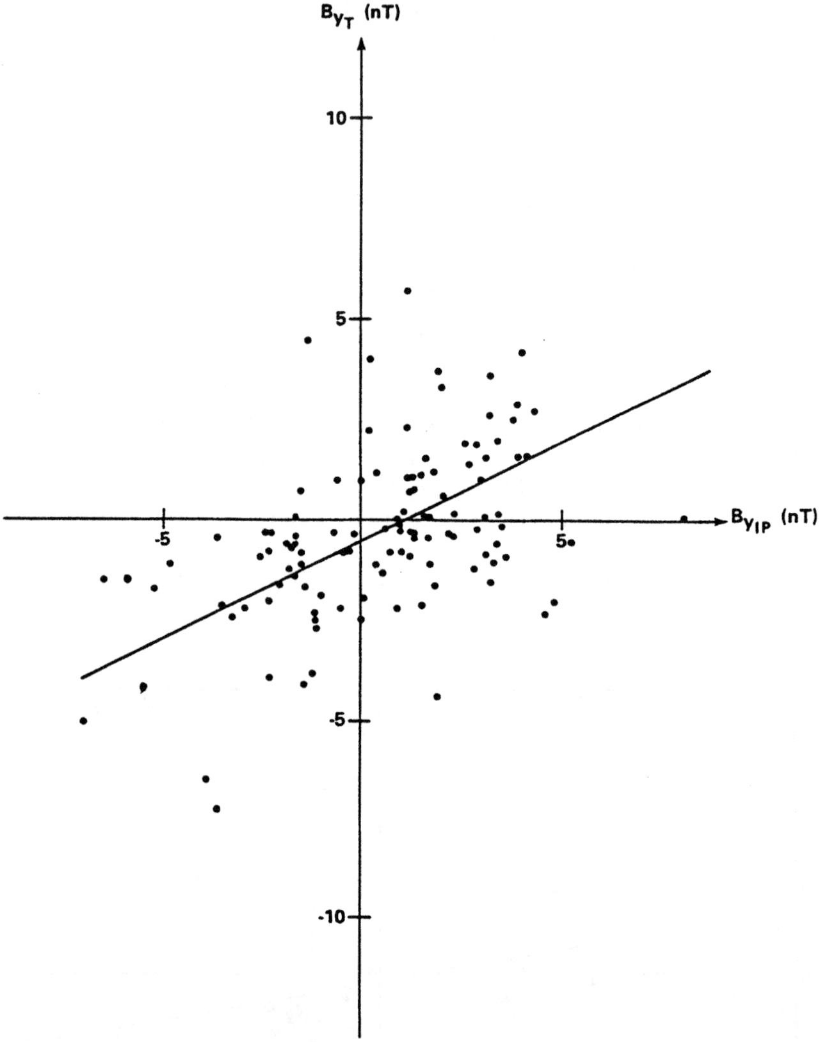

Fig. 4. A scatter plot of the B_y component in the tail at neutral sheet crossings versus the hourly averaged IMF B_y. The straight line is obtained from a least square fit to the points, suggesting that a partial penetration of about 50% of IMF B_y occurs in the neutral sheet region (after Lui, 1983).

with a correlation coefficient of 0.6. Similar results have been obtained by Fairfield (1979) and Sibeck *et al.* (1985) but with a lesser degree (~13%) of overall IMF B_y penetration when measurements in the entire tail region are included. The differences between these results may imply different degree of B_y penertration in different tail regions or the presence of field-aligned currents (Voigt, 1985).

In addition to the presence of dawn-dusk tail field related to the IMF B_y, asymmetric distribution of plasma have also been observed. Hardy *et al.* (1979) have reported on the existence of lobe plasma at the lunar distance. The characteristics of this plasma (in terms of number density, flow speed and temperature) are very similar to the plasma properties found in the plasma mantle and the magnetopause boundary layer; i.e., they are intermediate between the values for magnetosheath plasma and the hot plasma in the plasma sheet. The quadrants of the tail where the lobe plasma occurs predominantly are found to be related to the IMF B_y component. When the IMF B_y is positive, the lobe plasma is mostly detected in the dawn quadrant of the northern tail and in the dusk quadrant of the southern tail. When the IMF B_y becomes negative, the lobe plasma switches from the dawn to dusk quadrant in the northern tail and vice versa in the southern tail. Recently, a similar asymmetric plasma distribution is observed for the electron plasma in the distant tail lobe from ISEE-3 (GOSLING *et al.*, 1985). SIBECK and SISCOE (1984) have suggested that this distribution of lobe plasma is due to the advection of reconnected magnetic flux tubes (RMFT) into the tail. The RMFT are the tailward evolution of flux tubes in flux transfer events observed in the dayside as a result of sporadic interconnection between the interplanetary and geomagnetic field. Their model can also account for magnetopause boundary normals which have large components in the circumferential direction (i.e., tangent to the magnetopause and perpendicular to the tail axis) during some multiple magnetopause crossings. A readily testable prediction of this model is that the locations of multiple nightside magnetopause crossings should depend on the IMF orientation and should increase during intervals of southward IMF.

The electric field pattern in the tail appears to be affected by the IMF B_y component as well. This information has been inferred from ground magnetic perturbations as well as electric and magnetic field measurements made by low altitude satellite in the polar cap to where field lines in the tail lobe map (BERTHELIER *et al.*, 1974; FRIIS-CHRISTENSEN and WILHJELM, 1975; HEPPNER, 1972; MOZER *et al.*, 1974; MCDIARMID *et al.*, 1978a, b, 1979). For positive IMF B_y, the dawn-to-dusk convection electric field is strong in the dusk sector in the northern polar cap. In the southern polar cap, strong convection lies in the dawn sector. The sector preference for strong electric fields switches in both polar caps when the IMF B_y becomes negative. A similar trend is inferred from a study of the large-scale magnetic-field-aligned and ionospheric current systems (ZANETTI *et al.*, 1984).

COWLEY (1981) suggests that the IMF B_y component for an open magnetosphere could lead to two effects or a combination thereof. The first is a dawn–dusk asymmetric distribution of the normal field at the tail magnetopause which will lead to the dawn–dusk asymmetric plasma population and electric field pattern as discussed previously. The second effect is an IMF B_y induced torque on the tail magnetopause which will give rise to a twist of the tail midplane in the same sense as the torque exerted on the magnetopause. These two effects are schematically illustrated in Fig. 5 (after COWLEY, 1981). Evidence of tail twisting has recently been observed in the ISEE-3 distant tail measurements (SIBECK *et al.*, 1985).

2.5 *Plasma structures and magnetic field intensity*

In addition to giving rise to asymmetries, the IMF orientation also plays a role in determining the presence and thickness of plasma regions in the tail. For instance, the

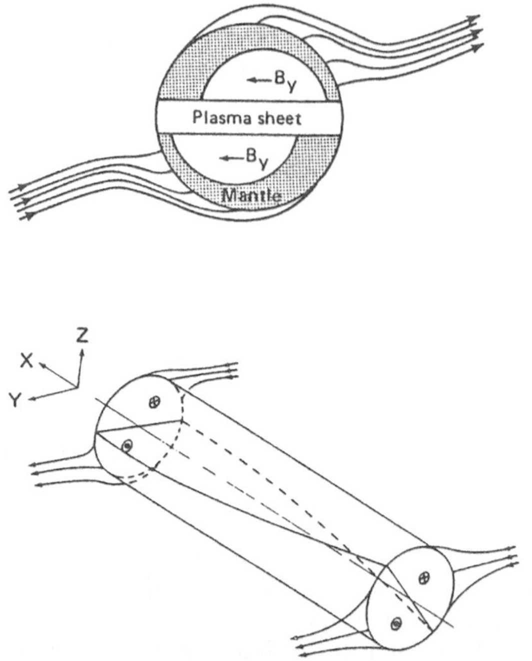

Fig. 5. A schematic diagram to illustrate two IMF B_y effects on the magnetotail. The top panel illustrates a partial penetration of IMF B_y in the tail as well as an asymmetric distribution of lobe plasma/boundary layer plasma between dawn-dusk sectors and north-south hemispheres. The bottom panel shows a twisting of the neutral sheet surface in the direction of the torque exerted on the nightside magnetopause as a result of interconnection of geomagnetic field and IMF (after Cowley, 1981).

plasma mantle, which is a broad region of magnetosheath-like plasma flowing downstream, varies in its thickness with IMF B_z. Figure 6 shows the correlation between the mantle thickness determined from Heos-2 and IMF B_z. The parameter $s_\perp$ is normalized to account for the obliqueness of the Heos-2 trajectory with respect to the magnetopause as well as the anticipated thickening of the plasma mantle with downstream distance as a result of convection towards the neutral sheet. The tendency for a thicker mantle during southward IMF is noticeable. This result is taken as an indication that the plasma mantle is a product of the dayside merging process.

The magnetic field strength in the tail lobe outside the plasma sheet is also found to be influenced by IMF B_z. From a comparison of simultaneous measurements of IMF and tail lobe field, Meng and Colburn (1974) have noticed that the tail lobe field strength increases about 10% after a long duration of southward IMF. They attributed this effect to a transfer of dayside flux into the tail lobe.

There has been recent interest in magnetospheric configuration and activity during northward IMF. Investigations of electric field patterns, Birkeland currents, and auroral forms in the polar region have suggested that the magnetotail may

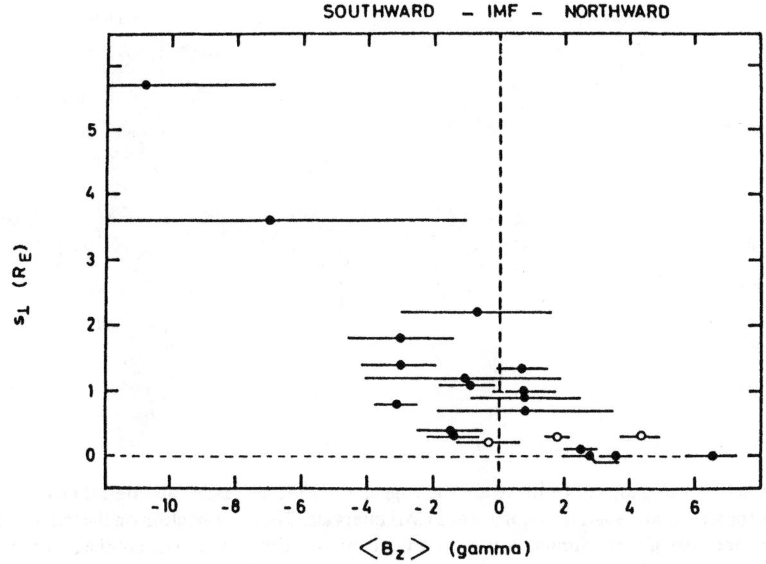

Fig. 6. A plot of the thickness of the plasma mantle as a function of IMF B_z. This result shows that the plasma mantle thickens progressively as the IMF becomes more negative (after SCKOPKE *et al.*, 1976).

undergo rather drastic configurational changes between southward and northward IMF conditions (FRANK *et al.*, 1983; POTEMRA *et al.*, 1984; CROOKER, 1977, 1979). Figure 7 is a schematic diagram illustrating the two tail topology (BYTHROW *et al.*, 1985). The cross- hatched region includes the plasma sheet and its boundary layer. The typical tail configuration is given on the left for southward or weakly northward IMF viewed from the tail towards the sun. When the IMF is strongly northward for a sustained period, there are indications that the tail lobe bifurcates, producing transpolar auroral arcs, sunward convection in the polar cap and NBZ current systems (ZANETTI *et al.*, 1984; POTEMRA *et al.*, 1984).

3. Dynamics

3.1 Vortices

Plasma measurements from ISEE-1 and 2 have shown the occasional presence of rotation of plasma flow directions primarily in the equatorial plane with a period of ~5 to 15 min (HONES *et al.*, 1981, 1983; SAUNDERS *et al.*, 1981, 1983a, b). This kind of flow rotation is sometimes observed up to 5 cycles and have been called plasma vortex events. The preferred rotation sense is clockwise in the morning sector and anticlockwise in the evening sector when viewed from north of the equatorial plane. Accompanying waves in the local magnetic field are seen although phase velocities and phase lags determined from plasma flow and magnetic field often differ (HONES *et al.*, 1983). From phase lags of the flow rotations between the two spacecraft ISEE-1 and 2, the wavelengths of the vortices have been deduced to be about one to many tens

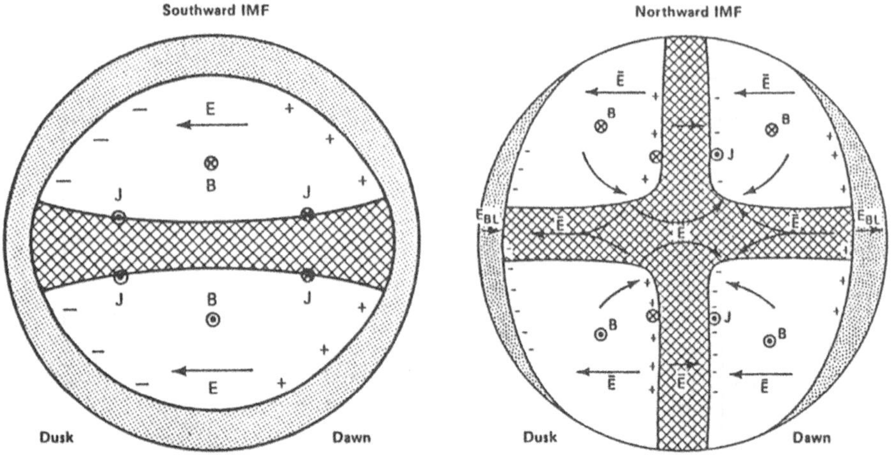

Fig. 7. A schematic diagram to illustrate the suggestion that a drastic topological change of the tail occurs for strong and sustained northward IMF intervals. The cross-section on the left of the diagram is representative of the common notion of the tail configuration. On the right of the diagram, the tail is bifurcated by the plasma convecting from the midnight region in the plasma sheet boundary towards the magnetopause. The asymmetry between the northern and the southern tail possibly associated with IMF B_y is not sketched here for simplicity. Field-aligned currents consistent with this geometry have also been reported (after Bythrow *et al.*, 1985).

of earth radii. However, wavelengths deduced from energetic particle data tend to be only about 1 R_e (Saunders *et al.*, 1983a, b). The energy flux associated with these vortex events tends to be pointing towards the central tail axis and tailward. The phase velocities are in the range of few tens to few hundred km/s, well below the Alfven and sound speeds. Although the vortices have been suggested as hydromagnetic waves driven by Kelvin-Helmholtz instability at the boundary between the magnetotail and the magnetosheath, there is not yet any study conducted to relate the occurrences of these events to interplanetary quantities.

3.2 *Storm-sudden-commencement effects*

Storm sudden-commencements (SSC's) are often triggered by the arrival of an interplanetary shock at the front of the magnetosphere. Armstrong and Tholen (1985) have recently completed a long term statistical study to examine the association between the occurrence of SSC's and energetic particle bursts in the tail. Based on 10 years (1973–1983) data from both IMP 7 and 8 covering over 300 SSC's, they have found that increase in intensity of both energetic protons (E>0.29 MeV) and energetic electrons (E>0.22 MeV) can be triggered by SSC occurrences. On the average, enhancement in energetic particle intensity are detected about 64% of the time following an SSC. This is an example of the magnetospheric response to an impulsive perturbation from the solar wind.

Another example of the tail response to an SSC is the compression of plasma sheet (Lui *et al.*, 1975). On September 29, 1969, an SSC was registered by the ground magnetic field record at 0452 UT. The Vela 5B spacecraft was at about 11 R_e in the tail at the time of the SSC. This SSC was accompanied immediately with a substorm expansion phase onset, which led to plasma sheet thinning recognizable in the plasma measurement as a decrease in plasma pressure. In spite of this thinning effect, an increase of pressure by a factor of about 5 was seen associated with the SSC.

3.3 *Magnetospheric substorm*

An age old question in magnetospheric research is whether the magnetospheric substorm is primarily an externally driven process or a spontaneous internal process. The intent of this section is not to provide a comprehensive review on this subject but to highlight the recent development.

Evidence supporting the idea that magnetic field energy is first stored in the magnetotail following southward turning of the IMF and subsequently released through explosive reconnection intitiated at the substorm expansion phase onset can be well represented in Fig. 8 (CAAN *et al.*, 1975). Superposed epoch analysis is done on

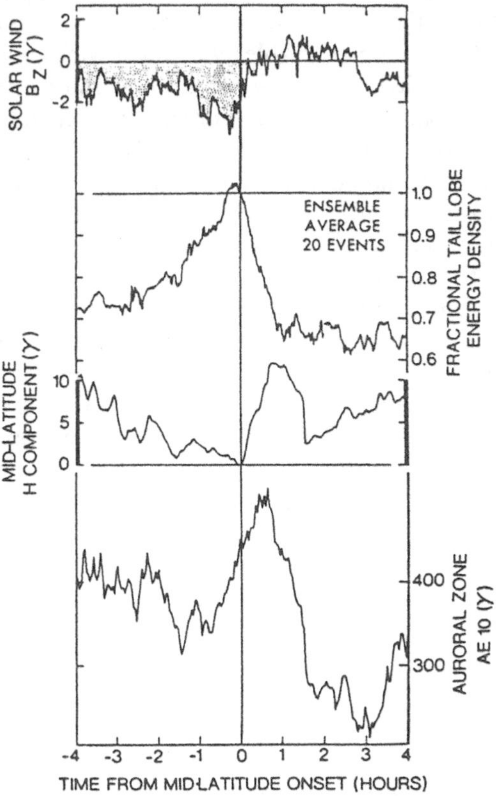

Fig. 8. Superposed analysis of the IMF B_z, tail lobe energy density, mid-latitude ground magnetic perturbations and the *AE* index. This result has often been quoted to support the idea that the substorm process is primarily an unloading process (after CAAN *et al.*, 1975).

20 substorm events. The plotted quantities are the average values after shifting all their times to line up with the onset time of the mid-latitude positive bay ($t=0$) for these 20 events. It is seen that prior to $t=0$, the averaged IMF B_z is negative. The IMF B_z begins to recover to positive values near $t=0$. The averaged tail lobe energy density increases about 2 hours prior to $t=0$ time from ~0.75 to about 1.0 and decreases rapidly in about 1 hour after $t=0$. The averaged AE index which monitors the magnetic substorm activity on the ground starts to rise at about 1 hour before $t=0$ and reaches the peak about 1 hour later. One feature which causes concern from researchers with a different view on the subject is that the *AE* index 1 hour prior to $t=0$ is already above 300 nT. This large value of the *AE* index suggests that the substorm expansion phase is probably already in progress about 1 hour prior to $t=0$ and that the mid-latitude positive bay onset is not a reliable timing mark for ordering observed quantities with respect to the onset of substorm expansion phase. Furthermore, individual case study has shown that the tail lobe energy density on some occasions can continue to increase subsequent to substorm expansion phase onset (FAIRFIELD *et al.*, 1981).

The linear prediction filtering technique has also been applied to substantiate the unloading nature of the substorm. Research along this line has shown that ~40% of the variance of the *AL* index is predictable by the solar wind condition (MCPHERRON *et al.*, 1984). From this result, it is suggested that an internal process uncorrelated with the solar wind condition must contribute significantly to the substorm energy output. This inference is however difficult to substantiate convincingly since the residuals from the prediction filter could be due to a large number of possible sources, such as the imperfectness of the *AL* index as a substorm output monitor or the non-global characterization of the solar wind parameters determined by a single spacecraft.

Evidence put forth in support of the idea that the substorm is primarily a directly driven process has been in relating the energy output rate in a substorm with the energy input rate from the solar wind (PERREAULT and AKASOFU, 1978; AKASOFU, 1981; KAN and LEE, 1979). It is emphasized from these studies that the general temporal profiles of the energy output rate resemble closely the temporal profiles of the energy input rate. The input function used (called the energy coupling function in Fig. 9) is the expression $vB^2\sin^4(\theta/2)\, l_0^2$ where v is the solar wind speed, B is the IMF strength, θ is the polar angle of the magnetic field projected on the YZ-plane, and l_0 is a constant taken to be ~7 R_e. The output function is the total energy output rate in the magnetosphere which includes contributions from ring current, auroral electrojet, and particle precipitation. A comparison of the two parameters in Fig. 9 seems to show encouraging similarities in terms of temporal profiles. The similarity in shape persists even when high time resolution data are used for the correlative study (AKASOFU *et al.*, 1982; BAKER *et al.*, 1983).

One feature which causes concern to researchers with the view of substorm being an unloading process is in the construction of the magnetospheric energy output rate. In general, the decay of the D_{st} index during a magnetic storm period is rapid in the early storm recovery phase and is more gradual in the later stage of the storm recovery. Due to this non-uniform decay time of the ring current in magnetic storms, the ring current particle lifetime used in the expression for the total magnetospheric

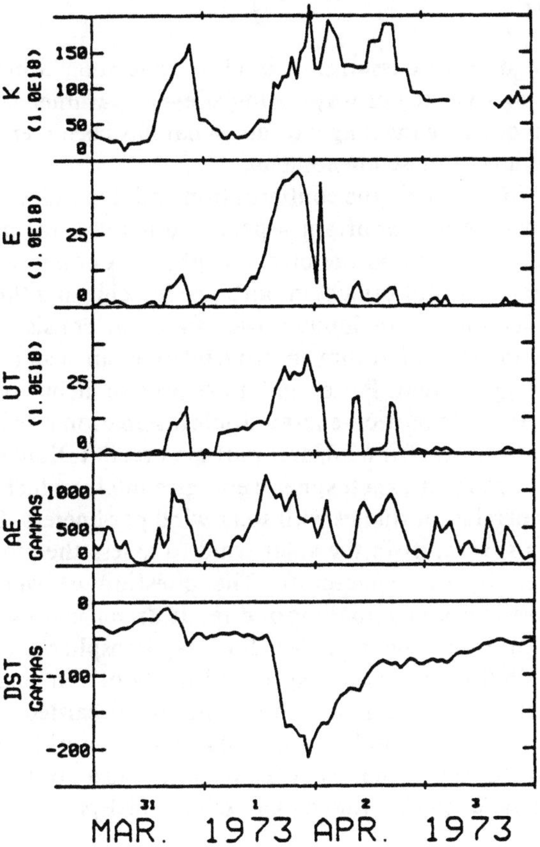

Fig. 9. Temporal profiles of the solar wind kinetic energy flux K, the solar wind-magnetosphere energy coupling function E, the total energy consumption rate UT, the AE index and the D_{st} index. The close resemblance of the temporal profiles between the energy coupling function and the energy output rate UT is used to substantiate the notion that the magnetospheric substorm is being directly-driven by the solar wind (after AKASOFU, 1981).

energy output rate by the correlative studies mentioned above is taken to be one of the two values depending on the energy input rate. In this way, the energy output rate depends to some extent on the energy input rate which could bring about the resemblance of these two parameters. Another point raised against the directly driven idea is the presence of a significant time lag (~15 to 45 min.) for the output power profile in comparison with the input power profile. The counter argument for this point is that the time lag reflects the large inductance inherent in the scale size of the magnetosphere and is not a modification of the input power profile by any internal process. It should, however, be noted that there does not seem to be any disagreement between the two schools of thought that some aspects of magnetospheric substorm activities can be attributed to an unloading process and some to a directly driven process.

4. Final Remark

From the foregoing discussions, it is clear that solar wind control on the magnetotail comes in a variety of ways. Almost every parameter in the solar wind plays some role in modifying the magnetotail. What has been mentioned is merely a rather sketchy overview of these phenomena.

The solar wind influence on the configuration and dynamics of the magnetotail has been and still will be an important aspect in our quest for understanding the relationships between the sun and the magnetosphere. A continuous acquisition of the solar wind parameters is of utmost importance in achieving this goal.

On the identification of an input power function consisting of solar wind quantities, an area in need of future improvement is an accurate assessment of magnetospheric energy content. Better and more precise means of determining the ring current energy, trapped particle energy, Joule dissipation rate in the ionosphere, as well as energy output rate from auroral particle precipitation are all essential in devising a reliable index for magnetospheric energy content, which can then be used with confidence in correlation studies with solar wind parameters. It is also desirable to have at least two spacecraft in the solar wind to assess the global nature of the parameters obtained by one spacecraft. The question of whether or not the magnetospheric substorm is an unloading or directly driven process can be addressed satisfactorily only after a reliable index for magnetospheric energy content is obtained. The possibility that both aspects (directly-driven and unloading) are important in magnetospheric substorm should not be discarded. Finally, if we can identify a power input expression based on solar wind quantities that is correlated well with certain substorm parameters or features, we must not forget to investigate what physical significance the identified expression conveys.

The author is grateful to D. G. Sibeck and B. H. Mauk for useful discussions on the manuscript. This research is supported by the Atmospheric Sciences Section of the National Science Foundation, Grant ATM83-05537 to the Johns Hopkins University.

REFERENCES

AKASOFU, S.-I., Energy coupling between the solar wind and the magnetosphere, *Space Sci. Rev.*, **28**, 121, 1981.

AKASOFU, S.-I., J. F. CARBARY, C.-I. MENG, J. P. SULLIVAN, and R. P. LEPPING, A high time resolution study of the solar wind-magnetosphere energy coupling function, *Planet. Space Sci.*, **30**, 537, 1982.

ARMSTRONG, T. P. and S. THOLEN, Triggering of magnetospheric particle bursts by SSC events, in *Proceedings of the AGU Chapman Conference on Solar Wind-Magnetosphere Coupling*, Feb. 12–15, 1985, Pasadena, California, 25, 1985.

AUBRY, M. P., M. G. KIVELSON, R. L. MCPHERRON, C. T. RUSSELL, and D. S. COLBURN, Outer magnetosphere near midnight at quiet and disturbed times, *J. Geophys. Res.*, **77**, 5487, 1972.

AXFORD, W. I. and C. O. HINES, A unifying theory of high-latitude geophysical phenomena and geomagnetic storms, *Can. J. Phys.*, **39**, 1433, 1961.

BAKER, D. N., R. D. ZWICKL, S. J. BAME, E. W. HONES, Jr., B. T. TSURUTANI, E. J. SMITH, and S.-I. AKASOFU, An ISEE 3 high time resolution study of interplanetary parameter correlations with magnetospheric activity, *J. Geophys. Res.*, **88**, 6230, 1983.

BAME, S. J., J. R. ASBRIDGE, H. E. FELTHAUSER, R. A. OLSON, and I. B. STRONG, Electrons in the plasma sheet of the earth's magnetic tail, *Phys. Rev. Lett.*, **16**, 138, 1966.

BAME, S. J., R. C. ANDERSON, J. R. ASBRIDGE, D. N. BAKER, W. C. FELDMAN, J. T. GOSLING, E. W. HONES, Jr., D. J. MCCOMAS, and R. D. ZWICKL, Plasma regimes in the deep geomagnetic tail: ISEE-3, *Geophys. Res. Lett.*, **10**, 912, 1983.

BEHANNON, K. W., Geometry of the geomagnetic tail, *J. Geophys. Res.*, **75**, 743, 1970.

BERTHELIER, A., J. J. BERTHELIER, and C. GUERIN, The effect of the east–west component of the interplanetary magnetic field on magnetospheric convection as deduced from magnetic perturbations at high latitudes, *J. Geophys. Res.*, **79**, 421, 1974.

BYTHROW, P. F., W. J. BURKE, T. A. POTEMRA, L. J. ZANETTI, and A. T. Y. LUI, Ionospheric evidence for irregular reconnection and turbulent plasma flows in the magnetotail during periods of northward IMF, *J. Geophys. Res.*, **90**, 5319, 1985.

CAAN, M. N., R. L. MCPHERRON, and C. T. RUSSELL, Substorm and interplanetary magnetic field effects on the geomagnetic tail lobes, *J. Geophys. Res.*, **80**, 191, 1975.

COWLEY, S. W. H., Magnetospheric asymmetries associated with the *y*-component of the IMF, *Planet. Space Sci.*, **29**, 79, 1981.

COWLEY, S. W. H., Asymmetry effects associated with the *x*-component of the IMF in a magnetically open magnetosphere, *Planet. Space Sci.*, **31**, 809, 1983.

CROOKER, N. U., The magnetospheric asymmetries associated with the *y*-component of the IMF, *Planet. Space Sci.*, **29**, 79, 1981.

CROOKER, N. U., Dayside merging and cusp geometry, *J. Geophys. Res.*, **84**, 951, 1979.

DALY, P. W., T. R. SANDERSON, and K. P. WENZEL, Survey of energetic (E>35 keV) ion anisotropies in the deep geomagnetic tail, *J. Geophys. Res.*, **89**, 10733, 1984.

DESSLER, A. J., length of the magnetospheric tail, *J. Geophys. Res.*, **69**, 3913, 1964.

DUNGEY, J. W., Interplanetary magnetic field and auroral zones, *Phys. Rev. Lett.*, **6**, 47, 1961.

DUNGEY, J. W., The length of the magnetospheric tail, *J. Geophys. Res.*, **70**, 1753, 1965.

EASTMAN, T. E., E. W. HONES, Jr., S. J. BAME, and J. R. ASBRIDGE, The magnetospheric boundary layer: site of plasma, momentum and energy transfer form the magnetosheath into the magnetosphere, *Geophys. Res. Lett.*, **3**, 685, 1976.

EASTMAN, T. E., L. A. FRANK, C. Y. HUANG, The boundary layers as the primary transport regions of the earth's magnetotail, *J. Geophys. Res.*, **90**, 9541, 1985.

FAIRFIELD, D. H., Simultaneous measurements on three satellites and the observation of the geomagnetic tail at 1000 R_e, *J. Geophys. Res.*, **73**, 6179, 1968.

FAIRFIELD, D. H., Average and unusual locations of the earth's magnetopause and bow shock, *J. Geophys. Res.*, **76**, 6700, 1971.

FAIRFIELD, D. H., On the average configuration of the geomagnetic tail, *J. Geophys. Res.*, **84**, 1950, 1979.

FAIRFIELD, D. H., A statistical determination of the shape and position of the geomagnetic neutral sheet, *J. Geophys. Res.*, **85**, 775, 1980.

FAIRFIELD, D. H. and N. F. NESS, Magnetic field measurements with the IMP 2 satellite, *J. Geophys. Res.*, **72**, 2379, 1967.

FAIRFIELD, D. H., R. P. LEPPING, E. W. HONES, Jr., S. J. BAME, and J. R. ASBRIDGE, Simultaneous measurements of magnetotail dynamics by IMP spacecraft, *J. Geophys. Res.*, **86**, 1396, 1981.

FELDMAN, W. C., S. J. SCHWARTZ, S. J. BAME, D. N. BAKER, J. BIRN, J. T. GOSLING, E. W. HONES, Jr., D. J. MCCOMAS, J. A. SLAVIN, E. J. SMITH, and R. D. ZWICKL, Evidence for slow-mode shocks in the deep geomagnetic tail, *Geophys. Res. Lett.*, **11**, 599, 1984.

FRANK, L. A., J. D. CRAVEN, C. Y. HUANG, K. L. ACKERSON, and E. J. SMITH, Observations pertaining to the character of theta auroras, *EOS Trans. AGU*, **64**, 300, 1983.

FRIIS-CHRISTENSEN, E. and J. WILHJELM, Polar cap currents for different directions of the interplanetary magnetic field in the *y–z* plane, *J. Geophys. Res.*, **80**, 1248, 1975.

GLOECKLER, G., M. SCHOLER, F. M. IPAVICH, D. HOVESTADT, B. KLECKER, and A. B. GALVIN, Abundances and spectra of suprathermal H^+, He^{++} and heavy ions in a fast moving plasma structure (plasmoid) in the distant geotail, *Geophys. Res. Lett.*, **11**, 603, 1984.

GOSLING, J. T., D. N. BAKER, S. J. BAME, W. C. FELDMAN, R. D. ZWICKL, E. J. SMITH, North–south and dawn-dusk plasma asymmetries in the distant tail lobes: ISEE-3, *J. Geophys. Res.*, **90**, 6354, 1985.

HARDY, D. A., P. H. REIFF, and W. J. BURKE, Response of magnetotail plasma at lunar distance to changes in the interplanetary magnetic field, the solar wind plasma, and substorm activity, *J. Geophys. Res.*, **84**, 1382, 1979.

HEIKKILA, W. J., Criticism of reconnection models of the magnetosphere, *Planet. Space Sci.*, **26**, 121 1978.

HEIKKILA, W. J., Impulsive plasma transport through the magnetopause, *Geophys. Res. Lett.*, **9**, 159, 1982.

HEIKKILA, W. J., Exit of boundary layer plasma from the distant magnetotail, *Geophys. Res. Lett.*, **10**, 218, 1983.

HEPPNER, J. P., N. F. NESS, T. L. SKILLMAN, and C. S. SCEARCE, Explorer 10 magnetic field measurements, *J. Geophys. Res.*, **68**, 1, 1963.

HEPPNER, J. P., Polar cap electric field distributions related to the interplanetary magnetic field direction, *J. Geophys. Res.*, **77**, 4877, 1972.

HONES, E. W., Jr., J. R. ASBRIDGE, S. J. BAME, M. D. MONTGOMERY, S. SINGER, and S.-I. AKASOFU, Measurements of magnetotail plasma flow made with Vela 4B, *J. Geophys. Res.*, **77**, 5503, 1972.

HONES, E. W., Jr., J. BIRN, J. R. ASBRIDGE, G. PASCHMANN, N. SCKOPKE, and G. HAERENDEL, Further determination of the characteristics of magnetospheric plasma vortices with ISEE 1 and 2, *Geophys. Res. Lett.*, **86**, 814, 1981.

HONES, E. W., Jr., J. BIRN, S. J. BAME, and C. T. RUSSELL, New observations of plasma vortices and insights into their interpretation, *Geophys. Res. Lett.*, **10**, 674, 1983.

HUANG, C. Y., L. A. FRANK, and T. E. EASTMAN, High-altitude observations of an intense inverted V events, *J. Geophys. Res.*, **89**, 7423, 1984.

INTRILIGATOR, D. S., J. H. WOLFE, D. D. MCKIBBIN, and H. R. COLLARD, Preliminary comparison of solar wind plasma observations in the geomagnetospheric wake at 1000 and 500 earth radii, *Planet. Space Sci.*, **17**, 321, 1969.

INTRILIGATOR, D. S., H. R. COLLARD, J. D. MIHALOV, O. L. VAISBERG, and J. H. WOLFE, Evidence for earth magnetospheric tail associated phenomena at 3100 R_e, *Geophys. Res. Lett.*, **6**, 585, 1979.

JOHNSON, F. S., The gross character of the geomagnetic tail in the solar wind, *J. Geophys. Res.*, **65**, 3049, 1960.

KAN, J. R. and L. C. LEE, Energy coupling function and solar-wind-magnetosphere dynamo, *Geophys. Res. Lett.*, **6**, 577, 1979.

KING, J. H., Interplanetary medium data book, supplement 1, 1975–1978, NSSDC/WDC-A-R&S 79-08, 1979.

KLECKER, B., M. SCHOLER, D. HOVESTADT, G. GLOECKLER, F. M. IPAVICH, E. J. SMITH, and B. T. TSURUTANI, Correlation between proton anisotropy and magnetic field direction in the distant geotail, *Geophys. Res. Lett.*, **11**, 1038, 1984.

LUI, A. T. Y., Observations on plasma sheet dynamics during magnetospheric substorms, in *Dynamics of Magnetosphere*, edited by S.-I. Akasofu, p. 563, D. Reidel Publ. Co., Dordrecht, Netherlands, 1980.

LUI, A. T. Y., Characteristics of the cross-tail current in the earth's magnetotail, in *Magnetospheric Currents, AGU monograph 28,*, p. 158, 1983.

LUI, A. T. Y. and S. M. KRIMIGIS, Association between energetic particle bursts and Birkeland currents in the geomagnetic tail, *J. Geophys. Res.*, **89**, 10741, 1984.

LUI, A. T. Y., E. W. HONES, Jr., D. VENKATESAN, S.-I. AKASOFU, and S. J. BAME, Complete plasma dropouts at Vela satellites during thinning of the plasma sheet, *J. Geophys. Res.*, **80**, 4649, 1975.

LUI, A. T. Y., C.-I. MENG, and S.-I. AKASOFU, Search for the magnetic neutral line in the near-earth plasma sheet, 1, Critical re-examination of earlier studies on magnetic field observations, *J. Geophys. Res.*, **81**, 5934, 1976.

LUI, A. T. Y., L. A. FRANK, K. L. ACKERSON, C.-I. MENG, and S.-I. AKASOFU, Plasma flows and magnetic field vectors in the plasma sheet during substorms, *J. Geophys. Res.*, **83**, 3849, 1978.

LUI, A. T. Y., D. J. WILLIAMS, T. E. EASTMAN, L. A. FRANK, and S.-I. AKASOFU, Streaming reversal of energetic particles in the magnetotail during a substorm, *J. Geophys. Res.*, **89**, 1540, 1984.

MAEZAWA, K., Magnetotail boundary motion associated with geomagnetic substorms, *J. Geophys. Res.*, **80**, 3543, 1975.

MAKITA, K., C.-I. MENG, and S.-I. AKASOFU, The shift of the auroral electron precipitation boundaries in the dawn-dusk section in association with geomagnetic activity and interplanetary magnetic field, *J. Geophys. Res.*, **88**, 7967, 1983.

MARIANI, F. and N. F. NESS, Observations of the geomagnetic tail at 500 earth radii by Pioneer 8, *J. Geophys. Res.*, **74**, 5033, 1969.

MCDIARMID, I. B., J. R. BURROWS, and M. D. WILSON, Comparison of magnetic field perturbations at high latitudes with charged particle and IMF measurements, *J. Geophys. Res.*, **83**, 681, 1978a.

MCDIARMID, I. B., J. R. BURROWS, and M. D. WILSON, Magnetic field perturbations in the dayside cleft and their relationship to the IMF, *J. Geophys. Res.*, **83**, 5753, 1978b.

MCDIARMID, I. B., J. R. BURROWS, and M. D. WILSON, Large-scale magnetic field perturbations and particle measurements at 1400 km on the dayside, *J. Geophys. Res.*, **84**, 1431, 1979.

MCPHERRON, R. L., R. A. FAY, C. R. GARRITY, L. F. BARGATZE, D. N. BAKER, C. R. CLAUER, and C. SEARLS, Coupling of the solar wind to measures of magnetic activity in *Achievements of the International Magnetospheric Study (IMS) ESA SP-217*, p. 161, 1984.

MENG, C.-I. and D. S. COLBURN, Magnetotail variations associated with the southward interplanetary magnetic field, *J. Geophys. Res.*, **79**, 1831, 1974.

MENG, C.-I. and K. A. ANDERSON, Magnetic field configuration in the magnetotail near 60 R_e, *J. Geophys. Res.*, **79**, 5143, 1974.

MENG, C.-I., A. T. Y. LUI, S. M. KRIMIGIS, S. ISMAIL, C. O. BOSTROM, and D. J. WILLIAMS, Spatial distribution of energetic particles in the distant magnetotail, *J. Geophys. Res.*, **86**, 5682, 1981.

MIHALOV, J. D., D. S. COLBURN, R. G. CURRIE, and C. P. SONNETT, Configuration and reconnection of the geomagnetic tail, *J. Geophys. Res.*, **73**, 943, 1968.

MOZER, F. S., W. D. GONZALES, F. BOGOTT, M. C. KELLEY, and S. SCHUTZ, High-latitude electric fields and the three-dimensional interaction between the interplanetary and terrestrial magnetic fields, *J. Geophys. Res.*, **79**, 56, 1974.

NESS, N. F., The earth's magnetic tail, *J. Geophys. Res.*, **70**, 2989, 1965.

NESS, N. F., C. S. SCEARCE, and S. CANTARANO, Probable observations of the geomagnetic tail at 10^3 earth radii by Pioneer 7, *J. Geophys. Res.*, **72**, 3769, 1967.

PARKS, G. K., D. LEAVER, C. S. LIN, K. A. ANDERSON, R. P. LIN, and H. REME, ISEE 1/2 particles observations of outer plasma sheet boundary, *J. Geophys. Res.*, **84**, 6471, 1979.

PASCHMANN, G., B. U. O. SONNERUP, I. PAPAMASTORAKIS, N. SCKOPKE, G. HAERENDEL, S. J. BAME, J. R. ASBRIDGE, J. T. GOSLING, C. T. RUSSELL, and R. C. ELPHIC, Plasma acceleration at the Earth's magnetopause: evidence for reconnection, *Nature*, **282**, 243, 1979.

PERREAULT, P. and S.-I. AKASOFU, A study of geomagnetic storms, *Geophys. J. R. Astron. Soc.*, **54**, 547, 1978.

PIDDINGTON, J. H., A theory of polar geomagnetic storms, *Geophys. J.*, **3**, 314, 1960.

POTEMRA, T. A., L. J. ZANETTI, P. F. BYTHROW, A. T. Y. LUI, and T. IIJIMA, B_y-dependent convection patterns during northward interplanetary magnetic field, *J. Geophys. Res.*, **89**, 9753, 1984.

REID, G. C. and H. H. SAUER, Evidence for non-uniformity of solar-proton precipitation over the polar caps, *J. Geophys. Res.*, **72**, 4383, 1967.

REIFF, P. H., R. W. SPIRO, and T. W. HILL, Dependence of the polar cap potential drop on interplanetary parameters, *J. Geophys. Res.*, **86**, 7639, 1981.

RUSSELL, C. T. and K. I. BRODY, Some remarks on the position and shape of the neutral sheet, *J. Geophys. Res.*, **72**, 6104, 1967.

SAUNDERS, M. A., D. J. SOUTHWOOD, E. W. HONES, Jr., and C. T. RUSSELL, A hydromagnetic vortex seen by ISEE-1 and 2, *J. Atm. Terr. Phys.*, **43**, 927, 1981.

SAUNDERS, M. A., D. J. SOUTHWOOD, T. A. FRITZ, and E. W. HONES, Jr., Hydromagnetic vortices—1, the 11 December, 1977 event, *Planet. Space Sci.*, **31**, 1099, 1983a.

SAUNDERS, M. A., D. J. SOUTHWOOD, and E. W. HONES, Jr., Hydromagnetic vortices—2, further dawnside events, *Planet Space Sci.*, **31**, 1117, 1983b.

SCKOPKE, N., G. PASCHMANN, H. ROSENBAUER, and D. H. FAIRFIELD, Influence of the interplanetary magnetic field on occurrence and thickness of the plasma mantle, *J. Geophys. Res.*, **81**, 2687, 1976.

SIBECK, D. G. and G. L. SISCOE, Downstream Properties of magnetic flux transfer events, *J. Geophys. Res.*, **89**, 10709, 1984.

SIBECK, D. G., G. L. SISCOE, J. A. SLAVIN, E. J. SMITH, B. T. TSURUTANI, and R. P. LEPPING, The distant magnetotail's response to a strong IMF B_y: twisting, flattening, and field line bending, *J. Geophys. Res.*, **90**, 4011, 1985.

SINGER, S. J., P. CONNER, W. D. EVANS, M. D. MONTGOMERY, and E. E. STOGSDILL, Plasma observations at 10^5 km, *Space Res.*, **5**, 546, 1965.

SLAVIN, J. A., B. T. TSURUTANI, E. J. SMITH, D. E. HONES, and D. G. SIBECK, Average configuration of the distant (<200 R_e) magnetotail: initial ISEE-3 magnetic field results, *Geophys. Res. Lett.*, **10**, 973, 1983.

SLAVIN, J. A., E. J. SMITH, D. G. SIBECK, D. N. BAKER, R. D. ZWICKL, and S.-I. AKASOFU, An ISEE-3 study of average and substorm conditions in the distant magnetotail, *J. Geophys. Res.*, **90**, 10875, 1985.

SMITH, E. J., A comparison of Explorer 6 and Explorer 10 magnetometer data, *J. Geophys. Res.*, **67**, 2045, 1962.

SPEISER, T. W. and N. F. NESS, The neutral sheet in the geomagnetic tail: its motion, equivalent currents, and field line connection through it, *J. Geopys. Res.*, **72**, 131, 1967.

SPREITER, J. R. and A. Y. ALKSNE, Plasma flow around the magnetosphere, *Rev. of Geophysics*, **7**, 11, 1969.

TOICHI, T. and T. MIYAZAKI, Flapping motions of the tail plasma sheet induced by the interplanetary magnetic field variations, *Planet. Space Sci.*, **24**, 147, 1976.

VAN ALLEN, J. A., J. F. FENNELL, and N. F. NESS, Asymmetric access of energetic solar protons to the earth's north and south polar caps, *J. Geophys. Res.*, **76**, 4262, 1971.

VILLANTE, U., An overview of Pioneer observations of the distant geomagnetic tail, *Space Sci. Rev.*, **20**, 123, 1977.

VOIGT, G.-H., The influence of the IMF B_y component on the magneto-hydrostatic magnetotail, in *Proceedings of the AGU Chapman Conference on Magnetotail Physics*, October 28–31, 1985, Laurel, Maryland, p. 6, 1985.

WALKER, R. C., U. VILLANTE, and A. J. LAZARUS, Pioneer 7 observations of plasma flow and field reversal regions in the distant geomagnetic tail, *J. Geophys. Res.*, **80**, 1238, 1975.

WOLFE, J. H., R. W. SILVA, D. D. MCKIBBIN, and R. H. MASON, Preliminary observations of a geomagnetospheric wake at 1000 earth radii, *J. Geophys. Res.*, **72**, 4577, 1967.

WU, C. C., The effects of dipole tilt on the structure of the magnetosphere, *J. Geophys. Res.*, **89**, 11048, 1984.

YEAGER, D. M. and L. A. FRANK, Low-energy electron intensities at large distances over the Earth's polar cap, *J. Geophys. Res.*, **81**, 3966, 1976.

ZANETTI, L. J., T. A. POTEMRA, T. IJIMA, W. BAUMJOHANN, and P. F. BYTHROW, Ionospheric and Birkeland current distributions for northward interplanetary magnetic field: inferred polar convection, *J. Geophys. Res.*, **89**, 7453, 1984.

ZWICKL, R. D., D. N. BAKER, S. J. BAME, W. C. FELDMAN, J. T. GOSLING, E. W. HONES, Jr., D. J. MCCOMAS, B. T. TSURUTANI, and J. A. SLAVIN, Evolution of the earth's distant magnetotail: ISEE-3 electron plasma results, *J. Geophys. Res.*, **89**, 11007, 1984.

Solar Wind-Magnetosphere Coupling, edited by Y. Kamide and J. A. Slavin, 691–695.

Observation of IMF-Associated Magnetic Field Perturbations in the GSM *X* Component at Geostationary Orbit

W. J. HUGHES* and S. W. H. COWLEY

The Blackett Laboratory, Imperial College of Science and Technology, London SW7 2BZ, U.K.

One year of magnetic field data from the geostationary spacecraft ATS 6 have been analyzed for the presence of perturbations associated with the IMF. A previous analysis showed the presence of effects in the GSM *Y* component associated with IMF B_y. It is shown here that strong perturbation fields in the GSM *X* component also occur, depending on both IMF B_x and B_y. The morphology is complex, depending strongly on the sign of IMF B_z, and also on ATS 6 local time.

1. Introduction

FAIRFIELD (1979) discovered the existence of a dawn–dusk directed perturbation magnetic field in near-Earth magnetotail data which is positively correlated with the external interplanetary magnetic field (IMF) GSM B_y component. More recently, TSURUTANI *et al.* (1984) have detected the same effect 150–220 R_E down-tail in ISEE-3 data. The nature of the perturbation field is such that, in effect, there is a partial penetration of the IMF B_y field into the Earth's magnetotail, amounting to some 10–20% of the external field. Theoretical work by COWLEY (1981a) showed that this phenomenon was consistent with other IMF B_y-associated asymmetry effects observed in ionospheric polar cap flows and in the plasma mantle in the magnetotail lobes, and that this suite of effects results directly from dawn-dusk stresses exerted on open flux tubes in the presence of IMF B_y, as initially suggested by ATKINSON (1972) and JØRGENSEN *et al.* (1972).

It was also pointed out by COWLEY (1981a) that some of the observed asymmetry effects also involve closed flux tubes mapping to the auroral zones (e.g., dawn–dusk shifts of the polar cap boundary), and that these could be understood if the Fairfield perturbation field occurs not only on open flux tubes, but in the region of closed quasi-dipolar flux tubes as well. Accordingly, COWLEY and HUGHES (1983) examined ATS 6 geostationary orbit magnetometer data for the occurrence of this effect and found that it was indeed present. On average 28% of the IMF B_y field appears at this location, though variations do occur depending upon local time, IMF B_z (or magnetic disturbance level) and season. It was also clearly established that the effect was indeed positively correlated with IMF B_y, as opposed to being negatively correlated with IMF B_x, for example, it being noted that IMF B_x and B_y are highly related in solar wind data via the average spiral direction of the field.

Following from the above considerations, COWLEY (1981b) went on to consider possible effects which might be associated with IMF B_x. It was suggested that a

*On leave from: Department of Astronomy, Boston University, Boston, MA 02215, U.S.A.

'partial penetration' effect might occur in this case also, again resulting from stresses exerted on open flux tubes. This paper demonstrates for the first time that IMF-related B_x perturbation fields do indeed occur in ATS 6 data, but that their morphology is considerably more complex than envisaged by COWLEY (1981b). The next section briefly describes the data set employed (essentially the same as that used in the Y-component studies by COWLEY and HUGHES (1983)), while results are presented in Section 3.

2. Data Analysis

The data used in this study consist of 4710 simultaneous 1-hour averages of the IMF and of the perturbation field observed at ATS 6. The IMF data were provided on tape by the US National Space Science Data Center (KING, 1977), while the ATS 6 data were provided by Dr. R. L. McPherron. The latter were obtained during the interval 1 June 1974 to 17 May 1975 when the spacecraft was stationed at the geographic equator at 94° W longitude, where the magnetic latitude is ~10°. The 1-hour ATS 6 perturbation vectors were derived by computing nine averaged 'standard' daily variations of the field at 1 min resolution (three seasons divided into three disturbance levels defined by K_p), subtracting each 1 min vector from the appropriate 'standard' value, rotating the resulting difference vector into GSM coordinates and then averaging over each UT hour. The procedure is identical to that described in detail by COWLEY and HUGHES (1983), with the exception that GSM coordinates are used here instead of the solar magnetic coordinates in the latter paper, it being noted however that the Y component studied by COWLEY and HUGHES (1983) is identical in the two systems. It should also be noted that by averaging the ATS 6 data over UT hours (to correspond with the IMF data) a 20 min mismatch is introduced between these averages and LT hour averages, since at 94° W longitude LT is UT minus 6 hr 20 min. Thus in this paper the "noon quadrant" means the 6 hr period from 15 to 21 UT corresponding to 0840 to 1440 LT.

The results presented in this paper are in the form of contour maps of the ATS 6 GSM X perturbation field (derived as above and denoted here by lower case b_x) plotted on the IMF B_x–B_y plane. The map for each subset of the data (defined by ATS 6 local time and sign of IMF B_z) was obtained by first averaging the ATS 6 b_x data into 1 nT×1 nT boxes in the IMF B_x–B_y plane (the perimeters defined by integer IMF B_x, B_y values), and then directly averaging four adjacent values so obtained to produce an average $\langle b_x \rangle$ value at each integer IMF point. Individual 1-hour data points contributing to each $\langle b_x \rangle$ value are thus located within overlapping 2 nT×2 nT IMF B_x–B_y boxes centred on the integer IMF points. This procedure was adopted so as to smooth the data before contouring, but without introducing biases which would be caused by the very non-uniform distribution of data points in the IMF B_x–B_y plane. Finally, no $\langle b_x \rangle$ value is derived at a particular IMF B_x, B_y point if one of the contributing 1 nT square boxes contains no data, this condition determining the perimeter of the contoured regions shown in the figures.

3. GSM *X* Perturbation Fields at Geostationary Orbit

Initial examination of the ATS 6 b_x perturbation fields showed immediately that strong IMF effects are indeed present, and that their form depends markedly on the sign of IMF B_z. On dividing the data into four IMF B_z ranges defined by the boundary values 6.0, 1.5, 0.0, −1.5 and −6.0 nT it was found that the effects are very similar in the two positive ranges and in the two negative ranges, but that the results for positive B_z differ considerably from those for negative B_z. Therefore, there must exist a very sharp transition in the response of the field as IMF B_z changes sign, and accordingly we so divide the data here and examine how the effects depend upon ATS 6 local time.

Figure 1 shows contour maps of ATS 6 $\langle b_x \rangle$ values for IMF $B_z>0$ split into

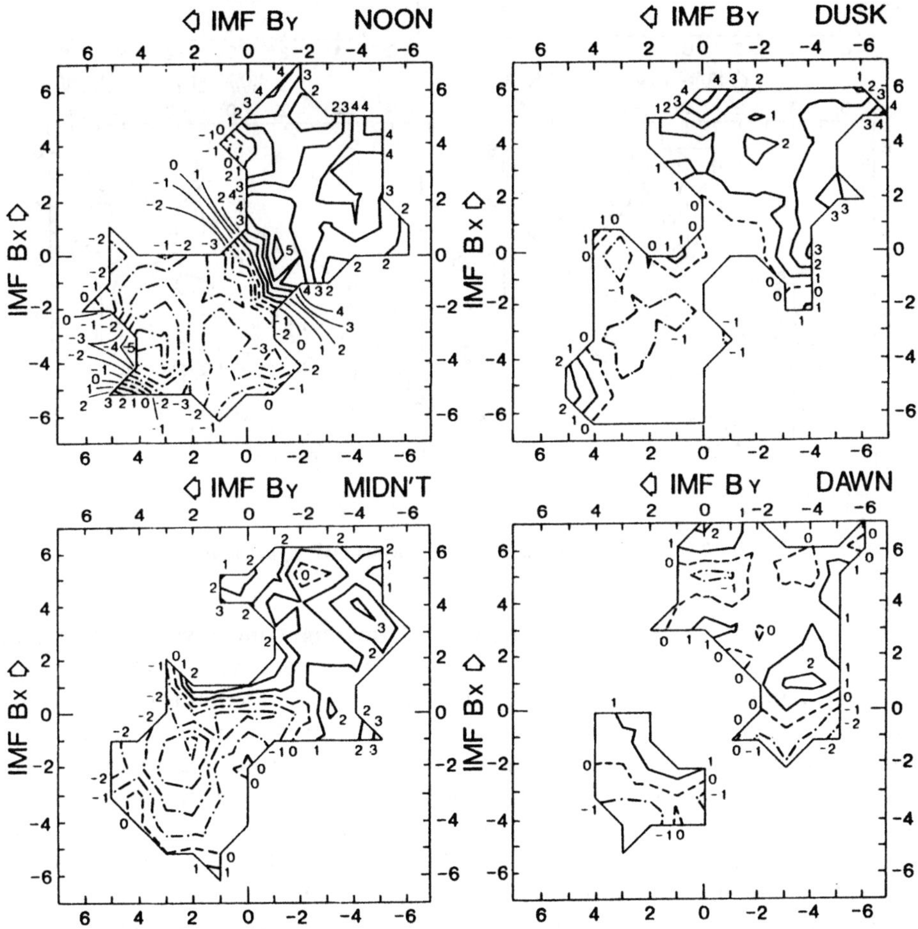

Fig. 1. Contour maps of averaged ATS 6 $\langle b_x \rangle$ magnetic field perturbations plotted on the IMF B_x–B_y plane for simultaneous IMF B_z positive, and various ATS 6 local time quadrants. Positive contours are given by the heavy solid lines, zeros by the dashed lines, and negative contours by the dot-dash lines.

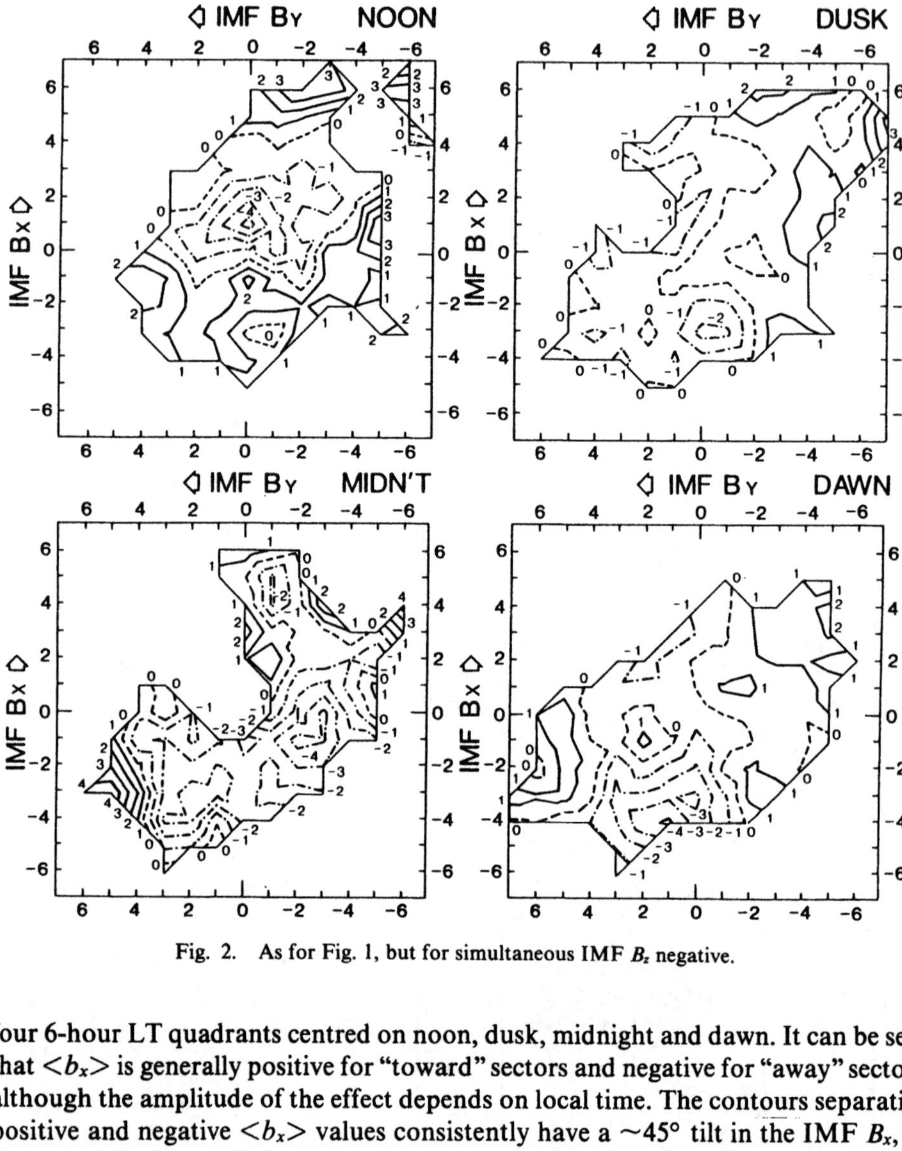

Fig. 2. As for Fig. 1, but for simultaneous IMF B_z negative.

four 6-hour LT quadrants centred on noon, dusk, midnight and dawn. It can be seen that $<b_x>$ is generally positive for "toward" sectors and negative for "away" sectors, although the amplitude of the effect depends on local time. The contours separating positive and negative $<b_x>$ values consistently have a ~45° tilt in the IMF B_x, B_y plane, and further analysis attempting to order the data by a single field component in the IMF B_x, B_y plane showed that the IMF component along the average spiral direction consistently orders the data best. Thus, although we know of no physical explanation for the effect, it appears that for IMF B_z>0 the B_x field at ATS 6 responds to the IMF component along the average 45° spiral direction, rather than to either IMF B_x or B_y components alone. The response has the form of a step function in the ATS 6 data (somewhat smoothed by the averaging in the contour maps), the magnitude of the step varying with local time. Overall the effect is largest at noon, the total step being ~8 nT (such that it contributes significantly to the total field variation in this regime), falling to ~4 nT at midnight, ~2 nT at dusk, and no consistent effect is observed in the dawn sector.

Figure 2 presents similar information for intervals when IMF $B_z<0$. Although strong perturbation fields are again present, the picture is obviously less well defined than for IMF $B_z>0$. At noon a negative correlation with IMF B_x appears to be present, while at midnight and dawn this reverses to a positive correlation. At dusk the perturbations seem better ordered by IMF B_y than by IMF B_x, though the amplitude is small and close to the noise level.

Finally, although not perhaps immediately apparent in the above figures, another effect is also present in the ATS 6 b_x data, for both IMF B_z positive and negative. This is the tendency for ATS 6 b_x to become increasingly positive as the magnitude of the IMF in the x–y plane increases, independent of the latter's orientation. This perturbation only starts to become noticeable above the other effects present for $|B_{IMF}|\gtrsim 5$ nT, and so is confined to the periphery of the maps in Figs. 1 and 2.

In summary, it has been demonstrated in this paper that IMF-associated perturbation fields in the GSM X component are present in geostationary orbit magnetic data, and that they have a complex morphology. The most remarkable finding is the distinctly different response which occurs for IMF B_z positive and negative, providing further evidence for the very different coupling which occurs between the solar wind and the Earth's magnetosphere in these two situations.

We would like to thank Karl Lam for writing the software which produced the contour plots and Sandra Chapman for their realization. Work at Boston University was partially supported by NSF grant number ATM 8209025.

REFERENCES

ATKINSON, G., Magnetospheric flows and substorms, in *Magnetosphere-Ionosphere Interactions*, edited by K. Folkestad, p. 203, Universitetsforlaget, Oslo, Norway, 1972.

COWLEY, S. W. H., Magnetospheric asymmetries associated with the *Y*-component of the IMF, *Planet. Space Sci.*, **29**, 79, 1981a.

COWLEY, S. W. H., Asymmetry effects associated with the *X*-component of the IMF in a magnetically open magnetosphere, *Planet. Space Sci.*, **29**, 809, 1981b.

COWLEY, S. W. H. and W. J. HUGHES, Observation of an IMF sector effect in the Y magnetic field component at geostationary orbit, *Planet. Space Sci.*, **31**, 73, 1983.

FAIRFIELD, D. H., On the average configuration of the geomagnetic tail, *J. Geophys. Res.*, **84**, 1950, 1979.

KING, J. H., Interplanetary medium data book—Appendix, *Rep. NSSDC/WDC-A R&S 77-04a*, NSSDC, NASA Goddard SFC, Greenbelt, MD 20771, USA, 1977.

JØRGENSEN, T. S., E. FRIIS-CHRISTENSEN, and J. WILHJELM, Interplanetary magnetic field direction and high latitude ionospheric currents, *J. Geophys. Res.*, **77**, 1976, 1972.

TSURUTANI, B. T., D. E. JONES, R. P. LEPPING, E. J. SMITH, and D. G. SIBECK, The relationship between the IMF B_y and the distant tail (150–238 R_E) lobe and plasma sheet B_y fields, *Geophys. Res. Lett.*, **11**, 1082, 1984.

Solar Wind-Magnetosphere Coupling, edited by Y. Kamide and J. A. Slavin, 697–705.

Ion Composition of the Bulk Ring Current during a Magnetic Storm: Observations with the CHEM-Instrument on AMPTE/CCE

W. STÜDEMANN[1], G. GLOECKLER[2], B. WILKEN[1], F. M. IPAVICH[2], G. KREMSER[1], D. C. HAMILTON[2], and D. HOVESTADT[3]

[1]*Max-Planck-Institut für Aeronomie, Katlenburg-Lindau, F.R.G.*
[2]*University of Maryland, College Park, Maryland 20742, U.S.A.*
[3]*Max-Planck-Institut für Extraterrestrische Physik, Garching, F.R.G.*

Measurements of the ion composition in the entire energy range of the ring current population are presented. The measurements were obtained by the University of Maryland/Max-Planck-Institut für Aeronomie CHEM instrument capable of determinig separately the ion mass and mass per charge in the energy band 1–300 keV/charge. We present energy spectra for all major constituents (H^+, He^+, He^{2+}, O^+, and O^{2+}) during the main phase of a magnetic storm. From these spectra the number densities, energy densities and the mean energies are calculated and displayed as radial profiles covering the range $3 \leq L \leq 9$ (R_E). Mean energies of He^{2+} are about twice that of H^+ and He^+ throughout this storm. When normalizing mean energies by the ionic charge, the time profiles for the mean energies of all major ions tend to bunch together. The largest variations of the energy densities and mean energies occur for O^+ ions. At $L=7$ R_E, for instance, the energy density of O^+ increases to ~50% of that of H^+, and the mean energy reaches a maximum of 50 keV compared to ~8% and ~15 keV at prestorm time, respectively. The importance of O^+ ions during a magnetic storm which was well established by earlier measurements below 20 keV/charge was manifested at higher energies. In one case the O^+ energy density equalled the proton energy density even up to the highest energy of 300 keV observed by this instrument.

1. Introduction

The lack of knowledge about the ion composition in the entire energy range of the ring current population is a well known fact and was emphasized by e.g., WILLIAMS (1980) and SCHULZ (1983). The only composition measurements performed were below 17 keV and above a few hundred keV. From these the dominating role of O^+ ions at energies below 17 keV during magnetic storms was first found by the Lockheed Group (e.g., SHELLEY *et al.*, 1972). Later on GEOS and PROGNOZ measurements added more details to the picture of a possible ionospheric source active during magnetic storms (e.g., BALSIGER *et al.*, 1983; LUNDIN *et al.*, 1980). But the mass-per-charge resolving instruments flown on these spacecraft were also limited to low energies. At higher energies (above 400 keV/Nuc) measurements e.g. on Exploree 45 and ATS-6 (FRITZ and SPJELDVIK, 1981; FRITZ and WILKEN, 1976) suggested the solar wind as a source for ring current particles. In the wide energy band between 20 keV/charge and 400 keV/Nuc where the ring current energy distribution has its maximum value, composition measurements remained an instrumental challenge.

2. Instrumentation and Data Analysis

One of the main objectives of the CHEM-instrument (Charge-, Energy-, Mass-spectrometer) on the CCE (Charge Composition Explorer) within the AMPTE programme was to fill this gap. The instrument employs an electrostatic analyzer, postacceleration and a time-of-flight/energy spectrometer. Each particle is identified by its energy/charge, velocity and residual energy. From these quantities the onboard processor calculates the ion's mass and mass/charge. The energy range covered by CHEM extends from 1 to 300 keV/charge. A comprehensive description of the instrument is given by GLOECKLER *et al.* (1985a).

For a number of ion species in a given charge state (H^+, He^+, He^{2+}, O^+, O^{2+} and others) the instrument returns count rates for each energy/charge step of the electrostatic analyzer. These data are spin averaged, and this corresponds to effectively pitch angle averaging due to the relative orientation of the satellite spin vector.

The following analysis has used these rate channels to construct energy spectra for H^+, He^+, He^{2+}, O^+, and O^{2+} ions.

The algorithms used to determine the number density and the energy density assume—for the time being—that the shapes of the pitch angle distributions for the various ion species are independent of mass as well as of energy. In addition the algorithms are based on the assumption of a $\sin^m\alpha$ pitch angle distribution. As long as m is above 3, there is no strong dependance on the actual value of m.

3. Observations

The data presented in the following sections are taken from the period September 3 through 6, 1984. The SSC on September 4, 1984 at 0746 UT triggered a moderate magnetic storm resulting in a minimum D_{ST} value of 120 nT on the morning of September 5, 1984 at about 07 UT. Two major particle injections occurred causing minima of the surface magnetic field on September 4 at ~16 UT and on September 5 at ~07 UT. The recovery phase of the storm followed the second injection continuing into the morning of September 6. An overview of this storm is given by WILLIAMS and SUGIURA (1985). Initial results from the CHEM instrument during this storm are reported by GLOECKLER *et al.* (1985b). Data from the same storm including all plasma composition instruments on the CCE spacecraft are also published by KRIMIGIS *et al.* 1985).

In the following section ion composition data will be presented from this magnetic storm starting with sample energy spectra of major magnetospheric ions during the main phase of the storm (Fig. 1). Next, a number of energy spectra of Hydrogen are presented for a full pass through the ring current region (Fig. 2). Using similar measurements for the other species a radial profile of number densities is obtained (Fig. 3). The ion composition in the region of the maximum energy density of the ring current (L=3.9) at about a time when the ring current has become strongest is shown by the energy densities of various ion species as a function of the upper

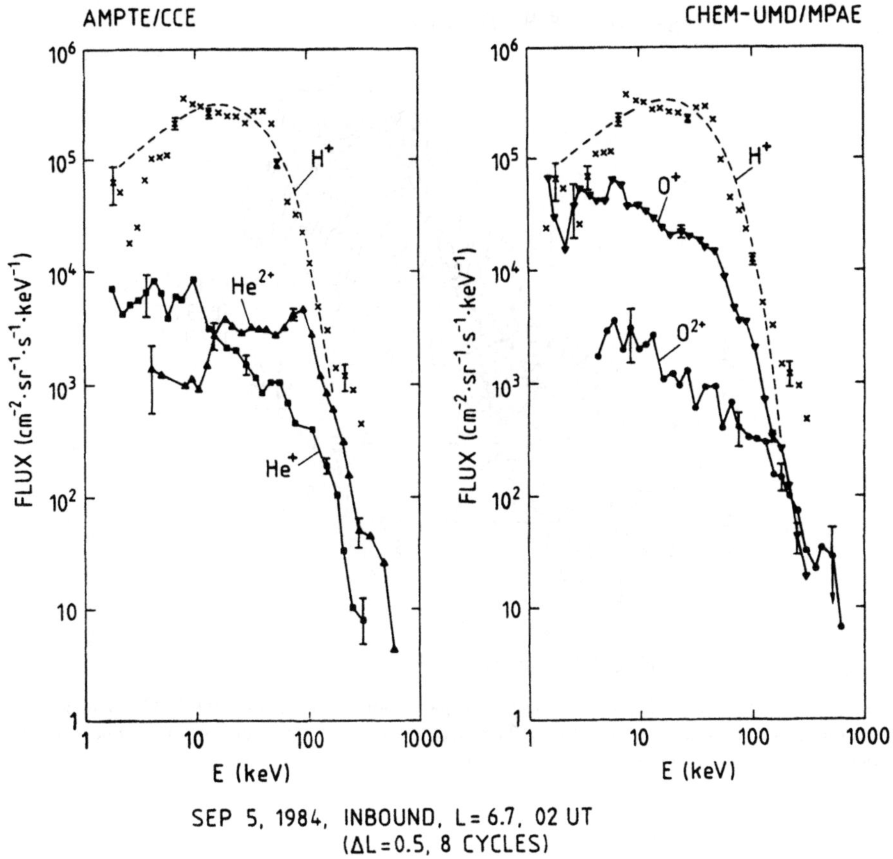

Fig. 1. Energy spectra of H^+, He^+, He^{2+}, O^+, and O^{2+} at the main phase of a magnetic storm on September 5, 1984. For a better presentation, energy spectra of He^+ and He^{2+} are on the left and O^+ and O^{2+} on the right side with H^+ included on both. Energy range is 1–300 keV/charge. The data are taken during the main phase of a magnetic storm in the outer region of the ring current. Local time is approximately noon.

energy limit (Fig. 4). Finally, Figure 5 depicts the overview over the entire magnetic storm.

Figure 1 shows the energy spectra of H^+, He^+, He^{2+}, O^+, and O^{2+} obtained in the outer region of the ring current (L=6.7) at about the time of the second injection on September 5. The various rates are summed up over 8 energy cycles corresponding to an integration time of about 25 min or a radial range of ΔL=0.5. Error bars in Fig. 1 indicate the statistical uncertainties of the count rates. For demonstration purposes the figure is split into two parts with the H^+ spectrum included on both.

A pseudo three dimensional representation of H^+ spectra for an entire pass through the region of the ring current on September 4, 1984 in the time period 15–19 UT is shown in Fig. 2. The D_{ST} index was recovering from a negative excursion peaking at -60γ at ~15 UT. The spectra are constructed at equal radial intervals.

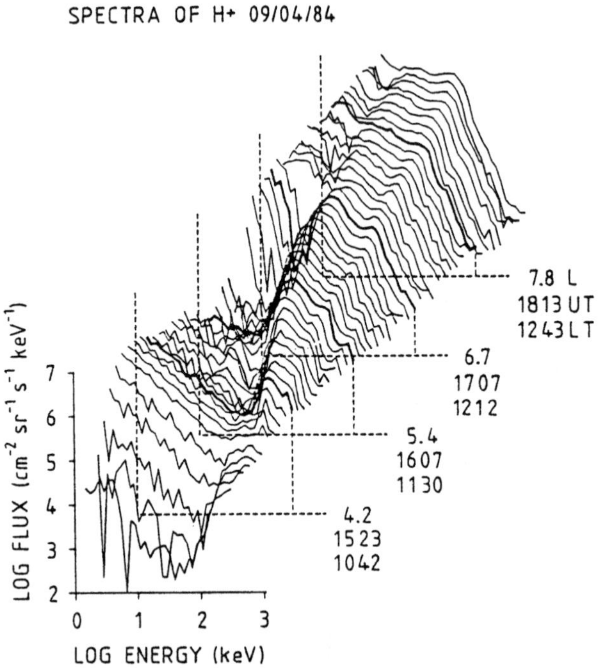

Fig. 2. Energy spectra of H^+ from the outbound pass about 8 hours after the SSC. Each spectrum is averaged over a spatial region of 0.1 R_E. The spectra will be used to determine the number- and energy densities and the mean energies (see next figure).

Each tenth spectrum is marked by a heavy line and the instantaneous radial distance, universal-and local times are given.

Inside about $L=6.7$ protons below an energy of a few tens of keV disappear completely from the distribution so that inside of about $L=4$ only high energy protons are present. It is obvious that further inside part of the energy distribution of the protons even exceeds the upper energy limit of the instrument. The power law spectral shape at low energies leading to the broad spatial maximum around $L=4$ is in part due to energetic penetrators (probably electrons) causing accidental coincidences in the time-of-flight sensor.

Radial profiles of the computed number- and energy densities for the same orbit as before are displayed in Fig. 3 for H^+, He^+, He^{2+}, O^+, and O^{2+}. As mentioned before these densities are integrated over the range 1 to 300 keV/charge. Number densities are almost constant between 4.5 and 7 R_E. He^{2+} in the outer magnetosphere amounts to about 5% of the protons (prestorm ~2%), and O^+ reaches effectively the number density of H^+ in a region outside $L=4.5$ where the contamination due to penetrators (see above) could be disregarded. Background rates caused by penetrating particles are monitored routinely by the instrument and for this particular pass show an increase inside $L=4.5$, but no correction is performed in this presentation. Also the

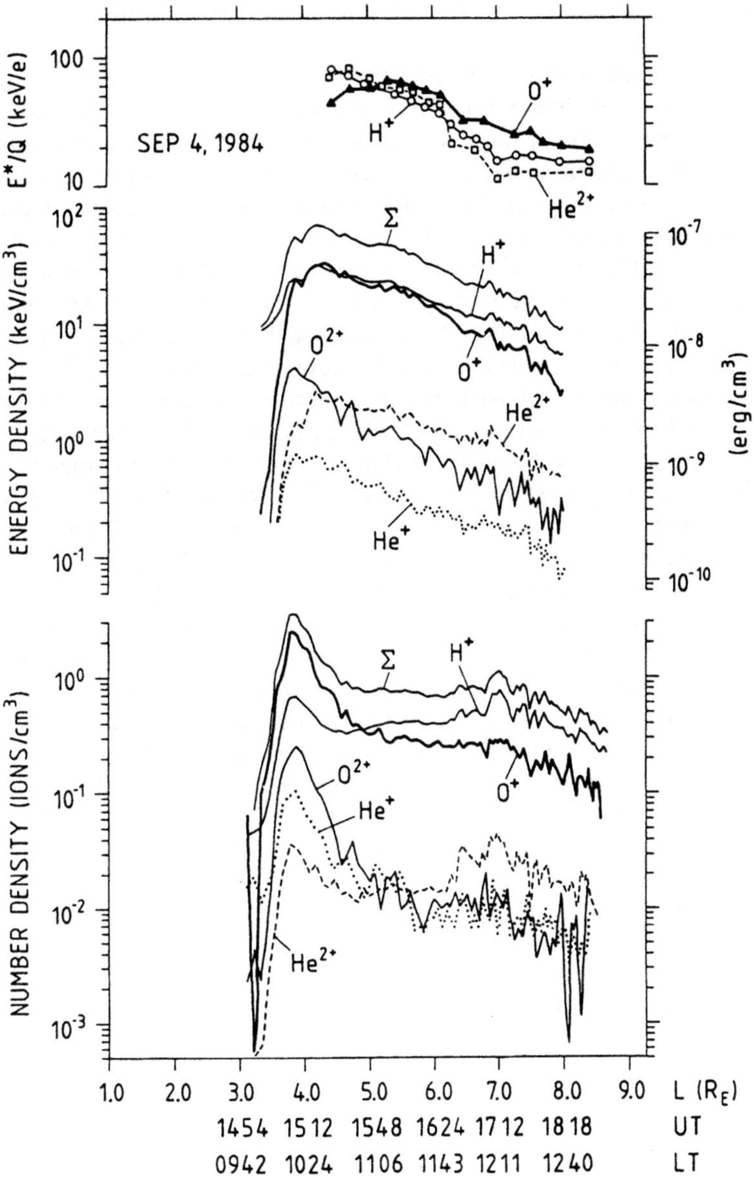

Fig. 3. Number- and energy-densities for H^+, He^+, He^{2+}, O^+ and O^{2+} together with the mean energies for the three most abundant species versus radial distances from the same outbound pass as Fig. 2. The mean energies are given as energy/charge. Units for the energy density are keV/cm^3 (left) as well as erg/cm^3 (right). Note that inside 6 R_E singly charged oxygen ions carry the same energy density as the protons up to an energy of 300 keV/charge.

steep decrease both in number densities and energy densities inside $L=3.8$ is not corrected for the high energy tail of ions extending beyond the upper energy threshold of CHEM.

From the energy- and number densities given in Fig. 3, the mean energy E^* can be obtained defined as their ratio. This is performed for H^+, O^+, and He^{2+} ions and displayed as mean energy/charge in the top panel of Fig. 3. Normalization of the mean energies by the ionic charge moves the curve for He^{2+} into close vicinity of the H^+ curve. All curves show a general trend of decresing mean energy with increasing radial distance. While the mean energy for O^+ reaches a maximum of about 63 keV at $L=5.3$ mean energies for He^{2+} and H^+ are still increasing with decreasing radial distance.

Out of the five in- and outbound passes which covered the period of this mangetic storm the particular pass of Fig. 3 was selected because of the surprisingly high contribution of singly charged Oxygen to the total energy density of the ring current. Inside a radial distance of ~6 R_E the energy density of O^+ equals that of the protons not only at small energies but also up to and including the highest energy channel of 300 keV/charge observable by the CHEM instrument. Since the number density of O^+ ions outside 5 R_E is slightly smaller than that of the protons, the high energy density is caused by the high value of the mean energy of the oxygen population.

Figure 4 contains the accumulated energy densities for H^+, He^+, He^{2+}, O^+, and O^{2+} below an energy E as a function of energy normalized to the total energy density of all ions. The data are taken at the maxima of the energy densities ($L=3.9$) on September 5 at about 0415 UT. For better visibility all values below 1% are in a logarithmic scale. While O^+ energy densities on the pass selected for the previous figure equalled those of H^+ even up to energies of 300 keV, energy densities of O^+ are dominant only below ~60 keV on this pass. The numbers in brackets denote each species' contribution to the total energy density. Mean energies of 75 and 29 keV are also indicated for H^+ and O^+, respectively. At the highest energy value the energy density of H^+ is still increasing indicative for the high energy tail of protons above 300 keV.

4. Summary and Discussion

Ion composition measurements are reported covering—for the first time—the entire energy range of the ring current population. The measurements clearly demonstrate the importance of singly charged Oxygen ions during magnetically active periods because of their high contribution to the ring current energy density.

Figure 5 summarizes the results obtained during this magnetic storm. It displays the temporal variations of energy densities (middle panel) and mean energies (top panel) for H^+, He^+, He^{2+}, O^+, and O^{2+} ions throughout the event. Snapshots are taken at the outer edge of the ring current at $L=7$ for each consecutive in-and outbound pass. Although variations of the ion composition are more dynamic at the maximum of the ring current around L ~4 a radial distance of 7 R_E was chosen as a precaution against the penetrating background radiation when uncorrected data are used. The bottom panel of Fig. 5 shows the magnetic field decrease on the ground represented by the hourly equatorial D_{ST}-values.

As was already mentioned with respect to Fig. 4, mean energies for ions of a

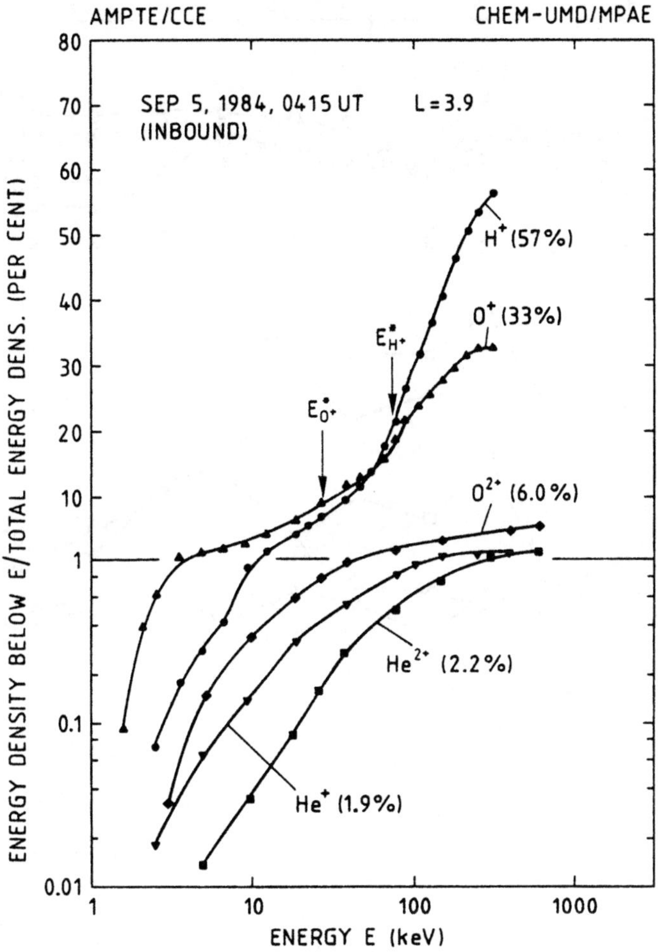

Fig. 4. Accumulated energy densities (in per cent) below the energy E as a function of energy for H^+, He^+, He^{2+}, O^+, O^{2+} ions normalized to the total energy density of all these ions below 300 keV/charge. The data are from the same satellite pass as the spectra of Fig. 1, but obtained from the maximum of the ring current energy density at L=3.9. $E^*_{O^+}$ and $E^*_{H^+}$ denote mean energies for O^+ and H^+ of 29 and 75 keV, respectively. The percentage numbers in brackets specify the contribution of each ion population at the upper energy limit of the CHEM instrument (300 keV/charge) to the total energy density. For better presentation the vertical axis is logarithmic below 1%.

higher charge state are in better proximity when normalized to energy/charge. This behaviour points to an energy/charge dependance of the particles' history.

At the time of the SSC the CCE-spacecraft was on an inbound pass close to apogee at L=8.8: Following the SSC the magnetosphere became compressed so that the satellite entered the magnetosheath. At a radial distance of 7.5 R_E it reentered the magnetosphere. This is about the time when the steep increase in the energy densities

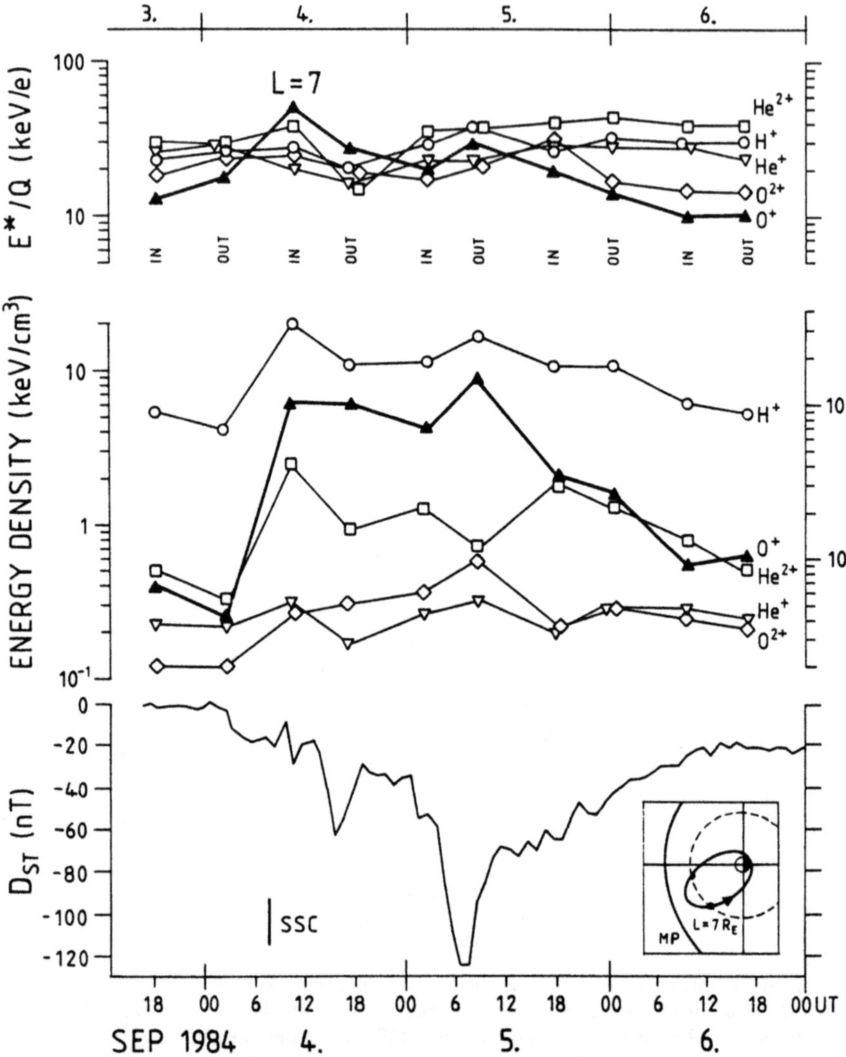

Fig. 5. Summary plot of energy densities and mean energies for H^+, He^+, He^{2+}, O^+ and O^{2+} ions at $L=7$ over the entire period of the magnetic storm from the evening of September 3 through September 6, 1984. The hourly equatorial D_{ST} values are given in the bottom panel for reference. Data were sampled whenever the AMPTE/CCE spacecraft passed a radial distance of 7 R_E on consecutive in- and outbound passes. As can be seen from the sketch on the bottom of the figure the crossing of $L=7$ on an inbound pass occurs around early afternoon compared to noon time on outbound passes. The SSC preceeding this storm is indicated on September 4 at 0746 UT.

of O^+, H^+, and He^{2+} are detected. There are different reasons for this increase for H^+ than there are for O^+. While the H^+ energy density increases due to an increase in number density only, the energy density of O^+ increases by a factor of ~9 due to the number density and another factor of ~3 due to the increase of the mean energy. The He^{2+} energy density behaves intermediately with increases by factors of 5 and 1.5 due to the number density and mean energy, respectively.

Following this early phase of the compression of the magnetosphere, there were two major injection events mentioned earlier which further caused the number densities to increase but led to a decrease of the mean energy of the O^+ ions. The peak value of the O^+ energy density occurred at the same time when the magnetic field depression reached its minimum.

We are grateful to many individuals at the University of Maryland (UMD), the Max-Planck-Institut für Aeronomie (MPAE) and the Technische Universität Braunschweig (TUB) for bringing the instrument to a success. We especially would like to thank R. A. Lundgren, E. O. Tums, and J. C. Cain (UMD), Dr. W. Weiss, H. P. Winterhoff, and H. Sommer (MPAE), and W. Rieck and H. Dinse (TUB). In addition we are indebted to F. Both of MPAE and L. Kistler, H. Weiss, and N. Lal of UMD for their assistance in data analysis. This work has been supported by the Bundesminister für Forschung und Technologie under grant N° 010M158 0 (W. Germany) and by NASA (USA) via JHU/APL contract 601527.

REFERENCES

Balsiger, H., J. Geiss, and D. T. Young, The composition of thermal and hot ions observed by the GEOS-1 and -2 spacecraft, in *Energetic Ion Composition in the Earth's Magnetosphere*, edited by R. G. Johnson, pp. 195–230, Terra Scientific Publishing Company, Tokyo, 1983.

Fritz, T. A. and W. N. Spjeldvik, Steady state observations of geomagnetically trapped energetic heavy ions and their implications for theory, *Planet. Space Sci.*, **29**, 1169, 1981.

Fritz, T. A. and B. Wilken, Substorm generated fluxes of heavy ions at the geostationary orbit, in *Magnetospheric Particles and Fields*, edited by B. M. McCormac, D. Reidel Publishing Company, Dordrecht, 1976.

Gloeckler, G., F. M. Ipavich, W. Stüdemann, B. Wilken, D. C. Hamilton, G. Kremser, D. Hovestadt, F. Gliem, R. A. Lundgren, W. Riek, E. O. Tums, J. C. Cain, L. S. MaSung, W. Weiss, and P. Winterhoff, The charge-energy-mass (CHEM) spectrometer for 0.3 to 300 keV/e ions on AMPTE-CCE, *IEEE Trans. Geosci. Rem. Sens.*, Vol. GE-23, 234, 1985a.

Gloeckler, G., B. Wilken, W. Stüdemann, F. M. Ipavich, D. Hovestadt, D. C. Hamilton, and G. Kremser, First composition measurements of the bulk of the storm time ring current (1 to 300 keV/e), *Geophys. Res. Lett.*, **12**, 325, 1985b.

Krimigis, S. M., G. Gloeckler, R. W. McEntire, T. A. Potemra, F. L. Scarf, and E. G. Shelley, Magnetic storm of September 4, 1984: A synthesis of ring current spectra and energy densities measured with AMPTE/CCE, *Geophys. Res. Lett.*, **12**, 329, 1985.

Lundin, R., L. R. Lyons, and N. Pissarenko, Observation of the ring current composition at $L<4$, *Geophys. Res. Lett.*, **7**, 425, 1980.

Schulz, M., Principles of magnetospheric ion composition, in *Energetic Ion Composition in the Earth's Magnetosphere*, edited by R. G. Johnson, pp. 1–21, Terra Scientific Publishing Company, Tokyo, 1983.

Shelley, E. G., R. G. Johnson, and R. D. Sharp, Satellite observations of energetic heavy ions during a geomagnetic storm, *J. Geophys. Res.*, **77**, 6104, 1972.

Williams, D. J., Ring current composition and sources, in *Dynamics of the Magnetosphere*, edited by S. I. Akasofu, p. 407, D. Reidel Publishing Company, Dordrecht, 1980.

Williams, D. J. and M. Sugiura, The Charge composition Explorer and the 4–17 September 1984 geomagnetic storm, *Geophys. Res. Lett.*, **12**, 305, 1985.

Solar Wind-Magnetosphere Coupling, edited by Y. Kamide and J. A. Slavin, 707–716.

Solar Wind Control of the Geomagnetic Mass Spectrometer

J. H. WAITE, Jr.,[1] M. LOCKWOOD,[2] T. E. MOORE,[1] M. O. CHANDLER,[3] J. L. HORWITZ,[3] and C. R. CHAPPELL[1]

[1]*Space Science Laboratory, NASA Marshall Space Flight Center, Huntsville, Alabama 35812, U.S.A.*
[2]*Rutherford Appleton Laboratory, Chilton, Didcot, OX11 OQX, England*
[3]*Department of Physics, The University of Alabama in Huntsville, Huntsville, Alabama 35899, U.S.A.*

A localized source of energization of ionospheric H^+, He^+, O^+, and N^+ ions in the polar cusp produces a substantial source of upwelling plasma that is transported into the magnetosphere. The roughly equal heating of the various ion species results in field-aligned flow velocities for the ions which are inversely proportional to the square root of their mass. The resulting velocity filter effect of solar-driven $\boldsymbol{E}\times\boldsymbol{B}$ ion convection provides a mass dispersion of ionospheric plasma in the polar cap that we call the Geomagnetic Mass Spectrometer. The control of the solar wind is twofold: (1) strong convection channels presumably associated with dayside magnetic merging appear to play a role in the energization of the upwelling plasma, and (2) ion transport of the plasma into the polar cap magnetosphere and plasmasheet is controlled by the strength and direction of the solar-wind induced $\boldsymbol{E}\times\boldsymbol{B}$ ion convection at high latitudes. The large strength of this low-energy plasma source ($>10^{25}$ ions s^{-1}) coupled with transport of this plasma into the nightside auroral zone and plasmasheet make this an extremely important source of magnetospheric plasma.

1. Introduction

Evidence for a poleward dispersion in mass, pitch angle, and energy of solar wind particles entering the cusp/magnetopause injection boundary is a well-documented phenomenon (cf., BURCH *et al.*, 1982; BURCH *et al.*, 1986, this volume). The ordered ion dispersion signature is the result of a localized cusp injection point near the magnetopause boundary and the subsequent time of flight spreading of the ions due to the action of high-latitude $\boldsymbol{E}\times\boldsymbol{B}$ ion convection. Ions with smaller field-aligned velocities (i.e., lower energies and/or larger pitch angles) take longer to travel from the point of injection to a given altitude and therefore are spread further poleward or further equatorward depending on the direction of $\boldsymbol{E}\times\boldsymbol{B}$ convection (i.e., the sign of the B_z component of the Interplanetary Magnetic Field (IMF)).

Recent measurements of low-energy ionospheric ions by the Dynamics Explorer/retarding ion mass spectrometer (DE/RIMS) have revealed similar dispersion, in both mass and energy, from the cusp region ionosphere that may, on occasion, extend across the polar cap and into the nightside auroral zone (WAITE *et al.*, 1985; MOORE *et al.*, 1985; LOCKWOOD *et al.*, 1985a). Both mass and energy dispersion of these ions have been observed, with the lighter, warmer particles being found closer to the dayside cusp region and the heavier, colder particles being found farther into the nightside polar cap. Thus the phenomenon has been aptly named the Geomagnetic Mass Spectrometer.

Evidence from the DE/RIMS data set for the Geomagnetic Mass Spectrometer is presented in the first section, followed by presentation of the ionospheric source characteristics and the influence of solar wind conditions on the outflow via the polar cap $\boldsymbol{E}\times\boldsymbol{B}$ convection pattern. In the final section, the implications of this new source of ionospheric ions for the magnetosphere are discussed.

2. Observations

The data presented in this paper were recorded by DE/RIMS between October 19, 1981, and October 18, 1983. The precession of the line of apsides of the orbit of DE 1 permits sampling of the polar cap between 1.11 and 4.65 earth radii geocentric distance. The RIMS instrument observes 0–50 eV ions with full pitch angle coverage for H^+, He^{++}, He^+, O^{++}, O^+, and N^+ ions with a cycle time of 16 s. A full description of the RIMS instrument and the DE 1 orbital characteristics is given by CHAPPELL *et al.* (1981).

2.1. The geomagnetic mass spectrometer

The Geomagnetic Mass Spectrometer has been observed in mass and energy dispersions of H^+, He^+, and O^+ ionospheric outflows at high altitudes over the polar cap. An example showing these dispersions is found in Fig. 1. At the top of the figure are energy-versus-time spectrograms measured by RIMS for the H^+, He^+, and O^+ ion species. The energy scale goes from 0 to 50 eV; the integral energy count rates are color-coded according to the color bar on the right and are proportional to flux with the corresponding approximate conversion ratios of count rates (per sample) to integral flux ($cm^{-2}s^{-1}sr^{-1}$) of 3.1×10^4, 1.9×10^4, and 4.4×10^4 for the H^+, He^+, and O^+ ions, respectively. In the middle of the figure is a plot of the DE 1 and DE 2 orbital segments over the polar cap during the sampling time. Note the poleward displacement of the heavier ions with respect to the lighter ions, particularly O^+ as compared to H^+. Note as well the decreasing characteristic energy of the outflow in each ion species. The colored lines for the different ion species represent a self-consistent mapping of the ions observed by RIMS at high altitudes to their common origin in the topside ionosphere using the ion convection trajectory model of HORWITZ (1984). In the lower portion of the figure, DE 2 data from the magnetometer, ion drift meter, and Langmuir probe (see *Space Sci. Instru.*, special DE issue, Volume 5, 1981, for instrument descriptions) indicate, respectively, a region of field-aligned currents, horizontal ion convection shears, and elevated electron temperatures that are associated with a topside ionospheric outflow region near the cusp. It appears from these data that the most plausible interpretation is that a localized ionospheric source region and species-independent convective flow lead to dispersion in parallel velocity that separates the ion species spatially in the observed energy range.

2.2 Source region characteristics

The precession of the DE 1 apogee to lower latitudes allowed DE 1 to sample the suggested source region in the winter and spring of 1982. The results of these findings are illustrated in Fig. 2 and are reported in detail by LOCKWOOD *et al.* (1985a) and MOORE *et al.* (1986). In the center of Fig. 2 the approximate orbital trajectory of DE 1 through the cusp source region is shown for an example on day 105 of 1982. The RIMS spin phase angle-versus-time spectrograms for the O^+ and H^+ ions are shown in the upper right quadrant of the figure. The data are plotted as a function of the spin phase angle of the detector with respect to the satellite orbital direction on the vertical axis; time is on the horizontal axis. The spin axis of the spacecraft is normal to the spacecraft orbit and allows sampling of the full pitch angle range of the ions twice per spin. The upward and downward field-aligned directions are marked on the plot by the dashed and dotted white traces, respectively. Note the sudden upward excursion of the O^+ and H^+ ion flux at 1120 UT with a dramatic equatorward cutoff at 1127 UT. These events, called "upwelling ion events," show large, upward field-aligned fluxes of all measured ion species (H^+, He^+, O^{++}, O^+, N^+ and sometimes NO^+, O_2^+, and N_2^+) in this region. This upwelling ion signature is consistently seen in the region of the polar cusp or polar cleft at altitudes between 0.4 and 1.2 earth radii as shown by the occurrence probabilities of these events also plotted as a function of invariant latitude and magnetic local time in the upper left quadrant of Fig. 2.

Escape from the ionosphere for heavy ions such as O^+, N^+, and molecular ions requires greater than 10 eV of energy to overcome the gravitational potential barrier. This energy is supplied by transverse heating of the ions which is illustrated in the lower right-hand quadrant of Fig. 2 by means of an O^+ ion velocity distribution function. The ram enhancement of the cold O^+ ions (due to the spacecraft motion) and the transverse heating of the O^+ ions (the $\boldsymbol{B}$ field direction is indicated by the white line through the ion distribution function) are clearly evident in this figure. All observed ion species appear to receive roughly equal amounts of heating ($\sim$10 eV).

The source of this heating is not fully understood at this time. However, a list of source region characteristics suggests that there are several potential sources of ion heating active in the region, such as ion-neutral collisional heating and wave-particle heating. The source region characteristics, as observed by DE 1, are illustrated in the bottom left quadrant of Fig. 2 and are listed below (see also MOORE *et al.*, 1986):

- Outfluxes of H^+ and O^+ ions of $\sim 10^9\ cm^{-2}s^{-1}$ and lesser outflows of He^+, O^{++}, N^+ and sometimes molecualr ions.
- Transverse ion heating of about 10 eV.
- Upward and downward field-aligned currents of $\sim$1 to $2\times10^{-6}\ A/m^2$.
- A horizontal convection jet of $\sim$10 Km s^{-1}.
- Low-frequency, broadband plasma wave emissions from <1 Hz which extend to well above the proton gyrofrequency.
- Upward field-aligned electron beams with 20 eV of energy.
- Elevated electron temperatures $\approx$10,000 K.

These source characteristics suggest extremely interesting, wide ranging and highly persistent phenomena associated with the upwelling ion source region. Since the thermal velocity of the observed heating is of the same order of magnitude as the convection jet speed, and both are much larger than the thermal speed of the surrounding ionosphere, it is likely that the heating is related to viscous dissipation of convection energy.

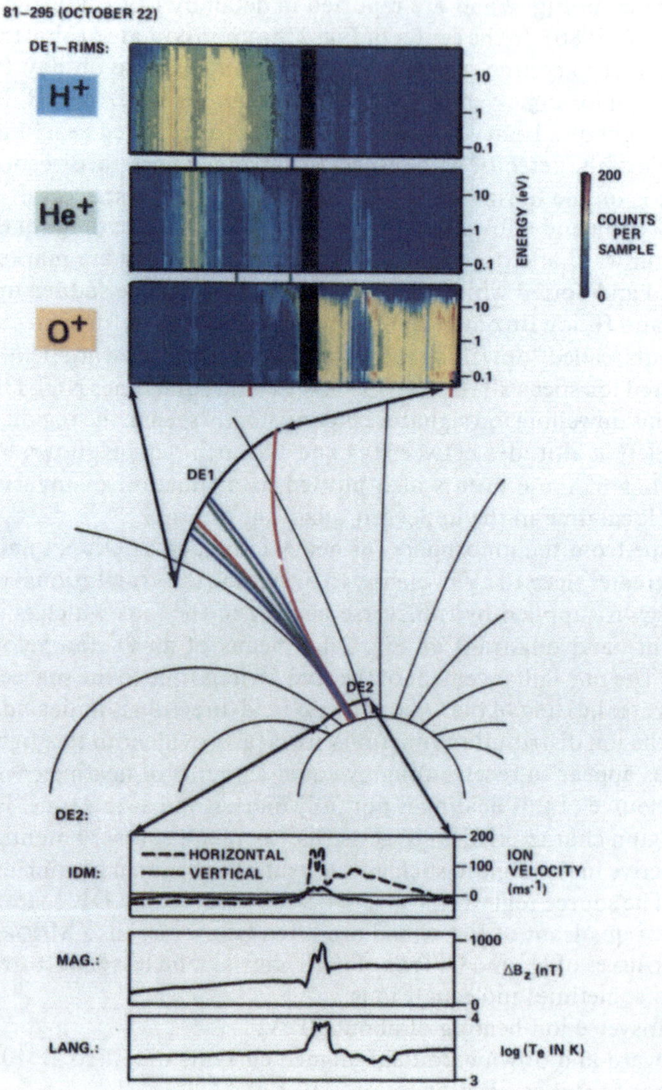

Fig. 1. Example of the Geomagnetic Mass Spectrometer at high latitudes. RIMS data show the poleward displacement of the heavy/colder ions with respect to the light/warmer ions (top), DE 1 and DE 2 orbital segments along with ion convection trajectories mapped using the RIMS high-altitude data down to a cusp source altitude (middle), and characteristics of the cusp source region as sampled by the DE 2 satellite (bottom).

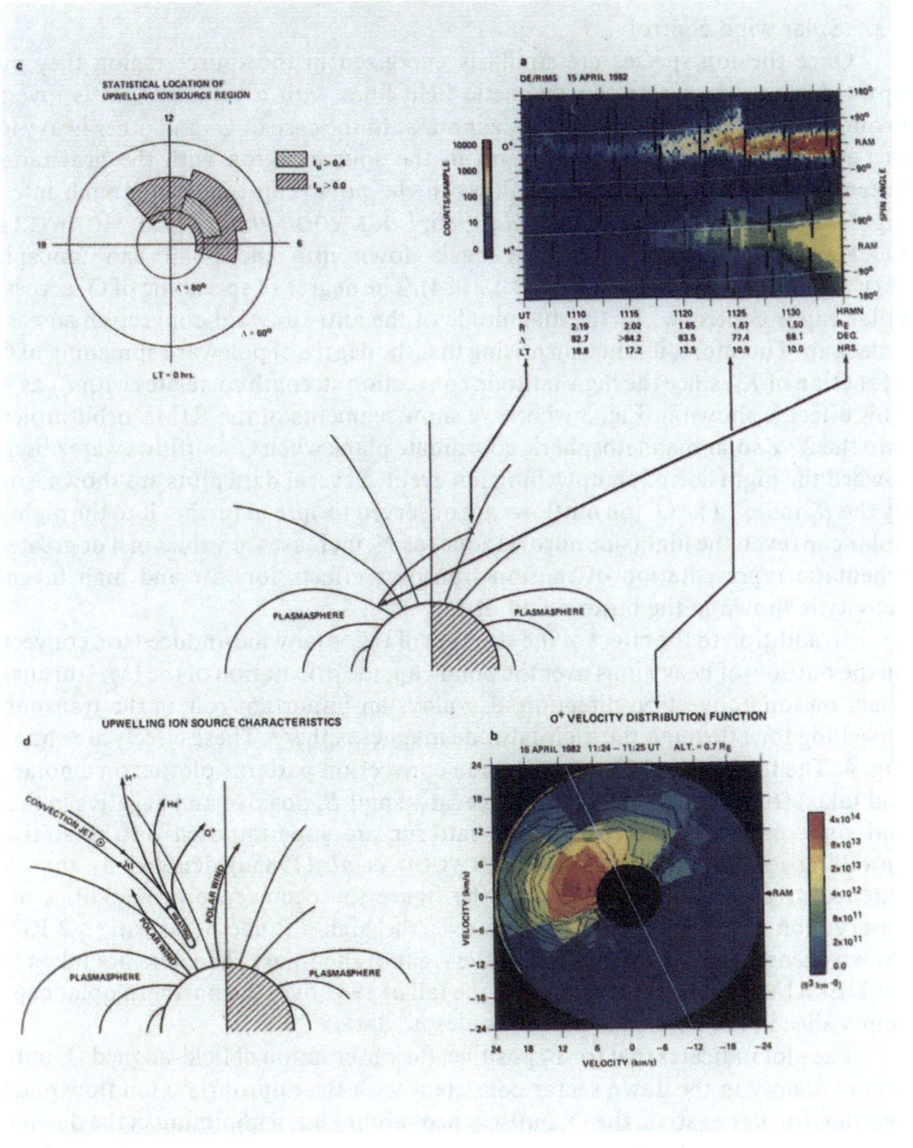

Fig. 2. Multi-panel figure giving information on the location and characteristics of the cusp/cleft upwelling ion source region. The middle of the figure shows the DE 1 orbital segment for the RIMS data samples shown in the top and bottom left-hand panels. a) Spin phase angle-versus-time spectrogram for O^+ and H^+ during an upwelling ion event on April 15, 1982; b) O^+ velocity distribution function for the center of the upwelling ion event on April 15, 1982, between 11:24 and 11:25 UT; c) plot of the occurrence probability of upwelling ion events on an invariant latitude-magnetic local time polar dial for over 2 years of DE/RIMS data; and d) some of the observed source region characteristics of upwelling ion events in the cusp/cleft topside ionosphere.

2.3. Solar wind control

Once the ion species are similarly energized in the source region they move upward along the polar cap magnetic field lines with a velocity that is inversely proportional to the square root of their mass. In the case of O^+ and other heavy ions, the near balance of the energization in the source region with the gravitational potential energy leads to O^+ ion flows in the polar cap that can stretch into the nightside auroral zone (WAITE *et al.*, 1985; LOCKWOOD *et al.*, 1985a; HORWITZ and LOCKWOOD, 1985) or even flow back down into the polar cap ionosphere (LOCKWOOD *et al.*, 1985b; HORWITZ, 1984). The degree of spreading of O^+ across the polar cap is controlled by the magnitude of the anti-sunward convection across the polar cap. Therefore, it is not surprising that the degree of poleward spreading of O^+ is a function of K_p, since the high-latitude convection strength correlates with K_p as well. This effect is shown in Fig. 3 where we show segments of the RIMS orbit projected into the X–Z solar magnetospheric coordinate plane when O^+ outflows were observed toward the nightside of an upwelling ion event. Several data plots are shown, sorted by the K_p index. The O^+ ion outflows are observed to spread further into the nightside polar cap (even the nightside auroral zone) as K_p increases to values of 4 or greater. A schematic representation of the ion transport effects for low and high magnetic activity is shown in the bottom half of Fig. 3.

In addition to the effect of the strength of the solar wind-induced ion convection on the outflow of heavy ions over the polar cap, the orientation of the IMF through its effect on ion convection direction also plays an important role in the transport of upwelling ions through the high-latitude magnetosphere. These effects are shown in Fig. 4. The top panel of Fig. 4 shows ion convection patterns plotted on a polar dial and taken from HEELIS (1984) for B_z negative and B_y positive and negative in the left and right panels, respectively. The patterns are superimposed on the statistical upwelling ion source region of LOCKWOOD *et al.* (1985a) denoted by the cross-hatched areas. In the lower half of the figure the occurrence probabilities of the observation of field-aligned O^+ flows over the high-altitude polar cap (>2 R_E) are shown according to the grey scale at the right of the figure. The data are taken from the DE/RIMS measurements during the fall of 1981 over the northern polar cap. An empty slice in the polar dial plot indicates no data.

The plot indicates that for B_y positive, the observation of field-aligned O^+ outflow occurs mainly in the dawn sector consistent with the appropriate ion flow patterns and that for B_y negative, the O^+ outflow probability has a minimum in the dawn side, also consistent with the corresponding ion convection patterns. Note that the asymmetry of the source region location appears to be reflected in the asymmetry of the O^+ outflow occurrence probability plots for the B_y positive and negative cases shown in the bottom panel of Fig. 4. For northward IMF (B_z>0) O^+ outflow occurrence is generally lower, but again reflects IMF dawn-dusk asymmetries, dependent on the sense of B_y.

3. Discussion and Conclusions

There are two essential elements in a magnetic sector mass spectrometer: a source

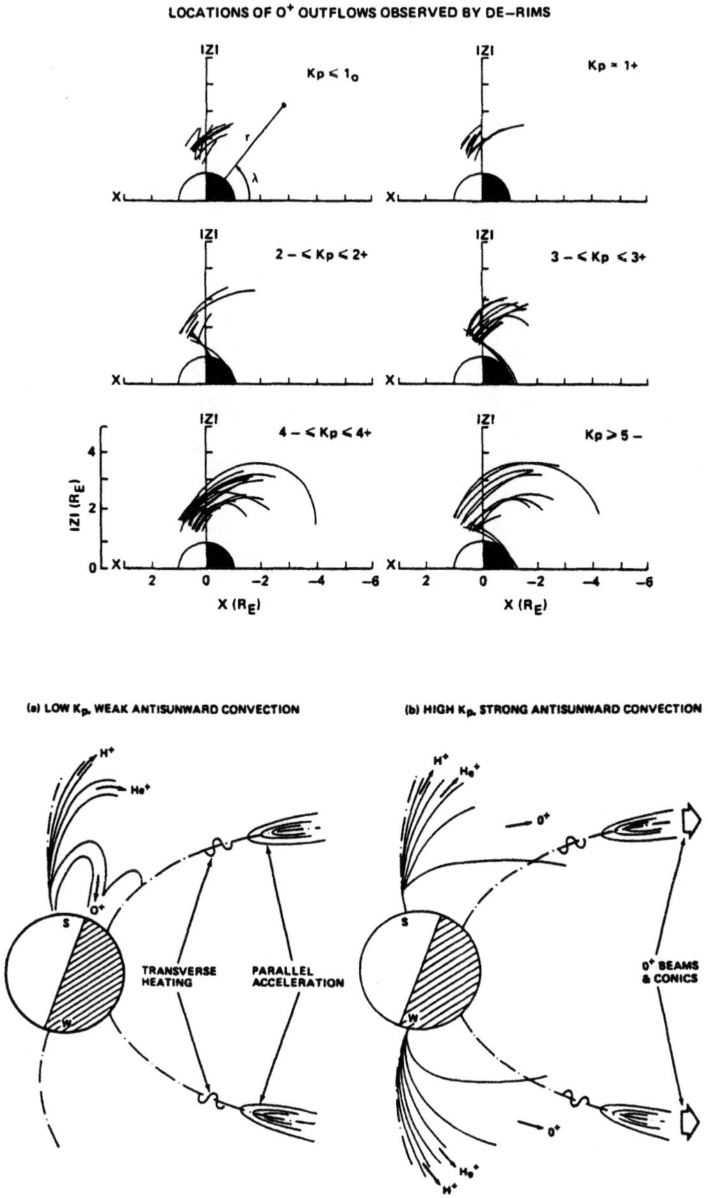

Fig. 3. Effects of magnetic activity on the transport of low-energy ions in the polar cap magnetosphere. Segments of the DE/RIMS orbit during which O^+ outflow was detected in the polar cap magnetosphere as a function of K_p (top six panels). Each orbital segment is projected into a SM $X-Z$ plane. Schematic ion trajectories for both low and high magnetic activity conditions are depicted in lower portion.

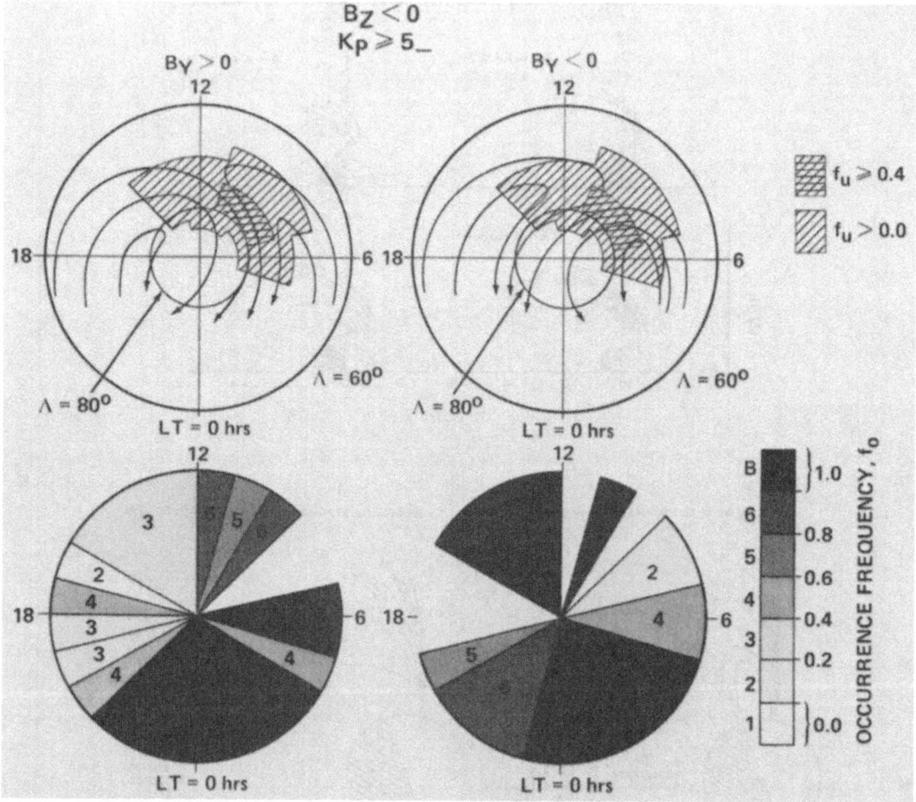

Fig. 4. Effects of changes in IMF B_y on ion transport in the polar cap magnetosphere; statistical upwelling ion source region on a polar dial with the appropriate convection patterns (HEELIS, 1984) for B_z southward and B_y positive and negative superimposed on the polar dial (top portion); occurrence probability for seeing O^+ ion outflow in the DE/RIMS data as a function of magnetic local time is shown in the polar cap for RIMS data during the time period October 19, 1981, to January 10, 1982 (bottom portion).

slit to collimate the incoming ions, and magnetic and electric fields that provide the dispersive separation of the source beam into its respective mass components during the ions' transit through the mass spectrometer. In the case of the Geomagnetic Mass Spectrometer, the important source slit is provided by a localized ion heating region in the polar cusp/cleft region where H^+, He^+, O^+, and N^+ ions and possibly molecular ions (CRAVEN *et al.*, 1985) are transversely heated. In this ion source region the various ion species receive roughly equal amounts of energy and therefore reach field-aligned flow velocities that are inversely proportional to the square root of their mass. The differences in parallel velocity, and hence time of flight, result in an antisunward mass and energy dispersion as a result of antisunward species-independent ion convection at high latitudes. The variation in the ion convection strength can therefore determine the degree of ion spreading across the polar cap with higher field strengths resulting in increased spreading of the ions across the polar cap. In addition, the source energization strength determines the magnitude of the field-aligned flow velocities and thus the proportion of ions that receive enough energy to escape into the magnetosphere. The energization strength may be dependent on the strength of convection jets associated with magnetic merging occurring near the cusp region. This effect is particularly important for the heavy ions with the degree of energization determining what fraction of the O^+ ions can flow upwards into the high-latitude polar magnetosphere (WAITE *et al.*, 1985) as compared to the fraction that falls back down into the polar cap as the result of gravity (HORWITZ and LOCKWOOD, 1985; LOCKWOOD *et al.*, 1985b). O^+ ions may also be reflected upward by the magnetic mirror force and execute "hopping" trajectories (see the bottom half of Fig. 3). The coupled effects of source energization strength and $\boldsymbol{E}\times\boldsymbol{B}$ convection strength, therefore, determine the ultimate destination of transported ions of the various ion species (i.e., magnetotail, nightside auroral zone, plasmasheet, etc.). Furthermore, the orientation of the IMF through its effect on the ion convection pattern controls the direction of ion transport with both B_z and B_y effects being clearly evident in the data. The combined influence on energization strength and ion transport due to solar wind processes demonstrates a high degree of solar wind control of the Geomagnetic Mass Spectrometer process which is an important source of heavy ions ($>10^{25}$ O^+ ions per second) for the nightside plasmasheet.

REFERENCES

BURCH, J. L., P. H. REIFF, R. A. HEELIS, J. D. WINNINGHAM, W. B. HANSON, C. GURGIOLO, J. D. MENIETTI, R. A. HOFFMAN, and J. N. BARFIELD, Plasma injection and transport in the mid-altitude polar cusp, *Geophys. Res. Lett.*, **9**, 921, 1982.

BURCH, J. L., J. D. MENIETTI, and J. N. BARFIELD, DE-1 Observations of solar wind-magnetosphere coupling processes in the polar cusp, this volume, 1986.

CHAPPELL, C. R., S. A. FIELDS, C. R. BAUGHER, J. H. HOFFMAN, W. B. HANSON, W. W. WRIGHT, and H. D. HAMMACK, The retarding ion mass spectrometer on Dynamics Explorer-A, *Space Sci. Instru.*, **5**, 477, 1981.

CRAVEN, P. D., R. C. OLSEN, C. R. CHAPPELL, and L. KAKANI, Observations of molecular ions in the Earth's magnetosphere, *J. Geophys. Res.*, **90**, 7599, 1985.

HEELIS, R. A., The effects of interplanetary magnetic field orientation on dayside high latitude ionospheric convection, *J. Geophys. Res.*, **89**, 2873, 1984.

HORWITZ, J. L., Features of ion trajectories in the polar magnetosphere, *Geophys. Res. Lett.*, **11**, 701, 1984.

HORWITZ, J. L. and M. LOCKWOOD, The cleft ion fountain: A two-dimensional kinetic model, *J. Geophys. Res.*, **90**, 9749, 1985.

LOCKWOOD, M., J. H. WAITE, Jr., T. E. MOORE, J. F. E. JOHNSON, and C. R. CHAPPELL, A new source of suprathermal O^+ ions near the dayside polar cap boundary, *J. Geophys. Res.*, **90**, 4099, 1985a.

LOCKWOOD, M., M. O. CHANDLER, J. L. HORWITZ, J. H. WAITE, Jr., T. E. MOORE, and C. R. CHAPPELL, The cleft ion fountain, *J. Geophys. Res.*, **90**, 9736, 1985b.

MOORE, T. E., C. R. CHAPPELL, M. LOCKWOOD, and J. H. WAITE, Jr., Superthermal ion signatures of auroral acceleration processes, *J. Geophys. Res.*, **90**, 1611, 1985.

MOORE, T. E., M. LOCKWOOD, M. O. CHANDLER, J. H. WAITE, Jr., C. R. CHAPPELL, A. PERSOON, and M. SUGUIRA, Upwelling O^+ ion source characteristics, *J. Geophys. Res.*, in press, 1986.

WAITE, J. H., Jr., T. NAGAI, J. F. E. JOHNSON, C. R. CHAPPELL, J. L. BURCH, T. L. KILLEEN, P. B. HAYS, G. R. CARIGNAN, W. K. PETERSON, and E. G. SHELLEY, Escape of suprathermal O^+ ions in the polar cap, *J. Geophys. Res.*, **90**, 1619, 1985.

Solar Wind-Magnetosphere Coupling, edited by Y. Kamide and J. A. Slavin, 717–730.

Solar Wind-Magnetosphere Coupling and the Distant Magnetotail: ISEE-3 Observations

J. A. SLAVIN[1], E. J. SMITH[1], D. G. SIBECK[2], D. N. BAKER[3], R. D. ZWICKL[3], S.-I. AKASOFU[4], and R. P. LEPPING[5]

[1]*Jet Propulsion Laboratory, California Institute of Technology, Pasadena, California 91109, U.S.A.*
[2]*Department of Atmospheric Sciences, University of California, Los Angeles, California 90024, U.S.A.*
[3]*Los Alamos National Laboratory, University of California, Los Alamos, New Mexico 87545, U.S.A.*
[4]*University of Alaska, Fairbanks, Alaska 99701, U.S.A.*
[5]*NASA Goddard Space Flight Center, Greenbelt, Maryland 20771, U.S.A.*

ISEE-3 Geotail observations are used to investigate the relationship between the interplanetary magnetic field, substorm activity, and the distant magnetotail. Magnetic field and plasma observations are used to present evidence for the existence of a quasi-permanent, curved reconnection neutral line in the distant tail. The distance to the neutral line varies from $|X|=120$–$140\ R_e$ near the center of the tail to beyond $|X|=200\ R_e$ at the flanks. Downstream of the neutral line the plasma sheet magnetic field is shown to be negative and directly proportional to negative B_z in the solar wind as observed by IMP-8. V_x in the distant plasma sheet is also found to be proportional to IMF B_z with southward IMF producing the highest anti-solar flow velocities. A global dayside reconnection efficiency of $20\pm5\%$ is derived from the ISEE-3/IMP-8 magnetic field comparisons. Substorm activity, as measured by the AL index, produces enhanced negative B_z and tailward V_x in the distant plasma sheet in agreement with the basic predictions of the reconnection-based models of substorms. The rate of magnetic flux transfer out of the tail as a function of AL is found to be consistent with previous near-earth studies. Similarly, the mass and energy fluxes carried by plasma sheet flow down the tail are consistent with theoretical mass and energy budgets for an open magnetosphere. In summary, the ISEE-3 Geotail observations appear to provide good support for reconnection models of solar wind-magnetosphere coupling and substorm energy rates.

1. Introduction

Most theories of solar wind-magnetosphere coupling are based upon the early works of DUNGEY (1961) and AXFORD and HINES (1961). In both cases the solar wind exerts normal and tangential stresses on the geomagnetic field which compress most of the field lines to form the dayside magnetosphere, but with some fraction being pulled downstream to produce the magnetotail. However, while Axford and Hines suggested that the tangential stress is due to quasi-viscous forces operating at the magnetopause, Dungey proposed that plasma processes exist which allow interplanetary and geomagnetic field lines to "merge" at the dayside magnetopause. Merging produces "open" field lines which have one foot rooted in the planet and the other in the solar wind. The existence of open field lines enables the solar wind to exert a drag on the magnetosphere via simple $\boldsymbol{J}\times\boldsymbol{B}$ forces. The open field lines are then pulled back into the tail where they eventually "reconnect" to produce closed geomagnetic field lines which convect back toward the dayside magnetosphere and interplanetary field lines which move tailward to rejoin the solar wind.

Two decades of magnetospheric research have produced considerable evidence that both the reconnection and viscous interaction processes are important for solar wind-magnetosphere coupling. However, it appears that reconnection plays the dominant role, particularly during storms and substorms (COWLEY, 1982). Given this picture of the interaction, it has been possible to conduct quantitative studies of the global magnetospheric energy budget (STERN, 1984). While knowledge of how magnetospheric energy is dissipated in the upper atmosphere/ionosphere has reached a fairly mature state (PERREAULT and AKASOFU, 1978), comparatively little is known about the downstream boundary condition on this system, the distant magnetotail. In this study observations from the recent ISEE-3 Geotail Mission are analyzed for the purposes of (1) testing the predictions of the reconnection models of solar wind-magnetosphere coupling, and (2) investigating the rates of mass and energy loss down the distant plasma sheet as a function of substorm activity. Overall, the results are generally consistent with the reconnection-based models of the magnetosphere and they indicate a close coupling between the solar wind, the near-earth magnetosphere, and the distant magnetotail.

2. Quasi-Permanent Distant Neutral Line

After four years at the forward sun-earth Lagrange point, ISEE-3 was moved into the magnetotail where it spent part of 1982 and most of 1983 (BAME *et al.*, 1983; SLAVIN *et al.*, 1983; SCHOLER *et al.*, 1983). Although the analysis of observations from this mission is still at an early stage, many studies of distant magnetotail properties and phenomena have already been published (e.g., COWLEY *et al.*, 1984; HONES *et al.*, 1984; SCHOLER *et al.*, 1984; BAKER *et al.*, 1984; SLAVIN *et al.*, 1984; TSURUTANI *et al.*, 1984; SIBECK *et al.*, 1984; FELDMAN *et al.*, 1984; SMITH *et al.*, 1984; DALY *et al.*, 1984). Briefly stated, ISEE-3 made two deep tail orbits with apogees in the 220–240 R_e range and a number of lower apogee orbits. For the purpose of this study, the 25 December 1982–20 April 1983 interval covering approximately the first half of the mission will be used. Detailed studies of the evolution of plasma and magnetic field parameters within the various regions of the tail have already been conducted (ZWICKL *et al.*, 1984; SLAVIN *et al.*, 1985) and should be referred to for more detailed information on the instruments, trajectory, and general properties of the tail.

The top panel of Fig. 1 displays a schematic view of the classical open magnetosphere. The two principal regions of the tail are the lobes and the plasma sheet. Magnetic merging at the dayside magnetopause creates open field lines which are pulled back to form the lobes. The low beta ($\beta = nk(T_e + T_i)/(B^2/8\pi)$) lobes are separated by a high beta region termed the plasma sheet. The particles in the plasma sheet are of solar wind and ionospheric origin with $\boldsymbol{E}\times\boldsymbol{B}$ drift toward the $X-Y$ plane concentrating them in the plasma sheet (HILL, 1974; PILIPP and MORFILL, 1978). The addition of particles to the plasma sheet is balanced in two ways: 1) particles entering the plasma sheet earthward of the neutral line are eventually convected back to the

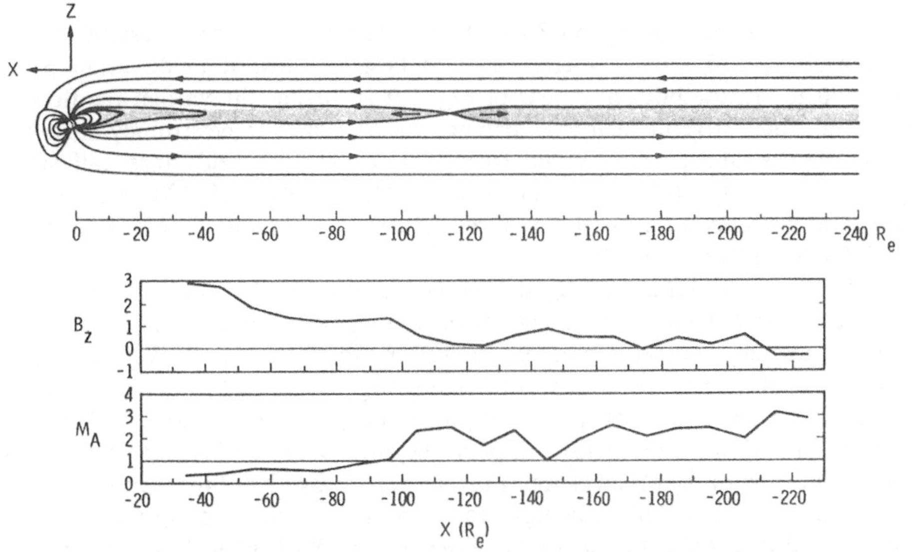

Fig 1. The top panel displays a cut through the center of the magnetosphere parallel to the GSM X– Y plane. The plasma sheet has been shaded and flow directions are indicated by arrows on either side of the distant neutral line. The bottom panels plot mean B_z and Alfvenic Mach number observed by ISEE-3 in the plasma sheet as a function of X. The close correspondence between where the flow speed becomes super-Alfvenic and B_z ceases to be strongly northward mark the average location of the distant neutral line, $|X| = 100$–$140\ R_e$.

forward magnetosphere where they are lost to precipitation or pass into the magnetosheath, and 2) particles entering the plasma sheet tailward of the neutral line are convected down the tail to rejoin the solar wind.

Similarly, the transport of magnetic field lines into the tail is, on average, balanced by reconnection at a neutral line in the nightside magnetosphere. Two lobe field lines reconnect to produce a closed field line which moves back toward the Earth and an interplanetary field line which returns to the solar wind. During substorms additional neutral lines are thought to form in the near-earth magnetotail (RUSSELL and MCPHERRON, 1973), but there is disagreement concerning site and dynamics of the merging process (e.g., FRANK *et al.*, 1976; HONES *et al.*, 1977).

The ISEE-3 Geotail observations have provided the first opportunity to confirm and quantify the nightside half of the open magnetosphere hypothesis with in situ measurements. For the purpose of identifying the average location of the distant neutral line all of the plasma sheet intervals in the ISEE-3 data set were identified as in SLAVIN *et al.* (1985). Two parameters, B_z in GSM coordinates and flow speed normalized to the local Alfven speed, i.e., M_A, have been averaged into ten earth radii bins and plotted against the X coordinate of spacecraft position. The magnetic field observations were made by the JPL VHM magnetometer and the bulk flow parameters were derived from the LANL electron plasma measurements as described by ZWICKL *et al.* (1984). BAME *et al.* (1984), COWLEY *et al.* (1984), SCHOLER *et al.*

(1984), DALY *et al.* (1984), and ZWICKL *et al.* (1984) have already established that plasma sheet bulk flow velocity becomes largely tailward in direction beyond $|X|=100$ R$_e$. In the bottom panels of Fig. 1 it is further shown that the flow also becomes super-Alfvenic near that distance and B_z ceases to be strongly northward. These results are all consistent with lack of closed field lines tailward of approximately $|X|=100$ R$_e$ (COWLEY *et al.*, 1984; ZWICKL *et al.*, 1984; SLAVIN *et al.*, 1985) and the reported observation of reconnection slow mode shocks (FELDMAN *et al.*, 1984; SMITH *et al.*, 1984).

However, the failure of B_z to become negative at the point where the flow becomes super-Alfvenic and tailward does represent a difficulty for the existence of a quasi-permanent neutral line across the full width of the tail at $|X|=100$ R$_e$. The behavior of the magnetic field is further investigated in Fig. 2 where B_z is plotted against the Y coordinate of spacecraft position for three ranges in $|X|$. The results indicate that B_z did indeed turn negative beyond $|X|=100$ R$_e$, but only near the east–west center of the plasma sheet. Furthermore, the width of the region of negative B_z appears to increase with growing distance from the Earth. The top panel shows the corresponding reconnection neutral line shape in the X–Y plane (SLAVIN *et al.*, 1985) where a parabola has been fit to the two observations of neutral line width. The curved nature of the resulting neutral line tends to focus the anti-solar flow tailward of the neutral line to produce the easily observable high speed flow away from the Earth. The flow direction is observed to be tailward across the entire width of the plasma sheet (SLAVIN *et al.*, 1985), and hence, we do not expect that the neutral line stretches across the entire width of the tail. Rather, the closed field lines being pulled downstream at the flanks of the tail must eventually break-off due to reconnection closer to the earth and probably near the central portion of the tail. The existence of a curved neutral line has been suggested previously on the basis of energetic particle observations (RUSSELL, 1977) and is generally attributed to the warping effect of closed field lines dragged down the flanks of the tail by the viscous component of the solar wind interaction with the magnetosphere. Finally, the bottom panel of Fig. 2 displays cuts through the tail parallel to the Y–Z plane at three downstream distances. The plasma sheet starts out on closed field lines, but is gradually consumed by a growing region of interplanetary field lines as the length of the neutral line in the Y direction increases. Eventually, the width of the region of reconnected field lines spreads across more than half of the tail and causes the averages as a function of X in Fig. 1 to go negative at $|X|>200$ R$_e$. In this model the entire region separating the two lobes is composed of interplanetary field lines and magnetospheric debris, such as plasmoids, by $|X|=300$–400 R$_e$. This flow of plasma and magnetic field out the back of the magnetosphere represents mass and energy derived from the solar wind that did not reach the Earth and must be taken into account in the magnetospheric energy budget.

3. Distant Tail Response to IMF B_z

Southward IMF upstream of the Earth results in dayside reconnection which transfers magnetic field lines to the tail as already discussed. Studies of the dayside

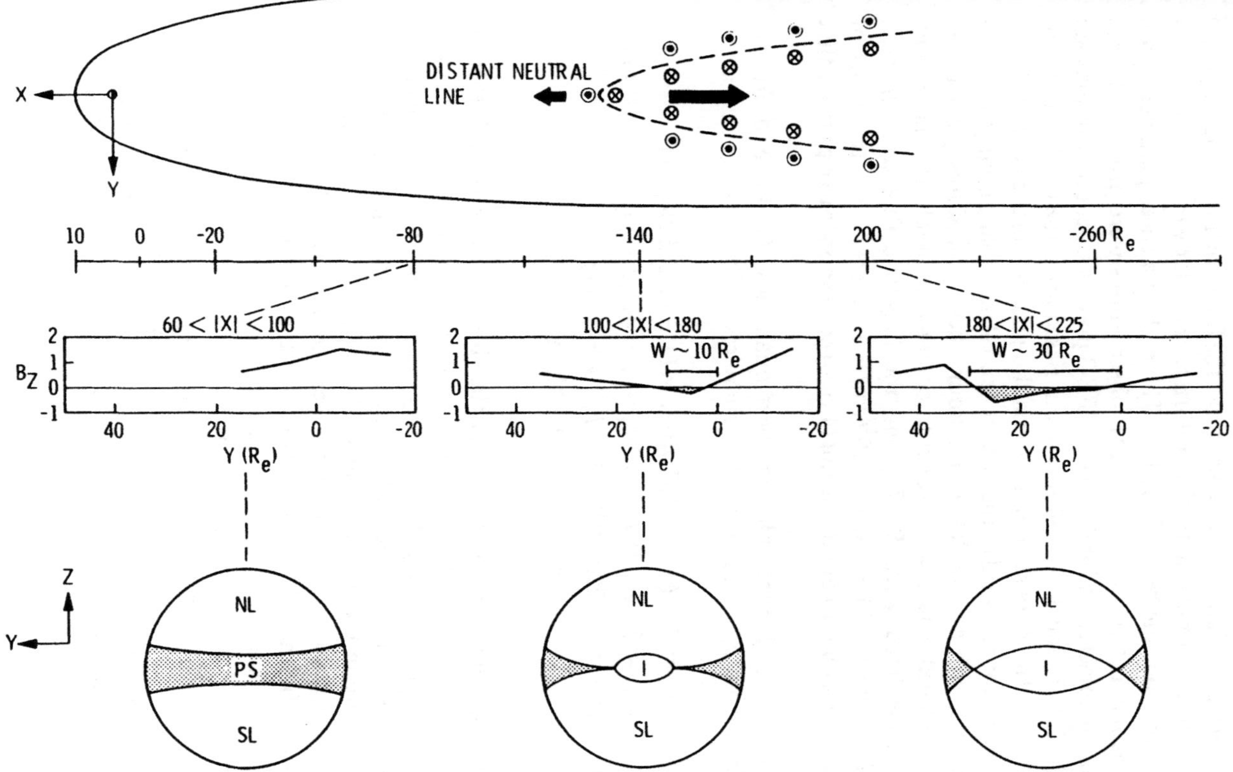

Fig. 2. The top panel displays a cut through the center of the magnetosphere parallel to the GSM X–Y plane. A dashed line indicates the shape of the distant neutral line suggested by the ISEE-3 magnetotail observations with arrows indicating flow direction. In the middle panel mean plasma sheet B_z is plotted as a function of GSM Y for three different ranges in X. Earthward of $|X| = 100\ R_e$ the magnetic field is northward everywhere. However, at greater distances B_z becomes negative in a region near the center of the tail that grows in Y as $|X|$ increases. Finally, the bottom panel displays cuts through the magnetotail parallel to the Y–Z at several downstream distances. The magnetic field topology implied by the curved neutral line model is indicated by stiple for closed field (i.e., plasma sheet) regions, NL (SL) for north (south) lobe open field lines, and I for interplanetary (i.e., disconnected) field lines moving to rejoin the solar wind.

magnetopause and the near tail have measured the rates at which magnetic flux is transferred to the tail and returned to the forward magnetosphere (FAIRFIELD and NESS, 1970; CAAN *et al.*, 1973; HOLZER and SLAVIN, 1978; SLAVIN and HOLZER, 1979). In this study we investigate, for the first time, the final portion of the magnetic flux transfer cycle; the rate of magnetic flux return to the solar wind via the flow of reconnected flux tubes out the rear of the tail. Three plasma sheet parameters are plotted against upstream B_z in Fig. 3. The IMF observations were provided by the GSFC fluxgate magnetometer on IMP-8. The ISEE-3 plasma sheet observations are limited to the $|X|>100$ R_e and $30>Y>0$ R_e region. Based upon the results shown in the previous two figures this data selection criteria should result in the analysis being limited mostly to the region tailward of the distant neutral line. Using hourly averaged IMP-8 measurements with a time delay of 1 hour, these procedures yielded 392 ISEE-3 5 minute averaged values with which to conduct the analysis. The plasma sheet parameters considered were B_z, V_x, and the magnetic flux transfer rate, $\Phi_T(\text{Mx/s})=B_z(\text{nT})\ V_x(\text{km/s})\ 30(R_e)\ 6380\times10^5(\text{cm}/R_e)$. Hence, a net transport of southward magnetic flux ($B_z<0$) out of the tail ($V_x<0$) corresponds to a positive flux transfer rate, Φ_T.

As shown in Fig. 3, all three plasma sheet quantities, B_z, V_x, and Φ_T, are correlated with solar wind B_z. The linear least square regressions in each case are:

$$\begin{aligned} B_z(\text{ISEE-3; nT}) &= 0.17(\pm.02)\ B_z(\text{IMP-8; nT}) - 0.05(\pm.06)\ (\text{nT}) \\ V_x(\text{ISEE-3; km/s}) &= 36(\pm4.8)\ B_z(\text{IMP-8; nT}) - 411(\pm13)\ (\text{km/s}) \\ \Phi_T(\text{ISEE-3; Mx/s}) &= -0.20(\pm.03)\times10^{13} B_z(\text{IMP-8; nT}) \\ &\quad + 0.24(\pm.07)\times10^{13}(\text{Mx/s}) \end{aligned} \tag{1}$$

The correlation coefficients are 0.4 in each of the three cases. Although there is considerable unexplained variance, these correlations are significant at the >99% confidence level (BEVINGTON, 1969). Accordingly, it appears that conditions in the distant plasma sheet exhibit the same strong dependence upon upstream B_z as has been observed elsewhere in the magnetosphere. In particular, if it is assumed that plasma sheet flow speed and solar wind flow were comparable, then an average dayside reconnection efficiency of 17±2% is implied by the regression of ISEE-3 B_z on IMP-8 B_z. However, if instead an average solar wind speed of 430 km/s is assumed, then the bottom panel of Fig. 3 indicates that the overall efficiency of the dayside merging process is 24±4%. Combining these results to get an average efficiency of 20±5% yields excellent agreement with previous estimates based upon studies of dayside cusp motion (BURCH, 1973), magnetopause erosion (HOLZER and SLAVIN, 1978), and magnetospheric dawn-dusk electric field (LEI *et al.*, 1981) and demonstrates strong observational internal consistency for the open model of the magnetosphere.

4. Distant Tail Response to Substorm Activity

Whether the energy is stored (CAAN *et al.*, 1973) or directly supplied (AKASOFU,

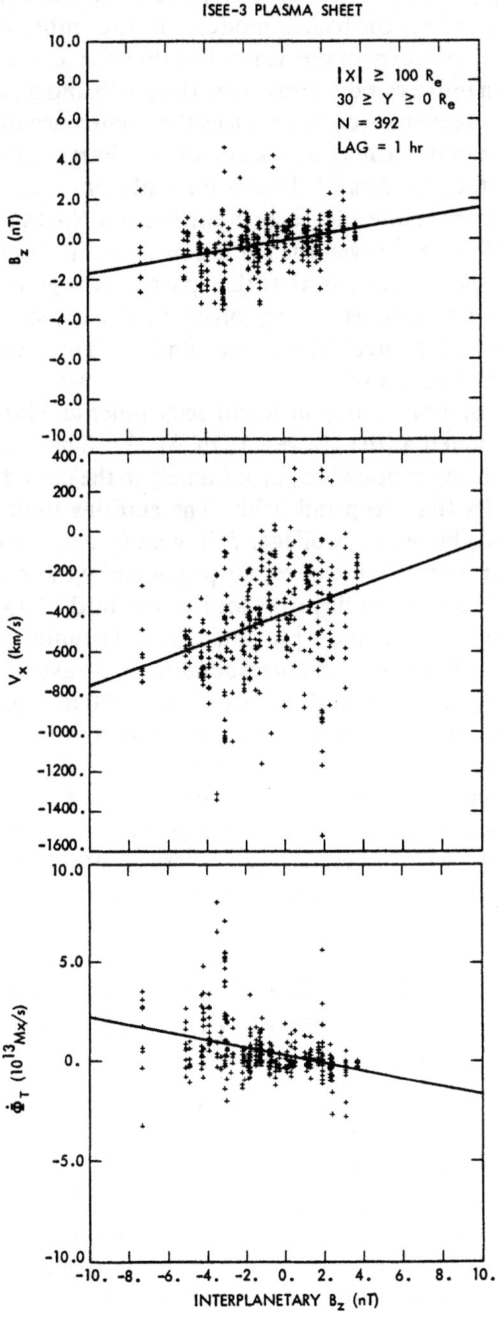

Fig. 3. Five minuted averaged B_z, V_x, and flux transfer rate, Φ_T, observed by ISEE-3 in the plasma sheet at $|X| > 100$ R$_e$ and $30 > Y > 0$ R$_e$ are plotted against 1 hour averaged IMP-8 GSM B_z with a time lag of 1 hour. The linear fits to the data are described in the text.

1983), overall magnetospheric power dissipation increases dramatically during substorms. The reconnection-based models of the substorm process require additional neutral lines to form in the near tail giving rise to magnetic islands termed plasmoids which then move rapidly away from the Earth and down the tail (e.g., BIRN and HONES, 1981). Under these circumstances the occurrence of substorm activity is expected to produce rapid plasma flow away from the Earth and southward magnetic fields in the distant tail. Studies of the cis-lunar plasma sheet following substorm onset have resulted in mixed results (e.g., NISHIDA and NAGAYAMA, 1973; CAAN *et al.*, 1979; LUI *et al.*, 1977). However, the ISEE-3 measurements at $|X|>100$ R_e offer an advantage for this type of study in that the spacecraft is generally tailward of the region(s) where reconnection is taking place as discussed earlier. Under these circumstances motion of the neutral line back and forth over the spacecraft will not interfere with the correlative studies.

For the purpose of monitoring substorm occurrence and intensity, a nine station *AE* index was constructed at the University of Alaska. The interval considered was days 1–55, 1983 which corresponded approximately to the period when ISEE-3 was at $|X|>200$ R_e during its first deep tail orbit. The stations used were Narssarssuaq, Abisko, Dixon Island, Tixie Bay, College, Yellowknife, Fort Churchill, Great Whale River, and Leirvogur. Overall, the coverage provided by these nine stations is quite good and they provide a reasonable proxy for the standard 12 station *AE* which is not yet available for 1983. Based upon previous ISEE-3 studies (BAKER *et al.*, 1984; SLAVIN *et al.*, 1985), a 30 minute lag corresponding to an average speed of 700 km/s was assumed (see the plasma sheet flow speeds for $B_z<0$ in Fig. 3). In Fig. 4 plasma sheet B_z, V_x, and Φ_T measured at ISEE-3 are displayed as a function of the $|AL|$ index. The *AL* index, a measure of westward electrojet intensity, was chosen because of its simpler interpretation relative to *AE* and past successes in relating this index to magnetospheric processes (e.g., SLAVIN and HOLZER, 1979). As with the IMP-8/ISEE-3 correlations in the previous figure, there is considerable scatter present in all three panels. However, clear dependences on *AL* are present with linear regression analyses yielding:

$$\begin{aligned} B_z(\text{ISEE-3 nT}) &= -0.12(\pm.03)\times 10^{-2}\,|AL|\ (\text{nT}) - 0.04(\pm.08)\ (\text{nT}) \\ V_x(\text{ISEE-3; km/s}) &= -0.63(\pm.05)\,|AL|\ (\text{nT}) - 326(\pm16)\ (\text{km/s}) \\ \Phi_T(\text{ISEE-3; Mx/s}) &= 0.13(\pm.03)\times 10^{11}\,|AL|\ (\text{nT}) \\ &\quad + 0.22(\pm.09)\times 10^{13}\ (\text{Mx/s}) \end{aligned} \tag{2}$$

with fair to poor correlation coefficients of 0.2, 0.5, and 0.2, respectively. In each case, increased substorm activity was followed by enhanced anti-sunward flows and negative B_z in the distant tail as predicted by the reconnection models. Furthermore, there is excellent agreement with previous near-earth studies as to the rate of magnetic flux transport into and out of the magnetotail as a function of *AL* (SLAVIN and HOLZER, 1979).

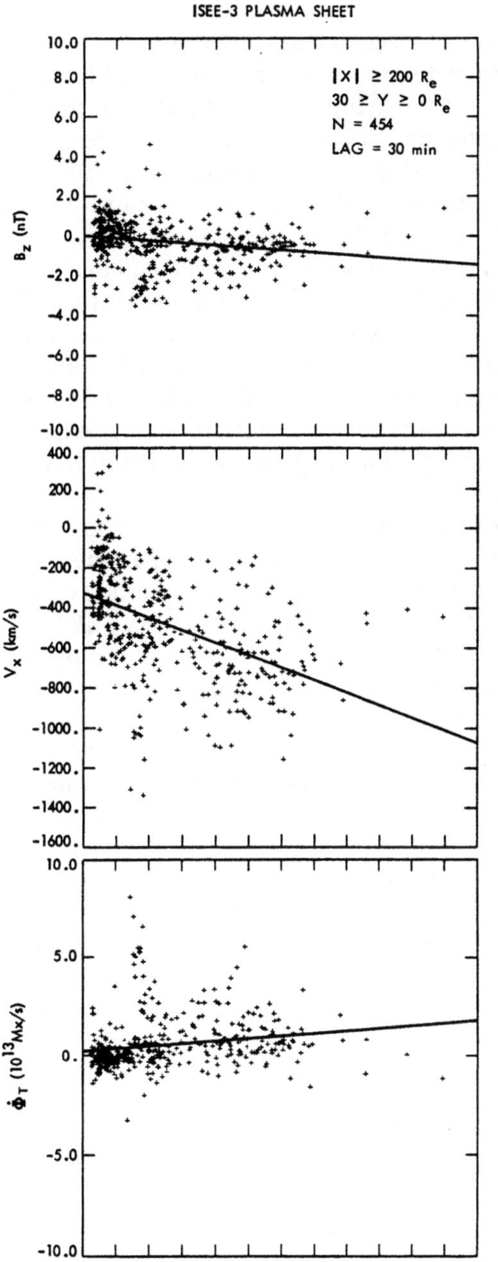

Fig. 4. Five minute averaged B_z, V_x, and flux transfer rate, $\dot{\Phi}_T$, observed by ISEE-3 in the plasma sheet at $|X| \geq 200$ R_e and $30 \geq Y \geq 0$ R_e are plotted against five minute averaged $|AL|$ with a time lag of 30 minutes. The linear fits to the data are described in the text.

5. Mass and Energy Loss Down the Tail

It has long been recognized that not all the mass and energy transferred from the solar wind to the magnetotail is eventually deposited in the ionosphere and upper atmosphere (SISCOE and CUMMINGS, 1969). In addition to earthward flow, nightside reconnection also generates the tailward flow considered in detail by this study. Energy budget calculations for the magnetosphere typically assume that the Poynting flux across the magnetopause earthward of the neutral line eventually reaches the Earth while energy entering downstream of the neutral line simply returns to interplanetary space. If this hypothesis is correct, and the distance to the neutral line in the central regions of the tail is 120–140 R_e as shown in Figs. 1 and 2, then by $|X|=200$ R_e the flux of energy returning to the solar wind may approach equality with the sunward flux earthward of the neutral line which is ultimately dissipated in the ionosphere and upper atmosphere (i.e., both are fed by the Poynting flux through an approximately 100 R_e segment of magnetotail). Similar arguments apply to the entry of solar wind plasma into the magnetosphere (PILIPP and MORFILL, 1978) as discussed earlier.

For the purpose of investigating magnetospheric losses due to plasma sheet flow away from the Earth, two additional plasma sheet parameters have been examined as a function of substorm activity. The first is simply the mass flux, nV. The second is the net energy flux down the tail, $\dot{U}_T = V_x(nm_p V^2 + B^2/8\pi + nkT_e)$. The thermal energy of the ions has been neglected because it is not measured by ISEE-3. As before, five minute averages of the ISEE-3 magnetic field and plasma measurements were used and the analysis was limited to the central portion (i.e., $30 > Y > 0$ R_e) of the distant (i.e., $|X| > 200$ R_e) plasma sheet. In this case, however, AE was used as the substorm index due to its being preferred by many studies considering total energy dissipation rates (PERREAULT and AKASOFU, 1978; WEI *et al.*, 1985).

The results in Fig. 5 show no dependence of mass flux on AE. At these distances the flux of particles away from the Earth is relatively constant near $10^7/\text{cm}^2$-s. Assuming a width of 30 R_e for the region of tailward flow and a plasma sheet thickness of 2 R_e, or slightly less than the 3 R_e observed at $|X|=60$ R_e by RICH *et al.* (1973), the total plasma sheet loss rate is 2.4×10^{26} particles/s. This number is roughly similar to that believed to be lost out of the plasma sheet earthward of the neutral line (HILL, 1974). Hence, the plasma sheet mass flux observations at $|X|=200$ R_e support the models where a neutral line with an average position of 120–140 R_e partitions the magnetosphere with respect to the mass budget.

The correlation between plasma sheet energy flux, $\dot{U}_T$, and AE is somewhat better:

$$\begin{aligned}\dot{U}_T(\text{ISEE-3; ergs/cm}^2\text{-s}) = {} & 0.25(\pm.08)\times10^{-4}AE(\text{nT}) \\ & + 0.24(\pm.02)\times10^{-1}(\text{ergs/cm}^2\text{-s})\end{aligned} \tag{3}$$

but with a correlation coefficient of only 0.2. Making the same assumptions as before regarding the cross section of the flow region, the total energy loss to plasma sheet flow away from the Earth is $6\text{–}12\times10^{17}$ ergs/s or 6– 12×10^{10} Watts. This rate of

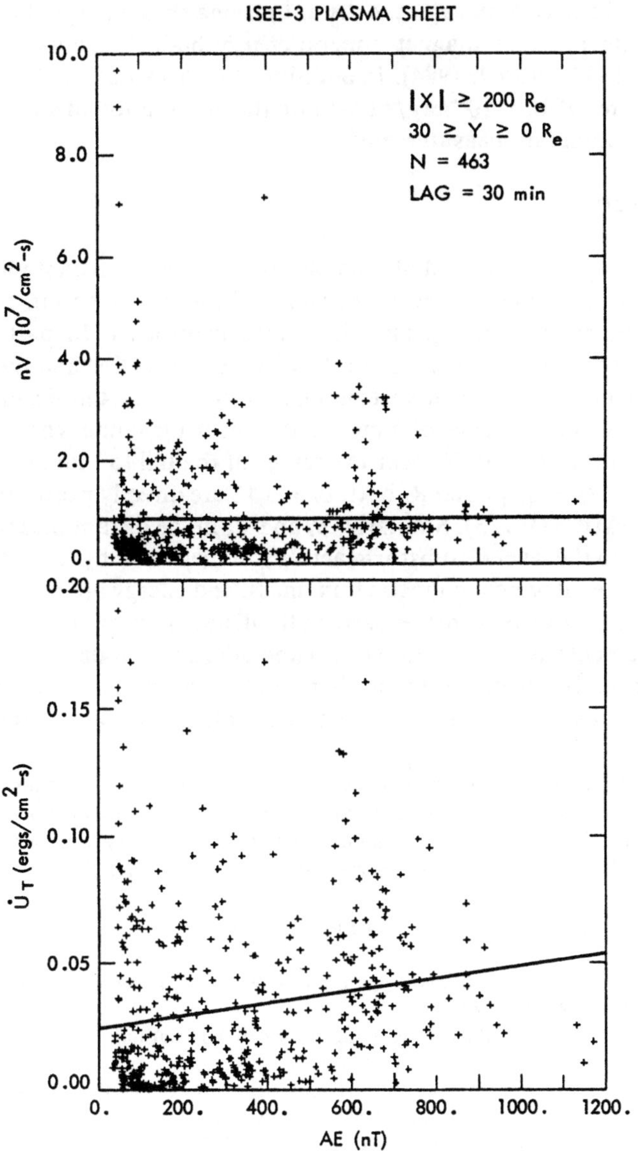

Fig. 5. Five minute averaged anti-solar mass and energy flux observed by ISEE-3 in the plasma sheet at $|X| \geq 200$ R_e and $30 \geq Y \geq 0$ R_e are plotted against five minute averaged AE with a 30 minute time lag. The linear fits to the data are described in the text.

energy loss to the solar wind is very similar to the ionospheric joule heating which takes place during AE levels of 200-1000 nT (i.e., $4-20 \times 10^{10}$ W; WEI *et al.*, 1985) and, therefore, the Poynting flux entering the magnetotail earthward of the distant neutral line. As with the mass flux case, the distant plasma sheet energy fluxes appear to be consistent with previous magnetospheric energy budget calculations (SISCOE and CUMMINGS, 1969; STERN, 1984). In addition, FELDMAN *et al.* (1984) have inferred dissipation rates of 10^{17}–10^{18} erg/s based on the observation of slow mode shocks in the ISEE-3 distant tail measurements.

6. Summary

ISEE-3 magnetic field and plasma observations in the distant plasma sheet have been used to test some of the predictions of the reconnection models of solar wind-magnetosphere coupling and the substorm process. In particular, we have investigated the position and shape of the distant neutral line and the dependence of conditions in the distant plasma sheet upon IMF B_z and the *AE*/*AL* index. The results have supported the existence of a curved distant neutral line whose distance to the Earth varies from 120–140 R_e near the center of the tail to more than 200 R_e at the flanks (Figs. 1–2). At $|X| > 100$ R_e both B_z and V_x are directly proportional to IMF B_z measured at IMP-8 (Fig. 3). Similarly, B_z and V_x in the distant plasma sheet are also proportional to the level of substorm activity as measured by the *AL* index (Fig. 4). Finally, the rates at which magnetic flux, mass, and energy are being returned to the solar wind via flow away from the Earth in the distant plasma sheet are all found to be in good agreement with expectation based upon the open model of the magnetosphere (Fig. 5). Hence, the results of this study provide good support for the reconnection-based models of solar wind-magnetosphere coupling and the substorm process.

The research described in this report was carried out at the Jet Propulsion Laboratory of the California Institute of Technology under contract to the National Aeronautics and Space Administration. The work at Los Alamos National Laboratory was performed under the auspices of the U.S. Department of Energy with support from NASA.

REFERENCES

AKASOFU, S.-I., Solar wind disturbances and the solar wind-magnetosphere energy coupling function, *Space Sci. Rev.*, **34**, 173, 1983.

AXFORD, W. I. and C. O. HINES, A unifying theory of high latitude geophysical phenomena and geomagnetic storms, *Can. J. Phys.*, **3**, 1433, 1961.

BAKER, D. N., S. J. BAME, R. D. BELIAN, W. C. FELDMAN, J. T. GOSLING, P. R. HIGBIE, E. W. HONES, Jr., D. J. MCCOMAS, and R. D. ZWICKL, Correlated dynamical changes in the near Earth and distant magnetotail regions: ISEE-3, *J. Geophys. Res.*, **89**, 3855, 1984.

BAME, S. J., R. C. ANDERSON, J. R. ASBRIDGE, D. N. BAKER, W. C. FELDMAN, J. T. GOSLING, E. W. HONES, Jr., D. J. MCCOMAS, and R. D. ZWICKL, Plasma regimes in the deep geomagnetic tail: ISEE-3, *Geophys. Rev. Lett.*, **10**, 912, 1983.

BIRN, J. and E. W. HONES, Jr., Three dimensional computer modeling of dynamic reconnection in the geomagnetic tail, *J. Geophys. Res.*, **86**, 6802, 1981.

BURCH, J. L., The rate of erosion of dayside magnetic flux based upon a quantitative study of the dependence of polar cusp latitude on the IMF, *Radio Sci.*, **8**, 995, 1973.

CAAN, M. N., R. L. MCPHERRON, and C. T. RUSSELL, Solar wind and substorm related changes in the lobes of the geomagnetic tail, *J. Geophys. Res.*, **78**, 8087, 1973.

CAAN, M. N., D. H. FAIRFIELD, and E. W. HONES, Jr., Magnetic fields in flowing magnetotail plasmas and their significance for magnetic reconnection, *J. Geophys. Res.*, **84**, 1971, 1979.

Cowley, S. W. H., The causes of convection in the Earth's magnetosphere: A review of recent developments during the IMS, *Rev. Geophys. Space Phys.*, **20**, 531, 1982.

Cowley, S. W. H., R. J. Hynds, I. G. Richardson, P. W. Daly, T. R. Sanderson, K.-P. Wenzel, J. A. Slavin, and B. T. Tsurutani, Energetic ion regimes in the deep geomagnetic tail: ISEE-3, *Geophys. Res. Lett.*, **11**, 275, 1984.

Daly, P. W., T. R. Sanderson, and K.-P. Wenzel, Survey of energetic (E>35 keV) ion anisotropies in the deep geomagnetic tail, *J. Geophys. Res.*, **89**, 19733, 1974.

Dungey, J. W., Interplanetary magnetic field and the auroral zones, *Phys. Rev. Lett.*, **6**, 47, 1961.

Fairfield, D. H. and N. F. Ness, Configuration of the geomagnetic tail during substorms, *J. Geophys. Res.*, **75**, 7032, 1970.

Feldman, W. C., D. N. Baker, S. J. Bame, J. Birn, E. W. Hones, Jr., S. J. Schwartz, and R. L. Tokar, Power dissipation at slow mode shocks in the distant geomagnetic tail, *Geophys. Res. Lett.*, **11**, 1058, 1984.

Frank, L. A., K. L. Ackerson, and R. P. Lepping, On hot tenuous plasmas, fireballs, and boundary layers in the Earth's magnetotail, *J. Geophys. Res.*, **81**, 5859, 1976.

Hill, T. W., Origin of the plasma sheet, *Rev. Geophys. Space Phys.*, **12**, 379, 1974.

Holzer, R. E. and J. A. Slavin, Magnetic flux transfer associated with expansions and contractions of the dayside magnetosphere, *J. Geophys. Res.*, **83**, 3831, 1978.

Hones, E. W., Jr., Substorm processes in the magnetotail: Comments on 'On hot tenuous plasmas fireballs, and boundary layers in the Earth's magnetotail' by L. A. Frank, K. L. Ackerson, and R. P. Lepping, *J. Geophys. Res.*, **82**, 5633, 1977.

Hones, E. W., Jr., D. N. Baker, S. J. Bame, W. C. Feldman, J. T. Gosling, D. J. McComas, R. D. Zwickl, J. A. Slavin, E. J. Smith, and B. T. Tsurutani, Structure of the magnetotail at 220 R_e and its response to geomagnetic activity, *J. Geophys. Res.*, **11**, 5, 1984.

Lei, W. R. Gendrin, B. Higel, and J. Berchem, Relationship between the solar wind electric field and the magnetospheric convection electric field, *Geophys. Res. Lett.*, **8**, 1099, 1981.

Lui, A. T. Y., C.-I. Meng, and S.-I. Akasofu, Search for the magnetic neutral line in the near Earth plasma sheet, 3: An extensive study of magnetic field observations at lunar distances, *J. Geophys. Res.*, **82**, 3603, 1977.

Nishida, A. and N. Nagayama, Synoptic survey of the neutral line in the magnetotail during substorm expansion phase, *J. Geophys. Res.*, **78**, 3782, 1973.

Perreault, P. and S.-I. Akasofu, A study of geomagnetic storms, *Geophys. J. R. Astron. Soc.*, **54**, 547, 1978.

Pilipp, W. G. and G. Morfill, The formation of the plasma sheet resulting from plasma mantle dynamics, *J. Geophys. Res.*, **83**, 5671, 1978.

Rich, F. J., D. L. Reasoner, and W. J. Burke, Plasma sheet at lunar distance: Characteristics and interactions with lunar surface, *J. Geophys. Res.*, **78**, 8097, 1973.

Russell, C. T. and R. L. McPherron, The magnetotail and substorms, *Space Sci. Rev.*, **15**, 205, 1973.

Russell, C. T., Some comments on the topology of the geomagnetic tail, *J. Geophys. Res.*, **82**, 1625, 1977.

Scholer, M., G. Gloeckler, D. Hovestadt, F. M. Ipavich, B. Klecker, and C. Y. Fan, Anisotropies and flows of suprathermal particles in the distant magnetotail: ISEE-3 observations, *Geophys. Res. Lett.*, **10**, 1203, 1983.

Scholer, M., G. Gloeckler, D. Hovestadt, B. Klecker, and F. M. Ipavich, Characteristics of plasmoid-like structures in the distant magnetotail, *J. Geophys. Res.*, **89**, 8872, 1984.

Sibeck, D. G., G. L. Siscoe, J. A. Slavin, E. J. Smith, S. J. Bame, and F. L. Scarf, Magnetotail flux ropes, *Geophys. Res. Lett.*, **11**, 1090, 1984.

Siscoe, G. L. and W. D. Cummings, On the cause of geomagnetic bays, *Planet. Space Sci.*, **17**, 1795, 1969.

Slavin, J. A. and R. E. Holzer, Empirical relationships between interplanetary conditions, magnetospheric flux transfer, and the *AL* index, in *Quantitative Modeling of Magnetospheric Processes*, edited by W. P. Olson, pp. 423–436, AGU, Washington, D.C., 1979.

Slavin, J. A., B. T. Tsurutani, E. J. Smith, D. E. Jones, and D. G. Sibeck, Average configuration of the distant (<220 R_e) magnetotail: Initial ISEE-3 magnetic field results, *Geophys. Res. Lett.*, **10**, 973, 1983.

SLAVIN, J. A., E. J. SMITH, B. T. TSURUTANI, D. G. SIBECK, H. G. SINGER, D. N. BAKER, J. T. GOSLING, E. W. HONES, Jr., and F. L. SCARF, Substorm associated traveling compression regions in the distant tail: ISEE-3 Geotail observations, *Geophys. Res. Lett.*, **11**, 657, 1984.

SLAVIN, J. A., E. J. SMITH, D. G. SIBECK, D. N. BAKER, R. D. ZWICKL, and S.-I. AKASOFU, An ISEE-3 study of average and substorm conditions in the distant magnetotail, *J. Geophys. Res.*, **90**, 10875, 1985.

SMITH, E. J., J. A. SLAVIN, B. T. TSURUTANI, W. C. FELDMAN, and S. J. BAME, Slow mode shocks in the Earth's magnetotail: ISEE-3, *Geophys. Res. Lett.*, **11**, 1054, 1984.

STERN, D. P., Energetics of the magnetosphere, *NASA Tech. Memorandum* 86063, 1984.

TSURUTANI, B. T., D. E. JONES, and D. G. SIBECK, The two lobed structure of the distant (X>200 R_e) magnetotail, *Geophys. Res. Lett.*, **11**, 1066, 1984.

WEI, S., B.-H. AHN, and S.-I. AKASOFU, The global joule heat production rate and the *AE* index, *Planet. Space Sci.*, **33**, 279, 1985.

ZWICKL, R. D., D. N. BAKER, S. J. BAME, W. C. FELDMAN, J. T. GOSLING, E. W. HONES, Jr., D. J. MCCOMAS, B. T. TSURUTANI, and J. A. SLAVIN, Evolution of the Earth's distant magnetotail: ISEE-3 electron plasma results, *J. Geophys. Res.*, **89**, 11007, 1984.

Solar Wind-Magnetosphere Coupling, edited by Y. Kamide and J. A. Slavin, 731–738.

Twisting of the Geomagnetic Tail

D. G. SIBECK[1,*], J. A. SLAVIN[2], E. J. SMITH[2], and B. T. TSURUTANI[2]

[1]*Department of Atmospheric Sciences, UCLA, Los Angeles, California 90024, U.S.A.*
[2]*Jet Propulsion Laboratory, California Institute of Technology, Pasadena, California 91109, U.S.A.*

Theory and models indicate that the IMF torques the magnetotail to produce an observable twist. The torque should bring the north lobe south of the GSM xy plane on one flank of the magnetotail and the south lobe north of the plane on the other flank. A comparison of distant ($|x| \geq 180$ R_E) magnetotail observations with the observed and inferred IMF B_y shows that twisting is the single most important factor accounting for observations of the north lobe south of the GSM xy plane and the south lobe north of the plane. The observed magnetotail twist about its aberrated axis at these distances is about 18°. If magnetopause boundary normals are defined so that north lobe normals have northward components and south lobe normals have southward components, then normals to the twisted low-latitude magnetotail magnetopause should point inward towards the other flank of the magnetotail, rather than outward as for an untwisted tail. We find that the twisting model successfully explains 75% of these inward-pointing normals. The required magnetotail magnetic field and circumferential velocity for twisting are small compared to observed lobe field strengths and magnetosheath Alfvén velocities.

1. Introduction

FAIRFIELD (1980) reviews models and observations of the near-earth magnetotail neutral sheet which show that its location above or below the solar magnetospheric (SM) xy plane depends on the tilt of the geomagnetic dipole axis and the position of an observer in the SM xy plane. Although the SM xy plane can serve as a useful approximation to the neutral sheet location, the warped near-earth neutral sheet lies above (below) the SM xy plane near the tail axis in summer (winter). Near the magnetotail flanks, the warped neutral sheet lies closer to, but on the opposite side of, the SM xy plane.

RUSSELL (1972, 1973) first suggested that the magnetotail neutral sheet could twist about the magnetotail axis in response to the interplanetary magnetic field (IMF), particularly when the IMF has a northward component. As discussed by COWLEY (1981, 1983), the asymmetric deposition of merged magnetosphere and magnetosheath field lines torques the magnetotail. When the IMF B_y is positive, the sense of the torque is to raise the south lobe above the duskside ecliptic plane and to lower the north lobe below the dawnside ecliptic plane. The torque, and twist, reverse when the IMF B_y is negative. One might also expect the twist to increase for a stronger IMF, a larger positive B_z, and with distance down the magnetotail. Twisting in numerical magnetotail models conforms to these expectations (BRECHT *et al.*, 1981;

*Now at Johns Hopkins University, Applied Physics Laboratory, Laurel, MD 20707, U.S.A.

Akasofu, personal communication, 1983; OGINO *et al.*, 1984). COWLEY (1981) provides a theroretical model for magnetotail twisting with expressions for the required magnetic and electric field components, currents, and velocities.

Twisting has been more difficult to verify observationally. It may cause the near-earth geotail neutral sheet motion reported by TOICHI and MIYAZAKI (1976) and partially account for the butterfly shape of the magnetotail plasma sheet (COWLEY, 1981). AKASOFU *et al.* (1982) invoked twisting to explain observations of the Saturnian plasma sheet. Jovian neutral sheet boundary normal directions oscillate in the y–z plane (BEHANNON *et al.*, 1981) and could be consistent with twisting. However, COWLEY (1984) has interpreted geotail neutral sheet boundary normals at $x = -1000\ R_E$ reported by VILLANTE (1976) as limiting the geotail twist to less than 20° about the geotail axis. Finally, observations of a B_y magnetic field component in the plasma sheet (TSURUTANI *et al.*, 1984) are consistent with magnetotail twisting (Voigt, unpublished manuscript, 1984).

2. Model

Figure 1 illustrates four tests for magnetotail twisting. The figure shows the magnetotail cross-section viewed from the earth for a positive IMF B_y and (with a reversal in the direction of the y axis) from the distant magnetotail for a negative IMF B_y. Test 1 requires comparing observations of the north lobe below the GSM xy plane and the south lobe above the GSM xy plane with models of a twisted tail based on simultaneous IMF observations. The unwarped neutral sheet of an untwisted tail lies approximately in the GSM xy plane, and its normals point in the z direction. Test 2 requires identifying neutral sheet crossings away from the GSM xy plane and comparing their locations with those of the neutral sheet predicted for a twisted tail. Test 3 requires comparing the computed neutral sheet normal directions with the neutral sheet tilt predicted by twisting models. Computing neutral sheet normals is rather difficult since the fields on either side of the neutral sheet are nearly anti-parallel.

Test 4 requires computing magnetopause boundary normals. If we consider an untwisted magnetotail and force the z component of north lobe normals to be positive and that of south lobe normals to be negative, we find that normals on the duskside of the tail point duskward and normals on the dawnside of the tail point dawnward. Twisted tail normals do not follow this pattern. Figure 1 shows that the normals to the south lobe magnetopause above the duskside ecliptic plane must point southward and consequently also point dawnward. Similary, north lobe normals below the dawnside ecliptic plane point northward and duskward. Thus normals observed pointing in unusual directions should be compared with the normals predicted for a twisted tail.

It is important to remember that variations in the solar wind flow direction as well as neutral sheet warping can shift the neutral sheet and portions of the lobes above or below the GSM xy plane, particularly at large distances from earth. In addition, the neutral sheet may be wavy (LUI *et al.*, 1978), causing its normals to vary. However, since none of these phenomena depend on the IMF B_y, finding phenomena in the sense correct for twisting that depend on the IMF B_y would constitute strong

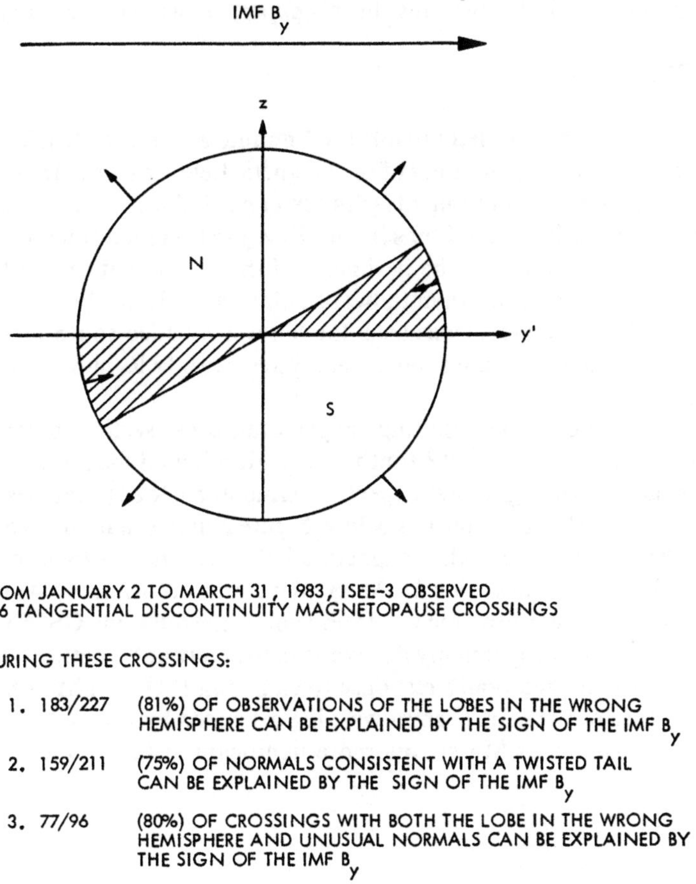

Fig. 1. The twisted magnetotail cross-section viewed from earth for a positive IMF B_y, and (with a reversal in the direction of the y axis) from the distant magnetotail. The north lobe and neutral sheet lie below the dawnside GSM xy plane, while the south lobe and neutral sheet lie above the duskside GSM xy plane. Magnetopause normals to these portions of the twisted tail (shown shaded) point towards the other side of the magnetotail. Normals to the neutral sheet (not shown) do not point in the z direction, but in the $-y$, $+z$ direction. The figure also depicts the twisted magnetotail cross-section viewed from the distant magnetotail when the IMF B_y is negative.

evidence for twisting.

The purpose of this paper is to compare simultaneous geotail and IMF B_y observations to determine if the magnetotail twists in response to the IMF B_y. We will use ISEE-3 geotail observations from the spacecraft's first distant ($|x| \geq 200$ R_E) magnetotail pass from January to March, 1983 across the low-latitude tail from dawn to dusk flanks. These observations will be compared with IMP 8 IMF observations supplemented by inferred IMF B_y polarities (COFFEY, 1983). TSURUTANI *et al.* (1984) and SIBECK *et al.* (1985b) present plots which determine the extent of all ISEE 3 distant magnetotail observations. In this paper, we will consider only tests 1 and 4.

3. Observations

We identify the magnetotail lobes in 5 minute averages of ISEE-3 observations as intervals when the x component of the magnetic field was more than twice as great as the combined y and z components (SIBECK *et al.*, 1985b). All lobe observations were then accumulated in $2 \times 2\ R_E^2$ bins in the GSM y–z (magnetotail cross-section) plane. A heavy solid line outlines the perimeter of these observations. If there were more observations of earthward than tailward pointing fields in a bin, the bin was labelled north lobe. If the converse was true, the bin was labelled south lobe. The bins were plotted in a crudely aberrated coordinate system with 12 R_E subtracted from the GSM y coordinate.

Figure 2 shows the resulting magnetotail cross-section pattern viewed from earth, with regions in which the south lobe predominated shaded and regions of north lobe unshaded. We might have expected to find north lobe polarities above the GSM xy plane and south lobe polarities below. Figure 2 shows that numerous exceptions to this guideline occur near the magnetotail flanks. The north lobe was frequently observed below the dawnside GSM xy plane and occasionally below the duskside GSM xy plane. The south lobe was found above the duskside GSM xy plane for days in a region extending over many R_E. We wish to determine whether these observations result from the magnetotail's response to changing IMF conditions.

The sense of the neutral sheet warping discussed by FAIRFIELD (1980) would be the same in January as March, but more significant in January when ISEE-3 was on

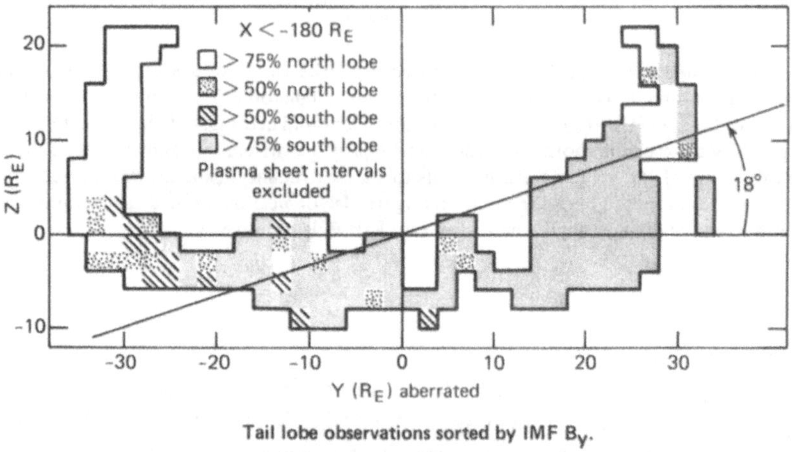

Fig. 2. ISEE-3 observations of the distant magnetotail during the first deep tail pass. There are 16658 five munute observations sorted into 252 bins. The number of samples per bin ranges from 1 to 610, with a mode of 20 and a median of 30–40. Each 2 R_E^2 bin is classified according to the percent of time during which North ($+B_x$) and South ($-B_x$) lobe fields were observed. The observations are plotted in aberrated GSM coordinates.

the dawnside of the aberrated magnetotail. At this time the north pole is tilted in the anti-sunward direction and the near-earth magnetotail neutral sheet is warped so that the north lobe should lie below the GSM *xy* plane near the tail center and the south lobe should lie above the plane at the flanks. There is little evidence for this in the distant magnetotail, since the north lobe is seldom observed at the tail center below the GSM *xy* plane, but is frequently seen below the plane at the flanks.

Another possibility is that solar wind flow was from the north when ISEE-3 was at the magnetotail flanks, and from the south when ISEE-3 observed south lobe above the GSM *xy* plane. This certainly could account for some observations near the GSM *xy* plane, but we doubt whether the solar wind flow could deviate so greatly from the radial direction for days.

Instead, we think the twisting model can explain the magnetotail observations. We sorted ISEE-3 tail observations using the observed and inferred IMF B_y polarity. Since the expected sense of the tail twist reverses when the polarity of the IMF B_y reverses, tail observations made during intervals of positive B_y were plotted where they occurred, but were plotted at a negative aberrated y coordinate during intervals of negative IMF B_y. The magnetotail observations were not plotted for the entire day when the observed or inferred IMF was not known, nor when only inferred IMF polarities were available and these indicated both a toward and away polarity within the same twelve hour interval. Since there is no inferred IMF for the second half of days, we used the inferred IMF polarity of the first half days. The result of this procedure should show the twist appropriate for a positive IMF B_y: north lobe below the dawnside GSM *xy* plane and south lobe above the duskside GSM *xy* plane.

Figure 3 shows that this is indeed the case. The sorting process brings nearly all the north lobe observations below the GSM *xy* plane to the dawnside of the tail, and nearly all observations of south lobe above the ecliptic plane to the duskside of the magnetotail. An average twist of about 18° is sufficient to explain the observations.

Figure 3 also shows that the excursions of the twisted lobes across the ecliptic plane are greatest near the magnetotail flanks. It is likely that the chance of observing north or south lobe just inside the low-latitude magnetopause on the magnetotail flanks depends on the IMF B_y polarity. More than 750 magnetopause crossings could be identified in high-resolution plots of the ISEE 3 plasma and magnetic field measurements during the first distant tail pass (SIBECK *et al.*, 1985a and b). We found 227 observations of the north lobe below and south lobe above the GSM *xy* plane at magnetopause crossings. Of these, 183 (81%) could be explained by the sense of tail twist predicted for the observed IMP 8 or inferred IMF B_y. If we cannot explain 19% of crossings, about 19% out of the 81% we can explain were fortuitous. Then our success rate for explaining crossings is 81 − 19%=62%. The comparison shows that at the magnetotail flanks, twisting is the most important factor responsible for observations of the north lobe below and the south lobe above the GSM *xy* plane.

Our next step is to look for normals on the dusk side of the tail which point dawnward and those on the dawnside of the tail which point duskward. We thought these would be consistent with twisting by Test 4 (See Fig. 1). To determine the boundary normals, we computed the cross-product of the five minute averaged magnetic fields just inside and outside the magnetopause as is appropriate for a

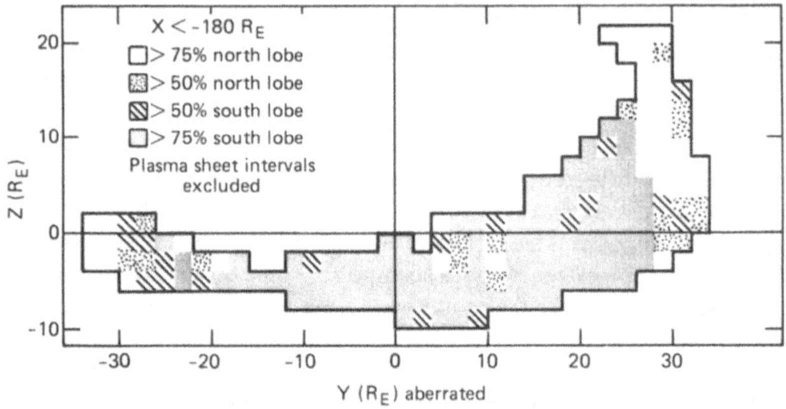

Tail lobe observations

Fig. 3. The magnetotail lobe observations shown in Fig. 2 have been sorted based on the polarity of the observed or inferred IMF B_y. The shading scheme remains the same, but there are now 5484 five minutes averages sorted into 187 bins. The number of samples per bin ranges from 1 to 255, with a median of 20–30 and a mode of 10–20. To produce this figure, observations shown in Fig. 2 were replotted at the same y coordinate when the IMF B_y was positive. They were replotted at a negative y coordinate when the IMF B_y was negative. They were not replotted when the sign of the IMF B_y was unknown. As predicted by the twisting model, north lobe is seen below the dawnside GSM xy plane, and south lobe above the duskside GSM xy plane.

tangential discontinuity magnetopause (SIBECK *et al.*, 1985a and b). We found 211 normals that pointed towards the opposite side of the magnetotail, suggesting twisting. Of these, 159 (75%) can be explained in terms of the twisting model in Fig. 1 based on the inferred and/or observed IMF B_y. We also examined a limited subset of the magnetopause crossings with both the lobe on the wrong side of the ecliptic plane and the unusual boundary normals. Of 96 such crossings, 77 (80%) could be explained by the sign of the observed or inferred IMF B_y. We conclude that twisting is the single greatest factor producing the unusual normals.

4. Theory

COWLEY (1981) provides formulae which determine the circumferential magnetotail velocity and magnetic field required to produce an observed twist of θ degrees. The required velocity is

$$V_\theta = \pi\theta R V_{sw}/180L \tag{1}$$

and the required magnetic field strength is

$$B_\theta = \pi\theta R B_x 180L, \tag{2}$$

where R is the tail radius, L the distance down the tail at which the twist is observed, V_{sw} the solar wind velocity, and B_x the lobe field strength. Taking $\theta=18°$, $R=30$ R_E, $L=200$ R_E, $V_{sw}=400$ km sec^{-1}, and $B_x=10$ nT, we obtain $V_\theta=18$ km sec^{-1} and $B_\theta=0.45$ nT. These values are quite reasonable since the twisting velocity is about one third of the magnetosheath Alfvén and sound speeds, and the magnetic field strength is on the order of that typically observed to perturb the x-oriented tail lobe fields (SLAVIN *et al.*, 1985).

5. Conclusion

In conclusion, we have discussed four tests for magnetotail twisting, of which two have been applied to the first ISEE-3 distant tail pass. We found that most observations of the north lobes below the GSM xy plane and south lobe above the GSM xy plane occurred in a sense predictable from the twisting model and the IMF B_y polarity. This study suggested that twisting should be most easily observed at the dawn and dusk flanks of the magnetotail. A study of magnetopause crossings showed that more than half of the crossings in which north lobe was observed below the GSM xy plane and south lobe above could be explained by the sign of the IMF B_y. Twisting should create regions of the dawnside magnetopause over which normals point duskward, and regions of the duskside magnetopause over which normals point dawnward. More than 75% of such boundary normals observed by ISEE-3 can be explained by twisting determined by the IMF B_y. Twisting does not require unusually large magnetic fields or velocities.

We would like to thank Professor C. T. Russell for helpful comments and suggestions. This work was performed at the Jet Propulsion Laboratory, California Institute of Technology, under contract with NASA. The work carried out at UCLA was supported in part by a grant from NSF (ATM81-20455).

REFERENCES

AKASOFU, S.-I., M. ROEDERER, and S. M. KRIMIGIS, Dawn-dusk asymmetry of the tail region of the magnetosphere of Saturn and the interplanetary magnetic field, *Planet. Space Sci.*, **30** 1061–1063, 1982.

BEHANNON, K. W., L. F. BURLAGA, and N. F. NESS, The Jovian magnetotail and its current sheet, *J. Geophys. Res.*, **86**, 8385–8401, 1981.

BRECHT, S. H., J. LYON, J. A. FEDDER, and K. HAIN, A simulation of east-west IMF effects on the magnetosphere, *Geophys. Res. Lett.*, **8**, 397–400, 1981.

COFFEY, H. E., Geomagnetic and solar data, *J. Geophys. Res.*, **88**, 4145, 5785–5786, 1983.

COWLEY, S. W. H., Magnetospheric asymmetries associated with the y-component of the IMF, *Planet. Space Sci.*, **29**, 79–96, 1981.

COWLEY, S. W. H., Interpretation of observed relations between solar wind characteristics and effects at ionospheric altitudes, in *High-latitude Space Plasma Physics*, ed. by B. Hultqvist and T. Hagfors, pp. 225–249, Plenum Press, New York, 1983.

COWLEY, S. W. H., The distant geomagnetic tail in theory and observations, in *Magnetic Reconnection in Space and Laboratory Plasmas*, edited by E. W. Hones, Jr., pp. 228–239, AGU, Washington D.C., 1984.

FAIRFIELD, D. H., A statistical determination of the shape and position of the geomagnetic neutral sheet, *J. Geophys. Res.*, **85**, 775–780, 1980.

LUI, A. T. Y., C.-I. MENG, S.-I. AKASOFU, Wavy nature of the magnetotail neutral sheet, *Geophys. Res. Lett.*, **5** 279–282, 1978.

OGINO, T., R. J. WALKER, and M. ASHOUR-ABDALLA, A magnetohydrodynamic simulation of the effects of the interplanetary magnetic field (IMF) B_y component on the interaction of the solar wind with the earth's magnetosphere during intervals with southward IMF, *Trans. Am. Geophys. Un.*, **65**, 1063, 1984.

RUSSELL, C. T., The configuration of the magnetosphere, in *Critical Problems of Magnetospheric Physics*, edited by E. R. Dyer, pp. 1–16, IUCSTP Secretariat, Nat. Acad. of Sci., Washington, D.C., 1972.

RUSSELL, C. T., Comments on the paper 'The internal structure of the geomagnetic neutral sheet' by K. Schindler and N. F. Ness, *J. Geophys. Res.*, **78** 7576–7579, 1973.

SIBECK, D. G., G. L. SISCOE, J. A. SLAVIN, E. J. SMITH, B. T. TSURUTANI, and R. P. LEPPING, The distant magnetotail's response to a strong interplanetary magnetic field B_y: Twisting, flattening and field line bending, *J. Geophys. Res.*, **90**, 4011–4019, 1985a.

SIBECK, D. G., G. L. SISCOE, J. A. SLAVIN, and R. P. LEPPING, Major flattening of the distant geotail, submitted to *J. Geophys. Res.*, 1985b.

SLAVIN, J. A., E. J. SMITH, D. G. SIBECK, D. N. BAKER, R. D. ZWICKL, and S.-I. AKASOFU, An ISEE-3 study of average and substorm conditions in the distant magnetotail, in press *J. Geophys. Res.*, 1985.

TOICHI, T. and T. MIYAZAKI, Flapping motions of the tail plasma sheet induced by the interplanetary magnetic field variations, *Planet., Space Sci.*, **24**, 147–159, 1976.

TSURUTANI, B. T., J. A. SLAVIN, E. J. SMITH, R. OKIDA, and D. E. JONES, Magnetic structure of the distant geotail from −60 to −220 R_E: ISEE-3, *Geophys. Res. Lett.*, **11**, 1–4, 1984.

VILLANTE, U., Neutral sheet observations at 1000 R_E, *J. Geophys. Res.*, **81**, 212–215, 1976.

Solar Wind-Magnetosphere Coupling, edited by Y. Kamide and J. A. Slavin, 739–745.

Observations of a Large-Scale Vortex-Like Structure in the Deep-Tail Plasma Sheet Boundary Layer

T. R. SANDERSON,[1] P. DALY,[1] K.-P. WENZEL,[1] E. W. HONES JR.,[2] and E. J. SMITH[3]

[1]*Space Science Dept. of ESA, ESTEC, Noordwijk, The Netherlands*
[2]*Los Alamos National Laboratory, Los Alamos, New Mexico 87545, U.S.A.*
[3]*Jet Propulsion Laboratory, California Institute of Technology, Pasadena, California 91109, U.S.A.*

We present ISEE-3 observations of a large-scale vortex-like structure in the deep tail of the magnetosphere at $X_{GSM}=-217$ R_e. The structure is characterised by two clockwise rotations of the energetic ion anisotropy vector. Variations in the magnetic field vector approximately 180° out-of-phase with the ion variations are observed. We believe this structure is most likely the signature within the magnetosphere of a surface wave at the magnetopause driven by a Kelvin-Helmholtz instability. Conditions inside and outside of the magnetosphere, as observed by ISEE-3 and −2 respectively, are examined, these conditions suggesting that the surface wave is most likely propagating in the slow mode.

1. Introduction

In this paper we present ISEE-3 observations of a large-scale vortex-like structure in the deep tail of the magnetosphere at $X_{GSM}=-217$ R_e, $Y_{GSM}=4.3$ R_e and $Z_{GSM}=-7.8$ R_e. We believe that this vortex-like structure is the signature within the tail of a surface wave on the magnetopause, driven by the shear due to the different flow conditions on either side of the magnetopause. Using measurements made outside of the magnetosphere but within the magnetosheath by ISEE-2, we make some simple assumptions to derive the conditions outside of the magnetosphere and adjacent to the position of ISEE-3, so that we are able to examine the conditions necessary for the instability to occur.

2. Instrumentation

The ion measurements presented here were obtained with the low energy proton instrument on ISEE-3, which consists of three identical telescopes oriented at 30°, 60° and 135° with respect to the spin axis. The instrument measures three-dimensional anisotropies in 8 logarithmically spaced channels from 35 to 1600 keV with a time resolution of 16 s. Each energy channel for each telescope records counts in 8 azimuthal sectors (except for the highest energy channel which has only 4 sectors). The geometrical factor for each telescope is 0.05 cm^2 sr. A concise description of this instrument and the dat analysis methods used can be found in SANDERSON *et al.* (1985). The plasma measurements used here were obtained with the Los Alamos plasma instrument on the same spacecraft. This instrument measures electrons from ~10 eV to ~1000 eV in 15 energy levels, interspersing sequences of two-dimensional measurements, each measurement made every 12s, with three-dimensional measurements, each measurement accomplished in 72s. A description of this instrument can be found in BAME *et al.* (1978). The magnetic field measurements were obtained with the vector helium magnetometer on the same spacecraft. The magnetometer measures the *X*, *Y* and *Z* components 6 times per second. The data used here, however, have been integrated over a time period corresponding to one spin of the spacecraft (3s) so as to conform with the spin-synchronous energetic ion sampling time. A further description can be found in FRANDSEN *et al.* (1978).

3. Observations

On January 29, 1983, whilst ISEE-3 was at $X=-217$ R_e during the first excursion into the distant geotail, the signature of a large scale vortex-like structure was observed in the energetic ions, plasma and magnetic field. Preliminary observations of this event have been reported by SANDERSON *et al.* (1984). Figure 1 shows combined energetic ion and magnetic field data for the 1-hour period commencing at 1145 UT on January 29, 1983. The top panel of the figure shows the spin averaged 35-56 keV ion intensity. The next three panels show the first order ion anisotropy amplitude A_1/A_0, and the direction (co-latitude θ_1 and azimuth ϕ_1), derived from a spherical harmonic fit to the data (SANDERSON *et al.*, 1985). The directions presented here represent flow directions. The remaining three panels show the magnetic field magnitude *B*, and direction θ and ϕ measured in the GSE coordinate system. This figure shows the conditions prevailing when ISEE-3 passed from the plasma sheet into the energetic ion boundry layer and then into the southern lobe. The spacecraft was in the energetic ion boundary layer from 1152 UT until 1229 UT. (This layer is identified by an energetic ion signature similar to that of the plasma sheet, and a magnetic field signature similar to that of the lobe, COWLEY *et al.*, 1984). During this period we observed the vortex-like structure, recognisable by the slow, almost monotonic decrease in the value of the azimuth angle of the ion anisotropy vector, ϕ_1, over a period of more than 30 minutes, representing a slow rotation of the vector through approximately 2 full revolutions in a clockwise direction. This rotation was only observed whilst the spacecraft was in the boundary layer. During the same period that we observed the vortex-like structure in the energetic ions, we observed variations in the magnetic field. To aid comparison between the magnetic field and the ion data, we have resolved the first-order ion anisotropy into three orthogonal components, A_x/A_0, A_y/A_0 and A_z/A_0, measured along each of the axes of the GSE coordinate system, and show these in Fig. 2 together with the 3 components of the magnetic field, B_x, B_y and B_z. The variations in the *x* and *y* components of the anisotropy are well correlated with the variations in the *x* and *y* components of the magnetic field, but are approximately 180° out of phase with the field. (The ion anisotropy components have been inverted to show this relationship.)

To compare the anisotropy measurements with the magnetic field measurements, we performed a covariance analysis on both the 1st order anisotropy and the magnetic field vectors. The results are shown in Table 1. The numerical values reveal a remarkable similarity between the variations in the anisotropy vector and the magnetic field vector, the only significant difference being the ratio of minimum to

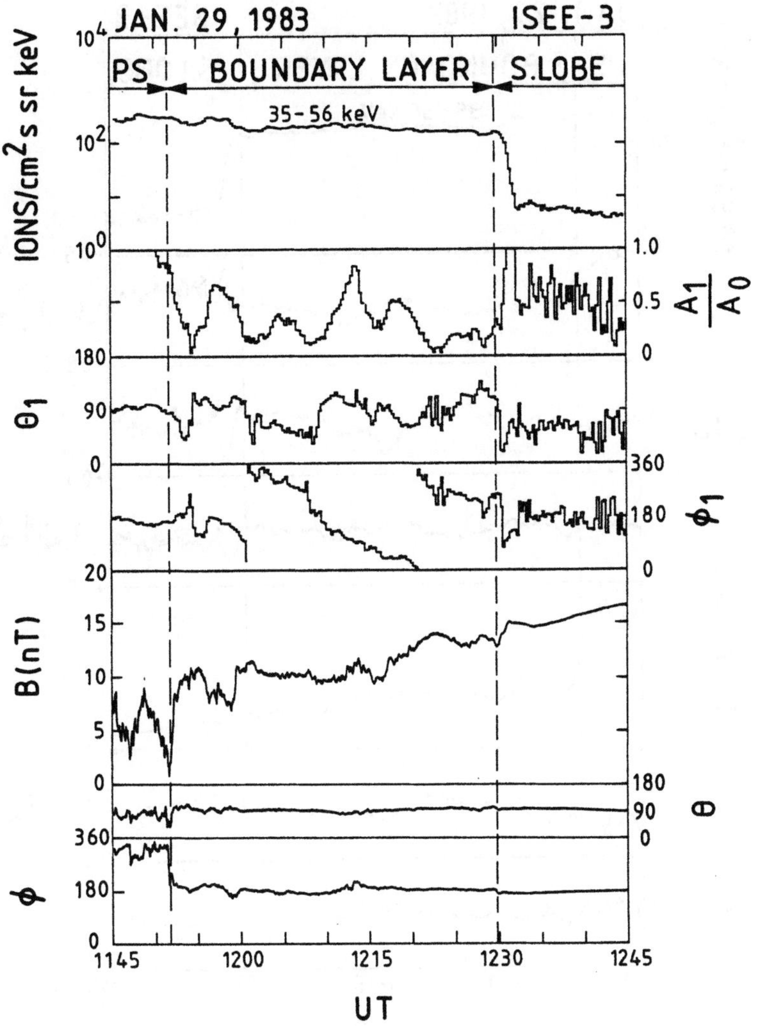

Fig 1. Energetic ion intensity, 3 dimensional vector anisotropy A_1/A_0, θ_1 and ϕ_1, and magnetic field magnitude B and direction θ and ϕ (GSE coordinates) measured during the passage of ISEE-3 from the plasma sheet (PS) through the boundary layer to the southern lobe, on January 29, 1983. The large scale vortex structure observed in the boundary layer is recognisable by the slow, monotonic change in the anisotoropy azimuth angle (4th panel from top).

maximum eigen values, λ_1/λ_3, signifying somewhat more variations along the directions of minimum variance for the field than for the ions. The directions of minimum variance ($\theta_1=19°$, $\phi_1=151°$ for the ions, $\theta=22°$, $\phi=157°$ for the magnetic field) show that both the rotation of the ion anisotropy vector, and the variations observed in the magnetic field are contained in the same plane. This plane is almost parallel to the plane of the ecliptic and therefore to the magnetotail plasma sheet.

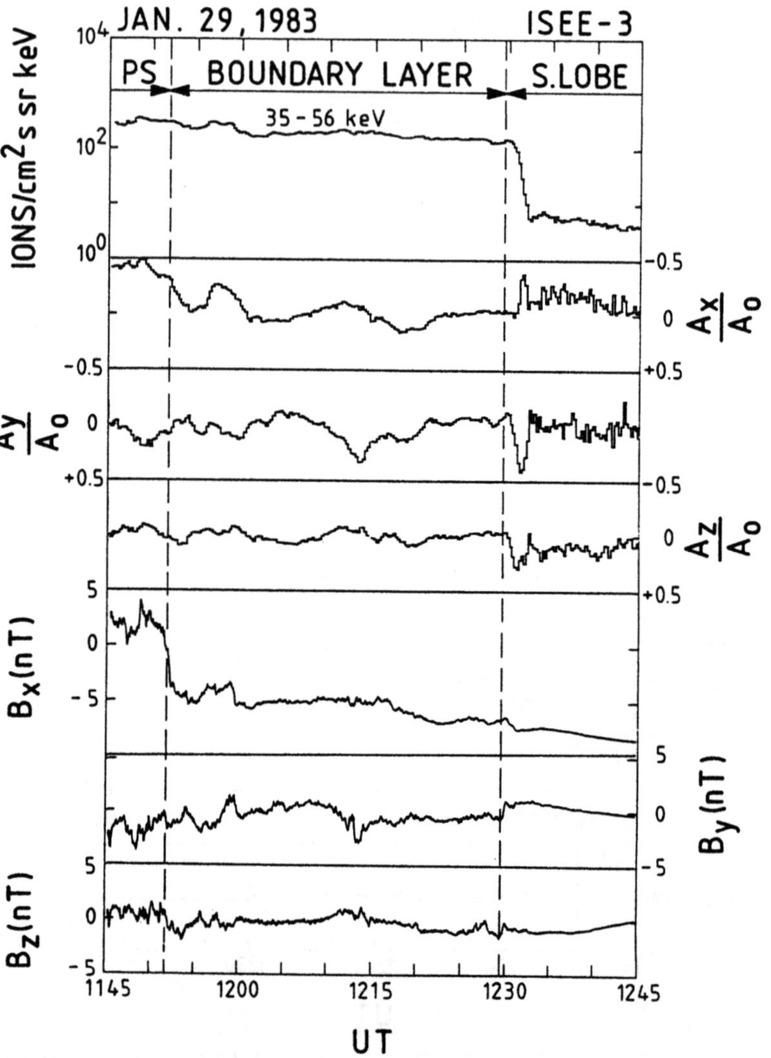

Fig. 2. Energetic ion intensity, 3-dimensional vector anisotropy components A_x/A_0, A_y/A_0 and A_z/A_0 and magnetic field components B_x, B_y and B_z (GSE coordinates) for the period shown in Fig. 1. The three anisotropy components are plotted inverted to show the 180° phase relationship between the anisotropy and the magnetic field components.

4. Discussion

Within a fluid we can observe the effects of surface waves at some distance from the surface by the vertical and horizontal oscillations that each particle performs, the oscillations appearing as either circular motion for an infinite fluid, or elliptical

Table 1. Covariance analysis of the ion anisotropy vector and the magnetic field vector.

Ions		Mag field	
$\frac{\lambda_1}{\lambda_2}$	= 1.39	$\frac{\lambda_1}{\lambda_2}$	= 1.36
$\frac{\lambda_1}{\lambda_3}$	= 10.6	$\frac{\lambda_1}{\lambda_3}$	= 4.11
$\theta_{1\ min}$	= 19°	θ_{min}	= 22°
$\phi_{1\ min}$	= 151°	ϕ_{min}	= 157°

motion (with the major axis parallel to the surface) for a fluid with finite depth, the size of the orbit decreasing with increasing distance from the surface. The vortex-like structure described here, which is characterised by the elliptical motion of the ions, is consistent with the presence of a surface wave at the magnetopause.

At the time of observation, ISEE-3 was within the boundary layer which separates the plasma sheet from the southern lobe of the tail, at $X_{GSM}=-217$, $Y_{GSM}=4.3$ and $Z_{GSM}=-7.8$ R_E. The position of the spacecraft, although on the dusk side of the sun-earth line, is on the dawn side of the magnetotail if the effects of aberration by the solar wind are taken into account. The negative value of the y component of the magnetic field (Fig. 2) is also consistent with a location on the dawn side of the tail. We assume that the boundary layer where we observed the vortex-like structure extends to the interface where the surface waves are being produced. The surface waves make their presence known within the boundary layer as a circular motion of the energetic ions (as shown in panel 4 of Fig. 1). The variations in the 3 components of the magnetic field vector are approximately 180° out of phase with the variations in the 3 components of the ion anisotropy vector, not unlike the observations closer to the earth of SAUNDERS *et al.* (1983). Closer to the earth, however, field line curvature plays an important rôle, giving rise to signatures consistent with coupling between the slow and Alfvén modes (SOUTHWOOD and SAUNDERS, 1985). The variations of the ion anisotropy and the magnetic field are such that both disturbance vectors rotate in a clockwise direction, which is the same sense as that observed by HONES *et al.* (1978). This sense of rotation is consistent with the surface wave propagating tailward on the dawn side of the tail. During the period when the spacecraft was in the boundary layer, we observed two rotations of the disturbance vectors, which together with the measured plasma flow velocity of ~50–100 km/s gives a wavelength of ~15–30 R_e. The value is similar to values of several R_e quoted by HONES *et al.* (1981, 1983), and BIRN *et al.* (1985) for vortices observed in the plasma sheet closer to the earth, and values quoted by LEPPING and BURLAGA (1979), and SCKOPKE *et al.* (1981), (3–8 R_e), for direct observations of surface waves on the flanks of the magnetopause. The small difference between the observations closer to the earth and our observations may be due to the increase in wavelength with distance downtail reported by BIRN *et al.* (1985).

Inside the tail, in the boundary layer, we have in-situ observations from ISEE-3.

For the measurements outside of the magnetotail, we have taken measurements from ISEE-2, which was in the magnetosheath at $X_{GSM}=-6.5$, $Y_{GSM}=-18.7$ and $Z_{GSM}=-7.6$ R_e on the same side of the magnetosphere as ISEE-3. If we assume that the magnetosheath plasma remains substantially unchanged during the 1 hour in which it propagates from ISEE-2 to a downstream position corresponding to the position of ISEE-3 at $X=-217$ R_e, we can use the observations at ISEE-2 to infer the conditions outside of the magnetosphere and adjacent to ISEE-3.

The most relevant study for our observations is the work of PU and KIVELSON (1983). PU and KIVELSON (1983) present growth rates and critical velocities for the onset of a Kelvin-Helmholtz instability as a function of $U=v/C_A$, where v is the relative velocity between the two adjacent media, and C_A is the Alfvén velocity within the magnetosphere. PU and KIVELSON (1983) present results for a ratio of magnetic fields, $B_1/B_2=1.5$, densities $n_2/n_1=10$, and plasma beta's $\beta_1=0.54$ and $\beta_2=2.5$ (the subscripts 1 and 2 refer to measurement in the magnetosphere and in the magnetosheath, respectively), for the configuration when the two magnetic fields, and the relative velocity v are all parallel. The ISEE-2 and -3 observations have a ratio $B_1/B_2=$ .83 which is similar to the parameters used in their model. We have no way of measuring β_1, and no measurement of β_2. Their predicitions show that a slow mode should be excited for $0.9<U<1.2$, and a fast mode for $1.3<U<2.4$, with the maximum growth rates occurring at $U=1.0$ and 1.7 for the slow and fast modes, respectively. Outside the magnetosphere, the ISEE-2 observations show that the plasma flow velocity is ~375–400 km/s (P.-A. Lindquist, R. D. Zwickl, private communications, 1985), whilst inside the magnetotail at ISEE-3, the flow velocity is ~50–100 km/s. Taking the relative velocity v as 300 km/s, and the measured value of C_A (480 km/s), gives $U=0.625$. This figure is somewhat lower than the critical value of 0.9, but bearing in mind the difference between the measured parameters and those used in the predictions, suggests that we are most likely observing within the magnetosphere the effect of a slow-mode surface wave at the magnetospause.

5. Conclusion

We have observed in the deep geotail at $X=-217$ R_e, a vortex-like structure in the energetic ions and magnetic field. At the location of the spacecraft we observe: (i) 2 clockwise rotations of the energetic ion anisotropy vector within a period of half an hour, (ii) a variation in the magnetic field vector approximately 180° out of phase with the ion and plasma variations.

We believe this is the signature of a surface wave on the magnetopause caused by the Kelvin-Helmholtz instability. We have inferred the conditions outside of the magnetosphere from ISEE-2 measurements upstream of the location of ISEE-3, and find that the difference between the flow velocity of the magnetosheath and the flow velocity of the magnetotail is consistent with that required to cause a slow mode surface wave at the magnetopause on the dawn side of the magnetosphere.

The portions of this study carried out at the Jet Propulsion Laboratory, California Institute of Technology, Pasadena, were supported under contract with the National

Aeronautics and Space Administration.

REFERENCES

BAME, S. J., J. R. ASHBRIDGE, H. E. FELTHAUSER, J. P. GLORE, H. L. HAWK, and J. CHAVEZ, ISEE-C Solar wind plasma experiment, *IEEE Trans.*, **GE-16**, 160, 1978.

BIRN, J., E. W. HONES Jr., S. J. BAME, and C. T. RUSSELL, Analysis of 16 plasma vortex events in the geomagnetic tail, *J. Geophys. Res.*, **90**, 7449, 1985.

COWLEY, S. W. H., R. J. HYNDS, I. G. RICHARDSON, P. W. DALY, T. R. SANDERSON, K.-P. WENZEL, J. A. SLAVIN, and B. T. TSURUTANI, Energetic ion regimes in the deep geomagnetic tail: ISEE-3, *Geophys. Res. Lett.*, **11**, 275, 1984.

FRANDSEN, A. M. A., B. V. CONNOR, J. VAN AMERSFOORT, and E. J. SMITH, The ISEE-C helium magnetometer, *IEEE Trans. Geosci. Electron.*, **GE-16**, 195, 1978.

HONES Jr., E. W., G. PASCHMANN, S. J. BAME, J. R. ASHBRIDGE, N. SCKOPKE, and K. SCHINDLER, Vortices in magnetospheric plasma flow, *Geophys. Res. Lett.*, **5**, 1059, 1978.

HONES Jr., E. W., J. BIRN, S. J. BAME, J. R. ASHBRIDGE, G. PASCHMANN, N. SCKOPKE, and G. HAERENDEL, Further determination of the characteristics of magnetospheric plasma vortices with ISEE 1 and 2, *J. Geophys. Res.*, **86**, 814, 1981.

HONES, E. W., Jr., J. BIRN, S. J. BAME, and C. T. RUSSELL, New observations of plasma vortices and insights into their interpretation, *Geophys. Res. Lett.*, **10**, 674, 1983.

LEPPING, R. P. and L. F. BURLAGA, Geomagnetopause surface fluctuations observed by Voyager 1, *J. Geophys. Res.*, **84**, 7099, 1979.

PU, Z. and M. G. KIVELSON, Kelvin-Helmholtz instability at the magnetopause: Solution for incompressible plasmas, *J. Geophys. Res.*, **88**, 841, 1983.

SANDERSON, T. R., P. W. DALY, K.-P. WENZEL, E. W. HONES, Jr., and S. J. BAME, Energetic ion observations of a large-scale vortex in the distant geotail, *Geophys. Res. Lett.*, **11**, 1094, 1984.

SANDERSON, T. R., R. REINHARD, P. VAN NES, and K.-P. WENZEL, Observations of three-dimensional anisotropies of 35- to 1000-keV protons associated with interplanetary shocks, *J. Geophys. Res.*, **90**, 19, 1985.

SAUNDERS, M. A., D. J. SOUTHWOOD, T. A. FRITZ, and E. W. HONES Jr., Hydromagnetic vortices-I. The 11 December 1977 event, *Planet. Space Sci.*, **31**, 1099, 1983.

SCKOPKE, N., G. PASCHMANN, G. HAERENDEL, B. U. Ö. SONNERUP, S. J. BAME, T. G. FORBES, E. W. HONES, Jr., and C. T. RUSSELL, Structure of the low-latitude boundary layer, *J. Geophys. Res.*, **86**, 1099, 1981.

SOUTHWOOD, D. J. and M. A. SAUNDERS, Curvature coupling of slow and Alfvén MHD waves in a magnetotail field configuration, *Planet. Space Sci.*, **33**, 127, 1985.

Solar Wind-Magnetosphere Coupling, edited by Y. Kamide and J. A. Slavin, 747–757.

Planetary Boundary Layers in the Solar Wind

H. PÉREZ-DE-TEJADA

Instituto de Geofísica, Universidad Nacional Autónoma de México, Ensenada, Baja California, México

A comparative study of the solar wind streaming in the vicinity of the effective obstacles of the earth, Venus and Mars, is presented. The available experimental evidence indicates that a notable difference exists in the velocity profile of the solar wind around these planets. At the earth, the velocity of the solar wind streaming by the flanks of the magnetopause remains nearly constant along directions normal to that boundary. In Venus and Mars the velocity profile exhibits, in that region, a strong transverse gradient which implies decreased values of the momentum flux of the solar wind. Such different boundary layers are interpreted as due to mass loading processes and charge exchange collisions as well as to the more efficient transport of momentum across the boundary of the effective obstacles of weakly magnetized planets.

1. Introduction

The observation of the solar wind in the plasma environment of the terrestrial planets has revealed a wide variety of conditions which reflect the different type of interactions that take place at the boundary of their effective obstacles. The presence of a strong intrinsic planetary magnetic field gives place to an effective obstacle very different from that presented by weakly magnetized planets. At the earth the geomagnetic field prevents the solar wind from reaching directly the upper layers of the atmosphere and, as a result, the solar wind is mostly inflicted by changes caused by its interaction with the planetary magnetic field. Venus represents, on the other hand, the extreme case of a very weakly magnetized planet in which the ionosphere controls almost completely the process of interaction with the solar wind. In this case the local ionosphere is compressed by the solar wind and confined below a boundary (ionopause) which is usually located a few hundred kilometers above the surface of the planet. An intermediate situation appears to characterize the conditions present in Mars where a small intrinsic magnetization may marginally act as the effective obstacle which deflects the solar wind around the planet.

As a result of the different origin of the effective planetary obstacles the character of the interaction process that takes place at their boundaries is also notably different. At the earth, the solar wind particles interact with the planetary magnetic field and produce a current layer which compresses and bounds the planetary field. At Venus, on the other hand, the outer boundary of the ionosphere is the site of an electrodynamic interaction in which the incident solar wind flow builds up a layer of enhanced magnetic fluxes. The development of such a layer (or barrier) of magnetic fluxes can be viewed as resulting from the accumulation of interplanetary magnetic field lines around the Venus ionosphere. Thus, the conditions at the boundary of the Venus ionosphere are in a sense opposite to those at the boundary of a planetary magnetosphere with a strong magnetic field present outside, and a weakly magnetized plasma inside the obstacle.

In the present report we will examine the existing experimental evidence regarding the behavior of the solar wind outside magnetized and weakly magnetized planets. We will show that the temperature and velocity profiles of the solar wind plasma streaming around their effective obstacles reflect the character of the interaction process that takes place in each case.

2. Observations

Over the years it has been recognized that a dominant feature of the earth's magnetosphere is the presence of an internal plasma flow, or magnetospheric boundary layer, directed in the antisolar direction adjacent to, and earthward of, the magnetopause (see, for example, EASTMAN, 1984). The origin of the motion of the plasma in that region has been traced to phenomena associated with magnetic reconnection processes (DUNGEY, 1961; SONNERUP, 1979); with a viscous transfer of momentum across the magnetopause (AXFORD and HINES, 1961; FAYE-PETERSEN and HECKMAN, 1968; CASSEN, 1970; LUNDIN and DUBININ, 1984); and with the turbulent diffusion of solar wind particles entering through the cusps of the magnetosphere (HAERENDEL *et al.*, 1978; JOHNSON, 1978). In these and other studies it has been shown that the flow within the magnetosphere is generally less dense and slower than that measured outside.

An important feature of the observed velocity profiles is the fact that significant velocity gradients occur primarily across the magnetospheric boundary layer. In the magnetosheath, on the other hand, the flow velocity remains nearly uniform when sampled along directions transverse to the magnetopause. Observed variations in density and pressure anisotropy are consistent with those expected from calculations of the solar wind around a blunt obstacle (see, for example, SPREITER *et al.*, 1966; CROOKER *et al.*, 1979). An example is shown in Fig. 1 which clearly reveals the different character of the velocity field within and outside the earth's magnetosphere. The left panel reproduces the density, temperature, velocity and magnetic field profiles reported by ROSENBAUER *et al.* (1975) from measurements conducted with the HEOS 2 as that spacecraft crossed the magnetopause on June 4, 1972. The region of the high-latitude magnetosphere covered in that pass is illustrated schematically in the right hand side panel of Fig. 1. In the outbound leg, the magnetopause is identified by the drastic change in the orientation of the magnetic field which occurs at 1100 UT. Prior to this time, while the vehicle was still within the magnetosphere, the velocity values detected are significantly smaller than those measured in the magnetosheath. In this latter region the velocity profiles seem to be rather uniform with no strong variations near the magnetopause. The fact that a well-defined decrease of the velocity occurs only near this boundary indicates that little or no changes occur in the momentum flux of the shocked solar wind (within the magnetosphere the boundary layer flow comprises only $\lesssim 2\%$ of the total particle flux diverted by the magnetospheric obstacle).

A behavior radically different has been reported, on the other hand, from the

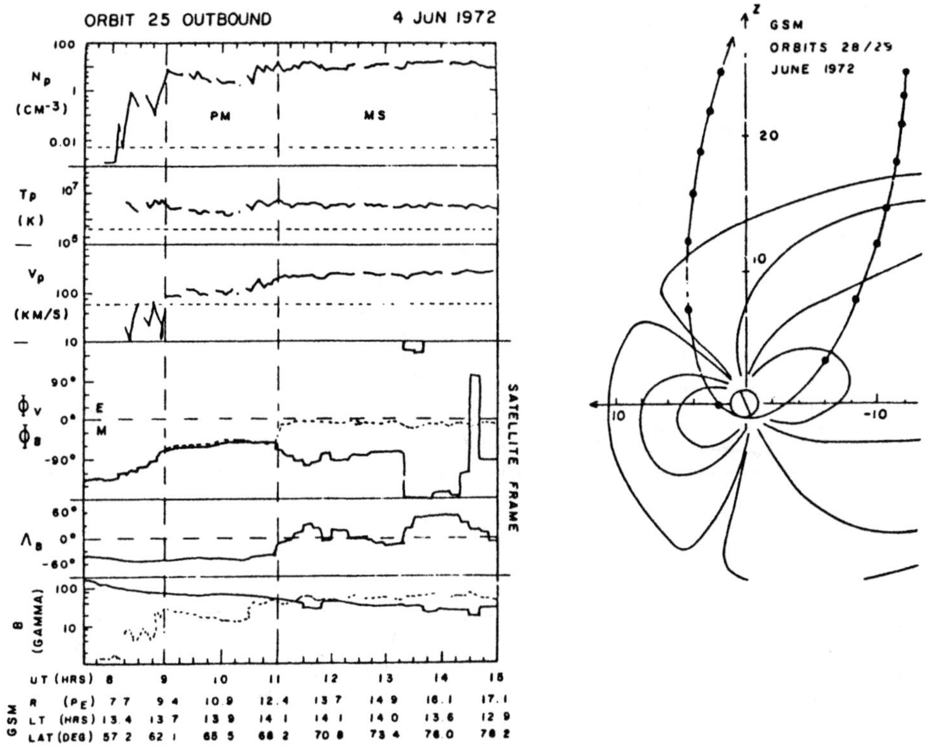

Fig. 1. Density, temperature, velocity and magnetic field profiles measured within and outside the earth's magnetosphere during the outbound leg of the HEOS-2 satellite on June 4, 1972. The trajectory of the spacecraft is indicated in the right hand side panel, on a plane perpendicular to the magnetic equator (from ROSENBAUER *et al.*, 1975).

in-situ exploration of the Venus plasma environment. Most notable is the detection of a region, exterior to the venusian ionosphere, where the flow velocity exhibits a severe decrease with respect to the values seen farther away from the planet (BRIDGE *et al.*, 1967; VERIGIN *et al.*, 1978; SHEFFER *et al.*, 1979; ROMANOV *et al.*, 1979). Figure 2 illustrates the velocity and temperature profiles reported in the Romanov *et al.* study from measurements conducted along the trajectory of the Venera 10 orbiter in the region exterior to the near wake of the planet on april 19, 1976. (The trajectory of the spacecraft is shown in the upper panel in a plane on which the vertical coordinate gives the distance to the Sun–Venus axis).

The velocity decrease initiated near 0200 MT is reminiscent of that shown in Fig. 1 for the plasma mantle of the earth's magnetosphere. Between both cases there is, however, a basic difference which reveals the different character of the effective obstacle that each planet produces in the solar wind. At the earth the velocity decrease occurs primarily within the magnetosphere and a nearly uniform profile is observed beyond the magnetopause. At Venus, on the other hand, the velocity decrease begins well outside of the downstream extension of the boundary of the obstacle. This latter

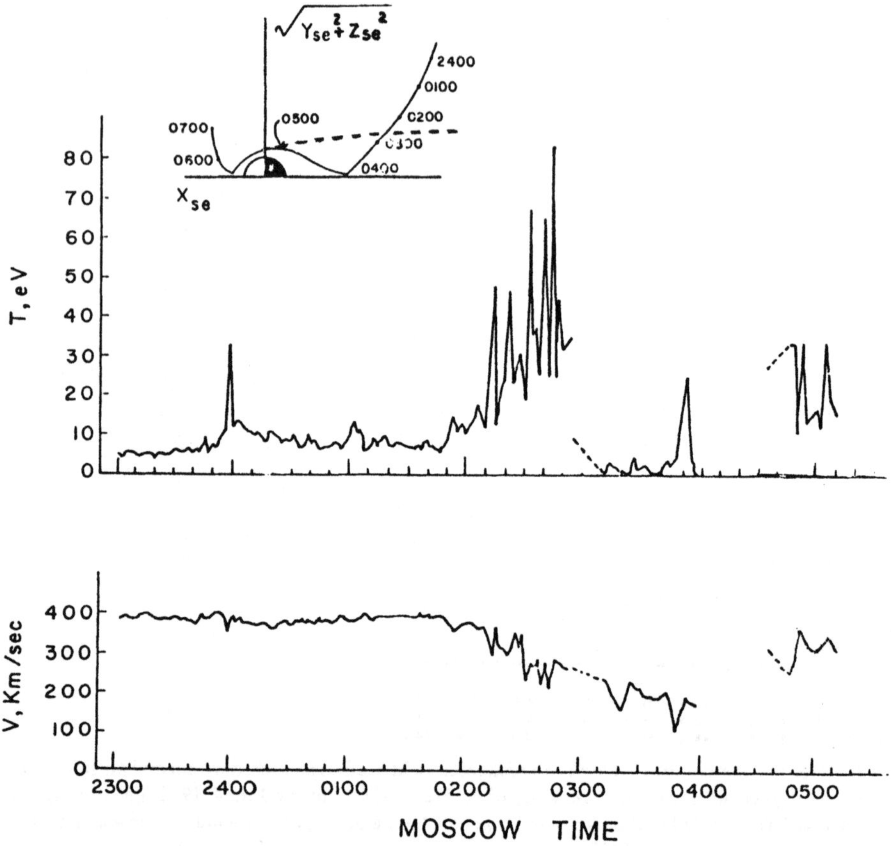

Fig. 2. Velocity and temperature profiles measured in the near Venus wake with instruments onboard the Venera 10 orbiter on April 19, 1976. The trajectory of the spacecraft is indicated in the upper panel on a plane in which the vertical axis is perpendicular to the Sun–Venus axis (from ROMANOV *et al.*, 1979).

transition (indicated by the dashed line in the upper panel of Fig. 2) was detected at ~0250 MT as a sudden change in the temperature and composition of the local plasma. The magnetic field data for this crossing (DOLGINOV *et al.*, 1977, see Fig. 4a) indicates that a certain increase of the B_x component (directed along the Sun–Venus axis) took place at this time; a behavior which suggests a more preferent tail-directed orientation of the magnetic field lines within the wake. (No significant changes in the direction, nor the intensity, of the magnetic field were detected before that time, across the region where the steady velocity decrease was measured).

In addition to the different location of the velocity profile, the data of Fig. 2 shows that, at Venus, the temperature of the plasma exterior to the obstacle is significantly enhanced. We note, in particular, that the initial increase of the plasma temperature seen near 0200 MT coincides with the initial decrease of the flow velocity; a situation which suggests that an effective transfer of kinetic energy into thermal energy takes place in the inner regions of the ionosheath. A different temperature field is seen, on the other hand, beyond the earth's magnetopause where, as shown in Fig. 1, the temperature remains fairly uniform throughout the region exterior to that boundary. As mentioned before, this observation indicates a different type of response of the shocked solar wind as it interacts with the effective obstacles of the earth and Venus.

The velocity decrease and the temperature increase seen in the inner regions of the Venus ionosheath are persistent features identified in most of the plasma data recorded in that region. Velocity and temperature profiles similar to those shown in Fig. 2 were also reported by SHEFFER *et al.*, (1979, see Fig. 2) from the plasma data of the Mariner 5, and by VERIGIN *et al.*, (1978, see Fig. 8) from the plasma data of the Venera 9 and 10 orbiters. These reports also contain density measurements which indicate that a steady density decrease is observed across the region adjacent to the boundary of the wake, where the flow velocity exhibits lower values. Such variations imply that the momentum flux of the plasma in that region is significantly smaller than that of the plasma streaming in the outer ionosheath; a result which, as discussed below, has important consequences in regard to the budget of the momentum flux of the incident solar wind.

The observed density, velocity and temperature profiles of the solar wind near Venus are also useful to interpret those reported from measurements conducted near Mars. Despite the apparent existence of an intrinsic magnetic field which is partially responsible for producing the effective obstacle of that planet, the available experimental evidence suggests that the solar wind streaming outside the obstacle exhibits a behavior similar to that seen at Venus. It should be stressed, however, that observations of plasma and magnetic fluxes near Mars have been scarce and circumstantial. Almost all of the data available has been obtained from Soviet spacecraft (Mars 2, 3 and 5) which orbited the planet but did not penetrate into the optical umbra. As a result, there are no magnetic field measurements in the inner regions of the wake, and it has not been possible to determine conclusively the actual strength of the martian magnetic field. The data available apply only to the region exterior to the optical umbra and has led to controversial interpretations of the relative role that the planetary magnetic field and the local ionosphere have in producing the deflection of the solar wind around the planet (RUSSELL, 1978; DOLGINOV, 1978).

The existence of a magnetopause at Mars (rather than of an ionopause) as boundary of the effective martian obstacle has been suggested by the apparent absence of an abrupt decrease of the ionospheric density with height as observed in Venus. In fact, most of the results of the radio-occultation experiments of the Soviet spacecraft indicate an electron profile which decreases gradually with height. Evidence for a sharp discontinuity is present only in the profiles derived from the Mars 6 (VASILIEV *et al.*, 1975), and Viking 2 (HANSON *et al.*, 1977) measurements. The existence of a magnetic obstacle in the Mars plasma environment has also been suggested by DOLGINOV (1976) from observations conducted with the Mars 5 spacecraft. A significant enhancement of the magnetic field strength was detected in the vicinity of the optical umbra in the near wake. These results, as well as the apparent insufficient plasma pressure of the martian ionosphere to stand off the solar wind (SLAVIN and HOLZER, 1982), have been the basis for suggesting that the effective obstacle is produced primarily by an appreciable internal magnetization (see also, INTRILIGATOR and SMITH, 1979; BREUS, 1980).

As noted above, however, there are also arguments which support a Venus-type interaction at the boundary of the effective martian obstacle. These refer to the occasional detection of an ionopause-like transition and, most important, to the observation of velocity and temperature profiles similar to those seen outside of the Venus ionosphere. Figure 3 shows (lower panel) density, temperature, velocity, and magnetic field values measured along the trajectory of the Mars 5 on 20/II/1974, as reported by VAISBERG *et al.* (1976). From the velocity profile it is apparent that the local flow exhibits a behavior similar to that seen near Venus. We note, in particular, the steady increase of the velocity between 01 40 MT (at position labeled MP) and 02 30 MT (at position labeled BL). As shown at the bottom of the lower panel of Fig. 3, these measurements were made outside of the region of enhanced magnetic field intensities (which fall between positions 2 and 3), and thus again differ from that present at the earth (Fig. 1) where notable velocity gradients are detected primarily within the magnetosphere. Figure 3 also shows that the temperature of the slower moving plasma seen near and within the martian obstacle is higher than that of the faster moving plasma detected farther outside. The inferred temperature profile is similar to that measured in the Venus plasma environment even though it does not seem to extend so far away from the obstacle. This latter characteristic may simply reflect the different width of the boundary layers of both planets; a circumstance which may be due to the presence of stronger magnetic fluxes at the boundary of the martian obstacle.

3. Discussion

The fact that the velocity of the solar wind streaming in the vicinity of the earth's magnetosphere does not exhibit a behavior similar to that seen near Venus and Mars reveals the different nature of the process of interaction that takes place in those planets. The strong reduction of the flow velocity seen outside of the Venus ionosphere indicates a severe loss of momentum flux of the incident solar wind. A significant loss of momentum is also present outside the martian effective obstacle but, as pointed out before, it does not seem to occur outside of the earth's magnetosphere where a uniform velocity profile is observed. In this latter case the solar wind appears to move without an appreciable deceleration in the region adjacent to the magnetopause; even though significant plasma fluxes of solar wind origin still penetrate the obstacle's boundary to produce the magnetospheric boundary layer.

Various processes have been suggested to explain the loss of momentum of the solar wind as it streams around the ionospheres of Venus and Mars. Charge exchange collisions and the mass loading of the incident plasma with planetary ions are mechanisms which are expected to contribute to the continuous removal of momentum of the incident plasma. The first process occurs when an oncoming solar wind proton captures the electron of a planetary hydrogen atom and becomes a fast moving hydrogen atom. In such a collision a planetary hydrogen ion is also formed

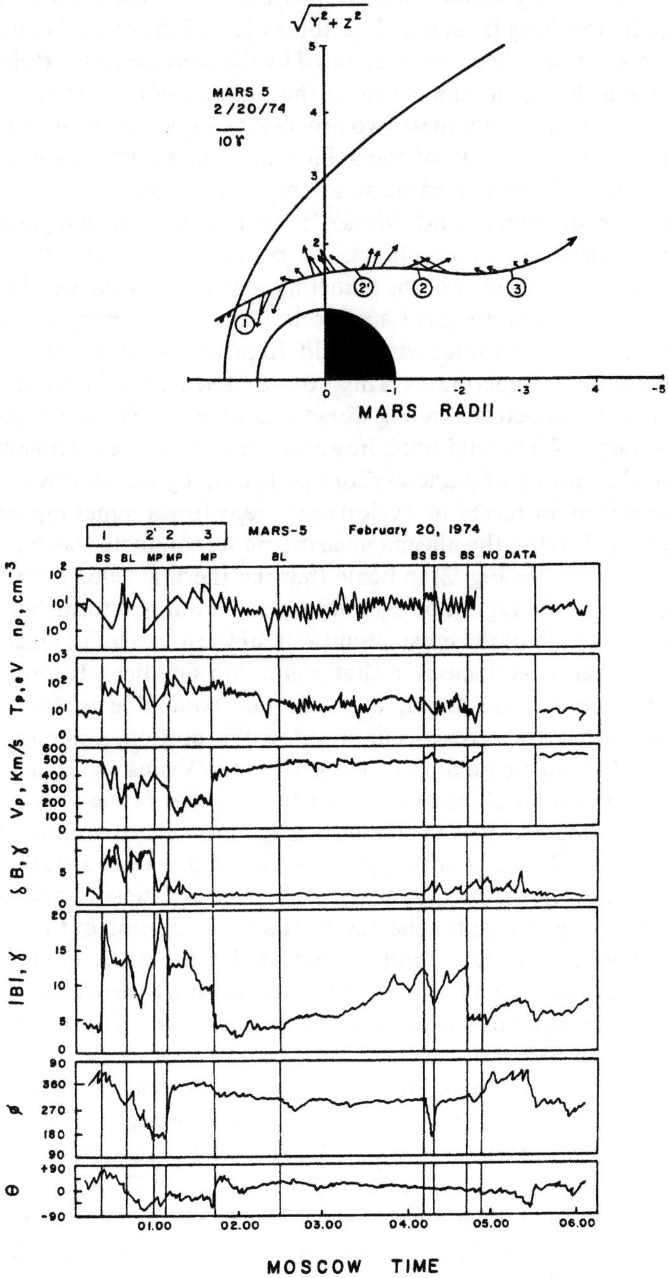

Fig. 3. Density, temperature, velocity and magnetic field profiles measured in the near Mars wake with instruments onboard the Mars 5 space probe on Feb. 20, 1974. The trajectory of the spacecraft is indicated in the upper panel on a plane in which the vertical axis is perpendicular to the Sun–Mars axis (from VAISBERG *et al.*, 1976).

but it carries only a very small momentum flux. The second process deals with ion-electron pairs resulting from the photoionization of the neutral exosphere in the solar wind. These particles become accelerated by the convective $\boldsymbol{V}\times\boldsymbol{B}$ electric field of the solar wind and thus gain momentum at the expense of the latter.

Even though it is clear that these two processes should contribute to the removal of some of the momentum flux of the solar wind, their participation may not be sufficient to explain the enhanced plasma temperatures seen in the vicinity of the effective obstacles of Venus and Mars. In fact, since the temperature of the exosphere/ionosphere of these planets is much lower than the solar wind temperature, the thermal energy of the planetary ions is much smaller than that of the solar wind particles. The subsequent acceleration of the planetary particles through the $\boldsymbol{V}\times\boldsymbol{B}$ electric field of the solar wind could, in principle, be viewed as resulting in high equivalent plasma temperatures. This is due to the fact that in cyclotronic motion the velocity of particles detected along directions transverse to the magnetic field is expected to be large. We should note, however, that there is experimental evidence indicating that the motion of planetary ions picked up by the solar wind may not be adequately described in terms of cyclotronic trajectories superimposed on $\boldsymbol{E}\times\boldsymbol{B}$ electric drifts only. Firstly, the plasma measurements conducted with the Mariner 5 (SHEFFER *et al.*, 1979, see Fig. 2) indicate that the thermal velocities of the plasma particles measured in the region of enhanced plasma temperatures (adjacent to the boundary of the wake) is, at the most, about 25% of the local (bulk) kinetic velocities. Even though the thermal velocities in that region are significantly larger than those measured in the outer ionosheath, they are not compable to the local kinetic velocities, as would be expected to occur in cyclotronic motion. Secondly, the analysis of the plasma and magnetic field data measured in the Venus wake has revealed that the direction of motion of planetary O^+ ions follows that of the bulk motion of the solar wind, and is uncorrelated to the local magnetic field orientation (PEREZ-DE-TEJADA *et al.*, 1982). This behavior suggests that other processes are also necessary to explain the displacement of the planetary ions. Wave-particle interactions may, in fact, be ultimately responsible for the deviations of the motion of the particles from simple cyclotronic trajectories superimposed on $\boldsymbol{V}\times\boldsymbol{B}$ electric drifts. Under such conditions it is not evident that the high plasma temperatures seen in the vicinity of the obstacles of weakly magnetized planets can be accounted for in terms of mass loading and charge exchange collisions.

An additional limitation of the mass loading processes is the fact that the measured concentration of planetary O^+ ions in the Venus wake is only $\sim 1\%$ of the local proton population (MIHALOV *et al.*, 1980; INTRILIGATOR, 1982). With such low concentrations the momentum flux that they carry is not sufficient to account for the deficiency of momentum flux implied by the velocity profile shown in Fig. 2. PEREZ-DE-TEJADA (1982) has estimated that in order to properly explain the fate of the "missing" momentum flux of the ionosheath flow, in terms of mass loading processes, it would be necessary that the local concentration of planetary O^+ ions were $\sim 15\%$ of the local proton population, or about one order of magnitude larger than that actually measured.

An alternative, and most likely supplementary, mechanism required to explain the velocity reduction and the enhanced plasma temperatures seen in the Venus inner ionosheath flow has been proposed by PEREZ-DE-TEJADA (1982) in terms of viscous effects at the ionopause. In this view the solar wind transfers viscously some of its momentum to the upper layers of the Venus dayside ionosphere, and heats up the local plasma through the onset of turbulent dissipative processes. Measurements conducted with instruments onboard the Pioneer Venus orbiter (KNUDSEN *et al.*, 1980) have revealed that a transterminator plasma flow does exist in the Venus upper ionosphere; and that its (height integrated) momentum flux is comparable to the (height integrated) deficiency of momentum flux exhibited by the shocked solar wind in the vicinity of the terminator (see PEREZ-DE-TEJADA, 1982).

This result is consistent with the concept that the latter quantity is primarily delivered to the upper ionospheric plasma, and that a strong coupling exists between the ionospheric and the solar wind plasmas. Current studies of viscous processes in collisionless plasmas (MIURA and SATO, 1978; KORZHOV *et al.*, 1984) indicate that an effective transfer of momentum between streaming flows with a velocity shear can result from the non-linear evolution of unstable density waves produced by Kelvin-Helmholtz oscillations. The existing treatments are not directly applicable, however, to the conditions expected at the Venus ionopause where the much denser ionospheric plasma may represent a significantly more viscous obstacle to the solar wind than a planetary magnetic field. Further studies are clearly required to examine the efficiency of the viscous processes for this case.

Similar arguments seem to be also applicable to the data of Fig. 3 for the Mars plasma environment. The observation of decelerated plasma fluxes outside of the martian magnetosphere (exterior to the region of enhanced magnetic fluxes) implies a certain deficit of momentum flux within the shocked solar wind. As noted earlier, however, the velocity boundary layer in that planet does not appear to be as extended as that present outside the Venus ionosphere. Thus, even though the data of Fig. 3 suggests an interaction similar to that seen at Venus, it is possible that the conditions present at the boundary of the martian obstacle are somewhat different. As indicated before, the presence of magnetic fluxes intrinsic to Mars may be the dominant factor determining the position and geometry of the martian effective obstacle. A situation which can be contemplated within this interpretation is that the solar wind interacts with abundant concentrations of ionospheric plasma contained within a small martian magnetosphere. Under such conditions it would still be possible that, despite the protective action of the planetary magnetic field, a viscous-like transport of momentum takes place across the magnetopause and that an ionospheric flow be generated near the terminator.

These concepts serve to illustrate the varied response that the shocked solar wind exhibits as it flows around the effective obstacles of magnetized and weakly magnetized planets and, at the same time, emphasize the necessity to further identify the degree in which mass loading, charge exchange collisions and viscosity participate in the interaction process that takes place in the vicinity of weakly magnetized planets. On the basis of what has been written here it is apparent, however, that the observation of large scale velocity gradients, as well as of increased plasma temperatures, in the region exterior to the obstacles of Venus and Mars suggests that the plasma-plasma interaction processes modify more severely the solar wind than those associated with the plasma-magnetic field interaction that takes place at the earth's magnetopause.

REFERENCES

AXFORD, W. I. and C. O. HINES, A unifying theory of high latitude geophysical phenomena, *Can. J. Physics*, **39**, 1433, 1961.

BRIDGE, H. S., A. J. LAZARUS, C. W. SNYDER, E. J. SMITH, L. DAVIES, P. J. COLEMAN, and D. E. JONES, Plasma and magnetic fields observed near Venus, *Science*, **158**, 1669, 1967.

BREUS, T. K., Peculiarities of the solar wind interaction with the upper atmospheres of Venus and Mars, *Geophys. International*, **19**, 17, 1980.

CASSEN, P. and J. SZABO, The viscous magnetopause, *Planet. Space Sci.*, **18**, 349, 1970.

CROOKER, N. U., T. E. EASTMAN, and G. S. STILES, Observations of plasma depletion in the magnetosheath at the dayside magnetopause, *J. Geophys. Res.*, **84**, 869, 1979.

DOLGINOV, Sh. Sh., The magnetic field of Mars according to the data from the Mars 3 and Mars 5, *J. Geophys. Res.*, **81**, 3353, 1976.

DOLGINOV, Sh. Sh., On the magnetic field of Mars: Mars 2 and 3 evidence, *Geophys. Res. Lett.*, **5**, 89, 1978; Mars 5 evidence, *Geophys. Res. Lett.*, **5**, 93, 1978.

DOLGINOV, Sh. Sh., L. N. ZHUZGOV, V. A. SHAROVA, V. B. BUZIN, and Ye. G. YEROSHENKO, The magnetosphere of Venus, *Inst. Zem. Magnet.*, Pub. I9(I93); Goddard Tech. Translation, GSFC-16, 1977.

DUNGEY, J. W., Interplanetary magnetic field and the auroral zones, *Phys. Rev. Lett.*, **6**, 47, 1961.

EASTMAN, T. E., Observations of the magnetospheric boundary layer, in *Proc. Conf. IMS*, Graz, Austria, June 1984 (ESA SP-217, September 1984, p. 77).

FAYE-PETERSON, R. and G. HECKMAN, Viscous magnetospheric boundary layer, *Ann. Geophys.*, **24**, 347, 1968.

GRINGAUZ, K. I., Interaction of the solar wind with Mars as seen by charged particle traps on Mars 2, 3 and 5 satellites, *Rev. Geophys. Space Phys.*, **14**, 391, 1976.

HAERENDEL, G., G. PASHMANN, N. SCKOPKE, H. ROSENBAUER, and P. C. HEDGECOCK, The front side boundary layer of the magnetosphere and the problem of reconnection, *J. Geophys. Res.*, **83**, 3195, 1978.

HANSON, W. B., S. SANATANI, and D. R. ZUCCARO, The martian ionosphere as observed by the viking retarding potential analyzer, *J. Geophys. Res.*, **82**, 4351, 1977.

JOHNSON, F. S., The driving force of magnetospheric convection, *Rev. Geophys. Space Phys.*, **16**, 161, 1978.

INTRILIGATOR, D. S., Observation of mass addition to the shocked solar wind of Venus ionosheath, *Geophys. Res. Lett.*, **9**, 727, 1982.

INTRILIGATOR, D. S. and E. J. SMITH, Mars in the solar wind, *J. Geophys. Res.*, **84**, 8427, 1979.

KNUDSEN, W. C., K. SPENNER, K. L. MILLER, and V. NOVAK, Transport of ionospheric O^+ ions across the Venus terminator and implications, *J. Geophys. Res.*, **85**, 7803, 1980.

KORZHOV, N. P., V. V. MISHIN, and V. M. TOMOZOV, On the role of plasma parameters and the Kelvin-Helmholtz instability in a viscous interaction of solar wind streams, *Planet. Space Sci.*, **32**, 1169, 1984.

LUNDIN, R. and R. M. DUBININ, Solar wind energy transfer regions inside the dayside magnetopause: Accelerated plasmaspheric ions as tracers for MHD processes in the dayside boundary layer, *Planet. Space Sci.*, **32**, 757, 1984.

MIHALOV, J. D., J. H. WOLFE, and D. S. INTRILIGATOR, Pioneer Venus plasma observations in the solar wind Venus interaction region, *J. Geophys. Res.*, **85**, 7613, 1980.

MIURA, A. and T. SATO, Shear instability: Auroral arc deformation and anomalous momentum transport, *J. Geophys. Res.*, **83**, 2109, 1978.

PEREZ-DE-TEJADA, H., Viscous dissipation at the Venus ionopause, *J. Geophys. Res.*, **87**, 7405, 1982.

PEREZ-DE-TEJADA, H., D. S. INTRILIGATOR, and C. T. RUSSELL, Orientation of plasma and magnetic field fluxes in the Venus wake, *Nature*, **299**, 325, 1982.

ROMANOV, S. A., V. N. SMIRNOV, and O. L. VAISBERG, On the nature of the solar wind-Venus interaction, *Cosmic Res.*, **16**, 603, 1979.

ROSENBAUER, H., H. GRUNWALDT, M. D. MONTGOMERY, G. PASHMANN, and N. SCKOPKE, Heos 2 plasma observations in the distant polar magnetosphere: The plasma mantle, *J. Geophys. Res.*, **80**, 2723, 1975.

RUSSELL, C. T., The magnetic field of Mars: Mars 3 evidence re-examined, *Geophys. Res. Lett.*, **5**, 81, 1978; Mars 5 evidence re-examined, *Geophys. Res. Lett.*, **5**, 85, 1978.

SHEFFER, R. E., A. J. LAZARUS, and H. S. BRIDGE, A re-examination of plasma measurements from the Mariner 5 Venus encounter, *J. Geophys. Res.*, **84**, 2109, 1979.

SLAVIN, J. A. and R. E. HOLZER, The solar wind interaction with Mars revisited, *J. Geophys. Res.*, **87**, 10285, 1982.

SONNERUP, B. U. O., Theory of the low latitude boundary layers, in *Magnetospheric Boundary Layers*, edited by B. Bathrick, 395 ESA, SP 148, Paris, 1979.

SPREITER, J. R., A. T. ALKSNE, and A. L. SUMMERS, Hydromagnetic flow around the magnetopause, *Planet. Space Sci.*, **14**, 223, 1966.

VAISBERG, O. L., A. V. BOGDANOV, V. N. SMIRNOV, and S. A. ROMANOV, On the nature of the solar wind Mars interaction, in *Solar Wind Interaction with the Planets Mercury, Venus and Mars*, edited by N. F. Ness, p. 21, NASA SP-397, 1976.

VASILYEV, M. B., A. S. VYSHLOV, M. A. KOLOSOV, N. A. SAVICH, V. A. SAMOVOL, L. M. SAMOZNAEV, A. I. SIDORENKO, Yu. N. ZAITSEV, G. M. PETROV, O. N. RZHIGA, D. Ya. SHTERN, and L. I. POMENOVA, Preliminary results of the two frequency radio transillumination of the martian ionosphere by means of the Mars automatic interplanetary stations in 1974, *Cosmic Res.*, **13**, 41, 1975.

VERIGIN, M. I., K. I. GRINGAUZ, T. GOMBOSI, T. K. BREUS, V. V. BEZRUKIKH, A. P. REMIZOV, and G. I. VOLKOV, Plasma near Venus from the Venera 9 and 10 wide angle Analyzer data, *J. Geophys. Res.*, **83**, 3721, 1978.

Solar Wind-Magnetosphere Coupling, edited by Y. Kamide and J. A. Slavin, 759–766.

Numerical Modeling on Magnetospheric Substorm

K. MIN, H. OKUDA, and T. SATO*

Plasma Physics Laboratory, Princeton University, P.O. Box 451, Princeton New Jersey 08540, U.S.A.

Magnetospheric substorm is studied numerically by means of a two-dimensional time-dependent nonlinear resistive MHD code. Our model includes magnetic reconnection driven by the entering solar wind plasmas into the magnetosphere as a mechanism of magnetospheric substorm. Two-dimensional dipole field is used as an initial magnetic field configuration. As a result of interaction between the solar wind and the dipole field, tail-like configuration of the magnetic field is realized. When the neutral sheet current density is enhanced enough, current-driven anomalous resistivity triggers magnetic reconnection. Strong plasma jetting, plasmoid formation and its tailward motion are observed during the substorm expansive phase. Recovery phase is also studied by reducing the entering solar wind energy flux. Magnetic field configuration is restored back to a tail-like one and the plasma sheet is broadened. The x-point of reconnection is observed to move in the tailward direction during the recovery phase.

1. Introduction

Although it is generally believed that the ultimate cause of the magnetospheric substorm is the solar wind energy flux entering the magnetosphere, possibly as a result of dayside reconnection, there are tow controversial opinions on the detailed mechanism of the magnetospheric substorm, namely the directly driven model and the unloading model. The driven model has its point on that the energy dissipation rate of the magnetosphere is directly related to the energy input rate from the solar wind (AKASOFU *et al.*, 1985). In the unloading model, on the other hand, magnetic energy is stored in the tail region before the explosive energy conversion occurs. The unloading model includes magnetic reconnection as a powerful energy conversion mechanism (HONES, 1984). Even in the unloading model, there are still controversial opinions on the magnetic reconnection process. One is the spontaneous reconnection model where the tearing instability triggers magnetic reconnection (SHINDLER, 1984). In this model, once the system with magnetic free energy is given initially, there is no place for the solar wind to take part in. The other is the driven reconnection model. SATO and HASEGAWA (1982) argued that the energy conversion is more powerful in the driven reconnection model than the spontaneous tearing model.

The magnetospheric substorm has also been studied intensively by using numerical simulation models (LEBOEUF *et al.*, 1978, 1981; LYON *et al.*, 1980, 1981; BIRN, 1980; BIRN and HONES, 1981; BRECHT *et al.*, 1982; SATO *et al.*, 1984).

*Permanently at Institute for Fusion Theory, Hiroshima University, Hiroshima 730, Japan.

However, these works are not complete by any means. Especially, the recovery phase has not been studied well.

In this paper we would like to study the magnetospheric substorm including the recovery phase by controlling the solar wind energy flux entering the magnetosphere. Our model involves magnetic reconnection driven by the solar wind plasmas, which enter the magnetosphere presumably as a result of dayside reconnection. The solar wind considered here is the plasma flow after its passage through the bow shock and the magnetopause, and thus, it is not supersonic.

2. Numerical Model

The magnetohydrodynamic equations including the plasma resistivity are used in the conservation forms. The variables are normalized by the following three characteristic quantities: B_0 (magnetic field), ρ_0 (density), and L (length). Accordingly, velocity, time, current density, pressure, electric field, resistivity, and plasma acceleration per unit volume (rate of change of the energy density) are normalized by $V_{A_0}=B_0/(\mu_0\rho_0)^{1/2}$, $\tau_A=L/V_{A_0}$, $B_0/(\mu_0 L)$, $B_0^2/(2\mu_0)$, $E_0=V_{A_0}B_0$, $\mu_0 L V_{A_0}$, $V_{A_0}B_0^2/(\mu_0 L)$, respectively. Numerically, $B_0=4\times10^{-4}$ Gauss, $\rho_0=1.67\times10^{-25}$ g/cm^3, $L=6400$ km, $V_{A_0}=2800$ km/sec, $\tau_A=2.3$ sec, and $E_0=1.1$ mV/cm. These normalized equations are transformed into the two-step Lax-Wendroff scheme.

Our simulation domain is located in the tail region, confined by four walls which are located at 5 R_E and at 85 R_E in the sun-earth direction (x-direction), and at 23 R_E and -23 R_E in the north-south direction (z-direction). The symmetry in the dawn dusk direction is assumed. Calculations are performed only in the northern half of the considered region. The simulation box is divided into 160$\times$77 grids so that the mesh size is 0.5 R_E in the x-direction and 0.3 R_E in the z-direction. The earth side boundary at $x=5$ is used as an input boundary through which the solar wind plasmas enter the simulation box. Since our simulation region is located inside the magnetosphere, the entering solar wind considered here is neither supersonic nor super-Alfvenic. The density, two components of the magnetic field, B_x and B_z, and the pressure are kept fixed at the initial values on this earth side boundary. The z-component of the velocity is also assumed to remain zero, while the x-component of the velocity is given by

$$V_x = \begin{cases} 0 & \text{for } 0 \le z \le 5; \\ V_{EX} & \text{for } 5 \le z \le 23, \end{cases}$$

where $V_{EX}=EB_z/B^2$ and E is the imposed electric field. Free boundary conditions are imposed on the boundaries at $z=23$ R_E and at $x=85$ R_E, and the mirror boundary conditions are imposed at $z=0$. It is found in the simulation that the plasmas also enter through the free boundary at $z=23$ in addition to the input boundary at $x=5$.

Our resistivity models is

$$\eta = \begin{cases} \alpha(J - J_c)^2 & \text{if } J \ge J_c; \\ 0 & \text{otherwise,} \end{cases}$$

where α is an adjustable parameter and J_c is a threshold value. In this calculation we choose $\alpha=100$ and $J_c=-0.01$. The dependence of the resistivity on the current density implies that our anomalous resistivity is assumed to be generated by the current driven instability (HUBA *et al.*, 1977; SATO, 1979; TANAKA and SATO, 1981a, 1981b). It is not clear whether or not such anomalous resistivity is really generated in the tail region of our concern. Thus, in this paper, we want to present one particular model as a possible mechanism for the magnetospheric substorm. However, it should be noted that the plasma acceleration is larger than the Joule heating by a factor of two even with this anomalous resistivity model (MIN *et al.*, 1985).

Two-dimensional dipole magnetic field is used as an initial configuration along with the non-moving plasma of constant and uniform density ($0.1/\mathrm{cm}^3$) and temperature (300 eV) as initial conditions. The resulting plasma β ratio (plasma pressure divided by the magnetic pressure) is larger than 1 everywhere except the central region near the earth side boundary ($x\leq35$, $-10\leq z\leq10$) where the dipole magnetic field is strong. The detailed information of our numerical model was given in our previous paper (MIN *et al.*, 1985).

3. Results

In our simulation model the energy flux entering the simulation domain is controlled by the electric field imposed on the earth-side boundary ($x=5$). This external electric field is increased smoothly from 0 to -1×10^{-3} in 100 τ_A so that plasmas can flow into the simulation domain according to the $\boldsymbol{E}\times\boldsymbol{B}$ drift. In our coordinate system dawn to dusk direction has the negative sign. It should be noted that the field aligned plasma entry is negligible in the present model. As the plasmas entering through the earth-side boundary as well as through the north and south boundaries move in the tailward direction, the initial dipole field is deformed into a tail-like configuration. Figure 1 shows the geomagnetic tail formation at time $T=4000\ \tau_A$. Formation of the plasma sheet and the neutral sheet current, and the earthward plasma flow in the near-earth region can be observed. Since we have started with the constant density and temperature everywhere, the formation of the plasma sheet is due to the entering solar wind. Thus, the static pressure gradient of the plasma sheet does not balance the solar wind pressure until the equilibrium is reached. In our simulation the equilibrium is not observed, and the plasma sheet becomes thinner and thinner due to the magnetic stress and the dynamic plasma pressure. Also, because of the high β ratio we have chosen, the plasma energy flux is larger than the associated Poynting flux initially. However, as the magnetotail develops, the magnetic field intensity in the tail lobe region increases and the Poynting flux through the north and south boundaries becomes comparable to the plasma energy flux at $T=4000$.

When the neutral sheet current density is enhanced enough, the current-driven anomalous resistivity is generated and it triggers magnetic reconnection at the point where the current density is peaked. In our model reconnection occurs at about $T=5000\ \tau_A$. Once reconnection starts, the fast energy conversion occurs and the plasmas are accelerated in the down-stream region (Fig. 2). In contrast to the

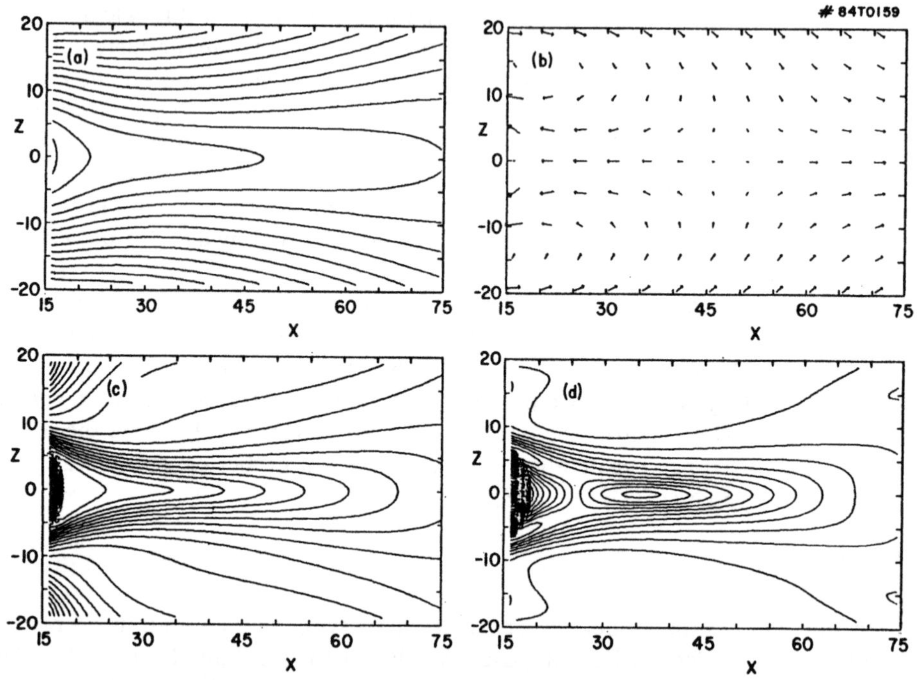

Fig. 1. Magnetotail formation at T=4000: (a) magnetic field lines, (b) velocity, (c) pressure, and (d) cross-tail current density.

pre-reconnection phase, the width of the plasma sheet does not change. The formation of plasmoid and its tailward motion are also observed. On the other hand, the x-point of reconnection does not move. Figure 2 shows the substorm expansive phase at $T=6500\ \tau_A$. The magnetic Reynolds number corresponding to the maximum resistivity is around 80 at $T=6500\ \tau_A$.

To study the recovery phase we reduce the imposed electric field on the earth side boundary from -1×10^{-3} to -1×10^{-4} in 1000 τ_A starting from $T=7000\ \tau_A$ when reconnection is developed well. Our system responds instantly to this change. Figure 3 shows the response of the system in terms of plasma acceleration (conversion of magnetic energy into the kinetic energy subtracted by Joule heating) and the electric field measured at $x=35$ R_E and $z=0$. Both the electric field (at $x=35$, $z=0$) and the plasma acceleration reach steady values at $T=9000\ \tau_A$. However, this does not mean that the system is in steady state after $T=9000\ \tau_A$. Instead, the recovery phase is still in its progress. The plot of electric field shows a characteristic time scale (or a slope) after $T=8000\ \tau_A$ when the transition period is over. During the recovery phase the x-point of reconnection moves in the tail-ward direction and the magnetic field configuration is restored back to a tail-like one in the earth-side region of the x-point. Plasma sheet becomes thick in the recovered region. Plasma flow in the earth-side region of the x-point is generally in the earthward direction. However, the x-point is not necessarily the point where the direction of plasma flow velocity reverses. Current density is still

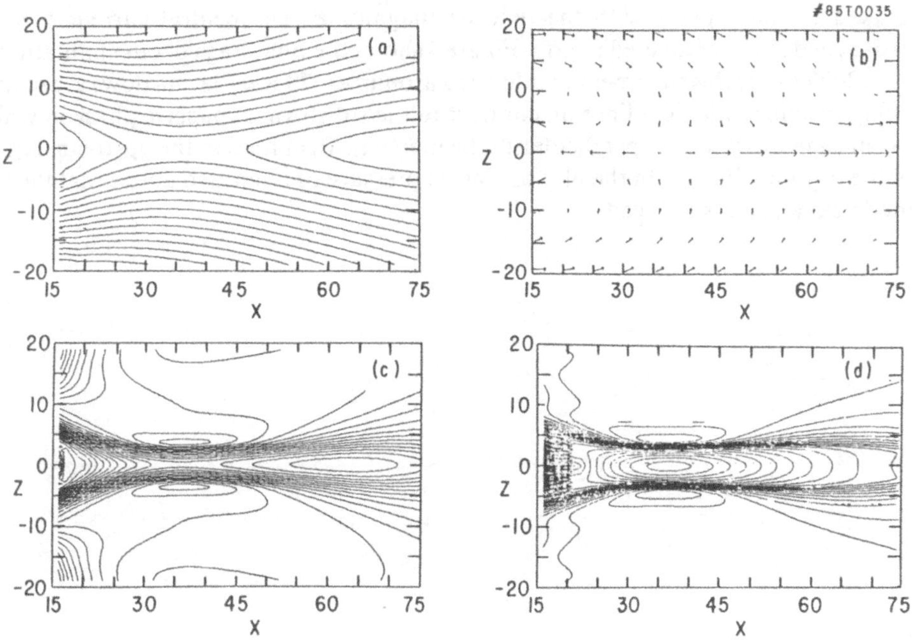

Fig. 2. Magnetic reconnection at $T=6500$: (a) magnetic field lines, (b) velocity, (c) pressure, and (d) cross-tail current density.

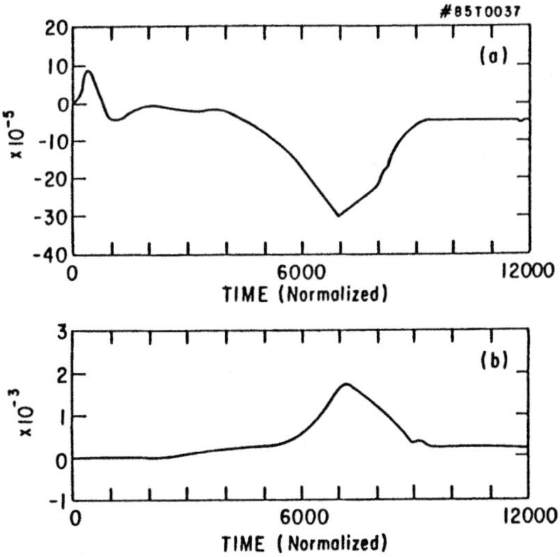

Fig. 3. Response of the magnetosphere to the change of solar wind condition: (a) electric field measured at $X=35$, $z=0$, and (b) plasma acceleration. The electric field imposed on the left boundary has been reduced from -1×10^{-3} to -1×10^{-4} in 1000 τ_A starting from $T=7000$.

peaked at the x-point with the reduced magnitude. The neutral current sheet is broadened in the recovered region. Figure 4 shows the final stage of our simulation at $T=12000\,\tau_A$. At this time x-point is located around $x=70$ R_E. Plasma flows are mainly along the magnetic field lines, in contrast to the substorm expansive phase in which the plasma flows are perpendicular to the magnetic field lines in the upstream region of the x-point (Fig. 2). In the plot of plasma pressure one can also notice that the fast mode expansion is reduced.

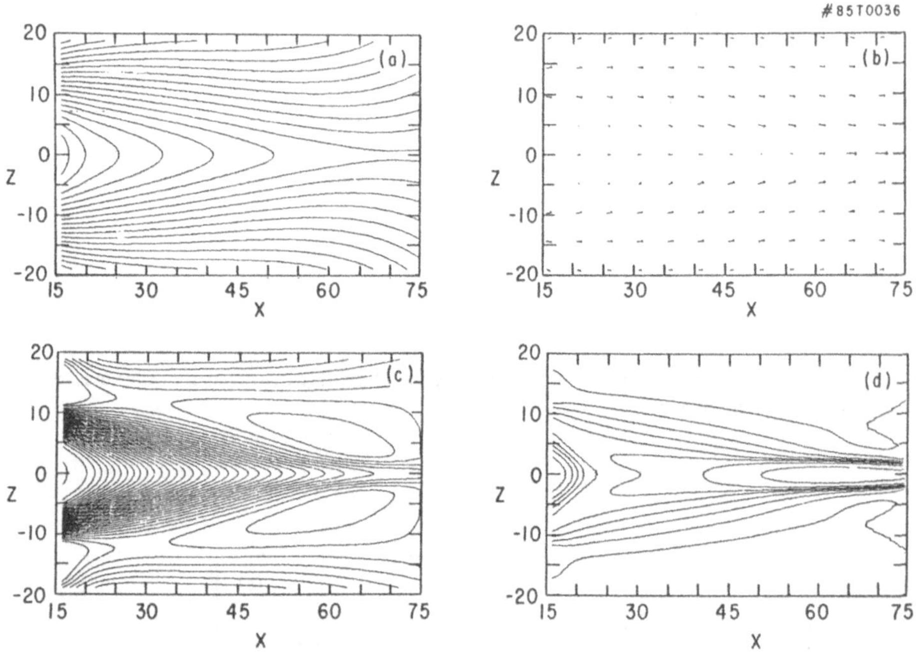

Fig. 4. Substorm recovery phase at $T=12000$: (a) magnetic field lines, (b) velocity, (c) pressure, and (d) cross-tail current density.

4. Discussion

As we have discussed in Section 3, our numerical model is quite similar to the phenomenological model suggested by Hones which is based on satellite observations (HONES, 1984). The results show the plasma sheet thinning in the growth phase, plasma jetting, plasmoid formation and its tailward motion in the expansive phase, and the restoration of the tail-like magnetic field configuration as well as the thickening of plasma sheet in the recovery phase. Although the dependence of the magnetospheric substorm on the entering solar wind energy flux is not discussed in this paper, our previous simulation result showed that the distance to the x-point from the earth and the time scale for the substorm are both decreased when the imposed electric field is increased (MIN *et al.*, 1985).

In the driven reconnection model, the interaction between the entering solar wind and the magnetosphere is important not only in the growth phase but also in the expansive phase as an ongoing process. The fact that the x-point is formed nearer to the earth when the intruding solar wind energy flux is larger (i.e., larger imposed electric field) implies that the solar wind pressure is the cause for the deformation of the magnetic field near the supposed x-point. For the earth side region of the x-point the dipole magnetic pressure is larger than the intruding solar wind pressure, and the magnetic field does not change much, while the tail side region of the supposed x-point where the magnetic pressure is relatively small undergoes a large change in the magnetic field configuration. That is, the position of the x-point of magnetic reconnection can be said to be determined by the pressure relationship (including the magnetic pressure) between the solar wind and the earth's dipole field (including the plasma sheet). During the expansive phase, the x-point does not move so long as the same pressure condition is maintained. However, when the intruding solar wind pressure is reduced, the pressure of the earth's dipole exceeds the solar wind pressure and the x-point is pushed in the tailward direction. In this way the recovery phase can be explained in the driven reconnection model, in contrast to the spontaneous reconnection model in which the final stage of reconnection is probably a nonlinear saturation state. Although our driven reconnection model has a feature of unloading process like a spontaneous tearing model since the pre-stored magnetic energy is converted explosively into the kinetic energy, it also has a characteristic of the driven process in that the system responds instantly to the change of external condition.

Another important fact is that the plasma flow velocity does not necessarily change its direction across the x-point in the downstream region. This is particularly true during the recovery phase in which the x-point moves tailward and the magnetic tension is reduced. Pressure grdient may also play an important role because the plasma β ratio is large. One should note that the plasmas are not accelerated abruptly at the x-point. Rather, the plasma jetting occurs due to the successive acceleration by the magnetic tension.

Since our model is two dimensional, the three dimensional effects such as the field aligned currents and the variation in the dawn dusk direction are not observed, although these effects are thought to be important in the real situation.

5. Summary

By studying two-dimensional MHD simulation, we were able to show that the driven reconnection model can explain the magnetospheric substorm phenomena including the recovery phase to a good extent. We observed the plasma sheet thinning in the growth phase, plasma jetting, pasmoid formation and its tailward motion in the expansive phase. By reducing the entering solar wind energy flux the substorm recovery phase was realized. Tailward motion of the x-point and the broadening of the plasma sheet were observed. These results well agree with the substorm model made by Hones inferred from the satellite observations. We believe that magnetic reconnection driven by the solar wind entering the magnetosphere can be a viable mechanism of the magnetospheric substorm.

This work was supported by the National Science Foundation grant ATM83-11102 and by the U.S. Department of Energy contract DE-AC02-76-CHO-3073.

REFERENCES

AKASOFU, S.-I., C. OLMSTED, E. J. SMITH, B. TSURUTANI, R. OKIDA, and D. N. BAKER, Solar wind variations and geomagnetic storms: A study of individual storms based on high time resolution ISEE 3 data, *J. Geophys. Res.*, **90**, 325, 1985.

BIRN, J., Computer studies of the dynamic evolution of the geomagnetic tail, *J. Geophys. Res.*, **85**, 1214, 1980.

BIRN, J. and E. W. HONES, Jr., Three dimensional computer modeling of dynamic reconnection in the geomagnetic tail, *J. Geophys. Res.*, **86**, 6802, 1981.

BRECHT, S. H., J. G. LYON, J. A. FEDDER, and K. HAIN, A time dependent three-dimensional simulation of the earth's magnetosphere: Reconnection events, *J. Geophys. Res.*, **87**, 6098, 1982.

HONES, E. W. Jr., Plasma sheet behavior during substorms, in *Magnetic Reconnection in Space and Laboratory Plasmas*, edited by E. W. Hones, Jr., AGU, Washington D.C., 1984.

HUBA, J. D., N. T. GLADD, and K. PAPADOPOULOS, The lower hybrid drift instability as a source of anomalous resistivity for magnetic field line reconnection, *Geophys. Res. Lett.*, **4**, 125, 1977.

LEBOEUF, J. N., T. TAJIMA, C. F. KENNEL, and J. M. DAWSON, Global simulation of the time-dependent magnetosphere, *Geophys. Res. Lett.*, **5**, 609, 1978.

LEBOEUF, J. N., T. TAJIMA, C. F. KENNEL, and J. M. DAWSON, Global simulations of the three-dimensional magnetospher, *Geophys. Res. Lett.*, **8**, 257, 1981.

LYON, J., S. H. BRECHT, J. A. FEDDER, and P. J. PALMADESSO, The effect on the earth's magnetotail from shocks in the solar wind, *Geophys. Res. Lett.*, **7**, 712, 1980.

LYON, J., S. H. BRECHT, J. D. HUBBA, J. A. FEDDER, and P. J. PALMADESSO, Computer simulation of a geomagnetic substorm, *Phys. Rev. Lett.*, **46**, 1038, 1981.

MIN, K., H. OKUDA, and T. SATO, Numerical studies on magnetotail formation and driven reconnection, *J. Geophys. Res.*, **90**, 4035, 1985.

SATO, T., Strong plasma acceleration by slow shocks resulting from magnetic reconnection, *J. Geophys. Res.*, **84**, 7177, 1979.

SATO, T. and A. HASEGAWA, Externally driven magnetic reconnection versus tearing mode instability, *Geophys. Res. Lett.*, **9**, 52, 1982.

SATO, T., R. J. WALKER, and M. ASHOUR-ABDALLA, Driven magnetic reconnection in three dimensions: energy conversion and field aligned current generation, *J. Geophys. Res.*, **89**, 9761, 1984.

SCHINDLER, K., Spontaneous reconnection, in *Magnetic Reconnection in Space and Laboratory Plasmas*, edited by E. W. Hones, Jr., AGU, Washington D.C., 1984.

TANAKA, M. and T. SATO, Simulations on lower hybrid drift instability and anomalous resistivity in the magnetic neutral sheet, *J. Geophys. Res.*, **86**, 5541, 1981a.

TANAKA, M. and T. SATO, Multiple excitation of lower hybrid drift waves in the neutral sheet, *Phys. Rev. Lett.*, **47**, 714, 1981b.

5. SUMMARY PANEL

Solar Wind-Magnetosphere Coupling, edited by Y. Kamide and J. A. Slavin, 769–807.

Coupling between the Solar Wind and the Earth's Magnetosphere: Summary Comments

R. A. WOLF[1], S.-I. AKASOFU[2], S. W. H. COWLEY[3], R. L. MCPHERRON[4], G. ROSTOKER[5], G. L. SISCOE[6], and B. U. Ö. SONNERUP[7]

[1]*Department of Space Physics and Astronomy, Rice University, Houston, Texas 77251, U.S.A.*
[2]*Geophysical Institute, University of Alaska, Fairbanks, Alaska 99701, U.S.A.*
[3]*Blackett Laboratory, Imperial College of Science and Technology London SW7 2BZ, U.K.*
[4]*Institute of Geophysics and Planetary Physics, University of California, Los Angeles, Los Angeles, California 90024, U.S.A.*
[5]*Department of Physics, University of Alberta, Edmonton, Alberta, Canada T6C 2J1*
[6]*Department of Atmospheric Sciences, University of California, Los Angeles, California 90024, U.S.A.*
[7]*Thayer School of Engineering, Dartmouth College, Hanover, New Hampshire 03755, U.S.A.*

Presented in the form of sections written by individual members of the summary panel for the AGU-Chapman Conference on Solar-Wind/Magnetosphere Coupling, this paper attempts to summarize the Conference. There is now abundant observational evidence that reconnection occurs at the magnetopause; however, the process occurs in a more complicated way than originally described by Dungey. Uncertainty remains with regard to the detailed physics of the process, to the way it occurs in the complicated three-dimensional geometry of the real magnetopause, and to its regulation by the solar wind. There is also evidence of the existence of a closed-field-line boundary layer, similar to what was originally proposed by Axford and Hines. Some quantitative arguments suggest that this closed-field-line boundary layer plays only a minor role in driving magnetospheric convection for conditions when convection is strong; however, this point is still highly controversial. In the AMPTE mission, attempts at magnetospheric detection of Li^+ or Ba^+ ions released in the magnetosheath were not successful. However, AMPTE is providing the first direct composition measurements of magnetospheric ions for the full range of ring-current energies. Near the peak of the one magnetic storm studied so far, O^+ ions constituted approximately 25–30% of the energy in ring-current particles, for the energy range 1–300 keV/e. With regard to the controversy between the "directly driven" picture of the substorm and the "loading/unloading" picture, there seems to be agreement that the substorm phenomenon is, to a large extent, driven by the solar wind, but that storage and release of energy in the magnetosphere is also significantly involved. The relative importance of the two processes must vary from substorm to substorm, and from one substorm phase to another. Disagreement remains as to which of these basic phenomena should be regarded as more important overall.

1. Introduction

This paper is a written version of the comments made by members of the summary panel at the end of the AGU Chapman Conference on Solar-Wind/Magnetosphere Coupling. The panel was charged with summarizing our understanding of solar-wind/magnetosphere coupling, in the light of the developments and ideas discussed at the conference.

Owing to the large scatter in the personalities and interests of the various panelists, the following set of flexible ground rules was decided upon:

1. Each panelist was asked to review a specific subject area; these subject areas corresponded generally, but not exactly, to conference sessions;
2. However, each panelist was also allowed to devote a substantial part of his presentation to a conference topic that he felt was particularly crucial, even if it lay partially or entirely outside his assigned subject area;
3. Panelists were not encouraged to try to mention every conference paper relating to his assigned subject area;
4. Each panelist was asked to write up his comments in the form of an individually signed section of this paper.

The resulting summary paper is, of course, uneven in style. We hope, however, that it conveys the spirit of the discussions at the conference. It should also serve to exemplify the present state of our understanding of solar-wind/magnetosphere coupling.

Section 2, by B. U. Ö. Sonnerup, summarizes the "Solar Wind and Magnetospheric Configuration" session of the conference, with emphasis on magnetic reconnection at the magnetopause. Section 3, by G. Rostoker, summarizes the "Solar Wind and Geomagnetic Activity" session, and argues that more serious consideration should be given to viscous-interaction processes at the magnetopause. Section 4, by R. L. McPherron, summarizes the "AMPTE Mission" session. The "Polar Cusp, Auroral Zone, and Polar Cap" session is summarized in Section 5, by R. A. Wolf, and in Section 6, by S.-I. Akasofu. Section 7, by R. L. McPherron, summarizes the "Magnetotail" session. Sections 6 and 7 together serve to summarize the present state of the argument between proponents of the "driven" and "unloading" views of the magnetospheric substorm; there is agreement that both driven and unloading processes occur in substorms; Akasofu argues that driven processes, which in the past were underemphasized, may in fact be dominant; McPherron addresses several confusions in substorm terminology and describes how the reconnection model combines driven and unloading aspects. Section 8, by S. W. H. Cowley, summarizes the status of theoretical work on solar-wind/magnetosphere processes, particularly with reference to the papers presented in the "Theoretical Modeling and Simulation" session.

In Section 9, G. L. Siscoe broadly summarizes the conference, and the present status of solar-terrestrial physics, by comparing it to the historical development of the older science of meteorology. Analogies are pointed out between key concepts in meteorology and corresponding concepts in magnetospheric physics. Siscoe emphasizes the relative lack of comprehensive *in situ* data as the central limitation to the rate of progress in magnetospheric research.

2. Reconnection at the Magnetopause

B. U. Ö. SONNERUP

The papers presented at this conference have provided a rich variety of direct and indirect evidence for the occurrence of magnetic coupling across the magnetopause. The direct evidence is in the form of flux transfer events (FTEs) and other reconnection events observed in plasma, magnetic field, and energetic particle data, as well as in the form of measured electric fields tangential to the magnetopause (LINDQVIST and MOZER, this conference, 1985; KNOTT and PEDERSEN, this conference, 1985). The indirect evidence is extraordinarily varied, ranging from the influence, not only of the interplanetary B_z component but also of the B_y and B_x components on magnetospheric convection and polar cap behavior, to the twisting of the tail neutral sheet (SIBECK *et al.*, this volume, 1986) and the organization of plasma entry across the tail magnetopause caused by B_y, and to the identification of polar-rain electrons with the solar-wind Strahl component (FAIRFIELD and SCUDDER, this volume, 1986). It is striking that simple geometrical models of the coupling, obtained by the superposition of vacuum or other simple magnetic fields (CROOKER, this volume, 1986), seem to be capable of predicting in a qualitative way many of the effects of the interplanetary magnetic field (IMF) on convection in the polar-cap ionosphere, including patterns obtained during northward B_z, as well as effects such as the partial "penetration" of B_y deep in the magnetosphere (HUGHES and COWLEY, this volume, 1986) and tail.

The evidence available to date suggests that most of the potential drop across the polar caps during southward B_z is a direct result of the magnetic coupling and that only 10–20% (BAUMJOHANN and HAERENDEL, this volume, 1986; REIFF and LUHMANN, this volume, 1986) may be associated with viscosity of one form or another (including momentum transfer resulting from turbulent mixing via the Kelvin-Helmholtz instability). Since the magnetic coupling, which involves some 10–20% of the total flux incident upon the magnetosphere is a direct result of reconnection (merging) operating somewhere on the dayside magnetopause, the single most important fundamental question relating to solar-wind magnetosphere coupling is:

How and where does reconnection occur at the magnetopause and which solar-wind parameters control its occurrence?

Reconnection in the geomagnetic tail is a problem of comparable importance and interest. However, it can be viewed as an unavoidable consequence of reconnection at the magnetopause, specifically as being required to avoid excessive accumulation of flux on open field lines in the tail lobes.

There is general agreement that reconnection is initiated by the tearing mode but it is not clear how this instability evolves three-dimensionally and/or in the nonlinear regime. A two-dimensional MHD simulation of tearing in the presence of stagnation point flow of the type expected near the subsolar point was presented by LEE (this volume, 1986). Depending on the boundary conditions, the calculations show, either the continual formation, nonlinear growth, and transport along the magnetopause of magnetic islands, or the establishment of a quasisteady standard reconnection configuration. LEE and FU (1985) have associated the former situation, which occurs when the system length along the magnetopause is much greater than the magnetopause thickness, with FTEs, although the actual process of establishing the interconnection between the helical ($B_y \neq 0$) fields in the islands and the magnetosheath and magnetospheric fields cannot be dealt with in a two-dimensional model.

The main difference between the Lee-Fu FTE configuration and that envisaged by RUSSELL and ELPHIC (1978) is that in the former model, the FTE flux tubes are the magnetic islands themselves, whereas in the latter they are holes in the magnetopause between islands (e.g., SONNERUP, 1984). In the former model, the FTE flux tube is embedded in, and forms part of the dayside magnetopause over a large longitude segment, whereas in the latter it crosses the magnetopause more or less directly. These geometrical diferences should be tested against a variety of observations both locally at the magnetopause and at the FTE "footprints" in the ionosphere (GOERTZ *et al.*, 1985).

Observations at the dayside magnetopause indicate that FTEs are far more frequent than quasisteady reconnection events. Data presented by SMITH *et al.* (this volume, 1986b) and by CROLEY *et al.* (this conference, 1985) at this conference also show that FTEs may have very small as well as extremely large magnetic amplitudes.

Another aspect of magnetopause reconnection, whether in a steady state or as FTEs, is the hypothesis of "antiparallel merging," advanced by CROOKER (1979; this volume, 1986). A possible physical basis for this idea is that a finite magnetic-field component along a reconnection line may allow effective short circuiting by the electrons of the reconnection electric field. The alternate hypothesis of "component merging" has its roots in MHD theories of reconnection in which a magnetic-field component along a reconnection line has little or, in incompressible models, no influence on the process. There is an urgent need to establish which of these assumptions more nearly represents the actual situation. The antiparallel merging hypothesis usually implies reconnection sites away from the subsolar stagnation point region, e.g., near the cusps. One potential difficulty with such sites is that a reconnection line would tend to be swept downstream by the prevailing plasma velocity component transverse to the line. Perhaps new reconnection lines are formed and shed continually at such locations (this may also be the case at the reconnection site in the distant tail current sheet where the flow is predominantly tailward).

To date, the search for key parameters in the solar wind which influence the occurrence of reconnection at the magnetopause has placed heavy emphasis on the sign and magnitude of the interplanetary B_z component (e.g., KING, this volume, 1986). However, the process must be governed by a set of nondimensional groups of which the angles defining the IMF relative to the earth's dipole are but two. Other potentially important parameters are the Alfvén Mach number, the plasma beta value and the pressure anisotropy, as well as the fluctuation level in the magnetosheath adjacent to the magnetopause. As discussed by WHANG *et al.* (this conference, 1985), solar-wind fluctuations are modified as they cross the bow shock and, in addition, the shock manufactures its own fluctuations (e.g., VENKATESAN *et al.*, this volume, 1986). The tearing mode, which normally evolves slowly, may be driven to rapid growth by the ponderomotive force of impinging magnetosonic fluctuations (SAKAI and WASHIMI, 1982), by ion pressure anisotropy ($p_\perp > p_\parallel$; CHEN and PALMADESSO, 1984) and by low plasma density (QUEST and CORONITI, 1981).

The solar wind also contains an assortment of large-scale discontinuities and other structures: the impact of interplanetary shocks on the magnetosphere has been studied extensively (e.g., SMITH *et al.*, this volume, 1986a; KORTH *et al.*, this volume, 1986; KING, this volume, 1986), but the interactions at the magnetopause resulting from the impact of solar-wind rotational and tangential discontinuities, including the sector boundaries, are less well understood. These interactions include the initiation and cessation of reconnection at various locations on the magnetopause.

In an average sense, the magnetospheric shape appears to be approximately that predicted by gasdynamic models which do not take into account reconnection or indeed any magnetic stresses, even though, in those models, such stresses become sufficiently large near the magnetopause to be capable of modifying the flow substantially. However, magnetopause reconnection is known to produce systematic large-scale configuration changes (RUSSELL, this volume, 1986), and the possiblity exists that large local deformations of the magnetopause may also occur, in particular during active solar conditions. For example, magnetosheath flux tubes may get hung up on the frontside magnetosphere or in the cusps. The resulting indentations could then serve as suitable reconnection sites and, when reconnection occurs, further deepening of the indentations and other large modifications of field and flow may be expected.

In summary, while much has been learned about magnetopause reconnection and its various effects in the magnetosphere, the basic physics of the process remains poorly understood. Continued studies of this and other magnetopause processes will require new high-resolution observations in the magnetopause region and continued high-quality information concerning the properties of the incident solar-wind plasma (KING, this volume, 1986; FREEMAN and LOPEZ, this volume, 1986), as well as new and more realistic computer simulations.

3. Solar-Magnetospheric Coupling and Geomagnetic Activity

Gordon ROSTOKER

Researchers studying the origin of geomagnetic activity are inevitably interested in the means by which solar wind energy can penetrate into the magnetosphere. At this meeting, it was evident that present views favor the concept that energy enters the magnetosphere through the connection of the interplanetary magnetic field (IMF) with frontside earth dipolar field lines with reconnection of the merged field lines occurring somewhere back in the magnetotail. That such field line reconnection should take place when the IMF is varying with time is evident from Faraday's Law

$$\frac{d\Phi_B}{dt} = -\oint E \cdot dl$$

from which it is clear that a non zero electric field exists integrated over the length of the separator line. If there is no time rate of change of magnetic field, then $\oint E \cdot dl = 0$ along the separator line; however, as pointed out by COWLEY (1984a), this does not rule out the possibility of reconnection, as E could have one sense of direction along part of the separator line and the opposite sense along another part of that line such that, overall, $\oint E \cdot dl = 0$. Thus, while time-dependent reconnection must take place in the magnetosphere, the existence and importance of steady-state reconnection depends on an appropriate variation of E along the separator line. Most reconnection studies deal only with steady state aspects and I found it interesting that, at this Chapman conference, magnetospheric effects emphasizing time-dependent reconnection were presented by both FEJER (this volume, 1986) (equatorial ionosphere effects) and HEIKKILA (this volume, 1986) (entry of solar wind plasma into the boundary layers). The need to pay more careful attention to time dependent reconnection has been emphasized by BRATENAHL and BAUM (1984), and I would suggest that their concern is well warranted. The situation is made more complex in that the magnetosphere has a response function of its own to impulsive changes such as solar wind discontinuities in field and/or plasma parameters. The papers on linear prediction filtering by MCPHERRON *et al.* (this volume, 1986) and BARGATZE *et al.* (this volume, 1986a, b) showed the complexity of the response function in terms of the behavior of the geomagnetic auroral electrojet index AL. The fact that a change in IMF conditions which takes place on a time scale of seconds can lead to changes in magnetospheric behavior extending over an interval of up to 2 hours begs the question as to whether or not the magnetosphere is ever in a steady state. Let me quantify that statement by presenting Fig. 1, which shows three groups of three months each (panels a, b, c) and the sum of all three (panel d) in which the variability of the IMF theta angle is monitored. The theta angle is the angle which the magnetic field vector makes with the ecliptic plane, with $\theta=0$ indicating the IMF lies completely within the ecliptic plane. The study started with a $\theta=0$ crossing and tabulated the amount of time between following consecutive $\theta=0$ crossings. Evidently, the probability of having 50 minutes during which the polarity of the IMF B_z is unchanged is very low, so that the magnetosphere rarely can complete its response to one IMF change before a new change occurs. The question of whether or not the magnetosphere adheres to physical processes governed by steady-state conditions or is typically in the midst of some response to a time-dependent effect warrants further consideration.

During the course of his presentation, RUSSELL (this volume, 1986) indicated that he believed that steady-state geomagnetic activity (the geomagnetic signature of which is the so-called convection bay) is also driven by reconnection, but that the process is envisioned to be steady state. The role of steady-state reconnection is, however, a little less clear than that of time-dependent reconnection, and I would like to suggest that the relative contributions of steady-state reconnection and the so-called "viscous interaction" at the magnetopause boundary is still not entirely certain.

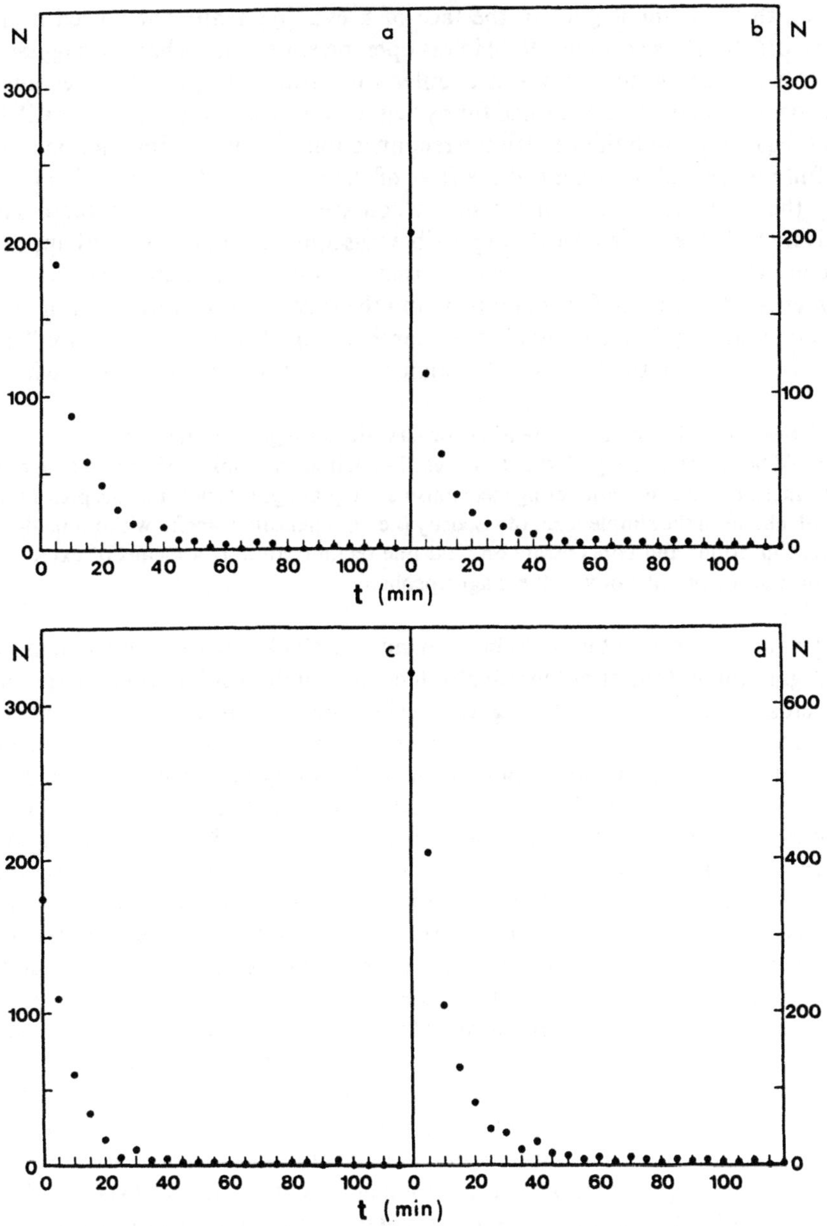

Fig. 1. Variability of the polarity of the north-south component of the interplanetary magnetic field (courtesy of NSSDC). Panels a, b, and c show three separate three-month data sets (Nov. 1978–Jan. 1979, May 1979–July 1979, and Nov. 1979– Jan. 1980, respectively) and panel d shows the sum of the three. Each plot shows the number of cases in which the IMF B_z component switched polarity less than 5 minutes after a prior polarity change, less than 10 minutes after a prior polarity change, ..., and less than 130 minutes after a prior polarity change. This set of results, compiled using IMF J data by T. D. Phan and G. Rostoker, demonstrates that it is not reasonable to assume that the magnetosphere is responding to steady-state conditions under most circumstances.

Making this statement goes in the face of prevailing sentiment voiced clearly by BAUMJOHANN (this volume, 1986) in his opening statement, where he suggested that 90% of geomagnetic activity was accounted for through the process of reconnection, with no more than 10% accounted for by "viscous" effects. One reason for still having doubt regarding this balance between reconnection and viscous interaction is the lack of a full theoretical treatment of the flow of the magnetized magnetosheath plasma along the magnetopause. For reasons which are not entirely clear, there has been relatively little theoretical work on possible "viscous" mechanisms. Perhaps the clear evidence which emerged in the late 1960s that the solar wind-magnetosphere coupling was regulated by the IMF B_z comonent and the success of the reconnection model in predicting this regulation dissuaded reasearchers from further studies of the "viscous" effect. In any event, the last word for some time went to AXFORD (1969) who stated:

> "However, although there are a few observations suggesting that the viscous effect could be important (e.g., FREEMAN *et al.*, 1968), it seems unlikely that we would be so unlucky to have two competing mechanisms of equal significance. For the present, we will abandon the simple form of viscosity (i.e., momentum transfer without field-line reconnection) and concentrate on field-line reconnection as a means of exerting a shear stress on the body of the magnetosphere."

I would suggest that it might be even more "unlucky" if we were not to more fully investigate the potential of "viscous" effects in coupling solar wind energy into the magnetosphere. I say this in the light of the fact there seems to be a prevalent ongoing *a priori* assumption that the "viscous" interation is *not* modulated by the IMF B_z and represents some weak residual capable of contributing no more than ~20 kV to the cross polar cap potential drop. An ~15–20 kV potential drop exists across the polar cap during intervals of northward IMF (WYGANT *et al.*, 1983), which researchers are apparently willing to attribute, in part, to the viscous interaction. However they assume this contribution will remain unchanged even if the IMF begins to turn towards the south. If, on the contrary, the "viscous" interaction is modulated by the IMF B_z component, the clearly prevalent view that reconnection is the only entry mechanism of import would warrant re-examination. The recent study by MIURA (1984) involving numerical simulation of the non-linear Kelvin-Helmholtz instability (KHI) in a velocity shear zone has shown that, even for unfavorable conditions involving northward IMF, the potential drop across the low latitude boundary layer would be ~10–15 kV. This result, stemming from numerical simulations, begs for re-examination of the behavior of momentum transport across velocity shear zones in the presence of magnetic shear (relevant for episodes of southward IMF B_z). While the growth of the KHI for various orientations of the IMF has been well studied (e.g., LEE and OLSON, 1980) the regulation of momentum transport by the orientation of the IMF B_z component has not yet been adequately addressed. It is highly likely that the presence or absence of magnetic shear across a velocity shear zone is an important factor although whether the presence of magnetic shear favors or militates against momentum transport from the magnetosheath to the boundary layers is not intuitively obvious. It warrants study and, at this Chapman Conference, an answer was not forthcoming to this question. At this point in time, we seem to have four mechanisms which migth contribute to the entry of magnetosheath plasma into the earth's low latitude boundary layer (viz. diffusion (EVIATAR and WOLF, 1968; TSURUTANI and THORNE, 1982), anomalous momentum transport through the action of the Kelvin-Helmholtz instability (MIURA, 1984),impulsive mass transport through the actions of inductive electric fields (HEIKKILA, 1982, and this volume, 1986), and gradient drift as espoused by OLSON and PFITZER (this conference, 1985)). In fact, all of these mechanisms may be operative to some extent and the total amount of energy entering the low latitude boundary layer may represent the sum of the individual components. The clear evidence that geomagnetic activity is modulated by the IMF B_z component, however, demands that the "viscous" interaction be similarly modulated if it is going to be able to vie with reconnection as a process which can be responsible for facilitating entry of the solar wind energy required to power magnetospheric activity. The only potential for that behavior seems to lie with the anomalous momentum transport process associated with non-linear development of the KHI.

Thus, the message I wish to convey in this summary is that the relative roles of magnetic field line reconnection and "viscouslike" processes in powering the conventional auroral processes is not yet a closed book. Resolution of the question of whether or not anomalous momentum transport across a velocity shear zone is affected by magnetic shear will go a long way to answer this question. Finally, the question of whether or not reconnection processes should be treated as time dependent, as opposed to a steady state remains open. The typically fluctuating IMF B_z component and the long ($\sim$2 hours) impulse response time of the magnetosphere calls for further theoretical research to investigate this problem.

4. AMPTE

R. L. MCPHERRON

The Active Magnetospheric Particle Tracer Explorer (AMPTE) mission consists of three spacecraft (MCENTIRE, this conference, 1985). These include the Ion Release Module (IRM), the United Kingdom Spacecraft (UKS), and the Charge Composition Explorer (CCE). In August 1984 these three spacecraft were launched into eccentric earth orbits, with the IRM and UKS having apogees at 18.7 R_E and the CCE at 9 R_E. Initially the apogees were located at 1430 LT, but subsequently precessed to earlier local times. By mid-September they reached local noon, and the first ion release was performed when the spacecraft were near apogee in the solar wind. Three months later a second ion release was performed in the dawn magnetosheath. A third and fourth release are currently planned in March and June 1985 in the center of the magnetotail and the dusk magnetosheath, respectively.

The scientific objectives of the AMPTE mission included four major areas of study. The first is the interaction between different plasmas. These studies are being accomplished by observing the consequences of lithium and barium ion releases in different plasmas such as the solar wind, magnetosheath, and plasma sheet. The IRM spacecraft, from which the ion modules are released, observes the plasmas in the region of release, while the UKS spacecraft observes the plasmas some tens of kilometers away. Additional observations are provided by optical observations from ground and aircraft.

The second, and prime, objective of the AMPTE mission is tracing ions released by the IRM outside the magnetosphere, or deep in the tail, to lower altitudes where they are hopefully detected by the CCE spacecraft. These experiments combine complex numerical models of ion trajectories with observations of ambient conditions to estimate the times of arrivals and fluxes of ions at the CCE spacecraft.

A third major objective is the study of natural ion populations. Instruments on the three spacecraft provide ionic composition measurements over the entire energy range from a few eV to several MeV. Simultaneous observations at two locations near and far from the earth provide information from which one deduces the probable source of magnetospheric ions and the processes that have brought about their energization.

The fourth major objective of the AMPTE mission is the study of natural boundaries. The bow shock, magnetopause, and plasma sheet boundary separate regions of distinct plasma. There is, however, a variety of processes which connect the two sides of the boundaries and cause the transfer of mass, momentum and energy from one side to the other. The high time resolution measurements at separations of tens of kilometers provided by the mother/daughter pair IRM and UKS facilitate studies of such boundaries.

Initial results of the AMPTE mission were presented at this meeting by several investigators. In the area of plasma interactions several points were made. First, both the September release of lithium in the solar wind, and the December release of barium in the magnetosheath produced the expected diamagnetic cavities (Cowley, personal communication). Also as expected, these cavities caused an upstream pile-up of solar wind magnetic field which eventually disappeared as the cavities were swept downstream. An interesting observation of a short-lived natural diamagnetic cavity was also reported just inside the magnetopause (LÜHR *et al.*, this conference, 1985). Electrostatic waves normally present in the plasma disappeared inside this cavity, as they did also inside the artificial cavities. The barium release in the dawn magnetosheath produced a "Christmas Comet" visible from the ground (MCENTIRE, this conference, 1985). This comet formed a double tail from which a piece eventually detached in conjunction with a change in the polarity of the ambient magnetic field.

The attempt to trace lithium ions released in the solar wind was not successful. Trajectory simulations show that a substantial fraction of the released ions should have reached the magnetopause. Estimates of their penetration and subsequent drift led to calculated fluxes at the CCE spacecraft which should have been observable However, none were observed above instrument thresholds.

Several detailed observations of natural ions were reported. On one occasion, an unusually strong solar wind moved the magnetopause sufficiently earthward that the CCE instrumentation was able to make the first measurement of the ratio of carbon to oxygen in the solar wind (IPAVICH *et al.*, this conference, 1985). On another occasion, a moderate magnetic storm made possible the first measurements of ring current composition (STÜDEMANN *et al.*, this volume, 1986). The major components include H^+, He^+, He^{++}, and O^+. The relative composition of these components varies significantly during a storm, with O^+ being most variable and constituting as much as 30% of the ring current energy. In addition to significant time variations in the sources and sinks for these ions, there also appears to be significant charge exchange between different charge states of a given ion. Many energetic heavy ions, probably of solar origin, are also present in the ring current (MCENTIRE *et al.*, this conference, 1985).

In the area of boundary studies, initial results concerning flux transfer events were reported (SMITH *et al.*, this volume, 1986b). It has been found that higher time resolution magnetic field data reveal many small FTE's. The spectrum of these events is very important because many small events can be as effective in flux transfer as a few large events.

5. The Polar Cusp and Auroral Zone

R. A. WOLF

The characteristics of the auroral and polar-cusp regions of the ionosphere, and their dependence on solar-wind conditions, have been studied for many years. Here we briefly comment on a few specific topics discussed at the Conference that seem particularly timely and crucial to understanding the solar-wind/magnetosphere interaction. In what follows, the discussion moves from the ionospheric region most directly driven by the solar wind (i.e., the polar cusp) to regions that are driven less and less directly.

5.1 Throat location when the IMF has an "ortho-gardenhose" orientation

One particularly interesting topic related to the cusp is the location of the "throat" through which convecting flux tubes flow poleward into the polar cap. Observers seem to agree that, in times of southward IMF, when the convection is well defined, the throat is usually located in the pre-noon local-time sector. Thus, the convection pattern, unlike the region-1 and region-2 Birkeland-current patterns (e.g., IIJIMA and POTEMRA, 1978), does not exhibit approximate symmetry about the noon-midnight meridian plane. As discussed by HEELIS (this conference, 1985), this somewhat unexpected asymmetry of the flow pattern may result from the fact that the interplanetary magnetic field is typically oriented near the gardenhose angle. That is, it might be that the throat would be located at local noon if the equatorial projection of the IMF were in the sunward or antisunward direction, or past noon if the IMF projection were nearly orthogonal to the gardenhose angle. HEELIS (this conference, 1985) also showed plasma-flow observations for a period when the IMF was close to this ortho-gardenhose orientation. Although those data suggest that the throat may indeed have been in the local-afternoon sector, more cases need to be examined before the result is conclusive. On the other hand, FRIIS-CHRISTENSEN (this volume, 1986) inferred flow patterns from ground magnetograms for periods when the IMF was near the ortho-gardenhose orientation, and found that the patterns did not differ importantly from those observed when the IMF was near the classic gardenhose angle. The Birkeland currents are a result of finite ionospheric conductance, i.e., they would be zero if the ionosphere were an insulator. If the Birkeland-current pattern generally reflects the approximate symmetry of the solar-wind flow velocity for all IMF orientations, but the ionospheric flow pattern is generally tilted toward earlier local times, then apparently the nonzero conductance of the ionosphere importantly affects the interaction between the solar wind and the magnetosphere at the dayside magnetopause.

5.2 *Plasma-flow, Birkeland-current, and large-scale precipitation patterns in the polar cusp and polar cap for periods of northward IMF*

In the past few years, a great deal of research has been devoted to investigating the dependence of these patterns on IMF, particularly through the analysis of Dynamics Explorer data. The polar caps have been found to exhibit more interesting structure for northward IMF—when the auroral zone is relatively quiet—than for southward IMF—when the auroral zone is active. The situation is too complex for sensible review here. In general, it appears that, although there are still arguments about significant details, some agreement seems to be gradually emerging among the various observers concerning the nature of those patterns, their dependence on the IMF, and the relationship between flows, currents, and aurora (see, e.g., BURCH *et al.*, this volume, 1986; REIFF and LUHMANN, this volume, 1986; ZANETTI and POTEMRA, this volume, 1986; HEELIS, this conference, 1985). Major disagreements persist with regard to the magnetospheric implications of the observed ionospheric pattern for times of northward IMF.

5.3 *Definition of the polar cap for northward IMF*

MENG and MAKITA (this volume, 1986) have made a key point with regard to interpretation of the dynamics of the high-latitude ionosphere in times of northward IMF. According to Meng and Makita, in periods of northward IMF, precipitation resembling a dilute plasma sheet persists to very high latitudes, with only a small high-latitude "island" free of plasma-sheet-like electron precipitation. (See also MENG, 1981, MURPHEE *et al.*, 1982; HARDY *et al.*, this volume, 1986.) Meng postulates that the high-latitude region of low-intensity plasma-sheet precipitation maps to closed plasma-sheet field lines, and he thus suggests that this region be defined as part of the "auroral oval"; the polar cap would then consist of just the small island that is free of electrons above a few hundred electron volts, the precipitation there consisting primarily of polar rain and related very soft precipitation. When combined with the standard open-model picture of the magnetosphere for southward IMF, Meng's picture implies that a turning of the IMF from southward to northward implies a conversion of many open tail-lobe flux tubes into closed, distant-plasma-sheet flux tubes.

5.4 *Substorm sequences of auroral photos*

FRANK (this conference, 1985) showed sequences of auroral images obtained with the DE-1 spacecraft at intervals of approximately twelve minutes. He pointed out that substorm activity seems to start from a localized brightening, an "ignition point", which is initially located near local midnight. This initial bright spot develops and moves westward as the substorm develops.

5.5 IMF-dependence of convection in the lower-latitude part of the auroral zone

The Millstone Hill incoherent-backscatter radar measures line-of-sight plasma velocities for magnetic latitudes of 50–72°, which usually corresponds to the lower-latitude part of the auroral zone, the midlatitude-trough region, and the outer plasmasphere. FOSTER *et al.* (this volume, 1986) reported a statistical analysis of the IMF dependence of the electric fields in this region, which is not magnetically connected to the solar wind or to a magnetopause boundary layer and thus is not directly driven by the solar wind. The large-scale form of the flow patterns observed in this region corresponds to classic convection for all orientations of the IMF, although the statistical analysis exhibits significant IMF dependence of various details of both the flow pattern and the electron-precipitation pattern (which was studied using data from the NOAA/TIROS satellites). The electric field in this lower-auroral/sub-auroral region is important for ring-current injection and the dynamics of the plasmasphere and radiation belts.

5.6 IMF effects at low latitudes

FEJER (this volume, 1986) summarized observational information on the dependence of low-latitude electric fields on the IMF. The low-latitude region is driven only indirectly by the solar wind. The solar-wind influence is attenuated greatly by internal magnetospheric processes, particularly the tendency of the inner edge of the plasma sheet to shield the region equatorward and earthward of it from the magnetospheric-convection electric field. Neutral winds driven by dayside solar heating are relatively large at low latitudes, which complicates the problem of identifying the smaller IMF-associated effects. Furthermore, the IMF affects low-latitude electric fields in two ways—through direct electrostatic penetration and through the effects of magnetosphere-driven neutral winds. HAREL (this conference, 1985) gave a preliminary report on the first serious effort at theoretical calculation of the combination of the latter two effects. According to FEJER (this volume, 1986), the equatorial ionosphere responds to both northward and southward shifts of the IMF. However, its response to a sudden northward turning is considerably stronger than its response to a sudden southward turning.

6. The Polar Caps and Substorms

Syun-Ichi AKASOFU

In the past we have often neglected upper atmospheric phenomena poleward of the auroral oval, because our main interests have traditionally been concentrated in understanding magnetospheric substorms along the auroral oval; substorms are due primarily to the southward component of the interplanetary magnetic field (IMF). After all, it is our recent finding that a large value of the IMF southward component provides perhaps the simplest situation. As the IMF B_z component becomes positive, however, a great variety of fascinating phenomena take place in the highest latitude region of the earth. The other two components (B_x, B_y) become equally or even more important than the IMF $B_z>0$ component in controlling various upper atmospheric phenomena in the polar region.

During the conference, a number of workers have stressed that a full understanding of the solar wind-magnetosphere interaction requires a detailed study of upper atmospheric phenomena in the highest latitude region under conditions of the IMF $B_z>0$, $B_y \gtreqless 0$ and $B_x \gtreqless 0$, as well as of $B_z<0$. Most theorists attempted to explain polar cap phenomena for $B_z>0$ in terms of reconnection, but WU (this conference, 1985) demonstrated by his MHD simulation study that some of the dynamical aspects of the polar cap phenomena can be reproduced without reconnection.

The area surrounded by the auroral oval was given special attention by FRANK (this conference, 1985) and MENG and MAKITA (this volume, 1986), since it might provide quantitative information on the total amount of open flux in the magnetotail, and its time variations. Noting that a weak luminosity might not be detectable by auroral imagers, Meng and Makita used DMSP electron data and demonstrated that the area of the polar cap reaches a maximum value at about the maximum epoch of a substorm (determined by the *AE* index). If this would indeed be the case for most substorms, magnetospheric substorms must *primarily* be due to a directly driven process and cannot simply be caused by unloading of the magnetic energy stored in the magnetotail prior to substorm onset. For details of the cases presented by MENG and MAKITA (this volume, 1986), see MAKITA *et al.* (1985).

FRANK (this conference, 1985) showed three examples of his study of changes of polar cap size during substorms. In the first example, there were two successive substorms. The polar cap size *increased* after the first onset and *decreased* after the second onset. Thus, the first substorm cannot simply be due to unloading of the magnetic energy, while the second one could well be. Further, the second one was associated with a northward turning of the IMF, so that the observed decrease of the polar cap size may partially be due to "switching-off" of the solar wind-magnetosphere dynamo. It is interesting to speculate that this "switching-off" can generate a voltage of $(100\text{H})\times(10^7\text{A})/(10^3\ \text{sec}) \sim 10^6$ volts, triggering a northward turning substorm, which are known to be a subset of substorms (Rostoker, private communication, 1985), comprising not more than 20% of all substorms. It is expected that the polar cap size will increase during a typical substorm.

There is a very unique phenomenon on the earth to monitor the *net* transfer of the magnetic flux. It is the dayside cusp. It has long been known that the cusp shifts equatorward as a substorm intensifies, indicating that more magnetic flux is transferred from the dayside magnetosphere to the magnetotail during the maximum epoch of a substorm in agreement with the observation by MAKITA *et al.* (1985).

A number of papers were presented on the correlation between solar wind/IMF quantities (V, B, VB_s, etc.) and magnetospheric quantities (AE, Dst, U_T, etc.). Figure 2 is a portion of the figure from CLAUER *et al.* (1981), indicating a high degree of solar wind-IMF control of the *AL* index. The fact that one can predict gross time variations of substorms with such high degree of accuracy indicates strongly that magnetospheric

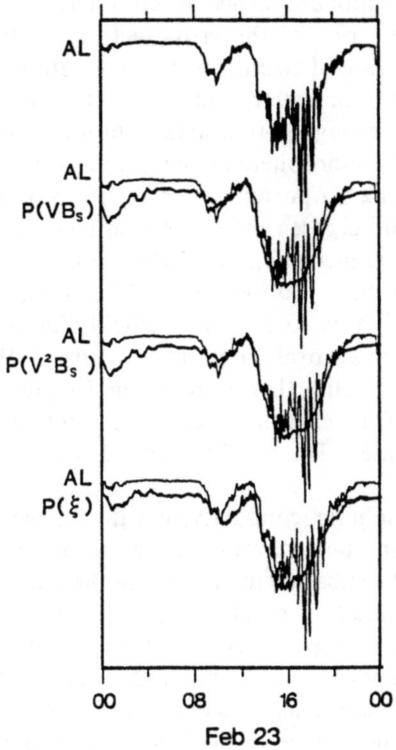

Fig. 2. The relationship between the geomagnetic *AL* index and the *AL* index predicted (*P*) on the basis of three solar wind parameters VB_s, V^2B_s and epsilon. Taken from CLAUER *et al.* (1981).

substorms are not a catastrophic, explosive and unpredictable phenomenon (as explosive reconnection theories imply), but should *primarily* be a directly driven phenomenon, superposed with secondary internal processes.

Note that the superposed high-frequency component consists of geomagnetic pulsations, which appear to be of internal origin, at least in this particular case. In spite of their appearance in the *AL* index, such pulsations have much less energy than the substorm current system.

There is no doubt that magnetospheric substorms draw energy from both the directly driven and unloading processes. However, in the past *only* unloading processes had been exclusively emphasized and *only* explosive reconnection had been considered as the unloading process. At this point in the history of substorm studies, it is important to make it clear that the directly driven process is at least equally important and is perhaps a primary process.

It is unfortunate that most theoretical studies on substorms still emphasize only explosive reconnection for mainly historical reasons, except for computer simulation studies of driven reconnection (SATO and HAYASHI, 1979; WALKER and SATO, 1984).

I believe that one could probably reconcile the driven aspect of the magnetospheric substorm with the unloading aspect by considering the magnetospheric circuit shown in Fig. 3. It shows one of the cross-sections of the magnetotail and the associated circuits, consisting of two parts, the magnetotail circuit and the auroral circuit which is connected to the polar ionosphere. One can consider a similar circuit for each cross-section of the magnetotail, so that the locus of the contact points of the auroral circuit on the polar ionosphere delineates the auroral oval.

Suppose that the dynamo power increases, from a very low value, say 10^{17} ergs/sec (10^{10} watts), to 10^{19} erg/sec and subsequently declines to below 10^{18} erg/sec. The currents in both the magnetotail loop and the polar circuit will increase. As the power of the dynamo reaches $\sim 10^{18}$ erg/sec (10^{11} watts), the field-aligned current density will become $\sim 10^{-6}$ A/m^2. As a result, a field-aligned potential drop of a few kilovolts will develop in the auroral potential structure, so that the current-carrying electrons are accelerated, enabling them to reach the E region of the ionosphere and to cause ionization there. The resulting sudden brightening of the aurora may be identified as substorm onset. The conductivity of the ionosphere will increase at substorm onset.

The increased ionospheric conductivity will change the proportion of the currents in the tail and the polar circuit, while the power is increasing from 10^{18} erg/sec to 10^{19} erg/sec. The subsequent changes of the polar cap size depend on how the power varies further, and how much the ionosphere draws the current along the auroral circuit. If the increasing power (from 10^{18} erg/sec to 10^{19} erg/sec) is sufficient to increase the currents in both the cross-tail and auroral circuits, both the open flux and the AE index will increase. There is a tendency for many magnetotail workers to identify an increase of the magnetic field B as a signature of "growth phase," regardless of what is in progress in the auroral ionosphere (even though an auroral substorm may be in progress). From the above considerations, there is no

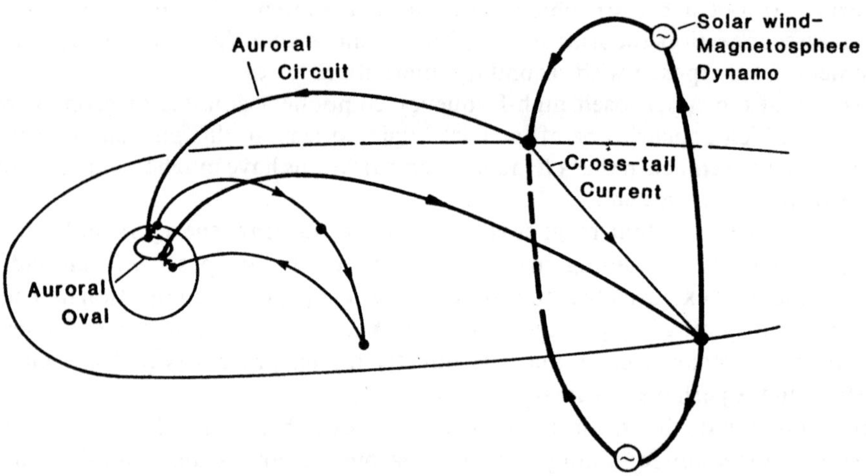

Fig. 3. Schematic illustration of the solar wind-magnetosphere dynamo and the two circuits, the cross-tail and auroral circuits.

contradiction in having simultaneously an increasing B in the magnetotail and an auroral substorm. In any case, a variety of cases are expected to occur, depending on time variations of the power and of the ionospheric conductivity. Another possibility is that some internal processes modulate the output when it is driven. Therefore, our attention should not be directed to the "directly-driven versus unloading" controversy, but to the relative importance between them.

One must be cautious in inferring the magnetic energy in the magnetotail on the basis of the magnetic field measurement in the magnetotail alone, since B changes may be affected by changes of the cross-section of the plasma sheet and of the magnetotail. Plasma sheet thinning alone could reduce B as much as 30% by assuming that the half-thickness of the plasma sheet decreased from 4 R_E to 0.5 R_E (assuming also the constant radius of the magnetotail cross-section to be 15 R_E), so that the argument presented by FAIRFIELD (this volume, 1986) requires further study.

We would expect that the magnetic energy in the cross-tail current is unloaded as soon as the power begins to decline. However, the majority of substorms seem to subside within an hour after the input power drops below 10^{18} erg/sec. Therefore, there is not much energy available in the magnetotail for substorms, unless the solar wind-magnetosphere dynamo power is above 10^{18} erg/sec. This point is illustrated clearly by Fig. 4, in which magnetospheric disturbances (measured in terms of the *AE* and *Dst* indices) subsided soon after the northward turning of the IMF, which caused the input power to drop below 10^{18} erg/sec. Substorm activity subsided quickly despite the long period of southward IMF (implying, to some workers, storage of a large amount of magnetic energy in the tail).

SLAVIN *et al.* (this volume, 1986) presented an interesting observation of substorm effects in the distant tail. One of the most interesting features in their study is that both the anti-sunward flow speed ($-V_x$) and the southward magnetic field ($-B_z$) are well correlated with the *AE* index. Since a high degree of the correlation between the *AE* index and the solar wind IMF quantities (B_s, VB_s, V^2B_s, etc.) was demonstrated during the conference, this result implies that the flow speed is well controlled by the solar wind/IMF quantities. Taking into account the observation by REIFF *et al.* (1981), both the sunward and anti-sunward flows from the reconnection region are indeed well controlled by the solar wind-magnetosphere dynamo power. If explosive aspects of substorms were dominant, the high correlation between *AE* and V_x would not be expected.

In summary: Like any natural phenomena, magnetospheric substorms are a multi-dimensional pheomenon. One aspect (dimension) will give us an impression that the substorm is an unloading process, while another aspect (dimension) will give us an impression that the substorm is a directly driven process. The issue is the relative importance between the two processes. The relative importance may change from substorm to substorm and also during individual substorms. However, my present view is that the directly driven process is the primary one.

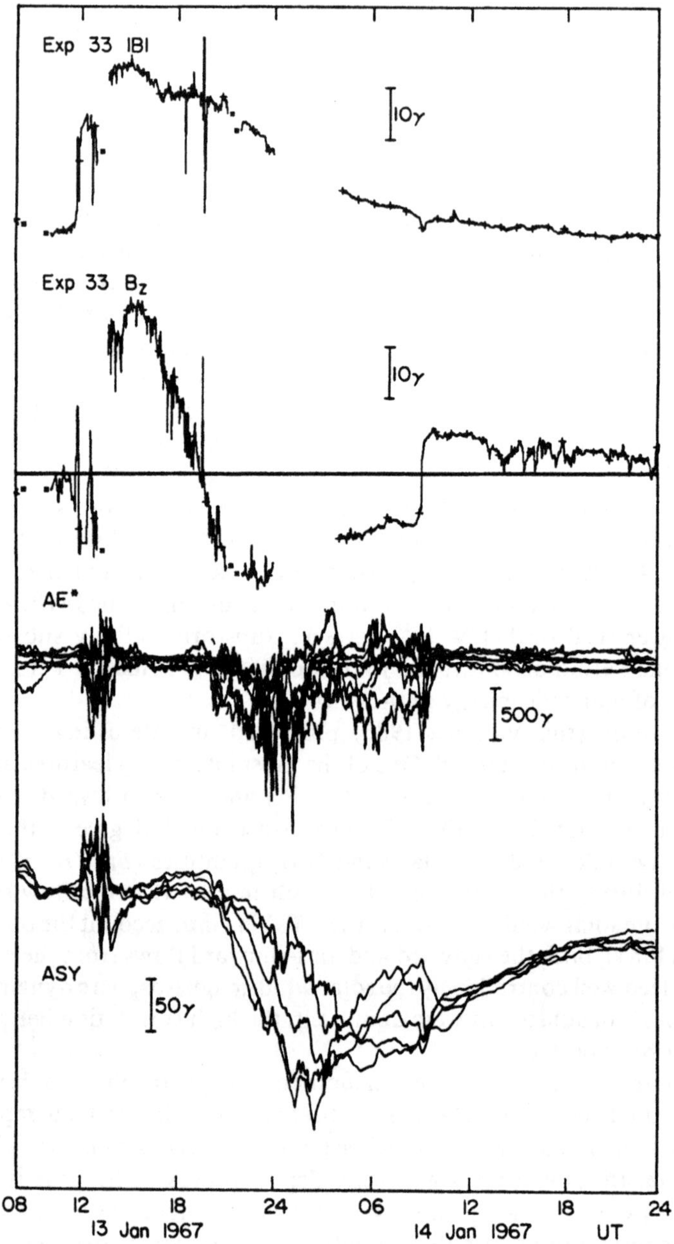

Fig. 4. The IMF magnitude B, the north–south component B_z and the combined high latitude (AE) and low latitude (ASY) magnetic records. Note that the storm practically ended soon after a sharp northward turning of the IMF at about 10 UT on January 14, 1967.

7. Comments on the Magnetotail and Substorms

R. L. MCPHERRON

In his review of magnetotail phenomena, LUI (this volume, 1986) concluded that the processes which are responsible for a substorm are still not understood and that the subject is highly controversial. FAIRFIELD (this volume, 1986) took issue with this conclusion, pointing out that advocates of reconnection were well aware of alternate models and had incorporated many of their features in their own models. Siscoe commented that the problem appeared to him to be a matter of semantics with the proponents of the two major models using the same words to describe different phenomena.

The foregoing remarks, which occurred during the magnetotail session, indicate that many researchers are not yet convinced that the reconnection model is a valid explanation of substorm phenomena. Furthermore, it would appear that they do not even agree on the definition of some fundamental terms. These definitions and their relation to the reconnection model are the topics I address here.

First, what is a magnetospheric substorm? As originally defined (CORONITI *et al.*, 1968), it was an extension of the concept of an auroral substorm (AKASOFU, 1964) to include the collection of phenomena that occur throughout the magnetosphere at the time of an auroral and magnetic substorm. Today these phenomena are recognized as being associated with two major processes; the driven process and the unloading process. The driven process is often referred to as magnetospheric convection, while the unloading process is referred to as explosive reconnection, near-earth neutral line, or plasmoid formation. The driven process is a direct manifestation of solar wind coupling to the magnetosphere, while the unloading process is a delayed manifestation.

In contrast to Siscoe's view, I believe there does exist a real controversy concerning the nature of substorms. In the driven model of substorms (AKASOFU, 1981), it has been suggested that most substorm phenomena can be explained as a consequence of a direct electrical connection between the solar wind and the ionosphere. In this model, loading and unloading of energy in the magnetotail either do not occur or are dynamically insignificant in terms of energy dissipation in the magnetosphere. Thus Akasofu has referred to the magnetospheric substorm as "...closer to a driven process than an unloading process" (AKASOFU, 1981). In contrast, the reconnection model incorporates both driven and unloading processes, and its proponents argue that their relative importance varies from substorm to substorm such that either may be dominant (BAKER *et al.*, 1984). However, although driven processes are contained in the reconnection model, most studies of substorms in the magnetotail emphasize the formation and evolution of the near-earth neutral line and how it is responsible for the release of energy stored inthe magnetotail lobes. This emphasis has led some observers to conclude that proponents of reconnection believe the magnetosphere is primarily an unloading system.

In the original definition of a substorm (AKASOFU, 1964) there were two phases of a substorm: expansion and recovery. I later altered this definition by including a growth phase (McPHERRON, 1970). In the reconnection model (McPHERRON, 1979), the growth phase corresponds to the driven process, storing energy in the magnetotail and developing conditions in the near-earth plasma sheet that allow the formation of a localized neutral line. The expansion phase corresponds to the evolution of this neutral line as it reconnects, first, plasma sheet flux, and later, lobe flux. The recovery phase corresponds to the retreat of this neutral line to the distant tail and the return to quiet conditions. In the driven model of substorms, there is no growth phase in the sense of establishing conditions for energy release. Furthermore, the expansion phase is no longer viewed as an explosive phenomenon, but instead is viewed as the movement of a slowly varying, localized, ionospheric current across one of a few sparsely distributed observatories. Sudden brightening of the aurora is attributed to an instability of field-aligned currents in the driven system. The recovery phase is explained as a reduction in the coupling currents.

Difficulties arise in the use of the concepts of substorm phases when magnetic activity becomes more complex. For example, it has been known for some time that energy may be stored in the tail at the same time as it is being released. Thus growth and expansion phases may overlap as they do during multiple-onset substorms. Similarly, it has been suggested that the magnetosphere may reach an equilibrium state in which energy stored in the tail is released through a previously formed near-earth neutral line as rapidly as it arrives. This is the so-called "convection bay" (PYTTE *et al.*, 1978), and in a very real sense represents a magnetosphere which is a driven system. There has been some confusion recently about the use of the terms "substorm" and "substorm onset". The term substorm usually refers to the entire sequence of events, from the first appearance of magnetic activity to the last. In the reconnection model this corresponds to a distinct interval of southward interplanetary magnetic field. During this interval there may be one expansion phase (the classic substorm), a sequence of isolated expansion phases (a substorm sequence), a set of overlapping expansion phases (multiple onsets or intensifications), or a steady-state expansion phase (convection bay). In contrast to this usage, some authors use the term substorm to refer to an individual expansion phase, implicitly neglecting the other phenomena which led up to the expansion, or which follow it. Such usage implies that the process responsible for an individual expansion is the dominant process in a substorm.

The term onset originally denoted the beginning of the expansion phase of a substorm. In the reconnection model it still does. However, it has been suggested that during complex activity only the beginning of the first expansion should be so designated, with later onsets referred to as intensifications (ROSTOKER *et al.*, 1980). In the driven model, substorm onset is the beginning of magnetic activity, since advocates of this model apparently believe sudden brightening or appearance of intense magnetic activity near midnight is not significant. These differing uses of the term onset have caused considerable confusion since such usage seems to define both the beginning of the growth phase and the beginning of the expansion phase as the same time. We personally advocate the use of the terms "beginning of a substorm" and "onset of expansion phase" to distinguish what is meant.

These remarks should not be construed as indicating that there are no questions remaining concerning the cause of geomagnetic activity. As apparent in the review of tail observations presented at this meeting, and also in the new results obtained in the deep tail, there are problems with the reconnection model. Perhaps the most significant problem is that what is thought to be the best indicator of a near-earth neutral line, southward magnetic field in the plasma sheet associated with tailward plasma flow, is subject to alternate interpretation. For example, tailward flow is often observed with northward field. Also, southward field and tailward flow are often seen in the plasma sheet boundary layer, where it is possible that the first is caused by waves, or field-aligned currents on the boundary, and the second by beams of oxygen ions streaming outward.

Despite such problems, the near-earth neutral line model has been remarkably successful in organizing observations in the deep tail. It should be remembered that this model has been developed from near-earth observations. The fact that the model successfully organizes other observations as well speaks well for its underlying validity. Alternative models, such as the driven model, have not attempted to explain the tail phenomena so closely correlated with the sudden onset of a substorm expansion phase and hence were incomplete explanations of geomagnetic activity. Recent changes in the driven model (Akasofu, preceding section) include the existence of energy storage and release, but Akasofu still argues that these processes are not the dominant ones. Clearly, an outstanding problem of magnetospheric physics is determining the relative importance of the directly driven and unloading processes, and why the relative importance changes in response to the solar wind.

8. Theory and Simulation

S. W. H. COWLEY

The aim of this section is to provide a brief overview of the current state of theoretical and simulation studies related to solar-wind magnetosphere coupling, setting into context some of the contributions made at this conference. We will also comment on the relative development of theoretical and observational work in various areas.

8.1 Solar wind-magnetosphere coupling at the magnetopause

It is now well established that during intervals of moderate to strong geomagnetic disturbance, magnetospheric dynamics is dominated by a large scale internal convective flow set up by coupling to the external flow of the solar wind. A centrally important question therefore concerns the nature of the boundary processes which result in momentum transport into the magnetosphere, and which hence initiate the convection cycle from which much of magnetospheric structure and dynamics ensues. An impressive body of evidence has by now been gathered in favor of Dungey's open magnetosphere, thus indicating that magnetic reconnection is the dominant coupling process. Current best estimates indicate that other processes probably generate no more than 5–10 kV voltage drops across the magnetosphere out of typical totals of 30–80 kV (e.g., COWLEY, 1984b; and references therein).

The theory of reconnection is thus of considerable importance in magnetospheric physics and has attracted a considerable amount of study, both analytic and numerical, which is discussed at length in the proceedings of another recent Chapman Conference (*Magnetic Reconnection in Space and Laboratory Plasmas*, edited by E. W. Hones, Jr., Geophysical Monograph 30, AGU, Washington, D.C., 1984). Steady-state reconnection is reasonably well-developed theoretically, particularly in the MHD approximation (e.g., HEYN *et al.*, 1985; and references therein), though this clearly has its limitations and, for example, does not describe plasma mixing along reconnected field lines threading the magnetopause. The theory of steady reconnection in a collision-free plasma has yet to reach a degree of development similar to the MHD models. Observationally magnetopause reconnection does on occasion appear to remain sufficiently steady that steady-state equilibria become applicable (SONNERUP *et al.*, 1981). However, flux transfer events (FTEs), believed to be the signature of unsteady, localized reconnection, occur much more frequently near the magnetopause, appearing essentially whenever the solar wind magnetic field has a southward component (RIJNBEEK *et al.*, 1984; BERCHEM and RUSSELL, 1984).

A major theoretical question thus concerns the reason why dayside reconnection appears usually to be unsteady, and the physics behind the formation of the large discrete open FTE flux tubes, whose transverse dimensions is typically $\lesssim 1$ R_E. One approach is via the tearing mode instability, initial results for which have been presented by LEE (this volume, 1986). The outcome of such studies should shed light also on the question of the preferred sites of reconnection at the magnetopause. This topic has been tackled theoretically in recent years by investigating the consequences of plausible assumptions concerning the circumstances under which magnetopause reconnection occurs (CROOKER, this volume, 1986). Experimental results against which such models can be tested are, however, far from complete. A thorough magnetopause survey over a wide range of latitudes and local times is required to address this question. Although dayside reconnection appears to be the dominant contributor to momentum coupling at the boundary, it clearly seems not to be the whole story as far as magnetopause boundary layer formation is concerned. It is known that substantial boundary layers can also exist when the interplanetary field is directed northward (e.g., PASCHMANN *et al.*, 1978), but these have been very little studied, either experimentally or theoretically. A systematic study of boundary layers and how their properties depend upon the direction of the IMF is urgently required. Theoretically, the topic has not progressed very much beyond the enumeration of possible mechanisms such as wave-induced diffusion, gradient drift effects etc. (see, e.g., VASYLIUNAS, this conference, 1985; OLSON and PFITZER, this conference, 1985). Even reconnection remains a possiblitity if the original DUNGEY (1963) mechanism for northward fields is operative, since this process results in the attachment of magnetosheath flux tubes to the dayside magnetosphere. The scenario described by HEIKKILA (this volume, 1986) suffers by requiring large field-aligned voltage drops on either side of the magnetopause in the initial equilibrium which are sustained by unspecified means (see the discussion in COWLEY (1984b)).

8.2 Flow in the polar cusp, polar cap and tail lobes

Given that momentum transfer does occur at the magnetopause the next question concerns the subsequent formation of the geomagnetic tail and the generation of flow in the polar cusp and polar cap ionosphere. Although considerable observational information now exists about this flow and how it responds to the vector direction of the IMF, the theory of this subject is very underdeveloped and has largely been confined to the drawing of cartoons. Ionospheric flows have been derived from given field-aligned current distributions, and from measured magnetic field perturbations, by solving the ionosheric current continuity equation with a given conductivity model (e.g., BARBOSA, this conference, 1985). Ideally, however, one wishes to solve for the flows which would be generated by the stresses arising from a given magnetopause reconnection rate with a given orientation of the IMF. A full solution of this problem will probably require the formulation of a global magnetospheric simulation having a degree of sophistication rather in excess of that which has been achieved to date.

At the present time virtually the only theoretical work done on this important problem has involved mapping solar wind flows along a model magnetic field consisting of the linear superposition of a dipole field with a uniform and steady IMF (STERN, 1973; LYONS, 1985). That this procedure has any validity is worthy of comment, given the fact that we are dealing with a highly conducting plasma medium and not a vacuum or dielectric. The reason why this idealization will give qualitatively correct results is that open flux tubes exert stresses on the plasma that they contain, in general causing the plasma to move such as to reduce the stress. The state with zero electromagnetic stress (zero currents) is the dipole plus uniform field superposition. This model therefore represents the idealized end state of the stress reduction process. A rather more sophisticated development of this idea has been presented by TOFFOLETTO and HILL (this volume, 1986), who impose an internal cross-magnetospheric magnetic field by specifying the normal magnetic flux over the boundary of a more complex magnetospheric model. Magnetosheath flows are then mapped into the magnetosphere along equipotential open flux tubes. This procedure is likely to produce rather more realistic results than the dipole plus uniform IMF idealization, but the results should be qualitatively similar to the latter provided the normal flux distribution over the magnetopause boundary is chosen appropriately. Simple kinematic models of fields and flows in the geomagnetic tail using a similar idea to that adopted by Toffoletto and Hill were previously presented by COWLEY (1981).

8.3 The plasma sheet and inner magnetosphere

Of all magnetospheric domains the most thoroughly studied and understood from a theoretical viewpoint is undoubtedly the inner quasi-dipolar magnetosphere. Following the single-particle studies of the late 1960s and early 1970s which gave a basic picture of ring current formation, the effects resulting from magnetosphere-ionosphere coupling have formed a subsequent focus of enquirey, this work reaching its highest development in the formulation of the Rice Convection Model. Although already sophisticated, this model is still capable of further development, however, through the inclusion, e.g., of neutral winds, precipitation losses of hot plasma to the ionosphere and magnetospheric magnetic field perturbations self-consistently determined from plasma currents (HAREL, this conference, 1985).

Turning finally to the plasma sheet, it seems that the basic picture derived twenty years ago by T. W. Speiser and co-workers from studies of single-particle motion in an electromagnetic field appropriate to reconnection, gives a good zeroth order account of the formation of the accelerated plasma populations in the tail when combined with stress-balance considerations (COWLEY, 1980; COWLEY and SOUTHWOOD, 1980). Provided the tail neutral line lies in a region where the down-tail tail-lobe flow is sub-Alfvénic, steady reconnection can occur. Tailward of the neutral line the lobe plasma is accelerated to form a thin high-speed jet traveling away from the Earth, while Earthward of the neutral line a similar jet will be formed traveling Earthward. Subsequent mirroring of the latter particles at the Earth and return to the tail results in isotropization and formation of the thick hot plasma sheet population proper in the near-Earth tail, with an Earthward flow layer located at its outer surface. Quasi-static equilibria of the former region and their stability have been extensively studied, principally by K. Schindler and co-workers.

While this overall picture of the tail appears to serve as a useful paradigm, interesting theoretical problems have arisen from the new information about the more distant tail which has recently been provided by the ISEE-3 geotail mission. Plasma sheet flow has been found to be predominantly tailward at distances beyond $\sim 100\ R_E$, indicating that the tail neutral line typically lies Earthward of this location on the basis of the simple steady-state picture described above. It would then be expected that the B_z field threading the plasma sheet should be negative on average beyond $\sim 100\ R_E$, whereas in fact observations show that this is only true in the tail center, while on the flanks the average B_z remains positive (SLAVIN *et al.*, this volume, 1986). Clearly these facts require theoretical explanation. One possibility is to consider what happens if the tail lobe flow were to become super-Alfvénic, as is most likely to occur on the tail flanks where the mantle plasma is fastest and densest. In this situation a steady state cannot occur if the neutral line lies in such a flow. Instead, both the neutral line and the Earth-connected closed flux tubes lying Earthward of it will be blown down-tail by the flow, leading to the occurrence of tailward plasma sheet flow in the presence of positive B_z as observed. The plasma sheet flow speed in this case is expected to be slower than the lobe plasma speed opposite to the case tailward of the neutral line where B_z is negative. Ultimately, however, these tailward-flowing closed tubes must be severed nearer the Earth by reconnection, and if the new neutral line again lies within the super-Alfvénic flow the sequence will repeat. The disconnection will result in the formation of a plasmoid containing both positive and negative B_z which will flow rapidly down-tail, replacing for short intervals the slower tailward flow of the previously Earth-connected plasma sheet. The time-averaged B_z in such a scenario would be positive, but the average B_z flux transport (average of V_xB_z) would nevertheless be expected to be smaller and negative. While somewhat speculative, this discussion should serve to indicate some of the interesting theoretical possibilities whose consideration has been stimulated by the ISEE-3 results.

Another important question concerning the nearer-Earth part of the plasma sheet is whether steady earthward convection can occur within it (given steady boundary conditions), or whether such flow will usually generate unstable configurations also leading to time-dependence. Observations seem to be in favor of the latter suggestion. Another result of the ISEE-3 geotail mission has been the observation of large substorm-associated plasmoids propagating down-tail, confirming earlier suggestions based on near-Earth data that a large part of the quiet-time plasma sheet becomes pinched off at substorm onset by rapid near-Earth reconnection (HONES *et al.*, 1984). Several recent simulation studies have recently appeared which attempt to model in some detail this major dynamical reconfiguration (e.g., BIRN, this conference, 1985; MIN *et al.*, this volume, 1986). Some of these computations are compromised by rather unrealistic boundary conditions, however, and all of them employ very simplistic assumptions about plasma conductivity. These remarks are not intended as criticism, but only to indicate that if a lot has already been done, then there is still a lot left to do. This then seems a suitable note on which to conclude this brief review.

9. Solar-Terrestrial Relations—1984

George L. SISCOE

I teach a course on meteorology history and have lived through much of the history of our field. Comparing how these sciences developed reveals striking similarities. They appear like sister disciplines, with meteorology much the older sister. Sitting through this conference has seemed to me like reading in some future history book the stages of life which various aspects of our field had reached at the end of 1984. It is therefore natural to ask when the corresponding aspects of meteorology had reached these same stages. I would like to try to fulfill my assignment to summarize this broad ranging conference by stating what I found in carrying out the exercise. Although this is a hazardous and very subjective understanding, I think it leads to an important conclusion with which we can all agree.

We gain another end in spotting our current epoch on the time line of meteorology. The problems we face often resemble problems meteorologists faced much earlier. Their insights, mistakes and solutions can teach us what to look for, what to avoid and what to expect.

Table 1 gives my assessment of where various components of our field now stand relative to the evolution of our older and larger sister science. The table begins with subjects of comparable development in the two fields and moves toward subjects with increasing lags in their development on our side.

Although this conference de-emphasized microphysics, I include it in the table because in this area advances in space plasma physics seem on a par with those in meteorology. The parity results from their nearness to large, well developed branches of physics, thermodynamics of phase transitions in one case and plasma kinetic and wave physics in the other. In both cases laboratory work importantly constrains and guides progress. Also both areas have benefited much from the local nature of microphysical processes. They have now reached the stage of development at which

Table 1. A Mapping of the present state of solar terrestrial relations onto corresponding stages in the development of meteorology.

Microphysics	1984
Numerical simulation and modeling	1960–1970
Synoptics-data gathering	
(A) Remote sensing	1970–1984
(B) In situ	1840–1910
Dynamics	1860–1900
Forecasting	1860

inhomogeneities and boundaries in the medium become important, noticeably slowing progress. I refer to the explanation of the observed polydispersed cloud droplet distribution in the one field and patchy, sporadic reconnection in the other.

The atmosphere and the magnetosphere defy detailed quantitative prediction by analytical means. Numerical atmospheric modeling began in 1956. In the United States four very large "research class" general circulation models (GCM's) for the atmosphere have been developed by groups at UCLA, Geophysical Fluid Dynamics Laboratory (Princeton), Goddard and NCAR. Many research and operational units have versions of these codes. A large array of special purpose, mesoscale and cloud dynamics codes also exist. Meteorologists extensively employ numerical simulation in research.

The Rice magnetospheric modeling effort began in the early 1970's and the UCLA and NRL efforts at the end of the same decade. In some respects present magnetospheric GCM's are very sophisticated, but because they are limited by our lack of complete knowledge of the magnetospheric energization process, they compare with atmospheric GCM's of the early 1960's. In 1963 the atmospheric energy input processes, radiative transfer and latent heat release, were made part of the GCM code, rather than being specified. To drive magnetospheric convection, the Rice model requires the transpolar potential as an input parameter. The solar wind energy coupling mechanism still needs to be determined definitively and quantitatively before it can be incorporated. The first atmospheric GCM in 1956 exhibited the instability that produces extratropical cyclones, the main mid-latitude weather systems. Substorms play their role in the magnetosphere. As yet no global magnetospheric GCM develops substorms in the sense of an instability given only solar wind conditions as input. the Los Alamos numerical tail models show that substorm behavior might result when a fully connected magnetospheric GCM emerges.

Under the heading "Synoptics" in the table, we have "Remote Sensing" and "In Situ" data gathering techniques. Though atmospheric radar remote sensing began around forty years ago, I see no evident lag in its use in our field. Both fields use Doppler radar to get morphological and wind information. Meteorologists use radar primarily to study the mesoscale structure of storms. We use the Millstone Hill and Sondrestrom facilities for synoptic studies of the auroral zone. Multispectral, high resolution atmospheric imaging and remote sensing from polar orbiting satellites began in 1960 with the Tiros series and in 1965 with the Nimbus series. Geostationary satellite atmospheric coverage began in 1966 with ATS. DMSP satellites give us auroral images from low altitude poalr orbits; but until DE, our field had no synoptic scale images of the auroral oval and no sequences showing a complete substorm life cycle. Our capability with respect to coverage, resolution and use of multispectral techniques probably compares with meteorology in the early 1970's. Without DE we would be at mid-1960's level.

Regarding synoptic studies using in situ data, we must distinguish traditional surface observations and observations from balloon and satellites. Magnetometers play the role of barometers in surface observations. The patterns of ionospheric currents inferred from surface magnetometer chains by modern inversion algorithms exhibit comparable spatial resolution relative to the features to be resolved as did the isobars and wind patterns on surface weather charts prepared in the 1870's. Meteorological research reached a high point in this decade. Synopticians (data analyzers) identified the main features and properties of cyclones while dynamicists (theorists) represented complete physical-mathematical models of these storm systems. Recognizing the crippling limitations of earth-bound observations, meteorologists began systematic attempts to obtain upper air data in the 1890's. They used manned and unmanned balloons, kites and mountain tops. As a result the troposphere was extensively sounded by 1900, and the stratosphere was discovered in 1902. The tropopause was mapped and its dynamical behavior in relation to seasons and storms was reasonably well determined by 1907. Crucial upper air observations taken to test cyclone models occurred in the 1890's. Some of you might be pleased to know that these observations forced meteorologists to abandon (with much acrimony) the cyclone energization process that had until then, for well considered reasons, been nearly universally accepted.

On our side the magnetopause was discovered in 1960 (but predicted earlier) and the tail in 1965 (also predicted). The magnetopause and the magnetosphere's internal structures were reasonably well mapped by 1970. ISEE-3's observations of plasmoids in the tail correspond roughly to the definitive 1890's upper air measurements in the cores of cyclones.

One finds other parallels in the synoptic discoveries in the two fields. In 1857 the Dutch meteorologist Buys Ballot discovered the law that now bears his name ("the only unchallenged law in meteorology"). From the wind direction, it tells in which direction the low pressure lies. The Svalgaard Mansurov "law" resembles it closely, giving the sign of $(B_y)_{\mathrm{IMF}}$ from the sign of vertical magnetic perturbation field in the polar cap. Both laws follow directly and simply from the dynamical principle operating in each case.

Perhaps the most instructive parallel concerns the famous Espy-Redfield controversy on the "Law of storms," which raged in the 1830's and 1840's. Espy discovered the role latent heat release plays in driving cumulus convection. By extrapolating this process to cyclones he produced a pure radial inflow storm model. Redfield studied data from storms at sea, mainly hurricanes, and reached the opinion that instead of flowing toward, winds circulate around a storm's center. By careful analysis of synoptic weather maps (which he introduced), Elias Loomis showed in 1846 that storms combine both wind fields (Fig. 5). Using our sparse, non-synchronous synoptic data alone we cannot do as Loomis did and resolve our controversy over directly driven versus spontaneous substorms. Yet, as in the Espy-Redfield case, given the arguments supporting the two positions, a resolution that omits either one entirely seems inconceivable. The historical precedent here suggests we should concentrate on formulating the appropriate synthesis.

The year 1919 ushered in a revolution in meteorology when Jacob Bjerknes, through painstaking and remarkably synoptic studies, discovered weather fronts and saw the cyclone life cyle as the evolution of an instability operating at the interface between cold polar and warm subtropical air. Bjerknes' now fully established model replaced the by-then discredited model, which began with Espy, attributing the energy of cyclones to latent heat of condensation. The cover of *ISIS* reproduced in Fig. 6 recognizes this milestone in the history of science. A high density of weather stations figured importantly in enabling Bjerknes' breakthrough (Fig. 7).

To be safe I propose we adopt the view that substorms are no less complicated and no less devious than cyclones; that the key to understanding substorms might be higher resolution synoptic data; that there is a useful lesson to us in the history of the development of understanding what a cylone is and how it works. Each significant stage in that development from Espy, Redfield and Loomis to upper air observations to Bjerknes can be attributed to observations in new regions or to increased spatial resolution using simultaneous measurements, that is exploratory or higher resolution

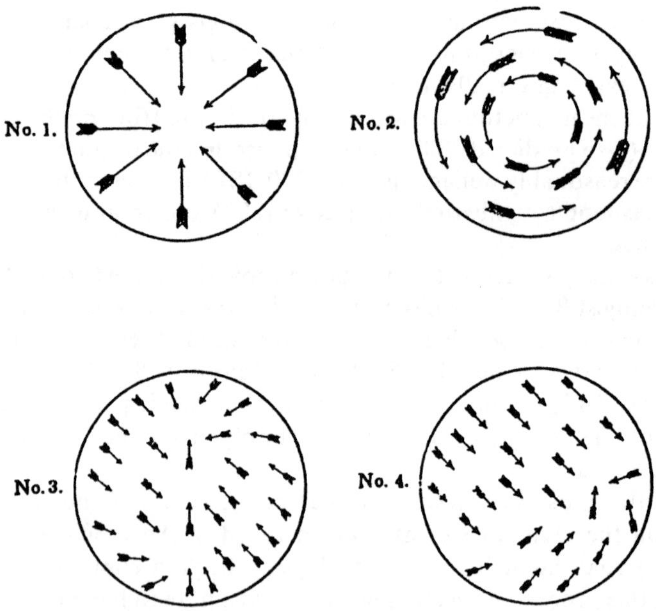

Fig. 5. Diagrams of surface winds in cyclones: No. 1, according to Espy; No. 2, according to Redfield; Nos. 3 and 4, according to Loomis' own investigations. From LOOMIS (1846).

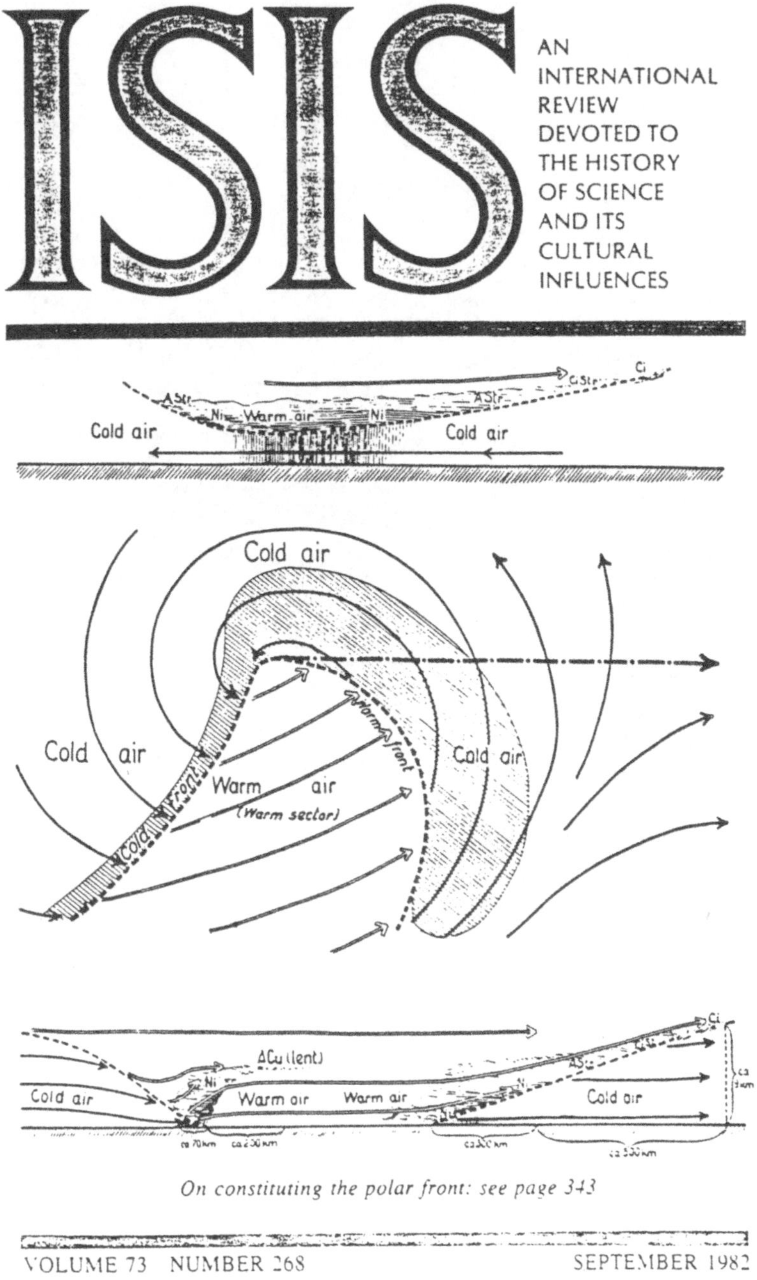

Fig. 6. Cover from the journal *ISIS*, showing Bjerknes' concept of the interaction between cold polar and warm subtropical air.

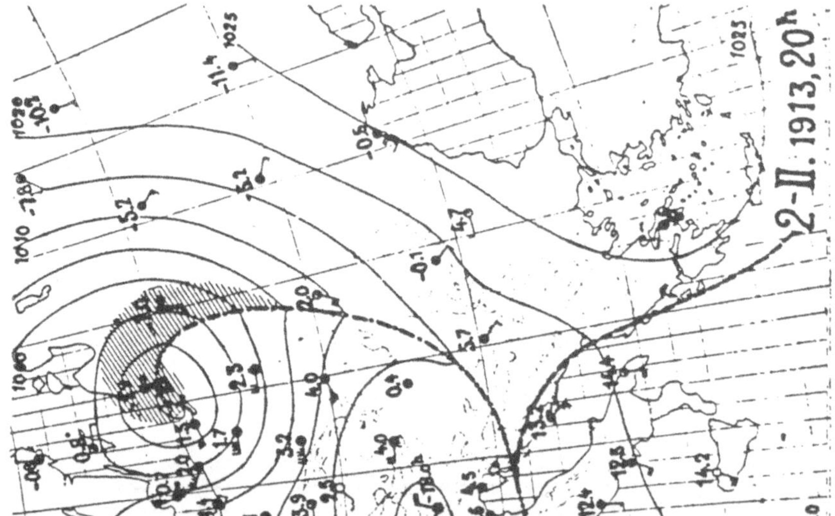

Fig. 7. A storm system studied by Jacob Bjerknes, illustrating the high density of simultaneous data that enabled the discovery of weather fronts.

synoptic studies. These were necessary but not sufficient. Each breakthrough also involved a deep understanding of the operative dynamics, which also evolved.

Before comparing magnetospheric and atmospheric dynamics, we should consider a final possible parallel between developments in meteorology and magnetospheric physics dependent on exploratory synoptic studies. Jet streams were discovered in 1944. Like weather fronts, they are thin, long features. Their relative inaccessiblity and narrowness delayed their discovery. They are now recognized as the main atmospheric conduits transporting energy poleward across the middle latitudes. Until jet streams were discovered, meteorologists attributed their inability to balance the global energy budget to inaccuracies in some of the terms entering the equation, rather than to a qualitative inadequacy of their model.

Chris Russell might like to draw an analogy between this story and the discovery of FTE's, an event which we all agree was important. But I worry that to do so might deflect us from recognizing it as a warning that we should not become sanguine about our understanding of solar wind energy transfer processes until we explore the high latitude dayside magnetopause. This region connects the atmosphere directly to the solar wind through the polar cusp. In it, one finds optimal merging geometries for all IMF orientations. Also some studies have identified it as the portal of entry of magnetosheath plasma. Yet, except for limited sampling by the HEOS spacecraft, this region is virtually terra incognita to us. It might be that reports showing that important things happen only where spacecraft have been testify more to their authors' cleverness than to their wisdom. Like the people in Plato's cave, we might be marveling at low latitude phantoms that are mere shadows of the processes taking place at high latitudes.

As for synoptics, we can draw parallels and lessons by comparing the development of dynamics in the two fields. Figure 8 shows stages in the advance of our ideas on atmospheric and magnetospheric circulation. In 1735 George Hadley modeled the general atmospheric circulation by an equator-to-pole, thermally driven meridional cell. He attributed the midlatitude westerlies and the easterly trades to the change in the rotational speed of the earth's surface as seen by air moving across latitudes. In 1961 Dungey and Axford and Hines formulated their magnetospheric circulation models at roughly similar levels of detail to Hadley's model. Though Axford and Hines considered some effects or rotation, its primary mangetospheric signature, like the westerlies and the trades in the atmosphere, is a convection-free region, the plasmasphere, described theoretically, simultaneously and independently by Brice and Nishida in 1966.

In the first decades of the 19th century it became generally clear that Hadley's cell extended from the equator only to about thirty degrees latitude, rather than to the pole. This property led in 1856 to a major milestone in meteorology. It motivated William Ferrel to discover the north-south component of the Coriolis force (and give its correct explanation). This allowed him to show how the westerlies and the trades pile up at thirty degrees, producing a high pressure there which sinks Hadley's cell. Similarly the westerlies pull away from higher latitudes producing the storm belt of low pressure at sixty degrees. Ferrel therefore introduced a three celled general circulation model, which has become standard. He also formulated for the first time

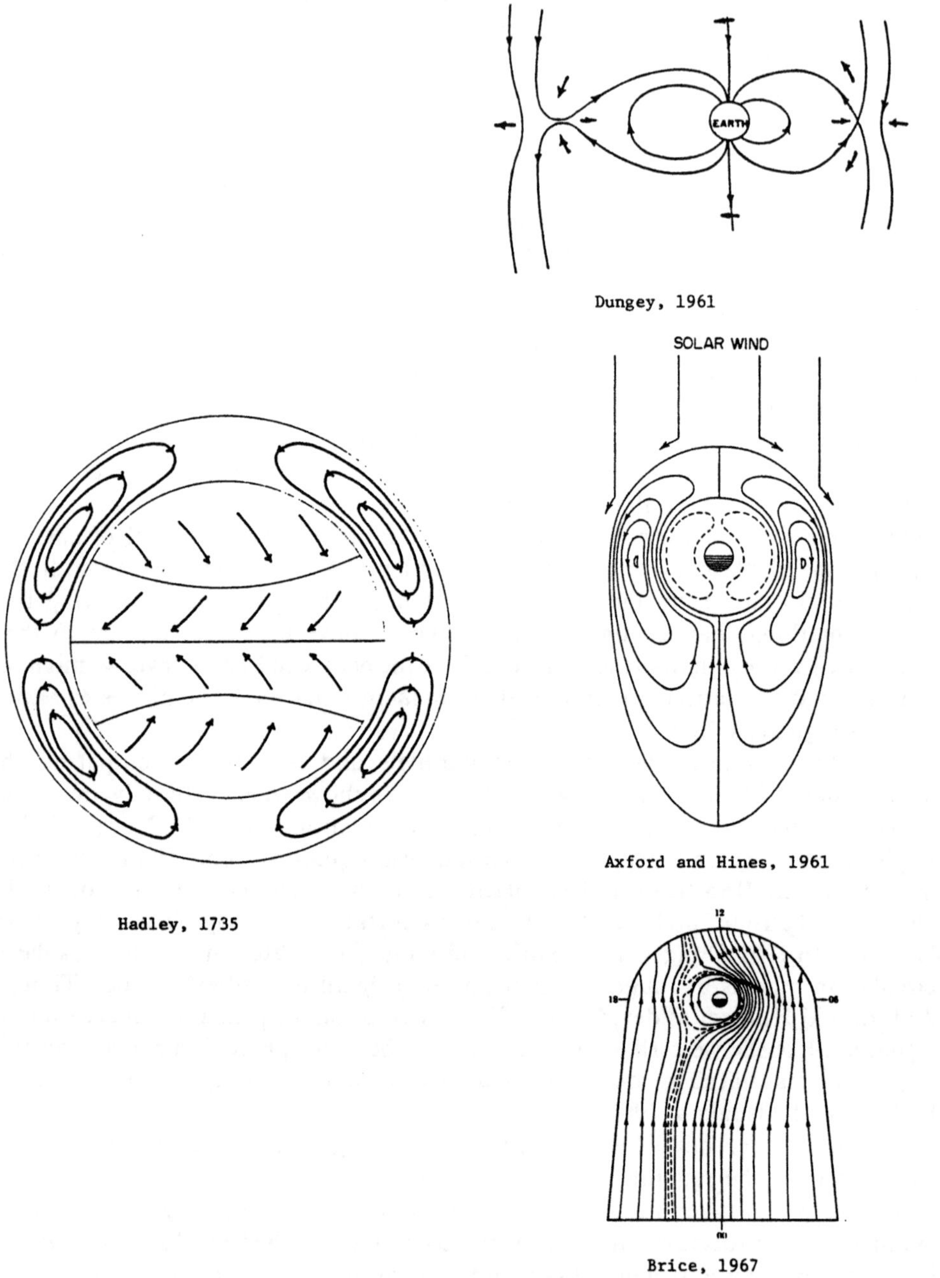

Fig. 8. Comparison of the development of leading ideas in dynamic meteorology and magnetospheric physics.

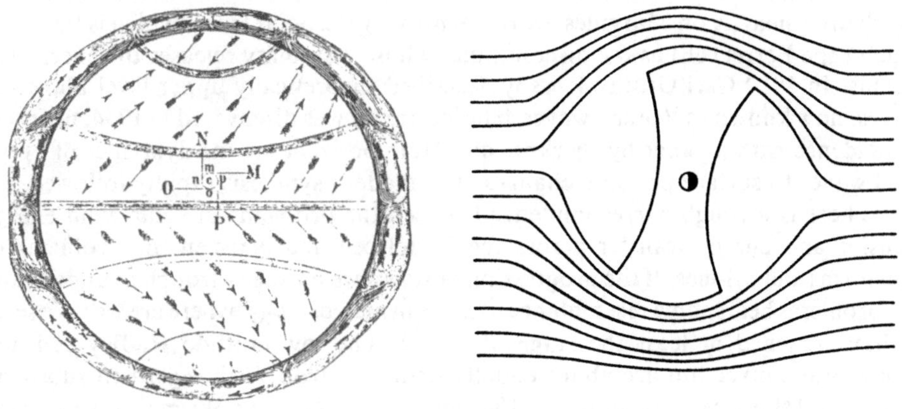

Fig. 8. (continued).

the equations of motion for an atmosphere on a rotating planet, permitting the mathematical investigation of atmospheric dynamics.

In our field, Ferrel's achievement corresponds to Vasyliunas' 1972 formulation of the equations of magnetosphere-ionosphere coupling. These led to the subsequent mathematical treatments of magnetospheric convection by Wolf, Southwood and others. Interestingly, the self-shielding of the midlatitude troposphere from Hadley convection has its analog in the self-shielding of the midlatitude ionosphere from magnetospheric convection.

In both fields the great dynamical-theoretical issues after the Ferrel-Vasyliunas stage became what energized the general circulation and how is it related to storms? In meteorology the midlatitude cell, now called the Ferrel cell, causes all of the problems. This cell circulated backwards. It rises on the cold side and sinks on the warm. We now know that the Ferrel cell is unstable. Atmospheric energy crosses midlatitudes mainly in the meandering horizontal motions of the jet streams. The dynamical equations show that the upper air motions necessarily diverge and converge along the poleward and equatorward legs of the meanders causing eastward migrating low pressure cyclones under the divergences paired with trailing high pressure anticyclones under the convergences. This model explains energy transfer across midlatitudes and ties storms intimately to the transfer process.

What were the major steps leading to the present model subsequent to Ferrel's work? In a series of papers between 1901 and 1906 Max Margules developed the equations of equilibrium at discontinuities in a thermally inhomogeneous rotating atmosphere and showed that such discontinuities represent potential energy that can be converted into kinetic energy. Bjerknes then proposed his cyclone model in 1919, which drew heavily on Margules' work, identifying the polar front (which is the polar edge of the Ferrel cell) as the discontinuity whose instability releases the energy for storms. In 1939 Carl Gustav Rossby described theoretically upper level planetary waves, now known as Rossby waves. The jet stream was discovered in 1944, nearly in coincidence with a paper by Bjerknes and Holmboe relating the dynamics of upper level waves to surface pressure changes. The modern synthesis rapidly followed.

There is a rough correspondence between the problems of solar wind energy transfer and energy transfer across the Ferrel cell and between the problems of substorms and cylones. If the progression of stages we must go through to understand our problems bears some resemblance to those in meteorology, where are we? We seem to have reached at least the stage of Ferrel. The low latitude Hadley cell and subauroral convection are about equally under control. With regard to substorm dynamics, I think we are around the 1900 level in meteorology. We may have seen the Margules stage in the tail equilibrium and instability studies carried out at Rice and Los Alamos. If so, we await our Bjerknes to give the first synthesis which identifies the essential features and describes how they evolve through a cooperative interaction of the tail and the ionosphere. Undoubtedly large pieces of that synthesis already exist, such as, probably, the plasmoid generation model. We can rightly expect a synthesis rather than a creation. The final stage must await the definitive identification and mathematical formulation of the solar wind energy transfer process, the jet stream-Rossby-Bjerknes and Holmboe stage. Then magnetospheric GCM's can accept solar wind parameters as input and generate, without empirical parameters such as ε, the full internal state of the magnetosphere including the substorm.

The item "Forecasting" completes the table. Here it is hard to compete with meteorology. Multispectral, high resolution, high altitude satellite images are good, frequent and readily available. The density of surface weather stations providing upper air soundings is high. The data processing facilities and forecast algorithms are powerful and advanced. Weather forecasting is a worldwide industry supported by university research and government laboratories and manned by a large force of career professionals. From the beginning forecasting has been the main goal of atmospheric research. We have the advantage that whereas meteorologists must forecast for thousands of points on the map, we deal with global events like substorms, geomagnetic storms, and magnetic disturbances generally.

Given that a forecast is a statement of the form "A storm is coming" rather than "A storm is here," and understanding that all of our magnetospheric weather is carried from the sun by the solar wind, it is evident that a forecast required information about the state of the solar wind before it reaches earth. As Dungey predicted in 1961 and Fairfield and Cahill confirmed in 1966, the best upstream predictor of geomagnetic activity is $(B_z)_{\mathrm{IMF}}$(in the proper coordinate system). The formula, ("B_z south-storm's acomin'," corresponds closely to "barometer's low-storm's acomin'." Barometer makers marked their instruments with the words Change, Rain, Much Rain and Stormy within forty years after Torricelli invented the device in 1643. When ISEE-3 was in its upstream station with a real time link to the Boulder forecast center, the B_z forecast formula could be applied with a one to two hour lead time. This level of data collection, sophistication of forecast formula, and forecast precision (storm or no storm anywhere) corresponds roughly to the meteorological situation in the 1860's. Then the invention of the telegraph enabled linking together many stations to produce synoptic charts in real time. In both fields the intensity of the storm could be predicted by the size of the depression in B_z or pressure. Buys Ballot was authorized by the Dutch government to telegraph storm warnings in 1860, and Admiral Fitzroy, then head of the British Meteorological Department, inaugurated a system of storm warnings in ports in 1861. In both cases the warnings were based on synoptic charts prepared from data telegraphed from remote stations. The Boulder forecasts were probably better than those of Buys Ballot and Fitzroy because many storms hitting the Netherlands and England came from the open ocean and did not forecast well.

The absence of the ISEE-3 weather station relocates solar-terrestrial forecasting capability on the meteorological time line. The only upstream data now are solar data. Visually inspecting the solar surface to make a solar-terrestrial forecast corresponds in kind and roughly in accuracy to weather forecasts based on the visual appearance of the sky. Modern attempts to forecast solar-terrestrial disturbances under the serious handicap of no upstream solar wind data of course use scientifically understood principles. Nonetheless, it emphasizes the seriousness of our problem to note that weather forecasting based on sky watching was an art developed already in Aristotle's time and reached a sort of culmination around 1600 with the widely circulated tract titled *The Shepherd of Banbury's Rules*. Even without this dramatization, it is painfully evident to all of us that an upstream solar wind monitor is indispensible not only for solar-terrestrial forecasting, but for any hope of seriously attacking our field's major problems.

I think the main lessons for us to learn from the experience of meteorology are these. The crucial lapses in our understanding, preventing our constructing a fully connected magnetospheric GCM requiring only solar wind input data, are the definitive identifications and mathematical formulations of the energy transfer process at the magnetopause and the substorm process in the tail. (Everyone knows this.) To achieve the desired understanding we should obtain high resolution in situ synoptic data in critical regions. (ISTP is essential.) Well coordinated synoptic studies of special events are highly valuable. (CDAW's are good things and should be continued.) Dynamicists and synopticians should focus on constructing substorm syntheses. We cannot be confident that we will ever understand the solar wind energy transfer process until we explore the high latitude dayside magnetopause.

Although meteorology took nearly fifty years to progress through these steps, science in general progresses faster now. Our pace will be set by how fast we can acquired high resolution in situ data, and I think that means ISTP.

REFERENCES

Akasofu, S.-I., The development of the auroral substorms, *Planet. Space Sci.*, **12**, 273, 1964.

Akasofu, S.-I., A newly emerging substorm model, *Planet. Space Sci.*, **29**, 1069, 1981.

Axford, W. I., Magnetospheric convection, *Rev. Geophys.*, **7**, 421–459, 1969.

BAKER, D. N., S.-I. AKASOFU, W. BAUMJOHANN, J. W. BIEBER, D. M. FAIRFIELD, E. W. HONES, Jr., B. MAUK, R. L. MCPHERRON, and T. E. MOORE, Substorms in the magnetosphere, in *Solar-Terrestrial Physics: Present and Future*, edited by D. M. Butler and K. Papadopoulos, NASA Reference Publ. 1120, 1984.

BARBOSA, D. D., Polar convection patterns under quiet conditions, this conference, 1985.

BARGATZE, L. F., D. N. BAKER, and R. L. MCPHERRON, Magnetospheric response to solar wind variations, this volume, 1986a.

BARGATZE, L. F., R. L. MCPHERRON, and D. N. BAKER, Solar wind-magnetosphere energy input functions, this volume, 1986b.

BAUMJOHANN, W., Merits and limitations of the use of geomagnetic indices in solar wind-magnetosphere coupling studies, this volume, 1986.

BAUMJOHANN, W. and G. HAERENDEL, Dayside convection, viscous interaction, and magnetic merging, this volume, 1986.

BERCHEM, J. and C. T. RUSSELL, Flux transfer events on the magnetopause: spatial distribution and controlling factors, *J. Geophys. Res.*, **89**, 6689, 1984.

BIRN J., On the role of driving boundary conditions in MHD simulations of magnetotail reconnection, this conference, 1985.

BRATANAHL, A. and P. J. BAUM, A plasmoid release mechanism that could explain the substorm's impulsive earthward diversion of cross tail current, in *Magnetic Reconnection in Space and Laboratory Plasmas*, edited by E. W. Hones, Jr., p. 347, Am. Geophys. Un., Washington, D.C., 1984.

BURCH, J. L., J. D. MENIETTI, and J. N. BARFIELD, DE-1 observations of solar-wind magnetosphere coupling processes in the polar cusp, this volume, 1986.

CHEN, J. and P. PALMADESSO, Tearing instability in an anisotropic neutral sheet, *Phys. Fluids*, **27**, 1198, 1984.

CLAUER, C. R., R. L. MCPHERRON, C. SEARLS, and M. G. KIVELSON, Solar wind control of auroral zone geomagnetic activity, *Geophys. Res. Lett.*, **8**, 915, 1981.

CORONITI, F. V., R. L. MCPHERRON, and G. K. PARKS, Studies of the magnetospheric substorm. 3. Concept of the magnetospheric substorm and its relation to electron precipitation and micropulsations, *J. Geophys. Res.*, **73**, 1715, 1968.

COWLEY, S. W. H., Plasma populations in a simple open model magnetosphere, *Space Sci. Rev.*, **26**, 217, 1980.

COWLEY, S. W. H., Magnetospheric asymmetries associated with the *y*-component of the IMF, *Planet. Space Sci.*, **29**, 79, 1981.

COWLEY, S. W. H., Evidence for the occurrence and importance of reconnection between the Earth's magnetic field and the interplanetary magnetic field, in *Magnetic Reconnection in Space and Laboratory Plasmas*, edited by E. W. Hones, Jr., p. 347, American Geophys. Un., Washington, D.C., 1984a.

COWLEY, S. W. H., Solar wind control of magnetospheric convection, in *Proc. Conf. Achievements of the IMS*, ESA SP-217, p. 483, 1984b.

COWLEY, S. W. H. and D. J. SOUTHWOOD, Some properties of the steady-state geomagnetic tail, *Geophys. Res. Lett.*, **7**, 833, 1980.

CROLEY, D. R., J. F. FENNELL, and B. G. LEDLEY, Equatorial magnetic field and charged particle data observed during magnetopause crossings at synchronous orbit, this conference, 1985.

CROOKER, N. U., The configuration of dayside merging, in *Dynamics of the Magnetosphere*, edited by S.-I. Akasofu, p. 101, D. Reidel, Dordrecht, Holland, 1979.

CROOKER, N. U., Global geometry of merging at the dayside magnetopause, this volume, 1986.

DUNGEY, J. W., The structure of the exosphere or adventures in velocity space, in *Geophysics of the Earth's Environment*, edited by C. DeWitt, J. Hieblot and L. Le Beau, p. 503, Gordon and Breach, New York, 1963.

EVIATAR, A. and R. A. WOLF, Transfer processes in the magnetopause, *J. Geophys. Res.*, **73**, 5561, 1968.

FAIRFIELD, D. H., Solar wind control of the magnetotail, this volume, 1986.

FAIRFIELD, D. H. and J. D. SCUDDER, Polar rain: solar coronal electrons in the Earth's magnetotail, this volume, 1986.

FEJER, B. G., Equatorial ionospheric electric fields associated with magnetospheric disturbances, this

volume, 1986.
FOSTER, J. C., J. M. HOLT, R. G. MUSGROVE, and D. S. EVANS, Solar wind dependence of high-latitude convection and precipitation, this volume, 1986.
FRANK, L. A., Solar wind-magnetospheric coupling as inferred from global auroral imaging, this conference, 1985.
FREEMAN, J. W., Jr., C. S. WARREN, and J. J. MAGUIRE, Plasma flow directions at the magnetopause on January 13 and 14, 1967, *J. Geophys. Res.*, **73**, 5719, 1968.
FREEMAN J. W. and R. LOPEZ, Solar cycle variations in the solar wind, this volume, 1986.
FRIIS-CHRISTENSEN, E., Solar wind control of the polar cusp, this volume, 1986.
GOERTZ, C. K., E. NIELSEN, A. KORTH, K. H. GLASSMEIER, C. HALDOUPIS, P. HOEG, and D. HAYWARD, Observations of a possible ground signature of flux transfer events, *J. Geophys. Res.*, **90**, 4069, 1985.
HARDY, D. A., M. S. GUSSENHOVEN, K. RIEHL, R. BURKHARDT, N. HEINEMANN, and T. L. SCHUMAKER, The characteristics of polar cap precipitations and their dependence on the interplanetary magnetic field and the solar wind, this volume, 1986.
HAREL, M., Coupling of the magnetosphere and ionosphere during substorms, this conference, 1985.
HEELIS, R. A., High latitude ionospheric convection patterns and their relationship to IMF orientation, this conference, 1985.
HEIKKILA, W. J., Impulsive plasma transport through the magnetopause, *Geophys. Res. Lett.*, **9**, 159, 1982.
HEIKKILA, W. J., Transport of plasma across the magnetopause, this volume, 1986.
HEYN, M. F., H. K. BIERNAT, V. S. SEMENOV, and I. V. KUBYSHKIN, Dayside magnetopause reconnection, *J. Geophys. Res.*, **90**, 1781, 1985.
HONES, E. W., Jr., D. N. BAKER, S. J. BAME, W. C. FELDMAN, J. T. GOSLING, D. J. MCCOMAS, R. D. ZWICKL, J. A. SLAVIN, E. J. SMITH, and B. T. TSURUTANI, Structure of the magnetotail at 220 R_E and its response to geomagnetic activity, *J. Geophys. Res.*, **11**, 5, 1984.
HUGHES, W. J. and S. W. H. COWLEY, Observation of IMF-associated magnetic field perturbations in the GSM *X*-component at geostationary orbit, this volume, 1986.
IIJIMA T. and T. A. POTEMRA, Large-scale characteristics of field-aligned currents associated with substorms, *J. Geophys. Res.*, **83**, 2852, 1978.
IPAVICH, F. M., G. GLOECKLER, D. C. HAMILTON, W. STUDEMAN, B. WILKEN, G. KREMSER, and D. HOVESTADT, First direct measurements of C^{+6} in the solar wind, this conference, 1985.
KING, J. H., Solar wind parameters and magnetosphere coupling studies, this volume, 1986.
KNOTT, K. and A. PEDERSEN, Magnetopause electric fields: observations in the geostationary orbit at times of magnetospheric compression, this conference, 1985.
KORTH, A., G. KREMSER, N. CORNILLEAU-WEHRLIN, and J. SOLOMON, Observations of energetic electrons and VLF waves at geostationary orbit during storm sudden commencements (SSC), this volume, 1986.
LEE, L. C., Magnetic flux transfer at the Earth's magnetopause, this volume, 1986.
LEE, L. C. and J. V. OLSON, Kelvin-Helmholtz instability and the variation of geomagnetic pulsation activity, *Geophys. Res. Lett.*, **7**, 777–780, 1980.
LEE, L. C. and Z. F. FU, A theory of magnetic flux transfer at the earth's magnetopause, *Geophys. Res. Lett.*, **12**, 105, 1985.
LINDQVIST, P.-A. and F. S. MOZER, The average tangential electric field at the noon magnetopause, this conference, 1985.
LOOMIS, E., On two storms which were experienced throughout the United States in the month of February, 1842, *Trans. Am. Phil. Soc.*, **9**, 1846.
LÜHR, H., N. KLOCKER, B. HÄUSLER, O. BAUER, R. TREUMANN, G. PASCHMANN, N. SCKOPKE, H. KOONS, R. HOLZWORTH, D. GURNETT, R. ANDERSON, and C. CARLSON, A natural magnetic cavity near the magnetopause, this conference, 1985.
LUI, A. T. Y., Solar wind influence on magnetotail configuration and dynamics, this volume, 1986.
LYONS, L. R., A simple model for polar cap convection patterns and generation of theta-auroras, *J. Geophys. Res.*, **90**, 1561, 1985.
MAKITA, K. C., C.-I. MENG, and S.-I. AKASOFU, Temporal and spatial variations of the polar cap dimension and its relation to the energy input rate and the *AE* index, *J. Geophys. Res.*, **90**, 2744, 1985.
MCENTIRE, R. W., Preliminary results from the AMPTE mission, this conference, 1985.

McENTIRE, R. W., A. T. Y. LUI, S. M. KRIMIGIS, and E. P. KEATH, Energetic heavy ions in the magnetosphere, this conference, 1985.
McPHERRON, R. L., Growth phase of magnetospheric substorms, *J. Geophys. Res.*, **75**, 5592, 1970.
McPHERRON, R. L., Magnetospheric substorms, *Rev. Geophys. Space Phys.*, **17**, 657, 1979.
McPHERRON, R. L., D. N. BAKER, and L. F. BARGATZE, Linear filters as a method of real time prediction of geomagnetic activity, this volume, 1986.
MENG, C.-I., Polar cap arcs and the plasma sheet, *Geophys. Res. Lett.*, **8**, 273, 1981.
MENG, C.-I. and K. MAKITA, Dynamic variations of the polar cap, this volume, 1986.
MIN, K. W., H. OKUDA, and T. SATO, Simulation on magnetospheric substorm, this volume, 1986.
MIURA, A., Anomalous transport by magnetohydrodynamic Kelvin-Helmholtz instabilities in the solar wind-magnetosphere interaction, *J. Geophys. Res.*, **89**, 801, 1984.
MURPHEE, J. S., C. D. ANGER, and L. L. COGGER, The instantaneous relationship between polar cap and oval auroras at times of northward interplanetary magnetic field, *Can. J. Phys.*, **60**, 349, 1982.
OLSON, W. P. and K. A. PFITZER, The coupling of the solar wind to the magnetosphere, this conference, 1985.
PASCHMANN, G., N. SCKOPKE, G. HAERENDEL, J. PAPAMASTORAKIS, S. J. BAME, J. R. ASBRIDGE, J. T. GOSLING, E. W. JONES, Jr., and E. R. TECH, ISEE plasma observations near the subsolar magnetopause, *Space Sci. Rev.*, **22**, 717, 1978.
PYTTE, T., R. L. McPHERRON, E. W. HONES, Jr., and H. I. WEST, Jr., Multiple-satellite studies of magnetospheric substorms: distinction between polar magnetic substorms and convection-driven negative bays, *J. Geophys. Res.*, **83**, 663, 1978.
QUEST, K. B. and F. V. CORONITI, Tearing at the dayside magnetopause, *J. Geophys. Res.*, **86**, 3289, 1981.
REIFF, P. H. and J. G. LUHMANN, Solar wind control of the polar cap electrostatic potential difference, this volume, 1986.
REIFF, P. H., R. W. SPIRO, and T. W. HILL, Dependence of polar cap potential drop on interplanetary parameters, *J. Geophys. Res.*, **86**, 7639, 1981.
RIJNBEEK, R. P., S. W. H. COWLEY, D. J. SOUTHWOOD, and C. T. RUSSELL, A survey of flux transfer events observed by ISEE 1 and 2 magnetometers, *J. Geophys. Res.*, **89**, 786, 1984.
ROSTOKER, G., S.-I. AKASOFU, J. FOSTER, R. A. GREENWALD, Y. KAMIDE, K. KAWASAKI, A. T. Y. LUI, R. L. McPHERRON, and C. T. RUSSELL, Magnetospheric substorms—definitions and signatures, *J. Geophys. Res.*, **85**, 1663, 1980.
RUSSELL, C. T., The configuration of the magnetosphere, in *Critical Problems in Magnetospheric Physics*, edited by E. R Dyer, p. 1, IUCSTP-NAS, Washington, D.C., 1972.
RUSSELL, C. T., Solar wind control of magnetospheric configuration, this volume, 1986.
RUSSELL, C. T. and R. C. ELPHIC, Initial ISEE magnetometer results: magnetopause observations, *Space Sci. Rev.*, **22**, 681, 1978.
SAKAI, J.-I. and H. WASHIMI, A triggering of a solar flare by magnetosonic waves in a neutral sheet plasma, *Astrophys. J.*, **258**, 823, 1982.
SATO, T. and T. HAYASHI, Externally driven magnetic reconnection and a powerful magnetic energy converter, *Phys. Fluids*, **22**, 1189, 1979.
SIBECK, D. G., J. A. SLAVIN E. J. SMITH, and B. T. TSURUTANI, Geomagnetotail twisting, this volume, 1986.
SLAVIN, J. A., E. J. SMITH, D. N. BAKER, R. D. ZWICKL, and S.-I. AKASOFU, Solar-wind-magnetosphere coupling and the distant magnetotail: ISEE-3 observations, this volume, 1986.
SMITH, E. J., J. A. SLAVIN, R. D. ZWICKL, and S. J. BAME, Shocks and storm sudden commencements, this volume, 1986a.
SMITH, M. F., D. J. RODGERS, R. P. RIJNBEEK, D. J. SOUTHWOOD, A. J. COATES, and A. D. JOHNSTONE, Flux transfer events can be miniaturized, this volume, 1986b.
SONNERUP, B. U. Ö., Magnetic field reconnection at the magnetopause: an overview, in *Magnetic Reconnection in Space and Laboratory*, edited by E. W. Hones, Jr., p. 92, Geophys. Monograph 30, Am. Geophys. U., Washington, D.C., 1984.
SONNERUP, B. U. Ö., G. PASCHMANN, I. PAPAMASTARAKIS, N. SCKOPKE, G. HAERENDEL, S. J. BAME, J. R. ASBRIDGE, J. T. GOSLING, and C. T. RUSSELL, Evidence for magnetic field reconnection at the Earth's magnetopause, *J. Geophys. Res.*, **86**, 10049, 1981.
STERN, D. P., A study of the electric field in an open magnetospheric model, *J. Geophys. Res.*, **78**, 7292,

1973.

STÜDEMAN, W., G. GLOECKLER, B. WILKEN, F. IPAVICH, G. KREMSER, D. HAMILTON, and D. HOVESTADT, Ion composition of the bulk ring current during a magnetic storm: observations with the CHEM-instrument on AMPTE/CCE, this volume, 1986.

TOFFOLETTO, F. R. and T. W. HILL, Quantitative model of magnetic coupling between solar wind and magnetosphere, this volume, 1986.

TSURUTANI B. T. and R. M. THORNE, Diffusion processes in the magnetopause boundary, *Geophys. Res. Lett.*, **9**, 1247, 1982.

VASYLIUNAS, V. M., Solar wind control of magnetospheric convection, this volume, 1986.

VENKATESAN, D., A. WOLFE, A. MELONI, L. J. LANZEROTTI, C. G. MACLENNAN, and J. F. BAMBER, Observations of hydromagnetic-band energy in the magnetosphere near $L{\sim}2$ and $L{\sim}3$ and solar wind wave and field conditions, this volume, 1986.

WALKER, R. J. and T. SATO, Externally driven magnetic reconnection, in *Magnetic Reconnection in Space and Laboratory Plasmas*, edited by E. W. Hones, Jr., p. 272, AGU Monograph series 30, AGU, Washington, D.C., 1984.

WHANG, Y. C., F. WEI, and H. DU, Jump of interplanetary fluctuations across Earth's bow shock, this conference, 1985.

WU, C. C., MHD modelling of the solar wind-magnetosphere interaction, this conference, 1985.

WYGANT, J. R., R. B. TORBERT, and F. S. MOZER, Comparison of S3-3 polar cap potential drops with the interplanetary magnetic field and models of magnetopause reconnection, *J. Geophys. Res.*, **88**, 5727-5735, 1983.

ZANETTI, L. J. and T. A. POTEMRA, The dependence of Birkeland current systems on the interplanetary magnetic field, this volume, 1986.

GPSR Compliance

The European Union's (EU) General Product Safety Regulation (GPSR) is a set of rules that requires consumer products to be safe and our obligations to ensure this.

If you have any concerns about our products, you can contact us on ProductSafety@springernature.com

In case Publisher is established outside the EU, the EU authorized representative is:

Springer Nature Customer Service Center GmbH
Europaplatz 3
69115 Heidelberg, Germany

Batch number: 10823769

Printed by Printforce, the Netherlands